Probability and Mathematical Statistics (Continued)

MATTHES, KERSTAN, and MECKE • Infi
Processes
MUIRHEAD • Aspects of Multivariate Statistical Theory
OLIVER and SMITH • Influence Diagrams, Belief Nets and Decision Analysis
PRESS • Bayesian Statistics: Principles, Models, and Applications
PURI and SEN • Nonparametric Methods in General Linear Models
PURI and SEN • Nonparametric Methods in Multivariate Analysis
PURI, VILAPLANA, and WERTZ • New Perspectives in Theoretical and Applied Statistics
RANDLES and WOLFE • Introduction to the Theory of Nonparametric Statistics
RAO • Asymptotic Theory of Statistical Inference
RAO • Linear Statistical Inference and Its Applications, *Second Edition*
RAO • Real and Stochastic Analysis
RAO and SEDRANSK • W.G. Cochran's Impact on Statistics
ROBERTSON, WRIGHT and DYKSTRA • Order Restricted Statistical Inference
ROGERS and WILLIAMS • Diffusions, Markov Processes, and Martingales, Volume II: Îto Calculus
ROHATGI • An Introduction to Probability Theory and Mathematical Statistics
ROHATGI • Statistical Inference
ROSS • Stochastic Processes
RUBINSTEIN • Simulation and the Monte Carlo Method
RUZSA and SZEKELY • Algebraic Probability Theory
SCHEFFE • The Analysis of Variance
SEBER • Linear Regression Analysis
SEBER • Multivariate Observations
SEBER and WILD • Nonlinear Regression
SEN • Sequential Nonparametrics: Invariance Principles and Statistical Inference
SERFLING • Approximation Theorems of Mathematical Statistics
SHORACK and WELLNER • Empirical Processes with Applications to Statistics
STOYANOV • Counterexamples in Probability
WHITTAKER • Graphical Models in Applied Multivariate Statistics

Applied Probability and Statistics

ABRAHAM and LEDOLTER • Statistical Methods for Forecasting
AGRESTI • Analysis of Ordinal Categorical Data
AGRESTI • Categorical Data Analysis
AICKIN • Linear Statistical Analysis of Discrete Data
ANDERSON and LOYNES • The Teaching of Practical Statistics
ANDERSON, AUQUIER, HAUCK, OAKES, VANDAELE, and WEISBERG • Statistical Methods for Comparative Studies
ARTHANARI and DODGE • Mathematical Programming in Statistics
ASMUSSEN • Applied Probability and Queues
*BAILEY • The Elements of Stochastic Processes with Applications to the Natural Sciences
BARNETT • Interpreting Multivariate Data
BARNETT and LEWIS • Outliers in Statistical Data, *Second Edition*
BARTHOLOMEW • Stochastic Models for Social Processes, *Third Edition*
BARTHOLOMEW and FORBES • Statistical Techniques for Manpower Planning
BATES and WATTS • Nonlinear Regression Analysis and Its Applications
BHATTACHARYA and WAYMIRE • Stochastic Processes with Applications
BECK and ARNOLD • Parameter Estimation in Engineering and Science
BELSLEY, KUH, and WELSCH • Regression Diagnostics: Identifying Influential Data and Sources of Collinearity

*Now available in a lower priced paperback edition in the Wiley Classics Library.

BHAT • Elements of Applied Stochastic Processes, *Second Edition*
BHATTACHARYA and WAYMIRE • Stochastic Processes with Applications
BLOOMFIELD • Fourier Analysis of Time Series: An Introduction
BOLLEN • Structural Equations with Latent Variables
BOX • R. A. Fisher, the Life of a Scientist
BOX and DRAPER • Empirical Model-Building and Response Surfaces
BOX and DRAPER • Evolutionary Operation: A Statistical Method for Process Improvement
BOX, HUNTER, and HUNTER • Statistics for Experimenters: An Introduction to Design, Data Analysis, and Model Building
BROWN and HOLLANDER • Statistics: A Biomedical Introduction
BUCKLEW • Large Deviation Techniques in Decision, Simulation, and Estimation
BUNKE and BUNKE • Nonlinear Regression, Functional Relations and Robust Methods: Statistical Methods of Model Building
BUNKE and BUNKE • Statistical Inference in Linear Models, Volume I
CHAMBERS • Computational Methods for Data Analysis
CHATTERJEE and HADI • Sensitivity Analysis in Linear Regression
CHATTERJEE and PRICE • Regression Analysis by Example
CHOW • Econometric Analysis by Control Methods
CLARKE and DISNEY • Probability and Random Processes: A First Course with Applications, *Second Edition*
COCHRAN • Sampling Techniques, *Third Edition*
COCHRAN and COX • Experimental Designs, *Second Edition*
CONOVER • Practical Nonparametric Statistics, *Second Edition*
CONOVER and IMAN • Introduction to Modern Business Statistics
CORNELL • Experiments with Mixtures, Designs, Models, and the Analysis of Mixture Data, *Second Edition*
COX • A Handbook of Introductory Statistical Methods
COX • Planning of Experiments
DANIEL • Applications of Statistics to Industrial Experimentation
DANIEL • Biostatistics: A Foundation for Analysis in the Health Sciences, *Fourth Edition*
DANIEL and WOOD • Fitting Equations to Data: Computer Analysis of Multifactor Data, *Second Edition*
DAVID • Order Statistics, *Second Edition*
DAVISON • Multidimensional Scaling
DEGROOT, FIENBERG and KADANE • Statistics and the Law
*DEMING • Sample Design in Business Research
DILLON and GOLDSTEIN • Multivariate Analysis: Methods and Applications
DODGE • Analysis of Experiments with Missing Data
DODGE and ROMIG • Sampling Inspection Tables, *Second Edition*
DOWDY and WEARDEN • Statistics for Research
DRAPER and SMITH • Applied Regression Analysis, *Second Edition*
DUNN • Basic Statistics: A Primer for the Biomedical Sciences, *Second Edition*
DUNN and CLARK • Applied Statistics: Analysis of Variance and Regression, *Second Edition*
ELANDT-JOHNSON and JOHNSON • Survival Models and Data Analysis
FLEISS • The Design and Analysis of Clinical Experiments
FLEISS • Statistical Methods for Rates and Proportions, *Second Edition*
FLURY • Common Principal Components and Related Multivariate Models
FOX • Linear Statistical Models and Related Methods
FRANKEN, KÖNIG, ARNDT, and SCHMIDT • Queues and Point Processes
GALLANT • Nonlinear Statistical Models
GIBBONS, OLKIN, and SOBEL • Selecting and Ordering Populations: A New Statistical Methodology

Continued on back end papers

*Now available in a lower priced paperback edition in the Wiley Classics Library.

Stochastic Processes with Applications

Stochastic Processes with Applications

RABI N. BHATTACHARYA

Department of Mathematics
Indiana University
Bloomington, Indiana

EDWARD C. WAYMIRE

Department of Mathematics
Oregon State University
Corvallis, Oregon

A Wiley-Interscience Publication
John Wiley & Sons, Inc.
New York / Chichester / Brisbane / Toronto / Singapore

Library of Congress Cataloging in Publication Data:

Bhattacharya, R. N. (Rabindra Nath), 1937–
Stochastic processes with applications / Rabi N. Bhattacharya, Edward C. Waymire.
p. cm.—(Wiley series in probability and mathematical statistics. Applied probability and statistics.)
Includes bibliographical references.
1. Stochastic processes. I. Waymire, Edward C. II. Title. III. Series.
QA274.B49 1990
519.5—dc20 89-28530
ISBN 0-471-84272-9 CIP

Printed in the United States of America

10 9 8 7 6 5 4 3 2 1

To
Gowri and Linda,
with love

Contents

Preface

This is a text on stochastic processes for graduate students in science and engineering, including mathematics and statistics. It has become somewhat commonplace to find growing numbers of students from outside of mathematics enrolled along with mathematics students in our graduate courses on stochastic processes. In this book we seek to address such a mixed audience. For this purpose, in the main body of the text the theory is developed at a relatively simple technical level with some emphasis on computation and examples. Sometimes to make a mathematical argument complete, certain of the more technical explanations are relegated to the end of the chapter under the label *theoretical complements*. This approach also allows some flexibility in instruction. A few sample course outlines have been provided to illustrate the possibilities for designing various types of courses based on this book. The theoretical complements also contain some supplementary results and references to the literature.

Measure theory is used sparingly and with explanation. The instructor may exercise control over its emphasis and use depending on the background of the majority of the students in the class. Chapter 0 at the end of the book may be used as a short course in measure theoretical probability for self study. In any case we suggest that students unfamiliar with measure theory read over the first few sections of the chapter early on in the course and look up standard results there from time to time, as they are referred in the text.

Chapter applications, appearing at the end of the chapters, are largely drawn from physics, computer science, economics, and engineering. There are many additional examples and applications illustrating the theory; they appear in the text and among the exercises.

Some of the more advanced or difficult exercises are marked by asterisks. Many appear with hints. Some exercises are provided to complete an argument or statement in the text. Occasionally certain well-known results are only a few steps away from the theory developed in the text. Such results are often cited in the exercises, along with an outline of steps, which can be used to complete their derivation.

Rules of cross-reference in the book are as follows. Theorem m.n, Proposition m.n, or Corollary m.n, refers to the nth such assertion in section m of the same chapter. Exercise n, or Example n, refers to the nth Exercise, or nth Example, of the same section. Exercise m.n (Example m.n) refers to Exercise n (Example n) of a different section m within the same chapter. When referring to a result or an example in a different chapter, the chapter number is always mentioned along with the label m.n to locate it within that chapter.

This book took a long time to write. We gratefully acknowledge research support from the National Science Foundation and the Army Research Office during this period. Special thanks are due to Wiley editors Beatrice Shube and Kate Roach for their encouragement and assistance in seeing this effort through.

RABI N. BHATTACHARYA
EDWARD C. WAYMIRE

Bloomington, Indiana
Corvallis, Oregon
February 1990

Sample Course Outlines

COURSE 1

Beginning with the Simple Random Walk, this course leads through Brownian Motion and Diffusion. It also contains an introduction to discrete/continuous-parameter Markov Chains and Martingales. More emphasis is placed on concepts, principles, computations, and examples than on complete proofs and technical details.

Chapter I	*Chapter II*	*Chapter III*
§1–7 (+ Informal Review of Chapter 0, §4) §13 (Up to Proposition 13.5)	§1–4 §5 (By examples) §11 (Example 2) §13	§1–3 §5

Chapter IV	*Chapter V*	*Chapter VI*
§1–7 (Quick survey by examples)	§1 §2 (Give transience/recurrence from Proposition 2.5) §3 (Informal justification of equation (3.4) only) §5–7 §10 §11 (Omit proof of Theorem 11.1) §12–14	§4

COURSE 2

The principal topics are the Functional Central Limit Theorem, Martingales, Diffusions, and Stochastic Differential Equations. To complete proofs and for supplementary material, the theoretical complements are an essential part of this course.

Chapter I	*Chapter V*	*Chapter VI*	*Chapter VII*
§1–4 (Quick survey) §6–10 §13	§1–3 §6–7 §11 §13–17	§4	§1–4

COURSE 3

This is a course on Markov Chains that also contains an introduction to Martingales. Theoretical complements may be used only sparingly.

Chapter I	*Chapter II*	*Chapter III*	*Chapter IV*	*Chapter VI*
§1–6 §13	§1–9 §11 §12 or 15 §13–14	§1 §5	§1–11	§1–2 §4–5

Stochastic Processes
with Applications

CHAPTER I

Random Walk and Brownian Motion

1 WHAT IS A STOCHASTIC PROCESS?

Denoting by X_n the value of a stock at an nth unit of time, one may represent its (erratic) evolution by a family of random variables $\{X_0, X_1, \ldots\}$ indexed by the *discrete-time parameter* $n \in \mathbb{Z}_+$. The number X_t of car accidents in a city during the time interval $[0, t]$ gives rise to a collection of random variables $\{X_t: t \geqslant 0\}$ indexed by the *continuous-time parameter* t. The velocity $X_{\mathbf{u}}$ at a point $\mathbf{u}$ in a turbulent wind field provides a family of random variables $\{X_{\mathbf{u}}: \mathbf{u} \in \mathbb{R}^3\}$ indexed by a *multidimensional spatial parameter* $\mathbf{u}$. More generally we make the following definition.

Definition 1.1. Given an index set I, a *stochastic process* indexed by I is a collection of random variables $\{X_\lambda: \lambda \in I\}$ on a probability space $(\Omega, \mathscr{F}, P)$ taking values in a set S. The set S is called the *state space* of the process.

In the above, one may take, respectively: (i) $I = \mathbb{Z}_+$, $S = \mathbb{R}_+$; (ii) $I = [0, \infty)$, $S = \mathbb{Z}_+$; (iii) $I = \mathbb{R}^3$, $S = \mathbb{R}^3$. For the most part we shall study stochastic processes indexed by a one-dimensional set of real numbers (e.g., time). Here the natural ordering of numbers coincides with the sense of evolution of the process. This order is lost for stochastic processes indexed by a multidimensional parameter; such processes are usually referred to as *random fields*. The state space S will often be a set of real numbers, finite, countable, (i.e., discrete) or uncountable. However, we also allow for the possibility of vector-valued variables. As a matter of convenience in notation the index set is often suppressed when the context makes it clear. In particular, we often write $\{X_n\}$ in place of $\{X_n: n = 0, 1, 2, \ldots\}$ and $\{X_t\}$ in place of $\{X_t: t \geqslant 0\}$.

For a stochastic process the values of the random variables corresponding to the occurrence of a sample point $\omega \in \Omega$ constitute a *sample realization* of the process. For example, a sample realization of the coin-tossing process corresponding to the occurrence of $\omega \in \Omega$ is of the form $(X_0(\omega), X_1(\omega), \ldots, X_n(\omega), \ldots)$. In this case $X_n(\omega) = 1$ or 0 depending on whether the outcome of the nth toss is a head or a tail. In the general case of a discrete-time stochastic process with state-space S and index set $I = \mathbb{Z}_+ = \{0, 1, 2, \ldots\}$, the sample realizations of the process are of the form $(X_0(\omega), X_1(\omega), \ldots, X_n(\omega), \ldots)$, $X_n(\omega) \in S$. In the case of a continuous-parameter stochastic process with state space S and index set $I = \mathbb{R}_+ = [0, \infty)$, the sample realizations are functions $t \to X_t(\omega) \in S$, $\omega \in \Omega$. Sample realizations of a stochastic process are also referred to as *sample paths* (see Figures 1.1a, b).

In the so-called *canonical* choice for Ω the sample points of Ω represent sample paths. In this way Ω is some set of functions ω defined on I taking values in S, and the value $X_t(\omega)$ of the process at time t corresponding to the outcome $\omega \in \Omega$ is simply the coordinate projection $X_t(\omega) = \omega_t$. Canonical representations of sample points as sample paths will be used often in the text.

Stochastic models are often specified by prescribing the probabilities of events that depend only on the values of the process at finitely many time points. Such events are called *finite-dimensional events.* In such instances the probability measure P is only specified on a subclass $\mathscr{C}$ of the events contained in a sigmafield $\mathscr{F}$. Probabilities of more complex events, for example events that depend on the process at infinitely many time points (*infinite-dimensional events*), are

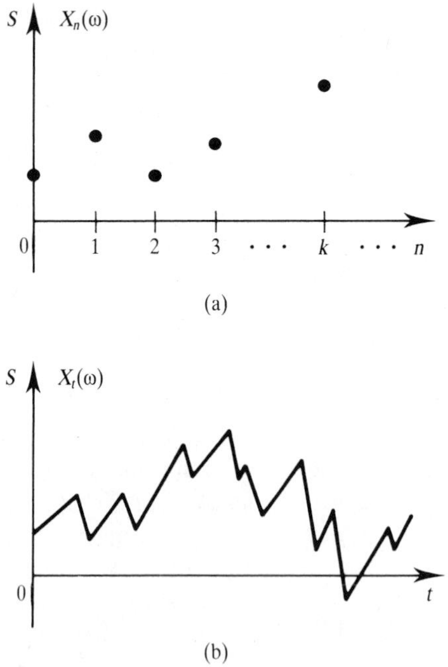

Figure 1.1

frequently calculated in terms of the probabilities of finite-dimensional events by passage to a limit.

The ideas contained in this section will be illustrated in the example and in exercises.

Example 1. The sample space Ω for repeated (and unending) tosses of a coin may be represented by the sequence space consisting of sequences of the form $\omega = (\omega_1, \omega_2, \ldots, \omega_n, \ldots)$ with $\omega_n = 1$ or $\omega_n = 0$. For this choice of Ω, the value of X_n corresponding to the occurrence of the sample point $\omega \in \Omega$ is simply the *nth coordinate projection* of ω; i.e., $X_n(\omega) = \omega_n$. Suppose that the probability of the occurrence of a head in a single toss is p. Since for any number n of tosses the results of the first $n - 1$ tosses have no effect on the odds of the nth toss, the random variables $X_1, \ldots, X_n$ are, for each $n \geqslant 1$, independent. Moreover, each variable has the same (*Bernoulli*) distribution. These facts are summarized by saying that $\{X_1, X_2, \ldots\}$ is a sequence of *independent and identically distributed* (*i.i.d.*) random variables with a common Bernoulli distribution. Let F_n denote the event that the specific outcomes $\varepsilon_1, \ldots, \varepsilon_n$ occur on the first n tosses respectively. Then

$$F_n = \{X_1 = \varepsilon_1, \ldots, X_n = \varepsilon_n\} = \{\omega \in \Omega\colon \omega_1 = \varepsilon_1, \ldots, \omega_n = \varepsilon_n\}$$

is a finite-dimensional event. By independence,

$$P(F_n) = p^{r_n}(1 - p)^{n - r_n} \tag{1.1}$$

where r_n is the number of 1's among $\varepsilon_1, \ldots, \varepsilon_n$. Now consider the singleton event G corresponding to the occurrence of a specific sequence of outcomes $\varepsilon_1, \varepsilon_2, \ldots, \varepsilon_n, \ldots$. Then

$$G = \{X_1 = \varepsilon_1, \ldots, X_n = \varepsilon_n, \ldots\} = \{(\varepsilon_1, \varepsilon_2, \ldots, \varepsilon_n, \ldots)\}$$

consists of the single outcome $\omega = (\varepsilon_1, \varepsilon_2, \ldots, \varepsilon_n, \ldots)$ in Ω. G is an infinite-dimensional event whose probability is easily determined as follows. Since $G \subset F_n$ for each $n \geqslant 1$, it follows that

$$0 \leqslant P(G) \leqslant P(F_n) = p^{r_n}(1 - p)^{n - r_n} \qquad \text{for each } n = 1, 2, \ldots. \tag{1.2}$$

Now apply a limiting argument to see that, for $0 < p < 1$, $P(G) = 0$. Hence the probability of every singleton event in Ω is zero.

2 THE SIMPLE RANDOM WALK

Think of a particle moving randomly among the integers according to the following rules. At time $n = 0$ the particle is at the origin. At time $n = 1$ it moves either one unit forward to $+1$ or one unit backward to -1, with respective probabilities p and $q = 1 - p$. In the case $p = \frac{1}{2}$, this may be accomplished by tossing a balanced coin and making the particle move forward or backward corresponding to the occurrence of a "head" or a "tail", respectively. Similar experiments can be devised for any fractional value of p. We may think of the experiment, in any case, as that of repeatedly tossing a coin that falls "head" with probability p and shows "tail" with probability $1 - p$. At time n the particle moves from its present position S_{n-1} by a unit distance forward or backward depending on the outcome of the nth toss. Suppose that X_n denotes the displacement of the particle at the nth step from its position S_{n-1} at time $n - 1$. According to these considerations the *displacement* (or *increment*) *process* $\{X_n\}$ associated with $\{S_n\}$ is an i.i.d. sequence with $P(X_n = +1) = p$, $P(X_n = -1) = q = 1 - p$ for each $n \geqslant 1$. The *position process* $\{S_n\}$ is then given by

$$S_n := X_1 + \cdots + X_n, \qquad S_0 = 0. \tag{2.1}$$

Definition 2.1. The stochastic process $\{S_n: n = 0, 1, 2, \ldots\}$ is called the *simple random walk*. The related process $S_n^x = S_n + x, n = 0, 1, 2, \ldots$ is called the *simple random walk starting at* x.

The simple random walk is often used by physicists as an approximate model of the fluctuations in the position of a relatively large solute molecule immersed in a pure fluid. According to Einstein's diffusion theory, the solute molecule gets kicked around by the smaller molecules of the fluid whenever it gets within the range of molecular interaction with fluid molecules. Displacements in any one direction (say, the vertical direction) due to successive collisions are small and taken to be independent. We shall return to this physical model in Section 7.

One may also think of X_n as a gambler's gain in the nth game of a series of independent and stochastically identical games: a negative gain means a loss. Then $S_0^x = x$ is the gambler's initial capital, and S_n^x is the capital, positive or negative, at time n.

The first problem is to calculate the distribution of S_n^x. To calculate the probability of $\{S_n^x = y\}$, count the number u of $+1$'s in a path from x to y in n steps. Since $n - u$ is then the number of -1's, one must have $u - (n - u) = y - x$, or $u = (n + y - x)/2$. For this, n and $y - x$ must be both even or both odd, and $|y - x| \leqslant n$. Hence

$$P(S_n^x = y) = \begin{cases} \dbinom{n}{\frac{n+y-x}{2}} p^{(n+y-x)/2} q^{(n-y+x)/2} & \text{if } |y - x| \leqslant n, \\ \text{and } y - x, n & \text{have the same parity,} \\ 0 & \text{otherwise.} \end{cases} \tag{2.2}$$

3 TRANSIENCE AND RECURRENCE PROPERTIES OF THE SIMPLE RANDOM WALK

Let us first consider the manner in which a particle escapes from an interval. Let T_y^x denote the first time that the process starting at x reaches y, i.e.

$$T_y^x := \min\{n \geqslant 0 : S_n^x = y\}. \tag{3.1}$$

To avoid trivialities, assume $0 < p < 1$. For integers c and d with $c < d$, denote

$$\phi(x) := P(T_d^x < T_c^x). \tag{3.2}$$

In other words, $\phi(x)$ is the probability that the particle starting at x reaches d before it reaches c. Since in one step the particle moves to $x + 1$ with probability p, or to $x - 1$ with probability q, one has

$$\phi(x) = p\phi(x+1) + q\phi(x-1) \tag{3.3}$$

so that

$$\begin{aligned} \phi(x+1) - \phi(x) &= \frac{q}{p}[\phi(x) - \phi(x-1)], \qquad c + 1 \leqslant x \leqslant d - 1 \\ \phi(c) &= 0, \\ \phi(d) &= 1. \end{aligned} \tag{3.4}$$

Thus, $\phi(x)$ is the solution to the *discrete boundary-value problem* (3.4). For $p \neq q$, Eq. 3.4 yields

$$\begin{aligned} \phi(x) &= \sum_{y=c}^{x-1} [\phi(y+1) - \phi(y)] = \sum_{y=c}^{x-1} \left(\frac{q}{p}\right)^{y-c} [\phi(c+1) - \phi(c)] \\ &= \phi(c+1) \sum_{y=c}^{x-1} \left(\frac{q}{p}\right)^{y-c} = \phi(c+1) \frac{1 - (q/p)^{x-c}}{1 - q/p}. \end{aligned} \tag{3.5}$$

To determine $\phi(c + 1)$ take $x = d$ in Eq. 3.5 to get

$$1 = \phi(d) = \phi(c+1) \frac{1 - (q/p)^{d-c}}{1 - q/p}.$$

Then

$$\phi(c+1) = \frac{1 - q/p}{1 - (q/p)^{d-c}}$$

so that

$$P(T_d^x < T_c^x) = \frac{1 - (q/p)^{x-c}}{1 - (q/p)^{d-c}} \qquad \text{for } c \leqslant x \leqslant d,\ p \neq q. \tag{3.6}$$

Now let

$$\psi(x) := P(T_c^x < T_d^x). \tag{3.7}$$

By symmetry (or the same method as above),

$$P(T_c^x < T_d^x) = \frac{1 - (p/q)^{d-x}}{1 - (p/q)^{d-c}} \qquad \text{for } c \leqslant x \leqslant d,\ p \neq q. \tag{3.8}$$

Note that $\phi(x) + \psi(x) = 1$, proving that the particle starting in the interior of $[c, d]$ will eventually reach the boundary (i.e., either c or d) with probability 1. Now if $c < x$, then (Exercise 3)

$$P(\{S_n^x\} \text{ will ever reach } c) = P(T_c^x < \infty) = \lim_{d \to \infty} \psi(x)$$

$$= \begin{cases} \lim\limits_{d \to \infty} \dfrac{-\left(\dfrac{p}{q}\right)^{-x}}{-\left(\dfrac{p}{q}\right)^{-c}}, & \text{if } p > \tfrac{1}{2} \\ 1, & \text{if } p < \tfrac{1}{2}, \end{cases}$$

$$= \begin{cases} \left(\dfrac{q}{p}\right)^{x-c}, & \text{if } p > \tfrac{1}{2} \\ 1, & \text{if } p < \tfrac{1}{2}. \end{cases} \tag{3.9}$$

By symmetry, or as above,

$$P(\{S_n^x\} \text{ will ever reach } d) = P(T_d^x < \infty) = \begin{cases} 1, & \text{if } p > \tfrac{1}{2} \\ \left(\dfrac{p}{q}\right)^{d-x}, & \text{if } p < \tfrac{1}{2}. \end{cases} \tag{3.10}$$

Observe that one gets from these calculations the (geometric) distribution function for the *extremes* $M^x = \sup_n S_n^x$ and $m^x = \inf_n S_n^x$ (Exercise 7).

Note that, by the strong law of large numbers (Chapter 0),

$$P\left(\frac{S_n^x}{n} = \frac{x + S_n}{n} \to p - q \text{ as } n \to \infty\right) = 1. \tag{3.11}$$

Hence, if $p > q$, then the random walk drifts to $+\infty$ (i.e., $S_n^x \to +\infty$) with probability 1. In particular, the process is certain to reach $d > x$ if $p > q$. Similarly, if $p < q$, then the random walk drifts to $-\infty$ (i.e., $S_n^x \to -\infty$), and starting at $x > c$ the process is certain to reach c if $p < q$. In either case, no matter what the integer y is,

$$P(S_n^x = y \quad \text{i.o.}) = 0, \qquad \text{if } p \neq q, \tag{3.12}$$

where i.o. is shorthand for "infinitely often." For if $S_n^x = y$ for integers $n_1 < n_2 < \cdots$ through a sequence going to infinity, then

$$\frac{S_{n_k}^x}{n_k} = \frac{y}{n_k} \to 0 \qquad \text{as } n_k \to \infty,$$

the probability of which is zero by Eq. 3.11.

Definition 3.1. A state y for which Eq. 3.12 holds is called *transient*. If all states are transient then the stochastic process is said to be a *transient process*.

In the case $p = q = \frac{1}{2}$, according to the boundary-value problem (3.4), the graph of $\phi(x)$ is along the line of constant slope between the points $(c, 0)$ and $(d, 1)$. Thus,

$$P(T_d^x < T_c^x) = \frac{x - c}{d - c}, \qquad c \leqslant x \leqslant d,\ p = q = \tfrac{1}{2}. \tag{3.13}$$

Similarly,

$$P(T_c^x < T_d^x) = \frac{d - x}{d - c}, \qquad c \leqslant x \leqslant d,\ p = q = \tfrac{1}{2}. \tag{3.14}$$

Again we have

$$\phi(x) + \psi(x) = 1. \tag{3.15}$$

Moreover, in this case, given any initial position $x > c$,

$$\begin{aligned} P(\{S_n^x\} \text{ will eventually reach } c) &= P(T_c^x < \infty) \\ &= \lim_{d \to \infty} P(\{S_n^x\} \text{ will reach } c \text{ before it reaches } d) \\ &= \lim_{d \to \infty} \frac{d - x}{d - c} = 1. \end{aligned} \tag{3.16}$$

Similarly, whatever the initial position $x < d$,

$$P(\{S_n^x\} \text{ will eventually reach } d) = P(T_d^x < \infty)$$

$$= \lim_{c \to -\infty} \frac{x - c}{d - c} = 1. \tag{3.17}$$

Thus, no matter where the particle may be initially, it will eventually reach any given state y with probability 1. After having reached y for the first time, it will move to $y + 1$ or to $y - 1$. From either of these positions the particle is again bound to reach y with probability 1, and so on. In other words (Exercise 4),

$$P(S_n^x = y \quad \text{i.o.}) = 1, \qquad \text{if } p = q = \tfrac{1}{2}. \tag{3.18}$$

This argument is discussed again in Example 4.1 of Chapter II.

Definition 3.2. A state y for which Eq. 3.18 holds is called *recurrent*. If all states are recurrent, then the stochastic process is called a *recurrent process*.

Let η_x denote the *time of the first return to x*,

$$\eta_x := \inf\{n \geqslant 1 : S_n^x = x\}. \tag{3.19}$$

Then, conditioning on the first step, it will follow (Exercise 6) that

$$P(\eta_x < \infty) = 2 \min(p, q). \tag{3.20}$$

4 FIRST PASSAGE TIMES FOR THE SIMPLE RANDOM WALK

Consider the random variable $T_y := T_y^0$ representing the first time the simple random walk starting at zero reaches the level (state) y. We will calculate the distribution of T_y by means of an analysis of the sample paths of the simple random walk. Let $F_{N,y} = \{T_y = N\}$ denote the event that the particle reaches state y for the first time at the Nth step. Then,

$$F_{N,y} = \{S_n \neq y \quad \text{for} \quad n = 0, 1, \ldots, N - 1, S_N = y\}. \tag{4.1}$$

Note that "$S_N = y$" means that there are $(N + y)/2$ plus 1's and $(N - y)/2$ minus 1's among $X_1, X_2, \ldots, X_N$ (see Eq. 2.1). Therefore, we assume that $|y| \leqslant N$ and $N + y$ is even. Now there are as many paths leading from $(0, 0)$ to (N, y) as there are ways of choosing $(N + y)/2$ plus 1's among $X_1, X_2, \ldots, X_N$, namely

$$\binom{N}{\frac{N+y}{2}}.$$

Each of these choices has the same probability of occurrence, specifically $p^{(N+y)/2}q^{(N-y)/2}$. Thus,

$$P(F_{N,y}) = Lp^{(N+y)/2}q^{(N-y)/2} \tag{4.2}$$

where L is the number of paths from $(0, 0)$ to (N, y) that do not touch or cross the level y prior to time N. To calculate L, consider the complementary number L' of paths that do reach y prior to time N,

$$L' = \binom{N}{\frac{N+y}{2}} - L. \tag{4.3}$$

First consider the case of $y > 0$. If a path from $(0, 0)$ to (N, y) has reached y prior to time N, then either (a) $S_{N-1} = y + 1$ (see Figure 4.1a) or (b) $S_{N-1} = y - 1$ and the path from $(0, 0)$ to $(N - 1, y - 1)$ has reached y prior to time $N - 1$ (see Figure 4.1b). The contribution to L' from (a) is

$$\binom{N-1}{\frac{N+y}{2}}.$$

We need to calculate the contribution to L' from (b).

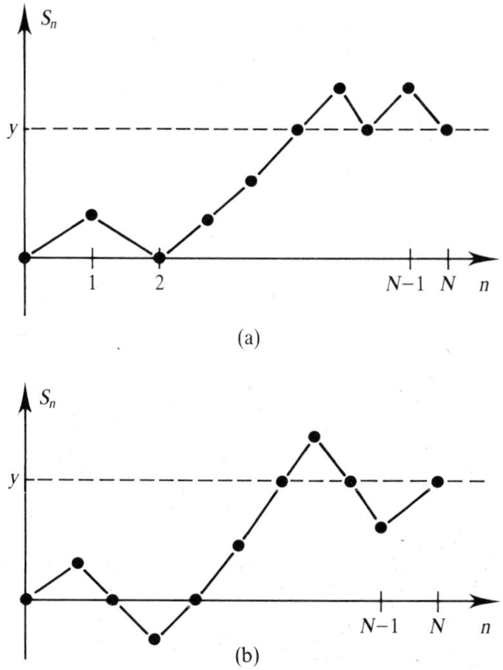

Figure 4.1

Proposition 4.1. (*A Reflection Principle*). Let $y > 0$. The collection of all paths from $(0, 0)$ to $(N - 1, y - 1)$ that touch or cross the level y prior to time $N - 1$ is in one-to-one correspondence with the collection of all possible paths from $(0, 0)$ to $(N - 1, y + 1)$.

Proof. Given a path γ from $(0, 0)$ to $(N - 1, y + 1)$, there is a first time τ at which the path reaches level y. Let γ' denote the path which agrees with γ up to time τ but is thereafter the mirror reflection of γ about the level y (see Figure 4.2). Then γ' is a path from $(0, 0)$ to $(N - 1, y - 1)$ that touches or crosses the level y prior to time $N - 1$. Conversely, a path from $(0, 0)$ to $(N - 1, y - 1)$ that touches or crosses the level y prior to time $N - 1$ may be reflected to get a path from $(0, 0)$ to $(N - 1, y + 1)$. This reflection transformation establishes the one-to-one correspondence. ■

It now follows from the reflection principle that the contribution to L' from (b) is

$$\binom{N-1}{\frac{N+y}{2}}.$$

Hence

$$L' = 2\binom{N-1}{\frac{N+y}{2}}. \tag{4.4}$$

Therefore, by (4.3), (4.2),

$$P(T_y = N) = P(F_{N,y}) = \left[\binom{N}{\frac{N+y}{2}} - 2\binom{N-1}{\frac{N+y}{2}}\right] p^{(N+y)/2} q^{(N-y)/2}$$

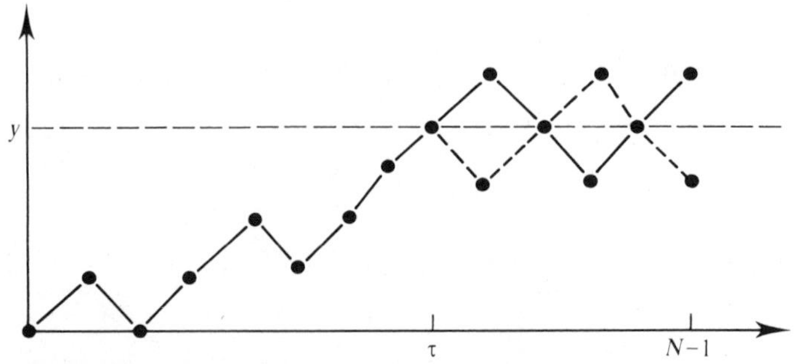

Figure 4.2

$$= \frac{|y|}{N} \binom{N}{\frac{N+y}{2}} p^{(N+y)/2} q^{(N-y)/2} \qquad \text{for } N \geqslant y,\ y + N \text{ even},\ y > 0 \tag{4.5}$$

To calculate $P(T_y = N)$ for $y < 0$, simply relabel H as T and T as H (i.e., interchange $+1$, -1). Using this new code, the desired probability is given by replacing y by $-y$ and interchanging p, q in (4.5), i.e.,

$$P(T_y = N) = -\frac{y}{N} \binom{N}{\frac{N+y}{2}} q^{(N-y)/2} p^{(N+y)/2}.$$

Thus, for all integers $y \neq 0$, one has

$$P(T_y = N) = \frac{|y|}{N} \binom{N}{\frac{N+y}{2}} p^{(N+y)/2} q^{(N-y)/2} = \frac{|y|}{N} P(S_N = y) \tag{4.6}$$

for $N = |y|, |y| + 2, |y| + 4, \ldots$. In particular, if $p = q = \frac{1}{2}$, then (4.6) yields

$$P(T_y = N) = \frac{|y|}{N} \binom{N}{\frac{N+y}{2}} \frac{1}{2^N} \qquad \text{for } N = |y|, |y| + 2, |y| + 4, \ldots. \tag{4.7}$$

However, observe that the *expected time to reach y* is infinite since by *Stirling's formula*, $k! = (2\pi k)^{1/2} k^k e^{-k}(1 + o(1))$ as $k \to \infty$, the *tail* of the *p.m.f.* of T_y is of the order of $N^{-3/2}$ as $N \to \infty$ (Exercise 10).

5 MULTIDIMENSIONAL RANDOM WALKS

The *k-dimensional unrestricted simple symmetric random walk* describes the motion of a particle moving randomly on the integer lattice $\mathbb{Z}^k$ according to the following rules. Starting at a site $\mathbf{x} = (x_1, \ldots, x_k)$ with integer coordinates, the particle moves to a neighboring site in one of the $2k$ coordinate directions randomly selected with probability $1/2k$, and so on, independently of previous displacements. The *displacement* at the nth step is a random variable $\mathbf{X}_n$ whose possible values are vectors of the form $\pm\mathbf{e}_i$, $i = 1, \ldots, k$, where the jth component of $\mathbf{e}_i$ is 1 for $j = i$ and 0 otherwise. $\mathbf{X}_1, \mathbf{X}_2, \ldots$ are i.i.d. with

$$P(\mathbf{X}_n = \mathbf{e}_i) = P(\mathbf{X}_n = -\mathbf{e}_i) = 1/2k \qquad \text{for } i = 1, \ldots, k. \tag{5.1}$$

The corresponding *position process* is defined by

$$\mathbf{S}_0^{\mathbf{x}} = \mathbf{x}, \qquad \mathbf{S}_n^{\mathbf{x}} = \mathbf{x} + \mathbf{X}_1 + \cdots + \mathbf{X}_n, \qquad n \geqslant 1. \tag{5.2}$$

The case $k = 1$ is that already treated in the preceding sections with $p = q = \frac{1}{2}$. In particular, for $k = 1$ we know that the simple symmetric random walk is *recurrent*.

Consider the coordinates of $\mathbf{X}_n = (X_n^1, \ldots, X_n^k)$. Although X_n^i and X_n^j are *not independent*, notice that they are *uncorrelated* for $i \neq j$. Likewise, it follows that the coordinates of the position vector $\mathbf{S}_n^{\mathbf{x}} = (S_n^{x,1}, \ldots, S_n^{x,k})$ are uncorrelated. In particular,

$$E\mathbf{S}_n^{\mathbf{x}} = \mathbf{x},$$

$$\operatorname{Cov}(S_n^{x,i}, S_n^{x,j}) = \begin{cases} n, & \text{if } i = j \\ 0, & \text{if } i \neq j. \end{cases} \tag{5.3}$$

Therefore the covariance matrix of $\mathbf{S}_n^{\mathbf{x}}$ is $n\mathbf{I}$ where $\mathbf{I}$ is the $k \times k$ identity matrix.

The problem of describing the recurrence properties of the simple symmetric random walk in k dimensions is solved by the following theorem of Pólya.

Theorem 5.1. [*Pólya*]. $\{\mathbf{S}_n^{\mathbf{x}}\}$ is recurrent for $k = 1, 2$ and transient for $k \geqslant 3$.

Proof. The result has already been obtained for $k = 1$. In general, let $\mathbf{S}_n = \mathbf{S}_n^{\mathbf{0}}$ and write

$$r_n = P(\mathbf{S}_n = \mathbf{0})$$

$$f_n = P(\mathbf{S}_n = \mathbf{0} \text{ for the first time after time 0 at } n), \qquad n \geqslant 1. \tag{5.4}$$

Then we get the convolution equation

$$r_n = \sum_{j=0}^{n} f_j r_{n-j} \qquad \text{for } n = 1, 2, \ldots,$$

$$r_0 = 1, \qquad f_0 = 0. \tag{5.5}$$

Let $\hat{r}(s)$ and $\hat{f}(s)$ denote the respective probability generating functions of $\{r_n\}$ and $\{f_n\}$ defined by

$$\hat{r}(s) = \sum_{n=0}^{\infty} r_n s^n, \qquad \hat{f}(s) = \sum_{n=0}^{\infty} f_n s^n \qquad (0 < s < 1). \tag{5.6}$$

The convolution equation (5.5) transforms as

$$\hat{r}(s) = 1 + \sum_{n=1}^{\infty} \sum_{j=0}^{n} f_j r_{n-j} s^j s^{n-j} = 1 + \sum_{j=0}^{\infty} \left(\sum_{m=0}^{\infty} r_m s^m \right) f_j s^j = 1 + \hat{f}(s)\hat{r}(s). \tag{5.7}$$

Therefore,

$$\hat{r}(s) = \frac{1}{1 - \hat{f}(s)}. \tag{5.8}$$

The probability of eventual return to the origin is given by

$$\gamma := \sum_{n=1}^{\infty} f_n = \hat{f}(1). \tag{5.9}$$

Note that by the Monotone Convergence Theorem (Chapter 0), $\hat{r}(s) \nearrow \hat{r}(1)$ and $\hat{f}(s) \nearrow \hat{f}(1)$ as $s \nearrow 1$. If $\hat{f}(1) < 1$, then $\hat{r}(1) = (1 - \hat{f}(1))^{-1} < \infty$. If $\hat{f}(1) = 1$, then $\hat{r}(1) = \lim_{s \nearrow 1}(1 - \hat{f}(s))^{-1} = \infty$. Therefore, $\gamma < 1$ (i.e., $\mathbf{0}$ *is transient*) if and only if $\beta := \hat{r}(1) < \infty$.

This criterion is applied to the case $k = 2$ as follows. Since a return to $\mathbf{0}$ is possible at time $2n$ if and only if the numbers of steps among the $2n$ in the positive horizontal and vertical directions equal the respective numbers of steps in the negative directions,

$$\begin{aligned} r_{2n} &= 4^{-2n} \sum_{j=0}^{n} \frac{(2n)!}{j!\,j!(n-j)!(n-j)!} = \frac{1}{4^{2n}} \binom{2n}{n} \sum_{j=0}^{n} \binom{n}{j}^2 \\ &= \frac{1}{4^{2n}} \binom{2n}{n} \sum_{j=0}^{n} \binom{n}{j}\binom{n}{n-j} = \frac{1}{4^{2n}} \binom{2n}{n}^2. \end{aligned} \tag{5.10}$$

The combinatorial identity used to get the last line of (5.10) follows by considering the number of ways of selecting samples of size n from a population of n objects of type 1 and n objects of type 2 (Exercise 2). Apply Stirling's formula to (5.10) to get $r_{2n} = O(1/n) > c/n$ for some $c > 0$. Therefore, $\beta = \hat{r}(1) = +\infty$ and so $\mathbf{0}$ is recurrent in the case $k = 2$.

In the case $k = 3$, similar considerations of "coordinate balance" give

$$\begin{aligned} r_{2n} &= 6^{-2n} \sum_{(j,m):j+m \leq n} \frac{(2n)!}{j!\,j!\,m!\,m!(n-j-m)!(n-j-m)!} \\ &= \frac{1}{2^{2n}} \binom{2n}{n} \sum_{j+m \leq n} \left\{ \frac{1}{3^n} \frac{n!}{j!\,m!(n-j-m)!} \right\}^2. \end{aligned} \tag{5.11}$$

Therefore, writing

$$p_{j,m} = \frac{n!}{j!\,m!(n-j-m)!} \frac{1}{3^n}$$

and noting that these are the probabilities for the *trinomial distribution*, we have that

$$r_{2n} = \frac{1}{2^{2n}}\binom{2n}{n} \sum_{j+m \leqslant n} (p_{j,m})^2 \tag{5.12}$$

is nearly an average of $p_{j,m}$'s (with respect to the $p_{j,m}$ distribution). In any case,

$$r_{2n} \leqslant \frac{1}{2^{2n}}\binom{2n}{n} \sum_{j+m \leqslant n} \left[\max_{j,m} p_{j,m}\right] p_{j,m} = \frac{1}{2^{2n}}\binom{2n}{n} \max_{j,m} p_{j,m}. \tag{5.13}$$

The maximum value of $p_{j,m}$ is attained at j and m nearest to $n/3$ (Exercise 5). Therefore, writing $[x]$ for the *integer part* of x,

$$r_{2n} \leqslant \frac{1}{2^{2n}}\binom{2n}{n} \frac{1}{3^n} \frac{n!}{\left[\frac{n}{3}\right]!\left[\frac{n}{3}\right]!\left[\frac{n}{3}\right]!}. \tag{5.14}$$

Apply Stirling's formula to get (see 5.19 below),

$$r_{2n} \leqslant \frac{C}{2^{2n}}\binom{2n}{n}\frac{1}{n} \leqslant \frac{C'}{n^{3/2}}, \qquad \text{for some } C' > 0. \tag{5.15}$$

In particular,

$$\sum_n r_n < \infty. \tag{5.16}$$

The general case, $r_{2n} \leqslant c_k n^{-k/2}$ for $k > 3$, is left as an exercise (Exercise 1). ■

The constants appearing in the estimate (5.15) are easily computed from the monotonicity of the ratio $n!/\{(2\pi n)^{1/2} n^n e^{-n}\}$; whose limit as $n \to \infty$ is 1 according to Stirling's formula. To see that the ratio is monotonically decreasing, simply observe that

$$\begin{aligned}
\log \frac{n!}{(2\pi n)^{1/2} n^n e^{-n}} &= \log n! - \tfrac{1}{2}\log n - n \log n + n - \log(2\pi)^{1/2} \\
&= \left\{\sum_{j=1}^{n} \log j - \tfrac{1}{2}\log n\right\} - \{n \log n - n\} - \log(2\pi)^{1/2} \\
&= \left\{\sum_{j=2}^{n} \frac{\log(j-1) + \log(j)}{2} - \int_1^n \log x \, dx\right\} + 1 - \log(2\pi)^{1/2}
\end{aligned} \tag{5.17}$$

where the integral term may be checked by integration by parts. The point is that the term defined by

$$T_n = \sum_{j=2}^{n} \frac{\log(j-1) + \log(j)}{2} \tag{5.18}$$

provides the inner trapezoidal approximation to the area under the curve $y = \log x$, $1 \leqslant x \leqslant n$. Thus, in particular, a simple sketch shows

$$0 \leqslant \int_1^n \log x \, dx - T_n$$

is monotonically increasing. So, in addition to the asymptotic value of the ratio, one also has

$$1 \leqslant \frac{n!}{(2\pi n)^{1/2} n^n e^{-n}} < \frac{e}{(2\pi)^{1/2}}, \qquad n = 1, 2, \ldots. \tag{5.19}$$

6 CANONICAL CONSTRUCTION OF STOCHASTIC PROCESSES

Often a stochastic process is defined on a *given* probability space as a sequence of functions of other *already constructed* random variables. For example, the simple random walk $\{S_n = X_1 + \cdots + X_n\}$, $S_0 = 0$ is defined in terms of the coin-tossing process $\{X_n\}$ in Section 2. At other times, a probability space is constructed specifically to define the stochastic process. For example, the probability space for the coin-tossing process was constructed starting from the specifications of the probabilities of finite sequences of heads and tails. This latter method, called the *canonical construction*, is elaborated upon in this section.

Consider the case that the state space is $\mathbb{R}^1$ (or a subset of it) and the parameter is discrete ($n = 1, 2, \ldots$). Take Ω to be the space of all sample paths; i.e., $\Omega := (\mathbb{R}^1)^\infty := \mathbb{R}^\infty$ is the space of all sequences $\omega = (\omega_1, \omega_2, \ldots)$ of real numbers. The appropriate sigmafield $\mathscr{F} := \mathscr{B}^\infty$ is then the smallest sigmafield containing all finite-dimensional sets of the form $\{\omega \in \Omega : \omega_1 \in B_1, \ldots, \omega_k \in B_k\}$, where $B_1, \ldots, B_k$ are Borel subsets of $\mathbb{R}^1$. The coordinate functions X_n are defined by $X_n(\omega) = \omega_n$.

As in the case of coin tossing, the underlying physical process sometimes suggests a specification of probabilities of finite-dimensional events defined by the values of the process at time points $1, 2, \ldots, n$ for each $n \geqslant 1$. That is, for each $n \geqslant 1$ a probability measure P_n is prescribed on $(\mathbb{R}^n, \mathscr{B}^n)$. The problem is that we require a probability measure P on $(\Omega, \mathscr{F})$ such that P_n is the distribution of $X_1, \ldots, X_n$. That is, for all Borel sets $B_1, \ldots, B_n$,

$$P(\omega \in \Omega : \omega_1 \in B_1, \ldots, \omega_n \in B_n) = P_n(B_1 \times \cdots \times B_n). \tag{6.1}$$

Equivalently,

$$P(X_1 \in B_1, \ldots, X_n \in B_n) = P_n(B_1 \times \cdots \times B_n). \tag{6.2}$$

Since the events $\{X_1 \in B_1, \ldots, X_n \in B_n, X_{n+1} \in \mathbb{R}^1\}$ and $\{X_1 \in B_1, \ldots, X_n \in B_n\}$ are identical subsets of $\mathscr{B}^\infty$, for there to be a well-defined probability measure P prescribed by (6.1) or (6.2) it is *necessary* that

$$P_{n+1}(B_1 \times \cdots \times B_n \times \mathbb{R}^1) = P_n(B_1 \times \cdots \times B_n) \tag{6.3}$$

for all Borel sets $B_1, \ldots, B_n$ in $\mathbb{R}^1$ and $n \geqslant 1$. *Kolmogorov's Existence Theorem* asserts that the *consistency* condition (6.3) *is* also *sufficient* for such a probability measure P to exist and that there is only one such P on $(\mathbb{R}^\infty, \mathscr{B}^\infty) = (\Omega, \mathscr{F})$ (theoretical complement 1). This holds more generally, for example, when the state space S is $\mathbb{R}^k$, a countable set, or any Borel subset of $\mathbb{R}^k$. A proof for the simple case of finite state processes is outlined in Exercise 3.

Example 1. Consider the problem of canonically constructing a sequence $X_1, X_2, \ldots$ of i.i.d. random variables having the common (marginal) distribution Q on $(\mathbb{R}^1, \mathscr{B}^1)$. Take $\Omega = \mathbb{R}^\infty$, $\mathscr{F} = \mathscr{B}^\infty$, and X_n the nth coordinate projection $X_n(\omega) = \omega_n$, $\omega \in \Omega$. Define, for each $n \geqslant 1$ and all Borel sets $B_1, \ldots, B_n$,

$$P_n(B_1 \times \cdots \times B_n) = Q(B_1) \cdots Q(B_n). \tag{6.4}$$

Since $Q(\mathbb{R}^1) = 1$, the consistency condition (6.3) follows immediately from the definition (6.4). Now one simply invokes the Kolmogorov Existence Theorem to get a probability measure P on $(\Omega, \mathscr{F})$ such that

$$\begin{aligned} P(X_1 \in B_1, \ldots, X_n \in B_n) &= Q(B_1) \cdots Q(B_n) \\ &= P(X_1 \in B_1) \cdots P(X_n \in B_n). \end{aligned} \tag{6.5}$$

The simple random walk can be constructed within the framework of the canonical probability space $(\Omega, \mathscr{F}, P)$ constructed for coin tossing, although this is a noncanonical probability space for $\{S_n\}$. Alternatively, a canonical construction can be made directly for $\{S_n\}$ (Exercise 2(i)). This, on the other hand, provides a noncanonical probability space for the displacement (coin-tossing) process defined by the differences $X_n = S_n - S_{n-1}$, $n \geqslant 1$.

Example 2. The problem is to construct a *Gaussian stochastic process* having prescribed means and covariances. Suppose that we are given a sequence of real numbers $\mu_1, \mu_2, \ldots$, and an array, σ_{ij}, $i, j = 1, 2, \ldots$, of real numbers satisfying

(*Symmetry*)

$$\sigma_{ij} = \sigma_{ji} \qquad \text{for all } i, j, \tag{6.6}$$

(*Non-negative Definiteness*)

$$\sum_{i,j=1}^{n} \sigma_{ij} x_i x_j \geqslant 0 \qquad \text{for all } n\text{-tuples } (x_1, \ldots, x_n) \text{ in } \mathbb{R}^n. \tag{6.7}$$

Property (6.7) is the condition that $\mathbf{D}_n = ((\sigma_{ij}))_{1 \leqslant i,j \leqslant n}$ be a nonnegative definite matrix for each n. Again take $\Omega = \mathbb{R}^\infty$, $\mathscr{F} = \mathscr{B}^\infty$, and $X_1, X_2, \ldots$ the respective coordinate projections. For each $n \geqslant 1$, let P_n be the n-dimensional Gaussian distribution on $(\mathbb{R}^n, \mathscr{B}^n)$ having mean vector $(\mu_1, \ldots, \mu_n)$ and covariance matrix $\mathbf{D}_n$. Since a linear transformation of a Gaussian random vector is also Gaussian, the consistency condition (6.3) can be checked by applying the coordinate projection mapping $(x_1, \ldots, x_{n+1}) \to (x_1, \ldots, x_n)$ from $\mathbb{R}^{n+1}$ to $\mathbb{R}^n$ (Exercise 1).

Example 3. Let S be a countable set and let $\mathbf{p} = ((p_{i,j}))$ be a matrix of nonnegative real numbers such that for each fixed i, $p_{i,j}$ is a probability distribution (sums to 1 over j in S). Let $\boldsymbol{\pi} = (\pi_i)$ be a probability distribution on S. By the Kolmogorov Existence Theorem there is a probability distribution $P_{\boldsymbol{\pi}}$ on the infinite sequence space $\Omega = S \times S \times \cdots \times S \times \cdots$ such that $P_{\boldsymbol{\pi}}(X_0 = j_0, \ldots, X_n = j_n) = \pi_{j_0} p_{j_0, j_1} \cdots p_{j_{n-1}, j_n}$, where X_n denotes the nth projection map (Exercise 2(ii)). In this case the process $\{X_n\}$, having distribution $P_{\boldsymbol{\pi}}$, is called a *Markov chain*. These processes are the subject of Chapter II.

7 BROWNIAN MOTION

Perhaps the simplest way to introduce the continuous-parameter stochastic process known as *Brownian motion* is to view it as the limiting form of an unrestricted random walk. To physically motivate the discussion, suppose a solute particle immersed in a liquid suffers, on the average, f collisions per second with the molecules of the surrounding liquid. Assume that a collision causes a small random displacement of the solute particle that is independent of its present position. Such an assumption can be justified in the case that the solute particle is much heavier than a molecule of the surrounding liquid. For simplicity, consider displacements in one particular direction, say the vertical direction, and assume that each displacement is either $+\Delta$ or $-\Delta$ with probabilities p and $q = 1 - p$, respectively. The particle then performs a one-dimensional random walk with step size Δ. Assume for the present that the vessel is very large so that the random walk initiated far away from the boundary may be considered to be unrestricted. Suppose at time zero the particle is at the position x relative to some origin. At time $t > 0$ it has suffered approximately $n = tf$ independent displacements, say $Z_1, Z_2, \ldots, Z_n$. Since f is extremely large (of the order of 10^{21}), if t is of the order of 10^{-10} second then n is very large. The position of the particle at time t, being x plus the sum of n independent Bernoulli random variables, is, by the central limit theorem, approximately Gaussian with mean $x + tf(p - q)\Delta$ and variance $tf4\Delta^2 pq$. To make the limiting argument firm, let

$$p = \frac{1}{2} + \frac{\mu}{2\sqrt{f}\,\sigma} \qquad \text{and} \qquad \Delta = \frac{\sigma}{\sqrt{f}}.$$

Here μ and σ are two fixed numbers, $\sigma > 0$. Then as $f \to \infty$, the mean displacement $tf(p - q)\Delta$ converges to $t\mu$ and the variance converges to $t\sigma^2$. In the limit, then, the position X_t of the particle at time $t > 0$ is *Gaussian* with probability density function (in y) given by

$$p(t; x, y) = \frac{1}{(2\pi\sigma^2 t)^{1/2}} \exp\left\{ -\frac{(y - x - t\mu)^2}{2\sigma^2 t} \right\}. \tag{7.1}$$

If $s > 0$ then $X_{t+s} - X_t$ is the sum of displacements during the time interval $(t, t + s]$. Therefore, by the argument above, $X_{t+s} - X_t$ is Gaussian with mean $s\mu$ and variance $s\sigma^2$, and it is independent of $\{X_u: 0 \leqslant u \leqslant t\}$. In particular, for every finite set of time points $0 < t_1 < t_2 < \cdots < t_n$ the random variables $X_{t_1}, X_{t_2} - X_{t_1}, \ldots, X_{t_m} - X_{t_{m-1}}$ are independent. A stochastic process with this last property is said to be a *process with independent increments.* This is the continuous-time analogue of random walks. From the physical description of the process $\{X_t\}$ as representing (a coordinate of) the path of a diffusing solute particle, one would expect that the sample paths of the process (i.e., the trajectories $t \to X_t(\omega) = \omega_t$) may be taken to be continuous. That this is indeed the case is an important mathematical result originally due to Norbert Wiener. For this reason, Brownian motion is also called the *Wiener process.* A complete definition of Brownian motion goes as follows.

Definition 7.1. A *Brownian motion with drift μ and diffusion coefficient σ^2* is a stochastic process $\{X_t: t \geqslant 0\}$ having continuous sample paths and independent Gaussian increments with mean and variance of an increment $X_{t+s} - X_t$ being $s\mu$ and $s\sigma^2$, respectively. If $X_0 = x$, then this Brownian motion is said to *start at x.* A Brownian motion with zero drift and diffusion coefficient of 1 is called the *standard Brownian motion.*

Families of random variables $\{X_t\}$ constituting Brownian motions arise in many different contexts on diverse probability spaces. The canonical model for Brownian motion is given as follows.

1. The sample space $\Omega := C[0, \infty)$ is the set of all real-valued continuous functions on the time interval $[0, \infty)$. This is the set of all possible trajectories (sample paths) of the process.
2. $X_t(\omega) := \omega_t$ is the value of the sample path ω at time t.
3. Ω is equipped with the smallest sigmafield $\mathscr{F}$ of subsets of Ω containing the class $\mathscr{F}_0$ of all finite-dimensional sets of the form $F = \{\omega \in \Omega: a_i < \omega_{t_i} \leqslant b_i, i = 1, 2, \ldots, k\}$, where $a_i < b_i$ are constants and $0 < t_1 < t_2 < \cdots < t_k$ are a finite set of time points. $\mathscr{F}$ is said to be *generated* by $\mathscr{F}_0$.

4. The existence and uniqueness of a probability measure P_x on $\mathscr{F}$, called the *Wiener measure starting at* x, as specified by Definition 7.1 is determined by the probability assignments of the form of (7.2) below.

For the set F above, $P_x(F)$ can be calculated as follows. Definition (7.1) gives the joint density of $X_{t_1}, X_{t_2} - X_{t_1}, \ldots, X_{t_k} - X_{t_{k-1}}$ as that of k independent Gaussian random variables with means $t_1\mu, (t_2 - t_1)\mu, \ldots, (t_k - t_{k-1})\mu$, respectively, and variances $t_1\sigma^2, (t_2 - t_1)\sigma^2, \ldots, (t_k - t_{k-1})\sigma^2$, respectively. Transforming this (product) joint density, say in variables $z_1, z_2, \ldots, z_k$, by the change of variables $z_1 = y_1, z_2 = y_2 - y_1, \ldots, z_k = y_k - y_{k-1}$ and using the fact that the Jacobian of this linear transformation is unity, one obtains

$$
\begin{aligned}
&P_x(a_i < X_{t_i} \leqslant b_i \text{ for } i = 1, 2, \ldots, k) \\
&= \int_{a_1}^{b_1} \cdots \int_{a_{k-1}}^{b_{k-1}} \int_{a_k}^{b_k} \frac{1}{(2\pi\sigma^2 t_1)^{1/2}} \exp\left\{-\frac{(y_1 - x - t_1\mu)^2}{2\sigma^2 t_1}\right\} \\
&\quad \times \frac{1}{(2\pi\sigma^2(t_2 - t_1))^{1/2}} \exp\left\{-\frac{(y_2 - y_1 - (t_2 - t_1)\mu)^2}{2\sigma^2(t_2 - t_1)}\right\} \\
&\quad \cdots \frac{1}{(2\pi\sigma^2(t_k - t_{k-1}))^{1/2}} \exp\left\{-\frac{(y_k - y_{k-1} - (t_k - t_{k-1})\mu)^2}{2\sigma^2(t_k - t_{k-1})}\right\} dy_k\, dy_{k-1} \cdots dy_1.
\end{aligned}
\tag{7.2}
$$

The joint density of $X_{t_1}, X_{t_2}, \ldots, X_{t_k}$ is the integrand in (7.2) and may be expressed, using (7.1), as

$$
p(t_1; x, y_1)p(t_2 - t_1; y_1, y_2) \cdots p(t_k - t_{k-1}; y_{k-1}, y_k). \tag{7.3}
$$

The probabilities of a number of infinite-dimensional events will be calculated in Sections 9–13, and in Chapter IV. Some further discussion of mathematical issues in this connection are presented in Section 8 also. The details of a construction of the Brownian motion and its Wiener measure distribution are given in the theoretical complements of Section 13.

If $\{X_t^{(j)}\}, j = 1, 2, \ldots, k$, are k independent standard Brownian motions, then the vector-valued process $\{\mathbf{X}_t\} = \{\{X_t^{(1)}, X_t^{(2)}, \ldots, X_t^{(k)}\}\}$ is called a *standard k-dimensional Brownian motion.* If $\{\mathbf{X}_t\}$ is a standard k-dimensional Brownian motion, $\boldsymbol{\mu} = (\mu^{(1)}, \ldots, \mu^{(k)})$ a vector in $\mathbb{R}^k$, and $\mathbf{A}$ a $k \times k$ nonsingular matrix, then the vector-valued process $\{\mathbf{Y}_t = \mathbf{A}\mathbf{X}_t + t\boldsymbol{\mu}\}$ has *independent increments*, the increment $\mathbf{Y}_{t+s} - \mathbf{Y}_t = \mathbf{A}(\mathbf{X}_{t+s} - \mathbf{X}_t) + (t + s - t)\boldsymbol{\mu}$ being *Gaussian* with mean vector $s\boldsymbol{\mu}$ and covariance (or dispersion) matrix $s\mathbf{D}$, where $\mathbf{D} = \mathbf{A}\mathbf{A}'$ and $\mathbf{A}'$ denotes the transpose of $\mathbf{A}$. Such a process $\mathbf{Y}$ is called a *k-dimensional Brownian motion* with *drift vector* $\boldsymbol{\mu}$ and *diffusion matrix*, or *dispersion matrix* $\mathbf{D}$.

8 THE FUNCTIONAL CENTRAL LIMIT THEOREM (FCLT)

The argument in Section 7 indicating Brownian motion (with zero drift parameter for simplicity) as the limit of a random walk can be made on the basis of the classical central limit theorem which applies to every i.i.d. sequence of increments $\{Z_m\}$ having finite mean and variance. While we can only obtain convergence of the finite-dimensional distributions by such considerations, much more is true. Namely, probabilities of certain infinite-dimensional events will also converge. The convergence of the full distributions of random walks to the full distribution of the Brownian motion process is informally explained in this section. A more detailed and precise discussion is given in the theoretical complements of Sections 8 and 13.

To state this limit theorem somewhat more precisely, consider a sequence of i.i.d. random variables $\{Z_m\}$ and assume for the present that $EZ_m = 0$ and $\operatorname{Var} Z_m = \sigma^2 > 0$. Define the random walk

$$S_0 = 0, \qquad S_m = Z_1 + \cdots + Z_m \qquad (m = 1, 2, \ldots). \tag{8.1}$$

Define, for each value of the *scale parameter* $n \geqslant 1$, the stochastic process

$$X_t^{(n)} = \frac{S_{[nt]}}{\sqrt{n}} \qquad (t \geqslant 0), \tag{8.2}$$

where $[nt]$ is the *integer part* of nt. Figure 8.1 plots the sample path of $\{X_t^{(n)}: t \geqslant 0\}$ up to time $t = 13/n$ if the successive displacements take values $Z_1 = -1$, $Z_2 = +1$, $Z_3 = +1$, $Z_4 = +1$, $Z_5 = -1$, $Z_6 = +1$, $Z_7 = +1$, $Z_8 = -1$, $Z_9 = +1$, $Z_{10} = +1$, $Z_{11} = +1$, $Z_{12} = -1$.

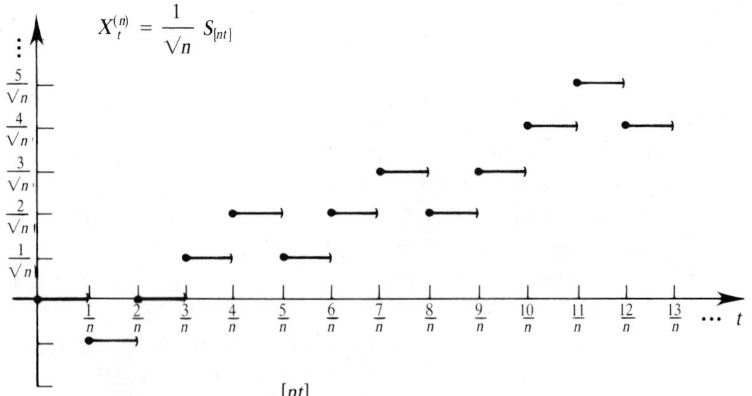

$$EX_t^{(n)} = 0, \quad \operatorname{Var} X_t^{(n)} = \frac{[nt]}{n} \sim t,$$

$$\operatorname{Cov}(X_{t+s}^{(n)}, X_t^{(n)}) = \frac{[ns]}{n} \sim s.$$

Figure 8.1

The process $\{S_{[nt]}: t \geqslant 0\}$ records the discrete-time random walk $\{S_m: m = 0, 1, 2, \ldots\}$ on a continuous *time scale* whose unit is n times that of the discrete time unit, i.e., S_m is plotted at time m/n. The process $\{X_t^{(n)}\} = \{(1/\sqrt{n})S_{[nt]}\}$ also *scales distance* by measuring distances on a scale whose unit is $\sqrt{n}$ times the unit of measurement used for the random walk. This is a convenient normalization, since

$$EX_t^{(n)} = 0, \qquad \text{Var } X_t^{(n)} = \frac{[nt]\sigma^2}{n} \simeq t\sigma^2 \qquad \text{for large } n. \tag{8.3}$$

In a time interval $(t_1, t_2]$ the overall "displacement" $X_{t_2}^{(n)} - X_{t_1}^{(n)}$ is the sum of a large number $[nt_2] - [nt_1] \simeq n(t_2 - t_1)$ of *small* i.i.d. random variables

$$\frac{1}{\sqrt{n}} Z_{[nt_1]+1}, \ldots, \frac{1}{\sqrt{n}} Z_{[nt_2]}.$$

In the case $\{Z_m\}$ is i.i.d. Bernoulli, this means reducing the step sizes of the random variables to $\Delta = 1/\sqrt{n}$. In a physical application, looking at $\{X_t^{(n)}\}$ means the following.

1. The random walk is observed at times $t_1 < t_2 < t_3 < \cdots$ sufficiently far apart to allow a large number of individual displacements to occur during each of the time intervals $(t_1, t_2], (t_2, t_3], \ldots$, and
2. Measurements of distance are made on a *"macroscopic"* scale whose unit of measurement is much larger than the average magnitude of the individual displacements. The normalizing *large parameter n* scales time and $n^{1/2}$ scales space coordinates.

Since the sample paths of $\{X_t^{(n)}\}$ have jumps (though small for large n) and are, therefore, discontinuous, it is technically more convenient to *linearly interpolate* the random walk between one jump point and the next, using the same space–time scales as used for $\{X_t^{(n)}\}$. The *polygonal process* $\{\tilde{X}_t^{(n)}\}$ is formally defined by

$$\tilde{X}_t^{(n)} = \frac{S_{[nt]}}{\sqrt{n}} + (nt - [nt])\frac{Z_{[nt]+1}}{\sqrt{n}}, \qquad t \geqslant 0. \tag{8.4}$$

In this way, just as for the limiting Brownian motion process, the paths of $\{\tilde{X}_t^{(n)}\}$ are continuous. Figure 8.2 plots the path of $\{\tilde{X}_t^{(n)}\}$ corresponding to the path of $\{X_t^{(n)}\}$ drawn in Figure 8.1. In a time interval $m/n \leqslant t < (m+1)/n$, $X_t^{(n)}$ is constant at level $1/\sqrt{n}\, S_m$, while $\tilde{X}_t^{(n)}$ changes linearly from $1/\sqrt{n}\, S_m$ at time $t = m/n$ to

$$\frac{1}{\sqrt{n}} S_{m+1} = \frac{S_m}{\sqrt{n}} + \frac{Z_{m+1}}{\sqrt{n}}, \qquad \text{at time } t = \frac{m+1}{n}.$$

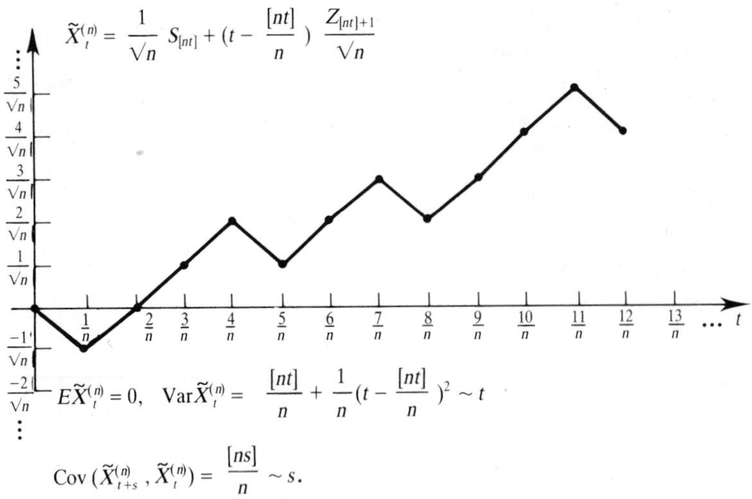

Figure 8.2

Thus, in any given interval $[0, T]$ the maximum difference between the two processes $\{X_t^{(n)}\}$ and $\{\tilde{X}_t^{(n)}\}$ does not exceed

$$\varepsilon_n(T) = \max\left(\frac{|Z_1|}{\sqrt{n}}, \frac{|Z_2|}{\sqrt{n}}, \ldots, \frac{|Z_{[nT]+1}|}{\sqrt{n}}\right).$$

To see that the difference between $\{X_t^{(n)}\}$ and $\{\tilde{X}_t^{(n)}\}$ is negligible for large n, consider the following estimate. For each $\delta > 0$,

$$\begin{aligned} P(\varepsilon_n(T) > \delta) &= 1 - P(\varepsilon_n(T) \leqslant \delta) \\ &= 1 - P\left(\frac{|Z_m|}{\sqrt{n}} \leqslant \delta \qquad \text{for all } m = 1, 2, \ldots, [nT] + 1\right) \\ &= 1 - (P(|Z_1| \leqslant \delta\sqrt{n}))^{[nT]+1} \\ &= 1 - (1 - P(|Z_1| > \delta\sqrt{n}))^{[nT]+1}. \end{aligned} \tag{8.5}$$

Assuming for simplicity that $E|Z_1|^3 < \infty$, Chebyshev's inequality yields $P(|Z_1| > \delta\sqrt{n}) \leqslant E|Z_1|^3/\delta^3 n^{3/2}$. Use this in (8.5) to get (Exercise 9)

$$\begin{aligned} P(\varepsilon_n(T) > \delta) &\leqslant 1 - \left(1 - \frac{E|Z_1|^3}{\delta^3 n^{3/2}}\right)^{[nT]+1} \\ &\simeq 1 - \exp\left\{-\frac{E|Z_1|^3 T}{\delta^3 n^{1/2}}\right\} \to 0 \end{aligned} \tag{8.6}$$

when n is large. Here $\simeq$ indicates that the difference between the two sides goes to zero. Thus, on any closed and bounded time interval the behaviors of $\{X_t^{(n)}\}$ and $\{\tilde{X}_t^{(n)}\}$ are the same in the large-n limit.

Note that given any finite set of time points $0 < t_1 < t_2 < \cdots < t_k$, the joint distribution of $(X_{t_1}^{(n)}, X_{t_2}^{(n)}, \ldots, X_{t_k}^{(n)})$ converges to the finite-dimensional distribution of $(X_{t_1}, X_{t_2}, \ldots, X_{t_k})$, where $\{X_t\}$ is a Brownian motion with zero drift and diffusion coefficient σ^2. To see this, note that $X_{t_1}^{(n)}, X_{t_2}^{(n)} - X_{t_1}^{(n)}, \ldots, X_{t_k}^{(n)} - X_{t_{k-1}}^{(n)}$ are independent random variables that by the classical central limit theorem (Chapter 0) converge in distribution to Gaussian random variables with zero means and variances $t_1\sigma^2, (t_2 - t_1)\sigma^2, \ldots, (t_k - t_{k-1})\sigma^2$. That is to say, the joint distribution of $(X_{t_1}^{(n)}, X_{t_2}^{(n)} - X_{t_1}^{(n)}, \ldots, X_{t_k}^{(n)} - X_{t_{k-1}}^{(n)})$ converges to that of $(X_{t_1}, X_{t_2} - X_{t_1}, \ldots, X_{t_k} - X_{t_{k-1}})$. By a linear transformation, one gets the desired *convergence of finite-dimensional distributions of* $\{X_t^{(n)}\}$ (and, therefore, of $\{\tilde{X}_t^{(n)}\}$) to those of the Brownian motion process $\{X_t\}$ (Exercise 1).

Roughly speaking, to establish the full convergence in distribution of $\{\tilde{X}_t^{(n)}\}$ to Brownian motion, one further looks at a finite set of time points comprising a fine subdivision of a bounded interval $[0, T]$ and shows that the fluctuations of the process $\{\tilde{X}_t^{(n)}\}$ on $[0, T]$ between successive points of this subdivision are sufficiently small in probability, a property called the *tightness* of the process. This control over fluctuations together with the convergence of $\{\tilde{X}_t^{(n)}\}$ evaluated at the time points of the subdivision ensures convergence in distribution to a *continuous process* whose finite-dimensional distributions are the same as those of Brownian motion (see theoretical complements for details). Since there is no process other than Brownian motion with continuous sample paths that has these limiting finite-dimensional distributions, it follows that the limit must be Brownian motion.

A precise statement of the *functional central limit theorem* (FCLT) is the following.

Theorem 8.1. (*The Functional Central Limit Theorem*). Suppose $\{Z_m\colon m = 1, 2, \ldots\}$ is an i.i.d. sequence with $EZ_m = 0$ and variance $\sigma^2 > 0$. Then as $n \to \infty$ the stochastic processes $\{X_t^{(n)}\colon t \geqslant 0\}$ (or $\{\tilde{X}_t^{(n)}\colon t \geqslant 0\}$) converge in distribution to a Brownian motion starting at the origin with zero drift and diffusion coefficient σ^2.

An important way in which to view the convergence asserted in the FCLT is as follows. First, the sample paths of the polygonal process $\{\tilde{X}_t^{(n)}\}$ belong to the space $\Omega = C[0, \infty)$ of all continuous real-valued function on $[0, \infty)$, as do those of the limiting Brownian motion $\{X_t\}$. This space $C[0, \infty)$ is a metric space with a natural notion of convergence of sequences $\{\omega^{(n)}\}$, say, being that "$\{\omega^{(n)}\}$ converges to ω in $C[0, \infty)$ as n tends to infinity if and only if $\{\omega^{(n)}(t)\colon a \leqslant t \leqslant b\}$ converges uniformly to $\{\omega(t)\colon a \leqslant t \leqslant b\}$ for all closed and bounded intervals $[a, b]$." Second, the distributions of the processes $\{\tilde{X}_t^{(n)}\}$ and $\{X_t\}$ are probability measures P_n and P on a certain class $\mathscr{F}$ of events of $C[0, \infty)$, called *Borel subsets*, which is generated by and therefore includes all of the finite-dimensional events. $\mathscr{F}$ includes as well various important infinite-dimensional events, e.g., the events $\{\max_{a \leqslant t \leqslant b} \tilde{X}_t^{(n)} > y\}$ and $\{\max_{a \leqslant t \leqslant b} \tilde{X}_t^{(n)} \leqslant x\}$ pertaining to extremes of the process. More generally, if f is a continuous function on $C[0, \infty)$ then the event $\{f(\{\tilde{X}_t^{(n)}\}) \leqslant x\}$ is also a Borel subset of $C[0, \infty)$ (Exercise 2). With events of this type in mind, a precise meaning of *convergence in distribution* (*or weak convergence*) of the probability measures P_n to P on this *infinite-dimensional* space $C[0, \infty)$ is that *the probability distributions of the real-valued (one dimensional) random variables* $f(\{\tilde{X}_t^{(n)}\})$ *converge (in distribution as described in Chapter 0) to the distribution of* $f(\{X_t\})$ *for each real-valued continuous function* f *defined on* $C[0, \infty)$. Since a number of important infinite-dimensional events can be expressed in terms of continuous functionals of the processes, this makes calculations of probabilities possible by taking limits; for examples of infinite dimensional events whose probabilities do *not* converge see Exercise 9.3(iv).

Because the limiting process, namely Brownian motion, is the same for all increments $\{Z_m\}$ as above, the limit Theorem 8.1 is also referred to as the *Invariance Principle*, i.e., invariance with respect to the distribution of the increment process.

There are two distinct types of applications of Theorem 8.1. In the first type it is used to calculate probabilities of infinite-dimensional events associated with Brownian motion by studying simple random walks. In the second type it (invariance) is used to calculate asymptotics of a large variety of partial-sum processes by studying simple random walks and Brownian motion. Several such examples are considered in the next two sections.

9 RECURRENCE PROBABILITIES FOR BROWNIAN MOTION

The first problem is to calculate, for a Brownian motion $\{X_t^x\}$ with drift $\mu = 0$ and diffusion coefficient σ^2, starting at x, the probability

$$P(\tau_c^x < \tau_d^x) = P(\{X_t^x\} \text{ reaches } c \text{ before } d) \qquad (c < x < d), \tag{9.1}$$

where

$$\tau_y^x := \inf\{t \geqslant 0\colon X_t^x = y\}. \tag{9.2}$$

Since $\{B_t = (X_t^x - x)/\sigma\}$ is a standard Brownian motion starting at zero,

$$P(\tau_c^x < \tau_d^x) = P\left(\{B_t\} \text{ reaches } \frac{c - x}{\sigma} \text{ before } \frac{d - x}{\sigma}\right). \tag{9.3}$$

Now consider the i.i.d. Bernoulli sequence $\{Z_m\colon m = 1, 2, \ldots\}$ with $P(Z_m = 1) = P(Z_m = -1) = \frac{1}{2}$, and the associated random walk $S_0 = 0$, $S_m = Z_1 + \cdots + Z_m$ $(m \geqslant 1)$. By the FCLT (Theorem 8.1), the polygonal process $\{\tilde{X}_t^{(n)}\}$ associated with this random walk converges in distribution to $\{B_t\}$. Hence (theoretical complement 2)

$$P(\tau_c^x < \tau_d^x) = \lim_{n\to\infty} P\left(\{\tilde{X}_t^{(n)}\} \text{ reaches } \frac{c-x}{\sigma} \text{ before } \frac{d-x}{\sigma}\right)$$

$$= \lim_{n\to\infty} P(\{S_m\} \text{ reaches } c_n \text{ before } d_n), \tag{9.4}$$

where

$$c_n = \left[\frac{c-x}{\sigma}\sqrt{n}\right],$$

and

$$d_n = \begin{cases} \dfrac{d-x}{\sigma}\sqrt{n} & \text{if } d_n = \dfrac{d-x}{\sigma}\sqrt{n} \text{ is an integer,} \\ \left[\dfrac{d-x}{\sigma}\sqrt{n}\right] + 1 & \text{if not.} \end{cases}$$

By relation (3.14) of Section 3, one has

$$P(\tau_c^x < \tau_d^x) = \lim_{n\to\infty} \frac{d_n}{d_n - c_n} = \lim_{n\to\infty} \frac{\dfrac{d-x}{\sigma}\sqrt{n}}{\dfrac{d-x}{\sigma}\sqrt{n} - \dfrac{c-x}{\sigma}\sqrt{n}}. \tag{9.5}$$

Therefore,

$$P(\tau_c^x < \tau_d^x) = \frac{d-x}{d-c} \qquad (c < x < d,\ \mu = 0). \tag{9.6}$$

Similarly, using relation (3.13) of Section 3 instead of (3.14), one gets

$$P(\tau_d^x < \tau_c^x) = \frac{x-c}{d-c} \qquad (c < x < d,\ \mu = 0). \tag{9.7}$$

Letting $d \to +\infty$ in (9.6) and $c \to -\infty$ in (9.7), one has

$$\begin{aligned} P(\tau_c^x < \infty) &= P(\{X_t^x\} \text{ ever reaches } c) = 1 \qquad (c < x,\ \mu = 0), \\ P(\tau_c^x < \infty) &= P(\{X_t^x\} \text{ ever reaches } d) = 1 \qquad (x < d,\ \mu = 0). \end{aligned} \tag{9.8}$$

The relations (9.8) mean that a *Brownian motion with zero drift is recurrent*, just as a simple symmetric random walk was shown to be in Section 3.

The next problem is to calculate the corresponding probabilities when the drift is a nonzero quantity μ. Consider, for each large n, the Bernoulli sequence

$$\{Z_{m,n}: m = 1, 2, \ldots\} \qquad \text{with} \begin{cases} P(Z_{m,n} = +1) = p_n = \dfrac{1}{2} + \dfrac{\mu}{2\sigma\sqrt{n}}, \\ \\ P(Z_{m,n} = -1) = q_n = \dfrac{1}{2} - \dfrac{\mu}{2\sigma\sqrt{n}}. \end{cases}$$

Write $S_{m,n} = Z_{1,n} + \cdots + Z_{m,n}$ for $m \geqslant 1$, $S_{0,n} = 0$. Then,

$$\begin{aligned} EX_t^{(n)} &= \frac{ES_{[nt],n}}{\sqrt{n}} = \frac{[nt]\dfrac{\mu}{\sigma\sqrt{n}}}{\sqrt{n}} \to \frac{t\mu}{\sigma}, \\ \operatorname{Var} X_t^{(n)} &= \frac{[nt] \operatorname{Var} Z_{1,n}}{n} = \frac{[nt]}{n}\left(1 - \left(\frac{\mu}{\sigma\sqrt{n}}\right)^2\right) \to t, \end{aligned} \tag{9.9}$$

and a slight modification of the FCLT, with no significant difference in proof, implies that $\{X_t^{(n)}\}$ and, therefore, $\{\tilde{X}_t^{(n)}\}$ converges in distribution to a Brownian motion with drift μ/σ and diffusion coefficient of 1 that starts at the origin. Let $\{X_t^x\}$ be a Brownian motion with drift μ and diffusion coefficient σ^2 starting at x. Then $\{W_t = (X_t^x - x)/\sigma\}$ is a Brownian motion with drift μ/σ and diffusion coefficient of 1 that starts at the origin. Hence, by using relation (3.8) of Section 3,

$$\begin{aligned} P(\tau_c^x < \tau_d^x) &= P(\{X_t^x\} \text{ reaches } c \text{ before } d) \\ &= P\left(\{W_t\} \text{ reaches } \frac{c-x}{\sigma} \text{ before } \frac{d-x}{\sigma}\right) \\ &= \lim_{n\to\infty} (\{S_{m,n}: m = 0, 1, 2, \ldots\} \text{ reaches } c_n \text{ before } d_n) \\ &= \lim_{n\to\infty} \frac{1 - (p_n/q_n)^{\frac{d-x}{\sigma}\sqrt{n}}}{1 - (p_n/q_n)^{\frac{d-x}{\sigma}\sqrt{n} - \frac{c-x}{\sigma}\sqrt{n}}} \\ &= \lim_{n\to\infty} \frac{1 - \left(\dfrac{1 + \dfrac{\mu}{\sigma\sqrt{n}}}{1 - \dfrac{\mu}{\sigma\sqrt{n}}}\right)^{\frac{d-x}{\sigma}\sqrt{n}}}{1 - \left(\dfrac{1 + \dfrac{\mu}{\sigma\sqrt{n}}}{1 - \dfrac{\mu}{\sigma\sqrt{n}}}\right)^{\frac{d-c}{\sigma}\sqrt{n}}} \end{aligned}$$

$$= \frac{1 - \dfrac{\exp\left\{\dfrac{d-x}{\sigma^2}\mu\right\}}{\exp\left\{-\dfrac{d-x}{\sigma^2}\mu\right\}}}{1 - \dfrac{\exp\left\{\dfrac{(d-c)}{\sigma^2}\mu\right\}}{\exp\left\{-\dfrac{d-c}{\sigma^2}\mu\right\}}}.$$

Therefore,

$$P(\tau_c^x < \tau_d^x) = \frac{1 - \exp\{2(d-x)\mu/\sigma^2\}}{1 - \exp\{2(d-c)\mu/\sigma^2\}} \qquad (c < x < d,\ \mu \neq 0). \tag{9.10}$$

If relation (3.6) of Section 3 is used instead of (3.8), then

$$P(\tau_d^x < \tau_c^x) = \frac{1 - \exp\{-2(x-c)\mu/\sigma^2\}}{1 - \exp\{-2(d-c)\mu/\sigma^2\}} \qquad (c < x < d,\ \mu \neq 0). \tag{9.11}$$

Letting $d \uparrow \infty$ in (9.10), one gets

$$\begin{aligned} P(\tau_c^x < \infty) &= \exp\left\{-\frac{2(x-c)\mu}{\sigma^2}\right\} \qquad && (c < x,\ \mu > 0), \\ P(\tau_c^x < \infty) &= 1 && (c < x,\ \mu < 0). \end{aligned} \tag{9.12}$$

Thus, in this case the extremal random variable $\min_{t \geq 0} X_t^0$ is exponentially distributed (Exercise 4). Letting $c \downarrow -\infty$ in (9.11) one obtains

$$\begin{aligned} P(\tau_d^x < \infty) &= 1 \qquad && (x < d,\ \mu > 0), \\ P(\tau_d^x < \infty) &= \exp\{2(d-x)\mu/\sigma^2\} && (x < d,\ \mu < 0). \end{aligned} \tag{9.13}$$

In particular it follows that $\max_{t \geq 0} X_t^0$ is exponentially distribute (Exercise 4). Relations (9.12), (9.13) imply that *a Brownian motion with a nonzero drift is transient*. This can also be deduced by an appeal to (a continuous time version of) the strong law of large numbers, just as in (3.11), (3.12) of Section 3 (Exercise 1).

10 FIRST PASSAGE TIME DISTRIBUTIONS FOR BROWNIAN MOTION

We have seen in Section 4, relation (4.7), that for a simple symmetric random walk starting at zero, the first passage time T_y to the state $y \neq 0$ has the distribution

$$P(T_y = N) = \frac{|y|}{N}\binom{N}{\frac{N+y}{2}}\frac{1}{2^N}, \qquad N = |y|, |y|+2, |y|+4, \ldots. \tag{10.1}$$

Now let $\tau_z := \tau_z^0$ be the first time a standard Brownian motion starting at the origin reaches z. Let $\{\tilde{X}_t^{(n)}\}$ be the polygonal process corresponding to the simple symmetric random walk. Considering the first time $\{\tilde{X}_t^{(n)}\}$ reaches z, one has by the FLCT (Theorem 8.1) and Eq. 10.1 (Exercise 1),

$$\begin{aligned} P(\tau_z > t) &= \lim_{n\to\infty} P(T_{[z\sqrt{n}]} > [nt]) \\ &= \lim_{n\to\infty} \sum_{N=[nt]+1}^{\infty} P(T_{[z\sqrt{n}]} = N) \\ &= \lim_{n\to\infty} \sum_{\substack{N=[nt]+1,\\ N-y \text{ even}}}^{\infty} \frac{|y|}{N}\binom{N}{\frac{N+y}{2}}\frac{1}{2^N} \qquad (y = [z\sqrt{n}]). \end{aligned} \tag{10.2}$$

Now according to *Stirling's formula*, for large integers M, we can write

$$M! = (2\pi)^{1/2} e^{-M} M^{M+1/2}(1 + \delta_M) \tag{10.3}$$

where $\delta_M \to 0$ as $M \to \infty$. Since $y = [z\sqrt{n}]$, $N > [nt]$, and $\frac{1}{2}(N \pm y) > \{[nt] - |[z\sqrt{n}]|\}/2$, both N and $N \pm y$ tend to infinity as $n \to \infty$. Therefore, for $N + y$ even,

$$\begin{aligned} \frac{|y|}{N}\binom{N}{\frac{N+y}{2}}2^{-N} &= \frac{|y|}{(2\pi)^{1/2}N} \frac{e^{-N}N^{N+\frac{1}{2}}2^{-N}}{e^{-(N+y)/2}\left(\frac{N+y}{2}\right)^{(N+y)/2+\frac{1}{2}} e^{-(N-y)/2}\left(\frac{N-y}{2}\right)^{(N-y)/2+\frac{1}{2}}} \\ &\quad \times (1 + \mathrm{o}(1)) \\ &= \frac{2|y|}{(2\pi)^{1/2}N^{3/2}}\left(1 + \frac{y}{N}\right)^{-(N+y)/2-\frac{1}{2}}\left(1 - \frac{y}{N}\right)^{-(N-y)/2-\frac{1}{2}}(1 + \mathrm{o}(1)) \\ &= \frac{2|y|}{(2\pi)^{1/2}N^{3/2}}\left(1 + \frac{y}{N}\right)^{-(N+y)/2}\left(1 - \frac{y}{N}\right)^{-(N-y)/2}(1 + \mathrm{o}(1)), \end{aligned} \tag{10.4}$$

where $\mathrm{o}(1)$ denotes a quantity whose magnitude is bounded above by a quantity $\varepsilon_n(t, z)$ that depends only on n, t, z and which goes to zero as $n \to \infty$. Also,

$$\log\left[\left(1+\frac{y}{N}\right)^{-(N+y)/2}\left(1-\frac{y}{N}\right)^{-(N-y)/2}\right] = -\frac{N+y}{2}\left[\frac{y}{N}-\frac{y^2}{2N^2}+O\left(\frac{|y|^3}{N^3}\right)\right]$$
$$+\frac{N-y}{2}\left[\frac{y}{N}+\frac{y^2}{2N^2}+O\left(\frac{|y|^3}{N^3}\right)\right]$$
$$= -\frac{y^2}{2N}+\theta(N,y), \tag{10.5}$$

where $|\theta(N, y)| \leqslant n^{-1/2}c(t, z)$ and $c(t, z)$ is a constant depending only on t and z. Combining (10.4) and (10.5), we have

$$\frac{|y|}{N}\binom{N}{\frac{N+y}{2}}2^{-N} = \sqrt{\frac{2}{\pi}}\frac{|y|}{N^{3/2}}\exp\left\{-\frac{y^2}{2N}\right\}(1+o(1))$$
$$= \sqrt{\frac{2}{\pi}}\frac{|z|\sqrt{n}}{N^{3/2}}\exp\left\{-\frac{nz^2}{2N}\right\}(1+o(1)), \tag{10.6}$$

where $o(1) \to 0$ as $n \to \infty$, uniformly for $N > [nt]$, $N - [z\sqrt{n}]$ even. Using this in (10.2), one obtains

$$P(\tau_z > t) = \lim_{n\to\infty} \sum_{\substack{N>[nt],\\ N-[z\sqrt{n}]\text{ even}}} \sqrt{\frac{2}{\pi}}\frac{|z|\sqrt{n}}{N^{3/2}}\exp\left\{-\frac{nz^2}{2N}\right\}. \tag{10.7}$$

Now either $[nt] + 1$ or $[nt] + 2$ is a possible value of N. In the first case the sum in (10.7) is over values $N = [nt] - 1 + 2r$ ($r = 1, 2, \ldots$), and in the second case over values $[nt] + 2r$ ($r = 1, 2, \ldots$). Since the differences in corresponding values of N/n are $2/n$, one may take $N = [nt] + 2r$ for the purpose of calculating (10.7). Thus,

$$P(\tau_z > t) = \lim_{n\to\infty}\sum_{r=1}^{\infty}\sqrt{\frac{2}{\pi}}\frac{|z|\sqrt{n}}{([nt]+2r)^{3/2}}\exp\left\{-\frac{nz^2}{2([nt]+2r)}\right\}$$
$$= \sqrt{\frac{2}{\pi}}|z|\lim_{n\to\infty}\sum_{r=1}^{\infty}\frac{1}{2}\left(\frac{2}{n}\right)\frac{1}{(t+2r/n)^{3/2}}\exp\left\{-\frac{z^2}{2(t+2r/n)}\right\}$$
$$= \sqrt{\frac{2}{\pi}}|z|\cdot\frac{1}{2}\int_t^{\infty}\frac{1}{u^{3/2}}\exp\left\{-\frac{z^2}{2u}\right\}du. \tag{10.8}$$

Now, by the change of variables $v = |z|/\sqrt{u}$, we get

$$P(\tau_z > t) = \sqrt{\frac{2}{\pi}}\int_0^{|z|/\sqrt{t}} e^{-v^2/2}\,dv. \tag{10.9}$$

The first passage time distribution for the more general case of Brownian motion $\{X_t\}$ with zero drift and diffusion coefficient $\sigma^2 > 0$, starting at the origin, is now obtained by applying (10.9) to the standard Brownian motion $\{(1/\sigma)X_t\}$. Therefore,

$$P(\tau_z > t) = \sqrt{\frac{2}{\pi}} \int_0^{|z|/\sigma\sqrt{t}} e^{-v^2/2}\, dv. \tag{10.10}$$

The probability density function $f_{\sigma^2}(t)$ of τ_z is obtained from (10.10) as

$$f_{\sigma^2}(t) = \frac{|z|}{(2\pi\sigma^2)^{1/2} t^{3/2}} e^{-z^2/2\sigma^2 t} \qquad (t > 0). \tag{10.11}$$

Note that for large t the *tail* of the p.d.f. $f_\sigma{}^2(t)$ is of the order of $t^{-3/2}$. Therefore, although $\{X_t^0\}$ will reach z in a finite time with probability 1, the *expected time* is infinite (Exercise 11).

Consider now a Brownian motion $\{X_t\}$ with a nonzero drift μ and diffusion coefficient σ^2 that starts at the origin. As in Section 9, the polygonal process $\{\tilde{X}_t^{(n)}\}$ corresponding to the simple random walk $S_{m,n} = Z_{1,n} + \cdots + Z_{m,n}$, $S_{0,n} = 0$, with $P(Z_{m,n} = 1) = p_n = \frac{1}{2} + \mu/(2\sigma\sqrt{n})$, converges in distribution to $\{W_t = X_t/\sigma\}$, which is a Brownian motion with drift μ/σ and diffusion coefficient 1. On the other hand, writing $T_{y,n}$ for the first passage time of $\{S_{m,n}: m = 0, 1, \ldots\}$ to y, one has, by relation (4.6) of Section 4,

$$\begin{aligned}
P(T_{y,n} = N) &= \frac{|y|}{N}\binom{N}{\frac{N+y}{2}} p_n^{(N+y)/2} q_n^{(N-y)/2} \\
&= \frac{|y|}{N}\binom{N}{\frac{N+y}{2}} 2^{-N}\left(1 + \frac{\mu}{\sigma\sqrt{n}}\right)^{(N+y)/2}\left(1 - \frac{\mu}{\sigma\sqrt{n}}\right)^{(N-y)/2} \\
&= \frac{|y|}{N}\binom{N}{\frac{N+y}{2}} 2^{-N}\left(1 - \frac{\mu^2}{\sigma^2 n}\right)^{N/2}\left(1 + \frac{\mu}{\sigma\sqrt{n}}\right)^{y/2}\left(1 - \frac{\mu}{\sigma\sqrt{n}}\right)^{-y/2}.
\end{aligned} \tag{10.12}$$

For $y = [w\sqrt{n}]$ for some given nonzero w, and $N = [nt] + 2r$ for some given $t > 0$ and all positive integers r, one has

$$\begin{aligned}
&\left(1 - \frac{\mu^2}{\sigma^2 n}\right)^{N/2}\left(1 + \frac{\mu}{\sigma\sqrt{n}}\right)^{y/2}\left(1 - \frac{\mu}{\sigma\sqrt{n}}\right)^{-y/2} \\
&\qquad = \left(1 - \frac{\mu^2}{\sigma^2 n}\right)^{nt/2 + r}\left(1 + \frac{\mu}{\sigma\sqrt{n}}\right)^{w\sqrt{n}/2}\left(1 - \frac{\mu}{\sigma\sqrt{n}}\right)^{-w\sqrt{n}/2}(1 + o(1))
\end{aligned}$$

$$= \exp\left\{-\frac{t\mu^2}{2\sigma^2}\right\}\left(1 - \frac{\mu^2}{\sigma^2 n}\right)^r \exp\left\{\frac{\mu w}{2\sigma}\right\} \exp\left\{\frac{\mu w}{2\sigma}\right\}(1 + o(1))$$

$$= \exp\left\{-\frac{t\mu^2}{2\sigma^2} + \frac{\mu w}{\sigma}\right\}\left[\left(1 - \frac{\mu^2}{\sigma^2 n}\right)^n\right]^{r/n}(1 + o(1))$$

$$\exp\left\{-\frac{t\mu^2}{2\sigma^2} + \frac{\mu w}{\sigma}\right\}\left[\exp\left\{-\frac{\mu^2}{\sigma^2}\right\} + \varepsilon_n\right]^{r/n}(1 + o(1)) \qquad (10.13)$$

where ε_n does not depend on r and goes to zero as $n \to \infty$, and o(1) represents a term that goes to zero uniformly for all $r \geqslant 1$ as $n \to \infty$. The first passage time τ_z to z for $\{X_t\}$ is the same as the first passage time to $w = z/\sigma$ for the process $\{W_t\} = \{X_t/\sigma\}$. It follows from (9.12), (9.13) that if

(i) $\mu < 0$ and $z > 0$ or
(ii) $\mu > 0$ and $z < 0$,

then there is a positive probability that the process $\{W_t\}$ will never reach $w = z/\sigma$ (i.e., $\tau_z = \infty$). On the other hand, the sum of the probabilities in (10.12) over $N > [nt]$ only gives the probability that the random walk reaches $[w\sqrt{n}]$ in a *finite time* greater than $[nt]$. By the FCLT, (10.6)–(10.8) and (10.12) and (10.13), we have

$$P(t < \tau_z < \infty) = \sqrt{\frac{2}{\pi}}|w| \exp\left\{-\frac{t\mu^2}{2\sigma^2} + \frac{w\mu}{\sigma}\right\} \lim_{n\to\infty} \sum_{r=1}^{\infty} \frac{1}{n(t + 2r/n)^{3/2}}$$

$$\times \exp\left\{-\frac{w^2}{2(t + 2r/n)}\right\}(e^{-\mu^2/\sigma^2} + \varepsilon_n)^{r/n}$$

$$= \sqrt{\frac{2}{\pi}}|w| \exp\left\{-\frac{t\mu^2}{2\sigma^2} + \frac{w\mu}{\sigma}\right\}\cdot\frac{1}{2}\int_t^\infty \frac{1}{v^{3/2}} \exp\left\{-\frac{w^2}{2v}\right\}$$

$$\times \left[\exp\left\{-\frac{\mu^2}{\sigma^2}\right\}\right]^{(v-t)/2} dv$$

$$= \frac{1}{(2\pi)^{1/2}}|w| \exp\left\{-\frac{t\mu^2}{2\sigma^2} + \frac{w\mu}{\sigma}\right\}\int_t^\infty \frac{1}{v^{3/2}}$$

$$\times \exp\left\{-\frac{w^2}{2v} - (v - t)\frac{\mu^2}{2\sigma^2}\right\} dv$$

$$= \frac{1}{(2\pi)^{1/2}}|w| \exp\left\{\frac{w\mu}{\sigma}\right\}\int_t^\infty \frac{1}{v^{3/2}} \exp\left\{-\frac{w^2}{2v} - \frac{\mu^2 v}{2\sigma^2}\right\} dv,$$

$$\text{for } w = z/\sigma. \qquad (10.14)$$

Therefore, for $t \geqslant 0$,

$$P(t < \tau_z < \infty) = \frac{1}{(2\pi)^{1/2}} \left|\frac{z}{\sigma}\right| \exp\left\{\frac{\mu z}{\sigma^2}\right\} \int_t^\infty \frac{1}{v^{3/2}} \exp\left\{-\frac{z^2}{2\sigma^2 v} - \frac{\mu^2 v}{2\sigma^2}\right\} dv. \quad (10.15)$$

Differentiating this with respect to t (and changing the sign) the probability density function of τ_z is given by

$$f_{\sigma^2,\mu}(t) = \frac{|z|}{(2\pi\sigma^2)^{1/2} t^{3/2}} \exp\left\{\frac{\mu z}{\sigma^2} - \frac{z^2}{2\sigma^2 t} - \frac{\mu^2}{2\sigma^2} t\right\}.$$

Therefore,

$$f_{\sigma^2,\mu}(t) = \frac{|z|}{(2\pi\sigma^2)^{1/2} t^{3/2}} \exp\left\{-\frac{1}{2\sigma^2 t}(z - \mu t)^2\right\} \qquad (t > 0). \quad (10.16)$$

In particular, letting $p(t; 0, y)$ denote the p.d.f. (7.1) of the distribution of the position X_t^0 at time t, (10.16) can be expressed as (see 4.6)

$$f_{\sigma^2,\mu}(t) = \frac{|z|}{t} p(t; 0, z). \quad (10.17)$$

As mentioned before, the integral of $f_{\sigma^2,\mu}(t)$ is less than 1 if either

(i) $\mu > 0$, $z < 0$ or
(ii) $\mu < 0$, $z > 0$.

In all other cases, (10.16) is a proper probability density function. By putting $\mu = 0$ in (10.16), one gets (10.11).

11 THE ARCSINE LAW

Consider a *simple symmetric random walk* $\{S_m\}$ starting at zero. The problem is to calculate the distribution of the last visit to zero by $S_0, S_1, \ldots, S_{2n}$. For this we first calculate the probability that the number of $+1$'s exceeds the number of -1's until time N and with a given positive value of the excess at time N.

Lemma 1. Let a, b be two integers, $0 \leqslant a < b$. Then

$$P(S_1 > 0, S_2 > 0, \ldots, S_{a+b-1} > 0, S_{a+b} = b - a)$$
$$= \left[\binom{a+b-1}{b-1} - \binom{a+b-1}{b}\right]\left(\frac{1}{2}\right)^{a+b} = \binom{a+b}{b}\frac{b-a}{a+b}\left(\frac{1}{2}\right)^{a+b}. \quad (11.1)$$

Proof. Each of the $\binom{a+b}{b}$ paths from $(0,0)$ to $(a+b, b-a)$ has probability $(\frac{1}{2})^{a+b}$. We seek the number M of those for which $S_1 = 1$, $S_2 > 0$, $S_3 > 0, \ldots,$ $S_{a+b-1} > 0$, $S_{a+b} = b - a$. Now the paths from $(1,1)$ to $(a+b, b-a)$ that cross or touch zero (the horizontal axis) are in one–one correspondence with those that go from $(1,-1)$ to $(a+b, b-a)$. This correspondence is set up by reflecting each path of the last type about zero (i.e., about the horizontal time axis) up to the first time after time zero that zero is reached, and leaving the path from then on unchanged. The reflected path leads from $(1,1)$ to $(a+b, b-a)$ and crosses or touches zero. Conversely, any path leading from $(1,1)$ to $(a+b, b-a)$ that crosses or touches zero, when reflected in the same manner, yields a path from $(1,-1)$ to $(a+b, b-a)$. But the number of all paths from $(1,-1)$ to $(a+b, b-a)$ is simply $\binom{a+b-1}{b}$, since it requires b plus 1's and $a-1$ minus 1's among $a+b-1$ steps to go from $(1,-1)$ to $(a+b, b-a)$. Hence

$$M = \binom{a+b-1}{b-1} - \binom{a+b-1}{b},$$

since there are altogether $\binom{a+b-1}{b-1}$ paths from $(1,1)$ to $(a+b, b-a)$. Now a straightforward simplification yields

$$M = \binom{a+b}{b}\frac{b-a}{a+b}.$$ ■

Lemma 2. For the simple symmetric random walk starting at zero we have,

$$P(S_1 \neq 0, S_2 \neq 0, \ldots, S_{2n} \neq 0) = P(S_{2n} = 0) = \binom{2n}{n}\left(\frac{1}{2}\right)^{2n}. \tag{11.2}$$

Proof. By symmetry, the leftmost side of (11.2) equals

$$\begin{aligned}
2P(S_1 > 0, S_2 > 0, \ldots, S_{2n} > 0)\\
&= 2\sum_{r=1}^{n} P(S_1 > 0, S_2 > 0, \ldots, S_{2n-2} > 0, S_{2n} = 2r)\\
&= 2\sum_{r=1}^{n}\left[\binom{2n-1}{n+r-1} - \binom{2n-1}{n+r}\right]\left(\frac{1}{2}\right)^{2n}\\
&= 2\binom{2n-1}{n}\left(\frac{1}{2}\right)^{2n} = \binom{2n}{n}\left(\frac{1}{2}\right)^{2n} = P(S_{2n} = 0),
\end{aligned}$$

where we have adopted the convention that $\binom{2n-1}{2n} = 0$ in writing the middle equality.

Theorem 11.1. Let $\Gamma^{(m)} = \max\{j: 0 \leqslant j \leqslant m, S_j = 0\}$. Then

$$\begin{aligned} P(\Gamma^{(2n)} = 2k) &= P(S_{2k} = 0)P(S_{2n-2k} = 0) \\ &= \binom{2k}{k}\left(\frac{1}{2}\right)^{2k}\binom{2n-2k}{n-k}\left(\frac{1}{2}\right)^{2n-2k} \\ &= \frac{(2k)!(2n-2k)!}{(k!)^2((n-k)!)^2}\left(\frac{1}{2}\right)^{2n} \qquad \text{for } k = 0, 1, 2, \ldots, n. \end{aligned} \tag{11.3}$$

Proof. By considering the conditional probability given $\{S_{2k} = 0\}$ one can easily justify that

$$\begin{aligned} P(\Gamma^{(2n)} = 2k) &= P(S_{2k} = 0, S_{2k+1} \neq 0, S_{2k+2} \neq 0, \ldots, S_{2n} \neq 0) \\ &= P(S_{2k} = 0)P(S_1 \neq 0, S_2 \neq 0, \ldots, S_{2n-2k} \neq 0) \\ &= P(S_{2k} = 0)P(S_{2n-2k} = 0). \end{aligned}$$

■

Theorem 11.1 has the following symmetry relation as a corollary.

$$P(\Gamma^{(2n)} = 2k) = P(\Gamma^{(2n)} = 2n - 2k) \qquad \text{for all } k = 0, 1, \ldots, n. \tag{11.4}$$

Theorem 11.2. (*The Arc Sine Law*). Let $\{B_t\}$ be a standard Brownian motion at zero. Let $\gamma = \sup\{t: 0 \leqslant t \leqslant 1, B_t = 0\}$. Then γ has the probability density function

$$f(x) = \frac{1}{\pi(x(1-x))^{1/2}}, \qquad 0 < x < 1. \tag{11.5}$$

$$P(\gamma \leqslant x) = \int_0^x f(y)\,dy = \frac{2}{\pi}\sin^{-1}\sqrt{x}. \tag{11.6}$$

Proof. Let $\{S_0 = 0, S_1, S_2, \ldots\}$ be a simple symmetric random walk. Define $\{\tilde{X}_t^{(n)}\}$ as in (8.4). By the FCLT (Theorem 8.1) one has

$$P(\gamma \leqslant x) = \lim_{n\to\infty} P(\gamma^{(n)} \leqslant x) \qquad (0 < x < 1),$$

where

$$\gamma^{(n)} = \sup\{t: 0 \leqslant t \leqslant 1, \tilde{X}_t^{(n)} = 0\} = \frac{1}{n}\sup\{m: 0 \leqslant m \leqslant n, S_m = 0\} = \frac{1}{n}\Gamma^{(n)}.$$

In particular, taking the limit over even integers, it follows that

$$P(\gamma \leqslant x) = \lim_{n\to\infty} P\left(\frac{1}{2n}\Gamma^{(2n)} \leqslant x\right) = \lim_{n\to\infty} P(\Gamma^{(2n)} \leqslant 2nx),$$

where $\Gamma^{(2n)}$ is defined in Theorem 11.1. By Theorem 11.1 and Stirling's approximation

$$\lim_{n\to\infty} P(\Gamma^{(2n)} \leqslant 2nx) = \lim_{n\to\infty} \sum_{k=0}^{[nx]} \frac{(2k)!(2n-2k)!}{(k!)^2((n-k)!)^2} 2^{-2n}$$

$$= \lim_{n\to\infty} \sum_{k=0}^{[nx]} \frac{(2\pi)^{1/2} e^{-2k}(2k)^{2k+\frac{1}{2}}}{((2\pi)^{1/2} e^{-k} k^{k+\frac{1}{2}})^2}$$

$$\times \frac{(2\pi)^{1/2} e^{-2(n-k)}(2(n-k))^{[2(n-k)+\frac{1}{2}]} 2^{-2n}}{((2\pi)^{1/2} e^{-(n-k)}(n-k)^{n-k+\frac{1}{2}})^2}$$

$$= \lim_{n\to\infty} \sum_{k=0}^{[nx]} \frac{1}{\pi\sqrt{k}\,(n-k)^{1/2}}$$

$$= \frac{1}{\pi} \lim_{n\to\infty} \sum_{k=0}^{[nx]} \frac{1}{n} \frac{1}{\left(\frac{k}{n}\left(1-\frac{k}{n}\right)\right)^{1/2}} = \frac{1}{\pi}\int_0^x \frac{1}{(y(1-y))^{1/2}}\,dy. \quad \blacksquare$$

The following (invariance) corollary follows by applying the FCLT.

Corollary 11.3. Let $\{Z_1, Z_2, \ldots\}$ be a sequence of i.i.d. random variables such that $EZ_1 = 0, EZ_1^2 = 1$. Then, defining $\{\tilde{X}_t^{(n)}\}$ as in (8.4) and $\gamma^{(n)}$ as above, one has

$$\lim_{n\to\infty} P(\gamma^{(n)} \leqslant x) = \frac{2}{\pi} \sin^{-1} \sqrt{x}. \tag{11.7}$$

From the arcsine law of the time of the last visit to zero it is also possible to get the distribution of the length of time in [0, 1] the standard Brownian motion spends on the positive side of the origin (i.e., an *occupation time law*) again as an arcsine distribution. This fact is recorded in the following corollary (Exercise 2).

Corollary 11.4. Let $U = |\{t \leqslant 1 : B_t \in \mathbb{R}_+\}|$, where $|\ \ |$ denotes Lebesgue measure and $\{B_t\}$ is standard Brownian motion starting at 0. Then,

$$P(U \leqslant x) = \frac{2}{\pi} \sin^{-1} \sqrt{x}. \tag{11.8}$$

12 THE BROWNIAN BRIDGE

Let $\{B_t\}$ be a standard Brownian motion starting at zero. Since $B_t - tB_1$ vanishes for $t = 0$ and $t = 1$, the stochastic process $\{B_t^*\}$ defined by

$$B_t^* := B_t - tB_1, \qquad 0 \leqslant t \leqslant 1, \tag{12.1}$$

is called the *Brownian bridge* or the *tied-down Brownian motion.*

Since $\{B_t\}$ is a Gaussian process with independent increments, it is simple to check that $\{B_t^*\}$ is a Gaussian process; i.e., its finite dimensional distributions are Gaussian. Also,

$$EB_t^* = 0, \tag{12.2}$$

and

$$\begin{aligned}\operatorname{Cov}(B_{t_1}^*, B_{t_2}^*) &= \operatorname{Cov}(B_{t_1}, B_{t_2}) - t_2 \operatorname{Cov}(B_{t_1}, B_1) - t_1 \operatorname{Cov}(B_{t_2}, B_1)\\ &\quad + t_1 t_2 \operatorname{Cov}(B_1, B_1)\\ &= t_1 - t_2 t_1 - t_1 t_2 + t_1 t_2 = t_1(1 - t_2), \qquad \text{for } t_1 \leqslant t_2. \end{aligned} \tag{12.3}$$

From this one can also write down the joint normal density of $(B_{t_1}^*, B_{t_2}^*, \ldots, B_{t_k}^*)$ for arbitrary $0 < t_1 < t_2 < \cdots < t_k < 1$ (Exercise 1).

The Brownian bridge arises quite naturally in the asymptotic theory of statistics. To explain this application, let us consider a sequence of real-valued i.i.d. random variables $Y_1, Y_2, \ldots$, having a (common) distribution function F. The nth *empirical distribution* is the discrete probability distribution on the line assigning a probability $1/n$ to each of the n values $Y_1, Y_2, \ldots, Y_n$. The corresponding distribution function F_n is called the (nth) *empirical distribution function,*

$$F_n(t) = \frac{1}{n} \# \{j: 1 \leqslant j \leqslant n, Y_j \leqslant t\}, \qquad -\infty < t < \infty, \tag{12.4}$$

where $\# A$ denotes the cardinality of the set A.

Suppose $Y_{(1)} \leqslant Y_{(2)} \leqslant \cdots \leqslant Y_{(n)}$ is the ordering of the first n observations. Figure 12.1 illustrates F_5. Note that $\{F_n(t): t \geqslant 0\}$ is for each n a stochastic

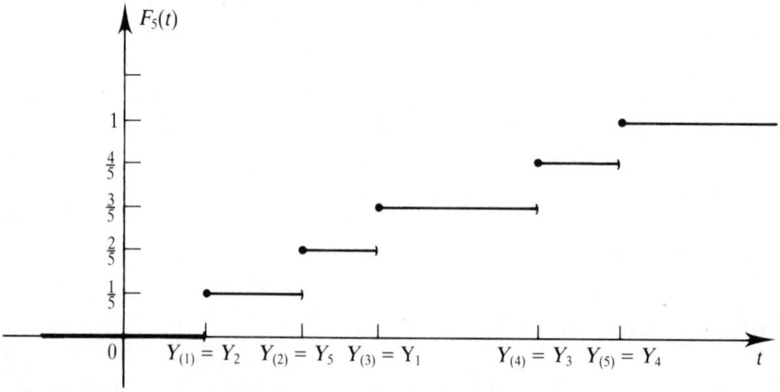

Figure 12.1

process. Now for each t the random variable

$$nF_n(t) = \sum_{j=1}^{n} \mathbf{1}_{\{Y_j \leqslant t\}} \tag{12.5}$$

is the sum of n i.i.d. Bernoulli random variables each taking the value 1 with probability $F(t) = P(Y_j \leqslant t)$ and the value 0 with probability $1 - F(t)$. Now $E(\mathbf{1}_{\{Y_j \leqslant t\}}) = F(t)$ and, for $t_1 \leqslant t_2$,

$$\operatorname{Cov}(\mathbf{1}_{\{Y_j \leqslant t_1\}}, \mathbf{1}_{\{Y_k \leqslant t_2\}}) = \begin{cases} 0, & \text{if } j \neq k \\ F(t_1)(1 - F(t_2)), & \text{if } j = k, \end{cases} \tag{12.6}$$

since in the case $j = k$,

$$\operatorname{Cov}(\mathbf{1}_{\{Y_j \leqslant t_1\}}, \mathbf{1}_{\{Y_k \leqslant t_2\}}) = E(\mathbf{1}_{\{Y_j \leqslant t_1\}}\mathbf{1}_{\{Y_j \leqslant t_2\}}) - E(\mathbf{1}_{\{Y_j \leqslant t_1\}})E(\mathbf{1}_{\{Y_j \leqslant t_2\}})$$

$$= F(t_1) - F(t_1)F(t_2) = F(t_1)(1 - F(t_2)).$$

It follows from the central limit theorem that

$$n^{-1/2}\left(\sum_{j=1}^{n} \mathbf{1}_{\{Y_j \leqslant t\}} - nF(t)\right) = \sqrt{n}(F_n(t) - F(t))$$

is asymptotically (as $n \to \infty$) Gaussian with mean zero and variance $F(t)(1 - F(t))$. For $t_1 < t_2 < \cdots < t_k$, the multidimensional central limit theorem applied to the i.i.d. sequence of k-dimensional random vectors $(\mathbf{1}_{\{Y_j \leqslant t_1\}}, \mathbf{1}_{\{Y_j \leqslant t_2\}}, \ldots, \mathbf{1}_{\{Y_j \leqslant t_k\}})$ shows that $(\sqrt{n}(F_n(t_1) - F(t_1)), \sqrt{n}(F_n(t_2) - F(t_2)), \ldots, \sqrt{n}(F_n(t_k) - F(t_k)))$ is asymptotically (k-dimensional) Gaussian with zero mean and dispersion matrix $\Sigma = ((\sigma_{ij}))$, where

$$\sigma_{ij} = \operatorname{Cov}(\mathbf{1}_{\{Y_j \leqslant t_i\}}, \mathbf{1}_{\{Y_j \leqslant t_j\}}) = F(t_i)(1 - F(t_j)), \qquad \text{for } t_i \leqslant t_j. \tag{12.7}$$

In the special case of observations from the uniform distribution on [0, 1], one has

$$F(t) = t, \qquad 0 \leqslant t \leqslant 1, \tag{12.8}$$

so that the finite-dimensional distributions of the stochastic process $\sqrt{n}(F_n(t) - F(t))$ converge to those of the Brownian bridge as $n \to \infty$. As in the case of the functional central limit theorem (FCLT), probabilities of many infinite-dimensional events of interest also converge to those of the Brownian bridge (theoretical complements 1 and 2). The precise statement of such a result is as follows.

Proposition 12.1. If $Y_1, Y_2, \ldots$ is an i.i.d. sequence having the uniform distribution on $[0, 1]$, then the normalized empirical process $\{\sqrt{n}(F_n(t) - t)\colon 0 \leqslant t \leqslant 1\}$ converges in distribution to the Brownian bridge as $n \to \infty$.

Let $Y_1, Y_2, \ldots$ be an i.i.d. sequence having a (common) distribution function F that is continuous on the real number line. Note that in the case that F is strictly increasing on an interval (a, b) with $F(a) = 0$, $F(b) = 1$, one has for $0 < t < 1$,

$$P(F(Y_k) \leqslant t) = P(Y_k \leqslant F^{-1}(t)) = F(F^{-1}(t)) = t, \tag{12.9}$$

so the sequence $U_1 = F(Y_1)$, $U_2 = F(Y_2), \ldots$ is i.i.d. uniform on $[0, 1]$. The same is true more generally (Exercise 2). Let F_n be the empirical distribution function of $Y_1, \ldots, Y_n$, and G_n that of $U_1, \ldots, U_n$. Then, since the proportion of Y_k's, $1 \leqslant k \leqslant n$, that do not exceed t coincides with the proportion of U_k's, $1 \leqslant k \leqslant n$, that do not exceed $F(t)$, we have

$$\sqrt{n}[F_n(t) - F(t)] = \sqrt{n}[G_n(F(t)) - F(t)], \qquad a \leqslant t \leqslant b. \tag{12.10}$$

If $a = -\infty$ $(b = +\infty)$, the index set $[a, b]$ for the process is to exclude a (b). Since $\sqrt{n}(G_n(t) - t), 0 \leqslant t \leqslant 1$, converges in distribution to the Brownian bridge, and since $t \to F(t)$ is increasing on (a, b), one derives the following extension of Proposition 12.1.

Proposition 12.2. Let $Y_1, Y_2, \ldots$ be a sequence of i.i.d. real-valued random variables with continuous distribution function F on (a, b) where $F(a) = 0$, $F(b) = 1$. Then the normalized empirical process $\sqrt{n}(F_n(t) - F(t))$, a $\leqslant t \leqslant b$, converges in distribution to the Gaussian process $\{Z_t\} := \{B^*_{F(t)}\colon a \leqslant t \leqslant b\}$, as $n \to \infty$.

It also follows from (12.10) that the *Kolmogorov–Smirnov statistic* defined by $D_n := \sup\{\sqrt{n}|F_n(t) - F(t)|\colon a \leqslant t \leqslant b\}$ satisfies

$$D_n = \sup_{a \leqslant t \leqslant b} \sqrt{n}|F_n(t) - F(t)| = \sup_{a \leqslant t \leqslant b} \sqrt{n}|G_n(F(t)) - F(t)| = \sup_{0 \leqslant t \leqslant 1} \sqrt{n}|G_n(t) - t|. \tag{12.11}$$

Thus, *the distribution of D_n is the same* (namely that obtained under the uniform distribution) *for all continuous F*. This common distribution has been tabulated for small and moderately large values of n (see theoretical complement 2). By Proposition 12.2, for large n, the distribution is approximately the same as that of the statistic defined by (also see theoretical complement 1)

$$D := \sup_{0 \leqslant t \leqslant 1} |B^*_t|. \tag{12.12}$$

A calculation of the distribution of D yields (theoretical complement 3 and Exercise 4*(iii))

$$P(D > d) = 2 \sum_{k=1}^{\infty} (-1)^{k-1} e^{-2k^2 d^2}, \qquad d > 0. \tag{12.13}$$

These facts are often used to test the *statistical hypothesis* that observations $Y_1, Y_2, \ldots, Y_n$ are from a specified distribution with a continuous distribution function F. If the observed value, say d, of D_n is so large that the probability (approximated by (12.13) for large n) is very small for a value of D_n as large as or larger than d to occur (under the assumption that $Y_1, \ldots, Y_n$ do come from F), then the hypothesis is rejected.

In closing, note that by the strong law of large numbers, $F_n(t) \to F(t)$ as $n \to \infty$, *with probability 1*. From the FCLT, it follows that

$$\sup_{-\infty < t < \infty} |F_n(t) - F(t)| \to 0 \qquad \text{in probability as } n \to \infty. \tag{12.14}$$

In fact, it is possible to show that the uniform convergence in (12.14) is also *almost sure* (Exercise 8). This stronger result is known as the *Glivenko–Cantelli lemma*.

13 STOPPING TIMES AND MARTINGALES

An extremely useful concept in probability is that of a *stopping time*, sometimes also called a *Markov time*. Consider a sequence of random variables $\{X_n: n = 0, 1, \ldots\}$, defined on some probability space $(\Omega, \mathcal{F}, P)$. Stopping times with respect to $\{X_n\}$ are defined as follows. Denote by $\mathcal{F}_n$ the sigmafield $\sigma\{X_0, \ldots, X_n\}$ comprising all events that depend only on $\{X_0, X_1, \ldots, X_n\}$.

Definition 13.1. A *stopping time* τ *for the process* $\{X_n\}$ is a random variable taking nonnegative integer values, including possibly the value $+\infty$, such that

$$\{\tau \leqslant n\} \in \mathcal{F}_n \qquad (n = 0, 1, \ldots). \tag{13.1}$$

Observe that (13.1) is equivalent to the condition

$$\{\tau = n\} \in \mathcal{F}_n \qquad (n = 0, 1, \ldots), \tag{13.2}$$

since $\mathcal{F}_n$ are increasing sigmafields (i.e., $\mathcal{F}_n \subset \mathcal{F}_{n+1}$) and τ is integer-valued.

Informally, (13.1) says that, using τ, the decision to stop or not to stop by time n depends only on the observations $X_0, X_1, \ldots, X_n$.

An important example of a stopping time is the *first passage time* τ_B to a (Borel) set $B \subset \mathbb{R}^1$,

$$\tau_B := \min\{n \geqslant 0: X_n \in B\}. \tag{13.3}$$

If X_n does not lie in B for any n, one takes $\tau_B = \infty$. Sometimes the minimum in (13.3) is taken over $\{n \geqslant 1, X_n \in B\}$, in which case we call it the *first return time* to B, denoted η_B.

A less interesting but useful example of a stopping time is a *constant time*,

$$\tau := m \tag{13.4}$$

where m is a fixed positive integer.

One may define, for every positive integer r, the rth *passage time* $\tau_B^{(r)}$ to B recursively, by

$$\begin{aligned} \tau_B^{(r)} &:= \min\{n > \tau_B^{(r-1)}: X_n \in B\} \qquad (r = 2, \ldots) \\ \tau_B^{(1)} &:= \tau_B. \end{aligned} \tag{13.5}$$

Again, if X_n does not lie in B for any $n > \tau_B^{(r-1)}$, take $\tau_B^{(r)} = \infty$. Also note that if $\tau_B^{(r)} = \infty$ for some r then $\tau_B^{(r')} = \infty$ for all $r' \geqslant r$. It is a simple exercise to check that each $\tau_B^{(r)}$ is a stopping time (Exercise 1).

The usefulness of the concept of stopping times will now be illustrated by a result that in gambling language says that *in a fair game the gambler has no winning strategy*. To be precise, consider a sequence $\{X_n: n = 0, 1, \ldots\}$ of independent random variables, $S_n = X_0 + X_1 + \cdots + X_n$. Obviously, if $EX_n = 0$, $n \geqslant 1$, then $ES_n = S_0$ for each n. We will now consider an extension of this property when n is replaced by certain stopping times. Since $\{X_0, \ldots, X_n\}$ and $\{S_0, S_1, \ldots, S_n\}$ determine each other, stopping times for $\{S_n\}$ are the same as those for $\{X_n\}$.

Theorem 13.1. Let τ be a stopping time for the process $\{S_n\}$. If

1. $EX_n = 0$ for $n \geqslant 1$, $E|X_0| \equiv E|S_0| < \infty$,
2. $P(\tau < \infty) = 1$,
3. $E|S_\tau| < \infty$, and
4. $E(S_m \mathbf{1}_{\{\tau > m\}}) \to 0$ as $m \to \infty$,

then

$$ES_\tau = ES_0. \tag{13.6}$$

Proof. First assume $\tau \leqslant m$ for some integer m. Then

$$\begin{aligned} ES_\tau &= E(S_0 \mathbf{1}_{\{\tau = 0\}}) + E(S_1 \mathbf{1}_{\{\tau = 1\}}) + \cdots + E(S_m \mathbf{1}_{\{\tau = m\}}) \\ &= E(X_0 \mathbf{1}_{\{\tau \geqslant 0\}}) + E(X_1 \mathbf{1}_{\{\tau \geqslant 1\}}) + \cdots + E(X_j \mathbf{1}_{\{\tau \geqslant j\}}) + \cdots + E(X_m \mathbf{1}_{\{\tau \geqslant m\}}) \\ &= EX_0 + E(X_1 \mathbf{1}_{\{\tau \geqslant 1\}}) + \cdots + E(X_j \mathbf{1}_{\{\tau \geqslant j\}}) + \cdots + E(X_m \mathbf{1}_{\{\tau \geqslant m\}}). \end{aligned} \tag{13.7}$$

Now $\{\tau \geqslant j\} = \{\tau < j\}^c = \{\tau \leqslant j - 1\}^c$ depends only on $X_0, X_1, \ldots, X_{j-1}$. Therefore,

$$E(X_j \mathbf{1}_{\{\tau \geqslant j\}}) = E[\mathbf{1}_{\{\tau \geqslant j\}} E(X_j \mid \{X_0, \ldots, X_{j-1}\})]$$
$$= E[\mathbf{1}_{\{\tau \geqslant j\}} E(X_j)] = 0 \qquad (j = 1, 2, \ldots, m), \tag{13.8}$$

so that (13.7) reduces to

$$ES_\tau = EX_0 = ES_0. \tag{13.9}$$

To prove the general result, define

$$\tau_m := \tau \wedge m = \min\{\tau, m\}, \tag{13.10}$$

and check that τ_m is a stopping time (Exercise 2). By (13.9),

$$ES_{\tau_m} = ES_0 \qquad (m = 1, 2, \ldots). \tag{13.11}$$

Since $\tau = \tau_m$ on the set $\{\tau \leqslant m\}$, one has

$$|ES_\tau - ES_{\tau_m}| = |E(S_\tau - S_{\tau_m})| = |E((S_\tau - S_m)\mathbf{1}_{\{\tau > m\}})|$$
$$\leqslant |E(S_\tau \mathbf{1}_{\{\tau > m\}})| + |E(S_m \mathbf{1}_{\{\tau > m\}})|. \tag{13.12}$$

The first term on the right side of the inequality in (13.12) goes to zero as $m \to \infty$, by assumptions (2), (3) (Exercise 3), while the second term goes to zero by assumption (4). ■

Assumptions (2), (3) ensure that ES_τ is well defined and finite. Assumption (4) is of a technical nature, but cannot be dispensed with. To demonstrate this, consider a simple symmetric random walk $\{S_n\}$ starting at zero (i.e., $S_0 = 0$). Write τ_y for $\tau_{\{y\}}$, the *first passage* time to the state y, $y \neq 0$. Then (1), (2), (3) are satisfied. But

$$ES_{\tau_y} = y \neq 0. \tag{13.13}$$

The reason (13.6) does not hold in this case is that assumption (4) is violated (see Exercise 4). If, on the other hand,

$$\tau = \min\{n: S_n = -a \text{ or } b\}, \tag{13.14}$$

where a and b are positive integers, then $P(\tau < \infty) = 1$. There are various ways of proving this last assertion. A more general result, namely, Proposition 13.4, is proved later in the section to take care of this condition. To check condition (3) of Theorem 13.1, note that $|S_\tau| \leqslant \max\{a, b\}$, so that $E|S_\tau| \leqslant \max\{a, b\}$. Also, on the set $\{\tau > m\}$ one has $-a < S_m < b$ and therefore

$$|E(S_m \mathbf{1}_{\{\tau > m\}})| \leqslant \max\{a, b\} E|\mathbf{1}_{\{\tau > m\}}|$$

$$= \max\{a, b\} P\{\tau > m\} \to 0 \qquad \text{as } m \to \infty. \tag{13.15}$$

Thus condition (4) is verified. Hence the conclusion (13.6) of Theorem 13.1 holds. This means

$$0 = ES_\tau = -aP(\tau_{-a} < \tau_b) + bP(\tau_{-a} > \tau_b)$$

$$= -a(1 - P(\tau_{-a} > \tau_b)) + bP(\tau_{-a} > \tau_b), \tag{13.16}$$

which may be solved for $P(\tau_{-a} > \tau_b)$ to yield

$$P(\tau_{-a} > \tau_b) = \frac{a}{a + b}, \tag{13.17}$$

a result that was obtained by a different method earlier (see Chapter I, Eq. 3.13).

To deal with the case $EX_n \neq 0$, as is the case with the simple asymmetric random walk, the following corollary to Theorem 13.1 is useful.

Corollary 13.2. (*Wald's Equation*). Let $\{Y_n: n = 1, 2, \ldots\}$ be a sequence of i.i.d. random variables with $EY_n = \mu$. Let $S'_n = Y_1 + \cdots + Y_n$, and let τ be a stopping time for the process $\{S'_n: n = 1, 2, \ldots\}$ such that

2′. $E\tau < \infty$,
3′. $E|S'_\tau| < \infty$, and
4′. $\lim_{m \to \infty} |E(S'_m \mathbf{1}_{\{\tau > m\}})| = 0$.

Then

$$ES'_\tau = (E\tau)\mu. \tag{13.18}$$

Proof. To prove this, simply set $X_0 = 0$, $X_n = Y_n - \mu$ $(n \geqslant 1)$, and $S_n = X_1 + \cdots + X_n = S'_n - n\mu$, and apply Theorem 13.1 to get

$$0 = ES_\tau = E(S'_\tau - \tau\mu) = ES'_\tau - E(\tau)\mu, \tag{13.19}$$

which yields (13.18). Note that $E|S_\tau| \leqslant E|S'_\tau| + (E\tau)|\mu| < \infty$, by (2′) and (3′). Also

$$|E(S_m \mathbf{1}_{\{\tau > m\}})| \leqslant |E(S'_m \mathbf{1}_{\{\tau > m\}})| + E|\tau\mu \mathbf{1}_{\{\tau > m\}}|$$

$$= |E(S'_m \mathbf{1}_{\{\tau > m\}})| + |\mu| E|\tau \mathbf{1}_{\{\tau > m\}}| \to 0. \qquad \blacksquare$$

For an application of Corollary 13.2, consider the case $Y_n = +1$ or -1 with probabilities p and $q = 1 - p$, respectively, $\frac{1}{2} < p < 1$. Let $\tau = \min\{n \geqslant 1:$

$S_n' = -a$ or $b\}$, where a, b are positive. Then (13.19) yields

$$-aP(\tau_{-a} < \tau_b) + bP(\tau_{-1} > \tau_b) = (E\tau)(p - q). \tag{13.20}$$

From this one gets

$$E\tau = \frac{(b + a)P(\tau_{-a} > \tau_b) - a}{p - q}. \tag{13.21}$$

Making use of the evaluation

$$P(\tau_{-a} > \tau_b) = \frac{1 - \left(\frac{q}{p}\right)^a}{1 - \left(\frac{q}{p}\right)^{a+b}},$$

(Chapter I, Eq. 3.6) one has

$$E\tau = \frac{b + a}{p - q}\left(\frac{1 - \left(\frac{q}{p}\right)^a}{1 - \left(\frac{q}{p}\right)^{a+b}}\right) - \frac{a}{p - q}. \tag{13.22}$$

Assumption (2′) for this case follows from Proposition 13.4 below, while (3′), (4′), follow exactly as in the case of the simple symmetric random walk (see Eq. 13.15).

In the proof of Theorem 13.1, the only property of the sequence $\{X_n\}$ that is made use of is the property

$$E(X_{n+1} \mid \{X_0, X_1, \ldots, X_n\}) = 0 \qquad (n = 0, 1, 2, \ldots). \tag{13.23}$$

Theorem 13.1 therefore remains valid if assumption (1) is replaced by (13.23), and (2)–(4) hold. No other assumption (such as independence) concerning the random variables is needed. This motivates the following definition.

Definition 13.2. A sequence $\{X_n : n = 0, 1, 2, \ldots\}$ satisfying (13.23) is called a sequence of *martingale differences*, and the sequence of their partial sums, $\{S_n\}$, is called a *martingale*.

Note that (13.23) is equivalent to

$$E(S_{n+1} \mid \{S_0, \ldots, S_n\}) = S_n \qquad (n = 0, 1, 2, \ldots), \tag{13.24}$$

since

$$\begin{aligned} E(S_{n+1} \mid \{S_0, S_1, \ldots, S_n\}) &= E(S_{n+1} \mid \{X_0, X_1, \ldots, X_n\}) \\ &= E(S_n + X_{n+1} \mid \{X_0, X_1, \ldots, X_n\}) \\ &= S_n + E(X_{n+1} \mid \{X_0, X_1, \ldots, X_n\}) = S_n. \end{aligned} \tag{13.25}$$

Conversely, if (13.24) holds for a sequence of random variables $\{S_n: n = 0, 1, 2, \ldots\}$, then the sequence $\{X_n: n = 0, 1, 2, \ldots\}$ defined by

$$X_n = S_n - S_{n-1} \qquad (n = 1, 2, 3, \ldots), \qquad X_0 = S_0, \tag{13.26}$$

satisfies (13.23).

Martingales necessarily have *constant expected values*. Likewise, if $\{X_n\}$ is a martingale difference sequence, then $EX_n = 0$ for each $n \geqslant 1$. Theorem 13.1, Corollary 13.2, and Theorem 13.3 below, assert that this constancy of expectations of a martingale continues to hold at appropriate stopping times.

In the gambling setting, the martingale property (13.24), or (13.23), is often taken as the definition of a fair game, since whatever be the outcomes of the first n plays, the expected net gain at the $(n + 1)$st play is zero. As an example of a strategy for the gambler, suppose that it is decided not to stop until an amount a is lost or an amount b is gained, whichever comes first. Under (13.23) and conditions (2)–(4) of Theorem 13.1, the expected gain at the end of the game is still zero. This conclusion holds for more general stopping times, as stated in Theorem 13.3 below. Before this result is stated, it would be useful to extend the definition of a martingale somewhat. To motivate this new definition, consider a sequence of i.i.d. random variables $\{Y_n: n = 1, 2, \ldots\}$ such that $EY_n = 0$, $EY_n^2 = \mathrm{Var}(Y_n) = 1$. Then $\{S_n^2 - n: n = 0, 1, 2, \ldots\}$ is a martingale, where S_0 is an arbitrary random variable independent of $\{Y_n: n = 1, 2, \ldots\}$ satisfying $ES_0^2 < \infty$.

To see this, form the difference sequence

$$\begin{aligned} &X_{n+1} := S_{n+1}^2 - (n + 1) - (S_n^2 - n) = Y_{n+1}^2 + 2S_n Y_{n+1} - 1 \qquad (n = 0, 1, 2, \ldots), \\ &X_0 := S_0^2. \end{aligned} \tag{13.27}$$

Then, writing $Y_0 = S_0$,

$$\begin{aligned} &E(X_{n+1} \mid \{Y_0, Y_1, Y_2, \ldots, Y_n\}) \\ &\qquad = E(Y_{n+1}^2 \mid \{Y_0, Y_1, \ldots, Y_n\}) + 2S_n E(Y_{n+1} \mid \{Y_0, Y_1, \ldots, Y_n\}) - 1 \\ &\qquad = E(Y_{n+1}^2) + 2S_n E(Y_{n+1}) - 1 = 1 + 0 - 1 = 0. \end{aligned} \tag{13.28}$$

Since $X_0, X_1, X_2, \ldots, X_n$ are determined by (i.e., are Borel measurable functions of) $Y_0, Y_1, Y_2, \ldots, Y_n$, it follows that

$$E(X_{n+1} \mid \{X_0, X_1, X_2, \ldots, X_n\}) = E[E(X_{n+1} \mid \{Y_0, Y_1, \ldots, Y_n\}) \mid \{X_0, X_1, X_2, \ldots, X_n\}] = 0, \quad (13.29)$$

by (13.28). Thus,

$$E(X_{n+1} \mid \{Y_0, Y_1, \ldots, Y_n\}) = 0 \qquad (n = 1, 2, \ldots) \quad (13.30)$$

implies that

$$E(X_{n+1} \mid \{X_0, X_1, \ldots, X_n\}) = 0 \qquad (n = 1, 2, \ldots). \quad (13.31)$$

In general, however, the converse is not true; namely, (13.31) does not imply (13.30). To understand this better, consider a sequence of random variables $\{Y_n: n = 0, 1, 2, \ldots\}$. Suppose that $\{X_n: n = 0, 1, 2, \ldots\}$ is another sequence of random variables such that, for every n, $X_0, X_1, X_2, \ldots, X_n$ can be expressed as functions of $Y_0, Y_1, Y_2, \ldots, Y_n$. Also assume $EX_n^2 < \infty$ for all n. The condition (13.31) implies that X_{n+1} is orthogonal to all square integrable functions of $X_0, X_1, \ldots, X_n$, while (13.30) implies that X_{n+1} is orthogonal to all square integrable functions of $Y_0, Y_1, \ldots, Y_n$ (Chapter 0, Eq. 4.20). The latter class of functions is larger than the former class. Property (13.30) is therefore stronger than property (13.31).

One may express (13.30) as

$$E(X_{n+1} \mid \mathscr{F}_n) = 0 \qquad (n = 0, 1, 2, \ldots), \quad (13.32)$$

where $\mathscr{F}_n := \sigma\{Y_0, Y_1, \ldots, Y_n\}$ is the sigmafield of all events determined by $\{Y_0, Y_1, \ldots, Y_n\}$. Note that X_n is $\mathscr{F}_n$-measurable, i.e., a Borel-measurable function of $Y_0, \ldots, Y_n$. Also, $\{\mathscr{F}_n\}$ is increasing, $\mathscr{F}_n \subset \mathscr{F}_{n+1}$. This motivates the following more general definition of a martingale.

Definition 13.3. Let $\{X_n\}$ be a sequence of random variables, and $\{\mathscr{F}_n\}$ an increasing sequence of sigmafields such that, for every n, $X_0, X_1, X_2, \ldots, X_n$ are $\mathscr{F}_n$-measurable. If $E|X_n| < \infty$ and (13.32) holds for all n, then $\{X_n: n = 0, 1, 2, \ldots\}$ is said to be *a sequence of* $\{\mathscr{F}_n\}$*-martingale differences*. The sequence of partial sums $\{Z_n = X_0 + \cdots + X_n: n = 0, 1, 2, \ldots\}$ is then said to be *a* $\{\mathscr{F}_n\}$*-martingale*.

Note that a martingale in this sense is also a martingale in the sense of Definition 13.2, since (13.30) implies (13.31). In order to state an appropriate generalization of Theorem 13.1 we need to extend the definition of stopping times given earlier.

Definition 13.4. Let $\{\mathscr{F}_n: n = 0, 1, 2, \ldots\}$ be an increasing sequence of sub-sigmafields of the basic sigmafield $\mathscr{F}$. A random variable τ with nonnegative integral values, including possibly the value $+\infty$, is a $\{\mathscr{F}_n\}$*-stopping time* if (13.1) (or, (13.2)) holds for all n.

Theorem 13.3. Let $\{X_n\}, \{Z_n\}, \{\mathscr{F}_n\}$ be as in Definition 13.3. Let τ be a $\{\mathscr{F}_n\}$-stopping time such that

1. $P(\tau < \infty) = 1$,
2. $E|Z_\tau| < \infty$, and
3. $\lim_{m \to \infty} E(Z_m \mathbf{1}_{\{\tau > m\}}) = 0$.

Then

$$EZ_\tau = EZ_0. \tag{13.33}$$

The proof of Theorem 13.3 is essentially identical to that of Theorem 13.1 (Exercise 6), except that (13.6) becomes

$$EZ_\tau = EZ_m = EZ_0 \tag{13.34}$$

which may not be zero.

For an application of Theorem 13.3 consider an i.i.d. Bernoulli sequence $\{Y_n : n = 1, 2, \ldots\}$:

$$P(Y_n = 1) = P(Y_n = -1) = \tfrac{1}{2}. \tag{13.35}$$

Let $Y_0 = 0$, $\mathscr{F}_n = \sigma\{Y_0, Y_1, \ldots, Y_n\}$. Then, by (13.27) and (13.28), the sequence of random variables

$$Z_n := S_n^2 - n \qquad (n = 0, 1, 2, \ldots) \tag{13.36}$$

is a $\{\mathscr{F}_n\}$-martingale sequence. Here $S_n = Y_0 + \cdots + Y_n$. Let τ be the first time the random walk $\{S_n : n = 0, 1, 2, \ldots\}$ reaches $-a$ or b, where a, b are two given positive integers. Then τ is a $\{\mathscr{F}_n\}$-stopping time. By Proposition 13.4 below, one has $P(\tau < \infty) = 1$ and $E\tau^k < \infty$ for all k. Moreover,

$$E|Z_\tau| = E|S_\tau^2 - \tau| \leqslant ES_\tau^2 + E\tau \leqslant \max\{a^2, b^2\} + E\tau < \infty, \tag{13.37}$$

and

$$\begin{aligned} E|Z_m \mathbf{1}_{\{\tau > m\}}| &\leqslant E((\max\{a^2, b^2\} + m)\mathbf{1}_{\{\tau > m\}}) \\ &= (\max\{a^2, b^2\} + m)P(\tau > m) \\ &= \max\{a^2, b^2\}P(\tau > m) + mP(\tau > m). \end{aligned} \tag{13.38}$$

Now $P(\tau > m) \to 0$ as $m \to \infty$ and, by Chevyshev's inequality,

$$mP(\tau > m) \leqslant \frac{m}{m^2} E\tau^2 \to 0. \tag{13.39}$$

Conditions (1)–(3) of Theorem 13.3 are therefore verified. Hence,

$$EZ_\tau = EZ_1 = ES_1^2 - 1 = 1 - 1 = 0, \tag{13.40}$$

i.e., using (13.17),

$$E\tau = ES_\tau^2 = a^2P(\tau_{-a} < \tau_b) + b^2P(\tau_{-a} > \tau_b) = \frac{a^2b}{a+b} + \frac{b^2a}{a+b} = ab. \tag{13.41}$$

By changing the starting position $Y_0 = S_0 = 0$ of the random walk to $Y_0 = S_0 = x$ one then gets

$$E\tau = (x - c)(d - x), \qquad c < x < d, \tag{13.42}$$

where $\tau = \min\{n: S_n^x = c \text{ or } d\}$.

The next result has been made use of in deriving (13.17), (13.22) and (13.42).

Proposition 13.4. Let $\{X_n: n = 1, 2, \ldots\}$ be an i.i.d. sequence of random variables such that $P(X_n = 0) < 1$. Let τ be the first escape time of the (general) random walk $\{S_n^x := x + X_1 + \cdots + X_n: n = 1, 2, \ldots\}$, $S_0^x = x$, from the interval $(-a, b)$, where a and b are positive numbers. Then $P(\tau < \infty) = 1$, and the moment generating function $\phi(z) = Ee^{z\tau}$ of τ is finite in a neighborhood of $z = 0$.

Proof. There exists an $\varepsilon > 0$ such that either $P(X_n > \varepsilon) > 0$ or $P(X_n < -\varepsilon) > 0$. Assume first that $\delta := P(X_n > \varepsilon) > 0$. Define

$$n_0 = \left[\frac{a+b}{\varepsilon}\right] + 1, \tag{13.43}$$

where $[(a + b)/\varepsilon]$ is the integer part of $(a + b)/\varepsilon$. No matter what the starting position $x \in (-a, b)$ of the random walk may be, if $X_n > \varepsilon$ for all $n = 1, 2, \ldots, n_0$, then $S_{n_0}^x = x + X_1 + \cdots + X_{n_0} > x + n_0\varepsilon > x + a + b \geqslant b$. Therefore,

$$P(\tau \leqslant n_0) \geqslant P(S_{n_0}^x \geqslant b) \geqslant P(X_n > \varepsilon \text{ for } n = 1, 2, \ldots, n_0) \geqslant \delta^{n_0} = \delta_0, \tag{13.44}$$

say. Now consider the events

$$\begin{aligned} A_k &= \{-a < S_n^x < b \text{ for } (k-1)n_0 < n \leqslant kn_0\} \qquad (k = 1, 2, 3, \ldots), \\ A_1 &= \{-a < S_n^x < b \text{ for } 1 \leqslant n \leqslant n_0\}. \end{aligned} \tag{13.45}$$

By (13.44),

$$P(A_1) = P(\tau > n_0) \leqslant 1 - \delta_0. \tag{13.46}$$

Next,

$$P(\tau > 2n_0) = E(\mathbf{1}_{A_1}\mathbf{1}_{A_2}) = E[\mathbf{1}_{A_1}E(\mathbf{1}_{A_2} \mid \{S_1^x, \ldots, S_n^x\})]. \tag{13.47}$$

Now

$$\begin{aligned} E(\mathbf{1}_{A_2} \mid \{S_1^x, \ldots, S_{n_0}^x\}) &= P(-a < S_{n_0+m}^x < b \text{ for } m = 1, \ldots, n_0 \mid \{S_1^x, \ldots, S_{n_0}^x\}) \\ &= P(-a < S_{n_0}^x + X_{n_0+1} + \cdots + X_{n_0+m} < b, \\ &\qquad \text{for } m = 1, \ldots, n_0 \mid \{S_1^x, \ldots, S_{n_0}^x\}). \end{aligned} \tag{13.48}$$

Since $X_{n_0+1} + \cdots + X_{n_0+m}$ is independent of $S_1^x, \ldots, S_{n_0}^x$, the last conditional probability may be evaluated as

$$\begin{aligned} P(-a < z + X_{n_0+1} + \cdots + X_{n_0+m} < b \text{ for } m = 1, \ldots, n_0)_{z = S_{n_0}^x} \\ = P(-a < z + X_1 + \cdots + X_m < b \text{ for } m = 1, \ldots, n_0)_{z = S_{n_0}^x}. \end{aligned} \tag{13.49}$$

The equality in (13.49) is due to the fact that the distribution of $(X_1, X_2, \ldots, X_{n_0})$ is the same as that of $(X_{n_0+1}, X_{n_0+2}, \ldots, X_{2n_0})$. Note that $S_{n_0}^x \in (-a, b)$ on the set A_1. Hence the last probability in (13.49) is not larger than $1 - \delta_0$, by (13.46). Therefore, (13.47) yields

$$P(A_1 \cap A_2) = P(\tau > 2n_0) \leqslant E[\mathbf{1}_{A_1}(1 - \delta_0)] = (1 - \delta_0)P(A_1) \leqslant (1 - \delta_0)^2. \tag{13.50}$$

By recursion it follows that

$$\begin{aligned} P(\tau > kn_0) &= E(\mathbf{1}_{A_1} \cdots \mathbf{1}_{A_k}) = E[\mathbf{1}_{A_1} \cdots \mathbf{1}_{A_{k-1}} E(\mathbf{1}_{A_k} \mid \{S_1^x, \ldots, S_{(k-1)n_0}^x\})] \\ &\leqslant E[\mathbf{1}_{A_1} \cdots \mathbf{1}_{A_{k-1}}(1 - \delta_0)] \leqslant (1 - \delta_0)P(A_1 \cap \cdots \cap A_{k-1}) \\ &= (1 - \delta_0)P(\tau > (k-1)n_0) \leqslant (1 - \delta_0)(1 - \delta_0)^{k-1} \\ &= (1 - \delta_0)^k \qquad (k = 1, 2, \ldots). \end{aligned} \tag{13.51}$$

Since $\{\tau = \infty\} \subset \{\tau > kn_0\}$ for every k, one has

$$P(\tau = \infty) \leqslant (1 - \delta_0)^k \qquad \text{for all } k, \tag{13.52}$$

and consequently, $P(\tau = \infty) = 0$. Finally,

$$\begin{aligned} Ee^{z\tau} &= \sum_{m=1}^{\infty} e^{mz} P(\tau = m) \leqslant \sum_{m=1}^{\infty} e^{m|z|} P(\tau = m) \\ &= \sum_{k=1}^{\infty} \left(\sum_{m=(k-1)n_0+1}^{kn_0} \right) e^{m|z|} P(\tau = m) \end{aligned}$$

$$\leqslant \sum_{k=1}^{\infty} e^{kn_0|z|}P(\tau > (k-1)n_0) \leqslant \sum_{k=1}^{\infty} e^{kn_0|z|}(1-\delta)^{k-1}$$

$$= e^{n_0|z|} \sum_{k=1}^{\infty} ((1-\delta)e^{n_0|z|})^{k-1} < \infty \qquad \text{for } |z| < \frac{-\log(1-\delta)}{n_0}. \tag{13.53}$$

One may proceed in an entirely analogous manner assuming $P(X_n < -\varepsilon) > 0$. ■

An immediate corollary to Proposition 13.4 is that $E\tau^k < \infty$ for all $k = 1, 2, \ldots$ (Exercise 18(i)).

It is easy to check that (Exercise 8), instead of the independence of $\{X_n\}$, it is enough to assume that for some $\varepsilon > 0$ there is a $\delta > 0$ such that

$$\begin{aligned} &P(X_{n+1} > \varepsilon \mid \{X_1, \ldots, X_n\}) \geqslant \delta > 0 \qquad \text{for all } n, \\ \text{or} \quad &P(X_{n+1} < -\varepsilon \mid \{X_1, \ldots, X_n\}) \geqslant \delta > 0 \qquad \text{for all } n. \end{aligned} \tag{13.54}$$

Thus, Proposition 13.4 has the following extension, which is useful in studying processes other than random walks.

Proposition 13.5. The conclusion of Proposition 13.4 holds if, instead of the assumption that $\{X_n\}$ is i.i.d., (13.54) holds for a pair of positive numbers ε, δ.

The next result in this section is *Doob's Maximal Inequality*. A maximal inequality relates the growth of the maximum of partial sums to the growth of the partial sums, and typically shows that the former does not grow much faster than the latter.

Theorem 13.6. (*Doob's Maximal Inequality*). Let $\{Z_0, \ldots, Z_n\}$ be a martingale, $M_n := \max\{|Z_0|, \ldots, |Z_n|\}$.

(a) For all $\lambda > 0$,

$$P(M_n \geqslant \lambda) \leqslant \frac{1}{\lambda} E(|Z_n| \mathbf{1}_{\{M_n \geqslant \lambda\}}) \leqslant E|Z_n|/\lambda. \tag{13.55}$$

(b) If $EZ_n^2 < \infty$ then, for all $\lambda > 0$,

$$P(M_n \geqslant \lambda) \leqslant \frac{1}{\lambda^2} E(Z_n^2 \mathbf{1}_{\{M_n \geqslant \lambda\}}) \leqslant EZ_n^2/\lambda^2, \tag{13.56}$$

and

$$EM_n^2 \leqslant 4EZ_n^2. \tag{13.57}$$

Proof. (a) Write $\mathscr{F}_k := \sigma\{Z_0, \ldots, Z_k\}$, the sigmafield of events determined by $Z_0, \ldots, Z_k$. Consider the events $A_0 := \{|Z_0| \geqslant \lambda\}$, $A_k := \{|Z_j| < \lambda \text{ for } 0 \leqslant j < k, |Z_k| \geqslant \lambda\}$ $(k = 1, \ldots, n)$. The events A_k are pairwise disjoint and

$$\bigcup_{0}^{n} A_k = \{M_n \geqslant \lambda\}.$$

Therefore,

$$P(M_n \geqslant \lambda) = \sum_{k=0}^{n} P(A_k). \tag{13.58}$$

Now $1 \leqslant |Z_k|/\lambda$ on A_k. Using this and the martingale property,

$$P(A_k) = E(\mathbf{1}_{A_k}) \leqslant \frac{1}{\lambda} E(\mathbf{1}_{A_k}|Z_k|) = \frac{1}{\lambda} E[\mathbf{1}_{A_k}|E(Z_n \mid \mathscr{F}_k)|] \leqslant \frac{1}{\lambda} E[\mathbf{1}_{A_k} E(|Z_n| \mid \mathscr{F}_k)]. \tag{13.59}$$

Since $\mathbf{1}_{A_k}$ is $\mathscr{F}_k$-measurable, $\mathbf{1}_{A_k} E(|Z_n| \mid \mathscr{F}_k) = E(\mathbf{1}_{A_k}|Z_n| \mid \mathscr{F}_k)$; and as the expectation of the conditional expectation of a random variable Z equals EZ (see Chapter 0, Theorem 4.4),

$$P(A_k) \leqslant \frac{1}{\lambda} E[E(\mathbf{1}_{A_k}|Z_n| \mid \mathscr{F}_k)] = \frac{1}{\lambda} E(\mathbf{1}_{A_k}|Z_n|). \tag{13.60}$$

Summing over k one arrives at (13.55).

(b) The inequality (13.56) follows in the same manner as above, starting with

$$P(A_k) = E(\mathbf{1}_{A_k}) \leqslant \frac{1}{\lambda^2} E(\mathbf{1}_{A_k} Z_k^2), \tag{13.61}$$

and then noting that

$$\begin{aligned} E(Z_n^2 \mid \mathscr{F}_k) &= E(Z_k^2 + (Z_n - Z_k)^2 - 2Z_k(Z_n - Z_k) \mid \mathscr{F}_k) \\ &= E(Z_k^2 + (Z_n - Z_k)^2 \mid \mathscr{F}_k) - 2Z_k E(Z_n - Z_k \mid \mathscr{F}_k) \\ &= E(Z_k^2 \mid \mathscr{F}_k) + E((Z_n - Z_k)^2 \mid \mathscr{F}_k) \geqslant E(Z_k^2 \mid \mathscr{F}_k). \end{aligned} \tag{13.62}$$

Hence,

$$E(\mathbf{1}_{A_k} Z_n^2) \geqslant E[\mathbf{1}_{A_k} E(Z_k^2 \mid \mathscr{F}_k)] = E[E(\mathbf{1}_{A_k} Z_k^2 \mid \mathscr{F}_k)] = E(\mathbf{1}_{A_k} Z_k^2). \tag{13.63}$$

Now use (13.63) in (13.61) and sum over k to get (13.56).

In order to prove (13.57), first express M_n^2 as $2 \int_0^{M_n} \lambda \, d\lambda$ and interchange the order of taking expectation and integrating with respect to Lebesgue measure

(Fubini's Theorem, Chapter 0, Theorem 4.2), to get

$$EM_n^2 = \int_\Omega \left(2 \int_0^{M_n} \lambda\, d\lambda\right) dP = 2 \int_\Omega \left(\int_0^\infty \lambda \mathbf{1}_{\{M_n \geq \lambda\}}\, d\lambda\right) dP$$

$$= 2 \int_0^\infty \lambda \left(\int_\Omega \mathbf{1}_{\{M_n \geq \lambda\}}\, dP\right) d\lambda = 2 \int_0^\infty \lambda P(M_n \geq \lambda)\, d\lambda. \quad (13.64)$$

Now use the first inequality in (13.55) to derive

$$EM_n^2 \leqslant 2 \int_0^\infty E(\mathbf{1}_{\{M_n \geq \lambda\}}|Z_n|)\, d\lambda = 2 \int_\Omega |Z_n| \left(\int_0^{M_n} d\lambda\right) dP$$

$$= 2 \int_\Omega |Z_n| M_n\, dP = 2E(|Z_n| M_n) \leqslant 2(EZ_n^2)^{1/2}(EM_n^2)^{1/2}, \quad (13.65)$$

using the Schwarz Inequality. Now divide the extreme left and right sides of (13.65) by $(EM_n^2)^{1/2}$ to get (13.57). ■

A well-known inequality of Kolmogorov follows as a simple consequence of (13.56).

Corollary 13.7. Let $X_0, X_1, \ldots, X_n$ be independent random variables, $EX_j = 0$ for $1 \leqslant j \leqslant n$, $EX_j^2 < \infty$ for $0 \leqslant j \leqslant n$. Write $S_k := X_0 + \cdots + X_k$, $M_n := \max\{|S_k|: 0 \leqslant k \leqslant n\}$. Then for every $\lambda > 0$,

$$P(M_n \geq \lambda) \leqslant \frac{1}{\lambda^2} \int_{\{M_n \geq \lambda\}} S_n^2\, dP \leqslant \frac{1}{\lambda^2} ES_n^2. \quad (13.66)$$

The main results in this section may all be extended to continuous-parameter martingales. For this purpose, consider a stochastic process $\{X_t: t \geqslant 0\}$ having right continuous sample paths $t \to X_t$. A *stopping time* τ for $\{X_t\}$ is a random variable with values in $[0, \infty]$, satisfying the property

$$\{\tau \leqslant t\} \in \mathscr{F}_t \qquad \text{for all } t \geqslant 0, \quad (13.67)$$

where $\mathscr{F}_t := \sigma\{X_u: 0 \leqslant u \leqslant t\}$ is the sigmafield of events that depend only on the ("past" of the) process up to time t. As in the discrete-parameter case, first passage times to finite sets are stopping times. Write for Borel sets $B \subset \mathbb{R}^1$,

$$\tau_B := \min\{t \geqslant 0: X_t \in B\}, \quad (13.68)$$

for the *first passage time to the set B*. If B is a singleton, $B = \{y\}$, τ_B is written simply as τ_y, the *first passage time to the state y*.

A stochastic process $\{Z_t\}$ is said to be a $\{\mathscr{F}_t\}$-*martingale* if

$$E(Z_t \mid \mathscr{F}_s) = Z_s \qquad (s < t). \tag{13.69}$$

For a Brownian motion $\{X_t\}$ with drift μ, the process $\{Z_t := X_t - t\mu\}$ is easily seen to be a martingale with respect to $\{\mathscr{F}_t\}$. For this $\{X_t\}$ another example is $\{Z_t := (X_t - t\mu)^2 - t\sigma^2\}$, where σ^2 is the diffusion coefficient of $\{X_t\}$.

The following is the continuous-parameter analogue of Theorem 13.6(b).

Proposition 13.8. Let $\{Z_t : t \geqslant 0\}$ be a right continuous square integrable $\{\mathscr{F}_t\}$-martingale, and let $M_t := \sup\{|Z_s| : 0 \leqslant s \leqslant t\}$. Then

$$P(M_t > \lambda) \leqslant EZ_t^2/\lambda^2 \qquad (\lambda > 0) \tag{13.70}$$

and

$$EM_t^2 \leqslant 4EZ_t^2. \tag{13.71}$$

Proof. For each n let $0 = t_{1,n} < t_{2,n} < \cdots < t_{n,n} = t$ be such that the sets $I_n := \{t_{j,n} : 1 \leqslant j \leqslant n\}$ are increasing (i.e., $I_n \subset I_{n+1}$) and $\bigcup_1^\infty I_n$ is dense in $[0, t]$. Write $M_{t,n} := \max\{|Z_{t_{j,n}}| : 1 \leqslant j \leqslant n\}$. By (13.56),

$$P(M_{t,n} \geqslant \lambda) \leqslant \frac{EZ_t^2}{\lambda^2}.$$

Letting $n \uparrow \infty$, one obtains (13.70) as the sets $F_n := \{M_{t,n} \geqslant \lambda\}$ increase to a set that contains $\{M_t > \lambda\}$ as $n \uparrow \infty$.

Next, from (13.57),

$$EM_{t,n}^2 \leqslant 4EZ_t^2.$$

Now use the Monotone Convergence Theorem (Chapter 0, Theorem 3.2), to get (13.71). ■

The next result is an analogue of Theorem 13.3.

Proposition 13.9. (*Optional Stopping*). Let $\{Z_t : t \geqslant 0\}$ be a right continuous square integrable $\{\mathscr{F}_t\}$-martingale, and τ a $\{\mathscr{F}_t\}$-stopping time such that

(i) $P(\tau < \infty) = 1$,
(ii) $E|Z_\tau| < \infty$, and
(iii) $EZ_{\tau \wedge r} \to EZ_\tau$ as $r \to \infty$.

Then

$$EZ_\tau = EZ_0. \tag{13.72}$$

Proof. Define, for each positive integer n, the random variables

$$\tau^{(n)} := \begin{cases} k2^{-n} & \text{on } \{(k-1)2^{-n} < \tau \leqslant k2^{-n}\} \quad (k = 0, 1, 2, \ldots), \\ \infty & \text{on } \{\tau = \infty\}. \end{cases} \tag{13.73}$$

It is simple to check that $\tau^{(n)}$ is a stopping time with respect to the sequence of sigmafields $\{\mathscr{F}_{k2^{-n}}: k = 0, 1, \ldots\}$, as is $\tau^{(n)} \wedge r$ for every positive integer r. Since $\tau^{(r)} \wedge r \leqslant r$, it follows from Theorem 13.3 that

$$EZ_{\tau^{(n)} \wedge r} = EZ_0. \tag{13.74}$$

Now $\tau^{(n)} \geqslant \tau$ for all n and $\tau^{(n)} \downarrow \tau$ as $n \uparrow \infty$. Therefore, $\tau^{(n)} \wedge r \downarrow \tau \wedge r$. By the right continuity of $t \to Z_t$, $Z_{\tau^{(n)} \wedge r} \to Z_{\tau \wedge r}$.

Also, $|Z_{\tau^{(n)} \wedge r}| \leqslant M_r := \sup\{|Z_t|: 0 \leqslant t \leqslant r\}$, and $EM_r \leqslant (EM_r^2)^{1/2} < \infty$ by (13.71). Applying Lebesgue's Dominated Convergence Theorem (Chapter 0, Theorem 3.4) to (13.74), one gets $EZ_{\tau \wedge r} = EZ_0$. The proof is now complete by assumption (iii). ■

One may apply (13.72) exactly as in the case of the simple symmetric random walk starting at zero (see (13.16) and (13.17)) to get, in the case $\mu = 0$,

$$P(\tau_{-a} > \tau_b) := P(\{X_t\} \text{ reaches } b \text{ before } -a) = \frac{a}{a+b}, \tag{13.75}$$

for arbitrary $a > 0, b > 0$. Similarly, applying (13.72) to $\{Z_t := (X_t - t\mu)^2 - t\sigma^2\}$, as in the case of the simple asymmetric random walk (see (13.20)–(13.22), and use (9.11)),

$$E\tau_{\{-a,b\}} = \frac{(b+a)P(\tau_{-a} > \tau_b) - a}{\mu} = \frac{\left(1 - \exp\left\{-\dfrac{2a\mu}{\sigma^2}\right\}\right)(b+a)}{\left(1 - \exp\left\{-\dfrac{2(b+a)\mu}{\sigma^2}\right\}\right)\mu} - \frac{a}{\mu}, \tag{13.76}$$

where $\{X_t\}$ is a Brownian motion with a nonzero drift μ, starting at zero.

14 CHAPTER APPLICATION: FLUCTUATIONS OF RANDOM WALKS WITH SLOW TRENDS AND THE HURST PHENOMENON

Let $\{Y_n: n = 1, 2, \ldots\}$ be a sequence of random variables representing the annual flows into a reservoir over a span of N years. Let $S_n = Y_1 + \cdots + Y_n$, $n \geqslant 1$,

$S_0 = 0$. Also let $\bar{Y}_N = N^{-1}S_N$. A variety of complicated natural processes (e.g., sediment deposition, erosion, etc.) constrain the life and capacity of a reservoir. However, a particular design parameter analyzed extensively by hydrologists, based on an idealization in which water usage and natural loss would occur at an annual rate estimated by $\bar{Y}_N$ units per year, is the (dimensionless) statistic defined by

$$\frac{R_N}{D_N} := \frac{M_N - m_N}{D_N}, \tag{14.1}$$

where

$$\begin{aligned} M_N &:= \max\{S_n - n\bar{Y}_N : n = 0, 1, \dots, N\} \\ m_N &:= \min\{S_n - n\bar{Y}_N : n = 0, 1, \dots, N\}, \end{aligned} \tag{14.2}$$

$$D_N := \left[\frac{1}{N}\sum_{n=1}^{N}(Y_n - \bar{Y}_N)^2\right]^{1/2}, \qquad \bar{Y}_N = \frac{1}{N}S_N. \tag{14.3}$$

The hydrologist Harold Edwin Hurst stimulated a tremendous amount of interest in the possible behaviors of R_N/D_N for large values of N. On the basis of data that he analyzed for regions of the Nile River, Hurst published a finding that his plots of $\log(R_N/D_N)$ versus $\log N$ are linear with slope $H \approx 0.75$. William Feller was soon to show this to be an anomaly relative to the standard statistical framework of i.i.d. flows $Y_1, \dots, Y_n$ having finite second moment. The precise form of Feller's analysis is as follows.

Let $\{Y_n\}$ be an i.i.d. sequence with

$$EY_n = d, \qquad \operatorname{Var} Y_n = \sigma^2 > 0. \tag{14.4}$$

First consider that, by the central limit theorem, $S_N = Nd + O(N^{1/2})$ in the sense that $(S_N - Nd)/\sqrt{N}$ is, for large N, distributed approximately like a Gaussian random variable with mean zero and variance σ^2. If one defines

$$\begin{aligned} \tilde{M}_N &= \max\{S_n - nd : 0 \leqslant n \leqslant N\}, \\ \tilde{m}_N &= \min\{S_n - nd : 0 \leqslant n \leqslant N\}, \\ \tilde{R}_N &= \tilde{M}_N - \tilde{m}_N, \end{aligned} \tag{14.5}$$

then by the functional central limit theorem (FCLT)

$$\left(\frac{\tilde{M}_N}{\sigma\sqrt{N}}, \frac{\tilde{m}_N}{\sigma\sqrt{N}}\right) \Rightarrow (\tilde{M}, \tilde{m}), \qquad \frac{\tilde{R}_N}{\sigma\sqrt{N}} \Rightarrow \tilde{R} \qquad \text{as } N \to \infty, \tag{14.6}$$

where $\Rightarrow$ denotes convergence in distribution of the sequence on the left to the distribution of the random variable(s) on the right. Here

$$\begin{aligned}\tilde{M} &:= \max\{B_t: 0 \leqslant t \leqslant 1\},\\ \tilde{m} &:= \min\{B_t: 0 \leqslant t \leqslant 1\},\\ \tilde{R} &:= \tilde{M} - \tilde{m},\end{aligned} \tag{14.7}$$

with $\{B_t\}$ a standard Brownian motion starting at zero. It follows that the magnitude of $\tilde{R}_N$ is $O(N^{1/2})$. Under these circumstances, therefore, one would expect to find a fluctuation between the maximum and minimum of partial sums, centered around the mean, over a period N to be of the order $N^{1/2}$. To see that this still remains for $R_N/(\sqrt{N}\,D_N)$ in place of $\tilde{R}_N/(\sqrt{N}\,\sigma)$, first note that, by the strong law of large numbers applied to Y_n and Y_n^2 separately, we have with probability 1 that $\bar{Y}_N \to d$, and

$$D_N^2 = \frac{1}{N}\sum_{n=1}^{N} Y_n^2 - \bar{Y}_N^2 \to EY_1^2 - d^2 = \sigma^2, \qquad \text{as } N \to \infty. \tag{14.8}$$

Therefore, with probability 1, as $N \to \infty$,

$$\frac{R_N}{\sqrt{N}\,D_N} \sim \frac{R_N}{\sqrt{N}\,\sigma} \tag{14.9}$$

where "$\sim$" indicates "asymptotic equality" in the sense that the ratio of the two sides goes to 1 as $N \to \infty$. This implies that the asymptotic distributions of the two sides of (14.9) are the same. Next notice that

$$\begin{aligned}\frac{M_N}{\sigma\sqrt{N}} &= \max_{0\leqslant n\leqslant N}\left(\frac{(S_n - nd) - n(\bar{Y}_N - d)}{\sigma\sqrt{N}}\right)\\ &= \max_{0\leqslant t\leqslant 1}\left(\frac{S_{[Nt]} - [Nt]d}{\sigma\sqrt{N}} - \frac{[Nt]}{N}\left(\frac{S_N - Nd}{\sigma\sqrt{N}}\right)\right)\\ &\Rightarrow \max_{0\leqslant t\leqslant 1}(B_t - tB_1) := M\end{aligned} \tag{14.10}$$

and

$$\frac{m_N}{\sigma\sqrt{N}} = \min_{0\leqslant t\leqslant 1}\left(\frac{S_{[Nt]} - [Nt]d}{\sigma\sqrt{N}} - \frac{[Nt]}{N}\left(\frac{S_N - Nd}{\sigma\sqrt{N}}\right)\right) \Rightarrow \min_{0\leqslant t\leqslant 1}(B_t - tB_1) := m,$$

and

$$\left(\frac{M_N}{\sigma\sqrt{N}}, \frac{m_N}{\sigma\sqrt{N}}\right) \Rightarrow (M, m). \tag{14.11}$$

Therefore,

$$\frac{R_N}{D_N\sqrt{N}} \sim \frac{R_N}{\sigma\sqrt{N}} \Rightarrow M - m = R, \tag{14.12}$$

where R is a strictly positive random variable. Once again then, R_N/D_N, the so-called *rescaled adjusted range* statistic, is of the order of $O(N^{1/2})$.

The basic problem raised by Hurst is to identify circumstances under which one may obtain an exponent $H > \frac{1}{2}$. The next major theoretical result following Feller was again somewhat negative, though quite insightful. Specifically, P. A. P. Moran considered the case of i.i.d. random variables $Y_1, Y_2, \ldots$ having "fat tails" in their distribution. In this case the re-scaling by D_N serves to compensate for the increased fluctuation in R_N to the extent that cancellations occur *resulting again in* $H = \frac{1}{2}$.

The first positive result was obtained by Mandelbrot and VanNess, who obtained $H > \frac{1}{2}$ under a stationary but strongly dependent model having moments of all orders, in fact Gaussian (theoretical complements 1, and theoretical complements 1.3, Chapter IV). For the present section we will consider the case of independent but nonstationary flows having finite second moments. In particular, it will now be shown that under an appropriately slow trend superimposed on a sequence of i.i.d. random variables the Hurst effect appears. We will say that the *Hurst exponent* is H if $R_N/(D_N N^H)$ converges in distribution to a nonzero real-valued random variable as N tends to infinity. In particular, this includes the case of convergence in probability to a positive constant.

Let $\{X_n\}$ be an i.i.d. sequence with $EX_n = d$ and $\text{Var}\, X_n = \sigma^2$ as above, and let $f(n)$ be an arbitrary real-valued function on the set of positive integers. We assume that the observations Y_n are of the form

$$Y_n = X_n + f(n). \tag{14.13}$$

The partial sums of the observations are

$$\begin{aligned} S_n &= Y_1 + \cdots + Y_n = X_1 + \cdots + X_n + f(1) + \cdots + f(n) \\ &= S_n^* + \sum_{j=1}^{n} f(j), \qquad S_0 = 0, \end{aligned} \tag{14.14}$$

where

$$S_n^* = X_1 + \cdots + X_n. \tag{14.15}$$

Introduce the notation D_N^* for the standard deviation of the X-values $\{X_n: 1 \leqslant n \leqslant N\}$,

$$D_N^{*2} := \frac{1}{N} \sum_{n=1}^{N} (X_n - \bar{X}_N)^2. \tag{14.16}$$

Then, writing $\bar{f}_N = \sum_{n=1}^{N} f(n)/N$,

$$
\begin{aligned}
D_N^2 &:= \frac{1}{N}\sum_{n=1}^{N} (Y_n - \bar{Y}_N)^2 \\
&= \frac{1}{N}\sum_{n=1}^{N} (X_n - \bar{X}_N)^2 + \frac{1}{N}\sum_{n=1}^{N} (f(n) - \bar{f}_N)^2 + \frac{2}{N}\sum_{n=1}^{N} (f(n) - \bar{f}_N)(X_n - \bar{X}_N) \\
&= D_N^{*2} + \frac{1}{N}\sum_{n=1}^{N} (f(n) - \bar{f}_N)^2 + \frac{2}{N}\sum_{n=1}^{N} (f(n) - \bar{f}_N)(X_n - \bar{X}_N). \qquad (14.17)
\end{aligned}
$$

Also write

$$
\begin{aligned}
M_N &= \max_{0 \leqslant n \leqslant N} \{S_n - n\bar{Y}_N\} = \max_{0 \leqslant n \leqslant N} \left\{S_n^* - n\bar{X}_N + \sum_{j=1}^{n} (f(j) - \bar{f}_N)\right\}, \\
m_N &= \min_{0 \leqslant n \leqslant N} \{S_n - n\bar{Y}_N\} = \min_{0 \leqslant n \leqslant N} \left\{S_n^* - n\bar{X}_N + \sum_{j=1}^{n} (f(j) - \bar{f}_N)\right\}, \qquad (14.18) \\
R_N &= M_N - m_N, \qquad R_N^* = \max_{0 \leqslant n \leqslant N} \{S_n^* - n\bar{X}_N\} - \min_{0 \leqslant n \leqslant N} \{S_n - n\bar{X}_N\}.
\end{aligned}
$$

For convenience, write

$$
\begin{aligned}
\mu_N(n) &:= \sum_{j=1}^{n} (f(j) - \bar{f}_N), \qquad \mu_N(0) = 0, \\
\Delta_N &:= \max_{0 \leqslant n \leqslant N} \mu_N(n) - \min_{0 \leqslant n \leqslant N} \mu_N(n).
\end{aligned} \qquad (14.19)
$$

Observe that

$$
\begin{aligned}
M_N &\leqslant \max_{0 \leqslant n \leqslant N} \mu_N(n) + \max_{0 \leqslant n \leqslant N} (S_n^* - n\bar{X}_N), \\
m_N &\geqslant \min_{0 \leqslant n \leqslant N} \mu_N(n) + \min_{0 \leqslant n \leqslant N} (S_n^* - n\bar{X}_N),
\end{aligned} \qquad (14.20)
$$

and

$$
\begin{aligned}
M_N &\geqslant \max_{0 \leqslant n \leqslant N} \mu_N(n) + \min_{0 \leqslant n \leqslant N} (S_n^* - n\bar{X}_N), \\
m_N &\leqslant \min_{0 \leqslant n \leqslant N} \mu_N(n) + \max_{0 \leqslant n \leqslant N} \{S_n^* - n\bar{X}_N).
\end{aligned} \qquad (14.21)
$$

From (14.20) one gets $R_N \leqslant \Delta_N + R_N^*$, and from (14.21), $R_N \geqslant \Delta_N - R_N^*$. In other words,

$$|R_N - \Delta_N| \leqslant R_N^*. \tag{14.22}$$

Note that in the same manner,

$$|R_N - R_N^*| \leqslant \Delta_N. \tag{14.23}$$

It remains to estimate D_N and Δ_N.

Lemma 1. If $f(n)$ converges to a finite limit, then D_N^2 converges to σ^2 with probability 1.

Proof. In view of (14.17), it suffices to prove

$$\frac{1}{N}\sum_{n=1}^{N}(f(n) - \bar{f}_N)^2 + \frac{2}{N}(f(n) - \bar{f}_N)(X_n - \bar{X}_N) \to 0 \qquad \text{as } N \to \infty. \tag{14.24}$$

Let α be the limit of $f(n)$. Then

$$\frac{1}{N}\sum_{n=1}^{N}(f(n) - \bar{f}_N)^2 = \frac{1}{N}\sum_{n=1}^{N}(f(n) - \alpha)^2 - (\bar{f}_N - \alpha)^2. \tag{14.25}$$

Now if a sequence $g(n)$ converges to a limit θ, then so do its *Caesaro means* $N^{-1}\sum_1^N g(n)$. Applying this to the sequences $(f(n) - \alpha)^2$ and $f(n)$, observe that (14.25) goes to zero as $N \to \infty$. Next

$$\frac{1}{N}\sum_{n=1}^{N}(f(n) - \bar{f}_N)(X_n - \bar{X}_N) = \frac{1}{N}\sum_{n=1}^{N}(f(n) - \alpha)(X_n - d) - (\bar{f}_N - \alpha)(\bar{X}_N - d). \tag{14.26}$$

The second term on the right clearly tends to zero as N increases. Also, by Schwarz inequality,

$$\left|\frac{1}{N}\sum_{n=1}^{N}(f(n) - \alpha)(X_n - d)\right| \leqslant \frac{1}{N}\left(\sum_{n=1}^{N}(f(n) - \alpha)^2\right)^{1/2}\left(\sum_{n=1}^{N}(X_n - d)^2\right)^{1/2}.$$

By the strong law of large numbers, $N^{-1}\sum_{n=1}^{N}(X_n - d)^2 \to E(X_1 - d)^2 = \sigma^2$ and the Caesaro means $N^{-1}\sum_{n=1}^{N}(f(n) - \alpha)^2$ go to zero as $N \to \infty$, since $(f(n) - \alpha)^2 \to 0$ as $n \to \infty$. ■

From (14.12), (14.22), and Lemma 1 we get the following result.

Theorem 14.1. If $f(n)$ converges to a finite limit, then for every $H > \frac{1}{2}$,

$$\left| \frac{R_N}{D_N N^H} - \frac{\Delta_N}{D_N N^H} \right| \to 0 \qquad \text{in probability as } N \to \infty. \tag{14.27}$$

In particular, the Hurst effect with exponent $H > \frac{1}{2}$ holds if and only if, for some positive number c',

$$\lim_{N \to \infty} \frac{\Delta_N}{N^H} = c'. \tag{14.28}$$

Example 1. Take

$$f(n) = \alpha + c(n + m)^\beta \qquad (n = 1, 2, \ldots), \tag{14.29}$$

where α, c, m, β are parameters, with $c \neq 0$, $m \geqslant 0$. The presence of m indicates the *starting point* of the trend, namely, m units of time before the time $n = 0$. Since the asymptotics are not affected by the particular value of m, we assume henceforth that $m = 0$ without essential loss of generality. For simplicity, also take $c > 0$. The case $c < 0$ can be treated in the same way.

First let $\beta < 0$. Then $f(n) \to \alpha$, and Theorem 14.1 applies. Recall that

$$\Delta_N = \max_{0 \leqslant n \leqslant N} \mu_N(n) - \min_{0 \leqslant n \leqslant N} \mu_N(n), \tag{14.30}$$

where

$$\begin{aligned} \mu_N(n) &= \sum_{j=1}^{n} (f(j) - \bar{f}_N) \qquad \text{for } 1 \leqslant n \leqslant N, \\ \mu_N(0) &= 0. \end{aligned} \tag{14.31}$$

Notice that, with $m = 0$ and $c > 0$,

$$\mu_N(n) - \mu_N(n-1) = c\left(n^\beta - \frac{1}{N} \sum_{j=1}^{N} j^\beta \right) \tag{14.32}$$

is positive for $n < (N^{-1} \sum_{j=1}^{N} j^\beta)^{1/\beta}$, and negative or zero otherwise. This shows that the *maximum* of $\mu_N(n)$ is attained at $n = n_0$ given by

$$n_0 = \left[\left(\frac{1}{N} \sum_{j=1}^{N} j^\beta \right)^{1/\beta} \right], \tag{14.33}$$

where $[x]$ denotes the *integer part* of x. The minimum value of $\mu_N(n)$ is *zero*, attained at $n = 0$ and $n = N$. Thus,

$$\Delta_N = \mu_N(n_0) = c \sum_{k=1}^{n_0} \left(k^\beta - \frac{1}{N} \sum_{j=1}^{N} j^\beta \right). \tag{14.34}$$

By a comparison with a Riemann sum approximation to $\int_0^1 x^\beta \, dx$, one obtains

$$\frac{1}{N} \sum_{j=1}^{N} j^\beta = N^\beta \sum_{j=1}^{N} \left(\frac{j}{N}\right)^\beta \frac{1}{N} \sim \begin{cases} (1+\beta)^{-1} N^\beta & \text{for } \beta > -1 \\ N^{-1} \log N & \text{for } \beta = -1 \\ N^{-1} \sum_{j=1}^{\infty} j^\beta & \text{for } \beta < -1. \end{cases} \tag{14.35}$$

By (14.33) and (14.35),

$$n_0 \sim \begin{cases} (1+\beta)^{-1/\beta} N & \text{for } \beta > -1 \\ N/\log N & \text{for } \beta = -1 \\ \left(\sum_1^\infty j^\beta \right)^{1/\beta} N^{1/(-\beta)} & \text{for } \beta < -1. \end{cases} \tag{14.36}$$

From (14.34)–(14.36) it follows that

$$\Delta_N \sim \begin{cases} cn_0 \left(\dfrac{n_0^\beta}{1+\beta} - \dfrac{N^\beta}{1+\beta} \right) \sim c_1 N^{1+\beta}, & \beta > -1, \\ cn_0 (n_0^{-1} \log n_0 - N^{-1} \log N) \sim c \log N, & \beta = -1, \\ c \sum_{j=1}^{\infty} j^\beta = c_2, & \beta < -1. \end{cases} \tag{14.37}$$

Here c_1, c_2 are positive constants depending only on β. Now consider the following cases.

CASE 1: $-\frac{1}{2} < \beta < 0$. In this case Theorem 14.1 applies with $H(\beta) = 1 + \beta > \frac{1}{2}$. Note that, by Lemma 1, $D_N \sim \sigma$ with probability 1. Therefore,

$$\frac{R_N}{D_N N^{1+\beta}} \to c' > 0 \quad \text{in probability as } N \to \infty, \qquad \text{if } -\tfrac{1}{2} < \beta < 0. \tag{14.38}$$

CASE 2: $\beta < -\frac{1}{2}$. Use inequality (14.23), and note from (14.37) that $\Delta_N = o(N^{1/2})$. Dividing both sides of (14.23) by $D_N N^{1/2}$ one gets, in probability as $N \to \infty$,

$$\frac{R_N}{D_N N^{1/2}} \sim \frac{R_N^*}{D_N N^{1/2}} \sim \frac{R_N^*}{\sigma N^{1/2}} \qquad \text{if } \beta < -\frac{1}{2}. \tag{14.39}$$

But $R_N^*/\sigma N^{1/2}$ converges in distribution to R by (14.11). Therefore, the Hurst exponent is $H(\beta) = \frac{1}{2}$.

CASE 3: $\beta = 0$. In this case the Y_n are i.i.d. Therefore, as proved at the outset, the Hurst exponent is $\frac{1}{2}$.

CASE 4: $\beta > 0$. In this case Lemma 1 does not apply, but a simple computation yields

$$D_N \sim c_3 N^\beta \qquad \text{with probability 1 as } N \to \infty, \qquad \text{if } \beta > 0. \tag{14.40}$$

Here c_3 is an appropriate positive number. Combining (14.40), (14.37), and (14.22) one gets

$$\frac{R_N}{N D_N} \to c_4 \qquad \text{in probability as } N \to \infty, \qquad \text{if } \beta > 0, \tag{14.41}$$

where c_4 is a positive constant. Therefore, $H(\beta) = 1$.

CASE 5: $\beta = -\dfrac{1}{2}$. A slightly more delicate argument than used above shows that

$$\frac{R_N}{D_N N^{1/2}} \Rightarrow \max_{0 \leqslant t \leqslant 1} (B_t - tB_1 - 2ct(1-t)), \tag{14.42}$$

where $\{B_t\}$ is a standard Brownian motion starting at zero. Thus, $H(-\frac{1}{2}) = \frac{1}{2}$. In this case one considers the process $\{Z_N(s)\}$ defined by

$$Z_N\left(\frac{n}{N}\right) = \frac{S_n - n\overline{Y}_N}{\sqrt{N}\, D_N} \qquad \text{for } n = 1, 2, \ldots, \text{N},$$

and linearly interpolated between n/N and $(n+1)/N$. Then $\{Z_N(s)\}$ converges in distribution to $\{B_s^* + 2c\sqrt{s}(1 - \sqrt{s})\}$, where $\{B_s^*\}$ is the Brownian bridge. In this case the asymptotic distribution of $R_N/(\sqrt{N}\, D_N)$ is the *nondegenerate* distribution of

$$\max_{0 \leqslant s \leqslant 1} B_s^* - \min_{0 \leqslant s \leqslant 1} B_s^*,$$

(Exercise 1).

The graph of $H(\beta)$ versus β in Figure 14.1 summarizes the results of the preceding cases 1 through 5.

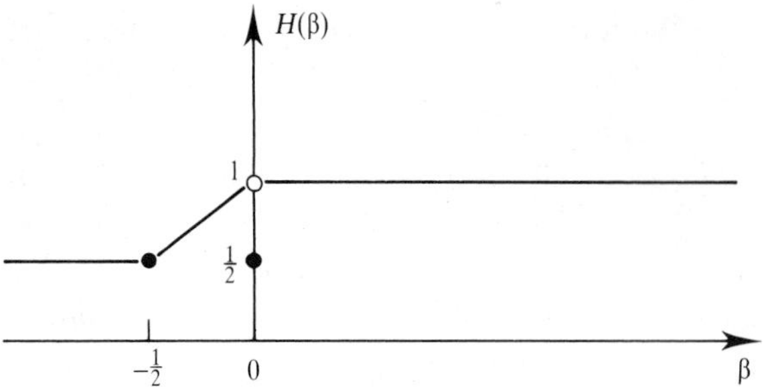

Figure 14.1

For purposes of data analysis, note that (14.38) implies

$$\log \frac{R_N}{D_N} - [\log c' + (1 + \beta) \log N] \to 0 \qquad \text{if } -\tfrac{1}{2} < \beta < 0. \qquad (14.43)$$

In other words, for large N the plot of $\log R_N/D_N$ against $\log N$ should be approximately linear with slope $H = 1 + \beta$, if $-\frac{1}{2} < \beta < 0$.

Under the i.i.d. model one would expect to find a fluctuation between the maximum and the minimum of partial sums, centered around the sample mean, over a period N to be of the order of $N^{1/2}$. One may then try to check the appropriateness of the model, i.e., the presumed i.i.d. nature of the observations, by taking successive (disjoint) blocks of Y_n values each of size N, calculating the difference between the maximum and minimum of partial sums in each block, and seeing whether this difference is of the order of $N^{1/2}$. In this regard it is of interest that many other geophysical data sets indicative of climatic patterns have been reported to exhibit the Hurst effect.

EXERCISES

Exercises for Section I.1

1. The following events refer to the coin-tossing model. Determine which of these events is finite-dimensional and calculate the probability of each event.
   (i) 1 appears for the first time on the 10th toss.
   (ii) 1 appears for the last time on the 10th toss.
   (iii) 1 appears on every other toss.
   (iv) The proportion of 1's stabilizes to a given value r as the number of tosses is increased without bound.
   (v) Infinitely many 1's occur.

2. For the coin-tossing experiment (Example 1) determine the probability that the *first head* is followed immediately by a tail.

3. Let L_n denote the length of the run of 1's starting at the nth toss in the coin-tossing model. Calculate the distribution of L_n.

4. Each integer lattice site of $\mathbb{Z}_+^d$ is independently colored red or green with probabilities p and $q = 1 - p$, respectively. Let E_m be the event that the number of green sites equals the number of red sites in the block of sites of side lengths $2m$ (sites per side) with a corner at the origin. Calculate $P(E_m \text{ i.o.})$ for $d \geqslant 3$. [*Hint*: Use the Borel–Cantelli Lemma, Chapter 0, Lemma 6.1.]

5. A die is repeatedly tossed and the number of spots is recorded at each stage. Fix j, $1 \leqslant j \leqslant 6$, and let p_n be the probability that j occurs among the first n tosses. Calculate p_n and the probability that j eventually occurs.

6. (*A Fair Coin Simulation*) Suppose that you are given a coin for which the probability of a head is p where $0 < p < 1$. At each unit of time toss the coin twice and at the nth such double toss record:

$$\begin{aligned} X_n &= 1 && \text{if a head followed by a tail occurs,} \\ X_n &= -1 && \text{if a tail followed by a head occurs,} \\ X_n &= 0 && \text{if the outcomes of the double toss coincide.} \end{aligned}$$

Let, $\tau = \min\{n \geqslant 1 \colon X_n = 1 \text{ or } -1\}$.
   (i) Verify that $P(\tau < \infty) = 1$.
   (ii) Calculate the distribution of $Y = X_\tau$.
   (iii) Calculate $E\tau$.

7. Show that the two probability distributions for unending independent tosses of a coin, corresponding to distinct probabilities $p_1 \neq p_2$ for a head in a single toss, assign respective total probabilities to mutually disjoint subsets of the coin-tossing sample space Ω. [*Hint*: Consider the density of 1's in the various possible sequences in Ω and use the SLLN (Chapter 0, Theorem 6.1). Such distributions are said to be *mutually singular*.]

8. Suppose that M particles can be in each of N possible states $s_1, s_2, \ldots, s_N$. Construct a probability space and calculate the distribution of $(X_1, \ldots, X_N)$, where X_i is the number of particles in state s_i, for each of the following schemes (i)–(iii).
   (i) (*Maxwell–Boltzmann*) The particles are distinguishable, say labeled $m_1, \ldots, m_M$, and are randomly assigned states in such a way that all possible distinct assignments are equally likely to occur. (Imagine putting balls (particles) into boxes (states).)
   (ii) (*Bose–Einstein*) The particles are not distinguishable but are randomly assigned states in such a way that all possible values of the numbers of particles in the various states are equally likely to occur.
   (iii) (*Fermi–Dirac*) The particles are not distinguishable but are randomly assigned states in such a way that there can be at most one particle in any one of the states and all possible values of the numbers of particles in various states under the exclusion principle are equally likely to occur.
   (iv) For each of the above distributions calculate the asymptotic distribution of X_i as M and $N \to \infty$ such that $M/N \to \rho$, where $\rho > 0$ is the asymptotic density of occupied states.

9. Suppose that the sample paths of $\{X_t\}$ solve the following problem, $dX/dt = -\beta X$, $X_0 = 1$, with probability 1, where β is a random parameter having a normal distribution.
   (i) Calculate EX_t.
   (ii) Calculate the solution $x(t)$ to the problem with β replaced by $E\beta$.
   (iii) How do EX_t and $x(t)$ compare for short times? What about in the long run?

(iv) Calculate the distribution of the process $\{X_t\}$.

*10. (i) Show that the sample space $\Omega = \{\omega = (\omega_1, \omega_2, \ldots): \omega_i = 1 \text{ or } 0\}$ for repeated tosses of a coin is uncountable.

(ii) Show that under binary expansion of numbers in the unit interval, the event $\{\omega \in \Omega: \omega_1 = \varepsilon_1, \omega_2 = \varepsilon_2, \ldots, \omega_n = \varepsilon_n\}$, $\varepsilon_i = 0$ or 1, is represented by an interval in $[0, 1]$ of length $1/2^n$.

*11. Suppose that $\Omega = [0, 1]$ and $\mathscr{F}$ is the Borel sigmafield of $[0, 1]$. Suppose that P is defined by the uniform p.d.f. $f(x) = 1$, $x \in [0, 1]$. Remove the middle one-third segment from $[0, 1]$ and let $J_{11} = [0, 1/3]$ and $J_{12} = [2/3, 1]$ be the remaining left and right segments, respectively. Let $I_2 = J_{11} \cup J_{12}$. Next remove the middle one-third from each of these and let $J_{111} = [0, 1/9]$ and $J_{112} = [2/9, 3/9]$ be the remaining left and right segments of J_{11}, and let $J_{121} = [6/9, 7/9]$ and $J_{122} = [8/9, 1]$ be the remaining left and right intervals of J_{12}. Let $I_3 = J_{111} \cup J_{112} \cup J_{121} \cup J_{122}$. Repeat this process for each of these segments, and so on. At the nth stage, I_n is the union of 2^{n-1} disjoint intervals each of length $1/3^{n-1}$. In particular, the probability (under f) that a randomly selected point belongs to I_n, i.e., the length of I_n, is $P(I_n) = 2^{n-1}/3^{n-1}$. The sets $I_1 \supset I_2 \supset \cdots \supset I_n \supset \cdots$ form a decreasing sequence. The *Cantor set* is the limiting set defined by $C = \bigcap I_n$.

(i) Show that C is in one-to-one correspondence with the interval $(0, 1]$.

(ii) Verify that C is a Borel set and calculate $P(C)$.

(iii) Construct a continuous c.d.f that does not have a p.d.f. [*Hint*: Define the *Cantor ternary function* F on $[0, 1]$ as follows. Let $F_0(x) = (2k - 1)/2^n$ for x in the kth interval from the left among those 2^{n-1} subintervals deleted for the first time at the nth stage. Then F_0 is well defined and has a continuous extension to a function F on all of $[0, 1]$ with $F(1) = 1$ and $F(0) = 0$.]

Exercises for Section I.2

1. Let $\{X_n: n \geqslant 1\}$ be an i.i.d. sequence with $EX_n^2 < \infty$, and define the general random walk starting at x by $S_0^x = x$, $S_n^x = x + X_1 + \cdots + X_n$. Calculate each of the following numerical characteristics of the distribution.

(i) ES_n^x

(ii) Var S_n^x

(iii) $\text{Cov}(S_n^x, S_m^x)$

2. (*A Mechanical Model*) Consider the scattering device depicted in Figure Ex.I.2, consisting of regularly spaced scatterers in a triangular array. Balls are successively dropped from the top, scattered by the pegs and finally caught in the bins along the bottom row.

(i) Calculate the probabilities that a ball will land in each of the respective bins if the tree has n levels.

(ii) What is the expected number of balls that will fall in each of the bins if N balls are dropped in succession?

*3. (*Dorfman's Blood Testing Scheme*) Prior to World War II the army made individual tests of blood samples for venereal disease. So N recruits entering a processing station required N tests. R. Dorfman suggested pooling the blood samples of m individuals and then testing the pooled sample. If the test is negative then only *one*

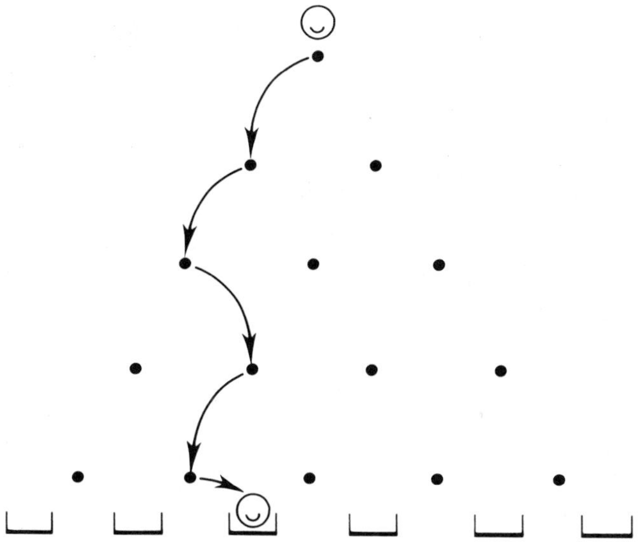

Figure Ex.I.2

test is needed. If the test of the pool is positive then at least one individual has the disease and each of the m persons must be retested individually, resulting in this event in $m + 1$ *tests*. Let $X_1, X_2, \ldots, X_N$ be an i.i.d. sequence of 0 or 1 valued random variables with $p = P(X_n = 1)$ for $n = 1, 2, \ldots$. Let the event $\{X_n = 1\}$ be used to indicate that the nth individual is infected; then the parameter p measures the *incidence* of the disease in the population. Let $S_n = X_1 + X_2 + \cdots + X_n$ denote the number of infected individuals among the first n individuals tested, $S_0 = 0$. Let T_k denote the number of tests required for the kth group of m individuals tested, $k = 1, 2, \ldots, [N/m]$. Thus, for $m \geqslant 2$, $T_k = m + 1$ if $S_{mk} - S_{m(k-1)} \neq 0$ and $T_k = 1$ if $S_{mk} - S_{m(k-1)} = 0$. The total number of tests (cost) for N individuals tested in groups of size m each is

$$C_N = \sum_{k=1}^{[N/m]} T_k + (N - m[N/m]), \qquad m \geqslant 2.$$

Find m such that, for given N (large) and p, the expected number of tests per person is minimal. [*Hint*: Consider the limit as $N \to \infty$ and show that the optimal m, if one exists, is the integer value of m that minimizes the function

$$c(m) = \lim_{N\to\infty} \frac{EC_N}{N} = 1 + \frac{1}{m} - (1 - p)^m, \qquad m \geqslant 2, \quad c(1) = 1.$$

Analyze the extreme values of the function $g(x) = (1/x) - (1 - p)^x$ for $x > 0$ (see D. W. Turner, F. E. Tidmore, D. M. Young (1988), *SIAM Review*, **30**, pp. 119–122).]

4. Let $\{S_n^x\}$ be the simple symmetric random walk starting at x and let

$u(n, y) = P(S_n^x = y)$. Verify that $u(n, y)$ satisfies the following initial value problem

$$\partial_n u = \tfrac{1}{2}\nabla_y^2 u, \qquad u(0, y) = \delta_{x,y},$$

where $\delta_{x,y}$ is Kronecker's delta and

$$\partial_n u(n, y) = u(n + 1, y) - u(n, y),$$
$$\nabla_y u(n, y) = u(n, y + \tfrac{1}{2}) - u(n, y - \tfrac{1}{2}).$$

5. Suppose that $\{S_n^{(1)}\}$ and $\{S_n^{(2)}\}$ are independent (general) random walks whose increments (step sizes) have the common distribution $\{p_x : x \in \mathbb{Z}\}$. Verify that $\{T_n = S_n^{(1)} - S_n^{(2)}\}$ is a random walk with step size distribution $\{\bar{p}_x : x \in \mathbb{Z}\}$ given by a *symmetrization* of $\{p_x : x \in \mathbb{Z}\}$, i.e. $\bar{p}_x = \bar{p}_{-x}$, $x \in \mathbb{Z}$. Express $\{\bar{p}_x\}$ in terms of $\{p_x : x \in \mathbb{Z}\}$.

Exercises for Section I.3

1. Write out a complete derivation of (3.3) by conditioning on X_1.
2. (i) Write out the symmetry argument for (3.8) using (3.6). [*Hint*: Look at the random walk $\{-S_n^x : n \geqslant 0\}$.]
   (ii) Verify that (3.6) can be expressed as

$$P(T_d^x < T_c^x) = \exp\{\gamma_p(d - x)\}\frac{\sinh\{\gamma_p(x - c)\}}{\sinh\{\gamma_p(d - c)\}}, \qquad \text{where } \gamma_p = \ln((q/p)^{1/2}).$$

3. Justify the use of limits in (3.9) using the *continuity properties* (Chapter 0, (1.1), (1.2)) of a probability measure.
4. Verify that $P(S_n^x \neq y \text{ for all } n \geqslant N) = 0$ for each $N = 1, 2, \ldots$ to establish (3.18).
5. (i) If $p < q$ and $x < d$, then give the symmetry argument to calculate, using (3.9), the probability that the simple random walk starting at x will eventually reach d in (3.10).
   (ii) Verify that (3.10) may be expressed as

$$P(T_d^x < \infty) = e^{-\alpha_p(d-x)} \qquad \text{for some } \alpha_p \geqslant 0.$$

6. Let η_x denote the "time of the first return to x" for the simple random walk $\{S_n^x\}$ starting at x. Show that

$$P(\eta_x < \infty) = P(\{S_n^x\} \text{ eventually returns to } x)$$
$$= 2\min(p, q).$$

[*Hint*: $P(\eta_x < \infty) = pP(T_x^{x+1} < \infty) + qP(T_x^{x-1} < \infty)$.]

7. Let $M \equiv M^0 := \sup_{n \geqslant 0} S_n$, $m \equiv m^0 := \inf_{n \geqslant 0} S_n$ be the (possibly infinite) *extreme* statistics for the simple random walk $\{S_n\}$ starting at 0.

   (i) Calculate the distribution function and probability mass function for M in the case $p < q$. [*Hint*: Use (3.10) and consider $P(M \geqslant t)$.]
   (ii) Do the calculations corresponding to (i) for m when $p > q$, and for M^x, and m^x.

8. Let $\{X_n\}$ be an arbitrary i.i.d. sequence of integer-valued displacement random variables for a general random walk on $\mathbb{Z}$ denoted by $S_n = X_1 + \cdots + X_n$, $n \geqslant 1$, and $S_0 = 0$. Show that the random walk is transient if EX_1 exists and is nonzero.

9. Say that an "equalization" occurs at time n in coin tossing if the number of heads acquired by time n coincides with the number of tails. Let E_n denote the event that an equalization occurs at time n.
   (i) Show that if $p \neq \frac{1}{2}$ then $P(E_{2n}) \sim c_1 r^n n^{-1/2}$, for some $0 < r < 1$, as $n \to \infty$. [*Hint*: Use Stirling's formula: $n! \sim (2\pi n)^{1/2} n^n e^{-n}$ as $n \to \infty$; see W. Feller (1968), *An Introduction to Probability Theory and its Applications*, Vol. I, 3rd ed., pp. 52–54, Wiley, New York.]
   (ii) Show that if $p = \frac{1}{2}$ then $P(E_{2n}) \sim c_2 n^{-1/2}$ as $n \to \infty$.
   (iii) Calculate $P(E_n \text{ i.o.})$ for $p \neq \frac{1}{2}$ by application of the Borel–Cantelli Lemma (Chapter 0, Lemma 6.1) and (i). Discuss why this approach fails for $p = \frac{1}{2}$. (Compare Exercise 1.4.)
   (iv) Calculate $P(E_n \text{ i.o.})$ for arbitrary values of p by application of the results of this section.

10. (*A Gambler's Ruin*) A gambler plays a game in which it is possible to win a unit amount with probability p or lose a unit amount with probability q at each play. What is the probability that a gambler with an initial capital x playing against an infinitely rich adversary will eventually go broke (i) if $p = q = \frac{1}{2}$, (ii) if $p > \frac{1}{2}$?

11. Suppose that two particles are initially located at integer points x and y, respectively. At each unit of time a particle is selected at random, both being equally likely to be selected, and is displaced one unit to the right or left with probabilities p and q respectively. Calculate the probability that the two particles will eventually meet. [*Hint*: Consider the evolution of the difference between the positions of the two particles.]

12. Let $\{X_n\}$ be a discrete-time stochastic process with state spaces $S = \mathbb{Z}$ such that state y is transient. Let τ_N denote the proportion of time spent in state y among the first N time units. Show that $\lim_{N\to\infty} \tau_N = 0$ with probability 1.

13. (*Range of Random Walk*) Let $\{S_n\}$ be a simple random walk starting at 0 and define the *Range* R_n in time n by

$$R_n = \#\{S_0 = 0, S_1, \ldots, S_n\}.$$

R_n represents the number of distinct states visited by the random walk in time 0 to n.
   (i) Show that $E(R_n/n) \to |p - q|$ as $n \to \infty$. [*Hint*: Write

$$R_n = \sum_{k=0}^{n} I_k \qquad \text{where } I_0 = 1,\ I_k = I_k(S_1, \ldots, S_k) \text{ is defined by}$$

$$I_k = \begin{cases} 1 & \text{if } S_k \neq S_j \text{ for all } j = 0, 1, \ldots, k-1, \\ 0 & \text{otherwise.} \end{cases}$$

Then

$$\begin{aligned} EI_k &= P(S_k - S_{k-1} \neq 0, S_k - S_{k-2} \neq 0, \ldots, S_k \neq 0) \\ &= P(S_j \neq 0, j = 1, 2, \ldots, k) \quad \text{(justify)} \end{aligned}$$

$$= 1 - P(\text{time of the "first return to 0"} \leqslant k)$$

$$= 1 - \sum_{j=1}^{k-1} \{P(T_0^1 = j)p + P(T_0^{-1} = j)q\} \to |p - q| \text{ as } k \to \infty. \quad]$$

(ii) Verify that $R_n/n \to 0$ in probability as $n \to \infty$ for the symmetric case $p = q = \frac{1}{2}$. [*Hint*: Use (i) and Chebyshev's (first moment) Inequality. (Chapter 0, (2.16).] For almost sure convergence, see F. Spitzer (1976), *Principles of Random Walk*, Springer-Verlag, New York.

14. (*I.I.D. Products*) Let $X_1, X_2, \ldots$ be an i.i.d. sequence of nonnegative random variables having a nondegenerate distribution with $EX_1 = 1$. Define $T_n = \prod_{k=1}^{n} X_k$, $n \geqslant 1$. Show that with probability unity, $T_n \to 0$ as $n \to \infty$. [*Hint*: If $P(X_1 = 0) > 0$ then

$$P(T_n = 0 \text{ for all } n \text{ sufficiently large})$$
$$= P(X_k = 0 \text{ for some } k) = \lim_{n \to \infty} (1 - [1 - P(X_1 = 0)]^n) = 1.$$

For the case $P(X_1 = 0) = 0$, take logarithms and apply Jensen's Inequality (Chapter 0, (2.7)) and the SLLN (Chapter 0, Section 0.6) to show $\log T_n \to -\infty$ a.s. Note the strict inequality in Jensen's Inequality by nondegeneracy.]

15. Let $\{S_n\}$ be a simple random walk starting at 0. Show the following.
   (i) If $p = \frac{1}{2}$, then $\sum_{n=0}^{\infty} P(S_n = 0)$ diverges.
   (ii) If $p \neq \frac{1}{2}$, then $\sum_{n=0}^{\infty} P(S_n = 0) = (1 - 4pq)^{-1/2} = |p - q|^{-1}$. [*Hint*: Apply the Taylor series generalization of the Binomial theorem to $\sum_{n=0}^{\infty} z^n P(S_n = 0)$ noting that

$$\binom{2n}{n}(pq)^n = (-1)^n(4pq)^n\binom{-\frac{1}{2}}{n}. \quad]$$

   (iii) Give a proof of the transience of 0 using (ii) for $p \neq \frac{1}{2}$. [*Hint*: Use the Borel–Cantelli Lemma (Chapter 0).]

16. Define the backward difference operator by

$$\nabla\phi(x) = \phi(x) - \phi(x - 1).$$

   (i) Show that the boundary-value problem (3.4) can be expressed as

$$\tfrac{1}{2}\sigma^2\,\nabla^2\phi + \mu\,\nabla\phi = 0 \text{ on } (c, d) \qquad \text{and} \qquad \phi(c) = 0, \quad \phi(d) = 1,$$

   where $\mu = p - q$ and $\sigma^2 = 2p$.
   (ii) Verify that if $p = q (\mu = 0)$, then ϕ satisfies the following averaging property on the interior of the interval.

$$\phi(x) = [\phi(x - 1) + \phi(x + 1)]/2 \qquad \text{for } c < x < d.$$

   Such a function ϕ is called a *harmonic function*.
   (iii) Show that a harmonic function on $[c, d]$ must take its maximum and its minimum on the boundary $\partial = \{c, d\}$.

(iv) Verify that if two harmonic functions agree on ∂, then they must coincide on all of $[c, d]$.

(v) Give an alternate proof of the fact that a symmetric simple random walk starting at x in (c, d) must eventually reach the boundary based on the above *maximum/minimum principle* for harmonic functions. [*Hint*: Verify that the sum of two harmonic functions is harmonic and use the above ideas to determine the minimum of the escape probability from $[c, d]$ starting at x, $c \leqslant x \leqslant d$.]

17. Consider the simple random walk with $p \leqslant q$ starting at 0. Let N_j denote the number of visits to $j > 0$ that occur prior to the first return to 0. Give an argument that $EN_j = (p/q)^j$. [*Hint*: The number of excursions to j before returning to 0 has a geometric distribution. Condition on the first displacement.]

Exercises for Section I.4

1. Let T^x denote the time to reach the boundary for a simple random walk $\{S_n^x\}$ starting at x in (c, d). Let $\mu = 2p - 1$, $\sigma^2 = 2p$.

(i) Verify that $ET^x < \infty$. [*Hint*: Take $x = 0$. Choose N such that $P(|S_N^0| > d - c) \geqslant \frac{1}{4}$. Argue that $P(T^0 > rN) \leqslant (\frac{3}{4})^r$, $r = 1, 2, \ldots$ using the fact that the r sums over $(jN - N, jN]$, $j = 1, \ldots, r$, are i.i.d. and distributed as S_N^0.]

(ii) Show that $m(x) = ET^x$ solves the boundary value problem

$$\frac{\sigma^2}{2}\nabla^2 m + \mu\,\nabla m = -1,$$

$$m(c) = m(d) = 0, \qquad \nabla m(x) := m(x) - m(x-1).$$

(iii) Find an analytic expression for the solution to the nonhomogeneous boundary value problem for the case $\mu = 0$. [*Hint*: $m(x) = -x^2$ is a particular solution and 1, x solve the homogeneous problem.]

(iv) Repeat (iii) for the case $\mu \neq 0$. [*Hint*: $m(x) = |q - p|^{-1}x$ is a particular solution and 1, $(q/p)^x$ solve the homogeneous problem.]

(v) Describe ET^x in the case $c = 0$ and $d \to \infty$.

2. Establish the following identities as a consequences of the results of this section.

(i) $$\sum_{\substack{N \geqslant |y|,\\ N+y \text{ even}}} \left\{ \frac{|y|}{N} \binom{N}{\frac{N+y}{2}} 2^{-N} \right\} = 1 \qquad \text{for all } y \neq 0.$$

(ii) For $p > q$,

$$\sum_{\substack{N \geqslant |y|,\\ N+y \text{ even}}} \left\{ \frac{|y|}{N} \binom{N}{\frac{N+1}{2}} p^{(N+y)/2} q^{(N-y)/2} \right\} = \begin{cases} 1 & \text{for } y > 0 \\ \left(\dfrac{p}{q}\right)^y & \text{for } y < 0. \end{cases}$$

(iii) Give the corresponding results for $p < q$.

3. (*A Reflection Property*) For the simple symmetric random walk $\{S_n\}$ starting at 0 show that, for $y > 0$,

$$P\left(\max_{n \leqslant N} S_n \geqslant y\right) = 2P(S_N \geqslant y) - P(S_N = y).$$

4. Let $S_n = X_1 + \cdots + X_n$, $S_0 = 0$, and suppose that $X_1, X_2, \ldots$ are independent random variables such that $S_n - S_1, S_n - S_2, \ldots$ have symmetric probability distributions.
   (i) Show that $P(\max_{n \leqslant N} S_n \geqslant y) = 2P(S_N \geqslant y) - P(S_N = y)$ for all $y > 0$.
   (ii) Show that $P(\max_{n \leqslant N} S_n \geqslant x, S_N = y) = P(S_N = y)$ if $y \geqslant x$ and $P(\max_{n \leqslant N} S_n \geqslant x, S_N = y) = P(S_N = 2x - y)$ if $y \leqslant x$.

5. (*A Gambler's Ruin*) A gambler wins or loses 1 unit with probabilities p and $q = 1 - p$, respectively, at each play of a game. The gambler has an initial capital of x units and the adversary has an initial capital of $d > x$ units. The game is played repeatedly until one of the players is broke.
   (i) Calculate the probability that the gambler will eventually go broke.
   (ii) What is the expected duration of the game?

6. (*Bertrand's Classical Ballot Problem*) Candidates A and B have probabilities p and $1 - p$ $(0 < p < 1)$ of winning any particular vote. If A scores m votes and B scores n votes, $m > n$, then what is the probability that A will maintain a lead throughout the process of sequentially counting all $m + n$ votes cast? [*Hint*: P (out of $m + n$ votes cast, A scores m votes, B scores n votes, and A maintains a lead throughout) $= P(T_0^{m-n} = m + n)$.]

7. Let $\{S_n\}$ be the simple random walk starting at 0 and let

$$M_N = \max\{S_n : n = 0, 1, 2, \ldots, N\},$$
$$m_N = \min\{S_n : n = 0, 1, 2, \ldots, N\}.$$

   (i) Calculate the distribution of M_N.
   (ii) Calculate the distribution of m_N.
   (iii) Calculate the joint distribution of M_N and S_N. [*Hint*: Let $a > 0, b$ be integers, $a \geqslant b$. Then

$$\begin{aligned} P(M_N \geqslant a, S_N \geqslant b) = P(T_a \leqslant N, S_N \geqslant b) &= \sum_{n=1}^{N} P(T_a = n, S_N \geqslant b) \\ &= \sum_{n=1}^{N} P(T_a = n, S_N - S_n \geqslant b - a) \\ &= \sum_{n=1}^{N} P(T_a = n)P(S_{N-n} \geqslant b - a). \quad] \end{aligned}$$

8. What percentage of the particles at y at time N are there for the first time in a dilute system of many noninteracting (i.e., independent) particles each undergoing a simple random walk starting at the origin?

*9. Suppose that the points of the state space $S = \mathbb{Z}$ are painted blue with probability

ρ or green with probability $1 - \rho$, $0 \leqslant \rho \leqslant 1$, independently of each other and of a simple random walk $\{S_n\}$ starting at 0. Let B denote the random set of states (integer sites) colored blue and let $N_n(\rho)$ denote the amount of time (*occupation time*) that the random walk spends in the set B prior to time n, i.e.,

$$N_n(\rho) = \sum_{k=0}^{n} I_B(S_k).$$

(i) Show that $EN_n(\rho) = (n+1)\rho$. [*Hint*: $EI_B(S_k) = E\{E[I_B(S_k) \mid S_k]\}$.]
(ii) Verify that

$$\lim_{n\to\infty} \operatorname{Var}\left\{\frac{N_n(\rho)}{\sqrt{n}}\right\} = \begin{cases} \infty & \text{for } p = \frac{1}{2} \\ \dfrac{\rho(1-\rho)}{|p-q|} & \text{for } p \neq \frac{1}{2}. \end{cases}$$

[*Hint*: Use Exercise 13.15.]
(iii) For $p \neq \frac{1}{2}$, use (ii), to show $N_n(\rho)/n \to \rho$ in probability as $n \to \infty$. [*Hint*: Use Chebyshev's Inequality.]
(iv) For $p = \frac{1}{2}$, show $N_n(\rho)/n \to \rho$ in probability. [*Hint*: Show $\operatorname{Var}(N_n(\rho)/n) \to 0$ as $n \to \infty$.]

10. Apply *Stirling's Formula*, ($k! = (2\pi k)^{1/2} k^k e^{-k}(1 + o(1))$ as $k \to \infty$), to show for the simple symmetric random walks starting at 0 that

(i) $P(T_y = N) \sim \dfrac{|y|}{(2\pi)^{1/2}} N^{-3/2}$ as $N \to \infty$.

(ii) $ET_y = \infty$.

Exercises for Section I.5

1. (i) Complete the proof of *Pólya's Theorem* for $k > 3$. (See Exercise 5 below.)
 (ii) Give an alternative proof of transience for $k \geqslant 3$ by an application of the Borel–Cantelli Lemma Part 1 (Chapter 0, (6.1)). Why cannot Part 2 of the lemma be directly applied to prove recurrence for $k = 1, 2$?

2. Show that

$$\sum_{k=0}^{n} \binom{n}{k}^2 = \binom{2n}{n}.$$

[*Hint*: Consider the number of ways in which n balls can be selected from a box of n black and n white balls.]

3. (i) Show that for the 2-dimensional simple symmetric random walk, the probability of a return to $(0, 0)$ at time $2n$ is the same as that for two independent walkers, one along the horizontal and the other along the vertical, to be at $(0, 0)$ at time $2n$. Also verify this by a geometric argument based on two independent walkers with step size $1/\sqrt{2}$ and viewed along the axes rotated by 45°.

(ii) Show that relations (5.5) hold for a general random walk on the integer lattice in any dimension. Use these to compute, for the simple symmetric random walk in dimension two, the probabilities f_j that the random walk returns to the origin at time j for the first time for $j = 1, \ldots, 8$. Similarly compute f_j in dimension three for $1 \leqslant j \leqslant 4$.

4. (i) Show that the method of Exercise 3(i) above does not hold in $k = 3$ dimensions.
   (ii) Show that the motion of three independent simple symmetric random walkers starting at $(0, 0, 0)$ in $\mathbb{Z}^3$ is transient.

5. Show that the trinomial coefficient

$$\binom{n}{j, k, n-j-k} = \frac{n!}{j!k!(n-j-k)!}$$

is largest for $j, k, n - j - k$, closest to $n/3$. [*Hint*: Suppose a maximum is attained for $j = J$, $k = K$. Consider the inequalities of the form

$$\binom{n}{j, k, n-j-k} \leqslant \binom{n}{J, K, n-J-K}$$

when j, k and/or $n - j - k$ differ from $J, K, n - J - K$, respectively, by ± 1. Use this to show $|n - J - 2K| \leqslant 1$, $|n - K - 2J| \leqslant 1$.]

6. Give a probabilistic interpretation to the relation (see Eq. 5.9) $\beta = 1/(1 - \gamma)$. [*Hint*: Argue that the number of returns to 0 is geometrically distributed with parameter γ.]

7. Show that a multidimensional random walk is transient when the (one-step) mean displacement is nonzero.

8. Calculate the probability that the simple symmetric k-dimensional random walk will return i.o. to a previously occupied site. [*Hint*: The conditional probability, given $\mathbf{S}_0, \ldots, \mathbf{S}_n$, that $\mathbf{S}_{n+1} \notin \{\mathbf{S}_0, \ldots, \mathbf{S}_n\}$ is at most $(2k - 1)/2k$. Check that

$$P(\mathbf{S}_{n+1}, \ldots, \mathbf{S}_{n+m} \in \{\mathbf{S}_0, \ldots, \mathbf{S}_n\}^c \leqslant \left(\frac{2k-1}{2k}\right)^m$$

for each $m \geqslant 1$.]

9. (i) Estimate (numerically) the expected number $(\beta - 1)$ of returns to the origin. [*Hint*: Estimate a bound for C in (5.15) and bound (5.16) with a Riemann integral.]
   (ii) Give a numerical estimate of the probability γ that the simple random walk in $k = 3$ dimensions will return to the origin. [*Hint*: Use (i).]

*10. Calculate the probability that a simple symmetric random walk in $k = 3$ dimensions will eventually hit a given line $\{(ra, rb, rc): r \in \mathbb{Z}\}$, where $(a, b, c) \neq (0, 0, 0)$ is a lattice point.

*11. (*A Finite Switching Network*) Let $F = \{x_1, x_2, \ldots, x_k\}$ be a finite set of k sites that can be either "on" (1) or "off" (0). At each instant of time a site is randomly selected and switched from its current state ε to $1 - \varepsilon$. Let $S = \{0, 1\}^F = \{(\varepsilon_1, \ldots, \varepsilon_k): \varepsilon_i = 0 \text{ or } 1\}$. Let $X_1, X_2, \ldots$ be i.i.d. S-valued random variables with

$P(X_n = e_i) = p_i$, $i = i, \ldots, k$, where $e_i \in S$ is defined by $e_i(j) = \delta_{i,j}$, and $p_i \geqslant 0$, $\sum p_i = 1$. Define a random walk on S, regarded as a group under coordinatewise addition mod 2, by

$$S_n = X_1 \oplus \cdots \oplus X_n, \qquad S_0 = (0, 0, \ldots, 0).$$

Show that the configuration in which all switches are off is recurrent in the cases $k = 1, 2$. The general case will follow from the methods and theory of Chapter II when $k < \infty$. The problem when $k = \infty$ has an interesting history: see F. Spitzer (1976), *Principles of Random Walk*, Springer Verlag, New York, and references therein.

*12. Use Exercise 11 above and the examples of random walks on $\mathbb{Z}$ to arrive at a general formulation of the notion of a random walk on a group. Describe a random walk on the unit circle in the complex plane as an illustration of your ideas.

13. Let $\{X_n\}$ denote a recurrent random walk on the 1-dimensional integer lattice. Show that

$$E_x(\text{number of visits to } x \text{ before hitting } 0) = +\infty, \qquad x \neq 0.$$

[*Hint*: Translate the problem by x and consider that starting from 0, the number of visits to 0 before hitting $-x$ is bounded below by the number of visits to 0 before leaving the (open) interval centered at 0 of length $|x|$. Use monotonicity to pass to the limit.]

Exercises for Section I.6

1. Let $\mathbf{X}$ be an $(n + 1)$-dimensional Gaussian random vector and $\mathbf{A}$ an arbitrary linear transformation from $\mathbb{R}^{n+1}$ to $\mathbb{R}^n$. Show that $\mathbf{AX}$ is a Gaussian random vector in $\mathbb{R}^n$.

2. (i) Determine the consistent specification of finite-dimensional distributions for the canonical construction of the simple random walk.
   (ii) Verify Kolmogorov consistency for Example 3.

*3. (*A Kolmogorov Extension Theorem: Special Case*) This exercise shows the role of *topology* in proving what otherwise seems a (purely) *measure-theoretical* assertion in the simplest case of Kolmogorov's theorem. Let $\Omega = \{0, 1\}^{\mathbb{N}}$ be the product space consisting of sequences of the form $\omega = (\omega_1, \omega_2, \ldots)$ with $\omega_i \in \{0, 1\}$. Give Ω the product topology for the discrete topology on $\{0, 1\}$. By *Tychonoff's Theorem* from topology, this makes Ω compact. Let X_n be the nth coordinate projection mapping on Ω. Let $\mathscr{F}$ be the Borel sigmafield for Ω.
   (i) Show that $\mathscr{F}$ coincides with the sigmafield for Ω generated by events of the form

$$F(\varepsilon_1, \ldots, \varepsilon_n) = \{\omega \in \Omega \colon \omega_i = \varepsilon_i \text{ for } i = 1, 2, \ldots, n\},$$

   for an arbitrarily prescribed sequence $\varepsilon_1, \ldots, \varepsilon_n$ of 1's and 0's. [*Hint*: $\rho(\omega, \eta) = \sum_{n=1}^{\infty} |\omega_n - \eta_n|/2^n$ metrizes the product topology on Ω. Consider the open balls of radii of the form $r = 2^{-N}$ centered at sequences which are 0 from some n onward and use separability.]

(ii) Let $\{P_n\}$ be a consistent family of probability measures, with P_n defined on $(\mathbb{R}^n, \mathscr{B}^n)$, and such that P_n is concentrated on $\Omega_n = \{0, 1\}^n$. Define a set function for events of the form $F = F(\varepsilon_1, \ldots, \varepsilon_n)$ in (i), by

$$P(F) = P_n(\{\omega_i = \varepsilon_i, i = 1, 2, \ldots, n\}).$$

Show that there is a unique probability measure P on the sigmafield $\mathscr{F}_n$ of cylinder sets of the form

$$C = \{\omega \in \Omega: (\omega_1, \ldots, \omega_n) \in B\},$$

where $B \subset \{0, 1\}^n$, which agrees with this formula for $F \in \mathscr{F}_n$, $n \geqslant 1$.

(iii) Show that $\mathscr{F}^0 := \bigcup_{n=1}^{\infty} \mathscr{F}_n$ is a field of subsets of Ω but not a sigmafield.

(iv) Show that P is a countably additive measure on $\mathscr{F}^0$. [*Hint*: Ω is compact and the cylinder sets are both open and closed for the product topology on Ω.]

(v) Show that P has a unique extension to a probability measure on $\mathscr{F}$. [*Hint*: Invoke the Carathéodory Extension Theorem (Chapter 0, Section 1) under (iii), (iv).]

(vi) Show that the above arguments also apply to any *finite-state* discrete-parameter stochastic process.

4. Let $(\Omega, \mathscr{F}, P)$ and $(S, \mathscr{L})$ represent measurable spaces. A function X defined on Ω and taking values in S is called *measurable* if $X^{-1}(B) \in \mathscr{F}$ for all $B \in \mathscr{L}$, where $X^{-1}(B) = \{X \in B\} = \{\omega \in \Omega: X(\omega) \in B\}$. This is the meaning of an S-valued random variable. The distribution of X is the induced probability measure Q on $\mathscr{L}$ defined by

$$Q(B) = P(X^{-1}(B)) = P(X \in B), \qquad B \in \mathscr{L}.$$

Let $(\Omega, \mathscr{F}, P)$ be the canonical model for nonterminating repeated tosses of a coin and $X_n(\omega) = \omega_n$, $\omega \in \Omega$. Show that $\{X_m, X_{m+1}, \ldots\}$, m an arbitrary positive integer, is a measurable function on $(\Omega, \mathscr{F}, P)$ taking values in $(\Omega, \mathscr{F})$ with the distribution P; i.e., $\{X_m, X_{m+1}, \ldots,\}$ is a *noncanonical model* for an infinite sequence of coin tossings.

5. Suppose that $\mathbf{D}^{(1)}$ and $\mathbf{D}^{(2)}$ are covariance matrices.

(i) Verify that $\alpha\mathbf{D}^{(1)} + \beta\mathbf{D}^{(2)}$, $\alpha, \beta \geqslant 0$, is a covariance matrix.

(ii) Let $\{\mathbf{D}^{(n)} = ((\sigma_{ij}^{(n)}))\}$ be a sequence of covariance matrices $(k \times k)$ such that $\lim_n \sigma_{ij}^{(n)} = \sigma_{ij}$ exists. Show that $\mathbf{D} = ((\sigma_{ij}))$ is a covariance matrix.

*6. Let $\hat{\mu}(t) = \int_{-\infty}^{\infty} e^{itx}\mu(dx)$ be the Fourier transform of a positive finite measure μ. (Chapter 0, (8.46)).

(i) Show that $(\hat{\mu}(t_i - t_j))$ is a nonnegative definite matrix for any $t_1 < \cdots < t_k$.

(ii) Show that $\hat{\mu}(t) = e^{-|t|}$ if μ is the *Cauchy distribution.*

*7. (*Pólya Criterion for Characteristic Functions*) Suppose that ϕ is a real-valued nonnegative function on $(-\infty, \infty)$ with $\phi(-t) = \phi(t)$ and $\phi(0) = 1$. Show that if ϕ is continuous and convex on $[0, \infty)$, then ϕ is the *Fourier transform* (characteristic function) of a probability distribution (in particular, for any $t_1 < t_2 < \cdots < t_k$, $k \geqslant 1$, $((\phi(t_i - t_j)))$ is *nonnegative definite* by Exercise 6), via the following steps.

(i) Check that

$$\hat{\gamma}(t) = \begin{cases} 1 - |t|, & |t| \leqslant 1, \quad t \in \mathbb{R}^1, \\ 0, & |t| > 1 \end{cases}$$

is the characteristic function of the (probability) measure with p.d.f. $\gamma(x) = (1 - \cos x)/\pi x^2$, $-\infty < x < \infty$. Use the Fourier inversion formula [(8.43), Chapter 0].

(ii) Check that the characteristic function of a convex combination (mixture) of probability distributions is the corresponding convex combination of characteristic functions.

(iii) Let $0 < a_1 < \cdots < a_n$ be real numbers, $p_i \geqslant 0$, $\sum_{i=1}^n p_i = 1$. Show that $\hat{v}(t) = p_1\hat{\gamma}(t/a_1) + \cdots + p_n\hat{\gamma}(t/a_n)$ is a characteristic function. Draw a graph of $\hat{v}(t)$ and check that the slope of the segment between a_k and a_{k+1} is $-[(p_k/a_k) + \cdots + (p_n/a_n)]$, $k = 1, 2, \ldots, n-1$. Interpret the numbers $p_1, p_1 + p_2, \ldots, p_1 + \cdots + p_n = 1$ along the vertical axis with reference to the polygonal graph of $\hat{v}(t)$.

(iv) Show that a function $\phi(t)$ satisfying Pólya's criterion can be approximated by a function of the form (iii) to arbitrary accuracy. [*Hint*: Approximate $\phi(t)$ by a polygonal path consisting of n segments of decreasing slopes.]

(v) Show that the (pointwise) limit of characteristic functions is a characteristic function.

*8. Show that the following define covariance functions for a Gaussian process.

(i) $\sigma_{ij} = e^{-|i-j|}, \qquad i, j = 0, 1, 2, \ldots.$

(ii) $\sigma_{ij} = \min(i, j) = (|i| + |j| - |i - j|)/2.$
[*Hint*: Use Exercises 6 and 7.]

Exercises for Section I.7

1. Verify that $(B_{t_1}, \ldots, B_{t_m})$ has an m-dimensional Gaussian distribution and calculate the mean and the variance–covariance matrix using the fact that the Brownian motion has independent Gaussian increments.

2. Let $\{X_t\}$ be a process with stationary and independent increments starting at 0 with $EX_s^2 < \infty$ for $s > 0$. Assume EX_t, EX_t^2 are continuous functions of t.

   (i) Show that $EX_t = mt$ for some constant m.
   (ii) Show that $\operatorname{Var} X_t = \sigma^2 t$ for some constant $\sigma^2 \geqslant 0$.
   (iii) Calculate the limiting distribution of

$$Y_t^{(n)} = \frac{(X_{nt} - mnt)}{\sqrt{n}}$$

   as $n \to \infty$, for $t > 0$, fixed.

3. (i) (*Diffusion Limit Scalings*) Let X_t be a random variable with mean $x + tf(p - q)\Delta$ and variance $tf\Delta^2 pq$. Give a direct calculation of p and $q = 1 - p$ in terms of f and Δ using only the requirements that the mean and variance of $X_t - x$ should stabilize to some limiting values proportional to t as $f \to \infty$ and $\Delta \to 0$.

   (ii) Verify convergence to the Gaussian distribution for the distribution of $(\sigma/\sqrt{n})S_{[nt]}$ as $n \to \infty$, where $\{S_n\}$ is the simple random walk with $p_n = (\mu/2\sqrt{n}) + \frac{1}{2}$, by application of Liapunov's CLT (Chapter 0, Corollary 7.3).

4. Let $\{X_t\}$ be a Brownian motion starting at 0 with diffusion coefficient $\sigma^2 > 0$ and zero drift.
   (ii) Show that the process has the following *scaling property*. For each $\lambda > 0$ the process $\{Y_t\}$ defined by $Y_t = \lambda^{-1/2} X_{\lambda t}$ is distributed exactly as the process $\{X_t\}$.
   (ii) How does (i) extend to k-dimensional Brownian motion?

5. Let $\{X_t\}$ be a stochastic process which has stationary and independent increments.
   (i) Show that the distribution of the increments must be *infinitely divisible*; i.e., for each integer n, the distribution of $X_t - X_s$ $(s < t)$ can be expressed as an n-fold convolution of a probability measure μ_n.
   (ii) Suppose that the increment $X_t - X_s$ has the Cauchy distribution with p.d.f. $(t - s)/\pi[(t - s)^2 + x^2]$ for $s < t$, $x \in \mathbb{R}^1$. Show that the Cauchy process so described is invariant under the rescaling $\{Y_t\}$ where $Y_t = \lambda^{-1} X_{\lambda t}$ for $\lambda > 0$; i.e., $\{Y_t\}$ has the same distribution as $\{X_t\}$. (This process can be constructed by methods of theoretical complements 1, 2 to Section IV.1.)

6. Let $\{X_t\}$ be a Brownian motion starting at 0 with zero drift and diffusion coefficient $\sigma^2 > 0$. Define $Y_t = |X_t|$, $t \geqslant 0$.
   (i) Calculate EY_t, Var Y_t.
   (ii) Is $\{Y_t\}$ a process with independent increments?

7. Let $R_t = X_t^2$ where $\{X_t\}$ is a Brownian motion starting at 0 with zero drift and diffusion coefficient $\sigma^2 > 0$. Calculate the distribution of R_t.

8. Let $\{B_t\}$ be a standard Brownian motion starting at 0. Define

$$V_n = \sum_{i=1}^{2^n} |B_{i/2^n} - B_{(i-1)/2^n}|.$$

   (i) Verify that $EV_n = 2^{n/2} E|B_1|$.
   (ii) Show that Var V_n = Var$|B_1|$.
   (iii) Show that with probability one, $\{B_t\}$ is not of bounded variation on $0 \leqslant t \leqslant 1$. [*Hint*: Show that $V_{n+1} \geqslant V_n$, $n \geqslant 1$, and, using Chebyshev's Inequality, $P(V_n > M) \to 1$ as $n \to \infty$ for any $M > 0$.]

9. The *quadratic variation* over $[0, t)$ of a function f on $[0, \infty)$ is defined by $V_t(f) = \lim_{n\to\infty} v_n(t, f)$, where

$$v_n(t, f) = \sum_{k=1}^{2^n} [f(kt/2^n) - f((k-1)t/2^n)]^2$$

provided the limit exists.

   (i) Show if f is continuous and of bounded variation then $V_t(f) = 0$.
   (ii) Show that $Ev_n(t, \{X_s\}) \to \sigma^2 t$ as $n \to \infty$ for Brownian motion $\{X_s\}$ with diffusion coefficient σ^2 and drift μ.
   (iii) Verify that $v_n(t, \{X_s\}) \to \sigma^2 t$ in probability as $n \to \infty$.
   (*iv) Show that the limit in (iii) holds almost surely. [*Hint*: Use Borel–Cantelli Lemma and Chebyshev Inequality.]

10. Let $\{\mathbf{X}_t^{\mathbf{x}}\}$ be the k-dimensional Brownian motion with drift $\boldsymbol{\mu}$ and diffusion coefficient matrix $\mathbf{D}$. Calculate the mean of $\mathbf{X}_t^{\mathbf{x}}$ and the variance–covariance matrix of $\mathbf{X}_t^{\mathbf{x}}$.

11. Let $\{X_t\}$ be any mean zero Gaussian process. Let $t_1 < t_2 < \cdots < t_n$.
   (i) Show that the characteristic function of $(X_{t_1}, \ldots, X_{t_n})$ is of the form $e^{-\frac{1}{2}Q(\xi)}$ for some quadratic form $Q(\xi) = \langle \mathbf{A}\xi, \xi \rangle$.
   (ii) Establish the *pair-correlation decomposition formula* for block correlations:

$$E\{X_{t_1} X_{t_2} \cdots X_{t_n}\} = \begin{cases} 0 & \text{if } n \text{ is odd} \\ \sum^* E\{X_{t_i} X_{t_j}\} \cdots E\{X_{t_m} X_{t_k}\} & \text{if } n \text{ is even,} \end{cases}$$

   where $\sum^*$ denotes the sum taken over all possible decompositions into all possible *disjoint* pairs $\{t_i, t_j\}, \ldots, \{t_m, t_k\}$ obtained from $\{t_1, \ldots, t_n\}$. [*Hint*: Use induction on derivatives of the (multivariate) characteristic function at $(0, 0, \ldots, 0)$ by first observing that $\partial e^{-\frac{1}{2}Q(\xi)}/\partial \xi_i = -\alpha_i e^{-\frac{1}{2}Q(\xi)}$ and $\partial \alpha_i / \partial \xi_j = a_{ij}$, where $\alpha_i = \sum_j a_{ij}\xi_j$ and $\mathbf{A} = ((a_{ij}))$.]

Exercises for Section I.8

1. Construct a matrix representation of the linear transformation on $\mathbb{R}^k$ of the increments $(x_{t_1}, x_{t_2} - x_{t_1}, \ldots, x_{t_k} - x_{t_{k-1}})$ to $(x_{t_1}, x_{t_2}, \ldots, x_{t_k})$.

*2. (i) Show that the functions $f\colon C[0, \infty) \to \mathbb{R}$ defined by

$$f(\omega) = \max_{a \leqslant t \leqslant b} \omega(t), \qquad g(\omega) = \min_{a \leqslant t \leqslant b} \omega(t)$$

   are continuous for the topology of uniform convergence on bounded intervals.
   (ii) Show that the set $\{f(\{\tilde{X}_t^{(n)}\}) \leqslant x\}$ is a Borel subset of $C[0, \infty)$ if f is continuous on $C[0, \infty)$.
   (iii) Let $\mathbf{f}\colon C[0, \infty) \to \mathbb{R}^k$ be continuous. Explain how it follows from the definition of weak convergence on $C[0, \infty)$ given after the statement of the FCLT (pages 23–24) that the random vectors $\mathbf{f}(\{\tilde{X}_t^{(n)}\})$ must converge in distribution to $\mathbf{f}(\{X_t\})$ too.
   (iv) Show that convergence in distribution on $C[0, \infty)$ implies convergence of the finite-dimensional distributions. Exercise 3 below shows that the converse is *not* true in general.

*3. Suppose that for each $n = 1, 2, \ldots$, $\{x_n(t), 0 \leqslant t \leqslant 1\}$, is the deterministic process whose sample path is the continuous function whose graph is given by Figure Ex.I.8.
   (i) Show that the finite-dimensional distributions converges to those of the a.s. identically zero process $\{z(t)\}$, i.e., $z(t) \equiv 0$, $0 \leqslant t \leqslant 1$.
   (ii) Check that $\max_{0 \leqslant t \leqslant 1} x_n(t)$ does not converge to $\max_{0 \leqslant t \leqslant 1} z(t)$ in distribution.

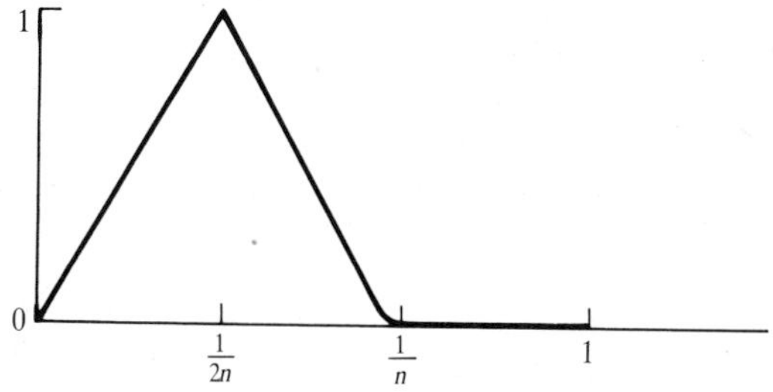

Figure Ex.I.8

*4. Give an example to demonstrate that it is *not* the case that the FCLT gives convergence of probabilities of all infinite-dimensional events in $C[0, \infty)$. [*Hint*: The polygonal process has finite total variation over $0 \leqslant t \leqslant 1$ with probability 1. Compare with Exercise 7.8.]

5. Verify that the probability density function $p(t; x, y)$ of the position at time t of the Brownian motion starting at x with drift μ and diffusion coefficient σ^2 solves the so-called Fokker–Planck equation (for fixed x) given by

$$\frac{\partial p}{\partial t} = \tfrac{1}{2}\sigma^2 \frac{\partial^2 p}{\partial y^2} - \mu \frac{\partial p}{\partial y}.$$

(i) Check that for fixed y, p also satisfies the adjoint equation

$$\frac{\partial p}{\partial t} = \tfrac{1}{2}\sigma^2 \frac{\partial^2 p}{\partial x^2} + \mu \frac{\partial p}{\partial x}.$$

(ii) Show that p is a symmetric function of x and y if and only if $\mu = 0$.
(iii) Let $c(t, y) = \int c_0(x)p(t, x, y)\, dx$, where c_0 is a positive bounded initial concentration smoothly distributed over a finite interval. Verify that $c(t, y)$ solves the Fokker–Planck equation with initial condition c_0, i.e.,

$$\frac{\partial c}{\partial t} = \tfrac{1}{2}\sigma^2 \frac{\partial^2 c}{\partial y^2} - \mu \frac{\partial c}{\partial y}, \qquad c(0^+, y) = c_0(y).$$

6. (*Collective Risk in Actuary Science*) Suppose that an insurance company has an initial reserve (total assets) of $X_0 > 0$ units. Policy holders are charged a (gross) risk premium rate α per unit time and claims are made at an average rate λ. The average claim amount is μ with variance σ^2. Discuss modeling the risk reserve process $\{X_t\}$ as a Brownian motion starting at x with drift coefficient of the form $\alpha - \mu\lambda$ and diffusion coefficient $\lambda\sigma^2$, on some scale.

7. (*Law of Proportionate Effect*) A material (e.g., pavement) is subject to a succession of random impacts or loads in the form of positive random variables $L_1, L_2, \ldots$ (e.g., traffic). It is assumed that the (measure of) material strength T_k after the kth impact is proportional to the strength T_{k-1} at the preceding stage through the applied load L_k, $k = 1, 2, \ldots$, i.e., $T_k = L_k T_{k-1}$. Assume an initial strength $T_0 \equiv 1$ as normalization, and that $E(\log L_1)^2 < \infty$. Describe conditions under which it is appropriate to consider the *geometric Brownian motion* defined by $\{\exp(\mu t + \sigma^2 B_t)\}$, where $\{B_t\}$ is standard Brownian motion, as a model for the strength process.

8. Let $X_1, X_2, \ldots$ be i.i.d. random variables with $EX_n = 0$, $\operatorname{Var} X_n = \sigma^2 > 0$. Let $S_n = X_1 + \cdots + X_n$, $n \geqslant 1$, $S_0 = 0$. Express the limiting distribution of each of the random variables defined below in terms of the distribution of the appropriate random variable associated with Brownian motion having drift 0 and diffusion coefficient $\sigma^2 > 0$.
(i) Fix $\theta > 0$, $Y_n = n^{-\theta/2} \max\{|S_k|^\theta : 1 \leqslant k \leqslant n\}$.
(ii) $Y_n = n^{-1/2} S_n$.
(iii) $Y_n = n^{-3/2} \sum_{k=1}^{n} S_k$. [*Hint*: Consider the integral of $t \to S_{[nt]}$, $0 \leqslant t \leqslant 1$.]

9. (i) Write $R_n(x) = |(1 + x/n)^n - e^x|$. Show that

$$R_n(x) \leqslant \sum_{r=2}^{n} \left\{1 - \left(1 - \frac{1}{n}\right) \cdots \left(1 - \frac{r-1}{n}\right)\right\} \frac{|x|^r}{r!} + \frac{|x|^{n+1}}{(n+1)!} e^{|x|},$$

$$\sup_{|x| \leqslant c} R_n(x) \to 0 \qquad \text{as } n \to \infty \qquad \text{(for every } c > 0).$$

(ii) Use (i) to prove (8.6). [*Hint*: Use Taylor's theorem for the inequality, and Lebesgue's Dominated Convergence Theorem (Chapter 0, Section 0.3).]

Exercises for Section I.9

1. (i) Use the SLLN to show that the Brownian motion with nonzero drift is transient.
   (ii) Extend (i) to the k-dimensional Brownian motion with drift.

2. Let $X_t = X_0 + vt$, $t \geqslant 0$, where v is a nonrandom constant-rate parameter and X_0 is a random variable.
   (i) Calculate the conditional distribution of X_t, given $X_s = x$, for $s < t$.
   (ii) Show that all states are transient if $v \neq 0$.
   (iii) Calculate the distribution of X_t if the initial state is normally distributed with mean μ and variance σ^2.

3. Let $\{X_t\}$ be a Brownian motion starting at 0 with diffusion coefficient $\sigma^2 > 0$ and zero drift.
   (i) Define $\{Y_t\}$ by $Y_t = tX_{1/t}$ for $t > 0$ and $Y_0 = 0$. Show that $\{Y_t\}$ is distributed as Brownian motion starting at 0. [*Hint*: Use the law of large numbers to prove sample path continuity at $t = 0$.]
   (ii) Show that $\{X_t\}$ has infinitely many zeros in every neighborhood of $t = 0$ with probability 1.
   (iii) Show that the probability that $t \to X_t$ has a right-hand derivative at $t = 0$ is zero.
   (iv) Use (iii) to provide another example to Exercise 8.4.

4. Show that the distribution of $\min_{t \geqslant 0} X_t^0$ is exponential if $\{X_t^0\}$ is Brownian motion starting at 0 with drift $\mu > 0$. Likewise, calculate the distribution of $\max_{t \geqslant 0} X_t^0$ when $\mu < 0$.

*5. Let $\{S_n\}$ denote the simple symmetric random walk starting at 0, and let

$$m_n = \min_{0 \leqslant k \leqslant n} S_k, \qquad M_n = \max_{0 \leqslant k \leqslant n} S_k, \qquad n = 1, 2, \ldots.$$

Let $\{B_t\}$ denote a standard Brownian motion and let $m = \min_{0 \leqslant t \leqslant 1} B_t$, $M = \max_{0 \leqslant t \leqslant 1} B_t$. Then, by the FCLT, $n^{-1/2}(m_n, M_n, S_n)$ converges in distribution to (m, M, B_1); for rigorous justification use theoretical complements I.8, I.9 noting that the functional $\omega \to (\min_{0 \leqslant t \leqslant 1} \omega_t, \max_{0 \leqslant t \leqslant 1} \omega_t, \omega_1)$ is a continuous map of the metric space $C[0, 1]$ into $\mathbb{R}^3$. For notational convenience, let

$$p_n(j) = P(S_n = j), \qquad p_n(u, v, y) = P(u < m_n \leqslant M_n < v, S_n = y),$$

for integers u, v, y such that $u \leqslant 0 \leqslant v$, $u < v$ and $u \leqslant y \leqslant v$. Also let $\Phi(a, b) = P(a < Z < b)$, where Z has the standard normal distribution. The following use of the reflection principle is taken from an exercise in P. Billingsley (1968),

Convergence of Probability Measures, Wiley, New York, p. 86. These results for Brownian motion are also obtained by other methods in Chapter V.

(i) $p_n(u, v, y) = p_n(y) - \pi(v, y) - \pi(u, y) + \pi(v, u, y) + \pi(u, v, y) - \pi(v, u, v, y) - \pi(u, v, u, y) + \cdots$, where for any fixed sequence of nonnegative integers $y_1, y_2, \ldots, y_k, y, k, n, \pi(y_1, y_2, \ldots, y_k, y)$ denotes the probability that an n-step random walk meets y_1 (at least once), then meets $-y_2$, then meets $y_3, \ldots$, then meets $(-1)^{k-1}y_k$, and ends at y.

(ii) $\pi(y_1, y_2, \ldots, y_k, y) = p_n(2y_1 + 2y_2 + \cdots + 2y_{k-1} + (-1)^{k+1}y)$ if $(-1)^{k+1}y \geq y_k$, $\pi(y_1, y_2, \ldots, y_k, y) = p_n(2y_1 + 2y_2 + \cdots + 2y_k - (-1)^{k+1}y)$ if $(-1)^{k+1}y \leq y_k$. [*Hint*: Use Exercise 4.4(ii), the reflection principle, and induction on k. Reflect through $(-1)^k y_{k-1}$ the part of the path to the right of the first passage through that point following successive passages through $y_1, -y_2, \ldots, (-1)^{k-1}y_{k-2}$.]

(iii) $$p_n(u, v, y) = \sum_{k=-\infty}^{\infty} p_n(y + 2k(v - u)) - \sum_{k=-\infty}^{\infty} p_n(2v - y + 2k(v - u)).$$

(iv) For integers, $u \leq 0 \leq v$, $u \leq y_1 < y_2 \leq v$,

$$P(u < m_n < v, y_1 < S_n < y_2) = \sum_{k=-\infty}^{\infty} P(y_1 + 2k(v-u) < S_n < y_2 + 2k(v-u))) - \sum_{k=-\infty}^{\infty} P(2v - y_2 + 2k(v-u) < S_n < 2v - y_1 + 2k(v-u)).$$

[*Hint*: Sum over y in (iii).]

(v) For real numbers $u \leq 0 \leq v$, $u \leq y_1 < y_2 \leq v$,

$$P(u < m \leq M < v, y_1 < B_1 < y_2) = \sum_{k=-\infty}^{\infty} \Phi(y_1 + 2k(v-u), y_2 + 2k(v-u)) - \sum_{k=-\infty}^{\infty} \Phi(2v - y_2 + 2k(v-u), 2v - y_1 + 2k(v-u)).$$

[*Hint*: Respectively substitute the integers $[u\sqrt{n}]$, $-[-u\sqrt{n}]$, $[y_1\sqrt{n}]$, $-[-y_2\sqrt{n}]$ into (iv) ([ ] denoting the greatest integer function). Use Scheffe's Theorem (Chapter 0) to justify the interchange of limit with summation over k.]

(vi) $$P(M < v, y_1 < B_1 < ky_2) = \Phi(y_1, y_2) - \Phi(2v - y_2, 2v - y_1).$$

[*Hint*: Take $u = -n - 1$ in (iv) and then pass to the limit.]

(vii) $$P(u < m \leq M < v) = \sum_{k=-\infty}^{\infty} (-1)^k \Phi(u + 2k(v-u), v + 2k(v-u)).$$

[*Hint*: Take $y_1 = u$, $y_2 = v$ in (v).]

(viii) $P(\sup|B_t| < v) = \sum_{k=-\infty}^{\infty} (-1)^k \Phi((2k-1)v, (2k+1)v)$.

[*Hint*: Take $u = -v$ in (vii).]

Exercises for Section I.10

*1. (i) Show that (10.2) holds at each point $z \leqslant 0$ ($\geqslant 0$) of continuity of the distribution function for $\min_{0 \leqslant s \leqslant t} X_s$ ($\max_{0 \leqslant s \leqslant t} X_s$). [*Hint*: These latter functionals are continuous.]
(ii) Use (i) and (10.9) to assert (10.2) for *all* z.

2. Calculate the probability that a Brownian motion with drift μ and diffusion coefficient $\sigma^2 > 0$ starting at x will reach $y \neq x$ in time t or less.

3. Suppose that solute particles are undergoing Brownian motion in the horizontal direction in a semi-infinite tube whose left end acts as an absorbing boundary in the sense that when a particle reaches the left end it is taken out of the flow. Assume that initially a proportion $\psi(x)\,dx$ of the particles are present in the element of volume between x and $x + dx$ from the left end, so that $\int_0^\infty \psi(x)\,dx = 1$. For a given drift μ away from the left end and diffusion coefficient $\sigma^2 > 0$, calculate the fraction of particles eventually absorbed. What if $\mu = 0$?

4. Two independent Brownian motions with drift μ_i and diffusion coefficient σ_i^2, $i = 1, 2$, are found at time $t = 0$ at positions x_i, $i = 1, 2$, with $x_1 < x_2$.
(i) Calculate the probability that the two particles will never meet.
(ii) Calculate the probability that the particles will meet before time $s > 0$.

5. (i) Calculate the distribution of the maximum value of the Brownian motion starting at 0 with drift μ and diffusion coefficient σ^2 over the time period $[0, t]$.
*(ii) For the case $\mu = 0$ give a geometric "reflection" argument that $P(\max_{0 \leqslant s \leqslant t} X_s \geqslant y) = 2P(X_t \geqslant y)$. Use (i) to verify this.

6. Calculate the distribution of the minimum value of a Brownian motion starting at 0 with drift μ and diffusion coefficient σ^2 over the time period $[0, t]$.

7. Let $\{B_t\}$ be standard Brownian motion starting at 0 and let $a, b > 0$.
(i) Calculate the probability that $-at < B_t < bt$ for all sufficiently large t.
(ii) Calculate the probability that $\{B_t\}$ last touches the line $y = -at$ instead of $y = bt$. [*Hint*: Consider the process $\{Z_t\}$ defined by $Z_0 = 0$, $Z_t = tB_{1/t}$ for $t > 0$, and Exercise 9.3(i).]

8. Let $\{(B_t^{(1)}, B_t^{(2)})\}$ be a two-dimensional standard Brownian motion starting at (0, 0) (see Section 7). Let $\tau_y = \inf\{t \geqslant 0\colon B_t^{(2)} = y\}$, $y > 0$. Calculate the distribution of $B_{\tau_y}^{(1)}$. [*Hint*: $\{B_t^{(1)}\}$ and $\{B_t^{(2)}\}$ are independent one-dimensional Brownian motions. Condition on τ_y. Evaluate the integral by substituting $u = (x^2 + y^2)/t$.]

9. Let $\{B_t\}$ be a standard Brownian motion starting at 0. Describe the geometric structure of sample paths for each of the following stochastic processes and calculate EY_t.

(i) (*Absorption*) $\begin{cases} Y_t = B_t & \text{if } \max_{0 \leqslant s \leqslant t} B_s < a \\ Y_t = a & \text{if } \max_{0 \leqslant s \leqslant t} B_s \geqslant a, \end{cases}$

where $a > 0$ is a constant.

(*ii) (*Reflection*) $\begin{cases} Y_t = B_t & \text{if } B_t \leqslant a \\ Y_t = 2a - B_t & \text{if } B_t \geqslant a, \end{cases}$

where $a > 0$ is a constant.

(*iii) (*Periodic*) $Y_t = B_t - [B_t]$,

where $[x]$ denotes the greatest integer less than or equal to x.

10. Let, for $\alpha > 0$,

$$f_\alpha(t) = \frac{\alpha}{(2\pi)^{1/2} t^{3/2}} e^{-\alpha^2/2t} \qquad (t > 0).$$

(i) Verify the convolution property

$$f_\alpha * f_\beta(t) = f_{\alpha+\beta}(t) \qquad \text{for any } \alpha, \beta > 0.$$

(ii) Verify that the distribution of τ_z is a *stable* law with *exponent* $\theta = 2$ (index $\frac{1}{2}$) in the sense that if $T_1, T_2, \ldots, T_n$ are i.i.d. and distributed as τ_z then $n^{-\theta}(T_1 + \cdots + T_n)$ is distributed as τ_z (see Eq. 10.2).

(iii) (*Scaling property*) τ_z is distributed as $z^2\tau_1$.

11. Let τ_z be the first passage time to z for a standard Brownian motion starting at 0 with zero drift.

(i) Verify that $E\tau_z$ is *not* finite.

(ii) Show that $Ee^{-\lambda\tau_z} = e^{-|z|\sqrt{2}\cdot\lambda^{1/2}}$, $\lambda > 0$. [*Hint*: Tedious integration will work.]

(iii) Use Laplace transforms to check that $(1/n)\tau_{[z\sqrt{n}]}$ converges in distribution to τ_z as $n \to \infty$.

12. Let $\{B_t\}$ be standard Brownian motion starting at 0. Let $s < t$. Show that the probability that $\{B_t\}$ has at least one zero in (s, t) is given by $(2/\pi)\cos^{-1}(s/t)^{1/2}$. [*Hint*: Let

$$\rho(x) = P(\{B_t\} \text{ has at least one zero in } (s, t) \mid B_s = x).$$

Then for $x > 0$,

$$\rho(x) = P\left(\min_{s \leqslant r \leqslant t} B_r \leqslant 0 \mid B_s = x\right) = P\left(\max_{s \leqslant r \leqslant t} B_r \geqslant 0 \mid B_s = -x\right)$$

$$= P\left(\max_{s \leqslant r \leqslant t} B_r \geqslant x \mid B_s = 0\right) = P(\tau_x \leqslant t - s).$$

Likewise for $x < 0$, $\rho(x) = P(\tau_{-x} \leqslant t - s)$. So the desired probability can be obtained by calculating

$$E\rho(|B_s|) = \int_0^\infty \rho(x)\left(\frac{2}{\pi s}\right)^{1/2} e^{-\frac{1}{2s}x^2}\,dx. \;]$$

Exercises for Section I.11

Throughout this set of exercises $\{S_n\}$ denotes the simple symmetric random walk starting at 0.

1. Show the following for $r \neq 0$.

(i) $P(S_1 \neq 0, S_2 \neq 0, \ldots, S_{2n-1} \neq 0, S_{2n} = 2r) = \binom{2n}{n+r}\frac{|r|}{n} 2^{-2n}$.

(ii) $P(\Gamma^{(2n)} = 2k, S_{2n} = 2r) = \binom{2k}{k}\binom{2n-2k}{n-k+r}\frac{|r|}{n-k} 2^{-2n}$.

(*iii) Calculate the joint distribution of (γ, B_1).

2. Let $U_n = \#\{k \leqslant n: S_{k-1}$ or $S_k > 0\}$ denote the amount of time spent (by the polygonal path) on the positive side of the state space during time 0 to n.

(i) Show that

$$P(U_{2n} = 2k) = P(\Gamma^{(2n)} = 2k) \sim \frac{1}{\pi(k(n-k))^{1/2}}.$$

[*Hint*: Check $k = 0$, $k = n$ first. Use mathematical induction on n and consider the conditional distribution of U_{2n} given the time of the first return to 0. Derive the last equality of (11.3) for U_{2n} using the induction hypothesis and (11.3).]

(ii) Prove Corollary 11.4 by calculating the distribution of the proportion of time spent above 0 in the limit as $n \to \infty$ for the random walk.

(iii) How does one reconcile the facts that $U_{2n}/2n \nrightarrow \frac{1}{2}$ as $n \to \infty$ and $P(S_n > 0) \to \frac{1}{2}$ as $n \to \infty$? [*Hint*: Consider the average length of time between returns to zero.]

*3. Let $\tau^{(n)} = \min\{k \leqslant n: S_k = \max_{0 \leqslant j \leqslant n} S_j\}$ denote the location of the first absolute maximum in time 0 to n. Then

$$\{\tau^{(n)} = k\} = \{S_k > 0, S_k > S_1, \ldots, S_k > S_{k-1}, S_k \geqslant S_{k+1}, \ldots, S_k \geqslant S_n\}$$

$$\text{for } k \geqslant 1,\ \{\tau^{(n)} = 0\} = \{S_k \leqslant 0, 1 \leqslant k \leqslant n\}.$$

(i) Show that

$$P(\tau^{(2n)} = 0) = P(\tau^{(2n-1)} = 0) = \binom{2n}{n} 2^{-2n}.$$

[*Hint*: Use (10.1) and induction.]

(ii) Show that

$$P(\tau^{(2n)} = 2n) = P(\tau^{(2n+1)} = 2n) = \frac{1}{2}\binom{2n}{n} 2^{-2n}.$$

[*Hint*: Consider the dual paths to $\{S_k: k = 0, 1, \ldots, n\}$ obtained by reversing the order of the displacements, $S_1' = S_n - S_{n-1}$, $S_2' = (S_n - S_{n-1}) + (S_{n-1} - S_{n-2})$,

$\ldots, S'_n = S_n$. This transformation corresponds to a rotation through 180 degrees. Use (11.2).]

(iii) Show that

$$P(\tau^{(2n)} = 2k) = P(\tau^{(2n)} = 2k + 1) = \frac{1}{2}\binom{2k}{k}2^{-2k}\binom{2(n-k)}{n-k}2^{-2(n-k)}$$

for $k = 1, \ldots, n$ in the first case and $k = 0, \ldots, n - 1$ in the second. [*Hint*: A path of length $2n$ with a maximum at $2k$ can be considered in two sections. Apply (i) and (ii) to each section.]

(iv) $\lim_{n\to\infty} P\left(\frac{\tau^{(n)}}{n} \leqslant t\right) = \frac{2}{\pi}\sin^{-1}\sqrt{t}, \qquad 0 < t < 1.$

*4. Let $\Gamma^{(n)}$, U_n be as defined in Exercise 2. Define $V_n = \#\{k \leqslant \Gamma^{(n)}: S_{k-1} \geqslant 0, S_k \geqslant 0\} = U_{T^{(n)}}$. Show that $P(V_{2n} = 2r \mid S_{2n} = 0) = 1/(n + 1)$, $r = 0, 1, \ldots, n$. [*Hint*: Use induction and Exercise 3(i) to show that $P(V_{2n} = 2r, S_{2n} = 0)$ does not depend on r, $0 \leqslant r \leqslant n$.]

Exercises for Section I.12

1. Show that the finite-dimensional distributions of the Brownian bridge are Gaussian.

2. Suppose that F is an arbitrary distribution function (not necessarily continuous). Define an inverse to F as $F^{-1}(y) = \inf\{x: F(x) > y\}$. Show that if Y is uniform on $[0, 1]$ then $X = F^{-1}(Y)$ has distribution function F.

3. Let $\{B_t\}$ be standard Brownian motion starting at 0 and let $B_t^* = B_t - tB_1, 0 \leqslant t \leqslant 1$.
   (i) Show that $\{B_t^*\}$ is independent of B_1.
   (ii) (*The Inverse Simulation*) Give a construction of standard Brownian motion from the Brownian bridge. [*Hint*: Use (i).]

*4. Let $\{B_t\}$ be a standard Brownian motion starting at 0 and let $\{B_t^*\}$ be the Brownian bridge.
   (i) Show that for time points $0 \leqslant t_1 < t_2 < \cdots < t_k \leqslant 1$,

$$\lim_{\varepsilon\to 0} P(B_{t_i} \leqslant x_i, i = 1, 2, \ldots, k \mid -\varepsilon < B_1 < \varepsilon) = P(B_{t_i}^* \leqslant x_i, i = 1, \ldots, k).$$

   Likewise, for conditioning on $B_1 \in D_\varepsilon = [0, \varepsilon)$ or $D_\varepsilon = [-\varepsilon, 0)$ the limit is unchanged; for the existence of $\{B_t^*\}$ as the limit distribution (tightness) as $\varepsilon \to 0$, see theoretical complement 3.
   (ii) Show that for $m^* = \inf_{0\leqslant t\leqslant 1} B_t^*$, $M^* = \sup_{0\leqslant t\leqslant 1} B_t^*$, $u < 0 < v$,

$$P(u < m^* \leqslant M^* \leqslant v) = \sum_{k=-\infty}^{\infty} \exp\{-2k^2(v-u)^2\} - \sum_{k=-\infty}^{\infty} \exp\{-2[v + k(v-u)]^2\}.$$

   [*Hint*: Express as a limit of the ratio of probabilities as in (i) and use Exercise 9.5(v). Also, $\Phi(x, x + \varepsilon) = \varepsilon/(2\pi)^{1/2}\exp(-x^2/2) + o(1)$ as $\varepsilon \to 0$.]

(iii) Prove

$$P\left(\sup_{0\leqslant t\leqslant 1} |B_t^*| \leqslant y\right) = 1 + 2\sum_{k=1}^{\infty} (-1)^k e^{-2k^2y^2}, \qquad y > 0.$$

[*Hint*: Take $u = -v$ in (ii).]

(iv) $P(M^* < v) = 1 - e^{-2v^2}$, $v > 0$. [*Hint*: Use Exercise 9.5(vi) for the ratio of probabilities described in (i).]

5. (*Random Walk Bridge*) Let $\{S_n\}$ denote the simple symmetric random walk starting at 0.

(i) Calculate $P(S_m = y \mid S_{2n} = 0)$, $0 \leqslant m < 2n$.

(*ii) Let $U_{2n} = \#\{k \leqslant 2n: S_{k-1} \geqslant 0, S_k \geqslant 0\}$. Calculate $P(U_{2n} = r \mid S_{2n} = 0)$. [*Hint*: See Exercise 11.4*.]

*6. (*Brownian Meander*) The Brownian meander $\{B_t^+\}$ is defined as the limiting distribution of the standard Brownian motion $\{B_t\}$ starting at 0, conditional on $\{m = \min_{0\leqslant t\leqslant 1} B_t > -\varepsilon\}$ as $\varepsilon \to 0$ (see theoretical complement 4 for existence). Let $m^+ = \min_{0\leqslant t\leqslant 1} B_t^+$, $M^+ = \max_{0\leqslant t\leqslant 1} B_t^+$. Prove the following:

(i) $$P(M^+ \leqslant x, B_1^+ \leqslant y) = \sum_{k=-\infty}^{\infty} [e^{-(2kx)2/2} - e^{-(2kx+y)^2/2}], \qquad 0 < y \leqslant x.$$

[*Hint*: Express as a limit of ratios of probabilities and use Exercise 9.5(v). Also $P(m > -\varepsilon) = (2/\pi)^{1/2}\varepsilon + o(1)$; see Exercise 10.5(ii) noting $\min(A) = -\max(-A)$ and symmetry. Justify interchange of limits with the Dominated Convergence Theorem (Chapter 0).]

(ii) $P(M^+ \leqslant x) = 1 + 2\sum_{k=1}^{\infty} (-1)^k \exp\{-(kx)^2/2\}$. [*Hint*: Consider (i) with $y = x$.]

(iii) $EM^+ = (2\pi)^{1/2} \log 2 = 1.7374\ldots.$ [*Hint*: Compute $\int_0^\infty P(M^+ > x)\,dx$ from (ii).]

(iv) (*Rayleigh Distribution*) $P(B_1^+ \leqslant x) = 1 - e^{-x^2/2}$, $x \geqslant 0$. [*Hint*: Consider (i) in the limit as $x \to \infty$.]

*7. (*Brownian Excursion*) The Brownian excursion $\{B_t^{*+}\}$ is defined by the limiting distribution of $\{B_t^+\}$ conditioned on $\{m^* > -\varepsilon\}$ as $\varepsilon \downarrow 0$ (see theoretical complement 4 for existence). Let $M^{*+} = \max_{0\leqslant t\leqslant 1} B_t^{*+}$. Prove the following:

(i) $$P(M^{*+} \leqslant x) = 1 + 2\sum_{k=1}^{\infty} [1 - (2kx)^2] \exp\{-(2kx)^2/2\}, \qquad x > 0.$$

[*Hint*: Write $P(M^{*+} < x)$ as the limit of ratios as $\varepsilon \to 0$. Multiply numerator and denominator by ε^{-2}, apply Exercise 4, and use l'Hospital's rule twice to evaluate the limit term by term. Check that the Dominated Convergence Theorem (Chapter 0) can be used to justify limit interchange.]

(ii) $EM^{*+} = (\pi/2)^{1/2}$. [*Hint*: Note that interchange of (limits) integral with sum in (i) to compute $EM^{*+} = \int_0^\infty P(M^{*+} > x)\,dx$ leads to an absurdity, since the values of the termwise integrals are zero. Express EM^{*+} as

$$2\lim_{\Delta\to 0} \int_\Delta^\infty \sum_{k=1}^{\infty} [(2kx)^2 - 1]\exp\{-\tfrac{1}{2}(2kx)^2\}\,dx$$

and note that for $k \geqslant \frac{1}{2}\Delta$ the integrand is nonnegative on $[\Delta, \infty)$. So Lebesgue's monotone convergence can be applied to interchange integral with sum over $k \geqslant 1/(2\Delta)$ to get zero for this. Thus, EM^{*+} is the limit as $\Delta \to 0$ of a finite sum over $k < \frac{1}{2}\Delta$ of an integral that can be evaluated (by parts). Note that this gives a Riemann sum limit for $2\int_0^\infty \exp(-\frac{1}{2}x^2)\,dx = (\pi/2)^{1/2}$.]

(iii) $$\mu^{*+}(r) := E(M^{*+})^r = \begin{cases} (\pi/2)^{1/2}, & \text{if } r = 1 \\ (2(2)^{1/2})^{-r}4(\pi)^{1/2}\dfrac{r!}{\Gamma\left(\dfrac{r-1}{2}\right)}\zeta(r), & \text{if } r = 2, 3, \ldots, \end{cases}$$

where $\zeta(r) = \sum_{k=1}^\infty k^{-r}$ is the Riemann Zeta function ($r \geqslant 2$). [*Hint*: The case $r = 1$ is given in (ii) above.] For the case $r \geqslant 2$, we have

$$\mu^{*+}(r) = r\int_0^\infty x^{r-1}(1 - F_0^+(x))\,dx = 2r\sum_{k=1}^\infty \frac{1}{k^r}\int_0^\infty t^{r-1}\{(2t)^2 - 1\}e^{-(2t)^2/2}\,dt,$$

where the interchange of limits is justified for $r \geqslant 2$ since

$$\sum_{k=1}^\infty \frac{1}{k^r}\int_0^\infty t^{r-1}|(2t)^2 - 1|e^{-(2t)^2/2}\,dt < \infty, \qquad r = 2, 3, \ldots.$$

In particular, letting Z denote a standard normal random variable,

$$\mu^{*+}(r) = 2^{-r}r\zeta(r)(2)^{1/2}(\pi)^{1/2}[E|Z|^{r+1} - E|Z|^{r-1}].$$

Consider the two cases whether r is odd or even.

8. Prove the *Glivenko–Cantelli Lemma* by justifying the following steps.
   (i) For each t, the event $\{F_n(t) \to F(t)\}$ has probability 1.
   (ii) For each t, the event $\{F_n(t^-) \to F(t^-)\}$ has probability 1.
   (iii) Let $\tau(y) = \inf\{t\colon F(t) \geqslant y\}$, $0 < y < 1$. Then $F(\tau(y)^-) \leqslant y \leqslant F(\tau(y))$.
   (iv) Let

$$D_{m,n} = \max_{1 \leqslant k \leqslant m}\{|F_n(\tau(k/m)) - F(\tau(k/m))|, |F_n(\tau(k/m)^-) - F(\tau(k/m)^-)|\}.$$

Then, by considering the cases

$$\tau\left(\frac{k-1}{m}\right) \leqslant t \leqslant \tau\left(\frac{k}{m}\right), \qquad t < \tau\left(\frac{1}{m}\right) \text{ and } t > \tau(1),$$

check that

$$\sup_t |F_n(t) - F(t)| \leqslant D_{m,n} + \frac{1}{m}.$$

   (v) $$C = \bigcup_{m=1}^\infty \bigcup_{k=1}^m \{\{F_n(\tau(k/m)) \nrightarrow F(\tau(k/m))\} \cup \{F_n(\tau(k/m)^-) \nrightarrow F(\tau(k/m)^-)\}$$

has probability zero, and for $\omega \in C^c$ and each $m \geqslant 1$

$$D_{m,n}(\omega) \to 0 \qquad \text{as } n \to \infty.$$

(vi) $\sup_t |F_n(t, \omega) - F(t)| \to 0 \quad$ as $n \to \infty \quad$ for $\omega \in C^c$.

9. (*The Gnedenko–Koroljuk Formula*) Let $(X_1, \ldots, X_n)$ and $(Y_1, \ldots, Y_n)$ be two independent i.i.d. random samples with continuous distribution functions F and G, respectively. To test the *null hypothesis* that both samples are from the same population (i.e., $F = G$), let F_n and G_n be the respective *empirical distribution functions* and consider the statistic $D_{n,n} = \sup_x |F_n(x) - G_n(x)|$. Under the null hypothesis, $X_1, \ldots, X_n, Y_1, \ldots, Y_n$ are $2n$ i.i.d. random variables with the common distribution F. Verify that under the null hypothesis:
   (i) the distribution of $D_{n,n}$ does not depend on F and can be explicitly calculated according to the formula

$$P\left(D_{n,n} < \frac{r}{n}\right) = P\left(\max_{0 \leqslant k \leqslant 2n} |S_k^{(2n)*}| < r\right),$$

   where $\{S_k^{(2n)*}: k = 0, 1, 2, \ldots, 2n\}$ is the *simple symmetric random walk bridge* (starting at 0 and tied down at $k = 2n$) as defined in Exercise 5. [*Hint*: Arrange $X_1, \ldots, X_n, Y_1, \ldots, Y_n$ in increasing order as $X_{(1)} < X_{(2)} < \cdots < X_{(2n)}$ and define the kth displacement of $\{S_k^{(2n)*}\}$ by

$$S_k^{(2n)*} - S_{k-1}^{(2n)*} = +1 \qquad \text{if } X_{(k)} \in \{X_1, \ldots, X_n\}$$

   and

$$S_k^{(2n)*} - S_{k-1}^{(2n)*} = -1 \qquad \text{if } X_{(k)} \in \{Y_1, \ldots, Y_n\}. \;]$$

   (ii) Find the analytic expression for the probability in (i). [*Hint*: Consider the event that the simple random walk with absorbing boundaries at $\pm r$ returns to 0 at time $2n$. First condition on the initial displacement.]
   (iii) Calculate the large-sample-theory (i.e., asymptotic as $n \to \infty$) limit distribution of $\sqrt{n}\, D_{n,n}$. See Exercise 4(iii).
   (iv) Show

$$P\left(\sup_x (F_n(x) - G_n(x)) < \frac{r}{n}\right) = 1 - \frac{\binom{2n}{n-r}}{\binom{2n}{n}}, \qquad r = 1, \ldots, n.$$

   [*Hint*: Only one absorbing barrier occurs in the random walk approach.]

Exercises for Section I.13

1. Prove that $\tau_B^{(r)}$ defined by (13.5) ($r = 1, 2, \ldots$) are stopping times.
2. If τ is a stopping time and $m \geqslant 0$, prove that $\tau \wedge m := \min\{\tau, m\}$ is a stopping time.
3. Prove $E(S_\tau \mathbf{1}_{\{\tau > m\}}) \to 0$ as $m \to \infty$ under assumptions (2) and (3) of Theorem 13.1.

4. For the simple symmetric random walk, starting at x, show that $E\{S_m \mathbf{1}_{\{\tau_y = r\}}\} = yP(\tau_y = r)$ for $r \leqslant m$.

5. Prove that EZ_n is independent of n (i.e., constant) for a martingale $\{Z_n\}$. Show also that $E(Z_n \mid \{Z_0, \ldots, Z_k\}) = Z_k$ for any $n > k$.

6. Write out a proof of Theorem 13.3 along the lines of that of Theorem 13.1.

7. Let $\{S_n\}$ be a simple symmetric random walk with $p \in (\frac{1}{2}, 1)$.
   (i) Prove that $\{(q/p)^{S_n}: n = 0, 1, 2, \ldots\}$ is a martingale.
   (ii) Let $c < x < d$ be integers, $S_0 = x$, and $\tau = \tau_c \wedge \tau_d := \min(\tau_c, \tau_d)$. Apply Theorem 13.3 to the martingale in (i) and τ to compute $P(\{S_n\}$ reaches c before $d)$.

8. Write out a proof of Proposition 13.5 along the lines of that of Proposition 13.4.

9. Under the hypothesis that the pth absolute moments are finite for some $p \geqslant 1$, derive the Maximal Inequality $P(M_n \geqslant \lambda) \leqslant E|Z_n|^p/\lambda^p$ in the context of Theorem 13.6.

10. (*Submartingales*) Let $\{Z_n: n = 0, 1, 2, \ldots\}$ be a finite or infinite sequence of integrable random variables satisfying $E(Z_{n+1} \mid \{Z_0, \ldots Z_n\}) \geqslant Z_n$ for all n. Such a sequence $\{Z_n\}$ is called a *submartingale*.
   (i) Prove that, for any $n > k$, $E(Z_n \mid \{Z_0, \ldots, Z_k\}) \geqslant Z_k$.
   (ii) Let $M_n = \max\{Z_0, \ldots, Z_n\}$. Prove the maximal inequality $P(M_n \geqslant \lambda) \leqslant EZ_n^2/\lambda^2$ for $\lambda > 0$. [*Hint*: $E(Z_k \mathbf{1}_{A_k}(Z_n - Z_k)) = E(Z_k \mathbf{1}_{A_k} E(Z_n - Z_k \mid \{Z_0, \ldots, Z_k\})) \geqslant 0$ for $n > k$, where $A_k := \{Z_0 < \lambda, \ldots, Z_{k-1} < \lambda, Z_k \geqslant \lambda\}$.]
   (iii) Extend the result of Exercise 9 to nonnegative submartingales.

11. Let $\{Z_n\}$ be a martingale. If $E|Z_n|^p < \infty$ then prove that $|Z_n|^p$ is a submartingale, $p \geqslant 1$. [*Hint*: Use Jensen's or Hölder's Inequality, Chapter 0, (2.7), (2.12).]

12. (*An Exponential Martingale*) Let $\{X_j: j \geqslant 0\}$ be a sequence of independent random variables having finite moment-generating functions $\phi_j(\xi) := E\exp\{\xi X_j\}$ for some $\xi \neq 0$. Define $S_n := X_1 + \cdots + X_n$, $Z_n = \exp\{\xi S_n\}/\prod_{j=1}^n \phi_j(\xi)$.
   (i) Prove that $\{Z_n\}$ is a martingale.
   (ii) Write $M_n = \max\{S_1, \ldots, S_n\}$. If $\xi > 0$, prove that

$$P(M_n \geqslant \lambda) \leqslant \exp\{-\xi\lambda\} \prod_{j=1}^{n} \phi_j(\xi) \qquad (\lambda > 0).$$

   (iii) Write $m_n = \min\{S_1, \ldots, S_n\}$. If $\xi < 0$, prove

$$P(m_n \leqslant -\lambda) \leqslant \exp\{\xi\lambda\} \prod_{j=1}^{n} \phi_j(\xi) \qquad (\lambda > 0).$$

13. Let $\{X_n: n \geqslant 1\}$ be i.i.d. Gaussian with mean zero and variance $\sigma^2 > 0$. Let $S_n = X_1 + \cdots + X_n$, $M_n = \max\{S_1, \ldots, S_n\}$. Prove the following for $\lambda > 0$.
   (i) $P(M_n \geqslant \lambda) \leqslant \exp\{-\lambda^2/(2\sigma^2 n)\}$. [*Hint*: Use Exercise 12(ii) and an appropriate choice of ξ.]
   (ii) $P(\max\{|S_j|: 1 \leqslant j \leqslant n\} \geqslant \lambda\sigma\sqrt{n}) \leqslant 2\exp\{-\lambda^2/2\}$.

14. Let τ_1, τ_2 be stopping times. Show the following assertions (i)–(v) hold.
   (i) $\tau_1 \vee \tau_2 := \max(\tau_1, \tau_2)$ is a stopping time.

(ii) $\tau_1 \wedge \tau_2 := \min(\tau_1, \tau_2)$ is a stopping time.
(iii) $\tau_1 + \tau_2$ is a stopping time.
(iv) $\alpha\tau_1$, where α is a positive integer, is a stopping time.
(v) If $\tau_1 < \tau_2$ a.s. then it need not be the case that $\tau_2 - \tau_1$ is a stopping time.
(vi) If τ is an even integer-valued stopping time, must $\frac{1}{2}\tau$ be a stopping time?

15. (*A Doob–Meyer Decomposition*)
   (i) Let $\{Y_n\}$ be an arbitrary *submartingale* (see Exercise 10) with respect to sigmafields $\mathscr{F}_0 \subset \mathscr{F}_1 \subset \mathscr{F}_2 \subset \cdots$. Show that there is a unique sequence $\{V_n\}$ such that:
      (a) $0 = V_0 \leqslant V_1 \leqslant V_2 \leqslant \cdots$ a.s.
      (b) V_n is $\mathscr{F}_{n-1}$-measurable.
      (c) $\{Z_n\} := \{Y_n - V_n\}$ is a *martingale* with respect to $\mathscr{F}_n$. [*Hint*: Define $V_n = V_{n-1} + E\{Y_n \mid \mathscr{F}_{n-1}\} - Y_{n-1}$, $n \geqslant 1$.]
   (ii) Calculate the $\{V_n\}$, $\{Z_n\}$ decomposition for $Y_n = S_n^2$, where $\{S_n\}$ is the simple symmetric random walk starting at 0. [*Note*: A sequence $\{V_n\}$ satisfying (b) is called a *predictable* sequence with respect to $\{\mathscr{F}_n\}$.]

16. Let $\{S_n\}$ be the simple random walk starting at 0. Let $\{G_n\}$ be a *predictable* sequence of nonnegative random variables with respect to $\mathscr{F}_n = \sigma\{X_1, \ldots, X_n\} = \sigma\{S_0, \ldots, S_n\}$, where $X_n = S_n - S_{n-1}$ $(n = 1, 2, \ldots)$; i.e., each G_n is $\mathscr{F}_{n-1}$-measurable. Assume each G_n to have finite first moment. Such a sequence $\{G_n\}$ will be called a *strategy*. Define

$$W_n = W_0 + \sum_{k=1}^{n} G_k(S_k - S_{k-1}), \qquad n \geqslant 1,$$

where W_0 is an integrable nonnegative random variable independent of $\{S_n\}$ (representing *initial capital*). Show that regardless of the strategy $\{G_n\}$ we have the following.
   (i) If $p = \frac{1}{2}$ then $\{W_n\}$ is a martingale.
   (ii) If $p > \frac{1}{2}$ then $\{W_n\}$ is a submartingale.
   (iii) If $p < \frac{1}{2}$ then $\{W_n\}$ is a *supermartingale* (i.e., $E|W_n| < \infty$, $E(W_{n+1} \mid \mathscr{F}_n) \leqslant W_n$, $n = 1, 2, \ldots$).
   (iv) Calculate EW_n, $n \geqslant 1$, in the case of the so-called *double-or-nothing strategy* defined by $G_n = 2^{S_{n-1}}\mathbf{1}_{\{S_{n-1} = n-1\}}$, $n \geqslant 1$.

17. Let $\{S_n\}$ be the simple symmetric random walk starting at 0. Let $\tau = \inf\{n \geqslant 0: S_n = 2 - n\}$.
   (i) Calculate $E\tau$ from the distribution of τ.
   (ii) Use the martingale stopping theorem to calculate $E\tau$.
   (*iii) How does this generalize to the cases $\tau = \inf\{n \geqslant 0: S_n = b - n\}$, where b is a positive integer? [*Hint*: Check that $n + S_n$ is *even* for $n = 0, 1, 2, \ldots$.]

18. (i) Show that if X is a random variable such that $g(z) = Ee^{zX}$ is finite in a neighborhood of $z = 0$, then $EX^k < \infty$ for all $k = 1, 2, \ldots$.
   (ii) For a Brownian motion $\{X_t\}$ with drift μ and diffusion coefficient σ^2, prove that $\exp\{\lambda X_t - \lambda t\mu - \lambda^2\sigma^2 t/2\}$ $(t \geqslant 0)$ is a martingale.

19. Consider an arbitrary Brownian motion with drift μ and diffusion coefficient $\sigma^2 > 0$.
   (i) Let $m(x) = ET^x$, where T^x is the time to reach the boundary $\{c, d\}$ starting at $x \in [c, d]$. Show that $m(x)$ solves the boundary-value problem

$$\tfrac{1}{2}\sigma^2 \frac{d^2m}{dx^2} + \mu \frac{dm}{dx} = -1, \qquad m(c) = m(d) = 0.$$

(ii) Let $r(x) = P_x(\tau_d < \tau_c)$ for $x \in [c, d]$. Verify that $r(x)$ solves the boundary value problem

$$\tfrac{1}{2}\sigma^2 \frac{d^2r}{dx^2} + \mu \frac{dr}{dx} = 0, \qquad r(c) = 0, \quad r(d) = 1.$$

Exercises for Section I.14

1. In the case $\beta = \frac{1}{2}$, consider the process $\{Z_N(s)\}$ defined by

$$Z_N\left(\frac{n}{N}\right) = \frac{S_n - n\bar{Y}_N}{\sqrt{N}\, D_N}, \qquad n = 1, 2, \ldots, N,$$

and linearly interpolate between n/N and $(n + 1)/N$.

(i) Show that $\{Z_N\}$ converges in distribution to $B_s^* + 2cs^{1/2}(1 - s^{1/2})$, where $\{B_s^*\} = \{B_s - sB_1 : 0 \leqslant s \leqslant 1\}$ is the Brownian Bridge.

(ii) Show that $R_N/(\sqrt{N}\, D_N)$ converges in distribution to $\max_{0 \leqslant s \leqslant 1} B_s^* - \min_{0 \leqslant s \leqslant 1} B_s^*$.

(iii) Show that the asymptotic distribution in (ii) is nondegenerate.

THEORETICAL COMPLEMENTS

Theoretical Complements to Section I.1

1. Events that can be specified by the values of $X_n, X_{n+1}, \ldots$ for each value of n (i.e., events that depend only on the long-run values of the sequences) are called *tail events*.

Theorem T.1.1. (*Kolmogorov Zero-One Law*). A tail event for a sequence of independent random variables has probability either zero or one. □

Proof. To see this one uses the general measure-theoretic fact that the probability of any event A belonging to the sigmafield $\mathscr{F} = \sigma\{X_1, X_2, \ldots, X_n, \ldots\}$ (generated by $X_1, X_2, \ldots, X_n, \ldots$) can be approximated by events $A_1, \ldots, A_n, \ldots$ belonging to the *field of events* $\mathscr{F}_0 = \bigcup_{n=1}^{\infty} \sigma\{X_1, \ldots, X_n\}$ in the sense that $A_n \in \sigma\{X_1, \ldots, X_n\}$ for each n and $P(A \Delta A_n) \to 0$ as $n \to \infty$, where Δ denotes the *symmetric difference* $A \Delta A_n = (A \cap A_n^c) \cup (A^c \cap A_n)$. Applying this approximation to a tail event A, one obtains that since $A \in \sigma\{X_{n+1}, X_{n+2}, \ldots\}$ for each n, A is *independent* of each event A_n. Thus, $0 = \lim_{n\to\infty} P(A \Delta A_n) = 2P(A)P(A^c) = 2P(A)(1 - P(A))$. The only solutions to the equation $x(1 - x) = 0$ are 0 and 1. ■

2. Let $S_n = X_1 + \cdots + X_n$, $n \geqslant 1$. *Events that depend on the tail of the sums are trivial* (i.e., *have probability 1 or 0*) *whenever the summands* $X_1, X_2, \ldots$ *are i.i.d.* This is a consequence of the following more general zero–one law for events that *symmetrically depend* on the terms $X_1, X_2, \ldots$ of an i.i.d. sequence of random variables (or vectors). Let $\mathscr{B}^\infty$ denote the sigmafield of subsets of $\mathbb{R}^\infty = \{(x_1, x_2, \ldots): x_i \in \mathbb{R}^1\}$ generated by events depending on finitely many coordinates.

Theorem T.1.2. (*Hewitt–Savage Zero–One Law*). Let $X_1, X_2, \ldots$ be an i.i.d. sequence of random variables. If an event $A = \{(X_1, X_2, \ldots) \in B\}$, where $B \in \mathscr{B}^\infty$, is invariant under finite permutations $(X_{i_1}, X_{i_2}, \ldots)$ of terms of the sequence $(X_1, X_2, \ldots)$, that is, $A = \{(X_{i_1}, X_{i_2}, \ldots) \in B\}$ for any finite permutation $(i_1, i_2, \ldots)$ of $(1, 2, \ldots)$, then $P(A) = 1$ or 0. □

As noted above, the symmetric dependence with respect to $\{X_n\}$ applies, for example, to tail events for the sums $\{S_n\}$.

Proof. To prove the Hewitt–Savage 0–1 law, proceed as in the Kolmogorov 0–1 law by selecting finite-dimensional approximants to A of the form $A_n = \{(X_1, \ldots, X_n) \in B_n\}$, $B_n \in \mathscr{B}^n$, such that $P(A \Delta A_n) \to 0$ as $n \to \infty$. For each fixed n, let $(i_1, i_2, \ldots)$ be the permutation $(2n, 2n-1, \ldots, 1, 2n+1, \ldots)$ and define $\tilde{A}_n = \{(X_{i_1}, \ldots, X_{i_n}) \in B_n\}$. Then $\tilde{A}_n$ and A_n are independent with $P(A_n \cap A_n) = P(A_n)P(\tilde{A}_n) = (P(A_n))^2 \to (P(A))^2$ as $n \to \infty$. On the other hand, $P(A \Delta \tilde{A}_n) = P(A \Delta A_n) \to 0$, so that $P(A_n \Delta \tilde{A}_n) \to 0$ and, in particular, therefore $P(A_n \cap \tilde{A}_n) \to P(A)$ as $n \to \infty$. Thus $x = P(A)$ satisfies $x = x^2$. ■

Theoretical Complements to Section I.3

1. **Theorem T.3.1.** If $\{X_n\}$ is an i.i.d. sequence of integer-valued random variables for a general random walk $S_n = X_1 + \cdots + X_n$, $n \geqslant 1$, $S_0 = 0$, on $\mathbb{Z}$ with $\mu \equiv EX_1 = 0$, then $\{S_n\}$ is recurrent. □

Proof. To prove this, first observe that $P(S_n = 0 \text{ i.o.})$ is 1 or 0 by the Hewitt–Savage zero–one law (theoretical complement 1.2). If $\sum_{n=1}^\infty \mathrm{P}(S_n = 0) < \infty$, then $P(S_n = 0 \text{ i.o.}) = 0$ by the Borel–Cantelli Lemma. If $\sum_{n=1}^\infty P(S_n = 0)$ is divergent (i.e., the expected number of visits to 0 is infinite), then we can show that $P(S_n = 0 \text{ i.o.}) = 1$ as follows. Using *independence* and the property that the *shifted* sequence $X_k, X_{k+1}, \ldots$ has the same distribution as $X_1, X_2, \ldots$, one has

$$\begin{aligned}1 \geqslant P(S_n = 0 \text{ finitely often}) &\geqslant \sum_n P(S_n = 0, S_m \neq 0, m > n)\\ &= \sum_n P(S_n = 0)P(S_m - S_n \neq 0, m > n)\\ &= \sum_n P(S_n = 0)P(S_m \neq 0, m \geqslant 1).\end{aligned}$$

Thus, if $\sum_n P(S_n = 0)$ diverges, then $P(S_m \neq 0, m \geqslant 1) = 0$ or equivalently $P(S_m = 0 \text{ for some } m \geqslant 1) = 1$. This may now be extended by induction to get that at least r visits to 0 is certain for each $r = 1, 2, \ldots$. One may also use the strong Markov property of Chapter II, Section 4, with the time of the rth visit to 0 as the stopping time.

From here the proof rests on showing $\sum_n P(S_n = 0)$ is divergent when $\mu = 0$. Consider the *generating function* of the sequence $P(S_n = 0)$, $n = 0, 1, 2, \ldots$, namely,

$$g(x) := \sum_{n=0}^{\infty} P(S_n = 0)x^n, \qquad |x| < 1.$$

The problem is to investigate the divergence of $g(x)$ as $x \to 1^-$. Note that $P(S_n = 0)$ is the 0th-term Fourier coefficient of the characteristic function (Fourier series)

$$Ee^{itS_n} = \sum_{k=-\infty}^{\infty} P(S_n = k)e^{itk}.$$

Thus,

$$P(S_n = 0) = \frac{1}{2\pi}\int_{-\pi}^{\pi} Ee^{itS_n}\,dt = \frac{1}{2\pi}\int_{-\pi}^{\pi} \varphi^n(t)\,dt$$

where $\varphi(t) = Ee^{itX_1}$. It follows that for $|x| < 1$,

$$g(x) = \frac{1}{2\pi}\int_{-\pi}^{\pi} \frac{dt}{1 - x\varphi(t)}.$$

Thus, recurrence or transience depends on the divergence or convergence, respectively, of this integral. Now, with $\mu = 0$, we have $\varphi(t) = 1 - o(|t|)$ as $t \to 0$. Thus, for any $\varepsilon > 0$ there is a $\delta > 0$ such that $|1 - \varphi_1(t)| \leqslant \varepsilon|t|$, $|\varphi_2(t)| \leqslant \varepsilon|t|$, for $|t| \leqslant \delta$, where $\varphi(t) = \varphi_1(t) + i\varphi_2(t)$ has real and imaginary parts φ_1, φ_2. Now, for $0 < x < 1$, noting that $g(x)$ is real valued,

$$\begin{aligned}
\int_{-\pi}^{\pi} \frac{dt}{1 - x\phi(t)} &= \int_{-\pi}^{\pi} \mathrm{Re}\left(\frac{1}{1 - x\varphi(t)}\right) dt = \int_{-\pi}^{\pi} \frac{1 - x\varphi_1(t)}{|1 - x\varphi(t)|^2}\,dt \geqslant \int_{-\delta}^{\delta} \frac{1 - x\varphi_1(t)}{|1 - x\varphi(t)|^2}\,dt \\
&= \int_{-\delta}^{\delta} \frac{1 - x\varphi_1(t)}{(1 - x\varphi_1(t))^2 + x^2\varphi_2^2(t)}\,dt \geqslant \int_{-\delta}^{\delta} \frac{1 - x}{(1 - x + x\varepsilon|t|)^2 + x^2\varepsilon^2 t^2}\,dt \\
&\geqslant \int_{-\delta}^{\delta} \frac{1 - x}{2(1 - x)^2 + 3x^2\varepsilon^2 t^2}\,dt \geqslant \int_{-\delta}^{\delta} \frac{1 - x}{3[(1 - x)^2 + \varepsilon^2 t^2]}\,dt \\
&= \frac{2}{3\varepsilon}\tan^{-1}\left(\frac{\varepsilon\delta}{1 - x}\right) \to \frac{\pi}{3\varepsilon} \qquad \text{as } x \to 1^-.
\end{aligned}$$

Since ε is arbitrary, this completes the argument. ■

The above argument is a special case of the so-called *Chung–Fuchs recurrence criterion* developed by K. L. Chung and W. H. J. Fuchs (1951), "On the Distribution of Values of Sums of Random Variables," *Mem. Amer. Math. Soc.*, No. 6.

Theoretical Complements to Section I.6

1. The *Kolmogorov Extension Theorem* holds for more general spaces S than the case $S = \mathbb{R}^1$ presented. In general it requires that S be homeomorphic to a *complete* and *separable* metric space. So, for example, it applies whenever S is $\mathbb{R}^k$ or a rectangle,

or when S is a finite or countable set. Assuming some background in analysis, a simple proof of Kolmogorov's theorem (due to Edward Nelson (1959), "Regular Probability Measures on Function Space," *Annals of Math.*, **69**, pp. 630–643) in the case of compact S can be made as follows. Define a linear functional l on the subspace of $C(S^I)$ consisting of continuous functions on S^I that depend on finitely many coordinates, by

$$l(f) := \int_{S^{\{i_1,\dots,i_k\}}} \bar{f}(x_{i_1}, \dots, x_{i_k}) \mu_{i_1}, \dots, \mu_{i_k}(dx_{i_1} \cdots dx_{i_k}),$$

where

$$f((x_i)_{i \in I}) = \bar{f}(x_{i_1}, \dots, x_{i_k}).$$

By consistency, l is a well-defined linear functional on a subspace of $C(S^I)$ that, by the *Stone–Weierstrass Theorem* of functional analysis, is *dense* in $C(S^I)$; note that S^I is *compact* for the product topology by *Tychonoff's Theorem* from topology (see H. L. Royden (1968), *Real Analysis*, 2nd ed., Macmillan, New York, pp. 174, 166). In particular, l has a natural extension to $C(S^I)$ that is linear and continuous. Now apply the *Riesz–Representation Theorem* to get a (probability) measure μ on $(S^I, \mathscr{F})$ such that for any $f \in C(S^I)$, $l(f) = \int_{S^I} f \, d\mu$ (see Royden, *loc. cit.*, p. 310). To make the proof in the noncompact but separable and complete case, one can use a fundamental (homeomorphic) embedding of S into $\mathbb{R}^\infty$ (see P. Billingsley (1968), *Convergence of Probability Measures*, Wiley, New York, p. 219); this is a special case of *Urysohn's Theorem* in topology (see H. L. Royden, *loc. cit.*, p. 149). Then, by making a two-point compactification of $\mathbb{R}^1$, S can be further embedded in the Hilbert cube $[0, 1]^\infty$, where the measure μ can be obtained as above. Consistency allows one to restrict μ back to S. □

Theoretical Complements to Section I.7

1. (*Brownian Motion and the Inadequacy of Kolmogorov's Extension Theorem*) Let $S = C[0, 1]$ denote the space of continuous real-valued functions on $[0, 1]$ equipped with the uniform metric

$$\rho(\mathbf{x}, \mathbf{y}) = \max_{0 \leqslant t \leqslant 1} |\mathbf{x}(t) - \mathbf{y}(t)|, \qquad \mathbf{x}, \mathbf{y} \in C[0, 1].$$

The Borel sigmafield $\mathscr{B}$ of $C[0, 1]$ for the metric ρ is the smallest sigmafield of subsets of $C[0, 1]$ that contains all finite-dimensional events of the form

$$\{\mathbf{x} \in C[0, 1] : a_i < \mathbf{x}(t_i) \leqslant b_i, i = 1, \dots, k\},$$
$$k \geqslant 1, \quad 0 \leqslant t_1 < \cdots < t_k \leqslant 1, \quad a_1, \dots, a_k, b_1, \dots, b_k \in \mathbb{R}^1.$$

The problem of constructing standard Brownian motion (starting at 0) on the time interval $0 \leqslant t \leqslant 1$ corresponds to constructing a process $\{X_t : 0 \leqslant t \leqslant 1\}$ having Gaussian finite-dimensional distributions with mean zero and variance–covariance matrix $\gamma_{ij} = \min(t_i, t_j)$ for the time points $0 \leqslant t_0 < \cdots < t_k \leqslant 1$, and having a.s. continuous sample paths. One can easily construct a process on the product space

$\Omega = \mathbb{R}^{[0,1]}$ (or $\mathbb{R}^{[0,\infty)}$) equipped with the sigmafield $\mathscr{F}$ generated by finite-dimensional events of the form $\{\mathbf{x} \in \mathbb{R}^{[0,1]}: a_i < \mathbf{x}(t_i) \leqslant b_i, i = 1, \ldots, k\}$ having the prescribed finite-dimensional distributions; in particular, consistency in the Kolmogorov extension theorem can be checked by noting that for any time points $t_1 < t_2 < \cdots < t_k$, the matrix $\gamma_{ij} = \min(t_i, t_j)$, $1 \leqslant i, j \leqslant k$, is symmetric and nonnegative definite, since for any real numbers $x_1, \ldots, x_k$, $t_0 = 0$,

$$\sum_{i,j} \gamma_{ij} x_i x_j = \sum_{i,j} x_i x_j \sum_{r=1}^{\min(i,j)} (t_r - t_{r-1}) = \sum_r (t_r - t_{r-1}) \left(\sum_{i=r}^{k} x_i \right)^2 \geqslant 0.$$

However, the problem with this construction of a probability space $(\Omega, \mathscr{F}, P)$ for $\{X_t\}$ is that events in $\mathscr{F}$ can only depend on specifications of values at countably many time points. Thus, the subset $C[0,1]$ of Ω is *not* measurable; i.e., $C[0,1] \notin \mathscr{F}$. This dilemma is resolved in the theoretical complement to Section I.13 by showing that there is a modification of the process $\{X_t\}$ that yields a process $\{B_t\}$ with sample paths in $C[0,1]$ and having the same finite-dimensional distributions as $\{X_t\}$; i.e., $\{B_t\}$ is the desired Brownian motion process. The basic idea for this modification is to show that almost all paths $q \to X_q$, $q \in D$, where D is a countable dense set of time points, are uniformly continuous. With this, one can then define $\{B_t\}$ by the continuous extension of these paths given by

$$B_t = \begin{cases} X_q & \text{if } t = q \in D \\ \lim\limits_{q \downarrow t} X_q & \text{if } t \notin D. \end{cases}$$

It is then a simple matter to check that the *finite-dimensional distributions* of $\{X_t\}$ carry over to $\{B_t\}$ under this extension. The distribution of $\{B_t\}$ defines the *Wiener measure* on $C[0,1]$. So, from the point of view of Kolmogorov's (canonical) construction, the main problem remains that of showing a.s. uniform continuity on a countable dense set. A solution is given in theoretical complement 13.1.

2. In general, if $\{X_t: t \in I\}$ is a stochastic process with an uncountable index set, for example $I = [0,1]$, then the measurability of events that depend on sample paths at uncountably many points $t \in I$ can be at issue. This issue goes beyond the finite-dimensional distributions, and is typically solved by exploiting properties of the sample paths of the model. For example, if I is discrete, say $I = \{0, 1, 2, \ldots\}$, then the event $\{\sup_{n \in I} X_n > x\}$ is the *countable union* $\bigcup_{n=0}^{\infty} \{X_n > x\}$ of measurable sets, which is, therefore, measurable. But, on the other hand, if I is uncountable, say $I = [0,1]$, then $\{\sup_{t \in I} X_t > x\}$ is an *uncountable union* of measurable sets. This of course does not make $\{\sup_{t \in I} X_t > x\}$ measurable. If, for example, it is known that $\{X_t\}$ has continuous sample paths, however, then, letting T denote the rationals in $[0,1]$, we have $\{\sup_{t \in I} X_t > x\} = \{\sup_{t \in T} X_t > x\} = \bigcup_{t \in T} \{X_t > x\}$. Thus $\{\sup_{t \in I} X_t > x\}$ is seen to be measurable under these circumstances. While it is not always possible to construct a stochastic process $\{X_t: t \geqslant 0\}$ having continuous sample paths for a given consistent specification of finite-dimensional distributions, it is often possible to construct a model $(\Omega, \mathscr{F}, P)$, $\{X_t: t \geqslant 0\}$ with the following property, called *separability. There is a countable dense subset T of $I = [0, \infty]$ and a set $D \in \mathscr{F}$ with $P(D) = 0$ such that for each $\omega \in \Omega - D$, and each $t \in I$ there is a sequence $t_1, t_2, \ldots$ in T (which may depend on ω) such that $t_n \to t$ and $X_{t_n}(\omega) \to X_t(\omega)$* (P. Billingsley (1986),

Probability and Measure, 2nd ed., Wiley, New York, p. 558). In theory, this is enough sample path regularity to make manageable most such measurability issues connected with processes at uncountably many time points. In practice though, one seeks to explicitly construct models with sufficient sample path regularity that such considerations are often avoidable. The latter is the approach of this text.

Theoretical Complements to Section I.8

1. Let $\{X_t^{(n)}\}$, $n = 1, 2, \ldots$ and $\{X_t\}$ be stochastic processes whose sample paths belong to a metric space S with metric ρ; for example, $S = C[0,1]$ with $\rho(\omega, \eta) = \sup_{0 \leqslant t \leqslant 1} |\omega_t - \eta_t|$, $\omega, \eta \in S$. Let $\mathscr{B}$ denote the Borel sigmafield for S. The distributions P of $\{X_t\}$ and P_n of $\{X_t^{(n)}\}$, respectively, are probability measures on $(S, \mathscr{B})$. Assume $\{X_t^{(n)}\}$ and $\{X_t\}$ are defined on a probability space $(\Omega, \mathscr{F}, Q)$. Then,

$$\begin{aligned} &\text{(i)} \qquad P(B) = Q(\{\omega \in \Omega\colon \{X_t(\omega)\} \in B\}) \\ &\text{(ii)} \qquad P_n(B) = Q(\{\omega \in \Omega\colon \{X_t^{(n)}(\omega)\} \in B\}), \qquad n \geqslant 1. \end{aligned} \tag{T.8.1}$$

Convergence in distribution of $\{X_t^{(n)}\}$ *to* $\{X_t\}$ *has been defined in the text to mean that the sequence of real-valued random variables* $Y_n := f(\{X_t^{(n)}\})$ *converges in distribution to* $Y := f(\{X_t\})$ *for each continuous (for the metric* ρ*) function* $f\colon S \to \mathbb{R}^1$. However, an equivalent condition is that for each *bounded and continuous* real-valued function $f\colon S \to \mathbb{R}^1$ one has,

$$\lim_{n \to \infty} Ef(\{X_t^{(n)}\}) = Ef(\{X_t\}). \tag{T.8.2}$$

To see the equivalence, first observe that for any continuous $f\colon S \to \mathbb{R}^1$, the functions $\cos(rf)$ and $\sin(rf)$ are, for each $r \in \mathbb{R}^1$, *continuous and bounded* functions on S. Therefore, assuming that condition (T.8.2) gives the convergence of the characteristic functions of the Y_n to that of Y for each continuous f on S. In particular, the Y_n must converge in distribution to Y. To go the other way, suppose that $f\colon S \to \mathbb{R}^1$ is continuous and bounded. Assume without loss of generality that $0 < f \leqslant 1$. Then, for each $N \geqslant 1$,

$$\sum_{k=1}^{N} \frac{k-1}{N} P\left(\frac{k-1}{N} < f \leqslant \frac{k}{N}\right) \leqslant \int_S f\,dP \leqslant \sum_{k=1}^{N} \frac{k}{N} P\left(\frac{k-1}{N} < f \leqslant \frac{k}{N}\right). \tag{T.8.3}$$

Equivalently, by rearranging terms,

$$\frac{1}{N} \sum_{k=1}^{N} P\left(f > \frac{k-1}{N}\right) - \frac{1}{N} \leqslant \int_S f\,dP \leqslant \frac{1}{N} \sum_{k=1}^{N} P\left(f > \frac{k-1}{N}\right). \tag{T.8.4}$$

Likewise, these inequalities hold with P_n in place of P. Therefore, by (T.8.4) applied to P,

$$\int_S f\,dP \leqslant \frac{1}{N} \sum_{k=1}^{N} P\left(f > \frac{k-1}{N}\right)$$

$$\leqslant \frac{1}{N} \sum_{k=1}^{N} \liminf_n P_n\left(f > \frac{k-1}{N}\right)$$

$$\leqslant \liminf_n \int_S f\,dP_n + \frac{1}{N}, \tag{T.8.5}$$

by (T.8.4) applied to P_n, and the fact that $\lim \operatorname{Prob}(Y_n > x) = \operatorname{Prob}(Y > x)$ for all points x of continuity of the d.f. of Y implies $\liminf \operatorname{Prob}(Y_n > y) \geqslant \operatorname{Prob}(Y > y)$ for *all* y. Letting $N \to \infty$ gives

$$\int_S f\,dP \leqslant \liminf_n \int_S f\,dP_n.$$

The same argument applied to $-f$ gives that

$$\int_S f\,dP \geqslant \limsup_n \int_S f\,dP_n.$$

Thus, in general,

$$\limsup_n \int_S f\,dP_n \leqslant \int_S f\,dP \leqslant \liminf_n \int_S f\,dP_n \tag{T.8.6}$$

which implies

$$\limsup_n \int_S f\,dP_n = \liminf_n \int_S f\,dP_n = \int_S f\,dP. \tag{T.8.7}$$

This is the desired condition (T.8.2). ■

With the above equivalence in mind we make the following general definition.

Definition. A sequence $\{P_n\}$ of probability measures on $(S, \mathscr{B})$ *converges weakly* (or *in distribution*) to a probability measure P on $(S, \mathscr{B})$ provided that $\lim_n \int_S f\,dP_n = \int_S f\,dP$ for all bounded and continuous functions $f: S \to \mathbb{R}^1$.

Weak convergence is sometimes denoted by $P_n \Rightarrow P$ as $n \to \infty$. Other equivalent notions of convergence in distribution are as follows. The proofs are along the lines of the arguments above. (P. Billingsley (1968), *Convergence of Probability Measures*, Theorem 9.1, pp. 11–14.)

Theorem T.8.1. (*Alexandrov*). $P_n \Rightarrow P$ as $n \to \infty$ if and only if

(i) $\lim_{n\to\infty} P_n(A) = P(A)$ for all $A \in \mathscr{B}$ such that $P(\partial A) = 0$, where ∂A denotes the boundary of A for the metric ρ.

(ii) $\limsup_n P_n(F) \leqslant P(F)$ for all closed sets $F \subset S$.

(iii) $\liminf_n P_n(G) \geqslant P(G)$ for all open sets $G \subset S$. □

2. The convergence of a sequence of probability measures is frequently established by an application of the following theorem due to Prohorov.

Theorem T.8.2. (*Prohorov*). Let $\{P_n\}$ be a sequence of probability measures on the metric space S with Borel sigmafield $\mathscr{B}$. If for each $\varepsilon > 0$ there is a compact set $K_\varepsilon \subset S$ such that

$$P_n(K_\varepsilon) \geqslant 1 - \varepsilon \qquad \text{for all } n = 1, 2, \ldots, \tag{T.8.8}$$

then $\{P_n\}$ has a subsequence weakly convergent to a probability measure Q on $(S, \mathscr{B})$. Moreover, if S is complete and separable then the condition (T.8.8) is also necessary. □

The condition (T.8.8) is referred to as *tightness* of the sequence of probability measures $\{P_n\}$. A proof of sufficiency of the tightness condition in the special case $S = \mathbb{R}^1$ is given in Chapter 0, Theorem 5.2. For the general result, consult Billingsley, *loc. cit.*, pp. 37–40. A version of (T.8.8) for processes with continuous paths is computed below in Theorem T.8.4.

3. In the case of probability measures $\{P_n\}$ on $S = C[0, 1]$, if the finite-dimensional distributions of P_n converge to those of P and if the sequence $\{P_n\}$ is *tight*, then it will follow from Prohorov's theorem that $\{P_n\}$ converges weakly to P. To check tightness it is useful to have the following characterization of (relatively) compact subsets of $C[0, 1]$ from real analysis (A. N. Kolmogorov and S. V. Fomin (1975), *Introductory Real Analysis*, Dover, New York, p. 102).

Theorem T.8.3. (*Arzela–Ascoli*). A subset A of functions in $C[0, 1]$ has compact closure if and only if

(i) $\sup_{\omega \in A} |\omega_0| < \infty$,

(ii) $\lim_{\delta \to 0} \sup_{\omega \in A} v_\omega(\delta) = 0$,

where $v_\omega(\delta)$ is the oscillation in $\omega \in C[0, 1]$ defined by $v_\omega(\delta) = \sup_{|s-t|<\delta} |\omega_s - \omega_t|$. □

The condition (ii) refers to the *equicontinuity* of the functions in A in the sense that given any $\varepsilon > 0$ there is a common $\delta > 0$ such that for all functions $\omega \in A$ we have $|\omega_t - \omega_s| < \varepsilon$ if $|t - s| < \delta$. Conditions (i) and (ii) together imply that A is *uniformly bounded* in the sense that there is a number B for which

$$\|\omega\| := \sup_{0 \leqslant t \leqslant 1} |\omega_t| \leqslant B \qquad \text{for all } \omega \in A.$$

This is because for N sufficiently large we have $\sup_{\omega \in A} v_\omega(1/N) < 1$ and, therefore, for each $0 \leqslant t \leqslant 1$

$$|\omega_t| \leqslant |\omega_0| + \sum_{i=1}^{N} |\omega_{it/N} - \omega_{(i-1)t/N}| \leqslant \sup_{\omega \in A} |\omega_0| + N \sup_{\omega \in A} v_\omega\left(\frac{1}{N}\right) = B.$$

4. Combining the Prohorov theorem (T.8.2) with the Arzela–Ascoli theorem (T.8.3) gives the following criterion for tightness of probability measures $\{P_n\}$ on $S = C[0, 1]$.

Theorem T.8.4. Let $\{P_n\}$ be a sequence of probability measures on $C[0, 1]$. Then $\{P_n\}$ is tight if and only if the following two conditions hold.

(i) For each $\eta > 0$ there is a number B such that

$$P_n(\{\omega \in C[0,1]: |\omega_0| > B\}) \leqslant \eta, \qquad n = 1, 2, \ldots.$$

(ii) For each $\varepsilon > 0$, $\eta > 0$, there is a $0 < \delta < 1$ such that

$$P_n(\{\omega \in C[0,1]: v_\omega(\delta) \geqslant \varepsilon\}) \leqslant \eta, \qquad n \geqslant 1. \qquad \square$$

Proof. If $\{P_n\}$ is tight, then given $\eta > 0$ there is a compact K such that $P_n(\mathrm{K}) > 1 - \eta$ for all n. By the Arzela–Ascoli theorem, if $B > \sup_{\omega \in K} |\omega_0|$ then

$$P_n(\{\omega \in C[0,1]: |\omega_0| \geqslant B\}) \leqslant P_n(\mathrm{K}^C) \leqslant 1 - (1 - \eta) = \eta.$$

Also given $\varepsilon > 0$ select $\delta > 0$ such that $\sup_{\omega \in K} v_\omega(\delta) < \varepsilon$. Then

$$P_n(\{\omega \in C[0,1]: v_\omega(\delta) \geqslant \varepsilon\}) \leqslant P_n(K^c) < \eta \qquad \text{for all } n \geqslant 1.$$

The converse goes as follows. Given $\eta > 0$, first select B using (i) such that $P_n(\{\omega: |\omega_0| \leqslant B\}) \geqslant 1 - \frac{1}{2}\eta$, for $n \geqslant 1$. Select δ_r using (ii) such that $P_n(\{\omega: v_\omega(\delta_r) < 1/r\}) \geqslant 1 - 2^{-(r+1)}\eta$ for $n \geqslant 1$. Now take K to be the closure of

$$\{\omega: |\omega_0| \leqslant B\} \cap \bigcap_{r=1}^{\infty} \left\{\omega: v_\omega(\delta_r) < \frac{1}{r}\right\}.$$

Then $P_n(K) > 1 - \eta$ for $n \geqslant 1$, and K is compact by the Arzela–Ascoli theorem. ∎

The above theorem taken with Prohorov's theorem is a cornerstone of weak convergence theory in $C[0,1]$. If one has proved convergence of the finite-dimensional distributions of $\{X_t^{(n)}\}$ to $\{X_t\}$ then the distributions of $\{X_0^{(n)}\}$ must be tight as probability measures on $\mathbb{R}^1$ so that condition (i) is implied. In view of this and the Prohorov theorem we have the following necessary and sufficient condition for weak convergence in $C[0,1]$ based on convergence of the finite-dimensional distributions and tightness.

Theorem T.8.5. Let $\{X_t^{(n)}: 0 \leqslant t \leqslant 1\}$ and $\{X_t: 0 \leqslant t \leqslant 1\}$ be stochastic processes on $(\Omega, \mathscr{F}, P)$ which have a.s. continuous sample paths and suppose that the finite-dimensional distributions of $\{X_t^{(n)}\}$ converge to those of $\{X_t\}$. Then $\{X_t^{(n)}\}$ converges weakly to $\{X_t\}$ if and only if for each $\varepsilon > 0$

$$\lim_{\delta \to 0} \sup_n P\left(\sup_{|s-t| \leqslant \delta} |X_t^{(n)} - X_s^{(n)}| > \varepsilon\right) = 0. \qquad \square$$

Corollary. For the last limit to hold it is sufficient that there be positive numbers α, β, M such that

$$E|X_t^{(n)} - X_s^{(n)}|^\alpha \leqslant M|t - s|^{1+\beta} \qquad \text{for all } s, t, n. \qquad \square$$

To prove the corollary, let D be the set of all *dyadic rationals* in $[0,1]$, i.e., numbers in $[0,1]$ of the form $j/2^m$ for integers j and m. By sample path continuity, the oscillation of the process over $|t - s| \leqslant \delta$ is given by

$$v_n(D, \delta) = \sup_{\substack{|t-s| \leqslant \delta, \\ s,t \in D}} |X_t^{(n)} - X_s^{(n)}| \quad \text{a.s.}$$

Take $\delta = 2^{-k+1}$. Then, taking suprema over positive integers i, j, n,

$$v_n(D, \delta) \leqslant 2 \sup_{j2^{-k} < i2^{-m} < (j+1)2^{-k}} |X_{i2^{-m}}^{(n)} - X_{j2^{-k}}^{(n)}|.$$

Now, for $j2^{-k} < i2^{-m} < (j+1)2^{-k}$, writing

$$i2^{-m} = j2^{-k} + \sum_{\nu=1}^{r} 2^{-m_\nu} \qquad \text{where } k < m_1 < m_2 < \cdots < m_r \leqslant m,$$

we have, writing $\alpha(\mu) = \sum_{\nu=1}^{\mu} 2^{-m_\nu}$,

$$X_{i2^{-m}}^{(n)} - X_{j2^{-k}}^{(n)} = \sum_{\mu=1}^{r} X_{j2^{-k}+\alpha(\mu)} - X_{j2^{-k}+\alpha(\mu-1)}\}.$$

Therefore,

$$v_n(D, \delta) \leqslant 2 \sum_{m=k+1}^{\infty} \sup_{0 \leqslant h \leqslant 2^m - 1} |X_{(h+1)2^{-m}}^{(n)} - X_{h2^{-m}}^{(n)}|.$$

Let $\varepsilon > 0$ and take $\delta = 2^{-k+1}$ so small (i.e., k so large) that $\sum_{m=k+1}^{\infty} 1/m^2 < \varepsilon/2$. Then

$$\begin{aligned} P(v_n(D, \delta) > \varepsilon) &\leqslant \sum_{m=k+1}^{\infty} P\left(\sup_{0 \leqslant h \leqslant 2^m - 1} |X_{(h+1)2^{-m}}^{(n)} - X_{h2^{-m}}^{(n)}| > \frac{1}{m^2} \right) \\ &\leqslant \sum_{m=k+1}^{\infty} \sum_{h=0}^{2^m - 1} P\left(|X_{(h+1)2^{-m}}^{(n)} - X_{h2^{-m}}^{(n)}| > \frac{1}{m^2} \right). \end{aligned}$$

Now apply Chebyshev's Inequality to get

$$\begin{aligned} P(v_n(D, \delta) > \varepsilon) &\leqslant \sum_{m=k+1}^{\infty} m^{2\alpha} \sum_{h=0}^{2^m - 1} E|X_{(h+1)2^{-m}}^{(n)} - X_{h2^{-m}}^{(n)}|^\alpha \\ &\leqslant \sum_{m=k+1}^{\infty} m^{2\alpha} 2^m M 2^{-m(1+\beta)} = M \sum_{m=k+1}^{\infty} \frac{m^{2\alpha}}{2^{m\beta}}. \end{aligned}$$

This bound does not depend on n and goes to zero as $\delta \to 0$ (i.e., $k = \log_2 \delta^{-1} + 1 \to +\infty$) since it is the tail of a convergent series. This proves the corollary. ∎

5. (*FCLT and a Brownian Motion Construction*) As an application of the above corollary one can obtain a proof of *Donsker's Invariance Principle* for i.i.d. summands having *finite fourth moments*. Moreover, since the simple random walk has moments of all orders, this approach can also be used to give an alternative rigorous construction of the Wiener measure based on Prohorov's theorem as the limiting distribution of random walks. (Compare theoretical complement 13.1 for another construction). Let $Z_1, Z_2, \ldots$ be i.i.d. random variables on a probability space $(\Omega, \mathscr{F}, P)$ having mean zero, variance one, and finite fourth moment $m_4 = EZ_1^4$. Define $S_0 = 0$, $S_n = Z_1 + \cdots + Z_n$, $n \geqslant 1$, and

$$\tilde{X}_t^{(n)} = n^{-1/2} S_{[nt]} + n^{-1/2}(nt - [nt]) Z_{[nt]+1}, \qquad 0 \leqslant t \leqslant 1.$$

We will show that there are positive numbers α, β and M such that

$$E|\tilde{X}_t^{(n)} - \tilde{X}_s^{(n)}|^\alpha \leqslant M|t - s|^{1+\beta} \qquad \text{for } 0 \leqslant s, t \leqslant 1, \quad n = 1, 2, \ldots. \tag{T.5.1}$$

By our corollary this will prove tightness of the distributions of the process $\{X_t^{(n)}\}$, $n = 1, 2, \ldots$. This together with the finite-dimensional CLT proves the FCLT under the assumption of finite fourth moments. One needs to calculate the probabilities of fluctuations described in the Arzela–Ascoli theorem more carefully to get the proof under finite second moments alone.

To establish (T.5.1), take $\alpha = 4$. First consider the case $s = (j/n) < (k/n) = t$ are at the grid points. Then

$$\begin{aligned} E\{\tilde{X}_t^{(n)} - \tilde{X}_s^{(n)}\}^4 &= n^{-2} E\{Z_{j+1} + \cdots + Z_k\}^4 \\ &= n^{-2} \sum_{i_1=j+1}^{k} \sum_{i_2=j+1}^{k} \sum_{i_3=j+1}^{k} \sum_{i_4=j+1}^{k} E\{Z_{i_1} Z_{i_2} Z_{i_3} Z_{i_4}\} \\ &= n^{-2}\left\{(k-j)EZ_1^4 + \binom{4}{2}\binom{k-j}{2}(EZ_1^2)^2\right\}. \end{aligned}$$

Thus, in this case,

$$\begin{aligned} E\{\tilde{X}_t^{(n)} - \tilde{X}_s^{(n)}\}^4 &= n^{-2}\{(k-j)m_4 + 3(k-j)(k-j-1)\} \\ &\leqslant n^{-2}\{(k-j)m_4 + 3(k-j)^2\} \leqslant (m_4 + 3)\left(\frac{k}{n} - \frac{j}{n}\right)^2 \\ &\leqslant (m_4 + 3)|t - s|^2 = c_1|t - s|^2, \qquad \text{where } c_1 = m_4 + 3. \end{aligned}$$

Next, consider the more general case $0 \leqslant s, t \leqslant 1$, but for which $|t - s| \geqslant 1/n$. Then, for $s < t$,

$$\begin{aligned} E\{\tilde{X}_t^{(n)} - \tilde{X}_s^{(n)}\}^4 &= n^{-2} E\left\{\sum_{j=[ns]+1}^{[nt]} Z_j + (nt - [nt]) Z_{[nt]+1} - ([ns] - ns) Z_{[ns]+1}\right\}^4 \\ &\leqslant n^{-2} 3^4 \left\{E\left(\sum_{j=[ns]+1}^{[nt]} Z_j\right)^4 + (nt - [nt])^4 EZ_{[nt]+1}^4 \right. \\ &\qquad \left. + (ns - [ns])^4 EZ_{[ns]+1}^4\right\} \\ &\leqslant n^{-2} 3^4 \{c_1([nt] - [ns])^2 + (nt - [nt])^2 m_4 + (ns - [ns])^2 m_4\} \end{aligned}$$

$$\leqslant n^{-2}3^4 c_1\{([nt]-[ns])^2+(nt-[nt])^2+(ns-[ns])^2\}$$
$$\leqslant n^{-2}3^4 c_1\{([nt]-[ns])+(nt-[nt])+(ns-[ns])\}^2$$
$$= n^{-2}3^4 c_1\{nt-ns+2(ns-[ns])\}^2$$
$$\leqslant n^{-2}3^4 c_1\{nt-ns+2(nt-ns)\}^2$$
$$= n^{-2}3^6 c_1(nt-ns)^2 = 3^6 c_1(t-s)^2.$$

In the above, we used the fact that $(a+b+c)^4 \leqslant 3^4(a^4+b^4+c^4)$ to get the first inequality. The analysis of the first (gridpoint) case was then used to get the second inequality. Finally, if $|t-s| < 1/n$, then either

(a) $\dfrac{k}{n} \leqslant s < t < \dfrac{k+1}{n}$ for some $0 \leqslant k \leqslant n-1$, or

(b) $\dfrac{k}{n} \leqslant s < \dfrac{k+1}{n}$ and $\dfrac{k+1}{n} \leqslant t < \dfrac{k+2}{n}$ for some $0 \leqslant k \leqslant n-1$.

In (a), $S_{[nt]} = S_{[ns]}$, so that, since $|nt - ns| \leqslant 1$,

$$E\{\tilde{X}_t^{(n)} - \tilde{X}_s^{(n)}\}^4 = n^{-2}E\{n(t-s)Z_{k+1}\}^4 = m_4 n^2(t-s)^4$$
$$= m_4(nt-ns)^2(t-s)^2 \leqslant m_4(t-s)^2,$$

In (b), $S_{[nt]} - S_{[ns]} = Z_{k+1}$, so that

$$E\{\tilde{X}_t^{(n)} - \tilde{X}_s^{(n)}\}^4 = n^{-2}E\left\{Z_{k+1} + n\left(t-\frac{k+1}{n}\right)Z_{k+2} - n\left(s-\frac{k}{n}\right)Z_{k+1}\right\}^4$$
$$= n^{-2}E\left\{n\left(t-\frac{k+1}{n}\right)Z_{k+2} - n\left(s-\frac{k+1}{n}\right)Z_{k+1}\right\}^4$$
$$\leqslant 2^4 n^2\left(t-\frac{k+1}{n}\right)^4 m_4 + \left(\frac{k+1}{n}-s\right)^4 m_4$$
$$= 2^4 n^2 m_4\left\{\left(t-\frac{k+1}{n}\right)^4 + \left(\frac{k+1}{n}-s\right)^4\right\}$$
$$\leqslant 2^4 n^2 m_4\left\{t-\frac{k+1}{n}+\frac{k+1}{n}-s\right\}^4 = 2^4 n^2 m_4(t-s)^4$$
$$= 2^4 m_4(nt-ns)^2(t-s)^2 \leqslant 2^4 m_4(t-s)^2.$$

Take $\beta = 1$, $M = \max\{2^4 m_4, 3^6(m_4+3)\} = 3^6(m_4+3)$ for $\alpha = 4$. ■

The FCLT (Theorem 8.1) is stated in the text for convergence in $S = C[0,\infty)$, when S has the topology of uniform convergence on compacts. One may take the metric to be $\rho(\omega,\omega') = \sum_{k=1}^{\infty} 2^{-k}d_k/(1+d_k)$, where $d_k = \max\{|\omega(t)-\omega'(t)| : 0 \leqslant t \leqslant k\}$. Since the above arguments apply to $[0,k]$ in place of $[0,1]$, the assertion of Theorem 8.1 follows (under the moment condition $m_4 < \infty$).

6. (*Measure Determining Classes*) Let (S, ρ) be a metric space, $\mathscr{B}(S)$ its Borel sigmafield. A class $\mathscr{C} \subset \mathscr{B}(S)$ is *measure-determining* if, for any two finite measures μ, ν, $\mu(C) = \nu(C)\ \forall C \in \mathscr{C}$ implies $\mu = \nu$. An example is the class $\mathscr{D}$ of all closed sets. To see this, consider the lambda class $\mathscr{A}$ of all sets A for which $\mu(A) = \nu(A)$. If this class contains $\mathscr{D}$ then by the Pi–Lambda Theorem (Chapter 0, Theorem 4.1) $\mathscr{A} \supset \sigma(\mathscr{D}) = \mathscr{B}(S)$. Similarly, the class $\mathscr{O}$ of all open sets is measure-determining. A class $\mathscr{G}$ of real-valued bounded Borel measurable functions on S is *measure-determining* if $\int f\,d\mu = \int f\,d\nu\ \forall g \in \mathscr{G}$ implies $\mu = \nu$. The class $C_b(S)$ of real-valued bounded continuous functions on S is measure-determining. To prove this, it is enough to show that for each $F \in \mathscr{D}$ there exists a sequence $\{f_n\} \subset C_b(S)$ such that $f_n \uparrow \mathbf{1}_F$ as $n \downarrow \infty$. For this, let $h_n(r) = 1 - nr$ for $0 \leqslant r \leqslant 1/n$, $h_n(r) = 0$ for $r \geqslant 1/n$. Then take $f_n(x) = h_n(\rho(x, F))$.

Theoretical Complements to Section I.9

1. If $f: C[0, \infty) \to \mathbb{R}^1$ is continuous, then the FCLT provides that the real-valued random variables $X_n = f(\{\tilde{X}_t^{(n)}\})$, $n \geqslant 1$, converge in distribution to $X = f(\{X_t\})$. Thus, if one checks by direct computation that the limit $F(a) = \lim_n P(X_n \leqslant a)$, or $F(a^-) = \lim_n P(X_n < a)$, exists for *all* a, then F is the d.f. of X. Applying this to $f(\{X_s\}) := \max\{X_s : 0 \leqslant s \leqslant t\} \equiv M_t$ we get (10.10), for $P(\tau_z > t) = P(M_t < z)$, if $z > 0$. The case $z < 0$ is similar.

Joint distributions of several functionals may be similarly obtained by looking at linear combinations of the functionals. Here is the precise statement.

Theorem T.9.1. If $\mathbf{f}: C[0, \infty) \to \mathbb{R}^k$ is continuous, say $\mathbf{f} = (f_1, \ldots, f_k)$ where $f_i: C[0, \infty) \to \mathbb{R}^1$, then the random vectors $\mathbf{X}_n = \mathbf{f}(\{\tilde{X}_t^{(n)}\})$, $n \geqslant 1$, converge in distribution to $\mathbf{X} = \mathbf{f}(\{X_t\})$.

Proof. For any $r_1, \ldots, r_k \in \mathbb{R}^1$, $\sum_{j=1}^k r_j f_j: C[0, \infty) \to \mathbb{R}^1$ is continuous so that $\sum_{j=1}^k r_j f_j(\{\tilde{X}_t^{(n)}\})$ converges in distribution to $\sum_{j=1}^k r_j f_j(\{X_t\})$ by the FCLT. Therefore, its characteristic function converges to $Ee^{i\langle \mathbf{r}, \mathbf{f}(\{\tilde{X}_t\})\rangle}$ as $n \to \infty$ for each $\mathbf{r} \in \mathbb{R}^k$. This means $\mathbf{f}(\{\tilde{X}_t^{(n)}\})$ converges in distribution to $\mathbf{f}(\{X_t\})$ as asserted. ■

2. (*Mann–Wald*) It is sufficient that $f: C[0, \infty) \to \mathbb{R}^1$ be only a.s. continuous with respect to the limiting distribution for the FCLT to apply, i.e., for the convergence of $f(\{\tilde{X}_t^{(n)}\})$ in distribution to $f(\{X_t\})$. That is,

Theorem T.9.2. If $\{\tilde{X}_t^{(n)}\}$ converges in distribution to $\{X_t\}$ and if $P(\{X_t\} \in D_f) = 0$, where

$$D_f = \{\mathbf{x} \in C[0, \infty): f \text{ is discontinuous at } \mathbf{x}\},$$

then $f(\{\tilde{X}_t^{(n)}\})$ converges in distribution to $f(\{X_t\})$. □

Proof. This can be proved using Alexandrov's Theorem (T.8.1(ii)), since for any closed set F, $f^{-1}(F) \subset \overline{f^{-1}(F)} = D_f \cup f^{-1}(F)$, where the overbar denotes the closure of the set. ■

In applications it is the requirement that the limit distribution assign probability zero to the event D_f that is sometimes nontrivial to check. As an example, consider (9.5) and (9.6). Let $F = \{\tau_c^x < \tau_d^x\}$. Recall that $\Omega \equiv C[0, \infty)$ is given the topology of uniform convergence on bounded subintervals of $[0, \infty)$. Consider an element $\omega \in \Omega$ that reaches a number *less* than c before reaching d. It is simple to check that ω belongs to the *interior* of F. On the other hand, if ω belongs to the closure of F, then either (i) $\omega \in F$, or (ii) ω neither reaches c nor d. Thus, if $\omega \in \partial F$, then either (ii) occurs, the probability of which is zero for a Brownian motion since $|X_t| \to \infty$ in probability, as $t \to \infty$, or (iii) in the interval $[\tau_c^x, \tau_d^x)$, ω never goes below c. By the strong Markov property (see Chapter V, Theorem 11.1), the latter event (iii) has the same probability as that of a Brownian motion, starting at c, never reaching below c before reaching d. In turn, the last probability is the same as that of a Brownian motion, starting at 0, never reaching below 0 before reaching $d - c$. This probability is zero, since $\tau_z \equiv \tau_z^0$ converges to zero in probability as $z \uparrow 0$ (see Eq. 10.10). By combining such arguments with those given in theoretical complement 1 above, one may also arrive at (10.15) for Brownian motions with nonzero drifts.

Theoretical Complements to Section I.12

1. A proof of Proposition 12.1 for the special case of infinite-dimensional events that depend on the empirical process through the functional ($\omega \to \sup_{0 \leqslant t \leqslant 1} |\omega_t|$) used to define the Kolmogorov–Smirnov statistic (12.11) is given below. This proof is based on a trick of M. D. Donsker (1952), "Justification and Extension of Doob's Heuristic Approach to the Kolmogorov–Smirnov Theorems," *Annals Math. Statist.*, **23**, pp. 277–281, which allows one to apply the FCLT as given in Section 8 (and proved in theoretical complements to Section I.8 under the assumption of finite fourth moments).

The key to Donsker's proof is the simple observation that *the distribution of the order statistic* $(Y_{(1)}, \ldots, Y_{(n)})$ *of n i.i.d. random variables* $Y_1, Y_2, \ldots$ *from the uniform distribution on* $[0, 1]$ *can also be obtained as the distribution of the ratios*

$$\left(\frac{S_1}{S_{n+1}}, \frac{S_2}{S_{n+1}}, \ldots, \frac{S_n}{S_{n+1}}\right),$$

where $S_k = T_1 + \cdots + T_k$, $k \geqslant 1$, *and* $T_1, T_2, \ldots$ *is an i.i.d. sequence of (mean* 1) *exponentially distributed random variables.* □

Intuitively, if the T_i are regarded as the successive times between occurrence of some phenomena, then S_{n+1} is the time to the $(n+1)$st occurrence and, in units of S_{n+1}, the occurrence times should be randomly distributed because of lack of memory and independence properties. A version of this simple fact is given in Chapter IV (Proposition 5.6) for the *Poisson process*. The calculations are essentially the same, so this is left as an exercise here.

The precise result that we will prove here is as follows. The symbol $\stackrel{d}{=}$ below denotes equality in distribution.

Proposition T.12.1. Let $Y_1, Y_2, \ldots$ be i.i.d. uniform on $[0, 1]$ and let, for each $n \geqslant 1$, $\{F_n(t)\}$ be the corresponding empirical process based on $Y_1, \ldots, Y_n$. Then $D_n = \sqrt{n} \sup_{0 \leqslant t \leqslant 1} |F_n(t) - t|$ converges in distribution to $\sup_{0 \leqslant t \leqslant 1} |B_t^*|$ as $n \to \infty$.

Proof. In the notation introduced above, we have

$$D_n := \sqrt{n} \sup_{0 \leqslant t \leqslant 1} |F_n(t) - t| = \sqrt{n} \max_{k \leqslant n} \left| Y_{(k)} - \frac{k}{n} \right|$$

$$\stackrel{d}{=} \sqrt{n} \max_{k \leqslant n} \left| \frac{S_k}{S_{n+1}} - \frac{k}{n} \right| = \frac{n}{S_{n+1}} \max_{k \leqslant n} \left| \frac{S_k - k}{\sqrt{n}} - \frac{k}{n} \frac{S_{n+1} - n}{\sqrt{n}} \right|$$

$$\stackrel{d}{=} \frac{n}{S_{n+1}} \sup_{0 \leqslant t \leqslant 1} |X_t^{(n)} - t X_1^{(n)}| + O(n^{-1/2}), \tag{T.12.1}$$

where

$$X_t^{(n)} = \frac{S_{[nt]} - [nt]}{\sqrt{n}}$$

and, by the SLLN, $n/(S_{n+1}) \to 1$ a.s. as $n \to \infty$. The result follows from the FCLT, (8.6), and the definition of Brownian bridge. ∎

This proposition is sufficient for the particular application to the large-sample distribution given in the text. A precise definition of weak convergence of the scaled empirical distribution function as well as a proof of Proposition 12.1 can be found in P. Billingsley (1968), *Convergence of Probability Measures*, Wiley, New York.

2. Tabulations of the distribution of the Kolmogorov–Smirnov statistic can be found in L. H. Miller (1956), "Table of Percentage Points of Kolmogorov Statistics," *J. Amer. Statist. Assoc.*, **51**, pp. 111–121.

3. Let $(X \mid A)$ denote the conditional distribution of a random variable or stochastic process X given an event A. As suggested by Exercise 12.4*, from the convergence of the finite-dimensional distributions it is possible to obtain the Brownian bridge as the limiting distribution of $(\{B_t\} \mid 0 \leqslant B_1 \leqslant \varepsilon)$ as $\varepsilon \to 0$, where $\{B_t\}$ denotes standard Brownian motion on $0 \leqslant t \leqslant 1$. To make this observation firm, one needs to check tightness of the family $\{(\{B_t\} \mid 0 \leqslant B_1 \leqslant \varepsilon) : \varepsilon > 0\}$. The following argument is used by P. Billingsley (1968), *Convergence of Probability Measures*, Wiley, New York, p. 84.

Let F be any closed subset of $C[0, 1]$. Then, since $\sup_{0 \leqslant t \leqslant 1} |B_t^* - B_t| = |B_1|$,

$$P(\{B_t\} \in F \mid 0 \leqslant B_1 \leqslant \varepsilon) \leqslant P(\{B_t^*\} \in F_\varepsilon \mid 0 \leqslant B_1 \leqslant \varepsilon),$$

where $F_\varepsilon := \{\omega \in C[0, 1] : \text{dist}(\omega, F) \leqslant \varepsilon\}$ for $\text{dist}(\omega, A) := \inf\{\rho(\omega, \gamma) : \gamma \in A\}$, $A \subset C[0, 1]$. But, starting with finite-dimensional sets and then using the monotone class argument (Chapter 0), one may check that the events $\{\{B_t^*\} \in F_\varepsilon\}$ and $\{0 \leqslant B_1 \leqslant \varepsilon\}$ are independent. Therefore, $P(\{B_t\} \in F \mid 0 \leqslant B_1 \leqslant \varepsilon) \leqslant P(\{B_t^*\} \in F_\varepsilon)$ for any $\varepsilon > 0$. Since F is closed, the events $\{\{B_t^*\} \in F_\varepsilon\}$ decrease to $\{\{B_t^*\} \in F\}$ as $\varepsilon \to 0$, and tightness follows from the continuity of the probability measure P and Alexandrov's Theorem T.8.1(ii).

4. A check of *tightness* is also required for the Brownian meander and Brownian excursion as described in Exercises 12.6 and 12.7, respectively. For this, consult R. T. Durrett, D. L. Iglehart, and D. R. Miller (1977), "Weak Convergence to Brownian Meander and Brownian Excursion," *Ann. Probab.*, **5**, pp. 117–129. The distribution of the extremal functionals outlined in the exercises can also be found in R. T. Durrett and D. L. Iglehart (1977), "Functionals of Brownian Meander and Brownian Excursion," *Ann. Probab.*, **5**, pp. 130–135; K. L. Chung (1976), "Excursions in Brownian Motion," *Ark. Mat.*, pp. 155–177; D. P. Kennedy (1976), "Maximum Brownian Excursion," *J. Appl. Probability*, **13**, 371–376. Durrett, Iglehart and Miller (1977) also show that the $*$ and $+$ commute in the sense that the Brownian excursion can be obtained either by a meander of the Brownian bridge (as done in Exercise 12.7) or as a bridge of the meander (i.e., conditioning the meander in the sense of theoretical complement 3 above). Brownian meander and Brownian excursion have been defined in a variety of other ways in work originating in the late 1940's with Paul Lévy; see P. Lévy (1965), *Processus Stochastiques et Mouvement Brownien*, Gauthier-Villars, Paris. The theory was extended and terminology introduced in K. Itô and H. P. McKean, Jr. (1965), *Diffusion Processes and Their Sample Paths*, Springer Verlag, New York. The general theory is introduced in D. Williams (1979), *Diffusions, Markov Processes, and Martingales*, Vol. 1, Wiley, New York. A much fuller theory is then given in L. C. G. Rogers and D. Williams (1987), *Diffusions, Markov Processes, Martingales*, Vol. II, Wiley, New York. Approaches from the point of view of Markov processes (see theoretical complement 11.2, Chapter V) having nonstationary transition law are possible. Another very useful approach is from the point of view of FCLTs for random walks conditioned on a late return to zero; see W. D. Kaigh (1976), "An Invariance Principle for Random Walk Conditioned by a Late Return to Zero," *Ann. Probab.*, **4**(1), pp. 115–121, and references therein. A connection with extreme values of branching processes is described in theoretical complement 11.2, Chapter V.

Theoretical Complements to Section I.13

1. (*Construction of Wiener Measure*) Let $\{X_t : t \geqslant 0\}$ be the process having the finite-dimensional distributions of the standard Brownian motion starting at 0 as constructed by the Kolmogorov Extension Theorem on the product probability space $(\Omega, \mathscr{F}, P)$. As discussed in theoretical complement 7.1, events pertaining to the behavior of the process at uncountably many time points (e.g., path continuity) cannot be represented in this framework. However, we will now see how to modify the model in such a way as to get around this difficulty.

Let D be the set of nonnegative dyadic rational numbers and let $J_{n,k} = [k/2^n, (k+1)/2^n]$. We will use the maximal inequality to check that

$$\sum_{n=1}^{\infty} P\left(\max_{0 \leqslant k < n2^n} \sup_{q \in J_{n,k} \cap D} |X_q - X_{k/2^n}| > \frac{1}{n} \right) < \infty. \tag{T.13.1}$$

By the Borel–Cantelli lemma we will get from this that with probability 1, for all n sufficiently large,

$$\max_{0 \leqslant k < n2^n} \sup_{q \in J_{n,k} \cap D} |X_q - X_{k/2^n}| \leqslant \frac{1}{n}. \tag{T.13.2}$$

In particular, it will follow that with probability 1, for every $t > 0$, $q \to X_q$ is uniformly continuous on $D \cap [0, t]$. Thus, almost all sample paths of $\{X_q : q \in D\}$ have a unique extension to continuous functions $\{B_t : t \geqslant 0\}$. That is, letting $C = \{\omega \in \Omega$: for each

$t \geq 0$, $q \to X_q(\omega)$ is uniformly continuous on $D \cap [0, t]\}$, define for $\omega \in C$,

$$B_t(\omega) = \begin{cases} X_q(\omega), & \text{if } t = q \in D, \\ \lim\limits_{q \to t} X_q(\omega), & \text{if } t \notin D, \end{cases} \tag{T.13.3}$$

where the limit is over dyadic rational q decreasing to t. By construction, $\{B_t: t \geq 0\}$ has continuous paths with probability 1. Moreover, for $0 < t_1 < \cdots < t_k$, with probability one, $(B_{t_1}, \ldots, B_{t_k}) = \lim_{n \to \infty} (X_{q_1^{(n)}}, \ldots, X_{q_k^{(n)}})$ for dyadic rational $q_1^{(n)}, \ldots, q_k^{(n)}$ decreasing to $t_1, \ldots, t_k$. Also, the random vector $(X_{q_1^{(n)}}, \ldots, X_{q_k^{(n)}})$ has the multivariate normal distribution with mean vector $\mathbf{0}$ and variance–covariance matrix $\gamma_{ij} = \min(t_i, t_j)$, $1 \leq i, j \leq k$ as a limiting distribution. It follows from these two facts that this must be the distribution of $(B_{t_1}, \ldots, B_{t_k})$. Thus, $\{B_t\}$ is a standard Brownian motion process.

To verify the condition (T.13.1) for the Borel–Cantelli lemma, just note that by the maximal inequality (see Exercises 4.3, 13.11),

$$\begin{aligned} P\left(\max_{i \leq 2^m} |X_{t+i\delta 2^{-m}} - X_t| \geq \alpha\right) &\leq 2P(|X_{t+\delta} - X_t| \geq \alpha) \\ &\leq \frac{2}{\alpha^4} E(X_{t+\delta} - X_t)^4 \\ &= \frac{6\delta^2}{\alpha^4}, \end{aligned} \tag{T.13.4}$$

since the increments of $\{X_t\}$ are independent and Gaussian with mean 0. Now since the events $\{\max_{i \leq 2^m} |X_{t+i\delta 2^{-m}} - X_t| \geq \alpha\}$ increase with m, we have, letting $m \to \infty$,

$$P\left(\sup_{0 \leq q \leq 1, q \in D} |X_{t+q\delta} - X_t| > \alpha\right) \leq \frac{6\delta^2}{\alpha^4}. \tag{T.13.5}$$

Thus,

$$\begin{aligned} P\left(\max_{0 \leq k < n2^n} \sup_{q \in J_{n,k} \cap D} |X_q - X_{k/2^n}| > \frac{1}{n}\right) &\leq \sum_{k=0}^{n2^n - 1} P\left(\sup_{q \in J_{n,k} \cap D} |X_q - X_{k/2^n}| > \frac{1}{n}\right) \\ &\leq n2^n \frac{6(2^{-n})^2}{(n^{-1})^4} = \frac{6n^5}{2^n}, \end{aligned} \tag{T.13.6}$$

which is summable.

Theoretical Complements to Section I.14

1. The treatment given in this section follows that in R. N. Bhattacharya, V. K. Gupta, and E. Waymire (1983), "The Hurst Effect Under Trends," *J. Appl. Probability*, **20**, pp. 649–662. The research on the Hurst effect is rather extensive. For the other related results mentioned in the text consult the following references:

W. Feller (1951), "The Asymptotic Distribution of the Range of Sums of Independent Random Variables," *Ann. Math. Statist.*, **22**, pp. 427–432.

P. A. P. Moran (1964), "On the Range of Cumulative Sums," *Ann. Inst. Statist. Math.*, **16**, pp. 109–112.

B. B. Mandelbrot and J. W. Van Ness (1968), "Fractional Brownian Motions, Fractional Noises and Applications," *SIAM Rev.*, **10**, pp. 422–437. Processes of the type considered by Mandelbrot and Van Ness are briefly described in theoretical complement 1.3 to Chapter IV of this book.

CHAPTER II

Discrete-Parameter Markov Chains

1 MARKOV DEPENDENCE

Consider a discrete-parameter stochastic process $\{X_n\}$. Think of $X_0, X_1, \ldots, X_{n-1}$ as "the past," X_n as "the present," and $X_{n+1}, X_{n+2}, \ldots$ as "the future" of the process relative to time n. The law of evolution of a stochastic process is often thought of in terms of the conditional distribution of the future given the present and past states of the process. In the case of a sequence of independent random variables or of a simple random walk, for example, this conditional distribution does not depend on the past. This important property is expressed by Definition 1.1.

Definition 1.1. A stochastic process $\{X_0, X_1, \ldots, X_n, \ldots\}$ has the *Markov property* if, for each n and m, the conditional distribution of $X_{n+1}, \ldots, X_{n+m}$ given $X_0, X_1, \ldots, X_n$ is the same as its conditional distribution given X_n alone. A process having the Markov property is called a *Markov process*. If, in addition, the state space of the process is countable, then a Markov process is called a *Markov chain*.

In view of the next proposition, it is actually enough to take $m = 1$ in the above definition.

Proposition 1.1. A stochastic process $X_0, X_1, X_2, \ldots$ has the Markov property if and only if for each n the conditional distribution of X_{n+1} given $X_0, X_1, \ldots, X_n$ is a function only of X_n.

Proof. For simplicity, take the state space S to be countable. The necessity of the condition is obvious. For sufficiency, observe that

$$\begin{aligned}
P(X_{n+1} = j_1, \dots, X_{n+m} = j_m \mid X_0 = i_0, \dots, X_n = i_n)& \\
&= P(X_{n+1} = j_1 \mid X_0 = i_0, \dots, X_n = i_n)\cdot \\
&\quad P(X_{n+2} = j_2 \mid X_0 = i_0, \dots, X_n = i_n, X_{n+1} = j_1)\cdots \\
&\quad P(X_{n+m} = j_m \mid X_0 = i_0, \dots, X_{n+m-1} = j_{m-1}) \\
&= P(X_{n+1} = j_1 \mid X_n = i_n)P(X_{n+2} = j_2 \mid X_{n+1} = j_1)\cdots \\
&\quad P(X_{n+m} = j_m \mid X_{n+m-1} = j_{m-1}). \qquad (1.1)
\end{aligned}$$

The last equality follows from the hypothesis of the proposition. Thus the conditional distribution of the future as a function of the past and present states $i_0, i_1, \dots, i_n$ depends only on the present state i_n. This is, therefore, the conditional distribution given $X_n = i_n$ (Exercise 1). ■

A Markov chain $\{X_0, X_1, \dots\}$ is said to have a *homogeneous* or *stationary transition law* if the distribution of $X_{n+1}, \dots, X_{n+m}$ given $X_n = y$ depends on the *state* at time n, namely y, but not on the *time* n. Otherwise, the transition law is called *nonhomogeneous*. An i.i.d. sequence $\{X_n\}$ and its associated random walk possess time-homogeneous transition laws, while an independent nonidentically distributed sequence $\{X_n\}$ and its associated random walk have nonhomogeneous transitions. Unless otherwise specified, by a Markov process (chain) we shall mean a Markov process (chain) with a *homogeneous* transition law.

The *Markov property* as defined above refers to a special type of *statistical dependence* among families of random variables indexed by a *linearly ordered parameter set*. In the case of a continuous parameter process, we have the following analogous definition.

Definition 1.2. A continuous-parameter stochastic process $\{X_t\}$ has the *Markov property* if for each $s < t$, the conditional distribution of X_t given $\{X_u, u \leqslant s\}$ is the same as the conditional distribution of X_t given X_s. Such a process is called a *continuous-parameter Markov process*. If, in addition, the state space is countable, then the process is called a *continuous-parameter Markov chain*.

2 TRANSITION PROBABILITIES AND THE PROBABILITY SPACE

An i.i.d. sequence and a random walk are merely two examples of Markov chains. To define a general Markov chain, it is convenient to introduce a *matrix* $\mathbf{p}$ to describe the probabilities of transition between successive states in the evolution of the process.

Definition 2.1. A *transition probability matrix* or a *stochastic matrix* is a square matrix $\mathbf{p} = ((p_{ij}))$, where i and j vary over a finite or denumerable set S, satisfying

(i) $p_{ij} \geqslant 0$ for all i and j,

(ii) $\sum_{j \in S} p_{ij} = 1$ for all i.

The set S is called the *state space* and its elements are *states*.

Think of a particle that moves from point to point in the state space according to the following scheme. At time $n = 0$ the particle is set in motion either by starting it at a fixed state i_0, called the *initial state*, or by randomly locating it in the state space according to a probability distribution $\boldsymbol{\pi}$ on S, called the *initial distribution*. In the former case, $\boldsymbol{\pi}$ is the distribution concentrated at the state i_0, i.e., $\pi_j = 1$ if $j = i_0$, $\pi_j = 0$ if $j \neq i_0$. In the latter case, the probability is π_i that at time zero the particle will be found in state i, where $0 \leqslant \pi_i \leqslant 1$ and $\sum_i \pi_i = 1$. Given that the particle is in state i_0 at time $n = 0$, a random trial is performed, assigning probability $p_{i_0 j'}$ to the respective states $j' \in S$. If the outcome of the trial is the state i_1, then the particle moves to state i_1 at time $n = 1$. A second trial is performed with probabilities $p_{i_1 j'}$ of states $j' \in S$. If the outcome of the second trial is i_2, then the particle moves to state i_2 at time $n = 2$, and so on.

A typical sample point of this experiment is a sequence of states, say $(i_0, i_1, i_2, \ldots, i_n, \ldots)$, representing a *sample path*. The set of all such sample paths is the *sample space* Ω. The position X_n at time n is a random variable whose value is given by $X_n = i_n$ if the sample path is $(i_0, i_1, \ldots, i_n, \ldots)$. The precise specification of the probability P_π on Ω for the above experiment is given by

$$P_\pi(X_0 = i_0, X_1 = i_1, \ldots, X_n = i_n) = \pi_{i_0} p_{i_0 i_1} p_{i_1 i_2} \cdots p_{i_{n-1} i_n}. \tag{2.1}$$

More generally, for finite-dimensional events of the form

$$A = \{(X_0, X_1, \ldots, X_n) \in B\}, \tag{2.2}$$

where B is an arbitrary set of $(n + 1)$-tuples of elements of S, the probability of A is specified by

$$P_\pi(A) = \sum_{(i_0, i_1, \ldots, i_n) \in B} \pi_{i_0} p_{i_0 i_1} \cdots p_{i_{n-1} i_n}. \tag{2.3}$$

By Kolmogorov's existence theorem, P_π extends uniquely as a probability measure on the smallest sigmafield $\mathscr{F}$ containing the class of all events of the form (2.2); see Example 6.3 of Chapter I. This probability space $(\Omega, \mathscr{F}, P_\pi)$ with $X_n(\omega) = \omega_n$, $\omega \in \Omega$, is a *canonical model for the Markov chain with transition probabilities* $((p_{ij}))$ *and initial distribution* $\boldsymbol{\pi}$ (the Markov property is established below at (2.7)–(2.10)). In the case of a Markov chain starting in state i, that is, $\pi_i = 1$, we write P_i in place of P_π.

To specify various joint distributions and conditional distributions associated with this Markov chain, it is convenient to use the notation of matrix multiplication. By definition the (i, j) element of the matrix $\mathbf{p}^2$ is given by

$$p_{ij}^{(2)} = \sum_{k \in S} p_{ik} p_{kj}. \tag{2.4}$$

The elements of the matrix $\mathbf{p}^n$ are defined recursively by $\mathbf{p}^n = \mathbf{p}^{n-1}\mathbf{p}$ so that the (i, j) element of $\mathbf{p}^n$ is given by

$$p_{ij}^{(n)} = \sum_{k \in S} p_{ik}^{(n-1)} p_{kj} = \sum_{k \in S} p_{ik} p_{kj}^{(n-1)}, \qquad n = 2, 3, \ldots. \tag{2.5}$$

It is easily checked by induction on n that the expression for $p_{ij}^{(n)}$ is given directly in terms of the elements of $\mathbf{p}$ according to

$$p_{ij}^{(n)} = \sum_{i_1, \ldots, i_{n-1} \in S} p_{ii_1} p_{i_1 i_2} \cdots p_{i_{n-2} i_{n-1}} p_{i_{n-1} j}. \tag{2.6}$$

Now let us check the Markov property of this probability model. Using (2.1) and summing over unrestricted coordinates, the joint distribution of $X_0, X_{n_1}, X_{n_2}, \ldots, X_{n_k}$, with $0 = n_0 < n_1 < n_2 < \cdots < n_k$, is given by

$$\begin{aligned} P_\pi(X_0 = i, X_{n_1} = j_1, X_{n_2} = j_2, \ldots, X_{n_k} = j_k) \\ = \sum_1 \sum_2 \cdots \sum_k (\pi_i p_{ii_1} p_{i_1 i_2} \cdots p_{i_{n_1 - 1} j_1})(p_{j_1 i_{n_1+1}} p_{i_{n_1+1} i_{n_1+2}} \cdots p_{i_{n_2-1} j_2}) \cdots \\ \times (p_{j_{k-1} i_{n_{k-1}+1}} p_{i_{n_{k-1}+1} i_{n_{k-1}+2}} \cdots p_{i_{n_k - 1} j_k}), \end{aligned} \tag{2.7}$$

where $\sum_r$ is the sum over the rth block of indices $i_{n_{r-1}+1}, \ldots, i_{n_r}$ $(r = 1, 2, \ldots, k)$. The sum $\sum_k$, keeping indices in all other blocks fixed, yields the factor $p_{j_{k-1} j_k}^{(n_k - n_{k-1})}$ using (2.6) for the last group of terms. Next sum successively over the $(k-1)$st, $\ldots$, second, and first blocks of factors to get

$$P_\pi(X_0 = i, X_{n_1} = j_1, X_{n_2} = j_2, \ldots, X_{n_k} = j_k) = \pi_i p_{ij_1}^{(n_1)} p_{j_1 j_2}^{(n_2 - n_1)} \cdots p_{j_{k-1} j_k}^{(n_k - n_{k-1})}. \tag{2.8}$$

Now sum over $i \in S$ to get

$$P_\pi(X_{n_1} = j_1, X_{n_2} = j_2, \ldots, X_{n_k} = j_k) = \left(\sum_{i \in S} \pi_i p_{ij_1}^{(n_1)}\right) p_{j_1 j_2}^{(n_2 - n_1)} \cdots p_{j_{k-1} j_k}^{(n_k - n_{k-1})}. \tag{2.9}$$

Using (2.8) and the elementary definition of conditional probabilities, it now follows that the conditional distribution of X_{n+m} given $X_0, X_1, \ldots, X_n$ is given by

$$\begin{aligned} P_\pi(X_{n+m} = j \mid X_0 = i_0, X_1 = i_1, \ldots, X_{n-1} = i_{n-1}, X_n = i) \\ = p_{ij}^{(m)} = P_\pi(X_{n+m} = j \mid X_n = i) \\ = P_\pi(X_m = j \mid X_0 = i), \qquad m \geqslant 1, \quad j \in S. \end{aligned} \tag{2.10}$$

Although by Proposition 1.1 the case $m = 1$ would have been sufficient to prove the Markov property, (2.10) justifies the terminology that $\mathbf{p}^m := ((p_{ij}^{(m)}))$ is the *m-step transition probability matrix*. Note that $\mathbf{p}^m$ is a stochastic matrix for all $m \geqslant 1$.

The calculation of the distribution of X_m follows from (2.10). We have,

$$P_\pi(X_m = j) = \sum_i P_\pi(X_m = j. X_0 = i) = \sum_i P_\pi(X_0 = i) P_\pi(X_m = j \mid X_0 = i)$$
$$= \sum_i \pi_i p_{ij}^{(m)} = (\boldsymbol{\pi}'\mathbf{p}^m)_j, \tag{2.11}$$

where $\boldsymbol{\pi}'$ is the transpose of the column vector $\boldsymbol{\pi}$, and $(\boldsymbol{\pi}'\mathbf{p}^m)_j$ is the jth element of the row vector $\boldsymbol{\pi}'\mathbf{p}^m$.

3 SOME EXAMPLES

The transition probabilities for some familiar Markov chains are given in the examples of this section. Although they are excluded from the general development of this chapter, examples of a non-Markov process and a Markov process having a nonhomogeneous transition law are both supplied under Example 8 below.

Example 1. (*Completely Deterministic Motion*). Let the only elements of $\mathbf{p}$, which may be either a finite or denumerable square matrix, be 0's and 1's. That is, for each state i there exists a state $h(i)$ such that

$$p_{ih(i)} = 1, \qquad p_{ij} = 0 \qquad \text{for } j \neq h(i) \qquad (i \in S). \tag{3.1}$$

This means that if the process is now in state i it must be in state $h(i)$ at the next instant. In this case, if the initial state X_0 is known then one knows the entire future. Thus, if $X_0 = i$, then

$$X_1 = h(i),$$
$$X_2 = h(h(i)) := h^{(2)}(i), \ldots,$$
$$X_n = h(h^{(n-1)}(i)) = h^{(n)}(i), \ldots.$$

Hence $p_{ij}^{(n)} = 1$ if $j = h^{(n)}(i)$ and $p_{ij}^{(n)} = 0$ if $j \neq h^{(n)}(i)$. Pseudorandom number generators are of this type (see Exercise 7).

Example 2. (*Completely Random Motion or Independence*). Let all the rows of $\mathbf{p}$ be identical, i.e., suppose p_{ij} is the same for all i. Write for the common row,

$$p_{ij} = p_j \qquad (i \in S, j \in S). \tag{3.2}$$

In this case, $X_0, X_1, X_2, \ldots$ forms a sequence of independent random variables. The distribution of X_0 is π while $X_1, \ldots, X_n, \ldots$ have a common distribution given by the probability vector $(p_j)_{j \in S}$. If we let $\pi = (p_j)_{j \in S}$, then $X_0, X_1, \ldots$ form a sequence of independent and identically distributed (i.i.d.) random variables. The coin-tossing example is of this kind. There, $S = \{0, 1\}$ (or $\{H, T\}$) and $p_0 = \frac{1}{2}$, $p_1 = \frac{1}{2}$.

Example 3. (*Unrestricted Simple Random Walk*). Here, $S = \{0, \pm 1, \pm 2, \ldots\}$ and $\mathbf{p} = ((p_{ij}))$ is given by

$$\begin{aligned} p_{ij} &= p \qquad \text{if } j = i + 1 \\ &= q \qquad \text{if } j = i - 1 \\ &= 0 \qquad \text{if } |j - i| > 1. \end{aligned} \tag{3.3}$$

where $0 < p < 1$ and $q = 1 - p$.

Example 4. (*Simple Random Walk with Two Reflecting Boundaries*). Here $S = \{c, c + 1, \ldots, d\}$, where c and d are integers, $c < d$. Let

$$\begin{aligned} p_{ij} &= p \qquad \text{if } j = i + 1 \text{ and } c < i < d \\ &= q \qquad \text{if } j = i - 1 \text{ and } c < i < d \\ p_{c,c+1} &= 1, \qquad p_{d,d-1} = 1. \end{aligned} \tag{3.4}$$

In this case, if at any point of time the particle finds itself in state c, then at the next instant of time it moves with probability 1 to $c + 1$. Similarly, if it is at d at any point of time it will move to $d - 1$ at the next instant. Otherwise (i.e., in the interior of $[c, d]$), its motion is like that of a simple random walk.

Example 5. (*Simple Random Walk with Two Absorbing Boundaries*). Here $S = \{c, c + 1, \ldots, d\}$, where c and d are integers, $c < d$. Let

$$p_{cc} = 1, \qquad p_{dd} = 1. \tag{3.5}$$

For $c < i < d$, p_{ij} is as defined by (3.3) or (3.4). In this case, once the particle reaches c (or d) it stays there forever.

Example 6. (*Unrestricted General Random Walk*). Take $S = \{0, \pm 1, \pm 2, \ldots\}$. Let Q be an arbitrary probability distribution on S, i.e.,

(i) $Q(i) \geq 0$ for all $i \in S$,
(ii) $\sum_{i \in S} Q(i) = 1$.

Define the transition matrix $\mathbf{p}$ by

$$p_{ij} = Q(j - i), \qquad i, j \in S. \tag{3.6}$$

One may think of this Markov chain as a partial-sum process as follows. Let X_0 have a distribution π. Let $Z_1, Z_2, \ldots$ be a sequence of i.i.d. random variables with common distribution Q and independent of X_0. Then,

$$X_n = X_0 + Z_1 + \cdots + Z_n, \qquad \text{for } n \geqslant 1, \tag{3.7}$$

is a Markov chain with the transition probability (3.6). Also note that Example 3 is a special case of Example 6, with $Q(-1) = q$, $Q(1) = p$, and $Q(i) = 0$ for $i \neq \pm 1$.

Example 7. (*Bienaymé–Galton–Watson Simple Branching Processes*). Particles such as neutrons or organisms such as bacteria can generate new particles or organisms of the same type. The number of particles generated by a single particle is a random variable with a probability function f; that is, $f(j)$ is the probability that a single particle generates j particles, $j = 0, 1, 2, \ldots$. Suppose that at time $n = 0$ there are $X_0 = i$ particles present. Let $Z_1, Z_2, \ldots, Z_i$ denote the numbers of particles generated by the first, second, ..., ith particle, respectively, in the initial set. Then each of $Z_1, \ldots, Z_i$ has the probability function f and it is assumed that $Z_1, \ldots, Z_i$ are independent random variables. The size of the *first generation* is $X_1 = Z_1 + \cdots + Z_i$, the total number generated by the initial set. The X_1 particles in the first generation will in turn generate a total of X_2 particles comprising the second generation in the same manner; that is, the X_1 particles generate new particles independently of each other and the number generated by each has probability function f. This goes on so long as offspring occur. Let X_n denote the size of the *nth generation*. Then, using the convolution notation for distributions of sums of independent random variables,

$$\begin{aligned} p_{ij} &= P(Z_1 + \cdots + Z_i = j) = f^{*i}(j), \qquad i \geqslant 1, \quad j \geqslant 0, \\ p_{00} &= 1, \qquad p_{0j} = 0 \qquad \text{if } j \neq 0. \end{aligned} \tag{3.8}$$

The last row says that "zero" is an *absorbing state*, i.e., if at any point of time $X_n = 0$, then $X_m = 0$ for all $m \geqslant n$, and *extinction* occurs.

Example 8. (*Pólya Urn Scheme and a Non-Markovian Example*). A box contains r red balls and b black balls. A ball is randomly selected from the box and its color is noted. The ball selected together with $c \geqslant 0$ balls of the same color are then placed back into the box. This process is repeated in successive trials numbered $n = 1, 2, \ldots$. Indicate the event that a red ball occurs at the nth trial by $X_n = 1$ and that a black ball occurs at the nth trial by $X_n = 0$. A straightforward induction calculation gives for $0 < \sum_{k=1}^{n} \varepsilon_k < n$,

$$P(X_1 = \varepsilon_1, \ldots, X_n = \varepsilon_n) = \frac{[r + (s_n - 1)c][r + (s_n - 2)c]\cdots r[b + (\tau_n - 1)c]\cdots b}{[r + b + (n-1)c][r + b + (n-2)c]\cdots[r+b]} \tag{3.9}$$

where

$$s_n = \sum_{k=1}^{n} \varepsilon_k, \qquad \tau_n = n - s_n. \tag{3.10}$$

In the case $s_n = n$ (i.e., $\varepsilon_1 = \cdots = \varepsilon_n = 1$),

$$P(X_1 = 1, \ldots, X_n = 1) = \frac{[r + (n-1)c]\cdots r}{[r + b + (n-1)c]\cdots[r+b]} \tag{3.11}$$

and if $s_n = 0$ (i.e., $\varepsilon_1 = \cdots = \varepsilon_n = 0$) then

$$P(X_1 = 0, \ldots, X_n = 0) = \frac{[b + (n-1)c]\cdots b}{[r + b + (n-1)c]\cdots[r+b]}. \tag{3.12}$$

In particular,

$$\begin{aligned} P(X_n = \varepsilon_n \mid X_1 = \varepsilon_1, \ldots, X_{n-1} = \varepsilon_{n-1}) &= \frac{P(X_1 = \varepsilon_1, \ldots, X_n = \varepsilon_n)}{P(X_1 = \varepsilon_1, \ldots, X_{n-1} = \varepsilon_{n-1})} \\ &= \frac{[r + (s_n - 1)c]\cdots r[b + (\tau_n - 1)c]\cdots b}{[r + (s_{n-1} - 1)c]\cdots r[b + (\tau_{n-1} - 1)c]\cdots b} \, \frac{[r + b + (n-2)c]\cdots[r+b]}{[r + b + (n-1)c]\cdots[r+b]} \\ &= \begin{cases} \dfrac{[r + s_{n-1}c]}{r + b + (n-1)c} & \text{if } \varepsilon_n = 1 \\[2ex] \dfrac{[b + \tau_{n-1}c]}{r + b + (n-1)c} & \text{if } \varepsilon_n = 0. \end{cases} \end{aligned} \tag{3.13}$$

It follows that $\{X_n\}$ is *non-Markov* unless $c = 0$ (in which case $\{X_n\}$ is i.i.d.). Note, however, that $\{X_n\}$ does have a distinctive symmetry property reflected in (3.9). Namely, the joint distribution is a function of $s_n = \sum_{k=1}^{n} \varepsilon_k$ only, and is therefore invariant under permutations of $\varepsilon_1 \cdots \varepsilon_n$. Such a stochastic process is called *exchangeable* (or *symmetrically dependent*). The Pólya urn model was originally introduced to illustrate a notion of "contagious disease" or "accident proneness' for actuarial mathematics. Although $\{X_n\}$ is non-Markov for $c \neq 0$, it is interesting to note that the partial-sum process $\{S_n\}$, representing the evolution of accumulated numbers of red balls sampled, does have the Markov property. From (3.13) one can also get that

$$P(S_n = s \mid S_1 = s_1, \ldots, S_{n-1} = s_{n-1}) = \begin{cases} \dfrac{r + cs_{n-1}}{r + b + (n-1)c} & \text{if } s = 1 + s_{n-1} \\ \dfrac{b + (n - s_{n-1})c}{r + b + (n-1)c} & \text{if } s = s_{n-1}. \end{cases} \tag{3.14}$$

Observe that the transition law (3.14) depends explicitly on the time point n. In other words, *the partial-sum process $\{S_n\}$ is a Markov process with a nonhomogeneous transition law*. A related continuous-time version of this Markov process, again usually called the *Pólya process* is described in Exercise 1 of Chapter IV, Section 4.1. An alternative model for contagion is also given in Example 1 of Chapter IV, Section 4, and that one has a homogeneous transition law.

4 STOPPING TIMES AND THE STRONG MARKOV PROPERTY

One of the most useful general properties of a Markov chain is that the Markov property holds even when the "past" is given up to certain types of *random* times. Indeed, we have tacitly used it in proving that the simple symmetric random walk reaches every state infinitely often with probability 1 (see Eq. 3.18 of Chapter I). These special random times are called *stopping times* or (less appropriately) *Markov times*.

Definition 4.1. Let $\{Y_n : n = 0, 1, 2, \ldots\}$ be a stochastic process having a countable state space and defined on some probability space $(\Omega, \mathscr{F}, P)$. A random variable τ defined on this space is said to be a *stopping time* if

(i) It assumes only nonnegative integer values (including, possibly, $+\infty$), and

(ii) For every nonnegative integer m the event $\{\omega: \tau(\omega) \leqslant m\}$ is determined by $Y_0, Y_1, \ldots, Y_m$.

Intuitively, if τ is a stopping time, then whether or not to stop by time m can be decided by observing the stochastic process up to time m. For an example, consider the *first time τ_y the process $\{Y_n\}$ reaches the state y*, defined by

$$\tau_y(\omega) = \inf\{n \geqslant 0: Y_n(\omega) = y\}. \tag{4.1}$$

If ω is such that $Y_n(\omega) \neq y$ whatever be n (i.e., if the process never reaches y), then take $\tau_y(\omega) = \infty$. Observe that

$$\{\omega: \tau_y(\omega) \leqslant m\} = \bigcup_{n=0}^{m} \{\omega: Y_n(\omega) = y\}. \tag{4.2}$$

Hence τ_y is a stopping time. The *rth return times* $\tau_y^{(r)}$ *of* y are defined recursively by

$$\begin{aligned}\tau_y^{(1)}(\omega) &= \inf\{n \geqslant 1: Y_n(\omega) = y\},\\ \tau_y^{(r)}(\omega) &= \inf\{n > \tau_y^{(r-1)}(\omega): Y_n(\omega) = y\},\end{aligned} \qquad \text{for } r = 2, 3, \ldots. \tag{4.3}$$

Once again, the infimum over an empty set is to be taken as ∞. Now whether or not the process has reached (or hit) the state y at least r times by the time m depends entirely on the values of $Y_1, \ldots, Y_m$. Indeed, $\{\tau_y^{(r)} \leqslant m\}$ is precisely the event that at least r of the variables $Y_1, \ldots, Y_m$ equal y. Hence $\tau_y^{(r)}$ *is a stopping time*. On the other hand, if η_y denotes *the last time the process reaches the state* y, then η_y is *not* a stopping time; for whether or not $\eta_y \leqslant m$ cannot in general be determined without observing the entire process $\{Y_n\}$.

Let S be a countable state space and $\mathbf{p}$ a transition probability matrix on S, and let P_π denote the distribution of the Markov process with transition probability $\mathbf{p}$ and initial distribution $\boldsymbol{\pi}$. It will be useful to identify the events that depend on the process up to time n. For this, let Ω denote the set of all sequences $\omega = (i_0, i_1, i_2, \ldots)$ of states, and let $Y_n(\omega)$ be the nth coordinate of ω (if $\omega = (i_0, i_1, \ldots, i_n, \ldots)$, then $Y_n(\omega) = i_n$). Let $\mathscr{F}_n$ denote the class of all events that depend only on $Y_0, Y_1, \ldots, Y_n$. Then the $\mathscr{F}_n$ form an increasing sequence of sigmafields of finite-dimensional events. The Markov property says that *given the "past"* $Y_0, Y_1, \ldots, Y_m$ *up to time* m, *or given* $\mathscr{F}_m$, *the conditional distribution of the "after-m" stochastic process* $Y_m^+ = \{(Y_m)_n^+\} := \{Y_{m+n}: n = 0, 1, \ldots\}$ *is* P_{Y_m}. In other words, if the process is re-indexed after time m with $m + n$ being regarded as time n, then this stochastic process is conditionally distributed as a Markov chain having transition probability $\mathbf{p}$ and initial state Y_m.

A WORD ON NOTATION. Many of the conditional distributions do not depend on the initial distribution. So the subscripts on P_π, P_i, etc., are suppressed as a matter of convenience in some calculations.

Suppose now that τ is the stopping time. *"Given the past up to time τ"* means given the values of τ and $Y_0, Y_1, \ldots, Y_\tau$. By the *"after-τ" process* we now mean *the stochastic process*

$$Y_\tau^+ = \{Y_{\tau+n}: n = 0, 1, 2, \ldots\},$$

which is well defined only on the set $\{\tau < \infty\}$.

Theorem 4.1. Every Markov chain $\{Y_n: n = 0, 1, 2, \ldots\}$ has the *strong Markov property*; that is, for every stopping time τ, the conditional distribution of the after-τ process $Y_\tau^+ = \{Y_{\tau+n}: n = 0, 1, 2, \ldots\}$, given the past up to time τ is P_{Y_τ} on the set $\{\tau < \infty\}$.

Proof. Choose and fix a nonnegative integer m and a positive integer k along with k time points $0 \leqslant m_1 < m_2 < \cdots < m_k$, and states $i_0, i_1, \ldots, i_m$,

$j_1, j_2, \ldots, j_k$. Then,

$$P(Y_{\tau+m_1} = j_1, Y_{\tau+m_2} = j_2, \ldots, Y_{\tau+m_k} = j_k \mid \tau = m, Y_0 = i_0, \ldots, Y_m = i_m)$$
$$= P(Y_{m+m_1} = j_1, Y_{m+m_2} = j_2, \ldots, Y_{m+m_k} = j_k \mid \tau = m, Y_0 = i_0, \ldots, Y_m = i_m). \tag{4.4}$$

Now if the event $\{\tau = m\}$ (which is determined by the values of $Y_0, \ldots, Y_m$) is not consistent with the event "$Y_0 = i_0, \ldots, Y_m = i_m$" then $\{\tau = m, Y_0 = i_0, \ldots, Y_m = i_m\} = \varnothing$ and the conditioning event is impossible. In that case the conditional probability may be defined arbitrarily (or left undefined). However, if $\{\tau = m\}$ is consistent with (i.e., implied by) "$Y_0 = i_0, \ldots, Y_m = i_m$" then $\{\tau = m, Y_0 = i_0, \ldots, Y_m = i_m\} = \{Y_0 = i_0, \ldots, Y_m = i_m\}$, and the right side of (4.4) becomes

$$P(Y_{m+m_1} = j_1, Y_{m+m_2} = j_2, \ldots, Y_{m+m_k} = j_k \mid Y_0 = i_0, \ldots, Y_m = i_m). \tag{4.5}$$

But by the Markov property, (4.5) equals

$$P_{i_m}(Y_{m_1} = j_1, Y_{m_2} = j_2, \ldots, Y_{m_k} = j_k)$$
$$= P_{Y_\tau}(Y_{m_1} = j_1, Y_{m_2} = j_1, Y_{m_2} = j_2, \ldots, Y_{m_k} = j_k), \tag{4.6}$$

on the set $\{\tau = m\}$. ■

We have considered just "future" events depending on only finitely many (namely k) time points. The general case (applying to infinitely many time points) may be obtained by passage to a limit. Note that the equality of (4.4) and (4.5) holds (in case $\{\tau = m\} \supset \{Y_0 = i_0, \ldots, Y_m = i_m\}$) for *all* stochastic processes and *all* events whose (conditional) probabilities may be sought. The equality between (4.5) and (4.6) is a consequence of the Markov property. Since the latter property holds for all future events, so does the corresponding result for stopping times τ.

Events determined by the past up to time τ comprise a sigmafield $\mathscr{F}_\tau$ called the *pre-τ sigmafield.* The strong Markov property is often expressed as: the *conditional distribution of Y_τ^+ given $\mathscr{F}_\tau$ is P_{Y_τ}* (on $\{\tau < \infty\}$). Note that $\mathscr{F}_\tau$ is the smallest sigmafield containing all events of the form $\{\tau = m, Y_0 = i_0, \ldots, Y_m = i_m\}$.

Example 1. In this example let us reconsider the validity of relation (3.18) of Chapter I in light of the strong Markov property. Let d and y be two integers. For the simple symmetric random walk $\{S_n: n = 0, 1, 2, \ldots\}$ starting at x, $\tau_y^{(1)}$ is an almost surely finite stopping time, for the probability ρ_{xy} that the random walk ever reaches y is 1 (see Chapter I, Eqs. 3.16–3.17). Denoting by E_x the expectation with respect to P_x, we have

$$
\begin{aligned}
P_x(\tau_y^{(2)} < \infty) &= P_x(S_{\tau_y^{(1)}+n} = y \text{ for some } n \geqslant 1) \\
&= E_x[P_x(S_{\tau_y^{(1)}+n} = y \text{ for some } n \geqslant 1 \mid \tau_y^{(1)}, S_0, \ldots, S_{\tau_y^{(1)}})] \\
&= E_x[P_y(S_n = y \text{ for some } n \geqslant 1)] \text{ (since } S_{\tau_y^{(1)}} = y) \\
&\qquad\qquad\qquad\qquad\qquad \text{(Strong Markov Property)} \\
&= P_y(S_n = y \text{ for some } n \geqslant 1) \\
&= P_y(S_1 = y - 1, S_n = y \text{ for some } n \geqslant 1) \\
&\quad + P_y(S_1 = y + 1, S_n = y \text{ for some } n \geqslant 1) \\
&= P_y(S_1 = y - 1)P_y(S_n = y \text{ for some } n \geqslant 1 \mid S_1 = y - 1) \\
&\quad + P_y(S_1 = y - 1)P_y(S_n = y \text{ for some } n \geqslant 1 \mid S_1 = y + 1) \\
&= \tfrac{1}{2}P_y(S_{1+m} = y \text{ for some } m \geqslant 0 \mid S_1 = y - 1) \\
&\quad + \tfrac{1}{2}P_y(S_{1+m} = y \text{ for some } m \geqslant 0 \mid S_1 = y + 1) \qquad (m = n - 1) \\
&= \tfrac{1}{2}P_{y-1}(S_m = y \text{ for some } m \geqslant 0) + \tfrac{1}{2}P_{y+1}(S_m = y \text{ for some } m \geqslant 0) \\
&\qquad\qquad\qquad\qquad\qquad \text{(Markov property)} \\
&= \tfrac{1}{2}\rho_{y-1,y} + \tfrac{1}{2}\rho_{y+1,y} = \tfrac{1}{2} + \tfrac{1}{2} = 1. \qquad (4.7)
\end{aligned}
$$

Now all the steps in (4.7) remain valid if one replaces $\tau_y^{(1)}$ by $\tau_y^{(r-1)}$ and $\tau_y^{(2)}$ by $\tau_y^{(r)}$ and assumes that $\tau_y^{(r-1)} < \infty$ almost surely. Hence, by induction, $P_x(\tau_y^{(r)} < \infty) = 1$ for all positive integers r. This is equivalent to asserting

$$
\begin{aligned}
1 &= P_x(\tau_y^{(r)} < \infty \text{ for all positive integers } r) \\
&= P_x(S_n = y \text{ for infinitely many } n). \qquad (4.8)
\end{aligned}
$$

The importance of the strong Markov property will be amply demonstrated in Sections 9–11.

5 A CLASSIFICATION OF STATES OF A MARKOV CHAIN

The unrestricted simple random walk $\{S_n\}$ is an example in which any state $i \in S$ can be reached from every state j in a finite number of steps with positive probability. If $\mathbf{p}$ denotes its transition probability matrix, then $\mathbf{p}^2$ is the transition probability matrix of $\{Y_n\} := \{S_{2n}: n = 0, 1, 2, \ldots\}$. However, for the Markov chain $\{Y_n\}$, transitions in a finite number of steps are possible from odd to odd integers and from even to even, but not otherwise. For $\{S_n\}$ one says that there is one class of *"essential"* states and for $\{Y_n\}$ that there are two classes of essential states.

A different situation occurs when the random walk has two absorbing boundaries on $S = \{c, c + 1, \ldots, d - 1, d\}$. The states c, d can be reached (with positive probability) from $c + 1, \ldots, d - 1$. However, $c + 1, \ldots, d - 1$ cannot be reached from c or d. In this case $c+1, \ldots, d-1$ are called *"inessential"* states while $\{c\}$, $\{d\}$ form two classes of essential states.

The term "inessential" refers to states that will not play a role in the long-run behavior of the process. If a chain has several essential classes, the process restricted to each class can be analyzed separately.

Definition 5.1. Write $i \to j$ and read it as either "*j is accessible from i*" or "the process can go from i to j" if $p_{ij}^{(n)} > 0$ for some $n \geqslant 1$.

Since

$$p_{ij}^{(n)} = \sum_{i_1, i_2, \ldots, i_{n-1} \in S} p_{ii_1} p_{i_1 i_2} \cdots p_{i_{n-1} j}, \tag{5.1}$$

$i \to j$ if and only if there exists one chain $(i, i_1, i_2, \ldots, i_{n-1}, j)$ such that $p_{ii_1}, p_{i_1 i_2}, \ldots, p_{i_{n-1} j}$ are strictly positive.

Definition 5.2. Write $i \leftrightarrow j$ and read "*i and j communicate*" if $i \to j$ and $j \to i$. Say "*i is essential*" if $i \to j$ implies $j \to i$ (i.e., if any state j is accessible from i, then i is accessible from that state). We shall let $\mathscr{E}$ denote the set of all essential states. States that are not essential are called *inessential*.

Proposition 5.1

(a) For every i there exists (at least one) j such that $i \to j$.
(b) $i \to j$, $j \to k$ imply $i \to k$.
(c) "i essential" implies $i \leftrightarrow i$.
(d) i essential, $i \to j$ imply "j is essential" and $i \leftrightarrow j$.
(e) On $\mathscr{E}$ the relation "$\leftrightarrow$" is an equivalence relation (i.e., reflexive, symmetric, and transitive).

Proof. (a) For each i, $\sum_{j \in S} p_{ij} = 1$. Hence there exists at least one j for which $p_{ij} > 0$; for this j one has $i \to j$.

(b) $i \to j$, $j \to k$ means that there exist $m \geqslant 1$, $n \geqslant 1$ such that $p_{ij}^{(m)} > 0$, $p_{jk}^{(n)} > 0$. Hence,

$$\begin{aligned} p_{ik}^{(m+n)} &= \sum_{l \in S} p_{il}^{(m)} p_{lk}^{(n)} \\ &= p_{ij}^{(m)} p_{jk}^{(n)} + \sum_{l \neq j} p_{il}^{(m)} p_{lk}^{(n)} \geqslant p_{ij}^{(m)} p_{jk}^{(n)} > 0. \end{aligned} \tag{5.2}$$

Hence, $i \to k$. Note that the first equality is a consequence of the relation $\mathbf{p}^{m+n} = \mathbf{p}^m \mathbf{p}^n$.

(c) Suppose i is essential. By (a) there exists j such that $p_{ij} > 0$. Since i is essential, this implies $j \to i$, i.e., there exists $m \geqslant 1$ such that $\mathbf{p}_{ji}^{(m)} > 0$. But then

$$p_{ii}^{(m+1)} = \sum_{l \in S} p_{il} p_{li}^{(m)} = p_{ij} p_{ji}^{(m)} + \sum_{l \neq j} p_{il} p_{li}^{(m)} > 0. \tag{5.3}$$

Hence $i \to i$ and, therefore, $i \leftrightarrow i$.

(d) Suppose i is essential, $i \to j$. Then there exist $m \geqslant 1$, $n \geqslant 1$ such that $p_{ij}^{(m)} > 0$ and $p_{ji}^{(n)} > 0$. Hence $i \leftrightarrow j$. Now suppose k is any state such that $j \to k$, i.e., there exists $m' \geqslant 1$ such that $p_{jk}^{(m')} > 0$. Then, by (b), $i \to k$. Since i is essential, one must have $k \to i$. Together with $i \to j$ this implies (again by (b)) $k \to j$. Thus, if any state k is accessible from j, then j is accessible from that state k, proving that j is essential.

(e) If $\mathscr{E}$ is empty (which is possible, as for example in the case $p_{i,i+1} = 1$, $i = 0, 1, 2, \ldots$), then, there is nothing to prove. Suppose $\mathscr{E}$ is nonempty. Then: (i) On $\mathscr{E}$ the relation "$\leftrightarrow$" is reflexive by (c). (ii) If i is essential and $i \leftrightarrow j$, then (by (d)) j is essential and, of course, $i \leftrightarrow j$ and $j \leftrightarrow i$ are equivalent properties. Thus "$\leftrightarrow$" is symmetric (on $\mathscr{E}$ as well as on S). (iii) If $i \leftrightarrow j$ and $j \leftrightarrow k$, then $i \to j$ and $j \to k$. Hence $i \to k$ (by (b)). Also, $k \to j$ and $j \to i$ imply $k \to i$ (again by (b)). Hence $i \leftrightarrow k$. This shows that "$\leftrightarrow$" is transitive (on $\mathscr{E}$ as well as on S). ■

From the proof of (e) the relation "$\leftrightarrow$" is seen to be symmetric and transitive on all of S (and not merely $\mathscr{E}$). However, it is not generally true that $i \leftrightarrow i$ (or, $i \to i$) for all $i \in S$. In other words, reflexivity may break down on S.

Example 1. (*Simple (Unrestricted) Random Walk*). $S = \{0, \pm 1, \pm 2, \ldots\}$. Assume, as usual, $0 < p < 1$. Then $i \to j$ for all states $i \in S, j \in S$. Hence $\mathscr{E} = S$.

Example 2. (*Simple Random Walk with Two Absorbing Boundaries*). Here $S = \{c, c+1, \ldots, d\}$, $\mathscr{E} = \{c, d\}$. Note that c is not accessible from d, nor is d accessible from c.

Example 3. (*Simple Random Walk with Two Reflecting Boundaries*). Here $S = \{c, c+1, \ldots, d\}$, and $i \to j$ for all $i \in S$ and $j \in S$. Hence $\mathscr{E} = S$.

Example 4. Let $S = \{1, 2, 3, 4, 5\}$ and let

$$\mathbf{p} = \begin{bmatrix} \frac{1}{5} & \frac{1}{5} & \frac{1}{5} & \frac{1}{5} & \frac{1}{5} \\ 0 & \frac{1}{3} & 0 & \frac{2}{3} & 0 \\ 0 & 0 & \frac{1}{4} & 0 & \frac{3}{4} \\ 0 & \frac{2}{3} & 0 & \frac{1}{3} & 0 \\ \frac{1}{3} & 0 & \frac{1}{3} & 0 & \frac{1}{3} \end{bmatrix}. \tag{5.4}$$

Note that $\mathscr{E} = \{2, 4\}$.

In Examples 1 and 3 above, there is one essential class and there are no inessential states.

Definition 5.3. A transition probability matrix **p** having one essential class and no inessential states is called *irreducible.*

Now fix attention on $\mathscr{E}$. Distinct subsets of essential states can be identified according to the following considerations. Let $i \in \mathscr{E}$. Consider the set $\mathscr{E}(i) = \{j \in \mathscr{E} : i \to j\}$. Then, by (d), $i \leftrightarrow j$ for all $j \in \mathscr{E}(i)$. Indeed, if $j, k \in \mathscr{E}(i)$, then $j \leftrightarrow k$ (for $j \to i$, $i \to k$ imply $j \to k$; similarly, $k \to j$). Thus, all members of $\mathscr{E}(i)$ communicate with each other. Let $r \in \mathscr{E}$, $r \notin \mathscr{E}(i)$. Then r is not accessible from a state in $\mathscr{E}(i)$ (for, if $j \in \mathscr{E}(i)$ and $j \to r$, then $i \to j$, $j \to r$ will imply $i \to r$ so that $r \in \mathscr{E}(i)$, a contradiction). Define $\mathscr{E}(r) = \{j \in \mathscr{E}, r \to j\}$. Then, as before, all states in $\mathscr{E}(r)$ communicate with each other. Also, no state in $\mathscr{E}(r)$ is accessible from any state in $\mathscr{E}(i)$ (for if $l \in \mathscr{E}(r)$, and $j \in \mathscr{E}(i)$ and $j \to l$, then $i \to l$; but $r \leftrightarrow l$, so that $i \to l$, $l \to r$ implying $i \to r$, a contradiction). In this manner, one decomposes $\mathscr{E}$ into a number of disjoint classes, each class being a maximal set of communicating states. No member of one class is accessible from any member of a different class. Also note that if $k \in \mathscr{E}(i)$, then $\mathscr{E}(i) = \mathscr{E}(k)$. For if $j \in \mathscr{E}(i)$, then $j \to i$, $i \to k$ imply $j \to k$; and since j is essential one has $k \to j$. Hence $j \in \mathscr{E}(k)$. The classes into which $\mathscr{E}$ decomposes are called *equivalence classes.* In the case of the unrestricted simple random walk $\{S_n\}$, we have $\mathscr{E} = S = \{0, \pm 1, \pm 2, \ldots\}$, and all states in $\mathscr{E}$ communicate with each other; only one equivalence class. While for $\{X_n\} = \{S_{2n}\}$, $\mathscr{E} = S$ consists of two disjoint equivalence classes, the odd integers and the even integers.

Our last item of bookkeeping concerns the role of possible cyclic motions within an essential class. In the unrestricted simple random walk example, note that $p_{ii} = 0$ for all $i = 0, \pm 1, \pm 2, \ldots$, but $p_{ii}^{(2)} = 2pq > 0$. In fact $p_{ii}^{(n)} = 0$ for all odd n, and $p_{ii}^{(n)} > 0$ for all even n. In this case, we say that the period of i is 2. More generally, if $i \to i$, then the *period of i* is the greatest common divisor of the integers in the set $A = \{n \geqslant 1 : p_{ii}^{(n)}\}$. If $d = d_i$ is the period of i, then $p_{ii}^{(n)} = 0$ whenever n is not a multiple of d and d is the largest integer with this property.

Proposition 5.2

(a) If $i \leftrightarrow j$ then i and j possess the same period. In particular "period" is constant on each equivalence class.

(b) Let $i \in \mathscr{E}$ have a period $d = d_i$. For each $j \in \mathscr{E}(i)$ there exists a unique integer r_j, $0 \leqslant r_j \leqslant d - 1$, such that $p_{ij}^{(n)} > 0$ implies $n = r_j \pmod{d}$ (i.e., either $n = r_j$ or $n = sd + r_j$ for some integer $s \geqslant 1$).

Proof. (a) Clearly,

$$p_{ii}^{(a+m+b)} \geqslant p_{ij}^{(a)} p_{jj}^{(m)} p_{ji}^{(b)} \tag{5.5}$$

for all positive integers a, m, b. Choose a and b such that $p_{ij}^{(a)} > 0$ and $p_{ji}^{(b)} > 0$. If $p_{jj}^{(m)} > 0$, then $p_{jj}^{(2m)} \geqslant p_{jj}^{(m)} p_{jj}^{(m)} > 0$, and

$$p_{ii}^{(a+m+b)} \geqslant p_{ij}^{(a)} p_{jj}^{(m)} p_{ji}^{(b)} > 0, \qquad p_{ii}^{(a+2m+b)} \geqslant p_{ij}^{(a)} p_{jj}^{(2m)} p_{ji}^{(b)} > 0. \tag{5.6}$$

Therefore, d (the period of i) divides $a + m + b$ and $a + 2m + b$, so that it divides the difference $m = (a + 2m + b) - (a + m + b)$. Hence, the period of i does not exceed the period of j. By the same argument (since $i \leftrightarrow j$ is the same as $j \leftrightarrow i$), the period of j does not exceed the period of i. Hence the period of i equals the period of j.

(b) Choose a such that $p_{ji}^{(a)} > 0$. If $p_{ij}^{(m)} > 0$, $p_{ij}^{(n)} > 0$, then $p_{ii}^{(m+a)} \geqslant p_{ij}^{(m)} p_{ji}^{(a)} > 0$, and $p_{ii}^{(n+a)} \geqslant p_{ij}^{(n)} p_{ji}^{(a)} > 0$. Hence d, the period of i, divides $m + a$, $n + a$ and, therefore, $m - n = m + a - (n + a)$. Since this is true for all m, n such that $p_{ij}^{(m)} > 0$, $p_{ij}^{(n)} > 0$, it means that the difference between any two integers in the set $A = \{n: p_{ij}^{(n)} > 0\}$ is divisible by d. This implies that there exists a *unique* integer r_j, $0 \leqslant r_j \leqslant d - 1$, such that $n = r_j \pmod d$ for all $n \in A$ (i.e., $n = sd + r_j$ for some integer $s \geqslant 0$ where s depends on n). ∎

It is generally *not true* that the period of an essential state i is $\min\{n \geqslant 1: p_{ii}^{(n)} > 0\}$. To see this consider the chain with state space $\{1, 2, 3, 4\}$ and transition matrix

$$\begin{bmatrix} 0 & 1 & 0 & 0 \\ 0 & 0 & 1 & 0 \\ 0 & \frac{1}{2} & 0 & \frac{1}{2} \\ 1 & 0 & 0 & 0 \end{bmatrix}.$$

Schematically, only the following one-step transitions are possible.

$$\begin{array}{l} \qquad\qquad\quad 4 \to 1 \\ \qquad\qquad\; \nearrow \\ 1 \to 2 \to 3 \\ \qquad\qquad\; \searrow \\ \qquad\qquad\quad 2 \end{array}.$$

Thus $p_{11}^{(2)} = 0$, $p_{11}^{(4)} > 0$, $p_{11}^{(6)} > 0$, etc., and $p_{11}^{(n)} = 0$ for all odd n. The states communicate with each other and their common period is 2, although $\min\{n: p_{11}^{(n)} > 0\} = 4$. Note that $\min\{n \geqslant 1: p_{ii}^{(n)} > 0\}$ is a multiple of d_i since d_i divides all n for which $p_{ii}^{(n)} > 0$. Thus, $d_i \leqslant \min\{n \geqslant 1: p_{ii}^{(n)} > 0\}$.

Proposition 5.3. Let $i \in \mathscr{E}$ have period $d > 1$. Let C_r be the set of $j \in \mathscr{E}(i)$ such that $r_j = r$, where r_j is the remainder term as defined in Proposition 5.2(b). Then

(a) $C_0, C_1, \ldots, C_{d-1}$ are disjoint, $\bigcup_{r=0}^{d-1} C_r = \mathscr{E}(i)$.

(b) If $j \in C_r$, then $p_{jk} > 0$ implies $k \in C_{r+1}$, where we take $r + 1 = 0$ if $r = d - 1$.

Proof. (a) Follows from Proposition 5.2(b).

(b) Suppose $j \in C_r$ and $p_{ij}^{(n)} > 0$. Then $n = sd + r$ for some $s \geqslant 0$. Hence, if $p_{jk} > 0$ then

$$p_{ik}^{(n+1)} \geqslant p_{ij}^{(n)} p_{jk} > 0, \tag{5.7}$$

which implies $k \in C_{r+1}$ (since $n + 1 = sd + r + 1 = r + 1 \pmod d$), by Proposition 5.2(b). ■

Here is what Proposition 5.3 means. Suppose i is an essential state and has a period $d > 1$. In one step (i.e., one time unit) the process can go from $i \in C_0$ only to some state in C_1 (i.e., $p_{ij} > 0$ only if $j \in C_1$). From states in C_1, in one step the process can go only to states in C_2. This means that in two steps the process can go from i only to states in C_2 (*i.e.*, $p_{ij}^{(2)} > 0$ only if $j \in C_2$), and so on. In d steps the process can go from i only to states in $C_{d+1} = C_0$, completing one cycle (of d steps). Again in $d + 1$ steps the process can go from i only to states in C_1, and so on. In general, in $sd + r$ steps the process can go from i only to states in C_r. Schematically, one has the picture in Figure 5.1 for the case $d = 4$ and a fixed state $i \in C_0$ of period 4.

Example 5.5. In the case of the unrestricted simple random walk, the period is 2 and all states are essential and communicate with each other. Fix $i = 0$. Then $C_0 = \{0, \pm 2, \pm 4, \ldots\}$, $C_1 = \{\pm 1, \pm 3, \pm 5, \ldots\}$. If we take i to be any even integer, then C_0, C_1 are as above. If, however, we start with i odd, then $C_0 = \{\pm 1, \pm 3, \pm 5, \ldots\}$, $C_1 = \{0, \pm 2, \pm 4, \ldots\}$.

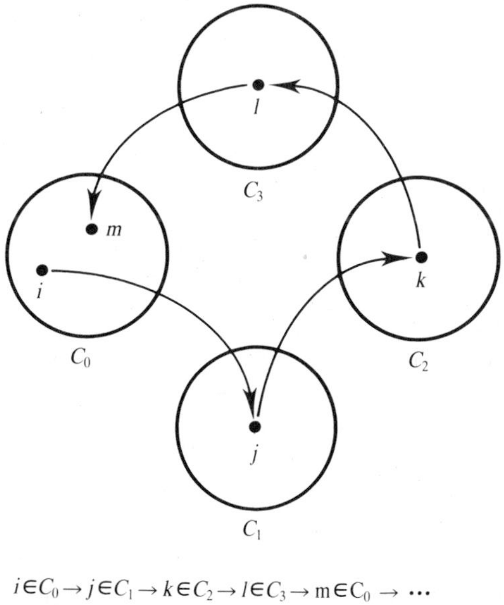

$i \in C_0 \to j \in C_1 \to k \in C_2 \to l \in C_3 \to m \in C_0 \to \cdots$

Figure 5.1

6 CONVERGENCE TO STEADY STATE FOR IRREDUCIBLE AND APERIODIC MARKOV PROCESSES ON FINITE SPACES

As will be demonstrated in this section, if the state space is finite, a complete analysis of the limiting behavior of $\mathbf{p}^n$, as $n \to \infty$, may be carried out by elementary methods that also provide sharp rates of convergence to the so-called *steady-state* or *invariant distributions.* Although later, in Section 9, the asymptotic behavior of general Markov chains is analyzed in detail, including the *law of large numbers* and the *central limit theorem* for Markov chains that admit unique steady-state distributions, the methods of the present section are also suited for applications to certain more general (nonfinite) state spaces (e.g., closed and bounded intervals). These latter extensions are outlined in the exercises.

First we consider what happens to the n-step transition law if all states are accessible from each other in one time step.

Proposition 6.1. Suppose S is finite and $p_{ij} > 0$ for all i, j. Then there exists a unique probability distribution $\boldsymbol{\pi} = \{\pi_j : j \in S\}$ such that

$$\sum_i \pi_i p_{ij} = \pi_j \qquad \text{for all } j \in S \tag{6.1}$$

and

$$|p_{ij}^{(n)} - \pi_j| \leqslant (1 - N\delta)^n \qquad \text{for all } i, j \in S, \quad n \geqslant 1, \tag{6.2}$$

where $\delta = \min\{p_{ij} : i, j \in S\}$ and N is the number of elements in S. Also, $\pi_j \geqslant \delta$ for all $j \in S$ and $\delta \leqslant 1/N$.

Proof. Let $M_j^{(n)}, m_j^{(n)}$ denote the maximum and the minimum, respectively, of the elements $\{p_{ij}^{(n)} : i \in S\}$ of the jth column of $\mathbf{p}^n$. Since $p_{ij} \geqslant \delta$ and $p_{ij} = 1 - \sum_{k \neq j} p_{ik} \leqslant 1 - (N-1)\delta$ for all i, one has

$$M_j^{(1)} - m_j^{(1)} \leqslant 1 - (N-1)\delta - \delta = 1 - N\delta. \tag{6.3}$$

Fix two states i, i' arbitrarily. Let $J = \{j \in S : p_{ij} > p_{i'j}\}$, $J' = \{j \in S : p_{ij} \leqslant p_{i'j}\}$. Then,

$$0 = 1 - 1 = \sum_j (p_{ij} - p_{i'j}) = \sum_{j \in J'} (p_{ij} - p_{i'j}) + \sum_{j \in J} (p_{ij} - p_{i'j}),$$

so that

$$\sum_{j \in J'} (p_{ij} - p_{i'j}) = -\sum_{j \in J} (p_{ij} - p_{i'j}), \tag{6.4}$$

and

$$\sum_{j\in J}(p_{ij}-p_{i'j}) = \sum_{j\in J} p_{ij} - \sum_{j\in J} p_{i'j}$$
$$= 1 - \sum_{j\in J'} p_{ij} - \sum_{j\in J} p_{i'j} \leqslant 1 - (\#J')\delta - (\#J)\delta = 1 - N\delta. \quad (6.5)$$

Therefore,

$$p_{ij}^{(n+1)} - p_{i'j}^{(n+1)} = \sum_k p_{ik}p_{kj}^{(n)} - \sum_k p_{i'k}p_{kj}^{(n)} = \sum_k (p_{ik}-p_{i'k})p_{kj}^{(n)}$$
$$= \sum_{k\in J}(p_{ik}-p_{i'k})p_{kj}^{(n)} + \sum_{k\in J'}(p_{ik}-p_{i'k})p_{kj}^{(n)}$$
$$\leqslant \sum_{k\in J}(p_{ik}-p_{i'k})M_j^{(n)} + \sum_{k\in J'}(p_{ik}-p_{i'k})m_j^{(n)}$$
$$= (M_j^{(n)} - m_j^{(n)})\left(\sum_{k\in J}(p_{ik}-p_{i'k})\right) \leqslant (1-N\delta)(M_j^{(n)} - m_j^{(n)}). \quad (6.6)$$

Letting i, i' be such that $p_{ij}^{(n+1)} = M_j^{(n+1)}$, $p_{i'j} = m_j^{(n+1)}$, one gets from (6.6),

$$M_j^{(n+1)} - m_j^{(N+1)} \leqslant (1-N\delta)(M_j^{(n)} - m_j^{(n)}). \quad (6.7)$$

Iteration now yields, using (6.3) as well as (6.7),

$$M_j^{(n)} - m_j^{(n)} \leqslant (1-N\delta)^n \qquad \text{for } n \geqslant 1. \quad (6.8)$$

Now

$$M_j^{(n+1)} = \max_i p_{ij}^{(n+1)} = \max_i\left(\sum_k p_{ik}p_{kj}^{(n)}\right) \leqslant \max_i\left(\sum_k p_{ik}M_j^{(n)}\right) = M_j^{(n)},$$
$$m_j^{(n+1)} = \min_i p_{ij}^{(n+1)} = \min_i\left(\sum_k p_{ik}p_{kj}^{(n)}\right) \geqslant \min_i\left(\sum_k p_{ik}m_j^{(n)}\right) = m_j^{(n)},$$

i.e., $M_j^{(n)}$ is nonincreasing and $m_j^{(n)}$ is nondecreasing in n. Since $M_j^{(n)}, m_j^{(n)}$ are bounded above by 1, (6.7) now implies that both sequences have the same limit, say π_j. Also, $\delta \leqslant m_j^{(1)} \leqslant m_j^{(n)} \leqslant \pi_j \leqslant M_j^{(n)}$ for all n, so that $\pi_j \geqslant \delta$ for all j and

$$m_j^{(n)} - M_j^{(n)} \leqslant p_{ij}^{(n)} - \pi_j \leqslant M_j^{(n)} - m_j^{(n)},$$

which, together with (6.8), implies the desired inequality (6.2).

Finally, taking limits on both sides of the identity

$$p_{ij}^{(n+1)} = \sum_k p_{ik}^{(n)}p_{kj} \quad (6.9)$$

one gets $\pi_j = \sum_k \pi_k p_{kj}$, proving (6.1). Since $\sum_j p_{ij}^{(n)} = 1$, taking limits, as $n \to \infty$, it follows that $\sum_j \pi_j = 1$. To prove uniqueness of the probability distribution $\boldsymbol{\pi}$ satisfying (6.1), let $\bar{\boldsymbol{\pi}} = \{\bar{\pi}_j : j \in S\}$ be a probability distribution satisfying $\bar{\boldsymbol{\pi}}'\mathbf{p} = \bar{\boldsymbol{\pi}}'$. Then by iteration it follows that

$$\bar{\pi}_j = \sum_i \bar{\pi}_i p_{ij} = (\bar{\boldsymbol{\pi}}'\mathbf{p})_j = (\bar{\boldsymbol{\pi}}'\mathbf{pp})_j = (\bar{\boldsymbol{\pi}}'\mathbf{p}^2)_j = \cdots = (\bar{\boldsymbol{\pi}}'\mathbf{p}^n)_j = \sum_i \bar{\pi}_i p_{ij}^{(n)}. \quad (6.10)$$

Taking limits as $n \to \infty$, one gets $\bar{\pi}_j = \sum_i \bar{\pi}_i \pi_j = \pi_j$. Thus, $\bar{\boldsymbol{\pi}} = \boldsymbol{\pi}$. ∎

For an irreducible aperiodic Markov chain on a finite state space S there is a positive integer ν such that

$$\delta' := \min_{i,j} p_{ij}^{(\nu)} > 0. \quad (6.11)$$

Applying Proposition 6.1 with $\mathbf{p}$ replaced by $\mathbf{p}^\nu$ one gets a probability $\boldsymbol{\pi} = \{\pi_j : j \in S\}$ such that

$$\max_{i,j} |p_{ij}^{(n\nu)} - \pi_j| \leqslant (1 - N\delta')^n, \qquad n = 1, 2, \ldots. \quad (6.12)$$

Now use the relations

$$|p_{ij}^{(n\nu+m)} - \pi_j| = \left|\sum_k p_{ik}^{(m)}(p_{kj}^{(n\nu)} - \pi_j)\right| \leqslant \sum_k p_{ik}^{(m)}(1 - N\delta')^n = (1 - N\delta')^n, \qquad m \geqslant 1, \quad (6.13)$$

to obtain

$$|p_{ij}^{(n)} - \pi_j| \leqslant (1 - N\delta')^{[n/\nu]}, \qquad n = 1, 2, \ldots, \quad (6.14)$$

where $[x]$ is the integer part of x. From here one obtains the following corollary to Proposition 6.1.

Corollary 6.2. Let $\mathbf{p}$ be an irreducible and aperiodic transition law on a state space S having N states. Then there is a unique probability distribution $\boldsymbol{\pi}$ on S such that

$$\sum_i \pi_i p_{ij} = \pi_j \qquad \text{for all } j \in S. \quad (6.15)$$

Also,

$$|p_{ij}^{(n)} - \pi_j| \leqslant (1 - N\delta')^{[n/\nu]} \qquad \text{for all } i, j \in S, \quad (6.16)$$

for some $\delta' > 0$ and some positive integer ν.

The property (6.15) is important enough to justify a definition.

Definition 6.1. A probability measure $\boldsymbol{\pi}$ satisfying (6.15) is said to be an *invariant or steady-state distribution* for $\mathbf{p}$.

Suppose that $\boldsymbol{\pi}$ is an invariant distribution for $\mathbf{p}$ and let $\{X_n\}$ be the Markov chain with transition law $\mathbf{p}$ starting with initial distribution $\boldsymbol{\pi}$. It follows from the Markov property, using (2.9), that

$$\begin{aligned} P_\pi(X_m = i_0, X_{m+1} = i_1, \ldots, X_{m+n} = i_n) &= (\boldsymbol{\pi}'\mathbf{p}^m)_{i_0} p_{i_0 i_1} p_{i_1 i_2} \cdots p_{i_{n-1} i_n} \\ &= \pi_{i_0} p_{i_0 i_1} p_{i_1 i_2} \cdots p_{i_{n-1} i_n} \\ &= P_\pi(X_0 = i_0, X_1 = i_1, \ldots, X_n = i_n), \end{aligned} \tag{6.17}$$

for any given positive integer n and arbitrary states $i_0, i_1, \ldots, i_n \in S$. In other words, the distribution of the process is invariant under time translation; i.e., $\{X_n\}$ is a *stationary* Markov process according to the following definition.

Definition 6.2. A stochastic process $\{Y_n\}$ is called a *stationary* stochastic process if for all n, m

$$P(Y_0 = i_0, Y_1 = i_1, \ldots, Y_n = i_n) = P(Y_m = i_0, Y_{1+m} = i_1, \ldots, Y_{n+m} = i_n). \tag{6.18}$$

Proposition 6.1 or its corollary 6.2 establish the existence and uniqueness of an invariant initial distribution that makes the process stationary. Moreover, the *asymptotic convergence rate* (*relaxation time*) may also be expressed in the form

$$|p_{ij}^{(n)} - \pi_j| \leqslant ce^{-n\lambda} \tag{6.19}$$

for some $c, \lambda > 0$.

Suppose that $\{X_n\}$ is a stochastic process with state space $J = [c, d]$ and having the Markov property. Suppose also that the conditional distribution of X_{n+1} given $X_n = x$ has a density $p(x, y)$ that is jointly continuous in (x, y) and does not depend on $n \geqslant 1$. Given $X_0 = x_0$, the (conditional on X_0) joint density of $X_1, \ldots, X_n$ is given by

$$f_{(X_1,\ldots,X_n \mid X_0 = x_0)}(x_1, \ldots, x_n) = p(x_0, x_1)p(x_1, x_2)\cdots p(x_{n-1}, x_n). \tag{6.20}$$

If X_0 has a density $\mu(x)$, then the joint density of $X_0, \ldots, X_n$ is

$$f_{(X_0,\ldots,X_n)}(x_0, \ldots, x_n) = \mu(x_0)p(x_0, x_1)\cdots p(x_{n-1}, x_n). \tag{6.21}$$

Now let

$$\delta = \min_{x,y \in [c,d]} p(x, y). \tag{6.22}$$

In a manner analogous to the proof of Proposition 6.1, one can also obtain the following result (Exercise 9).

Proposition 6.3. If $\delta = \min_{x,y \in [c,d]} p(x, y) > 0$ then there is a continuous probability density function $\pi(x)$ such that

$$\int_c^d \pi(x)p(x, y)\, dx = \pi(y) \qquad \text{for all } y \in (c, d) \tag{6.23}$$

and

$$|p^{(n)}(x, y) - \pi(y)| \leqslant [1 - \delta(d - c)]^{n-1}\theta \qquad \text{for all } x, y \in (c, d) \tag{6.24}$$

where

$$\theta = \max_{x,y,z \in [c,d]} \{p(x, y) - p(z, y)\}. \tag{6.25}$$

Here $p^{(n)}(x, y)$ is the n-step transition probability density function of X_n given $X_0 = x$.

Markov processes on general state spaces are discussed further in theoretical complements to this chapter.

Observe that if $\mathbf{A} = ((a_{ij}))$ is an $N \times N$ matrix with strictly positive entries, then one may readily deduce from Proposition 6.1 that if furthermore $\sum_{j=1}^N a_{ij} = 1.$ for each i, then the *spectral radius* of $\mathbf{A}$, i.e., the *magnitude* of the largest eigenvalue of $\mathbf{A}$, must be 1. Moreover, $\lambda = 1$ must be a simple eigenvalue (i.e., multiplicity 1) of $\mathbf{A}$. To see this let $\mathbf{z}$ be a (left) eigenvector of $\mathbf{A}$ corresponding to $\lambda = 1$. Then for t sufficiently large, $\mathbf{z} + t\boldsymbol{\pi}$ is also a positive eigenvector (and normalizable), where $\boldsymbol{\pi}$ is the invariant distribution (normalized positive eigenvector for $\lambda = 1$) of $\mathbf{A}$. Thus, uniqueness makes $\mathbf{z}$ a scalar multiple of $\boldsymbol{\pi}$. The following theorem provides an extension of these results to the case of arbitrary *positive matrices* $\mathbf{A} = ((a_{ij}))$, i.e., $a_{ij} > 0$ for $1 \leqslant i, j \leqslant N$. Use is not made of this result until Sections 11 and 12, so that it may be skipped on first reading. At this stage it may be regarded as an application of probability to analysis.

Theorem 6.4. [*Perron–Frobenius*]. Let $\mathbf{A} = ((a_{ij}))$ be a positive $N \times N$ matrix.

(a) There is a unique eigenvalue λ_0 of $\mathbf{A}$ that has largest magnitude. Moreover, λ_0 is positive and has a corresponding positive eigenvector.

(b) Let $\mathbf{x}$ be any nonnegative nonzero vector. Then

$$\mathbf{v} = \lim_{n \to \infty} \lambda_0^{-n} \mathbf{A}^n \mathbf{x} \tag{6.26}$$

exists and is an eigenvector of $\mathbf{A}$ corresponding to λ_0, unique up to a scalar multiple determined by $\mathbf{x}$, but otherwise independent of $\mathbf{x}$.

Proof. Define $\Lambda^+ = \{\lambda > 0: \mathbf{Ax} \geqslant \lambda\mathbf{x}$ for some nonnegative nonzero vector $\mathbf{x}\}$; here inequalities are to be interpreted componentwise. Observe that the set Λ^+ is nonempty and bounded above by $\|\mathbf{A}\| := \sum_{i=1}^N \sum_{j=1}^N a_{ij}$. Let λ_0 be the least upper bound of Λ^+. There is a sequence $\{\lambda_n: n \geqslant 1\}$ in Λ^+ with limit λ_0 as $n \to \infty$. Let $\{\mathbf{x}_n: n \geqslant 1\}$ be corresponding nonnegative vectors, normalized so that $\|\mathbf{x}_n\| := \sum_{i=1}^N x_i = 1$, $n \geqslant 1$, for which $\mathbf{Ax}_n \geqslant \lambda_n \mathbf{x}_n$. Then, since $\|\mathbf{x}_n\| = 1$, $n = 1, 2, \ldots$, $\{\mathbf{x}_n\}$ must have a convergent subsequence, with limit denoted $\mathbf{x}_0$, say. Therefore $\mathbf{Ax}_0 \geqslant \lambda_0 \mathbf{x}_0$ and hence $\lambda_0 \in \Lambda^+$. In fact, it follows from the least upper bound property of λ_0 that $A\mathbf{x}_0 = \lambda_0 \mathbf{x}_0$. For otherwise there must be a component with strict inequality, say $\sum_{j=1}^N a_{1j}x_j - \lambda_0 x_1 = \delta > 0$, where $\mathbf{x}_0 = (x_1, \ldots, x_N)'$, and $\sum_{j=1}^N a_{kj}x_j - \lambda_0 x_k \geqslant 0$, $k = 2, \ldots, N$. But then taking $\mathbf{y} = (x_1 + (\delta/\lambda_0), x_2, \ldots, x_N)'$ we get $\mathbf{Ay} > \lambda_0 \mathbf{y}$ with strict inequality in each component. This contradicts the maximality of λ_0. To prove that if λ is any other eigenvalue then $|\lambda| \leqslant \lambda_0$, let $\mathbf{z}$ be an eigenvector corresponding to λ and define $|\mathbf{z}| = (|z_1|, \ldots, |z_n|)$. Then $\mathbf{Az} = \lambda\mathbf{z}$ implies $\mathbf{A}|\mathbf{z}| \geqslant |\lambda||\mathbf{z}|$. Therefore, by definition of λ_0, we have $|\lambda| \leqslant \lambda_0$. To prove part (ii) of the Theorem we can apply Proposition 6.1 to the transition probability matrix

$$\left(\left(p_{ij} = \frac{a_{ij}x_j}{\lambda_0 x_i}\right)\right),$$

where $\mathbf{x}_0 = (x_1, \ldots, x_N)$ is a positive eigenvector corresponding to λ_0. In particular, noting that

$$p_{ij}^{(2)} = \sum_{k=1}^N p_{ik} p_{kj} = \sum_{k=1}^N \frac{a_{ik}x_k}{\lambda_0 x_i} \frac{a_{kj}x_j}{\lambda_0 x_k} = \frac{a_{ij}^{(2)}}{\lambda_0^2} \frac{x_j}{x_i},$$

and inductively

$$p_{ij}^{(n)} = \frac{a_{ij}^{(n)}}{\lambda_0^n} \frac{x_j}{x_i},$$

the result follows. ■

Corollary 6.5. Let $\mathbf{A} = ((a_{ij}))$ be an $N \times N$ matrix with positive entries and let $\mathbf{B}$ be the $(N-1) \times (N-1)$ matrix obtained from $\mathbf{A}$ by striking out an ith row and jth column. Then the spectral radius of $\mathbf{B}$ is strictly smaller than that of $\mathbf{A}$.

Proof. Let λ_{00} be the largest positive eigenvalue of $\mathbf{B}$ and without loss of generality take $\mathbf{B} = ((a_{ij}: i, j = 1, \ldots, N-1))$. Since, for some positive $\mathbf{x}$,

$$\sum_{j=1}^{N} a_{ij}x_j = \lambda_0 x_i, \qquad i = 1, \ldots, N, \quad x_i > 0,$$

we have

$$(\mathbf{Bx})_i = \sum_{j=1}^{N-1} a_{ij}x_j = \lambda_0 x_i - a_{iN}x_N < \lambda_0 x_i, \qquad i = 1, \ldots, N-1.$$

Thus, by the property (6.26) applied to λ_{00} we must have $\lambda_{00} < \lambda_0$. ■

Corollary 6.6. Let $\mathbf{A} = ((a_{ij}))$ be a matrix of strictly positive elements and let λ_0 be the positive eigenvalue of maximum magnitude (spectral radius). Then λ_0 is a simple eigenvalue.

Proof. Consider the characteristic polynomial $p(\lambda) = \det(\mathbf{A} - \lambda\mathbf{I})$. Differentiation with respect to λ gives $p'(\lambda) = -\sum_{k=1}^{N} \det(\mathbf{A}_k - \lambda\mathbf{I})$, where $\mathbf{A}_k$ is the matrix obtained from $\mathbf{A}$ by striking out the kth row and column. By Corollary 6.5 we have $\det(\mathbf{A}_k - \lambda_0\mathbf{I}) \neq 0$, since each $\mathbf{A}_k$ has smaller spectral radius than λ_0. Moreover, since each polynomial $\det(\mathbf{A}_k - \lambda\mathbf{I})$ has the same leading term, they all have the same sign at $\lambda = \lambda_0$. Thus $p'(\lambda_0) \neq 0$. ■

Another formula for the spectral radius is given in Section 14.

7 STEADY-STATE DISTRIBUTIONS FOR GENERAL FINITE-STATE MARKOV PROCESSES

In general, the transition law $\mathbf{p}$ may admit several essential classes, periodicities, or inessential states. In this section, we consider the asymptotics of $\mathbf{p}^n$ for such cases.

First suppose that S is finite and is, under $\mathbf{p}$, a single class of periodic essential states of period $d > 1$. Then the matrix $\mathbf{p}^d$, regarded as a (one-step) transition probability matrix of the process viewed every d time steps, admits d (equivalence) classes $C_0, C_1, \ldots, C_{d-1}$, each of which is aperiodic. Applying (6.16) to C_r and $\mathbf{p}^d$ (instead of S and $\mathbf{p}$) one gets, writing N_r for the number of elements in C_r,

$$|p_{ij}^{(nd)} - \pi_j| \leqslant (1 - N_r\delta_r)^{[n/\nu_r]} \qquad \text{for } i, j \in C_r, \tag{7.1}$$

where $\pi_j = \lim_{n\to\infty} p_{ij}^{(nd)}$, ν_r is the smallest positive integer such that $p_{ij}^{(\nu_r d)} > 0$

for all $i, j \in C_r$, and $\delta_r = \min\{p_{ij}^{(\nu_r d)}: i, j \in C_r\}$. Let $\delta = \min\{\delta_r: r = 0, 1, \ldots, d-1\}$. Then one has, writing $L = \min\{N_r: 0 \leqslant r \leqslant d-1\}$, $\nu = \max\{\nu_r: 0 \leqslant r < d-1\}$,

$$|p_{ij}^{(nd)} - \pi_j| \leqslant (1 - L\delta)^{[n/\nu]} \qquad \text{for } i, j \in C_r \quad (r = 0, 1, \ldots, d-1). \tag{7.2}$$

If $i \in C_r$ and $j \in C_s$ with $s = r + m$ (mod d), then one gets, using the facts that $p_{kj}^{(nd)} = 0$ if k is not in C_s, and that $\sum_{k \in C_s} p_{ik}^{(m)} = 1$,

$$|p_{ij}^{(nd+m)} - \pi_j| = \left| \sum_{k \in C_s} p_{ik}^{(m)}(p_{kj}^{(nd)} - \pi_j) \right| \leqslant (1 - L\delta)^{[n/\nu]} \qquad \text{for } i \in C_r, \quad j \in C_s. \tag{7.3}$$

Of course,

$$p_{ij}^{(nd+m')} = 0 \qquad \text{if } i \in C_r, j \in C_s, \text{ and } m' \neq s - r \text{ (mod } d). \tag{7.4}$$

Note that $\{\pi_j: j \in C_r\}$ is the unique invariant initial distribution on C_r for the restriction of $\mathbf{p}^d$ to C_r.

Now, in view of (7.4),

$$\sum_{t=1}^{nd} p_{ij}^{(t)} = \sum_{n'=0}^{n-1} p_{ij}^{(n'd+m)} \qquad \text{if } i \in C_r, j \in C_s, \text{ and } m = s - r \text{ (mod } d). \tag{7.5}$$

If $r = s$ then $m = 0$ and the index of the second sum in (7.5) ranges from 1 to n. By (7.3) one then has

$$\lim_{n \to \infty} \frac{1}{n} \sum_{t=1}^{nd} p_{ij}^{(t)} = \pi_j \qquad (i, j \in S). \tag{7.6}$$

This implies, on multiplying (7.6) by $1/d$,

$$\lim_{n \to \infty} \frac{1}{n} \sum_{t=1}^{n} p_{ij}^{(t)} = \frac{\pi_j}{d} \qquad (i, j \in S). \tag{7.7}$$

Now observe that $\boldsymbol{\pi} = \{\pi_j/d: j \in S\}$ is the unique invariant initial distribution of the periodic chain defined by $\mathbf{p}$ since

$$\begin{aligned} \sum_{k \in S} \frac{\pi_k}{d} p_{kj} &= \sum_{k \in S} \left(\lim_{n \to \infty} \frac{1}{n} \sum_{t=1}^{n} p_{ik}^{(t)} \right) p_{kj} \\ &= \lim_{n \to \infty} \frac{1}{n} \sum_{t=1}^{n} p_{ij}^{(t+1)} = \frac{\pi_j}{d} \qquad (j \in S). \end{aligned} \tag{7.8}$$

Moreover, if $\bar{\boldsymbol{\pi}} = \{\bar{\pi}_j: j \in S\}$ is another invariant distribution, then

$$\bar{\pi}_j = \sum_{k \in S} \bar{\pi}_k p_{kj} = \sum_{k \in S} \bar{\pi}_k p_{kj}^{(t)} \qquad (t = 1, 2, \ldots), \tag{7.9}$$

so that, on averaging over $t = 1, \ldots, n-1, n$,

$$\bar{\pi}_j = \frac{1}{n} \sum_{t=1}^{n} \sum_{k \in S} \bar{\pi}_k p_{kj}^{(t)} = \sum_{k \in S} \bar{\pi}_k \left(\frac{1}{n} \sum_{t=1}^{n} p_{kj}^{(t)} \right). \tag{7.10}$$

The right side converges to $\sum_k \bar{\pi}_k(\pi_j/d) = \pi_j/d$, as $n \to \infty$ by (7.7). Hence $\bar{\pi}_j = \pi_j/d$.

If the finite state space S consists of b equivalence classes of essential states ($b > 1$), say $\mathscr{E}_1, \mathscr{E}_2, \ldots, \mathscr{E}_b$, then the results of the preceding paragraphs may be applied to each $\mathscr{E}_i$, for the restriction of $\mathbf{p}$ to $\mathscr{E}_i$, separately. Let $\boldsymbol{\pi}^{(i)}$ denote the unique invariant initial distribution for the restriction of $\mathbf{p}$ to $\mathscr{E}_i$. Write $\bar{\boldsymbol{\pi}}^{(i)}$ for the probability distribution on S that assigns probability $\pi_j^{(i)}$ to states $j \in \mathscr{E}_i$ and zero probabilities to states not in $\mathscr{E}_i$. Then it is easy to check that $\sum_{i=1}^{b} \alpha_i \bar{\boldsymbol{\pi}}^{(i)}$ is an invariant initial distribution for $\mathbf{p}$, for every b-tuple of nonnegative numbers $\alpha_1, \alpha_2, \ldots, \alpha_b$ satisfying $\sum_{i=1}^{b} \alpha_i = 1$. The probabilistic interpretation of this is as follows. Suppose by a random mechanism the equivalence class $\mathscr{E}_i$ is chosen with probability α_i ($1 \leqslant i \leqslant b$). If $\mathscr{E}_i$ is the class chosen in this manner, then start the process with initial distribution $\bar{\boldsymbol{\pi}}^{(i)}$ on S. The resulting Markov chain $\{X_n\}$ is a *stationary process*.

It remains to consider the case in which S has some *inessential* states. In the next section it will be shown that if S is finite, then, no matter how the Markov chain starts, after a finite (possibly random) time the process will move only among essential states, so that its asymptotic behavior depends entirely on the restriction of $\mathbf{p}$ to the class of essential states. The main results of this section may be summarized as follows.

Theorem 7.1. Suppose S is finite.

(i) If $\mathbf{p}$ is a transition probability matrix such that there is only one essential class $\mathscr{E}$ of (communicating) states and these states are aperiodic, then there exists a unique invariant distribution $\boldsymbol{\pi}$ such that $\sum_{j \in S} \pi_j = 1$ and (6.16) holds for all $i, j \in \mathscr{E}$.

(ii) If $\mathbf{p}$ is such that there is only one essential class of states $\mathscr{E}$ and it is periodic with period d, then again there exists a unique invariant distribution $\bar{\boldsymbol{\pi}}$ such that $\sum_{j \in \mathscr{E}} \bar{\pi}_j = 1$ and (7.3), (7.4), (7.7) hold for $i, j \in \mathscr{E}$ and $\bar{\pi}_j = \pi_j/d$.

(iii) If there are b equivalence classes of essential states, then (i), (ii), as the case may be, apply to each class separately, and every convex combination of the invariant distributions of these classes (regarded as state spaces in their own right) is an invariant distribution for $\mathbf{p}$.

8 MARKOV CHAINS: TRANSIENCE AND RECURRENCE PROPERTIES

Let $\{X_n\}$ be a Markov chain with countable state space S and transition probability law $\mathbf{p} = ((p_{ij}))$. As in the case of random walks, the frequency of returns to a state is an important feature of the evolution of the process.

Definition 8.1. A state j is said to be *recurrent* if

$$P_j(X_n = j \text{ i.o.}) = 1, \tag{8.1}$$

and *transient* if

$$P_j(X_n = j \text{ i.o.}) = 0. \tag{8.2}$$

Introduce the *successive return times to the state j* as

$$\begin{aligned} \tau_j^{(0)} = 0, \qquad \tau_j^{(1)} &= \min\{n > 0: X_n = j\}, \\ \tau_j^{(r)} = \min\{n > \tau_j^{(r-1)}&: X_n = j\} \qquad (r = 1, 2, \ldots), \end{aligned} \tag{8.3}$$

with the convention that $\tau_j^{(r)}$ is infinite if there is no $n > \tau_j^{(r-1)}$ for which $X_n = j$. Write

$$\rho_{ji} = P_j(X_n = i \text{ for some } n \geqslant 1) = P_j(\tau_i^{(1)} < \infty). \tag{8.4}$$

Using the strong Markov property (Theorem 4.1) we get

$$\begin{aligned} P_j(\tau_i^{(r)} < \infty) &= P_j(\tau_i^{(r-1)} < \infty \text{ and } X_{\tau_i^{(r-1)}+n} = i \text{ for some } n \geqslant 1) \\ &= E_j(\mathbf{1}_{\{\tau_i^{(r-1)} < \infty\}} P_{X_{\tau_i^{(r-1)}}}(X_n = i \text{ for some } n \geqslant 1)) \\ &= E_j(\mathbf{1}_{\{\tau_i^{(r-1)} < \infty\}} P_i(X_n = i \text{ for some } n \geqslant 1)) \\ &= E_j(\mathbf{1}_{\{\tau_i^{(r-1)} < \infty\}} \rho_{ii}) = P_j(\tau_i^{(r-1)} < \infty)\rho_{ii}. \end{aligned} \tag{8.5}$$

Therefore, by iteration,

$$P_j(\tau_i^{(r)} < \infty) = P_j(\tau_i^{(1)} < \infty)\rho_{ii}^{r-1} = \rho_{ji}\rho_{ii}^{r-1} \qquad (r = 2, 3, \ldots). \tag{8.6}$$

In particular, with $i = j$,

$$P_j(\tau_j^{(r)} < \infty) = \rho_{jj}^r \qquad (r = 1, 2, 3, \ldots). \tag{8.7}$$

Now

$$P_j(X_n = j \text{ for infinitely many } n) = P_j(\tau_j^{(r)} < \infty \text{ for all } r)$$

$$= \lim_{r\to\infty} P_j(\tau_j^{(r)} < \infty) = \begin{cases} 1 & \text{if } \rho_{jj} = 1 \\ 0 & \text{if } \rho_{jj} < 1. \end{cases} \tag{8.8}$$

Further, write $N(j)$ for the *number of visits to the state j* by the Markov chain $\{X_n\}$, and denote its expected value by

$$G(i, j) = E_i N(j). \tag{8.9}$$

Now by (8.6) and summation by parts (Exercise 1), if $i \neq j$ then

$$E_i N(j) = \sum_{r=0}^{\infty} P_i(N(j) > r) = \sum_{r=0}^{\infty} P_i(\tau_j^{(r+1)} < \infty) = \rho_{ij} \sum_{r=0}^{\infty} \rho_{jj}^r \tag{8.10}$$

so that, if $i \neq j$,

$$G(i, j) = \begin{cases} 0 & \text{if } i \not\to j, \text{ i.e., if } \rho_{ij} = 0, \\ \rho_{ij}/(1 - \rho_{jj}) & \text{if } i \to j \text{ and } \rho_{jj} < 1, \\ \infty & \text{if } i \to j \text{ and } \rho_{jj} = 1. \end{cases} \tag{8.11}$$

This calculation provides two useful characterizations of recurrence; one is in terms of the long-run expected number of returns and the other in terms of the probability of eventual return.

Proposition 8.1

(a) Every state is either recurrent or transient. A state j is recurrent iff $\rho_{jj} = 1$ iff $G(j, j) = \infty$, and transient iff $\rho_{jj} < 1$ iff $G(j, j) < \infty$.

(b) If j is recurrent, $j \to i$, then i is recurrent, and $\rho_{ji} = \rho_{ij} = 1$. Thus, recurrence (or transience) is a class property. In particular, if all states communicate with each other, then either they are all recurrent, or they are all transient.

(c) Let j be recurrent, and $S(j) := \{i \in S : j \to i\}$ be the class of states which communicate with j. Let $\bar{\pi}$ be a probability distribution on $S(j)$. Then

$$P_{\bar{\pi}}(X_n \text{ visits every state in } S(j) \text{ i.o.}) = 1. \tag{8.12}$$

Proof. Part (a) follows from (8.8), (8.11). For part (b), suppose j is recurrent and $j \to i$ $(i \neq j)$. Let A_r denote the event that the Markov chain visits i between the rth and $(r + 1)$st visits to state j. Then under P_j, A_r $(r \geqslant 0)$ are independent events and have the same probability θ, say. Now $\theta > 0$. For if $\theta = 0$, then $P_j(X_n = i \text{ for some } n \geqslant 1) = P_j(\bigcup_{r \geqslant 0} A_r) = 0$, contradicting $j \to i$. It now follows from the second half of the Borel–Cantelli Lemma (Chapter 0, Lemma 6.1) that $P_j(A_r \text{ i.o.}) = 1$. This implies $G(j, i) = \infty$. Interchanging i and j in (8.11) one then obtains $\rho_{ii} = 1$. Hence i is recurrent. Also, $\rho_{ji} \geqslant P_j(A_r \text{ i.o.}) = 1$. By the same argument, $\rho_{ij} = 1$.

To prove part (c) use part (b) to get

$$P_{\bar{\pi}}(X_n \text{ visits } i \text{ i.o.}) = \sum_{k \in S(j)} \bar{\pi}_k P_k(X_n \text{ visits } i \text{ i.o.}) = \sum_{k \in S(j)} \bar{\pi}_k = 1. \quad (8.13)$$

Hence

$$P_{\bar{\pi}}\left(\bigcap_{i \in S(j)} \{X_n \text{ visits } i \text{ i.o.}\} \right) = 1. \quad (8.14)$$

■

Note that $G(i, i) = 1/(1 - \rho_{ii})$, i.e., replace ρ_{ij} by 1 in (8.10).

For the simple random walk with $p > \frac{1}{2}$, one has (see (3.9), (3.10) of Chapter I),

$$G(i, j) = \begin{cases} \left(\dfrac{q}{p}\right)^{i-j} \Big/ (1 - 2p) & \text{for } i > j, \\ 1/(1 - 2p) & \text{for } i \leqslant j. \end{cases} \quad (8.15)$$

Proposition (8.1) shows that the difference between recurrence and transience is quite dramatic. If j is recurrent, then $P_j(N(j) = \infty) = 1$. If j is transient, not only is it true that $P_j(N(j) < \infty) = 1$, but also $E_j(N(j)) < \infty$. Note also that every inessential state is transient (Exercise 7).

Example 1. (*Random Rounding*). A real number $x \geqslant 0$ is *truncated* to its (greatest) integer part by the function $[x] := \max\{n \in \mathbb{Z}: n \leqslant x\}$. On the other hand, $[x + 0.5]$ is the value of x *rounded-off* to the nearest whole number. While both operations are quite common in digital filters, electrical engineers have also found it useful to use *random switchings* between rounding and truncation in the hardware design of certain recursive digital multipliers. In random rounding, one applies the function $[x + u]$, where u is randomly selected from $[0, 1)$, to digitize. The objective in such designs is to remove spurious fixed points and limit cycles (theoretical complement 8.1).

For the underlying mathematical ideas, first consider the simple one-dimensional digital recursion (with deterministic round-off), $x_{n+1} = [ax_n + 0.5]$, $n = 0, 1, 2, \ldots$, where $|a| < 1$ is a fixed parameter. Note that $x_0 = 0$ is always a fixed point; however, there can be others in $\mathbb{Z}$ (depending on a). For example, if $a = \frac{1}{2}$, then $x_0 = 1$ is another. Let $U_1, U_2, \ldots$ be an i.i.d. sequence of random variables uniformly distributed on $[0, 1)$ and consider $X_{n+1} = [aX_n + U_{n+1}]$, $n \geqslant 0$ on the state space $\mathbb{Z}$. Again $X_0 = 0$ is a fixed point (*absorbing* state). However, for this case, as will now be shown, $X_n \to 0$ *as* $n \to \infty$ *with probability 1*. To see this, first note that 0 is an *essential* state and is *accessible* from any other state x. That 0 is accessible in the case $0 < a < 1$ goes as follows. For $x > 0$, $[ax + u] = [ax] \leqslant ax < x$ if $u \in [0, 1 - (ax - [ax]))$. If $x < 0$, $ax \notin \mathbb{Z}$, $|[ax + u]| = |[ax] + 1| \leqslant a|x|$ if $u \in (1 + ([ax] - ax), 1)$. In either case, these events have positive probability. If $ax \in \mathbb{Z}$, then $|[ax + u]| = |ax|$. Since 0 is absorbing, all states $x \neq 0$ must, therefore, be *inessential*, and thus *transient*. To obtain convergence, simply note that the state space of the process started at x is finite since, with probability 1, the process remains in the interval $[-|x|, |x|]$. The *finiteness* is critical for this argument, as one can readily see by considering for contrast the case of the simple asymmetric ($p > \frac{1}{2}$) random walk on the nonnegative integers with absorbing boundary at 0.

Consider now the k-dimensional problem $\mathbf{X}_{n+1} = [\mathbf{A}\mathbf{X}_n + \mathbf{U}_{n+1}]$, $n = 0, 1, 2, \ldots$, where $\mathbf{A}$ is a $k \times k$ real matrix and $\{\mathbf{U}_n\}$ is an i.i.d. sequence of random vectors uniformly distributed over the k-dimensional cube, $[0, 1)^k$ and $[\ ]$ is defined componentwise, i.e., $[\mathbf{x}] = ([x_1], \ldots, [x_k])$, $\mathbf{x} \in \mathbb{R}^k$. It is convenient to use the *norm* $\|\ \|_0$ defined by $\|\mathbf{x}\|_0 := \max\{|x_1|, \ldots, |x_k|\}$. The ball $B_r(\mathbf{0}) := \{\mathbf{x}\colon \|\mathbf{x}\|_0 \leqslant r\}$ (of radius r centered at $\mathbf{0}$) for this norm is the square of side length $2r$ centered at 0. Assume $\|\mathbf{A}\mathbf{x}\|_0 < \|\mathbf{x}\|_0$, for all $\mathbf{x} \neq \mathbf{0}$ (i.e., $\|\mathbf{A}\| := \sup_{\mathbf{x} \neq \mathbf{0}} \|\mathbf{A}\mathbf{x}\|_0 / \|\mathbf{x}\|_0 < 1$) as our *stability condition*. Once again we wish to show that $\mathbf{X}_n \to \mathbf{0}$ as $n \to \infty$ with probability 1. As in the one-dimensional case, $\mathbf{0}$ is an (absorbing) essential state that is accessible from every other state $\mathbf{x}$ because there is a subset $N(\mathbf{x})$ of $[0, 1)^k$ having positive volume such that for each $\mathbf{u} \in N(\mathbf{x})$, one has $\|[\mathbf{A}\mathbf{x} + \mathbf{u}]\|_0 \leqslant \|\mathbf{A}\mathbf{x}\|_0 < \|\mathbf{x}\|_0$. So each state $\mathbf{x} \neq \mathbf{0}$ is inessential and thus transient. The result follows since the process starting at $\mathbf{x} \in \mathbb{Z}^k$ does not escape $B_{\|\mathbf{x}\|_0}(\mathbf{0})$, since $\|[\mathbf{A}\mathbf{x} + \mathbf{u}]\|_0 \leqslant \|\mathbf{A}\mathbf{x}\|_0 + 1 < \|\mathbf{x}\|_0 + 1$ and, since $\|[\mathbf{A}\mathbf{x} + \mathbf{u}]\|_0$ and $\|\mathbf{x}\|_0$ are both integers, $\|[\mathbf{A}\mathbf{x} + \mathbf{u}]\|_0 \leqslant \|\mathbf{x}\|_0$.

Linear models $\mathbf{X}_{n+1} = \mathbf{A}\mathbf{X}_n + \boldsymbol{\varepsilon}_{n+1}$, with $\{\boldsymbol{\varepsilon}_n\}$ a sequence of i.i.d. random vectors, are systematically analyzed in Section 13.

9 THE LAW OF LARGE NUMBERS AND INVARIANT DISTRIBUTIONS FOR MARKOV CHAINS

The existence of an invariant distribution is intimately connected with the limiting frequency of returns to recurrent states. For a process in steady state one expects the equilibrium probability of a state j to coincide with the fraction of time spent on the average by the process in state j. To this effect a major goal of this section is to obtain the invariant distribution as a consequence of a (strong) law of large numbers.

Assume from now on, unless otherwise specified, that *S comprises a single class of (communicating) recurrent states under* $\mathbf{p}$ *for the Markov chain* $\{X_n\}$.

Let f be a real-valued function on S, and define the cumulative sums

$$S_n = \sum_{m=0}^{n} f(X_m) \qquad (n = 1, 2, \ldots). \tag{9.1}$$

For example, if $f(i) = 1$ for $i = j$ and $f(i) = 0$ for $i \neq j$, then $S_n/(n+1)$ is the average number of visits to j in time 0 to n. As in Section 8, let $\tau_j^{(r)}$ denote the time of the rth visit to state j. Write the contribution to the sum S_n from the rth *block of time* $(\tau_j^{(r)}, \tau_j^{(r+1)}]$ as,

$$Z_r = \sum_{m=\tau_j^{(r)}+1}^{\tau_j^{(r+1)}} f(X_m) \qquad (r = 0, 1, 2, \ldots). \tag{9.2}$$

By the strong Markov property, the conditional distribution of the process $\{X_{\tau_j^{(r)}}, X_{\tau_j^{(r)}+1}, \ldots, X_{\tau_j^{(r)}+n}, \ldots\}$, given the past up to time $\tau_j^{(r)}$, is P_j, which is the distribution of $\{X_0, X_1, \ldots, X_n, \ldots\}$ under $X_0 = j$. Hence, the conditional distribution of Z_r given the process up to time $\tau_j^{(r)}$ is that of $Z_0 = f(X_1) + \cdots + f(X_{\tau_j^{(1)}})$, given $X_0 = j$. This conditional distribution does not change with the values of $X_0, X_1, \ldots, X_{\tau_j^{(r)}}(=j), \tau_j^{(r)}$. Hence, Z_r is independent of all events that are determined by the process up to time $\tau_j^{(r)}$. In particular, Z_r is independent of $Z_1, \ldots, Z_{r-1}$. Thus, we have the following result.

Proposition 9.1. The sequence of random variables $\{Z_1, Z_2, \ldots\}$ is i.i.d., no matter what the initial distribution of $\{X_n: n = 0, 1, 2, \ldots\}$ may be.

This decomposition will be referred to as the *renewal decomposition*. The strong law of large numbers now provides that, with probability 1,

$$\lim_{r\to\infty} \frac{1}{r} \sum_{s=1}^{r} Z_s = EZ_1, \tag{9.3}$$

provided that $E|Z_1| < \infty$. In what follows we will make the stronger assumption that

$$E \sum_{m=\tau_j^{(1)}+1}^{\tau_j^{(2)}} |f(X_m)| < \infty. \tag{9.4}$$

Write N_n *for the number of visits to state* j *by time* n. Now,

$$N_n = \max\{r \geqslant 0: \tau_j^{(r)} \leqslant n\}. \tag{9.5}$$

Then (see Figure 9.1),

$$S_n = \sum_{m=0}^{\tau_j^{(1)}} f(X_m) + \sum_{r=1}^{N_n} Z_r - \sum_{n+1}^{\tau_j^{(N_n+1)}} f(X_m). \tag{9.6}$$

For each sample path there are a finite number, $\tau_j^{(1)} + 1$, of summands in the first sum on the right side, except for a set of sample paths having probability zero. Therefore,

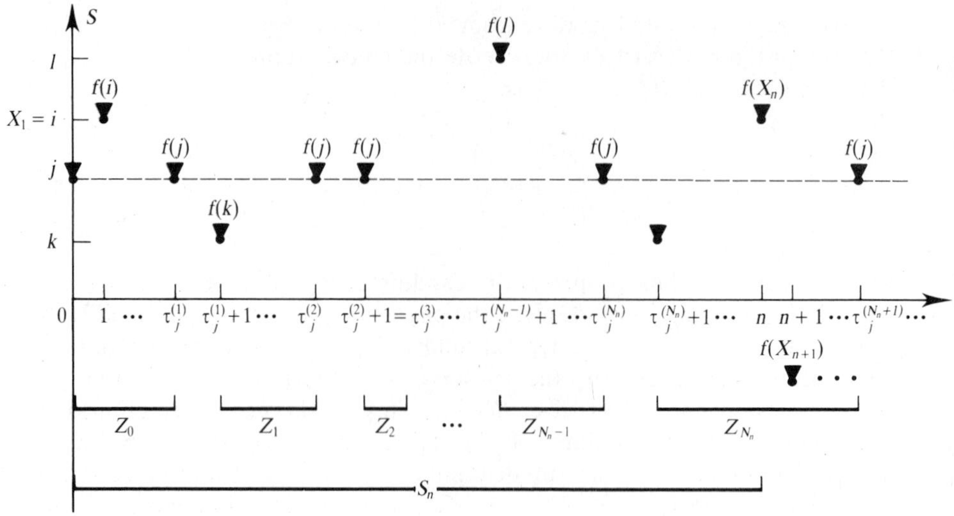

Figure 9.1

$$\lim_{n\to\infty} \frac{1}{n} \sum_{m=0}^{\tau_j^{(1)}} f(X_m) = 0, \qquad \text{with probability } 1. \tag{9.7}$$

The last sum on the right side of (9.6) has at most $\tau_j^{(N_n+1)} - \tau_j^{(N_n)}$ summands, this number being the time between the last visit to j by time n and the next visit to j. Although this sum depends on n, under the condition (9.4) we still have that (Exercise 1)

$$\left|\frac{1}{n} \sum_{m=n}^{\tau_j^{(N_n+1)}} f(X_m)\right| \leqslant \frac{1}{n} \sum_{m=\tau_j^{(N_n)}+1}^{\tau_j^{(N_n+1)}} |f(X_m)| \to 0 \quad \text{a.s.} \qquad \text{as } n\to\infty. \tag{9.8}$$

Therefore,

$$\frac{S_n}{n} = \frac{1}{n} \sum_{r=1}^{N_n} Z_r + R_n = \left(\frac{N_n}{n}\right) \frac{1}{N_n} \sum_{r=1}^{N_n} Z_r + R_n, \tag{9.9}$$

where $R_n \to 0$ as $n \to \infty$ with probability 1 under (9.4). Also, for each sample path outside a set of probability 0, $N_n \to \infty$ as $n \to \infty$ and therefore by (9.3) (taking limit over a subsequence)

$$\lim_{n\to\infty} \frac{1}{N_n} \sum_{r=1}^{N_n} Z_r = EZ_1 \tag{9.10}$$

if (9.4) holds. Now, replacing f by the constant function $f \equiv 1$ in (9.10), we have

$$\lim_{n\to\infty} \frac{\tau_j^{(N_n+1)} - \tau_j^{(1)}}{N_n} = E(\tau_j^{(2)} - \tau_j^{(1)}), \tag{9.11}$$

assuming that the right side is finite. Since

$$n - \tau_j^{(N_n)} \leqslant \tau_j^{(N_n+1)} - \tau_j^{(N_n)},$$

which is negligible compared to N_n as $n \to \infty$, one gets (Exercise 2)

$$\lim_{n\to\infty} \frac{n}{N_n} = E(\tau_j^{(2)} - \tau_j^{(1)}). \tag{9.12}$$

Note that the right side, $E(\tau_j^{(2)} - \tau_j^{(1)})$, is the average recurrence time of $j(= E_j\tau_j^{(1)})$, and the left side is the reciprocal of the asymptotic proportion of time spent at j.

Definition 9.1. A state j *is positive recurrent if*

$$E_j\tau_j^{(1)} < \infty. \tag{9.13}$$

Combining (9.9)–(9.11) gives the following result. Note that positive recurrence is a class property (see Theorem 9.4 and Exercise 4).

Theorem 9.2. Suppose j is a positive recurrent state under $\mathbf{p}$ and that f is a real-valued function on S such that

$$E_j\{|f(X_1)| + \cdots + |f(X_{\tau_j^{(1)}})|\} < \infty. \tag{9.14}$$

Then the following are true.

(a) With P_j-probability 1,

$$\lim_{n\to\infty} \frac{1}{n} \sum_{m=0}^{n} f(X_m) = E_j(f(X_1) + \cdots + f(X_{\tau_j^{(1)}}))/E_j\tau_j^{(1)}. \tag{9.15}$$

(b) If S comprises a single class of essential states (irreducible) that are all positive recurrent, then (9.15) holds with probability 1 regardless of the initial distribution.

Corollary 9.3. If S consists of a single positive recurrent class under $\mathbf{p}$, then for all $i, j \in S$,

$$\lim_{n\to\infty} \frac{1}{n} \sum_{m=1}^{n} p_{ij}^{(m)} = \frac{1}{E_j\tau_j^{(1)}}. \tag{9.16}$$

A WORD ON PROOFS. The calculations of limits in the following proofs require the use of certain standard results from analysis such as Lebesgue's Dominated Convergence, Fatou's Lemma, and Fubini's theorem. These results are given in Chapter 0; however, the reader unfamiliar with these theorems may proceed formally through the calculations to gain acquaintance with the statements.

Proof. Take f to be the indicator function of the state j,

$$\begin{aligned} f(j) &= 1 \\ f(k) &= 0 \end{aligned} \qquad \text{for all } k \neq j. \tag{9.17}$$

Then $Z_r \equiv 1$ for all $r = 0, 1, 2, \ldots$ since there is only *one* visit to state j in $(\tau_j^{(r)}, \tau_j^{(r+1)}]$. Hence, taking expectation on both sides of (9.15) under the initial state i, one gets (9.16) after interchanging the order of the expectation and the limit. This is permissible by Lebesgue's Dominated Convergence Theorem since $|(n+1)^{-1} \sum_0^n f(X_m)| \leqslant 1$. ■

It will now be shown that the quantities defined by

$$\pi_j = (E_j \tau_j^{(1)})^{-1} \qquad (j \in S) \tag{9.18}$$

constitute an invariant distribution if the states are all positive recurrent and communicate with each other. Let us first show that $\sum_j \pi_j = 1$ in this case. For this, introduce the random variables

$$T_i^{(r)} = \#\{m \in (\tau_j^{(r)}, \tau_j^{(r+1)}]: X_m = i\} \qquad (r = 0, 1, 2, \ldots), \tag{9.19}$$

i.e., *$T_i^{(r)}$ is the amount of time the Markov chain spends in the state i between the rth and* $(r+1)$*th passages to j*. The sequence $\{T_i^{(r)}: r = 1, 2, \ldots\}$ is i.i.d. by the strong Markov property. Write

$$\theta_j(i) = E_j T_i^{(r)}. \tag{9.20}$$

Then,

$$\sum_{i \in S} \theta_j(i) = \sum_i E_j T_i^{(1)} = E_j\left(\sum_i T_i^{(1)}\right) = E_j(\tau_j^{(2)} - \tau_j^{(1)}) = \frac{1}{\pi_j}. \tag{9.21}$$

Also, taking f to be the indicator function of $\{i\}$ and replacing j by i in Theorem 9.2, one obtains

$$\lim_{n \to \infty} \frac{\#\{m \leqslant n: X_m = i\}}{n} = \pi_i \qquad \text{with probability } 1. \tag{9.22}$$

On the other hand, by the strong law of large numbers applied to

$\{T_i^{(r)}: r = 1, 2, \ldots\}$, and by (9.12) and (9.18), the limit on the left side also equals

$$\lim_{n\to\infty} \frac{\sum_{r=1}^{N_n} T_i^{(r)}}{n} = \lim_{n\to\infty} \frac{N_n}{n}\left(\sum_{r=1}^{N_n} \frac{T_i^{(r)}}{N_n}\right) = \pi_j \theta_j(i). \tag{9.23}$$

Hence,

$$\pi_i = \pi_j \theta_j(i). \tag{9.24}$$

Adding over i, one has using (9.21),

$$\sum_i \pi_i = \pi_j \sum_i \theta_j(i) = \frac{\pi_j}{\pi_j} = 1. \tag{9.25}$$

By Scheffé's Theorem (Theorem 3.7 of Chapter 0) and (9.16),

$$\sum_{i\in S} \left| \frac{1}{n} \sum_{m=1}^{n} p_{ji}^{(m)} - \pi_i \right| \to 0 \qquad \text{as } n \to \infty. \tag{9.26}$$

Hence,

$$\left| \sum_{i\in S} \pi_i p_{ij} - \sum_{i\in S} \left(\frac{1}{n} \sum_{m=1}^{n} p_{ji}^{(m)} \right) p_{ij} \right| \leqslant \sum_{i\in S} \left| \pi_i - \frac{1}{n} \sum_{m=1}^{n} p_{ji}^{(m)} \right| \to 0 \qquad \text{as } n \to \infty. \tag{9.27}$$

Therefore,

$$\begin{aligned} \sum_{i\in S} \pi_i p_{ij} &= \lim_{n\to\infty} \sum_{i\in S} \frac{1}{n} \left(\sum_{m=1}^{n} p_{ji}^{(m)} \right) p_{ij} \\ &= \lim_{n\to\infty} \frac{1}{n} \sum_{m=1}^{n} \sum_{i\in S} p_{ji}^{(m)} p_{ij} \\ &= \lim_{n\to\infty} \frac{1}{n} \sum_{m=1}^{n} p_{jj}^{(m+1)} = \pi_j. \end{aligned} \tag{9.28}$$

In other words, $\boldsymbol{\pi}$ is invariant.

Finally, let us show that for a positive recurrent chain the probability distribution $\boldsymbol{\pi}$ above is the only invariant distribution. For this let $\bar{\boldsymbol{\pi}}$ be another invariant distribution. Then

$$\bar{\pi}_j = \sum_{i\in S} \bar{\pi}_i p_{ij}^{(m)} \qquad (m = 1, 2, \ldots). \tag{9.29}$$

Averaging over $m = 1, 2, \ldots, n$, one gets

$$\bar{\pi}_j = \sum_{i \in S} \bar{\pi}_i \left(\frac{1}{n} \sum_{m=1}^{n} p_{ij}^{(m)} \right). \tag{9.30}$$

The right side converges to $\sum_i \bar{\pi}_i \pi_j = \pi_j$ by Lebesgue's Dominated Convergence Theorem. Hence $\pi_j = \bar{\pi}_j$.

We have proved above that if all states of S are positive recurrent and communicate with each other, then there exists a unique invariant distribution given by (9.18). The next problem is to determine what happens if S comprises a single class of recurrent states that are *not* positive recurrent.

Definition 9.2. A recurrent state j is said to be *null recurrent* if

$$E_j \tau_j^{(1)} = \infty. \tag{9.31}$$

If j is a null recurrent state and $i \leftrightarrow j$, then for the Markov chain having initial state i the sequence $\{Z_r : r = 1, 2, \ldots\}$ defined by (9.2) with $f \equiv 1$ is still an i.i.d. sequence of random variables, but the common mean is infinity. It follows (Exercise 3) that, with P_i-probability 1,

$$\lim_{n \to \infty} \frac{\tau_j^{(N_n)}}{N_n} = \infty.$$

Since $n \geqslant \tau_j^{(N_n)}$, we have

$$\lim_{n \to \infty} \frac{n+1}{N_n} = \infty,$$

and, therefore,

$$\lim_{n \to \infty} \frac{N_n}{n+1} = 0 \qquad \text{with } P_i\text{-probability } 1. \tag{9.32}$$

Since $0 \leqslant N_n/(n+1) \leqslant 1$ for all n, Lebesgue's Dominated Convergence Theorem applied to (9.32) yields

$$\lim_{n \to \infty} E_i\left(\frac{N_n}{n+1} \right) = 0. \tag{9.33}$$

But

$$E_i N_n = E_i\left(\sum_{m=1}^{n} \mathbf{1}_{\{X_m = j\}} \right) = \sum_{m=1}^{n} p_{ij}^{(m)}, \tag{9.34}$$

and (9.33), (9.34) lead to

$$\lim_{n\to\infty} \frac{1}{n+1} \sum_{m=1}^{n} p_{ij}^{(m)} = 0 \tag{9.35}$$

if j is null-recurrent and $j \to i$ (or, equivalently, $j \leftrightarrow i$). In other words, (9.16) holds if S comprises either a single class of positive recurrent states, or a single class of null recurrent states.

In the case that j is transient, (8.11) implies that $G(i, j) < \infty$, i.e.,

$$\sum_{m=0}^{\infty} p_{ij}^{(m)} < \infty.$$

In particular, therefore,

$$\lim_{n\to\infty} p_{ij}^{(n)} = 0 \qquad (i \in S), \tag{9.36}$$

if j is a transient state.

The main results of this section may be summarized as follows.

Theorem 9.4. Assume that all states communicate with each other. Then one has the following results.

(a) Either all states are recurrent, or all states are transient.

(b) If all states are recurrent, then they are either all positive recurrent, or all null recurrent.

(c) There exists an invariant distribution if and only if all states are positive recurrent. Moreover, in the positive recurrent case, the invariant distribution $\boldsymbol{\pi}$ is unique and is given by

$$\pi_j = (E_j \tau_j^{(1)})^{-1} \qquad (j \in S). \tag{9.37}$$

(d) In case the states are positive recurrent, no matter what the initial distribution $\boldsymbol{\mu}$, if $E_\pi |f(X_1)| < \infty$, then

$$\lim_{n\to\infty} \frac{1}{n} \sum_{m=1}^{n} f(X_m) = \sum_{i\in S} \pi_i f(\text{i}) = E_\pi f(X_1) \tag{9.38}$$

with $P_{\boldsymbol{\mu}}$-probability 1.

Proof. Part (a) follows from Proposition 8.1(a).

To prove part (b), assume that j is positive recurrent and let $i \neq j$. Since $T_i^{(r)} \leqslant \tau_j^{(r+1)} - \tau_j^{(r)}$, one has $ET_i^{(r)} < \infty$. Hence (9.23) holds. Also, (9.22) holds with $\pi_i > 0$ if i is positive recurrent, and $\pi_i = 0$ if i is null recurrent (use (9.32) in the latter case). Thus (9.24) holds. Now $\theta_j(i) > 0$; for otherwise $T_i^{(r)} = 0$ with probability 1 for all $r \geqslant 0$, implying $\rho_{ji} = 0$, which is ruled out by the assumption. The relation (9.24) implies $\pi_i > 0$, since $\pi_j > 0$ and $\theta_j(i) > 0$. Therefore, all states are positive recurrent.

For part (c), it has been proved above that there exists a unique invariant probability distribution $\boldsymbol{\pi}$ given by (9.35) if all states are positive recurrent. Conversely, suppose $\bar{\boldsymbol{\pi}}$ is an invariant probability distribution. We need to show that all states are positive recurrent. This is done by elimination of other possibilities. If the states are transient, then (9.36) holds; using this in (9.29) (or (9.30)) one would get $\bar{\pi}_j = 0$ for all j, which is a contradiction. Similarly, null recurrence implies, by (9.30) and (9.35), that $\bar{\pi}_j = 0$ for all j. Therefore, the states are all positive recurrent. Part (d) will follow from Theorem 9.2 if:

(i) The hypothesis (9.14) holds whenever

$$\sum_{i \in S} \pi_i |f(i)| < \infty, \tag{9.39}$$

and

(ii)
$$E_j Z_0 = E_j(f(X_1) + \cdots + f(X_{\tau_j^{(1)}})) = \frac{1}{\pi_j} \sum_{i \in S} \pi_i f(i). \tag{9.40}$$

To verify (i) and (ii), first note that by the definition of $T_i^{(1)}$,

$$\sum_{m=\tau_j^{(1)}+1}^{\tau_j^{(2)}} |f(X_m)| = \sum_{i \in S} |f(i)| \, T_i^{(1)}. \tag{9.41}$$

Since $ET_i^{(1)} = \theta_j(i) = \pi_i/\pi_j$ (see Eq. 9.24), taking expectations in (9.41) yields

$$E\left(\sum_{m=\tau_j^{(1)}+1}^{\tau_j^{(2)}} |f(X_m)| \right) = E\left(\sum_{i \in S} |f(i)| \, T_i^{(1)} \right) = \sum_{i \in S} |f(i)| \frac{\pi_i}{\pi_j}. \tag{9.42}$$

The last equality follows upon interchanging the orders of summation and expectation, which by Fubini's theorem is always permissible if the summands are nonnegative. Therefore (9.14) follows from (9.39). Now as in (9.42),

$$\begin{aligned} E_j Z_0 = EZ_1 &= E\left(\sum_{m=\tau_j^{(1)}+1}^{\tau_j^{(2)}} f(X_m) \right) = E\left(\sum_{i \in S} f(i) T_i^{(1)} \right) \\ &= \sum_{i \in S} f(i) ET_i^{(1)} = \sum_{i \in S} f(i) \frac{\pi_i}{\pi_j}, \end{aligned} \tag{9.43}$$

where this time the interchange of the orders of summation and expectation is justified again using Fubini's theorem by finiteness of the double "integral." ■

If the assumption that "all states communicate with each other" in Theorem 9.4 is dropped, then S can be decomposed into a set $\mathscr{I}$ of inessential states and (disjoint) classes $S_1, S_2, \ldots, S_t$ of essential states. The transition probability matrix $\mathbf{p}$ may be restricted to each one of the classes $S_1, \ldots, S_t$ and the conclusions of Theorem 9.2 will hold individually for each class. If more than one of these classes is positive recurrent, then more than one invariant distribution exist, and they are supported on disjoint sets. Since any convex combination of invariant distributions is again invariant, an infinity of invariant distributions exist in this case. The following result takes care of the set $\mathscr{I}$ of inessential states in this connection (also see Exercise 4).

Proposition 9.5

(a) If j is inessential then it is transient.

(b) Every invariant distribution assigns zero probability to inessential, transient, and null recurrent states.

Proof. (a) If j is inessential then there exist $i \in S$ and $m \geqslant 1$ such that

$$p_{ji}^{(m)} > 0 \qquad \text{and} \qquad p_{ij}^{(n)} = 0 \qquad \text{for all } n \geqslant 1. \tag{9.44}$$

Hence

$$\begin{aligned} P_j(N(j) < \infty) &\geqslant P_j(X_m = i,\ X_n \neq j \text{ for } n > m) \\ &= p_{ji}^{(m)} P_i(X_n \neq j \text{ for } n > 0) = p_{ji}^{(m)} > 0. \end{aligned} \tag{9.45}$$

By Proposition 8.1, j is transient, since (9.45) says j is not recurrent.

(b) Next use (9.36), (9.35) and (9.30) and argue as in the proof of part (c) of Theorem 9.2 to conclude that $\pi_j = 0$ if j is either transient or null recurrent. ■

Corollary 9.6. If S is finite, then there exists at least one positive recurrent state, and therefore at least one invariant distribution $\boldsymbol{\pi}$. This invariant distribution is unique if and only if all positive recurrent states communicate.

Proof. Suppose it possible that all states are either transient or null recurrent. Then

$$\lim_{n \to \infty} \frac{1}{n+1} \sum_{m=0}^{n} p_{ij}^{(m)} = 0 \qquad \text{for all } i, j \in S. \tag{9.46}$$

Since $(n+1)^{-1} \sum_{m=1}^{n} p_{ij}^{(m)} \leqslant 1$ for all j, and there are only finitely many states j, by Lebesgue's Dominated Convergence Theorem,

$$\sum_{j\in S}\lim_{n\to\infty}\left(\frac{1}{n+1}\sum_{m=1}^{n}p_{ij}^{(m)}\right)=\lim_{n\to\infty}\sum_{j\in S}\left(\frac{1}{n+1}\sum_{m=0}^{n}p_{ij}^{(m)}\right)$$
$$=\lim_{n\to\infty}\left(\frac{1}{n+1}\sum_{m=0}^{n}\sum_{j\in S}p_{ij}^{(m)}\right)$$
$$=\lim_{n\to\infty}\left(\frac{1}{n+1}\sum_{m=0}^{n}1\right)=\lim_{n\to\infty}\frac{n+1}{n+1}=1. \quad (9.47)$$

But the first term in (9.47) is zero by (9.46). We have reached a contradiction. Thus, there exists at least one positive recurrent state. The rest follows from Theorem 9.2 and the remark following its proof. ■

10 THE CENTRAL LIMIT THEOREM FOR MARKOV CHAINS

The same method as used in Section 9 to obtain the law of large numbers may be used to derive a *central limit theorem* for $S_n = \sum_{m=0}^{n} f(X_m)$, where f is a real-valued function on the state space S. Write

$$\mu = E_\pi f(X_0) = \sum_{i\in S}\pi_i f(i), \tag{10.1}$$

and assume that, for $Z_r = \sum_{m=\tau_j^{(r)}+1}^{\tau_j^{(r+1)}} f(X_m)$, $r = 0, 1, 2, \ldots$,

$$E_j(Z_0 - E_j Z_0)^2 < \infty. \tag{10.2}$$

Now replace f by $\bar{f} = f - \mu$, and write

$$\bar{S}_n = \sum_{m=0}^{n}\bar{f}(X_m) = \sum_{m=0}^{n}(f(X_m)-\mu),$$
$$\bar{Z}_r = \sum_{m=\tau_j^{(r)}+1}^{\tau_j^{(r)}}\bar{f}(X_m) \qquad (r = 0, 1, 2, \ldots). \tag{10.3}$$

Then by (9.40),

$$E_j\bar{Z}_r = (E_j\tau_j^{(1)})E_\pi\bar{f}(X_0) = 0 \qquad (r = 0, 1, 2, \ldots). \tag{10.4}$$

Thus $\{\bar{Z}_r: r = 1, 2, \ldots\}$ is an i.i.d. sequence with mean zero and finite variance

$$\sigma^2 = E_j\bar{Z}_0^2. \tag{10.5}$$

Now apply the classical central limit theorem to this sequence. As $r \to \infty$, $(1/\sqrt{r})\sum_{k=1}^{r}\bar{Z}_k$ converges in distribution to the Gaussian law with mean zero and variance σ^2. Now express $\bar{S}_n$ as in (9.6) with f replaced by $\bar{f}$, S_n by $\bar{S}_n$, and

Z_r by $\bar{Z}_r$, to see that the limiting distribution of $(1/\sqrt{n})\bar{S}_n$ is the same as that of (Exercise 1)

$$\frac{1}{\sqrt{n}}\sum_{r=1}^{N_n}\bar{Z}_r = \left(\frac{N_n}{n}\right)^{1/2}\frac{1}{\sqrt{N_n}}\sum_{r=1}^{N_n}\bar{Z}_r. \tag{10.6}$$

We shall need an extension of the central limit theorem that applies to sums of random numbers of i.i.d. random variables. We can get such a result as an extension of Corollary 7.2 in Chapter 0 as follows.

Proposition 10.1. Let $\{X_j: j \geqslant 1\}$ be i.i.d., $EX_j = 0$, $0 < \sigma^2 := EX_j^2 < \infty$. Let $\{\nu_n: n \geqslant 1\}$ be a sequence of nonnegative integer-valued random variables with

$$\lim_{n\to\infty}\frac{\nu_n}{n} = \alpha \qquad \text{in probability} \tag{10.7}$$

for some constant $\alpha > 0$. Then $\sum_{j=1}^{\nu_n} X_j/\sqrt{\nu_n}$ converges in distribution to $N(0, \sigma^2)$.

Proof. Without loss of generality, let $\sigma = 1$. Write $S_n := X_1 + \cdots + X_n$. Choose $\varepsilon > 0$ arbitrarily. Then,

$$P(|S_{\nu_n} - S_{[n\alpha]}| \geqslant \varepsilon([n\alpha])^{1/2}) \leqslant P(|\nu_n - [n\alpha]| \geqslant \varepsilon^3[n\alpha]) + P\left(\max_{\{m:|m-[n\alpha]|<\varepsilon^3[n\alpha]\}} |S_m - S_{[n\alpha]}| \geqslant \varepsilon([n\alpha])^{1/2}\right). \tag{10.8}$$

The first term on the right goes to zero as $n \to \infty$, by (10.7). The second term is estimated by Kolmogorov's Maximal Inequality (Chapter I, Corollary 13.7), as being no more than

$$(\varepsilon([n\alpha])^{1/2})^{-2}(\varepsilon^3[n\alpha]) = \varepsilon. \tag{10.9}$$

This shows that

$$\frac{S_{\nu_n} - S_{[n\alpha]}}{([n\alpha])^{1/2}} \to 0 \qquad \text{in probability.} \tag{10.10}$$

Since $S_{[n\alpha]}/([n\alpha])^{1/2}$ converges in distribution to $N(0, 1)$, it follows from (10.10) that so does $S_{\nu_n}/([n\alpha])^{1/2}$. The desired convergence now follows from (10.7). ■

By Proposition 10.1, $N_n^{-1/2}\sum_{r=1}^{N_n}\bar{Z}_r$ is asymptotically Gaussian with mean zero and variance σ^2. Since N_n/n converges to $(E_j\tau_j^{(1)})^{-1}$, it follows that the expression in (10.6) is asymptotically Gaussian with mean zero and variance

$(E_j\tau_j^{(1)})^{-1}\sigma^2$. This is then the asymptotic distribution of $n^{-1/2}\bar{S}_n$. Moreover, defining, as in Chapter I,

$$W_n(t) = \frac{\bar{S}_{[nt]}}{\sqrt{n+1}}$$
$$\tilde{W}_n(t) = W_n(t) + \frac{1}{\sqrt{n+1}}(nt - [nt])X_{[nt]+1} \qquad (t \geqslant 0), \tag{10.11}$$

all the finite-dimensional distributions of $\{W_n(t)\}$, as well as $\{\tilde{W}_n(t)\}$, converge in distribution to those of Brownian motion with zero drift and diffusion coefficient

$$D = (E_j\tau_j^{(1)})^{-1}\sigma^2 \tag{10.12}$$

(Exercise 2). In fact, convergence of the full distribution may also be obtained by consideration of the above renewal argument. The precise form of the functional central limit theorem (FCLT) for Markov chains goes as follows (see theoretical complement 1).

Theorem 10.2. (*FCLT*). If S is a positive recurrent class of states and if (10.2) holds then, as $n \to \infty$, $W_n(1) = (n+1)^{-1/2}\bar{S}_n$ converges in distribution to a Gaussian law with mean zero and variance D given by (10.12) and (10.5). Moreover, the stochastic process $\{W_n(t)\}$ (or $\{\tilde{W}_n(t)\}$) converges in distribution to Brownian motion with zero drift and diffusion coefficient D.

Rather than using (10.12), a more convenient way to compute D is sometimes the following. Write E_π, Var_π, Cov_π for expectation, variance, and covariance, respectively, when the initial distribution is $\boldsymbol{\pi}$ (the unique invariant distribution). The following computation is straightforward. First write, for any two functions h, g that are square summable with respect to $\boldsymbol{\pi}$,

$$\langle h, g\rangle_\pi = \sum_{i \in S} h(i)g(i)\pi_i = E_\pi[h(X_0)g(X_0)]. \tag{10.13}$$

Then,

$$\begin{aligned}
\text{Var}_\pi((n+1)^{-1/2}\bar{S}_n) &= E_\pi\left(\sum_{m=0}^{n} \bar{f}(X_m)\right)^2 \Big/ (n+1) \\
&= E_\pi\left[\sum_{m=0}^{n} \bar{f}^2(X_m) + 2\sum_{m=1}^{n}\sum_{m'=0}^{m-1} \bar{f}(X_{m'})\bar{f}(X_m)\right] \Big/ (n+1) \\
&= \frac{1}{n+1}\sum_{m=0}^{n} E_\pi \bar{f}^2(X_m)
\end{aligned}$$

$$+\frac{2}{n+1}\sum_{m=1}^{n}\sum_{m'=0}^{m-1} E_\pi[\bar{f}(X_{m'})E_\pi(\bar{f}(X_m)\mid X_{m'})]$$

$$= E_\pi \bar{f}^2(X_0) + \frac{2}{n+1}\sum_{m=1}^{n}\sum_{m'=0}^{m-1} E_\pi[\bar{f}(X_{m'})(\mathbf{p}^{m-m'}\bar{f})(X_{m'})]$$

$$= E_\pi \bar{f}^2(X_0) + \frac{2}{n+1}\sum_{m=1}^{n}\sum_{m'=0}^{m-1} E_\pi[\bar{f}(X_0)(\mathbf{p}^{m-m'}\bar{f})(X_0)]$$

$$= E_\pi \bar{f}^2(X_0) + \frac{2}{n+1}\sum_{m=1}^{n}\sum_{m'=0}^{m-1} \langle \bar{f}, \mathbf{p}^{m-m'}\bar{f}\rangle_\pi$$

$$= E_\pi \bar{f}^2(X_0) + 2\left\langle \bar{f}, \frac{1}{n+1}\sum_{m=1}^{n}\sum_{k=1}^{m} \mathbf{p}^k\bar{f}\right\rangle_\pi \qquad (k = m - m') \tag{10.14}$$

Now assume that the limit

$$\gamma = \lim_{m\to\infty}\sum_{k=1}^{m}\langle \bar{f}, \mathbf{p}^k\bar{f}\rangle_\pi \tag{10.15}$$

exists and is finite. Then it follows from (10.14) that

$$D = \lim_{n\to\infty} \mathrm{Var}_\pi((n+1)^{-1/2}\bar{S}_n) = E_\pi \bar{f}^2(X_0) + 2\gamma = \sum_{j\in S}\pi_j \bar{f}^2(j) + 2\gamma. \tag{10.16}$$

Note that

$$\langle \bar{f}, \mathbf{p}^k\bar{f}\rangle_\pi = \mathrm{Cov}_\pi\{f(X_0), f(X_k)\}, \tag{10.17}$$

and

$$\sum_{k=1}^{m}\langle \bar{f}, \mathbf{p}^k\bar{f}\rangle_\pi = \sum_{k=1}^{m}\mathrm{Cov}\{f(X_0), f(X_k)\}.$$

The condition (10.15) that γ exists and is finite is the condition that the correlation decays to zero at a sufficiently rapid rate for time points k units apart as $k \to \infty$.

11 ABSORPTION PROBABILITIES

Suppose that $\mathbf{p}$ is a transition probability matrix for a Markov chain $\{X_n\}$ starting in state $i \in S$. Suppose that $j \in S$ is an *absorbing state* of $\mathbf{p}$, i.e., $p_{jj} = 1$, $p_{jk} = 0$ for $k \neq j$. Let τ_j denote the time required to reach j,

$$\tau_j = \inf\{n: X_n = j\}. \tag{11.1}$$

To calculate the distribution of τ_j, consider that

$$P_i(\tau_j > m) = \sum{}^* p_{ii_1} p_{i_1 i_2} \cdots p_{i_{m-1} i_m}, \tag{11.2}$$

where $\sum^*$ denotes summation over all m-tuples $(i_1, i_2, \dots, i_m)$ of elements from $S - \{j\}$. Now let $\mathbf{p}^0$ denote the matrix obtained by deleting the jth row and jth column from $\mathbf{p}$,

$$\mathbf{p}^0 = ((p_{ik}: i, k \in S - \{j\})). \tag{11.3}$$

Then, by definition of matrix multiplication, the calculation (11.2) may be expressed as

$$P_i(\tau_j > m) = \sum_k p_{ik}^{0(m)}, \tag{11.4}$$

and, therefore,

$$P_i(\tau_j = m) = \sum_k p_{ik}^{0(m-1)} - \sum_k p_{ik}^{0(m)}, \qquad j \neq i. \tag{11.5}$$

Observe that the above idea can be applied to the calculation of the first passage time to any state $j \in S$ or, for that matter, to any nonempty set A of states such that $i \notin A$. Moreover, it follows from Theorem 6.4 and its corollaries that the rate of absorption is, therefore, furnished by the *spectral radius* of $\mathbf{p}^0$. This will be amply demonstrated in examples of this section.

Proposition 11.1. Let $\mathbf{p}$ be a transition probability matrix for a Markov chain $\{X_n\}$ starting in state i. Let A be a nonempty subset of S, $i \notin A$. Let

$$\tau_A = \inf\{n \geqslant 0: X_n \in A\}. \tag{11.6}$$

Then,

$$P_i(\tau_A \leqslant m) = 1 - \sum_k p_{ik}^{0(m)}, \qquad m = 1, 2, \dots, \tag{11.7}$$

where $\mathbf{p}^0$ is the matrix obtained by deleting the rows and columns of $\mathbf{p}$ corresponding to the states in A.

In general, the matrix $\mathbf{p}^0$ is not a proper transition probability matrix since the row sums may be strictly less than 1 upon the removal of certain columns from $\mathbf{p}$. However, if each of the rows in $\mathbf{p}$ corresponding to states $j \in A$ is replaced by e_j' having 1 in the jth place and 0 elsewhere, then the resulting matrix $\hat{\mathbf{p}}$, say, is a transition probability matrix and

$$P_i(\tau_A \leqslant m) = 1 - \sum_{k \notin A} \hat{p}_{ik}^{(m)}. \tag{11.8}$$

The reason (11.8) holds is that up to the first passage time τ_A the distribution of Markov chains having transition probability matrices $\mathbf{p}$ and $\hat{\mathbf{p}}$ (starting at i) are the same. In particular,

$$\hat{p}_{ik}^{(m)} = p_{ik}^{0(m)} \qquad \text{for } i, k \notin A. \tag{11.9}$$

In the case that there is more than one state in A, an important problem is to determine the distribution of X_{τ_A}, starting from $i \notin A$. Of course, if $P_i(\tau_A < \infty) < 1$ then X_{τ_A} is a *defective* random variable under P_i, being defined on the set $\{\tau_A < \infty\}$ of P_i-probability less than 1.

Write

$$a_j(i) := P_i(\{\tau_A < \infty, X_{\tau_A} = j\}) \qquad (j \in A, i \in S). \tag{11.10}$$

By the Markov property, (conditional on X_1 in (11.10)),

$$a_j(i) = \sum_k \hat{p}_{ik} a_j(k) \qquad (j \in A, i \in S). \tag{11.11}$$

Denoting by $\mathbf{a}_j$ the vector $(a_j(i): i \in S)$, viewed as a column vector, one may express (11.10) as

$$\mathbf{a}_j = \hat{\mathbf{p}}\mathbf{a}_j \qquad (j \in A). \tag{11.12}$$

Alternatively, (11.11) or (11.12) may be replaced by

$$\begin{aligned} &a_j(i) = \sum_k p_{ik} a_j(k) \qquad \text{for } i \notin A, \\ &a_j(i) = \begin{cases} 1 & \text{if } i = j \\ 0 & \text{if } i \in A, \text{ but } i \neq j \end{cases} \qquad (j \in A). \end{aligned} \tag{11.13}$$

A function (or, vector) $\mathbf{a} = (a(i): i \in S)$ is said to be $\mathbf{p}$-harmonic on $B(\subset S)$ if

$$a(i) = (\mathbf{pa})(i) \qquad \text{for } i \in B. \tag{11.14}$$

Hence $\mathbf{a}_j$ is $\mathbf{p}$-harmonic on A^c and has the (boundary-)values on A,

$$a_j(i) = \begin{cases} 1 & \text{if } i = j \\ 0 & \text{if } i \in A, i \neq j. \end{cases} \tag{11.15}$$

We have thus proved part (a) of the following proposition.

Proposition 11.2. Let $\mathbf{p}$ be a transition probability matrix and A a nonempty subset of S.

(a) Then the distribution of X_{τ_A}, as defined by (11.10), satisfies (11.13).
(b) This is the unique bounded solution if and only if

$$P_i(\tau_A < \infty) = 1 \qquad \text{for all } i \in S. \tag{11.16}$$

Proof. (b) Let $i \in A^c$. Then

$$\sum_{k \in A^c} \hat{p}_{ik}^{(n)} = P_i(\tau_A > n) \downarrow P_i(\tau_A = \infty) \qquad \text{as } n \uparrow \infty. \tag{11.17}$$

Hence, if (11.16) holds, then

$$\lim_{n \to \infty} \hat{p}_{ik}^{(n)} = 0 \qquad \text{for all } i, k \in A^c. \tag{11.18}$$

On the other hand,

$$\hat{p}_{ik}^{(n)} = P_i(\tau_A \leqslant n, X_{\tau_A} = k) \uparrow P_i(\tau_A < \infty, X_{\tau_A} = k) = a_k(i) \tag{11.19}$$

if $i \in A^c$, $k \in A$, and

$$\hat{p}_{ik}^{(n)} = \delta_{ik} \qquad \text{for all } n, \text{ if } i \in A,\ k \in S.$$

Now let $\mathbf{a}$ be another solution of (11.13), besides $\mathbf{a}_j$. Then $\mathbf{a}$ satisfies (11.12), which on iteration yields $\mathbf{a} = \hat{\mathbf{p}}^n\mathbf{a}$. Taking the limit as $n \uparrow \infty$, and using (11.18), (11.19), one gets

$$a(i) = \lim_{n \to \infty} \sum_k \hat{p}_{ik}^{(n)} a(k) = \sum_{k \in A} a_k(i) a(k) = a_j(i) \tag{11.20}$$

for all $i \in A^c$, since $a(k) = 0$ for $k \in A \backslash \{j\}$ and $a(j) = 1$. Hence $\mathbf{a}_j$ is the unique solution of (11.13).

Conversely, if $P_i(\tau_A < \infty) < 1$ for some $i \in A^c$, then the function $\mathbf{h} = (h(i){:}\ i \in S)$ defined by

$$h(i) := 1 - P_i(\tau_A < \infty) = P_i(\tau_A = \infty) \qquad (i \in S), \tag{11.21}$$

may be shown to be $\mathbf{p}$-harmonic in A^c with (boundary-)value *zero* on A. The harmonic property is a consequence of the Markov property (Exercise 5),

$$\begin{aligned} h(i) = P_i(\tau_A = \infty) &= \sum_k p_{ik} P_k(\tau_A = \infty) \\ &= \sum_k p_{ik} h(k) = \sum_k \hat{p}_{ik} h(k) \qquad (i \in A^c). \end{aligned} \tag{11.22}$$

Since $P_i(\tau_A = 0) = 1$ for $i \in A$, $h(i) = 0$ for $i \in A$. It follows that both $\mathbf{a}_j$ and $\mathbf{a}_j + \mathbf{h}$ satisfy (11.13). Since $\mathbf{h} \neq \mathbf{0}$, the solution of (11.13) is not unique. ■

Example 1. (*A Random Replication Model*). The simple Wright–Fisher model originated in genetics as a model for the evolution of gene frequencies. Here the mathematical model will be described in a somewhat different physical context. Consider a collection of $2N$ individuals, each one of which is either in *favor of* or *against* some issue. Let X_n denote the number of individuals in favor of the issue at time $n = 0, 1, 2, \ldots$. In the evolution, each one of the individuals will randomly re-decide his or her position under the influence of the current overall opinion as follows. Let $\theta_n = X_n/2N$ denote the proportion in favor of the issue at time n. Then given $X_0, X_1, \ldots, X_n$, each of the $2N$ individuals, independently of the choices of the others, elects to be in favor with probability θ_n or against the issue with probability $1 - \theta_n$. That is,

$$P(X_{n+1} = k \mid X_0, X_1, \ldots, X_n) = \binom{2N}{k}\theta_n^k(1 - \theta_n)^{2N-k}, \tag{11.23}$$

for $k = 0, 1, \ldots, 2N$. So $\{X_n\}$ is a Markov chain with state space $S = \{0, 1, \ldots, 2N\}$ and one-step transition matrix $\mathbf{p} = ((p_{ij}))$, where

$$p_{ij} = \binom{2N}{j}\left(\frac{i}{2N}\right)^j\left(1 - \frac{i}{2N}\right)^{2N-j}, \qquad i, j = 0, 1, \ldots, 2N. \tag{11.24}$$

Notice that $\{X_n\}$ is an aperiodic Markov chain. The "boundary" states $\{0\}$ and $\{2N\}$ form closed classes of essential states. The set of states $\{1, 2, \ldots, 2N - 1\}$ constitute an inessential class. The model has a special conservation property of the form of the following martingale property,

$$E(X_{n+1} \mid X_0, X_1, \ldots, X_n) = E(X_{n+1} \mid X_n) = \left(\frac{X_n}{2N}\right)2N = X_n, \tag{11.25}$$

for $n = 0, 1, 2, \ldots$. In particular, therefore,

$$EX_{n+1} = E\{E(X_{n+1} \mid X_0, X_1, \ldots, X_n)\} = EX_n, \tag{11.26}$$

for $n = 0, 1, 2, \ldots$. However, since S is finite we know that in the long run $\{X_n\}$ is certain to be absorbed in state 0 or $2N$, i.e., the population is certain to eventually come to a *unanimous opinion*, be it pro or con. It is of interest to calculate the absorption probabilities as well as the rate of absorption. Here, with $A = \{0, 2N\}$, one has $\mathbf{p} = \hat{\mathbf{p}}$.

Let $a_j(i)$ denote the probability of ultimate absorption at $j = 0$ or at $j = 2N$ starting from state $i \in S$. Then,

$$\begin{cases} a_{2N}(i) = \sum_k p_{ik} a_{2N}(k) & \text{for } 0 < i < 2N, \\ a_{2N}(2N) = 1, \quad a_{2N}(0) = 0, \end{cases} \tag{11.27}$$

and $a_0(i) = 1 - a_{2N}(i)$. In view of (11.26) and (11.19) we have

$$i = E_i X_0 = E_i X_n = \sum_{k=0}^{2N} k p_{ik}^{(n)} \to 0 a_0(i) + 2N a_{2N}(i) = 2N a_{2N}(i)$$

for $i = 1, \ldots, 2N - 1$. Therefore,

$$a_{2N}(i) = \frac{i}{2N}, \qquad i = 0, 1, 2, \ldots, 2N, \tag{11.28}$$

$$a_0(i) = \frac{2N - i}{2N}, \qquad i = 0, 1, 2, \ldots, 2N. \tag{11.29}$$

Check also that (11.29) satisfies (11.27).

In order to estimate the rate at which fixation of opinion occurs, we shall calculate the eigenvalues of $\hat{\mathbf{p}}(= \mathbf{p}$ here).

Let $\mathbf{v} = (v_0, \ldots, v_{2N})'$ and consider the eigenvalue problem

$$\sum_{j=0}^{2N} p_{ij} v_j = \lambda v_i, \qquad i = 0, 1, \ldots, 2N. \tag{11.30}$$

The *rth factorial moment* of the binomial distribution $(p_{ij}: 0 \leqslant j \leqslant 2N)$ is

$$\begin{aligned} \sum_{j=0}^{2N} j(j-1)\cdots(j-r+1) p_{ij} &= \left(\frac{i}{2N}\right)^r (2N)\cdots(2N-r+1) \sum_{j=r}^{2N} \binom{2N-r}{j-r} \\ &\quad \times \left(\frac{i}{2N}\right)^{j-r} \left(1 - \frac{i}{2N}\right)^{2N-j} \\ &= \left(\frac{i}{2N}\right)^r (2N)(2N-1)\cdots(2N-r+1), \end{aligned} \tag{11.31}$$

for $r = 1, 2, \ldots, 2N$. Equation (11.31) contains a transformation between "factorial powers" and "ordinary powers" that deserves to be examined for connections with (11.30). The "factorial powers" $(j)_r := j(j-1)\cdots(j-r+1)$ are simply polynomials in the "ordinary powers" and can be expressed as

$$j(j-1)\cdots(j-r+1) = \sum_{k=1}^{r} s_k^r j^k. \tag{11.32}$$

Likewise, "ordinary powers" j^r can be expressed as polynomials in the "factorial powers" as

$$j^r = \sum_{k=0}^{r} S_k^r (j)_k, \tag{11.33}$$

with the convention $(j)_0 = 1$. Note that $S_r^r = 1$ for all $r \geqslant 0$. The coefficients $\{s_k^r\}$, $\{S_k^r\}$ are commonly referred to as *Stirling coefficients of the first and second kinds*, respectively.

Now every vector $\mathbf{v} = (v_0, \ldots, v_{2N})'$ may be represented as the successive values of a unique (factorial) polynomial of degree $2N$ evaluated at $0, 1, \ldots, 2N$ (Exercise 7), i.e.,

$$v_j = \sum_{r=0}^{2N} \alpha_r (j)_r \qquad \text{for } j = 0, 1, \ldots, 2N. \tag{11.34}$$

According to (11.24), (11.32),

$$\sum_{j=0}^{2N} p_{ij} v_j = \sum_{r=0}^{2N} \alpha_r \sum_{j=0}^{2N} p_{ij} (j)_r = \sum_{r=0}^{2N} \alpha_r \frac{(2N)_r}{(2N)^r} i^r = \sum_{r=0}^{2N} \alpha_r \frac{(2N)_r}{(2N)^r} \sum_{n=0}^{r} S_n^r (i)_n$$

$$= \sum_{n=0}^{2N} \left(\sum_{r=n}^{2N} \alpha_r \frac{(2N)_r}{(2N)^r} S_n^r \right) (i)_n. \tag{11.35}$$

It is now clear that (11.30) holds if and only if

$$\sum_{r=n}^{2N} \alpha_r \frac{(2N)_r}{(2N)^r} S_n^r = \lambda \alpha_n. \tag{11.36}$$

In particular, taking $n = 2N$ and noting $S_r^r = 1$, we see that (11.36) holds if

$$\lambda = \lambda_{2N} = \frac{(2N)_{2N}}{(2N)^{2N}}, \qquad \alpha_{2N} = \alpha_{2n}^{(2N)} = 1, \tag{11.37}$$

and $\alpha_r = \alpha_r^{(2N)}$, $r = 2N - 1, \ldots, 0$, are solved recursively from (11.36). Next take $\alpha_{2N}^{(2N-1)} = 0$, $\alpha_{2N-1}^{(2N-1)} = 1$ and solve for

$$\lambda = \lambda_{2N-1} = \frac{(2N)_{2N-1}}{(2N)^{2N-1}}, \qquad \text{etc.}$$

Then,

$$\lambda_r = \frac{(2N)_r}{(2N)^r} = \left(1 - \frac{1}{2N}\right) \cdots \left(1 - \frac{r-1}{2N}\right), \qquad 0 \leqslant r \leqslant 2N. \tag{11.38}$$

Notice that $\lambda_0 = \lambda_1 = 1$. The next largest eigenvalue is $\lambda_2 = 1 - (1/2N)$. Let $\mathbf{V} = ((v_{ij}))$ be the matrix whose columns are the eigenvectors obtained above. Writing $\mathbf{D} = \mathbf{diag}(\lambda_0, \lambda_1, \ldots, \lambda_{2N})$, this means

$$\mathbf{pV} = \mathbf{VD}, \tag{11.39}$$

or

$$\mathbf{p} = \mathbf{VDV}^{-1}, \tag{11.40}$$

so that

$$\mathbf{p}^m = \mathbf{VD}^m\mathbf{V}^{-1}. \tag{11.41}$$

Therefore, writing $\mathbf{V}^{-1} = ((v^{ij}))$, we have from (11.8)

$$P_i(\tau_{\{0,2N\}} > m) = \sum_{j=1}^{2N-1} p_{ij}^{(m)} = \sum_{j=1}^{2N-1} \sum_{k=0}^{2N} \lambda_k^m v_{ik} v^{kj} = \sum_{k=0}^{2N} \left(\sum_{j=1}^{2N-1} v_{ik} v^{kj} \right) \lambda_k^m. \tag{11.42}$$

Since the left side of (11.42) must go to zero as $m \to \infty$, the coefficients of $\lambda_0^m \equiv 1$ and $\lambda_1^m \equiv 1$ must be zero. Thus,

$$\begin{aligned} P_i(\tau_{\{0,2N\}} > m) &= \sum_{k=0}^{2N} \sum_{j=1}^{2N-1} v_{ik} v^{kj} \lambda_k^m \\ &= \lambda_2^m \left[\left(\sum_{j=1}^{2N-1} v_{i2} v^{2j} \right) + \sum_{k=3}^{2N} \left(\sum_{j=1}^{2N-1} v_{ik} v^{kj} \right) \left(\frac{\lambda_k}{\lambda_2} \right)^m \right] \\ &\sim (\text{const.}) \lambda_2^m \qquad \text{for large } m. \end{aligned} \tag{11.43}$$

Example 2. (*Bienaymé–Galton–Watson Simple Branching Process*). A simple branching process was introduced as Example 3.7. The state X_n of the process at time n represents the total number of offspring in the nth generation of a population that evolves by i.i.d. *random replications* of parent individuals. If the offspring distribution is denoted by f then, as explained in Section 3, the one-step transition probabilities are given by

$$p_{ij} = \begin{cases} f^{*i}(j), & \text{if } i \geqslant 1, j \geqslant 0, \\ 1 & \text{if } i = 0, j = 0, \\ 0 & \text{if } i = 0, j \neq 0. \end{cases} \tag{11.44}$$

According to the values of p_{ij} at the boundary, the state $i = 0$ is *absorbing* (permanent *extinction*). Write ρ_{i0} for the probability that eventually extinction occurs given $X_0 = i$. Also write $\rho = \rho_{10}$. Then $\rho_{i0} = \rho^i$, since each of the i sequences of generations arising from the i initial particles has the same chance

ρ of extinction, and the i sequences evolving independently must all be extinct in order that there may be eventual extinction, given $X_0 = i$.

If $f(0) = 0$, then $\rho = \rho_{10} = 0$ and extinction is impossible. If $f(0) = 1$, then $\rho_{10} \geqslant p_{10}^{(1)} = 1$ and extinction is certain (no matter what X_0 is). To avoid these and other trivialities, we assume (unless otherwise specified)

$$0 < f(0) < 1, \qquad f(0) + f(1) < 1. \tag{11.45}$$

Introduce the *probability generating function* of f:

$$\phi(z) = \sum_{j=0}^{\infty} f(j)z^j = f(0) + \sum_{j=1}^{\infty} f(j)z^j \qquad (|z| \leqslant 1). \tag{11.46}$$

Since a power series can be differentiated term by term within its radius of convergence, one has

$$\phi'(z) \equiv \frac{d}{dz}\phi(z) = \sum_{j=1}^{\infty} jf(j)z^{j-1} \qquad (|z| < 1). \tag{11.47}$$

If the *mean* μ *of the number of particles* generated by a single particle is finite, i.e., if

$$\mu \equiv \sum_{j=1}^{\infty} jf(j) < \infty, \tag{11.48}$$

then (11.47) holds even for the left-hand derivative at $z = 1$, i.e.,

$$\mu = \phi'(1). \tag{11.49}$$

Since $\phi'(z) > 0$ for $0 < z < 1$, ϕ is strictly increasing. Also, since $\phi''(z)$ (which exists and is finite for $0 \leqslant z < 1$) satisfies

$$\phi''(z) \equiv \frac{d^2}{dz^2}\phi(z) = \sum_{j=2}^{\infty} j(j-1)f(j)z^{j-2} > 0 \qquad \text{for } 0 < z < 1, \tag{11.50}$$

the function ϕ is *strictly convex* on $[0,1]$. In other words, the line segment joining any two points on the curve $y = \phi(z)$ lies strictly above the curve (except at the two points joined). Because $\phi(0) = f(0) > 0$ and $\phi(1) = \sum_{j=0}^{\infty} f(j) = 1$, the graph of ϕ looks like that of Figure 11.1 (curve a or b).

The maximum of $\phi'(z)$ is μ, which is attained at $z = 1$. Hence, in the case $\mu > 1$, the graph of $y = \phi(z)$ must lie below that of $y = z$ near $z = 1$ and, because $\phi(0) = f(0) > 0$, must cross the line $y = z$ at a point z_0, $0 < z_0 < 1$. Since the slope of the curve $y = \phi(z)$ continuously increases as z increases in $(0, 1)$, z_0 is the unique solution of the equation $z = \phi(z)$ that is smaller than 1.

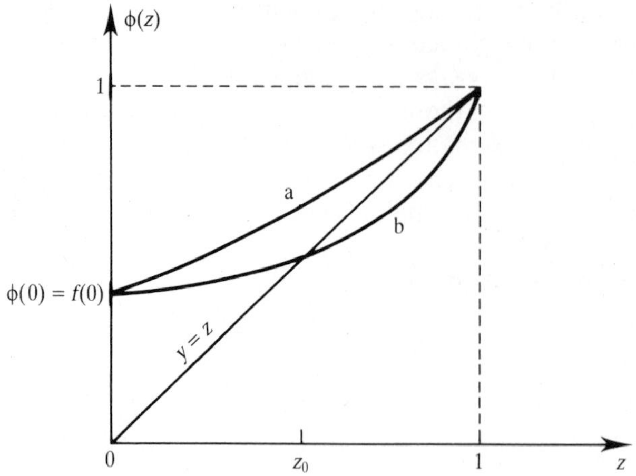

Figure 11.1

In case $\mu \leqslant 1$, $y = \phi(z)$ must lie strictly above the line $y = z$, except at $z = 1$. For if it meets the line $y = z$ at a point $z_0 < 1$, then it must go under the line in the immediate vicinity to the right of z_0, since its slope falls below that of the line (i.e., unity). In order to reach the height $\phi(1) = 1$ (also reached by the line at the same value $z = 1$) its slope then must exceed 1 somewhere in $(z_0, 1]$; this is impossible since $\phi'(z) \leqslant \phi'(1) = \mu \leqslant 1$ for all z in $[0, 1]$. Thus, the only solution of the equation $z = \phi(z)$ is $z = 1$.

Now observe

$$\rho = \rho_{10} = \sum_{j=0}^{\infty} P(X_1 = j \mid X_0 = 1)\rho_{j0} = \sum_{j=0}^{\infty} f(j)\rho^j = \phi(\rho), \qquad (11.51)$$

thus *if* $\mu \leqslant 1$, *then* $\rho = 1$ *and extinction is certain.* On the other hand, suppose $\mu > 1$. Then ρ is either z_0 or 1. We shall now show that $\rho = z_0 (< 1)$. For this, consider the quantities

$$q_n := P(X_n = 0 \mid X_0 = 1) = p_{10}^{(n)} \qquad (n = 1, 2, \ldots). \qquad (11.52)$$

That is, q_n is the probability that the sequence of generations originating from a single particle is extinct at time n. As n increases, $q_n \uparrow \rho$; for clearly, $\{X_n = 0\} \subset \{X_m = 0\}$ for all $m \geqslant n$, so that $q_n \leqslant q_m$ if $n \leqslant m$. Also

$$\left\{\lim_{n\to\infty} X_n = 0\right\} = \bigcup_{n=0}^{\infty} \{X_n = 0\} = \{\text{extinction occurs}\}.$$

Now, by independence of the generations originating from different particles,

$$P(X_n = 0 \mid X_0 = j) = q_n^j \qquad (j = 0, 1, 2, \ldots),$$

$$\begin{aligned} q_{n+1} &= P(X_{n+1} = 0 \mid X_0 = 1) = P(X_1 = 0 \mid X_0 = 1) \\ &\quad + \sum_{j=1}^{\infty} P(X_1 = j, X_{n+1} = 0 \mid X_0 = 1) \\ &= f(0) + \sum_{j=1}^{\infty} P(X_1 = j \mid X_0 = 1) P(X_{n+1} = 0 \mid X_0 = 1, X_1 = j) \\ &= f(0) + \sum_{j=1}^{\infty} f(j) q_n^j = \phi(q_n) \qquad (n = 1, 2, \ldots). \end{aligned} \tag{11.53}$$

Since $q_1 = f(0) = \phi(0) < \phi(z_0) = z_0$ (recall that $\phi(z)$ is strictly increasing in z for $0 < z < 1$), one has using (11.53) with $n = 1$, $q_2 = \phi(q_1) < \phi(z_0) = z_0$, and so on. Hence, $q_n < z_0$ for all n. Therefore, $\rho = \lim_{n \to \infty} q_n \leqslant z_0$. This proves $\rho = z_0$.

If $f(0) + f(1) = 1$ and $0 < f(0) < 1$, then $\phi''(z) = 0$ for all z, and the graph of $\phi(z)$ is the line segment joining $(0, f(0))$ and $(1, 1)$. Hence, $\rho = 1$ in this case.

Let us now compute the average size of the nth generation. One has

$$\begin{aligned} E(X_{n+1} \mid X_0 = 1) &= \sum_{k=1}^{\infty} k p_{1k}^{(n+1)} = \sum_{k=1}^{\infty} k \left(\sum_{j=1}^{\infty} p_{1j}^{(n)} p_{jk}^{(1)} \right) \\ &= \sum_{k=1}^{\infty} \sum_{j=1}^{\infty} k p_{1j}^{(n)} p_{jk}^{(1)} = \sum_{j=1}^{\infty} p_{1j}^{(n)} \left(\sum_{k=1}^{\infty} k p_{jk}^{(1)} \right) \\ &= \sum_{j=1}^{\infty} p_{1j}^{(n)} E(X_1 \mid X_0 = j) = \sum_{j=1}^{\infty} p_{1j}^{(n)} j E(X_1 \mid X_0 = 1) \\ &= \sum_{j=1}^{\infty} p_{1j}^{(n)} j\mu = \mu \sum_{j=1}^{\infty} j p_{1j}^{(n)} = \mu E(X_n \mid X_0 = 1). \end{aligned} \tag{11.54}$$

Continuing in this manner, one obtains

$$\begin{aligned} E(X_{n+1} \mid X_0 = 1) &= \mu E(X_n \mid X_0 = 1) = \mu^2 E(X_{n-1} \mid X_0 = 1) \\ &= \cdots = \mu^n E(X_1 \mid X_0 = 1) = \mu^{n+1}. \end{aligned} \tag{11.55}$$

It follows that

$$E(X_n \mid X_0 = j) = j\mu^n. \tag{11.56}$$

Thus, in the case $\mu < 1$, the expected size of the population at time n decreases to zero exponentially fast as $n \to \infty$. If $\mu = 1$, then the expected size at time n does not depend on n (i.e., it is the same as the initial size). If $\mu > 1$, then the expected size of the population increases to infinity exponentially fast.

12 ONE-DIMENSIONAL NEAREST-NEIGHBOR GIBBS STATE

The notion of a Gibbs distribution has its origins in statistical physics as a probability distribution with respect to which bulk thermodynamic properties of materials in equilibrium can be expressed as expected values (called *phase averages*). The thrust of Gibbs' idea is that a theoretically convenient way in which to view materials at the microscopic scale is as a large system composed of randomly distributed but interacting components, such as the positions and momenta of molecules comprising the material.

To arrive at an appropriate probability distribution for computing large-scale averages, Gibbs argued that the probability of a given configuration of component values in equilibrium should be inversely proportional to an exponential function of the total energy of the configuration. In this way the lowest-energy configurations (*ground states*) are the most likely (*modes*) to occur. Moreover, the *additivity* of the combined total energy of two noninteracting systems is reflected in the *multiplication* rule for (exponential) probabilities of independent values for such a specification. It is a tribute to the genius of Gibbs that this approach leads to the correct thermodynamics at the bulk material scale. Perhaps because of the great success these ideas have enjoyed in physics, Gibbs' probability distributions have also been introduced to represent systems having a large number of interacting components in a wide variety of other contexts, ranging from both genetic and automata codes to economic and sociological systems.

For purposes of orientation one may think of random values $\{X_n: n \in \Lambda\}$ (states) from a finite set S distributed over a finite set of sites. The set of possible configurations is then represented by the cartesian product $\Omega := \{\omega = (\sigma_n: n \in \Lambda) \mid n \to \sigma_n \in S,\ n \in \Lambda, \text{ is a function on } \Lambda\}$. For Λ and S finite, Ω is also a finite set and a (*free-boundary*) *Gibbs distribution* on Ω can be described by a probability mass function (p.m.f.) of the form

$$P(X_n = \sigma_n, n \in \Lambda) := Z^{-1} \exp\{-\beta U(\omega)\}, \qquad \omega = (\sigma_n) \in \Omega \tag{12.1}$$

where $U(\omega)$ is a real-valued function on Ω, referred to as *total potential energy* of configurations $\omega \in \Omega$, β is a positive parameter called *inverse temperature*, and

$$Z := \sum_{\omega \in \Omega} \exp\{-\beta U(\omega)\}, \tag{12.2}$$

is the normalization constant, referred to as the *partition function*.

Observe that if P is any probability distribution on a finite set Ω that assigns strictly positive probability $p(\omega)$ to each $\omega \in \Omega$, then P can trivially be expressed in the form (12.1) with $U(\omega) = -\beta^{-1} \log\{p(\omega)\}$. In physics, however, one regards the total potential energy as a sum of energies at individual sites plus the sum of interaction energies between pairs of sites plus the sum of energies between triples, etc., and the probability distribution is specified for various types of such interactions. For example, if $U(\omega) = \sum_{n \in \Lambda} \varphi_1(\sigma_n)$ for

$\omega = (\sigma_n : n \in \Lambda)$ is a sum of single-site energies, higher-order interactions being zero, then the distribution (12.1) is that of *independent* components; similarly for the case of *infinite temperature* ($\beta = 0$).

It is both natural and quite important to consider the problem of extending the above formulation to the so-called *infinite volume* setting in which Λ is a countably infinite set of sites. We will consider this problem here for one-dimensional systems on $\Lambda = \mathbb{Z}$ consisting of interacting components taking values in a finite set S and for which the nonzero interaction energies contributing to U are at most between *pairs* of *nearest-neighbor* (i.e., *adjacent*) integer sites.

Let $\Omega = S^{\mathbb{Z}}$ for a finite set S and let $\mathscr{F}$ be the sigmafield generated by finite-dimensional cylinder sets of the form

$$C = \{\omega = (\sigma_n) \in \Omega : \sigma_k = s_k \text{ for } k \in \Lambda\}, \tag{12.3}$$

where Λ is a finite subset of $\mathbb{Z}$ and $s_k \in S$ is specified for each $k \in \Lambda$. Although it would be enough to consistently prescribe the probabilities for such events C, except for the independent case it is not quite obvious how to do this starting with energy considerations. First consider then the independent case. For single-site energies $\varphi_1(s, n)$, $s \in S$, at the respective sites $n \in \mathbb{Z}$, the probabilities of cylinder events can be specified according to the formula

$$P(C) = Z_\Lambda^{-1} \exp\left\{-\beta \sum_{k \in \Lambda} \varphi_1(s_k, k)\right\}, \tag{12.4}$$

where Z_Λ appropriately normalizes P for each set of sites Λ. In the *homogeneous* (or *translation-invariant*) case we have $\varphi_1(s, n) = \varphi_1(s)$, $s \in S$, depending on the site n only through the component values s at n.

Now suppose that we have in addition to single-site energies $\varphi_1(s)$, *pairwise nearest-neighbor interactions* represented by $\varphi_2(s, n; t, m)$ for $|n - m| = 1, s, t \in S$. For translation invariance of interactions, take $\varphi_2(s, n; t, m) = \varphi_2(s, t)$ if $|n - m| = 1$ and 0 otherwise. Now because values inside of Λ should be dependent on values across boundaries of Λ it is less straightforward to *consistently* write down expressions for $P(C)$ than in (12.4). In fact, in this connection it is somewhat more natural to consider a (local) specification of *conditional probabilities* of finite-dimensional events of the form C, given information about a configuration outside of Λ. Take $\Lambda = \{n\}$ and consider a specification of the conditional probability of $C = \{X_n = s\}$ given an event of the form $\{X_k = s_k \text{ for } k \in D_n \backslash \{n\}\}$, where D_n is a finite set of sites that includes n and the two neighboring sites $n - 1$ and $n + 1$, as follows

$$\begin{aligned} P(X_n = s \mid X_k = s_k, k \in D_n \backslash \{n\}) \\ = Z_{n,D_n}^{-1} \exp\{-\beta[\varphi_1(s) + \varphi_2(s, s_{n-1}) + \varphi_2(s, s_{n+1})]\}, \end{aligned} \tag{12.5}$$

where

$$Z_{n,D_n} = \sum_{s \in S} \exp\{-\beta[\varphi_1(s) + \varphi_2(s, s_{n-1}) + \varphi_2(s, s_{n+1})]\}. \tag{12.6}$$

That is, the state at n depends on the given states at sites in $D_n\backslash\{n\}$ only through the neighboring values at $n-1, n+1$. One would like to know that there is a probability distribution having these conditional probabilities. For the one-dimensional case at hand, we have the following basic result.

Theorem 12.1. Let S be an arbitrary finite set, $\Omega = S^{\mathbb{Z}}$, and let $\mathscr{F}$ be the sigmafield of subsets of Ω generated by finite-dimensional cylinder sets. Let $\{X_n\}$ be the coordinate projections on Ω. Suppose that P is a probability measure on Ω with the following properties.

(i) $P(C) > 0$ for every finite-dimensional cylinder set $C \in \mathscr{F}$.

(ii) For arbitrary $n \in \mathbb{Z}$ let $\partial(n) = \{n-1, n+1\}$ denote the boundary of $\{n\}$ in $\mathbb{Z}$. If f is any S-valued function defined on a finite subset D_n of $\mathbb{Z}$ that contains $\{n\} \cup \partial(n)$ and if $a \in S$ is arbitrary, then the conditional probability

$$P(X_n = a \mid X_m = f(m), m \in D_n\backslash\{n\})$$

depends only on the values of f on $\partial(n)$.

(iii) The value of the conditional probability in (ii) is invariant under translation by any amount $\Delta \in \mathbb{Z}$.

Then P is the distribution of a (unique) stationary (i.e., translation-invariant) Markov chain having strictly positive transition probabilities and, conversely, every stationary Markov chain with strictly positive transition probabilities satisfies (i), (ii), and (iii).

Proof. Let

$$g_{b,c}(a) = P(X_n = a \mid X_{n-1} = b, X_{n+1} = c). \tag{12.7}$$

The family of conditional probabilities $\{g_{b,c}(a)\}$ is referred to as the *local structure* of the probability distribution. We will prove the converse statement first and along the way we will calculate the local structure of P in terms of the transition probabilities. So assume that P is a stationary Markov chain with strictly positive transition matrix $((p(a, b): a, b \in S))$ and marginal distribution $(\pi(a): a \in S)$. Consider cylinder set probabilities given by

$$P(X_n = a_0, X_{n+1} = a_1, \ldots, X_{n+h} = a_h) = \pi(a_0)p(a_0, a_1)\cdots p(a_{h-1}, a_h). \tag{12.8}$$

So, in particular, the condition (i) is satisfied; also see (2.9), (2.6), *Prop. 6.1.* For m and $n \geqslant 1$ it is a straightforward computation to verify

$$\begin{aligned} P(X_0 = a \mid X_{-m} = b_{-m}, \ldots, X_{-1} = b_{-1}, X_1 = b_1, \ldots, X_n = b_n) \\ = P(X_0 = a \mid X_{-1} = b_{-1}, X_1 = b_1) \\ = p(b_{-1}, a)p(a, b_1)/p^{(2)}(b_{-1}, b_1) = g_{b_{-1},b_1}(a). \end{aligned} \tag{12.9}$$

Therefore, the condition (ii) holds for P, since condition (iii) also holds because P is the distribution of a stationary Markov chain. Next suppose that P is a probability distribution satisfying (i), (ii), and (iii). We must show that P is the distribution of a stationary Markov chain. Fix an arbitrary element of S, denoted as θ, say. Let the local structure of P be as defined in (12.7). Observe that for each $b, c \in S$, $g_{b,c}(\ )$ is a probability measure on S. Outlined in Exercise 1 are the steps required to show that

$$g_{b,c}(a) = \frac{q(b, a)q(a, c)}{q^{(2)}(b, c)}, \tag{12.10}$$

where

$$q(b, c) = \frac{g_{\theta,c}(b)}{g_{\theta,c}(\theta)} \tag{12.11}$$

and

$$q^{(n+1)}(b, c) = \sum_{a \in S} q^{(n)}(b, a)q(a, c). \tag{12.12}$$

An induction argument, as outlined in Exercise 2, can be used to check that

$$P(X_{k+h} = a_k, 1 \leqslant k \leqslant r \mid X_{h+1-n} = a, X_{h+r+n} = b)$$
$$= \frac{q^{(n)}(a, a_1)q(a_1, a_2)\cdots q^{(n)}(a_r, b)}{q^{(r+2n-1)}(a, b)}, \tag{12.13}$$

for $h \in \mathbb{Z}$, $a_i, a, b \in S$, $n, r \geqslant 1$. We would like to conclude that P coincides with the (uniquely determined) stationary Markov chain having transition probabilities $((\hat{q}(b, c)))$ defined by normalizing $((q(b, c)))$ to a transition probability matrix according to the special formula

$$\hat{q}(b, c) = \frac{q(b, c)\mu(c)}{\sum_{a \in S} q(b, a)\mu(a)} = \frac{q(b, c)\mu(c)}{\lambda\mu(b)}, \tag{12.14}$$

where $\boldsymbol{\mu} = (\mu(a))$ is a positive eigenvector of $\mathbf{q}$ corresponding to the largest (in magnitude) eigenvalue $\lambda = \lambda_{\max}$ of $\mathbf{q}$. Note that $\hat{\mathbf{q}}$ is a strictly positive transition probability matrix with unique invariant distribution $\boldsymbol{\pi}$, say, and such that (Exercise 3)

$$\hat{q}^{(n)}(b, c) = \frac{q^{(n)}(b, c)\mu(c)}{\lambda^n \mu(b)}. \tag{12.15}$$

Let Q be the probability distribution of this Markov chain. It is enough to show that P and Q agree on the cylinder events. Using (12.13) we have for any $n \geqslant 1$,

$$\begin{aligned}
P(X_0 &= a_0, \dots, X_r = a_r) \\
&= \sum_{a \in S} \sum_{b \in S} P(X_{-n} = a, X_{n+r} = b) P(X_0 = a_0, \dots, X_r = a_r \mid X_{-n} = a, X_{n+r} = b) \\
&= \sum_{a \in S} \sum_{b \in S} P(X_{-n} = a, X_{n+r} = b) \frac{q^{(n)}(a, a_0) q(a_0, a_1) \cdots q(a_{r-1}, a_r) q^{(n)}(a_r, b)}{q^{(r+2n)}(a, b)} \\
&= \sum_{a \in S} \sum_{b \in S} P(X_{-n} = a, X_{n+r} = b) \frac{\hat{q}^{(n)}(a, a_0) \hat{q}(a_0, a_1) \cdots \hat{q}(a_{r-1}, a_r) \hat{q}^{(n)}(a_r, b)}{\hat{q}^{(r+2n)}(a, b)}.
\end{aligned} \tag{12.16}$$

Now let $n \to \infty$ to get by the fundamental convergence result of Proposition 6.1 (Exercise 4),

$$\begin{aligned}
P(X_0 = a_0, \dots, X_r = a_r) &= \pi(a_0) \hat{q}(a_0, a_1) \cdots \hat{q}(a_{r-1}, a_r) \\
&= Q(X_0 = a_0, \dots, X_r = a_r).
\end{aligned} \tag{12.17}$$

■

Probability distributions on $\mathbb{Z}$ that satisfy (i)–(iii) of Theorem 12.1 are also referred to as *one-dimensional Markov random fields* (MRF). This definition has a natural extension to probability distributions on $\mathbb{Z}^d$ called the *d-dimensional Markov random field.* For that matter, the important element of the definition of a MRF on a countable set Λ is that Λ have a *graph structure* to accommodate the notion of *nearest-neighbor* (or *adjacent*) sites. The result here shows that one-dimensional Markov random fields can be locally specified and are in fact stationary Markov chains. While existence of a probability distribution satisfying (i)–(iii) can be proved for any dimension d (for finite S), the probability distribution need not be unique. This interesting phenomenon is known as a *phase transition.* Existence may fail in the case that S is countably infinite too (see theoretical complement 1).

13 A MARKOVIAN APPROACH TO LINEAR TIME SERIES MODELS

The canonical construction of Markov chains on the space of trajectories has been explained in Chapter I, Section 6, using Kolmogorov's Existence Theorem. In the present section and the next, another widely used general method of construction of Markov processes on arbitrary state spaces is illustrated. Markovian models in this form arise naturally in many fields, and they are often easier to analyze in this noncanonical representation.

Example 1. *(The Linear Autoregressive Model of Order One, or the* **AR(1)** *Model).* Let b be a real number and $\{\varepsilon_n : n \geqslant 1\}$ an i.i.d. sequence of real-valued random variables defined on some probability space $(\Omega, \mathscr{F}, P)$. Given an initial random variable X_0 independent of $\{\varepsilon_n\}$, define recursively the sequence of random variables $\{X_n: n \geqslant 0\}$ as follows:

$$X_0, X_1 := bX_0 + \varepsilon_1, \qquad X_{n+1} := bX_n + \varepsilon_{n+1} \qquad (n \geqslant 0). \tag{13.1}$$

As $X_0, X_1, \ldots, X_n$ are determined by $\{X_0, \varepsilon_1, \ldots, \varepsilon_n\}$, and ε_{n+1} is independent of the latter, one has, for all Borel sets C,

$$\begin{aligned} P(X_{n+1} \in C \mid \{X_0, X_1, \ldots, X_n\}) &= [P(bx + \varepsilon_{n+1} \in C)]_{x = X_n} \\ &= [P(\varepsilon_{n+1} \in C - bx)]_{x = X_n} = Q(C - bX_n), \end{aligned} \tag{13.2}$$

where Q is the common distribution of the random variables ε_n. Thus $\{X_n: n \geqslant 0\}$ is a Markov process on the state space $S = \mathbb{R}^1$, having the *transition probability* (of going from x to C in one step)

$$p(x, C) := Q(C - bx), \tag{13.3}$$

and *initial distribution* given by the distribution of X_0. The analysis of this Markov process is, however, facilitated more by its representation (13.1) than by an analytical study of the asymptotics of n-step transition probabilities. Note that successive iteration in (13.1) yields

$$\begin{aligned} &X_1 = bX_0 + \varepsilon_1, \qquad X_2 = bX_1 + \varepsilon_2 = b^2X_0 + b\varepsilon_1 + \varepsilon_2 \qquad \cdots \\ &X_n = b^nX_0 + b^{n-1}\varepsilon_1 + b^{n-2}\varepsilon_2 + \cdots + b\varepsilon_{n-1} + \varepsilon_n \qquad (n \geqslant 1). \end{aligned} \tag{13.4}$$

The distribution of X_n is, therefore, the same as that of

$$Y_n := b^nX_0 + \varepsilon_1 + b\varepsilon_2 + b^2\varepsilon_3 + \cdots + b^{n-1}\varepsilon_n \qquad (n \geqslant 1). \tag{13.5}$$

Assume now that

$$|b| < 1 \tag{13.6}$$

and $|\varepsilon_n| \leqslant c$ with probability 1 for some constant c. Then it follows from (13.5) that

$$Y_n \to \sum_{n=0}^{\infty} b^n\varepsilon_{n+1} \quad \text{a.s.}, \tag{13.7}$$

regardless of X_0. Let π denote the distribution of the random variable on the right side in (13.7). Then Y_n converges in distribution to π as $n \to \infty$ (Exercise 1). Because the distribution of X_n is the same as that of Y_n, it follows that X_n converges in distribution to π. Therefore, π is the unique invariant distribution for the Markov process $\{X_n\}$, i.e., for $p(x, dy)$ (Exercise 1).

The assumption that the random variable ε_1 is bounded can be relaxed. Indeed, it suffices to assume

$$\sum_{n=1}^{\infty} P(|\varepsilon_1| > c\delta^n) < \infty \qquad \text{for some } \delta < \frac{1}{|b|}, \text{ and some } c > 0. \tag{13.8}$$

For (13.8) is equivalent to assuming $\sum P(|\varepsilon_{n+1}| > c\delta^n) < \infty$ so that, by the Borel–Cantelli Lemma (see Chapter 0, Section 6),

$$P(|\varepsilon_{n+1}| \leqslant c\delta^n \text{ for all but finitely many } n) = 1.$$

This implies that, with probability 1, $|b^n\varepsilon_{n+1}| \leqslant c(|b|\delta)^n$ for all but finitely many n. Since $|b|\delta < 1$, the series on the right side of (13.7) is convergent and is the limit of Y_n.

It is simple to check that (13.8) holds if $|b| < 1$ and (Exercise 3)

$$E|\varepsilon_1|^r < \infty \qquad \text{for some } r > 0. \tag{13.9}$$

The conditions (13.6) and (13.8) (or (13.9)) are therefore sufficient for the existence of a unique invariant probability π and for the convergence of X_n in distribution to π.

Next, Example 1 is extended to multidimensional state space.

Example 2. (*General Linear Time Series Model*). Let $\{\boldsymbol{\varepsilon}_n: n \geqslant 1\}$ be a sequence of i.i.d. random vectors with values in $\mathbb{R}^m$ and common distribution Q, and let $\mathbf{B}$ be an $m \times m$ matrix with real entries b_{ij}. Suppose $\mathbf{X}_0$ is an m-dimensional random vector independent of $\{\boldsymbol{\varepsilon}_n\}$. Define recursively the sequence of random vectors

$$\mathbf{X}_0, \mathbf{X}_{n+1} := \mathbf{B}\mathbf{X}_n + \boldsymbol{\varepsilon}_{n+1} \qquad (n = 0, 1, 2, \ldots). \tag{13.10}$$

As in (13.2), (13.3), $\{\mathbf{X}_n\}$ is a Markov process with state space $\mathbb{R}^m$ and transition probability

$$p(\mathbf{x}, C) := Q(C - \mathbf{B}\mathbf{x}) \qquad (\text{for all Borel sets } C \subset \mathbb{R}^m). \tag{13.11}$$

Assume that

$$\|\mathbf{B}^{n_0}\| < 1 \qquad \text{for some positive integer } n_0. \tag{13.12}$$

Recall that the *norm of a matrix* $\mathbf{H}$ is defined by

$$\|\mathbf{H}\| := \sup_{|\mathbf{x}|=1} |\mathbf{H}\mathbf{x}|, \tag{13.13}$$

where $|\mathbf{x}|$ denotes the Euclidean length of $\mathbf{x}$ in $\mathbb{R}^m$. For a positive integer $n > n_0$ write $n = jn_0 + j'$, where $0 \leqslant j' < n_0$. Then using the fact $\|\mathbf{B}_1\mathbf{B}_2\| \leqslant \|\mathbf{B}_1\|\,\|\mathbf{B}_2\|$ for arbitrary $m \times m$ matrices $\mathbf{B}_1$, $\mathbf{B}_2$ (Exercise 2), one gets

$$\|\mathbf{B}^n\| = \|\mathbf{B}^{jn_0}\mathbf{B}^{j'}\| \leqslant \|\mathbf{B}^{n_0}\|^j\,\|\mathbf{B}^{j'}\| \leqslant c\,\|\mathbf{B}^{n_0}\|^j, \qquad c := \max\{\|\mathbf{B}^r\| : 0 \leqslant r < n_0\}. \tag{13.14}$$

From (13.12) and (13.14) it follows, as in Example 1, that the series $\sum \mathbf{B}^n \boldsymbol{\varepsilon}_{n+1}$ converges a.s. in Euclidean norm if (Exercise 3), for some $c > 0$,

$$\sum_{n=1}^{\infty} P(|\boldsymbol{\varepsilon}_1| > c\delta^n) < \infty \qquad \text{for some } \delta < \frac{1}{\|\mathbf{B}^{n_0}\|^{1/n_0}}. \tag{13.15}$$

Write, in this case,

$$\mathbf{Y} := \sum_{n=0}^{\infty} \mathbf{B}^n \boldsymbol{\varepsilon}_{n+1}. \tag{13.16}$$

It also follows, as in Example 1 (see Exercise 1), that no matter what the initial distribution (i.e., the distribution of $\mathbf{X}_0$) is, $\mathbf{X}_n$ converges in distribution to the distribution π of $\mathbf{Y}$. Therefore, π is the unique invariant distribution for $p(\mathbf{x}, d\mathbf{y})$.

For purposes of application it is useful to know that the assumption (13.12) holds if the maximum modulus of eigenvalues of $\mathbf{B}$, also known as the *spectral radius* $r(\mathbf{B})$ *of* $\mathbf{B}$, is less than 1. This fact is implied by the following result from linear algebra.

Lemma. Let $\mathbf{B}$ be an $m \times m$ matrix. Then the spectral radius $r(\mathbf{B})$ satisfies

$$r(\mathbf{B}) \geqslant \overline{\lim_{n\to\infty}}\, \|\mathbf{B}^n\|^{1/n}. \tag{13.17}$$

Proof. Let $\lambda_1, \ldots, \lambda_m$ be the eigenvalues of $\mathbf{B}$. This means $\det(\mathbf{B} - \lambda\mathbf{I}) = (\lambda_1 - \lambda)(\lambda_2 - \lambda)\cdots(\lambda_m - \lambda)$, where det is shorthand for determinant and $\mathbf{I}$ is the identity matrix. Let λ_m have the maximum modulus among the λ_i, i.e., $|\lambda_m| = r(\mathbf{B})$. If $|\lambda| > |\lambda_m|$ then $\mathbf{B} - \lambda\mathbf{I}$ is invertible, since $\det(\mathbf{B} - \lambda\mathbf{I}) \neq 0$. Indeed, by the definition of the inverse, each element of the inverse of $\mathbf{B} - \lambda\mathbf{I}$ is a polynomial in λ (of degree $m - 1$ or $m - 2$) divided by $\det(\mathbf{B} - \lambda\mathbf{I})$. Therefore, one may write

$$(\mathbf{B} - \lambda\mathbf{I})^{-1} = (\lambda_1 - \lambda)^{-1}\cdots(\lambda_m - \lambda)^{-1}(\mathbf{B}_0 + \lambda\mathbf{B}_2 + \cdots + \lambda^{m-1}\mathbf{B}_{m-1}) \qquad (|\lambda| > |\lambda_m|), \tag{13.18}$$

where $\mathbf{B}_j (0 \leqslant j \leqslant m - 1)$ are $m \times m$ matrices that do not involve λ. Writing $z = 1/\lambda$, one may express (13.18) as

$$(\mathbf{B} - \lambda\mathbf{I})^{-1} = (-\lambda)^{-m}(1 - \lambda_1/\lambda)^{-1} \cdots (1 - \lambda_m/\lambda)^{-1}\lambda^{m-1} \sum_{j=0}^{m-1} (1/\lambda)^{m-1-j}\mathbf{B}_j$$

$$= (-1)^m z(1 - \lambda_1 z)^{-1} \cdots (1 - \lambda_m z)^{-1} \sum_{j=0}^{m-1} z^{m-1-j}\mathbf{B}_j$$

$$= \left(z \sum_{n=0}^{\infty} a_n z^n\right) \sum_{j=0}^{m-1} z^{m-1-j}\mathbf{B}_j, \qquad (|z| < |\lambda_m|^{-1}). \tag{13.19}$$

On the other hand,

$$(\mathbf{B} - \lambda\mathbf{I})^{-1} = -z(\mathbf{I} - z\mathbf{B})^{-1} = -z \sum_{k=0}^{\infty} z^k \mathbf{B}^k \qquad \left(|z| < \frac{1}{\|\mathbf{B}\|}\right). \tag{13.20}$$

To see this, first note that the series on the right is convergent in norm for $|z| < 1/\|\mathbf{B}\|$, and then check that term-by-term multiplication of the series $\sum z^k \mathbf{B}^k$ by $\mathbf{I} - z\mathbf{B}$ yields the identity $\mathbf{I}$ after all cancellations. In particular, writing $b_{ij}^{(k)}$ for the (i, j) element of $\mathbf{B}^k$, the series

$$-z \sum_{k=0}^{\infty} z^k b_{ij}^{(k)} \tag{13.21}$$

converges absolutely for $|z| < 1/\|\mathbf{B}\|$. Since (13.21) is the same as the (i, j) element of the series (13.19), at least for $|z| < 1/\|\mathbf{B}\|$, their coefficients coincide (Exercise 4) and, therefore, the series in (13.21) is absolutely convergent for $|z| < |\lambda_m|^{-1}$ (as (13.19) is).

This implies that, for each $\varepsilon > 0$,

$$|b_{ij}^{(k)}| < (|\lambda_m| + \varepsilon)^k \qquad \text{for all sufficiently large } k. \tag{13.22}$$

For if (13.22) is violated, one may choose $|z|$ sufficiently close to (but less than) $1/|\lambda_m|$ such that $|z^{(k')} b_{ij}^{(k')}| \to \infty$ for a subsequence $\{k'\}$, contradicting the requirement that the terms of the convergent series (13.21) must go to zero for $|z| < 1/|\lambda_m|$.

Now $\|\mathbf{B}^k\| \leqslant m^{1/2} \max\{|b_{ij}^{(k)}| : 1 \leqslant i, j \leqslant m\}$ (Exercise 2). Since $m^{1/2k} \to 1$ as $k \to \infty$, (13.22) implies (13.17). ■

Two well-known time series models will now be treated as special cases of Example 2. These are the *pth order autoregressive* (or AR(p)) *model*, and the *autoregressive moving-average model* ARMA(p, q).

Example 2(a). (*AR*(p) *Model*). Let $p > 1$ be an integer, $\beta_0, \beta_1, \ldots, \beta_{p-1}$ real constants. Given a sequence of i.i.d. real-valued random variables $\{\eta_n : n \geqslant p\}$, and p other random variables $U_0, U_1, \ldots, U_{p-1}$ independent of $\{\eta_n\}$, define recursively

$$U_{n+p} := \sum_{i=0}^{p-1} \beta_i U_{n+i} + \eta_{n+p} \qquad (n \geqslant 0). \tag{13.23}$$

The sequence $\{U_n\}$ is not in general a Markov process, but the sequence of p-dimensional random vectors

$$\mathbf{X}_n := (U_n, U_{n+1}, \ldots, U_{n+p-1})' \qquad (n \geqslant 0) \tag{13.24}$$

is Markovian. Here the prime (′) denotes transposition, so $\mathbf{X}_n$ is to be regarded as a column vector in matrix operations. To prove the Markov property, consider the sequence of p-dimensional i.i.d. random vectors

$$\boldsymbol{\varepsilon}_n := (0, 0, \ldots, 0, \eta_{n+p-1})' \qquad (n \geqslant 1), \tag{13.25}$$

and note that

$$\mathbf{X}_{n+1} = \mathbf{B}\mathbf{X}_n + \boldsymbol{\varepsilon}_{n+1} \tag{13.26}$$

where $\mathbf{B}$ is the $p \times p$ matrix

$$\mathbf{B} := \begin{bmatrix} 0 & 1 & 0 & 0 & \cdots & 0 & 0 \\ 0 & 0 & 1 & 0 & \cdots & 0 & 0 \\ \cdot & \cdot & \cdot & \cdot & \cdots & \cdot & \cdot \\ 0 & 0 & 0 & 0 & \cdots & 0 & 1 \\ \beta_0 & \beta_1 & \beta_2 & \beta_3 & \cdots & \beta_{p-2} & \beta_{p-1} \end{bmatrix}. \tag{13.27}$$

Hence, arguing as in (13.2), (13.3), or (13.11), $\{\mathbf{X}_n\}$ is a Markov process on the state space $\mathbb{R}^p$. Write

$$\mathbf{B} - \lambda\mathbf{I} = \begin{bmatrix} -\lambda & 1 & 0 & 0 & \cdots & 0 & 0 \\ 0 & -\lambda & 1 & 0 & \cdots & 0 & 0 \\ \cdot & \cdot & \cdot & \cdot & \cdots & \cdot & \cdot \\ 0 & 0 & 0 & 0 & \cdots & -\lambda & 1 \\ \beta_0 & \beta_1 & \beta_2 & \beta_3 & \cdots & \beta_{p-2} & \beta_{p-1} - \lambda \end{bmatrix}.$$

Expanding $\det(\mathbf{B} - \lambda\mathbf{I})$ by its last row, and using the fact that the determinant of a matrix in *triangular form* (i.e., with all zero off-diagonal elements on one side of the diagonal) is the product of its diagonal elements (Exercise 5), one gets

$$\det(\mathbf{B} - \lambda\mathbf{I}) = (-1)^{p+1}(\beta_0 + \beta_1\lambda + \cdots + \beta_{p-1}\lambda^{p-1} - \lambda^p). \tag{13.28}$$

Therefore, the eigenvalues of $\mathbf{B}$ are the roots of the equation

$$\beta_0 + \beta_1\lambda + \cdots + \beta_{p-1}\lambda^{p-1} - \lambda^p = 0. \tag{13.29}$$

Finally, in view of (13.17), the following proposition holds (see (13.15) and Exercise 3).

Proposition 13.1. Suppose that the roots of the polynomial equation (13.29) are all strictly inside the unit circle in the complex plane, and that the common distribution G of $\{\eta_n\}$ satisfies

$$\sum_{n=1}^{\infty} G(\{x \in \mathbb{R}^1 : |x| > c\delta^n\}) < \infty \qquad \text{for some } \delta < \frac{1}{|\lambda_m|} \tag{13.30}$$

where $|\lambda_m|$ is the maximum modulus of the roots of (13.29). Then (i) there exists a unique invariant distribution π for the Markov process $\{\mathbf{X}_n\}$, and (ii) no matter what the initial distribution, $\mathbf{X}_n$ converges in distribution to π.

Once again it is simple to check that (13.30) holds if G has a finite absolute moment of some order $r > 0$ (Exercise 3).

An immediate consequence of Proposition 13.1 is that the time series $\{U_n : n \geqslant 0\}$ converges in distribution to a steady state π_U given, for all Borel sets $C \subset \mathbb{R}^1$, by

$$\pi_U(C) := \pi(\{\mathbf{x} \in \mathbb{R}^p : x^{(1)} \in C\}). \tag{13.31}$$

To see this, simply note that U_n is the first coordinate of $\mathbf{X}_n$, so that $\mathbf{X}_n$ converges to π in distribution implies U_n converges to π_U in distribution.

Example 2(b). (*ARMA*(p, q) *Model*). The *autoregressive moving-average model of order* (p, q), in short ARMA(p, q), is defined by

$$U_{n+p} := \sum_{i=0}^{p-1} \beta_i U_{n+i} + \sum_{j=1}^{q} \delta_j \eta_{n+p-j} + \eta_{n+p} \qquad (n \geqslant 0), \tag{13.32}$$

where p, q are positive integers, β_i $(0 \leqslant i \leqslant p-1)$ and δ_j $(1 \leqslant j \leqslant q)$ are real constants, $\{\eta_n : n \geqslant p - q\}$ is an i.i.d. sequence of real-valued random variables, and U_i $(0 \leqslant i \leqslant p-1)$ are arbitrary initial random variables independent of $\{\eta_n\}$. Consider the sequence $\{\mathbf{X}_n\}$, $\{\boldsymbol{\varepsilon}_n\}$ of $(p + q)$-dimensional vectors

$$\begin{aligned} \mathbf{X}_n &:= (U_n, \ldots, U_{n+p-1}, \eta_{n+p-q}, \ldots, \eta_{n+p-1})', \\ \boldsymbol{\varepsilon}_n &:= (0, 0, \ldots, 0, \eta_{n+p-1}, 0, \ldots, 0, \eta_{n+p-1})' \end{aligned} \qquad (n \geqslant 0), \tag{13.33}$$

where η_{n+p-1} occurs as the pth and $(p + q)$th elements of $\boldsymbol{\varepsilon}_n$.

$$\mathbf{X}_{n+1} = \mathbf{H}\mathbf{X}_n + \boldsymbol{\varepsilon}_{n+1} \qquad (n \geqslant 0), \tag{13.34}$$

where $\mathbf{H}$ is the $(p+q) \times (p+q)$ matrix

$$\mathbf{H} := \begin{pmatrix} b_{11} & \cdot & \cdots & b_{1p} & 0 & \cdot & \cdot & \cdots & 0 & 0 \\ \cdot & \cdot & \cdots & \cdot & \cdot & \cdot & \cdot & \cdots & \cdot & \cdot \\ b_{p1} & \cdot & \cdots & b_{pp} & \delta_q & \delta_{q-1} & \cdot & \cdots & \delta_2 & \delta_1 \\ 0 & \cdot & \cdots & 0 & 0 & 1 & 0 & \cdots & 0 & 0 \\ 0 & \cdot & \cdots & 0 & 0 & 0 & 1 & \cdots & 0 & 0 \\ \cdot & \cdot & \cdots & \cdot & \cdot & \cdot & \cdot & \cdots & \cdot & \cdot \\ 0 & 0 & \cdots & \cdot & \cdot & 0 & 0 & \cdots & 0 & 1 \\ 0 & 0 & \cdots & \cdot & \cdot & 0 & 0 & \cdots & 0 & 0 \end{pmatrix},$$

the first p rows and p columns of $\mathbf{H}$ being the matrix $\mathbf{B}$ in (13.27).

Note that $U_0, \ldots, U_{p-1}, \eta_{p-q}, \ldots, \eta_{p-1}$ determine $\mathbf{X}_0$, so that $\mathbf{X}_0$ is independent of η_p and, therefore, of $\boldsymbol{\varepsilon}_1$. It follows by induction that $\mathbf{X}_n$ and $\boldsymbol{\varepsilon}_{n+1}$ are independent. Hence $\{\mathbf{X}_n\}$ is a Markov process on the state space $\mathbb{R}^{p+q}$.

In order to apply the Lemma above, expand $\det(\mathbf{H} - \lambda\mathbf{I})$ in terms of the elements of its pth row to get (Exercise 5)

$$\det(\mathbf{H} - \lambda\mathbf{I}) = \det(\mathbf{B} - \lambda\mathbf{I})(-\lambda)^q. \tag{13.35}$$

Therefore, the eigenvalues of $\mathbf{H}$ are q zeros and the roots of (13.29). Thus, one has the following proposition.

Proposition 13.2. Under the hypothesis of Proposition 13.1, the ARMA(p, q) process $\{\mathbf{X}_n\}$ has a unique invariant distribution π, and $\mathbf{X}_n$ converges in distribution to π no matter what the initial distribution is.

As a corollary, the time series $\{U_n\}$ converges in distribution to π_U given for all Borel sets $C \subset \mathbb{R}^1$ by

$$\pi_U(C) := \pi(\{\mathbf{x} \in \mathbb{R}^{p+q} : x^{(1)} \in C\}), \tag{13.36}$$

no matter what the distribution of $(U_0, U_1, \ldots, U_{p-1})$ is, provided the hypothesis of Proposition 13.2 is satisfied.

In the case that $\boldsymbol{\varepsilon}_n$ is Gaussian, it is simple to check that under the hypothesis (13.12) in Example 2 the random vector $\mathbf{Y}$ in (13.16) is Gaussian. Therefore, π is Gaussian, so that the stationary vector-valued process $\{\mathbf{X}_n\}$ with initial distribution π is Gaussian (Exercise 6). In particular, if η_n are Gaussian in Example 2(a), and the roots of the polynomial equation (13.29) lie inside the unit circle in the complex plane, then the stationary process $\{U_n\}$, obtained when $(U_0, U_1, \ldots, U_{p-1})$ have distribution π in Example 2(a), is Gaussian. A similar assertion holds for Example 2(b).

14 MARKOV PROCESSES GENERATED BY ITERATIONS OF I.I.D. MAPS

The method of construction of Markov processes illustrated for linear time series models in Section 13 extends to more general Markov processes. The present section is devoted to the construction and analysis of some nonlinear models. Before turning to these models, note that one may regard the process $\{X_n\}$ in Example 13.1 (see (13.1)) to be generated by *successive iterations of* an i.i.d. sequence of *random maps* $\boldsymbol{\alpha}_1, \boldsymbol{\alpha}_2, \ldots, \boldsymbol{\alpha}_n, \ldots$ defined by

$$x \to \boldsymbol{\alpha}_n x = bx + \varepsilon_n \qquad (n \geqslant 1),$$

$\{\varepsilon_n : n \geqslant 1\}$ being a sequence of i.i.d. real-valued random variables. Each $\boldsymbol{\alpha}_n$ is random (affine linear) map on the state space $\mathbb{R}^1$ into itself. The Markov sequence $\{X_n\}$ is defined by

$$X_n = \boldsymbol{\alpha}_n \cdots \boldsymbol{\alpha}_1 X_0 \qquad (n \geqslant 1), \tag{14.1}$$

where the initial X_0 is a real-valued random variable independent of the sequence of random maps $\{\boldsymbol{\alpha}_n : n \geqslant 1\}$. A similar interpretation holds for the other examples of Section 13. Indeed it may be shown, under a very mild condition on the state space, that every Markov process in discrete time may be represented as (14.1) (see theoretical complement 1). Thus the method of the last section and the present one is truly a general device for constructing and analyzing Markov processes on general state spaces.

Example 1. (*Iterations of I.I.D. Increasing Maps*). Let the state space be an interval J, finite or infinite. On some probability space $(\Omega, \mathscr{F}, P)$ is given a sequence of i.i.d. continuous and increasing random maps $\{\boldsymbol{\alpha}_n : n \geqslant 1\}$ on J into itself. This means first of all that for each $\omega \in \Omega$, $\boldsymbol{\alpha}_n(\omega)$ is a continuous and increasing (i.e., nondecreasing) function on J into J. Second, there exists a set Γ of continuous increasing functions on J into J such that $P(\boldsymbol{\alpha}_n \in \Gamma) = 1$ for all n; Γ has a sigmafield $\mathscr{B}(\Gamma)$ generated by sets of the form $\{\gamma \in \Gamma : a < \gamma x < b\}$ where $a < b$ and x are arbitrary elements of J, and γx denotes the value of γ at x. The maps $\boldsymbol{\alpha}_n$ on Ω into Γ are measurable, i.e., $F_n := \{\omega \in \Omega : \boldsymbol{\alpha}_n(\omega) \in D\} \in \mathscr{F}$ for every $D \in \mathscr{B}(\Gamma)$. Also, $P(F_n)$ is the same for all n. Finally, $\{\boldsymbol{\alpha}_n : n \geqslant 1\}$ are independent, i.e., events $\{\boldsymbol{\alpha}_n \in D_n\}$ is an independent sequence for every given sequence $\{D_n\} \subset \mathscr{B}(\Gamma)$.

For any finite set of functions $\gamma_1, \gamma_2, \ldots, \gamma_k$ in Γ, one defines the *composition* $\gamma_1\gamma_2 \cdots \gamma_k$ in the usual manner. For example, $\gamma_1\gamma_2 x = \gamma_1(\gamma_2 x)$, the value of γ_1 at the point $\gamma_2 x$.

For each $x \in J$ define the sequence of random variables

$$X_0(x) := x, \qquad X_n(x) := \boldsymbol{\alpha}_n X_{n-1}(x) = \boldsymbol{\alpha}_n \boldsymbol{\alpha}_{n-1} \cdots \boldsymbol{\alpha}_1 x \qquad (n \geqslant 1). \quad (14.2)$$

In view of the independence of $\{\boldsymbol{\alpha}_n : n \geqslant 1\}$, $\{X_n : n \geqslant 0\}$ is a Markov process (Exercise 1) on J, starting at x and having the (one-step) transition probability

$$p(y, C) := P(\boldsymbol{\alpha}_n y \in C) = \mu(\{\gamma \in \Gamma : \gamma y \in C\}) \qquad (C \text{ Borel subset of } J), \quad (14.3)$$

where μ is the common distribution of $\boldsymbol{\alpha}_n$.

It will be shown now that the following condition guarantees the existence of a unique invariant probability π as well as *stability*, i.e., convergence of $X_n(x)$ in distribution to π for every initial state x. Assume

$$\delta_1 := P(X_{n_0}(x) \leqslant z_0 \forall x) > 0 \quad \text{and} \quad \delta_2 := P(X_{n_0}(x) \geqslant z_0 \forall x) > 0$$
$$\text{for some } z_0 \in J \text{ and some integer } n_0. \quad (14.4)$$

Define

$$\Delta_n := \sup_{x,y,z \in J} |P(X_n(x) \leqslant z) - P(X_n(y) \leqslant z)|. \quad (14.5)$$

For the existence of a unique invariant probability π and for stability it is enough to show that $\Delta_n \to 0$ as $n \to \infty$. For this implies $P(X_n(x) \leqslant z)$ converges, uniformly in $z \in J$, to a distribution function (of a probability measure on J). To see this last fact, observe that $X_{n+m}(x) \equiv \boldsymbol{\alpha}_{n+m} \cdots \boldsymbol{\alpha}_1 x$ has the same distribution as $\boldsymbol{\alpha}_n \cdots \boldsymbol{\alpha}_1 \boldsymbol{\alpha}_{n+m} \cdots \boldsymbol{\alpha}_{n+1} x$, so that

$$|P(X_{n+m}(x) \leqslant z) - P(X_n(x) \leqslant z)|$$
$$= |P(X_n(\boldsymbol{\alpha}_{n+m} \cdots \boldsymbol{\alpha}_{n+1} x) \leqslant z) - P(X_n(x) \leqslant z)| \leqslant \Delta_n,$$

by comparing the conditional probabilities given $\boldsymbol{\alpha}_{n+m}, \ldots, \boldsymbol{\alpha}_{n+1}$ first. Thus, if $\Delta_n \to 0$, then the sequence of distribution functions $\{P(X_n(x) \leqslant z)\}$ is a Cauchy sequence with respect to uniform convergence for $z \in J$. It is simple to check that the limiting function of this sequence is a distribution function of a probability measure π on J (Exercise 2). Further, $\Delta_n \to 0$ implies that this limit π does not depend on the initial state x, showing that P is the unique invariant probability (Exercise 13.1).

In order to prove $\Delta_n \to 0$, the first step is to establish, under the assumption (14.4), the inequality

$$\Delta_{n_0} \leqslant \delta := \max\{1 - \delta_1, 1 - \delta_2\}. \quad (14.6)$$

For this fix $x, y \in J$ and first take $z < z_0$. On the set $F_2 := \{X_{n_0}(x) \geqslant z_0 \forall x\}$ the events $\{X_{n_0}(x) \leqslant z\}$, $\{X_{n_0}(y) \leqslant z\}$ are both empty. Hence, by the second condition in (14.4),

$$|P(X_{n_0}(x) \leqslant z) - P(X_{n_0}(y) \leqslant z)| = |E(\mathbf{1}_{\{X_{n_0}(x) \leqslant z\}} - \mathbf{1}_{\{X_{n_0}(y) \leqslant z\}})| \leqslant P(F_2^c) = 1 - \delta_2, \tag{14.7}$$

since the difference between the two indicator functions in (14.7) is zero on F_2. Similarly, if $z > z_0$ then on the set $F_1 := \{X_{n_0}(x) \leqslant z_0 \forall x\}$ the two indicator functions both equal 1, so that their difference vanishes and one gets

$$|P(X_{n_0}(x) \leqslant z) - P(X_{n_0}(y) \leqslant z)] \leqslant P(F_1^c) = 1 - \delta_1. \tag{14.8}$$

Combining (14.7) and (14.8) one gets

$$|P(X_{n_0}(x) \leqslant z) - P(X_{n_0}(y) \leqslant z)| \leqslant \delta \tag{14.9}$$

for all $z \neq z_0$. But the function on the left is right-continuous in z. Therefore, letting $z \downarrow z_0$, (14.9) holds also for $z = z_0$. In other words, (14.6) holds.

Next note that Δ_n is monotonically decreasing,

$$\Delta_{n+1} \leqslant \Delta_n. \tag{14.10}$$

For,

$$|P(X_{n+1}(x) \leqslant z) - P(X_{n+1}(y) \leqslant z)|$$
$$= |P(\boldsymbol{\alpha}_{n+1} \cdots \boldsymbol{\alpha}_2 \boldsymbol{\alpha}_1 x \leqslant z) - P(\boldsymbol{\alpha}_{n+1} \cdots \boldsymbol{\alpha}_2 \boldsymbol{\alpha}_1 y \leqslant z)| \leqslant \Delta_n,$$

by comparing the conditional probabilities given $\boldsymbol{\alpha}_1$.

The final step in proving $\Delta_n \to 0$ is to show

$$\Delta_n \leqslant \delta^{[n/n_0]} \tag{14.11}$$

where $[r]$ is the integer part of r. In view of (14.10), it is enough to prove

$$\Delta_{jn_0} \leqslant \delta^j \qquad (j = 1, 2, \ldots). \tag{14.12}$$

Suppose that this is true for some $j \geqslant 1$. Then,

$$|P(X_{(j+1)n_0}(x) \leqslant z) - P(X_{(j+1)n_0}(y) \leqslant z)|$$
$$= |E(\mathbf{1}_{\{\boldsymbol{\alpha}_{(j+1)n_0} \cdots \boldsymbol{\alpha}_{jn_0+1} X_{jn_0}(x) \leqslant z\}} - \mathbf{1}_{\{\boldsymbol{\alpha}_{(j+1)n_0} \cdots \boldsymbol{\alpha}_{jn_0+1} X_{jn_0}(y) \leqslant z\}})|$$
$$= |E(\mathbf{1}_{\{X_{jn_0}(x) \in (\boldsymbol{\alpha}_{(j+1)n_0} \cdots \boldsymbol{\alpha}_{jn_0+1})^{-1}(-\infty, z]\}} - \mathbf{1}_{\{X_{jn_0}(y) \in (\boldsymbol{\alpha}_{(j+1)n_0} \cdots \boldsymbol{\alpha}_{jn_0+1})^{-1}(-\infty, z]\}})|. \tag{14.13}$$

Let

$$F_3 := \{\boldsymbol{\alpha}_{(j+1)n_0} \cdots \boldsymbol{\alpha}_{jn_0+1} x \leqslant z_0 \forall x\}, \qquad F_4 := \{\boldsymbol{\alpha}_{(j+1)n_0} \cdots \boldsymbol{\alpha}_{jn_0+1} x \geqslant z_0 \forall x\}.$$

Take $z < z_0$ first. Then the inverse image of $(-\infty, z]$ in (14.13) is empty on F_4, so that the difference between the two indicator functions vanishes on F_4. On the complement of F_4, the inverse image of $(-\infty, z]$ under the continuous increasing map $\boldsymbol{\alpha}_{(j+1)n_0} \cdots \boldsymbol{\alpha}_{jn_0+1}$ is an interval $(-\infty, Z'] \cap J$, where Z' is a random variable. Therefore, (14.13) leads to

$$|P(X_{(j+1)n_0}(x) \leqslant z) - P(X_{(j+1)n_0}(y) \leqslant z)| = |E\mathbf{1}_{F_4^c}(\mathbf{1}_{\{X_{jn_0}(x) \leqslant Z'\}} - \mathbf{1}_{\{X_{jn_0}(y) \leqslant Z'\}})|. \tag{14.14}$$

As F_4^c and Z' are determined by $\boldsymbol{\alpha}_{(j+1)n_0}, \ldots, \boldsymbol{\alpha}_{jn_0+1}$ and the latter are independent of $X_{jn_0}(x)$, $X_{jn_0}(y)$ one gets, by taking conditional expectation given $\{\boldsymbol{\alpha}_{(j+1)n_0}, \ldots, \boldsymbol{\alpha}_{jn_0+1}\}$,

$$\begin{aligned} &|P(X_{(j+1)n_0}(x) \leqslant z) - P(X_{(j+1)n_0}(y) \leqslant z)| \\ &\qquad \leqslant |E\mathbf{1}_{F_4^c}\Delta_{jn_0}| = (1 - \delta_2)\Delta_{jn_0} \leqslant \delta\Delta_{jn_0} \leqslant \delta^{j+1}. \end{aligned} \tag{14.15}$$

Similarly, if $z > z_0$, the inverse image in (14.13) is J on F_3. Therefore, the difference between the two indicator functions in (14.13) vanishes on F_3, and one has (14.14), (14.15) with F_4, δ_2 replaced by F_3 and δ_1. As (14.12) holds for $j = 1$ (see (14.6)), the induction is complete and (14.12) holds for all $j \geqslant 1$. Since $\delta < 1$, it follows that under the hypothesis (14.4), $\Delta_n \to 0$ exponentially fast as $n \to \infty$ and, therefore, there exists a unique invariant probability.

If J is a closed bounded interval $[a, b]$, then the condition (14.4) is essentially *necessary* for stability. To see this, define

$$Y_0(x) \equiv x, \qquad Y_n(x) := \boldsymbol{\alpha}_1\boldsymbol{\alpha}_2 \cdots \boldsymbol{\alpha}_n x \qquad (n \geqslant 1). \tag{14.16}$$

Then $Y_n(x)$ and $X_n(x)$ have the same distribution. Also,

$$\begin{aligned} Y_1(a) \geqslant a, \qquad Y_2(a) &= Y_1(\boldsymbol{\alpha}_2 a) \geqslant Y_1(a), \ldots \\ Y_{n+1}(a) &= Y_n(\boldsymbol{\alpha}_{n+1}a) \geqslant Y_n(a), \ldots \end{aligned}$$

i.e., the sequence of random variables $\{Y_n(a): n \geqslant 0\}$ is increasing. Similarly, $\{Y_n(b): n \geqslant 0\}$ is decreasing. Let the limits of these two sequences be $\underline{Y}$, $\overline{Y}$, respectively. As $Y_n(a) \leqslant Y_n(b)$ for all n, $\underline{Y} \leqslant \overline{Y}$. If $P(\underline{Y} < \overline{Y}) > 0$, then $\underline{Y}$ and $\overline{Y}$ cannot have the same distribution. In other words, $Y_n(a)$ (and, therefore, $X_n(a)$) and $Y_n(b)$ (and, therefore, $X_n(b)$) converge in distribution to different limits π_1, π_2 say. On the other hand, if $\underline{Y} = \overline{Y}$ a.s., then these limiting distributions are the same. Also, $Y_n(a) \leqslant Y_n(x) \leqslant Y_n(b)$ for all x, so that $Y_n(x)$ converges in distribution to the same limit π, whatever x. Therefore, π is the unique invariant probability. Assume that π does not assign all its mass at a single point. That is, rule out the case that with probability 1 all γ's in Γ have a common fixed point. Then there exist $m < M$ such that $P(\overline{Y} < m) > 0$ and $P(\underline{Y} > M) > 0$. There exists n_0 such that $P(Y_{n_0}(b) < m) > 0$ and $P(Y_{n_0}(a) > M) > 0$. Now any $z_0 \in [m, M]$ satisfies (14.4).

As an application of Example 1, consider the following example from economics.

Example 1(a). (*A Descriptive Model of Capital Accumulation*). Consider an economy that has a single producible good. The economy starts with an initial stock $X_0 = x > 0$ of this good which is used to produce an output Y_1 in period 1. The *output* Y_1 is not a deterministic function of the *input* x. In view of the randomness of the state of nature, Y_1 takes one of the values $f_r(x)$ with probability $p_r > 0$ $(1 \leqslant r \leqslant N)$. Here f_r are *production functions* having the following properties:

(i) f_r is twice continuously differentiable, $f_r'(x) > 0$ and $f_r''(x) < 0$ for all $x > 0$.

(ii) $\lim_{x \downarrow 0} f_r(x) = 0$, $\lim_{x \downarrow 0} f_r'(x) > 1$, $\lim_{x \uparrow \infty} f_r'(x) = 0$.

(iii) If $r > r'$, then $f_r(x) > f_{r'}(x)$ for all $x > 0$.

The strict concavity of f_r in (i) reflects a *law of diminishing returns*, while (iii) assumes an ordering of the technologies or production functions f_r, from the least productive f_1 to the most productive f_N.

A fraction β $(0 \leqslant \beta < 1)$ of the output Y_1 is *consumed*, while the rest $(1 - \beta)Y_1$ is invested for the production in the next period. The total stock X_1 at hand for investment in period 1 is $\theta X_0 + (1 - \beta)Y_1$. Here $\theta < 1$ is the rate of *depreciation* of capital used in production. This process continues indefinitely, each time with an independent choice of the production function (f_r with probability p_r, $1 \leqslant r \leqslant N$). Thus, the capital X_{n+1} at hand in period $n + 1$ satisfies

$$X_{n+1} = \theta X_n + (1 - \beta)\varphi_{n+1}(X_n) \qquad (n \geqslant 0), \tag{14.17}$$

where φ_n is the random production function in period n,

$$P(\varphi_n = f_r) = p_r \qquad (1 \leqslant r \leqslant N),$$

and the φ_n $(n \geqslant 1)$ are independent. Thus the Markov process $\{X_n(x)\colon n \geqslant 0\}$ on the *state space* $(0, \infty)$ may be represented as

$$X_n(x) = \boldsymbol{\alpha}_n \cdots \boldsymbol{\alpha}_1 x,$$

where, writing

$$g_r(x) := \theta x + (1 - \beta) f_r(x), \qquad 1 \leqslant r \leqslant N, \tag{14.18}$$

one has

$$P(\boldsymbol{\alpha}_n = g_r) = p_r \qquad (1 \leqslant r \leqslant N). \tag{14.19}$$

Suppose, in addition to the assumptions already made, that

$$\theta + (1-\beta)\lim_{x\downarrow 0} f_r'(x) > 1 \qquad (1 \leqslant r \leqslant N), \tag{14.20}$$

i.e., $\lim_{x\downarrow 0} g_r'(x) > 1$ for all r. As $\lim_{x\to\infty} g_r'(x) = \theta + (1-\beta)\lim_{x\to\infty} f_r'(x) = \theta < 1$, it follows from the strictly increasing and strict concavity properties of g_r that each g_r has a *unique fixed point* a_r (see Figure 14.1)

$$g_r(a_r) = a_r \qquad (1 \leqslant r \leqslant N). \tag{14.21}$$

Note that by property (iii) of f_r, $a_1 < a_2 < \cdots < a_N$. If $y \geqslant a_1$, then $g_r(y) \geqslant g_r(a_1) \geqslant g_1(a_1) = a_1$, so that $X_n(x) \geqslant a_1$ for all $n \geqslant 0$ if $x \geqslant a_1$. Similarly, if $y \leqslant a_N$ then $g_r(y) \leqslant g_r(a_N) \leqslant g_N(a_N) = a_N$, so that $X_n(x) \leqslant a_N$ for all $n \geqslant 0$ if $x \leqslant a_N$. As a consequence, if the initial state x is in $[a_1, a_N]$, then the process $\{X_n(x): n \geqslant 0\}$ remains in $[a_1, a_N]$ forever. In this case, one may take $J = [a_1, a_N]$ to be the effective state space. Also, if $x \geqslant a_1$ then the nth iterate of g_1, namely $g_1^{(n)}(x)$, decreases as n increases. For if $x \geqslant a_1$, then $g_1(x) \leqslant x$, $g_1^{(2)}(x) = g_1(g_1(x)) \leqslant g_1(x)$, etc. The limit of this decreasing sequence is a fixed point of g_1 (Exercise 3) and, therefore, must be a_1. Similarly, if $x \leqslant a_N$ then $g_N^{(n)}(x)$ increases, as n increases, to a_N. In particular,

$$\lim_{n\to\infty} g_1^{(n)}(a_N) = a_1, \qquad \lim_{n\to\infty} g_N^{(n)}(a_1) = a_N.$$

Thus, there exists an integer n_0 such that

$$g_1^{(n_0)}(a_N) < g_N^{(n_0)}(a_1). \tag{14.22}$$

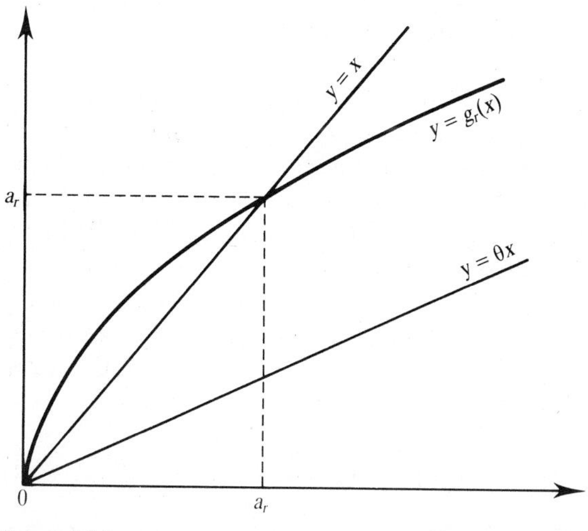

Figure 14.1

This means that if $z_0 \in [g_1^{(n_0)}(a_N), g_N^{(n_0)}(a_1)]$, then

$$P(X_{n_0}(x) \leqslant z_0 \forall x \in [a_1, a_N]) \geqslant P(\boldsymbol{\alpha}_n = g_1 \text{ for } 1 \leqslant n \leqslant n_0) = p_1^{n_0} > 0,$$

$$P(X_{n_0}(x) \geqslant z_0 \forall x \in [a_1, a_N]) \geqslant P(\boldsymbol{\alpha}_n = g_N \text{ for } 1 \leqslant n \leqslant n_0) = p_N^{n_0} > 0.$$

Hence, the condition (14.4) of Example 1 holds, and there exists a unique invariant probability π, if the state space is taken to be $[a_1, a_N]$.

Next fix the initial state x in $(0, a_1)$. Then $g_1^{(n)}(x)$ increases, as n increases. The limit must be a fixed point and, therefore, a_1. Since $g_r(a_1) > a_1$ for $r = 2, \ldots, N$, there exists $\varepsilon > 0$ such that $g_r(y) > a_1$ $(2 \leqslant r \leqslant N)$ if $y \in [a_1 - \varepsilon, a_1]$. Now find n_ε such that $g_1^{(n_\varepsilon)}(x) \geqslant a_1 - \varepsilon$. If $\tau_1 := \inf\{n \geqslant 1: X_n(x) \geqslant a_1\}$, then it follows from the above that

$$P(\tau_1 > n_\varepsilon + k) \leqslant p_1^k \qquad (k \geqslant 1),$$

because $\tau_1 > n_\varepsilon + k$ implies that the last k among the first $n_\varepsilon + k$ function $\boldsymbol{\alpha}_n$ are g_1. Since p_1^k goes to zero as $k \to \infty$, it follows from this that τ_1 is a.s. finite. Also $X_{\tau_1}(x) \leqslant a_N$ as $g_r(y) \leqslant g_r(a_N) \leqslant g_N(a_N) = a_N$ $(1 \leqslant r \leqslant N)$ for $y \leqslant a_1$, so that in a single step it is not possible to go from a state less than a_1 to a state larger than a_N. By the strong Markov property, and the result in the preceding paragraph on the existence of a unique invariant distribution and stability on $[a_1, a_N]$, it follows that $X_{\tau_1 + m}(x)$ converges in distribution to π, as $m \to \infty$ (Exercise 5). From this, one may show that $p^{(n)}(x, dy)$ converges weakly to $\pi(dy)$ for all x, as $n \to \infty$, so that π is the unique invariant probability on $(0, \infty)$ (Exercise 5).

In the same manner it may be checked that $X_n(x)$ converges in distribution to π if $x > a_N$. Thus, no matter what the initial state x is, $X_n(x)$ converges in distribution to π. Therefore, on the state space $(0, \infty)$ there exists a unique invariant distribution π (assigning probability 1 to $[a_1, a_N]$), and stability holds. In analogy with the case of Markov chains, one may call the set of states $\{x; 0 < x < a_1 \text{ or } x > a_N\}$ *inessential.*

The study of the existence of unique invariant probabilities and stability is relatively simpler for those cases in which the transition probabilities $p(x, dy)$ have a density $p(x, y)$, say, with respect to some reference measure $\mu(dy)$ on the state space. In the case of Markov chains this measure may be taken to be the counting measure, assigning mass 1 to each singleton in the state space. For a class of simple examples with an uncountable state space, let $S = \mathbb{R}^1$ and f a bounded measurable function on $\mathbb{R}^1$, $a \leqslant f(x) \leqslant b$. Let $\{\varepsilon_n\}$ be an i.i.d. sequence of real-valued random variables whose common distribution has a strictly positive continuous density φ with respect to Lebesgue measure on $\mathbb{R}^1$. Consider the Markov process

$$X_{n+1} := f(X_n) + \varepsilon_{n+1} \qquad (n \geqslant 0), \tag{14.23}$$

with X_0 arbitrary (independent of $\{\varepsilon_n\}$). Then the transition probability $p(x, dy)$

has the density

$$p(x, y) := \varphi(y - f(x)). \tag{14.24}$$

Note that

$$\varphi(y - f(x)) \geqslant \psi(y) \qquad \text{for all } x \in \mathbb{R}^1, \tag{14.25}$$

where

$$\psi(y) := \min\{\varphi(y - z): a \leqslant z \leqslant b\} > 0.$$

Then (see theoretical complement 6.1) it follows that this Markov process has a unique invariant probability with a density $\pi(y)$ and that the distribution of X_n converges to $\pi(y)\,dy$, whatever the initial state.

The following example illustrates the dramatic difference between the cases when a density exists and when it does not.

Example 2. Let $S = [-2, 2]$ and consider the Markov process $X_{n+1} = f(X_n) + \varepsilon_{n+1}$ $(n \geqslant 0)$, X_0 independent of $\{\varepsilon_n\}$, where $\{\varepsilon_n\}$ is an i.i.d. sequence with values in $[-1, 1]$, and

$$f(x) = \begin{cases} x + 1 & \text{if } -2 \leqslant x \leqslant 0, \\ x - 1 & \text{if } \ \ 0 < x \leqslant 2. \end{cases} \tag{14.26}$$

First let ε_n be *Bernoulli*, $P(\varepsilon_n = 1) = \frac{1}{2} = P(\varepsilon_n = -1)$. Then, with $X_0 \equiv x \in (0, 2]$,

$$X_1(x) = \begin{cases} x - 2 & \text{with probability } \frac{1}{2}, \\ x & \text{with probability } \frac{1}{2}, \end{cases}$$

and $X_1(x - 2)$ has the same distribution as $X_1(x)$. It follows that

$$P(X_2(x) = x - 2 \mid X_1(x) = x) = \tfrac{1}{2} = P(X_2(x) = x - 2 \mid X_1(x) = x - 2),$$
$$P(X_2(x) = x \mid X_1(x) = x) = \tfrac{1}{2} = P(X_2(x) = x \mid X_1(x) = x - 2).$$

In other words, $X_1(x)$ and $X_2(x)$ are independent and have the same two-point distribution π_x. It follows that $\{X_n(x): n \geqslant 1\}$ is i.i.d. with common distribution π_x. In particular, π_x is an invariant initial distribution. If $x \in [-2, 0]$, then $\{X_n(x): n \geqslant 1\}$ is i.i.d. with common distribution π_{x+2}, assigning probabilities $\frac{1}{2}$ and $\frac{1}{2}$ to $\{x + 2\}$ and $\{x\}$. Thus, there is an uncountable family of invariant initial distributions $\{\pi_x: 0 < x < 1\} \cup \{\pi_{x+2}: -1 \leqslant x \leqslant 0\}$.

On the other hand, suppose ε_n is uniform on $[-1, 1]$, i.e., has the density $\frac{1}{2}$ on $[-1, 1]$ and zero outside. Check that (Exercise 6) $\{X_{2n}(x): n \geqslant 1\}$ is an i.i.d.

sequence whose common distribution does not depend on x and has a density

$$\pi(y) := \frac{2 - |y|}{4}, \qquad -2 \leqslant y \leqslant 2. \tag{14.27}$$

Thus, $\pi(y)\,dy$ is the unique invariant probability, and stability holds.

The final example deals with absorption probabilities.

Example 3. (*Survival Probability of an Economic Agent*). Suppose that in a one-good economy the agent starts with an initial stock x. A fixed amount $c > 0$ is consumed ($x > c$) and the remainder $x - c$ is invested for production in the next period. The stock produced in period 1 is $X_1 = \varepsilon_1(X_0 - c) = \varepsilon_1(x - c)$, where ε_1 is a nonnegative random variable. Again, after consumption, $X_1 - c$ is invested in production of a stock of $\varepsilon_2(X_1 - c)$, provided $X_1 > c$. If $X_1 \leqslant c$, the agent is *ruined*. In general,

$$X_0 = x, \qquad X_{n+1} = \varepsilon_{n+1}(X_n - c) \qquad (n \geqslant 0), \tag{14.28}$$

where $\{\varepsilon_n : n \geqslant 1\}$ is an i.i.d. sequence of nonnegative random variables. The state space may be taken to be $[0, \infty)$ with *absorption* at 0. The *probability of survival* of the economic agent, starting with an initial stock $x > c$, is

$$\rho(x) := P(X_n > c \text{ for all } n \geqslant 0 \mid X_0 = x). \tag{14.29}$$

If $\delta := P(\varepsilon_1 = 0) > 0$, then it is simple to check that $P(\varepsilon_n = 0$ for some $n \geqslant 0) = 1$ (Exercise 8), so that $\rho(x) = 0$ for all x. Therefore, assume

$$P(\varepsilon_1 > 0) = 1. \tag{14.30}$$

From (14.28) one gets, by successive iteration,

$$X_{n+1} > c \text{ iff } X_n > c + \frac{c}{\varepsilon_{n+1}} \text{ iff } X_{n-1} > c + \frac{c + \dfrac{c}{\varepsilon_{n+1}}}{\varepsilon_n} = c + \frac{c}{\varepsilon_n} + \frac{c}{\varepsilon_n \varepsilon_{n+1}} \cdots$$

$$\text{iff } X_0 \equiv x > c + \frac{c}{\varepsilon_1} + \frac{c}{\varepsilon_1\varepsilon_2} + \cdots + \frac{c}{\varepsilon_1\varepsilon_2\cdots\varepsilon_{n+1}}.$$

Hence, on the set $\{\varepsilon_n > 0 \text{ for all } n\}$,

$$\{X_n > c \text{ for all } n\} = \left\{x > c + \frac{c}{\varepsilon_1} + \frac{c}{\varepsilon_1\varepsilon_2} + \cdots + \frac{1}{\varepsilon_1\varepsilon_2\cdots\varepsilon_n} \text{ for all } n\right\}$$

$$= \left\{x \geqslant c + c\sum_{n=1}^{\infty} \frac{1}{\varepsilon_1\varepsilon_2\cdots\varepsilon_n}\right\} = \left\{\sum_{n=1}^{\infty} \frac{1}{\varepsilon_1\varepsilon_2\cdots\varepsilon_n} \leqslant \frac{x}{c} - 1\right\}.$$

In other words,

$$\rho(x) = P\left\{\sum_{n=1}^{\infty} \frac{1}{\varepsilon_1\varepsilon_2\cdots\varepsilon_n} \leqslant \frac{x}{c} - 1\right\}. \tag{14.31}$$

This formula will be used to determine conditions on the common distribution of ε_n under which (1) $\rho(x) = 0$, (2) $\rho(x) = 1$, (3) $\rho(x) < 1$ $(x > c)$. Suppose first that $E \log \varepsilon_1$ exists and $E \log \varepsilon_1 < 0$. Then, by the Strong Law of Large Numbers,

$$\frac{1}{n}\sum_{r=1}^{n} \log \varepsilon_n \xrightarrow[\text{a.s.}]{} E \log \varepsilon_1 < 0,$$

so that $\log \varepsilon_1\varepsilon_2\cdots\varepsilon_n \to -\infty$ a.s., or $\varepsilon_1\varepsilon_2\cdots\varepsilon_n \to 0$ a.s. This implies that the infinite series in (14.31) diverges a.s., that is,

$$\rho(x) = 0 \qquad \text{for all } x, \text{ if } E \log \varepsilon_1 < 0. \tag{14.32}$$

Now by Jensen's Inequality (Chapter 0, Section 2), $E \log \varepsilon_1 \leqslant \log E\varepsilon_1$, with strict inequality unless ε_1 is degenerate. Therefore, if $E\varepsilon_1 < 1$, or $E\varepsilon_1 = 1$ and ε_1 is nondegenerate, then $E \log \varepsilon_1 < 0$. If ε_1 is degenerate and $E\varepsilon_1 = 1$, then $P(\varepsilon_1 = 1) = 1$, and the infinite series in (14.31) diverges. Therefore, (14.32) implies

$$\rho(x) = 0 \qquad \text{for all } x, \text{ if } E\varepsilon_1 \leqslant 1. \tag{14.33}$$

It is not true, however, that $E \log \varepsilon_1 > 0$ implies $\rho(x) = 1$ for large x. To see this and for some different criteria, define

$$m := \inf\{z \geqslant 0 \colon P(\varepsilon_1 \leqslant z) > 0\}. \tag{14.34}$$

Let us show that

$$\rho(x) < 1 \qquad \text{for all } x, \text{ if } m \leqslant 1. \tag{14.35}$$

For this fix $A > 0$, however large. Find n_0 such that

$$n_0 > A \prod_{r=1}^{\infty}\left(1 + \frac{1}{r^2}\right). \tag{14.36}$$

This is possible, as $\prod (1 + 1/r^2) < \exp\{\sum 1/r^2\} < \infty$. If $m \leqslant 1$ then $P(\varepsilon_1 \leqslant 1 + 1/r^2) > 0$ for all $r \geqslant 1$. Hence,

$$\begin{aligned}
&0 < P(\varepsilon_r \leqslant 1 + 1/r^2 \text{ for } 1 \leqslant r \leqslant n_0)\\
&\leqslant P\left(\sum_{r=1}^{n_0} \frac{1}{\varepsilon_1\cdots\varepsilon_r} \geqslant \sum_{r=1}^{n_0} \frac{1}{\prod_{j=1}^{r}\left(1 + \frac{1}{j^2}\right)}\right) \leqslant P\left(\sum_{r=1}^{n_0} \frac{1}{\varepsilon_1\cdots\varepsilon_r} \geqslant \frac{n_0}{\prod_{j=1}^{\infty}\left(1 + \frac{1}{j^2}\right)}\right)\\
&\leqslant P\left(\sum_{r=1}^{n_0} \frac{1}{\varepsilon_1\cdots\varepsilon_r} > A\right) \leqslant P\left(\sum_{r=1}^{\infty} \frac{1}{\varepsilon_1\cdots\varepsilon_r} > A\right).
\end{aligned}$$

Because A is arbitrary, (14.31) is less than 1 for all x, proving (14.35).

One may also show that, if $m > 1$, then

$$\rho(x) = \begin{cases} < 1 & \text{if } x < c\left(\dfrac{m}{m-1}\right), \\ = 1 & \text{if } x \geqslant c\left(\dfrac{m}{m-1}\right); \end{cases} \qquad (m > 1). \tag{14.37}$$

To prove this, observe that

$$\sum_{n=1}^{\infty} \frac{1}{\varepsilon_1 \cdots \varepsilon_n} \leqslant \sum_{n=1}^{\infty} \frac{1}{m^n} = \frac{1}{m-1},$$

with probability 1 (if $m > 1$). Therefore, (14.31) implies the second relation in (14.37). In order to prove the first relation in (14.37), let $x < cm/(m-1) - c\delta$ for some $\delta > 0$. Then $x/c - 1 < 1/(m-1) - \delta$. Choose $n(\delta)$ such that

$$\sum_{r=n(\delta)}^{\infty} \frac{1}{m^r} < \frac{\delta}{2}, \tag{14.38}$$

and then choose $\delta_r > 0$ $(1 \leqslant r \leqslant n(\delta) - 1)$ such that

$$\sum_{r=1}^{n(\delta)-1} \frac{1}{(m+\delta_1)\cdots(m+\delta_r)} > \sum_{r=1}^{n(\delta)-1} \frac{1}{m^r} - \frac{\delta}{2}. \tag{14.39}$$

Then

$$0 < P(\varepsilon_r < m + \delta_r \text{ for } 1 \leqslant r \leqslant n(\delta) - 1) \leqslant P\left(\sum_{r=1}^{n(\delta)-1} \frac{1}{\varepsilon_1 \cdots \varepsilon_r} > \sum_{r=1}^{n(\delta)-1} \frac{1}{m^r} - \frac{\delta}{2}\right)$$

$$\leqslant P\left(\sum_{r=1}^{\infty} \frac{1}{\varepsilon_1 \varepsilon_2 \cdots \varepsilon_r} > \sum_{r=1}^{\infty} \frac{1}{m^r} - \delta\right) = P\left(\sum_{r=1}^{\infty} \frac{1}{\varepsilon_1 \cdots \varepsilon_r} > \frac{1}{m-1} - \delta\right).$$

If $\delta > 0$ is small enough, the last probability is smaller than $P(\sum 1/(\varepsilon_1 \cdots \varepsilon_r) > x/c - 1)$, provided $x/c - 1 < 1/(m-1)$, i.e., if $x < cm/(m-1)$. Thus for such x one has $1 - \rho(x) > 0$, proving the first relation in (14.37).

15 CHAPTER APPLICATION: DATA COMPRESSION AND ENTROPY

The mathematical theory of information storage and retrieval rests largely on foundations established by Claude Shannon. In the present section we will consider one aspect of the general theory. We will suppose that "text" is constructed from symbols in a finite alphabet $S = \{a_1, a_2, \ldots, a_M\}$. The term "text" may be interpreted rather broadly and need not be restricted to the text of ordinary human language; other usages occur in genetic sequences of DNA, computer data storage, music, etc. However, applications to linguistics have played a central role in the historical development as will be discussed below. An encoding algorithm is a transformation in which finite sequences of text are replaced by sequences of code symbols in such a way that it must be possible to uniquely reconstruct (decode) the original text from the encoded text. For simplicity we shall consider compression codes in which the same alphabet is available for code symbols. Any sequence of symbols is referred to as a *word* and the number of symbols as its *length*. A word of length t will be encoded into a code-word of length s by the encoding algorithm. To compress the text one would use a short code for frequently occurring words and reserve the longer code-words for the more rarely occurring sequences. It is in this connection that the statistical structure of text (i.e., word frequencies) will play an important role. We consider a case in which the symbols occur in sequence according to a Markov chain having a stationary transition law $((p_{ij}: i\,j = 1, 2, \ldots, M))$. Shannon has suggested the following scheme for generating a Markov approximation to English text. Open a book and select a letter at random, say T. Next skip a few lines and read until a T is encountered, and select the letter that follows this T (we observed an H in our trial). Next skip a few more lines, and read until an H is encountered and take the next letter (we observed an A), and so on to generate a sample of text that should be more closely Markovian than text composed according to the usual rules of English grammar. Moreover, one expects this to resemble more closely the structure of the English language than independent samples of randomly selected single letters. Accordingly, one may also consider higher-order Markov approximations to the structure of English text, for example, by selecting letters according to the two preceding letters. Likewise, as was also done by Shannon and others, one may generate text by using "linguistic words" as the basic alphabetic symbols (see theoretical complement 1 for references). It is of significant historical notice that, in spite of a modern-day widespread utility in diverse physical sciences, Markov himself developed his ideas on dependence with linguistic applications in mind.

Let $X_0, X_1, \ldots$ denote the Markov chain with state space S having the stationary transition law $((p_{ij}))$ and a unique invariant initial distribution $\pi = (\pi_i)$. Then $\{X_n\}$ is a stationary process and the word $\alpha = (a_{i_1}, a_{i_2}, \ldots, a_{i_t})$ of length t has probability $\pi_{i_1} p_{i_1,i_2} \cdots p_{i_{t-1},i_t}$ of occurring. Suppose that under the coding transformation the word $\alpha = (a_{i_1}, \ldots, a_{i_t})$ is encoded to a word of length $s = c(a_{i_1}, \ldots, a_{i_t})$. Let

$$\mu_t = E[c(X_1, \ldots, X_t)]/t. \tag{15.1}$$

The quantity μ_t is referred to as the *average compression* for words of length t. The optimal extent to which a given (statistical) type of text can be compressed by a code is measured by the so-called *compression coefficient* defined by

$$\mu = \limsup_{t \to \infty} \mu_t. \tag{15.2}$$

The problem for our consideration here is to calculate the compression coefficient in terms of the parameters of the given Markov structure of the text and to construct an optimum compression code. We will show that the optimal compression coefficient is given by

$$\mu = \frac{H}{\log M} = \frac{\sum_i \pi_i H_i}{\log M} = -\frac{\sum_i \pi_i \left[\sum_j p_{ij} \log p_{ij}\right]}{\log M}, \tag{15.3}$$

in the sense that the compression coefficient of a code is never smaller than this, although there are codes whose coefficient is arbitrarily close to it.

The parameter $H_i = -\sum_j p_{ij} \log p_{ij}$ is referred to as the *entropy* of the transition distribution from state i and is a measure of the information obtained when the Markov chain moves one step ahead out of state i (Exercises 1 and 2). The quantity $H = \sum_i \pi_i H_i$ is called the *entropy* of the Markov chain. Observe that, given the transition law of the Markov chain, the optimal compression coefficient may easily be computed from (15.3) once the invariant initial distribution is determined.

For a word α of length t let $p_t(\alpha) = P((X_0, \dots, X_{t-1}) = \alpha)$. Then,

$$-\log p_t((X_0, \dots, X_{t-1})) = -\log \pi(X_0) + \sum_i (-\log p_{X_i, X_{i+1}}) = Y_0 + \sum_i Y_i \tag{15.4}$$

where $Y_1, Y_2, \dots$ is a stationary sequence of bounded random variables. By the law of large numbers applied to the stationary sequence $Y_1, Y_2, \dots$, we have by Theorems 9.2 and 9.3 (see Exercise 10.5),

$$-\frac{\log p_t((X_0, \dots, X_{t-1}))}{t} \to EY_1 = H \tag{15.5}$$

as $t \to \infty$ with probability 1 (Exercise 4); i.e., for almost all sample realizations, for large t the probability of the sequence $X_0, X_1, \dots, X_{t-1}$ is approximately $\exp\{-tH\}$. The result (15.5) is quite remarkable. It has a natural generalization that applies to a large class of stationary processes so long as a law of large numbers applies (Exercise 4).

An important consequence of (15.5) that will be used below is obtained by considering for each t the M^t words of length t arranged as $\alpha_{(1)}, \alpha_{(2)}, \dots$ in order of decreasing probability. For any positive number $\varepsilon < 1$, let

$$N_t(\varepsilon) = \min\left\{N : \sum_{i=1}^{N} p_t(\alpha_{(i)}) \geqslant \varepsilon\right\}. \tag{15.6}$$

Proposition 15.1. For any $0 < \varepsilon < 1$,

$$\lim_{t \to \infty} \frac{\log N_t(\varepsilon)}{t} = H. \tag{15.7}$$

Proof. Since almost sure convergence implies convergence in probability, it follows from (15.5) that for arbitrarily small positive numbers γ and δ, $\delta < \max\{\varepsilon, 1 - \varepsilon\}$, for all sufficiently large t we have

$$P\left(\left| -\frac{\log p_t(X_0, \ldots, X_{t-1})}{t} - H \right| < \gamma \right) > 1 - \delta.$$

In particular, for all sufficiently large t, say $t \geqslant T$,

$$\exp\{-t(H + \gamma)\} < p_t(X_0, \ldots, X_{t-1}) < \exp\{-t(H - \gamma)\}$$

with probability at least $1 - \delta$. Let R_t denote the set consisting of all words α of length t such that $e^{-t(H+\gamma)} < p_t(\alpha) < e^{-t(H-\gamma)}$. Fix t larger than T. Let $S_t = \{\alpha_{(1)}, \alpha_{(2)}, \ldots, \alpha_{(N_t(\varepsilon))}\}$. The sum of the probabilities of the $M_t(\varepsilon)$, say, words α of length t in R_t that are counted among the $N_t(\varepsilon)$ words in S_t equals $\sum_{\alpha \in S_t \cup R_t} p_t(\alpha) > \varepsilon - \delta$ by definition of $N_t(\varepsilon)$. Therefore,

$$N_t(\varepsilon) e^{-t(H-\gamma)} \geqslant M_t(\varepsilon) e^{-t(H-\gamma)} \geqslant \sum_{\alpha \in S_t \cup R_t} p_t(\alpha) > \varepsilon - \delta. \tag{15.8}$$

Also, none of the elements of S_t has probability less than $\exp\{-t(H + \gamma)\}$, since the set of all α with $p_t(\alpha) > \exp\{-t(H + \gamma)\}$ contains R_t and has total probability larger than $1 - \delta > \varepsilon$. Therefore,

$$N_t(\varepsilon) e^{-t(H+\gamma)} < 1. \tag{15.9}$$

Taking logarithms, we have

$$\frac{\log N_t(\varepsilon)}{t} < H + \gamma.$$

On the other hand, by (15.8),

$$N_t(\varepsilon) e^{-t(H-\gamma)} > \varepsilon - \delta. \tag{15.10}$$

Again taking logarithms and now combining this with (15.9), we get

$$H - \gamma + o(1) < \frac{\log N_t(\varepsilon)}{t} < H + \gamma, \qquad t \to \infty.$$

Since γ and δ are arbitrarily small the proof is complete. ∎

Returning to the problem of calculating the compression coefficient, first let us show that $\mu \geqslant H/\log M$. Let $\delta > 0$ and let $H' = H - 2\delta < H$. For an arbitrary given code, let

$$J_t = \{\alpha \mid \alpha \text{ is a word of length } t \text{ and } c(\alpha) < tH'/\log M\}. \qquad (15.11)$$

Then

$$\# J_t \leqslant M + M^2 + \cdots + M^{[tH'/\log M]} \leqslant M^{(tH'/\log M)}\{1 + 1/M + \cdots\} = \frac{e^{tH'}M}{M-1}, \qquad (15.12)$$

since the number of code-words of length k is M^k. Now observe that

$$t\mu_t = Ec(X_1, \ldots, X_t) \geqslant \frac{tH'}{\log M} P\{(X_1, \ldots, X_t) \in J_t^c\}$$
$$= \frac{tH'}{\log M}[1 - P\{(X_1, \ldots, X_t) \in J_t\}]. \qquad (15.13)$$

Therefore,

$$\mu = \limsup_t \mu_t \geqslant \frac{H'}{\log M} \limsup_t [1 - P\{(X_1, \ldots, X_t) \in J_t\}]. \qquad (15.14)$$

Now observe that for any positive number $\varepsilon < 1$, for the probability $P(\{(X_1, \ldots, X_t) \in J_t\})$ to exceed ε requires that $N_t(\varepsilon)$ be smaller than $\# J_t$. In view of (15.12) this means that

$$N_t(\varepsilon) < \frac{M}{M-1}\exp\{t(H - 2\delta)\} \qquad (15.15)$$

or

$$\frac{\log N_t(\varepsilon)}{t} < o(1) + H - 2\delta. \qquad (15.16)$$

Now by Proposition 15.1 for any given ε this can hold for at most finitely many values of t. In other words, we must have that the probability $P(\{(X_1, \ldots, X_t) \in J_t\})$ tends to 0 in the limit as t grows without bound. Therefore, (15.14) becomes

$$\mu \geqslant \frac{H'}{\log M} = \frac{H - 2\delta}{\log M}, \qquad (15.17)$$

and since $\delta > 0$ is arbitrary we get $\mu \geqslant H/\log M$ as desired.

To prove the reverse inequality, and therefore (15.3), again let δ be an arbitrarily small positive number. We shall construct a code whose compression coefficient μ does not exceed $(H + \delta)/\log M$. For arbitrary positive numbers γ and ε, we have

$$N_t(1 - \varepsilon) < e^{t(H+\gamma)} = M^{t(H+\gamma)/\log M}, \tag{15.18}$$

for all sufficiently large t. That is, the number of (relatively) high-probability words of length t, the sum of whose probabilities exceeds $1 - \varepsilon$, is no greater than the number $M^{t(H+\gamma)/\log M}$ of words of length $t(H + \gamma)/\log M$. Therefore, there are enough distinct sequences of length $t(H + \gamma)/\log M$ to code the $N_t(1 - \varepsilon)$ words "most likely to occur." For the lower-probability words, the sum of whose probabilities does not exceed $1 - (1 - \varepsilon) = \varepsilon$, just code each one as itself. To ensure uniqueness for *decoding*, one may put *one* of the previously unused sequences of length $t(H + \gamma)/\log M$ in front of each of the self-coded terms. The length $c(X_0, X_1, \ldots, X_{t-1})$ of code-words for such a code is then either $t(H + \gamma)/\log M$ or $t + t(H + \gamma)/\log M$, the latter occurring with probability at most ε. Therefore,

$$Ec(X_0, X_1, \ldots, X_{t-1}) \leqslant \frac{t(H+\gamma)}{\log M} + \left[t + \frac{t(H+\gamma)}{\log M}\right]\varepsilon = \frac{t(H+\delta)}{\log M} \tag{15.19}$$

where $\delta = \varepsilon H + \varepsilon \log M + \varepsilon\gamma + \gamma$. The desired inequality now follows. ∎

EXERCISES

Exercises for Section II.1

1. Verify that the conditional distribution of X_{n+1} given $X_0, X_1, \ldots, X_n$ is the conditional distribution of X_{n+1} given X_n if and only if the conditional distribution of X_{n+1} given $X_0, X_1, \ldots, X_n$ is a (measurable) function of X_n alone. [*Hint*: Use properties of conditional expectations, Section 0.4.]
2. Show that the simple random walk has the Markov property.
3. Show that every discrete-parameter stochastic process with independent increments is a Markov process.
4. Let A, B, C be events with $C, B \cap C$ having positive probabilities. Verify that the following are equivalent versions of the conditional independence of A and B given C: $P(A \cap B \mid C) = P(A \mid C)P(B \mid C)$ if and only if $P(A \mid B \cap C) = P(A \mid C)$.
5. (i) Let $\{X_n\}$ be a sequence of random variables with denumerable state space S. Call $\{X_n\}$ rth *order Markov-dependent* if

$$\begin{aligned} &P(X_{n+1} = j \mid X_0 = i_0, \ldots, X_n = i_n) \\ &\quad = P(X_{n+1} = j \mid X_{n-r+1} = i_{n-r+1}, \ldots, X_n = i_n) \qquad \text{for } i_0, \ldots, i_n, j \in S,\ n \geqslant r. \end{aligned}$$

Show that $Y_n = (X_n, X_{n+1}, \ldots, X_{n+r-1})$, $n = 0, 1, 2, \ldots$ is a (first-order) Markov chain under these circumstances.

(ii) Let $V_n = X_{n+1} - X_n$, $n = 0, 1, 2, \ldots$. Show that if $\{X_n\}$ is a Markov chain, then so is $\{(X_n, V_n)\}$. [*Hint*: Consider first $\{(X_n, X_{n+1})\}$ and then apply a one-to-one transformation.]

6. Show that a necessary and sufficient condition on the correlations of a discrete-parameter *stationary* Gaussian stochastic process $\{X_n\}$ to have a Markov property is $\text{Cov}(X_n, X_{n+m}) = \sigma^2\rho_1^m$ for some $\sigma^2 > 0$, $|\rho_1| < 1$, $m, n = 0, 1, 2, \ldots$. [A stochastic process $\{X_n: n \geqslant 0\}$ is said to be *stationary* if, for all n, m, the distribution of $(X_0, \ldots, X_n)$ is the same as that of $(X_m, X_{m+1}, \ldots, X_{m+n})$.]

7. Let $\{Y_n\}$ be an i.i.d. sequence of ± 1-valued Bernoulli random variables with parameter $0 < p < 1$. Define a new stochastic process by $X_n = (Y_n + Y_{n-1})/2$, for $n = 1, 2, \ldots$. Show that $\{X_n\}$ does *not* have the Markov property.

8. Let $\{S_n\}$ denote the simple symmetric random walk starting at the origin and let $R_n = |S_n|$. Show that $\{R_n\}$ is a Markov chain.

9. (*Random Walk in Random Scenery*) Let $\{Y_n: n \in \mathbb{Z}\}$ be a symmetric i.i.d. sequence of ± 1-valued random variables indexed by the set of integers $\mathbb{Z}$. Let $\{S_n\}$ be the simple symmetric random walk on the state space $\mathbb{Z}$ starting at $S_0 = 0$. The *random walk* $\{S_n\}$ is assumed to be independent of the *random scenery* $\{Y_n\}$. Define a new process $\{X_n\}$ by noting down the scenery at each integer site upon arrival in the course of the walk. That is, $X_n = Y_{S_n}$, $n = 0, 1, 2, \ldots$.
   (i) Calculate EX_n. [*Hint*: $X_n = \sum_{m=-n}^{n} Y_m \mathbf{1}_{\{S_n = m\}}$.]
   (*ii) Show that $\{X_n\}$ is stationary. [See Exercise 6 for a definition of stationarity.]
   (iii) Is $\{X_n\}$ Markovian?
   (iv) Show that $\text{Cov}(X_n, X_{n+m}) \sim (2\pi)^{1/2} m^{-1/2}$ for large even m, and zero for odd m. (For an analysis of the "long-range dependence" in this example, see H. Kesten and F. Spitzer (1979), "A Limit Theorem Related to a New Class of Self-Similar Processes," *Z. Wahr. Verw. Geb.*, **50**, 5–25.)

10. Let $\{Z_n: n = 0, 1, 2, \ldots\}$ be i.i.d. ± 1-valued with $P(Z_n = 1) = p \neq \frac{1}{2}$. Define $X_n = Z_n Z_{n+1}$, $n = 0, 1, 2, \ldots$. Show that for $k \leqslant n - 1$, $P(X_{n+1} = j \mid X_k = i) = P(X_{n+1} = j)$, i.e., X_{n+1} and X_k are independent for each $k = 0, \ldots, n-1$, $n \geqslant 1$. Is $\{X_n\}$ a Markov chain?

Exercises for Section II.2

1. (i) Show that the transition matrix for a sequence of *independent* integer-valued random variables is characterized by the property that its rows are identical; i.e., $p_{ij} = p_j$ for all $i, j \in S$.
   (ii) Under what further condition is the Markov chain an i.i.d. sequence?

2. (i) Let $\{Y_n\}$ be a Markov chain with a one-step transition matrix $\mathbf{p}$. Suppose that the process $\{Y_n\}$ is viewed only at every mth time step (m fixed) and let $X_n = Y_{nm}$, for $n = 0, 1, 2, \ldots$. Show that $\{X_n\}$ is a Markov chain with *one-step* transition law given by $\mathbf{p}^m$.
   (ii) Suppose $\{X_n\}$ is a Markov chain with transition probability matrix $\mathbf{p}$. Let $n_1 < n_2 < \cdots < n_k$. Prove that

$$P(X_{n_k} = j \mid X_{n_1} = i_1, \ldots, X_{n_{k-1}} = i_{k-1}) = p_{i_{k-1}j}^{(n_k - n_{k-1})}.$$

3. (*Random Walks on a Group*) Let G be a finite group with group operation denoted by $\oplus$. That is, G is a nonempty set and $\oplus$ is a well-defined binary operation for G such that (i) if $x, y \in G$ then $x \oplus y \in G$; (ii) if $x, y, z \in G$ then $x \oplus (y \oplus z) = (x \oplus y) \oplus z$; (iii) there is an $e \in G$ such that $x \oplus e = e \oplus x = x$ for all $x \in G$; (iv) for each $x \in G$ there is an element in G, denoted $-x$, such that $x \oplus (-x) = (-x) \oplus x = e$. If $\oplus$ is commutative, i.e., $x \oplus y = y \oplus x$ for all $x, y \in G$, then G is called *abelian*. Let $X_1, X_2, \ldots$ be i.i.d. random variables taking values in G and having the common probability distribution $Q(g) = P(X_n = g)$, $g \in G$.
   (i) Show that the *random walk on* G defined by $S_n = X_0 \oplus X_1 \oplus \cdots \oplus X_n$, $n \geqslant 0$, is a Markov chain and calculate its transition probability matrix. Note that it is *not* necessary for G to be abelian for $\{S_n\}$ to be Markov.
   (ii) (*Top-In Card Shuffles*) Construct a model for *card shuffling* as a Markov chain on a (nonabelian) permutation group on N symbols in which the top card of the deck is inserted at a randomly selected location in the deck at each shuffle.
   (iii) Calculate the transition probability matrix for $N = 3$. [*Hint*: Shuffles are of the form $(c_1, c_2, c_3) \to (c_2, c_1, c_3)$ or (c_2, c_3, c_1) only.] Also see Exercise 4.5.

4. An individual with a highly contagious disease enters a population. During each subsequent period, either the carrier will infect a new person or be discovered and removed by public health officials. A carrier is discovered and removed with probability $q = 1 - p$ at each unit of time. An unremoved infected individual is sure to infect someone in each time unit. The time evolution of the number of infected individuals in the population is assumed to be a Markov chain $\{X_n: n = 0, 1, 2, \ldots\}$. What are its transition probabilities?

5. The price of a certain commodity varies over the values 1, 2, 3, 4, 5 units depending on supply and demand. The price X_n at time n determines the demand D_n at time n through the relation $D_n = N - X_n$, where N is a constant larger than 5. The supply C_n at time n is given by $C_n = N - 3 + \varepsilon_n$ where $\{\varepsilon_n\}$ is an i.i.d. sequence of equally likely ± 1-valued Bernoulli random variables. Price changes are made according to the following policy:

$$
\begin{aligned}
X_{n+1} - X_n &= +1 \qquad \text{if } D_n - C_n > 0,\\
X_{n+1} - X_n &= -1 \qquad \text{if } D_n - C_n < 0,\\
X_{n+1} - X_n &= 0 \qquad \text{if } D_n - C_n = 0.
\end{aligned}
$$

   (i) Fix $X_0 = i_0$. Show that $\{X_n\}$ is a Markov chain with state space $S = \{1, 2, 3, 4, 5\}$.
   (ii) Compute the transition probability matrix of $\{X_n\}$.
   (iii) Calculate the two-step transition probabilities.

6. A reservoir has finite capacity of h units, where h is a positive integer. The daily inputs are i.i.d. integer-valued random variables $\{J_n: n = 1, 2, \ldots\}$ with the common p.m.f. $\{g_j = P(J_n = j), j = 0, 1, 2, \ldots\}$. One unit of water is released through the dam at the end of each day provided that the reservoir is not empty or does not exceed its capacity. If it is empty, there is no release. If it exceeds capacity, then the excess water is released. Let X_n denote the amount of water left in the reservoir on the nth day after release of water. Compute the transition matrix for $\{X_n\}$.

7. Suppose that at each unit of time each particle located in a fixed region of space has probability p, independently of the other particles present, of leaving the region. Also,

at each unit of time a random number of new particles having Poisson distribution with parameter λ enter the region independently of the number of particles already present at time n. Let X_n denote the number of particles in the region at time n. Calculate the transition matrix of the Markov chain $\{X_n\}$.

8. We are given two boxes A and B containing a total of N labeled balls. A ball is selected at random (all selections being equally likely) at time n from among the N balls and then a box is selected at random. Box A is selected with probability p and B with probability $q = 1 - p$ independently of the ball selected. The selected ball is moved to the selected box, unless the ball is already in it. Consider the Markov evolution of the number X_n of balls in box A. Calculate its transition matrix.

9. Each cell of a certain organism contains N particles, some of which are of type A and the others type B. The cell is said to be in state j if it contains exactly j particles of type A. Daughter cells are formed by cell division as follows: Each particle replicates itself and a daughter cell inherits N particles chosen at random from the $2j$ particles of type A and the $2N - 2j$ particles of type B present in the parental cell. Calculate the transition matrix of this Markov chain.

Exercises for Section II.3

1. Let $\mathbf{p}$ be the transition matrix for a completely random motion of Example 2. Show that $\mathbf{p}^n = \mathbf{p}$ for all n.

2. Calculate $p_{ij}^{(n)}$ for the unrestricted simple random walk.

3. Let $\mathbf{p} = ((p_{ij}))$ denote the transition matrix for the unrestricted general random walk of Example 6.
   (i) Calculate $p_{ij}^{(2)}$ in terms of the increment distribution Q.
   (ii) Show that $p_{ij}^{(n)} = Q^{*n}(j - i)$, where the *n-fold convolution* is defined recursively by

$$Q^{*n}(j) = \sum_k Q^{*(n-1)}(k)Q(j - k), \qquad Q^{*(1)} = Q.$$

4. Verify each of the following for the Pólya urn model in Example 8.
   (i) $P(X_n = 1) = r/(r + b)$ for each $n = 1, 2, 3, \dots$.
   (ii) $P(X_1 = \varepsilon_1, \dots, X_n = \varepsilon_n) = P(X_{1+h} = \varepsilon_1, \dots, X_{n+h} = \varepsilon_n)$, for any $h = 0, 1, 2, \dots$.
   (*iii) $\{X_n\}$ is a martingale (see Definition 13.2, Chapter I).

5. Describe the motion represented by a Markov chain having transition matrix of the following forms:

   (i) $\mathbf{p} = \begin{bmatrix} 1 & 0 \\ 0 & 1 \end{bmatrix}$,

   (ii) $\mathbf{p} = \begin{bmatrix} 0 & 1 \\ 1 & 0 \end{bmatrix}$,

   (iii) $\mathbf{p} = \begin{bmatrix} \frac{2}{5} & \frac{3}{5} \\ \frac{2}{5} & \frac{3}{5} \end{bmatrix}$.

(iv) Use the probabilistic description to write down $\mathbf{p}^n$ without algebraically performing the matrix multiplications. Generalize these to m-state Markov chains.

6. (*Length of a Queue*) Suppose that items arrive at a shop for repair on a daily basis but that it takes one day to repair each item. New arrivals are put on a waiting list for repair. Let A_n denote the number of arrivals during the nth day. Let X_n be the length of the waiting list at the end of the nth day. Assume that $A_1, A_2, \ldots$ is an i.i.d. nonnegative integer-valued sequence of random variables with $a(x) = P(A_n = x)$, $x = 0, 1, 2, \ldots$. Assume that A_{n+1} is independent of $X_0, \ldots, X_n$ $(n \geqslant 0)$. Calculate the transition probabilities for $\{X_n\}$.

7. (*Pseudo Random Number Generator*) The *linear congruential method* of generating integer values in the range 0 to $N - 1$ is to calculate $h(x) = (ax + c) \bmod(N)$ for some choice of integer coefficients $0 \leqslant a, c < N$ and an initial seed value of x. More generally, polynomials with integer coefficients can be used in place of $ax + c$. Note that these methods cycle after N iterations.

   (i) Show that the iterations may be represented by a Markov chain on a circle.
   (ii) Calculate the transition probabilities in the case $N = 5, a = 1, c = 2$.
   (iii) Calculate the transition probabilities in the case $h(x) = (x^2 + 2) \bmod(5)$.

   [See D. Knuth (1981), *The Art of Computer Programming*, Vol. II, 2nd ed., Addison-Wesley, Menlo Park, for extensive treatments.]

8. (*A Renewal Process*) A system requires a certain device for its operation that is subject to failure. Inspections for failure are made at regular points in time, so that an item that fails during the nth period of time between $n - 1$ and n is replaced at time n by a device of the same type having an independent service life. Let p_n denote the probability that a device will fail during the nth period of its use. Let X_n be the age (in number of periods) of the item in use at time n. A new item is started at time $n = 0$, and $X_n = 0$ if an item has just been replaced at time n. Calculate the transition matrix of the Markov chain $\{X_n\}$.

Exercises for Section II.4

1. A balanced six-sided die is rolled repeatedly. Let Z denote the smallest number of rolls for the occurrence of all six possible faces. Let $Z_1 = 1$, $Z_j =$ smallest number of tosses to obtain the jth new face after $j - 1$ distinct faces have occurred. Then $Z = Z_1 + \cdots + Z_6$.

   (i) Give a direct proof that $Z_1, \ldots, Z_6$ are independent random variables.
   (ii) Give a proof of (i) using the strong Markov property. [*Hint*: Define stopping times τ_j denoting the first time after τ_{j-1} that X_n is not among $X_1, \ldots, X_{\tau_{j-1}}$ where $X_1, X_2, \ldots$ are the respective outcomes on the successive tosses.]
   (iii) Calculate the distributions of $Z_2, \ldots, Z_6$.
   (iv) Calculate EZ and Var Z.

2. Let $\{S_n\}$ denote the two-dimensional simple symmetric random walk on the integer lattice starting at the origin. Define $\tau_r = \inf\{n: \|S_n\| = r\}$, $r = 1, 2, \ldots$, where $\|(a, b)\| = |a| + |b|$. Describe the distribution of the process $\{S_{\tau_r + n}: n = 0, 1, 2, \ldots\}$ in the two cases $r = 1$ and $r = 2$.

3. (*Coupon Collector's Problem*) A box contains N balls labeled $0, 1, 2, \ldots, N-1$. Let $T \equiv T_N$ be the number of selections (with replacement) required until each ball is sampled at least once. Let T_j be the number of selections required to sample j *distinct* balls. Show that

   (i) $T = (T_N - T_{N-1}) + (T_{N-1} - T_{N-2}) + \cdots + (T_2 - T_1) + T_1,$

   where $T_1 = 1, T_2 - T_1, \ldots, T_{j+1} - T_j, \ldots, T_N - T_{N-1}$ are independent geometrically distributed with parameters $(N-j)/N$, respectively.

   (ii) Let τ_j be the number of selections to get ball j. Then τ_j is geometrically distributed.

   (iii) $P(T > m) \leqslant Ne^{-m/N}$. [*Hint*: $P(T > m) \leqslant \sum_{j=1}^{N} P(\tau_j > m)$].

   (iv) $$P(T > m) = \sum_{k=1}^{N} (-1)^{k+1} \binom{N}{k} \left(1 - \frac{k}{N}\right)^m.$$

   [*Hint*: Use inclusion–exclusion on $\{T > m\} = \bigcup_{i=1}^{N} \{\tau_j > m\}$.]

   (v) Let $X_1, X_2, \ldots$ be the respective numbers on the balls selected. Is T a stopping time for $\{X_n\}$?

4. Let $\{X_n\}$ and $\{Y_n\}$ be independent Markov chains with common transition probability matrix $\mathbf{p}$ and starting in states i and j respectively.

   (i) Show that $\{(X_n, Y_n)\}$ is a Markov chain on the state space $S \times S$.

   (ii) Calculate the transition law of $\{(X_n, Y_n)\}$.

   (iii) Let $T = \inf\{n: X_n = Y_n\}$. Show that T is a stopping time for the process $\{(X_n, Y_n)\}$.

   (iv) Let $\{Z_n\}$ be the process obtained by watching $\{X_n\}$ up until time T and then switching to $\{Y_n\}$ after time T; i.e., $Z_n = X_n$, $n \leqslant T$, and $Z_n = Y_n$, for $n > T$. Show that $\{Z_n\}$ is a Markov chain and calculate its transition law.

5. (*Top-In Card Shuffling*) Suppose that a deck of N cards is shuffled by repeatedly taking the top card and inserting it into the deck at a random location. Let G_N be the (nonabelian) group of permutations on N symbols and let $X_1, X_2, \ldots$ be i.i.d. G_N-valued random variables with

   $$P(X_k = \langle i, i-1, \ldots, 1 \rangle) = 1/N \qquad \text{for } i = 1, 2, \ldots, N,$$

   where $\langle i, i-1, \ldots, 1 \rangle$ is the permutation in which the ith value moves to $i-1$, $i-1$ to $i-2, \ldots$ 2 to 1, and 1 to i. Let S_0 be the identity permutation and let $S_n = X_1 \cdots X_n$, where the group operation is being expressed multiplicatively. Let T denote the first time the original bottom card arrives at the top and is inserted back into the deck (cf. Exercise 2.3). Then

   (i) T is a stopping time.

   (ii) T has the additional property that $P(T = k, S_k = g)$ does not depend on $g \in G_N$. [*Hint*: Show by induction on N that at time $T - 1$ the $(N-1)!$ arrangements of the cards beneath the top card are equally likely; see Exercise 2.3(iii).]

   (iii) Property (ii) is equivalent to $P(S_k = g \mid T = k) = 1/|G_N|$; i.e., the deck is mixed at time T. This property is referred to as the *strong uniform time property* by D. Aldous and P. Diaconis (1986), "Shuffling Cards and Stopping Times," *Amer. Math. Monthly*, **93**, pp. 333–348, who introduced this example and approach.

(iv) Show that

$$\max_A \left| P(S_n \in A) - \frac{|A|}{|G_N|} \right| \leqslant P(T > n) \leqslant Ne^{-n/N}.$$

[*Hint*: Write

$$P(S_n \in A) = P(S_n \in A, T \leqslant n) + P(S_n \in A, T > n)$$

$$= \frac{|A|}{|G_N|} P(T \leqslant n) + P(S_n \in A \mid T > n)P(T > n) = \frac{|A|}{|G_N|} + rP(T > n),$$

$0 \leqslant r \leqslant 1$. For the rightmost upper bound, compare Exercise 4.3.]

6. Suppose that for a Markov chain $\{X_n\}$, $\rho_{x,y} := P_x(X_n = y \text{ for some } n \geqslant 1) = 1$. Prove that $P_x(X_n = y \text{ for infinitely many } n) = 1$.

7. (*Record Times*) Let $X_1, X_2, \ldots$ be an i.i.d. sequence of nonnegative random variables having a continuous distribution (so that the probability of a tie is zero). Define $R_1 = 1$, $R_k = \inf\{n \geqslant R_{k-1} + 1 : X_n \geqslant \max(X_1, \ldots, X_{n-1})\}$, for $k = 2, 3, \ldots$.
   (i) Show that $\{R_n\}$ is a Markov chain and calculate its transition probabilities. [*Hint*: All $i_k!$ rankings of $(X_1, X_2, \ldots, X_{i_k})$ are equally likely. Consider the event $\{R_1 = 1, R_2 = i_2, \ldots, R_k = i_k\}$ and count the number of rankings of $(X_1, X_2, \ldots, X_{i_k})$ that correspond to its occurrence.]
   (ii) Let $T_n = R_{n+1} - R_n$. Is $\{T_n\}$ a Markov chain? [*Hint*: Compute $P(T_3 = 1 \mid T_2 = 1, T_1 = 1)$ and $P(T_3 = 1 \mid T_2 = 1)$.]

8. (*Record Values*) Let $X_1, X_2, \ldots$ be an i.i.d. sequence of nonnegative random variables having a discrete distribution function. Define the record times $R_1 = 1$, $R_2, R_3, \ldots$ as in Exercise 7. Define the *record values* by $V_k = X_{R_k}$, $k = 1, 2, \ldots$.
   (i) Show that each R_k is a stopping time for $\{X_n\}$.
   (ii) Show that $\{V_k\}$ is a Markov process and calculate its transition probabilities.
   (iii) Extend (ii) to the case when the distribution function of X_k is continuous.

Exercises for Section II.5

1. Construct a finite-state Markov chain such that
   (i) There is only one inessential state.
   (ii) The set $\mathscr{E}$ of essential states decomposes into two equivalence classes with periods $d = 1$ and $d = 3$.

2. (i) Give an example of a transition matrix for which all states are inessential.
   (ii) Show that if S is finite then there is at least one essential state.

3. Classify all states for $\mathbf{p}$ given below into essential and inessential subsets. Decompose the set of all essential states into equivalence classes of communicating states.

$$\begin{bmatrix} \frac{1}{3} & 0 & 0 & 0 & \frac{2}{3} & 0 & 0 \\ 0 & 0 & 0 & \frac{1}{3} & 0 & 0 & \frac{2}{3} \\ \frac{1}{6} & \frac{1}{6} & \frac{1}{6} & \frac{1}{6} & \frac{1}{6} & \frac{1}{6} & 0 \\ 0 & \frac{1}{2} & 0 & 0 & 0 & \frac{1}{2} & 0 \\ \frac{2}{5} & 0 & 0 & 0 & \frac{3}{5} & 0 & 0 \\ 0 & 0 & 0 & \frac{5}{6} & 0 & 0 & \frac{1}{6} \\ 0 & \frac{1}{4} & 0 & 0 & 0 & \frac{3}{4} & 0 \end{bmatrix}.$$

4. Suppose that S comprises a single essential class of aperiodic states. Show that there is an integer v such that $p_{ij}^{(v)} > 0$ for all $i, j \in S$ by filling in the details of the following steps.
   (i) For a fixed (i, j), let $B_{ij} = \{v \geqslant 1: p_{ij}^{(v)} > 0\}$. Then for each state j, B_{jj} is closed under addition.
   (ii) (*Basic Number Theory Lemma*) If B is a set of positive integers having greatest common divisor 1 and if B is closed under addition, then there is an integer b such that $n \in B$ for all $n \geqslant b$. [*Hints*:
      (a) Let G be the smallest additive subgroup of $\mathbb{Z}$ that contains B. Then argue that $G = \mathbb{Z}$ since if d is the smallest positive integer in G it will follow that if $n \in B$, then, since $n = qd + r$, $0 \leqslant r < d$, one obtains $r = n - qd \in G$ and hence $r = 0$, i.e., d divides each $n \in B$ and thus $d = 1$.
      (b) If $1 \in B$, then each $n = 1 + 1 + \cdots + 1 \in B$. If $1 \notin B$, then by (a), $1 = \alpha - \beta$ for $\alpha, \beta \in B$. Check $b = (\alpha + \beta)^2 + 1$ suffices; for if $n > (\alpha + \beta)^2$, then, writing $n = q(\alpha + \beta) + r$, $0 \leqslant r < \alpha + \beta$, $n = q(\alpha + \beta) + r(\alpha - \beta) = (q + r)\alpha + (q - r)\beta$, and in particular $n \in B$ since $q + r > 0$ and $q - r > 0$ by virtue of $n > (r + 1)(\alpha + \beta)$.]
   (iii) For each (i, j) there is an integer b_{ij} such that $v \geqslant b_{ij}$ implies $v \in B_{ij}$. [*Hint*: Obtain b_{jj} from (ii) applied to (i) and then choose k such that $p_{ij}^{(k)} > 0$. Check that $b_{ij} = k + b_{jj}$ suffices.]
   (iv) Check that $v = \max\{b_{ij}: i, j \in S\}$ suffices for the statement of the exercise.

5. Classify the states in Exercises 2.4, 2.5, 2.6, 2.8 as essential and inessential states. Decompose the essential states into their respective equivalence classes.

6. Let $\mathbf{p}$ be the transition matrix on $S = \{0, 1, 2, 3\}$ defined below.

$$\begin{bmatrix} 0 & \frac{1}{2} & 0 & \frac{1}{2} \\ \frac{1}{2} & 0 & \frac{1}{2} & 0 \\ 0 & \frac{1}{2} & 0 & \frac{1}{2} \\ \frac{1}{2} & 0 & \frac{1}{2} & 0 \end{bmatrix}.$$

   Show that S is a single class of essential states of period 2 and calculate $\mathbf{p}^n$ for all n.

7. Use the strong Markov property to prove that if j is inessential then $P_\pi(X_n = j$ for infinitely many $n) = 0$.

8. Show by induction on N that all states communicate in the Top-In Card Shuffling example of Exercises 2.3(ii) and 4.5.

Exercises for Section II.6

1. (i) Check that "deterministic motion going one step to the right" on $S = \{0, 1, 2, \ldots\}$ provides a simple example of a homogeneous Markov chain for which there is no invariant distribution.
   (ii) Check that the "static evolution" corresponding to the identity matrix provides a simple example of a 2-state Markov chain with more than one invariant distribution (see Exercise 3.5(i)).
   (iii) Check that "deterministic cyclic motion" of oscillations between states provides a simple example of a 2-state Markov chain having a unique invariant distribution $\boldsymbol{\pi}$ that is *not* the limiting distribution $\lim_{n\to\infty} P_{\boldsymbol{\mu}}(X_n = j)$ for any other initial distribution $\boldsymbol{\mu} \neq \boldsymbol{\pi}$ (see Exercise 3.5(ii)).
   (iv) Show by *direct calculation* for the case of a 2-state Markov chain having strictly positive transition probabilities that there is a unique invariant distribution that is the limiting distribution for any initial distribution. Calculate the precise exponential rate of convergence.

2. Calculate the invariant distribution for Exercise 2.5. Calculate the so-called *equilibrium price* of the commodity, $E_\pi X_n$.

3. Calculate the invariant distribution for Exercise 2.8.

4. Calculate the invariant distribution for Exercise 2.3(ii) (see Exercise 5.8).

5. Suppose that n states $a_1, a_2, \ldots, a_n$ are arranged counterclockwise in a circle. A particle jumps one unit in the clockwise direction with probability p, $0 \leqslant p \leqslant 1$, or one unit in the counterclockwise direction with probability $q = 1 - p$. Calculate the invariant distribution.

6. (i) (*Coupling Bound*) If X and Y are arbitrary real-valued random variables and J an arbitrary interval then show that $|P(X \in J) - P(Y \in J)| \leqslant P(X \neq Y)$.
   (ii) (*Doeblin's Coupling Method*) Let $\mathbf{p}$ be an aperiodic irreducible finite-state transition matrix with invariant distribution $\boldsymbol{\pi}$. Let $\{X_n\}$ and $\{Y_n\}$ be independent Markov chains with transition law $\mathbf{p}$ and having respective initial distributions $\boldsymbol{\pi}$ and $\boldsymbol{\delta}_{\{i\}}$. Let T denote the first time n that $X_n = Y_n$.
   (a) Show $P(T < \infty) = 1$. [*Hint*: Let $\nu = \min\{n: p_{ij}^{(n)} > 0$ for all $i, j \in S\}$. Argue that

$$P(T > k\nu) \leqslant P(X_{k\nu} \neq Y_{k\nu} \mid X_{(k-1)\nu} \neq Y_{(k-1)\nu}, \ldots, X_\nu \neq Y_\nu) \cdots$$

$$P(X_{2\nu} \neq Y_{2\nu} \mid X_\nu \neq Y_\nu) P(X_\nu \neq Y_\nu) \leqslant (1 - N\delta^2)^k,$$

   where $\delta = \min_{i,j} p_{ij}^{(\nu)} > 0$, $N = |S|$.]
   (b) Show that $\{(X_n, Y_n)\}$ is a Markov chain on $S \times S$ and T is a stopping time for this Markov chain.
   (c) Define $\{Z_n\}$ to be the process obtained by observing $\{Y_n\}$ until it meets $\{X_n\}$ and then watching $\{X_n\}$ from then on. Show that $\{Z_n\}$ is a Markov chain with transition law $\mathbf{p}$ and invariant initial distribution $\boldsymbol{\pi}$.
   (d) Show $|P(Z_n = j) - P(X_n = j)| \leqslant P(T \geqslant n)$.
   (e) Show $|p_{ij}^{(n)} - \pi_j| \leqslant P(T \geqslant n) \leqslant (1 - N\delta^2)^{(n/\nu) - 1}$.

7. Let $\mathbf{A} = ((a_{ij}))$ be an $N \times N$ matrix. Suppose that $\mathbf{A}$ is a transition probability matrix with strictly positive entries a_{ij}.
   (i) Show that the *spectral radius*, i.e., the *magnitude* of the largest eigenvalue of $\mathbf{A}$,

must be 1. [*Hint*: Check first that λ is an *eigenvalue* of $\mathbf{A}$ (in the sense $\mathbf{Ax} = \lambda\mathbf{x}$ has a nonzero solution $\mathbf{x} = (x_1 \cdots x_n)'$) if and only if $\mathbf{zA} = \lambda\mathbf{z}$ has a nonzero solution $\mathbf{z} = (z_1 \cdots z_n)$; recall that $\det(B) = \det(B')$ for any $N \times N$ matrix B.]

(ii) Show that $\lambda = 1$ must be a *simple* eigenvalue of A (i.e., geometric multiplicity 1). [*Hint*: Suppose $\mathbf{z}$ is any (left) eigenvector corresponding to $\lambda = 1$. By the results of this section there must be an invariant distribution (positive eigenvector) $\boldsymbol{\pi}$. For t sufficiently large $\mathbf{z} + t\boldsymbol{\pi}$ is also positive (and normalizable).]

*8. Let $\mathbf{A} = ((a_{ij}))$ be a $N \times N$ matrix with positive entries. Show that the spectral radius is also given by $\min\{\lambda > 0: \mathbf{Ax} \leqslant \lambda\mathbf{x}$ for some positive $\mathbf{x}\}$. [*Hint*: $\mathbf{A}$ and its transpose $\mathbf{A}'$ have the same eigenvalues (why?) and therefore the same spectral radius. $\mathbf{A}'$ is *adjoint* to $\mathbf{A}$ with respect to the usual (dot) inner product in the sense $(\mathbf{Ax}, \mathbf{y}) = (\mathbf{x}, \mathbf{A}'\mathbf{y})$ for all $\mathbf{x}, \mathbf{y}$, where $(\mathbf{u}, \mathbf{v}) = \sum_{i=1}^{N} u_i v_i$. Apply the maximal property to the spectral radius of $\mathbf{A}'$.]

9. Let $p(x, y)$ be a continuous function on $[c, d] \times [c, d]$ with $c < d$. Assume that $p(x, y) > 0$ and $\int p(x, y)\, dy = 1$. Let Ω denote the space of all sequences $\omega = (x_0, x_1, \ldots)$ of numbers $x_i \in [c, d]$. Let $\mathscr{F}_0$ denote the class of all finite-dimensional sets A of the form $A = \{\omega = (x_0, x_1, \ldots) \in \Omega: a_i < x_i < b_i, i = 0, 1, \ldots, n\}$, where $c \leqslant a_i < b_i \leqslant d$ for each i. Define $P_x(A)$ for such a set A by

$$P_x(A) = \int_{a_n}^{b_n} \cdots \int_{a_1}^{b_1} p(x, y_1)p(y_1, y_2) \cdots p(y_{n-1}, y_n)\, dy_n \cdots dy_1 \qquad \text{for } x \in [a_0, b_0].$$

Define $P_x(A) = 0$ if $x < a_0$ or $x > b_0$. The Kolmogorov Extension Theorem assures us that P_x has a unique extension to a probability measure defined for all events in the smallest sigmafield $\mathscr{F}$ of subsets of Ω that contains $\mathscr{F}_0$. For any nonnegative integrable function γ with integral 1, define

$$P_\gamma(A) = \int P_x(A)\gamma(x)\, dx, \qquad A \in \mathscr{F}.$$

Let X_n denote the nth coordinate projection mapping on Ω. Then $\{X_n\}$ is said to be a *Markov chain on the state space* $S = [c, d]$ with *transition density* $p(x, y)$ and *initial density* γ under P_γ. Under P_x the process is said to have *initial state* x.

(i) Prove the Markov property for $\{X_n\}$; i.e., the conditional distribution of X_{n+1} given $X_0, \ldots, X_n$ is $p(X_n, y)\, dy$.

(ii) Compute the distribution of X_n under P_γ.

(iii) Show that under P_γ the conditional distribution of X_n given $X_0 = x_0$ is $p^{(n)}(x_0, y)\, dy$, where

$$p^{(n)}(x, y) = \int p^{(n-1)}(x, z)p(z, y)\, dz \qquad \text{and} \qquad p^{(1)} = p.$$

(iv) Show that if $\delta = \inf\{p(x, y): x, y \in [c, d]\} > 0$, then

$$\int |p(x, y) - p(z, y)|\, dy < 2[1 - \delta(d - c)]$$

by breaking the integral into two terms involving y such that $p(x, y) > p(z, y)$ and those y such that $p(x, y) \leqslant p(z, y)$.

(v) Show that there is a continuous strictly positive function $\pi(y)$ such that

$$\max\{|p^{(n)}(x, y) - \pi(y)|: c \leqslant x, y \leqslant d\} < [1 - \delta(d - c)]^{n-1}\rho$$

where $\rho = \max\{|p(x, y) - p(z, y)|: c \leqslant x, y, z \leqslant d\} < \infty$.

(vi) Prove that π is an invariant distribution and moreover that under the present conditions π is unique.

(vii) Show that $P_y(X_n \in (a, b) \text{ i.o.}) = 1$, for any $c < a < b < d$. [*Hint*: Show that $P_y(X_n \notin (a, b)$ for $m < n \leqslant M) \leqslant [1 - \delta(b - a)]^{M-m}$. Calculate $P_y(X_n \notin (a, b)$ for all $n > m)$.]

10. (*Random Walk on the Circle*) Let $\{X_n\}$ be an i.i.d. sequence of random variables taking values in $[0, 1]$ and having a continuous p.d.f. $f(x)$. Let $\{S_n\}$ be the process on $[0, 1)$ defined by

$$S_n = x + X_1 + \cdots + X_n \qquad \text{mod } 1,$$

where $x \in [0, 1)$.

(i) Show that $\{S_n\}$ is a Markov chain and calculate its transition density.

(ii) Describe the time asymptotic behavior of $p^{(n)}(x, y)$.

(iii) Describe the invariant distribution.

11. (*The Umbrella Problem*) A person who owns r umbrellas distributes them between home and office according to the following routine. If it is raining upon departure from either place, an event that has probability p, say, then an umbrella is carried to the other location if available at the location of departure. If it is not raining, then an umbrella is not carried. Let X_n denote the number of *available umbrellas* at whatever place the person happens to be departing on the nth trip.

(i) Determine the transition probability matrix and the invariant distribution (equilibrium).

(ii) Let $0 < \alpha < 1$. How many umbrellas should the person own so that the probability of getting wet under the equilibrium distribution is at most α against a climate (p)? What number works against all possible climates for the probability α?

Exercises for Section II.7

1. Calculate the invariant distributions for Exercise 2.6.

2. Calculate the invariant distribution for Exercise 5.6.

3. A transition matrix is called *doubly-stochastic* if its transpose $\mathbf{p}'$ is also a transition matrix; i.e., if the elements of each column add to 1.

(i) Show that the vector consisting entirely of 1's is invariant under $\mathbf{p}$ and can be normalized to a probability distribution if S is finite.

(ii) Under what additional conditions is this distribution the unique invariant distribution?

4. (i) Suppose that $\boldsymbol{\pi} = (\pi_i)$ is an invariant distribution for $\mathbf{p}$. The distribution P_π of

the process is called *time-reversible* if $\pi_i p_{ij} = \pi_j p_{ji}$ for all $i, j \in S$ [$\boldsymbol{\pi}$ is often said to be *time-reversible* (with respect to $\mathbf{p}$) as well]. Show that if S is finite and $\mathbf{p}$ is doubly stochastic, then the (discrete) uniform distribution makes the process time-reversible if and only if $\mathbf{p}$ is symmetric.

(*ii) Suppose that $\{X_n\}$ is a Markov chain with invariant distribution $\boldsymbol{\pi}$ and started in $\boldsymbol{\pi}$. Then $\{X_n\}$ is a stationary process and therefore has an extension backward in time to $n = 0, \pm 1, \pm 2, \ldots$. [Use Kolmogorov's Extension Theorem.] Define the time-reversed process by $Y_n = X_{-n}$. Show that the reversed process $\{Y_n\}$ is a Markov chain with 1-step transition probabilities $q_{ij} = \pi_j p_{ji}/\pi_i$.

(iii) Show that under the time-reversibility condition (i), the processes in (ii), $\{Y_n\}$ and $\{X_n\}$, have the same distribution; i.e., *in equilibrium a movie of the evolution looks the same statistically whether run forward or backward in time.*

(iv) Show that an irreducible Markov chain on a state space S with an invariant initial distribution $\boldsymbol{\pi}$ is time-reversible if and only if (*Kolmogorov Condition*):

$$p_{ii_1} p_{i_1 i_2} \cdots p_{i_k i} = p_{ii_k} p_{i_k i_{k-1}} \cdots p_{i_1 i} \qquad \text{for all } i, i_1, \ldots, i_k \in S, \quad k \geqslant 1.$$

(v) If there is a $j \in S$ such that $p_{ij} > 0$ for all $i \neq j$ in (iv), then for time-reversibility it is both necessary and sufficient that $p_{ij} p_{jk} p_{ki} = p_{ik} p_{kj} p_{ji}$ for all i, j, k.

5. (*Random Walk on a Tree*) A *tree graph* on n vertices $v_1, v_2, \ldots, v_r$ is a *connected graph* that contains *no cycles.* [That is, there is given a collection of *unordered pairs of distinct vertices* (called *edges*) with the following property: Any two distinct vertices $u, v \in S$ are uniquely connected in the sense that there is a unique sequence $e_1, e_2, \ldots, e_n$ of edges $e_i = \{v_{k_i}, v_{m_i}\}$ such that $u \in e_1$, $v \in e_n$, $e_i \cap e_{i+1} \neq \varnothing$, $i = 1, \ldots, n-1$.] The *degree* ν_i of the vertex v_i represents the number of vertices *adjacent* to v_i, where $u, v \in S$ are called adjacent if there is an edge $\{u, v\}$. By a *tree random walk* on a given tree graph we mean a Markov chain on the state space $S = \{v_1, v_2, \ldots, v_r\}$ that at each time step n changes its state v_i to one of its ν_i randomly selected adjacent states, with equal probabilities and independently of its states prior to time n.

(i) Explain that such a Markov chain must have a unique invariant distribution.

(ii) Calculate the invariant distribution in terms of the vertex degrees $\nu_i, i = 1, \ldots, r$.

(iii) Show that the invariant distribution makes the tree random walk time-reversible.

6. Let $\{X_n\}$ be a Markov chain on S and define $Y_n = (X_n, X_{n+1})$, $n = 0, 1, 2, \ldots$.

(i) Show that $\{Y_n\}$ is a Markov chain on $S' = \{(i, j) \in S \times S : p_{ij} > 0\}$.

(ii) Show that if $\{X_n\}$ is irreducible and aperiodic then so is $\{Y_n\}$.

(iii) Show that if $\{X_n\}$ has invariant distribution $\boldsymbol{\pi} = (\pi_i)$ then $\{Y_n\}$ has invariant distribution $(\pi_i p_{ij})$.

7. Let $\{X_n\}$ be an irreducible Markov chain on a finite state space S. Define a graph G having states of S as vertices with edges joining i and j if and only if either $p_{ij} > 0$ or $p_{ji} > 0$.

(i) Show that G is connected; i.e., for any two sites i and j there is a path of edges from i to j.

(ii) Show that if $\{X_n\}$ has an invariant distribution $\boldsymbol{\pi}$ then for any $A \subset S$,

$$\sum_{i \in A} \sum_{j \in S \setminus A} \pi_i p_{ij} = \sum_{i \in A} \sum_{j \in S \setminus A} \pi_j p_{ji}$$

(i.e., the net probability flux across a *cut* of S into complementary subsets A, $S\backslash A$ is in balance).

(iii) Show that if G contains no cycles (i.e., is a tree graph in the sense of Exercise 5) then the process is time-reversible started with $\boldsymbol{\pi}$.

Exercises for Section II.8

1. Verify (8.10) using summation by parts as indicated. [*Hint*: Let Z be nonnegative integer-valued. Then

$$\sum_{r=0}^{\infty} P(Z > r) = \sum_{r=0}^{\infty} \sum_{n=r+1}^{\infty} P(Z = n).$$

Now interchange the order of summation.]

2. Classify the states in Examples 3.1–3.7 as transient or recurrent.

3. Show that if j is transient and $i \to j$ then $\sum_{n=0}^{\infty} p_{ij}^{(n)} < \infty$ and, in particular, $p_{ij}^{(n)} \to 0$ as $n \to \infty$. [*Hint*: Represent $N(j)$ as a sum of indicator variables and use (8.11).]

4. Classify the states for the models in Exercises 2.6, 2.7, 2.8, 2.9 as transient or recurrent.

5. Classify the states for $\{R_n\} = \{|S_n|\}$, where S_n is the simple symmetric random walk starting at 0 (see Exercise 1.8).

6. Show that inessential states are transient.

7. (*A Birth or Collapse Model*) Let

$$p_{i,i+1} = \frac{1}{i+1}, \qquad p_{i,0} = \frac{i}{i+1}, \qquad i = 0, 1, 2, \dots.$$

Determine whether $\mathbf{p}$ is transient or recurrent.

8. Solve Exercise 7 when

$$p_{i,0} = \frac{1}{i+1}, \qquad p_{i,i+1} = \frac{i}{i+1}, \qquad i \geqslant 1, \quad p_{0,1} = 1.$$

9. Let $p_{i,i+1} = p$, $p_{i,0} = q$, $i = 0, 1, 2, \dots$. Classify the states of $S = \{0, 1, 2, \dots\}$ as transient or recurrent $(0 < p < 1,\ q = 1 - p)$.

10. Fix $i, j \in S$. Write

$$r_n = P_i(X_n = j) \equiv p_{ij}^{(n)} \qquad (n \geqslant 1), \quad r_0 = 1,$$
$$f_n = P_i(X_m \neq j \text{ for } m < n,\ X_n = j) \qquad (n \geqslant 1).$$

(i) Prove (see (5.5) of Chapter I) that $r_n = \sum_{m=1}^{n} f_m r_{n-m}$ $(n \geqslant 1)$.

(ii) Sum (i) over n to give an alternative proof of (8.11).

(iii) Use (i) to indicate how one may compute the distribution of the first visit to state j (after time zero), starting in state i, in terms of $p_{ij}^{(n)}$ $(n \geqslant 1)$.

11. (i) Show that if $\|\cdot\|$ and $\|\cdot\|_0$ are any two *norms* on $\mathbb{R}^k$, then there are positive constants c_1, c_2 such that

$$c_1\|\mathbf{x}\| \leqslant \|\mathbf{x}\|_0 \leqslant c_2\|\mathbf{x}\| \qquad \text{for all } \mathbf{x} \in \mathbb{R}^k.$$

A norm on $\mathbb{R}^k$ is a nonnegative function $\mathbf{x} \to \|\mathbf{x}\|$ on $\mathbb{R}^k$ with the properties that
(a) $\|c\mathbf{x}\| = |c|\,\|\mathbf{x}\|$ for $c \in R$, $\mathbf{x} \in \mathbb{R}^k$,
(b) $\|\mathbf{x}\| = 0$, $\mathbf{x} \in \mathbb{R}^k$, iff $\mathbf{x} = \mathbf{0}$,
(c) $\|\mathbf{x} + \mathbf{y}\| \leqslant \|\mathbf{x}\| + \|\mathbf{y}\|$ for all $\mathbf{x}, y \in \mathbb{R}^k$.
[*Hint*: Use compactness of the unit ball in the case of the Euclidean norm, $\|\mathbf{x}\| = (x_1^2 + \cdots + x_k^2)^{1/2}$, $\mathbf{x} = (x_1, \ldots, x_k)$.]

(ii) Show that the stability condition given in Example 1 implies that $X_n \to 0$ in every norm.

Exercises for Section II.9

1. Let $Y_1, Y_2, \ldots$ be i.i.d. random variables.
   (i) Show that if $E|Y_i| < \infty$ then $\{\max(Y_1, \ldots, Y_n)\}/n \to 0$ a.s. as $n \to \infty$.
   (ii) Verify (9.8) under the assumption (9.4). [*Hint*: Show that $P(|Y_n| > \varepsilon \text{ i.o.}) = 0$ for every $\varepsilon > 0$.]

2. Verify for (9.12) that

$$\lim_{n\to\infty} \frac{n}{N_n} = E(\tau_j^{(2)} - \tau_j^{(1)}) = E_j\tau_j^{(1)},$$

provided $E_j(\tau_j^{(1)}) < \infty$.

3. Prove

$$\lim_{n\to\infty} \frac{\tau_j^{(N_n)}}{N_n} = \infty$$

P_i—a.s. for a null-recurrent state j such that $i \leftrightarrow j$.

4. Show that positive and null recurrence are class properties.

5. Let $\{X_n\}$ be an irreducible aperiodic positive recurrent Markov chain having transition matrix $\mathbf{p}$ on a denumerable state space S. Define a transition matrix $\mathbf{q}$ on $S \times S$ by

$$q_{(i,j),(k,l)} = p_{ik}\,p_{jl}, \qquad (i, j), (k, l) \in S \times S.$$

Let $\{Z_n\}$ be a Markov chain on $S \times S$ with transition law $\mathbf{q}$. Define $T_D = \inf\{n \colon Z_n \in D\}$, where $D = \{(i, i) \colon i \in S\} \subset S \times S$. Show that if $P_{(i,j)}(T_D < \infty) = 1$ for all i, j then for all $i, j \in S$, $\lim_{n\to\infty} p_{ij}^{(n)} = \pi_j$, where $\{\pi_j\}$ is the unique invariant distribution. [*Hint*: Use the coupling method described in Exercise 6.6 for finite state space.]

6. (*General Birth–Collapse*) Let $\mathbf{p}$ be a transition probability matrix on $S = \{0, 1, 2, \ldots\}$ of the form $p_{i,i+1} = p_i$, $p_{i,0} = 1 - p_i$, $i = 0, 1, 2, \ldots$, $0 < p_i < 1$, $i \geqslant 1$, $p_0 = 1$. Show:
   (i) All states are recurrent

$$\text{iff} \quad \lim_{k \to \infty} \prod_{j=1}^{k} p_j = 0 \qquad \text{iff} \quad \sum_{j=1}^{\infty} (1 - p_j) = \infty .$$

   (ii) If all states are recurrent, then positive recurrence holds

$$\text{iff} \quad \sum_{k=1}^{\infty} \prod_{j=1}^{k} p_j < \infty .$$

   (iii) Calculate the invariant distribution in the case $p_j = 1/(j + 2)$.

7. Let $\{S_n\}$ be the simple symmetric random walk starting at the origin. Define $f\colon \mathbb{Z} \to \mathbb{R}^1$ by $f(n) = 1$ if $n \geqslant 0$ and $f(n) = 0$ if $n < 0$. Describe the behavior of $[f(S_0) + \cdots + f(S_{n-1})]/n$ as $n \to \infty$.

8. Calculate the invariant distribution for the *Renewal Model* of Exercise 3.8, in the case that $p_n = p^{n-1}(1 - p)$, $n = 0, 1, 2, \ldots$ where $0 < p < 1$.

9. (*One-Dimensional Nearest-Neighbor Ising Model*) The one-dimensional nearest-neighbor Ising model of magnetism consists of a random distribution of ± 1-valued random variables (*spins*) at the sites of the integers $n = 0, \pm 1, \pm 2, \ldots$. The parameters of the model are the *inverse temperature* $\beta = \dfrac{1}{kT} > 0$ where T is temperature and k is a universal constant called *Boltzmann's constant*, an *external field parameter* H, and an *interaction parameter* (*coupling constant*) J. The spin variables X_n, n = $0, \pm 1, \pm 2, \pm 3, \ldots$, are distributed according to a stochastic process on $\{-1, 1\}$ indexed by $\mathbb{Z}$ with the *Markov property* and having stationary transition law given by

$$P(X_{n+1} = \eta \mid X_n = \sigma) = \frac{\exp\{\beta J \sigma \eta + \beta H \eta\}}{2 \cosh(\beta H + \beta J \sigma)}$$

for $\sigma, \eta \in \{+1, -1\}$, $n = 0, \pm 1, \pm 2, \ldots$; by the Markov property is meant that the conditional distribution of X_{n+1} given $\{X_k\colon k \leqslant n\}$ does not depend on $\{X_k, k \leqslant n - 1\}$.
   (i) Calculate the unique invariant distribution π for $\mathbf{p}$.
   (ii) Calculate the *large-scale magnetization* (i.e., ability to pick up nails), defined by

$$M_N = [X_{-N} + \cdots + X_N]/(2N + 1)$$

   in the so-called bulk (thermodynamic) limit as $N \to \infty$.
   (iii) Calculate and plot the graph (i.e., *magnetic isotherm*) of EX_0 as a function of H for *fixed temperature*. Show that in the limit as $H \to 0^+$ or $H \to 0^-$ the bulk magnetization EX_0 tends to 0; i.e., there is no (zero) *residual magnetization* remaining when H is turned off at any temperature.

(iv) Determine when the process (in equilibrium) is reversible for the invariant distribution; modify Exercise 7.4 accordingly.

10. Show that if $\{X_n\}$ is an irreducible positive-recurrent Markov chain then the condition (iv) of Exercise 7.4 is necessary and sufficient for *time-reversibility* of the stationary process started with distribution $\boldsymbol{\pi}$.

*11. An *invariant measure* for a transition matrix $((p_{ij}))$ is a sequence of nonnegative numbers (m_i) such that $\sum_i m_i p_{ij} = m_j$ for all $j \in S$. An invariant measure may or may not be *normalizable* to a probability distribution on S.

(i) Let $p_{i,i+1} = p_i$ and $p_{i,0} = 1 - p_i$ for $i = 0, 1, 2, \ldots$. Show that there is a unique invariant measure (up to multiples) if and only if $\lim_{n\to\infty} \prod_{k=1}^{n} p_k = 0$; i.e., if and only if the chain is recurrent, since the product is the probability of no return to the origin.

(ii) Show that invariant measures exist for the unrestricted simple random walk but are not unique in the transient case, and is unique (up to multiples) in the (null) recurrent case.

(iii) Let $p_{00} = p_{01} = \frac{1}{2}$ and $p_{i,i-1} = p_{i,i} = 2^{-i-2}$, and $p_{i,i+1} = 1 - 2^{-i-1}$, $i = 1, 2, 3, \ldots$. Show that the probability of not returning to 0 is positive (i.e., transience), but that there is a unique invariant measure.

12. Let $\{Y_n\}$ be any sequence of random variables having finite second moments and let $\gamma_{nm} = \text{Cov}(Y_n, Y_m)$, $\mu_n = EY_n$, $\sigma_n^2 = \text{Var}(Y_n) = \gamma_{nn}$, and $\rho_{nm} = \gamma_{nm}/\sigma_n\sigma_m$.

(i) Verify that $-1 \leqslant \rho_{nm} \leqslant 1$ for all n and m. [*Hint*: Use the Schwarz Inequality.]

(ii) Show that if $\rho_{n,m} \leqslant f(|n - m|)$, where f is a nonnegative function such that $n^{-2} \sum_{k=0}^{n-1} f(k) \sum_{k=1}^{n} \sigma_k^2 \to 0$ as $n \to \infty$, then the WLLN holds for $\{Y_n\}$.

(iii) Verify that if $\rho_{n,m} = \rho_{0,|n-m|} = \rho(|n - m|) \geqslant 0$, then it is sufficient that $n^{-1} \sum_{k=0}^{n-1} \rho(k) \to 0$ as $n \to \infty$ for the WLLN.

(iv) Show that in the case of nonpositive correlations it is sufficient that $n^{-1} \sum_{k=1}^{n} \sigma_k^2 \to 0$ as $n \to \infty$ for the WLLN.

13. Let $\mathbf{p}$ be the transition probability matrix for the asymmetric random walk on $S = \{0, 1, 2, \ldots\}$ with 0 absorbing and $p_{i,i+1} = p > \frac{1}{2}$ for $i \geqslant 1$. Explain why for fixed $i > 0$,

$$\mu_n(\{j\}) := \frac{1}{n} \sum_{m=1}^{n} p_{ij}^{(m)}, \qquad j \in S,$$

does *not* converge to the invariant distribution $\delta_0(\{j\})$ (as $n \to \infty$). How can this be modified to get convergence?

14. (*Iterated Averaging*)

(i) Let a_1, a_2, a_3 be three numbers. Define $a_4 = (a_1 + a_2 + a_3)/3$, $a_5 = (a_2 + a_3 + a_4)/3, \ldots$. Show that $\lim_{n\to\infty} a_n = (a_1 + 2a_2 + 3a_3)/6$.

(ii) Let $\mathbf{p}$ be an irreducible positive recurrent transition law and let $a_1, a_2, \ldots$ be any bounded sequence of numbers. Show that

$$\lim_{n\to\infty} \sum_j p_{ij}^{(n)} a_j = \sum_j a_j \pi_j,$$

where (π_j) is the invariant distribution of $\mathbf{p}$. Show that the result of (i) is a special case.

Exercises for Section II.10

1. (i) Let $Y_1, Y_2, \ldots$ be i.i.d. with $EY_1^2 < \infty$. Show that $\max(Y_1, \ldots, Y_n)/\sqrt{n} \to 0$ a.s. as $n \to \infty$. [*Hint*: Show that $P(Y_n^2 > n\varepsilon \text{ i.o.}) = 0$ for every $\varepsilon > 0$.]
   (ii) Verify that $n^{-1/2}\bar{S}_n$ has the same limiting distribution as (10.6).

2. Let $\{W_n(t): t \geq 0\}$ be the path process defined in (10.7). Let $t_1 < t_2 < \cdots < t_k$, $k \geq 1$, be an arbitrary finite set of time points. Show that $(W_n(t_1), \ldots, W_n(t_k))$ converges in distribution as $n \to \infty$ to the multivariate Gaussian distribution with mean zero and variance–covariance matrix $((D \min\{t_i, t_j\}))$, where D is defined by (10.12).

3. Suppose that $\{X_n\}$ is Markov chain with state space $S = \{1, 2, \ldots, r\}$ having unique invariant distribution (π_j). Let

$$N_n(i) = \#\{k \leq n: X_k = i\}, \qquad i \in S.$$

Show that

$$\sqrt{n}\left(\frac{N_n(1)}{n} - \pi_1, \ldots, \frac{N_n(r)}{n} - \pi_r\right)$$

is asymptotically Gaussian with mean 0 and variance–covariance matrix $\Gamma = ((\gamma_{ij}))$ where

$$\gamma_{ij} = \delta_{ij} - \pi_i\pi_j + \sum_{k=1}^{\infty} (p_{ij}^{(k)} - \pi_i\pi_j) + \sum_{k=1}^{\infty} (p_{ji}^{(k)} - \pi_j\pi_i), \qquad \text{for } 1 \leq i, j \leq r.$$

4. For the one-dimensional nearest-neighbor Ising model of Exercise 9.9 calculate the following:
   (i) The pair correlations $\rho_{n,m} = \text{Cov}(X_n, X_m)$.
   (ii) The large-scale variance (*magnetic susceptibility*) parameter $\text{Var}(X_0)$.
   (iii) Describe the distribution of the fluctuations in the (bulk limit) magnetization (cf. Exercise 9.9(ii)).

5. Let $\{X_n\}$ be a Markov chain on S and define $Y_n = (X_n, X_{n+1})$, $n = 0, 1, 2, \ldots$. Let $\mathbf{p} = ((p_{ij}))$ be the transition matrix for $\{X_n\}$.
   (i) Show that $\{Y_n\}$ is a Markov chain on the state space defined by $S' = \{(i, j) \in S \times S: p_{ij} > 0\}$.
   (ii) Show that if $\{X_n\}$ is irreducible and aperiodic then so is $\{Y_n\}$.
   (iii) Suppose that $\{X_n\}$ has invariant distribution $\boldsymbol{\pi} = (\pi_i)$. Calculate the invariant distribution of $\{Y_n\}$.
   (iv) Let $(i, j) \in S'$ and let T_n be the number of one-step transitions from i to j by $X_0, X_1, \ldots, X_n$ started with the invariant distribution of (iii). Calculate $\lim_{n\to\infty}(T_n/n)$ and describe the fluctuations about the limit for large n.

6. (*Large-Sample Consistency in Statistical Parameter Estimation*) Let $X_n = 1$ or 0 according to whether the nth day at a specified location is *wet* (rain) or *dry*. Assume $\{X_n\}$ is a two-state Markov chain with parameters $\beta = P(X_{n+1} = 1 \mid X_n = 0)$ and $\delta = P(X_{n+1} = 0 \mid X_n = 1)$, $n = 0, 1, 2, \ldots$, $0 < \beta < 1$, $0 < \delta < 1$. Suppose that $\{X_n\}$ is in *equilibrium* with the invariant initial distribution $\boldsymbol{\pi} = (\pi_1, \pi_0)$. Define *statistics* based on the sample $X_0, X_1, \ldots, X_n$ to estimate β, π_1, respectively, by $\hat{\pi}_1^{(n)} = S_n/(n+1)$

and $\hat{\beta}^{(n)} = T_n/n$, where $S_n = X_0 + \cdots + X_n$ is the number of wet days and $T_n = \sum_{k=0}^{n-1} \mathbf{1}_{\{(X_k, X_{k+1}) = (0,1)\}}$ is the number of dry-to-wet transitions. Calculate $\lim_{n \to \infty} \hat{\pi}_1^{(n)}$ and $\lim_{n \to \infty} \hat{\beta}^{(n)}$.

7. Use the result of Exercise 1.5 to describe an extension of the SLLN and the CLT to certain rth-order dependent Markov chains.

Exercises for Section II.11

1. Let $\{X_n\}$ be a two-state Markov chain on $S = \{0, 1\}$ and let τ_0 be the first time $\{X_n\}$ reaches 0. Calculate $P_1(\tau_0 = n)$, $n \geqslant 1$, in terms of the parameters p_{10} and p_{01}.

2. Let $\{X_n\}$ be a three-state Markov chain on $S = \{0, 1, 2\}$ where 0, 1, 2 are arranged counterclockwise on a circle, and at each time a transition occurs one unit clockwise with probability p or one unit counterclockwise with probability $1 - p$. Let τ_0 denote the time of the first return to 0. Calculate $P(\tau_0 > n)$, $n \geqslant 1$.

3. Let τ_0 denote the first time starting in state 2 that the Markov chain in Exercise 5.6 reaches state 0. Calculate $P_2(\tau_0 > n)$.

4. Verify that the Markov chains starting at i having transition probabilities $\mathbf{p}$ and $\hat{\mathbf{p}}$, and viewed up to time τ_A have the same distribution by calculating the probabilities of the event $\{X_0 = i, X_1 = i_1, \ldots, X_m = i_m, \tau_A = m\}$ under each of $\mathbf{p}$ and $\hat{\mathbf{p}}$.

5. Write out a detailed explanation of (11.22).

6. Explain the calculation of (11.28) and (11.29) as given in the text using earlier results on the long-term behavior of transition probabilities.

7. (*Collocation*) Show that there is a unique polynomial $p(x)$ of degree k that takes prescribed (given) values $v_0, v_1, \ldots, v_k$ at any prescribed (given) distinct points $x_0, x_1, \ldots, x_k$, respectively; such a polynomial is called a *collocation polynomial*. [*Hint*: Write down a linear system with the coefficients $a_0, a_1, \ldots, a_k$ of $p(x)$ as the unknowns. To show the system is nonsingular, view the determinant as a polynomial and identify all of its zeros.]

*8. (*Absorption Rates and the Spectral Radius*) Let $\mathbf{p}$ be a transition probability matrix for a finite-state Markov chain and let τ_j be the time of the first visit to j. Use (11.4) and the results of the *Perron–Frobenius Theorem* 6.4 and its corollary to show that exponential rates of convergence (as obtained in (11.43)) can be anticipated more generally.

9. Let $\mathbf{p}$ be the transition probability matrix on $S = \{0, \pm 1, \pm 2, \ldots\}$ defined by

$$p_{ij} = \begin{cases} \dfrac{1}{i} & \text{if } i > 0, j = 0, 1, 2, \ldots, i-1 \\[2ex] \dfrac{1}{|i|} & \text{if } i < 0, j = 0, -1, -2, \ldots, -i+1 \\[2ex] 1 & \text{if } i = 0, j = 0 \\ 0 & \text{if } i = 0, j \neq 0. \end{cases}$$

(i) Calculate the absorption rate.

(ii) Show that the mean time to absorption starting at $i > 0$ is given by $\sum_{k=1}^{i} (1/k)$.

10. Let $\{X_n\}$ be the simple branching process on $S = \{0, 1, 2, \ldots\}$ with offspring distribution $\{f_j\}$, $\sum_{j=0}^{\infty} jf_j \leqslant 1$.
   (i) Show that all nonzero states in S are transient and that $\lim_{n\to\infty} P_1(X_n = k) = 0$, $k = 1, 2, \ldots$.
   (ii) Describe the unique invariant probability distribution for $\{X_n\}$.

11. (i) Suppose that in a certain society each parent has exactly two children, and both males and females are equally likely to occur. Show that passage of the family surname to descendants of males eventually stops.
   (ii) Calculate the extinction probability for the male lineage as in (i) if each parent has exactly three children.
   (iii) Prompted by an interest in the survival of family surnames, A. J. Lotka (1939), "Théorie Analytique des Associations Biologiques II," *Actualités Scientifiques et Industrielles*, (N.780), Hermann et Cie, Paris, used data for white males in the United States in 1920 to estimate the probability function f for the number of male children of a white male. He estimated $f(0) = 0.4825$, $f(j) = (0.2126)(0.5893)^{j-1}$ $(j = 1, 2, \ldots)$.
      (a) Calculate the mean number of male offspring.
      (b) Calculate the probability of survival of the family surname if there is only one male with the given name.
      (c) What is the survival probability for j males with the given name under this model.
   (iv) (*Maximal Branching*) Consider the following modification of the simple branching process in which if there are k individuals in the nth generation, and if $X_{n,1}, X_{n,2}, \ldots, X_{n,k}$ are independent random variables representing their respective numbers of offspring, then the $(n + 1)$st generation will contain $Z_{n+1} = \max\{X_{n,1}, X_{n,2}, \ldots, X_{n,k}\}$ individuals. In terms of the survival of family names one may assume that only the son providing the largest number of grandsons is entitled to inherit the family title in this model (due to J. Lamperti (1970), "Maximal Branching Processes and Long-Range Percolation," *J. Appl. Probability*, 7, pp. 89–98).
      (a) Calculate the transition law for the successive size of the generations when the offspring distribution function is $F(x)$, $x = 0, 1, 2, \ldots$.
      (b) Consider the case $F(x) = 1 - (1/x)$, $x = 1, 2, 3, \ldots$. Show that

$$\lim_{k\to\infty} P(Z_{n+1} \leqslant kx \mid Z_n = k) = e^{-x^{-1}}, \qquad x > 0.$$

12. Let f be the offspring distribution function for a simple branching process having finite second moment. Let $\mu = \sum_k kf(k)$, $\sigma^2 = \sum_k (k - \mu)^2 f(k)$. Show that, given $X_0 = 1$,

$$\operatorname{Var} X_n = \begin{cases} \sigma^2\mu^{n-1}(\mu^n - 1)/(\mu - 1) & \text{if } \mu \neq 1 \\ n\sigma^2 & \text{if } \mu = 1. \end{cases}$$

13. Each of the following distributions below depends on a single parameter. Construct graphs of the nonextinction probability and the expected sizes of the successive generations as a function of the parameter.

(i) $f(j) = \begin{cases} p & \text{if } j = 2 \\ q & \text{if } j = 0 \\ 0 & \text{otherwise;} \end{cases}$

(ii) $f(j) = qp^j, \quad j = 0, 1, 2, \dots;$

(iii) $f(j) = \dfrac{\lambda^j}{j!} e^{-\lambda}, \quad j = 0, 1, 2, \dots.$

14. (*Electron Multiplier*) A weak current of electrons may be amplified by a device consisting of a series of plates. Each electron, as it strikes a plate, gives rise to a random number of electrons, which go on to strike the next plate to produce more electrons, and so on. Use the simple branching process with a Poisson offspring distribution for the numbers of electrons produced at successive plates.
   (i) Calculate the mean and variance in the amplification of a single electron at the nth plate (see Exercise 12).
   (ii) Calculate the survival probability of a single electron in an infinite succession of plates if $\mu = 1.01$.

15. (*A Generalized Gambler's Ruin*) Gamblers 1 and 2 have respective initial capitals $i > 0$ and $c - i > 0$ (whole numbers) of dollars. The gamblers engage in a series of *fair* bets (in whole numbers of dollars) that stops when and only when one of the gamblers goes broke. Let X_n denote gambler 1's capital at the nth play ($n \geq 1$), $X_0 = i$. Gambler 1 is allowed to select a different game (to bet) at each play subject to the condition that it be *fair* in the sense that $E(X_n \mid X_{n-1}) = X_{n-1}$, $n = 1, 2, \dots$, and that the amounts wagered be covered by the current capitals of the respective gamblers. Assume that gambler 1's selection of a game for the $(n + 1)$st play ($n \geq 0$) depends only on the current capital X_n so that $\{X_n\}$ is a Markov chain with stationary transition probabilities. Also assume that $P(X_{n+1} = i \mid X_n = i) < 1$, $0 < i < c$, although it may be possible to break even in a play. Calculate the probability that gambler 1 will eventually go broke. How does this compare to classical gamblers ruin (win or lose $1 bets placed on outcomes of fair coin tosses)? [*Hint*: Check that $a_c(i) = i/c$, $0 < i < c$, $a_0(i) = (c - i)/c$, $0 < i \leq c$ (*uniquely*) solves (11.13) using the equation $E(X_n \mid X_{n-1}) = X_{n-1}$.]

Exercises for Section II.12

1. (i) Show that for (12.10) it is sufficient to check that

$$\frac{g_{b,c}(a)}{g_{b,c}(\theta)} = \frac{q(b, a)q(a, c)}{q(b, \theta)q(\theta, c)}, \qquad a, b, c \in S. \qquad \left[\textit{Hint}: \sum_a g_{b,c}(a) = 1,\ \forall b, c \in S.\right]$$

   (ii) Use (12.11), (12.12) to show that this condition can be expressed as

$$\frac{g_{b,c}(a)}{g_{b,c}(\theta)} = \frac{g_{\theta,a}(b)}{g_{\theta,a}(\theta)} \frac{g_{\theta,\theta}(\theta)}{g_{\theta,\theta}(b)} \frac{g_{\theta,c}(a)}{g_{\theta,c}(\theta)}, \qquad a, b, c \in S.$$

(iii) Consider four adjacent sites on $\mathbb{Z}$ in states α, β, a, b, respectively. For notational convenience, let $[\alpha, \beta, a, b] = P(X_0 = \alpha, X_1 = \beta, X_2 = a, X_3 = b)$. Use the condition (ii) of Theorem 12.1 to show

$$[\alpha, \beta, a, b] = P(X_0 = \alpha, X_2 = a, X_3 = b)g_{\alpha,a}(\beta)$$

and, therefore,

$$\frac{[\alpha, \beta, a, b]}{[\alpha, \beta', a, b]} = \frac{g_{\alpha,a}(\beta)}{g_{\alpha,a}(\beta')}.$$

(iv) Along the lines of (iii) show that also

$$\frac{[\alpha, \beta, a, b]}{[\alpha, \beta, a', b]} = \frac{g_{\beta,b}(a)}{g_{\beta,b}(a')}.$$

Use the "substitution scheme" of (iii) and (iv) to verify (12.10) by checking (ii).

2. (i) Verify (12.13) for the case $n = 1$, $r = 2$. [*Hints*: Without loss of generality, take $h = 0$, and note,

$$P(X_1 = \beta, X_2 = \gamma \mid X_0 = \alpha, X_3 = b) = \frac{[\alpha, \beta, \gamma, b]}{\sum_u \sum_v [\alpha, u, v, b]},$$

and using Exercise 1(iii)–(iv) and (12.10),

$$\frac{[\alpha, u, v, b]}{[\alpha, u', v', b]} = \frac{g_{\alpha,v}(u)g_{u',b}(v)}{g_{\alpha,v}(u')g_{u',b}(v')} = \frac{q(\alpha, u)q(u, v)q(v, b)}{q(\alpha, u')q(u', v')q(v', b)}.$$

Sum over $u, v \in S$ and then take $u' = \beta$, $v' = \gamma$.]

(ii) Complete the proof of (12.13) for $n = 1$, $r \geqslant 2$ by induction and then $n \geqslant 2$.

3. Verify (12.15).

4. Justify the limiting result in (12.17) as a consequence of Proposition 6.1. [*Hint*: Use Scheffe's Theorem, Chapter 0.]

*5. Take $S = \{0, 1\}$, $g_{1,1}(1) = \mu$, $g_{1,0}(1) = \nu$. Find the transition matrix $\mathbf{q}$ and the invariant initial distribution for the Markov random field viewed as a Markov chain.

Exercises for Section II.13

1. (i) Let $\{\mathbf{Y}_n: n \geqslant 0\}$ be a sequence of random vectors with values in $\mathbb{R}^k$ which converge a.s. to a random vector Y. Show that the distribution Q_n of $\mathbf{Y}_n$ converges weakly to the distribution Q of Y.

(ii) Let $p(\mathbf{x}, d\mathbf{y})$ be a *transition probability* on the state space $S = \mathbb{R}^k$ (i.e., (a) for each $\mathbf{x} \in S$, $p(\mathbf{x}, d\mathbf{y})$ is a probability measure on $(\mathbb{R}^k, B^k)$ and (b) for each $B \in \mathscr{B}^k$, $\mathbf{x} \to p(\mathbf{x}, B)$ is a Borel-measurable function on $\mathbb{R}^k$). The n-step transition probability $p^{(n)}(\mathbf{x}, d\mathbf{y})$ is defined recursively by

$$p^{(1)}(\mathbf{x}, d\mathbf{y}) = p(\mathbf{x}, d\mathbf{y}), \qquad p^{(n+1)}(\mathbf{x}, B) = \int_S p(\mathbf{y}, B)p^{(n)}(\mathbf{x}, d\mathbf{y}).$$

Show that, if $p^{(n)}(\mathbf{x}, d\mathbf{y})$ converges weakly to the same probability measure $\pi(d\mathbf{y})$ for all $\mathbf{x}$, and $\mathbf{x} \to p(\mathbf{x}, d\mathbf{y})$ is *weakly continuous* (i.e., $\int_S f(\mathbf{y})p(\mathbf{x}, d\mathbf{y})$ is a continuous function of $\mathbf{x}$ for every bounded continuous function f on S), then π is the unique *invariant* probability for $p(\mathbf{x}, d\mathbf{y})$, i.e., $\int_B p(\mathbf{x}, B)\pi(d\mathbf{x}) = \pi(B)$ for all $B \in \mathscr{B}^k$. [*Hint*: Let f be bounded and continuous. Then

$$\int f(\mathbf{y})p^{(n+1)}(\mathbf{x}, d\mathbf{y}) \to \int f(\mathbf{y})\pi(d\mathbf{y}),$$

$$\int f(\mathbf{y})p^{(n+1)}(\mathbf{x}, d\mathbf{y}) = \int\left(\int f(\mathbf{y})p(\mathbf{z}, d\mathbf{y})\right)p^{(n)}(\mathbf{x}, d\mathbf{z}) \to \int\left(\int f(\mathbf{y})p(\mathbf{z}, d\mathbf{y})\right)\pi(d\mathbf{z}). \text{]}$$

(iii) Extend (i) and (ii) to arbitrary metric space S, and note that it suffices to require convergence of $n^{-1}\sum_{m=0}^{n-1}\int f(\mathbf{y})p^{(m)}(\mathbf{x}, d\mathbf{y})$ to $\int f(\mathbf{y})\pi(\mathbf{y})\,d\mathbf{y}$ for all bounded continuous f on S.

2. (i) Let $\mathbf{B}_1, \mathbf{B}_2$ be $m \times m$ matrices (with real or complex coefficients). Define $\|\mathbf{B}\|$ as in (13.13), with the supremum over unit vectors in $\mathbb{R}^m$ or $\mathbb{C}^m$. Show that

$$\|\mathbf{B}_1\mathbf{B}_2\| \leqslant \|\mathbf{B}_1\|\,\|\mathbf{B}_2\|.$$

(ii) Prove that if $\mathbf{B}$ is an $m \times m$ matrix then

$$\|\mathbf{B}\| \leqslant m^{1/2}\max\{|b_{ij}|: 1 \leqslant i, j \leqslant m\}.$$

(iii) If $\mathbf{B}$ is an $m \times m$ matrix and $\|\mathbf{B}\|$ is defined to be the supremum over unit vectors in $\mathbb{C}^m$, show that $\|\mathbf{B}^n\| \geqslant r^n(\mathbf{B})$. Use this together with (13.17) to prove that $\lim\|\mathbf{B}^n\|^{1/n}$ exists and equals $r(\mathbf{B})$. [*Hint*: Let λ_m be an eigenvalue such that $|\lambda_m| = r(\mathbf{B})$. Then there exists $\mathbf{x} \in \mathbb{C}^m$, $\|\mathbf{x}\| = 1$, such that $\mathbf{B}\mathbf{x} = \lambda_m\mathbf{x}$.]

3. Suppose ε_1 is a random vector with values in $\mathbb{R}^k$.
(i) Prove that if $\delta > 1$ and $c > 0$, then

$$\sum_{n=1}^{\infty} P(|\varepsilon_1| > c\delta^n) \leqslant \varepsilon E|Z| - 1, \qquad \text{where } Z = \frac{\log|\varepsilon_1| - \log c}{\log \delta}.$$

$$\left[\textit{Hint}: \sum_{n=1}^{\infty} P(|Z| > n) = \sum_{n=1}^{\infty} nP(n < |Z| \leqslant n+1).\ \right]$$

(ii) Show that if (13.15) holds then (13.16) converges. [*Hint*:

$$\|\delta^n\mathbf{B}^n\| \leqslant d\|\delta^{n_0}\mathbf{B}^{n_0}\|^{[n/n_0]}, \qquad \text{where } d = \max\{\delta^r\|\mathbf{B}\|^r: 0 \leqslant r \leqslant n_0\}.\ ]$$

(iii) Show that (13.15) holds, if it holds for some $\delta < 1/r(\mathbf{B})$. [*Hint*: Use the Lemma.]

4. Suppose $\sum a_n z^n$ and $\sum b_n z^n$ are absolutely convergent and are equal for $|z| < r$, where r is some positive number. Show that $a_n = b_n$ for all n. [*Hint*: Within its radius of

convergence a power series is infinitely differentiable and may be repeatedly differentiated term by term.]

5. (i) Prove that the determinant of an $m \times m$ matrix in triangular form equals the product of its diagonal elements.
   (ii) Check (13.28) and (13.35).

6. (i) Prove that under (13.15), $\mathbf{Y}$ in (13.16) is Gaussian if $\boldsymbol{\varepsilon}_n$ are Gaussian. Calculate the mean vector and the dispersion matrix of $\mathbf{Y}$ in terms of those of $\boldsymbol{\varepsilon}_n$.
   (ii) Apply (i) to Examples 2(a) and 2(b).

7. (i) In Example 1 show that $|b| < 1$ is necessary for the existence of a unique invariant probability.
   (ii) Show by example that $|b| < 1$ is not sufficient for the existence of a unique invariant probability. [*Hint*: Find a distribution Q of the noise ε_n with an appropriately heavy tail.]

8. In Example 1, assume $E\varepsilon_n^2 < \infty$, and write $a = E\varepsilon_n$, $X_{n+1} = a + bX_n + \theta_{n+1}$, where $\theta_n = \varepsilon_n - a$ $(n \geqslant 1)$. The *least squares estimates* of a, b are $\hat{a}_N, \hat{b}_N$, which minimize $\sum_{n=0}^{n-1} (X_{n+1} - a - bX_n)^2$ with respect to a, b.
   (i) Show that $\hat{a}_N = \bar{Y} - \hat{b}_N\bar{X}$, $\hat{b}_N = \sum_0^{N-1} X_{n+1}(X_n - \bar{X})/\sum_1^N (X_n - \bar{X})^2$, where $\bar{X} = N^{-1}\sum_0^{N-1} X_n$, $\bar{Y} = N^{-1}\sum_1^N X_n$. [*Hint*: Reparametrize to write $a + bX_n = a_1 + b(X_n - \bar{X})$.]
   (ii) In the case $|b| < 1$, prove that $\hat{a}_N \to a$ and $\hat{b}_N \to b$ a.s. as $N \to \infty$.

9. (i) In Example 2 let $m = 2$, $b_{11} = -4$, $b_{12} = 5$, $b_{21} = -10$, $b_{22} = 3$. Assume ε_1 has a finite absolute second moment. Does there exists a unique invariant probability?
   (ii) For the AR(2) or ARMA(2, q) models find a sufficient condition in terms of β_0 and β_1 for the existence of a unique invariant probability, assuming that η_n has a finite rth moment for some $r > 0$.

Exercises for Section II.14

1. Prove that the process $\{X_n(x): n \geqslant 0\}$ defined by (14.2) is a Markov process having the transition probability (14.3). Show that this remains true if the initial state x is replaced by a random variable X_0 independent of $\{\boldsymbol{\alpha}_n\}$.

2. Let $F_n(z) := P(X_n \leqslant z)$, $n \geqslant 0$, be a sequence of distribution functions of random variables X_n taking values in an interval J. Prove that if $F_{n+m}(z) - F_n(z)$ converges to zero uniformly for $z \in J$, as n and m go to ∞, then $F_n(z)$ converges uniformly (for all $z \in J$) to the distribution function $F(z)$ of a probability measure on J.

3. Let g be a continuous function on a metric space S (with metric ρ) into itself. If, for some $x \in S$, the iterates $g^{(n)}(x) \equiv g(g^{(n-1)}(x))$, $n \geqslant 1$, converge to a point $x^* \in S$ as $n \to \infty$, then show that x^* is a fixed point of g, i.e., $g(x^*) = x^*$.

4. Extend the strong Markov property (Theorem 4.1) to Markov processes $\{X_n\}$ on an interval J (or, on $\mathbb{R}^k$).

5. Let τ be an a.s. finite stopping time for the Markov process $\{X_n(x): n \geqslant 0\}$ defined by (14.2). Assume that $X_\tau(x)$ belongs to an interval J with probability 1 and $p^{(n)}(y, dz)$ converges weakly to a probability measure $\pi(dz)$ on J for all $y \in J$. Assume also that $p(y, J) = 1$ for all $y \in J$.

(i) Prove that $p^{(n)}(x, dz)$ converges weakly to $\pi(dz)$. [*Hint*: $p^{(k)}(x, J) \to 1$ as $k \to \infty$, $\int f(y)p^{(k+r)}(x, dy) = \int (\int f(z)p^{(r)}(y, dz))p^{(k)}(x, dy)$.]

(ii) Assume the hypothesis above for all $x \in J$ (with J not depending on x). Prove that π is the unique invariant probability.

6. In Example 2, if ε_n are i.i.d. uniform on $[-1, 1]$, prove that $\{X_{2n}(x): n \geqslant 1\}$ is i.i.d. with common p.d.f. given by (14.27) if $x \in [-2, 2]$.

7. In Example 2, modify f as follows. Let $0 < \delta < \frac{1}{2}$. Define $f_\delta(x) := f(x)$ for $-2 \leqslant x < -\delta$, and $\delta \leqslant x \leqslant 2$, and linearly interpolate between $(-\delta, \delta)$, so that f_δ is continuous.

   (i) Show that, for $x \in [\delta, 1]$ (or, $x \in [-1, -\delta]$) $\{X_n(x): n \geqslant 1\}$ is i.i.d. with common distribution π_x (or, π_{x+2}).

   (ii) For $x \in (1, 2]$ (or, $[-2, -1)$) $\{X_n(x): n \geqslant 2\}$ is i.i.d. with common distribution π_x (or, π_{x+2}).

   (iii) For $x \in (-\delta, \delta)$ $\{X_n(x): n \geqslant 1\}$ is i.i.d. with common distribution π_{-x+1}.

8. In Example 3, assume $P(\varepsilon_1 = 0) > 0$ and prove that $P(\varepsilon_n = 0$ for some $n \geqslant 0) = 1$.

9. In Example 3, suppose $E \log \varepsilon_i > 0$.

   (i) Prove that $\sum_{n=1}^{\infty} \{1/(\varepsilon_1 \cdots \varepsilon_n)\}$ converges a.s. to a (finite) nonnegative random variable Z.

   (ii) Let $d_1 := \inf\{z \geqslant 0: P(Z \leqslant z) > 0\}$, $d_2 := \sup\{z \geqslant 0: P(Z \geqslant z) > 0\}$. Show that

$$\rho(x) \begin{cases} = 0 & \text{if } x < c(d_1 + 1), \\ \in (0, 1) & \text{if } c(d_1 + 1) < x < c(d_2 + 1), \\ = 1 & \text{if } x > c(d_2 + 1). \end{cases}$$

10. In Example 3, define $M := \sup\{z \geqslant 0: P(\varepsilon_1 \geqslant z) > 0\}$.

    (i) Suppose $1 < M < \infty$. Then show that $\rho(x) = 0$ if $x < cM/(M - 1)$.

    (ii) If $M \leqslant 1$, then show that $\rho(x) = 0$ for all x.

    (iii) Let m be as in (14.34) and M as above. Let d_1, d_2 be as defined in Exercise 9(ii). Show that

$$\text{(a) } d_1 = \sum_{n=1}^{\infty} m^{-n} \qquad \left(= \frac{1}{m-1} \text{ if } m > 1, = \infty \text{ if } m \leqslant 1\right), \qquad \text{and}$$

$$\text{(b) } d_2 = \sum_{n=1}^{\infty} M^{-n} \qquad \left(= \frac{1}{m-1} \text{ if } M > 1, = \infty \text{ if } M \leqslant 1\right).$$

Exercises for Section II.15

1. Let $0 < p \leqslant 1$ and suppose $h(p)$ represents a *measure of uncertainty* regarding the occurrence of an event having probability p. Assume

   (a) $h(1) = 0$.

   (b) $h(p)$ is strictly decreasing for $0 < p \leqslant 1$.

   (c) h is continuous on $(0, 1]$.

   (d) $h(pr) = h(p) + h(r)$, $0 < p \leqslant 1$, $0 < r \leqslant 1$.

Intuitively, condition (d) says that the total uncertainty in the joint occurrence of two independent events is the cumulative uncertainty for each of the events. Verify that h must be of the form $h(p) = -c \log_2 p$ where $c = h(\frac{1}{2}) > h(1) = 0$ is a positive constant. Standardizing, one may take

$$h(p) = -\log_2 p.$$

2. Let $\mathbf{f} = (f_j)$ be a probability distribution on $S = \{1, 2, \ldots, M\}$. Define the *entropy* in $\mathbf{f}$ by the "average uncertainty," i.e.,

$$H(\mathbf{f}) = -\sum_{\alpha=1}^{M} f_j \log_2 f_j \qquad (0 \log 0 := 0).$$

   (i) Show that $H(\mathbf{f})$ is *maximized* by the uniform distribution on S.
   (ii) If $\mathbf{g} = (g_i)$ is another probability distribution on S then

$$H(\mathbf{f}) \leqslant -\sum_i f_i \log_2 g_i$$

   with equality if and only if $f_i = g_i$ for all $i \in S$.

3. Suppose that X is a random variable taking values $a_1, \ldots, a_M$ with respective probabilities $p(a_1), \ldots, p(a_M)$. Consider an arbitrary *binary coding* of the respective symbols a_i, $1 \leqslant i \leqslant M$, by a string $\phi(a_i) = (\varepsilon_1^{(i)}, \ldots, \varepsilon_{n_i}^{(i)})$ of 0's and 1's, such that no string $(\varepsilon_1^{(j)}, \ldots, \varepsilon_{n_j}^{(j)})$ can be obtained from a shorter code $(\varepsilon_1^{(i)}, \ldots, \varepsilon_{n_i}^{(i)})$, $n_i \leqslant n_j$, by adding more terms; such codes will be called *admissible*. The number n_i of *bits* is called the *length* of the code-word $\phi(a_i)$.
   (i) Show that an admissible binary code ϕ having respective lengths n_i exists if and only if

$$\sum_{i=1}^{M} (\tfrac{1}{2})^{n_i} \leqslant 1.$$

   [*Hint*: Let $n_1, \ldots, n_M$ be positive integers and let

$$\rho_k = \#\{i: n_i = k\}, \qquad k = 1, 2, \ldots.$$

   Then it is necessary and sufficient for an admissible code that $\rho_1 \leqslant 2$, $\rho_2 \leqslant 2^2 - 2\rho_1, \ldots, \rho_k \leqslant 2^k - \rho_1 2^{k-1} - \cdots - \rho_{k-1}2$, $k \geqslant 1$.]
   (ii) (*Noiseless Coding Theorem*) For any admissible binary code ϕ of $a_1, \ldots, a_M$ having respective lengths $n_1, \ldots, n_M$, the average length of code-words cannot be made smaller than the entropy of the distribution $\mathbf{p}$ of $a_1, \ldots, a_M$, i.e.,

$$\sum_{i=1}^{M} n_i p(a_i) \geqslant -\sum_{i=1}^{M} p(a_i) \log_2 p(a_i).$$

   [*Hint*: Use Exercise 2(ii) with $f_i = p(a_i)$, $g_i = 2^{-n_i}$ to show

$$H(\mathbf{p}) \leqslant \sum_{i=1}^{M} n_i p(a_i) + \log_2\left(\sum_{k=1}^{M} 2^{-n_k}\right).$$

   Then apply Exercise 3(i) to get the result.]

(iii) Show that it is always possible to construct an admissible binary code of $a_1, \ldots, a_M$ such that

$$H(\mathbf{p}) \leqslant \sum_{i=1}^{M} n_i\, p(a_i) \leqslant H(\mathbf{p}) + 1.$$

[*Hint*: Select n_i such that

$$-\log_2 p(a_i) \leqslant n_i \leqslant -\log_2 p(a_i) + 1 \qquad \text{for } 1 \leqslant i \leqslant M,$$

and apply 3(i).]

(iv) Verify that there is not a more efficient (admissible) encoding (i.e. minimal average number of bits) of the symbols a_1, a_2, a_3 for the distribution $p(a_1) = \frac{1}{2}$, $p(a_2) = p(a_3) = \frac{1}{4}$, than the code $\phi(a_1) = (0)$, $\phi(a_2) = (1, 0)$, $\phi(a_3) = (1, 1)$.

4. (i) Show that $Y_1, Y_2, \ldots$ in (15.4) satisfies the law of large numbers.
(ii) Show that for (15.5) to hold it is sufficient that $Y_1, Y_2, \ldots$ satisfy the law of large numbers.

THEORETICAL COMPLEMENTS

Theoretical Complements to Section II.6

1. Let S be an arbitrary state space equipped with a sigmafield $\mathscr{S}$ of events. Let $\mu(dx)$ be a sigmafinite measure on $(S, \mathscr{S})$ and let $p(x, y)$ be a transition probability density with respect to μ; i.e., p is a nonnegative measurable function on $(S \times S, \mathscr{S} \times \mathscr{S})$ such that for each fixed x, $p(x, y)$ is a μ-integrable function of y with total mass 1. Let Ω denote the space of all sequences $\omega = (x_0, x_1, \ldots)$ of states $x_i \in S$. Let $\mathscr{F}_0$ denote the class of all finite-dimensional sets A of the form

$$A = \{\omega = (x_0, x_1, \ldots) \in \Omega \colon x_i \in B_i,\ i = 0, 1, \ldots, n\},$$

where $B_i \in \mathscr{S}$ for each i. Define $P_x(A)$ for such a set A, $x \in B_0$, by

$$P_x(A) = \int_{B_n} \cdots \int_{B_1} p(x, y_1)p(y_1, y_2)\cdots p(y_{n-1}, y_n)\mu(dy_n)\cdots\mu(dy_1) \tag{T.6.1}$$

for $x \in S$. Define $P_x(A) = 0$ if $x \notin B_0$. The Kolmogorov Extension Theorem assures us that P_x has a unique extension to a probability measure defined for all events in the smallest sigmafield $\mathscr{F}$ of subsets of Ω that contain $\mathscr{F}_0$. For any probability measure γ on $(S, \mathscr{S})$, define

$$P_\gamma(A) = \int P_x(A)\gamma(dx), \qquad A \in \mathscr{F}. \tag{T.6.2}$$

Let X_n denote the nth coordinate projection mapping on Ω. Then $\{X_n\}$ is said to be a *Markov chain on the state space S with transition density $p(x, y)$ with respect to μ and initial distribution, γ under P_γ*. The results of Exercise 6.9 can be extended to this setting as follows. *Suppose that there is a positive integer r and a μ-integrable function ρ on S such that $\int_S \rho(x)\mu(dx) > 0$ and $p^{(r)}(x, y) > \rho(y)$ for all x, y in S. Then there is a unique invariant distribution π such that*

$$\sup_B \left| \int_B p^{(n)}(x, y)\mu(dy) - \pi(B) \right| \leqslant (1 - \alpha)^{n'},$$

where $\alpha = \int_S \rho(x)\mu(dx)$, $n' = [n/r]$, and the sup is over $B \in \mathscr{S}$. □

Proof. To see this define

$$\begin{aligned} M_n(B) &:= \sup_{u \in S} \left\{ \int_B p^{(n)}(u, y)\mu(dy) \right\}, \\ m_n(B) &:= \inf_{u \in S} \left\{ \int_B p^{(n)}(u, y)\mu(dy) \right\}, \\ \Delta_n &:= \sup_B \{M_n(B) - m_n(B)\}. \end{aligned} \tag{T.6.3}$$

Then

(i) $\Delta_n = \dfrac{1}{2} \sup_{x,z} \displaystyle\int_S |p^{(n)}(x, y) - p^{(n)}(z, y)| \mu(dy) \leqslant 1.$

(ii) $\left| \displaystyle\int_B p^{((k+1)r)}(x, y)\mu(dy) - p^{((k+1)r)}(z, y)\mu(dy) \right| \leqslant (1 - \alpha)[M_{kr}(B) - m_{kr}(B)].$

(iii) The probability measure π given by

$$\pi(B) = \lim_{n \to \infty} \int_B p^{(n)}(x, y)\mu(dy) \tag{T.6.4}$$

is well defined and

$$\sup_B \left| \int_B p^{(n)}(x, y)\mu(dy) - \pi(B) \right| \leqslant (1 - \alpha)^{n'}. \tag{T.6.5}$$

■

Also, the following facts are simple to check.

(iv) π is absolutely continuous with respect to μ and therefore by the Radon–Nykodym Theorem (Chapter 0) has a density $\pi(y)$ with respect to μ.

(v) For $B \in \mathscr{S}$ with $\pi(B) > 0$, one has $P_y(X_n \in B \text{ i.o.}) = 1$.

2. (*Doeblin Condition*) A well-known condition ensuring the existence of an invariant distribution is the following. *Suppose there is a finite measure φ and an integer $r \geqslant 1$, and $\varepsilon > 0$, such that $p^{(r)}(x, B) \leqslant 1 - \varepsilon$ whenever $\varphi(B) \leqslant \varepsilon$. Under this condition there is a decomposition of the state space as $S = \bigcup_{i=1}^m S_i$ such that*

(i) *Each S_i is closed in the sense that $p(x, S_i) = 1$ for $x \in S_i$ $(1 \leqslant i \leqslant m)$.*

(ii) *p restricted to S_i has a unique invariant distribution π_i.*

(iii) $\frac{1}{n}\sum_{r=1}^{n} p^{(r)}(x, B) \to \pi_i(B)$ *as* $n \to \infty$ *for* $x \in S_i$.

Moreover, the convergence is exponentially fast and uniform in x and B.

It is easy to check that the condition of theoretical complement 1 above implies the Doeblin condition with $\varphi(dx) = \rho(x)\mu(dx)$. The more general connections between the Doeblin condition with the $\varphi(dx) = \rho(x)\mu(dx)$ condition in theoretical complement 1 can be found in detail in J. L. Doob (1953), *Stochastic Processes*, Wiley, New York, pp. 190.

3. (*Lengths of Increasing Runs*) Here is an example where the more general theory applies.

Let $X_1, X_2, \ldots$ be i.i.d. uniform on $[0, 1]$. *Successive increasing runs* among the values of the sequence $X_1, X_2, \ldots$ are defined by placing a marker at 0 and then between X_j and X_{j+1} whenever X_j exceeds X_{j+1}, e.g., $|0.20, 0.24, 0.60|0.50|0.30, 0.70|0.20 \ldots$. Let Y_n denote the initial (smallest) value in the nth run, and let L_n denote the length of the nth run, $n = 1, 2, \ldots$.

(i) $\{Y_n\}$ is a Markov chain on the state space $S = [0, 1]$ with transition density

$$p(x, y) = \begin{cases} e^{1-x} & \text{if } y < x \\ e^{1-x} - e^{y-x} & \text{if } y > x. \end{cases}$$

(ii) Applying theoretical complement 1, $\{Y_n\}$ has a unique invariant distribution π. Moreover, π has a density given by $\pi(y) = 2(1 - y)$, $0 \leqslant y \leqslant 1$.

(iii) The limit in distribution of the length L_n as $n \to \infty$ may also be calculated from that of $\{Y_n\}$, since

$$P(L_n \geqslant m) = \int_0^1 P(L_n \geqslant m \mid Y_n = y)P(Y_n \in dy) = \int_0^1 \frac{(1-y)^{m-1}}{(m-1)!} P(Y_n \in dy),$$

and therefore,

$$P(L_\infty \geqslant m) := \lim_{n\to\infty} P(L_n \geqslant m) = \int_0^1 \frac{(1-y)^{m-1}}{(m-1)!} 2(1-y)\,dy.$$

Note from this that

$$EL_\infty = \sum_{m=1}^{\infty} P(L_\infty \geqslant m) = 2.$$

Theoretical Complement to Section 11.8

1. We learned about the *random rounding problem* from Andreas Weisshaar (1987), "Statistisches Runden in rekursiven, digitalen Systemen 1 und 2," *Diplomarbeit erstellt am Institut für Netzwerk- und Systemtheorie*, Universität Stuttgart. The stability condition of Example 8.1 is new, however computer simulations by Weisshaar suggest that the result is more generally true if the spectral radius is less than 1. However, this is not known rigorously. For additional background on the applications of this technique to the design of digital filters, see R. B. Kieburtz, V. B. Lawrance, and K. V. Mina (1977), "Control of Limit Cycles in Recursive Digital Filters by Randomized Quantization," *IEEE Trans. Circuits and Systems*, **CAS-24**(6), pp. 291–299, and references therein.

Theoretical Complements to Section II.9

1. The lattice case of *Blackwell's Renewal Theorem* (theoretical complement 2 below) may be used to replace the large-scale *Caesaro type convergence* obtained for the transition probabilities in (9.16) by the stronger elementwise convergence described as follows.

(i) If $j \in S$ is recurrent with period 1, then for any $i \in S$,

$$\lim_{n\to\infty} p_{ij}^{(n)} = \rho_{ij}(E_j\tau_j^{(1)})^{-1}$$

where ρ_{ij} is the probability $P_i(\tau_j^{(1)} < \infty)$ (see Eq. 8.4).

(ii) If $j \in S$ is recurrent with period $d > 1$, then by regarding $\mathbf{p}^d$ as a one-step transition law,

$$\lim_{n\to\infty} p_{ij}^{(nd)} = \rho_{ij}d(E_j\tau_j^{(1)})^{-1}.$$

To obtain these from the general *renewal theory* described below, take as the *delay* Z_0 the time to reach j for the first time starting at i. The *durations* of the subsequent replacements $Z_1, Z_2, \ldots$ represent the lengths of times between returns to j.

2. (*Renewal Theorems*) Let $Z_0, Z_1, \ldots$ be independent nonnegative random variables such that $Z_1, Z_2, \ldots$ are i.i.d. with common (nondegenerate) distribution function F, and Z_0 has distribution function G. In the customary framework of renewal theory, components subject to failure (e.g. lightbulbs) are instantly replaced upon failure, and $Z_1, Z_2, \ldots$ represent the random durations of the successive replacements. The delay random variable Z_0 represents the length of time remaining in the life of the initial component with respect to some specified time origin. For example, if the initial component has age a relative to the placement of the time origin, then one may take

$$G(x) = \frac{F(x+a) - F(a)}{1 - F(a)}.$$

Let

$$S_n = Z_0 + Z_1 + \cdots + Z_n, \qquad n \geqslant 0, \tag{T.9.1}$$

and let

$$N_t = \inf\{n \geqslant 0: S_n > t\}, \qquad t \geqslant 0. \tag{T.9.2}$$

We will use the notation S_n^0, N_t^0 sometimes to identify cases when $Z_0 = 0$ a.s. Then S_n is the time of the $(n+1)$st renewal and N_t counts the number of renewals up to and including time t. In the case that $Z_0 = 0$ a.s., the stochastic (counting) process $\{N_t\} \equiv \{N_t^0\}$ is called the *(ordinary) renewal process*. Otherwise $\{N_t\}$ is called a *delayed renewal process*.

For simplicity, first restrict attention to the case of the ordinary renewal process. Let $\mu = EZ_1 < \infty$. Then $1/\mu$ is called the *renewal rate*. The interpretation as an average rate of renewals is reasonable since

$$\frac{S_{N_t - 1}}{N_t} \leqslant \frac{t}{N_t} \leqslant \frac{S_{N_t}}{N_t} \tag{T.9.3}$$

and $N_t \to \infty$ as $t \to \infty$, so that by the *strong law of large numbers*

$$\frac{N_t}{t} \to \frac{1}{\mu} \quad \text{a.s.} \qquad \text{as } t \to \infty. \tag{T.9.4}$$

Since N_t is a stopping time for $\{S_n\}$ (for fixed $t \geqslant 0$), it follows from *Wald's Equation* (Chapter I, Corollary 13.2) that

$$ES_{N_t} = \mu EN_t. \tag{T.9.5}$$

In fact, the so-called *Elementary Renewal Theorem* asserts that

$$\frac{EN_t}{t} \to \frac{1}{\mu} \qquad \text{as } t \to \infty. \tag{T.9.6}$$

To deduce this from the above, simply observe that $\mu EN_t = ES_{N_t} \geqslant t$ and therefore

$$\liminf_{t\to\infty} \frac{EN_t}{t} \geqslant \frac{1}{\mu}.$$

On the other hand, assuming first that $Z_n \leqslant C$ a.s. for each $n \geqslant 1$, where C is a positive constant, gives $\mu EN_t \leqslant t + C$ and therefore for this case, $\limsup_{t\to\infty}(EN_t/t) \leqslant 1/\mu$. More generally, since truncations of the Z_n at the level C would at most decrease the S_n, and therefore at most increase N_t and EN_t, this last inequality applied to the truncated process yields

$$\limsup_{t\to\infty} \frac{EN_t}{t} \leqslant \frac{1}{CP(Z_1 \geqslant C) + \int_0^C xF(dx)} \to \frac{1}{\mu} \qquad \text{as } C \to \infty. \tag{T.9.7}$$

The above limits (T.9.4) and (T.9.6) also hold in the case that $\mu = \infty$, under the convention that $1/\infty$ is 0, by the SLLN and (T.9.7). Moreover, these asymptotics can now be applied to the delayed renewal process to get precisely the same conclusions for any given (initial) distribution G of delay.

With the special choice of $G = F_\infty$ defined by

$$F_\infty(x) = \frac{1}{\mu}\int_0^x P(Z_1 > u)\,du, \qquad x \geqslant 0, \tag{T.9.8}$$

the corresponding delayed renewal process N^∞, called the *equilibrium renewal process*, has the property that

$$EN^\infty(t, t+h] = h/\mu, \qquad \text{for any } h, \quad t \geqslant 0, \tag{T.9.9}$$

where

$$N^\infty(t, t+h] = N^\infty_{t+h} - N^\infty_t. \tag{T.9.10}$$

To prove this, define the *renewal function* $m(t) = EN_t$, $t \geqslant 0$, for $\{N_t\}$. Then for the general (delayed) process we have

$$\begin{aligned} m(t) = EN_t &= \sum_{n=1}^{\infty} P(N_t \geqslant n) = G(t) + \sum_{n=1}^{\infty} P(N_t \geqslant n+1) \\ &= G(t) + \sum_{n=1}^{\infty} \int_0^t P(N_{t-u} \geqslant n) F(du) = G(t) + \int_0^t m(t-u) F(du) \\ &= G(t) + m * F(t), \end{aligned} \tag{T.9.11}$$

where $*$ denotes the convolution defined by

$$m * F(t) = \int_{\mathbb{R}} m(t-s) F(ds). \tag{T.9.12}$$

Observe that $g(t) = t/\mu$, $t \geqslant 0$, solves the *renewal equation* (T.9.11) with $G = F_\infty$; i.e.,

$$\begin{aligned} \frac{t}{\mu} = \frac{1}{\mu}\int_0^t (1 - F(u))\,du + \frac{1}{\mu}\int_0^t F(u)\,du = F_\infty(t) + \frac{1}{\mu}\int_0^t \int_0^u F(ds)\,du \\ = F_\infty(t) + \frac{1}{\mu}\int_0^t \int_s^t du\, F(ds) = F_\infty(t) + \int_0^t \frac{t-s}{\mu} F(ds). \end{aligned} \tag{T.9.13}$$

To finish the proof of (T.9.9), observe that $g(t) = m^\infty(t) := EN^\infty_t$, $t \geqslant 0$, *uniquely* solves (T.9.11), with $G = F_\infty$, among functions that are bounded on finite intervals. For if $r(t)$ is another such function, then by iterating we have

$$\begin{aligned} r(t) &= F_\infty(t) + \int_0^t r(t-u) F(du) \\ &= F_\infty(t) + \int_0^t \left\{ F_\infty(t-u) + \int_0^{t-u} r(t-u-s) F(ds) \right\} F(du) \\ &= P(N^\infty_t \geqslant 1) + P(N^\infty_t \geqslant 2) + \int_0^t r(t-v) P(S^0_2 \in dv) \\ &= P(N^\infty_t \geqslant 1) + P(N^\infty_t \geqslant 2) + \cdots + P(N^\infty_t \geqslant n) + \int_0^t r(t-v) P(S^0_n \in dv). \end{aligned} \tag{T.9.14}$$

Thus,

$$r(t) = \sum_{n=1}^{\infty} P(N_t^{\infty} \geqslant n) = m^{\infty}(t)$$

since

$$\int_0^t |r(t-v)| P(S_n^0 \in dv) \leqslant \sup_{s \leqslant t} |r(s)| P(S_n^0 \leqslant t) \to 0 \qquad \text{as } n \to \infty;$$

i.e.,

$$P(S_n^0 \leqslant t) = P(N_t^0 \geqslant n) \to 0 \qquad \text{as } n \to \infty \qquad \text{since } \sum_{n=1}^{\infty} P(N_t^0 \geqslant n) = EN_t^0 < \infty.$$

Let d be a positive real number and let $L_d = \{0, d, 2d, 3d, \ldots\}$. The common distribution F of the durations $Z_1, Z_2, \ldots$ is said to be a *lattice distribution* if there is a number $d > 0$ such that $P(Z_1 \in L_d) = 1$. The largest such d is called the *period* of F.

Theorem T.9.1. (*Blackwell's Renewal Theorem*)

(i) If F is not lattice, then for any $h > 0$,

$$EN(t, t+h] = \sum_{k=1}^{\infty} P(t < S_k \leqslant t+h) \to \frac{h}{\mu} \qquad \text{as } t \to \infty, \tag{T.9.15}$$

where $N(t, t+h] := N_{t+h} - N_t$ is the number of renewals in time t to $t+h$.

(ii) If F is lattice with period d, then

$$EN^{(n)} = \sum_{k=1}^{\infty} P(S_k = nd) \to \frac{d}{\mu} \qquad \text{as } n \to \infty, \tag{T.9.16}$$

where

$$N^{(n)} = \sum_{k=0}^{\infty} \mathbf{1}_{\{S_k = nd\}} \tag{T.9.17}$$

is the number of renewals at the time nd. In particular, if $P(Z_1 = 0) = 0$, then $EN^{(n)}$ is simply the probability of a renewal at time nd. □

Note that assuming that the limit exists for each $h > 0$, the value of the limit in (i), likewise (ii), of Blackwell's theorem can easily be identified from the elementary renewal theorem (T.9.6) by noting that $\varphi(h) := \lim_{t \to \infty} EN(t, t+h]$ must then be linear in h, $\varphi(0) = 0$, and

$$\begin{aligned}
\varphi(1) &= \lim_{n \to \infty} \{EN_{n+1} - EN_n\} \\
&= \lim_{n \to \infty} \frac{[\{EN_1 - EN_0\} + \{EN_2 - EN_1\} + \cdots + \{EN_n - EN_{n-1}\}]}{n} \\
&= \lim_{n \to \infty} \frac{EN_n}{n} = \frac{1}{\mu}.
\end{aligned}$$

Recently, *coupling constructions* have been successfully used to prove Blackwell's renewal theorem. The basic idea, analogous to that outlined in Exercise 6.6 for another convergence problem, is to watch both the given delayed renewal and an equilibrium renewal with the same renewal distribution F until their renewal times begin (approximately) to agree. After sufficient agreement is established, then one expects the statistics of the given delayed renewal process to resemble more and more those of the equilibrium process from that time onwards. The details for this argument follow. A somewhat stronger equilibrium property than (T.9.9) will be invoked, but not verified until Chapter IV.

Proof. To make the coupling idea precise for the case of Blackwell's theorem with $\mu < \infty$, let $\{Z_n: n \geqslant 1\}$ and $\{\tilde{Z}_n: n \geqslant 1\}$ denote two independent sequences of renewal lifetime random variables with common distribution F, and let Z_0 and $\tilde{Z}_0$ be independent delays for the two sequences having distributions G and $\tilde{G} = F_\infty$, respectively. The tilde ($\tilde{}$) will be used in reference to quantities associated with the latter (equilibrium) process. Let $\varepsilon > 0$ and define,

$$\nu(\varepsilon) = \inf\{n \geqslant 0: |S_n - \tilde{S}_{\tilde{n}}| \leqslant \varepsilon \text{ for some } \tilde{n}\}, \tag{T.9.18}$$

$$\tilde{\nu}(\varepsilon) = \inf\{\tilde{n} \geqslant 0: |S_n - \tilde{S}_{\tilde{n}}| \leqslant \varepsilon \text{ for some } n\}. \tag{T.9.19}$$

Suppose we have established that (ε-recurrence) $P(\nu(\varepsilon) < \infty) = 1$ (i.e., the coupling will occur). Since the event $\{\nu(\varepsilon) = n, \tilde{\nu}(\varepsilon) = \tilde{n}\}$ is determined by $Z_0, Z_1, \ldots, Z_n$ and $\tilde{Z}_0, \tilde{Z}_1, \ldots, \tilde{Z}_{\tilde{n}}$, the sequence of lifetimes $\{\tilde{Z}_{\tilde{\nu}+k}: k \geqslant 1\}$ may be replaced by the sequence $\{Z_{\nu+k}: k \geqslant 1\}$ without changing the distributions of $\{\tilde{S}_n\}$, $\{\tilde{N}_t\}$, etc. Then, after such a modification for $\varepsilon < h/2$, observe with the aid of a simple figure that

$$\tilde{N}(t + \varepsilon, t + h - \varepsilon]1_{\{S_{\nu(\varepsilon)} \leqslant t\}} \leqslant \tilde{N}(t, t + h]1_{\{S_{\nu(\varepsilon)} \leqslant t\}} \leqslant \tilde{N}(t - \varepsilon, t + h + \varepsilon]1_{\{S_{\nu(\varepsilon)} \leqslant t\}}. \tag{T.9.20}$$

Therefore,

$$\begin{aligned}
&\tilde{N}(t + \varepsilon, t + h - \varepsilon] - \tilde{N}(t, t + h]1_{\{S_{\nu(\varepsilon)} > t\}} \\
&\qquad = \tilde{N}(t + \varepsilon, t + h - \varepsilon]1_{\{S_{\nu(\varepsilon)} \leqslant t\}} + (\tilde{N}(t + \varepsilon, t + h - \varepsilon] - \tilde{N}(t, t + h])1_{\{S_{\nu(\varepsilon)} > t\}} \\
&\qquad \leqslant \tilde{N}(t + \varepsilon, t + h - \varepsilon]1_{\{S_{\nu(\varepsilon)} \leqslant t\}} \\
&\qquad \leqslant N(t, t + h]1_{\{S_{\nu(\varepsilon)} \leqslant t\}} \\
&\qquad \leqslant N(t, t + h]1_{\{S_{\nu(\varepsilon)} \leqslant t\}} + N(t, t + h]1_{\{S_{\nu(\varepsilon)} > t\}}(= N(t, t + h]) \\
&\qquad \leqslant \tilde{N}(t - \varepsilon, t + h + \varepsilon]1_{\{S_{\nu(\varepsilon)} \leqslant t\}} + N(t, t + h]1_{\{S_{\nu(\varepsilon)} > t\}} \\
&\qquad \leqslant \tilde{N}(t - \varepsilon, t + h + \varepsilon] + N(t, t + h]1_{\{S_{\nu(\varepsilon)} > t\}}.
\end{aligned} \tag{T.9.21}$$

Taking expected values and noting the first, fifth, and seventh lines, we have the following *coupling inequality*,

$$\begin{aligned}
&E\tilde{N}(t + \varepsilon, t + h - \varepsilon] - E(\tilde{N}(t, t + h]1_{\{S_{\nu(\varepsilon)} > t\}}) \\
&\qquad \leqslant EN(t, t + h] \leqslant E\tilde{N}(t - \varepsilon, t + h + \varepsilon] + E(N(t, t + h]1_{\{S_{\nu(\varepsilon)} > t\}}).
\end{aligned} \tag{T.9.22}$$

Using (T.9.9),

$$E\tilde{N}(t+\varepsilon, t+h-\varepsilon] = \frac{h-2\varepsilon}{\mu} \qquad \text{and} \qquad E\tilde{N}(t-\varepsilon, t+h+\varepsilon] = \frac{h+2\varepsilon}{\mu}.$$

Therefore,

$$\left| EN(t, t+h] - \frac{h}{\mu} \right| \leqslant \frac{2\varepsilon}{\mu} + E(N(t, t+h]1_{\{S_{\nu(\varepsilon)} > t\}}). \tag{T.9.23}$$

Since $A = 1_{\{S_{\nu(\varepsilon)} > t\}}$ is independent of $\{Z_{N_t+k}: k \geqslant 1\}$, we have

$$E(N(t, t+h]1_{\{S_{\nu(\varepsilon)}>t\}}) \leqslant E_0 N_h P(S_{\nu(\varepsilon)} > t) = m(h)P(S_{\nu(\varepsilon)} > t), \tag{T.9.24}$$

where E_0 denotes expected value for the process N_h under zero-delay. More precisely, because $(t, t+h] \subset (t, S_{N_t} + h]$ and there are no renewals in (t, S_{N_t}), we have $N(t, t+h] \leqslant \inf\{k \geqslant 0: Z_{N_t+k} > h\}$. In particular, noting (T.9.2), this upper bound by an ordinary (zero-delay) renewal process with renewal distribution F, is independent of the event A, and furnishes the desired estimate (T.9.24).

Now from (T.9.23) and (T.9.24) we have the estimate

$$\left| EN(t, t+h] - \frac{h}{\mu} \right| \leqslant m(h)P(S_{\nu(\varepsilon)} > t) + \frac{2\varepsilon}{\mu}, \tag{T.9.25}$$

which is enough, since $\varepsilon > 0$ is arbitrary, provided that the initial ε-recurrence assumption, $P(\nu(\varepsilon) < \infty) = 1$, can be established. So, the bulk of the proof rests on showing that the coupling will eventually occur. The probability $P(\nu(\varepsilon) < \infty)$ can be analyzed separately for each of the two cases (i) and (ii) of Theorem T.9.1.

First take the lattice case (ii) with lattice spacing (period) d. Note that for $\varepsilon < d$,

$$\nu(\varepsilon) = \nu(0) = \inf\{n \geqslant 0: S_n - \tilde{S}_{\tilde{n}} = 0 \text{ for some } \tilde{n}\}. \tag{T.9.26}$$

Also, by recurrence of the mean-zero random walk on the integers (theoretical complement 3.1 of Chapter I), we have $P(\nu(0) < \infty) = 1$. Moreover, $S_{\nu(0)}$ is a.s. a finite integral multiple of d. Taking $t = nd$ and $h = d$ with $\varepsilon = 0$, we have

$$EN^{(n)} = EN(nd, nd+h] \to \frac{h}{\mu} = \frac{d}{\mu} \qquad \text{as } n \to \infty. \tag{T.9.27}$$

For case (i), observe by the Hewitt–Savage zero–one law (theoretical complement 1.2 of Chapter I) applied to the i.i.d. sequence $(Z_1, \tilde{Z}_1), (Z_2, \tilde{Z}_2), (Z_3, \tilde{Z}_3), \ldots$, that

$$P(R_n < \varepsilon \text{ i.o.} \mid \tilde{Z}_0 = z) = 0 \quad \text{or} \quad 1,$$

where $R_n = \min\{\tilde{S}_{\tilde{n}} - S_n: \tilde{S}_{\tilde{n}} - S_n \geqslant 0, \tilde{n} \geqslant 0\} = \tilde{S}_{\tilde{N}_{S_n}} - S_n$.

Now, the distribution of $\{\tilde{S}_{\tilde{N}_t+i} - t\}_i$ does not depend on t (Exercise 7.5, Chapter

IV). This, independence of $\{\tilde{Z}_j\}$ and $\{S_n\}$, and the fact that $\{S_{k+n} - S_k : n \geqslant 0\}$ does not depend on k, make $\{R_{n+k}\}$ also have distribution independent of k. Therefore, the probability $P(R_n < \varepsilon$ for some $n \geqslant k)$, does not depend on k, and thus

$$\{R_n < \varepsilon \text{ i.o.}\} = \bigcap_{k=0}^{\infty} \{R_n < \varepsilon \text{ for some } n \geqslant k\} \tag{T.9.28}$$

implies $P(R_n < \varepsilon \text{ i.o.}) = P(R_n < \varepsilon \text{ for some } n \geqslant 0) \leqslant P(\nu(\varepsilon) < \infty)$. Now,

$$\begin{aligned}\int P(R_n < \varepsilon \text{ i.o.} \mid \tilde{Z}_0 = z) P(\tilde{Z}_0 \in dz)& \\ &= P(R_n < \varepsilon \text{ i.o.}) = P(R_n < \varepsilon \text{ for some } n) \\ &= \int P(R_n < \varepsilon \text{ for some } n \mid \tilde{Z}_0 = z) P(\tilde{Z}_0 \in dz). \qquad \text{(T.9.29)}\end{aligned}$$

The proof that $P(R_n < \varepsilon$ for some $n \mid \tilde{Z}_0 = z) > 0$ (and therefore is 1) in (T.9.29) follows from a final technical lemma given below on "points of increase" of distribution functions of sums of i.i.d. nonlattice positive random variables; a point x is called a *point of increase* of a distribution function F if $F(b) - F(a) > 0$ whenever $a < x < b$.

Lemma. Let F be a nonlattice distribution function on $(0, \infty)$. The set Σ of points of increase of the functions $F, F^{*2}, F^{*3}, \ldots$ is "asymptotically dense at ∞" in the sense that for any $\varepsilon > 0$ and x sufficiently large, $\Sigma \cap (x, x + \varepsilon) \neq \varnothing$, i.e., the interval $(x, x + \varepsilon)$ meets Σ for x sufficiently large. □

The following proof follows that in W. Feller (1971), *An Introduction to Probability Theory and Its Applications*, 2nd ed., Wiley, New York, p. 147.

Proof. Let $a, b \in \Sigma$, $0 < a < b$, such that $b - a < \varepsilon$. Let $I_n = (na, nb]$. For $a < n(b - a)$, the interval I_n properly contains $(na, (n + 1)a)$, and therefore each $x > a^2/(b - a)$ belongs to some I_n, $n \geqslant 1$. Since Σ is easily checked to be closed under addition, the $n + 1$ points $na + k(b - a)$, $k = 0, 1, \ldots, n$, belong to Σ and partition I_n into n subintervals of length $b - a < \varepsilon$. Thus each $x > a^2/(b - a)$ is at a distance $\leqslant (b - a)/2 < \varepsilon/2$ of Σ. If for some $\varepsilon > 0$, $b - a \geqslant \varepsilon$ for all $a, b \in \Sigma$ then F must be a lattice distribution. To see this say, without loss of generality, $\varepsilon \leqslant b - a < 2\varepsilon$ for some $a, b \in \Sigma$. Then $\Sigma \cap I_n \subset \{na + k(b - a) : k = 0, 1, \ldots, n\}$. Since $(n + 1)a \in \Sigma \cap I_n$ for $a < n(b - a)$, $\Sigma \cap I_n$ must consist of multiples of $(b - a)$. Thus, if $c \in \Sigma$ then $c + k(b - a) \in I_n \cap \Sigma$ for n sufficiently large. Thus c is a multiple of $(b - a)$. ■

Coupling approaches to the renewal theorem on which the preceding is based can be found in the papers of H. Thorisson (1987), "A Complete Coupling Proof of Blackwell's Renewal Theorem," *Stoch. Proc. Appl.*, **26**, pp. 87–97; K. Athreya, D. McDonald, P. Ney (1978), "Coupling and the Renewal Theorem," *Amer. Math. Monthly*, **851**, pp. 809–814; T. Lindvall (1977), "A Probabilistic Proof of Blackwell's Renewal Theorem," *Ann. Probab.*, **5**, pp. 482–485.

3. (*Birkhoff's Ergodic Theorem*) Suppose $\{X_n : n \geqslant 0\}$ is a stochastic process on $(\Omega, \mathscr{F}, P)$ with values in $(S, \mathscr{S})$. The process $\{X_n\}$ is (*strictly*) *stationary* if for every pair of integers $m \geqslant 0$, $r \geqslant 1$, the distribution of $(X_0, X_1, \ldots, X_m)$ is the same as that

of $(X_r, X_{1+r}, \ldots, X_{m+r})$. An equivalent definition is: $\{X_n\}$ is stationary if the distribution μ, say, of $\mathbf{X} := (X_0, X_1, X_2, \ldots)$ is the same as that of $T^r\mathbf{X} := (X_r, X_{1+r}, X_{2+r}, \ldots)$ for all $r \geqslant 0$. Recall that the distribution of $(X_r, X_{r+1}, \ldots)$ is the probability measure induced on $(S^\infty, \mathscr{S}^{\otimes\infty})$ by the map $\omega \to (X_r(\omega), X_{1+r}(\omega), X_{2+r}(\omega), \ldots)$. Here S^∞ is the space of all sequences $\mathbf{x} = (x_0, x_1, x_2, \ldots)$ with $x_i \in S$ for all i, and $\mathscr{S}^{\otimes\infty}$ is the smallest sigmafield containing the class of all sets of the form $C = \{\mathbf{x} \in S^\infty : x_i \in B_i$ for $0 \leqslant i \leqslant n\}$ where $n \geqslant 0$ and $B_i \in \mathscr{S}$ $(0 \leqslant i \leqslant n)$ are arbitrary. The *shift transformation* T is defined by $T\mathbf{x} := (x_1, x_2, \ldots)$ on S^∞, so that $T^r\mathbf{x} = (x_r, x_{1+r}, x_{2+r}, \ldots)$.

Denote by $\mathscr{G}$ the sigmafield generated by $\{X_n : n \geqslant 0\}$. That is, $\mathscr{G}$ is the class of all sets of the form $G = \mathbf{X}^{-1}C = \{\omega \in \Omega : \mathbf{X}(\omega) \in C\}$, $C \in \mathscr{S}^{\otimes\infty}$. For a set G of this form, write $T^{-1}G := \{\omega \in \Omega : T\mathbf{X}(\omega) \in C\} = \{(X_1, X_2, \ldots) \in C\} = \{\mathbf{X} \in T^{-1}C\}$. Such a set G is said to be *invariant* if $P(G \,\Delta T^{-1}G) = 0$, where Δ denotes the symmetric difference. By iteration it follows that if $G = \{\mathbf{X} \in C\}$ is invariant then $P(G \,\Delta T^{-r}G) = 0$ for all $r \geqslant 0$, where $T^{-r}G = \{(X_r, X_{1+r}, X_{2+r}, \ldots) \in C\}$. Let f be a real-valued measurable function on $(S^\infty, \mathscr{S}^{\otimes\infty})$. Then $\varphi(\omega) := f(\mathbf{X}(\omega))$ is $\mathscr{G}$-measurable and, conversely, all $\mathscr{G}$-measurable functions are of this form. Such a function $\varphi = f(\mathbf{X})$ is *invariant* if $f(\mathbf{X}) = f(T\mathbf{X})$ a.s. Note that $G = \{\mathbf{X} \in C\}$ is invariant if and only if $\mathbf{1}_G = \mathbf{1}_C(\mathbf{X})$ is invariant. Again, by iteration, if $f(\mathbf{X})$ is invariant then $f(\mathbf{X}) = f(T^r\mathbf{X})$ a.s. for all $r \geqslant 1$.

Given any $\mathscr{G}$-measurable real-valued function $\varphi = f(\mathbf{X})$, the functions (extended real-valued)

$$\bar{f}(X) := \overline{\lim_{n\to\infty}}\, n^{-1}(f(\mathbf{X}) + f(T\mathbf{X}) + \cdots + f(T^{n-1}\mathbf{X}))$$

and

$$\underline{f}(\mathbf{X}) := \underline{\lim}_{n\to\infty}\, n^{-1}(f(\mathbf{X}) + \cdots + f(T^{n-1}\mathbf{X}))$$

are invariant, and the set $\{\bar{f}(\mathbf{X}) = \underline{f}(\mathbf{X})\}$ is invariant.

The class $\mathscr{I}$ of all invariant sets (in $\mathscr{G}$) is easily seen to be a sigmafield, which is called the *invariant sigmafield*. The invariant sigmafield $\mathscr{I}$ is said to be *trivial* if $P(G) = 0$ or 1 for every $G \in \mathscr{I}$.

Definition T.9.1. The process $\{X_n : n \geqslant 0\}$ and the shift transformation T are said to be *ergodic* if $\mathscr{I}$ is trivial.

The next result is an important generalization of the classical strong law of large numbers (Chapter 0, Theorem 6.1).

Theorem T.9.2. (*Birkhoff's Ergodic Theorem*). Let $\{X_n : n \geqslant 0\}$ be a stationary sequence on the state space S (having sigmafield $\mathscr{S}$). Let $f(\mathbf{X})$ be a real-valued $\mathscr{G}$-measurable function such that $E|f(\mathbf{X})| < \infty$. Then

(a) $n^{-1}\sum_{j=0}^{n-1} f(T^j\mathbf{X})$ converges a.s. and in L^1 to an invariant random variable $g(\mathbf{X})$, and

(b) $g(\mathbf{X}) = Ef(\mathbf{X})$ a.s. if $\mathscr{I}$ is trivial. □

We first need an inequality whose derivation below follows A. M. Garcia (1965), "A Simple Proof of E. Hopf's Maximal Ergodic Theorem," *J. Math. Mech.*, **14**, pp. 381–382. Write

$$M_n(f) := \max\{0, f(\mathbf{X}), f(\mathbf{X}) + f(T\mathbf{X}), \ldots, f(\mathbf{X}) + \cdots + f(T^{n-1}\mathbf{X})\},$$

$$M_n(f \circ T) = \max\{0, f(T\mathbf{X}), f(T\mathbf{X}) + f(T^2\mathbf{X}), \ldots, f(T\mathbf{X}) + \cdots + f(T^n\mathbf{X})\},$$

$$M(f) := \lim_{n\to\infty} M_n(f) = \sup_{n\geq 1} M_n(f). \tag{T.9.30}$$

Proposition T.9.3. (*Maximal Ergodic Theorem*). Under the hypothesis of Theorem T.9.2,

$$\int_{\{M(f)>0\}\cap G} f(\mathbf{X})\, dP \geqslant 0 \qquad \forall G \in \mathscr{I}. \tag{T.9.31}$$

Proof. Note that $f(\mathbf{X}) + M_n(f \circ T) = M_{n+1}(f)$ on the set $\{M_{n+1}(f) > 0\}$. Since $M_{n+1}(f) \geqslant M_n(f)$ and $\{M_n(f) > 0\} \subset \{M_{n+1}(f) > 0\}$, it follows that $f(\mathbf{X}) \geqslant M_n(f) - M_n(f \circ T)$ on $\{M_n(f) > 0\}$. Also, $M_n(f) \geqslant 0$, $M_n(f \circ T) \geqslant 0$. Therefore,

$$\begin{aligned}
\int_{\{M_n(f)>0\}\cap G} f(\mathbf{X})\, dP &\geqslant \int_{\{M_n(f)>0\}\cap G} (M_n(f) - M_n(f \circ T))\, dP \\
&= \int_G M_n(f)\, \mathrm{dP} - \int_{\{M_n(f)>0\}\cap G} M_n(f \circ T)\, dP \\
&\geqslant \int_G M_n(f)\, \mathrm{dP} - \int_G M_n(f \circ T)\, dP \\
&= 0,
\end{aligned}$$

where the last equality follows from the invariance of G and the stationarity of $\{X_n\}$. Thus, (T.9.31) holds with $\{M_n(f) > 0\}$ in place of $\{M(f) > 0\}$. Now let $n \uparrow \infty$. ■

Now consider the quantities

$$A_n(f) := \max\left\{f(\mathbf{X}), \tfrac{1}{2}(f(\mathbf{X}) + f(T\mathbf{X})), \ldots, \frac{1}{n}\sum_{r=0}^{n-1} f(T^r\mathbf{X})\right\},$$

$$A(f) := \lim_{n\to\infty} A_n(f) = \sup_{n\geq 1} A_n(f).$$

The following is a consequence of Proposition T.9.3.

Corollary T.9.4. (*Ergodic Maximal Inequality*). Under the hypothesis of Theorem T.9.1 one has, for every $c \in \mathbb{R}^1$,

$$\int_{\{A(f)>c\}\cap G} f(\mathbf{X})\, dP \geqslant cP(\{A(f) > c\} \cap G) \qquad \forall G \in \mathscr{I}. \tag{T.9.32}$$

Proof. Apply Proposition T.9.3 to the function $f - c$ to get

$$\int_{\{M(f-c)>0\}\cap G} f(\mathbf{X})\,dP \geqslant cP(\{M(f-c)>0\}\cap G).$$

But $\{M_n(f-c)>0\} = \{A_n(f-c)>0\} = \{A_n(f)>c\}$, and $\{M(f-c)>0\} = \{A(f)>c\}$. ■

We are now ready to prove Theorem T.9.2, using (T.9.31).

Proof of Theorem T.9.2. Write

$$\bar{f}(\mathbf{X}) := \overline{\lim_{n\to\Omega}}\,\frac{1}{n}\sum_{r=0}^{n-1} f(T^r\mathbf{X}), \qquad \underline{f}(\mathbf{X}) := \underline{\lim}_{n\to\infty}\,\frac{1}{n}\sum_{r=0}^{n-1} f(T^r\mathbf{X}), \tag{T.9.33}$$

$$G_{c,d}(f) := \{\bar{f}(\mathbf{X}) > c, \underline{f}(\mathbf{X}) < d\} \qquad (c, d \in \mathbb{R}^1).$$

Since $G_{c,d}(f) \in \mathscr{I}$ and $G_{c,d}(f) \subset \{A(f) > c\}$, (T.9.32) leads to

$$\int_{G_{c,d}(f)} f(\mathbf{X})\,dP \geqslant cP(G_{c,d}(f)). \tag{T.9.34}$$

Now take $-f$ in place of f and note that $\overline{(-f)} = -\underline{f}$, $\underline{(-f)} = -\bar{f}$, $G_{-d,-c}(-f) = G_{c,d}(f)$ to get from (T.9.34) the inequality

$$-\int_{G_{c,d}(f)} f(\mathbf{X})\,dP \geqslant -dP(G_{c,d}(f)),$$

i.e.,

$$\int_{G_{c,d}(f)} f(\mathbf{X})\,dP \leqslant dP(G_{c,d}(f)). \tag{T.9.35}$$

Now if $c > d$, then (T.9.34) and (T.9.35) cannot both be true unless $P(G_{c,d}(f)) = 0$. Thus, if $c > d$, then $P(G_{c,d}(f)) = 0$. Apply this to all pairs of rationals $c > d$ to get $P(\bar{f}(\mathbf{X}) > \underline{f}(\mathbf{X})) = 0$. In other words, $(1/n)\sum_{r=0}^{n-1} f(T^r\mathbf{X})$ converges a.s. to $g(\mathbf{X}) := \bar{f}(\mathbf{X})$.

To complete the proof of part (a), it is enough to assume $f \geqslant 0$, since $n^{-1}\sum_0^{n-1} f^+(T^r\mathbf{X}) \to \overline{f^+}(\mathbf{X})$ a.s. and $n^{-1}\sum_0^{n-1} f^-(T^r\mathbf{X}) \to \overline{f^-}(\mathbf{X})$ a.s., where $f^+ = \max\{f, 0\}$, $-f^- = \min\{f, 0\}$. Assume then $f \geqslant 0$. First, by Fatou's Lemma and stationarity of $\{X_n\}$,

$$E\bar{f}(\mathbf{X}) \leqslant \underline{\lim}_{n\to\infty} E\left(\frac{1}{n}\sum_{r=0}^{n-1} f(T^r(\mathbf{X}))\right) = Ef(\mathbf{X}) < \infty.$$

To prove the L^1-convergence, it is enough to prove the uniform integrability of the sequence $\{(1/n)S_n(f): n \geqslant 1\}$, where $S_n(f) := \sum_0^{n-1} f(T^r\mathbf{X})$. Now since $f(\mathbf{X})$ is nonnegative and integrable, given $\varepsilon > 0$ there exists a constant N_ε such that

$\|f(\mathbf{X}) - f_\varepsilon(\mathbf{X})\|_1 < \varepsilon$ where $f_\varepsilon(\mathbf{X}) := \min\{f(\mathbf{X}), N_\varepsilon\}$. Then

$$\int_{\{\frac{1}{n}S_n(f) > \lambda\}} \frac{1}{n} S_n(f)\, dP \leqslant \int \frac{1}{n} S_n(f - f_\varepsilon)\, dP + \int_{\{\frac{1}{n}S_n(f) > \lambda\}} \frac{1}{n} S_n(f_\varepsilon)\, dP$$
$$\leqslant \varepsilon + N_\varepsilon P\left(\left\{\frac{1}{n} S_n(f) > \lambda\right\}\right)$$
$$\leqslant \varepsilon + N_\varepsilon Ef(\mathbf{X})/\lambda. \tag{T.9.36}$$

It follows that the left side of (T.9.36) goes to zero as $\lambda \to \infty$, uniformly for all n.
Part (b) is an immediate consequence of part (a). ■

Notice that part (a) of Theorem T.9.2 also implies that $g(\mathbf{X}) = E(f(\mathbf{X}) \mid \mathscr{I})$.
Theorem T.9.2 is generally stated for any transformation T on a probability space $(\Omega, \mathscr{G}, \mu)$ satisfying $\mu(T^{-1}G) = \mu(G)$ for all $G \in \mathscr{G}$. Such a transformation is called *measure-preserving*. If in this case we take $\mathbf{X}$ to be the identity map: $\mathbf{X}(\omega) = \omega$, then parts (a) and (b) hold without any essential change in the proof.

Theoretical Complements to Section II.10

1. To prove the FCLT for Markov-dependent summands as asserted in Theorem 10.2, first consider

$$X_t^{(n)} := \frac{1}{\sigma\sqrt{n}} (\bar{Z}_1 + \cdots + \bar{Z}_{[nt]}).$$

Since $\bar{Z}_1, \bar{Z}_2, \ldots$ are i.i.d. with finite second moment, the FCLT of Chapter I provides that $\{X_t^{(n)}\}$ converges in distribution to standard Brownian motion. The corresponding result for $\{W_n(t)\}$ follows by an application of the Maximal Inequality to show

$$\sup_{0 \leqslant t \leqslant 1} |X_t^{(n)} - W_{[nE_j\tau]}(t)| \to 0 \quad \text{in probability as } n \to \infty, \tag{T.10.1}$$

where τ is the first return time to j.

Theoretical Complements to Section II.12

1. There are specifications of local structure that are defined in a natural manner but for which there are *no* Gibbs states having the given structure when, for example, $\Lambda = \mathbb{Z}$, but S is not finite. As an example, one can take $\mathbf{q}$ to be the transition matrix of a (general) random walk on $S = \mathbb{Z}$ such that $q_{ij} = q_{|i-j|} > 0$ for all i, j. In this case *no* probability distribution on S^Λ exists having the local structure furnished by (12.10). For proofs, refer to the papers of F. Spitzer (1974), "Phase Transition in One-Dimensional Nearest-Neighbor Systems," *J. Functional Analysis*, **20**, pp. 240–254; H. Kesten (1975), "Existence and Uniqueness of Countable One-Dimensional Markov Random Fields," *Ann. Probab.*, **4**, pp. 557–569. The treatment here follows F. Spitzer (1971), "Random Fields and Interacting Particle Systems," *MAA Lecture Notes*, Washington, D.C.

Theoretical Complements to Section II.14

1. (*Markov Processes and Iterations of I.I.D. Maps*) Let $p(x; dy)$ denote a *transition probability* on a state space $(S, \mathscr{S})$; that is, (1) for each $x \in S$, $p(x; dy)$ is a probability measure on $(S, \mathscr{S})$, and (2) for each $B \in \mathscr{S}$, $x \to p(x; B)$ is $\mathscr{S}$-measurable. We will assume that S is a Borel subset of a complete separable metric space, and $\mathscr{S}$ its Borel sigmafield $\mathscr{B}(S)$. It may be shown that S may be "relabeled" as a Borel subset C of $[0, 1]$, with $\mathscr{B}(C)$ as the relabeling of $\mathscr{B}(S)$. (See H. L. Royden (1968), *Real Analysis*, 2nd ed., Macmillan, New York, pp. 326–327). Therefore, without any essential loss of generality, we take S to be a Borel subset of $[0, 1]$. For each $x \in S$, let $F_x(\cdot)$ denote the distribution function of $p(x; dy)$: $F_x(y) := p(x; S \cap (-\infty, y])$. Define $F_x^{-1}(t) := \inf\{y \in \mathbb{R}^1: F_x(y) > t\}$. Let U be a random variable defined on some probability space $(\Omega, \mathscr{F}, P)$, whose distribution is uniform on $(0, 1)$. Then it is simple to check that $P(F_x^{-1}(U) \leqslant y) \geqslant P(F_x(y) > U) = P(U < F_x(y)) = F_x(y)$, and $P(F_x^{-1}(U) \leqslant y) \leqslant P(F_x(y) \geqslant U) = P(U \leqslant F_x(y)) = F_x(y)$. Therefore, $P(F_x^{-1}(U) \leqslant y) = F_x(y)$, that is, the distribution of $F_x^{-1}(U)$ is $p(x; dy)$. Now let $U_1, U_2, \ldots$ be a sequence of i.i.d. random variables on $(\Omega, \mathscr{F}, P)$, each having the uniform distribution on $(0, 1)$. Let X_0 be a random variable with values in S, independent of $\{U_n\}$. Define $X_{n+1} = f(X_n, U_{n+1})$ $(n \geqslant 0)$, where $f(\mathrm{x}, \mathrm{u}) := F_x^{-1}(u)$. It then follows from the above that $\{X_n: n \geqslant 0\}$ is a Markov process having transition probability $p(x; dy)$, and initial distribution that of X_0.

   Of course, this type of representation of a Markov process having a given transition probability and a given initial distribution is not unique.

   For additional information, see R. M. Blumenthal and H. K. Corson (1972), "On Continuous Collections of Measures," *Proc. 6th Berkeley Symposium on Math. Stat. and Prob.*, Vol. 2, pp. 33–40.

2. Example 1 is essentially due to L. E. Dubins and D. A. Freedman (1966), "Invariant Probabilities for Certain Markov Processes," *Ann. Math. Statist.*, **37**, pp. 837–847. The assumption of continuity of the maps is not needed, as shown in J. A. Yahav (1975), "On a Fixed Point Theorem and Its Stochastic Equivalent," *J. Appl. Probability*, **12**, pp. 605–611. An extension to multidimensional state space with an application to time series models may be found in R. N. Bhattacharya and O. Lee (1988), "Asymptotics of a Class of Markov Processes Which Are Not in General Irreducible," *Ann. Probab.*, **16**, pp. 1333–1347. Example 1(a) may be found in L. J. Mirman (1980), "One Sector Economic Growth and Uncertainty: A Survey," *Stochastic Programming* (M. A. H. Dempster, ed.), Academic Press, New York.

   It is shown in theoretical complement 3 below that the existence of a unique invariant probability implies ergodicity of a stationary Markov process. The SLLN then follows from Birkhoff's Ergodic Theorem (see Theorem T.9.2). Central limit theorems for normalized partial sums may be derived for appropriate functions on the state space, by Theorem T.13.3 in the theoretical complements of Chapter V. Also see, Bhattacharya and Lee, *loc. cit.*

   Example 3 is due to M. Majumdar and R. Radner, unpublished manuscript.

   K. S. Chan and H. Tong (1985), "On the Use of Deterministic Lyapunov Function for the Ergodicity of Stochastic Difference Equations," *Advances in Appl. Probability*, **17**, pp. 666–678, consider iterations of i.i.d. piecewise linear maps.

3. (*Irreducible Markov Processes*) A transition probability $p(x; dy)$ on the state space $(S, \mathscr{S})$ is said to be *φ-irreducible* with respect to a sigmafinite nonzero measure φ if, for each $x \in S$ and $B \in \mathscr{S}$ with $\varphi(B) > 0$, there exists an integer $n = n(x, B)$ such that

$p^{(n)}(x, B) > 0$. There is an extensive literature on the asymptotics of φ-irreducible Markov processes. We mention in particular, N. Jain and B. Jamison (1967), "Contributions to Doeblin's Theorem of Markov Processes," *Z. Wahrscheinlichkeitstheorie und Verw Gebiete*, **8**, pp. 19–40; S. Orey (1971), *Limit Theorems for Markov Chain Transition Probabilities*, Van Nostrand, New York; R. L. Tweedie (1975), "Sufficient Conditions for Ergodicity and Recurrence of Markov Chains on a General State Space," *Stochastic Process Appl.*, **3**, pp. 385–403. Irreducible Markov chains on countable state spaces S are the simplest examples of φ-irreducible processes; here φ is the counting measure, $\varphi(B) := $ number of points in B. Some other examples are given in theoretical complements to Section II.6.

There is no general theory that applies if $p(x; dy)$ is not φ-irreducible, for any sigmafinite φ. The method of iterated maps provides one approach, when the Markov process arises naturally in this manner. A simple example of a nonirreducible p is given by Example 2. Another example, in which p admits a unique invariant probability, is provided by the simple linear model: $X_{n+1} = \frac{1}{2}X_n + \varepsilon_{n+1}$, where ε_n are i.i.d., $P(\varepsilon_n = \frac{1}{2}) = P(\varepsilon_n = -\frac{1}{2}) = \frac{1}{2}$.

4. (*Ergodicity, SLLN, and the Uniqueness of Invariant Probabilities*) Suppose $\{X_n : n \geqslant 0\}$ is a stationary Markov process on a state space $(S, \mathscr{S})$, having a transition probability $p(x; dy)$ and an invariant initial distribution π. We will prove the following result: *The process $\{X_n\}$ is ergodic if and only if there does not exist an invariant probability π' that is absolutely continuous with respect to π and different from π.*

The crucial step in the proof is to show that every (shift) invariant bounded measurable function $h(\mathbf{X})$ is a.s. equal to a random variable $g(X_0)$ where g is a measurable function on $(S, \mathscr{S})$. Here $\mathbf{X} := (X_0, X_1, X_2, \ldots)$, and we let T denote the shift transformation and $\mathscr{I}$ the (shift) invariant sigmafield (see theoretical complement 9.3). Now if $h(\mathbf{X})$ is invariant, $h(\mathbf{X}) = h(T^n\mathbf{X})$ a.s. for all $n \geqslant 1$. Then, by the Markov property, $E(h(\mathbf{X}) \mid \sigma(X_0, \ldots, X_n\}) = E(h(T^n\mathbf{X}) \mid \sigma\{X_0, \ldots, X_n\}) = E(h(T^n\mathbf{X}) \mid \sigma\{X_n\}) = g(X_n)$, where $g(x) = E(h(X_0, X_1, \ldots) \mid X_0 = x)$. By the Martingale Convergence Theorem (see theoretical complement 5.1 to Chapter IV, Theorem T.5.2), applied to the martingale $g(X_n) = E(h(\mathbf{X}) \mid \sigma\{X_0, \ldots, X_n\})$, $g(X_n)$ converges a.s., and in L^1, to $E(h(\mathbf{X}) \mid \sigma\{X_0, X_1, \ldots\}) = h(\mathbf{X})$. But $g(X_n) - h(\mathbf{X}) = g(X_n) - h(T^n\mathbf{X})$ has the same distribution as $g(X_0) - h(\mathbf{X})$ for all $n \geqslant 1$. Therefore, $g(X_0) - h(\mathbf{X}) = 0$ a.s., since the limit of $g(X_n) - h(\mathbf{X})$ is zero a.s. In particular, if $G \in \mathscr{I}$ then there exists $B \in \mathscr{S}$ such that $\{X_0 \in B\} = G$ a.s. This implies $\pi(B) = P(X_0 \in B) = P(G)$. If $\{X_n\}$ is *not ergodic*, then there exists $G \in \mathscr{I}$ such that $0 < P(G) < 1$ and, therefore, $0 < \pi(B) < 1$ for a corresponding set $B \in \mathscr{S}$ as above. But the probability measure π_B defined by: $\pi_B(A) = \pi(A \cap B)/\pi(B)$, $A \in \mathscr{S}$, is invariant. To see this observe that $\int p(x; A)\pi_B(dx) = \int_B p(x; A)\pi(dx)/\pi(B) = P(X_0 \in B, X_1 \in A)/\pi(B) = P(X_1 \in B, X_1 \in A)/\pi(B)$ (since $\{X_0 \in B\}$ is invariant) $= P(X_0 \in A \cap B)/\pi(B)$ (by stationarity) $= \pi(A \cap B)/\pi(B) = \pi_B(A)$. Since $\pi_B(B) = 1 > \pi(B)$, and π_B is absolutely continuous with respect to π, one half of the italicized statement is proved.

To prove the other half, suppose $\{X_n\}$ is ergodic and π' is also invariant and absolutely continuous with respect to π. Fix $A \in \mathscr{S}$. By Birkhoff's Ergodic Theorem, and conditioning on X_0, $(1/n)\sum_{r=0}^{n-1} p^{(r)}(x; A)$ converges to $\pi(A)$ for all x outside a set of zero π-measure. Now the invariance of π' implies $\int (1/n)\sum_0^{n-1} p^{(r)}(x; A)\pi'(dx) = \pi'(A)$ for all n. Therefore, $\pi'(A) = \pi(A)$. Thus $\pi' = \pi$, completing the proof.

As a very special case, the following *strong law of large numbers* (SLLN) for Markov processes on general state spaces is obtained: *If $p(x; dy)$ admits a unique invariant probability π, and $\{X_n : n \geqslant 0\}$ is a Markov process with transition probability* *p and initial distribution* π, *then* $(1/n)\sum_0^{n-1} f(X_r)$ *converges to* $\int f(x)\pi(dx)$ *a.s. provided that* $\int |f(x)|\pi(dx) < \infty$. This also implies, by conditioning on X_0, that this almost sure convergence holds under all initial states x outside a set of zero π-measure.

5. (*Ergodic Decomposition of a Compact State Space*) Suppose S is a compact metric space and $S = \mathscr{B}(S)$ its Borel sigmafield. Let $p(x; dy)$ be a transition probability on $(S, \mathscr{B}(S))$ having the *Feller property*: $x \to p(x; dy)$ is weakly continuous on S into $\mathscr{P}(S)$—the set of all probability measures on $(S, \mathscr{B}(S))$. Let T^* denote the map on $\mathscr{P}(S)$ into $\mathscr{P}(S)$ defined by: $(T^*\mu)(B) = \int p(x; B)\mu(dx)$ $(B \in \mathscr{B}(S))$. Then T^* is *weakly continuous*. For if probability measures μ_n converge weakly to μ then, for every real-valued bounded continuous f on S, $\int f\, d(T^*\mu_n) = \int (\int f(y)p(x; dy))\mu_n(dx)$ converges to $\int (\int f(y)p(x; dy))\mu(dx) = \int f\, d(T^*\mu)$, since $x \to \int f(y)p(x;dy)$ is continuous by the Feller property of p.

Let us show that under the above hypothesis there exists at least one invariant probability for p. Fix $\mu \in \mathscr{P}(S)$. Consider the sequence of probability measures

$$\mu_n := \frac{1}{n}\sum_{r=0}^{n-1} T^{*r}\mu \qquad (n \geqslant 1),$$

where

$$T^{*0}\mu = \mu, \qquad T^{*1}\mu = T^*\mu, \qquad \text{and} \qquad T^{*(r+1)}\mu = T^*(T^{*r}\mu) \qquad (r \geqslant 1).$$

Since S is compact, by Prohorov's Theorem (see theoretical complement 8.2 of Chapter I), there exists a subsequence $\{\mu_{n'}\}$ of $\{\mu_n\}$ such that $\mu_{n'}$ converges weakly to a probability measure π, say. Then $T^*\mu_{n'}$ converges weakly to $T^*\pi$. On the other hand,

$$\left|\int f\, d\mu_{n'} - \int f\, d(T^*\mu_{n'})\right| = \frac{1}{n'}\left|\int f\, d\mu - \int f\, d(T^{*n'}\mu)\right| \leqslant (\sup\{|f(x)|: x \in S\})(2/n') \to 0,$$

as $n' \to \infty$. Therefore, $\{\mu_{n'}\}$ and $\{T^*\mu_{n'}\}$ converge to the same limit. In other words, $\pi = T^*\pi$, or π is invariant. This also shows that on a compact metric space, and with p having the Feller property, if there exists a unique invariant probability π then $T^*\mu_n := (1/n)\sum_0^{n-1} T^{*r}\mu$ converges weakly to π, no matter what (the initial distribution) μ is.

Next, consider the set $\mathscr{M} = \mathscr{M}_p$ of all invariant probabilities for p. This is a *convex and* (weakly) *compact subset* of $\mathscr{P}(S)$. Convexity is obvious. Weak compactness follows from the facts (i) $\mathscr{P}(S)$ is weakly compact (by Prohorov's Theorem), and (ii) T^* is continuous for the weak topology on $\mathscr{P}(S)$. For, if $\mu_n \in \mathscr{M}$ and μ_n converges weakly to μ, then $\mu_n = T^*\mu_n$ converges weakly to $T^*\mu$. Therefore, $T^*\mu = \mu$. Also, $\mathscr{P}(S)$ is a metric space (see, e.g., K. R. Parthasarathy (1967), *Probability Measures on Metric Spaces*, Academic Press, New York, p. 43). It now follows from the *Krein–Milman Theorem* (see H. L. Royden (1968), *Real Analysis*, 2nd ed., Macmillan, New York, p. 207) that $\mathscr{M}$ is the closed convex hull of its extreme points. Now if $\{X_n\}$ is not ergodic under an invariant initial distribution π, then, by the construction given in theoretical complement 4 above, there exists $B \in \mathscr{B}(S)$ such that $0 < \pi(B) < 1$ and $\pi = \pi(B)\pi_B + \pi(B^c)\pi_{B^c}$, with π_B and π_{B^c} mutually singular invariant probabilities. In other words, the set K, say, of extreme points of $\mathscr{M}$ comprises those π such that $\{X_n\}$ with initial distribution π is ergodic. Every $\pi \in \mathscr{M}$ is a (weak) limit of convex combinations of the form $\sum_1^n \lambda_i^{(n)}\mu_i^{(n)}$ $(n \to \infty)$, where $0 < \lambda_i^{(n)} < 1$, $\sum_i \lambda_i^{(n)} = 1$, $\mu_i^{(n)} \in K$.

Therefore, the limit π may be expressed uniquely as $\pi = \int_K \mu m(d\mu)$, where m is a probability measure on $(K, \mathscr{B}(K))$. This means, for every real-valued bounded continuous f, $\int_S f\,d\pi = \int_K (\int_S f\,d\mu) m(d\mu)$.

Theoretical Complements to Section II.15

1. For some of Claude Shannon's applications of information theory to language structure, see C. E. Shannon (1951), "Prediction and Entropy of Printed English," *Bell System Tech. J.*, **30**(1), pp. 50–64. The basic ideas originated in C. E. Shannon (1948), "A Mathematical Theory of Communication," *Bell System Tech. J.*, **27**, pp. 379–423, 623–656. There are a number of excellent textbooks and references devoted to this and other problems of information theory. A few standard references are: C. E. Shannon and W. Weaver (1949), *The Mathematical Theory of Communications*, University of Illinois Press, Urbana; and N. Abramson (1963), *Information Theory and Coding*, McGraw-Hill, New York.

CHAPTER III

Birth–Death Markov Chains

1 INTRODUCTION TO BIRTH–DEATH CHAINS

Each of the simple random walk examples described in Section 1.3 has the special property that it does not skip states in its evolution. In this vein, we shall study time-homogeneous Markov chains called *birth–death chains* whose transition law takes the form

$$p_{ij} = \begin{cases} \beta_i & \text{if } j = i + 1 \\ \delta_i & \text{if } j = i - 1 \\ \alpha_i & \text{if } j = i \\ 0 & \text{otherwise,} \end{cases} \tag{1.1}$$

where $\alpha_i + \beta_i + \delta_i = 1$. In particular, the *displacement* probabilities may depend on the state in which the process is located.

Example 1. (*The Bernoulli–Laplace Model*). A simple model to describe the mixing of two incompressible liquids in possibly different proportions can be obtained by the following considerations. Consider two containers labeled box I and box II, respectively, each having N balls. Among the total of $2N$ balls, there are $2r$ red and $2w$ white balls, $1 \leqslant r \leqslant w$. At each instant of time, a ball is randomly selected from each of the boxes, and moved to the other box. The state at each instant is the number of red balls in box I.

In this example, the state space is $S = \{0, 1, \ldots, 2r\}$ and the evolution is a Markov chain on S with transition probabilities given by

for $1 \leqslant i \leqslant 2r - 1$,

$$p_{i,i+1} = \frac{(w+1-i)(2r-i)}{(w+r)^2}$$

$$p_{ii} = \frac{i(2r-1)}{(2+r)^2} + \frac{(2r-i)(w+r-i)}{(w+r)^2} \tag{1.2}$$

$$p_{i,i-1} = \frac{i(w-r+i)}{(w+r)^2}$$

and

$$p_{00} = p_{2r,2r} = \frac{w-r}{w+r}$$

$$p_{01} = p_{2r,2r-1} = \frac{2r}{w+r}. \tag{1.3}$$

Just as the simple random walk is the discrete analogue of Brownian motion, the birth–death chains are the discrete analogues of the diffusions studied in Chapter V.

Most of this chapter may be read independently of Chapter II.

2 TRANSIENCE AND RECURRENCE PROPERTIES

The long-run behavior of a birth–death chain depends on the nature of its (local) transition probabilities $p_{i,i+1} = \beta_i$, $p_{i,i-1} = \delta_i$ at interior states i as well as on its transitions at boundaries, when present. In this section a case-by-case computation of recurrence properties will be made according to the presence and types of boundaries.

CASE I. Let $\{X_n\}$ be an *unrestricted* birth–death chain on $S = \{0, \pm 1, \pm 2, \ldots\} = \mathbb{Z}$. The transition probabilities are

$$p_{i,i+1} = \beta_i, \qquad p_{i,i-1} = \delta_i, \qquad p_{i,i} = 1 - \beta_i - \delta_i \tag{2.1}$$

with

$$0 < \beta_i < 1, \qquad 0 < \delta_i < 1, \qquad \beta_i + \delta_i \leqslant 1. \tag{2.2}$$

Let $c, d \in S$, $c < d$, and write

$$\psi(i) = P(\{X_n\} \text{ reaches } c \text{ before } d \mid X_0 = i) = P_i(T_c < T_d) \qquad (c \leqslant i \leqslant d), \tag{2.3}$$

where T_r denotes the first time the chain reaches r. Now,

$$\psi(i) = (1 - \beta_i - \delta_i)\psi(i) + \beta_i\psi(i+1) + \delta_i\psi(i-1),$$

or equivalently,

$$\beta_i(\psi(i+1)-\psi(i)) = \delta_i(\psi(i)-\psi(i-1)) \qquad (c+1 \leqslant i \leqslant d-1). \tag{2.4}$$

The *boundary conditions* for ψ are

$$\psi(c) = 1, \qquad \psi(d) = 0. \tag{2.5}$$

Rewrite (2.4) as

$$\psi(i+1)-\psi(i) = \frac{\delta_i}{\beta_i}(\psi(i)-\psi(i-1)), \tag{2.6}$$

for $(c+1 \leqslant i \leqslant d-1)$. Iteration now yields

$$\psi(x+1)-\psi(x) = \frac{\delta_x}{\beta_x}\frac{\delta_{x-1}}{\beta_{x-1}}\cdots\frac{\delta_{c+1}}{\beta_{c+1}}(\psi(c+1)-\psi(c)) \tag{2.7}$$

for $c+1 \leqslant x \leqslant d-1$. Summing (2.7) over $x = y, y+1, \ldots, d-1$, one gets

$$\psi(d)-\psi(y) = \sum_{x=y}^{d-1}\frac{\delta_x\delta_{x-1}\cdots\delta_{c+1}}{\beta_x\beta_{x-1}\cdots\beta_{c+1}}(\psi(c+1)-\psi(c)). \tag{2.8}$$

Let $y = c+1$ and use (2.5) to get

$$\psi(c+1) = \frac{\displaystyle\sum_{x=c+1}^{d-1}\frac{\delta_x\delta_{x-1}\cdots\delta_{c+1}}{\beta_x\beta_{x-1}\cdots\beta_{c+1}}}{1+\displaystyle\sum_{x=c+1}^{d-1}\frac{\delta_x\delta_{x-1}\cdots\delta_{c+1}}{\beta_x\beta_{x-1}\cdots\beta_{c+1}}}. \tag{2.9}$$

Using this in (2.8) (and using $\psi(d) = 0$, $\psi(c) = 1$) one gets

$$\psi(y) = \frac{\displaystyle\sum_{x=y}^{d-1}\frac{\delta_x\delta_{x-1}\cdots\delta_{c+1}}{\beta_x\beta_{x-1}\cdots\beta_{c+1}}}{1+\displaystyle\sum_{x=c+1}^{d-1}\frac{\delta_x\delta_{x-1}\cdots\delta_{c+1}}{\beta_x\beta_{x-1}\cdots\beta_{c+1}}} \qquad (c+1 \leqslant y \leqslant d-1). \tag{2.10}$$

Let ρ_{yc} denote the probability that starting at y the process eventually reaches c after time 0, i.e.,

$$\rho_{yc} = P_y(X_n = c \text{ for some } n \geqslant 1). \tag{2.11}$$

Then (Exercise 1),

$$\rho_{yc} = \lim_{d\uparrow\infty} \psi(y) = 1 \qquad \text{if } \sum_{x=c+1}^{\infty} \frac{\delta_x\delta_{x-1}\cdots\delta_{c+1}}{\beta_x\beta_{x-1}\cdots\beta_{c+1}} = \infty,$$

$$< 1 \qquad \text{if } \sum_{x=c+1}^{\infty} \frac{\delta_x\delta_{x-1}\cdots\delta_{c+1}}{\beta_x\beta_{x-1}\cdots\beta_{c+1}} < \infty \qquad (c < y). \tag{2.12}$$

Since, for $c + 1 \leqslant 0$,

$$\sum_{x=c+1}^{\infty} \frac{\delta_x\delta_{x-1}\cdots\delta_{c+1}}{\beta_x\beta_{x-1}\cdots\beta_{c+1}} = \sum_{x=c+1}^{0} \frac{\delta_{c+1}\delta_{c+2}\cdots\delta_x}{\beta_{c+1}\beta_{c+2}\cdots\beta_x} + \frac{\delta_{c+1}\delta_{c+2}\cdots\delta_0}{\beta_{c+1}\beta_{c+2}\cdots\beta_0} \sum_{x=1}^{\infty} \frac{\delta_1\delta_2\cdots\delta_x}{\beta_1\beta_2\cdots\beta_x} \tag{2.13}$$

and a similar equality holds for $c + 1 > 0$, (2.12) may be stated as

$$\rho_{yc} = 1 \qquad \text{for all } y > c \text{ iff } \sum_{x=1}^{\infty} \frac{\delta_1\delta_2\cdots\delta_x}{\beta_1\beta_2\cdots\beta_x} = \infty,$$

$$< 1 \qquad \text{for all } y > c \text{ iff } \sum_{x=1}^{\infty} \frac{\delta_1\delta_2\cdots\delta_x}{\beta_1\beta_2\cdots\beta_x} < \infty. \tag{2.14}$$

By relabeling the states i as $-i$ ($i = 0, \pm 1, \pm 2, \ldots$), one gets (Exercise 2)

$$\rho_{yd} = 1 \qquad \text{for all } y < d \text{ iff } \sum_{x=-\infty}^{0} \frac{\beta_x\beta_{x+1}\cdots\beta_0}{\delta_x\delta_{x+1}\cdots\delta_0} = \infty,$$

$$< 1 \qquad \text{for all } y < d \text{ iff } \sum_{x=-\infty}^{0} \frac{\beta_x\beta_{x+1}\cdots\beta_0}{\delta_x\delta_{x+1}\cdots\delta_0} < \infty. \tag{2.15}$$

By the Markov property, conditioning on X_1 (Exercise 3),

$$\rho_{yy} = \delta_y \rho_{y-1,y} + \beta_y \rho_{y+1,y} + (1 - \delta_y - \beta_y). \tag{2.16}$$

If both sums in (2.14) *and* (2.15) *diverge*, then $\rho_{y-1,y} = 1$, $\rho_{y+1,y} = 1$, so that (2.16) implies

$$\rho_{yy} = 1, \qquad \text{for all } y. \tag{2.17}$$

In other words, *all states are recurrent.*

If one of the sums (2.14) or (2.15) is convergent, say (2.14), then by (2.16) we get

$$\rho_{yy} < 1, \qquad y \in S. \tag{2.18}$$

A state $y \in S$ satisfying (2.18) is called a *transient state*; since (2.18) holds for all $y \in S$, the *birth–death chain is transient*. Just as in the case of a simple asymmetric random walk, the *strong Markov property* may be applied to see that with probability 1 each state occurs at most finitely often in a transient birth–death Markov chain.

CASE II. The next case is that of *two reflecting boundaries*. For this take $S = \{0, 1, 2, \ldots, N\}$, $p_{00} = 1 - \beta_0$, $p_{01} = \beta_0$, $p_{N,N-1} = \delta_N$, $p_{N,N} = 1 - \delta_N$; and $p_{i,i+1} = \beta_i$, $p_{i,i-1} = \delta_i$, $p_{i,i} = 1 - \beta_i - \delta_i$ for $1 \leqslant i \leqslant N - 1$. If one takes $c = 0$, $d = N$ in (2.3), then $\psi(y)$ gives the probability that the process starting at y reaches 0 before reaching N. The probability $\phi(y)$, for the process to reach N before 0 starting at y, may be obtained in the same fashion by changing the boundary conditions (2.5) to $\phi(0) = 0$, $\phi(N) = 1$ to get that $\phi(y) = 1 - \psi(y)$. Alternatively, check that $\phi(y) \equiv 1 - \psi(y)$ satisfies the equation (2.6) (with ϕ replacing ψ) and the boundary conditions $\phi(0) = 0$, $\phi(N) = 1$, and then argue that such a solution is necessarily unique (Exercise 4). All states are recurrent, by Corollary 9.6 (see Exercise 5 for an alternative proof).

CASE III. For the case of *one absorbing boundary*, say at 0, take $S = \{0, 1, 2, \ldots\}$, $p_{00} = 1$, $p_{i,i+1} = \beta_i$, $p_{i,i-1} = \delta_i$, $p_{i,i} = 1 - \beta_i - \delta_i$ for $i > 0$; $\beta_i, \delta_i > 0$ for $i > 0$, $\beta_i + \delta_i \leqslant 1$. For $c, d \in S$, the probability $\psi(y)$ is given by (2.10) and the probability ρ_{y0}, which is also interpreted as the *probability of eventual absorption* starting at $y > 0$, is given by

$$\rho_{y0} = \lim_{d \uparrow \infty} \frac{\displaystyle\sum_{x=y}^{d-1} \frac{\delta_x \delta_{x-1} \cdots \delta_1}{\beta_x \beta_{x-1} \cdots \beta_1}}{1 + \displaystyle\sum_{x=1}^{d-1} \frac{\delta_x \delta_{x-1} \cdots \delta_1}{\beta_x \beta_{x-1} \cdots \beta_1}}$$

$$= 1 \quad \text{iff} \ \sum_{x=1}^{\infty} \frac{\delta_1 \delta_2 \cdots \delta_x}{\beta_1 \beta_2 \cdots \beta_x} = \infty \qquad (\text{for } y > 0). \tag{2.19}$$

Whether or not the last series diverges,

$$\rho_{y0} \geqslant \delta_y \delta_{y-1} \cdots \delta_1 > 0, \qquad \text{for all } y > 0 \tag{2.20}$$

and

$$\begin{aligned} \rho_{yd} &\leqslant 1 - \delta_y \delta_{y-1} \cdots \delta_1 < 1, && \text{for } d > y > 0, \\ \rho_{0d} &= 0, && \text{for all } d > 0. \end{aligned} \tag{2.21}$$

By (2.16) it follows that

$$\rho_{yy} < 1 \qquad (y > 0). \tag{2.22}$$

Thus, all nonzero states y are *transient*.

CASE IV. As a final illustration of transience–recurrence conditions, take the case of *one reflecting boundary* at 0 with $S = \{0, 1, 2, 3, \ldots\}$ and $p_{00} = 1 - \beta_0$, $p_{01} = \beta_0$, $p_{i,i+1} = \beta_i$, $p_{i,i-1} = \delta_i$, $p_{i,i} = 1 - \beta_i - \delta_i$ for $i > 0$; $\beta_i > 0$ for all i, $\delta_i > 0$ for $i \geqslant 1$, $\beta_i + \delta_i \leqslant 1$. Let us now see that all states are *recurrent if and only if the infinite series (2.19) diverges, i.e., if and only if* $\rho_{y0} = 1$.

First assume that the infinite series in (2.19) diverges, i.e., $\rho_{y0} = 1$ for all $y > 0$. Then condition on X_1 to get

$$\rho_{00} = (1 - \beta_0) + \beta_0 \rho_{10}, \tag{2.23}$$

so that

$$\rho_{00} = 1. \tag{2.24}$$

Next look at (see Eq. 2.16)

$$\rho_{11} = \delta_1 \rho_{01} + \beta_1 \rho_{21} + (1 - \delta_1 - \beta_1). \tag{2.25}$$

Since $\rho_{20} = 1$ and the process does not skip states, $\rho_{22} = 1$. Also, $\rho_{01} = 1$ (Exercise 6). Thus, $\rho_{11} = 1$ and, proceeding by induction,

$$\rho_{yy} = 1, \qquad \text{for each } y \geqslant 0. \tag{2.26}$$

On the other hand, if the series in (2.19) converges then $\rho_{y0} < 1$ for all $y > 0$. In particular, from (2.23), we see $\rho_{00} < 1$. Convergence of the series in (2.19) also gives $\rho_{yc} < 1$ for all $c < y$ by (2.12). Now apply (2.16) to get

$$\rho_{yy} < 1, \qquad \text{for all } y \tag{2.27}$$

whenever the series in (2.19) converges. That is, *the birth–death chain is transient*.

The various remaining cases, for example, two absorbing, or one absorbing and one reflecting boundary, are left to the Exercises.

3 INVARIANT DISTRIBUTIONS FOR BIRTH–DEATH CHAINS

Suppose that there is a probability distribution $\boldsymbol{\pi}$ on S such that

$$\boldsymbol{\pi}'\mathbf{p} = \boldsymbol{\pi}'. \tag{3.1}$$

Then $\boldsymbol{\pi}'\mathbf{p}^n = \boldsymbol{\pi}'$ for each time $n = 1, 2, \ldots$. That is, $\boldsymbol{\pi}$ is *invariant* under the transition law $\mathbf{p}$. Note that if $\{X_n\}$ is started with an invariant initial distribution $\boldsymbol{\pi}$ then X_n has distribution $\boldsymbol{\pi}$ at each successive time point n. Moreover, $\{X_n\}$ is *stationary* in the sense that the P_π-distribution of $(X_0, X_1, \ldots, X_m)$ is for each $m > 1$ invariant under all time shifts, i.e., for all

$k \geqslant 1$,

$$P_\pi(X_0 = i_0, \ldots, X_m = i_m) = P_\pi(X_k = i_0, \ldots, X_{m+k} = i_m). \tag{3.2}$$

For a birth–death process on $S = \{0, 1, 2, \ldots, N\}$ with two reflecting boundaries, the invariant distribution $\boldsymbol{\pi}$ is easily obtained by solving $\boldsymbol{\pi}'\mathbf{p}' = \boldsymbol{\pi}'$, i.e.,

$$\begin{aligned} \pi_0(1 - \beta_0) + \pi_1\delta_1 &= \pi_0, \\ \pi_{j-1}\beta_{j-1} + \pi_j(1 - \beta_j - \delta_j) + \pi_{j+1}\delta_{j+1} &= \pi_j \qquad (j = 1, 2, \ldots, N-1), \end{aligned} \tag{3.3}$$

or

$$\pi_{j-1}\beta_{j-1} - \pi_j(\beta_j + \delta_j) + \pi_{j+1}\delta_{j+1} = 0. \tag{3.4}$$

The solution, subject to $\pi_j \geqslant 0$ for all j and $\sum_j \pi_j = 1$, is given by

$$\begin{aligned} \pi_j &= \frac{\beta_0 \cdots \beta_{j-1}}{\delta_1 \cdots \delta_j}\pi_0 \qquad (1 \leqslant j \leqslant N), \\ \pi_0 &= \left(1 + \sum_{j=1}^{N} \frac{\beta_0\beta_1 \cdots \beta_{j-1}}{\delta_1\delta_2 \cdots \delta_j}\right)^{-1}. \end{aligned} \tag{3.5}$$

For a birth–death process on $S = \{0, 1, 2, \ldots\}$ with 0 as a reflecting boundary, the system of equations $\boldsymbol{\pi}'\mathbf{p} = \boldsymbol{\pi}'$ are

$$\begin{aligned} \pi_0(1 - \beta_0) + \pi_1\delta_1 &= \pi_0, \\ \pi_{j-1}\beta_{j-1} + \pi_j(1 - \beta_j - \delta_j) + \pi_{j+1}\delta_{j+1} &= j \qquad (j \geqslant 1). \end{aligned} \tag{3.6}$$

The solution in terms of π_0 is

$$\pi_j = \frac{\beta_0\beta_1 \cdots \beta_{j-1}}{\delta_1\delta_2 \cdots \delta_j}\pi_0 \qquad (j \geqslant 1). \tag{3.7}$$

In order that this may be a probability distribution one must have

$$\sum_{j=1}^{\infty} \frac{\beta_0\beta_1 \cdots \beta_{j-1}}{\delta_1\delta_2 \cdots \delta_j} < \infty. \tag{3.8}$$

In this case one must take

$$\pi_0 = \left(1 + \sum_{j=1}^{\infty} \frac{\beta_0\beta_1 \cdots \beta_{j-1}}{\delta_1\delta_2 \cdots \delta_j}\right)^{-1}. \tag{3.9}$$

For an unrestricted birth–death process on $S = \{0, \pm 1, \pm 2, \ldots\}$ the equations $\boldsymbol{\pi}'\mathbf{p}' = \boldsymbol{\pi}'$ are

$$\pi_{j-1}\beta_{j-1} + \pi_j(1 - \beta_j - \delta_j) + \pi_{j+1}\delta_{j+1} = \pi_j \qquad (j = 0, \pm 1, \pm 2, \ldots) \tag{3.10}$$

which are solved (in terms of π_0) by

$$\pi_j = \begin{cases} \dfrac{\beta_0\beta_1\cdots\beta_{j-1}}{\delta_1\delta_2\cdots\delta_j}\,\pi_0 & (j \geqslant 1), \\[2ex] \dfrac{\delta_{j+1}\delta_{j+2}\cdots\delta_0}{\beta_j\beta_{j+1}\cdots\beta_{-1}}\,\pi_0 & (j \leqslant -1). \end{cases} \tag{3.11}$$

This is a probability distribution if and only if

$$\sum_{j \leqslant -1} \frac{\delta_{j+1}\delta_{j+2}\cdots\delta_0}{\beta_j\beta_{j+1}\cdots\beta_{-1}} < \infty, \qquad \sum_{j \geqslant 1} \frac{\beta_0\beta_1\cdots\beta_{j-1}}{\delta_1\delta_2\cdots\delta_j} < \infty, \tag{3.12}$$

in which case

$$\pi_0 = \left(1 + \sum_{j \leqslant -1} \frac{\delta_{j+1}\delta_{j+2}\cdots\delta_0}{\beta_j\beta_{j+1}\cdots\beta_{-1}} + \sum_{j \geqslant 1} \frac{\beta_0\beta_1\cdots\beta_{j-1}}{\delta_1\delta_2\cdots\delta_j}\right)^{-1}. \tag{3.13}$$

Notice that the convergence of the series in (3.12) implies the divergence of the series in (2.14), (2.15). In other words, the existence of an equilibrium distribution for the chain implies its recurrence. The same remark applies to the birth–death chain with one or two reflecting boundaries.

Example 1. (*Equilibrium for the Bernoulli–Laplace Model*). For the Bernoulli–Laplace model described in Section 1, the invariant distribution $\boldsymbol{\pi} = (\pi_i\colon i = 0, 1, \ldots, 2r)$ is the hypergeometric distribution calculated from (3.5) as

$$\pi_j = \frac{\beta_0\cdots\beta_{j-1}}{\delta_1\cdots\delta_j}\,\pi_0 = \frac{2r(w+r)}{j(2-r+j)} \prod_{i=1}^{j-1} \frac{(w+r-i)(2r-i)}{i(w-r+i)} = \frac{\dbinom{2r}{j}\dbinom{2w}{w+r-j}}{\dbinom{2w+2r}{w+r}}. \tag{3.14}$$

The assertions concerning positive recurrence contained in Theorem 3.1 below rely on the material in Section 2.9 and may be omitted on first reading.

Recall from Theorem 9.2(c) of Chapter I that in the case that all states communicate with each other, existence of an invariant distribution is equivalent to positive recurrence of all states.

Theorem 3.1

(a) For a birth–death chain on $S = \{0, 1, \ldots, N\}$ with both boundaries reflecting, the states are all positive recurrent and the unique invariant distribution is given by (3.5).

(b) For a birth–death chain on $S = \{0, 1, 2, \ldots\}$ with 0 a reflecting boundary, all states are recurrent or transient according as the series

$$\sum_{1}^{\infty} \frac{\delta_1 \delta_2 \cdots \delta_x}{\beta_1 \cdots \beta_x}$$

diverges or converges. All states are positive recurrent if and only if the series (3.8) converges. In the case that (3.8) converges, the unique invariant distribution is given by (3.7), (3.9).

(c) For an unrestricted birth–death chain on $S = \{0, \pm 1, \pm 2, \ldots\}$ all states are transient if and only if at least one of the series in (2.14) and (2.15) is convergent. All states are positive recurrent if and only if (3.12) holds; if (3.12) holds, then the unique invariant distribution is given by (3.11), (3.13).

4 CALCULATION OF TRANSITION PROBABILITIES BY SPECTRAL METHODS

We will apply the spectral theorem to calculate $\mathbf{p}^n$, for $n = 1, 2, \ldots$, in the case that $\mathbf{p}$ is the transition law for a birth–death chain.

First consider the case of a birth–death chain on $S = \{0, 1, \ldots, N\}$ with reflecting boundaries at 0 and N. Then the *invariant distribution* $\boldsymbol{\pi}$ is given by (3.5) as

$$\pi_1 = \frac{1}{\delta_1}\pi_0, \qquad \pi_j = \frac{\beta_1 \cdots \beta_{j-1}}{\delta_1 \cdots \delta_j}\pi_0 \qquad (2 \leqslant j \leqslant N). \tag{4.1}$$

It is straightforward to check that

$$\pi_i p_{i,i-1} = \pi_i \delta_i = \pi_{i-1}\beta_{i-1} = \pi_{i-1} p_{i-1,i}, \qquad i = 1, 2, \ldots, N, \tag{4.2}$$

from which it follows that

$$\pi_i p_{ij} = \pi_j p_{ji}, \qquad \text{for all } i, j. \tag{4.3}$$

In the applied sciences the symmetry property (4.3) is often referred to as *detailed balance* or *time reversibility*. Introduce the following *inner product* $(\cdot, \cdot)_{\boldsymbol{\pi}}$ in the vector space $\mathbb{R}^{N+1}$,

$$(\mathbf{x}, \mathbf{y})_\pi = \sum_{i=0}^{N} x_i y_i \pi_i, \qquad \mathbf{x} = (x_0, x_1, \ldots, x_N)', \qquad \mathbf{y} = (y_0, y_1, \ldots, y_N)', \tag{4.4}$$

so that the "length" $\|\mathbf{x}\|_\pi$ of a vector $\mathbf{x}$ is given by

$$\|\mathbf{x}\|_\pi = \left(\sum_{i=0}^{N} x_i^2 \pi_i\right)^{1/2}. \tag{4.5}$$

With respect to this inner product the linear transformation $\mathbf{x} \to \mathbf{px}$ is *symmetric* since by (4.3) we have

$$\begin{aligned}(\mathbf{px}, \mathbf{y})_\pi &= \sum_{i=0}^{N}\left(\sum_{j=0}^{N} p_{ij}x_j\right)y_i\pi_i = \sum_{i=0}^{N}\sum_{j=0}^{N} p_{ji}x_j y_i \pi_j \\ &= \sum_{i=0}^{N}\left(\sum_{j=0}^{N} p_{ji}\, y_i\right)x_j\pi_j \\ &= \sum_{i=0}^{N}\left(\sum_{j=0}^{N} p_{ij}\, y_j\right)x_i\pi_i = (\mathbf{py}, \mathbf{x})_\pi = (\mathbf{x}, \mathbf{py})_\pi. \end{aligned} \tag{4.6}$$

Therefore, by the spectral theorem, $\mathbf{p}$ has $N + 1$ real eigenvalues $\alpha_0, \alpha_1, \ldots, \alpha_N$ (not necessarily distinct) and corresponding eigenvectors $\boldsymbol{\phi}_0, \boldsymbol{\phi}_1, \ldots, \boldsymbol{\phi}_N$, which are of unit length and mutually orthogonal with respect to $(\cdot,\cdot)_\pi$. Therefore, the linear transformation $\mathbf{x} \to \mathbf{px}$ has the *spectral representation*

$$\begin{aligned}\mathbf{p} &= \sum_{k=0}^{N} \alpha_k \mathbf{E}_k, \\ \mathbf{px} &= \sum_{k=0}^{N} \alpha_k (\boldsymbol{\phi}_k, \mathbf{x})_\pi \boldsymbol{\phi}_k,\end{aligned} \tag{4.7}$$

where $\mathbf{E}_k$ denotes orthogonal projection with respect to $(\cdot,\cdot)_\pi$ onto the one-dimensional subspace spanned by $\boldsymbol{\phi}_k$: $E_k x = (\boldsymbol{\phi}_k, \mathbf{x})_\pi \boldsymbol{\phi}_k$. It follows that

$$\begin{aligned}\mathbf{p} &= \sum_{k=0}^{N} \alpha_k \mathbf{E}_k, \\ \mathbf{p}^n\mathbf{x} &= \sum_{k=0}^{N} \alpha_k^n (\boldsymbol{\phi}_k, \mathbf{x})_\pi \boldsymbol{\phi}_k.\end{aligned} \tag{4.8}$$

Letting $\mathbf{x} = \mathbf{e}_j$ denote the vector with 1 in the jth coordinate and zeros elsewhere, one gets

$$\begin{aligned}p_{ij}^{(n)} &= i\text{th element of } \mathbf{p}^n\mathbf{e}_j \\ &= \sum_{k=0}^{N} \alpha_k^n (\boldsymbol{\phi}_k, \mathbf{e}_j)_\pi \phi_{ki} = \sum_{k=0}^{N} \alpha_k^n \phi_{ki}\phi_{kj}\pi_j.\end{aligned} \tag{4.9}$$

Without loss of generality, one may take $\alpha_0 = 1$ and $\phi_0 = 1$ throughout. We now consider two special birth–death chains as examples.

Example 1. (*Simple Symmetric Random Walk with Two Reflecting Boundaries*)

$$S = \{0, 1, \ldots, N\}, \qquad p_{i,i+1} = p_{i,i-1} = \tfrac{1}{2}, \qquad \text{for } 1 \leqslant i \leqslant N-1,$$
$$p_{01} = 1 = p_{N,N-1}.$$

In this case the invariant initial distribution is given by

$$\pi_j = \frac{1}{N} \qquad (1 \leqslant j \leqslant N-1), \qquad \pi_0 = \pi_N = \frac{1}{2N}. \tag{4.10}$$

If α is an eigenvalue of $\mathbf{p}$, then a corresponding eigenvector $\mathbf{x} = (x_0, x_1, \ldots, x_N)'$ satisfies the equation

$$\tfrac{1}{2}(x_{j-1} + x_{j+1}) = \alpha x_j \qquad (1 \leqslant j \leqslant N-1), \tag{4.11}$$

along with "boundary conditions"

$$x_1 = \alpha x_0, \qquad x_{N-1} = \alpha x_N. \tag{4.12}$$

As a trial solution of (4.11) consider $x_j = \theta^j$ for some nonzero θ. Since all vectors of $\mathbb{C}^{N+1}$ may be expressed as unique linear combinations of functions $j \to \exp\{(2r\pi/N+1)ji\} = \theta^j$, with $i = (-1)^{1/2}$, one expects to arrive at the right combination in this manner. Then (4.11) yields $\frac{1}{2}(\theta^{j-1} + \theta^{j+1}) = \alpha\theta^j$, i.e.,

$$\theta^2 - 2\alpha\theta + 1 = 0, \tag{4.13}$$

whose two roots are

$$\theta_1 = \alpha + i\sqrt{1-\alpha^2}, \qquad \theta_2 = \alpha - i\sqrt{1-\alpha^2}. \tag{4.14}$$

The equation (4.11) is *linear* in $\mathbf{x}$, i.e., if $\mathbf{x}$ and $\mathbf{y}$ are both solutions of (4.11) then so is $ax + by$ for arbitrary numbers a and b. Therefore, every linear combination

$$x_j = A(\alpha)\theta_1^j + B(\alpha)\theta_2^j \qquad (0 \leqslant j \leqslant N) \tag{4.15}$$

satisfies (4.11). We now apply the boundary conditions (4.12) to fix $A(\alpha)$, $B(\alpha)$, up to a constant multiplier. Since every scalar multiple of a solution of (4.11) and (4.12) is also a solution, let us fix $x_0 = 1$. Note that $x_0 = 0$ implies $x_j = 0$ for all j. Letting $j = 0$ in (4.15), one has

$$A(\alpha) + B(\alpha) = 1, \qquad B(\alpha) = 1 - A(\alpha). \tag{4.16}$$

The first boundary condition, $x_1 = \alpha x_0 = \alpha$, then becomes

$$A(\alpha)(\theta_1 - \theta_2) + \theta_2 = \alpha, \tag{4.17}$$

or,

$$2A(\alpha)(1 - \alpha^2)^{1/2} i = (1 - \alpha^2)^{1/2} i,$$

i.e., at least for $\alpha \neq 1$,

$$A(\alpha) = \tfrac{1}{2}, \qquad B(\alpha) = \tfrac{1}{2}. \tag{4.18}$$

The second boundary condition $x_{N-1} = \alpha x_N$ may then be expressed as

$$\frac{1}{2}(\theta_1^{N-1} + \theta_2^{N-1}) = \frac{\alpha}{2}(\theta_1^N + \theta_2^N). \tag{4.19}$$

Now write $\theta_1 = e^{i\phi}$, $\theta_2 = e^{-i\phi}$, where ϕ is the *unique* angle in $[0, \pi]$ such that $\cos \phi = \alpha$. Note that cosine is strictly decreasing in $[0, \pi]$ and assumes its entire range of values $[-1, 1]$ on $[0, \pi]$. Note also that this is consistent with the requirement $\sin \phi = 1 - \alpha^2 \geqslant 0$. Then (4.19) becomes

$$\cos(N - 1)\phi = \alpha \cos N\phi = \cos \phi \cos N\phi, \tag{4.20}$$

i.e.,

$$\sin N\phi \sin \phi = 0, \tag{4.21}$$

whose only solutions in $[0, \pi]$ are

$$\phi = \cos \frac{k\pi}{N} \qquad (k = 0, 1, 2, \dots, N). \tag{4.22}$$

Thus, there are $N + 1$ distinct (and, therefore, simple) eigenvalues

$$\alpha_k = \cos \frac{k\pi}{N} \qquad (k = 0, 1, 2, \dots, N), \tag{4.23}$$

and corresponding eigenvectors $x^{(k)}$ $(k = 0, 1, \dots, N)$:

$$x_j^{(k)} = \tfrac{1}{2}(\theta_1^j + \theta_2^j) = \cos \frac{k\pi}{N} j \qquad (j = 0, 1, \dots, N). \tag{4.24}$$

Now,

$$\|\mathbf{x}^{(k)}\|^2 = \sum_{j=0}^{N} \pi_j \cos^2\left(\frac{k\pi j}{N}\right) = \frac{1}{2N} + \frac{1}{N}\sum_{j=1}^{N-1} \cos^2\left(\frac{k\pi j}{N}\right) + \frac{1}{2N}$$

$$= \frac{1}{N}\sum_{j=0}^{N-1} \cos^2\left(\frac{k\pi j}{N}\right) = \frac{1}{N}\sum_{j=0}^{N-1} \frac{1+\cos(2k\pi j/N)}{2}$$

$$= \begin{cases} 1 & \text{if } k = 0 \text{ or } N \\ \frac{1}{2} & \text{if } k = 1, 2, \ldots, N-1. \end{cases} \tag{4.25}$$

Thus, the normalized eigenvectors are

$$\boldsymbol{\phi}_0 = (1, 1, \ldots, 1)', \qquad \boldsymbol{\phi}_N = (1, -1, +1, -1, \ldots)';$$
$$\boldsymbol{\phi}_k = \sqrt{2}\,\mathbf{x}^{(k)}, \qquad \phi_{kj} = \sqrt{2}\cos\frac{k\pi j}{N} \qquad (1 \leqslant k \leqslant N-1). \tag{4.26}$$

Now use (4.9), (4.23), and (4.26) to get, for $0 \leqslant i, j \leqslant N$,

$$p_{ij}^{(n)} = \sum_{k=0}^{N} \alpha_k^n \phi_{ki} \phi_{kj} \pi_j$$

$$= \pi_j + 2\pi_j \sum_{k=1}^{N-1} \cos^n\left(\frac{k\pi}{N}\right)\cos\left(\frac{k\pi i}{N}\right)\cos\left(\frac{k\pi j}{N}\right) + (-1)^{n+j-i}\pi_j. \tag{4.27}$$

Thus, for $1 \leqslant j \leqslant N-1$, $0 \leqslant i \leqslant N$,

$$p_{ij}^{(n)} = \frac{1}{N} + \frac{2}{N}\sum_{k=1}^{N-1} \cos^n\left(\frac{k\pi}{N}\right)\cos\left(\frac{k\pi i}{N}\right)\cos\left(\frac{k\pi j}{N}\right) + (-1)^{n+j-i}\frac{1}{N}.$$

For $0 \leqslant i \leqslant N$,

$$p_{i0}^{(n)} = \frac{1}{2N} + \frac{1}{N}\sum_{k=1}^{N-1} \cos^n\left(\frac{k\pi}{N}\right)\cos\left(\frac{k\pi i}{N}\right) + (-1)^{n-1}\frac{1}{2N}.$$

For $0 \leqslant i \leqslant N$,

$$p_{iN}^{(n)} = \frac{1}{2N} + \frac{1}{N}\sum_{k=1}^{N-1} \cos^n\left(\frac{k\pi}{N}\right)\cos\left(\frac{k\pi i}{N}\right)\cos\left(\frac{k\pi}{N}\right) + (-1)^{n+N-i}\frac{1}{2N}. \tag{4.28}$$

Note that when n and $j-i$ have the same parity, say $n = 2m$ and $j - i$ is even, then

$$\left(p_{ij}^{(2m)} - \frac{2}{N}\right) = \frac{4}{N}\left[\cos\left(\frac{\pi}{N}\right)\right]^{2m}\cos\left(\frac{\pi i}{N}\right)\cos\left(\frac{\pi j}{N}\right)[1 + o(1)] \tag{4.29}$$

as $m \to \infty$. This establishes the precise rate of exponential convergence to the steady state. One may express this as

$$\lim_{m \to \infty} \left(p_{ij}^{(2m)} - \frac{2}{N} \right) e^{2m\lambda} = \frac{4}{N} \cos\left(\frac{\pi i}{N}\right) \cos\left(\frac{\pi j}{N}\right), \tag{4.30}$$

where $\lambda = \log \alpha_1$ (see (4.23)).

Example 2. (*Simple Symmetric Random Walk with One Reflecting Boundary*)

$$S = \{0, 1, 2, \ldots\}, \qquad p_{01} = 1 \qquad \text{and} \qquad p_{i,k-1} = \tfrac{1}{2} = p_{i,i+1}$$

for all $i \geqslant 1$. Note that $p_{ij}^{(n)}$ is the same as in the case of a random walk with two reflecting boundaries 0 and N, provided $N > n + i$, since the random walk cannot reach N in n steps (or fewer) starting from i if $N > n + i$. Hence for all i, j, n, $p_{ij}^{(n)}$ is obtained by taking the limit in (4.28) as $N \to \infty$, i.e.,

$$\begin{aligned} p_{ij}^{(n)} &= 2 \int_0^1 \cos^n(\pi\theta) \cos(i\pi\theta) \cos(j\pi\theta)\, d\theta \\ &= \frac{2}{\pi} \int_0^\pi \cos^n(\theta) \cos(i\theta) \cos(j\theta)\, d\theta \qquad (j \geqslant 1,\ i \geqslant 0), \\ p_{i0}^{(n)} &= \frac{1}{\pi} \int_0^\pi \cos^n(\theta) \cos(i\theta)\, d\theta \qquad (i \geqslant 0). \end{aligned} \tag{4.31}$$

An alternative calculation of (4.31) can be made by first noting that the condition (4.3) is valid for the sequence of weights $\{\pi_j\}$ given in (4.1) with $\pi_0 = 1$. This provides an inner product (sequence) space on which $\mathbf{p}$ is bounded self-adjoint linear transformation. The spectral theory extends to such settings as well.

An example in which the birth–death parameters are state-space dependent (i.e., nonconstant) is given in the chapter application.

5 CHAPTER APPLICATION: THE EHRENFEST MODEL OF HEAT EXCHANGE

The Ehrenfest model illustrates the process of heat exchange between two bodies that are in contact and insulated from the outside. The temperatures are assumed to change in steps of one unit and are represented by the numbers of balls in two boxes. The two boxes are marked I and II and there are $2d$ balls labeled $1, 2, \ldots, 2d$. Initially some of these balls are in box I and the remainder in box II. At each step a ball is chosen at random (i.e., with equal probabilities among ball numbers $1, 2, \ldots, 2d$) and moved from its box to the other box. If there are i balls in box I, then there are $2d - i$ balls in box II. Thus there is no overall heat loss or gain. Let X_n denote the number of balls in box I after the nth trial. Then $\{X_n: n = 0, 1, \ldots\}$ is a Markov chain with state space $S = \{0, 1, 2, \ldots, 2d\}$ and transition probabilities

$$\begin{aligned} p_{i,i-1} &= \frac{i}{2d}, \qquad p_{i,i+1} = 1 - \frac{i}{2d}, \qquad \text{for } i = 1, 2, \ldots, 2d-1, \\ p_{01} &= 1, \qquad p_{2d,2d-1} = 1, \\ p_{ij} &= 0, \qquad \text{otherwise.} \end{aligned} \tag{5.1}$$

This is a *birth–death chain with two reflecting boundaries* at 0 and $2d$. The transition probabilities are such that the mean change in temperature, in box I, say, at each step is propostional to the negative of the existing temperature gradient, or temperature difference, between the two bodies. We will first see that the model yields *Newton's law of cooling* at the level of the evolution of the averages. Assume that initially there are i balls in box I. Let $Y_n = X_n - d$, the excess of the number of balls in box I over d. Writing $e_n = E_i(Y_n)$, the expected value of Y_n given $X_0 = i$, one has

$$\begin{aligned} e_n &= E_i(X_n - d) = E_i[X_{n-1} - d + (X_n - X_{n-1})] \\ &= E_i(X_{n-1} - d) + E_i(X_n - X_{n-1}) = e_{n-1} + E_i\left(\frac{2d - X_{n-1}}{2d} - \frac{X_{n-1}}{2d}\right) \\ &= e_{n-1} + E_i\left(\frac{d - X_{n-1}}{d}\right) = e_{n-1} - \frac{e_{n-1}}{d} = \left(1 - \frac{1}{d}\right)e_{n-1}. \end{aligned}$$

Note that in evaluating $E_i(X_n - X_{n-1})$ we first calculated the conditional expectation of $X_n - X_{n-1}$ given X_{n-1} and then took the expectation of this conditional mean. Now, by successive applications of the relation $e_n = (1 - 1/d)e_{n-1}$,

$$e_n = \left(1 - \frac{1}{d}\right)^n e_0 = \left(1 - \frac{1}{d}\right)^n E_i(X_0 - d) = (i - d)\left(1 - \frac{1}{d}\right)^n. \tag{5.2}$$

Suppose in the physical model the frequency of transitions is τ per second. Then in time t there are $n = t\tau$ transitions. Write $v = -\log[(1 - (1/d))]^{\tau}$. Then

$$e_n = (i - d)e^{-vt}, \tag{5.3}$$

which is *Newton's law of cooling*.

The *equilibrium distribution* for the Ehrenfest model is easily seen, using (3.5), to be

$$\pi_j = \binom{2d}{j} 2^{-2d}, \qquad j = 0, 1, \ldots, 2d. \tag{5.4}$$

That is, $\boldsymbol{\pi} = (\pi_j: j \in S)$ is binomial with parameters $2d, \frac{1}{2}$. Note that $d = E_\pi X_n$ is the (constant) mean temperature under equilibrium in (5.3).

The physicists P. and T. Ehrenfest in 1907, and later Smoluchowski in 1916, used this model in order to explain an apparent paradox that at the turn of the century threatened to wreck Boltzmann's *kinetic theory of matter*. In the kinetic theory, heat exchange is a random process, while in *thermodynamics* it is an orderly irreversible progression toward equilibrium. In the present context, thermodynamic equilibrium would be achieved when the temperatures of the two bodies became equal, or at least approximately or macroscopically equal. But if one uses a kinetic model such as the one described above, from the state $i = d$ of thermodynamical equilibrium the system will eventually pass to a state of extreme disequilibrium (e.g., $i = 0$) owing to *recurrence*. This would contradict irreversibility of thermodynamics. However, one of the main objectives of kinetic theory was to explain thermodynamics, a largely phenomenological macroscopic-scale theory, starting from the molecular theory of matter.

Historically it was Poincaré who first showed that statistical-mechanical systems have the *recurrence property* (theoretical complement 2). A scientiest named Zermelo then forcefully argued that recurrence contradicted irreversibility.

Although Boltzmann rightly maintained that the time required by the random process to pass from the equilibrium state to a state of macroscopic nonequilibrium would be so large as to be of no physical significance, his reasoning did not convince other physicists. The Ehrenfests and Smoluchowski finally resolved the dispute by demonstrating how large the passage time may be from $i = d$ to $i = 0$ in the present model.

Let us now present in detail a method of calculating the mean first passage time $m_i = E_i T_0$, where $T_0 = \inf\{n \geqslant 0: X_n = 0\}$. Since the method is applicable to general birth–death chains, consider a state space $S = \{0, 1, 2, \ldots, N\}$ and a reflecting chain with parameters $\beta_i, \delta_i = 1 - \beta_i$ such that $0 < \beta_i < 1$ for $1 \leqslant i \leqslant N - 1$ and $\beta_0 = \delta_N = 1$. Then,

$$\begin{aligned} m_i &= 1 + \beta_i m_{i+1} + \delta_i m_{i-1} \qquad (1 \leqslant i \leqslant N-1),\\ m_0 &= 0, \qquad m_N = 1 + m_{N-1}. \end{aligned} \tag{5.5}$$

Relabel the states by $i \to u_i$ so that u_i is increasing with i,

$$u_0 = 0, \qquad u_1 = 1, \tag{5.6}$$

and, for all $x \in S$,

$$\psi(x) \equiv P_x(\{X_n\} \text{ reaches 0 before } N) = \frac{u_N - u_x}{u_N - u_0}. \tag{5.7}$$

In other words, in this new scale the probability of reaching the relabeled boundary $u = 0$, before u_N, starting from u_x (inside), is proportional to the distance from the boundary u_N. This scale is called the *natural scale*. The difference equations (5.5) when written in this scale assume a simple form, as will presently be shown. First let us determine u_x from (5.6) and (5.7) and the difference equation

$$\begin{gathered} \psi(x) = \beta_x \psi(x+1) + \delta_x \psi(x-1), \qquad 0 < x < N, \\ \psi(x) = 1, \qquad \psi(N) = 0, \end{gathered} \tag{5.8}$$

which may also be expressed as

$$\begin{gathered} \psi(x+1) - \psi(x) = \frac{\delta_x}{\beta_x}[\psi(x) - \psi(x-1)], \qquad 0 < x < N, \\ \psi(0) = 1, \qquad \psi(N) = 0. \end{gathered} \tag{5.9}$$

Equations (5.6), (5.7), and (5.9) yield

$$\begin{aligned} u_{x+1} - u_x &= \frac{\delta_x}{\beta_x}(u_x - u_{x-1}) \\ &= \frac{\delta_1 \delta_2 \cdots \delta_x}{\beta_1 \beta_2 \cdots \beta_x}(u_1 - u_0) = \frac{\delta_1 \delta_2 \cdots \delta_x}{\beta_1 \beta_2 \cdots \beta_x} \qquad (1 \leqslant x \leqslant N-1), \end{aligned} \tag{5.10}$$

or

$$u_{x+1} = 1 + \sum_{i=1}^{x} \frac{\delta_1 \delta_2 \cdots \delta_i}{\beta_1 \beta_2 \cdots \beta_i} \qquad (1 \leqslant x \leqslant N-1). \tag{5.11}$$

Now write

$$m(u_x) \equiv m_x. \tag{5.12}$$

Then (5.5) becomes

$$\begin{gathered} [m(u_{i+1}) - m(u_i)]\beta_i - [m(u_i) - m(u_{i-1})]\delta_i = -1 \qquad (1 \leqslant i \leqslant N-1), \\ m(u_0) = m(0) = 0, \qquad m(u_N) - m(u_{N-1}) = 1. \end{gathered} \tag{5.13}$$

One may rewrite this, using (5.10), as

$$\frac{m(u_{i+1}) - m(u_i)}{u_{i+1} - u_i} - \frac{m(u_i) - m(u_{i-1})}{u_i - u_{i-1}} = -\frac{\beta_0 \beta_1 \cdots \beta_{i-1}}{\delta_1 \delta_2 \cdots \delta_i} \qquad (1 \leqslant i \leqslant N-1)$$

or, summing over $i = x, x+1, \ldots, N-1$ and using the last boundary condition in (5.13), one has

$$\frac{1}{u_N - u_{N-1}} - \frac{m(u_x) - m(u_{x-1})}{u_x - u_{x-1}} = -\sum_{i=x}^{N-1} \frac{\beta_0\beta_1\cdots\beta_{i-1}}{\delta_1\delta_2\cdots\delta_i} \qquad (1 \leqslant x \leqslant N-1). \tag{5.14}$$

Relations (5.10) and (5.14) lead to

$$m(u_x) - m(u_{x-1}) = \frac{\beta_x\beta_{x+1}\cdots\beta_{N-1}}{\delta_x\delta_{x+1}\cdots\delta_{N-1}} + \sum_{i=x}^{N-1} \frac{\beta_x\cdots\beta_{i-1}\beta_i}{\delta_x\cdots\delta_i\beta_i} \qquad (1 \leqslant x \leqslant N-1). \tag{5.15}$$

The factor β_i/β_i is introduced in the last summands to take care of the summand corresponding to $i = x$ (this summand is actually $1/\delta_x$). Sum (5.15) over $x = 1, 2, \ldots, y$ to finally get, using $m(u_0) = 0$,

$$m(u_y) = \sum_{x=1}^{y} \frac{\beta_x\beta_{x+1}\cdots\beta_{N-1}}{\delta_x\delta_{x+1}\cdots\delta_{N-1}} + \sum_{x=1}^{y}\sum_{i=x}^{N-1} \frac{\beta_x\cdots\beta_{i-1}\beta_i}{\delta_x\cdots\delta_i\beta_i} \qquad (1 \leqslant y \leqslant N-1). \tag{5.16}$$

In particular, for the Ehrenfest model one gets

$$\begin{aligned} m_d = m(u_d) &= \sum_{x=1}^{d} \frac{(2d-x)\cdots 2\cdot 1}{x(x+1)\cdots(2d-1)} + \sum_{x=1}^{d}\sum_{i=x}^{2d-1} \frac{(2d-x)\cdots(2d-i)}{x(x+1)\cdots i(2d-i)} \\ &= \sum_{x=1}^{d} \frac{(2d-x)!(x-1)!}{(2d-1)!} + \sum_{x=1}^{d}\sum_{i=x}^{2d-1} \frac{(2d-x)!(x-1)!}{(2d-i)!\,i!} \\ &= \frac{1}{2d}2^{2d}\left(1 + O\left(\frac{1}{d}\right)\right). \end{aligned} \tag{5.17}$$

Next let us calculate

$$\bar{m}_i \equiv E_i T_d \qquad (0 \leqslant i \leqslant d), \tag{5.18}$$

where

$$T_d = \inf\{n \geqslant 0: X_n = d\}. \tag{5.19}$$

Writing $\bar{m}(u_i) = \bar{m}_i$, one obtains the same equations as (5.13) for $1 \leqslant i \leqslant d-1$, and boundary conditions

$$\bar{m}(u_0) = 1 + \bar{m}(u_1), \qquad \bar{m}(u_d) = 0. \tag{5.20}$$

As before, summing the equations over $i = 1, 2, \ldots, x$,

$$\frac{\bar{m}(u_{x+1}) - \bar{m}(u_x)}{u_{x+1} - u_x} = \frac{\bar{m}(u_1) - \bar{m}(u_0)}{u_1 - u_0} - \sum_{i=1}^{x} \frac{\beta_0\beta_1 \cdots \beta_{i-1}}{\delta_1\delta_2 \cdots \delta_i}$$

$$= -1 - \sum_{i=1}^{x} \frac{\beta_0\beta_1 \cdots \beta_{i-1}}{\delta_1\delta_2 \cdots \delta_i}, \tag{5.21}$$

where $\beta_0 = 1$. Therefore,

$$\bar{m}(u_{x+1}) - \bar{m}(u_x) = -\frac{\delta_1\delta_2 \cdots \delta_x}{\beta_1\beta_2 \cdots \beta_x} - \sum_{i=1}^{x} \frac{\delta_{i+1} \cdots \delta_x\delta_{x+1}}{\beta_i \cdots \beta_x\delta_{x+1}}, \tag{5.22}$$

which, on summing over $x = 1, 3, \ldots, d - 1$ and using (5.20), leads to

$$-\bar{m}(u_0) = -1 - \sum_{x=1}^{d-a} \frac{\delta_1\delta_2 \cdots \delta_x}{\beta_1\beta_2 \cdots \beta_x} - \sum_{x=1}^{d-1} \sum_{i=1}^{x} \frac{\delta_{i+1} \cdots \delta_x\delta_{x+1}}{\beta_i \cdots \beta_x\delta_{x+1}}. \tag{5.23}$$

For the present example this gives

$$\bar{m}_0 = 1 + \sum_{x=1}^{d-1} \frac{x!}{(2d-1) \cdots (2d-x)} + \sum_{x=1}^{d-1} \sum_{i=1}^{x} \left(\frac{(x+1)x \cdots (i+2)(i+1)}{(2d-i) \cdots (2d-x)(x+1)} \right)$$

$$\leqslant 1 + \sum_{x=1}^{d-1} \frac{x!}{(2d-1) \cdots (2d-x)} + \sum_{x=1}^{d-1} \frac{2d}{2d-x} \sum_{i=1}^{x} \left(\frac{x}{2d-x} \right)^{x-i}$$

$$\leqslant 1 + \sum_{x=1}^{d-1} \frac{x!}{(2d-1) \cdots (2d-x)} + \sum_{x=1}^{d-1} \frac{2d}{2(d-x)}$$

$$\leqslant 1 + \sum_{x=1}^{d-1} \frac{x!}{(2d-1) \cdots (2d-x)} + d(\log d + 1). \tag{5.24}$$

Since the sum in the last expression goes to zero as $d \to \infty$,

$$\bar{m}_0 \leqslant d + d \log d + O(1), \qquad \text{as } d \to \infty. \tag{5.25}$$

For $d = 10\,000$ balls and rate of transition one ball per second, it follows that

$$\bar{m}_0 \leqslant 102\,215 \text{ seconds} < 29 \text{ hours}, \qquad m_d = 10^{6000} \text{ years}. \tag{5.26}$$

It takes only about a day on the average for the system to reach equilibrium from a state farthest from equilibrium, but takes an average time inconceivably large, even compared to cosmological scales, for the system to go back to that state from equilibrium.

The original calculations by the Ehrenfests concerned the mean recurrence times. Using Theorem 9.2(c) of Chapter II it is possible to get these results quite simply as follows. Let $\tau_i^{(1)} = \min\{n \geqslant 1: X_n = i\}$. Then, the *mean recurrence time of the state i* is

$$E_i\tau_i^{(1)} = \frac{1}{\pi_i} = \frac{i!(2d-i)!}{2d!}2^{2d}. \tag{5.27}$$

For $d = 10\,000$ one gets, using Stirling's approximation for the second estimate,

$$E_0\tau_0^{(1)} \sim 2^{20\,000}, \qquad E_d\tau_d^{(1)} \simeq 100\sqrt{\pi}. \tag{5.28}$$

Thus, within time scales over which applications of thermodynamics make sense, one would not observe a passage from equilibrium to a (macroscopic) nonequilibrium state. Although Boltzmann did not live to see it, this vindication of his theory ended a rather spirited debate on its validity and contributed in no small measure to its eventual acceptance by physicists.

The spectral representation for $\mathbf{p}$ can be obtained by precisely the same steps as those outlined in Exercise 4.3. The $2d$ eigenvalues that one obtains are given by $\alpha_j = j/d$, $j = \pm 1, \pm 2, \ldots, \pm d$.

EXERCISES

Exercises for Section III.1

1. (*Artificial Intelligence*) The amplitudes of pure noise in a signal-detection device has p.d.f. $f_0(x)$, while the amplitude is distributed as $f_1(x)$ when a signal is present with the noise. A detection procedure is designed as follows. Select a threshold value $\theta_0 = r\delta$ for integer r. If the first amplitude observed, X_1, is larger than the initial threshold value θ_0, then decide that the signal is present. Otherwise decide that the signal is absent. Suppose that a signal is being sent with probability $p = \frac{1}{2}$ and that upon making a decision the observer learns whether or not the decision was correct. The observer keeps the same threshold value if the decision is correct, but if the decision was incorrect the threshold is increased or decreased by an amount δ depending on the type of error committed. The rule governing the learning process of the observer is then

$$\theta_{n+1} = \theta_n + \delta[\mathbf{1}_{(0,\infty)}(X_{n+1} - \theta_n) - S_n]$$

where S_n is 1 or 0 depending on whether a signal is sent or not. The signal transmission processes $\{S_n\}$ and $\{X_n\}$ are i.i.d.

   (i) Show that the threshold adjustment process $\{\theta_n\}$ is a birth–death Markov chain and identify the state space.
   (ii) Calculate the transition probabilities.

2. Suppose that balls labeled $1, \ldots, N$ are initially distributed between two boxes labeled I and II. The *state* of the system represents the number of balls in box I. Determine the one-step transition probabilities for each of the following rules of motion in the state space.
   (i) At each time step a ball is randomly (uniformly) selected from the numbers $1, 2, \ldots, N$. Independently of the ball selected, box I or II is selected with respective probabilities p_1 and $p_2 = 1 - p_1$. The ball selected is placed in the box selected.
   (ii) At each time step a ball is randomly (uniformly) selected from the numbers in box I with probability p_1 or from those in II with probability $p_2 = 1 - p_1$. A box is then selected with respective probabilities in proportion to current box sizes. The ball selected is placed in the box selected.
   (iii) At each time step a ball is randomly (uniformly) selected from the numbers in box I with probability proportional to the current size of I or from those in II with the complementary probability. A box is also selected with probabilities in proportion to current box size. The ball selected is placed in the box selected.

Exercises for Section III.2

1. Let A_d be the set $\{\omega: X_0(\omega) = y, \{X_n(\omega): n \geqslant 0\}$ reaches c before $d\}$, where $y > c$. Show that $A_d \uparrow A = \{\omega: X_0(\omega) = y, \{X_n(\omega): n \geqslant 0\}$ ever reaches $c\}$.

2. Prove (2.15) by using (2.14) and looking at $\{-X_n: n \geqslant 0\}$.

3. Prove (2.4), (2.16), and (2.23) by conditioning on X_1 and using the Markov property.

4. Suppose that $\varphi(i)(c \leqslant i \leqslant d)$ satisfy the equations (2.4) and the boundary conditions $\varphi(c) = 0$, $\varphi(d) = 1$. Prove that such a φ is unique.

5. Consider a birth–death chain on $S = \{0, 1, \ldots, N\}$ with both boundaries reflecting.
   (i) Prove that $P_i(T_j \geqslant mN) \leqslant (1 - \delta_N \delta_{N-1} \cdots \delta_1)^m$ if $i > j$, and $\leqslant (1 - \beta_0 \beta_1 \cdots \beta_{N-1})^m$ if $i < j$. Here $T_j = \inf\{n \geqslant 1: X_n = j\}$.
   (ii) Use (i) to prove that $\rho_{ij} \equiv P_i(T_j < \infty) = 1$ for all i, j.

6. Consider a birth–death chain on $S = \{0, 1, \ldots\}$ with 0 reflecting. Argue as in Exercise 5 to show that $\rho_{0y} = 1$ for all y.

7. Consider a birth–death chain on $S = \{0, 1, \ldots, N\}$ with $0, N$ absorbing. Calculate

$$\lim_{n \to \infty} n^{-1} \sum_{m=1}^{n} p_{ij}^{(m)}, \qquad \text{for all } i, j.$$

8. Let 0 be a reflecting boundary for a birth–death chain on $S = \{\ldots, -3, -2, -1, 0\}$. Derive the necessary and sufficient condition for recurrence.

9. If 0 is absorbing, and N reflecting, for a birth–death chain on $S = \{0, 1, \ldots, N\}$, then show that 0 is recurrent and all other states are transient.

10. Let $\mathbf{p}$ be the transition probability matrix of a birth–death chain on $S = \{0, 1, 2, \ldots\}$ with

$$\beta_j = \frac{j+2}{2(j+1)}, \qquad \delta_j = \frac{j}{2(j+1)}, \qquad j = 0, 1, 2, \ldots.$$

(i) Are the states transient or recurrent?
(ii) Compute the probability of reaching c before d, $c < d$, starting from state i, $c \leqslant i \leqslant d$.

11. Suppose $\mathbf{p}$ is the transition matrix of a birth–death chain on $S = \{0, 1, 2, \ldots\}$ such that $\beta_0 = 1$, $\beta_j \leqslant \delta_j$ for $j = 1, 2, \ldots$. Show that all states must be recurrent.

Exercises for Section III.3

1. Let $\{X_n\}$ be the asymmetric simple random walk on $S = \{0, 1, 2, \ldots\}$ with $\beta_j = p < \frac{1}{2}$, $j = 1, 2, \ldots$ and (partial) reflection at 0 with $p_{0,0} = p_{0,1} = \frac{1}{2}$.
   (i) Calculate the invariant initial distribution $\boldsymbol{\pi}$.
   (ii) Calculate $E_\pi X_n$ as a function of $p < \frac{1}{2}$.

2. (*A Birth–Death Queue*) During each unit of time either 0 or 1 customer arrives for service and joins a single line. The probability of one customer arriving is λ, and no customer arrives with probability $1 - \lambda$. Also during each unit of time, independently of new arrivals, a single service is completed with probability p or continues into the next period with probability $1 - p$. Let X_n be the total number of customers (waiting in line or being serviced) at the nth unit of time.
   (i) Show that $\{X_n\}$ is a birth–death chain on $S = \{0, 1, 2, \ldots\}$.
   (ii) Discuss transience, recurrence, positive-recurrence.
   (iii) Calculate the invariant initial distribution when $\lambda < p$.
   (iv) Calculate $E_\pi X_n$ when $\lambda < p$, where $\boldsymbol{\pi}$ is the invariant initial distribution.

3. Calculate the invariant distribution for Exercise 1.2(i) where

$$p_{i,j} = \begin{cases} \dfrac{N-i}{N}\,p_1, & \text{if } j = i+1, \\[2ex] \dfrac{i}{N}\,p_1 + \dfrac{(N-i)}{N}\,p_2, & \text{if } j = i,\ i = 0, 1, \ldots, N, \\[2ex] \dfrac{i}{N}\,p_2, & \text{if } j = i-1, \\[2ex] 0, & \text{otherwise}. \end{cases}$$

   Discuss the situation for Exercise 2(ii) and (iii).

4. (*Time Reversal*) Let $\{X_n\}$ be a (stationary) irreducible birth–death chain with invariant initial distribution $\boldsymbol{\pi}$. Show that $P_\pi(X_n = j \mid X_{n+1} = i) = P_\pi(X_{n+1} = j \mid X_n = i)$.

Exercises for Section III.4

1. Calculate $p_{ij}^{(n)}$ for the birth–death queue of Exercise 3.2.

2. Let T be a self-adjoint linear transformation on a finite-dimensional inner product space V. Show
   (i) All eigenvalues of T must be real.
   (ii) Eigenvectors of T associated with distinct eigenvalues are orthogonal.

3. Calculate the transition probabilities $\mathbf{p}^n$ for $n \geqslant 1$ by the spectral method in the case of Exercise 1.2(i) and $p_1 = p_2 = \frac{1}{2}$ according to the following steps.
   (i) Consider the eigenvalue problem for the transpose $\mathbf{p}'$. Write out difference equations for $\mathbf{p}'\mathbf{x} = \alpha\mathbf{x}$.
   (ii) Replace the system of equations in (i) by the infinite system

$$\frac{1}{2}x_0 + \frac{1}{2N}x_1 = \alpha x_0, \qquad \frac{N-i}{2N}x_i + \frac{1}{2}x_{i+1} + \frac{i+2}{2N}x_{i+2} = \alpha x_{i+1},$$

   $i = 0, 1, 2, \ldots$. Show that if for some α there is a nonzero solution $\mathbf{x} = (x_0, x_1, x_2, \ldots)'$ of this infinite system with $x_{N+1} = 0$, then α must be an eigenvalue of $\mathbf{p}'$ with corresponding eigenvector $(x_0, x_1, \ldots, x_N)'$. [*Hint*: Show that $x_{N+1} = 0$ implies $x_i = 0$ for all $i \geqslant N + 1$.]
   (iii) Introduce the generating function $\varphi(z) = \sum_{i=0}^{\infty} x_i z^i$ for the infinite system in (ii). Note that for the desired solutions satisfying $x_{N+1} = 0$, $\varphi(z)$ will be a polynomial of degree $\leqslant N$. Show that

$$\varphi'(z) = \frac{N(2\alpha - 1 - z)}{1 - z^2}\varphi(z), \qquad \varphi(0) = x_0.$$

   [*Hint*: Multiply both sides of the second equation in (ii) by z^{i+1} and sum over $i \geqslant 0$.]
   (iv) Show that (iii) has the unique solution

$$\varphi(z) = x_0(1 - z)^{N(1-\alpha)}(1 + z)^{N\alpha}.$$

   (v) Show that for $\alpha_j = j/N$, $j = 0, 1, \ldots, N$, $\varphi(z)$ is a polynomial of degree N and therefore, by (ii) and (iii), $\alpha_j = j/N, j = 0, 1, \ldots, N$, are the eigenvalues of $\mathbf{p}'$ and, therefore, of $\mathbf{p}$.
   (vi) Show that the eigenvector $\mathbf{x}^{(j)} = (x_0^{(j)}, \ldots, x_N^{(j)})'$ corresponding to $\alpha_j = j/N$ is given with $x_0^{(j)} = 1$, by $x_k^{(j)} =$ coefficient of z^k in $(1 - z)^{N-j}(1 + z)^j$.
   (vii) Write $\mathbf{B}$ for the matrix with columns $\mathbf{x}^{(0)}, \ldots, \mathbf{x}^{(N)}$. Then,

$$\mathbf{p}^n = (\mathbf{B}')^{-1}\,\text{diag}(\alpha_0^n, \ldots, \alpha_N^n)\mathbf{B}'$$

   where

$$(\mathbf{B}')^{-1} = \mathbf{B}\,\text{diag}\left(\frac{\pi_0}{\|\mathbf{x}^{(0)}\|_\pi^2}, \ldots, \frac{\pi_N}{\|\mathbf{x}^{(N)}\|_\pi^2}\right)$$

   and the (invariant) distribution $\boldsymbol{\pi}$ is binomial with $p = \frac{1}{2}$, N; see Exercise 1.2(i). [*Hint*: Use the definitions of eigenvectors and matrix multiplication to write $(\mathbf{p}')^n\mathbf{B} = \mathbf{B}\,\mathbf{diag}(\alpha_0^n, \ldots, \alpha_N^n)$. Multiply both sides by $\mathbf{B}^{-1}$ and take the transpose to get the spectral representation of $\mathbf{p}^n$. Note that the columns of $(\mathbf{B}')^{-1}$ are the eigenvectors of $\mathbf{p}$ since $\mathbf{p}(\mathbf{B}')^{-1} = (\mathbf{B}')^{-1}\,\mathbf{diag}(\alpha_0, \ldots, \alpha_N)$. To compute $(\mathbf{B}')^{-1}$ use orthogonality of the eigenvectors with respect to $(\cdot,\cdot)_\pi$ to first write $\mathbf{B}'\,\mathbf{diag}(\pi_0, \ldots, \pi_N)\mathbf{B} = \mathbf{diag}(\|\mathbf{x}^{(0)}\|_\pi^2, \ldots, \|\mathbf{x}^{(N)}\|_\pi^2)$. The formula for $(\mathbf{B}')^{-1}$ follows.]

4. (*Relaxation and Correlation Length*) Let $\mathbf{p}$ be the transition matrix for a finite state stationary birth–death chain $\{X_n\}$ on $S = \{0, 1, \ldots, N\}$ with reflecting boundaries at 0 and N. Show that

$$\sup_{f,g} \{\mathrm{Corr}_\pi(f(X_n), g(X_0))\} = e^{-\lambda_1 n},$$

where

$$\mathrm{Corr}_\pi(f(X_n), g(X_0)) = \frac{E_\pi\{[f(X_n) - Ef(X_n)][g(X_0) - Eg(X_0)]\}}{(\mathrm{Var}_\pi f(X_n))^{1/2}(\mathrm{Var}_\pi g(X_0))^{1/2}},$$

λ_1 is the largest nontrivial (i.e., $\neq 1$) eigenvalue of $\mathbf{p}$, and the supremum is over real-valued functions f and g on S. The parameter $\tau = 1/\lambda_1$ is called the *relaxation time* or *correlation length* in applications. [*Hint*: Use the *self-adjointness* of $\mathbf{p}$ (i.e., *time-reversibility*) with respect to $(\cdot,\cdot)_\pi$ to obtain an orthonormal basis $\{\varphi_n\}$ of eigenvectors of $\mathbf{p}$. Check equality in the case $f = g = \varphi_1$ is the eigenvector corresponding to λ_1. Restrict attention to f and g such that $E_\pi f = E_\pi g = 0$ and $\|f\|_\pi = \|g\|_\pi = 1$, and expand as

$$f = \sum_n (f, \varphi_n)_\pi \varphi_n, \qquad g = \sum_n (g, \varphi_n)_\pi \varphi_n.$$

Use the inequality $ab \leqslant (a^2 + b^2)/2$ to show $|\mathrm{Corr}_\pi(f(X_n), g(X_0))| \leqslant e^{-\lambda_1 n}$.]

5. (i) (*Simple Random Walk with Periodic Boundary*) States $0, 1, 2, \ldots, N-1$ are arranged clockwise in a circle. A transition occurs either one unit clockwise or one unit counterclockwise with respective probabilities p and $q = 1 - p$. Show that

$$p_{jk}^{(n)} = \frac{1}{N} \sum_{r=0}^{N-1} \theta^{r(j-k)} \{p\theta^r + q\theta^{(N-1)r}\}^n$$

where $\theta = e^{(2\pi i)/N}$ is an *Nth root of unity* (all Nth roots of unity being $1, \theta, \theta^2, \ldots, \theta^{N-1}$).

(*ii) (*General Random Walk with Periodic Boundary*) Suppose that for the arrangement in (i), a transition k units clockwise (equivalently, $N - k$ units counterclockwise) occurs with probability p_k, $k = 0, 1, \ldots, N-1$. Show that

$$p_{jk}^{(n)} = \frac{1}{N} \sum_{r=0}^{N-1} \theta^{r(j-k)} \left(\sum_{s=0}^{N-1} p_s \theta^{rs} \right)^n$$

where $\theta = e^{(2\pi i)/N}$ is an Nth root of unity.

THEORETICAL COMPLEMENTS

Theoretical Complements to Section III.5

1. Conservative dynamical systems consisting of one or many degrees of freedom (components, particles) are often represented by one-parameter families of

transformations $\{\mathbf{T}_t\}$ acting on points $\mathbf{x} = (x_1, \ldots, x_n)$ of $\mathbb{R}^n$ (position–momentum *phase space*). The transformations are obtained from the differential equations (*Newton's Law*) that govern the evolution started at arbitrary states $\mathbf{x} \in \mathbb{R}^n$; i.e., $\mathbf{T}_t\mathbf{x}$ is the state at time t when initially the state is $\mathbf{T}_0\mathbf{x} = \mathbf{x}$, $\mathbf{x} \in \mathbb{R}^n$. The mathematical theory described below applies to phenomena where the physics provide a law of evolution of the form

$$\frac{\partial \mathbf{T}_t\mathbf{x}}{\partial t} = \mathbf{f}(\mathbf{T}_t\mathbf{x}), \qquad t > 0, \qquad T_0\mathbf{x} = \mathbf{x}, \tag{T.5.1}$$

such that $\mathbf{f} = (f_1, \ldots, f_n) : \mathbb{R}^n \to \mathbb{R}^n$ uniquely determines the solution at *all* times $t > 0$ for each initial state $\mathbf{x}$ by (T.5.1).

Example 1. Consider a mass m on a spring initially displaced to a location x_1 (relative to rest position at $x_1 = 0$) and with initial momentum x_2. Then Hooke's law provides the force (acting along the gradient of the potential curve $U(x_1) = \frac{1}{2}kx_1^2$), according to which $\mathbf{f}(\mathbf{x}) \equiv (f_1(\mathbf{x}), f_2(\mathbf{x})) = ((1/m)x_2, -kx_1)$, where $k > 0$ is the spring constant. In particular, it follows that $\mathbf{T}_t\mathbf{x} = \mathbf{x}\mathbf{A}(t)$, where

$$\mathbf{A}(t) = \begin{bmatrix} \cos(\gamma t) & -m\gamma \sin(\gamma t) \\ \dfrac{1}{m\gamma}\sin(\gamma t) & \cos(\gamma t) \end{bmatrix}, \qquad t \geqslant 0, \qquad \text{where } \gamma = \sqrt{\frac{k}{m}} > 0.$$

Notice that areas (2-dimensional phase-space volume) are preserved under $\mathbf{T}_t$ since $\det \mathbf{A}(t) = 1$. The motion is obviously periodic in this case.

Example 2. A standard model in statistical mechanics is that of a system having k (generalized) position coordinates $q_1, \ldots, q_k$ and k corresponding (generalized) momentum coordinates $p_1, \ldots, p_k$. The law of evolution is usually cast in Hamiltonian form:

$$\frac{dq_i}{dt} = \frac{\partial H}{\partial p_i}, \qquad \frac{dp_i}{dt} = -\frac{\partial H}{\partial q_i}, \qquad i = 1, \ldots, k, \tag{T.5.2}$$

where $H \equiv H(q_1, \ldots, q_k, p_1, \ldots, p_k)$ is the Hamiltonian function representing the *total energy* (kinetic energy plus potential energy) of the system. Example 1 is of this form with $k = 1$, $H(q, p) = p^2/2m + \frac{1}{2}kq^2$. Writing $n = 2k$, $x_1 = q_1, \ldots, x_k = q_k$, $x_{k+1} = p_1, \ldots, x_{2k} = p_k$, this is also of the form (T.5.1) with

$$\mathbf{f}(\mathbf{x}) = (f_1(\mathbf{x}), \ldots, f_{2k}(\mathbf{x})) = \left(\frac{\partial H}{\partial x_{k+1}}, \ldots, \frac{\partial H}{\partial x_{2k}}, -\frac{\partial H}{\partial x_1}, \ldots, -\frac{\partial H}{\partial x_k}\right). \tag{T.5.3}$$

Observe that for H sufficiently smooth, the flow in phase space is generally *incompressible*. That is,

$$\operatorname{div} f(\mathbf{x}) \equiv \operatorname{trace}\left(\frac{\partial f_i}{\partial x_j}\right) = \sum_{i=1}^{2k} \frac{\partial f_i}{\partial x_i} = \sum_{i=1}^{k} \left[\frac{\partial}{\partial x_i}\left(\frac{\partial H}{\partial x_{k+i}}\right) + \frac{\partial}{\partial x_{k+i}}\left(-\frac{\partial H}{\partial x_i}\right)\right]$$

$$= 0 \qquad \text{for all } \mathbf{x}. \tag{T.5.4}$$

Liouville first noticed the important fact that incompressibility gives the volume preserving property of the flow in phase space.

Liouville Theorem T.5.1. Suppose that $\mathbf{f}(\mathbf{x})$ in (T.5.1) is such that $\operatorname{div} \mathbf{f}(\mathbf{x}) = 0$ for all $\mathbf{x}$. Then for each bounded (measurable) set $D \subset \mathbb{R}^n$, $|T_t D| = |D|$ for all $t \geqslant 0$, where $|\cdot|$ denotes n-dimensional volume (Lebesgue measure). □

Proof. By the uniqueness condition stated at the outset we have $\mathbf{T}_{t+h} = \mathbf{T}_t \mathbf{T}_h$ for all $t, h \geqslant 0$. So, by the change of variable formula,

$$|\mathbf{T}_{t+h} D| = \int_{\mathbf{T}_t D} \det\left(\frac{\partial \mathbf{T}_h \mathbf{x}}{\partial \mathbf{x}}\right) d\mathbf{x}.$$

To calculate the Jacobian, first note from (T.5.1) that

$$\left(\frac{\partial \mathbf{T}_h \mathbf{x}}{\partial \mathbf{x}}\right) = I + \frac{\partial \mathbf{f}}{\partial \mathbf{x}} h + \mathrm{O}(h^2) \qquad \text{as } h \to 0.$$

But, expanding the determinant and collecting terms, one sees for any matrix M that

$$\det(I + hM) = 1 + h \operatorname{trace}(M) + \mathrm{O}(h^2) \qquad \text{as } h \to 0.$$

Thus, since $\operatorname{trace}(\partial \mathbf{f}/\partial \mathbf{x}) = \operatorname{div} \mathbf{f}(\mathbf{x}) = 0$,

$$\det\left(\frac{\partial \mathbf{T}_h \mathbf{x}}{\partial \mathbf{x}}\right) = 1 + \mathrm{O}(h^2) \qquad \text{as } h \to 0.$$

It follows that for each $t \geqslant 0$

$$|\mathbf{T}_{t+h} D| = |\mathbf{T}_t D| + \mathrm{O}(h^2) \qquad \text{as } h \to 0,$$

or

$$\frac{d}{dt}|T_t D| = 0 \qquad \text{and} \qquad |T_0 D| = |D|,$$

i.e., $t \to |T_t D|$ is constant with constant value $|D|$. ■

2. Liouville's theorem becomes especially interesting when considered along side the following theorem of Poincaré.

Poincaré's Recurrence Theorem T.5.2. Let T be any volume preserving continuous one-to-one mapping of a bounded (measurable) region $D \subset \mathbb{R}^n$ onto itself. Then for each neighborhood Δ of any point $\mathbf{x}$ in D and every n, however large there is a subset B of Δ having positive volume such that for all $\mathbf{y} \in B$ $T^r \mathbf{y} \in \Delta$ for some $r \geqslant n$. □

Proof. Consider $\Delta, T^{-n}\Delta, T^{-2n}\Delta, \ldots$. Then there are distinct times i, j such that $|T^{-in}\Delta \cap T^{-jn}\Delta| \neq 0$; for otherwise

$$|D| \geq \left| \bigcup_{j=0}^{\infty} T^{-jn}\Delta \right| = \sum_{j=0}^{\infty} |T^{-jn}\Delta| = \sum_{j=0}^{\infty} |\Delta| = +\infty.$$

It follows that

$$|\Delta \cap T^{-n|j-i|}\Delta| \neq 0.$$

Take $B = \Delta \cap T^{-n|j-i|}\Delta$, $r = n|j-i|$. ■

3. S. Chandrasekhar (1943), "Stochastic Problems in Physics and Astronomy", *Reviews in Modern Physics*, **15**, 1–89, contains a discussion of Boltzmann and Zermello's classical analysis together with other applications of Markov chains to physics. More complete references on alternative derivations as well as the computation of the mean recurrence time of a state can be found in M. Kac (1947), "Random Walk and the Theory of Brownian Motion", *American Mathematical Monthly*, **54**, 369–391; also see E. Waymire (1982), "Mixing and Cooling from a Probabilistic Point of View", *SIAM Review*, **24**, 73–75.

CHAPTER IV

Continuous-Parameter Markov Chains

1 INTRODUCTION TO CONTINUOUS-TIME MARKOV CHAINS

Suppose that $\{X_t: t \geqslant 0\}$ is a *continuous-parameter* stochastic process with a finite or denumerably infinite state space S. Just as in the discrete-parameter case, the Markov property here also refers to the property that the conditional distribution of the future, given past and present states of the process, does not depend on the past. In terms of finite-dimensional events, the Markov property requires that for arbitrary time points $0 \leqslant s_0 < s_1 < \cdots < s_k < s < t < t_1 < \cdots < t_n$ and states $i_0, \ldots, i_k, i, j, j_1, \ldots, j_n$ in S

$$P(X_t = j, X_{t_1} = j_1, \ldots, X_{t_n} = j_n \mid X_{s_0} = i_0, \ldots, X_{s_k} = i_k, X_s = i)$$
$$= P(X_t = j, X_{t_1} = j_1, \ldots, X_{t_n} = j_n \mid X_s = i). \quad (1.1)$$

In other words, for any sequence of time points $0 \leqslant t_0 < t_1 < \cdots$, the discrete parameter process $Y_0 := X_{t_0}, Y_1 := X_{t_1}, \ldots$ is a Markov chain as described in Chapter II. The conditional probabilities $p_{ij}(s, t) = P(X_t = j \mid X_s = i), 0 \leqslant s < t$, are collectively referred to as the *transition probability law* for the process. In the case $p_{ij}(s, t)$ is a function of $t - s$, the transition law is called *time-homogeneous*, and we write $p_{ij}(s, t) = p_{ij}(t - s)$.

Simple examples of continuous-parameter Markov chains are the *continuous-time random walks*, or *processes with independent increments* on countable state space. Some others are described in the examples below.

Example 1. (*The Poisson Process*). The *Poisson process with intensity function* ρ is a process with state space $S = \{0, 1, 2, \ldots\}$ having independent increments distributed as

$$P(X_t - X_s = j) = \frac{\left(\int_s^t \rho(u)\,du\right)^j}{j!} \exp\left(-\int_s^t \rho(u)\,du\right), \tag{1.2}$$

for $j = 0, 1, 2, \ldots$, $s < t$, where $\rho(u)$, $u \geqslant 0$, is a continuous nonnegative function. Just as with the simple random walk in Chapter II, the Markov property for the Poisson process is a consequence of the *independent increments* property (Exercise 2). Moreover,

$$\begin{aligned} p_{ij}(s, t) &= P(X_t = j \mid X_s = i) \\ &= \frac{P(X_t = j, X_s = i)}{P(X_s = i)} \\ &= \frac{P(X_t - X_s = j - i)P(X_s = i)}{P(X_s = i)} \\ &= \begin{cases} \dfrac{\left(\int_s^t \rho(u)\,du\right)^{j-i}}{(j-i)!} \exp\left(-\int_s^t \rho(u)\,du\right) & \text{for } j \geqslant i \\ 0 & \text{if } j < i. \end{cases} \end{aligned} \tag{1.3}$$

In the case that $\rho(u) \equiv \lambda$ (constant) for each $u \geqslant 0$,

$$p_{ij}(s, t) = \begin{cases} \dfrac{[\lambda(t-s)]^{j-i}}{(j-i)!} e^{-\lambda(t-s)}, & i \leqslant j \\ 0, & j > i \end{cases} \tag{1.4}$$

is a function $p_{ij}(t - s)$ of s and t through $t - s$; i.e., the transition law is time-homogeneous. In this case the process is referred to as the *Poisson process with parameter* λ.

Example 2. (*The Compound Poisson Process*). Let $\{N_t\}$ be a Poisson process with parameter $\lambda > 0$ starting at 0, and let $Y_1, Y_2, \ldots$ be i.i.d. integer-valued random variables, independent of the process $\{N_t\}$, having a common probability mass function f. The process $\{X_t\}$ is defined by

$$X_t = \sum_{n=0}^{N_t} Y_n, \tag{1.5}$$

where Y_0 is independent of the process $\{N_t\}$ and of $Y_1, Y_2, \ldots$. The stochastic process $\{X_t\}$ is called a *Compound Poisson Process*. The process has independent increments and is therefore Markovian (Exercise 4). As a consequence of the independence of increments,

$$p_{ij}(s, t) = P(X_t = j \mid X_s = i) = P(X_t - X_s = j - i \mid X_s = i)$$
$$= P(X_t - X_s = j - i). \tag{1.6}$$

Therefore,

$$p_{ij}(s, t) = E\{P(X_t - X_s = j - i \mid N_t - N_s)\}$$
$$= \sum_{k=0}^{\infty} f^{*k}(j - i) \frac{[\lambda(t - s)]^k}{k!} e^{-\lambda(t-s)}, \tag{1.7}$$

where f^{*k} is the k-fold convolution of f with $f^{*0}(0) = 1$. In particular, $\{X_t\}$ has a time-homogeneous transition law. The *continuous-time simple random walk* is defined as the special case with $f(+1) = P(Y_n = +1) = p$, $f(-1) = P(Y_n = -1) = q$, $0 \leqslant p, q \leqslant 1$, $p + q = 1$.

Another popular continuous-parameter Markov chain with *nonhomogeneous transition law*, the *Pólya process* is provided in Exercise 6 of the next section as a limiting approximation to the (nonhomogeneous) discrete-parameter partial sum process in Example 3.8, Chapter II. However, unless stated otherwise, we shall generally restrict our attention to the study of Markov chains with a *time-homogeneous transition law*.

2 KOLMOGOROV'S BACKWARD AND FORWARD EQUATIONS

We continue to denote the finite or denumerable state space by S. To construct a Markov process in discrete time (i.e., to find all possible joint distributions) it was enough to specify a one-step transition matrix $\mathbf{p}$ along with an initial distribution $\boldsymbol{\pi}$. In the continuous-parameter case, on the other hand, the specification of a single transition matrix $\mathbf{p}(t_0) = ((p_{ij}(t_0)))$, where $p_{ij}(t_0)$ gives the probability that the process will be in state j at time t_0 if it is initially at state i, together with an initial distribution $\boldsymbol{\pi}$, is not adequate. For a single time point t_0, $\mathbf{p}(t_0)$ together with $\boldsymbol{\pi}$ will merely specify joint distributions of $X_0, X_{t_0}, X_{2t_0}, \ldots, X_{nt_0}, \ldots$; for example,

$$P_\pi(X_0 = i_0, X_{t_0} = i_1, \ldots, X_{nt_0} = i_n) = \pi_{i_0} p_{i_0 i_1}(t_0) p_{i_1 i_2}(t_0) \cdots p_{i_{n-1} i_n}(t_0). \tag{2.1}$$

Here $\mathbf{p}(t_0)$ takes the place of $\mathbf{p}$, and t_0 is treated as the unit of time. Events that depend on the process at time points that are not multiples of t_0 are excluded. Likewise, specifying transition matrices $\mathbf{p}(t_0), \mathbf{p}(t_1), \ldots, \mathbf{p}(t_k)$ for an arbitrary finite set of time points $t_0, t_1, \ldots, t_k$, will not be enough.

On the other hand, if one specifies all transition matrices $\mathbf{p}(t)$ of a time-homogeneous Markov chain for values of t in a time interval $0 < t \leqslant t_0$ for some $t_0 > 0$, then, regardless of how small $t_0 > 0$ may be, all other transition probabilities may be constructed from these. To understand this basic fact, first assume transition matrices $\mathbf{p}(t)$ to be given for *all* $t > 0$, together with an initial distribution $\boldsymbol{\pi}$. Then for any finite set of time points $0 < t_1 < t_2 < \cdots < t_n$, the joint distribution of $X_0, X_{t_1}, \ldots, X_{t_n}$ is given by

$$P_\pi(X_0 = i_0, X_{t_1} = i_1, X_{t_2} = i_2, \ldots, X_{t_{n-1}} = i_{n-1}, X_{t_n} = i_n)$$
$$= \pi_{i_0} p_{i_0 i_1}(t_1) p_{i_1 i_2}(t_2 - t_1) \cdots p_{i_{n-1} i_n}(t_n - t_{n-1}). \quad (2.2)$$

Specializing to $\pi_i = 1$, $t_1 = t$, $t_2 = t + s$, it follows that

$$P_i(X_t = j, X_{t+s} = k) = p_{ij}(t) p_{jk}(s),$$

and

$$P_i(X_{t+s} = k) = p_{ik}(t + s). \quad (2.3)$$

But $\{X_{t+s} = k\}$ is the countable union $\bigcup_{j \in S} \{X_t = j, X_{t+s} = k\}$ of pairwise disjoint events. Therefore,

$$P_i(X_{t+s} = k) = \sum_{j \in S} P_i(X_t = j, X_{t+s} = k). \quad (2.4)$$

The relations (2.3) and (2.4) provide the *Chapman–Kolmogorov equations*, namely,

$$p_{ik}(t + s) = \sum_{j \in S} p_{ij}(t) p_{jk}(s) \qquad (i, k \in S; s > 0, t > 0), \quad (2.5)$$

which may also be expressed in matrix notation by the following so-called *semigroup property*

$$\mathbf{p}(t + s) = \mathbf{p}(t)\mathbf{p}(s) \qquad (s > 0, t > 0). \quad (2.6)$$

Therefore, the *transition matrices* $\mathbf{p}(t)$ *cannot be chosen arbitrarily*. They must be so chosen as to satisfy the Chapman–Kolmogorov equations.

It turns out that (2.5) is the only restriction required for *consistency* in the sense of prescribing finite-dimensional distributions as in Section 6, Chapter I. To see this, take an arbitrary initial distribution $\boldsymbol{\pi}$ and time points $0 < t_1 < t_2 < t_3$. For arbitrary states i_0, i_1, i_2, i_3, one has from (2.2) that

$$P_\pi(X_0 = i_0, X_{t_1} = i_1, X_{t_2} = i_2, X_{t_3} = i_3)$$
$$= \pi_{i_0} p_{i_0 i_1}(t_1) p_{i_1 i_2}(t_2 - t_1) p_{i_2 i_3}(t_3 - t_2), \quad (2.7)$$

as well as

$$P_\pi(X_0 = i_0, X_{t_1} = i_1, X_{t_3} = i_3) = \pi_{i_0} p_{i_0 i_1}(t_1) p_{i_1 i_3}(t_3 - t_1). \quad (2.8)$$

But consistency requires that (2.8) be obtained from (2.7) by summing over i_2. This sum is

$$\sum_{i_2} \pi_{i_0} p_{i_0 i_1}(t_1) p_{i_1 i_2}(t_2 - t_1) p_{i_2 i_3}(t_3 - t_2)$$

$$= \pi_{i_0} p_{i_0 i_1}(t_1) \sum_{i_2} p_{i_1 i_2}(t_2 - t_1) p_{i_2 i_3}(t_3 - t_2). \quad (2.9)$$

By the Chapman–Kolmogorov equations (2.5), with $t = t_2 - t_1$, $s = t_3 - t_2$, one has

$$\sum_{i_2} p_{i_1 i_2}(t_2 - t_1) p_{i_2 i_3}(t_3 - t_2) = p_{i_1 i_3}(t_3 - t_1),$$

showing that the right sides of (2.8) and (2.9) are indeed equal. Thus, if (2.5) holds, then (2.2) defines joint distributions *consistently*, i.e., the joint distribution at any finite set of points as specified by (2.2) equals the probability obtained by summing successive probabilities of a joint distribution (like (2.2)) involving a larger set of time points, over states belonging to the additional time points.

Suppose now that $\mathbf{p}(t)$ is given for $0 < t \leqslant t_0$, for some $t_0 > 0$, and the transition probability matrices satisfy (2.6). Since any $t > t_0$ may be expressed uniquely as $t = rt_0 + s$, where r is a positive integer and $0 \leqslant s < t_0$, by (2.6) we have

$$\mathbf{p}(t) = \mathbf{p}(rt_0 + s) = \mathbf{p}(t_0)\mathbf{p}((r-1)t_0 + s)$$

$$= \mathbf{p}^2(t_0)\mathbf{p}((r-2)t_0 + s) = \cdots = \mathbf{p}^r(t_0)\mathbf{p}(s).$$

Thus, it is enough to specify $\mathbf{p}(t)$ on any interval $0 < t \leqslant t_0$, however small $t_0 > 0$ may be. In fact, we will see that under certain further conditions $\mathbf{p}(t)$ is determined by its values for infinitesimal times; i.e., in the limit as $t_0 \to 0$.

From now on we shall assume that

$$\lim_{t \downarrow 0} p_{ij}(t) = \delta_{ij}, \quad (2.10)$$

where *Kronecker's delta* is given by $\delta_{ij} = 1$ if $i = j$, $\delta_{ij} = 0$ if $i \neq j$. This condition is very reasonable in most circumstances. Namely, it requires that with probability 1, the process spends a positive (but variable) amount of time in the initial state i before moving to a different state j. The relations (2.10) are also expressed as

$$\lim_{t \downarrow 0} \mathbf{p}(t) = \mathbf{I}, \quad (2.11)$$

where $\mathbf{I}$ is the *identity matrix*, with 1's along the diagonal and 0's elsewhere.

We shall also write,

$$\mathbf{p}(0) = \mathbf{I}. \tag{2.12}$$

Then (2.11) expresses the fact that $\mathbf{p}(t)$, $0 \leqslant t < \infty$, is (componentwise) continuous at $t = 0$ as a function of t. It may actually be shown that owing to the rich additional structure reflected in (2.6), continuity implies that $\mathbf{p}(t)$ is in fact differentiable in t, i.e., *$p'_{ij}(t) = d(p_{ij}(t))/dt$ exist for all pairs (i, j) of states, and all $t \geqslant 0$*. At $t = 0$, of course, "derivative" refers to the right-hand derivative. In particular, the parameters q_{ij} given by

$$q_{ij} = \lim_{t \downarrow 0} \frac{p_{ij}(t) - p_{ij}(0)}{t} = \lim_{t \downarrow 0} \frac{p_{ij}(t) - \delta_{ij}}{t}, \tag{2.13}$$

are well defined. Instead of *proving* differentiability from continuity for transition probabilities, which is nontrivial, *we shall assume from now on that $p_{ij}(t)$ has a finite derivative for all* (i, j) as part of the required structure. Also, we shall write

$$\mathbf{Q} = ((q_{ij})), \tag{2.14}$$

for q_{ij} defined in (2.13). The quantities q_{ij} are referred to as the *infinitesimal transition rates* and $\mathbf{Q}$ the (*formal*) *infinitesimal generator*. Note that (2.13) may be expressed equivalently as

$$p_{ij}(\Delta t) = \delta_{ij} + q_{ij}\Delta t + o(\Delta t) \qquad \text{as } \Delta t \to 0. \tag{2.15}$$

Suppose for the time being that S *is finite*. Since the derivative of a finite sum equals the sum of the derivatives, it follows by differentiating both sides of (2.5) with respect to t and setting $t = 0$ that

$$p'_{ik}(s) = \sum_{j \in S} p'_{ij}(0) p_{jk}(s) = \sum_{j \in S} q_{ij} p_{jk}(s), \qquad i, k \in S, \tag{2.16}$$

or, in matrix notation after relabeling s as t in (2.16) for notational convenience,

$$\mathbf{p}'(t) = \mathbf{Q}\mathbf{p}(t) \qquad (t \geqslant 0). \tag{2.17}$$

The system of equations (2.16) or (2.17) is called *Kolmogorov's backward equations*.

One may also differentiate both sides of (2.5) with respect to s and then set $s = 0$ to get *Kolmogorov's forward equations* for a finite state space S,

$$p'_{ik}(t) = \sum_{j \in S} p_{ij}(t) q_{jk}, \qquad i, k \in S, \tag{2.18}$$

or, in matrix notation,

$$\mathbf{p}'(t) = \mathbf{p}(t)\mathbf{Q}. \tag{2.19}$$

Since $p_{ij}(t)$ are transition probabilities,

$$\sum_{j\in S} p_{ij}(t) = 1 \qquad \text{for all } i \in S. \tag{2.20}$$

Differentiating (2.20) term by term and setting $t = 0$,

$$\sum_{j\in S} q_{ij} = 0. \tag{2.21}$$

Note that

$$\begin{aligned} q_{ij} &:= p'_{ij}(0) \geqslant 0 \qquad \text{for } i \neq j, \\ q_{ii} &:= p'_{ii}(0) \leqslant 0, \end{aligned} \tag{2.22}$$

in view of the fact $p_{ij}(t) \geqslant 0 = p_{ij}(0)$ for $i \neq j$, and $p_{ii}(t) \leqslant 1 = p_{ii}(0)$.

In the general case of a *countable* state space S, the term-by-term differentiation used to derive Kolmogorov's equations may not always be justified. Conditions are given in the next two sections for the validity of these equations for transition probabilities on denumerable state spaces. However, regardless of whether or not the differential equations are valid for given transition probabilities $\mathbf{p}(t)$, we shall refer to the equations in general as Kolmogorov's backward and forward equations, respectively.

Example 1. (*Compound Poisson*). From (1.7),

$$\begin{aligned} p_{ij}(t) &= \sum_{k=0}^{\infty} f^{*k}(j-i)\frac{\lambda^k t^k}{k!}e^{-\lambda t}, & t \geqslant 0, \\ &= \delta_{ij}e^{-\lambda t} + f(j-i)\lambda t e^{-\lambda t} + o(t), & \text{as } t \downarrow 0. \end{aligned} \tag{2.23}$$

Therefore,

$$q_{ij} = p'_{ij}(0) = \lambda f(j-i), \qquad i \neq j,$$

and

$$q_{ii} = p'_{ii}(0) = -\lambda(1 - f(0)). \tag{2.24}$$

3 SOLUTIONS TO KOLMOGOROV'S EQUATIONS IN EXPONENTIAL FORM

We saw in Section 2 that if $\mathbf{p}(t)$ is a transition probability law on a finite state space S with $\mathbf{Q} = \mathbf{p}'(0)$, then $\mathbf{p}$ satisfies Kolmogorov's backward equation

$$p'_{ij}(t) = \sum_k q_{ik} p_{kj}(t), \qquad i, j \in S, \quad t \geqslant 0, \tag{3.1}$$

and Kolmogorov's forward equation

$$p'_{ij}(t) = \sum_k p_{ik}(t) q_{kj}, \qquad i, j \in S, \quad t \geqslant 0, \tag{3.2}$$

where $\mathbf{Q} = ((q_{ij}))$ satisfies the conditions in (2.21) and (2.22).

The important problem is, however, to construct transition probabilities $\mathbf{p}(t)$ having prescribed infinitesimal transition rates $\mathbf{Q} = ((q_{ij}))$ satisfying

$$\begin{gathered} q_{ij} \geqslant 0 \quad \text{for } i \neq j, \quad q_{ii} \leqslant 0, \\ -q_{ii} = \sum_{j \neq i} q_{ij}. \end{gathered} \tag{3.3}$$

In the case that S is finite it is known from the theory of ordinary differential equations that, subject to the initial condition $\mathbf{p}(0) = \mathbf{I}$, the unique solution to (3.1) is given by

$$\mathbf{p}(t) = e^{t\mathbf{Q}}, \qquad t \geqslant 0, \tag{3.4}$$

where the matrix $e^{t\mathbf{Q}}$ is defined by

$$e^{t\mathbf{Q}} = \sum_{n=0}^{\infty} \frac{(t\mathbf{Q})^n}{n!} = \mathbf{I} + \sum_{n=1}^{\infty} \frac{t^n}{n!} \mathbf{Q}^n. \tag{3.5}$$

Example 1. Consider the case $S = \{0, 1\}$ for a general two-state Markov chain with rates

$$-q_{00} = q_{01} = \beta, \qquad -q_{11} = q_{10} = \delta. \tag{3.6}$$

Then, observing that $\mathbf{Q}^2 = -(\beta + \delta)\mathbf{Q}$ and iterating, we have

$$\mathbf{Q}^n = (-1)^{n-1}(\beta + \delta)^{n-1}\mathbf{Q}, \qquad \text{for } n = 1, 2, \ldots. \tag{3.7}$$

Therefore,

$$\begin{aligned} \mathbf{p}(t) = e^{t\mathbf{Q}} &= \mathbf{I} - \frac{1}{\beta + \delta}(e^{-t(\beta+\delta)} - 1)\mathbf{Q} \\ &= \frac{1}{\beta + \delta}\begin{bmatrix} \delta + \beta e^{-(\beta+\delta)t} & \beta - \beta e^{-(\beta+\delta)t} \\ \delta - \delta e^{-(\beta+\delta)t} & \beta + \delta e^{-(\beta+\delta)t} \end{bmatrix}. \end{aligned} \tag{3.8}$$

It is also simple, however, to solve the (forward) equations directly in this case (Exercise 3).

In the case that S is countably infinite, results analogous to those for the finite case can be obtained under the following fairly restrictive condition,

$$\Lambda := \sup_{i,j} |q_{ij}| < \infty. \tag{3.9}$$

In view of (3.3) the condition (3.9) is equivalent to

$$\sup_{i} |q_{ii}| < \infty. \tag{3.10}$$

The solution $\mathbf{p}(t)$ is still given by (3.4), i.e., for each $j \in S$,

$$p_{ij}(t) = \delta_{ij} + tq_{ij} + \frac{t^2}{2!} q_{ij}^{(2)} + \cdots + \frac{t^n}{n!} q_{ij}^{(n)} + \cdots \tag{3.11}$$

for all $i \in S$. For, by (3.9) and induction (Exercise 1), we have

$$|q_{ij}^{(n)}| \leqslant \Lambda(2\Lambda)^{n-1}, \qquad \text{for } i, j \in S, \quad n \geqslant 1, \tag{3.12}$$

so that the series on the right in (3.11) converges to a function $r_{ij}(t)$, say, absolutely for all t. By term-by-term differentiation of this series for $r_{ij}(t)$, which is an analytic function of t, one verifies that $r_{ij}(t)$ satisfies the Kolmogorov backward equation and the correct initial condition. Uniqueness under (3.9) follows by the same estimates typically used in the finite case (Exercise 2).

To verify the Chapman–Kolmogorov equations (2.6), note that

$$\begin{aligned}
\mathbf{p}(t+s) = e^{(t+s)\mathbf{Q}} &= \mathbf{I} + (t+s)\mathbf{Q} + \frac{(t+s)^2}{2!}\mathbf{Q}^2 + \cdots + \frac{(t+s)^m}{m!}\mathbf{Q} + \cdots \\
&= \left[\mathbf{I} + t\mathbf{Q} + \frac{t^2}{2!}\mathbf{Q}^2 + \cdots + \frac{t^m}{m!}\mathbf{Q}^m + \cdots\right] \\
&\quad \times \left[\mathbf{I} + s\mathbf{Q} + \frac{s^2}{2!}\mathbf{Q}^2 + \cdots + \frac{s^n}{n!}\mathbf{Q}^n + \cdots\right] \\
&= e^{t\mathbf{Q}} e^{s\mathbf{Q}} = \mathbf{p}(t)\mathbf{p}(s).
\end{aligned} \tag{3.13}$$

The third equality is a consequence of the identity

$$\sum_{m+n=r} \frac{t^m}{m!}\mathbf{Q}^m \frac{s^n}{n!}\mathbf{Q}^n = \left(\sum_{m+n=r} \frac{t^m s^n}{m!\,n!}\right)\mathbf{Q}^r = \frac{(t+s)^r}{r!}\mathbf{Q}^r. \tag{3.14}$$

Also, the functions $H_i(t) := \sum_{j \in S} p_{ij}(t)$ satisfy

$$H_i'(t) = \sum_{j \in S} p_{ij}'(t) = \sum_{j \in S} \sum_{k \in S} q_{ik} p_{kj}(t) = \sum_{k \in S} q_{ik} H_k(t) \qquad (t \geqslant 0),$$

with initial conditions $H_i(0) = \sum_{j \in S} \delta_{ij} = 1$ $(i \in S)$. Since $H_i(t) := 1$ (for all $t \geqslant 0$, all $i \in S$) clearly satisfy these equations, one has $H_i(t) = 1$ for all t by *uniqueness* of such solutions. Thus, the solutions (3.11) have been shown to satisfy all conditions for being transition probabilities except for nonnegativity (Exercise 5). Nonnegativity will also follow as a consequence of a more general method of construction of solutions given in the next section. When it applies, the exponential form (3.4) (equivalently, (3.11)) is especially suitable for calculations of transition probabilities by spectral methods as will be seen in Section 9.

Example 2. (*Poisson Process*). The Poisson process with parameter $\lambda > 0$ was introduced in Example 1.1. Alternatively, the process may be regarded as a Markov process on the state space $S = \{0, 1, 2, \ldots\}$ with prescribed infinitesimal transition rates of the form

$$\begin{aligned} q_{ij} &= \lambda, && \text{for } j = i + 1, \quad i = 0, 1, 2, \ldots \\ q_{ii} &= -\lambda, && i = 0, 1, 2, \ldots \\ q_{ij} &= 0, && \text{otherwise}. \end{aligned} \tag{3.15}$$

By induction it will follow that

$$q_{ij}^{(n)} = \begin{cases} \dbinom{n}{j-i}(-1)^{n+j-i}\lambda^n, & \text{if } 0 \leqslant j - i \leqslant n \\ 0, & \text{otherwise}. \end{cases} \tag{3.16}$$

Therefore, the exponential formula (3.4) [(3.11)] gives for $j \geqslant i$, $t > 0$,

$$\begin{aligned} p_{ij}(t) &= \delta_{ij} + tq_{ij} + \frac{t^2}{2!} q_{ij}^{(2)} + \cdots \\ &= \sum_{k=0}^{\infty} \frac{t^{j-i+k}}{(j-i+k)!} q_{ij}^{(j-i+k)} = \sum_{k=0}^{\infty} \frac{t^{j-i+k}}{(j-i+k)!} \binom{j-i+k}{j-i} (-1)^k \lambda^{j-i+k} \\ &= \frac{(\lambda t)^{j-i}}{(j-i)!} \sum_{k=0}^{\infty} \frac{(-\lambda t)^k}{k!} = \frac{(\lambda t)^{j-i}}{(j-i)!} e^{-\lambda t}. \end{aligned} \tag{3.17}$$

Likewise,

$$p_{ij}(t) = 0 \qquad \text{for } j < i. \tag{3.18}$$

Thus, the transition probabilities coincide with those given in Example 1.

4 SOLUTIONS TO KOLMOGOROV'S EQUATIONS BY SUCCESSIVE APPROXIMATIONS

For the general problem of constructing transition probabilities $((p_{ij}(t)))$ having a prescribed set of infinitesimal transition rates given by $\mathbf{Q} = ((q_{ij}))$, where

$$-q_{ii} = \sum_{j \neq i} q_{ij}, \qquad q_{ij} \geqslant 0 \qquad \text{for } i \neq j, \tag{4.1}$$

the *method of successive approximations* will be used in this section. The main result provides a solution to the backward equations

$$p'_{ij}(t) = \sum_{k} q_{ik}\, p_{kj}(t), \qquad i, j \in S \tag{4.2}$$

under the initial condition

$$p_{ij}(0) = \delta_{ij}, \qquad i, j \in S. \tag{4.3}$$

The precise statement of this result is the following.

Theorem 4.1. Given any $\mathbf{Q}$ satisfying (4.1) there exists a smallest nonnegative solution $\mathbf{p}(t)$ of the backward equations (4.2) satisfying (4.3). This solution satisfies

$$\sum_{j \in S} p_{ij}(t) \leqslant 1 \qquad \text{for all } i \in S, \quad \text{all } t \geqslant 0. \tag{4.4}$$

In case equality holds in (4.4) for all $i \in S$ and $t \geqslant 0$, there does not exist any other nonnegative solution of (4.2) that satisfies (4.3) and (4.4).

Proof. Write $\lambda_i = -q_{ii}$ $(i \in S)$. Multiplying both sides of the backward equations (4.2) by the "integrating factor" $e^{\lambda_i s}$ one obtains

$$e^{\lambda_i s} p'_{ik}(s) = e^{\lambda_i s} \sum_{j \in S} q_{ij}\, p_{jk}(s),$$

or

$$\frac{d}{ds}\left(e^{\lambda_i s} p_{ik}(s)\right) = \lambda_i e^{\lambda_i s} p_{ik}(s) + e^{\lambda_i s} \sum_{j \in S} q_{ij}\, p_{jk}(s) = \sum_{j \neq i} e^{\lambda_i s} q_{ij}\, p_{jk}(s).$$

On integration between 0 and t one has, remembering that $p_{ik}(0) = \delta_{ik}$,

$$e^{\lambda_i t} p_{ik}(t) = \delta_{ik} + \sum_{j \neq i} \int_0^t e^{\lambda_i s} q_{ij}\, p_{jk}(s)\, ds,$$

or

$$p_{ik}(t) = \delta_{ik}e^{-\lambda_i t} + \sum_{j \neq i} \int_0^t e^{-\lambda_i(t-s)} q_{ij} p_{jk}(s)\, ds \qquad (t \geqslant 0;\, i, k \in S). \tag{4.5}$$

Reversing the steps shows that (4.2) together with (4.3) follow from (4.5). Thus (4.2) and (4.3) are equivalent to the *system of integral equations* (4.5). To solve the system (4.5) start with the *first approximation*

$$p_{ik}^{(0)} = \delta_{ik}e^{-\lambda_i t} \qquad (i, k \in S,\, t \geqslant 0) \tag{4.6}$$

and compute *successive approximations*, recursively, by

$$p_{ik}^{(n)}(t) = \delta_{ik}e^{-\lambda_i t} + \sum_{j \neq i} \int_0^t e^{-\lambda_i(t-s)} q_{ij} p_{jk}^{(n-1)}(s)\, ds \qquad (n \geqslant 1). \tag{4.7}$$

Since $q_{ij} \geqslant 0$ for $i \neq j$, it is clear that $p_{ik}^{(1)}(t) \geqslant p_{ik}^{(0)}(t)$. It then follows from (4.7) by induction that $p_{ik}^{(n+1)}(t) \geqslant p_{ik}^{(n)}(t)$ for all $n \geqslant 0$. Thus, $\bar{p}_{ik}(t) = \lim_{n \to \infty} p_{ik}^{(n)}(t)$ exists. Taking limits on both sides of (4.7) yields

$$\bar{p}_{ik}(t) = \delta_{ik}e^{-\lambda_i t} + \sum_{j \neq i} \int_0^t e^{-\lambda_i(t-s)} q_{ij} \bar{p}_{jk}(s)\, ds. \tag{4.8}$$

Hence, $\bar{p}_{ik}(t)$ satisfy (4.5). Also, $\bar{p}_{ik}(t) \geqslant p_{ik}^{(0)}(t) \geqslant 0$. Further, $\sum_{k \in S} p_{ik}^{(0)}(t) \leqslant 1$ for all $t \geqslant 0$ and all i. Assuming, as induction hypothesis, that $\sum_{k \in S} p_{ik}^{(n-1)}(t) \leqslant 1$ for all $t \geqslant 0$ and all i, it follows from (4.7) that

$$\begin{aligned}
\sum_{k \in S} p_{ik}^{(n)}(t) &= e^{-\lambda_i t} + \sum_{j \neq i} \int_0^t e^{-\lambda_i(t-s)} q_{ij} \left(\sum_{k \in S} p_{jk}^{(n-1)}(s) \right) ds \\
&\leqslant e^{-\lambda_i t} + \sum_{j \neq i} \int_0^t e^{-\lambda_i(t-s)} q_{ij}\, ds \\
&= e^{-\lambda_i t} + \lambda_i e^{-\lambda_i t} \int_0^t e^{\lambda_i s}\, ds = e^{-\lambda_i t} + \lambda_i e^{-\lambda_i t} \left[\frac{e^{\lambda_i t} - 1}{\lambda_i} \right] = 1.
\end{aligned}$$

Hence, $\sum_{k \in S} p_{ik}^{(n)}(t) \leqslant 1$ for all n, all $t \geqslant 0$, and the same must be true for

$$\sum_{k \in S} \bar{p}_{ik}(t) = \lim_{n \uparrow \infty} \sum_{k \in S} p_{ik}^{(n)}(t).$$

We now show that $\bar{\mathbf{p}}(t)$ is the smallest nonnegative solution of (4.5). Suppose $\mathbf{p}(t)$ is any other nonnegative solution. Then obviously $p_{ik}(t) \geqslant \delta_{ik}e^{-\lambda_i t} = p_{ik}^{(0)}(t)$ for all i, k, t. Assuming, as induction hypothesis, $p_{ik}(t) \geqslant p_{ik}^{(n-1)}(t)$ for all $i, k \in S$, $t \geqslant 0$, it follows from the fact that $p_{ik}(t)$ satisfies (4.5) that

$$p_{ik}(t) \geqslant \delta_{ik} e^{-\lambda_i t} + \sum_{j \neq i} \int_0^t e^{-\lambda_i(t-s)} q_{ij} p_{jk}^{(n-1)}(s)\, ds = p_{ik}^{(n)}(t).$$

Hence, $p_{ik}(t) \geqslant p_{ik}^{(n)}(t)$ for all $n \geqslant 0$ and, therefore, $p_{ik}(t) \geqslant \bar{p}_{ik}(t)$ for all $i, k \in S$ and all $t \geqslant 0$. The last assertion of the theorem is almost obvious. For if equality holds in (4.4) for $\bar{\mathbf{p}}(t)$, for all i and all $t \geqslant 0$, and $\mathbf{p}(t)$ is another transition probability matrix, then, by the above $p_{ik}(t) \geqslant \bar{p}_{ik}(t)$ for all i, k, and $t \geqslant 0$. If strict inequality holds for some $t = t_0$ and $i = i_0$ then summing over k one gets

$$\sum_{k \in S} p_{ik}(t_0) > \sum_{k \in S} \bar{p}_{ik}(t_0) = 1,$$

contradicting the hypothesis that $\mathbf{p}(t_0)$ is a transition probability matrix. ■

Note that we have not proved that $\bar{\mathbf{p}}(t)$ satisfies the Chapman–Kolmogorov equation (2.6). This may be proved by using Laplace transforms (Exercise 6). It is also the case that the forward equations (2.18) (or (2.19)) always hold for the minimal solution $\bar{\mathbf{p}}(t)$ (Exercise 5).

In the case that (3.9) holds, i.e., *the bounded rates condition*, there is only *one solution* satisfying the backward equations and the initial conditions and, therefore, $\bar{p}_{ik}(t)$ is given by exponential representation on the right side of (3.11). Of course, the solution may be unique even otherwise. We will come back to this question and the probabilistic implications of nonuniqueness in the next section.

Finally, the upshot of all this is that the Markov process is under certain circumstances specified by an initial distribution $\boldsymbol{\pi}$ and a matrix $\mathbf{Q}$, satisfying (4.1). In any case, the minimal solution always exists, although the total mass may be less than 1.

Example 1. Another simple model of *"contagion"* or *"accident proneness"* for actuarial mathematics, this one having a homogeneous transition law, is obtained as follows. Suppose that the probability of an accident in t to $t + \Delta t$ is $(\nu + \lambda r)\,\Delta t + o(\Delta t)$ given that r accidents have occurred previous to time t, for $\nu, \lambda > 0$. We have a pure birth process on $S = \{0, 1, 2, \ldots\}$ having infinitesimal parameters $q_{ii} = -(\nu + i\lambda)$, $q_{i,i+1} = \nu + i\lambda$, $q_{ik} = 0$ if $k < i$ or if $k > i + 1$. The forward equations yield

$$\begin{aligned} p'_{0n}(t) &= -(\nu + n\lambda)p_{0n}(t) + (\nu + (n-1)\lambda)p_{0,n-1}(t), \\ p'_{00}(t) &= -\nu p_{00}(t) \qquad (n \geqslant 1). \end{aligned} \tag{4.9}$$

Clearly,

$$\begin{aligned} p_{00}(t) &= e^{-\nu t}, \\ p'_{01}(t) &= -(\nu + \lambda)p_{01}(t) + \nu p_{00}(t) = -(\nu + \lambda)p_{01}(t) + \nu e^{-\nu t}. \end{aligned} \tag{4.10}$$

This equation can be solved with the aid of an integrating factor as follows. Let $g(t) = e^{(v+\lambda)t}p_{01}(t)$. Then (4.10) may be expressed as

$$\frac{dg(t)}{dt} = ve^{\lambda t},$$

or

$$g(t) = g(0) + \int_0^t ve^{\lambda u}\, du = \int_0^t ve^{\lambda u}\, du = \frac{v}{\lambda}(e^{\lambda t} - 1).$$

Therefore,

$$p_{01}(t) = e^{-(v+\lambda)t}g(t) = \frac{v}{\lambda}(e^{-vt} - e^{-(v+\lambda)t}) = \frac{v}{\lambda}e^{-vt}(1 - e^{-\lambda t}). \tag{4.11}$$

Next

$$\begin{aligned} p'_{02}(t) &= -(v + 2\lambda)p_{02}(t) + (v + \lambda)p_{01}(t) \\ &= -(v + 2\lambda)p_{02}(t) + (v + \lambda)\frac{v}{\lambda}e^{-vt}(1 - e^{-\lambda t}). \end{aligned} \tag{4.12}$$

Therefore, as in (4.10) and (4.11),

$$\begin{aligned} p_{02}(t) &= e^{-(v+2\lambda)t}\left[\int_0^t e^{(v+2\lambda)u}\frac{(v+\lambda)v}{\lambda}e^{-vu}(1 - e^{-\lambda u})\, du\right] \\ &= e^{-(v+2\lambda)t}\left[\frac{v(v+\lambda)}{\lambda}\int_0^t (e^{2\lambda u} - e^{\lambda u})\, du\right] \\ &= e^{-(v+2\lambda)t}\left[\frac{v(v+\lambda)}{\lambda}\right]\left[\frac{e^{2\lambda t} - 1}{2\lambda} - \frac{e^{\lambda t} - 1}{\lambda}\right] \\ &= \frac{v(v+\lambda)}{2\lambda^2}[e^{-vt} - 2e^{-(v+\lambda)t} + e^{-(v+2\lambda)t}] \\ &= \frac{v(v+\lambda)}{2\lambda^2}e^{-vt}[1 - e^{-\lambda t}]^2. \end{aligned}$$

Assume, as induction hypothesis, that

$$p_{0n}(t) = \frac{v(v+\lambda)\cdots(v+(n-1)\lambda)}{n!\,\lambda^n}e^{-vt}[1 - e^{-\lambda t}]^n.$$

Then,

$$p'_{0,n+1}(t) = -(v + (n+1)\lambda)p_{0,n+1}(t) + (v + n\lambda)p_{0n}(t),$$

yields

$$p_{0,n+1}(t) = e^{-(\nu+(n+1)\lambda)t} \int_0^t e^{(\nu+(n+1)\lambda)u}(\nu + n\lambda)p_{0n}(u)\, du$$

$$= \frac{\nu(\nu+\lambda)\cdots(\nu+n\lambda)}{n!\lambda^n} e^{-(\nu+(n+1)\lambda)t} \int_0^t e^{(n+1)\lambda u}[1 - e^{-\lambda u}]^n\, du. \tag{4.13}$$

Now, setting $x = e^{\lambda u}$,

$$\int_0^t e^{(n+1)\lambda u}[1 - e^{-\lambda u}]^n\, du = \frac{1}{\lambda}\int_1^{e^{\lambda t}} x^n\left(1 - \frac{1}{x}\right)^n dx = \frac{1}{\lambda}\int_1^{e^{\lambda t}} (x-1)^n\, dx$$

$$= \frac{1}{\lambda}\left[\frac{(x-1)^{n+1}}{n+1}\right]_1^{e^{\lambda t}} = \frac{1}{\lambda}\frac{(e^{\lambda t}-1)^{n+1}}{n+1}.$$

Hence,

$$p_{0,n+1}(t) = \frac{\nu(\nu+\lambda)\cdots(\nu+n\lambda)}{(n+1)!\lambda^{n+1}} e^{-\nu t}(1 - e^{-\lambda t})^{n+1}. \tag{4.14}$$

5 SAMPLE PATH ANALYSIS AND THE STRONG MARKOV PROPERTY

Let $\mathbf{Q} = ((q_{ij}))$ be transition rates satisfying (4.1) and such that the corresponding Kolmogorov backward equation admits a unique (transition probability semigroup) solution $\mathbf{p}(t) = ((p_{ij}(t)))$. Given an initial distribution $\boldsymbol{\pi}$ on S there is a Markov process $\{X_t\}$ with transition probabilities $\mathbf{p}(t)$, $t \geq 0$, and initial distribution $\boldsymbol{\pi}$ having right-continuous sample paths. Indeed, the process $\{X_t\}$ may be constructed as coordinate projections on the space Ω of right-continuous step functions on $[0, \infty)$ with values in S (theoretical complement 5.3).

Our purpose in the present section is to analyze the probabilistic nature of the process $\{X_t\}$. First we consider the distribution of the time spent in the initial state.

Proposition 5.1. Let the Markov chain $\{X_t: 0 \leq t < \infty\}$ have the initial state i and let $T_0 = \inf\{t > 0: X_t \neq i\}$. Then T_0 has an exponential distribution with parameter $-q_{ii}$. In the case $q_{ii} = 0$, the degeneracy of the exponential distribution can be interpreted to mean that i is an absorbing state, i.e., $P_i(T_0 = \infty) = 1$.

Proof. Choose and fix $t > 0$. For each integer $n \geq 1$ define the finite-dimensional event

$$A_n = \{X_{(m/2^n)t} = i \text{ for } m = 0, 1, \dots, 2^n\}.$$

The events A_n are decreasing as n increases and

$$\begin{aligned} A = \lim_{n\to\infty} A_n &:= \bigcap_{n=1}^{\infty} A_n \\ &= \{X_u = i \text{ for all } u \text{ in } [0, t] \text{ which is a binary rational multiple of } t\} \\ &= \{T_0 > t\}. \end{aligned}$$

To see why the last equality holds, first note that $\{T_0 > t\} = \{X_u = i$ for all u in $[0, t]\} \subset A$. On the other hand, since the sample paths are step functions, if a sample path is not in $\{T_0 > t\}$ then there occurs a jump to state j, different from i, at some time t_0 $(0 < t_0 \leqslant t)$. The case $t_0 = t$ may be excluded, since it is not in A. Because each sample path is a right-continuous step function, there is a time point $t_1 > t_0$ such that $X_u = j$ for $t_0 \leqslant u < t_1$. Since there is some u of the form $u = (m/2^n)t \leqslant t$ in every nondegenerate interval, it follows that $X_u = j$ for some u of the form $u = (m/2^n)t < t$; this implies that this sample path is not in A_n and, hence, not in A. Therefore, $\{T_0 > t\} \supset A$. Now note by (2.2)

$$\begin{aligned} P_i(T_0 > t) = P_i(A) = \lim_{n\uparrow\infty} P_i(A_n) &= \lim_{n\uparrow\infty}\left[p_{ii}\left(\frac{t}{2^n}\right)\right]^{2^n} \\ &= \lim_{n\uparrow\infty}\left[1 + \frac{t}{2^n}q_{ii} + o\left(\frac{1}{2^n}\right)\right]^{2^n} = e^{tq_{ii}}, \end{aligned} \tag{5.1}$$

proving that T_0 is exponentially distributed with parameter $-q_{ii}$. If $q_{ii} = 0$, the above calculation shows that $P_i(T_0 > t) = e^0 = 1$ for all $t \geqslant 0$. Hence $P_i(T_0 = \infty) = 1$. ■

The following random times are basic to the description of the evolution of continuous time Markov chains.

$$\begin{gathered} \tau_0 = 0, \qquad \tau_1 = \inf\{t > 0: X_t \neq X_0\}, \qquad \tau_n = \inf\{t > \tau_{n-1}: X_t \neq X_{\tau_{n-1}}\}, \\ T_0 = \tau_1, \qquad T_n = \tau_{n+1} - \tau_n \qquad \text{for } n \geqslant 1. \end{gathered} \tag{5.2}$$

Thus, T_0 is the *holding time* in the initial state, T_1 is the holding time in the state to which the process jumps first time, and so on. Generally, T_n is the holding time in the state to which the process jumps at its nth transition.

As usual, P_i denotes the distribution of the process $\{X_t\}$ under $X_0 = i$. As might be guessed, given the past up to and including time T_0, the process evolves from time T_0 onwards as the original Markov process would with initial state X_{T_0}. More generally, *given the sample path of the process up to time* $\tau_n = T_0 + T_1 + \cdots + T_{n-1}$, *the conditional distribution of* $\{X_{\tau_n+t}: t \geqslant 0\}$ *is* $P_{X_{\tau_n}}$, *depends only on the (present) state* X_{τ_n} *and on nothing else in the past.*

Although this seems intuitively clear from the Markov property, the time T_0 is not a constant, as in the Markov property, but a random variable.

The italicized statement above is a case of an extension of the Markov property known as the *strong Markov property*. To state this property we introduce a class of random times called *stopping times* or *Markov times*. A *stopping time* τ is a random time, i.e., a random variable with values in $[0, \infty]$, with the property that for every fixed time s, the occurrence or nonoccurrence of the event $\{\tau \leqslant s\}$ can be determined by a knowledge of $\{X_u: 0 \leqslant u \leqslant s\}$. For example, if one cannot decide whether or not the event $\{\tau \leqslant 10\}$ has happened by observing only $\{X_u: 0 \leqslant u \leqslant 10\}$, then τ is *not* a stopping time. The random variables $\tau_1 = T_0$, $\tau_2 = T_0 + T_1, \dots, \tau_n$ are stopping times, but T_1, $T_0 + T_2$ are not (Exercise 1).

Proposition 5.2. On the set $\{\omega \in \Omega: \tau(\omega) < \infty\}$, the conditional distribution of $\{X_{\tau+t}: t \geqslant 0\}$ given the past up to time τ is P_{X_τ}, if τ is a stopping time.

Stated another way, Proposition 5.2 says that on $\{\tau < \infty\}$, given the past of the process $\{X_t\}$ up to time τ, the future is (conditionally) distributed as the Markov chain starting at X_τ and having the same transition probabilities as those of $\{X_t\}$. This is the *strong Markov property*. The proof of Proposition 5.3 in the case that τ is discrete is similar to that already given in Section 4, Chapter II. The proof in the general case follows by approximating τ by discrete stopping times. For a detailed proof see Theorem 11.1 in Chapter V.

The strong Markov property will now be used to obtain a vivid probabilistic description of the Markov chain $\{X_t\}$. For this, let us write

$$\left.\begin{aligned} \lambda_i = -q_{ii}, \qquad & k_{ii} = 0 \\ & k_{ij} = \frac{q_{ij}}{-q_{ii}} \quad \text{for } i \neq j \end{aligned}\right\} \quad \text{if } q_{ii} \neq 0,$$
$$k_{ii} = 1 \quad \text{and} \quad k_{ij} = 0 \quad \text{for } i \neq j \quad \text{if } q_{ii} = 0. \tag{5.3}$$

Proposition 5.3. If the initial state is i, and $q_{ii} \neq 0$ then T_0 and X_{T_0} are independent and the distribution of X_{T_0} is given by

$$P_i(X_{T_0} = j) = k_{ij}, \qquad j \in S. \tag{5.4}$$

Proof. For s, $\Delta > 0$, and $j \neq i$ observe that

$$\begin{aligned} P_i(s < T_0 \leqslant s + \Delta, X_{s+\Delta} = j) \\ &= P_i(X_u = i \text{ for } 0 \leqslant u \leqslant s, X_{s+\Delta} = j) \\ &= P_i(X_u = i \text{ for } 0 \leqslant u \leqslant s) P_i(X_{s+\Delta} = j \mid X_u = i \text{ for } 0 \leqslant u \leqslant s) \\ &= P_i(T_0 > s) p_{ij}(\Delta) = e^{q_{ii}s} p_{ij}(\Delta). \end{aligned} \tag{5.5}$$

Dividing the first and last expressions of (5.5) by Δ and letting $\Delta \downarrow 0$ one gets the joint density–mass function of T_0 and X_{T_0} at s and j, respectively (Exercise 2),

$$f_{T_0, X_{T_0}}(s, j) = e^{q_{ii}s} p'_{ij}(0) = e^{q_{ii}s} q_{ij} = \lambda_i e^{-\lambda_i s} \frac{q_{ij}}{-q_{ii}}, \tag{5.6}$$

where $\lambda_i = -q_{ii}$. ■

Now use Propositions 5.1 and 5.3 and the strong Markov property to check the following computation.

$$\begin{aligned}
&P_{i_0}(T_0 \leqslant s_0, X_{T_0} = i_1, T_1 \leqslant s_1, X_{\tau_1} = i_2, \ldots, T_n \leqslant s_n, X_{\tau_n} = i_{n+1}) \\
&= \int_{s=0}^{s_0} P_{i_0}(T_1 \leqslant s_1, X_{\tau_1} = i_2, \ldots, T_n \leqslant s_n, X_{\tau_n} = i_{n+1} \mid T_0 = s, X_{T_0} = i_1)\lambda_{i_0} \\
&\qquad\qquad \times e^{-\lambda_{i_0}s} k_{i_0 i_1}\, ds \\
&= \int_{s=0}^{s_0} P_{i_1}(T_0 \leqslant s_1, X_{T_0} = i_2, \ldots, T_{n-1} \leqslant s_n, X_{\tau_n} = i_{n+1})\lambda_{i_0} e^{-\lambda_{i_0}s} k_{i_0 i_1}\, ds \\
&= \left(\int_{s=0}^{s_0} \lambda_{i_0} e^{-\lambda_{i_0}s}\, ds\right) k_{i_0 i_1} \left(\int_{s=0}^{s_1} \lambda_{i_1} e^{-\lambda_{i_1}s}\, ds\right) k_{i_1 i_2} \\
&\times P_{i_2}(T_0 \leqslant s_2, X_{T_0} = i_3, \ldots, T_{n-2} \leqslant s_n, X_{\tau_{n-2}} = i_{n+1}) \\
&= \cdots = \left(\prod_{j=0}^{n} \int_{s=0}^{s_j} \lambda_{i_j} e^{-\lambda_{i_j}s}\, ds\right)\left(\prod_{j=0}^{n} k_{i_j i_{j+1}}\right).
\end{aligned} \tag{5.7}$$

Note that

$$\prod_{j=0}^{n} k_{i_j i_{j+1}} = P_{i_0}(Y_1 = i_1, Y_2 = i_2, \ldots, Y_{n+1} = i_{n+1}),$$

where $\{Y_m : m = 0, 1, \ldots\}$ is a discrete-parameter Markov chain with initial state i_0 and transition matrix $\mathbf{K} = ((k_{ij}))$, while

$$\prod_{j=0}^{n} \int_{s=0}^{s_j} \lambda_{i_j} e^{-\lambda_{i_j}s}\, ds$$

is the joint distribution function of $n + 1$ independent exponential random variables with parameters $\lambda_{i_0}, \lambda_{i_1}, \ldots, \lambda_{i_n}$. We have, therefore, proved the following.

Theorem 5.4

(a) Let $\{X_t\}$ be a Markov chain having the infinitesimal generator $\mathbf{Q}$ and initial state i_0. Then $\{Y_n := X_{\tau_n} : n = 0, 1, 2, \ldots\}$ is a discrete parameter

Markov chain having the transition probability matrix $\mathbf{K} = ((k_{ij}))$ given by (5.3). Also, conditionally given $\{Y_n\}$, the holding times $T_0, T_1, T_2, \ldots$ are independent exponentially distributed random variables with parameters $\lambda_{i_0}, \lambda_{i_1}, \lambda_{i_2}, \ldots$.

(b) Conversely, suppose $\{Y_n\}$ is a discrete-parameter Markov chain having transition probability matrix $\mathbf{K}$ and initial state i_0, and for each sequence of states $i_0, i_1, i_2, \ldots$, let $T_0, T_1, T_2, \ldots$, be a sequence of independent exponentially distributed random variables with parameters $\lambda_{i_0}, \lambda_{i_1}, \lambda_{i_2}, \ldots$, respectively. Assume also that the sequences $\{T_0, T_1, T_2, \ldots\}$, corresponding to all distinct sequences of states, comprise an independent family. Define

$$\begin{aligned} X_t &= Y_0, \qquad 0 \leqslant t < T_0, \\ X_t &= Y_{k+1} \qquad \text{for } T_0 + T_1 + \cdots + T_k \leqslant t < T_0 + T_1 + \cdots + T_{k+1}. \end{aligned} \tag{5.8}$$

Then the process $\{X_t : t \geqslant 0\}$ so constructed is Markovian with initial state i_0 and infinitesimal generator $\mathbf{Q}$.

Part (b) of Theorem 5.4 provides a noncanonical construction of a Markov chain with the transition rates given by $\mathbf{Q} = ((q_{ij}))$.

By randomizing over the initial state, one easily extends the theorem to an arbitrary initial distribution $\boldsymbol{\pi}$.

The content of the system of integral equations (4.5) is now easy to understand. Let us write (4.5) as

$$p_{il}(t) = \delta_{il} e^{-\lambda_i t} + \sum_{j \neq i} \int_0^t \lambda_i e^{-\lambda_i s} k_{ij} p_{jl}(t-s)\, ds. \tag{5.9}$$

If $j \neq i$, then $k_{ij} p_{jl}(t-s)$ is the conditional probability, given $T_0 = s$, that $X_{T_0} = j$ and in an additional time $t - s$ the process will be in state l. Adding over all $j \neq i$, and then multiplying by the density of T_0 and integrating, one gets for the case $l \neq i$,

$$\begin{aligned} p_{il}(t) = P_i(X_t = l) &= \sum_{j \neq i} \int_0^t P_i(X_{T_0} = j, X_t = l \mid T_0 = s) \lambda_i e^{-\lambda_i s}\, ds \\ &= \sum_{j \neq i} \int_0^t \lambda_i e^{-\lambda_i s} k_{ij} p_{jl}(t-s)\, ds. \end{aligned} \tag{5.10}$$

If $l = i$, an extra term, namely, $e^{-\lambda_i t}$ must be added as the probability that no jump occurs in the time interval $(0, t]$.

Example 1. (*Homogeneous Poisson Process*). As an immediate consequence of Theorem 5.4 we get the following corollary.

Corollary 5.5. For a homogeneous Poisson process with parameter λ the successive holding times are i.i.d. exponential with parameter λ.

An interesting and useful property of the holding times of a Poisson process concerns the conditional distribution of the successive arrival times in the period from 0 to t given the number of arrivals in $[0, t]$.

Proposition 5.6. Let $T_0, T_0 + T_1, \ldots, T_0 + T_1 + \cdots + T_j, \ldots$, denote successive arrival times of the Poisson process $\{Y_t\}$ with parameter λ. Then the conditional distribution of $(T_0, T_0 + T_1, \ldots, T_0 + T_1 + \cdots + T_{k-1})$ given $\{X_t - X_0 = k\}$ is the same as that of k increasingly ordered independent random variables each having the uniform distribution on $(0, t]$.

Proof. Let $U_0, U_1, \ldots, U_{k-1}$ be k i.i.d. random variables each uniformly distributed on $(0, t]$. Let $U_{(0)}$ be the smallest of $U_0, U_1, \ldots, U_{k-1}$, $U_{(1)}$ the next smallest, etc., so that with probability one $U_{(0)} < U_{(1)} < \cdots < U_{(k-1)}$. Since each realization of $(U_{(0)}, U_{(1)}, \ldots, U_{(k-1)})$ corresponds to $k!$ permutations of $(U_0, U_1, \ldots, U_{k-1})$, each with the same density $1/t^k$, the joint density of $(U_{(0)}, U_{(1)}, \ldots, U_{(k-1)})$ is given by

$$g(s_0, \ldots, s_{k-1}) = \begin{cases} \dfrac{k!}{t^k} & \text{if } 0 < s_0 < s_1 < \cdots < s_{k-1} \leqslant t \\ 0 & \text{otherwise.} \end{cases} \tag{5.11}$$

Now, by Corollary 5.5, $T_0, T_1, \ldots, T_k$ are i.i.d. with joint density given by

$$f(x_0, x_1, \ldots, x_k) = \begin{cases} \lambda^{k+1} \exp\left\{-\lambda \sum_{i=0}^{k} x_i\right\} & \text{if } x_i > 0 \quad \text{for all } i, \\ 0 & \text{otherwise.} \end{cases} \tag{5.12}$$

Since the Jacobian of the transformation

$$(x_0, x_1, \ldots, x_k) \to (x_0, x_0 + x_1, \ldots, x_0 + x_1 + \cdots + x_k)$$

is 1, the joint density of $T_0, T_0 + T_1, \ldots, T_0 + T_1 + \cdots + T_k$ is given by

$$f_1(t_0, t_1, \ldots, t_k) = \lambda^{k+1} e^{-\lambda t_k} \qquad \text{for } 0 < t_0 < t_1 < t_2 < \cdots < t_k. \tag{5.13}$$

The density of the conditional distribution of $T_0, T_0 + T_1, \ldots, T_0 + T_1 + \cdots + T_k$ given $\{T_0 + T_1 + \cdots + T_{k-1} \leqslant t < T_0 + T_1 + \cdots + T_k\}$ is

$$f_2(s_0, s_1, \ldots, s_k) = \begin{cases} \dfrac{\lambda^{k+1} e^{-\lambda s_k}}{P(T_0 + T_1 + \cdots + T_{k-1} \leqslant t < T_0 + T_1 + \cdots + T_k)} \\ \qquad \text{if } 0 < s_0 < s_1 < \cdots < s_k \text{ and } s_{k-1} \leqslant t < s_k, \\ 0 \qquad \text{otherwise.} \end{cases} \tag{5.14}$$

But the denominator in (5.14) equals $P(Y_t = k)$, where $Y_t = X_t - X_0$. Therefore,

$$f_2(s_0, s_1, \ldots, s_k) = \begin{cases} \dfrac{\lambda^{k+1} e^{-\lambda s_k}}{e^{-\lambda t} \dfrac{(\lambda t)^k}{k!}} = \dfrac{\lambda k! e^{-\lambda s_k}}{e^{-\lambda t} t^k} \\ \qquad \text{if } 0 < s_0 < s_1 < \cdots < s_k \text{ and } s_{k-1} \leqslant t < s_k, \\ 0 \qquad \text{otherwise}. \end{cases} \tag{5.15}$$

Integrating this over s_k we get the conditional density of $T_0, T_0 + T_1, \ldots, T_0 + T_1 + \cdots + T_{k-1}$ given $\{Y_t = k\}$ as

$$f_3(s_0, s_1, \ldots, s_{k-1}) = \begin{cases} \dfrac{(\lambda)k!}{t^k e^{-\lambda t}} \displaystyle\int_t^{\infty} e^{-\lambda s_k}\, ds_k = \dfrac{k!}{e^{-\lambda t} t^k} e^{-\lambda t} = \dfrac{k!}{t^k} \\ \qquad \text{if } 0 < s_0 < s_1 < \cdots < s_{k-1} \leqslant t, \\ 0 \qquad \text{otherwise}. \end{cases} \tag{5.16}$$

Thus, as asserted, f_3 is the same as g. ■

Example 2. (*Continuous-Parameter Markov Branching Process*). This is the continuous-parameter version of the Bienaymé–Galton–Watson branching process discussed in Chapter II. It is another basic random replication scheme that is used to describe processes ranging from the growth of populations to the branching of river networks. Consider a population of $X_0 = i_0 \geqslant 1$ particles at time 0. Each particle, independently of the others, waits an exponentially distributed length of time with parameter $\lambda > 0$ and then splits into a random number k of particles with probability f_k (independently of the waiting time), where $f_k \geqslant 0$, $f_0 > 0$, $0 < f_0 + f_1 < 1$, $\sum_{k=0}^{\infty} f_k = 1$. The progeny continue to split according to this same process. Let X_t denote the number of particles present at time t. An equivalent description of the process $\{X_t\}$ is as follows. Since the minimum of i_0 i.i.d. exponential random variables with parameter $\lambda > 0$ is exponential with parameter $i_0\lambda$, the holding time T_0 in state i_0 is exponential with parameter $i_0\lambda$. At time T_0 the initial population of $X_0 = i_0$ particles becomes a population of $X_{T_0} = j \geqslant i_0 - 1$ particles with probability f_{j-i_0+1}. If $i_0 = 0$, then $X_t = 0$ for all $t \geqslant 0$. In view of the *lack of memory property* of the i.i.d. exponentially distributed splitting times of the particles and their progeny, the holding time T_1 to the next transition given $\{X_{T_0} = j, T_0\}$ $(j \geqslant 1)$ is exponential with parameter $\lambda_j = j\lambda$ (Exercise 13). Thus, $\{X_t\}$ may be described as a continuous-parameter Markov chain with state space $S = \{0, 1, 2, \ldots\}$ with absorbing state at 0 and having infinitesimal transition rates that can be calculated as follows. First note that the probability of k of i particles splitting within time $h > 0$ is

$$\binom{i}{k}(e^{-\lambda h})^{i-k}(1 - e^{-\lambda h})^k = O(h^k) = o(h) \qquad \text{for } i \geqslant k \geqslant 2.$$

Now,

$$p_{ii}(h) = e^{-i\lambda h} + f_1\binom{i}{1}\left[\int_0^h \lambda e^{-\lambda s}e^{-\lambda(h-s)}\,ds\right](e^{-\lambda h})^{i-1} + O(h^2)$$
$$= 1 - i\lambda(1 - f_1)h + o(h) \qquad \text{as } h \to 0. \tag{5.17}$$

Likewise, for $j \geqslant i + 1$,

$$p_{ij}(h) = \binom{i}{1} f_{j-i+1}\left[\int_0^h \lambda e^{-\lambda s}e^{-(j-i+1)\lambda(h-s)}\,ds\right](e^{-\lambda h})^{i-1} + o(h)$$
$$= i f_{j-i+1}\lambda h + o(h) \qquad \text{as } h \to 0. \tag{5.18}$$

For $j = i - 1$, we have

$$p_{i,i-1}(h) = \binom{i}{1} f_0\left[\int_0^h \lambda e^{-\lambda s}\,ds\right](e^{-\lambda h})^{i-1} + o(h)$$
$$= i\lambda f_0 h + o(h) \qquad \text{as } h \to 0. \tag{5.19}$$

Therefore, the transition rates are given by

$$q_{ij} = \begin{cases} i\lambda f_{j-i+1}, & j \geqslant i-1, \quad i \geqslant 1, \quad j \neq i, \\ 0, & i = 0, \quad j \geqslant 0 \quad \text{or} \quad j < i-1, \quad i \geqslant 1, \\ -i\lambda(1 - f_1), & i = j \geqslant 1. \end{cases} \tag{5.20}$$

The branching evolution is characteristically *multiplicative* and as such clearly does not possess (additive) independent increments. The branching mechanism is distinctively displayed in terms of the probability-generating functions of the (minimal) transition probabilities by the following considerations. Let

$$g_t^{(i)}(r) = \sum_{j=0}^{\infty} p_{ij}(t) r^j, \qquad i = 0, 1, 2, \ldots. \tag{5.21}$$

Since each of the i initial particles, independently of the others, has produced a random number of progeny at time t with distribution $(p_{1j}(t): j = 0, 1, \ldots)$, the distribution of the total progeny of the i initial particles at time t is the i-fold convolution $(p_{1j}^{*i}(t): j = 0, 1, 2, \ldots)$. Therefore,

$$g_t^{(i)}(r) = (g_t^{(1)}(r))^i, \qquad i = 0, 1, 2, \ldots. \tag{5.22}$$

The Chapman–Kolmogorov equations for the branching evolution take the (*transformed*) form

$$g_{t+s}^{(i)}(r) = g_t^{(i)}(g_s^{(1)}(r)), \tag{5.23}$$

since

$$\sum_{j=0}^{\infty} p_{ij}(t+s)r^j = \sum_{j=0}^{\infty} \sum_k p_{ik}(t)p_{kj}(s)r^j$$
$$= \sum_k p_{ik}(t)(g_s^{(1)}(r))^k = g_t^{(i)}(g_s^{(1)}(r)).$$

Fix $\tau > 0$ and consider the discrete parameter stochastic process $\{\tilde{X}_n\}$ defined by

$$\tilde{X}_n = X_{n\tau}, \qquad n = 0, 1, 2, \ldots. \tag{5.24}$$

The process $\{\tilde{X}_n\}$ is clearly a discrete-parameter Markov chain. In fact, from (5.24) one can see that $\{\tilde{X}_n\}$ has stationary transition probabilities $\tilde{p}_{ij} = p_{ij}(\tau)$. *This makes $\{\tilde{X}_n\}$ a discrete parameter Bienaymé–Galton–Watson branching process with offspring distribution $\tilde{f}_j = \tilde{p}_{1j} = p_{1j}(\tau)$.*

Observe that the event that $\{X_t\}$ eventually becomes extinct is equivalent to the event that $\{\tilde{X}_n\}$ eventually becomes extinct. It follows from the result obtained in Example 11.2 of Chapter II for the discrete parameter case that the probability ρ of eventual extinction of an initial single particle is the smallest nonnegative (fixed-point) r of the equation

$$g_\tau^{(1)}(r) = r. \tag{5.25}$$

In particular, observe that this root cannot depend on $\tau > 0$. One may therefore expect to be able to compute ρ from the generating function of the infinitesimal transition rates. Define

$$h^{(1)}(r) = \sum_{j=0}^{\infty} q_{1j}r^j = -\lambda r + \sum_{j=0}^{\infty} \lambda f_j r^j. \tag{5.26}$$

Proposition 5.7. The extinction probability ρ for the continuous-parameter branching process is the smallest nonnegative root of the equation

$$h^{(1)}(r) = 0.$$

Moreover, if $\sum_{j=0}^{\infty} jf_j \leqslant 1$ then $\rho = 1$.

Proof. Observe that the Kolmogorov backward equation for the (minimal) branching evolution transforms as follows.

$$\frac{\partial g_t^{(1)}(r)}{\partial t} = \sum_j \frac{\partial p_{1j}(t)}{\partial t} r^j = \sum_j \sum_k q_{1k} p_{kj}(t) r^j$$
$$= \sum_k q_{1k} g_t^{(k)}(r) = \sum_k q_{1k}(g_t^{(1)}(r))^k.$$

Thus,

$$\frac{\partial g_t^{(1)}(r)}{\partial t} = h^{(1)}(g_t^{(1)}(r)), \tag{5.27}$$

$$g_0^{(1)}(r) = \sum_{j=0}^{\infty} p_{1j}(0)r^j = r. \tag{5.28}$$

In particular, since $g_t^{(1)}(\rho) = \rho$ for all $t > 0$, it follows that ρ must satisfy

$$h^{(1)}(\rho) = \frac{\partial g_t^{(1)}(\rho)}{\partial t} = 0. \tag{5.29}$$

Since $h^{(1)}(0) = \lambda f_0 > 0$, $h^{(1)}(1) = \sum_{j=0}^{\infty} q_{1j} = 0$, and

$$\frac{d^2 h^{(1)}(r)}{dr^2} = \lambda \sum_{j=2}^{\infty} j(j-1)f_j r^{j-2} \geqslant 0, \qquad 0 \leqslant r < 1,$$

it follows that $h^{(1)}(r)$ can have at most one zero in the open interval $0 < r < 1$. If $\sum_{j=0}^{\infty} jf_j \leqslant 1$ then

$$\frac{d}{dr} h^{(1)}(r) = -\lambda + \sum_{j=1}^{\infty} \lambda j f_j r^{j-1}$$

is nonpositive for all $0 \leqslant r \leqslant 1$. Thus $\rho = 1$ in this case. ■

As a corollary it follows that since λ appears as a simple positive factor in (5.26), the extinction probability is the smallest nonnegative fixed point of the probability-generating function of the offspring distribution $\{f_j\}$. In particular, therefore, extinction is completely determined by the embedded discrete parameter Markov chain $\{Y_n\}$. In particular, if $\sum_j jf_j > 1$, then $\rho < 1$ (see Example 11.2 of Chapter II).

Example 3. (*Continuous-Parameter Binary Branching or Birth–Death*). The Bienaymé–Galton–Walton binary branching Markov process $\{X_t\}$ is a simple birth–death process on $\{0, 1, 2, \ldots\}$ with $X_0 = 1$, absorption at 0, having linear birth–death rates $q_{i,i+1} = q_{i,i-1} = i\lambda/2$, $i \geqslant 1$, and having right continuous sample paths a.s. Define

$$L := \inf\{t \geqslant 0\colon X_t = 0\}, \tag{5.30}$$

$$M_t := \#\{0 \leqslant s \leqslant t\colon X_s - X_{s-} = -1\}. \tag{5.31}$$

Imagine a binary tree graph emanating from a single root vertex (see Figure 5.1) such that starting from a single root particle (vertex), a particle lives for

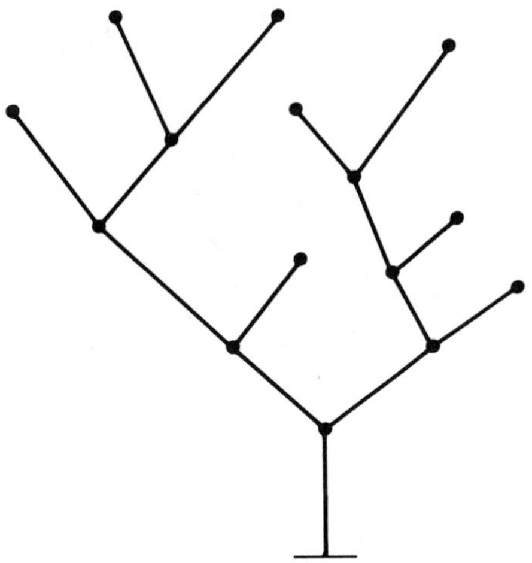

Figure 5.1

an exponentially distributed duration with mean λ^{-1} and then either splits into two particles or terminates with equal probabilities and independently of the holding times. Each new particle is again independently subjected to the same rules of evolution as its predecessors. Then L represents the *time to extinction* of this process, and M_t is the total number of deaths by time t. According to Proposition 5.7, L is finite with probability 1. In particular, M_L represents the number of degree-one vertices in the tree (excluding the root). In the context of random flow networks, such vertices represent *sources*, and the extreme value statistic L is the *main channel length*. We will consider the problem of computing the (conditional) mean of L given $M_L = n$. The process $\{(X_t, M_t)\}$ is a Markov process on $Z_+ \times Z_+$ having rates $\mathbf{A} = ((a_{\mathbf{ij}}))$, where

$$a_{\mathbf{i},\mathbf{j}} = \begin{cases} \frac{1}{2}i_1\lambda, & \mathbf{j} = \mathbf{i} + (1,0) = (i_1 + 1, i_2) \\ \frac{1}{2}i_1\lambda, & \mathbf{j} = \mathbf{i} + (-1,1) = (i_1 - 1, i_2 + 1) \\ -i_1\lambda, & \mathbf{j} = \mathbf{i} \\ 0, & \text{otherwise}. \end{cases} \tag{5.32}$$

Justification of the use of the forward equations below is deferred to Section 8 (Example 8.2, Exercise 8.14).

Lemma. Let $g(t, u, v) = E_{(1,0)}\{u^{X_t}v^{M_t}\}$, $0 \leqslant u, v \leqslant 1$, $t \geqslant 0$. Then,

$$g(t, u, v) = \frac{C_1(u - C_2) + C_2(C_1 - u)e^{\frac{1}{2}\lambda(C_1 - C_2)t}}{(u - C_2) - (u - C_1)e^{\frac{1}{2}\lambda(C_1 - C_2)t}}, \tag{5.33}$$

where $C_1 = C_1(v) = 1 - (1 - v)^{1/2}$, $C_2 = C_2(v) = 1 + (1 - v)^{1/2}$.

Proof. It follows from Kolmogorov's forward equation that for $\mathbf{i}_0 = (1, 0)$,

$$\begin{aligned}\frac{\partial}{\partial t} E_{\mathbf{i}_0}\{u^{X_t} v^{M_t}\} &= E_{\mathbf{i}_0}\left\{\sum_{(j_1, j_2)} u^{j_1} v^{j_2} a_{(X_t, M_t),(j_1, j_2)}\right\} \\ &= E_{\mathbf{i}_0}\{\tfrac{1}{2}\lambda X_t u^{X_t+1} v^{M_t} + \tfrac{1}{2}\lambda X_t u^{X_t-1} v^{M_t+1} - \lambda X_t u^{X_t} v^{M_t}\} \\ &= \tfrac{1}{2}\lambda u^2 \frac{\partial}{\partial u} E_{\mathbf{i}_0}\{u^{X_t} v^{M_t}\} + \tfrac{1}{2}\lambda v \frac{\partial}{\partial u} E_{\mathbf{i}_0}\{u^{X_t} v^{M_t}\} - \lambda u \frac{\partial}{\partial u} E_{\mathbf{i}_0}\{u^{X_t} v^{M_t}\}.\end{aligned} \tag{5.34}$$

Therefore, one has

$$\frac{\partial}{\partial t} g(t, u, v) = \frac{\lambda}{2}(u^2 + v - 2u)\frac{\partial}{\partial u} g(t, u, v) \tag{5.35}$$

with the initial condition,

$$g(0, u, v) = u. \tag{5.36}$$

Letting $\gamma(t)$ be the characteristic curve defined by

$$\gamma'(t) = -\frac{\lambda}{2}(\gamma^2(t) - 2\gamma(t) + v) = -\frac{\lambda}{2}(\gamma(t) - C_1)(\gamma(t) - C_2),$$

with $C_1 \equiv C_1(v)$ and $C_2 \equiv C_2(v)$ as defined in the statement of the lemma, (5.35) and (5.36) take the form

$$\frac{d}{dt} g(t, \gamma(t), v) = 0, \qquad g(0, \gamma(0), v) = \gamma(0).$$

These are easily integrated to get (5.33). ∎

From the lemma it follows that

$$\sum_{n=1}^{\infty} v^n P(M_L = n) = \lim_{t\to\infty} g(t, 1, v) = 1 - (1 - v)^{1/2} = \sum_{n=1}^{\infty} \binom{\frac{1}{2}}{n}(-1)^{n+1} v^n, \quad 0 \leqslant v < 1. \tag{5.37}$$

In particular, the classic formula of Cayley for the distribution of M_L follows as (see also Exercise 11)

$$P(M_L = n) = \binom{\frac{1}{2}}{n}(-1)^{n-1} = \frac{1}{2n-1}\binom{2n-1}{n} 2^{-2n+1}, \qquad n = 1, 2, \ldots. \tag{5.38}$$

Thus, by the Stirling formula one has,

$$P(M_L = n) \sim \frac{1}{2\sqrt{\pi}} n^{-3/2} \qquad \text{as } n \to \infty . \tag{5.39}$$

The lemma provides a way to compute the conditional moments as follows:

$$E\{L^r \mid M_L = n\} = r \int_0^\infty t^{r-1} P(L > t \mid M_L = n)\, dt \tag{5.40}$$

and

$$P(L > t \mid M_L = n) = \frac{P(L > t, M_L = n)}{P(M_L = n)} = \frac{P(M_L = n) - P(L \leqslant t, M_L = n)}{P(M_L = n)}$$

$$= \frac{P(M_L = n) - P(X_t = 0, M_L = n)}{P(M_L = n)} . \tag{5.41}$$

Now, letting

$$h_n = P(M_L = n) E(L^r \mid M_L = n), \qquad n \geqslant 1, \tag{5.42}$$

we have for $0 \leqslant v < 1$, using (5.33), (5.37), (5.40) and (5.41),

$$\hat{h}(v) := \sum_{n=1}^{\infty} h_n v^n = r \int_0^\infty t^{r-1} \{(1 - (1 - v)^{1/2}) - g(t, 0, v)\}\, dt$$

$$= r \int_0^\infty t^{r-1} \left\{ C_1 - \frac{C_1 C_2 - C_2 C_1 e^{\frac{1}{2}\lambda(C_1 - C_2)t}}{C_2 - C_1 e^{\frac{1}{2}\lambda(C_1 - C_2)t}} \right\} dt$$

$$= r \int_0^\infty t^{r-1} \left\{ \frac{C_1 (C_2 - C_1) e^{\frac{1}{2}\lambda(C_1 - C_2)t}}{C_2 - C_1 e^{\frac{1}{2}\lambda(C_1 - C_2)t}} \right\} dt . \tag{5.43}$$

Expanding the integrand as a geometric series, one obtains

$$\hat{h}(v) = r \frac{C_1 (C_2 - C_2)}{C_2} \int_0^\infty t^{r-1} \sum_{n=0}^{\infty} \left(\frac{C_1}{C_2} \right)^n e^{(n+1)\frac{\lambda}{2}(C_1 - C_2)t}\, dt$$

$$= \frac{2^r r}{\lambda^r} (C_2 - C_1)^{-r+1} \sum_{n=1}^{\infty} \left(\frac{C_1}{C_2} \right)^n n^{-r} \int_0^\infty s^{r-1} e^{-s}\, ds$$

$$= 2 \frac{r!}{\lambda^r} (1 - v)^{-(r-1)/2} \sum_{n=1}^{\infty} \left(\frac{C_1}{C_2} \right)^n n^{-r}. \tag{5.44}$$

In the case $r = 1$,

$$\hat{h}(v) = -\frac{2}{\lambda} \log\left(1 - \frac{C_1(v)}{C_2(v)} \right) = -\left(\frac{2}{\lambda} \right) \log\left(\frac{2(1 - v)^{1/2}}{1 + (1 - v)^{1/2}} \right). \tag{5.45}$$

To invert $\hat{h}(v)$ in the case $r = 1$, consider that $v\hat{h}'(v)$ is the generating function of $\{nh_n\}$. Thus, differentiating one gets that

$$v\hat{h}'(v) = \lambda^{-1}(1 - v)^{-1}[1 - (1 - v)^{-1/2}] = \lambda^{-1}\{(1 - v)^{-1} - (1 - v)^{-1/2}\} \tag{5.46}$$

and, therefore, after expanding $(1 - v)^{-1}$ and $(1 - v)^{-1/2}$ in a Taylor series about $v = 0$ and equating coefficients in the left and rightmost expansions in (5.46), it follows that

$$nP(M_L = n)E(L \mid M_L = n) = \lambda^{-1}\left\{1 + \binom{-\frac{1}{2}}{n}(-1)^{n-1}\right\}. \tag{5.47}$$

In particular, using (5.39) and Stirling's formula, it follows that

$$E(n^{-1/2}\lambda L \mid M_L = n) \sim 2\sqrt{\pi} \qquad \text{as } n \to \infty. \tag{5.48}$$

The asymptotic behavior of higher moments $E(\{n^{-1/2}\lambda L\}^r \mid M_L = n), r \geqslant 2$, may also be determined from (5.44) (theoretical complement 4). Moreover, it may be shown that as $n \to \infty$, the conditional distribution of $n^{-1/2}\lambda L$ given $M_L = n$ has a limiting distribution that coincides with the distribution of the *maximum* of a (suitably scaled) *Brownian excursion process* as described in Exercise 12.7, Chapter I (see theoretical complement 4).

6 THE MINIMAL PROCESS AND EXPLOSION

The Markov process with infinitesimal generator $\mathbf{Q}$ as constructed in Theorem 5.4 is called the *minimal process* with infinitesimal generator $\mathbf{Q}$ because it gives the process only up to the time

$$\zeta \equiv T_0 + T_1 + T_2 + \cdots = \sum_{n=0}^{\infty} T_n. \tag{6.1}$$

The time ζ is called the *explosion time*.

What we have shown in general is that *every* Markov chain with infinitesimal generator $\mathbf{Q}$ and initial distribution $\boldsymbol{\pi}$ is given, up to the explosion time, by the process $\{X_t\}$ described in Theorem 5.4. If $\zeta = \infty$ with probability 1, then the minimal process is the *only one* with infinitesimal generator $\mathbf{Q}$ and we have *uniqueness*. In the case $P_i(\zeta = \infty) < 1$, i.e., if explosion does take place with positive probability, then there are various ways of continuing the minimal process for $t \geqslant \zeta$ so as to preserve the Markov property and the backward equations (2.16) for the continued process $\{\tilde{X}_t\}$. One such way is to fix an arbitrary probability distribution $\boldsymbol{\psi}$ on S and let $\tilde{X}_\zeta = j$ with probability ψ_j $(j \in S)$. For $t > \zeta$ the continued process then evolves as a new (minimal) process with initial j and infinitesimal generator $\mathbf{Q}$ until a second explosion occurs, at which time $\{\tilde{X}_t\}$ is assigned a state according to $\boldsymbol{\psi}$ independently of the assignment at the time of the first explosion ζ, and so on. The transition probabilities $\tilde{p}_{ik}(t)$, say, of $\{\tilde{X}_t\}$ clearly satisfy (5.9), the integral version of the backward equations (2.16), since the derivation (5.10) is based on conditioning with respect to T_0, which is surely less than ζ, and on the Markov property. Unfortunately, the forward equations (2.19) do not hold for $\tilde{\mathbf{p}}$, since the conditioning in the corresponding integral equations is with respect to the *final jump* before time t and one or more explosions may have already occurred by time t.

We have already shown in Section 3 that if $\{q_{ii}: i \in S\}$ is bounded, the backward equations have a unique solution. This solution $p_{ik}(t)$ gives rise to a unique Markov process. Therefore we have the following elementary criterion for nonexplosion.

Proposition 6.1. If $\sup_{i\in S} \lambda_i$ $(= \sup_i |q_{ii}| = \sup_{i,j}|q_{ij}|)$ is finite, then $P_i(\zeta = \infty) = 1$ for all $i \in S$, and the minimal process having infinitesimal generator $\mathbf{Q}$ and an arbitrary initial distribution is the only Markov process with infinitesimal generator $\mathbf{Q}$.

Definition 6.1. A Markov process $\{X_t\}$ for which $P_i(\zeta = \infty) = 1$ for all $i \in S$ is called *conservative* or *nonexplosive*.

Compound Poisson processes are *conservative* as may be checked from

$$\sum_{j\in S} p_{ij}(t) = \sum_{n=0}^{\infty} \sum_{j\in S} e^{-\lambda t} \frac{(\lambda t)^n}{n!} f^{*n}(j-i) = \sum_{n=0}^{\infty} e^{-\lambda t} \frac{(\lambda t)^n}{n!} \left(\sum_{j\in S} f^{*n}(j-i) \right)$$
$$= \sum_{n=0}^{\infty} e^{-\lambda t} \frac{(\lambda t)^n}{n!} = 1. \tag{6.2}$$

Example 1. (*Pure Birth Process*). A pure birth process has state space $S = \{0, 1, 2, \ldots\}$, and its generator $\mathbf{Q}$ has elements $q_{ii} = -\lambda_i$, $q_{i,i+1} = \lambda_i$, $q_{ij} = 0$ for $j > i + 1$ or $j < i$, $i \in S$. Note that this has the same degenerate (i.e., completely deterministic) embedded spatial transition matrix as the Poisson process. However, the parameters λ_i of the exponential holding times are *not* assumed to be equal (spatially constant) here.

Fix an initial state i. Then the embedded chain $\{Y_n: n = 0, 1, \ldots\}$ has only one possible realization, namely, $Y_n = n + i$ $(n \geqslant 0)$. Therefore, the holding times $T_0, T_1, \ldots,$ are (unconditionally) independent and exponentially distributed with parameters $\lambda_i, \lambda_{i+1}, \ldots, \lambda_{i+n}, \ldots$. Consider the explosion time

$$\zeta = T_0 + T_1 + \cdots = \sum_{m=0}^{\infty} T_m.$$

First assume

$$\sum_{n=0}^{\infty} \frac{1}{\lambda_n} = \infty. \tag{6.3}$$

For $s \geqslant 0$, consider

$$\begin{aligned}\phi(s) = E_i e^{-s\zeta} &= E e^{-s\sum_0^\infty T_m} = \prod_{m=0}^{\infty} E_i e^{-sT_m}\\ &= \prod_{m=0}^{\infty} \frac{\lambda_{m+i}}{\lambda_{m+i}+s} = \prod_{m=0}^{\infty} \frac{1}{1+s/\lambda_{m+i}}.\end{aligned}$$

Choose and fix $s > 0$. Then,

$$\begin{aligned}\log \phi(s) &= -\sum_{m=0}^{\infty} \log\left(1 + \frac{s}{\lambda_{m+i}}\right) \leqslant -\sum_{m=0}^{\infty} \frac{s/\lambda_{m+i}}{1+s/\lambda_{m+1}}\\ &= -s\sum_{m=0}^{\infty} \frac{1}{\lambda_{m+i}+s} = -\infty,\end{aligned} \tag{6.4}$$

since $\log(1+x) \geqslant x/(1+x)$ for $x \geqslant 0$ (Exercise 1). Hence $\log \phi(s) = -\infty$, i.e., $\phi(s) = 0$. This is possible only if $P_i(\zeta = \infty) = 1$ since $e^{-s\zeta} > 0$ whenever $\zeta < \infty$. Thus *if* (6.3) *holds, then explosion is impossible.*

Next suppose that

$$\sum_{n=0}^{\infty} \frac{1}{\lambda_n} < \infty. \tag{6.5}$$

Then

$$E\zeta = \sum_{m=0}^{\infty} \frac{1}{\lambda_{m+i}} = \sum_{n=i}^{\infty} \frac{1}{\lambda_n} < \infty.$$

This implies $P_i(\zeta = \infty) = 0$, i.e., $P_i(\zeta < \infty) = 1$. In other words, *if* (6.5) *holds, then explosion is certain.*

In the case of explosion, the minimal process $\{X_t\colon 0 \leqslant t < \zeta\}$ can be continued to a conservative Markov process on the time interval $0 \leqslant t < \infty$ in a variety of ways. For illustration consider the continuation $\{\tilde{X}_t\}$ of an explosive pure birth process $\{X_t\}$ obtained by an instantaneous jump to state 0 at time $t = \zeta$, after which it evolves as a minimal process starting at 0 until the time of the second explosion, and so on. Let $\tilde{p}_{ij}(t)$ denote the transition probabilities for $\{\tilde{X}_t\}$. Likewise, $\tilde{P}_i$ denotes the distribution of $\{\tilde{X}_t\}$ given $\tilde{X}_0 = i$. Consider, for simplicity, $\tilde{p}_{00}(t)$. One has

$$\tilde{p}_{00}(t) = \tilde{P}_0(\tilde{X}_t = 0) = \sum_{n=0}^{\infty} \tilde{P}_0(\tilde{X}_t = 0, N_t = n) \tag{6.6}$$

where N_t is the number of explosions by time t. Since the event $\{\tilde{X}_t = 0, N_t = 0\}$ is the event that the minimal process holds at zero until time t,

$$\tilde{P}_0(X_t = 0, N_t = 0) = p_{00}(t) = e^{-\lambda_0 t}. \tag{6.7}$$

More generally, the event $\{\tilde{X}_t = 0, N_t = n\}$ occurs if and only if the nth explosion occurs at some time prior to t and the minimal process remains at 0 from this time until t. Let f_0 denote the probability density function of the first explosion time ζ for the process $\{\tilde{X}_t\}$ starting at 0. Since the times between successive explosions are i.i.d. with p.d.f. f_0, the p.d.f. of the nth explosion time is given by the n-fold convolution f_0^{*n}. Therefore,

$$\tilde{P}_0(\tilde{X}_t = 0, N_t = n) = \int_0^t p_{00}(t - s) f_0^{*n}(s)\, ds. \tag{6.8}$$

Using (6.7) and (6.8) in (6.6) we get

$$\tilde{p}_{00}(t) = e^{-\lambda_0 t}\left[1 + \sum_{n=1}^{\infty} \int_0^t e^{\lambda_0 s} f_0^{*n}(s)\, ds\right]. \tag{6.9}$$

Differentiating (6.9) gives

$$\tilde{p}'_{00}(t) = -\lambda_0 \tilde{p}_{00}(t) + \sum_{n=1}^{\infty} f_0^{*n}(t). \tag{6.10}$$

In particular, (6.10) shows that the forward equation does not hold for $\tilde{p}_{ij}(t)$.

Since the time of the first explosion starting at 0 is the holding time at 0 plus the remaining time for explosion starting at state 1, we have by the strong Markov property,

$$f_0 = e_{\lambda_0} * f_1 \tag{6.11}$$

where f_1 is the p.d.f. of the first explosion time starting in state 1 and e_{λ_0} is the exponential p.d.f. with parameter λ_0. Therefore,

$$\sum_{n=1}^{\infty} f_0^{*n}(t) = \sum_{n=1}^{\infty} e_{\lambda_0} * f_1 * f_0^{*(n-1)}(t). \tag{6.12}$$

Since,

$$e_{\lambda_0}(t) = \lambda_0 e^{-\lambda_0 t} = \lambda_0 p_{00}(t), \tag{6.13}$$

we can write (6.12) as

$$\begin{aligned}\sum_{n=1}^{\infty} f_0^{*n}(t) &= \lambda_0 \sum_{n=1}^{\infty} p_{00} * f_1 * f_0^{*(n-1)}(t) \\ &= \lambda_0 \int_0^t f_1(s) p_{00}(t-s)\, ds + \lambda_0 \sum_{n=1}^{\infty} \int_0^t f_1 * f_0^{*n}(s) p_{00}(t-s)\, ds \\ &= \lambda_0 \tilde{p}_{10}(t).\end{aligned} \tag{6.14}$$

Now (6.10) takes the form

$$\tilde{p}'_{00}(t) = -\lambda_0 \tilde{p}_{00}(t) + \lambda_0 \tilde{p}_{10}(t). \tag{6.15}$$

Equation (6.15) is the (0, 0) term in the backward equation for $\tilde{p}_{ij}(t)$. As remarked earlier (Exercise 2) the backward equations hold for all terms $\tilde{p}_{ij}(t)$.

Another method of constructing a conservative process from an explosive minimal process is to adjoin a new state Δ to S and define $X_t = \Delta$ for $t \geqslant \zeta$. Then Δ is an absorbing state and the transition probabilities are

$$p_{ij}^{\Delta}(t) = \begin{cases} p_{ij}(t) & \text{for } i, j \in S \\ 1 - \sum_{k \in S} p_{ik}(t) & \text{for } i \in S, j = \Delta \\ 1 & \text{for } i = j = \Delta \\ 0 & \text{for } i = \Delta, j \in S. \end{cases} \tag{6.16}$$

It is simple to check that (6.16) defines a transition probability (Exercise 3) and satisfies the backward as well as the forward equations for $i, j \in S$ (Exercise 3).

7 SOME EXAMPLES

Example 1. (*The Compound Poisson Process*). The state space for $\{X_t\}$ is $S = \{0, \pm 1, \pm 2, \ldots\}$. Let f be a given probability mass function on S with $f(0) < 1$, and let $\lambda > 0$. Define the **Q**-matrix by

$$\begin{aligned}\lambda_i &= -q_{ii} = \lambda(1 - f(0)), \\ k_{ij} &= \frac{q_{ij}}{\lambda_i} = \frac{f(j-i)}{1 - f(0)} \qquad \text{if } j \neq i.\end{aligned} \tag{7.1}$$

Since the λ_i do not depend on i, it follows from Theorem 5.1 that the embedded chain $\{Y_n\}$, corresponding to the transition matrix $\mathbf{K}$ and some initial distribution, is independent of the i.i.d. exponentially distributed sequence of holding times $\{T_0, T_1, \ldots\}$. Also, $\{Y_n\}$ is a random walk that may be represented as $Y_0 = X_0$, $Y_1 = X_0 + Z_1$, $Y_2 = X_0 + Z_1 + Z_2, \ldots, Y_n = X_0 + Z_1 + Z_2 + \cdots + Z_n, \ldots$, where $\{Z_1, Z_2, \ldots, Z_n, \ldots\}$ is a sequence of i.i.d. random variables with common probability function f, independent of X_0. It follows from this description that the transition probabilities are given as follows.

Let $\bar{\lambda} = \lambda(1 - f(0))$. Then,

$$\begin{aligned}
p_{ij}(t) &= \sum_{n=0}^{\infty} P_i \quad (n \text{ transitions occur in } (0, t], \text{ and } X_t = j) \\
&= \delta_{ij} e^{-\bar{\lambda}t} + \sum_{n=1}^{\infty} P_i(T_0 + T_1 + \cdots + T_{n-1} \leqslant t < T_0 + T_1 + \cdots + T_n, Y_n = j) \\
&= \delta_{ij} e^{-\bar{\lambda}t} + \sum_{n=1}^{\infty} P_i(T_0 + T_1 + \cdots + T_{n-1} \leqslant T < T_0 + T_1 + \cdots + T_n) \\
&\qquad \times P_i(Y_n = j) \\
&= \delta_{ij} e^{-\bar{\lambda}t} + \sum_{n=1}^{\infty} e^{-\bar{\lambda}t} \frac{(\bar{\lambda}t)^n}{n!} f^{*n}(j - i) \\
&= \sum_{n=0}^{\infty} e^{-\bar{\lambda}t} \frac{(\bar{\lambda}t)^n}{n!} f^{*n}(j - i),
\end{aligned} \tag{7.2}$$

with the convention that $f^{*0}(k) = \delta_{k0}$.

The rates given in (7.1) are the same as those that are obtained in (2.24).

Example 2. (*The Yule Linear Growth Process*). This process is sometimes used as a model for the growth of bacteria, the fission of neutrons, and other random replication processes. It is a special case of Example 2 in Section 5 with $f_0 = 0, f_2 = 1$. However, to keep this example self-contained, consider a situation in which there are initially i members of a population, each splitting into two identical particles after an exponentially distributed amount of time having parameter λ and independently of the others. Let X_t denote the size of the population at time t. Then $\{X_t\}$ is a Markov process on $S = \{1, 2, \ldots\}$ (Exercise 1), whose transition rates can be calculated explicitly as follows.

$$p_{ii}(h) = (e^{-\lambda h})^i = e^{-\lambda i h} = 1 - i\lambda h + o(h). \tag{7.3}$$

Similarly,

$$p_{i,i+1}(h) = \binom{i}{1}\left[\int_0^h \lambda e^{-\lambda s} e^{-2\lambda(h-s)}\, ds\right][e^{-\lambda h}]^{i-1} = i\lambda h + o(h) \tag{7.4}$$

and for $j \geqslant i + 2$,

$$0 \leqslant p_{ij}(h) \leqslant (1 - e^{-\lambda h})^{j-i} = o(h). \tag{7.5}$$

In particular, therefore, the infinitesimal rates are given by

$$q_{ii} = -i\lambda, \qquad q_{i,i+1} = i\lambda, \qquad q_{ij} = 0 \qquad \text{if } j > i + 1 \text{ or if } j < i. \tag{7.6}$$

Note that this makes $\{X_t\}$ a pure birth process. Since

$$\sum_{n=1}^{\infty} \frac{1}{\lambda_n} = \sum_{n=1}^{\infty} \frac{1}{n\lambda} = \frac{1}{\lambda} \sum_{n=1}^{\infty} \frac{1}{n} = \infty,$$

it follows from Example 6.1 that the process is *conservative*. We will use the backward equations to compute $p_{ik}(t)$ for all i, k. We have

$$\begin{aligned} p'_{ik}(t) &= 0 \qquad \text{for } k < i \quad (\text{indeed, } p_{ik}(t) = 0), \\ p'_{ii}(t) &= -i\lambda p_{ii}(t), \\ p'_{ik}(t) &= -i\lambda p_{ik}(t) + i\lambda p_{i+1,k}(t) \qquad (k \geqslant i + 1). \end{aligned} \tag{7.7}$$

Clearly,

$$p_{ii}(t) = e^{-\lambda_i t} = e^{-i\lambda t} \qquad (i = 1, 2, \ldots).$$

Substituting this into the last equation of (7.7) (with $k = i + 1$), one gets

$$p'_{i,i+1}(t) = -i\lambda p_{i,i+1}(t) + i\lambda e^{-(i+1)\lambda t}. \tag{7.8}$$

Multiplying both sides by $e^{i\lambda t}$ one gets

$$\begin{aligned} \frac{d}{dt}(e^{i\lambda t} p_{i,i+1}(t)) &= i\lambda e^{i\lambda t} p_{i,i+1}(t) + e^{i\lambda t} p'_{i,i+1}(t) \\ &= i\lambda e^{i\lambda t} e^{-(i+1)\lambda t} = i\lambda e^{-\lambda t} \end{aligned}$$

or

$$e^{i\lambda t} p_{i,i+1}(t) = \int_0^t i\lambda e^{-\lambda s}\, ds + e^{i\lambda} p_{i,i+1}(0) = i(1 - e^{-\lambda t}).$$

Therefore,

$$p_{i,i+1}(t) = ie^{-i\lambda t}(1 - e^{-\lambda t}). \tag{7.9}$$

Substituting this into (7.7) with $k = i + 2$, one gets

$$p'_{i,i+2}(t) = -i\lambda p_{i,i+2}(t) + i\lambda[(i+1)e^{-(i+1)\lambda t}(1 - e^{-\lambda t})],$$

which as above, on multiplying both sides by $e^{i\lambda t}$, yields

$$\begin{aligned} p_{i,i+2}(t) &= e^{-i\lambda t} \int_0^t i\lambda(i+1)e^{-\lambda s}(1 - e^{-\lambda s})\, ds \\ &= i(i+1)e^{-i\lambda t}[(1 - e^{-\lambda t}) + \tfrac{1}{2}(e^{-2\lambda t} - 1)] \\ &= \frac{i(i+1)}{2} e^{-i\lambda t}(1 - e^{-\lambda t})^2. \end{aligned} \tag{7.10}$$

By induction (a basis for which is laid out in (7.9)–(7.10)), one may prove (Exercise 2)

$$p_{ik}(t) = \frac{i(i+1)\cdots(k-1)}{(k-i)!} e^{-i\lambda t}(1 - e^{-\lambda t})^{k-i} \qquad \text{for } k > i. \tag{7.11}$$

Example 3. (*N-Server Queues*). Think of arrivals of customers as a Poisson process with parameter α. There are N servers. Let X_t denote the number of customers being served or waiting at time t. A customer arriving at time t is immediately served if X_t is less than N. On the other hand, if X_t exceeds or equals N, then the customer must stand in a queue and await a turn. Assume that the service time for each customer, counted from the moment of arrival at a free counter until the moment of departure from the counter, is exponentially distributed with parameter β, and assume that the service times for the different customers are independent. The process $\{X_t\}$ is Markovian (Exercise 4). The infinitesimal conditions may be found as follows. The state space is $S = \{0, 1, 2, \ldots\}$ and the infinitesimal probabilities are given by

$$p_{ii}(h) = \begin{cases} e^{-\alpha h}e^{-i\beta h} + O(h^2) & \text{if } 0 \leqslant i \leqslant N, \\ e^{-\alpha h}e^{-N\beta h} + O(h^2) & \text{if } i > N, \end{cases}$$

$$p_{i,i+1}(h) = \begin{cases} e^{-\alpha h}\alpha h e^{-i\beta h} + O(h^2) & \text{if } 0 \leqslant i \leqslant N, \\ e^{-\alpha h}\alpha h e^{-N\beta h} + O(h^2) & \text{if } i > N, \end{cases}$$

$$p_{i,i-1}(h) = \begin{cases} e^{-\alpha h}i(1 - e^{-\beta h})e^{-(i-1)\beta h} + O(h^2) & \text{if } 1 \leqslant i \leqslant N, \\ e^{-\alpha h}N(1 - e^{-\beta h})e^{-(N-1)\beta h} + O(h^2) & \text{if } i > N, \end{cases}$$

$$p_{ij}(h) = O(h^2) \qquad \text{if } |i - j| > 1.$$

Therefore,

$$q_{ii} = -(\alpha + i\beta) \quad \text{if } 0 \leqslant i \leqslant N, \qquad q_{ii} = -(\alpha + N\beta) \quad \text{if } i > N;$$
$$q_{i,i+1} = \alpha \quad \text{if } i \geqslant 0; \tag{7.12}$$
$$q_{i,i-1} = i\beta \quad \text{if } 1 \leqslant i \leqslant N, \qquad q_{i,i-1} = N\beta \quad \text{if } i > N.$$

It is difficult to calculate explicitly the transition probabilities by either of the methods of the two preceding examples. In the next example, we consider the case of infinitely many servers and zero waiting time for mathematical convenience. The *N-server queue* is a special case of the *continuous-parameter birth–death Markov process* to be analyzed in Example 1 of Section 8.

Example 4. (*Queues with Infinitely Many Servers*). In the physical model, X_t denotes the number of customers being served at time t. Suppose that the number Y_t of arrivals of customers is a Poisson process with parameter α. Each customer is upon arrival immediately attended to by a server, the service time being an exponential random variable with parameter β. The service times of different customers are independent. Suppose that initially there are $X_0 = i$ customers who are being served. Because of the *"lack of memory"* property of the exponential distribution, the remaining service times for these i customers are still independent exponential random variables with parameter β. Of the i customers served at time zero, let Z_t be the number of those who remain at the counters (being served) at time t. Then Z_t is a binomial random variable with parameters i and $e^{-\beta t}$. During the time interval $(0, t]$, however, there will be Y_t new arrivals. Suppose W_t of them remain at the counters at time t. Then $X_t = Z_t + W_t$. Of course, Z_t and W_t are independent random variables. To determine the distribution of X_t, given $X_0 = i$, we therefore need to compute the distribution of W_t given $Y_t = k$. Given that there are k arrivals in a Poisson process in $(0, t]$, the k arrival times, say $\tau_1, \tau_1 + \tau_2, \ldots, \tau_1 + \tau_2 + \cdots + \tau_k$, may be thought of as k independent observations $U_1, U_2, \ldots, U_k$ from the uniform distribution on $(0, t]$ arranged in increasing order of magnitude (see Proposition 5.6). Since the probability that an individual who arrives at time u will not leave the counter by time t is $e^{-\beta(t-u)}$, the probability that a random arrival in $(0, t]$ will not leave by time t is

$$\frac{1}{t}\int_0^t e^{-\beta(t-u)}\,du = \frac{1 - e^{-\beta t}}{\beta t}. \tag{7.13}$$

Therefore, the conditional distribution of W_t given $Y_t = k$ is binomial with parameters k and $(1 - e^{-\beta t})/\beta t$. Unconditionally, then,

$$
\begin{aligned}
P(W_t = j) &= \sum_{k=j}^{\infty} P(Y_t = k) P(W_t = j \mid Y_t = k) \\
&= \sum_{k=j}^{\infty} e^{-\alpha t} \frac{(\alpha t)^k}{k!} \binom{k}{j} \left(\frac{1 - e^{-\beta t}}{\beta t} \right)^j \left(1 - \frac{1 - e^{-\beta t}}{\beta t} \right)^{k-j} \\
&= \frac{e^{-\alpha t}}{j!} \left(\frac{1 - e^{-\beta t}}{\beta t} \right)^j (\alpha t)^j \sum_{k=j}^{\infty} \frac{(\alpha t)^{k-j}}{(k-j)!} \left(1 - \frac{1 - e^{-\beta t}}{\beta t} \right)^{k-j} \\
&= \frac{e^{-\alpha t}}{j!} \left(\frac{1 - e^{-\beta t}}{\beta t} \right)^j (\alpha t)^j e^{\alpha t[1 - (1 - e^{-\beta t})/\beta t]} \\
&= e^{-(\alpha/\beta)(1 - e^{-\beta t})} \left(\frac{[(\alpha/\beta)(1 - e^{-\beta t})]^j}{j!} \right) \qquad (j \geqslant 0). \qquad (7.14)
\end{aligned}
$$

In other words, W_t *is Poisson with parameter* $(\alpha/\beta)(1 - e^{-\beta t})$. Since Z_t is binomial, it follows from (7.14) that

$$
\begin{aligned}
p_{ik}(t) &= P(X_t = k \mid X_0 = i) = P(Z_t + W_t = k \mid X_0 = i) \\
&= \sum_{r=0}^{\min\{i,k\}} \binom{i}{r} (e^{-\beta t})^r (1 - e^{-\beta t})^{i-r} e^{-(\alpha/\beta)(1 - e^{-\beta t})} \frac{[(\alpha/\beta)(1 - e^{-\beta t})]^{k-r}}{(k-r)!} \\
&= e^{-(\alpha/\beta)(1 - e^{-\beta t})} (1 - e^{-\beta t})^i [(\alpha/\beta)(1 - e^{-\beta t})]^k i! \sum_{r=0}^{\min\{i,k\}} \frac{1}{r!(i-r)!(k-r)!} \left(\frac{\alpha}{\beta} \right)^{-r} \\
&\quad \times [e^{\beta t} + e^{-\beta t} - 2]^{-r} \qquad \text{for } k = 0, 1, \ldots . \qquad (7.15)
\end{aligned}
$$

From (7.15) one can directly check the infinitesimal conditions $q_{ii} = -(\alpha + i\beta)$ for $i \geqslant 0$, $q_{i,i+1} = \alpha$ and $q_{i,i-1} = i\beta$ for $i \geqslant 1$, $q_{0,1} = \alpha$. However, it is simpler to use the probabilistic description to arrive at these. For example, $p_{ii}(h)$ is the sum of the probabilities of the events that there are no arrivals in $(0, h]$ and none of the $X_0 = i$ customers leaves during $(0, h]$, or one customer arrives in $(0, h]$ *and* one of $i + 1$ customers leaves during $(0, h]$, etc. Therefore,

$$
p_{ii}(h) = e^{-\alpha h} (e^{-\beta h})^i + O(h^2). \qquad (7.16)
$$

More generally, since the chance of two or more occurrences in a small interval of length h is $O(h^2)$, one has

$$
\begin{aligned}
p_{ii}(h) &= e^{-(\alpha + i\beta)h} + O(h^2) = 1 - (\alpha + i\beta)h + O(h^2), \\
p_{i,i+1}(h) &= \left(\int_0^h \alpha e^{-\alpha s} e^{-\beta(h-s)} \, ds \right) e^{-i\beta h} + O(h^2)
\end{aligned}
$$

$$\begin{aligned}
&= e^{-\beta h - i\beta h} \int_0^h \alpha e^{-(\alpha-\beta)s}\, ds + O(h^2) = e^{-(i+1)\beta h}\alpha h + O(h^2) \\
&= \alpha h - \alpha(i+1)\beta h^2 + O(h^2) = \alpha h + O(h^2),
\end{aligned} \tag{7.17}$$

$$\begin{aligned}
p_{i,i-1}(h) &= (e^{-\alpha h}) i(1 - e^{-\beta h}) e^{-(i-1)\beta h} \\
&= i(1 - \alpha h)\beta h(1 - (i-1)\beta h) + O(h^2) = i\beta h + O(h^2).
\end{aligned}$$

Therefore,

$$\begin{aligned}
q_{ii} &= \lim_{h \downarrow 0} \frac{p_{ii}(h) - 1}{h} = -(\alpha + i\beta), \\
q_{i,i+1}(h) &= \lim_{h \downarrow 0} \frac{p_{i,i+1}(h) - 0}{h} = \alpha, \qquad q_{i,i-1} = i\beta.
\end{aligned} \tag{7.18}$$

For completeness one needs to argue that $\{X_t\}$ is a Markov process. For this note that X_{s+t} is a function of

(i) X_s and the *additional* service times up to time $s + t$ required by the X_s customers present at time s, and
(ii) the numbers and times of arrivals of those customers arriving during $(s, s + t]$ and their service times.

But the latter (i.e., (ii)) are stochastically independent of $\{X_u: 0 \leqslant u \leqslant s\}$ since $Y_{s+t} - Y_s$ is independent of Y_u for $0 \leqslant u \leqslant s$, and the service times of *all* customers are independent of each other as well as of the process $\{Y_u: u \geqslant 0\}$. Also, the conditional distribution of the *additional service times* of the X_s customers who are being served at time s, given $\{X_u, 0 \leqslant u \leqslant s\}$ (or, for that matter, given $\{Y_u, 0 \leqslant u \leqslant s\}$, as well as the service times of all these arrivals in $[0, s]$ already spent in $[0, s]$) is still that of X_s i.i.d. exponential random variables each having parameter β ("lack of memory" property). Hence $P(X_{s+t} = k \mid X_u: 0 \leqslant u \leqslant s) = P(X_{s+t} = k \mid X_s)$.

Example 5. (*Monte Carlo Approaches to the Telegrapher's Equation*). Monte Carlo methods broadly refer to numerical approximations based on averages of simulations of suitably designed random processes. Modern-day computing speed and precision make the random-number generator an important tool for numerical calculations. The results and methods of Chapter I indicate how discrete time and space *Monte Carlo numerical solutions* to initial and boundary-value problems associated with the heat equation (or diffusion equation)

$$\frac{\partial u}{\partial t} = D\frac{\partial^2 u}{\partial x^2} + c\frac{\partial u}{\partial x}$$

might be obtained from density profiles computed from simulated random walks. A connection between general *linear parabolic* (*diffusion*) *partial differential equations* and discrete space and/or time *Markov birth–death chains* is described in Section 4 of Chapter V.

In the present example we will consider a probabilistic approach to a special *hyperbolic* (*wave*) *equation* known as the *telegrapher's equation*. Namely,

$$\beta \frac{\partial^2 u}{\partial t^2} - v^2 \frac{\partial^2 u}{\partial x^2} + 2\alpha \frac{\partial u}{\partial t} = 0, \tag{7.19}$$

with the initial conditions of the form $u(x, 0) = \varphi(x)$ and $(\partial u/\partial t)\big|_{t=0} = 0$ for a suitable initial profile $\varphi(x)$. The telegrapher's equation arises in connection with the spatiotemporal evolution of both the voltage and the charge density in one-dimensional electrical transmission lines. In this framework the coefficients β, α, and v^2 are all nonnegative real numbers; in terms of the physical parameters $\beta/v^2 = LC$ and $2\alpha/v^2 = RC$, where R is the electrical resistance, L is the inductance, and C is the capacitance (see Exercise 9). However, these quantities play no particular role in the probabilistic derivation.

After dividing through by $\beta > 0$ and adjusting the other parameters accordingly, we may take $\beta = 1$. Partition the state space as a one-dimensional lattice of sites spaced $\Delta > 0$ units apart and partition time into units of $0, \tau, 2\tau, 3\tau, \ldots$, with $\Delta = v\tau$. A particle is started at $x = 0$ and in the first unit of time τ moves with speed $v > 0$, with some probability π^+ in the positive direction to $\Delta(= v\tau)$, or moves with probability $\pi^- = 1 - \pi^+$ in the negative direction at speed v to $-\Delta$. Subsequently, at each successive time step the particle will continue to travel at the rate v but at each time step, independently of the previous choice(s), may choose to *reverse* its previous direction with probability $p = \alpha\tau$. Notice that, given the initial velocity, the time step until the particle makes a reversal in direction is geometrically distributed with mean $1/p = 1/\alpha\tau$ and variance $q/p^2 = (1 - \alpha\tau)/(\alpha\tau)^2$. The reversal mechanism provides a kind of *stochastic oscillation* in the motion.

The position process $\{S_n\}$ is represented in terms of successive displacements by

$$S_0 = 0, \qquad S_n = X_1 + X_2 + \cdots + X_n, \qquad n \geqslant 1, \tag{7.20}$$

where $\{X_n\}$ is a two-state Markov chain with initial distribution $P(X_1 = \Delta) = \pi^+ = 1 - P(X_1 = -\Delta)$ and homogeneous one-step transition law determined by

$$P(X_{n+1} = \Delta \mid X_n = \Delta) = P(X_{n+1} = -\Delta \mid X_n = -\Delta) = 1 - p, \qquad n \geqslant 1. \tag{7.21}$$

Let us now consider the average profile after n time steps and viewed from the position of the particles initially moving to the right, starting from the initial position $m\Delta$. Let

$$u^+(n\tau, m\Delta) = E(\varphi(m\Delta + S_n) \mid X_1 = \Delta), \tag{7.22}$$

and let

$$u^-(n\tau, m\Delta) = E(\varphi(m\Delta + S_n) \mid X_1 = -\Delta). \tag{7.23}$$

Note that the conditional distribution of $\{S_n\}$ given $X_1 = -\Delta$ coincides with the conditional distribution of $\{-S_n\}$ given $X_1 = \Delta$. Therefore,

$$u^-(n\tau, m\Delta) = E(\varphi(m\Delta - S_n) \mid X_1 = \Delta). \tag{7.24}$$

Conditioning on X_2 in (7.22) we can write

$$\begin{aligned} u^+(n\tau, m\Delta) &= E(E(\varphi(m\Delta + S_n) \mid X_2) \mid X_1 = \Delta) \\ &= E(\varphi(m\Delta - S_{n-1} + \Delta) \mid X_1 = \Delta)p \\ &\quad + E(\varphi(m\Delta + S_{n-1} + \Delta) \mid X_1 = \Delta)(1 - p) \\ &= u^-((n-1)\tau, (m+1)\Delta)\alpha\tau + u^+((n-1)\tau, (m+1)\Delta)(1 - \alpha\tau). \end{aligned} \tag{7.25}$$

Conditioning likewise in (7.23) gives

$$u^-(n\tau, m\Delta) = u^+((n-1)\tau, (m-1)\Delta)\alpha\tau + u^-((n-1)\tau, (m-1)\Delta)(1 - \alpha\tau). \tag{7.26}$$

Note that taking $\pi^+ = \pi^- = \frac{1}{2}$, $u(n\tau, m\Delta) := E\varphi(m\Delta + S_n) = \frac{1}{2}(u^+ + u^-)$. We now have the first-order equations for u^+ and u^- given by

$$\begin{aligned} \frac{u^+(n\tau, x) - u^+((n-1)\tau, x)}{\tau} &= \frac{u^+((n-1)\tau, x + v\tau) - u^+((n-1)\tau, x)}{\tau} \\ &\quad - \alpha u^+((n-1)\tau, x + v\tau) + \alpha u^-((n-1)\tau, x + v\tau), \end{aligned} \tag{7.27}$$

$$\begin{aligned} \frac{u^-(n\tau, x) - u^-((n-1)\tau, x)}{\tau} &= \frac{u^-((n-1)\tau, x - v\tau) - u^-((n-1)\tau, x)}{\tau} \\ &\quad - \alpha u^-((n-1)\tau, x - v\tau) + \alpha u^+((n-1)\tau, x - v\tau), \end{aligned} \tag{7.28}$$

where $x = m\Delta$, $\Delta = v\tau$. In the limit as $\tau \to 0$, $\Delta = v\tau \to 0$, $n, m \to \infty$ such that $x = m\Delta$, $t = n\tau$, these equations go over to

$$\frac{\partial u^+}{\partial t} = v\frac{\partial u^+}{\partial x} - \alpha u^+ + \alpha u^-, \tag{7.29}$$

$$\frac{\partial u^-}{\partial t} = -v\frac{\partial u^-}{\partial x} - \alpha u^- + \alpha u^+. \tag{7.30}$$

Let, for the solutions u^+, u^- to these equations,

$$u = \frac{u^+ + u^-}{2} \quad \text{and} \quad w = \frac{u^+ - u^-}{2}. \tag{7.31}$$

By combining (7.29) and (7.30) we get the so-called *transmission line equations* for u and w,

$$\frac{\partial u}{\partial t} = v \frac{\partial w}{\partial x}, \tag{7.32}$$

$$\frac{\partial w}{\partial t} = v \frac{\partial u}{\partial x} - 2\alpha w. \tag{7.33}$$

Now w can be eliminated by differentiating (7.32) with respect to t and (7.33) with respect to x and combining the equations. This results in the *telegrapher's equation* for u. The initial conditions can also be checked by passage to the limit.

The natural Monte Carlo simulation for solving the telegrapher's equation suggested by the above analysis is to start a large number of noninteracting particles in motion according to the above scheme, say half of them starting to the right and the other half going to the left initially. One then approximates $u(t, x)$, $t = n\tau$, $x = m\Delta$, by the arithmetic mean $N^{-1} \sum_{k=1}^{N} \varphi(x + S_n^{(k)})$, where N is the number of particles. However, there is a practical issue that makes the approach *unfeasible*. Namely, for a small time–space grid, the probability $p = \alpha\tau$ of a reversal will also be very small; the smaller the grid size the larger the mean time to reversal and its variance. This means that an extremely large number of particle evolutions will have to be simulated in order to keep the fluctuations in the average down. Mark Kac first suggested that for this problem it is possible, and more practical, to use *continuous-time* simulations. The idea is to consider directly the time to velocity reversal. In the above scheme this time is τN_τ, where N_τ, the number of time steps until reversal, is geometrically distributed with parameter $p = \alpha\tau$. In the limit as $\tau \to 0$ the time therefore converges in distribution to the exponential distribution with parameter α. So it at least seems reasonable to consider the continuous-parameter position process $\{Y_t\}$ defined by the motion of a particle that, starting from the origin, travels at a rate v for an exponentially distributed length of time, reverses its direction, and then travels again for an (independent) exponentially distributed length of time at the speed v before again reversing its direction, and so on. The Poisson process of reversal times can be accurately simulated.

To make the above ideas firm, let $\{N_t\}$ be the continuous-parameter Poisson process with parameter α. Let ε be a ± 1-valued random variable independent of $\{N_t\}$ with $P(\varepsilon = 1) = \frac{1}{2}$. The *velocity process* is the two-state Markov process $\{V_t\}$ defined by

$$V_t = v\varepsilon(-1)^{N_t}, \qquad t \geqslant 0. \tag{7.34}$$

The *position process*, starting from x, is therefore given by

$$Y_t = x + \int_0^t V_s\,ds = x + v\varepsilon \int_0^t (-1)^{N_s}\,ds. \tag{7.35}$$

Although $\{Y_t\}$ is not a Markov process, the joint evolution $\{(Y_t, V_t)\}$ of position and velocity is a Markov process, as is the velocity process $\{V_t\}$ alone. This is seen as follows.

$$\begin{aligned}
P(a \leqslant Y_{t+s} &\leqslant b, V_{t+s} = w \mid \{Y_u : 0 \leqslant u \leqslant t\}, \{V_t : 0 \leqslant u \leqslant t\}) \\
&= P\left(a \leqslant Y_t + \int_0^s V_{t+\tau}\,d\tau \leqslant b, V_{t+s} = w \,\middle|\, \{Y_0 = x\}, \{V_u : 0 \leqslant u \leqslant t\}\right) \\
&= P(a \leqslant Y_s \leqslant b, V_s = w \mid Y_0 = y, V_0 = w')\big|_{y = Y_t, w' = V_t}.
\end{aligned} \tag{7.36}$$

One expects from the earlier considerations that the solution to the telegrapher's equation with the prescribed initial conditions is given by

$$\begin{aligned}
u(t, x) &= E\varphi(x + Y_t) \\
&= \frac{1}{2} E\left[\varphi\left(x + v\int_0^t (-1)^{N_s}\,ds\right) + \varphi\left(x - v\int_0^t (-1)^{N_s}\,ds\right)\right].
\end{aligned} \tag{7.37}$$

To verify that (7.37) indeed solves the problem for a reasonably large class of initial profiles, let $f(x, w) = \varphi(x)$, for $w = v, -v$. Define

$$g(t, x, w) = E\{f(Y_t, V_t) \mid V_0 = w, Y_0 = x\} = T_t f(x, w). \tag{7.38}$$

Let $g^+(t, x) = g(t, x, v)$, and $g^-(t, x) = g(t, x, -v)$. Since $\{(Y_t, V_t)\}$ is a Markov process, g satisfies the backward equation

$$\frac{\partial g}{\partial t} = Ag(t, x, w) = w\frac{\partial g}{\partial x} + \alpha[g(t, x, -w) - g(t, x, w)]. \tag{7.39}$$

In order to check that the infinitesimal generator A is of this form, let $x \to f(x, w)$ be, for $w = v$ and $w = -v$, a bounded twice-differentiable function with bounded derivatives. Then, as $s \downarrow 0$,

$$\begin{aligned}
g(s, x, w) &= E(f(Y_s, V_s) \mid Y_0 = x, V_0 = w) \\
&= E\left[f(x, V_s) + (Y_s - x)\frac{\partial f(x, V_s)}{\partial x} \,\middle|\, Y_0 = x, V_0 = w\right] + o(s) \\
&= (f(x, w)e^{-\alpha s} + \alpha s e^{-\alpha s} f(x, -w) + o(s))
\end{aligned}$$

$$
\begin{aligned}
&+ E\left((Y_s - x)\frac{\partial f(x, w)}{\partial x} \,\middle|\, Y_0 = x, V_0 = w \right) + o(s) \\
&= f(x, w) - \alpha s(f(x, w) - f(x, -w)) \\
&\quad + \frac{\partial f(x, w)}{\partial x} w \int_0^s (E(-1)^{N_\tau})\, d\tau + o(s) \\
&= f(x, w) + \alpha s(f(x, -w) - f(x, w)) \\
&\quad + w \frac{\partial f(x, w)}{\partial x} \int_0^s (e^{-\alpha\tau} + o(\tau))\, d\tau + o(s) \\
&= f(x, w) + \alpha s(f(x, -w) - f(x, w)) + w \frac{\partial f(x, w)}{\partial x} s + o(s). \qquad (7.40)
\end{aligned}
$$

Taking $w = v$ and $w = -v$ in (7.39), we get the two first-order equations (7.29) and (7.30) satisfied by g^+ and g^-, from which the corresponding transmission line equations and then the telegrapher's equation follow from (7.39) in the same way as before.

8 ASYMPTOTIC BEHAVIOR OF CONTINUOUS-TIME MARKOV CHAINS

Let $\{X_t\}$ be a Markov chain on a countable state space S having a transition probability matrix $\mathbf{p}(t)$, $t \geqslant 0$.

As in the case of discrete time, write $i \to j$ if $p_{ij}(t) > 0$ for some $t > 0$. States i and j *communicate*, denoted $i \leftrightarrow j$, if $i \to j$ and $j \to i$. Say "i is *essential*" if $i \to j$ implies $j \to i$ for all j, otherwise say "i is *inessential*." The following analog of Proposition 5.1 of Chapter II is proved in exactly the same manner, replacing $p_{ij}^{(m)}$, $p_{ji}^{(n)}$, etc., by $p_{ij}(t)$, $p_{ji}(s)$, etc. Let $\mathscr{E}$ denote the set of all essential states.

Proposition 8.1

(a) For every i there exists at least one j such that $i \to j$.

(b) If $i \to j$, $j \to k$ then $i \to k$.

(c) If i is essential then $i \leftrightarrow i$.

(d) If i is essential and $i \to j$ then "j is essential" and $i \leftrightarrow j$.

(e) On $\mathscr{E}$ the relation "$\leftrightarrow$" is an equivalence relation, i.e., reflexive, symmetric, and transitive.

By (e) of Proposition 8.1, $\mathscr{E}$ decomposes into disjoint equivalence classes of states such that members of a class communicate with each other. If i and j belong to different classes then $i \not\leftrightarrow j$, $j \not\leftrightarrow i$.

A significant departure from the discrete-parameter case is that states in *continuous parameter chains are not periodic*. More precisely, one has the following proposition.

Proposition 8.2

(a) For each state i, $p_{ii}(t) > 0$ for all $t > 0$.

(b) For each pair i, j of distinct states, either $p_{ij}(t) = 0$ for all $t > 0$ or $p_{ij}(t) > 0$ for all $t > 0$.

Proof. (a) If there is a positive $t_0 > 0$ such that $p_{ii}(t_0) = 0$, then since

$$\left(p_{ii}\left(\frac{t_0}{n}\right)\right)^n = p_{ii}\left(\frac{t_0}{n}\right)p_{ii}\left(\frac{t_0}{n}\right)\cdots p_{ii}\left(\frac{t_0}{n}\right) \leqslant p_{ii}(t_0),$$

$$p_{ii}\left(\frac{t_0}{n}\right) = 0$$

for all positive integers n. Taking the limit as $n \to \infty$, one gets by continuity that $\lim_{n\to\infty} p_{ii}(t_0/n) = 0$, which is a contradiction to continuity at $t = 0$ and the requirement $p_{ii}(0) = 1$.

(b) Let i, j be two distinct states. Suppose t_0 is a positive number such that $p_{ij}(t_0) = 0$. Since $p_{ij}(t_0) \geqslant p_{ii}(t_0 - s)p_{ij}(s)$ for all s, $0 \leqslant s < t_0$, and since $p_{ii}(t_0 - s) > 0$ it follows that $p_{ij}(s) = 0$ for all $s \leqslant t_0$. Then $q_{ij} = p'_{ij}(0) = 0$, and $k_{ij} = 0$. It follows from (5.4) that $P_i(X_{T_0} = j) = 0$. This implies that $k_{j'j} = 0$ for every j' such that $k_{ij'} > 0$. For, if $k_{ij'} > 0$ and $k_{j'j} > 0$, then no matter how small t is ($t > 0$), one has, denoting by $\{Y_n: n = 0, 1, 2, \ldots\}$ the embedded discrete parameter Markov chain,

$$\begin{aligned} p_{ij}(t) &\geqslant P_i(X_{T_0} = j', X_{T_0+T_1} = j, T_0 + T_1 \leqslant t < T_0 + T_1 + T_2) \\ &= P_i(Y_1 = j', Y_2 = j, T_0 + T_1 \leqslant t < T_0 + T_1 + T_2) \\ &= k_{ij'}k_{j'j} \iiint_{\{t_i + t_{j'} \leqslant t < t_i + t_{j'} + t_j\}} \lambda_i e^{-\lambda_i t_i}\lambda_{j'}e^{-\lambda_{j'}t_{j'}}\lambda_j e^{-\lambda_j t_j}\,dt_i\,dt_{j'}\,dt_j > 0 \end{aligned}$$

$$\text{if } \lambda_j > 0. \quad (8.1)$$

If $\lambda_j = 0$ then, given $Y_2 = j$, $T_2 = \infty$ with probability 1. Thus, the triple integral may be replaced by

$$\iint_{\{t_i + t_{j'} \leqslant t\}} \lambda_i \lambda_{j'} e^{-\lambda_i t_i + \lambda_{j'} t_{j'}}\,dt_i\,dt_{j'} > 0.$$

This contradicts "$p_{ij}(s) = 0$ for all $s \leqslant t_0$." Thus, $k_{ij}^{(2)} = 0$. In this manner one may prove that $k_{ij}^{(n)} = 0$ for all n. This implies $p_{ij}(t) = 0$ for all t. ■

A state i is *recurrent* if

$$P_i(\sup\{t \geqslant 0: X_t = i\} = \infty) = 1. \tag{8.2}$$

A state i is *transient* if

$$P_i(\sup\{t \geqslant 0: X_t = i\} < \infty) = 1. \tag{8.3}$$

Define the *time of first return to i* by

$$\eta_i = \inf\{t \geqslant T_0: X_t = i\}. \tag{8.4}$$

Also denote by ρ_{ij} the probability

$$\rho_{ij} = P_i(\eta_j < \infty). \tag{8.5}$$

Thus, if $j \neq i$, then ρ_{ij} is the probability that the particle starting at i will ever visit j. Note that ρ_{ii} is the probability that the particle will ever return to i after having visited another state.

Theorem 8.3

(a) A state i is recurrent or transient for $\{X_t\}$ according as i is recurrent or transient for the embedded discrete parameter chain $\{Y_n\}$ having transition matrix $\mathbf{K}$.

(b) i is recurrent if and only if $\rho_{ii} = 1$.

(c) Recurrence (or transience) is a class property.

Proof. (a) Suppose i is recurrent. This clearly implies that $P_i(Y_n = i$ for infinitely many integers $n) = 1$. Conversely, suppose i is a recurrent state of the discrete-parameter embedded chain $\{Y_n\}$. Let

$$\eta_i = \tau_i^{(1)} = \inf\{t \geqslant T_0: X_t = i\}, \qquad \tau_i^{(r)} = \inf\{t > \tau_i^{(r-1)}: X_t = i\}, \tag{8.6}$$

for $r = 2, 3, \ldots$, denote the times of successive returns to state i. Under P_i the random variables $\tau_i^{(1)}, \tau_i^{(2)} - \tau_i^{(1)}, \ldots, \tau_i^{(r)} - \tau_i^{(r-1)}$ are i.i.d. and positive. Hence, the nth partial sum $\tau_i^{(n)}$ converges to ∞ with probability 1 as $n \to \infty$ (Exercise 1). Therefore (8.2) holds.

If i is transient for $\{X_t\}$, then i cannot be recurrent for $\{Y_n\}$ since this would imply i is recurrent for $\{X_t\}$. This implies that i is transient for $\{Y_n\}$, since, for a discrete-parameter chain a state is either recurrent or transient.

(b) Since transitions occur only at times $T_0, T_0 + T_1, T_0 + T_1 + T_2, \ldots$, it follows that

$$\rho_{ij} = P_i(Y_n = j \text{ for some } n \geqslant 1). \tag{8.7}$$

In other words, the ρ_{ij} defined by (8.5) for $\{X_t\}$ are the same as those defined for the embedded discrete-parameter chain. Hence, (b) follows from (a) and Proposition 8.1 of Chapter II.

(c) i is recurrent and $i \leftrightarrow j$ for $\{X_t\}$ implies "i is recurrent and $i \leftrightarrow j$ for $\{Y_n\}$," which implies that j is recurrent for $\{Y_n\}$, which, in turn, implies that j is recurrent for $\{X_t\}$. ∎

A state i is *positive recurrent* if

$$\mu_i := E_i \eta_i < \infty. \tag{8.8}$$

Note that if (8.8) holds then $\rho_{ii} = 1$. Therefore, positive recurrence implies recurrence.

If a recurrent state i is such that $E_i \eta_i = \infty$, it is called *null recurrent*.

Suppose that S is a single essential class of positive recurrent states, under $\mathbf{p}(t)$. By the *strong law of large numbers* one has with probability 1,

$$\lim_{r \to \infty} \frac{\tau_i^{(r)}}{r} = \lim_{r \to \infty} \frac{\sum_{r'=1}^{r} \tau_i^{(r'+1)} - \tau_i^{(r')}}{r} = E(\tau_i^{(2)} - \tau_i^{(1)}) = E_i \tau_i^{(1)}. \tag{8.9}$$

Similarly, if f is a real-valued function on S such that

$$E_i \left| \int_0^{\tau_i^{(1)}} f(X_s)\, ds \right| < \infty, \tag{8.10}$$

then applying the strong law of large numbers to the i.i.d. sequence

$$Z_r = \int_{\tau_i^{(r)}}^{\tau_i^{(r+1)}} f(X_s)\, ds \qquad (r = 1, 2, \ldots), \tag{8.11}$$

it follows that, with probability 1,

$$\lim_{r \to \infty} \frac{1}{r} \sum_{r'=1}^{r} Z_{r'} = E_i \int_0^{\tau_i^{(1)}} f(X_s)\, ds. \tag{8.12}$$

Writing v_t = number of visits to state i during $(0, t]$, one has

$$\frac{1}{t} \int_0^t f(X_s)\, ds = \frac{1}{t} \int_0^{\tau_i^{(1)}} f(X_s)\, ds + \frac{1}{t} \sum_{r'=1}^{v_t} Z_{r'} + \frac{1}{t} \int_{\tau_i^{(v_t)}}^t f(X_s)\, ds. \tag{8.13}$$

The first and third terms on the right-hand side of (8.13) go to zero as $t \to \infty$ with probability 1 (Exercise 2). By (8.9), $(v_t/t) \to (E_i \tau_i^{(1)})^{-1}$ with probability 1 as $t \to \infty$. Therefore we have the following theorem.

Theorem 8.4. Suppose S comprises a single essential class of positive recurrent states. If f satisfies (8.10) for some i, then with probability 1,

$$\lim_{t\to\infty} \frac{1}{t}\int_0^t f(X_s)\,ds = (E_i\tau_i^{(1)})^{-1}E_i\int_0^{\tau_i^{(1)}} f(X_s)\,ds, \tag{8.14}$$

regardless of the initial distribution.

The following analog of Theorem 9.2 of Chapter II may be proved in exactly the same manner using Theorem 8.4.

Corollary 8.5. Suppose S consists of a single positive recurrent class under $\mathbf{p}(t)$. Then:

(a) For all $i, j \in S$

$$\lim_{t\to\infty} \frac{1}{t}\int_0^t p_{ij}(s)\,ds = (E_j\tau_j^{(1)})^{-1}. \tag{8.15}$$

(b) The transition probability $\mathbf{p}(t)$ admits a unique invariant initial distribution $\boldsymbol{\pi}$ given by

$$\pi_j = (E_j\tau_j^{(1)})^{-1}, \qquad j \in S.$$

(c) If (8.10) holds, then the limit in (8.14) also equals $E_\pi f(X_0) = \sum_j \pi_j f(j)$.

To state the analog of Theorem 10.2 of Chapter II, write $\bar{f}(j) = f(j) - E_\pi f(X_0)$ and

$$W_T(t) = \frac{1}{\sqrt{T}}\int_0^{Tt} \bar{f}(X_s)\,ds \qquad (t \geqslant 0). \tag{8.16}$$

Theorem 8.6. Assume, in addition to the hypothesis of Theorem 8.4, that

$$\sigma^2 := E_j\left(\int_0^{\tau_j^{(1)}} \bar{f}(X_s)\,ds\right)^2 < \infty. \tag{8.17}$$

Then, as $T \to \infty$, the stochastic process $\{W_T(t): t \geqslant 0\}$ converges in distribution to Brownian motion with drift zero and diffusion coefficient

$$\delta^2 = (E_j\tau_j^{(1)})^{-1}\sigma^2. \tag{8.18}$$

A complete analog of Theorem 9.4 of Chapter II follows from Theorems 8.3, 8.4, Corollary 8.5, if one observes that in the case S comprises a single communicating class, the Markov chain $\{X_t\}$ is positive recurrent if and only if it admits an invariant probability distribution. The "only if" part of the last statement follows from Corollary 8.5(b). For the sufficiency see Exercise 12.

In order to find an invariant probability $\boldsymbol{\pi}$, one may try to differentiate the right side of the equation $\boldsymbol{\pi}'\mathbf{p}(t) = \boldsymbol{\pi}'$ to get

$$\boldsymbol{\pi}'\mathbf{Q} = \mathbf{0}. \tag{8.19}$$

A sufficient condition for the validity of term-by-term differentiation is given in Exercise 11.

Example 1. (*Birth–Death Process with One Reflecting Boundary*)

$$S = \{0, 1, 2, \ldots\}, \qquad q_{01} = \lambda_0, \qquad q_{00} = -\lambda_0;$$
$$q_{i,i-1} = \delta_i\lambda_i, \qquad q_{i,i+1} = \beta_i\lambda_i, \qquad q_{ii} = -\lambda_i \qquad (i \geqslant 1).$$

Here $\beta_i, \delta_i > 0$ and $\beta_i + \delta_1 = 1$. By Theorem 8.3 and Section 2 of Chapter III, we know that this Markov chain is recurrent if and only if

$$\sum_{x=1}^{\infty} \frac{\delta_1\delta_2\cdots\delta_x}{\beta_1\beta_2\cdots\beta_x} = \infty. \tag{8.20}$$

Also, the equations (8.19) become

$$\begin{aligned} -\pi_0\lambda_0 + \pi_1\lambda_1\delta_1 &= 0, \\ \pi_{j-1}\beta_{j-1}\lambda_{j-1} - \pi_j\lambda_j + \pi_{j+1}\delta_{j+1}\lambda_{j+1} &= 0 \qquad \text{for } j \geqslant 1. \end{aligned} \tag{8.21}$$

Solving these successively in terms of π_0 one gets, with $\beta_0 = 1$,

$$\pi_1 = \frac{\lambda_0}{\lambda_1\delta_1}\pi_0, \qquad \pi_j = \left(\frac{\lambda_0}{\lambda_j}\right)\frac{\beta_0\beta_1\beta_2\cdots\beta_{j-1}}{\delta_1\delta_2\cdots\delta_j}\pi_0 \qquad \text{for } j \geqslant 2. \tag{8.22}$$

For the chain to be positive recurrent requires $\sum \pi_j < \infty$, i.e.,

$$\sum_{j=1}^{\infty}\left(\frac{\lambda_0}{\lambda_j}\right)\frac{\beta_0\beta_1\beta_2\cdots\beta_{j-1}}{\delta_1\delta_2\cdots\delta_j} < \infty. \tag{8.23}$$

If this holds, take

$$\pi_0 = \left(1 + \sum_{j=1}^{\infty}\left(\frac{\lambda_0}{\lambda_j}\right)\frac{\beta_0\beta_1\beta_2\cdots\beta_{j-1}}{\delta_1\delta_2\cdots\delta_j}\right)^{-1} \tag{8.24}$$

in (8.22) to get the unique invariant initial distribution. Thus, *the necessary and sufficient conditions for positive recurrence* (or for the existence of a steady state) *are* (8.20) *and* (8.23).

The *N-server queue* described in Example 7.3 is a continuous-parameter birth–death chain on $S = \{0, 1, 2, \ldots\}$ with infinitesimal rate parameters given by (7.12). In particular, the process has a reflecting boundary at zero so that the criterion for recurrence and positive recurrence apply. The series (8.20) diverges or converges according to the divergence or convergence of

$$\sum_{y=N}^{\infty} \frac{\delta_1\delta_2\cdots\delta_y}{\beta_1\beta_2\cdots\beta_y} = \sum_{y=N}^{\infty} \frac{\beta(2\beta)\cdots[(N-1)\beta](N\beta)^{y-N+1}}{\alpha^y}$$
$$= \frac{N!}{N^{N-1}} \sum_{y=N}^{\infty} \left(\frac{N\beta}{\alpha}\right)^y. \tag{8.25}$$

Thus, the process is recurrent if and only if $\beta \geqslant \alpha/N$. Similarly, the series (8.23) converges or diverges according as the following series converges or diverges,

$$\sum_{y=N}^{\infty} \frac{\lambda_0}{\lambda_y} \frac{\beta_1\beta_2\cdots\beta_{y-1}}{\delta_1\delta_2\cdots\delta_y} = \sum_{y=N}^{\infty} \frac{\alpha^{y-1}}{\beta^y N! N^{y-N+1}} \left(\frac{\alpha}{\alpha + N\beta}\right)$$
$$= \frac{N^{N-1}}{N!} \left(\sum_{y=N}^{\infty} \left(\frac{\alpha}{N\beta}\right)^y\right)(\alpha + N\beta)^{-1}. \tag{8.26}$$

Thus, the process is positive recurrent if and only if $\beta > \alpha/N$. *It is null recurrent if and only if* $\beta = \alpha/N$. In the case $\beta > \alpha/N$, the invariant initial distribution is given by

$$\pi_1 = \frac{\alpha}{\beta} \frac{1}{(\alpha + \beta)} \pi_0, \qquad \pi_j = \frac{1}{j!}\left(\frac{\alpha}{\beta}\right)^j \frac{1}{(\alpha + j\beta)} \pi_0 \qquad (\text{for } 2 \leqslant j \leqslant N),$$
$$\pi_j = \left(\frac{\alpha}{N\beta}\right)^j \frac{N^N}{N!} \frac{1}{(\alpha + N\beta)} \pi_0 \qquad (j > N), \tag{8.27}$$
$$\pi_0 = \left[1 + \sum_{j=1}^{N} \frac{1}{j!}\left(\frac{\alpha}{\beta}\right)^j \frac{1}{(\alpha + j\beta)} + \sum_{j=N+1}^{\infty} \frac{1}{(\alpha + N\beta)} \left(\frac{\alpha}{N\beta}\right)^j \frac{N^N}{N!}\right]^{-1}.$$

Example 2. (*Birth–Death Process with One Absorbing Boundary*). Here

$$S = \{0, 1, 2, \ldots\}, \qquad -q_{ii} = \lambda_i > 0 \quad \text{for } i \geqslant 1, \qquad q_{00} = 0,$$
$$q_{i,i+1} = \beta_i\lambda_i \quad \text{and} \quad q_{i,i-1} = \delta_i\lambda_i \quad \text{for } i \geqslant 1,$$

where

$$0 < \beta_i < 1 \quad \text{and} \quad \beta_i + \delta_i = 1 \quad \text{for } i \geqslant 1.$$

Thus, $k_{00} = 1$, $k_{ii} = 0$ if $i \geqslant 1$, $k_{i,i+1} = \beta_i$, $k_{i,i-1} = \delta_i$ for $i \geqslant 1$. The embedded discrete-parameter chain is a birth–death chain with absorption at 0. Therefore, "0" is an absorbing state for $\{X_t\}$, and it is the only absorbing state. Given two states c and d with $c < d$, let $\psi(x)$ denote the probability that $\{X_t\}$ reaches c before it reaches d starting from x. Note that this probability is the same for $\{X_t\}$ as it is for the embedded chain $\{Y_n\}$. Therefore by Eq. 2.10 of Chapter III,

$$\psi(x) = \frac{\displaystyle\sum_{y=x}^{d-1} \frac{\delta_{c+1}\delta_{c+2}\cdots\delta_y}{\beta_{c+1}\beta_{c+2}\cdots\beta_y}}{\left(\displaystyle\sum_{y=c+1}^{d-1} \frac{\delta_{c+1}\delta_{c+2}\cdots\delta_y}{\beta_{c+1}\beta_{c+2}\cdots\beta_y}\right) + 1} \qquad (c + 1 \leqslant x \leqslant d - 1). \tag{8.28}$$

Letting $d \to \infty$, one gets

$$\rho_{xc} = P_x(\{X_t\} \text{ ever reaches } c) = \lim_{d\to\infty} \frac{\displaystyle\sum_{y=x}^{d-1} \frac{\delta_{c+1}\delta_{c+2}\cdots\delta_y}{\beta_{c+1}\beta_{c+2}\cdots\beta_y}}{\left(\displaystyle\sum_{y=c+1}^{d-1} \frac{\delta_{c+1}\delta_{c+2}\cdots\delta_y}{\beta_{c+1}\beta_{c+2}\cdots\beta_y}\right) + 1} \qquad \text{for } x > c. \tag{8.29}$$

Thus, for $x > 0$, $\rho_{x0} = 1$ and, more generally, $\rho_{xc} = 1$ for $x > c$, if and only if

$$\sum_{y=1}^{\infty} \frac{\delta_1\delta_2\cdots\delta_y}{\beta_1\beta_2\cdots\beta_y} = \infty. \tag{8.30}$$

But if $\rho_{x0} = 1$ for all $x > 0$ then $P_x(\zeta = \infty) = 1$ for all $x \geqslant 0$, since the holding time at 0 is infinite. So there is no explosion if (8.30) holds. If the series in (8.30) is convergent, then $\rho_{x0} < 1$ for $x > 0$, and $\rho_{xc} < 1$ for $x > c$. However, this does not necessarily mean explosion. For example, let $\beta_x = p$, $\delta_x = 1 - p$ for $x > 0$ with $\frac{1}{2} < p < 1$, also take $\lambda_x = \lambda > 0$ for all $x > 0$, $q_{00} = 0$. Then by Proposition 6.1, the process is conservative and, therefore, nonexplosive.

Example 3. (*Chemical Reaction Kinetics*). The study of the time evolution of molecular concentrations that undergo chemical reactions is known as *chemical reaction kinetics*. The interest in reaction kinetics is not limited to chemistry. Ecologists and biologists are interested in the development of populations of various organic species that are involved in "reactions" through predation, births, deaths, immigration, emigration, etc.

The simplest view of chemical reactions is as an evolution that takes place through a succession of stages called *elementary reactions*. For example, $2A + B \to C$ is the notation used by chemists to denote an elementary reaction in which two molecules of A react with a molecule of B to produce a molecule of C. The notation $A \to B + C$ represents the dissociation of a molecule of A into molecules of types B and C. The reaction $A \to B$ represents a transformation of a molecule of A into a molecule of B (for example, as a result of radioactive emissions). The term *molecularity* is used to refer to the number of *reactant* molecules involved in a given elementary reaction. So, A → B is called *uni-molecular* while A + B → C is *bi-molecular*. Double arrows are used to denote *reversible* reactions. For example A ⇄ B signifies that either the reaction A → B or the reverse reaction B → A may occur at an elementary reaction stage.

As an illustration, consider a chemical solution that undergoes elementary reactions of the form C + A ⇄ C + B where C denotes a catalyst whose concentration is held constant. For example, if the concentration of C is large relative to the concentrations of A and B then, although C does react with A and B, the relative variation in its concentration is so small that it may be viewed as constant. In describing the evolution of such a process, it is convenient to disregard the concentrations of substances that are constant throughout the evolution. So we are left with the reversible unimolecular reaction A ⇄ B. Let X_t denote the number of molecules (*concentration*) of A at time t in a volume of substance held constant. The total number N of molecules A and B is constant throughout the reaction process. The *state space* of the process $\{X_t\}$ is $S = \{0, 1, 2, \ldots, N\}$.

The *holding time in state i* represents the time required for an elementary reaction (either A → B or B → A) to occur when there are i molecules of species A (and $N - i$ of species B). The transition (jump) from state i to state $i - 1$ corresponds to the reaction A → B by one of the A molecules with the catalyst and the transition from i to $i + 1$ corresponds to the reverse reaction. Let us assume that $\{X_t\}$ is a time-homogeneous *Markov process* such that each molecule of A has probability $r_A\,\Delta t + o(\Delta t)$ of reacting with the catalyst in time Δt to produce a molecule of B (as $\Delta t \to 0$). Suppose also that the reverse reaction has probability $r_B\,\Delta t + o(\Delta t)$ as $\Delta t \to 0$. In addition, suppose that the probability of occurrence of more than one elementary reaction in time Δt has order $o(\Delta t)$. This makes $\{X_t\}$ a continuous parameter (Ehrenfest type) birth–death chain on $S = \{0, 1, 2, \ldots, N\}$ with two reflecting boundaries at $i = 0$ and $i = N$ and having the infinitesimal transition rates q_{ij} given by

$$q_{ij} = \begin{cases} \beta_i\lambda_i = (N - i)r_B & \text{for } j = i + 1,\ 0 \leqslant i \leqslant N - 1 \\ \delta_i\lambda_i = ir_A & \text{for } j = i - 1,\ 1 \leqslant i \leqslant N \\ -\lambda_i = -[ir_A + (N - i)r_B] & \text{for } j = i,\ 0 \leqslant i \leqslant N \\ 0 & \text{otherwise.} \end{cases} \tag{8.31}$$

Observe that *uni-molecularity* is reflected in the birth–death property $p_{ij}(\Delta t) = o(\Delta t)$ as $\Delta t \to 0$ for $|j - i| > 1$. In case of higher-order molecularities, for example, 2A + C → B, the concentration of A would *skip* states. Also notice that the *conservation of mass* is obtained in the boundary conditions.

It follows from an application of Corollary 8.5 that there is a unique invariant initial distribution $\boldsymbol{\pi}$ that is also the limiting distribution regardless of the initial state. In fact, one may obtain $\boldsymbol{\pi}$ algebraically from the equation $\boldsymbol{\pi}'\mathbf{Q} = \mathbf{0}$.

However, it is difficult to explicitly solve the Kolmogorov equations (2.17) and (2.19) for the distribution at finite $t > 0$. By considerations of the *evolution of the generating functions* of the distribution under (2.17) or (2.19), one may obtain *moments* and other useful information at times $t < \infty$, when the coefficients are, as in the present case and in the branching case of Example 3 in Section 5, (affine) linear in the state variable j.

The generating function of X_t given $X_0 = i$, is

$$\varphi_i(z, t) = E_i(z^{X_t}) = \sum_{j=0}^{N} z^j p_{ij}(t) \tag{8.32}$$

so that, differentiating both sides with respect to t, using (2.19), and regarding $X \to z^X$ as a function of X for fixed z,

$$\begin{aligned}
\frac{\partial \varphi_i(z, t)}{\partial t} &= \frac{\partial}{\partial t} E_i(z^{X_t}) = E_i\left(\sum_j q_{X_t, j} z^j\right) \\
&= E_i(\mathbf{1}_{\{1 \leqslant X_t \leqslant N-1\}}[r_A X_t z^{X_t - 1} + (N - X_t) r_B z^{X_t + 1} - (r_A X_t + (N - X_t) r_B) z^{X_t}] \\
&\quad + \mathbf{1}_{\{X_t = 0\}}[N r_B z - N r_B] + \mathbf{1}_{\{X_t = N\}}[r_A X_t z^{X_t - 1} - N r_A z^N]) \\
&= r_A E_i(X_t z^{X_t - 1}) + N r_B E_i(z^{X_t + 1}) - r_B E_i(X_t z^{X_t + 1}) - r_A E_i(X_t z^{X_t}) \\
&\quad - N r_B E(z^{X_t}) + r_B E(X_t z^{X_t}) \\
&= r_A \frac{\partial \varphi_i}{\partial z}(z, t) + N r_B z \varphi_i(z, t) - z^2 r_B \frac{\partial \varphi_i}{\partial z}(z, t) - r_A z \frac{\partial \varphi_i}{\partial z}(z, t) \\
&\quad - N r_B \varphi_i(z, t) + r_B z \frac{\partial \varphi_i}{\partial z}(z, t).
\end{aligned} \tag{8.33}$$

Collecting similar terms in the last expression, one gets

$$\frac{\partial \varphi_i(z, t)}{\partial t} = (1 - z)(r_B z + r_A) \frac{\partial \varphi_i(z, t)}{\partial z} - N r_B (1 - z) \varphi_i(z, t). \tag{8.34}$$

This is a *first-order partial differential equation* subject to the "initial condition"

$$\varphi_i(z, 0) = E_i z^{X_0} = z^i. \tag{8.35}$$

A standard method of solving this equation is by *Lagrange's method of characteristics*. This method simply recognizes that (8.34) in essence says that the directional derivative

$$\left(\frac{\partial}{\partial t} - (1 - z)(r_B z + r_A) \frac{\partial}{\partial z}\right) \varphi_i(z, t) = -N r_B (1 - z) \varphi_i(z, t).$$

Introduce a real parameter θ and in the (z, t) plane define the (family of) parametric curve(s) $\theta \to (z(\theta), t(\theta))$, by

$$\frac{dz(\theta)}{d\theta} = -(1-z)(r_B z + r_A), \qquad \frac{dt(\theta)}{d\theta} = 1. \tag{8.36}$$

Now note that along such a curve the above directional derivative is none other than $d[\varphi_i(z(\theta), t(\theta))]/d\theta$, so that one has

$$\frac{d\varphi_i(z(\theta), t(\theta))}{d\theta} = -Nr_B(1-z)\varphi_i(z(\theta), t(\theta)), \tag{8.37}$$

whose general solution is

$$\varphi_i(z(\theta), t(\theta)) = \exp\left\{-Nr_B \int (1 - z(\theta))\, d\theta + d\right\}, \tag{8.38}$$

where d is a constant of integration. Now the solution of (8.36) is

$$t(\theta) = \theta, \qquad z(\theta) = -\frac{r_A}{r_B} + \frac{r_A + r_B}{r_B}(1 + ce^{(r_A + r_B)\theta})^{-1}. \tag{8.39}$$

Different values of c give rise to different curves of this parametric family. Each point (z, t) in the planar region $\{(z, t)\colon -1 < z < 1, t \geqslant 0\}$ lies on one and only one curve of this family. Thus, if we fix (z, t), then this point corresponds to c obtained by solving for c and θ with $t(\theta) = t$, $z(\theta) = z$ in (8.39). That is,

$$\theta = t, \qquad c = e^{-(r_A + r_B)t}\left[\left(\frac{r_A + r_B}{r_B}\right)\left(z + \frac{r_A}{r_B}\right)^{-1} - 1\right]. \tag{8.40}$$

Next note that, by (8.36),

$$-\int (1 - z)\, d\theta = \int \frac{dz}{r_B z + r_A} = \frac{1}{r_B} \log\left(z + \frac{r_A}{r_B}\right), \tag{8.41}$$

so that

$$\varphi_i(z(\theta), t(\theta)) = d'\left(z + \frac{r_A}{r_B}\right)^N, \tag{8.42}$$

where d' is to be obtained from the initial condition

$$\varphi_i(z(0), t(0)) = \varphi_i(z(0), 0) = z^i(0). \tag{8.43}$$

Specifically,

$$d' = z^i(0)\left(z(0) + \frac{r_A}{r_B}\right)^{-N} = z^{(0)}\left(\frac{r_A + r_B}{r_B}\right)^{-N}(1+c)^{-N}$$
$$= \left[-\frac{r_A}{r_B} + \frac{r_A + r_B}{r_B}(1+c)^{-1}\right]^i \left(\frac{r_A + r_B}{r_B}\right)^{-N}(1+c)^{-N}, \tag{8.44}$$

c being given by (8.40). Thus, writing $c(t, z)$ for c in (8.40),

$$\varphi_i(z, t) = \left[(1 + c(t, z))^{-1}\frac{r_A + r_B}{r_B} - \frac{r_A}{r_B}\right]^i \left(\frac{r_A + r_B}{r_B}\right)^{-N}(1 + c(t, z))^{-N}\left(z + \frac{r_A}{r_B}\right)^N. \tag{8.45}$$

From (8.45) and (8.40), by appropriate differentiation, one may calculate $E_i X_t$ and $\text{Var}_i\, X_t$. Moreover, as $t \to \infty$, $c(t, z) \to 0$. Hence,

$$\lim_{t\to\infty} \varphi_i(z, t) = \left(\frac{r_B}{r_A + r_B}\right)^N \left(z + \frac{r_A}{r_B}\right)^N, \tag{8.46}$$

which shows that, no matter what the initial state i may be, as $t \to \infty$, the limiting distribution of X_t is binomial with parameters N and $p = r_B/(r_A + r_B)$.

9 CALCULATION OF TRANSITION PROBABILITIES BY SPECTRAL METHODS

We are interested here in the computation of $e^{t\mathbf{Q}}$, where $\mathbf{Q}$ is an $(m+1) \times (m+1)$ matrix with eigenvalues $\alpha_0, \alpha_1, \ldots, \alpha_m$, possibly with repetitions, and corresponding eigenvectors

$$\mathbf{x}_0 = \begin{pmatrix} x_{00} \\ \vdots \\ x_{m0} \end{pmatrix}, \quad \ldots, \quad \mathbf{x}_m = \begin{pmatrix} x_{0m} \\ \vdots \\ x_{mm} \end{pmatrix}$$

which are linearly independent. Let

$$\mathbf{B} = (\mathbf{x}_0 \mathbf{x}_1 \cdots \mathbf{x}_m) = \begin{pmatrix} x_{00} & x_{01} & \cdots & x_{0m} \\ x_{10} & x_{11} & \cdots & x_{1m} \\ \vdots & \vdots & \vdots\vdots\vdots & \vdots \\ x_{m0} & x_{m1} & \cdots & x_{mm} \end{pmatrix}. \tag{9.1}$$

Then

$$\mathbf{QB} = (\alpha_0\mathbf{x}_0\alpha_1\mathbf{x}_1\cdots\alpha_m\mathbf{x}_m),$$
$$\mathbf{Q}^2\mathbf{B} = \mathbf{Q}(\alpha_0\mathbf{x}_0\alpha_1\mathbf{x}_1\cdots\alpha_m\mathbf{x}_m) = (\alpha_0^2\mathbf{x}_0\alpha_1^2\mathbf{x}_1\cdots\alpha_m^2\mathbf{x}_m), \ldots,$$
$$\mathbf{Q}^n\mathbf{B} = (\alpha_0^n\mathbf{x}_0\alpha_1^n\mathbf{x}_1\cdots\alpha_m^n\mathbf{x}_m).$$

Hence

$$e^{t\mathbf{Q}}\mathbf{B} = \left(\sum_{n=0}^{\infty}\frac{t^n}{n!}\mathbf{Q}^n\right)\mathbf{B} = \sum_{n=0}^{\infty}\frac{t^n}{n!}(\alpha_0^n\mathbf{x}_0\alpha_1^n\mathbf{x}_1\cdots\alpha_m^n\mathbf{x}_m)$$
$$= (e^{t\alpha_0}\mathbf{x}_0e^{t\alpha_1}\mathbf{x}_1\cdots e^{t\alpha_m}\mathbf{x}_m) = \mathbf{B}\,\mathbf{diag}(e^{t\alpha_0}, \ldots, e^{t\alpha_m}). \tag{9.2}$$

Therefore

$$e^{t\mathbf{Q}} = \mathbf{B}\begin{pmatrix} e^{t\alpha_0} & & & \\ & e^{t\alpha_1} & & 0 \\ 0 & & \ddots & \\ & & & e^{t\alpha_m} \end{pmatrix}\mathbf{B}^{-1}. \tag{9.3}$$

Example 1. (*Birth–Death Process with Reflecting Boundaries*). Let

$$S = \{0, 1, 2, \ldots, N\}, \qquad q_{00} = -\lambda_0, \qquad q_{01} = \lambda_0,$$
$$q_{NN} = -\lambda_N, \qquad q_{N,N-1} = \lambda_N; \qquad q_{i,i+1} = \beta_i\lambda_i, \qquad q_{i,i-1} = (1-\beta_i)\lambda_i = \delta_i\lambda_i$$

with $0 < \beta_i < 1$ for $1 \leqslant i \leqslant N-1$. The invariant distribution $\boldsymbol{\pi}$, which exists and is unique, satisfies $\boldsymbol{\pi}'\mathbf{p}(t) = \boldsymbol{\pi}'$. Differentiation at $t = 0$ yields

$$\boldsymbol{\pi}'\mathbf{Q} = \mathbf{0}, \tag{9.4}$$

or, in detail,

$$-\pi_0\lambda_0 + \pi_1\lambda_1\delta_1 = 0,$$
$$\pi_{j-1}\beta_{j-1}\lambda_{j-1} - \pi_j\lambda_j + \pi_{j+1}\delta_{j+1}\lambda_{j+1} = 0, \qquad \text{for } 1 \leqslant j \leqslant N-1, \tag{9.5}$$
$$\pi_{N-1}\beta_{N-1}\lambda_{N-1} - \pi_N\lambda_N = 0.$$

The solution is unique and given by

$$\pi_1 = \frac{\lambda_0}{\lambda_1\delta_1}\pi_0,$$
$$\pi_2 = \frac{1}{\delta_2\lambda_2}(\pi_1\lambda_1 - \pi_0\lambda_0) = \frac{1}{\delta_2\lambda_2}\left(\frac{\lambda_0}{\delta_1} - \lambda_0\right)\pi_0 = \left(\frac{\lambda_0}{\lambda_2}\right)\frac{\beta_1}{\delta_1\delta_2}\pi_0,$$
$$\pi_3 = \frac{1}{\delta_3\lambda_3}(\pi_2\lambda_2 - \pi_1\beta_1\lambda_1) = \frac{1}{\delta_3\lambda_3}\left(\frac{\lambda_0\beta_1}{\delta_1\delta_2} - \frac{\lambda_0\beta_1}{\delta_1}\right)\pi_0 = \left(\frac{\lambda_0}{\lambda_3}\right)\frac{\beta_1\beta_2}{\delta_1\delta_2\delta_3}\pi_0,$$

$$\pi_j = \left(\frac{\lambda_0}{\lambda_j}\right) \frac{\beta_1 \beta_2 \cdots \beta_{j-1}}{\delta_1 \delta_2 \cdots \delta_j} \pi_0 \qquad \text{for } 2 \leqslant j \leqslant N-1,$$

$$\pi_N = \frac{\beta_{N-1}\lambda_{N-1}}{\lambda_N} \pi_{N-1} = \left(\frac{\lambda_0}{\lambda_N}\right) \frac{\beta_1 \beta_2 \cdots \beta_{N-1}}{\delta_1 \delta_2 \cdots \delta_{N-1}} \pi_0. \tag{9.6}$$

Since $\sum \pi_i = 1$, one gets

$$\pi_0 = \left[1 + \frac{\lambda_0}{\lambda_1 \delta_1} + \sum_{j=2}^{N} \left(\frac{\lambda_0}{\lambda_j}\right) \frac{\beta_1 \cdots \beta_{j-1}}{\delta_1 \cdots \delta_j}\right]^{-1}. \tag{9.7}$$

In (9.5) we have used the convention $\delta_N = 1$. Now observe that $\mathbf{Q}$ is a symmetric linear transformation with respect to the inner product defined by

$$(\mathbf{x}, \mathbf{y})_\pi = \sum_{i=0}^{N} x_i y_i \pi_i. \tag{9.8}$$

That is,

$$(\mathbf{Qx}, \mathbf{y})_\pi = (\mathbf{x}, \mathbf{Qy})_\pi. \tag{9.9}$$

To show this, note that $\delta_j \lambda_j \pi_j = \beta_{j-1} \lambda_{j-1} \pi_{j-1}$, so that

$$\begin{aligned}
(\mathbf{Qx}, \mathbf{y})_\pi &= (-\lambda_0 x_0 + \lambda_0 x_1) y_0 \pi_0 + \sum_{j=1}^{N-1} (\delta_j \lambda_j x_{j-1} - \lambda_j x_j + \beta_j \lambda_j x_{j+1}) y_j \pi_j \\
&\quad + (\lambda_N x_{N-1} - \lambda_N x_N) y_N \pi_N \\
&= -\lambda_0 x_0 y_0 \pi_0 - \lambda_N x_N y_N \pi_N + \sum_{j=1}^{N} \delta_j \lambda_j x_{j-1} y_j \pi_j \\
&\quad - \sum_{j=1}^{N-1} \lambda_j x_j y_j \pi_j + \sum_{j=0}^{N-1} \beta_j \lambda_j x_{j+1} y_j \pi_j \\
&= -\lambda_0 x_0 y_0 \pi_0 - \lambda_N x_N y_N \pi_N - \sum_{j=1}^{N-1} \lambda_j x_j y_j \pi_j \\
&\quad + \sum_{j=1}^{N} \beta_{j-1} \lambda_{j-1} x_{j-1} y_j \pi_{j-1} + \sum_{j=1}^{N} \beta_{j-1} \lambda_{j-1} x_j y_{j-1} \pi_{j-1},
\end{aligned} \tag{9.10}$$

which is symmetric in $\mathbf{x}$ and $\mathbf{y}$. Hence $\mathbf{Q}$ has $N+1$ real eigenvalues $\alpha_0, \alpha_1, \ldots, \alpha_N$ possibly with repetitions, and corresponding mutually orthogonal eigenvectors $\mathbf{x}_0, \mathbf{x}_1, \ldots, \mathbf{x}_N$ of unit length. By orthogonality of the eigenvectors with respect to $(\cdot, \cdot)_\pi$,

$$(\mathbf{x}_i, \mathbf{x}_k)_\pi = \sum_{j=0}^{N} x_{ji} x_{jk} \pi_j = \delta_{ik}. \tag{9.11}$$

Equivalently,

$$\mathbf{B}' \, \mathbf{diag}(\pi_0, \ldots, \pi_N)\mathbf{B} = \mathbf{I}. \tag{9.12}$$

Therefore,

$$\mathbf{B}^{-1} = \mathbf{B}' \, \mathbf{diag}(\pi_0, \ldots, \pi_N). \tag{9.13}$$

Using (9.12) in (9.3) we get

$$\mathbf{p}(t) = e^{t\mathbf{Q}} = \mathbf{B} \, \mathbf{diag}(e^{t\alpha_0}, \ldots, e^{t\alpha_N})\mathbf{B}' \, \mathbf{diag}(\pi_0, \ldots, \pi_N). \tag{9.14}$$

From (9.14) we have

$$p_{ij}(t) = \sum_{k=0}^{N} x_{ik} e^{t\alpha_k} x_{jk} \pi_j. \tag{9.15}$$

As a special case consider the *continuous-parameter simple symmetric random walk* given by

$$\lambda_0 = \lambda_1 = \cdots = \lambda_N = 1 \qquad \text{and} \qquad \beta_i = \tfrac{1}{2} = \delta_i \qquad \text{for } 1 \leqslant i \leqslant N - 1.$$

If α is an eigenvalue of $\mathbf{Q}$, then a corresponding eigenvector $\mathbf{y}$ satisfies the equations

$$\begin{gathered} \tfrac{1}{2} y_{j-1} - y_j + \tfrac{1}{2} y_{j+1} = \alpha y_j \qquad (1 \leqslant j \leqslant N - 1), \\ -y_0 + y_1 = \alpha y_0, \qquad y_{N-1} - y_N = \alpha y_N. \end{gathered} \tag{9.16}$$

One may check that the eigenvalues $\alpha_0, \ldots, \alpha_N$ are given by

$$\alpha_k = \cos \frac{k\pi}{N} - 1 \qquad (k = 0, 1, 2, \ldots, N), \tag{9.17}$$

and the corresponding eigenvectors $\mathbf{x}_0, \ldots, \mathbf{x}_N$ are

$$\begin{gathered} \mathbf{x}_0 = (1, 1, \ldots, 1)', \qquad \mathbf{x}_N = (1, -1, 1, \ldots)', \\ x_{jk} = \sqrt{2} \cos\left(\frac{k\pi j}{N}\right), \qquad 1 \leqslant k \leqslant N - 1, \quad 0 \leqslant j \leqslant N. \end{gathered} \tag{9.18}$$

A systematic method for calculating these eigenvalues and eigenvectors is given in Example 4.1 of Chapter III. The eigenvectors here are the same as there, but the eigenvalues differ by 1 because the matrices differ by an identity matrix. Also from (9.6),

$$\pi_0 = \pi_N = \frac{1}{2N}, \qquad \pi_j = \frac{1}{N} \qquad \text{for } 1 \leqslant j \leqslant N-1.$$

Hence, by (9.15),

$$\begin{aligned} p_{ij}(t) &= \sum_{k=0}^{N} e^{(\cos(k\pi/N)-1)t} x_{ik} x_{jk} \pi_j \\ &= \pi_j \left[1 + \sum_{k=1}^{N-1} e^{(\cos(k\pi/N)-1)t} 2 \cos\left(\frac{k\pi i}{N}\right) \cos\left(\frac{k\pi j}{N}\right) + e^{-2t}\theta(i,j) \right], \end{aligned} \tag{9.19}$$

where $\theta(i, j) = x_{iN}x_{jN} = +1$ or -1 according as $|j - i|$ is even or odd. Note that $p_{ij}(t)$ converges to π_j exponentially fast as $t \to \infty$.

10 ABSORPTION PROBABILITIES

Suppose that P_i is the distribution of a continuous-parameter Markov chain $\{X_t\}$ starting at $i \in S$ and having homogeneous transition probabilities $\mathbf{p}(t) = ((p_{ij}(t)))$ with corresponding infinitesimal rates furnished by $\mathbf{Q} = \mathbf{p}'(0)$. For a nonempty set A of states, let $\hat{\mathbf{Q}}$ denote the matrix obtained from $\mathbf{Q}$ by replacing all rows corresponding to states in A by zeros. Let $\hat{P}_i$ denote the distribution of the process $\{X_t\}$ so transformed.

For an initial state $i \notin A$, all states belonging to A are *absorbing states* under $\hat{P}_i$. Moreover, in view of Theorem 5.4, the distributions under P_i and $\hat{P}_i$ of the process $\{X_t: 0 \leqslant t < \tau_A\}$, where

$$\tau_A = \inf\{t \geqslant 0: X_t \in A\} \tag{10.1}$$

is the *first passage time to* A, must coincide. Therefore, absorption probabilities starting from $i \notin A$ can be calculated directly in terms of the transition probabilities under the transformed distribution $\hat{P}_i$ as follows.

Proposition 10.1. Let $\mathbf{p}(t)$ and $\hat{\mathbf{p}}(t)$ denote the transition probabilities for the distributions P_i and $\hat{P}_i$ generated by $\mathbf{Q}$ and $\hat{\mathbf{Q}}$, respectively. Then, for $i \notin A$,

$$P_i(\tau_A > t) = \sum_{j \notin A} \hat{p}_{ij}(t). \tag{10.2}$$

Proof. Let $\{Y_n\}$ be the discrete-parameter Markov chain starting in state i under P_i obtained in accordance with Theorem 5.4 and let $T_0, T_1, T_2, \ldots$ denote the corresponding exponentially distributed holding times. Writing

$$\nu_A = \inf\{n \geqslant 0: Y_n \in A\}, \tag{10.3}$$

we have

$$P_i(t < \tau_A < \infty) = \sum_{m=1}^{\infty} \sum{}^* P_i(Y_0 = i, Y_1 = i_1, \ldots, Y_{m-1} = i_{m-1}, Y_m = j, \nu_A = m,$$
$$T_0 + \cdots + T_{m-1} > t)$$

$$= \sum_{m=1}^{\infty} \sum{}^* k_{ii_1} k_{i_1 i_2} \cdots k_{i_{m-1} j} G(t; \lambda_i, \lambda_{i_1}, \ldots, \lambda_{i_{m-1}}), \tag{10.4}$$

where $\sum^*$ denotes summation over all m-tuples $(i_1, \ldots, i_{m-1}, j)$ of states $i_1, i_2, \ldots, i_{m-1} \in S \backslash A$, $j \in A$, and $G(t; \lambda_i, \lambda_{i_1}, \ldots, \lambda_{i_{m-1}})$ is the probability that a sum of m independent exponentially distributed variables with parameters $\lambda_i, \lambda_{i_1}, \ldots, \lambda_{i_{m-1}}$ exceeds the value t; recall that $\lambda_k = -q_{kk}$. The corresponding probability under $\hat{P}_i$ is obtained on replacing $k_{ii_1} k_{i_1 i_2} k_{i_1 i_2} \cdots k_{i_{m-1} j}$ by $\hat{k}_{ii_1} \hat{k}_{i_1 i_2} \cdots \hat{k}_{i_{m-1} j}$ and G by $\hat{G}$.

Now the transition probabilities for the discrete-parameter Markov chains corresponding to $\mathbf{Q}$ and $\hat{\mathbf{Q}}$ are the same for all sequences of states in $\sum^*$. Therefore, for such sequences $G(t; \lambda_{i_1}, \ldots, \lambda_{i_{m-1}}) = \hat{G}(t; \lambda_{i_1}, \ldots, \lambda_{i_{m-1}})$, and

$$P_i(\tau_A > t) = \hat{P}_i(\tau_A > t) = \sum_{j \notin A} \hat{p}_{ij}(t). \qquad \blacksquare$$

Observe that since

$$\hat{p}_{lk}(t) = 0 \qquad \text{for all } l \in A,\ k \in S \backslash A, \tag{10.5}$$

the backward equations for $\hat{p}_{ij}(t)$, with $i, j \in S \backslash A$, take the form

$$\mathbf{p}^{0\prime}(t) = \mathbf{Q}^0 \mathbf{p}^0(t), \tag{10.6}$$

where $\mathbf{p}^0$ and $\mathbf{Q}^0$ are obtained from $\mathbf{p}$ and $\mathbf{Q}$ by *deleting* all rows and columns corresponding to states in the set A.

Example 1. (*Biomolecular Fixation*). The surface of a bacterium is assumed to consist of molecular sites that are susceptible to invasion by foreign molecules. Molecules of a particular composition, termed *acceptable*, will be permanently affixed to a site upon arrival, to the exclusion of further attachments by other molecules. Molecules that do not possess the acceptable composition remain at the site for some positive length of time but are eventually rejected. The problem here is to analyze the rate at which fixation occurs.

Suppose that foreign molecules arrive at a fixed site according to the occurrences of a Poisson process with parameter $\lambda > 0$. A proportion $\alpha > 0$ of the molecules that arrive at the site are acceptable. Other "unacceptable" molecules remain at the site for an exponentially distributed length of time with parameter $\beta > 0$. The calculation of the distribution of the length of time until fixation occurs is an absorption probability problem.

At any particular time t, the site is in one of three possible states a, b, or c, say, depending on whether the site is (a) occupied by an acceptable molecule, (b) occupied by an unacceptable molecule, or (c) unoccupied. The evolution of states at the given site is, according to the above assumptions, a continuous-time Markov chain $\{X_t\}$ with state space $S = \{a, b, c\}$ starting at c and having infinitesimal rates given by

$$\mathbf{Q} = \begin{array}{c} \\ a \\ b \\ c \end{array}\!\!\begin{array}{c} \begin{array}{ccc} a & b & c \end{array} \\ \begin{bmatrix} 0 & 0 & 0 \\ 0 & -\beta & \beta \\ \alpha\lambda & (1-\alpha)\lambda & -\lambda \end{bmatrix} \end{array}. \tag{10.7}$$

According to Proposition 10.1, the distribution of the first passage time τ_a to fixation (a) is given by

$$P_c(\tau_a > t) = p^0_{cc}(t) + p^0_{cb}(t), \tag{10.8}$$

where by (10.6), $\mathbf{p}^0(t)$ is determined by

$$\mathbf{p}^{0\prime}(t) = \mathbf{Q}^0\mathbf{p}^0(t), \qquad \mathbf{p}^0(0) = \mathbf{I}, \tag{10.9}$$

and

$$\mathbf{Q}^0 = \begin{array}{c} \\ b \\ c \end{array}\!\!\begin{array}{c} \begin{array}{cc} b & c \end{array} \\ \begin{bmatrix} -\beta & \beta \\ (1-\alpha)\lambda & -\lambda \end{bmatrix} \end{array}. \tag{10.10}$$

$\mathbf{Q}^0$ has distinct eigenvalues r_1, r_2 given by the zeros of the characteristic polynomial

$$\det(\mathbf{Q}^0 - r\mathbf{I}) = r^2 + (\lambda + \beta)r + \alpha\beta\lambda. \tag{10.11}$$

In particular,

$$r_1 = -\tfrac{1}{2}(\lambda + \beta) + \tfrac{1}{2}(\lambda^2 + 2(1 - 2\alpha)\beta\lambda + \beta^2)^{1/2}, \tag{10.12}$$

$$r_2 = -\tfrac{1}{2}(\lambda + \beta) - \tfrac{1}{2}(\lambda^2 + 2(1 - 2\alpha)\beta\lambda + \beta^2)^{1/2}. \tag{10.13}$$

Two corresponding linearly independent eigenvectors are

$$\mathbf{x}_1 = \begin{bmatrix} \beta \\ r_1 + \beta \end{bmatrix}, \qquad \mathbf{x}_2 = \begin{bmatrix} \beta \\ r_2 + \beta \end{bmatrix}. \tag{10.14}$$

Linear independence is easily checked since

$$\det \mathbf{B} = \det\begin{bmatrix} \beta & \beta \\ r_1 + \beta & r_2 + \beta \end{bmatrix} = \beta(r_2 - r_1) \neq 0. \tag{10.15}$$

The solution to the system (10.9) is given as in Section 9 by

$$\mathbf{p}^0(t) = e^{\mathbf{Q}^0 t} = \mathbf{B}\,\mathbf{diag}(e^{tr_1}, e^{tr_2})\mathbf{B}^{-1}$$

$$= \frac{1}{\beta(r_2 - r_1)}\begin{bmatrix} \beta\{(r_2+\beta)e^{r_1t} - (r_1+\beta)e^{r_2t}\} & \beta^2\{e^{r_2t} - e^{r_1t}\} \\ (r_1+\beta)(r_2+\beta)\{e^{r_1t} - e^{r_2t}\} & \beta\{(r_2+\beta)e^{r_2t} - (r_1+\beta)e^{r_1t}\} \end{bmatrix}. \tag{10.16}$$

Therefore, from (10.16) and (10.8) we obtain

$$P_c(\tau_a > t) = \frac{1}{r_2 - r_1}\left[r_2\left(1 + \frac{r_1}{\beta}\right)e^{r_1t} - r_1\left(1 + \frac{r_2}{\beta}\right)e^{r_2t}\right]. \tag{10.17}$$

The *average rate of molecular fixation* is given by

$$E_c\tau_a = \int_0^\infty P_c(\tau_a > t)\,dt = -\frac{r_1r_2 + \beta(r_1 + r_2)}{\beta r_1 r_2} = \frac{\beta + \lambda(1-\alpha)}{\alpha\beta\lambda}. \tag{10.18}$$

Example 2. (*Continuous-Parameter Markov Branching*). Let $\{X_t\}$ be the continuous-parameter Markov branching process defined in Example 5.2. The generator $\mathbf{Q} = ((q_{ij}))$ is given by

$$q_{ij} = \begin{cases} i\lambda f_{j-i+1}, & j \neq i, j = i-1, i+1, \ldots, i \geqslant 1, \\ -i\lambda(1 - f_1), & i = j \geqslant 1, \\ 0, & i = 0, j \geqslant 0 \text{ or } i \geqslant 1, j < i - 1. \end{cases} \tag{10.19}$$

The probability ρ of eventual extinction for an initial single particle was calculated in Section 5. Let

$$h(s) = h^{(1)}(s) = \sum_{j=0}^{\infty} q_{1j}s^j. \tag{10.20}$$

Assume that the offspring distribution $\{f_j\}$ has finite second moment with mean $\mu < 1$. Then $h'(1) < 0$ and $h''(1) < \infty$. In particular it follows from Proposition 5.7 that $\rho = 1$. We will exploit the special Markov branching structure to compute the tail probability $P_1(\tau_{\{0\}} > t)$ for large values of t under these assumptions. Let

$$\kappa_1 = h'(1) = \lambda(\mu - 1) < 0, \qquad \kappa_2 = h''(1) = \lambda \sum_{j=1}^{\infty} j(j-1)f_j < \infty. \tag{10.21}$$

We have,

$$P_1(\tau_{\{0\}} > t\} = \sum_{j=1}^{\infty} p_{1,j}(t) = 1 - p_{1,0}(t) = 1 - g_t^{(1)}(0), \tag{10.22}$$

where $g_t^{(1)}(r) = \sum_{j=1}^{\infty} p_{1,j}(t) r^j$. As in the proof of Proposition 5.7 at (5.27), the *Kolmogorov backward equation* for branching transforms according to

$$\frac{\partial g_t^{(1)}(r)}{\partial t} = h(g_t^{(1)}(r)), \tag{10.23}$$

with $g_0^{(1)}(r) = r$. Thus, for each $t > 0$, $0 \leqslant r < 1$,

$$\int_r^{g_t^{(1)}(r)} \frac{ds}{h(s)} = t. \tag{10.24}$$

Now,

$$h(1) = 0 \qquad \text{and} \qquad h(s) = h'(1)(s-1) + \tfrac{1}{2}\xi(s)(s-1)^2 \qquad \text{for } s \leqslant 1,$$

where $\xi(1^-) = h''(1) < \infty$. Thus, under the assumptions for (10.21),

$$\frac{1}{h(s)} = \frac{1}{\kappa_1(s-1) + \frac{1}{2}\xi(s)(s-1)^2} = \frac{1}{\kappa_1(s-1)\left[1 + \dfrac{\xi(s)(s-1)}{2\kappa_1}\right]}$$

$$= \frac{1}{\kappa_1(s-1)}\left\{1 - \frac{\dfrac{\xi(s)(s-1)}{2\kappa_1}}{1 + \dfrac{\xi(s)(s-1)}{2\kappa_1}}\right\} = \frac{1}{\kappa_1(s-1)} + \varphi(s), \tag{10.25}$$

where

$$\varphi(s) = \frac{-\dfrac{\xi(s)}{2\kappa_1^2}}{1 + \dfrac{\xi(s)(s-1)}{2\kappa_1}}.$$

In particular, note that

$$\frac{1}{h(s)} - \frac{1}{\kappa_1(s-1)} = \varphi(s)$$

is *bounded* for $0 \leqslant s < 1$. Define for $0 \leqslant x < 1$,

$$H(x) = -\int_x^1 \left(\frac{1}{h(s)} - \frac{1}{\kappa_1(s-1)}\right) ds + \frac{1}{\kappa_1}\log(1-x). \tag{10.26}$$

Then $H'(x) = 1/h(x) > 0$ for $0 \leqslant x < 1$ and (10.24) may be expressed as

$$H(g_t^{(1)}(r)) - H(r) = t, \qquad 0 \leqslant r < 1, \quad t \geqslant 0. \tag{10.27}$$

Since H is strictly increasing on $[0, 1)$, this can be solved to get

$$g_t^{(1)}(r) = H^{-1}(t + H(r)), \qquad 0 \leqslant r < 1, \quad t \geqslant 0. \tag{10.28}$$

Now, from (10.26), we have for $x \to 1^-$,

$$\begin{aligned} H(x) &= -\int_x^1 \varphi(s)\,ds + \frac{1}{\kappa_1}\log(1-x) \\ &= -\left(\frac{1}{1-x}\int_x^1 \varphi(s)\,ds\right)(1-x) + \frac{1}{\kappa_1}\log(1-x) \\ &= -\varphi(1^-)(1-x) + o(1-x) + \frac{1}{\kappa_1}\log(1-x) \\ &= \frac{\kappa_2}{2\kappa_1^2}(1-x) + \frac{1}{\kappa_1}\log(1-x) + o(1-x). \end{aligned} \tag{10.29}$$

Equivalently, solving this for $\log(1-x)$ we have

$$\log(1-x) = \kappa_1 H(x) - \frac{\kappa_2}{2\kappa_1}(1-x) + o(1-x). \tag{10.30}$$

Therefore, for $x \to 1^-$,

$$\begin{aligned} 1 - x &= e^{\kappa_1 H(x)} e^{-(\kappa_2/2\kappa_1)(1-x)}(1 + o(1-x)) \\ &= e^{\kappa_1 H(x)}\left(1 - \frac{\kappa_2}{2\kappa_1}(1-x) + o(1-x)\right). \end{aligned} \tag{10.31}$$

Now, by (10.22), (10.28), and (10.31) we have for $t \to \infty$, $y_t := H^{-1}(t + H(0))$,

$$\begin{aligned} P_1(\tau_{\{0\}} > t) &= 1 - g_t^{(1)}(0) = 1 - H^{-1}(t + H(0)) = 1 - y_t \\ &= e^{\kappa_1 H(y_t)}\left(1 - \frac{\kappa_2}{2\kappa_1}(1-y_t) + o(1-y_t)\right), \end{aligned} \tag{10.32}$$

where $y_t = H^{-1}(t + H(0)) \to 1$ as $t \to \infty$ is used in applying (10.31). Thus,

$$\begin{aligned} P_1(\tau_{\{0\}} > t) &= e^{\kappa_1(t+H(0))}\left(1 - \frac{\kappa_2}{2\kappa_1}(1 - y_t) + o(1 - y_t)\right) \\ &= e^{\kappa_1 t}e^{\kappa_1 H(0)}\left(1 - \frac{\kappa_2}{2\kappa_1}(1 - g_t^{(1)}(0)) + o(1 - g_t^{(1)}(0))\right) \\ &\sim ce^{-\lambda(1-\mu)t} \qquad \text{as } t \to \infty, \end{aligned} \tag{10.33}$$

where

$$-\kappa_1 = \lambda(1 - \mu) > 0, \qquad c = e^{\kappa_1 H(0)} > 0.$$

By consideration of the p.g.f. $k(z, t)$ of the conditional distribution of X_t given $\{X_t > 0, X_0 = 1\}$, one may obtain the existence of a nondegenerate limit distribution in the limit as $t \to \infty$ having p.g.f. of the form (Exercise 4*),

$$k(z, \infty) := \lim_{t\to\infty} k(z, t) = 1 - \exp\left(\kappa_1 \int_0^z \frac{ds}{h(s)}\right). \tag{10.34}$$

11 CHAPTER APPLICATION: AN INTERACTING SYSTEM THE SIMPLE SYMMETRIC VOTER MODEL

Although their interpretations vary, the *voter model* was independently introduced by P. Clifford and A. Sudbury and by R. Holley and T. Liggett. In either case, one considers a distribution of ± 1's at the point of the d-dimensional integer lattice $\mathbb{Z}^d$ at time $t = 0$. In the demographic interpretation one imagines the sites of $\mathbb{Z}^d$ as the locations of a species of one of the two types, $+1$ or -1. In the course of time, the type of species occupying a particular site can change owing to *invasion* by the opposition. The invasion is by occupants of neighboring sites and occurs at a rate proportional to the number of neighboring sites maintained by the opposing species. In the sociopolitical version, the values ± 1 represent opposing opinions (pro or con) held by occupants of locations indexed by $\mathbb{Z}^d$. As time evolves, a voter may change opinion on the issue. The rate at which the voters' position on the issue changes is proportional to the number of neighbors who hold the opposing opinion. This model is also related to a tumor-growth model introduced by T. Williams and R. Bjerknes, which, in fact, is now often referred to as the biased voter model. If one assumes the mechanism for cell division to be the same for abnormal as for normal cells in the Williams–Bjerknes model (i.e., no "carcinogenic advantage" in their model), then one obtains the voter model discussed here (see theoretical complements 4 for reference). While mathematical methods and theories for this and the more general models of this type are relatively recent, quite a bit can be learned about the voter model by applying some of the basic results of this chapter.

A sample configuration of ± 1-values is represented by points $\boldsymbol{\sigma} = (\sigma_{\mathbf{n}}: \mathbf{n} \in \mathbb{Z}^d)$ in the (uncountable) product space $S = \{1, -1\}^{\mathbb{Z}^d}$. The evolution of configurations for the voter model will be defined by a Markov process in S. The desired transition rates are such that in time t to $t + \Delta t$, the configuration may change from $\boldsymbol{\sigma}$, by a flip at some site $\mathbf{m}$, to the configuration $\boldsymbol{\sigma}^{(\mathbf{m})}$, where $\sigma_{\mathbf{n}}^{(\mathbf{m})} = -\sigma_{\mathbf{n}}$ if $\mathbf{n} = \mathbf{m}$ and $\sigma_{\mathbf{n}}^{(\mathbf{m})} = \sigma_{\mathbf{n}}$ otherwise. This occurs, with probability $c_{\mathbf{m}}(\boldsymbol{\sigma})\,\Delta t + o(\Delta t)$, where $c_{\mathbf{m}}(\boldsymbol{\sigma})$ is the number of *neighbors* $\mathbf{n}$ of $\mathbf{m}$ such that $\sigma_{\mathbf{n}} \neq \sigma_{\mathbf{m}}$; two sites that differ by one unit in one (and only one) of the coordinate directions are defined to be *neighbors*. More complicated changes involving a flip at two or more sites are to occur with probability $o(\Delta t)$ as $\Delta t \to 0$.

It is instructive to look closely at the flip-rates for the one-dimensional case. For a configuration $\boldsymbol{\sigma} \in S$ a flip at site m occurs at unit rate $c_m(\boldsymbol{\sigma}) = 1$ if the neighboring sites $m - 1$ and $m + 1$ have opposite values (opinions); i.e., $\sigma_{m-1}\sigma_{m+1} = -1$. If, on the other hand, the values at $m - 1$ and $m + 1$ agree mutually, but disagree with the value at m, then the flip-rate (probability) at m is proportionately increased to $c_m(\boldsymbol{\sigma}) = 2$. If both neighboring values agree mutually as well as with the value at m, then the flip-rate mechanism is turned off, $c_m(\boldsymbol{\sigma}) = 0$. Thus there is a *local tendency* toward consensus within the system. Observe that $c_m(\boldsymbol{\sigma})$ may be expressed as a local averaging via

$$c_m(\boldsymbol{\sigma}) = 1 - \tfrac{1}{2}\sigma_m(\sigma_{m-1} + \sigma_{m+1}). \tag{11.1}$$

More generally, in d dimensions the rates $c_{\mathbf{m}}(\boldsymbol{\sigma})$ may be expressed likewise as

$$\begin{aligned} c_{\mathbf{m}}(\boldsymbol{\sigma}) &= d\left(1 - \sigma_{\mathbf{m}} \sum_{\mathbf{n} \notin \mathbb{Z}^d} p_{\mathbf{mn}}\sigma_{\mathbf{n}}\right) \\ &= 2d \sum_{\{\mathbf{n}: \sigma_{\mathbf{n}} \neq \sigma_{\mathbf{m}}\}} p_{\mathbf{mn}}, \end{aligned} \tag{11.2}$$

where $\mathbf{p} = ((p_{\mathbf{mn}}))$ is the transition probability matrix on $\mathbb{Z}^d$ of the *simple symmetric random walk* on $\mathbb{Z}^d$, i.e.,

$$p_{\mathbf{mn}} = \begin{cases} \dfrac{1}{2d}, & \text{if } \mathbf{m} \text{ and } \mathbf{n} \text{ are neighbors,} \\ 0, & \text{otherwise.} \end{cases}$$

This helps explain the terminology *simple symmetric voter model*.

The first issue one must contend with is the existence of a Markov evolution $\{\boldsymbol{\sigma}(t): t \geqslant 0\}$ on the uncountable state space S having the prescribed infinitesimal transition rates. In general this itself can be a mathematically nontrivial matter when it comes to describing infinite interacting systems. However, for the special flip rates desired for the voter model, a relatively simple graphical construction of the process is possible; called the *percolation construction* because of the "fluid flow" interpretation described below.

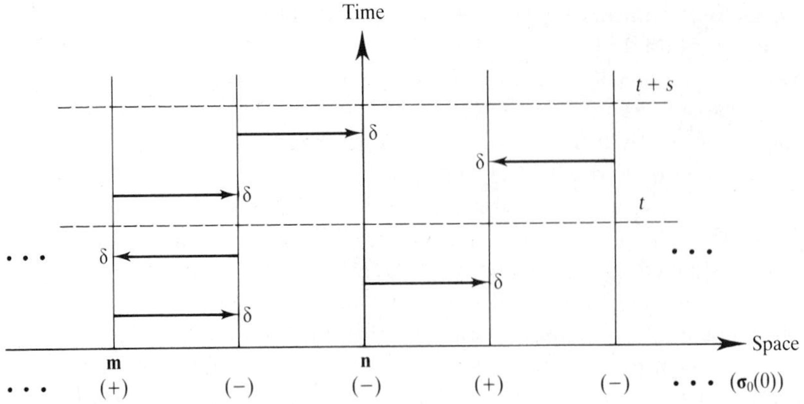

Figure 11.1

Consider a space–time diagram of $\mathbb{Z}^d \times [0, \infty)$ in which a "vertical time axis" is located at each site $\mathbf{m} \in \mathbb{Z}^d$, and imagine the points of $\mathbb{Z}^d$ as being laid out along a "horizontal axis"; see Figure 11.1. The basic idea is this. Imagine each vertical time axis as a wire that transports negative charge (opinion) upward from the (initially) negatively charged sites $\mathbf{m}$ such that $\sigma_{\mathbf{m}}(0) = -1$. For a certain random length of time (wire), the voter opinion at $\mathbf{m}$ will coincide with that of $\sigma_{\mathbf{m}}(0)$, but then this influence will be blocked, and the voter will receive the opinion of a randomly selected neighbor. Place a δ at the blockage time, and draw an arrow from the wire at the randomly selected site to the wire at $\mathbf{m}$ at this location of the blockage. δ stands for "death of influence from (directly) below." If the neighbor is conducting negative flow from some point below to this time, then it will be transferred across the arrow and up. So, while the source of influence at $\mathbf{m}$ might have changed, the opinion has not. However, if the selected neighbor is not conducting negative flow by this time, then a portion of the wire at $\mathbf{m}$ above this time will be given positive charge (opinion). Likewise, at certain times positive (plus) portions of the wire at a site can become negatively charged by the occurrence of an arrow from a randomly selected neighbor conducting negative flow across the arrow. This "space–time" flow of influence being qualitatively correct, it is now a matter of selecting the distribution of occurrence times of δ's and arrows with the Markov property of the evolution in mind. The specification of the average density of occurrences of δ's and arrows will then furnish the rate-parameter values.

Let $\{N_{\mathbf{m}}(t): t \geqslant 0\}$, $\mathbf{m} \in \mathbb{Z}^d$, be *independent* Poisson processes with intensity $2d$. The construction of a Poisson process having right-continuous unit jump sample functions is equivalent to the construction of a sequence $\{T_{\mathbf{m}}(k): k \geqslant 1\}$ of i.i.d. exponential inter-arrival times (see Section 5). The construction of countably many independent versions is made possible by Kolmogorov's existence theorem (theoretical complement 6.1 of Chapter I). Let $\{\mathbf{U}_{\mathbf{m}}(k): k \geqslant 1\}$, $\mathbf{m} \in \mathbb{Z}^d$, be independent i.i.d. sequences and independent of the processes

$\{N_{\mathbf{m}}(t)\}$, $\mathbf{m} \in \mathbb{Z}^d$, where $P(\mathbf{U}_{\mathbf{m}}(k) = \mathbf{n}) = p_{\mathbf{mn}}$. At the kth occurrence time $\tau_{\mathbf{m}}(k) := T_{\mathbf{m}}(1) + \cdots + T_{\mathbf{m}}(k)$ of the Poisson process at $\mathbf{m} \in \mathbb{Z}^d$, a neighboring site of $\mathbf{m}$, represented by $U_{\mathbf{m}}(k)$, is randomly selected. One may also consider independent Poisson processes $\{N_{\mathbf{mn}}(t): t \geqslant 0\}$, $\mathbf{m}, \mathbf{n} \in \mathbb{Z}^d$, with respectively intensity parameters $2dp_{\mathbf{mn}}$ representing those (Poisson) times at $\mathbf{m}$ when the neighboring site $\mathbf{n}$ is picked. Notice that $2dp_{\mathbf{mn}}$ is either 1 or 0 according to whether $\mathbf{m}$ and $\mathbf{n}$ are neighbors or not.

At the kth event time $\tau = \tau_{\mathbf{m}}(k)$ of the process $\{N_{\mathbf{m}}(t)\}$, let $\mathbf{n} = U_{\mathbf{m}}(k)$ be the corresponding neighbor selected. Place an arrow from $(\mathbf{n}, \tau)$ to $(\mathbf{m}, \tau)$, and attach the symbol δ to the point $(\mathbf{m}, \tau)$ at the arrowhead. For a given initial configuration $\boldsymbol{\sigma}(0) = (\sigma_{\mathbf{m}}(0))$, define a *flow* to be initiated at the sites $\mathbf{m}$ such that $\sigma_{\mathbf{m}}(0) = -1$. The flow passes vertically until blocked by the occurrence of a δ above, and also passes horizontally across arrows in the direction of the arrows and is stopped at the occurrence of a δ from moving upward. The configuration at time t is defined by

$$\begin{aligned} \sigma_{\mathbf{n}}(t) &= -1 \qquad \text{if the (negative) flow reaches } (\mathbf{n}, t) \text{ from some initial point } (\mathbf{m}, 0) \\ &= +1 \qquad \text{otherwise.} \end{aligned} \tag{11.3}$$

More precisely, the *flow is defined to reach* $(\mathbf{n}, t)$ *from* $(\mathbf{m}, 0)$ *if there are times* $0 = \tau_0 < \tau_1 < \cdots < \tau_{k+1} = t$ and *sites* $\mathbf{m} = \mathbf{n}_0, \mathbf{n}_1, \ldots, \mathbf{n}_k = \mathbf{n}$ *such that an arrow occurs from* $(\mathbf{n}_{i-1}, \tau_i)$ *to* $(\mathbf{n}_i, \tau_i)$ *and there are no* δ's *along the half-open vertical intervals (excluding lower end points) joining* $(\mathbf{n}_i, \tau_i)$ *to* $(\mathbf{n}_i, \tau_{i+i})$, $i = 1, \ldots, k$. In this case $(\mathbf{n}, t)$ and $(\mathbf{m}, 0)$ are said to be connected by a *continuous-time nearest-neighbor percolation.*

The transition from $\sigma_{\mathbf{m}}(t-) = -1$ to $\sigma_{\mathbf{m}}(t) = +1$ occurs if and only if $t = \tau$ is an event time of the process $\{N_{\mathbf{mn}}(t)\}$ for some $\mathbf{n}$ such that $\sigma_{\mathbf{n}}(t-) = +1$. Therefore the flipping rate from -1 to $+1$ is

$$\sum_{\mathbf{n}} 2dp_{\mathbf{mn}}(1 + \sigma_{\mathbf{n}})/2 = d\left(1 - \sigma_{\mathbf{m}} \sum_{\mathbf{n}} p_{\mathbf{mn}}\sigma_{\mathbf{n}}\right) = c_{\mathbf{m}}(\boldsymbol{\sigma}) \qquad \text{for } \sigma_{\mathbf{m}} = -1.$$

The transition from $\sigma_{\mathbf{m}}(t-) = +1$ to $\sigma_{\mathbf{m}}(t) = -1$ occurs if and only if $t = \tau$ is an event time of the process $\{N_{\mathbf{mn}}(t)\}$ for some $\mathbf{n}$ such that $\sigma_{\mathbf{n}}(t-) = -1$. Therefore the flipping rate from $+1$ to -1 is

$$\sum 2dp_{\mathbf{mn}}(1 - \sigma_{\mathbf{n}})/2 = d\left(1 - \sigma_{\mathbf{m}} \sum_{n} p_{\mathbf{mn}}\sigma_{\mathbf{n}}\right), \qquad \text{for } \sigma_{\mathbf{m}} = +1.$$

This (noncanonical) construction provides the existence of a *Markov process* with the desired infinitesimal rates. To check the Markov property, it is enough to consider the distribution of $\sigma_{\mathbf{n}}(t + s)$ for some finitely many sites $\mathbf{n} \in \mathbb{Z}^d$, and fixed $s, t \geqslant 0$, conditionally given the (entire) configurations $\boldsymbol{\sigma}(u)$, $u \leqslant t$. Working backwards through the space–time diagram from $\sigma_{\mathbf{n}_1}(t + s), \ldots, \sigma_{\mathbf{n}_k}(t + s)$, note that the value $\sigma_{\mathbf{n}_1}(t+s)$ must coincide with $\sigma_{\mathbf{m}_i}(t)$ for some $\mathbf{m}_i \in \mathbb{Z}^d$, where $\mathbf{m}_i$ depends only on $\boldsymbol{\sigma}(u)$, $u \geqslant t$, but *not* on $\boldsymbol{\sigma}(u)$, $u < t$. In the case of Figure 11.1 the history of $\sigma_{\mathbf{n}}(t+s)$ given $\boldsymbol{\sigma}(u)$, $u \leqslant t$, shows that the voter at $\mathbf{n}$ at time $t+s$ is copying the voter at $\mathbf{m}$ at time t (see theoretical complement 1).

The two configurations $\boldsymbol{\sigma}^+$ and $\boldsymbol{\sigma}^-$ representing *total consensus*, $\sigma_{\mathbf{n}}^+ = +1$ and $\sigma_{\mathbf{n}}^- = -1$ for all $\mathbf{n}$, are *absorbing*. In particular this makes each probability distribution of the form $p\delta_{\boldsymbol{\sigma}^+} + (1-p)\delta_{\boldsymbol{\sigma}^-}$, for $0 \leqslant p \leqslant 1$, an *invariant equilibrium distribution* for the system. To understand conditions under which it is also possible to have other equilibrium distributions of *persistent disagreement* requires some further analysis.

Probability distribution on S are defined for events belonging to the sigmafield $\mathscr{F}$ generated by events in S that depend on the values of configurations at finitely many sites. The state space S has a natural *metric space* structure for which $\mathscr{F}$ coincides with its Borel sigmafield (theoretical complement 2). Moreover, S is a *compact* metric space with this metric (theoretical complement 2). Consequently, any collection of probability measures on $(S, \mathscr{F})$ is necessarily *tight* (theoretical complement 8.2 of Chapter I). This is quite significant since it implies that *convergence in distribution* (*weak convergence*) *coincides with convergence of finite-dimensional distributions in this setting.* Special advantage is taken of this fact in the use of a method of *finite-dimensional* (multivariate) moments to study the long-run behavior of the distribution of the system.

Let f be a real-valued function on S that depends on a fixed set of finitely many coordinates of configurations $\boldsymbol{\sigma} \in S$. Then f must be bounded. Define linear operators $\mathbf{T}_t$, $(t \geqslant 0)$, for such functions f by

$$\mathbf{T}_t f(\boldsymbol{\sigma}) = E_{\boldsymbol{\sigma}} f(\boldsymbol{\sigma}(t)), \qquad \boldsymbol{\sigma} \in S. \tag{11.4}$$

Then,

$$\begin{aligned}\mathbf{T}_t f(\boldsymbol{\sigma}) - f(\boldsymbol{\sigma}) &= E_{\boldsymbol{\sigma}}\{f(\boldsymbol{\sigma}(t)) - f(\boldsymbol{\sigma})\} \\ &= \sum_{\mathbf{m}} \{f(\boldsymbol{\sigma}^{(\mathbf{m})}) - f(\boldsymbol{\sigma})\} c_{\mathbf{m}}(\boldsymbol{\sigma}) t + o(t),\end{aligned} \tag{11.5}$$

uniformly as $t \to 0^+$. Accordingly, for functions f on S depending on finitely many coordinates we have for $t \geqslant 0$, $\boldsymbol{\sigma} \in S$,

$$\frac{\partial \mathbf{T}_t f(\boldsymbol{\sigma})}{\partial t} = \mathbf{T}_t(\mathbf{A}f)(\boldsymbol{\sigma}) = E_{\boldsymbol{\sigma}}\{\mathbf{A}f(\boldsymbol{\sigma}(t))\}, \tag{11.6}$$

where, for such functions f,

$$\mathbf{A}f(\boldsymbol{\sigma}) = \lim_{t \to 0^+} \frac{\mathbf{T}_t f(\boldsymbol{\sigma}) - f(\boldsymbol{\sigma})}{t}$$

$$= \lim_{t \to 0^+} \frac{E_{\boldsymbol{\sigma}}\{f(\boldsymbol{\sigma}(t)) - f(\boldsymbol{\sigma})\}}{t}$$

$$= \sum_{\mathbf{m}} \{f(\boldsymbol{\sigma}^{(\mathbf{m})}) - f(\boldsymbol{\sigma})\} c_{\mathbf{m}}(\boldsymbol{\sigma}). \tag{11.7}$$

Let $\boldsymbol{\mu}$ be an arbitrary initial probability distribution on $(S, \mathscr{F})$ and let $\boldsymbol{\mu}_t$ denote the corresponding distribution of $\boldsymbol{\sigma}(t)$ with $\boldsymbol{\mu}_0 = \boldsymbol{\mu}$. The (*spatial*) *block correlations* of the distribution $\boldsymbol{\mu}_t$ over the finite set of sites $D = \{\mathbf{n}_1, \ldots, \mathbf{n}_r\}$, say, are the *multivariate moments* of $\sigma_{\mathbf{n}_1}(t), \ldots, \sigma_{\mathbf{n}_r}(t)$ defined by

$$\varphi_{\boldsymbol{\mu}}(t, D) = E_{\boldsymbol{\mu}}\{\sigma_{\mathbf{n}_1}(t) \times \cdots \times \sigma_{\mathbf{n}_r}(t)\}. \tag{11.8}$$

Using standard *inclusion–exclusion* calculations, one can verify that a probability distribution on $(S, \mathscr{F})$ is uniquely determined by its block correlations. Let $\varphi_{\boldsymbol{\sigma}}(t, D) = \varphi_{\delta_{\boldsymbol{\sigma}}}(t, D)$. Applying equation (11.6) to the function $f(\boldsymbol{\sigma}) := \sigma_{\mathbf{n}_1} \times \cdots \times \sigma_{\mathbf{n}_r}$, $\boldsymbol{\sigma} \in S$, we get an equation describing the evolution of the block correlations of the distributions as follows:

$$\frac{\partial \varphi_{\boldsymbol{\sigma}}(t, D)}{\partial t} = E_{\boldsymbol{\sigma}}\left\{\sum_{\mathbf{m}} \left[\prod_{\mathbf{k} \in D} \sigma_{\mathbf{k}}^{(\mathbf{m})}(t) - \prod_{\mathbf{k} \in D} \sigma_{\mathbf{k}}(t)\right] c_{\mathbf{m}}(\boldsymbol{\sigma}(t))\right\}$$

$$= -2E_{\boldsymbol{\sigma}}\left\{\sum_{\mathbf{m} \in D} \left[\prod_{\mathbf{k} \in D} \sigma_{\mathbf{k}}(t)\right] c_{\mathbf{m}}(\boldsymbol{\sigma}(t))\right\}$$

$$= -2d \sum_{\mathbf{m} \in D} E_{\boldsymbol{\sigma}}\left\{\left[\prod_{\mathbf{k} \in D} \sigma_{\mathbf{k}}(t)\right]\left[1 - \sigma_{\mathbf{m}}(t) \sum_{n} p_{\mathbf{mn}} \sigma_{\mathbf{n}}(t)\right]\right\}$$

$$= -2d|D|\varphi_{\sigma}(t, D) + 2d \sum_{\mathbf{m} \in D} \sum_{\mathbf{n}} p_{\mathbf{mn}} \varphi_{\sigma}(t, (D \setminus \{\mathbf{m}\}) \Delta \{\mathbf{n}\}), \tag{11.9}$$

where Δ is *symmetric difference*, $A \,\Delta B := A \cap B^c \cup A^c \cap B$, $A, B \subset \mathbb{Z}^d$, and $|D|$ is *cardinality* of D. Taking expected values, we get from (11.9) that

$$\frac{\partial \varphi_{\boldsymbol{\mu}}(t, D)}{\partial t} = -2d|D|\varphi_{\boldsymbol{\mu}}(t, D) + 2d \sum_{\mathbf{m} \in D} \sum_{\mathbf{n}} p_{\mathbf{mn}} \varphi_{\boldsymbol{\mu}}(t, (D \setminus \{\mathbf{m}\}) \Delta \{\mathbf{n}\})$$

$$= \sum_{\mathbf{m} \in D} \sum_{\mathbf{n}} \{\varphi_{\boldsymbol{\mu}}(t, (D \setminus \{\mathbf{m}\}) \Delta \{\mathbf{n}\}) - \varphi_{\boldsymbol{\mu}}(t, D)\} 2p_{\mathbf{mn}}. \tag{11.10}$$

As a warm-up to the equations (11.10) take $D = \{\mathbf{m}\}$ and suppose that $\boldsymbol{\mu}$ is an invariant (equilibrium) distribution. Then $E_{\boldsymbol{\mu}} \sigma_{\mathbf{m}} = \varphi_{\boldsymbol{\mu}}(0, \{\mathbf{m}\})$ is a *harmonic function* for the random walk; i.e., the left-hand side of (11.10) is zero so that the equations show that $\varphi_{\boldsymbol{\mu}}$ has the *averaging property* (harmonicity)

$$\varphi_{\boldsymbol{\mu}}(0, \{\mathbf{m}\}) = \sum_{\mathbf{n}} \varphi_{\boldsymbol{\mu}}(0, \{\mathbf{n}\}) p_{\mathbf{mn}}. \tag{11.11}$$

Now, for the *simple symmetric random walk* on $\mathbb{Z}^d$, one can show that the only bounded solutions to (11.11) are constants (theoretical complement 3). Therefore, the distribution of $\sigma_{\mathbf{m}}$ under the invariant equilibrium distribution is independent of $\mathbf{m}$ in this case. From here out we will restrict our consideration to the long-time behavior of various *translation invariant initial distributions*, with say,

$$E_\mu \sigma_{\mathbf{m}} = 2p - 1, \qquad 0 \leqslant p \leqslant 1, \quad \mathbf{m} \in \mathbb{Z}^d. \tag{11.12}$$

In particular, we will consider what happens in the long run when voters initially are, independently, $+1$ or -1 with probabilities p, $1 - p$, respectively.

The equation (11.9) or (11.10) suggests consideration of the following *finite-particle system*. A particle is initially placed at each site of the finite set $D_0 = D$. Each of these particles then undergoes a *continuous-time random walk* according to $\mathbf{p} = ((p_{\mathbf{mn}}))$ with exponential holding times with parameter $2d$, independently of the others until two particles meet. If two of the particles meet, then they are *mutually annihilated*. Let D_t denote the collection of sites occupied by particles at time $t \geqslant 0$. The evolution of the stochastic process $\{D_t\}$ takes place in the (*denumerable*) state space $S = \mathscr{L}^{(d)}$, consisting of all finite subsets of $\mathbb{Z}^d$. The empty set is *absorbing* for the process. If $D_t \neq \varnothing$ then in time t to $t + \Delta t$ a change to $(D_t \backslash \{\mathbf{m}\}) \Delta \{\mathbf{n}\}$, for some $\mathbf{m} \in D_t$ and $\mathbf{n} \in \mathbb{Z}^d$, occurs with probability $2dp_{\mathbf{mn}} \Delta t + o(\Delta t)$ as $\Delta t \to 0+$. Other types of changes in the configuration have probability $o(\Delta t)$ as $\Delta t \to 0+$.

Now consider that for fixed $\boldsymbol{\sigma} \in S$ we have by (11.9) that

$$\frac{\partial \varphi_{\boldsymbol{\sigma}}(t, D)}{\partial t} = \mathbf{A}^* \varphi_{\boldsymbol{\sigma}}(t, \cdot)(D), \tag{11.13}$$

together with

$$\varphi_{\boldsymbol{\sigma}}(0, D) = \prod_{\mathbf{m} \in D} \sigma_{\mathbf{m}}, \tag{11.14}$$

where for $D \in \mathscr{L}^{(d)}$,

$$\mathbf{A}^* f(D) = \sum_{\mathbf{m} \in D} \sum_{\mathbf{n}} \{ f((D \backslash \{\mathbf{m}\}) \Delta \{\mathbf{n}\}) - f(D) \} 2dp_{\mathbf{mn}}, \tag{11.15}$$

for bounded real-valued functions f on $\mathscr{L}^{(d)}$. Since $\mathbf{A}^*$ is the infinitesimal generator for $T_t^* f(D) = E_D f(D_t)$, $t \geqslant 0$, $D \in \mathscr{L}^{(d)}$, the *annihilating random walk*, the solution to (11.13), (11.14) is given by

$$\varphi_{\boldsymbol{\sigma}}(t, D) = E_D \varphi_{\boldsymbol{\sigma}}(0, D_t) = E_D \left\{ \prod_{\mathbf{m} \in D_t} \sigma_{\mathbf{m}} \right\}. \tag{11.16}$$

The representation (11.16) is known as the *duality equation* and is the basis for the proof of the following major result.

Theorem 11.1. Let $\mathbf{p} = ((p_{\mathbf{mn}}))$ be the transition probability matrix of the simple symmetric random walk on $\mathbb{Z}^d$ associated with the simple symmetric voter model on $\mathbb{Z}^d$. For an initial probability distribution $\boldsymbol{\mu}$ satisfying (11.12) we have the following:

(a) If $d \leqslant 2$, then $\boldsymbol{\mu}_t$ converges in distribution to $p\delta_{\sigma+} + (1-p)\delta_{\sigma-}$ as $t \to \infty$.

(b) If $d \geqslant 3$, then for each $p \in (0, 1)$ there is a distinct translation-invariant equilibrium distribution $\mathbf{v}^{(p)}$ on $(S, \mathscr{F})$, which is not a mixture of $\boldsymbol{\delta}_{\sigma+}$ and $\boldsymbol{\delta}_{\sigma-}$, such that $E_{\mathbf{v}^{(p)}}\sigma_{\mathbf{m}} = 2p - 1$. Moreover, if the initial distribution $\boldsymbol{\mu}$ is that of independent ± 1-valued Bernoulli random variables with probability parameter determined by (11.12), then $\boldsymbol{\mu}_t$ converges in distribution to $\mathbf{v}^{(p)}$ as $t \to \infty$.

Proof. To prove (a) first consider that

$$\begin{aligned} E_{\boldsymbol{\mu}}\sigma_{\mathbf{m}}(t) = \varphi_{\boldsymbol{\mu}}(t, \{m\}) &= \int_S \varphi_{\boldsymbol{\sigma}}(t, \{m\})\boldsymbol{\mu}(d\boldsymbol{\sigma}) \\ &= \int_S E_{\{\mathbf{m}\}}\left[\prod_{\mathbf{k}\in D_t} \sigma_{\mathbf{k}}\right]\boldsymbol{\mu}(d\boldsymbol{\sigma}) \\ &= \sum_{\mathbf{k}} p_{\mathbf{mk}}(t)E_{\boldsymbol{\mu}}\sigma_{\mathbf{k}} = 2p - 1 \qquad \text{for all } t \geqslant 0, \end{aligned} \tag{11.17}$$

where $p_{\mathbf{mk}}(t)$ is the transition law for the continuous-time random walk. Also for distinct sites $\mathbf{n}$ and $\mathbf{m}$ in $\mathbb{Z}^d$ we have

$$\begin{aligned} \varphi_{\boldsymbol{\mu}}(t, \{\mathbf{n}, \mathbf{m}\}) &= P_{\boldsymbol{\mu}}(\sigma_{\mathbf{m}}(t) = \sigma_{\mathbf{n}}(t)) - P_{\boldsymbol{\mu}}(\sigma_{\mathbf{m}}(t) \neq \sigma_{\mathbf{n}}(t)) \\ &= 1 - 2P_{\boldsymbol{\mu}}(\sigma_{\mathbf{m}}(t) \neq \sigma_{\mathbf{n}}(t)). \end{aligned} \tag{11.18}$$

Therefore, it is enough to show that $\varphi_{\boldsymbol{\mu}}(t, \{\mathbf{n}, \mathbf{m}\}) \to 1$ as $t \to \infty$ to prove (a). Now since the difference of the two independent simple symmetric random walks starting at $\mathbf{m}$ and $\mathbf{n}$ evolves as a (symmetrized) continuous-parameter simple symmetric random walk, it follows from the recurrence when $d \leqslant 2$, that $D_t \to \varnothing$ a.s. as $t \to \infty$; i.e., the particles will eventually meet. Using the duality relation and the Lebesgue Dominated Convergence Theorem we have, therefore,

$$\lim_{t\to\infty} \varphi_{\boldsymbol{\mu}}(t, \{\mathbf{n}, \mathbf{m}\}) = \lim_{t\to\infty} \int_S E_{\{\mathbf{n},\mathbf{m}\}}\left[\prod_{\mathbf{k}\in D_t} \sigma_{\mathbf{k}}\right]\boldsymbol{\mu}(d\boldsymbol{\sigma}) = 1. \tag{11.19}$$

To prove (b), on the other hand, take for $\boldsymbol{\mu}$ the Bernoulli product distribution of independent values subject to (11.12) and consider the duality relation

$$\varphi_\mu(t, D) = E_\mu\left\{E_D \prod_{\mathbf{m}\in D_t} \sigma_{\mathbf{m}}\right\} = E_D\left\{E_\mu \prod_{\mathbf{m}\in D_t} \sigma_{\mathbf{m}}\right\} = E_D \prod_{\mathbf{m}\in D_t} E_\mu \sigma_{\mathbf{m}}$$
$$= E_D(2p-1)^{|D_t|}, \tag{11.20}$$

where $|D_t|$ denotes the cardinality of D_t. Now since $|D_t|$ is a.s. nonincreasing, the limit, denoted $|D_\infty|$, exists a.s. and is positive with nonzero probability by *transience*. The limit distribution $\mathbf{v}^{(p)}$ is defined through its block correlation accordingly. That is,

$$\int_S \left\{\prod_{\mathbf{m}\in D} \sigma_{\mathbf{m}}\right\} \mathbf{v}^{(p)}(d\boldsymbol{\sigma}) = E_D(2p-1)^{|D_\infty|}. \tag{11.21}$$

Certainly it follows that $\mathbf{v}^{(p)}$ is translation-invariant and, being a time-asymptotic distribution of the evolution, one expects it to be invariant under the (further) evolution. More precisely, for any (bounded) continuous real-valued function f on S we have by the *Chapman–Kolmogorov equations* (semigroup property) that

$$\int_S f(\boldsymbol{\eta})\mathbf{v}_t^{(p)}(d\boldsymbol{\eta}) = \int_S \int_S f(\boldsymbol{\eta})p(t; \boldsymbol{\sigma}, d\boldsymbol{\eta})\mathbf{v}^{(p)}(d\boldsymbol{\sigma})$$
$$= \lim_{s\to\infty} \int_S \int_S f(\boldsymbol{\eta})p(t; \boldsymbol{\sigma}, d\boldsymbol{\eta})\boldsymbol{\mu}_s(d\boldsymbol{\sigma}), \tag{11.22}$$

since for fixed (continuous) f and $t > 0$, the (bounded) function $\boldsymbol{\sigma} \to \int_S f(\boldsymbol{\eta})p(t; \boldsymbol{\sigma}, d\boldsymbol{\eta})$ is continuous (theoretical complement) and, as noted earlier, by compactness of S $\boldsymbol{\mu}_s$ converges to $\mathbf{v}^{(p)}(d\boldsymbol{\sigma})$ in distribution (i.e., weak convergence) as $s \to \infty$. Now, from (11.22) and the Chapman–Kolmogorov equations again, one has

$$\int_S f(\boldsymbol{\eta})\mathbf{v}_t^{(p)}(d\boldsymbol{\eta}) = \lim_{s\to\infty} \int_S f(\boldsymbol{\eta})\boldsymbol{\mu}_{t+s}(d\boldsymbol{\eta}) = \int_S f(\boldsymbol{\eta})\mathbf{v}^{(p)}(d\boldsymbol{\eta}). \tag{11.23}$$

Thus, since the integrals of continuous functions on S with respect to the distributions $\mathbf{v}_t^{(p)}(d\boldsymbol{\eta})$ and $\mathbf{v}^{(p)}(d\boldsymbol{\eta})$ coincide, the two measures must be the same (see theoretical complement 8.6 of Chapter I). ■

The property that, for each bounded continuous function f on S, the function $\boldsymbol{\sigma} \to E_{\boldsymbol{\sigma}} f(\boldsymbol{\sigma}(t)) = \int_S f(\boldsymbol{\eta})p(t; \boldsymbol{\sigma}, d\boldsymbol{\eta})$ is also a bounded continuous function, is called the *Feller property*. As illustrated in the above proof (Eq. 11.22), this property is essential for confirming one's intuition about *invariance properties* of long-time limits under weak convergence. The other important role played by topology in the above was in the use of compactness to, in fact, get weak convergence from finite-dimensional calculations.

EXERCISES

Exercises for Section IV.1

1. Show that in order to establish the Markov property it is enough to check Eq. 1.1 for only one "future" time point t, for arbitrary $t > s$.

2. Show that a continuous-parameter process with independent increments has the Markov property. Show also that such a process has a homogeneous transition law if and only if, for every $h > 0$, the distribution of $X_{t+h} - X_t$ is the same for all $t \geqslant 0$.

3. Show that, given a Poisson process $\{X_t\}$ as in Example 1 with $\int_0^\infty \rho(u)\,du = \infty$, there exists an increasing transformation $\varphi: [0, \infty) \to [0, \infty)$ such that the process $\{Y_t := X_{\varphi(t)}\}$ is homogeneous with parameter $\lambda = 1$.

4. Prove that the compound Poisson process (Example 2) has independent increments.

5. Consider a compound Poisson process $\{X_t\}$ with state space $\mathbb{R}^1$ and an arbitrary jump distribution $\mu(dx)$ on $\mathbb{R}^1$.
   (ii) Show that $\{X_t\}$ is a Markov process and compute its transition probability $p(t; x, B) := P(X_t \in B \mid X_0 = x)$.
   (ii) Compute the characteristic function of X_t.

6. (*Doubly Stochastic Poisson or Cox Process*) Suppose that the parameter λ (mean rate) of a homogeneous Poisson process $\{X_t\}$ is random with distribution $\mu(d\lambda) = f(\lambda)\,d\lambda$ on $(0, \infty)$. In other words, conditionally given $\lambda = \lambda_0$, $\{X_t\}$ is a Poisson process with parameter λ_0.
   (i) Show that $\{X_t\}$ is not a process with independent increments.
   (ii) Show that $\{X_t\}$ is (generally) not a Markov process.
   (iii) Compute the distribution of X_t for arbitrary but fixed $t > 0$.
   (iv) Compute $\mathrm{Cov}(X_s, X_t)$.

7. Generalize Exercise 6 to compound Poisson processes.

8. The lifetimes of elements of a certain type are independent and exponentially distributed with parameter $\lambda > 0$. At time $t = 0$ there are $X_0 = n$ living elements present. Let X_t denote the number alive at time t. Show that $\{X_t\}$ is a Markov process and calculate its transition probabilities.

9. Let $\{X_t: t \geqslant 0\}$ be a process starting at x with (stationary) independent increments, and $EX_t^2 < \infty$. Assuming EX_t and EX_t^2 are continuous, prove the following.
   (i) $EX_t = mt + x$, $\mathrm{Var}\, X_t = \sigma^2 t$ for some constants m and σ^2.
   (ii) $(X_t - mt)/\sqrt{t}$ converges in distribution to the Gaussian distribution with mean 0 and variance σ^2, as $t \to \infty$.

10. (i) Show that the sum of a finite number of independent real-valued stochastic processes, each having independent increments is also a process with independent increments.
    (ii) Let $\{N_t^{(i)}\}$ $(i = 1, 2, \ldots, k)$ be independent Poisson processes with mean

parameter λ_i ($i = 1, 2, \ldots, k$). If $c_1, c_2, \ldots, c_k$ are arbitrary distinct positive constants, show that $\{\sum_1^k c_i N_t^{(i)}\}$ is a compound Poisson process and compute its jump distribution and the (Poisson) mean rate of occurrences.

(*iii) Prove that every compound Poisson process is a limit in distribution of superpositions of independent Poisson processes as described in (ii). [*Hint*: Compute the characteristic function of respective increments.]

Exercises for Section IV.2

1. Check the Kolmogorov consistency condition for P_π defined by Eq. 2.2, assuming the Chapman–Kolmogorov condition (2.5).

2. There are n identical components in a system that operate independently. When a component fails, it undergoes repair, and after repair is placed back into the system. Assume that for a component the operating times between successive failures are i.i.d. exponential with mean $1/\lambda$, and that these are independent of the successive repair times, which are i.i.d. exponential with mean $1/\mu$. The state of the system is the number of components in operation. Determine the infinitesimal generator of this Markov process.

3. For the process $\{X_t\}$ in Exercise 1.8, give the corresponding infinitesimal generator and Kolmogorov's backward and forward equations.

4. Let $\{X_t\}$ be a birth–death process on $S = \{0, 1, 2, \ldots\}$ with $q_{i,i+1} = i\beta$, $q_{i,i-1} = i\delta$, $i \geqslant 0$, where $\beta, \delta > 0$, $q_{ij} = 0$ if $|j - i| > 1$, $q_{01} = 0$. Let

$$m_i(t) = E_i X_t, \qquad s_i(t) = E_i X_t^2.$$

(i) Use the foward equation to show $m_i'(t) = (\beta - \delta)m_i(t)$, $m_i(0) = i$.
(ii) Show $m_i(t) = ie^{(\beta-\delta)t}$.
(iii) Show $s_i'(t) = 2(\beta - \delta)s_i(t) + (\beta + \delta)m_i(t)$.
(iv) Show that

$$s_i(t) = \begin{cases} ie^{2(\beta-\delta)t}\left[i + \dfrac{\beta+\delta}{\beta-\delta}(1 - e^{-(\beta-\delta)t})\right], & \text{if } \beta \neq \delta \\ i(i + 2\beta t), & \text{if } \beta = \delta. \end{cases}$$

(v) Calculate $\operatorname{Var}_i X_t$.

*5. Consider a compound Poisson process $\{X_t\}$ on $\mathbb{R}^1$ with an arbitrary jump distribution $\mu(dx)$.

(i) For a given bounded continuous function f on $\mathbb{R}^1$ compute

$$u(t, x) := E(f(X_t) \mid X_0 = x),$$

and show that

$$\frac{\partial}{\partial t} u(t, x) = -\lambda u(t, x) + \lambda \int u(t, x + y)\mu(dy) = \lambda \int (u(t, x + y) - u(t, x))\mu(dy).$$

(ii) Show that the limit of the last expression is

$$\lim_{t \downarrow 0} \frac{\partial u(t, x)}{\partial t} = (\mathbf{Q}f)(x),$$

where $\mathbf{Q}$ is the *integral operator* $(\mathbf{Q}f)(x) = \lambda \int (f(x + y) - f(x))\mu(dy)$.

(iii) Write the (backward) equation (i) above for $u(t, x)$ in terms of $\mathbf{Q}$.

6. Let $\{Y_t\}$ be a nonhomogeneous Markov chain with transition probabilities $p_{ij}(s, t) = P(Y_t = j \mid Y_s = i)$, continuous for $0 \leqslant s \leqslant t$, with $p_{ij}(s, s) = \delta_{ij}$, $i, j \in S$, and such that

$$\lim_{t \to s^+} \frac{p_{ij}(s, t) - \delta_{ij}}{t - s} = q_{ij}(s)$$

exists and is finite for each s.

(i) Show that the Chapman–Kolmogorov equations take the form,

$$p_{ik}(s, t) = \sum_j p_{ij}(s, r)p_{jk}(r, t), \qquad s < r < t.$$

(ii) For finite S show that the backward and forward equations, respectively, take the forms below:

$$(\textit{backward}) \qquad \frac{\partial p_{ik}(s, t)}{\partial s} = -\sum_j q_{ij}(s)p_{jk}(s, t),$$

$$(\textit{forward}) \qquad \frac{\partial p_{ik}(s, t)}{\partial t} = \sum_j p_{ij}(s, t)q_{jk}(t).$$

7. Consider a collection of particles that act independently in giving rise to succeeding generations of particles. Suppose that each particle, from the time it appears, waits a random length of time having an exponential distribution with parameter λ, and then either splits into two particles with probability p or disappears with probability $q = 1 - p$. Find the generator of this Markov process on the state space $S = \{0, 1, 2, 3, \ldots\}$, the state being the number of particles present.

8. In Exercise 7 above, suppose that new particles immigrate into the system (independently of particles present) at random times that form a Poisson process with parameter μ, and then give rise to succeeding generations as described in Exercise 7. Compute the generator of this Markov process.

*9. Let $\{X_t\}$ be a Markov process with state space an arbitrary measurable space $(S, \mathscr{S})$. Let $\mathbf{B}(S)$ be the space of (Borel-measurable) bounded real-valued functions on S with the uniform norm, $\|f\| = \sup_{x \in S}|f(x)|$, $f \in \mathbf{B}(S)$. Then $(\mathbf{B}(S), \|\cdot\|)$ is a *Banach space*. Define $\mathbf{T}_t f(x) = E_x f(X_t)$, $t \geqslant 0$, $x \in S$, for $f \in \mathbf{B}(S)$. Also, for $f \in \mathbf{B}(S)$ such that $\lim_{t \to 0^+}\{(\mathbf{T}_t f - f)/t\}$ exists in $(\mathbf{B}(S), \|\cdot\|)$, say that f belongs to the *domain* of $\mathbf{Q}$ and define $\mathbf{Q}f = \lim_{t \to 0^+}\{(\mathbf{T}_t f - f)/t\}$. Show that if the domain of $\mathbf{Q}$ is all of $\mathbf{B}(S)$, then $\mathbf{Q}$ must be a *bounded* linear operator on $(\mathbf{B}(S), \|\cdot\|)$; i.e., $\mathbf{Q}$ is *continuous* on $(\mathbf{B}(S), \|\cdot\|)$. [*Hint*: $\mathbf{T}_t f(x) = f(x) + \int_0^t \mathbf{T}_s \mathbf{Q}f(x)\, ds$, $t \geqslant 0$, $x \in S$, $f \in \mathbf{B}(S)$. Apply the *closed graph theorem* from functional analysis.]

10. (*Continuous-Parameter Pólya Process*) Fix $\tau > 0$ and consider a box containing $r = r(\tau)$ red balls and $b = b(\tau)$ black balls. Every τ units of time a ball is randomly selected, its color is noted, and together with $c = c(\tau)$ balls of the same color it is placed in the box. Let $S_{n\tau}$ denote the number of red balls sampled by the time $n\tau$, $n = 0, 1, 2, \ldots, S_0 = 0$. As in Eq. 3.14 of Chapter II, $\{S_{n\tau}\}$ is a discrete-parameter nonhomogeneous Markov chain with one-step transition probabilities

$$p_{i,i}(n\tau, (n+1)\tau) = 1 - p_{i,i+1}(n\tau, (n+1)\tau),$$

and

$$p_{i,i+1}(n\tau, (n+1)\tau) = P(S_{(n+1)\tau} = i + 1 \mid S_{n\tau} = i) = \frac{r + ci}{r + b + nc}, \qquad i = 0, 1, \ldots.$$

Let $p = p(\tau) = r/(r + b)$, $\gamma = \gamma(\tau) = c/(r + b)$, $t = n\tau$. Then the probability of a transition from i to $i + 1$ in time t to $t + \tau$ is given by

$$p_{i,i+1}(t, t + \tau) = \frac{p + \gamma i}{1 + \dfrac{\gamma}{\tau} t}.$$

Note that $p = r/(r + b)$ is the probability of selecting a red ball at the nth trial and $np = (p/\tau)t$ is the expected number of red balls sampled by time $t = n\tau$, i.e., p/τ is the mean sampling rate of red balls. Suppose that $p/\tau = p(\tau)/\tau \to 1$ and $\gamma/\tau = \gamma(\tau)/\tau \to \gamma_0 > 0$ as $\tau \to 0$.

(i) For fixed $\tau > 0$, use a combinatorial argument to show that the distribution of $S_{n\tau}$ is given by

$$P(S_{n\tau} = j \mid S_0 = 0) = \binom{n}{j} \frac{b(b + c)\cdots(b + (n - j)c - c)r(r + c)\cdots(r + jc - c)}{(b + r)(b + r + c)\cdots(b + r + nc - c)}.$$

(ii) Show that in the limit as $\tau \to 0$, the distribution in (i) converges to

$$f_j(t) = t^j(1 + \gamma_0 t)^{-j-\gamma_0^{-1}} \frac{((j - 1)\gamma_0 + 1)\cdots(2\gamma_0 + 1)(\gamma_0 + 1)}{j!}, \qquad j = 0, 1, 2, \ldots,$$

which is a *negative binomial* p.m.f.

(iii) Show that

$$\sum_{j=0}^{\infty} j f_j(t) = t \qquad \text{and} \qquad \sum_{j=0}^{\infty} (j - t)^2 f_j(t) = \gamma_0 t^2 + t.$$

(iv) The continuous-parameter Pólya process is often defined as a *nonhomogeneous* (pure birth) Markov chain $\{Y_t\}$ starting at 0 on $S = \{0, 1, 2, \ldots\}$ with transition probabilities denoted by $P(Y_t = j \mid Y_s = i) = p_{ij}(s, t), j, i = 0, 1, 2, \ldots, 0 \leqslant s \leqslant t$, and satisfying

$$p_{ij}(t, t + \Delta t) = \delta_{ij} + q_{ij}(t)\,\Delta t + o(\Delta t) \qquad \text{as } \Delta t \to 0,$$

where

$$q_{i,i+1}(t) = \frac{1 + i\gamma_0}{1 + t\gamma_0} = -q_{ii}(t), \qquad q_{ij}(t) = 0, \qquad j \neq i, i+1.$$

Check that $P(S_{(n+1)\tau} = j \mid S_{n\tau} = i) = p_{ij}(t, t + \tau) + o(\tau)$ as $\tau \to 0, n \to \infty, t = n\tau$.

Exercises for Section IV.3

1. Prove the inequality (3.12).
2. Prove that the solution $\mathbf{p}(t) = \exp\{t\mathbf{Q}\}$ is the unique bounded solution of Kolmogorov's backward (as well as forward) equations, under the hypothesis (3.9).
3. Solve the forward equations directly in Example 1.
4. Solve the Kolmogorov equations for a Markov process with three states and infinitesimal generator

$$\mathbf{Q} = \begin{bmatrix} -\lambda & \lambda & 0 \\ 0 & -\mu & \mu \\ 0 & 0 & 0 \end{bmatrix},$$

where λ, μ are positive.

5. Let $\mathbf{Q} = ((q_{ij}))$ such that $\sup_{i,j}|q_{ij}| < \infty$. Show that all elements of $e^{\mathbf{Q}t}$ are nonnegative for all $t \geqslant 0$ if $q_{ij} \geqslant 0$ for $i \neq j$. [*Hint*: Consider

$$r_{ij}(t) = \delta_{ij} + q_{ij}t + q_{ij}^{(2)}\frac{t^2}{2!} + \cdots$$

for small $t > 0$, and use the fact that the product of matrices with all entries nonnegative has itself all entries nonnegative together with the property $e^{t\mathbf{Q}} = (e^{(t/n)\mathbf{Q}})^n$.]

6. Each organism of a system, independently of the others, lives for an exponentially distributed time with parameter λ and is then replaced by two replicas that independently undergo the same replication process. Let X_t denote the total number of organisms at time t. Show that the bounded rates condition is *not* satisfied for the generator $\mathbf{Q}$ of $\{X_t\}$.
7. Calculate the transition probabilities for Exercise 2.2 in the case $n = 2$.
8. In radioactive transformations an unstable atom randomly disintegrates to an unstable atom of a new chemical and physical structure. Successive atomic states in the chain are labeled $0, 1, 2, \ldots, N$. Atoms in state i are transformed to state $i + 1$, $0 \leqslant i < N$ at the rate $q_{i,i+1} = \lambda_i$, where $\lambda_N = 0$. Calculate the distribution of the state at time t of an atom initially in state 0.

Exercises for Section IV.4

1. Calculate the successive approximations $p_{ik}^{(n)}(t)$ in the case of the (homogeneous) Poisson process with parameter $\lambda > 0$.

2. Write out the third iterate (of Eq. 4.7) $p_{ij}^{(3)}$ for the case $S = \{1, 2, 3\}$, $q_{1,1} = -\lambda$, $q_{1,2} = \lambda$, $q_{2,2} = -\mu$, $q_{2,3} = \mu$, $q_{3,3} = -\gamma$, $q_{3,2} = \gamma$, $q_{ij} = 0$ otherwise for $i, j \in S$.

3. (*A Pure Death Process*) Let $S = \{0, 1, 2, \ldots\}$ and let $q_{i,i-1} = \delta > 0$, $q_{ii} = -\delta, i \geqslant 1$, $q_{ij} = 0$, otherwise.
   (i) Calculate $p_{ij}(t)$, $t \geqslant 0$ using successive approximations.
   (ii) Calculate $E_i X_t$ and $\text{Var}_i X_t$.

4. Calculate approximations $\mathbf{p}^{(2)}(t)$ for Exercises 2.7 and 2.8.

5. Show that the minimal solution $\bar{\mathbf{p}}(t)$ also satisfies the forward equation. [*Hint*: Consider the integral version.]

6. Show that the minimal solution $\bar{\mathbf{p}}(t)$, $t \geqslant 0$, is a semigroup (i.e., that it satisfies the Chapman–Kolmogorov equations) by the following procedure.
   (i) Let f be a continuous bounded function on $[0, \infty)$ with Laplace transform

$$\hat{f}(v) = \int_0^\infty e^{-vt} f(t)\, dt.$$

   Define $F(s, t) = f(s + t)$, $s, t \geqslant 0$, and let

$$\hat{F}(v, \mu) = \int_0^\infty \int_0^\infty e^{-(vs + \mu t)} F(s, t)\, ds\, dt \qquad (\mu > 0,\ v > 0)$$

   be the bivariate Laplace transform of F. Show that $\hat{F}$ satisfies the *resolvent equation*:

$$\hat{F}(v, \mu) = -\frac{\hat{f}(v) - \hat{f}(\mu)}{v - \mu}, \qquad v \neq \mu.$$

   [*Hint*: Write $u = s + t$, $v = -s + t$, $0 \leqslant u < \infty$, $-u \leqslant v \leqslant u$, in the integral defining $\hat{F}$.]
   (ii) Let $\hat{\bar{\mathbf{p}}}(v) = ((\hat{\bar{\mathbf{p}}}_{ij}(v)))$ denote the matrix of transformed entries of the minimal solution. Show that $\bar{\mathbf{p}}(s + t) = \bar{\mathbf{p}}(s)\bar{\mathbf{p}}(t)$, $s, t \geqslant 0$ if and only if

$$-\frac{\hat{\bar{\mathbf{p}}}(v) - \hat{\bar{\mathbf{p}}}(\mu)}{v - \mu} = \hat{\bar{\mathbf{p}}}(\mu)\hat{\bar{\mathbf{p}}}(v), \qquad v \neq \mu.$$

   (iii) Show that the backward and forward equations (see also Exercise 5) transform, respectively, as

$$[\hat{B}]: \quad v\hat{\bar{\mathbf{p}}}(v) = \mathbf{I} + \mathbf{Q}\hat{\bar{\mathbf{p}}}(v),$$
$$[\hat{F}]: \quad v\hat{\bar{\mathbf{p}}}(v) = \mathbf{I} + \hat{\bar{\mathbf{p}}}(v)\mathbf{Q}.$$

   (iv) Use the backward equation to show

$$\mu\hat{\bar{\mathbf{p}}}(v) = \mathbf{A} + \mathbf{Q}\hat{\bar{\mathbf{p}}}(v)$$

   for $\mathbf{A} = \mathbf{I} + (\mu - v)\hat{\bar{\mathbf{p}}}(v)$. Use nonnegativity of $\hat{\bar{\mathbf{p}}}(v)$ and $\mathbf{A}$ for $\mu > v$, to show by induction, using (4.7), that $\hat{\hat{\mathbf{p}}}(\nu) \geqslant \hat{\mathbf{p}}^{(n)}(\mu)\mathbf{A}$ for $n = 0, 1, 2, \ldots$, and therefore

$$\hat{\hat{\mathbf{p}}}(\nu) \geqslant \hat{\hat{\mathbf{p}}}(\mu)\mathbf{A} = \hat{\hat{\mathbf{p}}}(\mu) + (\mu - \nu)\hat{\mathbf{p}}(\mu)\hat{\hat{\mathbf{p}}}(\nu), \qquad \mu > \nu.$$

(v) Use $[\hat{F}]$ to prove the reverse inequality and hence (ii). [*Hint*: Check $\mathbf{r}(\nu) = \hat{\hat{\mathbf{p}}}(\mu) + (\mu - \nu)\hat{\mathbf{p}}(\mu)\hat{\hat{\mathbf{p}}}(\nu)$ solves $[\hat{F}]$ and use minimality.]

7. Compute $p_{ik}(t)$ for all i, k in Example 1.

Exercises for Section IV.5

1. Show that the *holding time* T_n $(n \geqslant 1)$ is not a stopping time.

2. Let $A(\Delta), B(\Delta), B(0)$ denote the events $\{s < T_0 \leqslant s + \Delta\}$, $\{X_{s+\Delta} = j\}$, $\{X_{T_0} = j\}$, respectively.
   (i) Prove that $P_i(A(\Delta) \cap B(\Delta) \cap B^c(0)) = o(\Delta)$ and $P_i(A(\Delta) \cap B^c(\Delta) \cap B(0)) = o(\Delta)$ as $\Delta \downarrow 0$ (for $i \neq j$). [*Hint*: Apply the strong Markov property with respect to T_0 in order to show $P_i(s < T_0 < T_0 + T_1 \leqslant s + \Delta) = o(\Delta)$.]
   (ii) Use (i) to derive (5.6) from (5.5).

3. Let $U_1, U_2, \ldots, U_n$ be i.i.d. uniform on $[0, t]$, and order them as $U_{(1)} < U_{(2)} < \cdots < U_{(n)}$. Define $D_j = U_{(j)} - U_{(j-1)}$ $(j = 1, 2, \ldots, n+1)$ with $U_{(0)} = 0$, $U_{(n+1)} = t$.
   (i) Show that the joint distribution of $D_1, \ldots, D_{n+1}$ is the same as the conditional distribution of $n + 1$ i.i.d. exponential random variables $Y_1, Y_2, \ldots, Y_{n+1}$, given $Y_1 + \cdots + Y_{n+1} = t$.
   (ii) Show that D_k has p.d.f. given by $n(1 - x)^{n-1}$, $0 \leqslant x \leqslant 1$, $k = 1, \ldots, n + 1$.

4. Derive an extension of Proposition 5.6 for a nonhomogeneous Poisson process with intensity function $\rho(t)$ (see Example 1.1).

5. Suppose that a pure birth process on $S = \{0, 1, 2, \ldots\}$ has infinitesimal parameters $q_{kk} = -\lambda_k$.
   (i) If $\sum_0^\infty \lambda_k^{-1} = \infty$ and $\sum_0^\infty \lambda_k^{-2} < \infty$, then show that the variance of $\tau_n = T_0 + T_1 + \cdots + T_{n-1}$ (the time to reach n, starting from 0) goes to a finite limit as $n \to \infty$. Use Kolmogorov's zero–one law (see theoretical complements 1.1, Chapter I) to show that as $n \to \infty$, $\tau_n - \sum_0^{n-1} \lambda_k^{-1}$ converges (for all sample paths, outside a set of probability zero) to a finite random variable η.
   (ii) If $\sum_0^\infty \lambda_k^{-1} = \infty = \sum_0^\infty \lambda_k^{-2}$, but $\sum_0^\infty \lambda_k^{-3} < \infty$, show that

$$\left(\tau_n - \sum_0^{n-1} \lambda_k^{-1}\right) \Bigg/ \left(\sum_0^{n-1} \lambda_k^{-2}\right)^{1/2}$$

   is asymptotically (as $n \to \infty$) Gaussian with mean 0 and variance 1. [*Hint*: Use Liapounov's central limit theorem, Chapter 0.]

6. Messages arrive at a telegraph office according to a Poisson process with mean rate of occurrence of 4 messages per hour.
   (i) What is the probability that no message will have arrived during the *morning hours* (8 to 12)?
   (ii) What is the distribution of the time at which the *second afternoon message* arrives?

7. Let $\{X_t\}$ and $\{Y_t\}$ be independent Poisson processes with parameters λ and ρ respectively. Show that the probability of n occurrences of $\{Y_t\}$ within the time interval from the first to the $(r + 1)$st occurrence of $\{X_t\}$ is given by

$$\binom{n+r-1}{r-1}\left(\frac{\rho}{\lambda+\rho}\right)^n\left(\frac{\lambda}{\lambda+\rho}\right)^r, \qquad n = 0, 1, 2, \ldots .$$

8. Let $\{X_t\}$ be a Poisson process with parameter $\lambda > 0$. Let T_0 denote the time of the first arrival.
   (i) Let $N = X_{2T_0} - X_{T_0}$ and calculate $\text{Cov}(N, T_0)$.
   (ii) Calculate the (conditional) expected value of the time $T_0 + \cdots + T_{r-1}$ of the rth arrival given $X_t = n > r$.

9. Suppose that two colonies, 1 and 2, start as single units and independently undergo growth by pure birth processes with rates β_1, β_2, respectively. Calculate the expected size of colony 1 at the time when the first offspring is produced in colony 2.

10. Consider a *pure-death* process with rates $-q_{ii} = q_{i,i-1} = \lambda > 0,\ i \geqslant 1,\ q_{ij} = 0$ otherwise.
    (i) Calculate $p_{ij}(t)$.
    (ii) If initially there are n_1 particles of type 1 and n_2 particles of type 2 that independently undergo pure death at rates δ_1, δ_2, respectively, calculate the expected number of type 1 particles at the time of extinction of the type 2 particles.

11. Let $\{N_t\}$ be a homogeneous Poisson process with parameter $\lambda > 0$. Define $X_t = (-1)^{N_t}$, $t \geqslant 0$. Show that $\{X_t\}$ is a Markov process and compute its transition probabilities.

12. Consider all rooted binary tree graphs having n *sources* (i.e., degree-one vertices excluding the root). Call any edge incident to a source vertex an *external edge* and call the others *internal* (see Fig. Ex.IV.5). [By a *rooted* tree is meant a tree graph in which a vertex is singled out as the root. Other graph-theory terminology is described in Exercise 7.5 of Chapter II.]

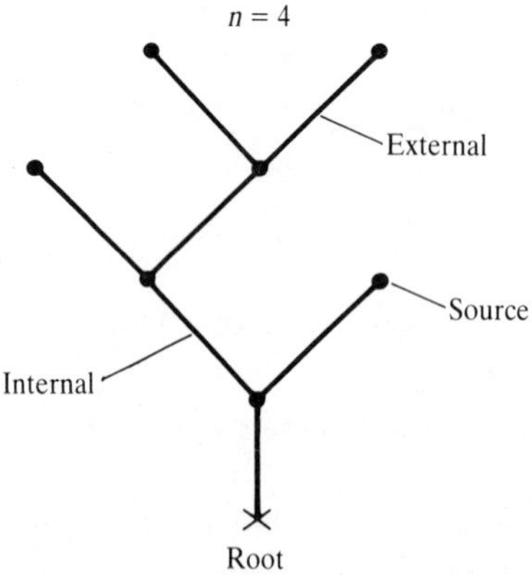

Figure Ex. IV.5

(i) Show that a rooted binary tree with n sources has n external edges and $n-1$ internal edges, $n \geqslant 1$. In particular, the total number of edges is $2n-1$, which is also the total number of vertices excluding the root.

(ii) Show that the following code establishes a one-to-one correspondence between the collection of all rooted binary tree graphs and the collection of all simple polygonal paths from $(0, 0)$ to $(2n-1, -1)$ in steps of ± 1 that do not touch or cross the line $y = -1$ prior to $x = 2n-1$. Starting with the edge incident to the root, traverse the tree along the leftmost path until reaching the leftmost source. Then follow back until reaching a junction leading to the next leftmost source, and so on, recording, on the first (and only on the first) traverse of an edge, + if it is internal and − if it is external. The path $2n-1$ long of (+)s and (−)s furnishes the coding of the tree in the form of a *"random walk excursion."*

(iii) Use (ii) and the reflection principle (Section 4, Chapter I) to calculate the distribution of M_L in Example 3.

(iv) Let $g(x) = Ex^{M_L}$ denote the probability-generating function of M_L. Use the recursive structure of the tree to establish the quadratic equation $g(x) = \frac{1}{2}x + \frac{1}{2}g^2(x)$.

(v) Use (iv) to give another derivation of the distribution of M_L.

13. Show that the holding time in state j for Example 2 is exponentially distributed with parameter $\lambda_j = j\lambda$ ($j \geqslant 1$).

Exercises for Section IV.6

1. Show that $(1 + x)\log(1 + x) \geqslant x$ for all $x \geqslant 0$.

2. Show that (i) the backward equation holds for $\tilde{p}_{ij}(t)$, and (ii) $\{\tilde{p}_{ij}(t): t > 0\}$ are transition probabilities.

3. Show that $\{\mathbf{p}_{ij}^{\Delta}(t): t > 0\}$ are transition probabilities on $S \cup \{\Delta\}$ satisfying the backward and forward equations.

4. (i) Consider an initial mass of size x_0 that grows (deterministically) to a size x_t at time t at a rate that is dependent upon size; say, $x_t' = f(x_t)$, $t \geqslant 0$. Give an example of a growth-rate function $f(x)$, $x \geqslant 0$, such that the mass will grow without bound within a finite time $\xi > 0$, i.e., $\lim_{t \to \xi} x_t = \infty$. [*Hint*: Consider a case in which each "element" of mass grows at a rate proportional to the total mass, so that the *total mass* itself grows at a rate proportional to the mass squared.]

   (ii) Let X_t denote the number of reactions that have occurred on or before time t in a chain reaction process. Suppose that $\{X_t\}$ is a pure birth process starting at 0. Show that the expected number of reactions that occur in a finite time interval $[0, t]$ need *not* be finite.

5. (*A Bus Stop Problem*) A passenger regularly travels from home to office either by bus or by walking. The travel time by walking is a constant t_w, whereas the travel time by bus from the stop to work is a random variable, independent of the bus arrival time, with a continuous distribution having mean $t_b < t_w$. Buses arrive randomly at the home stop according to a Poisson process with intensity parameter λ. The passenger uses the following *strategy* to decide whether to walk or ride. If the bus arrives within c time units, then ride the bus; if the wait reaches c, then walk. Determine (optimality) conditions on c that minimize the average travel time in terms

of t_b, t_w, λ. [*Note*: The solutions $c = 0$ (always walk), $c = \infty$ (always ride) are permitted.]

6. Consider a single telephone line that is either free (state 0) or busy (state 1). Suppose incoming calls form a Poisson process with a mean rate (intensity) λ per minute. The successive durations of calls are i.i.d. exponential random variables with parameter μ (mean duration of a call being μ^{-1} minutes), independent of the Poisson process of incoming calls. If a call arrives at a time when the line is busy, the call is simply lost.
   (i) Give the corresponding prescription of Kolmogorov's backward and forward equations for the state of the line.
   (ii) Solve the forward equation.

Exercises for Section IV.7

1. Show that the process $\{X_t\}$ described noncanonically in Example 2 is a Markov process.

2. Prove Eq. 7.11.

3. Let $\{X_t\}$ be the Yule linear growth process of Example 2. Show that
   (i) $E_i X_t = ie^{\lambda t}$.
   (ii) $\text{Var}_i X_t = ie^{\lambda t}(e^{\lambda t} - 1) = 2ie^{3\lambda t/2} \sinh(\frac{1}{2}\lambda t)$.

4. Show that the N-server queue process $\{X_t\}$ is a Markov process.

*5. (*Renewal Age Process*) Let $T_1, T_2, \ldots$ be an i.i.d. sequence of positive random variables with a continuous (*lifetime*) distribution function F. Let $S_0 = 0$, $S_n = T_1 + T_2 + \cdots + T_n$, $n \geqslant 1$, and for $A_0 = 0$, define for $t \geqslant 0$, the *age at time t* by $A_t := t - \max\{S_k : S_k \leqslant t, k \geqslant 0\}$. Then $\{a + A_t\}$ is the process starting at $a \geqslant 0$.
   (i) Show that $\{A_t\}$ has the Markov property: i.e., for arbitrary time points $0 \leqslant s_0 < s_1 < \cdots < s_k < s < t < t_1 < \cdots < t_n$, the conditional distribution of $A_t, A_{t_1}, \ldots, A_{t_n}$ given $A_{s_0}, \ldots, A_{s_k}, A_s$ does not depend on $A_{s_0}, \ldots, A_{s_k}$.
   (ii) The *failure rate* (also called *hazard rate or force of mortality*) for the objects being renewed is defined by $h(t) = f(t)/[1 - F(t)]$, where f is the p.d.f. (assumed to exist) of F. For continuously differentiable functions g having bounded derivatives show, for $a \geqslant 0$,

$$\mathbf{Q}g(a) \colon \lim_{t \to 0} E_a\left(\frac{g(A_t) - g(a)}{t}\right) = h(a)\{g(0) - g(a)\} + \frac{dg}{da}, \qquad a \geqslant 0.$$

   (iii) Show that $\mu(a) = Z^{-1} \exp\{-\int_0^a h(u)\,du\}$, $a \geqslant 0$, where Z is the normalization constant, solves $\int_0^\infty \mu(a)\mathbf{Q}g(a)\,da = 0$. In particular, show that μ is the density of an invariant probability for $\{A_t\}$.
   (iv) Show (i) holds also for the *residual lifetime* $\{S_{N_t} - t\}$, where $N_t = \inf\{n \geqslant 0 : S_n > t\}$. For $ET_1 < \infty$, show that $f_\infty(t) = (1 - F(t))/ET_1$, $t \geqslant 0$, is the density of an invariant probability. [The above has an interesting generalization by

F. Spitzer (1986), "A Multidimensional Renewal Theorem," in *Probability, Statistical Mechanics, and Number Theory*, Advances in Mathematics Supplementary Studies, Vol. 9 (G. C. Rota, ed.), pp. 147–155.]

6. (*Thinned Poisson*) Let $\{\tau_n\}$ be the sequence of occurrence times for a Poisson process $\{N_t\}$ with parameter λ. Let $\{\varepsilon_n\}$ be an i.i.d. sequence, independent of the Poisson process, of Bernoulli 0–1-valued random variables such that $P(\varepsilon_n = 0) = p$, $0 < p < 1$. Define the thinned process by

$$Y_t := \sum_{n=0}^{N_t} 1_{(0,t]}(\varepsilon_n \tau_n), \qquad t > 0, \qquad Y_0 = 0,$$

where $\{N_t = \max\{n: \tau_n \leqslant t\}\}$ is the Poisson counting process. Show that $\{Y_t\}$ is a Poisson process with intensity parameter $(1-p)\lambda$.

7. (i) (*Vibrating String*) For Example 5, check that in the (deterministic) case $p = \alpha\tau = 0$, $u(x, t) = \frac{1}{2}\{\varphi(x + vt) + \varphi(x - vt)\}$. In particular, that $u(x, t)$ solves

$$\frac{\partial^2 u}{\partial t^2} = v^2 \frac{\partial^2 u}{\partial x^2}, \qquad u(x, 0) = \varphi(x), \qquad \frac{\partial u(x, t)}{\partial t} = 0 \qquad \text{at } t = 0.$$

(ii) Check that there is a *randomization of time* represented by a nonnegative nondecreasing stochastic process $\{T_t\}$ such that

$$u(x, t) = \tfrac{1}{2}\{E\varphi(x + vT_t) + E\varphi(x - vT_t)\}.$$

[*Hint*: Simply consider (7.37).]

8. (*Diffusion limit*) Check that in the limit $\alpha \to \infty$, $v \to \infty$, $2\alpha/v^2 = D^{-1} > 0$, the diffusion equation

$$\frac{\partial u}{\partial t} = D \frac{\partial^2 u}{\partial x^2},$$

is consistent with the telegrapher's equation.

9. The flow of electricity through a coaxial cable is typically described by the telegrapher's equation (7.19) or (7.32), (7.33), where $u(x, t)$ represents the (instantaneous) voltage and $w(x, t)$ the current at a distance x from the sending end of the cable. The parameters α, β, and v^2 can be interpreted in terms of the electrical properties of the cable as outlined in this exercise. If one ignores leakage (conductance due to inadequate insulation), then the circuit diagram for the segment of cable from x to $x + \Delta x$ may be depicted as in Figure Ex.IV.7 below.

   Here the parameters R, L, and C are the *resistance*, *inductance*, and *capacitance per unit length*. These parameters are defined in accordance with certain physical principles. For example, *Ohm's law* says that the ratio of the voltage drop across a resistor to the current through the resistor is a constant (called the resistance) given here by $R\,\Delta x$. Thus, the voltage drop across the resistor is $R\,\Delta x\, w(x, t)$. Likewise,

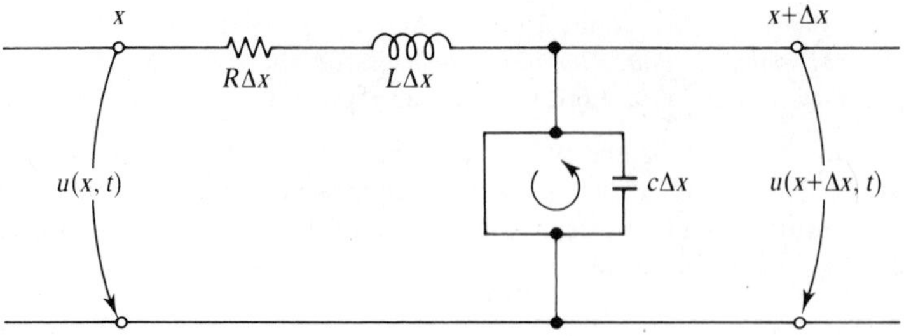

Figure Ex. IV.7

when there is a change in the flow of current $\partial w/\partial t$ in an inductor then there is a corresponding voltage drop of $L\,\Delta x\,\partial w/\partial t$. According to *Kirchhoff's second law*, the sum of potential drops (as measured by a voltmeter) around a closed loop in an electric network is zero. Thus, for Δx small, one has

$$u(x + \Delta x, t) - u(x, t) + R\,\Delta x\,w(x, t) + L\,\Delta x\,\frac{\partial w}{\partial t} = 0.$$

The nature of a capacitor is such that the ratio of the charge stored to the voltage drop is the capacitance $C\,\Delta x$. Thus, the capacitor current, being the time rate of change of charge, is given by $C\,\Delta x\,\partial u/\partial t$. According to *Kirchhoff's first law*, the sum of the currents flowing toward any point in an electrical network is zero (i.e., charge is conserved). Thus, for Δx small, one has

$$w(x + \Delta x, t) - w(x, t) + C\,\Delta x\,\frac{\partial u}{\partial t} = 0.$$

(i) Use the above to complete the derivation of the transmission line equations and the telegrapher's equation (for twice-continuously differentiable functions) and give the corresponding electrical interpretation of the parameters α, β, and v^2.
(ii) In the case when leakage is present there is an additional parameter (loss factor) G, called the *conductance per unit length*, such that the leakage current is proportional to the voltage $G\,\Delta x\,u(x, t)$. In this case the term $G\,\Delta x\,u(x, t)$ is added to the left-hand side in Kirchhoff's first law and one sees that the voltage and current satisfy transmission line equations of exactly the same general forms. Show that precisely the same equation is satisfied by both voltage and current.

10. In Example 4, suppose that the service time distribution is arbitrary, with distribution function F. Determine the distribution of W_t.

Exercises for Section IV.8

1. Let Z_i $(i \geqslant 1)$ be a sequence of i.i.d. nonnegative random variables, $P(Z_i > 0) > 0$. Prove that $\sum_{i=1}^{n} Z_i \to \infty$ almost surely.
2. Prove that the first and third terms on the right side of Eq. 8.13 go to zero (almost surely) as $t \to \infty$, provided (8.10) holds.

3. When is a compound Poisson process (i) a pure birth process, (ii) a birth–death process? Write down q_{ij}, k_{ij} in these cases.

4. Let $\{X_t\}$ be a birth–death chain, and let $c < x < d$ be three states.
   (i) Compute $P_x(\{X_t\}$ reaches c before $d)$ in terms of the infinitesimal rates (i.e., $\lambda_i, \beta_i, \delta_i$).
   (ii) Calculate $\rho_{xc} \equiv P_x(\{X_t\}$ ever reaches $c)$ in the case $\{X_t\}$ is nonexplosive. Briefly comment on what may happen in the case that explosion may occur with positive P_x-probability.

5. Given a birth–death chain $\{X_t\}$:
   (i) Derive a difference equation for $m(x) := E_x(\tau)$ where τ is the first time $\{X_t\}$ reaches c or d $(c \leqslant x \leqslant d)$.
   (ii) For the special case $q_{xx} = -\lambda$, $\beta_x = \delta_x = \frac{1}{2}$ $(c < x < d)$, compute $m(x)$.
   (*iii) Compute $m(x)$ for general birth–death chains.

6. Let $\mathbf{p}(t) = ((p_{ij}(t)))$ be transition probabilities for a Markov chain on a finite state space S. Adapt Proposition 6.1 of Chapter II to show that $\lim_{t\to\infty} p_{ij}(t)$ exists for all $i, j \in S$, and the convergence is exponentially fast. [*Hint*: Fix $t > 0$ and consider the discrete-parameter one-step transition probability matrix $\mathbf{q} = ((p_{ij}(t)))$. Then $\mathbf{q}^n = ((p_{ij}(nt))$, $n = 0, 1, 2, \ldots$. Use the semigroup property to show that $\lim_{n\to\infty} q_{ij}^{(n)}$ (exists) does not depend on $t > 0$.]

7. Suppose that $\{X_t\}$ is irreducible and positive-recurrent. Let $\{Y_n\}$ be the embedded discrete-parameter Markov chain with transition matrix $\mathbf{K}$. Let $\boldsymbol{\pi}$ be the invariant distribution for $\{X_t\}$ such that $\boldsymbol{\pi}'\mathbf{Q} = \mathbf{0}$. Calculate the invariant distribution for $\{Y_n\}$.

8. Let $\{X_t\}$ be a positive-recurrent birth–death process on $S = \{0, 1, 2, \ldots\}$ with birth–death rates $\beta_i\lambda_i$ and $\delta_i\lambda_i$ respectively. Show that

$$E_\pi(\delta_{X_t}\lambda_{X_t}) = E_\pi(\beta_{X_t}\lambda_{X_t}),$$

where π is the invariant initial distribution.

9. Consider the N-server queue under the invariant initial distribution (8.27).
   (i) Show that the expected number of customers waiting to be served (excluding those being served) is given by

$$\frac{\rho(N\rho)^N}{(1-\rho)^2 N!}\pi_0, \qquad \text{where } \rho = \frac{\alpha}{N\beta}\,(<1)$$

   is referred to as the *traffic intensity* parameter; i.e., the mean number of arrivals within the average service time β^{-1}/N.
   (ii) Show that the average length of time a customer must wait for service is

$$\frac{(N\rho)^N}{N(1-\rho)^2\beta N!}\pi_0.$$

   (iii) A particular hospital ward receives patients according to a Poisson process at an average rate of $\alpha = 2$ per day. The average length of stay is 5 days. Assuming the length of stay to be exponentially distributed, how many beds are necessary in order to achieve an equilibrium distribution for the total number of patients admitted and waiting admission? What will be the average waiting time for admission?

10. Let $\{X_t\}$ be the continuous-parameter Markov (binary) branching process with offspring distribution $f(0) = \delta$, $f(2) = \beta$, $\beta + \delta = 1$, and parameter λ.
   (i) Show that $\{X_t\}$ is a birth–death process with linear rates.
   (ii) Let $g_t^{(i)} \equiv g_t^{(i)}(r) := \sum_j P_i(X_t = j)r^j$ denote the p.g.f. of the P_i-distribution of X_t. Show that the forward equations transform as

$$\frac{\partial g_t^{(i)}}{\partial t} = [\bar{\beta}(r^2 - r) + \bar{\delta}(1 - r)] \frac{\partial g_t^{(i)}}{\partial r}, \qquad \bar{\beta} = \lambda\beta, \quad \bar{\delta} = \lambda\delta.$$

   (iii) Solve the equation for $g_t^{(i)}$.

11. Suppose that $\{X_t\}$ has transition probabilities $p_{ij}(t)$ and infinitesimal transition rates given by $\mathbf{Q} = ((q_{ij}))$. Suppose that $\boldsymbol{\pi}$ is a probability distribution satisfying $\pi'\mathbf{Q} = 0$. If $\sum_i \lambda_i \pi_i < \infty$, where $\lambda_i = -q_{ii}$, then show that $\boldsymbol{\pi}$ is an invariant initial distribution. [*Hint*: Differentiation of $\sum_i \pi_i p_{ij}(t)$ term by term is allowed.]

12. If S comprises a single communicating class and $\boldsymbol{\pi}$ is an invariant probability, then all states are positive recurrent. [*Hint*: Consider the discrete time chain X_n, $n \geqslant 0$. Use Theorem 9.4 of Chapter II. Note that η_i defined by (8.4) is smaller than the second passage time to state i for the discrete parameter chain.]

13. Show that positive recurrence is a class property. [*Hint*: Use arguments analogous to those used in the discrete-time case.]

14. Verify the forward equations for Example 5.3. [*Hint*: Use Exercise 4.5 and Example 8.2.]

Exercises for Section IV.9

1. Calculate the spectral representation of the transition probabilities $p_{ij}(t)$ for the continuous-parameter simple symmetric random walk on $S = \{0, 1, 2, \ldots\}$ with one reflecting boundary at 0. [*Hint*: Use (9.19).]

2. (*Time Reversibility*) Let $\{X_t\}$ be an irreducible positive recurrent Markov chain with invariant initial distribution $\boldsymbol{\pi}$.
   (i) Show that $\{X_t\}$ is a stationary process; i.e., for any $h > 0$, $0 \leqslant t_1 < t_2 < \cdots < t_k$ $(X_{t_1}, \ldots, X_{t_k})$ and $(X_{t_1 + h}, \ldots, X_{t_k + h})$ have the same distribution.
   (ii) Show that $\{X_t\}$ is *time-reversible*, in the sense that for any $0 \leqslant t_1 < t_2 < \cdots < t_k < T$ $(X_{t_1}, \ldots, X_{t_k})$ and $(X_{T - t_k}, \ldots, X_{T - t_1})$ have the same distribution, if and only if

$$\pi_i p_{ij}(t) = \pi_j p_{ji}(t) \qquad \text{for all } i, j \in S, \quad t \geqslant 0.$$

   This last property is sometimes referred to as *time-reversibility of* $\boldsymbol{\pi}$.
   (iii) Show that $\{X_t\}$ is time-reversible if and only if $\pi_i q_{ij} = \pi_j q_{ji}$ for all $i, j \in S$.
   (iv) Show that $\{X_t\}$ is time-reversible if and only if the discrete-parameter embedded chain is time-reversible; see Exercise 7.4 of Chapter II.
   (v) Show that the chemical reaction kinetics example is time-reversible under the invariant initial distribution.

3. (*Relaxation and Maximal Correlation*) Let $\{X_t\}$ be an irreducible finite-state Markov chain on S with transition probabilities $\mathbf{p}(t) = ((p_{ij}(t)))$, $t \geqslant 0$, and infinitesimal rates $\mathbf{Q} = ((q_{ij}))$. Let $\boldsymbol{\pi}$ be the unique invariant distribution and assume the distribution to be time-reversible as defined in Exercise 2(ii) above. Define the *maximal correlation* by $\rho(t) = \sup_{f,g} \operatorname{Corr}_\pi(f(X_t), g(X_0))$, where the supremum is over all real-valued functions f, g on S and

$$\operatorname{Corr}_\pi(f(X_t), g(X_0)) = \frac{E\{[f(X_t) - E_\pi f(X_t)][g(X_0) - E_\pi g(X_0)]\}}{\{\operatorname{Var}_\pi f(X_t)\}^{1/2}\{\operatorname{Var}_\pi g(X_0)\}^{1/2}}.$$

Show that

$$\rho(t) = e^{\lambda_1 t}, \qquad t \geqslant 0$$

where $\lambda_1 < 0$ is the largest nontrivial eigenvalue of $\mathbf{Q}$. The parameter $\gamma = -1/\lambda_1$ is called the *relaxation time* or *correlation length* parameter in this context. [*Hint*: For f, g such that $E_\pi f = E_\pi g = 0$, $\|f\|_\pi = \|g\|_\pi = 1$ $\operatorname{Corr}_\pi(f(X_t), g(X_0)) = (\mathbf{p}(t)f, g)_\pi$. So there is an orthonormal basis $\{\varphi_n\}$ of eigenvectors of $\mathbf{p}(t)$. Since

$$\mathbf{p}(t) = e^{\mathbf{Q}t} = \sum_{k=0}^{\infty} \frac{\mathbf{Q}^k t^k}{k!},$$

the eigenvalues of $\mathbf{p}(t)$ are of the form $e^{\lambda_n t}$ where λ_n are eigenvalues of $\mathbf{Q}$. Take $f = g = \varphi_1$ to get $\langle \mathbf{p}(t)f, g\rangle_\pi = e^{\lambda_1 t}$ and, for the general case (centered and scaled), expand f and g in terms of $\{\varphi_n\}$, i.e., $f = \sum_n (f, \varphi_n)_\pi \varphi_n$, $g = \sum_n (g, \varphi_n)_\pi \varphi_n$ to show $|\operatorname{Corr}_\pi(f(X_t), g(X_0))| \leqslant e^{\lambda_1 t}$. Use the simple inequality $ab \leqslant (a^2 + b^2)/2$.]

4. Let $\{X_t\}$ be an irreducible positive recurrent finite-state Markov chain with initial distribution $\boldsymbol{\mu}$. Let $\mu_j(t) = P_{\boldsymbol{\mu}}(X_t = j)$, $j \in S$, and let $\boldsymbol{\pi}$ denote the invariant initial distribution. Also let $\mathbf{Q} = ((q_{ij}))$ be the matrix of infinitesimal rates for $\{X_t\}$. Show that

$$\text{(i)} \quad \frac{d\mu_j(t)}{dt} = \sum_{i \in S} \{\mu_i(t) q_{ij} - \mu_j(t) q_{ji}\}, \qquad t \geqslant 0, \quad j \in S,$$

$$\mu_j(0) = \mu_j, \qquad j \in S.$$

(ii) Suppose that $\boldsymbol{\pi}$ is a *time-reversible* invariant distribution and define

$$\gamma_{ij} = \frac{1}{\pi_i q_{ij}} = \frac{1}{\pi_j q_{ji}}$$

(possibly infinite). Show that

$$\frac{d\mu_j(t)}{dt} = \sum_{k \in S} \frac{1}{\gamma_{jk}} \left(\frac{\mu_k(t)}{\pi_k} - \frac{\mu_j(t)}{\pi_j} \right), \qquad j \in S.$$

Consider an electrical network in which the states of S are the *nodes*, and γ_{jk} is the resistance in a wire connecting j and k. Suppose also that each node j carries a capacitance π_j. Then these equations are *Kirchhoff's equations* for the spread of an initial "electrical charge" $\mu_j(0) = \mu_j, j \in S$, with time (see also Exercise 7.9). The potential energy stored in the capacitors at the nodes at time t, when the initial distribution is $\boldsymbol{\mu}(0) = \boldsymbol{\mu}$, is given by

$$U(t) \equiv U(\boldsymbol{\mu}(t)) = -\frac{1}{2} \sum_{j \in S} \frac{(\mu_j(t))^2}{\pi_j^2}.$$

One expects that as time progresses energy will be dissipated as heat in the wires.

(iii) Show that if $\boldsymbol{\mu} \neq \boldsymbol{\pi}$, then U is strictly decreasing as a function of time.

(iv) Calculate $U(\boldsymbol{\mu}(t))$ for the two-state (flip-flop) Example 3.1 in the cases $\boldsymbol{\mu}(0) = \delta_{\{i\}}$, $i = 0, 1$.

5. As in Exercise 4 above, let $\{X_t\}$ be an irreducible positive recurrent finite-state Markov chain with initial distribution $\boldsymbol{\mu}$. Let $\mu_j(t) = P_{\boldsymbol{\mu}}(X_t = j)$, $j \in S$, and let $\boldsymbol{\pi}$ denote the invariant initial distribution. Let h be a strictly concave function on $[0, \infty)$ and define the *h-entropy* of the distribution at time t by

$$H(t) \equiv H(\boldsymbol{\mu}(t)) = \sum_{j \in S} h\left(\frac{\mu_j(t)}{\pi_j}\right)\pi_j, \qquad t \geqslant 0.$$

(i) Show for $\boldsymbol{\mu} \neq \boldsymbol{\pi}$, $H(t)$ is strictly increasing as a function of t and $\lim_{t \to \infty} H(\boldsymbol{\mu}(t)) = H(\boldsymbol{\pi})$.

(ii) Statistical entropy is defined by taking $h(x) = -x \log x$ in (i). Calculate $H(\boldsymbol{\mu}(t))$ for the two-state (flip-flop) Example 3.1 when $\boldsymbol{\mu}(0) = \boldsymbol{\delta}_{\{i\}}$, $i = 0, 1$.

Exercises for Section IV.10

1. Calculate the distribution $G(t; \lambda_i, \lambda_{i_1}, \ldots, \lambda_{i_{m-1}})$ appearing in Eq. 10.4. [*Hint*: Apply the *partial fractions* decomposition to the characteristic function of the sum.]

2. Calculate the distribution of the time until absorption at $j = 0$ for the *pure death* process on $S = \{0, 1, 2, \ldots\}$ starting at i, with infinitesimal rates $q_{ij} = i\delta$ if $j = i - 1$, $i \geqslant 1$, $q_{ii} = -i\delta$, $q_{ij} = 0$ otherwise. What is the average time to absorption?

3. A *zero-seeking particle* in state $i > 0$ waits for an exponentially distributed time with parameter λ and then jumps to a position uniformly distributed over $0, 1, 2, \ldots, i - 1$. If in state $i < 0$, it holds for an exponential time with parameter λ and then jumps to a position uniformly distributed over $i + 1, i + 2, \ldots, -1, 0$. Once in state 0 it stays there. Calculate the average time to reach zero, starting in state i.

*4. (i) Under the conditions stated in Example 2, show that conditioned on non-extinction at time t, X_t has a limit distribution as $t \to \infty$ with p.g.f. given by Eq. 10.34. [*Hint*: Check that $\sum_{j=0}^{\infty} P_1(X_t = j \mid \tau_{\{0\}} > t)r^j$ may be expressed as $\{g_t^{(1)}(r) - g_t^{(1)}(0)\}/(1 - g_t^{(1)}(0))$. Apply (10.28) to get

$$\frac{H^{-1}(t + H(r)) - H^{-1}(t + H(0))}{1 - H^{-1}(t + H(0))}$$

which by (10.31) is, asymptotically as $t \to \infty$, given by $1 - \exp\{\kappa_1(H(r) - H(0))\} \times (1 + o(1))$. Finally, apply (10.26).]

(ii) Determine the precise form of the distribution of the limit in the binary case $f_2 = p < q = f_0$, $p + q = 1$.

THEORETICAL COMPLEMENTS

Theoretical Complements to Section IV.1

1. (*Independent Increments and Infinite Divisibility*) A probability measure Q on $(\mathbb{R}^1, \mathscr{B}^1)$ is called *infinitely divisible* if for each $n \geq 1$, Q can be factored as an n-fold convolution of a probability distribution Q_n on $(\mathbb{R}^1, \mathscr{B})$, i.e., $Q = Q_n * \cdots * Q_n$ (n-fold), $n \geq 1$. Familiar examples are the Normal, Gamma, Poisson and Cauchy distributions, as well as the first passage time distribution of the Brownian motion (Chapter I). *A basic property of infinitely divisible distributions is that the characteristic function* $\hat{Q}(\xi) = \int_{\mathbb{R}^1} e^{i\xi x} Q(dx)$, $\xi \in \mathbb{R}^1$, *is never zero.* To see this, observe that by considering $\varphi(\xi) = |\hat{Q}^2(\xi)|$ if necessary, one may assume to be given a real nonnegative characteristic function which, for any $n \geq 1$, factors into an n-fold product of real nonnegative characteristic functions φ_n (of some probability distributions μ_n, say). Since $\varphi(0) = 1$ and φ is continuous, there is an interval $[-\tau, \tau]$ on which $\varphi(\xi)$ is strictly positive. Now, taking the unique version of $\log \varphi(\xi)$ which makes $\xi \to \log \varphi(\xi)$ continuous and $\log \varphi(0) = 0$, $\log \varphi(\xi) = n \log \varphi_n(\xi)$ and, expanding the log, $\log \varphi(\xi) = n \log\{1 - [1 - \varphi_n(\xi)]\} = -n[1 - \varphi_n(\xi)] - n[1 - \varphi_n(\xi)]^2/2 - \cdots$. In particular, one sees that $n[1 - \varphi_n(\xi)]$ is a bounded sequence for $\xi \in [-\tau, \tau]$. But, also

$$
\begin{aligned}
n[1 - \varphi_n(2\xi)] &= n\left[1 - \int_{\mathbb{R}^1} e^{2i\xi x}\mu_n(dx)\right] = n\int_{\mathbb{R}^1} (1 - e^{2i\xi x})\mu_n(dx) \\
&= n\int_{\mathbb{R}^1} [1 - \cos(2\xi x)]\mu_n(dx) = 2n\int_{\mathbb{R}^1} [1 - \cos^2(\xi x)]\mu_n(dx) \\
&\leq 4n\int_{\mathbb{R}^1} [1 - \cos(\xi x)]\mu_n(dx) = 4n[1 - \varphi_n(\xi)].
\end{aligned}
$$

So, the sequence $n[1 - \varphi_n(2\xi)]^k$ is bounded on $[-\tau, \tau]$. This means that each $\varphi_n(\xi)$, and therefore, $\varphi(\xi)$ must be nonzero on $[-2\tau, 2\tau]$. In particular, there could be no largest such τ and this proves φ has no real zeros.

If $\{X_t\}$ is a stochastic process with state space $S \subset \mathbb{R}^1$ having *stationary independent increments*, then for any $s < t$, the distribution of $X_t - X_s$ is clearly infinitely divisible since for each integer $n \geq 1$,

$$
X_t - X_s = \sum_{k=1}^{n} (X_{t_k} - X_{t_{k-1}}), \qquad t_k = s + (t - s)\frac{k}{n}, \qquad k = 0, 1, \ldots, n.
$$

Conversely, given an infinitely divisible distribution Q there is a family of probability measures Q_t, $t \geq 0$, such that $Q_1 = Q$ and $Q_t * Q_s = Q_{t+s}$, obtained, for example, though their respective characteristic functions by taking $\varphi_t(\xi) = (\varphi(\xi))^t := \exp\{t \log \varphi_1(\xi)\}$, since $\varphi_1(\xi) \neq 0$, for $\xi \in \mathbb{R}^1$. Thus one obtains a *consistent* specification of the finite-dimensional distributions of a stochastic process $\{X_t\}$ having stationary independent increments starting at 0 such that $X_t - X_s$ has distribution Q_{t-s}, $0 \leq s < t$.

The process $\{X_t\}$ with stationary independent increments is a Markov process on $\mathbb{R}^1$ starting at 0 having stationary transition probabilities given by

$$p(t; x, B) = P(X_{t+s} \in B \mid X_s = x) = Q_t(B - x), \qquad t > 0, \quad B \in \mathscr{B}^1,$$

where $B - x := \{y - x : y \in B\}$ is a translate of B.

2. (*Lévy–Khinchine Representation*) The Brownian motion process and the compound Poisson process are simple examples of processes having stationary independent increments. While the Brownian motion is the only one of these that is *a.s. continuous*, both are *continuous in probability* (called *stochastic continuity*). That is, for any $t_0 \geqslant 0$, $X_t \to X_{t_0}$ in probability as $t \to t_0$ since (i) for the Brownian motion, a.s. convergence implies convergence in probability and (ii) for the compound Poisson process, $P(|X_t - X_s| > \varepsilon) \leqslant 1 - e^{-\lambda(t-s)} = 0(|t - s|)$ for $\varepsilon > 0$. Although the sample paths of the Poisson and compound Poisson processes have jump discontinuities, stochastic continuity means that there are *no fixed discontinuities*.

Stochastic processes $\{X_t\}$ and $\{Y_t\}$ defined on the same probability space and having the same index set are said to be *stochastically equivalent* if $P(X_t = Y_t) = 1$ for each t. Stochastically equivalent processes must have the same finite-dimensional distributions. As an application of the fundamental theorem on the sample path regularity of stochastically continuous submartingales given in theoretical complement 5.2, it will follow that *a stochastically continuous process $\{X_t\}$ with independent increments is equivalent to a stochastic process $\{Y_t\}$ having the property that almost all of its sample paths are right-continuous and have left-hand limits at each t, i.e., have at most jump discontinuities of the first kind. Moreover, the process $\{Y_t\}$ is unique in the sense that if $\{Y'_t\}$ is any other such process, then $P(Y_t = Y'_t$ for every $t) = 1$.* Without loss of generality we may assume the given process $\{X_t\}$ with stationary independent increments to have jump discontinuities of the first kind. Such a process is called a (*homogeneous*) *Lévy process*. The prefix *homogeneous* refers only to stationarity of the increments.

Theorem T.1.1. If $\{X_t\}$ is a (nondegenerate) homogeneous Lévy process having a.s. continuous sample paths, then $\{X_t\}$ must be Brownian motion. □

Proof. Let $s < t$. By sample path continuity, for any $\varepsilon > 0$ there is a number $\delta = \delta(\varepsilon) > 0$ such that

$$P(|X_u - X_v| < \varepsilon \text{ whenever } |u - v| < \delta,\ s < u, v \leqslant t) > 1 - \varepsilon.$$

Let $\{\varepsilon_n\}$ be a sequence of positive numbers decreasing to zero and partition $(s, t]$ into subintervals $s = t_0^{(n)} < t_1^{(n)} < \cdots < t_{k_n}^{(n)} = t$ of lengths less than $\delta_n = \delta(\varepsilon_n)$. Then

$$X_t - X_s = \sum_{j=1}^{k_n} (X_{t_j^{(n)}} - X_{t_{j-1}^{(n)}}) := \sum_{j=1}^{k_n} X_j^{(n)}$$

where $X_1^{(n)}, \ldots, X_{k_n}^{(n)}$ are independent (*triangular array*). Observe that the truncated random variables

$$\tilde{X}_j^{(n)} := X_j^{(n)} \mathbf{1}_{\{|X_j^{(n)}| < \varepsilon_n\}}, \qquad j = 1, \ldots, k_n,$$

are also independent and $\sum_{j=1}^{k_n} \tilde{X}_j^{(n)} \to X_t - X_s$ in probability as $n \to \infty$ since

$$P\left(X_t - X_s = \sum_{j=1}^{k_n} \tilde{X}_j^{(n)}\right) > 1 - \varepsilon_n.$$

The result now follows by an application of the *Lindeberg CLT* (Chapter 0, Theorem 7.1). ■

Within this same context, the other extreme is represented by the following.

Theorem T.1.2. Let $\{X_t\}$ be a homogeneous Lévy process almost all of whose sample paths are step functions with unit jumps. Then $\{X_t\}$ is a Poisson process. □

The proof of Theorem T.1.2 will be based on the following basic *coupling lemma.*

Lemma. (*Coupling Bound*). Let X and Y be arbitrary random variables. Then for any (Borel) set B,

$$|P(X \in B) - P(Y \in B)| \leqslant P(X \neq Y).$$

Proof. First consider the case $P(X \in B) > P(Y \in B)$. Then

$$0 \leqslant P(X \in B) - P(X \in B, Y \in B) = P(X \in B, Y \notin B) \leqslant P(X \neq Y).$$

The argument applies symmetrically to the other case. ■

The coupling lemma can be used to get the following Poisson approximations which, while important in their own right, will be used in the proof of Theorem T.1.2.

Lemma. (*Poisson Approximations*). Let $Y_1, \ldots, Y_n$ be independent Bernoulli 0–1-valued random variables with $p_i = P(Y_i = 1)$, $i = 1, \ldots, n$. Then for any $\lambda > 0$ and $J \subseteq \{0, 1, 2, \ldots\}$,

(i) $$\left|P\left(\sum_{i=1}^{n} Y_i \in J\right) - F_{\rho_n}(J)\right| \leqslant \sum_{i=1}^{n} p_i^2;$$

(ii) $$\left|P\left(\sum_{i=1}^{n} Y_i \in J\right) - F_{\lambda}(J)\right| \leqslant \left|\lambda - \sum_{i=1}^{n} p_i\right| + \left(\max_i p_i\right) \sum_{i=1}^{n} p_i,$$

where

$$F_\lambda(J) = \sum_{m \in J} \frac{\lambda^m}{m!} e^{-\lambda}, \qquad \rho_n = \sum_{i=1}^{n} p_i.$$

Proof. Let Q be an arbitrary (discrete) probability distribution on $\{0, 1, 2, \ldots\}$ with p.m.f. $q_i = Q(\{i\})$, $i = 0, 1, 2, \ldots$. A *simulation random variable* having distribution Q based on the value of a random variable U uniformly distributed over $(0, 1]$ (i.e., a function of U) can be defined by

$$X = \tilde{Q}(U), \tag{T.1.1}$$

where $\tilde{Q}(x) \in \{0, 1, 2, \ldots\}$ is uniquely defined for each $0 < x < 1$ by the condition that (an empty sum being zero)

$$\sum_{j=0}^{\tilde{Q}(x)-1} q_j < x \leqslant \sum_{j=0}^{\tilde{Q}(x)} q_j. \tag{T.1.2}$$

Clearly, $X = \tilde{Q}(U)$ so constructed has distribution Q since

$$P(\tilde{Q}(U) = i) = P\left(\sum_{j=0}^{i-1} q_j < U \leqslant \sum_{j=0}^{i} q_j\right) = q_i.$$

Now let $U_1, U_2, \ldots, U_n$ be i.i.d. uniform over $(0, 1]$ and let $G_i \equiv G_{p_i}$ and $F_i \equiv F_{p_i}$ be the Bernoulli and Poisson distributions with parameter p_i, respectively, $i = 1, \ldots, n$. That is,

$$G_i(\{1\}) = p_i = 1 - G_i(\{0\}), \qquad F_i(\{k\}) = \frac{p_i^k}{k!} e^{-p_i}, \qquad k = 0, 1, 2, \ldots.$$

Then, by the coupling lemma, letting $\rho_n = \sum_{i=1}^n p_i$ and letting $\tilde{}$ indicate the corresponding "simulation" function as defined above, we have

$$\left|P\left(\sum_{i=1}^n Y_i \in J\right) - F_{\rho_n}(J)\right| \leqslant P\left(\sum_{i=1}^n \tilde{G}_i(U_i) \neq \sum_{i=1}^n \tilde{F}_i(U_i)\right), \tag{T.1.3}$$

since the distribution of the sum of independent Poisson random variables with parameters $p_1, \ldots, p_n$ is Poisson with parameter $\sum_{i=1}^n p_i$. Now from (T.1.3) we get

$$\begin{aligned}\left|P\left(\sum_{i=1}^n Y_i \in J\right) - F_{\rho_n}(J)\right| &\leqslant P\left(\bigcup_{i=1}^n \{\tilde{G}_i(U_i) \neq \tilde{F}_i(U_i)\}\right)\\ &\leqslant \sum_{i=1}^n P(\tilde{G}_i(U_i) \neq \tilde{F}_i(U_i)).\end{aligned} \tag{T.1.4}$$

Now, for fixed i, observe from Figure TC.IV.1 that

$$P(\tilde{G}_i(U_i) \neq \tilde{F}_i(U_i)) = (e^{-p_i} - (1 - p_i)) + (1 - e^{-p_i}(1 + p_i)) \leqslant p_i^2 \tag{T.1.5}$$

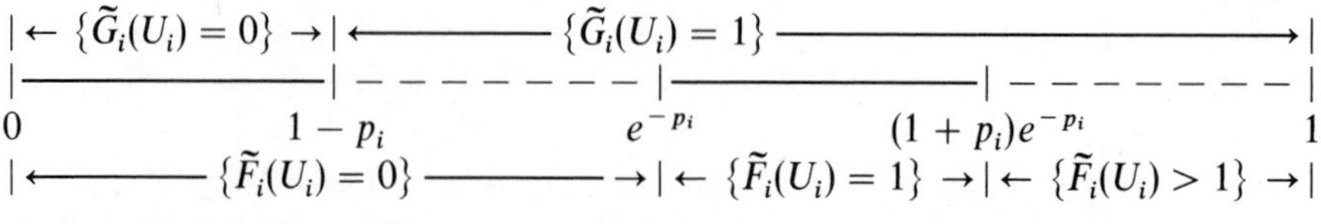

Figure TC.IV.1

since $1 - e^{-p_i} \leqslant p_i$. Thus, (T.1.4) and (T.1.5) prove (i). To derive the estimate (ii), let $\lambda > 0$ be arbitrary and use the triangle inequality to write

$$\left|P\left(\sum_{i=1}^{n} Y_i \in J\right) - F_\lambda(J)\right| \leqslant \left|P\left(\sum_{i=1}^{n} Y_i \in J\right) - F_{\rho_n}(J)\right| + |F_{\rho_n}(J) - F_\lambda(J)|$$

$$\leqslant \sum_{i=1}^{n} p_i^2 + |F_{\rho_n}(J) - F_\lambda(J)|$$

$$\leqslant \max_{1\leqslant i\leqslant n} p_i \sum_{i=1}^{n} p_i + |F_{\rho_n}(J) - F_\lambda(J)|. \tag{T.1.6}$$

To complete the proof, we simply need to show

$$|F_{\rho_n}(J) - F_\lambda(J)| \leqslant \left|\lambda - \sum_{i=1}^{n} p_i\right|.$$

For this, first suppose $\rho_n = \sum_{i=1}^{n} p_i > \lambda$ and let Z_1, Z_2 be independent Poisson random variables with parameters λ and $\sum_{i=1}^{n} p_i - \lambda$. Again use the coupling bound lemma to bound the last term in (T.1.6) by

$$P(Z_1 \neq Z_1 + Z_2) = P(Z_2 \neq 0) = 1 - \exp\left[-\left(\sum_{i=1}^{n} p_i - \lambda\right)\right] \leqslant \left|\lambda - \sum_{i=1}^{n} p_i\right|. \tag{T.1.7}$$

The symmetrical argument works for $\sum_{i=1}^{n} p_i < \lambda$ also. ■

Proof of Theorem T.1.2. It is enough to show that for each $t > 0$, X_t has a Poisson distribution with $EX_t = \lambda t$ for some $\lambda > 0$. Partition $(0, t]$ into n intervals of the form $(t_{i-1}, t_i]$, $i = 1, \dots, n$ having equal lengths $\Delta = t/n$. Let $A_i^{(n)} = \{X_{t_i} - X_{t_{i-1}} \geqslant 1\}$ and $S_n = \sum_{i=1}^{n} \mathbf{1}_{A_i}^{(n)}$ (the number of time intervals with at least one jump occurrence). Let D denote the shortest distance between jumps in the path X_s, $0 < s \leqslant t$. Then $P(X_t \neq S_n) \leqslant P(0 < D \leqslant t/n) \to 0$ as $n \to \infty$, since $\{X_t \neq S_n\}$ implies that there is at least one interval containing two or more jumps.

Now, by the Poisson approximation lemma (ii), taking $J = \{m\}$, we have

$$\left|P(S_n = m) - \frac{\lambda^m}{m!} e^{-\lambda}\right| \leqslant |np_n - \lambda| + \frac{(np_n)^2}{n}, \qquad \text{where } p_n = p_i^{(n)} = P(A_i^{(n)}),$$

$i = 1, \dots, n$, and $\lambda > 0$ is arbitrary. Thus, using the triangle inequality and the coupling bound lemma,

$$\left|P(X_t = m) - \frac{\lambda^m}{m!} e^{-\lambda}\right| \leqslant |P(X_t = m) - P(S_n = m)| + \left|P(S_n = m) - \frac{\lambda^m}{m!} e^{-\lambda}\right|$$

$$\leqslant P(X_t \neq S_n) + |np_n - \lambda| + np_n^2$$

$$= o(1) + |np_n - \lambda| + np_n^2. \tag{T.1.8}$$

The proof will be completed by determining λ such that $|np_n - \lambda| \to 0$, at least along a subsequence of $\{np_n\}$. For then we also have $np_n^2 = np_n P(A_i^{(n)}) \to 0$ as $n \to \infty$. But, since $P(X_t = 0) = P(\bigcap_{i=1}^n A_i^{(n)c}) = (1 - p_n)^n$, it follows that $P(X_t = 0) > 0$; for otherwise $P(A_i^{(n)}) = 1$ for each i and therefore $X_t \geqslant n$ for all n, which is not possible under the assumed sample path regularity. Therefore $np_n \leqslant -n \log(1 - p_n) = -\log P(X_t = 0)$ for all n. Since $\{np_n\}$ is a bounded sequence of positive numbers, there is at least one limit point $\lambda \geqslant 0$. This provides the desired λ. ■

The remarkable fact which we wish to record here (with a sketch of the proof and examples) is that *every (homogeneous) Lévy process may be represented as sum of a (possibly degenerate) Brownian motion and a limit of independent superpositions of compound Poisson processes with varying jump sizes.* Observe that if the sample paths of $\{X_t\}$ are step function of a fixed jump size $y \neq 0$, then $\{y^{-1}X_t\}$ is a Poisson process by Theorem T.1.2. The idea is that by removing (subtracting) the jumps of various sizes from $\{X_t\}$ one arrives at an independent homogeneous Lévy process with continuous sample paths. According to Theorem T.1.1, therefore, this is a Brownian motion process. More precisely, let $\{X_t\}$ be a homogeneous Lévy process and let $S = [0, T) \times \mathbb{R}^1$ for fixed $T > 0$. Let $\mathscr{B}(S)$ be the Borel sigmafield of S, and let $\mathscr{B}^+(S) = \{A \in \mathscr{B}(S): \rho(A, [0, T) \times \{0\}) > 0\}$, where $\rho(A, B) = \inf\{|x - y|: x \in A, y \in B\}$. The *space–time jump set* of $\{X_t\}$ is defined by

$$J = J(\omega) = \{(t, y) \in S: y = X_t(\omega) - X_{t^-}(\omega) \neq 0, 0 \leqslant t < T\},$$

for $\omega \in \Omega$. Define a *random counting measure* v on $\mathscr{B}(S)$ by

$$v(A) = v(\omega, A) = \# A \cap J(\omega), \qquad \omega \in \Omega, \quad A \in \mathscr{B}(S).$$

For fixed $\omega \in \Omega$, v is a *measure* on the sigmafield $\mathscr{B}(S)$, and for each $A \in \mathscr{B}(S)$, $v(A)$ is a random variable (i.e., $\omega \to v(\omega, A)$ is measurable). Moreover, for fixed $A \in \mathscr{B}(S)$, the process $v(A \cap [0, t] \times \mathbb{R}^1)$, $t \geqslant 0$, is a Poisson process. To prove this, take A an open subset of S and show that the process is a Lévy process with unit jump step function sample paths and apply Theorem T.1.2.

Define $\mu(A) = Ev(A)$ for $A \in \mathscr{B}^+(S)$. Then v is a measure on $\mathscr{B}^+(S)$; countable additivity can be proved from countable additivity of v by Lebesgue's Monotone convergence theorem. Now observe that for each fixed $B \in \mathscr{B}(\mathbb{R}^1 \backslash \{0\})$, the set function $\mu(C) = v(C \times B)$ is translation invariant on $\mathscr{B}([0, T))$. Therefore, $v(C \times B)$ is a multiple (dependent on B) of the Lebesgue measure $|C|$ of C, i.e., $v(C \times B) = K(B)|C|$, where $K(B)$ is the multiplying factor. Notice now that since μ is a measure, K must also be a measure on $\mathscr{B}(\mathbb{R}^1 \backslash \{0\})$. Define

$$v_t(B) = v([0, t] \times B), \qquad B \in \mathscr{B}(\mathbb{R}^1 \backslash \{0\}).$$

Since there are at most a finite number of jumps in $[0, t]$ of size $1/n$ or larger, $\int_{\{|y| > 1/n\}} y v_t(dy)$ is well-defined. However, in general, its limit may not exist pathwise. By appropriate centering, a limit may be shown to exist in L_2. Finally, one may prove that this process built up from jumps is independent of the Brownian motion component. In this way one obtains the following.

Theorem T.1.3. Let $\{X_t\}$ be a homogeneous Lévy process. Then there is a standard Brownian motion $\{B_t\}$ independent of $\{v_t(B): B \in \mathscr{B}(\mathbb{R}^1 \backslash \{0\}), 0 \leqslant t < T\}$ such that

$$X_t = mt + \sqrt{\sigma^2}\, B_t + \int_{-\infty}^{\infty} \left\{ y\nu_t(dy) - \frac{yt}{1+y^2} K(dy) \right\}$$

in distribution, where $m \in \mathbb{R}^1$, $\sigma^2 \geqslant 0$, and for fixed $B \in \mathscr{B}(\mathbb{R}^1 \setminus \{0\})$, $\nu_t(B)$ is a Poisson process in t.

Corollary. A probability distribution Q on $\mathbb{R}^1$ is infinitely divisible if and only if the characteristic function $\hat{Q}(\xi) = \int_{\mathbb{R}^1} e^{i\xi x} Q(dx)$, $\xi \in \mathbb{R}^1$, can be represented as

$$\hat{Q}(\xi) = \exp\left\{ imt\xi - \frac{\sigma^2 t}{2}\xi^2 + t\int_{-\infty}^{\infty} \left[e^{i\xi y} - 1 - \frac{i\xi y}{1+y^2} \right] K(dy) \right\},$$

for some $m \in \mathbb{R}^1$, $\sigma^2 \geqslant 0$, where K is a nonnegative measure on $\mathbb{R}^1$.

Example 1. $K \equiv 0$. Then $\{X_t\}$ is Brownian motion with drift m and diffusion coefficient σ^2.

Example 2

$$r = \lim_{\varepsilon \downarrow 0} \int_{|y| > \varepsilon} \frac{y}{1+y^2} K(dy)$$

exists. Then,

$$X_t = (m + r)t + \sqrt{\sigma^2}\, B_t + \int_{-\infty}^{\infty} y\nu_t(dy).$$

In this case

$$Ee^{i\xi X_t} = \exp\left\{ it(m + r)\xi - t\frac{\sigma^2}{2}\xi^2 + t\int_{-\infty}^{\infty} [e^{i\xi y} - 1] K(dy) \right\}.$$

Example 3. If $m + r = 0$, $\sigma^2 = 0$, then

$$X_t = \lim_{\varepsilon \downarrow 0} \int_{\{|y| > \varepsilon\}} y\nu_t(dy) \qquad \text{and} \qquad Ee^{i\xi X_t} = \exp\left\{ \lim_{\varepsilon \downarrow 0} t \int_{\{|y| > \varepsilon\}} (e^{i\xi y} - 1) K(dy) \right\}.$$

Example 4. If $\lambda = K(\mathbb{R}^1) < \infty$, $m + r = 0$, $\sigma^2 = 0$, then

$$X_t = \int_{-\infty}^{\infty} y\nu_t(dy), \qquad Ee^{i\xi X_t} = \exp\left\{ \lambda t \int_{-\infty}^{\infty} (e^{i\xi y} - 1)\lambda^{-1} K(dy) \right\}.$$

An important special class of (nondegenerate) homogeneous Lévy processes $\{X_t\}$ have the following basic *scaling property*: *For each* $\lambda > 0$, *there is a positive constant* $c_\lambda > 0$ *such that* $\{c_\lambda X_{\lambda t}\}$ *and* $\{X_t\}$ *have the same distribution.* These are the *Lévy stable processes*. In view of the Lévy–Khinchine formula we have the following theorem.

Theorem T.1.4. If $\{X_t\}$ is a homogeneous Lévy stable process, then

$$Ee^{i\xi X_t} = e^{it\xi m} \qquad \text{or} \qquad e^{-\alpha t\xi^2} \qquad \text{or} \qquad \exp\left\{\left(-\alpha t + i\frac{\xi\beta t}{|\xi|}\right)|\xi|^\theta\right\}$$

for some $\alpha > 0$, $\beta \in \mathbb{R}^1$, $0 < \theta < 2$.

Observe that in the first case $X_t = m$, $t \geqslant 0$, is a.s. constant and $c_\lambda = \lambda^0 = 1$. In the second case $\{X_t\}$ is Brownian motion with zero drift and $c_\lambda = \lambda^{-\frac{1}{2}}$. In the third case we have $c_\lambda = \lambda^{-1/\theta}$ and examples are the Cauchy process ($\theta = 1$) and the first-passage-time process for Brownian motion ($\theta = \frac{1}{2}$).

Thorough treatments of processes with independent increments may be found in K. Itô (1984), *Lectures on Stochastic Processes*, Tata Institute of Fundamental Research, distributed by Springer-Verlag, New York, and in I. I. Gikhman and A. V. Skorokhod (1969), *Introduction to the Theory of Random Processes*, W. B. Saunders, Philadelphia. The coupling arguments follow T. C. Brown (1984), "Poisson Approximations and the Definition of the Poisson Process," *Amer. Math. Monthly*, **91**, pp. 116–123.

3. (*Fractional Brownian Motion and Self-Similarity*) The *scaling* property (or *self-similarity*) of Lévy processes was extended to more general processes (including diffusions) in a classic paper by John Lamperti (1962), "Semistable Stochastic Processes," *Trans. Amer. Math. Soc.*, **104**, pp. 62–78. Lamperti used the term *semistable* to describe the class of processes $\{X_t\}$ with the property that for each $\lambda > 0$ there is a positive scale coefficient c_λ and a location parameter b_λ such that $\{c_\lambda X_{\lambda t} + b_\lambda\}$ and $\{X_t\}$ have the same distribution. The defining property is often referred to as simple scaling or self-similarity. Stochastic processes with this property are of substantial current interest in many branches of mathematics and the physical sciences (see references below as well as references therein). One family of such processes of some notoriety is the class of fractional Brownian motions defined as follows. Let $0 < \beta < 1$ be a fixed parameter and define, for each $s, t \geqslant 0$,

$$\rho_\beta(s, t) = \{|s|^{2\beta} + |t|^{2\beta} - |s - t|^{2\beta}\}/2. \tag{T.1.9}$$

Then ρ_β is a positive-definite symmetric function. To see this, one need only check that there is a family of square-integrable functions $\gamma_\beta(t,\cdot)$ on the real-number line such that for each s and t, $\rho_\beta(s, t)$ may be represented as an L_2 inner product $\langle\gamma_\beta(s,\cdot), \gamma_\beta(t,\cdot)\rangle$, namely,

$$\gamma_\beta(t, x) = |t - x|^{\beta - 1/2}\,\text{sign}(t - x) + |x|^{\beta - 1/2}\,\text{sign}(x). \tag{T.1.10}$$

One may then check the positive-definiteness using the bilinearity of the inner product; see, for example, M. Ossiander and E. Waymire (1989), "On Certain Positive Definite Kernels," *Proc. Amer. Math. Soc.*, **107**, 487–492.

The *fractional Brownian motion* $\{Z_t\}$ with exponent β may now be most simply defined as a mean-zero Gaussian process starting at 0 such that Z_s and Z_t have covariance given by $\rho_\beta(s, t)$. This class of processes was studied by Benoit Mandelbrot and others in connection with the Hurst effect problem of Section 1.14; see B. Mandelbrot (1982), *The Fractal Geometry of Nature*, Freeman, San Francisco (and references therein) and J. Feder (1988), *Fractals*, Plenum Press, New York.

Additional reference: M. Taqqu (1986), "A Bibliographical Guide to Self-Similar Processes and Long Range Dependence," in *Dependence in Statistics and Probability* (E. Eberlein, M. Taqqu, eds.), Birkhäuser, Boston.

Theoretical Complements to Section IV.5

1. (*The Upcrossing Inequality and (Sub)Martingale Convergence*) A fundamental property of *monotone* sequences of real numbers is that boundedness of the sequence implies the existence of a limit. An analogous property of submartingales also holds as will be developed below.

Consider a sequence $\{Z_n\}$ of real-valued random variables and sigmafields $\mathscr{F}_1 \subset \mathscr{F}_2 \subset \cdots$, such that Z_n is $\mathscr{F}_n$-measurable. Let $a < b$ be an arbitrary pair of real numbers. An *upcrossing* of the interval (a, b) by $\{Z_n\}$ is a passage to a value equal to or exceeding b from an earlier value equal to or below a. It is convenient to look at the corresponding process $\{X_n := (Z_n - a)^+\}$, where $(Z_n - a)^+ = \max\{(Z_n - a), 0\}$. The upcrossings of (a, b) by $\{Z_n\}$ are the upcrossings of $(0, b - a)$ by $\{X_n\}$. The successive upcrossing times η_{2k} $(k = 1, 2, \ldots)$ of $\{X_n\}$ are defined by $\eta_1 := \inf\{n \geqslant 1: X_n = 0\}$, $\eta_2 := \inf\{n \geqslant \eta_1: X_n \geqslant b - a\}$, $\eta_{2k+1} := \inf\{n \geqslant \eta_{2k}: X_n = 0\}$, $\eta_{2k+2} := \inf\{n \geqslant \eta_{2k+1}: X_n \geqslant b - a\}$. Then the η_k are $\{\mathscr{F}_n\}$-stopping times. Fix a positive integer N and define $\tau_k := \eta_k \wedge N = \min\{\eta_k, N\}$, $k = 1, 2, \ldots$. Then the τ_k are stopping times. Also, $\tau_k \equiv N$ for $k > [N/2]$ (the integer part of $N/2$) so that $X_{\tau_k} = X_N$ for $k > [N/2]$.

Let $U_N \equiv U_N(a, b)$ denote the number of upcrossings of (a, b) of $\{Z_n\}$ by time N. That is,

$$U_N(a, b) := \sup\{k \geqslant 1: \eta_{2k} \leqslant N\}, \tag{T.5.1}$$

with the convention that the supremum over an empty set is 0. Thus U_N is also the number of upcrossings of $(0, b - a)$ by $\{X_n\}$ in time N.

Since $X_{\tau_k} = X_N$ for $k > [N/2]$, one may write (setting $\tau_0 := 1$)

$$X_N - X_1 = \sum_{k=1}^{[N/2]} (X_{\tau_{2k-1}} - X_{\tau_{2k-2}}) + \sum_{k=1}^{[N/2]} (X_{\tau_{2k}} - X_{\tau_{2k-1}}). \tag{T.5.2}$$

To relate (T.5.2) to the number U_N, let v denote the largest k such that $\eta_k \leqslant N$, i.e., v is the last time $\leqslant N$ for an upcrossing or a downcrossing. Notice that $U_N = [v/2]$. If v is *even*, then $X_{\tau_{2k}} - X_{\tau_{2k-1}} \geqslant b - a$ if $2k \leqslant v$, and $= X_N - X_N = 0$ if $2k > v$. Now suppose v is *odd*. Then $X_{\tau_{2k}} - X_{\tau_{2k-1}} \geqslant b - a$ if $2k - 1 < v$, and $= 0$ if $2k - 1 > v$, and $= X_{\tau_{2k}} - 0 \geqslant 0$ if $2k - 1 = v$. Hence in every case

$$\sum_{k=1}^{[N/2]} (X_{\tau_{2k}} - X_{\tau_{2k-1}}) \geqslant [v/2](b - a) = U_N(b - a). \tag{T.5.3}$$

As a consequence,

$$X_N - X_1 \geqslant \sum_{k=1}^{[N/2]} (X_{\tau_{2k-1}} - X_{\tau_{2k-2}}) + (b - a)U_N. \tag{T.5.4}$$

Observe that (T.5.4) is true for an arbitrary sequence of random variables (or real numbers) $\{Z_n\}$. Assume now that $\{Z_n\}$ is a $\{\mathscr{F}_n\}$-submartingale. Then $\{X_n\}$ is a $\{\mathscr{F}_n\}$-submartingale by convexity. One may show, using the method of proof of Theorem 13.3 of Chapter I, that $\{X_{\tau_k}: k \geqslant 1\}$ is a submartingale. Hence EX_{τ_k} is increasing in k, so that

$$E\left[\sum_{k=1}^{[N/2]} (X_{\tau_{2k-1}} - X_{\tau_{2k-2}})\right] \geqslant 0.$$

Applying this to (T.5.2) one obtains the following.

Proposition T.5.1. (*Upcrossing Inequality*). Let $\{Z_n\}$ be a $\{\mathscr{F}_n\}$-submartingale. For each pair $a < b$ the expected number of upcrossings of (a, b) by $Z_1, \ldots, Z_N$ satisfies the inequality

$$EU_N(a, b) \leqslant \frac{E(Z_N - a)^+ - E(Z_1 - a)^+}{b - a} \leqslant \frac{E|Z_N| + |a|}{b - a}. \tag{T.5.5}$$

As an important consequence of this result we get the following theorem.

Theorem T.5.2. (*Submartingale Convergence Theorem*). Let $\{Z_n\}$ be a submartingale such that $E|Z_n|$ is bounded. Then $\{Z_n\}$ converges a.s. to an integrable random variable Z_∞. □

Proof. Let $U(a, b)$ denote the total number of upcrossings of (a, b) by $\{Z_n\}$. Then $U_N(a, b) \uparrow U(a, b)$ as $N \uparrow \infty$. Therefore, by the Monotone Convergence Theorem (Chapter 0)

$$EU(a, b) = \lim_{N \uparrow \infty} EU_N(a, b) \leqslant \sup_N \frac{E|Z_N| + |a|}{b - a} < \infty. \tag{T.5.6}$$

In particular, $U(a, b) < \infty$ almost surely, so that

$$P(\liminf Z_n < a < b < \limsup Z_n) = 0. \tag{T.5.7}$$

Since this holds for every pair $a, b = a + 1/m$ with a rational and m a positive integer, and the set of all such pairs is countable, one must have $\liminf Z_n = \limsup Z_n$ almost surely. If $\lim Z_n$ is not finite with probability 1, $|Z_n| \to \infty$ with positive probability so that $E \lim |Z_n| = \infty$. Then, by Fatou's Lemma (Chapter 0), $\lim E|Z_n| = \infty$, which contradicts the hypothesis of the theorem. The integrability of $|Z_\infty|$ follows from Fatou's Lemma. ■

Corollary. A nonnegative martingale $\{Z_n\}$ converges almost surely to a finite limit Z_∞. Also, $EZ_\infty \leqslant EZ_1$.

Proof. For a nonnegative martingale $\{Z_n\}$, $|Z_n| = Z_n$ and therefore, $\sup E|Z_n| = \sup EZ_n = EZ_1 < \infty$. By Fatou's Lemma (Chapter 0), $EZ_\infty = E(\lim Z_n) \leqslant \lim EZ_n = EZ_1$. ■

Convergence properties of *supermartingales* $\{X_n\}$ are obtained from the submartingale results applied to $\{-X_n\}$.

2. (*Martingales and Sample Path Regularity*) The control over fluctuations available through Doob's Inequality (Section 1.13) and the Upcrossing Inequality (theoretical complement 1) for (sub/super) martingales make it possible to find versions of such processes with very regular sample path behavior. In this connection the martingale property is of fundamental importance for the construction of models of stochastic processes having manageable sample path properties (cf. theoretical complements 1.1 and 5.3). The basic result is the following. Recall the meaning of *continuity in probability* from theoretical complement 1.2.

Theorem T.5.3. Let $\{X_t\}$ be a submartingale or supermartingale with respect to an increasing family of sigmafields $\{\mathscr{F}_t\}$. Suppose that $\{X_t\}$ is continuous in probability at each $t \geqslant 0$. Then there is a process $\{\tilde{X}_t\}$ such that:

(i) (*Stochastic Equivalence*) $\{\tilde{X}_t\}$ is equivalent to $\{X_t\}$ in the sense that $P(X_t = \tilde{X}_t) = 1$ for each $t \geqslant 0$.

(ii) (*Sample Path Regularity*) With probability 1 the sample paths of $\{\tilde{X}_t\}$ are bounded on compact intervals $a \leqslant t \leqslant b$, $(a, b \geqslant 0)$, and are right-continuous and have left-hand limits at each $t > 0$. □

Proof. Fix $T > 0$ and let Q_T denote the set of rational numbers in $[0, T]$. Write $Q_T = \bigcup_{n=1}^{\infty} R_n$, where each R_n is a *finite* subset of $[0, T]$ and $T \in R_1 \subset R_2 \subset \cdots$. By Doob's Maximal Inequality (Section 1.13) we have

$$P\left(\max_{t \in R_n} |X_t| > \lambda\right) \leqslant \frac{E|X_T|}{\lambda}, \qquad n = 1, 2, \ldots.$$

Therefore,

$$P\left(\sup_{t \in Q_T} |X_t| > \lambda\right) \leqslant \lim_{n \to \infty} P\left(\max_{t \in R_n} |X_t| > \lambda\right) \leqslant \frac{E|X_T|}{\lambda}.$$

In particular, the paths of $\{X_t : t \in Q_T\}$ are *bounded* with probability 1. Let (c, d) be any interval in $\mathbb{R}$ and let $U^{(T)}(c, d)$ denote the number of upcrossings of (c, d) by the process $\{X_t : t \in Q_T\}$. Then $U^{(T)}(c, d)$ is the limit of the number $U^{(n)}(c, d)$ of upcrossings of (c, d) by $\{X_t : t \in R_n\}$ as $n \to \infty$. By the upcrossing inequality (theoretical complement 1), one has

$$EU^{(n)}(c, d) \leqslant \frac{E|X_T| + |c|}{d - c}.$$

Since, the $U^{(n)}(c, d)$ are nondecreasing with n, it follows that $U^{(T)}(c, d)$ is a.s. finite. Taking unions over all (c, d), with c, d rational, it follows that with probability one $\{X_t : t \in Q_T\}$ has only finitely many upcrossings of any interval. In particular, therefore, left- and right-hand limits must exist at each $t < T$ a.s. To construct a right-continuous version of $\{X_t\}$, define $\tilde{X}_t = \lim_{s \to t^+, s \in Q_T} X_s$ for $t < T$. That $\{\tilde{X}_t\}$ is in fact stochastically equivalent to $\{X_t\}$ now follows from continuity in probability; i.e., $\tilde{X}_t = \lim_{s \to t^+} X_s = X_t$ since a.s. limits and limits in probability must a.s. coincide. Since T is arbitrary, the proof is complete. ■

3. (*Markov Processes and Sample Path Regularity*) The martingale regularity theorem of theoretical complement 2 can be applied to the problem of constructing Markov processes with regular sample path properties.

Theorem T.5.4. Let $\{X_t : t \geqslant 0\}$ be a Markov process with a compact metric state space S and transition probability function

$$p(t; x, B) = P(X_{t+s} \in B \mid X_s = x), \qquad x \in S, \quad s, t \geqslant 0, \quad B \in \mathscr{B}(S),$$

such that

(i) (*Uniform Stochastic Continuity*) For each $\varepsilon \geqslant 0$, $p(t; x, B_\varepsilon^c(x)) = o(1)$ *as* $t \to 0^+$ uniformly for $x \in S$, where $B_\varepsilon(x)$ is the ball centered at x of radius $\varepsilon > 0$.

(ii) (*Feller Property*) For each (bounded) continuous function f, the function $x \to \int_S f(y)p(t; x, dy) = E_x f(X_t)$ is continuous. Then there is a version $\{\tilde{X}_t\}$ of $\{X_t\}$ a.s. having right continuous sample paths with left-hand limits at each t. □

Proof. It is enough to prove that $\{X_t\}$ a.s. has left- and right-hand limits at each t, for then $\{X_t\}$ can be modified as $\tilde{X}_t = \lim_{s \to t^+} X_s$ which by stochastic continuity will provide a version of $\{X_t\}$. It is not hard to check stochastic continuity from (i) and (ii). Let $f \in C(S)$ be an arbitrary continuous function on S. Consider the semigroup $\{T_t\}$ acting on $C(S)$ by $T_t f(x) = E_x f(X_t)$, $x \in S$, $t \geqslant 0$. For $\lambda > 0$, write

$$R_\lambda f(x) = \int_0^\infty e^{-\lambda s} T_s f(x)\, ds, \qquad x \in S;$$

R_λ ($\lambda > 0$) is called the *resolvent* (Laplace transform) of the semigroup $\{T_t\}$. A basic property of the resolvent is that λR_λ behaves like the identity operator on $C(S)$ for λ large in the sense that

$$\begin{aligned}\|f - \lambda R_\lambda f\| &:= \sup_{x \in S} |f(x) - \lambda R_\lambda f(x)| = \sup_{x \in S} \left| \int_0^\infty e^{-\lambda s} \lambda (f(x) - T_s f(x))\, ds \right| \\ &\leqslant \int_0^\infty e^{-\lambda s} \lambda \|f - T_s f\|\, ds = \int_0^\infty e^{-\tau} \|f - T_{\lambda^{-1}\tau}\|\, d\tau. \end{aligned} \tag{T.5.8}$$

For each $\tau \geqslant 0$, by the properties (i) and (ii), $\|f - T_{\lambda^{-1}\tau}\| \to 0$ as $\lambda \to \infty$, and the integrand is bounded by $2e^{-\tau}\|f\|$ since $\|T_t f\| \leqslant \|f\|$ for any $t \geqslant 0$. By Lebesgue's Dominated Convergence Theorem (Chapter 0) it follows that $\|f - \lambda R_\lambda f\| \to 0$ as $\lambda \to \infty$; i.e., $\lambda R_\lambda f \to f$ uniformly on S as $\lambda \to \infty$. With this property of the resolvent in mind, consider the process $\{Y_t\}$ defined by

$$Y_t = e^{-\lambda t} R_\lambda f(X_t), \qquad t \geqslant 0, \tag{T.5.9}$$

where $f \in C(S)$ is fixed but arbitrary *nonnegative* function on S. Then $\{Y_t\}$ is a supermartingale with respect to $\mathscr{F}_t = \sigma\{X_s: s \leqslant t\}$, $t \geqslant 0$, since $E|Y_t| < \infty$, Y_t is $\mathscr{F}_t$-measurable, one has $T_t f(x) \geqslant 0$ ($x \in S$, $t \geqslant 0$), and

$$\begin{aligned} E\{Y_{t+h} \mid \mathscr{F}_t\} &:= e^{-\lambda(t+h)} E\{R_\lambda f(X_{t+h}) \mid \mathscr{F}_t\} = e^{-\lambda(t+h)} T_h R_\lambda f(X_t) \\ &= e^{-\lambda t} \int_0^\infty e^{-\lambda(s+h)} T_{s+h} f(X_t)\, ds = e^{-\lambda t} \int_h^\infty e^{-\lambda s} T_s f(X_t)\, ds \\ &\leqslant e^{-\lambda t} \int_0^\infty e^{-\lambda s} T_s f(X_t)\, ds = Y_t. \end{aligned} \tag{T.5.10}$$

Applying the martingale regularity result of Theorem T.5.3, we obtain a version $\{\tilde{Y}_t\}$ of $\{Y_t\}$ whose sample paths are a.s. right continuous with left-hand limits at each t.

Thus, the same is true for $\{\lambda e^{\lambda t}Y_t\} = \{\lambda R_\lambda f(X_t)\}$. Since $\lambda R_\lambda f \to f$ uniformly as $\lambda \to \infty$, the process $\{f(X_t)\}$ must, therefore, a.s. have left and right-hand limits at each t. The same will be true for any $f \in C(S)$ since one can write $f = f^+ - f^-$ with f^+ and f^- continuous *nonnegative* functions on S. So we have that for each $f \in C(S)$, $\{f(X_t)\}$ is a process whose left- and right-hand limits exist at each t (with probability 1). As remarked at the outset, it will be enough to argue that this means that the process $\{X_t\}$ will a.s. have left- and right-hand limits. This is where compactness of S enters the argument. Since S is a compact metric space, it has a countable dense subset $\{x_n\}$. The functions $f_n: S \to \mathbb{R}^1$ defined by $f_n(x) = \rho(x_n, x)$, $x \in S$, are continuous for the metric ρ, and separate points of S in the sense that if $x \neq y$, then for some n, $f_n(x) \neq f_n(y)$. In view of the above, for each n, $\{f_n(X_t)\}$ is a process whose left- and right-hand limits exist at each t with probability 1. Thus, the countable union of events of probability 0 having probability 0, it follows that, with probability 1, for all n the left- and right-hand limits exist at each t for $\{f_n(X_t)\}$. But this means that with probability one the left- and right-hand limits exist at each t for $\{X_t\}$ since the f_n's separate points; i.e., if either limit, say left, fails to exist at some t', then, by compactness of S, the sample path $t \to X_t$ must have at least two distinct limit points as $t \to t'^-$, contradicting the corresponding property for all the processes $\{f_n(X_t)\}$, $n = 1, 2, \ldots$. ■

In the case that S is locally compact, one may adjoin a point at infinity, denoted $\Delta(\notin S)$, to S. The topology of the one point compactification on $\bar{S} = S \cup \{\Delta\}$ defines a neighborhood system for Δ by complements of compact subsets of S. Let $\bar{\mathscr{B}}(S)$ be the sigmafield generated by $\{\mathscr{B}, \{\Delta\}\}$. The transition probability function $p(t; x, B)$, $(t \geqslant 0, x \in S, B \in \mathscr{B}(S))$ is extended to $\bar{p}(t; x, B)$ $(t \geqslant 0, x \in \bar{S}, B \in \bar{\mathscr{B}}(\bar{S}))$ by making Δ an absorbing state; i.e., $\bar{p}(t; \Delta, B) = 1$ if and only if $\Delta \in B$, $B \in \bar{\mathscr{B}}$, otherwise $\bar{p}(t; \Delta, B) = 0$. If the conditions of Theorem T.5.4 are fulfilled for $\bar{p}$, then one obtains a regular process $\{\bar{X}_t\}$ with state space $\bar{S}$. Defining

$$\tau_\Delta = \inf\{t > 0; \bar{X}_t = \Delta\},$$

the basic Theorem T.5.4 provides a process $\{\tilde{X}_t: t < \tau_\Delta\}$ with state space S whose sample paths are right continuous with left-hand limits at each $t < \tau_\Delta$. A detailed treatment of this case can be found in K. L. Chung (1982), *Lectures from Markov Processes to Brownian Motion*, Springer-Verlag, New York.

While the one-point compactification is natural on analytic grounds, it is not always probabilistically natural, since in general a given process may escape the state space in a variety of manners. For example, in the case of birth–death processes with state space $S = \mathbb{Z}$ one may want to consider escapes through the positive integers as distinct from escapes through the negative integers. These matters are beyond the present scope.

4. (*Tauberian Theorem*) According to (5.44), in the case $r \geqslant 2$ the generating function $h(v)$ can be expressed in the form

$$h(v) \sim (1 - v)^{-\rho} K((1 - v)^{-1}) \qquad \text{as } v \to 1^-, \tag{T.5.12}$$

where $\rho \geqslant 0$ and $K(x)$ is a *slowly varying* function as $x \to \infty$; that is, $K(tx)/K(t) \to 1$ as $t \to \infty$ for each $x > 0$. Examples of slowly varying functions at infinity are constants, various powers of $|\log x|$, and the coefficient appearing in (5.44) as a function of $x = (1 - v)^{-1}$. In the case $r = 1$, the generating function $vh'(v)$ for $\{nh_n\}$ is also of this form asymptotically with $\rho = 1$. Likewise, in the case of (5.37) one can differentiate to get that the generating function of $\{(n+1)P(M_L = n+1)\}$ is asymptotically of the form $\frac{1}{2}(1-v)^{-1/2}$ as $v \to 1^-$.

Let $\hat{a}(v)$ be the generating function for a sequence $\{a_n\}$ of nonnegative real numbers and suppose

$$\hat{a}(v) = \sum_{k=0}^{\infty} a_k v^k, \tag{T.5.13}$$

converges for $0 \leqslant v < 1$. The *Tauberian theorem* provides the asymptotic growth of the sums as

$$\sum_{k=0}^{n} a_k \sim (1/\rho\Gamma(\rho))n^\rho \qquad \text{as } n \to \infty, \tag{T.5.14}$$

from that of $\hat{a}(v)$ of the form

$$\hat{a}(v) \sim (1-v)^{-\rho}K((1-v)^{-1}) \qquad \text{as } v \to 1^- \tag{T.5.15}$$

as in (T.5.12). Under additional *regularity* (e.g., monotonicity) of the terms $\{a_n\}$ it is often possible to deduce the asymptotic behavior of the terms (i.e., differenced sums) from this as

$$a_n \sim (1/\Gamma(\rho))n^{\rho-1} \qquad \text{as } n \to \infty. \tag{T.5.16}$$

An especially simple proof can be found in W. Feller (1971), *An Introduction to Probability Theory and Its Applications*, Vol. II, Wiley, New York, pp. 442–447.

Using the Tauberian theorem one can compute the asymptotic form of the rth moments ($r \geqslant 1$) of $\lambda n^{-1/2}L$ given $M_L = n$ as $n \to \infty$. In fact, one obtains that for $r \geqslant 1$,

$$E\{\lambda^r n^{-r/2}L^r \mid M_L = n\} \sim (2\sqrt{2})^r \mu^{*+}(r) \qquad \text{as } n \to \infty, \tag{T.5.17}$$

where $\mu^{*+}(r)$ is the rth moment of the *maximum of the Brownian excursion* process given in Exercise 12.8(iii) of Chapter I. Moreover, in view of the last equation in the hint following that exercise, one can check that the moments $\mu^{*+}(r)$ uniquely determine the distribution function by checking that the moment-generating function with these coefficients has an infinite radius of convergence. From this one obtains convergence in distribution to that of the maximum of a (appropriately scaled) Brownian excursion. This result, which was motivated by considerations of the main channel length as an extreme value of a river network, can be found in V. K. Gupta, O. Mesa and E. Waymire (1990), "Tree Dependent Extreme Values: The Exponential Case," *J. Appl. Probability*, in press. A more comprehensive treatment of this problem is given by R. Durrett, H. Kesten and E. Waymire (1989), "Random Heights of Weighted Trees," MSI Report, Cornell University, Ithaca. Problems of this type also occur in the analysis of tree search algorithms in computer science (see P. Flajolet and A. M. Odlyzko (1982), "The Average Height of Binary Trees and Other Simple Trees," *J. Comput. System Sci.*, **25**, pp. 171–213).

Theoretical Complements to Section IV.11

1. A (noncanonical) probability space for the voter model is furnished by the graphical percolation construction. This approach was created by T. E. Harris (1978), "Additive Set-Valued Markov Processes and Percolation Methods," *Ann. Probab.*, **6**, pp. 355–378. To construct Ω, first, for each $\mathbf{m} \in \mathbb{Z}^d$, $\mathbf{n}$ belonging to the boundary set $\partial(\mathbf{m})$ of nearest neighbors to $\mathbf{m}$, let $\boldsymbol{\Omega}_{\mathbf{m},\mathbf{n}}$ denote the collection of right-continuous nondecreasing *unit jump step functions* $\omega_{\mathbf{m},\mathbf{n}}(t)$, $t \geqslant 0$, such that $\omega_{\mathbf{m},\mathbf{n}}(t) \to \infty$ as $t \to \infty$. By the term "step function," it is implied that there are at most finitely many jumps in any bounded interval (nonexplosive). Let $P_{\mathbf{m},\mathbf{n}}$ be the (canonical) Poisson probability distribution on $(\Omega_{\mathbf{m},\mathbf{n}}, \mathscr{G}_{\mathbf{m},\mathbf{n}})$ with intensity 1, where $\mathscr{G}_{\mathbf{m},\mathbf{n}}$ is the sigmafield generated by events of the form $\{\omega \in \Omega_{\mathbf{m},\mathbf{n}}: \omega(t) \leqslant k\}$, $t \geqslant 0$, $k = 0, 1, 2, \ldots$. Define $(\Omega', \mathscr{G}', P')$ as the product probability space $\Omega' = \prod \Omega_{\mathbf{m},\mathbf{n}}$, $\mathscr{G}' = \prod \mathscr{G}_{\mathbf{m},\mathbf{n}}$, $P' = \prod P_{\mathbf{m},\mathbf{n}}$, where the products $\prod$ are over $\mathbf{m} \in \mathbb{Z}^d$, $\mathbf{n} \in \partial(\mathbf{m})$. To get Ω, remove from Ω' any $\omega' = ((\omega_{\mathbf{m},\mathbf{n}}(t): t \geqslant 0): \mathbf{m} \in \mathbb{Z}^d, \mathbf{n} \in \partial(\mathbf{m}))$ such that two or more $\omega_{\mathbf{m},\mathbf{n}}(t)$ have a jump at the same time t. Then $\mathscr{G}(= \Omega \cap \mathscr{G}')$ and P are obtained by the corresponding restriction to Ω (and measure-theoretical completion). The percolation flow structure can now be defined on $(\Omega, \mathscr{G}, P)$ as in the text, but sample pointwise for each $\omega = ((\omega_{\mathbf{m},\mathbf{n}}(t): t \geqslant 0): \mathbf{m} \in \mathbb{Z}^d, \mathbf{n} \in \partial(\mathbf{n})) \in \Omega$. For a given initial configuration $\boldsymbol{\eta} \in S$ let $D = D(\boldsymbol{\eta}) := \{\mathbf{m} \in \mathbb{Z}^d: \eta_{\mathbf{m}} = -1\}$. A sample path of the process started at $\boldsymbol{\eta}$, denoted $\boldsymbol{\sigma}^{D(\boldsymbol{\eta})}(t, \omega) = (\sigma_{\mathbf{n}}^{D(\boldsymbol{\eta})}(t, \omega): \mathbf{n} \in \mathbb{Z}^d)$, $t \geqslant 0$, $\omega \in \Omega$, is given in terms of the percolation flow on ω by

$$\sigma_{\mathbf{n}}^{D}(t, \omega) = \begin{cases} -1 & \text{if the flow on } \omega \text{ reaches } (\mathbf{n}, t) \text{ from some } (\mathbf{m}, 0),\ \mathbf{m} \in D, \\ +1 & \text{otherwise}. \end{cases} \tag{T.11.1}$$

The *Markov property* follows from the following basic property of the construction:

$$\sigma^{D(\eta)}(t + s, \omega) = \boldsymbol{\sigma}^{D(\boldsymbol{\sigma}^{D(\eta)}(s,\omega))}(t, U_s\omega), \tag{T.11.2}$$

where, for $\omega \in \Omega$ as above, $U_s: \Omega \to \Omega$, is given by

$$U_s(\omega) = ((\omega_{\mathbf{m},\mathbf{n}}(t + s): t \geqslant 0): \mathbf{m} \in \mathbb{Z}^d, \mathbf{n} \in \partial(\mathbf{n})). \tag{T.11.3}$$

The Markov property follows from (T.11.2) as a consequence of the *invariance* of P under the map $U_s: \Omega \to \Omega$ for each s; this is easily checked for the Poisson distribution with constant intensity and, because of independence, it is enough.

2. The distribution of the process $\{\boldsymbol{\sigma}(t)\}$ started at $\boldsymbol{\eta} \in S = \{-1, 1\}^{\mathbb{Z}^d}$ is the induced probability measure $P_{\boldsymbol{\eta}}$ on $(S, \mathscr{F})$ defined by

$$P_{\boldsymbol{\eta}}(B) = P(\{\boldsymbol{\sigma}^{D(\boldsymbol{\eta})}(t)\} \in B), \qquad B \in \mathscr{F}. \tag{T.11.4}$$

A metric for S, which gives it the so-called product space topology, is defined by

$$\rho_{(\boldsymbol{\sigma},\boldsymbol{\eta})} = \sum_{\mathbf{n} \in \mathbb{Z}^d} \frac{|\sigma_{\mathbf{n}} - \eta_{\mathbf{n}}|}{2^{|\mathbf{n}|}}, \tag{T.11.5}$$

where $|\mathbf{n}| \equiv |(n_1, \ldots, n_d)| := \max(|n_1|, \ldots, |n_d|)$. Note that this metric is possible largely because of the denumerability of $\mathbb{Z}^d$. In any case, this makes the fact that, (i) $\mathscr{F}$ is the *Borel sigmafield* and (ii) S is *compact*, rather straightforward exercises for this metric (topology). Compactness is in fact true for arbitrary products of compact spaces under the product topology, but this is a much deeper result (called *Tychonoff's Theorem*).

To prove the *Feller property*, it is sufficient to consider continuity of the mappings of the form $\boldsymbol{\sigma} \to P_{\boldsymbol{\sigma}}(\sigma_{\mathbf{n}}(t) = -1, \mathbf{n} \in F)$ at $\boldsymbol{\eta} \in S$, for, fixed $t > 0$ and finite sets $F \subset \mathbb{Z}^d$. Inclusion–exclusion principles can then be used to get the continuity of $\boldsymbol{\sigma} \to P_{\boldsymbol{\sigma}}(B)$ for all finite-dimensional sets $B \in \mathscr{F}$. The rest will follow from compactness (tightness).

For simplicity, first consider the map $\boldsymbol{\sigma} \to P_{\boldsymbol{\sigma}}(\sigma_{\mathbf{n}}(t) = -1)$, for $F = \{\mathbf{n}\}$. Open neighborhoods of $\boldsymbol{\eta} \in S$ are provided by sets of the form,

$$R_\Lambda(\boldsymbol{\eta}) = \{\boldsymbol{\sigma} \in S: \sigma_{\mathbf{m}} = \eta_{\mathbf{m}} \text{ for all } \mathbf{m} \in \Lambda\}, \tag{T.11.6}$$

where Λ is a *finite* subset of $\mathbb{Z}^d$, say

$$\Lambda = \{\mathbf{m} = (m_1, \ldots, m_d) \in \mathbb{Z}^d: |m_i| < r\}, \qquad r > 0. \tag{T.11.7}$$

Now, in view of the simple percolation structure for the voter model, *regardless of the initial configuration* $\boldsymbol{\eta}$, one has

$$\sigma_{\mathbf{n}}^{D(\boldsymbol{\eta})}(t) = \sigma_{\mathbf{m}(\mathbf{n},t)}^{D(\boldsymbol{\eta})}(0) \quad \text{a.s.}, \tag{T.11.8}$$

where $\mathbf{m}(\mathbf{n}, t)$ is some (random) site, which does not depend on the initial configuration, obtained by following backwards through the percolation diagram down and against the direction of arrows. Thus, if $\{\boldsymbol{\eta}_k\}$ is a sequence of configurations in Ω that converges to $\boldsymbol{\eta}$ in the metric ρ as $k \to \infty$, then $\sigma_{\mathbf{n}}^{\boldsymbol{\eta}_k}(t) \to \sigma_{\mathbf{m}(\mathbf{n},t)}^{\boldsymbol{\eta}}(t)$ a.s. as $k \to \infty$. Thus, by Lebesgue's Dominated Convergence Theorem (Chapter 0), one has

$$1 - 2P_{\boldsymbol{\eta}_k}(\sigma_n(t) = -1) = E\sigma_{\mathbf{m}(\mathbf{n},t)}^{\boldsymbol{\eta}_k}(t) = 1 - 2P_{\boldsymbol{\eta}}(\sigma_n(t) = -1).$$

Thus, $P_{\boldsymbol{\eta}_k}(\sigma_n(t) = -1) \to P_{\boldsymbol{\eta}}(\sigma_n(t) = -1)$ as $k \to \infty$.

3. A real-valued function $\varphi(\mathbf{m})$, $\mathbf{m} \in \mathbb{Z}^d$, such that for each $\mathbf{m} \in \mathbb{Z}^d$,

$$\frac{1}{2d} \sum_{\mathbf{n} \in \partial(\mathbf{m})} \varphi(\mathbf{n}) = \varphi(\mathbf{m}), \tag{T.11.9}$$

where $\partial(\mathbf{m})$ denotes the set of nearest neighbors of $\mathbf{m}$, is said to be *harmonic* (with respect to the discrete Laplacian on $\mathbb{Z}^d$). Note that (T.11.9) may be expressed as

$$\varphi(\mathbf{Z}_0) = E_{\mathbf{m}}\varphi(\mathbf{Z}_1) = E_{\mathbf{m}}\varphi(\mathbf{Z}_k), \qquad k \geqslant 1, \tag{T.11.10}$$

where $\{\mathbf{Z}_k\}$ is the simple symmetric random walk starting at $\mathbf{Z}_0 = \mathbf{m}$.

Theorem T.11.1. (*Boundedness Principle*). Let φ be a real-valued bounded harmonic function on $\mathbb{Z}^d$. Then φ is a constant function. □

Proof. Suppose that $\{(\mathbf{Z}_k', \mathbf{Z}_k'')\}$ is a *coupling* of two copies of $\{\mathbf{Z}_k\}$; i.e., a Markov chain on $S \times S$ such that the marginal processes $\{\mathbf{Z}_k'\}$ and $\{\mathbf{Z}_k''\}$ are each Markov chains on S with the same transition law as $\{Z_k\}$. Then, if $\tau = \inf\{k \geqslant 1 : \mathbf{Z}'_k = \mathbf{Z}''_k\} < \infty$ a.s. one may define $\mathbf{Z}'_k = \mathbf{Z}''_k$ for all $k \geqslant \tau$ without disrupting the property that the process $\{(\mathbf{Z}'_k, \mathbf{Z}''_k)\}$ is a coupling of $\{\mathbf{Z}_k\}$. With this as the case, by the boundedness of φ and (T.11.10), one obtains

$$\begin{aligned}|\varphi(\mathbf{n}) - \varphi(\mathbf{m})| &= |E_{\mathbf{n}}\varphi(\mathbf{Z}_k) - E_{\mathbf{m}}\varphi(\mathbf{Z}_k)| = |E_{(\mathbf{n},\mathbf{m})}\{\varphi(\mathbf{Z}'_k) - \varphi(\mathbf{Z}''_k)\}| \\ &\leqslant E_{(\mathbf{n},\mathbf{m})}|\varphi(\mathbf{Z}'_k) - \varphi(\mathbf{Z}''_k)| \leqslant 2BP_{(\mathbf{n},\mathbf{m})}(\mathbf{Z}'_k \neq \mathbf{Z}''_k) \\ &= 2BP_{(\mathbf{n},\mathbf{m})}(\tau > k), \end{aligned} \tag{T.11.11}$$

where $B = \sup_{\mathbf{x}}|\varphi(\mathbf{x})|$. Letting $k \to \infty$ it would then follow that $\varphi(\mathbf{n}) = \varphi(\mathbf{m})$. The success of this approach rests on the construction of a coupling $\{(\mathbf{Z}'_k, \mathbf{Z}''_k)\}$ with $\tau < \infty$ a.s.; such a coupling is called a *successful coupling*.

The independent coupling is the simplest to try. To see how it works, take $d = 1$ and let $\{Z'_k\}$ and $\{Z''_k\}$ be independent simple symmetric random walks on $\mathbb{Z}$ starting at $n, m, n - m$ even. Then $\{(Z'_k, Z''_k)\}$ is a successful coupling since $\{Z'_k - Z''_k\}$ is easily checked to be a recurrent random walk using the results of Chapters II and IV or theoretical complement to Section 3 of Chapter I. This would also work for $d = 2$, but it fails for $d \geqslant 3$ owing to transience. In any case, here is another coupling that is easily checked to be successful for any d. Let $\{(\mathbf{Z}'_k, \mathbf{Z}''_k)\}$ be the Markov chain on $\mathbb{Z}^d \times \mathbb{Z}^d$ starting at $(\mathbf{n}, \mathbf{m})$ and having the stationary one-step transition probabilities furnished by the following rules of motion. At each time, first select a (common) coordinate axis at random (each having probability $1/d$). From $(\mathbf{n}, \mathbf{m})$ such that the coordinates of $\mathbf{n}, \mathbf{m}$ *differ* along the selected axis, independently select $\pm$ directions (each having probability $\frac{1}{2}$) along the selected axis for displacements of the components $\mathbf{n}, \mathbf{m}$. If, on the other hand, the coordinates of $\mathbf{n}, \mathbf{m}$ agree along the selected axis, then randomly select a *common* $\pm$ direction along the axis for displacement of both components $\mathbf{n}, \mathbf{m}$. Then the process $\{(\mathbf{Z}'_k, \mathbf{Z}''_k)\}$ with this transition law is a coupling. That it is successful for all $(\mathbf{n}, \mathbf{m})$, whose coordinates all have the same parity, follows from the recurrence of the simple symmetric random walk on $\mathbb{Z}^1$, since it guarantees with probability 1 that each of the d coordinates will eventually line up. Thus, one obtains $\varphi(\mathbf{n}) = \varphi(\mathbf{m})$ for all $\mathbf{n}, \mathbf{m}$ whose coordinates are each of the same parity. This is enough by (T.11.9) and the maximum principle for harmonic functions described below. ■

The following two results provide fundamental properties of harmonic functions given on a suitable *domain D*. Their proofs are obtained by iterating the averaging property out to the boundary, just as in the one-dimensional case of Exercise 3.16, Chapter I. First some preliminaries about the domain. We require that $D = D^0 \cup \partial D$, where $D^0, \partial D$ are *finite, disjoint* subsets of $\mathbb{Z}^d$ such that

(i) $\partial(\mathbf{n}) \subset D$ for each $\mathbf{n} \in D^0$, where $\partial(\mathbf{n})$ denotes the set of nearest neighbors of $\mathbf{n}$;
(ii) $\partial(\mathbf{m}) \cap D^0 \neq \varnothing$ for each $\mathbf{m} \in \partial D$;
(iii) for each $\mathbf{n}, \mathbf{m} \in D$ there is a path of respective nearest neighbors $\mathbf{n}_1, \mathbf{n}_2, \ldots, \mathbf{n}_k$ in D^0 such that $\mathbf{n}_1 \in \partial(\mathbf{n})$, $\mathbf{n}_k \in \partial(\mathbf{m})$.

Theorem T.11.2. (*Maximum Principle*). A real-valued function φ on a domain D, as defined above, that is harmonic on D^0, i.e.,

$$\varphi(\mathbf{m}) = \frac{1}{2d} \sum_{\mathbf{n} \in \partial(\mathbf{m})} \varphi(\mathbf{n}), \qquad \mathbf{m} \in D^0,$$

takes its maximum and minimum values on ∂D. □

In particular, it follows from the maximum principle that if φ also takes an extreme value on D^0, then it must be constant on D^0. This can be used to complete the proof of Theorem T.11.1 by suitably constructing a domain D with any of the 2^d coordinate parities on D^0 and ∂D desired.

Theorem T.11.3. (*Uniqueness Principle*). If φ_1 and φ_2 are harmonic functions on D, as defined in Theorem T.11.2, and if $\varphi_1 = \varphi_2$ on ∂D, then $\varphi_1 = \varphi_2$ on D. □

Proof. To prove the uniqueness principle, simply note that $\varphi_1 - \varphi_2$ is harmonic with zero boundary values and hence, by the Maximum Principle, zero extreme values. ■

The coupling described in Theorem T.11.1 can be found in T. Liggett (1985), *Interacting Particle Systems*, Springer-Verlag, New York, pp. 67–69, together with another coupling to prove the boundedness principle (*Choquet–Deny theorem*) for the more general case of an irreducible random walk on $\mathbb{Z}^d$.

4. *Additional references*: The voter model was first considered in the papers by P. Clifford and A. Sudbury (1973), "A Model for Spatial Conflict," *Biometrika*, **60**, pp. 581–588, and independently by R. Holley and T. M. Liggett (1975), "Ergodic Theorems for Weakly Interacting Infinite Systems and the Voter Model," *Ann. Probab.*, **3**, pp. 643–663. The approach in section 11 essentially follows F. Spitzer (1981), "Infinite Systems with Locally Interacting Components," *Ann. Probab.*, **9**, 349–364. The so-called biased voter (tumor-growth) model originated in T. Williams and R. Bjerknes (1972), "Stochastic Model for Abnormal Clone Spread Through Epithelial Basal Layer," *Nature*, **236**, pp. 19–21.

Much of the modern interest in the mathematical theory of infinite systems of interacting components from the point of view of continuous-time Markov evolutions was inspired by the fundamental papers of Frank Spitzer (1970), "Interaction of Markov Processes," *Advances in Math.*, **5**, pp. 246–290, and by R. L. Dobrushin (1971), "Markov Processes With a Large Number of Locally Interacting Components," *Problems Inform. Transmission*, **7**, pp. 149–164, 235–241. Since then several books and monographs have been written on the subject, the most comprehensive being that of T. Liggett (1985), *loc. cit.* Other modern books and monographs on the subject are those of F. Spitzer (1971), *Random Fields and Interacting Particle Systems*, MAA, Summer Seminar Notes; D. Griffeath (1979), *Additive and Cancellative Interacting Particle Systems*, Lecture Notes in Math., No. 724, Springer-Verlag, New York; R. Kindermann and J. L. Snell (1980), *Markov Random Fields and Their Applications*, Contemporary Mathematics Series, Vol. 1, AMS, Providence, RI; R. Durrett (1988), *Lecture Notes on Particle Systems and Percolation*, Wadsworth, Brooks/Cole, San Francisco.

CHAPTER V

Brownian Motion and Diffusions

1 INTRODUCTION AND DEFINITION

One-dimensional unrestricted diffusions are Markov processes in continuous time with state space $S = (a, b)$, $-\infty \leqslant a < b \leqslant \infty$, having continuous sample paths. The simplest example is Brownian motion $\{Y_t\}$ with drift coefficient μ and diffusion coefficient $\sigma^2 > 0$ introduced in Chapter I as a limit of simple random walks. In this case, $S = (-\infty, \infty)$. The transition probability distribution $p(t; x, dy)$ of Y_{s+t} given $Y_s = x$ has a density given by

$$p(t; x, y) = \frac{1}{(2\pi\sigma^2 t)^{1/2}} e^{-\frac{1}{2\sigma^2 t}(y - x - \mu t)^2}. \tag{1.1}$$

Because $\{Y_t\}$ has independent increments, the Markov property follows directly. Notice that (1.1) does not depend on s; i.e., $\{Y_t\}$ is a *time-homogeneous* Markov process. As before, by a Markov process we mean one having a *time-homogeneous transition law* unless stated otherwise.

One may imagine a Markov process $\{X_t\}$ that has continuous sample paths but that is *not* a process with independent increments. Suppose that, given $X_s = x$, for (infinitesimal) small times t, the displacement $X_{s+t} - X_s = X_{s+t} - x$ has mean and variance approximately $t\mu(x)$ and $t\sigma^2(x)$, respectively. Here $\mu(x)$ and $\sigma^2(x)$ are functions of the state of x, and not constants as in the case of $\{Y_t\}$. The distinction between $\{Y_t\}$ and $\{X_t\}$ is analogous to that between a simple random walk and a birth–death chain. More precisely, suppose

$$\begin{aligned} E(X_{s+t} - X_s \mid X_s = x) &= t\mu(x) + o(t), \\ E((X_{s+t} - X_s)^2 \mid X_s = x) &= t\sigma^2(x) + o(t), \\ E(|X_{s+t} - X_s|^3 \mid X_s = x) &= o(t), \end{aligned} \tag{1.2}$$

hold, as $t \downarrow 0$, for every $x \in S$.

Note that (1.2) holds for Brownian motions (Exercise 1). A more general formulation of the existence of infinitesimal mean and variance parameters, which does not require the existence of finite moments, is the following. For every $\varepsilon > 0$ assume that

$$\begin{aligned} E((X_{s+t} - X_s)\mathbf{1}_{\{|X_{s+t}-X_s| \leqslant \varepsilon\}} \mid X_s = x) &= t\mu(x) + o(t), \\ E((X_{s+t} - X_s)^2\mathbf{1}_{\{|X_{s+t}-X_s| \leqslant \varepsilon\}} \mid X_s = x) &= t\sigma^2(x) + o(t), \\ P(|X_{s+t} - X_s| > \varepsilon \mid X_s = x) &= o(t), \end{aligned} \tag{1.2$'$}$$

hold as $t \downarrow 0$.

It is a simple exercise to show that (1.2) implies (1.2)′ (Exercise 2). However, there are many Markov processes with continuous sample paths for which (1.2)′ hold, but not (1.2).

Example 1. (*One-to-One Functions of Brownian Motion*). Let $\{B_t\}$ be a standard one-dimensional Brownian motion and φ a twice continuously differentiable strictly increasing function of $(-\infty, \infty)$ onto (a, b). Then $\{X_t\} := \{\varphi(B_t)\}$ is a time-homogeneous Markov process with continuous sample paths. Take $\varphi(x) = e^{x^3}$. Now check that $E_x|X_t| = \infty$ for all $t > 0$, so that (1.2) does not hold. On the other hand, (1.2)′ does hold (as explained more generally in Section 3).

Definition 1.1. A Markov process $\{X_t\}$ on the state space $S = (a, b)$ is said to be a *diffusion with drift coefficient* $\mu(x)$ *and diffusion coefficient* $\sigma^2(x) > 0$, if

(i) it has continuous sample paths, and
(ii) relations (1.2)′ hold for all x.

If the transition probability distribution $p(t; x, dy)$ has a *density* $p(t; x, y)$, then, for (Borel) subsets B of S,

$$p(t; x, B) = \int_B p(t; x, y)\, dy. \tag{1.3}$$

It is known that a strictly positive and continuous density exists under the Condition (1.1) below, in the case $S = (-\infty, \infty)$ (see theoretical complement 1). Since any open interval (a, b) can be transformed into $(-\infty, \infty)$ by a strictly increasing smooth map, Condition (1.1) may be applied to $S = (a, b)$ after transformation (see Section 3).

Below, $\sigma(x)$ denotes either the positive square root of $\sigma^2(x)$ for all x or the negative square root for all x.

Condition (1.1). The functions $\mu(x)$, $\sigma(x)$ are continuously differentiable, with bounded derivatives on $S = (-\infty, \infty)$. Also, σ'' exists and is continuous, and

$\sigma^2(x) > 0$ for all x. If $S = (a, b)$, then assume the above conditions for the relabeled process under some smooth and strictly increasing transformation onto $(-\infty, \infty)$.

Although the results presented in this chapter are true under (1.2)′, in order to make the calculations less technical we will assume (1.2). It turns out that Condition (1.1) guarantees (1.2). From the Markov property the joint density of $X_{t_1}, X_{t_2}, \ldots, X_{t_n}$ for $0 < t_1 < t_2 < \cdots < t_n$ is given by the product $p(t_1; x, y_1)p(t_2 - t_1; y_1, y_2) \cdots p(t_n - t_{n-1}; y_{n-1}, y_n)$. Therefore, for an initial distribution π,

$$P_\pi(X_{t_1} \in B_1, \ldots, X_{t_n} \in B_n)$$
$$= \int_S \int_{B_1} \cdots \int_{B_n} p(t_1; x, y) \cdots p(t_n - t_{n-1}; y_{n-1}, y_n)\, dy_n \cdots dy_1\, \pi(dx). \quad (1.4)$$

As usual P_π denotes the distribution of the process $\{X_t\}$ for the initial distribution π. In the case $\pi = \delta_x$ we write P_x in place of P_{δ_x}. Likewise, E_π, E_x are the corresponding expectations.

Example 2. (*Ornstein–Uhlenbeck Process*). Let $\mathbf{V}_t$ denote the random velocity of a large solute molecule immersed in a liquid at rest. For simplicity, let V_t denote the vertical component of the velocity. The solute molecule is subject to two forces: *gravity* and *friction*. It turns out that the gravitational force is negligible compared to the frictional force exerted on the solute molecule by the liquid. The frictional force is directed oppositely to the direction of motion and is proportional to the velocity in magnitude. In the absence of statistical fluctuations, one would therefore have $m(dV_t/dt) = -\beta V_t$, where m is the *mass* of the solute molecule and β is the constant of proportionality known as the *coefficient of friction*. However, this frictional force $-\beta V$ may be thought of as the *mean* of a large number of random molecular collisions. Assuming that the central limit theorem applies to the superposition of displacements due to a large number of such collisions, the change in momentum $m\, dV_t$ over a time interval $(t, t + dt)$ is approximately Gaussian, provided dt is such that the mean number of collisions during $(t, t + dt)$ is large. The mean and variance of this Gaussian distribution are both proportional to the number of collisions, i.e., to dt. Therefore, one may take the (local) mean of V_t to be $-(\beta/m)V_t\, dt$ and the (local) variance to be $\sigma^2\, dt$.

$$\begin{aligned} E(V_{t+dt} - V_t \mid V_t = V) &= -(\beta/m)V\, dt + o(dt), \\ E((V_{t+dt} - V_t)^2 \mid V_t = V) &= \sigma^2\, dt + o(dt). \end{aligned} \quad (1.5)$$

Therefore, a reasonable model for the velocity process is a diffusion with drift $(-\beta/m)V$ and diffusion coefficient $\sigma^2 > 0$, called the *Ornstein–Uhlenbeck process*.

An important problem is to determine the transition probabilities for diffusions having given drift and diffusion coefficients. Various approaches to this problem will be developed in this chapter, but in the meantime for the Ornstein–Uhlenbeck process we can check that the transition function given by

$$p(t; x, y) = [\pi\sigma^2\beta^{-1}m(1 - \exp\{-2\beta m^{-1}t\})]^{-1/2} \exp\left\{-\frac{\beta m^{-1}(y - xe^{-\beta m^{-1}t})^2}{\sigma^2(1 - e^{-2\beta m^{-1}t})}\right\},$$

$$(t > 0, \ -\infty < x, y < \infty), \quad (1.6)$$

furnishes the solution. In other words, given $V_s = x$, V_{s+t} is *Gaussian* with *mean* $xe^{-\beta m^{-1}t}$ and *variance* $\frac{1}{2}\sigma^2(1 - e^{-2\beta m^{-1}t})\beta^{-1}m$. From this one may check (1.5) directly.

Example 3. (*A Nonhomogeneous Diffusion*). Suppose $\{X_t\}$ is a diffusion with drift $\mu(x)$ and diffusion coefficient $\sigma^2(x)$. Let f be a continuous function on $[0, \infty)$. Define

$$Z_t = X_t + \int_0^t f(u)\,du, \qquad t \geqslant 0. \quad (1.7)$$

Then, since f is a deterministic function, the process $\{Z_t\}$ inherits the Markov property from $\{X_t\}$. Moreover, $\{Z_t\}$ has continuous sample paths. Now,

$$E\{Z_{s+t} - Z_s \mid Z_s = z\} = E\left\{X_{s+t} - X_s \,\middle|\, X_s = z - \int_0^s f(u)\,du\right\} + \int_s^{s+t} f(u)\,du$$
$$= \mu\left(z - \int_0^s f(u)\,du\right)t + f(s)t + o(t). \quad (1.8)$$

Also,

$$E\{(Z_{s+t} - Z_s)^2 \mid Z_s = z\} = E\left\{(X_{s+t} - X_s)^2 \,\middle|\, X_s = z - \int_0^s f(u)\,du\right\}$$
$$+ \left(\int_s^{s+t} f(u)\,du\right)^2 + o(t)$$
$$= \sigma^2\left(z - \int_0^s f(u)\,du\right)t + o(t). \quad (1.9)$$

Similarly,

$$E\{|Z_{s+t} - Z_s|^3 \mid Z_s = z\} = o(t). \quad (1.10)$$

In this case $\{Z_t\}$ is referred to as a diffusion with (nonhomogeneous) drift and diffusion coefficients, respectively, given by

$$\mu(s, z) = \mu\left(z - \int_0^s f(u)\,du\right) + f(s),$$
$$\sigma^2(s, z) = \sigma^2\left(z - \int_0^s f(u)\,du\right). \tag{1.11}$$

Note that if $\{X_t\}$ is a Brownian motion with drift μ and diffusion coefficient σ^2, and if $f(t) := \nu$ is constant, then $\{Z_t\}$ is Brownian motion with drift $\mu + \nu$ and diffusion coefficient σ^2.

A rigorous construction of *unrestricted diffusions* by the method of *stochastic differential equations* is given in Chapter VII. An alternative method, similar to that of Chapter IV, Sections 3 and 4, is to solve Kolmogorov's backward or forward equations for the transition probability density. The latter is then used to construct probability measures P_π, P_x on the space of all continuous functions on $[0, \infty)$. The Kolmogorov equations, derived in the next section, may be solved by the methods of *partial differential equations* (PDE) (see theoretical complement 1). Although the general PDE solution is not derived in this book, the method is illustrated by two examples in Section 5. A third method of construction of diffusions is based on *approximation by birth–death chains*. This method is outlined in Section 4.

Diffusions with boundary conditions are constructed from unrestricted ones in Sections 6 and 7 by a probabilistic method. The *PDE method*, based on eigenfunction expansions, is described and illustrated in Section 8.

The aim of the present chapter is to provide a systematic development of some of the most important aspects of diffusions. Computational expressions and examples are emphasized. The development proceeds in a manner analogous to that pursued in previous chapters, and, as before, there is a focus on large-time behavior.

2 KOLMOGOROV'S BACKWARD AND FORWARD EQUATIONS, MARTINGALES

In Section 2 of Chapter IV, Kolmogorov's equations were derived for continuous-parameter Markov chains. The backward equations for diffusions are derived in Subsection 2.1 below, together with an important connection between Markov processes and martingales. The forward, or Fokker–Planck, equation is obtained in Subsection 2.2. Although the latter plays a less important mathematical role in the study of diffusions, it arises more naturally than the backward equation in physical applications.

2.1 The Backward Equation and Martingales

Suppose $\{X_t\}$ is a Markov process on a metric space S, having right-continuous sample paths and a transition probability $p(t; x, dy)$. Continuous-parameter

Markov chains on countable state spaces and diffusions on $S = (a, b)$ are examples of such processes. On the *set* $\mathbf{B}(S)$ *of all real-valued, bounded, Borel measurable function* f *on* S define the *transition operator*

$$(\mathbf{T}_t f)(x) := E(f(X_t) \mid X_0 = x) = \int f(y) p(t; x, dy), \qquad (t > 0). \tag{2.1}$$

Then $\mathbf{T}_t$ is a bounded linear operator on $\mathbf{B}(S)$ when the latter is given the *sup norm* defined by

$$\| f \| := \sup\{|f(y)|: y \in S\}. \tag{2.2}$$

Indeed $\mathbf{T}_t$ is a *contraction*, i.e., $\|\mathbf{T}_t f\| \leqslant \|f\|$ for all $f \in \mathbf{B}(S)$. For,

$$|(\mathbf{T}_t f)(x)| \leqslant \int |f(y)|\, p(t;, x, dy) \leqslant \|f\|. \tag{2.3}$$

The family of transition operators $\{\mathbf{T}_t : t > 0\}$ has the *semigroup property*,

$$\mathbf{T}_{s+t} = \mathbf{T}_s \mathbf{T}_t, \tag{2.4}$$

where the right side denotes the composition of two maps. This relation follows from

$$\begin{aligned}(\mathbf{T}_{s+t} f)(x) &= E(f(X_{s+t}) \mid X_0 = x) = E[E(f(X_{s+t}) \mid X_s) \mid X_0 = x] \\ &= E[(\mathbf{T}_t f)(X_s) \mid X_0 = x] = \mathbf{T}_s(\mathbf{T}_t f)(x).\end{aligned} \tag{2.5}$$

The relation (2.4) also implies that the transition operators *commute*,

$$\mathbf{T}_s \mathbf{T}_t = \mathbf{T}_t \mathbf{T}_s. \tag{2.6}$$

As discussed in Section 2 of Chapter IV for the case of continuous-parameter Markov chains, the semigroup relation (2.4) implies that the behavior of $\mathbf{T}_t$ near $t = 0$ completely determines the semigroup. Observe also that if $f \in C_b(S)$, the *set of all real bounded continuous functions on* S, then $(\mathbf{T}_t f)(x) \equiv E(f(X_t) \mid X_0 = x) \to f(x)$ as $t \downarrow 0$, by the right continuity of the sample paths. It then turns out, as in Chapter IV, that the derivative (operator) of the function $t \to \mathbf{T}_t$ at $t = 0$ determines $\{\mathbf{T}_t : t > 0\}$ (see theoretical complement 3).

Definition 2.1. The *infinitesimal generator* $\mathbf{A}$ *of* $\{\mathbf{T}_t : t > 0\}$, or of *the Markov process* $\{X_t\}$, is the linear operator $\mathbf{A}$ defined by

$$(\mathbf{A} f)(x) = \lim_{s \downarrow 0} \frac{(\mathbf{T}_s f)(x) - f(x)}{s}, \tag{2.7}$$

for all $f \in \mathbf{B}(S)$ such that the right side converges to some function uniformly in x. The class of all such f comprises the *domain* $\mathscr{D}_{\mathbf{A}}$ *of* $\mathbf{A}$.

In order to determine $\mathbf{A}$ explicitly in the case of a diffusion $\{X_t\}$, let f be a bounded twice continuously differentiable function. Fix $x \in S$, and a $\delta > 0$, however small. Find $\varepsilon > 0$ such that $|f''(x) - f''(y)| \leqslant \delta$ for all y such that $|y - x| \leqslant \varepsilon$. Write

$$(\mathbf{T}_t f)(x) = E(f(X_t)\mathbf{1}_{\{|X_t - x| \leqslant \varepsilon\}} \mid X_0 = x) + E(f(X_t)\mathbf{1}_{\{|X_t - x| > \varepsilon\}} \mid X_0 = x). \quad (2.8)$$

Now by a Taylor expansion of $f(X_t)$ around x,

$$\begin{aligned} E(f(X_t)\mathbf{1}_{\{|X_t - x| \leqslant \varepsilon\}} \mid X_0 = x) = E[\{f(x) + (X_t - x)f'(x) + \tfrac{1}{2}(X_t - x)^2 f''(x) \\ + \tfrac{1}{2}(X_t - x)^2(f''(\xi_t) - f''(x))\}\mathbf{1}_{\{|X_t - x| \leqslant \varepsilon\}} \mid X_0 = x], \end{aligned} \quad (2.9)$$

where ξ_t lies between x and X_t. By the first two relations in (1.2)′, the expectations of the first three terms on the right add up to

$$f(x) + t\mu(x)f'(x) + t\tfrac{1}{2}\sigma^2(x)f''(x) + \mathrm{o}(t), \qquad (t \downarrow 0). \quad (2.10)$$

The expectation of the remainder term in (2.9) is less than

$$\tfrac{1}{2}\delta E((X_t - x)^2 \mathbf{1}_{\{|X_t - x| \leqslant \varepsilon\}} \mid X_0 = x) = \tfrac{1}{2}\delta\sigma^2(x)t + \mathrm{o}(t), \qquad (t \downarrow 0). \quad (2.11)$$

The relations (2.8)–(2.11) lead to

$$\begin{aligned} \overline{\lim_{t \downarrow 0}} \left| \frac{(\mathbf{T}_t f)(x) - f(x)}{t} - \{\mu(x)f'(x) + \tfrac{1}{2}\sigma^2(x)f''(x)\} \right| \\ \leqslant \tfrac{1}{2}\delta\sigma^2(x) + \overline{\lim_{t \downarrow 0}} \frac{\|f\|}{t} E(\mathbf{1}_{\{|X_t - x| > \varepsilon\}} \mid X_0 = x). \end{aligned} \quad (2.12)$$

The $\overline{\lim}$ on the right is zero by the last relation in (1.2)′. As $\delta > 0$ is arbitrary it now follows that

$$\lim_{t \downarrow 0} \frac{(\mathbf{T}_t f)(x) - f(x)}{t} = \mu(x)f'(x) + \tfrac{1}{2}\sigma^2(x)f''(x). \quad (2.13)$$

Does the above computation prove that a bounded twice continuously differentiable f belongs to $\mathscr{D}_{\mathbf{A}}$? Not quite, for the limit (2.13) has not been shown to be uniform in x. There are three sources of nonuniformity. One is that the $\mathrm{o}(t)$ terms in (1.2)′ need not be uniform in x. The second is that the $\mathrm{o}(t)$ term in (2.10) may not be uniform in x, even if those in (1.2)′ are; for $\mu(x)f'(x)$, $\sigma^2(x)f''(x)$ may not be bounded. The third source of nonuniformity arises from the fact that given $\delta > 0$ there may not exist an ε independent of x such that $|f(y) - f(x)| < \delta$ for all x, y satisfying $|x - y| < \varepsilon$. The third source is removed by requiring that f'' be uniformly continuous. Assume $\mu(x)f'(x)$, $\sigma^2(x)f''(x)$ are bounded to take care of the second. The first source is intrinsic. One example where the errors in (1.2)′ are $o(t)$ uniformly in x, is a Brownian motion (Exercise 4). In the case the Brownian motion has a zero drift, the second source is removed if f'' is bounded. Thus, for a Brownian motion with zero drift, every bounded f having a uniformly continuous and bounded f'' is in $\mathscr{D}_{\mathbf{A}}$. Indeed, it may be shown that $\mathscr{D}_{\mathbf{A}}$ comprises precisely this class of functions. Similarly, if the Brownian motion has a nonzero drift, then $f \in \mathscr{D}_{\mathbf{A}}$ if f, f', f'' are all bounded and f'' is uniformly continuous (Exercise 4). In general one may ensure uniformity of $o(t)$ in (1.2)′ by restricting to a compact subset of S. Thus, all twice continuously differentiable f, vanishing outside a closed and bounded subinterval of S, belong to $\mathscr{D}_{\mathbf{A}}$. We have sketched a proof of the following result (see theoretical complements 1, 3 and Exercise 3.12 of Chapter VII).

Proposition 2.1. Let $\{X_t\}$ be a diffusion on $S = (a, b)$. Then, all twice continuously differentiable f, vanishing outside a closed bounded subinterval of S, belong to $\mathscr{D}_{\mathbf{A}}$, and for such f

$$(\mathbf{A}f)(x) = \mu(x)f'(x) + \tfrac{1}{2}\sigma^2(x)f''(x). \tag{2.14}$$

In what follows, the symbol $\mathbf{A}$ will stand, in the case of a diffusion $\{X_t\}$, for the second-order differential operator (2.14), and it will be applied to all twice-differentiable f, whether or not such an f is in $\mathscr{D}_{\mathbf{A}}$.

Turning to the derivation of the backward equation, note that the arguments leading to (2.13) hold for all bounded f that are twice continuously differentiable. If the function $x \to (\mathbf{T}_t f)(x)$ is bounded and twice continuously differentiable, then applying (2.13) to this function one gets

$$\frac{\partial}{\partial t}(\mathbf{T}_t f)(x) = \mathbf{A}(\mathbf{T}_t f)(x), \qquad (t > 0, x \in S). \tag{2.15}$$

This is *Kolmogorov's backward equation* for the function $(t, x) \to (\mathbf{T}_t f)(x)$ $(t > 0, x \in S)$. It will be shown below that if $f \in \mathscr{D}_{\mathbf{A}}$ then $\mathbf{T}_t f \in \mathscr{D}_{\mathbf{A}}$. The following proposition establishes this fact for general Markov processes.

Proposition 2.2. Let $\{\mathbf{T}_t\}$ be the family of transition operators for a Markov process on a metric space S. If $f \in \mathscr{D}_{\mathbf{A}}$ then the backward equation (2.15) holds.

Proof. If $f \in \mathscr{D}_{\mathbf{A}}$ then using (2.6) and (2.3),

$$\left\| \frac{\mathbf{T}_s(\mathbf{T}_t f) - \mathbf{T}_t f}{s} - \mathbf{T}_t(\mathbf{A}f) \right\| = \left\| \mathbf{T}_t\left(\frac{\mathbf{T}_s f - f}{s} - \mathbf{A}f \right) \right\|$$
$$\leqslant \left\| \frac{\mathbf{T}_s f - f}{s} - \mathbf{A}f \right\| \to 0 \qquad \text{as } s \downarrow 0, \tag{2.16}$$

which shows that $\mathbf{T}_t f \in \mathscr{D}_{\mathbf{A}}$ and, therefore, by the definition of $\mathbf{A}$, (2.15) holds. ■

The relation (2.16) also shows that

$$\mathbf{A}(\mathbf{T}_t f) = \mathbf{T}_t(\mathbf{A}f), \tag{2.17}$$

i.e., $\mathbf{T}_t$ *and* $\mathbf{A}$ *commute on* $\mathscr{D}_{\mathbf{A}}$. This fact is made use of in the proof of the following important result.

Theorem 2.3. Let $\{X_t\}$ be a right-continuous Markov process on a metric space S, and $f \in \mathscr{D}_{\mathbf{A}}$. Assume that $s \to (\mathbf{A}f)(X_s)$ is right-continuous for all sample paths. Then the process $\{Z_t\}$, where

$$Z_t := f(X_t) - \int_0^t (\mathbf{A}f)(X_s)\, ds \qquad (t \geqslant 0), \tag{2.18}$$

is a $\{\mathscr{F}_t\}$-martingale, with $\mathscr{F}_t := \sigma\{X_u : 0 \leqslant u \leqslant t\}$.

Proof. The assumption of right continuity of $s \to (\mathbf{A}f)(X_s)$ ensures the integrability and $\mathscr{F}_t$-measurability of the integral in (2.18) (theoretical complement 2). Now

$$E(Z_{s+t} \mid \mathscr{F}_s) = E(f(X_{s+t}) \mid \mathscr{F}_s) - \int_0^s (\mathbf{A}f)(X_u)\, du - E\left(\int_s^{s+t} (\mathbf{A}f)(X_u)\, du \,\Big|\, \mathscr{F}_s \right)$$
$$= (\mathbf{T}_t f)(X_s) - \int_0^s (\mathbf{A}f)(X_u)\, du - E\left(\int_0^t (\mathbf{A}f)(X_s^+)_u\, du \,\Big|\, \mathscr{F}_s \right), \tag{2.19}$$

where X_s^+ is the *after-s or shifted process* $(X_s^+)_u = X_{s+u}$ $(u \geqslant 0)$. By the Markov property,

$$E\left(\int_0^t (\mathbf{A}f)(X_s^+)_u\, du \,\Big|\, \mathscr{F}_s \right) = \left[E_x \int_0^t (\mathbf{A}f)(X_u)\, du \right]_{x = X_u}$$
$$= \left[\int_0^t (E_x \mathbf{A}f)(X_u)\, du \right]_{x = X_s}$$
$$= \left[\int_0^t \mathbf{T}_u(\mathbf{A}f)(x)\, du \right]_{x = X_s}, \tag{2.20}$$

interchanging the order of integration with respect to Lebesgue measure and taking expectation for the second equality (Fubini's Theorem, Section 4 of Chapter 0). Now replace the last integrand $\mathbf{T}_u(\mathbf{A}f)$ by $\mathbf{A}(\mathbf{T}_u f)$ (see (2.17)) to get

$$E\left(\int_0^t (\mathbf{A}f)(X_s^+)_u\,du \mid \mathscr{F}_s\right) = \int_0^t \mathbf{A}(\mathbf{T}_u f)(X_s)\,du. \tag{2.21}$$

Substituting (2.21) in (2.19) obtain

$$E(Z_{s+t} \mid \mathscr{F}_s) = (\mathbf{T}_t f)(X_s) - \int_0^t \mathbf{A}(\mathbf{T}_u f)(X_s)\,du - \int_0^s (\mathbf{A}f)(X_u)\,du. \tag{2.22}$$

By the backward equation (2.15), which holds by Proposition 2.2, and the fundamental theorem of calculus,

$$(\mathbf{T}_t f)(X_s) = f(X_s) + \int_0^t \frac{\partial}{\partial u}(\mathbf{T}_u f)(X_s)\,du = f(X_s) + \int_0^t \mathbf{A}(\mathbf{T}_u f)(X_s)\,du. \tag{2.23}$$

Use this in (2.22) to derive the desired result,

$$E(Z_{s+t} \mid \mathscr{F}_s) = f(X_s) - \int_0^s (\mathbf{A}f)(X_u)\,du = Z_s. \qquad \blacksquare$$

Combining Theorem 2.3 and Proposition 2.1, the following corollary is immediate.

Corollary 2.4. If Condition (1.1) holds for a diffusion $\{X_t\}$, then for every twice continuously differentiable f vanishing outside a compact subset of S, the process $\{Z_t\}$ in (2.18) is a $\{\mathscr{F}_t\}$-martingale.

It will be shown in Section 3 of Chapter VII by direct probabilistic arguments that $\{Z_t\}$ is a $\{\mathscr{F}_t\}$-martingale for a much wider class of twice-differentiable functions f than prescribed by the corollary. As a simple example, take $\{X_t\}$ to be a Brownian motion with drift μ and diffusion coefficient σ^2. Then

$$\mathbf{A} = \tfrac{1}{2}\sigma^2 \frac{d^2}{dx^2} + \mu \frac{d}{dx}.$$

If one takes $f(x) = x$ in (2.18), then $Z_t = X_t - t\mu$ $(t \geqslant 0)$, which is a martingale, even though f is unbounded. In the case $\mu = 0$, one may take $f(x) = x^2$ to get $Z_t = X_t^2 - t\sigma^2$ $(t \geqslant 0)$, which is also a martingale. Still, one can do a lot under the assumption of Corollary 2.4. One application is the following.

Proposition 2.5. Let $\{X_t\}$ be a diffusion on $S = (a, b)$, whose coefficients satisfy Condition (1.1). Let $[c, d] \subset (a, b)$. Then

$$P_x(\{X_t\} \text{ reaches } c \text{ before } d) = \frac{\displaystyle\int_x^d \exp\left\{-\int_c^z \frac{2\mu(y)}{\sigma^2(y)}\,dy\right\} dz}{\displaystyle\int_c^d \exp\left\{-\int_c^z \frac{2\mu(y)}{\sigma^2(y)}\,dy\right\} dz}, \qquad c \leqslant x \leqslant d. \tag{2.24}$$

Proof. Define a twice continuously differentiable function f that equals the right side of (2.24) for $c \leqslant x \leqslant d$, and vanishes outside $[c - \varepsilon, d + \varepsilon] \subset (a, b)$ for some $\varepsilon > 0$. This is always possible as f'' is continuous at c, d (Exercise 3). By Corollary 2.4,

$$Z_t := f(X_t) - \int_0^t (\mathbf{A}f)(X_s)\,ds \qquad (t \geqslant 0), \tag{2.25}$$

is a $\{\mathscr{F}_t\}$-martingale. Define the $\{\mathscr{F}_t\}$-stopping time (see Chapter I (13.68)),

$$\tau = \inf\{t \geqslant 0: X_t = c \text{ or } d\}. \tag{2.26}$$

Let $x \in (c, d)$. By the optional stopping result Proposition 13.9 of Chapter I

$$E_x Z_\tau = E_x Z_0. \tag{2.27}$$

But $Z_0 = f(X_0) \equiv f(x)$ under P_x, so that $E_x Z_0$ is the right side of (2.24). Now check that $\mathbf{A}f(x) = 0$ for $c < x < d$, so that $(\mathbf{A}f)(X_s) = 0$ for $0 \leqslant s < \tau$ if $x \in (c, d)$. Therefore,

$$Z_\tau = f(X_\tau) = \begin{cases} f(c) = 1 & \text{on } \{X_\tau = c\}, \\ f(d) = 0 & \text{on } \{X_\tau = d\}, \end{cases} \tag{2.28}$$

so that $E_x Z_\tau$ is simply the left side of (2.24). ■

As in Section 9 of Chapter I and Section 2 of Chapter III, one may derive criteria for transience and recurrence based on Corollary 2.4. This will be pursued later in Sections 4, 9, and 14. The martingale method will be more fully explored in Chapter VII.

2.2 The Fokker–Planck Equation

Consider a diffusion $\{X_t\}$ on $S = (a, b)$ whose coefficients satisfy Condition (1.1). Let $p(t; x, y)$ be the transition probability density of $\{X_t\}$. Letting $f = \mathbf{1}_B$ in

(2.5) leads to

$$\int_B p(s+t; x, y)\,dy = \int_S \left(\int_B p(t; z, y)\,dy\right) p(s; x, z)\,dz$$
$$= \int_B \left(\int_S p(t; z, y) p(s; x, z)\,dz\right) dy, \tag{2.29}$$

for all Borel sets $B \subset S$. This implies the so-called *Chapman–Kolmogorov equation*

$$p(s+t; x, y) = \int_S p(t; z, y) p(s; x, z)\,dz = (\mathbf{T}_s f)(x), \tag{2.30}$$

where $f(z) := p(t; z, y)$. A somewhat informal derivation of the backward equation for p may be based on (2.13) and (2.30) as follows. Suppose one may apply the computation (2.13) to the function f in (2.30). Then

$$(\mathbf{A}f)(x) = \lim_{s \downarrow 0} \frac{(\mathbf{T}_s f)(x) - f(x)}{s}. \tag{2.31}$$

But the right side equals $\partial p(t; x, y)/\partial t$, and the left side is $\mathbf{A}p(t; x, y)$. Therefore,

$$\frac{\partial p(t; x, y)}{\partial t} = \mu(x) \frac{\partial p(t; x, y)}{\partial x} + \tfrac{1}{2}\sigma^2(x) \frac{\partial^2 p(t; x, y)}{\partial x^2}, \tag{2.32}$$

which is the desired *backward equation for p*.

In applications to physical sciences the equation of greater interest is *Kolmogorov's forward equation* or the *Fokker–Planck equation* governing the probability density function of X_t, when X_0 has an arbitrary initial distribution π. Suppose for simplicity that π has a density g. Then the density of X_t is given by

$$(\mathbf{T}_t^* g)(y) := \int_a^b g(x) p(t; x, y)\,dx. \tag{2.33}$$

The operator $\mathbf{T}_t^*$ transforms a probability density g into another probability density. More generally, it transforms any integrable g into an integrable function $\mathbf{T}_t^* g$. It is *adjoint* (*transpose*) to $\mathbf{T}_t$ in the sense that

$$\langle \mathbf{T}_t^* g, f \rangle := \int_S (\mathbf{T}_t^* g)(y) f(y)\,dy = \int_S g(x)(\mathbf{T}_t f)(x)\,dx = \langle g, \mathbf{T}_t f \rangle. \tag{2.34}$$

Here $\langle u, v \rangle = \int u(x)v(x)\,dx$. If f is twice continuously differentiable and vanishes outside a compact subset of S, then one may differentiate with respect to t in

(2.34) and interchange the orders of integration and differentiation to get, using (2.17),

$$\left\langle \frac{\partial}{\partial t}\mathbf{T}_t^* g, f\right\rangle = \left\langle g, \frac{\partial}{\partial t}\mathbf{T}_t f\right\rangle = \langle g, \mathbf{T}_t\mathbf{A}f\rangle$$

$$= \langle \mathbf{T}_t^* g, \mathbf{A}f\rangle \equiv \int_S (\mathbf{T}_t^* g)(y)(\mathbf{A}f)(y)\,dy. \tag{2.35}$$

Now, assuming that f, h are both twice continuously differentiable and that f vanishes outside a finite interval, integration by parts yields

$$\langle h, \mathbf{A}f\rangle = \int_S h(y)[\mu(y)f'(y) + \tfrac{1}{2}\sigma^2(y)f''(y)]\,dy$$

$$= \int_S \left[-\frac{d}{dy}(\mu(y)h(y)) + \frac{d^2}{dy^2}\left(\frac{1}{2}\sigma^2(y)h(y)\right)\right]f(y)\,dy = \langle \mathbf{A}^* h, f\rangle, \tag{2.36}$$

where $\mathbf{A}^*$ is the *formal adjoint* of $\mathbf{A}$ defined by

$$(\mathbf{A}^* h)(y) = -\frac{d}{dy}(\mu(y)h(y)) + \frac{d^2}{dy^2}(\tfrac{1}{2}\sigma^2(y)h(y)). \tag{2.37}$$

Applying (2.36) in (2.35), with $\mathbf{T}_t^* g$ in place of h, one gets

$$\left\langle \frac{\partial}{\partial t}\mathbf{T}_t^* g, f\right\rangle = \langle \mathbf{A}^*\mathbf{T}_t^* g, f\rangle. \tag{2.38}$$

Since (2.38) holds for sufficiently many functions f, all infinitely differentiable functions vanishing outside some closed, bounded interval contained in S for instance, we get (Exercise 1),

$$\frac{\partial}{\partial t}(\mathbf{T}_t^* g)(y) = \mathbf{A}^*(\mathbf{T}_t^* g)(y). \tag{2.39}$$

That is,

$$\int_S \frac{\partial p(t; x, y)}{\partial t} g(x)\,dx = \int_S \left[-\frac{\partial}{\partial y}(\mu(y)p(t; x, y)) + \frac{1}{2}\frac{\partial^2}{\partial y^2}(\sigma^2(y)p(t; x, y))\right]g(x)\,dx. \tag{2.40}$$

Since (2.40) holds for sufficiently many functions g, all twice continuously differentiable functions vanishing outside a closed bounded interval contained in S for instance, we get *Kolmogorov's forward equation for the transition probability density* p (Exercise 1),

$$\frac{\partial p(t; x, y)}{\partial t} = -\frac{\partial}{\partial y}(\mu(y)p(t; x, y)) + \frac{\partial^2}{\partial y^2}(\tfrac{1}{2}\sigma^2(y)p(t; x, y)), \qquad (t > 0). \tag{2.41}$$

For a physical interpretation of the forward equation (2.41), consider a dilute concentration of solute molecules diffusing along one direction in a possibly nonhomogeneous fluid. An individual molecule's position, say in the x-direction, is a Markov process with drift $\mu(x)$ and diffusion coefficient $\sigma^2(x)$. Given an initial concentration $c_0(x)$, the concentration $c(t, x)$ at x at time t is given by

$$c(t, x) = \int_S c_0(z)p(t; z, x)\, dz, \tag{2.42}$$

where $p(t; z, x)$ is the transition probability density of the position process of an individual solute molecule. Therefore, $c(t, x)$ satisfies Kolmogorov's forward equation

$$\frac{\partial c(t, x)}{\partial t} = \mathbf{A}^* c(t, x) = -\frac{\partial}{\partial x} J(t, x), \tag{2.43}$$

with J given by

$$J(t, x) = -\frac{\partial}{\partial x}[\tfrac{1}{2}\sigma^2(x)c(t, x)) + \mu(x)c(t, x)]. \tag{2.44}$$

The Kolmogorov forward equation is also referred to as the Fokker–Planck equation in this context. Now the increase in the amount of solute in a small region $[x, x + \Delta x]$ during a small time interval $[t, t + \Delta t]$ is approximately

$$\frac{\partial c(t, x)}{\partial t} \Delta t\, \Delta x. \tag{2.45}$$

On the other hand, if $v(t, x)$ denotes the velocity of the solute at x at time t, moving as a continuum, then a fluid column of length approximately $v(t, x)\,\Delta t$ flows into the region at x during $[t, t + \Delta t]$. Hence the amount of solute that flows into the region at x during $[t, t + \Delta t]$ is approximately $v(t, x)c(t, x)\,\Delta t$, while the amount passing out at $x + \Delta x$ during $[t, t + \Delta t]$ is approximately $v(t, x + \Delta x)c(t, x + \Delta x)\,\Delta t$. Therefore, the increase in the amount of solute in $[x, x + \Delta x]$ during $[t, t + \Delta t]$ is approximately

$$[v(t, x)c(t, x) - v(t, x + \Delta x)c(t, x + \Delta x)]\,\Delta t. \tag{2.46}$$

Equating (2.45) and (2.46) and dividing by $\Delta t\,\Delta x$, one gets

$$\frac{\partial c(t,x)}{\partial t} = -\frac{\partial}{\partial x}(v(t,x)c(t,x)). \tag{2.47}$$

Equation (2.47) is generally referred to as the *equation of continuity* or the *equation of mass conservation.* The quantity $v(t,x)c(t,x)$ is called the *flux of the solute*, which is seen to be the rate per unit time at which the solute passes out at x (at time t) in the positive x-direction. In the present case, therefore, the flux is given by (2.44).

3 TRANSFORMATION OF THE GENERATOR UNDER RELABELING OF THE STATE SPACE

In many applications the state space of the diffusion of interest is a finite or a semi-infinite open interval. For example, in an economic model the quantity of interest, say price, demand, supply, etc., is typically nonnegative; in a genetic model the gene frequency is a proportion in the interval $(0,1)$. It is often the case that the end points or boundaries of the state space in such models cannot be reached from the interior. In other words, owing to some built-in mechanism the boundaries are *inaccessible.* A simple way to understand these processes is to think of them as strictly monotone functions of diffusions on $S = R^1$. Let

$$\mathbf{A} = \tfrac{1}{2}\sigma^2(x)\frac{d^2}{dx^2} + \mu(x)\frac{d}{dx} \tag{3.1}$$

be the generator of a diffusion on $S = (-\infty, \infty)$. That is to say, we consider a diffusion with coefficients $\mu(x)$, $\sigma^2(x)$. Let ϕ be a continuous one-to-one map of S onto $\tilde{S} = (a, b)$, where $-\infty \leqslant a < b \leqslant \infty$. It is simple to check that if $\{X_t\}$ is a diffusion on S, $\{Z_t\} := \{\phi(X_t)\}$ is a Markov process on $\tilde{S}$ having continuous sample paths (Exercise 1.3). If ϕ is smooth, then $\{Z_t\}$ is a diffusion whose drift and diffusion coefficients are given by Proposition 3.1 below.

First we need a lemma.

Lemma. If $\{X_t\}$ is a diffusion satisfying (1.2)$'$ for all $\varepsilon > 0$, then for every $r > 2$ one has

$$E(|X_{s+t} - X_s|^r \mathbf{1}_{\{|X_{s+t}-X_s| \leqslant \varepsilon'\}} \mid X_s = x) = o(t) \qquad \text{as } t \downarrow 0, \tag{3.2}$$

for all $\varepsilon' > 0$.

Proof. Let $r > 2$, $\varepsilon' > 0$ be given. Fix $\theta > 0$. We will show that the left side of (3.2) is less than θt for all sufficiently small t. For this, write

$\delta = (\theta/(2\sigma^2(x)))^{1/(r-2)}$. Then the left side of (3.2) is no more than

$$\begin{aligned}E(|X_{s+t} - X_s|^r \mathbf{1}_{\{|X_{s+t}-X_s|\leqslant\delta\}} \mid X_s = x) \\ + E(|X_{s+t} - X_s|^r \mathbf{1}_{\{|X_{s+t}-X_s|\leqslant\varepsilon', |X_{s+t}-X_s|>\delta\}} \mid X_s = x) \\ \leqslant \delta^{r-2} E(|X_{s+t} - X_s|^2 \mathbf{1}_{\{|X_{s+t}-X_s|\leqslant\delta\}} \mid X_s = x) + \varepsilon'^r P(|X_{s+t} - X_s| > \delta \mid X_s = x) \\ \leqslant \frac{\theta}{2\sigma^2(x)}(\sigma^2(x)t + o(t)) + (\varepsilon')^r o(t) = \frac{\theta}{2}t + o(t).\end{aligned} \tag{3.3}$$

The last inequality uses the fact that (1.2)′ holds for all $\varepsilon > 0$. Now the term $o(t)$ is smaller than $\theta t/2$ for all sufficiently small t. ∎

Proposition 3.1. Let $\{X_t\}$ be a diffusion on $S = (c, d)$ having drift and diffusion coefficients $\mu(x), \sigma^2(x)$, respectively. If ϕ is a three times continuously differentiable function on (c, d) onto (a, b) such that ϕ' is either strictly positive or strictly negative, then $\{Z_t := \phi(X_t)\}$ is a diffusion on (a, b) whose drift $\tilde{\mu}(\cdot)$ and diffusion coefficient $\tilde{\sigma}^2(\cdot)$ are given by

$$\begin{aligned}\tilde{\mu}(z) &= \phi'(\phi^{-1}(z))\mu(\phi^{-1}(z)) + \tfrac{1}{2}\phi''(\phi^{-1}(z))\sigma^2(\phi^{-1}(z)), \\ \tilde{\sigma}^2(z) &= (\phi'(\phi^{-1}(z)))^2\sigma^2(\phi^{-1}(z)),\end{aligned} \qquad z \in (a, b). \tag{3.4}$$

Proof. By a Taylor expansion,

$$\begin{aligned}Z_{s+t} - Z_s &= \phi(X_{s+t}) - \phi(X_s) \\ &= (X_{s+t} - X_s)\phi'(X_s) + \frac{1}{2!}(X_{s+t} - X_s)^2\phi''(X_s) + \frac{1}{3!}(X_{s+t} - X_s)^3\phi'''(\xi),\end{aligned} \tag{3.5}$$

where ξ lies between X_s and X_{s+t}. Fix $\varepsilon > 0$, $z \in (a, b)$. There exist positive constants $\delta_1(\varepsilon), \delta_2(\varepsilon)$ such that ϕ^{-1} maps the interval $[z - \varepsilon, z + \varepsilon]$ onto $[\phi^{-1}(z) - \delta_1(\varepsilon), \phi^{-1}(z) + \delta_2(\varepsilon)]$. Write $x = \phi^{-1}(z)$, and let $\delta_m = \min\{\delta_1(\varepsilon), \delta_2(\varepsilon)\}$, $\delta_M = \max\{\delta_1(\varepsilon), \delta_2(\varepsilon)\}$. Then

$$\begin{aligned}&\mathbf{1}_{\{|Z_{s+t}-z|\leqslant\varepsilon\}} = \mathbf{1}_{\{x-\delta_1(\varepsilon)\leqslant X_{s+t}\leqslant x+\delta_2(\varepsilon)\}}, \\ &\mathbf{1}_{\{|X_{s+t}-x|\leqslant\delta_m\}} \leqslant \mathbf{1}_{\{|Z_{s+t}-z|\leqslant\varepsilon\}} \leqslant \mathbf{1}_{\{|X_{s+t}-x|\leqslant\delta_M\}}.\end{aligned} \tag{3.6}$$

Therefore,

$$\begin{aligned}E((X_{s+t} - X_s)\mathbf{1}_{\{|Z_{s+t}-z|\leqslant\varepsilon\}} \mid Z_s = z) &= E((X_{s+t} - x)\mathbf{1}_{\{|Z_{s+t}-z|\leqslant\varepsilon\}} \mid X_s = x) \\ &= E((X_{s+t} - x)\mathbf{1}_{\{|X_{s+t}-x|\leqslant\delta_m\}} \mid X_s = x) \\ &\quad + E((X_{s+t} - x)\mathbf{1}_{\{|X_{s+t}-x|>\delta_m, |Z_{s+t}-z|\leqslant\varepsilon\}} \mid X_s = x).\end{aligned} \tag{3.7}$$

In view of the last inequality in (3.6), the last expectation is bounded in magnitude by

$$\delta_M P(|X_{s+t} - x| > \delta_m \mid X_s = x),$$

which is of the order o(t) as $t \downarrow 0$, by the last relation in (1.2)'. Also, by the first relation in (1.2)',

$$E((X_{s+t} - X_s)\mathbf{1}_{\{|X_{s+t}-x| \leqslant \delta_m\}} \mid X_s = x) = \mu(x)t + o(t).$$

Therefore,

$$E((X_{s+t} - X_s)\mathbf{1}_{\{|Z_{s+t}-z| \leqslant \varepsilon\}} \mid Z_s = z) = \mu(x)t + o(t). \tag{3.8}$$

In the same manner, one has

$$\begin{aligned} E((X_{s+t} - X_s)^2\mathbf{1}_{\{|Z_{s+t}-Z_s| \leqslant \varepsilon\}} \mid Z_s = z) &= E((X_{s+t} - X_s)^2\mathbf{1}_{\{|X_{s+t}-X_s| \leqslant \delta_m\}} \mid X_s = x) \\ &\quad + O(\delta_M^2 P(|X_{s+t} - x| > \delta_m \mid X_s = x) \\ &= \sigma^2(x)t + o(t). \end{aligned} \tag{3.9}$$

Also, by the Lemma above,

$$\begin{aligned} &E(|X_{s+t} - X_s|^3|\phi'''(\xi)|\mathbf{1}_{\{|Z_{s+t}-Z_s| \leqslant \varepsilon\}} \mid Z_s = z) \\ &\qquad \leqslant cE(|X_{s+t} - X_s|^3\mathbf{1}_{\{|X_{s+t}-X_s| \leqslant \delta_M\}} \mid X_s = x) = o(t). \end{aligned} \tag{3.10}$$

Using (3.8)–(3.10), and (3.5), one gets

$$E((Z_{s+t} - Z_s)\mathbf{1}_{\{|Z_{s+t}-Z_s| \leqslant \varepsilon\}} \mid Z_s = z) = \phi'(x)\mu(x)t + \tfrac{1}{2}\sigma^2(x)\phi''(x)t + o(t), \tag{3.11}$$

so that the drift coefficient of $\{Z_t\}$ is

$$\phi'(\phi^{-1}(z))\mu(\phi^{-1}(z)) + \tfrac{1}{2}\sigma^2(\phi^{-1}(z))\phi''(\phi^{-1}(z)).$$

In order to compute the diffusion coefficient of $\{Z_t\}$, square both sides of (3.5) to get

$$\begin{aligned} &E((Z_{s+t} - Z_s)^2\mathbf{1}_{\{|Z_{s+t}-Z_s| \leqslant \varepsilon\}} \mid Z_s = z) \\ &\qquad = (\phi'(x))^2 E((X_{s+t} - X_s)^2\mathbf{1}_{\{|Z_{s+t}-Z_s| \leqslant \varepsilon\}} \mid X_s = x) + R_t, \end{aligned} \tag{3.12}$$

where each summand in R_t is bounded by a term of the form

$$\begin{aligned} &cE(|X_{s+t} - X_s|^r\mathbf{1}_{\{|Z_{s+t}-Z_s| \leqslant \varepsilon\}} \mid X_s = x) \\ &\qquad \leqslant cE(|X_{s+t} - X_s|^r\mathbf{1}_{\{|X_{s+t}-X_s| \leqslant \delta_M\}} \mid X_s = x), \end{aligned}$$

where $r > 2$. Hence by the Lemma above $R_t = o(t)$ and we get, using (3.9) in (3.12),

$$E((Z_{s+t} - Z_s)^2 \mathbf{1}_{\{|Z_{s+t} - Z_s| \leqslant \varepsilon\}} \mid Z_s = z) = (\phi'(x))^2 \sigma^2(x) t + o(t). \quad (3.13)$$

Finally,

$$P(|Z_{s+t} - Z_s| > \varepsilon \mid Z_s = z) \leqslant P(|X_{s+t} - X_s| > \delta_m \mid X_s = x) = o(t), \quad (3.14)$$

by the last condition in (1.2)′. ■

Example 1. (*Geometric Brownian Motion*). Let $S = (-\infty, \infty)$, $\tilde{S} = (0, \infty)$, $\phi(x) = e^x$. If $\mathbf{A}$ is given by (3.1), then the coefficients of $\tilde{\mathbf{A}}$ are

$$\begin{aligned} \tilde{\mu}(z) &= z\mu(\log z) + \tfrac{1}{2} z\sigma^2(\log z), \\ \tilde{\sigma}^2(z) &= z^2 \sigma^2(\log z), \end{aligned} \qquad (0 < z < \infty). \quad (3.15)$$

In particular, a Brownian motion with drift μ and diffusion coefficient σ^2 becomes, under the transformation $x \to e^x$, a diffusion with state space $\tilde{S} = (0, \infty)$ and generator

$$\tilde{A} = \tfrac{1}{2}\sigma^2 z^2 \frac{d^2}{dz^2} + \left(\mu + \frac{\sigma^2}{2}\right) z \frac{d}{dz}. \quad (3.16)$$

In other words, the mean rate of growth as well as the mean rate of fluctuation in growth at z is proportional to the *size* z. Note that one has

$$Z_t = e^{X_t}. \quad (3.17)$$

But $\{X_t\}$ may be represented as $X_t = X_0 + t\mu + \sigma B_t$, $t \geqslant 0$, where $\{B_t\}$ is a standard one-dimensional Brownian motion starting at zero and independent of X_0. Then (3.17) becomes

$$Z_t = Z_0 e^{t\mu + \sigma B_t}, \qquad Z_0 = e^{X_0}. \quad (3.18)$$

The process $\{Z_t\}$ is sometimes referred to as the *geometric Brownian motion.*

Example 2. (*Geometric Ornstein–Uhlenbeck Process*). The Ornstein–Uhlenbeck process with generator (see Example 1.2)

$$\mathbf{A} = \tfrac{1}{2}\sigma^2 \frac{d^2}{dx^2} - \gamma x \frac{d}{dx}, \qquad (-\infty < x < \infty), \quad (3.19)$$

is transformed by the transformation $x \to e^x$ into a diffusion on $(0, \infty)$ with generator

$$\tilde{A} = \tfrac{1}{2}\sigma^2 z^2 \frac{d^2}{dz^2} - \left(\gamma z \log z - \frac{z}{2}\sigma^2\right)\frac{d}{dz}, \qquad (0 < z < \infty). \tag{3.20}$$

Example 3. $S = (-\infty, \infty)$, $\tilde{S} = (0, 1)$, $\phi(x) = e^x/(e^x + 1)$. For this case, first transform by $x \to e^x$, then by $y \to y/(1 + y)$. Thus, one needs to apply the transformation $\gamma(y)$: $y \to y/(1 + y)$ to the operator with coefficients (3.15). Since the inverse of γ is $\gamma^{-1}(z) = z/(1 - z)$, one may use (3.15) to obtain the transformed operator (Exercise 1)

$$\begin{aligned}\tilde{\mathbf{A}} &= \tfrac{1}{2}z^2(1-z)^2\sigma^2\left(\log\frac{z}{1-z}\right)\frac{d^2}{dz^2}\\ &\quad + \left[z(1-z)\mu\left(\log\frac{z}{1-z}\right) + \frac{z(1-z)}{2}\sigma^2\left(\log\frac{z}{1-z}\right)\right.\\ &\qquad\qquad \left. - z^2(1-z)\sigma^2\left(\log\frac{z}{1-z}\right)\right]\frac{d}{dz}\\ &= \tfrac{1}{2}z(1-z)\left[z(1-z)\sigma^2\left(\log\frac{z}{1-z}\right)\frac{d^2}{dz^2}\right.\\ &\quad \left. + \left\{2\mu\left(\log\frac{z}{1-z}\right) + (1-2z)\sigma^2\left(\log\frac{z}{1-z}\right)\right\}\frac{d}{dz}\right], \qquad 0 < z < 1. \end{aligned} \tag{3.21}$$

In particular, a Brownian motion is transformed into a diffusion on $(0, 1)$ with generator

$$\tilde{\mathbf{A}} = \tfrac{1}{2}z(1-z)\left[z(1-z)\sigma^2\frac{d^2}{dz^2} + (2\mu + (1-2z)\sigma^2)\frac{d}{dz}\right], \qquad 0 < z < 1, \tag{3.22}$$

and an Ornstein–Uhlenbeck process is transformed into a diffusion on $(0, 1)$ with generator

$$\tilde{\mathbf{A}} = \tfrac{1}{2}z(1-z)\left[z(1-z)\sigma^2\frac{d^2}{dz^2} - \left\{2\gamma\log\frac{z}{1-z} - (1-2z)\sigma^2\right\}\frac{d}{dz}\right], \qquad 0 < z < 1. \tag{3.23}$$

Example 4. Let $S = (-\infty, \infty)$, $\tilde{S} = (0, \infty)$, $\phi(x) = e^{x^3}$. If

$$\mathbf{A} = \tfrac{1}{2}\sigma^2\frac{d^2}{dx^2}$$

then

$$\tilde{\mathbf{A}} = \tfrac{1}{2}\sigma^2(9z^2(\log z)^{4/3})\frac{d^2}{dz^2} + \tfrac{1}{2}\sigma^2 z\{6(\log z)^{1/3} + 9(\log z)^{4/3}\}\frac{d}{dz}. \quad (3.24)$$

In this example the transformed process

$$Z_t = e^{X_t^3}, \qquad t \geqslant 0 \quad (3.25)$$

(where $\{X_t\}$ is a Brownian motion) *does not have a finite first moment* (Exercise 2).

4 DIFFUSIONS AS LIMITS OF BIRTH–DEATH CHAINS

We now show how one may arrive at diffusions as limits of discrete-parameter birth–death chains with decreasing step size and increasing frequencies. Because the underlying Markov chains do not in general have independent increments, the limiting diffusions will not have this property either. Indeed, *the Brownian motions are the only Markov processes with continuous sample paths that have independent increments* (see Theorem T.1.1 of Chapter IV).

Suppose we are given two real-valued functions $\mu(x)$, $\sigma^2(x)$ on $\mathbb{R}^1 = (-\infty, \infty)$ that satisfy Condition (1.1). Throughout this section, also assume that $\mu(x)$, $\sigma^2(x)$ are bounded and write

$$\sigma_0^2 = \sup_x \sigma^2(x). \quad (4.1)$$

Consider a discrete-parameter birth–death chain on $S = \{0, \pm\Delta, \pm 2\Delta, \ldots\}$ with step size $\Delta > 0$, having transition probabilities p_{ij} of going from $i\Delta$ to $j\Delta$ in one step given by,

$$p_{i,i-1} = \delta_i^{(\Delta)} := \frac{\sigma^2(i\Delta)\varepsilon}{2\Delta^2} - \frac{\mu(i\Delta)\varepsilon}{2\Delta}, \qquad p_{i,i+1} = \beta_i^{(\Delta)} := \frac{\sigma^2(i\Delta)\varepsilon}{2\Delta^2} + \frac{\mu(i\Delta)\varepsilon}{2\Delta},$$
$$p_{ii} = 1 - \frac{\sigma^2(i\Delta)\varepsilon}{\Delta^2} = 1 - \beta_i^{(\Delta)} - \delta_i^{(\Delta)}, \quad (4.2)$$

with the parameter ε given by

$$\varepsilon = \frac{\Delta^2}{\sigma_0^2}. \quad (4.3)$$

Note that under Condition (1.1) and boundedness of $\mu(x)$, $\sigma^2(x)$, the quantities $\beta_i^{(\Delta)}$, $\delta_i^{(\Delta)}$, and $1 - \beta_i^{(\Delta)} - \delta_i^{(\Delta)}$ are nonnegative for sufficiently small Δ. We shall let ε be the actual time in between two successive transitions. Note that, given that the process is at $x = i\Delta$, the mean displacement in a single step in time ε is

$$\Delta\,\beta_i^{(\Delta)} - \Delta\,\delta_i^{(\Delta)} = \mu(i\Delta)\varepsilon = \mu(x)\varepsilon. \tag{4.4}$$

Hence, the instantaneous rate of mean displacement per unit time, when the process is at x, is $\mu(x)$. Also, the mean squared displacement in a single step is

$$\Delta^2\,\beta_i^{(\Delta)} + (-\Delta)^2\,\delta_i^{(\Delta)} = \sigma^2(i\Delta)\varepsilon = \sigma^2(x)\varepsilon. \tag{4.5}$$

Therefore, $\sigma^2(x)$ is the instantaneous rate of mean squared displacement per unit time.

Conversely, in order that (4.4), (4.5) may hold one must have the birth–death parameters (4.2) (Exercise 1). The choice (4.3) of ε guarantees the nonnegativity of the transition probabilities $p_{ii}, p_{i,i-1}, p_{i,i+1}$.

Just as the simple random walk approximates Brownian motion under proper scaling of states and time, the above birth–death chain approximates the diffusion with coefficients $\mu(x)$ and $\sigma^2(x)$. Indeed, the approximation of Brownian motion by simple random walk described in Section 8 of Chapter I follows as a special case of the following result, which we state without proof (see theoretical complement 1). Below $[r]$ is the integer part of r.

Theorem 4.1. Let $\{Y_n^{(\Delta)}: n = 0, 1, 2, \ldots\}$ be a discrete-parameter birth–death chain on $S = \{0, \pm\Delta, \pm 2\Delta, \ldots\}$ with one-step transition probabilities (4.2) and with $Y_0^{(\Delta)} = [x_0/\Delta]\Delta$ where x_0 is a fixed number. Define the stochastic process

$$\{X_t^{(\Delta)}\} := \{Y_{[t/\varepsilon]}^{(\Delta)}\} \qquad (t \geqslant 0). \tag{4.6}$$

Then, as $\Delta \downarrow 0$, $\{X_t^{(\Delta)}\}$ converges in distribution to a diffusion $\{X_t\}$ with drift $\mu(x)$ and diffusion coefficient $\sigma^2(x)$, starting at x_0.

As a consequence of Theorem 4.1 it follows that for any bounded continuous function f we have

$$\lim_{\Delta\to\infty} E\left\{f(X_t^{(\Delta)}) \mid X_0^{(\Delta)} = \left[\frac{x}{\Delta}\right]\Delta\right\} = E_x f(X_t). \tag{4.7}$$

By (2.15), in the case of a smooth initial function f, the function $u(t, x) := E_x f(X_t)$ solves the initial value problem

$$\frac{\partial u}{\partial t} = \tfrac{1}{2}\sigma^2(x)\frac{\partial^2 u}{\partial x^2} + \mu(x)\frac{\partial u}{\partial x}, \qquad \lim_{t\downarrow 0} u(t, x) = f(x). \tag{4.8}$$

The expectation on the left side of (4.7) may be expressed as

$$u^{(\Delta)}(n, i) := \sum_j f(j\Delta)p_{ij}^{(n)} \tag{4.9}$$

where $n = [t/\varepsilon]$, $i = [x/\Delta]$, and $p_{ij}^{(n)}$ are the n-step transition probabilities. It is illuminating to check that the function $(n, i) \to u^{(\Delta)}(n, i)$ satisfies a *difference equation* that is a discretized version of the differential equation (4.8). To see this, note that

$$\begin{aligned} p_{ij}^{(n+1)} &= p_{ii}\,p_{ij}^{(n)} + p_{i,i+1}\,p_{i+1,j}^{(n)} + p_{i,i-1}\,p_{i-1,j}^{(n)} \\ &= (1 - \beta_i^{(\Delta)} - \delta_i^{(\Delta)})p_{ij}^{(n)} + \beta_i^{(\Delta)}p_{i+1,j}^{(n)} + \delta_i^{(\Delta)}p_{i-1,j}^{(n)} \\ &= p_{ij}^{(n)} + \beta_i^{(\Delta)}(p_{i+1,j}^{(n)} - p_{ij}^{(n)}) - \delta_i^{(\Delta)}(p_{ij}^{(n)} - p_{i-1,j}^{(n)}), \end{aligned} \tag{4.10}$$

or,

$$\begin{aligned} \frac{p_{ij}^{(n+1)} - p_{ij}^{(n)}}{\varepsilon} &= \frac{\mu(i\Delta)}{2\Delta}\{(p_{i+1,j}^{(n)} - p_{ij}^{(n)}) + (p_{ij}^{(n)} - p_{i-1,j}^{(n)})\} \\ &\quad + \frac{1}{2}\frac{\sigma^2(i\Delta)}{\Delta^2}(p_{i+1,j}^{(n)} - 2p_{ij}^{(n)} + p_{i-1,j}^{(n)}). \end{aligned} \tag{4.11}$$

Summing over j one gets a corresponding equation for $u^{(\Delta)}(n, i)$.

As $\Delta \downarrow 0$ the state space $S = \{0, \pm\Delta, \pm 2\Delta, \ldots\}$ approximates $\mathbb{R}^1 = (-\infty, \infty)$, provided we think of the state $j\Delta$ as representing an interval of width Δ around $j\Delta$. Accordingly, think of spreading the probability $p_{ij}^{(n)}$ over this interval. Thus, one introduces the *approximate density* $y \to p^{(\Delta)}(t; x, y)$ at time $t = n\varepsilon$ for states $x = i\Delta$, $y = j\Delta$, by

$$p^{(\Delta)}(n\varepsilon; i\Delta, j\Delta) := \frac{p_{ij}^{(n)}}{\Delta}. \tag{4.12}$$

By dividing both sides of (4.11) by Δ, one then arrives at the difference equation

$$\begin{aligned} &(p^{(\Delta)}((n+1)\varepsilon; i\Delta, j\Delta) - p^{(\Delta)}(n\varepsilon; i\Delta, j\Delta))/\varepsilon \\ &\quad = \mu(i\Delta)(p^{(\Delta)}(n\varepsilon; (i+1)\Delta, j\Delta) - p^{(\Delta)}(n\varepsilon; (i-1)\Delta, j\Delta))/2\Delta \\ &\qquad + \tfrac{1}{2}\sigma^2(i\Delta)(p^{(\Delta)}(n\varepsilon; (i+1)\Delta, j\Delta) - 2p^{(\Delta)}(n\varepsilon; i\Delta, j\Delta) \\ &\qquad + p^{(\Delta)}(n\varepsilon; (i-1)\Delta, j\Delta))/\Delta^2. \end{aligned} \tag{4.13}$$

This is a difference-equation version of the partial differential equation

$$\frac{\partial p(t; x, y)}{\partial t} = \mu(x)\frac{\partial p(t; x, y)}{\partial x} + \tfrac{1}{2}\sigma^2(x)\frac{\partial^2 p(t; x, y)}{\partial x^2}, \quad \text{for } t > 0,\ -\infty < x, y < \infty, \tag{4.14}$$

at *grid points* $(t, x, y) = (n\varepsilon, i\Delta, j\Delta)$.

Thus, the transition probability density $p(t; x, y)$ of the diffusion $\{X_t\}$ may be approximately computed by computing $p_{ij}^{(n)}$. The latter computation only involves raising a matrix to the nth power, a fairly standard computational task. In addition, Theorem 4.1 may be used to derive the probability $\psi(x)$ that $\{X_t\}$ reaches c before d, starting at $x \in (c, d)$, by taking the limit, as $\Delta \downarrow 0$, of the corresponding probability for the approximating birth–death chain $\{Y_n^{(\Delta)}\}$. In other words (see Section 2 of Chapter III, relation (2.10)),

$$\psi(x) = \lim_{\Delta \downarrow 0} \frac{\displaystyle\sum_{r=i_0}^{j-1} \frac{\delta_r^{(\Delta)}\delta_{r-1}^{(\Delta)} \cdots \delta_{i+1}^{(\Delta)}}{\beta_r^{(\Delta)}\beta_{r-1}^{(\Delta)} \cdots \beta_{i+1}^{(\Delta)}}}{1 + \displaystyle\sum_{r=i+1}^{j-1} \frac{\delta_r^{(\Delta)} \cdots \delta_{i+1}^{(\Delta)}}{\beta_r^{(\Delta)} \cdots \beta_{i+1}^{(\Delta)}}}, \tag{4.15}$$

where

$$i_0 = \left[\frac{x}{\Delta}\right], \qquad i = \left[\frac{c}{\Delta}\right], \qquad j = \left[\frac{d}{\Delta}\right]. \tag{4.16}$$

The limit in (4.15) is (Exercise 2)

$$\begin{aligned} \psi(x) &= \lim_{\Delta \downarrow 0} \frac{\displaystyle\sum_{r=i_0}^{j-1} \exp\left\{-\int_c^{r\Delta} (2\mu(y)/\sigma^2(y))\,dy\right\}}{1 + \displaystyle\sum_{r=i+1}^{j-1} \exp\left\{-\int_c^{r\Delta} (2\mu(y)/\sigma^2(y))\,dy\right\}} \\ &= \frac{\displaystyle\int_x^d \exp\left\{-\int_c^z (2\mu(y)/\sigma^2(y))\,dy\right\} dz}{\displaystyle\int_c^d \exp\left\{-\int_c^z (2\mu(y)/\sigma^2(y))\,dy\right\} dz}, \end{aligned} \tag{4.17}$$

which confirms the computation (2.24) given in Section 2. This leads to necessary and sufficient conditions for transience and recurrence of diffusions (Exercise 3). This is analogous to the derivation of the corresponding probabilities for Brownian motion given in Section 9 of Chapter I.

Alternative derivations of (4.17) are given in Section 9 (see Eq. 9.23 and Exercise 9.2), in addition to Section 2 (Eq. 2.24).

5 TRANSITION PROBABILITIES FROM THE KOLMOGOROV EQUATIONS: EXAMPLES

Under Condition (1.1) the Kolmogorov equations uniquely determine the transition probabilities. However, solutions are generally not obtainable in closed form, although numerical methods based on the scheme described in

Section 4 sometimes provide practical approximations. Two examples for which the solutions can be obtained explicitly from the Kolmogorov equations alone are given here.

Example 1. (*Brownian Motion*). Brownian motion is a diffusion with constant drift and diffusion coefficients. First assume that the drift is zero. Let the diffusion coefficient be σ^2. Then the forward, or Fokker–Planck, equation for $p(t; x, y)$ is

$$\frac{\partial p(t; x, y)}{\partial t} = \tfrac{1}{2}\sigma^2 \frac{\partial^2 p(t; x, y)}{\partial y^2} \qquad (t > 0, -\infty < x < \infty, -\infty < y < \infty). \tag{5.1}$$

Let the Fourier transform of $p(t; x, y)$ as a function of y be denoted by

$$\hat{p}(t; x, \xi) = \int_{-\infty}^{\infty} e^{i\xi y} p(t; x, y)\, dy. \tag{5.2}$$

Then (5.1) becomes

$$\frac{\partial \hat{p}}{\partial t} = -\frac{\sigma^2}{2}\xi^2 \hat{p}, \tag{5.3}$$

or,

$$\frac{\partial}{\partial t}\left(\hat{p}e^{t\frac{\sigma^2}{2}\xi^2}\right) = 0, \tag{5.4}$$

whose general solution is

$$\hat{p}(t; x, \xi) = c(x, \xi)\exp\{-\tfrac{1}{2}\sigma^2\xi^2 t\}. \tag{5.5}$$

Now $\hat{p}$ is the characteristic function of the distribution of X_t given $X_0 = x$, and as $t \downarrow 0$ this distribution converges to the distribution of X_0 that is degenerate at x. Therefore,

$$c(x, \xi) = \lim_{t \downarrow 0} \hat{p}(t; x, \xi) = E(e^{i\xi X_0}) = e^{i\xi x}, \tag{5.6}$$

and we obtain

$$\hat{p}(t; x, \xi) = \exp\{i\xi x - \tfrac{1}{2}\sigma^2\xi^2 t\}. \tag{5.7}$$

But the right side is the characteristic function of the normal distribution with mean x and variance $t\sigma^2$. Therefore,

$$p(t; x, y) = \frac{1}{(2\pi\sigma^2 t)^{1/2}} \exp\left\{-\frac{(y-x)^2}{2t\sigma^2}\right\} \qquad (t > 0, -\infty < x, y < \infty). \tag{5.8}$$

The transition probability density of a Brownian motion with nonzero drift may be obtained in the same manner as above (Exercise 1).

Example 2. (*The Ornstein–Uhlenbeck Process*). $S = (-\infty, \infty)$, $\mu(x) = -\gamma x$, $\sigma^2(x) := \sigma^2 > 0$. Here γ is a (positive) constant. Fix an initial state x. As a function of t and y, $p(t; x, y)$ satisfies the forward equation

$$\frac{\partial p}{\partial t} = \frac{\sigma^2}{2}\frac{\partial^2 p}{\partial y^2} + \gamma \frac{\partial}{\partial y}(yp(t; x, y)). \tag{5.9}$$

Let $\hat{p}$ be the Fourier transform of p as a function of y,

$$\hat{p}(t; \xi) = \int_{-\infty}^{\infty} e^{i\xi y} p(t; x, y)\, dy. \tag{5.10}$$

Then, upon integration by parts,

$$\begin{aligned}
\left(\frac{\partial p}{\partial y}\right)^{\wedge}(t; \xi) &= \int_{-\infty}^{\infty} e^{i\xi y} \frac{\partial p}{\partial y}(t; x, y)\, dy \\
&= -\int_{-\infty}^{\infty} i\xi e^{i\xi y} p(t; x, y)\, dy = -i\xi\hat{p}, \\
\left(\frac{\partial^2 p}{\partial y^2}\right)^{\wedge}(t; \xi) &= \int_{-\infty}^{\infty} e^{i\xi y} \frac{\partial^2 p}{\partial y^2}(t; x, y)\, dy = -i\xi \int_{-\infty}^{\infty} e^{i\xi y} \frac{\partial p}{\partial y}\, dy \\
&= (-i\xi)^2 \int_{-\infty}^{\infty} e^{i\xi y} p(t; x, y)\, dy = -\xi^2 \hat{p}, \\
\left(y\frac{\partial p}{\partial y}\right)^{\wedge}(t; \xi) &= \int_{-\infty}^{\infty} e^{i\xi y} y \frac{\partial p(t; x, y)}{\partial y} = \frac{1}{i}\frac{\partial}{\partial \xi}\int_{-\infty}^{\infty} e^{i\xi y}\frac{\partial p}{\partial y}\, dy \\
&= \frac{1}{i}\frac{\partial}{\partial \xi}\left(\frac{\partial p}{\partial y}\right)^{\wedge} = \frac{1}{i}\frac{\partial}{\partial \xi}(-i\xi\hat{p}) = -\hat{p} - \xi\frac{\partial \hat{p}(t; \xi)}{\partial \xi}.
\end{aligned} \tag{5.11}$$

Here we have assumed that $y(\partial p/\partial y)$, $\partial^2 p/\partial y^2$ are integrable and go to zero as $|y| \to \infty$. Taking Fourier transforms on both sides of (5.9) one has

$$\frac{\partial \hat{p}(t; \xi)}{\partial t} = -\frac{\sigma^2}{2}\xi^2\hat{p} + \gamma\hat{p} + \gamma\left(-\hat{p} - \xi\frac{\partial \hat{p}}{\partial \xi}\right) = -\frac{\sigma^2}{2}\xi^2\hat{p} - \gamma\xi\frac{\partial \hat{p}}{\partial \xi}.$$

Therefore,

$$\frac{\partial \hat{p}(t;\xi)}{\partial t} + \gamma\xi\frac{\partial \hat{p}}{\partial \xi} = -\frac{\sigma^2}{2}\xi^2\hat{p}. \tag{5.12}$$

Thus, we have reduced the second-order partial differential equation (5.9) to the first-order equation (5.12). The left side of (5.12) is the directional derivative of $\hat{p}$ along the vector $(1, \gamma\xi)$ in the (t, ξ)-plane. Let $\alpha(t) = d\exp\{\gamma t\}$. Then (5.12) yields

$$\frac{d}{dt}\hat{p}(t;\alpha(t)) = -\frac{\sigma^2}{2}\alpha^2(t)\hat{p}(t;\alpha(t)) \tag{5.13}$$

or

$$\begin{aligned}\hat{p}(t;\alpha(t)) &= c(x,d)\exp\left\{-\frac{\sigma^2}{2}\int_0^t \alpha^2(s)\,ds\right\}\\ &= c(x,d)\exp\left\{-\frac{\sigma^2 d^2}{4\gamma}(e^{2\gamma t}-1)\right\}.\end{aligned}$$

For arbitrary t and ξ one may choose $d = \xi e^{-\gamma t}$, so that $\alpha(t) = \xi$, and get

$$\hat{p}(t;\xi) = c(x, \xi e^{-\gamma t})\exp\left\{-\frac{\sigma^2}{4\gamma}\xi^2(1-e^{-2\gamma t})\right\}. \tag{5.14}$$

As $t \downarrow 0$, one then has

$$\lim_{t\downarrow 0}\hat{p}(t;\xi) = c(x,\xi). \tag{5.15}$$

On the other hand, as $t \downarrow 0$, $\hat{p}(t;\xi)$ converges to the Fourier transform of δ_x, which is $\exp\{i\xi x\}$. Thus, $c(x,\xi) = \exp\{i\xi x\}$ for every real ξ, so that

$$\hat{p}(t;\xi) = \exp\left\{i\xi x e^{-\gamma t} - \frac{\sigma^2}{4\gamma}(1-e^{-2\gamma t})\xi^2\right\}, \tag{5.16}$$

which is the Fourier transform of a Gaussian density with mean $xe^{-\gamma t}$ and variance $(\sigma^2/2\gamma)(1-e^{-2\gamma t})$. Therefore,

$$p(t;x,y) = \frac{1}{(2\pi)^{1/2}}\left(\frac{\sigma^2}{2\gamma}(1-e^{-2\gamma t})\right)^{-1/2}\exp\left\{\frac{-\gamma(y-xe^{-\gamma t})^2}{\sigma^2(1-e^{-2\gamma t})}\right\}. \tag{5.17}$$

Note that the above derivation does not require that γ be a positive parameter. However, observe that if $\gamma = \beta/m > 0$ then letting $t \to \infty$ in (5.16) we obtain the characteristic function of the Gaussian distribution with mean 0 and variance $\sigma^2/2\gamma = m\sigma^2/2\beta$ (the *Maxwell–Boltzmann velocity distribution*). It follows that

$$\pi(v) = \frac{1}{(\pi\sigma^2\gamma^{-1})^{1/2}} \exp\{-\gamma v^2/\sigma^2\}, \qquad -\infty < v < \infty,$$

is the p.d.f. of the invariant initial distribution of the process. With π as the initial distribution, $\{V_t\}$ is a stationary Gaussian process. The stationarity may be viewed as an "equilibrium" status in which energy exchanges between the particle and the fluid by thermal agitation and viscous dissipation have reached a balance (on the average). Observe that the average kinetic energy is given by $E_\pi(\frac{1}{2}mV_t^2) = (m^2/4\beta)\sigma^2$.

6 DIFFUSIONS WITH REFLECTING BOUNDARIES

So far we have considered unrestricted diffusions on $S = (-\infty, \infty)$, or on open intervals. End points of the state space of a diffusion that cannot be reached from the interior are said to be *inaccessible*. In this section and the next we look at diffusions restricted to subintervals of S with one or more end points *accessible* from the interior. In order to continue the process after it reaches a boundary point, one must specify some boundary behavior, or *boundary condition*, consistent with the requirement that the process be Markovian. One such boundary condition, known as the *reflecting boundary condition* or the *Neumann boundary condition*, is discussed in this section. Subsection 6.2 may be read independently of Subsection 6.1.

6.1 Reflecting Diffusions as Limits of Birth–Death Chains

As an aid to intuition let us first see, in the spirit of Section 4, how a reflecting diffusion on $S = [0, \infty)$ may be viewed as a limit of reflecting birth–death chains. Let $\mu(x)$ and $\sigma^2(x)$ satisfy Condition (1.1) on $[0, \infty)$ and let $\mu(x)$ be bounded. For sufficiently small $\Delta > 0$ one may consider a discrete-parameter birth–death chain on $S = \{0, \Delta, 2\Delta, \dots\}$ with one-step transition probabilities

$$\begin{aligned}
p_{i,i+1} &= \beta_i^{(\Delta)} = \frac{\sigma^2(i\Delta)\varepsilon}{2\Delta^2} + \frac{\mu(i\Delta)\varepsilon}{2\Delta},\\
p_{i,i-1} &= \delta_i^{(\Delta)} = \frac{\sigma^2(i\Delta)\varepsilon}{2\Delta^2} - \frac{\mu(i\Delta)\varepsilon}{2\Delta}, \qquad (i \geqslant 1), \qquad (6.1)\\
p_{ii} &= 1 - (\beta_i^{(\Delta)} + \delta_i^{(\Delta)}) = 1 - \frac{\sigma^2(i\Delta)\varepsilon}{\Delta^2}
\end{aligned}$$

and

$$p_{01} = \beta_0^{(\Delta)} = \frac{\sigma^2(0)\varepsilon}{2\Delta^2} + \frac{\mu(0)\varepsilon}{2\Delta}, \qquad p_{00} = 1 - \beta_0^{(\Delta)} = 1 - \frac{\sigma^2(0)\varepsilon}{2\Delta^2} - \frac{\mu(0)\varepsilon}{2\Delta}. \tag{6.2}$$

Exactly as in Section 4, the backward difference equations (4.11) or (4.13) are obtained for $i \geqslant 1, j \geqslant 0$. These are the discretized difference equations for

$$\frac{\partial p(t; x, y)}{\partial t} = \mu(x)\frac{\partial p(t; x, y)}{\partial x} + \tfrac{1}{2}\sigma^2(x)\frac{\partial^2 p(t; x, y)}{\partial x^2}, \qquad t > 0, \quad x > 0, \quad y > 0. \tag{6.3}$$

The backward boundary condition for the diffusion is obtained in the limit as $\Delta \downarrow 0$ from the corresponding equations for the chain for $i = 0, j \geqslant 0$,

$$\begin{aligned} p_{0j}^{(n+1)} &= p_{01}\, p_{1j}^{(n)} + p_{00}\, p_{0j}^{(n)} \\ &= \left(\frac{\sigma^2(0)\varepsilon}{2\Delta^2} + \frac{\mu(0)\varepsilon}{2\Delta}\right)p_{1j}^{(n)} + \left(1 - \frac{\sigma^2(0)\varepsilon}{2\Delta^2} - \frac{\mu(0)\varepsilon}{2\Delta}\right)p_{0j}^{(n)} \qquad (j \geqslant 0), \end{aligned}$$

or,

$$\frac{p_{0j}^{(n+1)} - p_{0j}^{(n)}}{\varepsilon} = \frac{\sigma^2(0)}{2\Delta^2}\,(p_{1j}^{(n)} - p_{0j}^{(n)}) + \frac{\mu(0)}{2\Delta}\,(p_{1j}^{(n)} - p_{0j}^{(n)}) \qquad (j \geqslant 0). \tag{6.4}$$

In the notation of (4.12) this leads to,

$$\begin{aligned} \frac{p^{(\Delta)}((n+1)\varepsilon; 0, j\Delta) - p^{(\Delta)}(n\varepsilon; 0, j\Delta)}{\varepsilon}\,\Delta &= \frac{\sigma^2(0)}{2\Delta}\,(p^{(\Delta)}(n\varepsilon; \Delta, j\Delta) - p^{(\Delta)}(n\varepsilon; 0, j\Delta)) \\ &\quad + \frac{\mu(0)}{2}\,(p^{(\Delta)}(n\varepsilon; \Delta, j\Delta) - p^{(\Delta)}(n\varepsilon; 0, j\Delta))\Delta. \end{aligned} \tag{6.5}$$

Fix $y \geqslant 0, t > 0$. For $n = [t/\varepsilon], j = [y/\Delta]$, the left side of (6.5) is approximately

$$\Delta\left(\frac{\partial p(t; x, y)}{\partial t}\right)_{x=0},$$

while the right side is approximately

$$\tfrac{1}{2}\sigma^2(0)\left(\frac{\partial p(t; x, y)}{\partial x}\right)_{x=0} + \Delta\tfrac{1}{2}\mu(0)\left(\frac{\partial p(t; x, y)}{\partial x}\right)_{x=0}.$$

Letting $\Delta \downarrow 0$, one has

$$\left(\frac{\partial p(t; x, y)}{\partial x}\right)_{x=0} = 0, \qquad (t > 0, y \geqslant 0). \tag{6.6}$$

The equations (6.3) and (6.6), together with an initial condition $p(0; x, \cdot) = \delta_x$ determine a transition probability density of a Markov process on $[0, \infty)$ having continuous sample paths and satisfying the infinitesimal conditions (1.2) on the interior $(0, \infty)$. This Markov process is called the *diffusion on* $[0, \infty)$ *with reflecting boundary at* 0 *and drift and diffusion coefficients* $\mu(x), \sigma^2(x)$. The equations (6.3) and (6.6) are called the *Kolmogorov backward equation and backward boundary condition*, respectively, for this diffusion. This particular boundary condition (6.6) is also known as a *Neumann boundary condition*. The precise nature of the approximation of the reflecting diffusion by the corresponding reflecting birth–death chain is the same as described in Theorem 4.1.

6.2 Sample Path Construction of Reflecting Diffusions

Independently of the above heuristic considerations, we shall give in the rest of this section complete probabilistic descriptions of diffusions reflecting at one or two boundary points, with a treatment of the periodic boundary along the way. The general method described here is sometimes called the *method of images*.

ONE-POINT BOUNDARY CASE. (*Diffusion on* $S = [0, \infty)$ *with "0" as a Reflecting Boundary*). First we consider a special case for which the probabilistic description of "reflection" is simple. Let $\mu(x), \sigma^2(x)$ be defined on $S = [0, \infty)$ and satisfy Condition (1.1) on S. Assume also that

$$\mu(0) = 0. \tag{6.7}$$

Now extend the coefficients $\mu(\cdot), \sigma^2(\cdot)$ on $\mathbb{R}^1$ by setting

$$\mu(-x) = -\mu(x), \qquad \sigma^2(-x) = \sigma^2(x), \qquad (x > 0). \tag{6.8}$$

Although $\mu(\cdot), \sigma^2(\cdot)$ so obtained may no longer be twice-differentiable on $\mathbb{R}^1$, they are Lipschitzian, and this suffices for the construction of a diffusion with these coefficients (Chapter VII).

Theorem 6.1. Let $\{X_t\}$ denote a diffusion on $\mathbb{R}^1$ with the extended coefficients $\mu(\cdot), \sigma^2(\cdot)$ defined above. Then $\{|X_t|\}$ is a Markov process on the state space $S = [0, \infty)$, whose transition probability density $q(t; x, y)$ is given by

$$q(t; x, y) = p(t; x, y) + p(t; x, -y) \qquad (x, y \in [0, \infty)), \tag{6.9}$$

where $p(t; x, y)$ is the transition probability density of $\{X_t\}$. Further, q satisfies the backward equation

$$\frac{\partial q(t; x, y)}{\partial t} = \tfrac{1}{2}\sigma^2(x)\frac{\partial^2 q}{\partial x^2} + \mu(x)\frac{\partial q}{\partial x} \qquad (t > 0; x > 0, y \geqslant 0), \tag{6.10}$$

and the backward boundary condition

$$\left.\frac{\partial q(t; x, y)}{\partial x}\right|_{x=0} = 0 \qquad (t > 0;\ y \geqslant 0). \tag{6.11}$$

Proof. First note that the two Markov processes $\{X_t\}$ and $\{-X_t\}$ on $\mathbb{R}^1$ have the same drift and diffusion coefficients (use Proposition 3.1, or see Exercise 1). Therefore, they have the same transition probability density function p, so that the conditional density $p(t; x, y)$ of X_t at y given $X_0 = x$ is the same as the conditional density of $-X_t$ at y given $-X_0 = x$; but the latter is the conditional density of X_t at $-y$, given $X_0 = -x$. Hence,

$$p(t; x, y) = p(t; -x, -y). \tag{6.12}$$

In order to show that $\{Y_t := |X_t|\}$ is a Markov process on $S = [0, \infty)$, consider an arbitrary real-valued bounded (Borel measurable or continuous) function g on $[0, \infty)$, and write $h(x) = g(|x|)$. Then, as usual, writing P_x for the distribution of $\{X_t\}$ starting at x and E_x for the corresponding expectations, one has for $E_x(g(Y_{s+t}) \mid \{Y_u, 0 \leqslant u \leqslant s\})$,

$$\begin{aligned}
&E_x(g(|X_{s+t}|) \mid \{|X_u| : 0 \leqslant u \leqslant s\}) \\
&\quad = E_x[E(g(|X_{s+t}|) \mid \{X_u : 0 \leqslant u \leqslant s\}) \mid \{|X_u| : 0 \leqslant u \leqslant s\}] \\
&\quad = E_x[E(h(X_{s+t}) \mid \{X_u : 0 \leqslant u \leqslant s\}) \mid \{|X_u| : 0 \leqslant u \leqslant s\}] \\
&\quad = E_x\left[\left(\int_{-\infty}^{\infty} h(y)p(t; x', y)\,dy\right)_{x'=X_s} \middle| \{|X_u| : 0 \leqslant u \leqslant s\}\right] \\
&\quad = E_x\left[\left(\int_0^{\infty} g(z)(p(t; x', z) + p(t; x', -z))\,dz\right)_{x'=X_s} \middle| \{|X_u| : 0 \leqslant u \leqslant s\}\right] \\
&\quad = E_x\left[\left(\int_0^{\infty} g(z)(p(t; x', z) + p(t; -x', z))\,dz\right)_{x'=X_s} \middle| \{|X_u| : 0 \leqslant u \leqslant s\}\right] \\
&\quad = E_x\left[\int_0^{\infty} g(z)q(t; |X_s|, z)\,dz \middle| \{|X_u| : 0 \leqslant u \leqslant s\}\right] \\
&\quad = \int_0^{\infty} g(z)q(t; |X_s|, z)\,dz,
\end{aligned} \tag{6.13}$$

where

$$q(t; x, y) = p(t; x, y) + p(t; -x, y). \tag{6.14}$$

This proves $\{|X_t|\}$ is a time-homogeneous Markov process with transition probability density q. By differentiating both sides of (6.9) with respect to t, and making use of the backward equation for p, one arrives at (6.10). Since the right side of (6.14) is an *even* differentiable function of x, (6.11) follows. ∎

Definition 6.1. A Markov process on $S = [a, \infty)$ that has continuous sample paths and whose transition probability density satisfies equations (6.10), (6.11), with 0 replaced by a, is called a *reflecting diffusion* on $[a, \infty)$ having drift and diffusion coefficients $\mu(x)$, $\sigma^2(x)$. The point a is then called a *reflecting boundary point*.

Note that in this definition $\mu(0)$ is not required to be zero. It is possible to give a description of such a general diffusion with a reflecting boundary, similar to that given in Theorem 6.1 for the case $\mu(0) = 0$ (see theoretical complement 1).

Example 1. Consider a reflecting diffusion on $S = [0, \infty)$ with $\mu(\cdot) := 0$, $\sigma^2(x) := \sigma^2 > 0$. This is called the *reflecting Brownian motion* on $[0, \infty)$. Its transition probability density is, by (6.9) or (6.14),

$$q(t; x, y) = (2\pi\sigma^2 t)^{-1/2}\left[\exp\left\{-\frac{(y-x)^2}{2\sigma^2 t}\right\} + \exp\left\{-\frac{(y+x)^2}{2\sigma^2 t}\right\}\right] \qquad (t > 0;\ x, y \geqslant 0). \quad (6.15)$$

Example 2. Consider the Ornstein–Uhlenbeck process (Example 5.2) $\{X_t\}$ on $\mathbb{R}^1$ with $\mu(x) = -\gamma x$, $\sigma^2(\cdot) := \sigma^2 > 0$. By Theorem 6.1, $\{|X_t|\}$ is the reflecting diffusion on $S = [0, \infty)$, with transition probability density (see (5.17) and (6.9))

$$q(t; x, y) = \left(\frac{\pi\sigma^2}{\gamma}(1 - e^{-2\gamma t})\right)^{-1/2}\left[\exp\left\{\frac{-\gamma(y - xe^{-\gamma t})^2}{\sigma^2(1 - e^{-2\gamma t})}\right\} + \exp\left\{\frac{-\gamma(y + xe^{-\gamma t})^2}{\sigma^2(1 - e^{-2\gamma t})}\right\}\right], \qquad (t > 0;\ x, y \geqslant 0). \quad (6.16)$$

By arguments similar to those given in Section 2, we shall now derive the forward equation for a reflecting diffusion on $[a, \infty)$. By considering twice continuously differentiable functions f, g vanishing outside a finite interval, and g vanishing in a neighborhood of a as well, one derives the *forward equation* exactly as in (2.33)–(2.41) (Exercise 6),

$$\frac{\partial p(t; x, y)}{\partial t} = \frac{\partial^2}{\partial y^2}\left(\tfrac{1}{2}\sigma^2(y)p(t; x, y)\right) - \frac{\partial}{\partial y}\left(\mu(y)p(t; x, y)\right), \qquad (t > 0;\ x \geqslant a,\ y > a). \quad (6.17)$$

To derive the forward boundary condition, differentiate both sides of the following equation with respect to t,

$$1 = \int_{[a,\infty]} p(t; x, y)\, dy, \tag{6.18}$$

to obtain, using (6.17),

$$\begin{aligned} 0 = \int_{[a,\infty)} \frac{\partial p(t; x, y)}{\partial t}\, dy &= \int_{[a,\infty)} \frac{\partial}{\partial y}\left(-\mu(y)p + \frac{\partial}{\partial y}\left(\tfrac{1}{2}\sigma^2(y)p\right)\right) dy \\ &= \mu(a)p(t; x, a) - \left(\frac{\partial}{\partial y}\left(\tfrac{1}{2}\sigma^2(y)p(t; x, y)\right)\right)_{y=a}. \end{aligned}$$

Hence the *forward boundary condition* is

$$\frac{\partial}{\partial y}\left(\tfrac{1}{2}\sigma^2(y)p(t; x, y)\right)_{y=a} - \mu(a)p(t; x, a) = 0. \tag{6.19}$$

PERIODIC BOUNDARY CASE. (*Diffusions on a Circle: Periodic Boundary Conditions*). Let $\mu(\cdot)$, $\sigma^2(\cdot)$ be defined on $\mathbb{R}^1$ and satisfy Condition (1.1). Assume that both are periodic functions of *period d*.

Theorem 6.2. Let $\{X_t\}$ be a diffusion on $\mathbb{R}^1$ with periodic coefficients $\mu(\cdot)$, $\sigma^2(\cdot)$ as above. Define

$$\{Z_t\} := \{X_t (\text{mod } d)\}. \tag{6.20}$$

Then $\{Z_t\}$ is a Markov process on $[0, d)$ whose transition probability density function $q(t; z, z')$ is given by

$$q(t; z, z') = \sum_{m=-\infty}^{\infty} p(t; z, z' + md) \qquad (t > 0;\ z, z' \in [0, d)) \tag{6.21}$$

where $p(t; x, y)$ is the transition probability density function of $\{X_t\}$.

Proof. Let f be a real-valued bounded continuous function on $[0, d]$ with $f(0) = f(d)$. Let g be the periodic extension of f on $\mathbb{R}^1$ defined by $g(x + md) = f(x)$ for $x \in [0, d]$ and $m = 0, \pm 1, \pm 2, \ldots$. We need to prove

$$E[f(Z_{s+t}) \mid \{Z_u : 0 \leqslant u \leqslant s\}] = \left[\int_0^d f(z')q(t; z, z')\, dz'\right]_{z'=Z_s}. \tag{6.22}$$

Now, since $f(Z_t) = g(X_t)$, and $\{X_t\}$ is Markovian,

$$E[f(Z_{s+t}) \mid \{X_u: 0 \leqslant u \leqslant s\}] = E[g(X_{s+t}) \mid \{X_u: 0 \leqslant u \leqslant s\}]$$
$$= \left[\int_{-\infty}^{\infty} g(y)p(t; x, y)\,dy\right]_{x=X_s}$$
$$= \sum_{m=-\infty}^{\infty} \left[\int_{md}^{(m+1)d} g(y)p(t; x, y)\,dy\right]_{x=X_s} \tag{6.23}$$

But the periodicity of $\mu(\cdot)$ and $\sigma^2(\cdot)$ implies that the Markov processes $\{X_t\}$ and $\{X_t - md\}$ have the same drift and diffusion coefficients (Proposition 3.1 or Exercise 2) and, therefore, the same transition probability densities. This means

$$p(t; x + md, y + md) = p(t; x, y) \qquad (t > 0;\ x, y \in [0, d);\ m = 0, \pm 1, \pm 2, \ldots). \tag{6.24}$$

Using (6.24) in (6.23), and using the fact that $g(md + z') = f(z')$ for $z \in [0, d)$, one gets

$$E[f(Z_{s+t}) \mid \{X_u: 0 \leqslant u \leqslant s\}] = \sum_{m=-\infty}^{\infty} \int_0^d g(md + z')p(t; X_s, md + z')$$
$$= \int_0^d f(z')\left[\sum_{m=-\infty}^{\infty} p(t; X_s, md + z')\right] dz' \tag{6.25}$$

Since $Z_s = X_s(\text{mod } d)$, there exists a unique integer $m_0 = m_0(X_s)$ such that $X_s = m_0 d + Z_s$. Then

$$p(t; X_s, md + z') = p(t; m_0 d + Z_s, md + z') = p(t; Z_s, (m - m_0)d + z'),$$

by (6.24). Hence (6.25) reduces to (taking $m' = m - m_0$)

$$E[f(Z_{s+t}) \mid \{X_u: 0 \leqslant u \leqslant s\}] = \int_0^d f(z')\left[\sum_{m'=-\infty}^{\infty} p(t; Z_s, m'd + z')\,dz'\right]$$
$$= \int_0^d f(z')q(t; Z_s, z')\,dz'. \tag{6.26}$$

The desired relation (6.22) is obtained by taking conditional expectations of extreme left and right sides of (6.26) with respect to $\{Z_u: 0 \leqslant u \leqslant s\}$, noting that $\{Z_u: 0 \leqslant u \leqslant s\}$ is determined by $\{X_u: 0 \leqslant u \leqslant s\}$ (i.e., $\sigma\{Z_u: 0 \leqslant u \leqslant s\} \subset \sigma\{X_u: 0 \leqslant u \leqslant s\}$). ■

Since $0(\text{mod } d) = 0 = d(\text{mod } d)$, the state space of $\{Z_t\}$ may be regarded as the compact set $[0, d]$ with 0 and d identified. Actually, the state space is best thought of as a circle of circumference d, with z as the arc length measured counterclockwise along the circle starting from a fixed point on the circle. It may also be identified with the *unit circle* by $z \leftrightarrow \exp\{i2\pi z/d\}$.

Definition 6.5. Let $\{X_t\}$ be a diffusion on $\mathbb{R}^1$ with periodic drift and diffusion coefficients with period d. Then the Markov process $\{Z_t\} := \{X_t(\text{mod } d)\}$, is called a *diffusion on* $[0, d)$ *with periodic boundary* or a *diffusion on the circle.*

Example 3. Let $\{X_t\}$ be a diffusion on $\mathbb{R}^1$ with drift and diffusion coefficients μ and $\sigma^2 > 0$. Then $\{Z_t := X_t(\text{mod } 2\pi)\}$ is the *circular Brownian motion*, so named since Z_t can be identified with e^{iZ_t}. The transition probability density of this diffusion is

$$q(t; z, z') = \sum_{m=-\infty}^{\infty} (2\pi\sigma^2 t)^{-1/2} \exp\left\{\frac{(2m\pi + z' - z - t\mu)^2}{2\sigma^2 t}\right\}. \tag{6.27}$$

Underlying the proofs of Theorems 6.1 and 6.2 is an important principle that may be stated as follows (Exercise 5 and theoretical complement 2).

Proposition 6.3. (*A General Principle*). Suppose $\{X_t\}$ is a Markov process on S having transition operators $\mathbf{T}_t$, $t \geqslant 0$. Let φ be an arbitrary (measurable) function on S. Then $\{\varphi(X_t)\}$ is a Markov process on $S' := \varphi(S)$ if $\mathbf{T}_t(f \circ \varphi)$ is a function of φ, for every bounded (measurable) f on S'.

TWO-POINT BOUNDARY CASE. (*Diffusions on* $S = [0, 1]$ *with* "0" *and* "1" *Reflecting*). To illustrate the ideas for the general case, let us first see how to obtain a probabilistic construction of a Brownian motion on $[0, 1]$ with zero drift, and both boundary points 0, 1 reflecting. This leads to another application of the above general principle 6.3 in the following example.

Example 4. Let $\{X_t\}$ be an unrestricted Brownian motion with zero drift, starting at $x \in [0, 1]$. Define $Z_t^{(1)} := X_t(\text{mod } 2)$. By Theorem 6.2, $\{Z_t^{(1)}\}$ is a diffusion on $[0, 2]$ (with "0" and "2" identified). Hence $\{Z_t^{(2)} := Z_t^{(1)} - 1\}$ is a diffusion on $[-1, 1]$ whose transition probability density is (see (6.27))

$$q^{(2)}(t; z, z') = \sum_{m=-\infty}^{\infty} (2\pi\sigma^2 t)^{-1/2} \exp\left\{-\frac{(2m + z' - z)^2}{2\sigma^2 t}\right\} \quad \text{for } -1 \leqslant z', z \leqslant 1. \tag{6.28}$$

In particular,

$$q^{(2)}(t; z, z') = q^{(2)}(t; -z, -z'). \tag{6.29}$$

It now follows, exactly as in the proof of Theorem 6.1, that $\{Z_t\} = \{\varphi(Z_t^{(2)})\} := \{|Z_t^{(2)}|\}$ is a Markov process on $[0, 1]$. For, if f is a continuous function on

[0, 1] then

$$\begin{aligned}
E[f(Z_t) \mid Z_0^{(2)} = z] &= E[f(|Z_t^{(2)}|) \mid Z_0^{(2)} = z] \\
&= \int_{-1}^{1} f(|z'|) q^{(2)}(t; z, z')\, dz' \\
&= \int_{-1}^{0} f(|z'|) q^{(2)}(t; z, z')\, dz' + \int_{0}^{1} f(|z'|) q^{(2)}(t; z, z')\, dz' \\
&= \int_{0}^{1} f(|z'|)(q^{(2)}(t; z, -z') + q^{(2)}(t; z, z'))\, dz' \\
&= \int_{0}^{1} f(|z'|)(q^{(2)}(t; -z, z') + q^{(2)}(t; z, z'))\, dz' \quad \text{(by (6.29))} \\
&= \int_{0}^{1} f(|z'|) q(t; |z|, z')\, dz', \quad \text{say}, \qquad (6.30)
\end{aligned}$$

which is a function of $\varphi(z) = |z|$. Hence, $\{Z_t\}$ is a Markov process on [0, 1] whose transition probability density is

$$\begin{aligned}
q(t; x, y) &= q^{(2)}(t; -x, y) + q^{(2)}(t; x, y) \\
&= \sum_{m=-\infty}^{\infty} (2\pi\sigma^2 t)^{-1/2}\left[\exp\left\{-\frac{(2m + y - x)^2}{2\sigma^2 t}\right\} + \exp\left\{-\frac{(2m + y + x)^2}{2\sigma^2 t}\right\}\right] \\
&\qquad \text{for } x, y \in [0, 1]. \qquad (6.31)
\end{aligned}$$

Note that

$$\begin{aligned}
\frac{\partial q(t; x, y)}{\partial t} &= \frac{\partial q^{(2)}(t; -x, y)}{\partial t} + \frac{\partial q^{(2)}(t; x, y)}{\partial t} \\
&= \sum_{m=-\infty}^{\infty} \left[\frac{\partial}{\partial t}(p(t; -x, 2m + y) + p(t; x, 2m + y))\right], \qquad (6.32)
\end{aligned}$$

where $p(t; x, y)$ is the transition probability density of a Brownian motion with zero drift and diffusion coefficient σ^2. Hence

$$\frac{\partial q(t; x, y)}{\partial t} = \tfrac{1}{2}\sigma^2 \frac{\partial^2 q(t; x, y)}{\partial x^2}, \qquad (x \in (0, 1), y \in [0, 1]). \qquad (6.33)$$

Also, the first equality in (6.31) shows that

$$\left.\frac{\partial q(t; x, y)}{\partial x}\right|_{x=0} = 0, \qquad (y \in [0, 1]). \qquad (6.34)$$

Since $q^{(2)}(t; -x, y) + q^{(2)}(t; x, y)$ is symmetric about $x = 1$ (see (6.21)) one has

$$\left.\frac{\partial q(t; x, y)}{\partial x}\right|_{x=1} = 0, \qquad (y \in [0, 1]). \tag{6.35}$$

Thus, q satisfies Kolmogorov's backward equation (6.33), and backward boundary conditions (6.34), (6.35).

In precisely the same manner as in Example 4, we may arrive at a more general result. In order to state it, consider $\mu(\cdot)$, $\sigma^2(\cdot)$ satisfying Condition (1.1) on $[0, 1]$. Assume

$$\mu(0) = 0 = \mu(1). \tag{6.36}$$

Extend $\mu(\cdot)$, $\sigma^2(\cdot)$ to $(-\infty, \infty)$ as follows. First set

$$\mu(-x) = -\mu(x), \qquad \sigma^2(-x) = \sigma^2(x) \qquad \text{for } x \in [0, 1] \tag{6.37}$$

and then set

$$\mu(x + 2m) = \mu(x), \qquad \sigma^2(x + 2m) = \sigma^2(x) \qquad \text{for } x \in [-1, 1],$$
$$m = 0, \pm 1, \pm 2, \ldots. \tag{6.38}$$

Theorem 6.4. Let $\mu(\cdot)$, $\sigma^2(\cdot)$ be extended as above, and let $\{X_t\}$ be a diffusion on $S = (-\infty, \infty)$ having these coefficients. Define $\{Z_t^{(1)}\} := \{X_t(\text{mod } 2)\}$, and $\{Z_t\} := \{|Z_t^{(1)} - 1|\}$. Then $\{Z_t\}$ is a diffusion with coefficients $\mu(\cdot)$ and $\sigma^2(\cdot)$ on $[0, 1]$, and reflecting boundary points $0, 1$.

The condition (6.36) guarantees continuity of the extended coefficients. However, if the given $\mu(\cdot)$ on $[0, 1]$ does not satisfy this condition, one may modify $\mu(x)$ at $x = 0$ and $x = 1$ as in (6.36). Although this makes $\mu(\cdot)$ discontinuous, Theorem 6.4 may be shown to go through (see theoretical complements 1 and 2).

7 DIFFUSIONS WITH ABSORBING BOUNDARIES

Diffusions with absorbing boundaries are rather simple to describe. Upon arrival at a boundary point (state) the process is to remain in that state for all times thereafter. In particular this entails jumps in the transition probability distribution at absorbing states.

7.1 One-Point Boundary Case (Diffusions on $S = [a, \infty)$ with Absorption at a)

Let $\mu(\cdot)$, $\sigma^2(\cdot)$ be defined on S and satisfy Condition (1.1). Extend $\mu(\cdot)$, $\sigma^2(\cdot)$ on all of $\mathbb{R}^1$ in some (arbitrary) manner such that Condition (1.1) holds on $\mathbb{R}^1$

for this extension. Let $\{X_t\}$ denote a diffusion on $\mathbb{R}^1$ having these coefficients, and starting at $x \in [a, \infty)$. Define a new stochastic process $\{\tilde{X}_t\}$ by

$$\tilde{X}_t := \begin{cases} X_t & \text{if } t < \tau_a \\ a = X_{\tau_a} & \text{if } t \geqslant \tau_a, \end{cases} \tag{7.1}$$

where τ_a is the *first passage time to a* defined by

$$\tau_a := \inf\{t \geqslant 0 \colon X_t = a\}. \tag{7.2}$$

Note that $\tau_a \equiv \tau_a(\{X_t\})$ is a function of the process $\{X_t\}$.

Theorem 7.1. The process $\{\tilde{X}_t\}$ is a time-homogeneous Markov process.

Proof. Let B be a Borel subset of (a, ∞). Then

$$P(\tilde{X}_{s+t} \in B, \text{ and } \tau_a \leqslant s \mid \{X_u \colon 0 \leqslant u \leqslant s\}) = 0. \tag{7.3}$$

On the other hand, introducing the *shifted process* $\{(X_s^+)_t := X_{s+t} \colon t \geqslant 0\}$ and using the Markov property of $\{X_t\}$ one gets, *on the set* $\{\tau_a > s\}$,

$$\begin{aligned} P(\tilde{X}_{s+t} \in B, \tau_a > s \mid \{X_u \colon 0 \leqslant u \leqslant s\}) \\ &= P(X_{s+t} \in B, \tau_a > s, \text{ and } \tau_a > s + t \mid \{X_u \colon 0 \leqslant u \leqslant s\}) \\ &= \mathbf{1}_{\{\tau_a > s\}} P((X_s^+)_t \in B, \text{ and } \tau_a(X_s^+) > t \mid \{X_u \colon 0 \leqslant u \leqslant s\}) \\ &= \mathbf{1}_{\{\tau_a > s\}} P(X_t \in B, \text{ and } \tau_a > t \mid \{X_0 = y\})|_{y = X_s} \\ &= \mathbf{1}_{\{\tau_a > s\}} P(\tilde{X}_t \in B \mid \{X_0 = y\})|_{y = X_s}. \end{aligned} \tag{7.4}$$

For the second equality in (7.4), we have used the fact that if $\tau_a > s$ then $\tau_a = s + \tau_a(X_s^+)$, being the first passage time to a for X_s^+. Also, $\{\tau_a > s\}$ is determined by $\{X_u \colon 0 \leqslant u \leqslant s\}$, so that $\mathbf{1}_{\{\tau_a > s\}}$ may be taken outside the conditional probability. Combining (7.3) and (7.4) one gets, for $B \subset (a, \infty)$,

$$\begin{aligned} P(\tilde{X}_{s+t} \in B \mid \{X_u \colon 0 \leqslant u \leqslant s\}) &= P(\tilde{X}_t \in B \mid \{X_0 = y\})|_{y = X_s} \mathbf{1}_{\{\tau_a > s\}} \\ &= P(\tilde{X}_t \in B \mid \{\tilde{X}_0 = y\})|_{y = \tilde{X}_s} \mathbf{1}_{\{\tau_a > s\}} \\ &= P(\tilde{X}_t \in B \mid \{\tilde{X}_0 = y\})|_{y = \tilde{X}_s}. \end{aligned} \tag{7.5}$$

The second equality holds, since $\tau_a > s$ implies $X_s = \tilde{X}_s$ (and, always, $X_0 = \tilde{X}_0$). The last equality holds, since $\tau_a \leqslant s$ implies $\tilde{X}_s = a$, so that for B contained in (a, ∞), $P(\tilde{X}_t \in B \mid \{\tilde{X}_0 = y\})|_{y = \tilde{X}_s} = 0$. Now the sigmafield $\sigma\{\tilde{X}_u \colon 0 \leqslant u \leqslant s\}$ is contained in $\sigma\{X_u \colon 0 \leqslant u \leqslant s\}$, since $\tilde{X}_u$ is determined by $\{X_v \colon 0 \leqslant v \leqslant u\}$. Hence $P(\tilde{X}_{s+t} \in B \mid \{\tilde{X}_u \colon 0 \leqslant u \leqslant s\})$ may be obtained by taking the conditional expectation of the left side of (7.5), given $\{\tilde{X}_u \colon 0 \leqslant u \leqslant s\}$. But the last expression in (7.5) is already a function of $\tilde{X}_s$. Hence

$$P(\tilde{X}_{s+t} \in B \mid \{\tilde{X}_u : 0 \leqslant u \leqslant s\}) = P(\tilde{X}_t \in B \mid \tilde{X}_0 = y)\big|_{y=\tilde{X}_s}. \tag{7.6}$$

For $B = \{a\}$, (7.6) may be checked by taking $B = (a, \infty)$, and by complementation. This establishes the Markov property of $\{\tilde{X}_t\}$. ■

Definition 7.1. The Markov process $\{\tilde{X}_t\}$ is called a *diffusion* on $[a, \infty)$ *having drift and diffusion coefficients* $\mu(\cdot), \sigma^2(\cdot)$, and *an absorbing boundary at* a.

The transition probability $\tilde{p}(t; x, dy)$ of $\{\tilde{X}_t\}$ is given by

$$\begin{aligned}\tilde{p}(t; x, B) &:= P(\tilde{X}_t \in B \mid \{\tilde{X}_0 = x\}) \\ &= \begin{cases} P(X_t \in B, \text{ and } \tau_a > t \mid \{X_0 = x\}) & \text{if } B \subset (a, \infty) \\ P(\tau_a \leqslant t \mid \{X_0 = x\}) & \text{if } B = \{a\}. \end{cases}\end{aligned} \tag{7.7}$$

In particular, the transition probability distribution will have a jump (positive probability) at the boundary point $\{a\}$ if a is accessible.

Let $p(t; x, y)$ denote the transition probability density of $\{X_t\}$. Since, for $B \subset (a, \infty)$ and $x > a$,

$$\begin{aligned}\tilde{p}(t; x, B) &= P(X_t \in B, \text{ and } \tau_a > t \mid \{X_0 = x\}) \\ &\leqslant P(X_t \in B \mid \{X_0 = x\}) = p(t; x, B), \qquad (B \subset (a, \infty)),\end{aligned} \tag{7.8}$$

$\tilde{p}(t; x, dy)$ is given by a density $p^0(t; x, y)$, say, on (a, ∞) with $p^0(t; x, y) \leqslant p(t; x, y)$ (for $x > a, y > a$) (Exercise 1). One may then rewrite (7.7) as

$$\tilde{p}(t; x, B) = \begin{cases} \displaystyle\int_B p^0(t; x, y)\, dy & \text{if } B \subset (a, \infty) \text{ and } x > a, \\ P_x(\tau_a \leqslant t) & \text{if } B = \{a\}, \end{cases} \tag{7.9}$$

where P_x denotes the distribution of $\{X_t\}$ starting at x.

Thus for the analytical determination of $\tilde{p}(t; x, dy)$ one needs to find the density $p^0(t; x, y)$ and the function $(t, x) \to P_x(\tau_a \leqslant t)$ for $x > a$. It is shown in Section 15 that $p^0(t; x, y)$ satisfies the same *backward equation* as does $p(t; x, y)$ (also see Exercise 2).

$$\frac{\partial p^0(t; x, y)}{\partial t} = \tfrac{1}{2}\sigma^2(x)\frac{\partial^2 p^0}{\partial x^2} + \mu(x)\frac{\partial p^0}{\partial x} \qquad (t > 0;\ x > a,\ y > a), \tag{7.10}$$

and the *Dirichlet boundary condition*

$$\lim_{x \downarrow a} p^0(t; x, y) = 0. \tag{7.11}$$

Indeed, (7.11) may be derived from the relation (see (7.9))

$$\int_{(a,\infty)} p^0(t; x, y)\,dy = P_x(\tau_a > t), \qquad (x > a, t > 0), \tag{7.12}$$

noting that as $x \downarrow a$ the probability on the right side goes to zero. If one assumes that $p^0(t; x, y)$ has a limit as $x \downarrow a$, then this limit must be zero.

By the same method as used in the derivation of (6.17) it will follow (Exercise 3) that p^0 satisfies the *forward equation*

$$\frac{\partial p^0(t; x, y)}{\partial t} = \frac{\partial^2}{\partial y^2}\left(\tfrac{1}{2}\sigma^2(y)p^0\right) - \frac{\partial}{\partial y}(\mu(y)p^0), \qquad (t > 0; x > a, y > a), \tag{7.13}$$

and the *forward boundary condition*

$$\lim_{y \downarrow a} p^0(t; x, y) = 0, \qquad (t > 0; x > a). \tag{7.14}$$

For a Brownian motion on $[0, \infty)$ with an absorbing boundary zero, p^0 is calculated in Section 8 analytically. A purely probabilistic derivation of p^0 is sketched in Exercises 11.5, 11.6, and 11.11.

7.2 Two-Point Boundary Case (Diffusions on $S = [a, b]$ with Two Absorbing Boundary Points a, b)

Let $\mu(\cdot), \sigma^2(\cdot)$ be defined on $[a, b]$. Extend these to $\mathbb{R}^1$ in any manner such that Condition (1.1) is satisfied. Let $\{X_t\}$ be a diffusion on $\mathbb{R}^1$ having these extended coefficients, starting at a point in $[a, b]$. Define the stopped process $\{\tilde{X}_t\}$ by,

$$\tilde{X}_t := X_{t \wedge \tau} = \begin{cases} X_t & \text{if } t < \tau, \\ X_\tau & \text{if } t \geqslant \tau, \end{cases} \tag{7.15}$$

where τ is the *first passage time to the boundary*,

$$\tau := \inf\{t \geqslant 0 \colon X_t = a \text{ or } X_t = b\}. \tag{7.16}$$

Virtually the same proof as given for Theorem 7.1 applies to show that $\{\tilde{X}_t\}$ is a Markov process on $[a, b]$.

Definition 7.2. The process $\{\tilde{X}_t\}$ in (7.15) is called a *diffusion* on $[a, b]$ *with coefficients* $\mu(\cdot), \sigma^2(\cdot)$ *and* with two *absorbing boundaries* a, b.

Once again, the transition probability $\tilde{p}(t; x, dy)$ of $\{\tilde{X}_t\}$ is given by a density

$p^0(t; x, y)$ when restricted to the interior (a, b),

$$\tilde{p}(t; x, B) = \int_B p^0(t; x, y)\,dy \qquad (t > 0;\ a < x < b,\ B \subset (a, b)). \quad (7.17)$$

This density has total mass less than 1,

$$\int_{(a,b)} p^0(t; x, y)\,dy = P_x(\tau > t), \qquad (t > 0;\ x \in (a, b)), \quad (7.18)$$

where P_x is the distribution of $\{X_t\}$. Also, by the same argument as in the one-point boundary case,

$$\lim_{y \downarrow a} p^0(t; x, y) = 0, \quad (7.19)$$

$$\lim_{y \uparrow b} p^0(t; x, y) = 0; \qquad (t > 0;\ x \in (a, b)). \quad (7.20)$$

Unlike the one-point boundary case, however, p^0 does not completely determine $\tilde{p}(t; x, dy)$ in the present case. For this, one also needs to calculate the probabilities

$$\tilde{p}(t; x, \{a\}) := P_x(\tau \leqslant t, X_\tau = a), \qquad \tilde{p}(t; x, \{b\}) := P_x(\tau \leqslant t, X_\tau = b). \quad (7.21)$$

In order to calculate these probabilities, let A denote the event that the diffusion $\{X_t\}$, starting at x, reaches a before reaching b, and let D_t be the event that by time t the process does not reach either a or b but eventually reaches a before b. Then $D_t \subset A$ and

$$\tilde{p}(t; x, \{a\}) = P_x(A \backslash D_t) = P_x(A) - P_x(D_t) = \psi(x) - P_x(D_t), \quad (7.22)$$

where $\psi(x)$ is the probability that starting at x the diffusion $\{X_t\}$ reaches a before b. Conditioning on X_t, one gets for $t > 0$, $a < x < b$,

$$P_x(D_t) = \int_{(a,b)} P_x(D_t \mid X_t = y) p^0(t; x, y)\,dy = \int_{(a,b)} \psi(y) p^0(t; x, y)\,dy. \quad (7.23)$$

Substituting (7.22) in (7.23) yields

$$\tilde{p}(t; x, \{a\}) = \psi(x) - \int_{(a,b)} \psi(y) p^0(t; x, y)\,dy, \qquad (t > 0,\ a < x < b). \quad (7.24)$$

Therefore we have the following result.

Proposition 7.2. Let $\mu(x)$ and $\sigma^2(x)$ satisfy Condition (1.1) on $(-\infty, \infty)$. Let $S = [a, b]$, for some $a < b$. Then the probability $\tilde{p}(t; x, \{a\})$ is given by

$$\tilde{p}(t; x, \{a\}) = \psi(x) - \int_a^b \psi(y) p^0(t; x, y)\, dy \qquad (t > 0, a < x < b), \quad (7.25)$$

where $\psi(y) = P_y(X_\tau = a)$, P_y being the distribution of the unrestricted process $\{X_t\}$ starting at y.

The function $\psi(x)$ in Proposition 7.2 is given by (2.24) as well as (4.17). It is also calculated in Section 8, where it is shown that $\psi(x)$ can be obtained as the solution to a boundary-value problem that is the continuous (or, differential equation) analog of the discrete (or difference equation) boundary-value problem (Chapter III, Eqs. 2.4–2.5) for the birth–death process. It is given by

$$\psi(x) = \frac{\displaystyle\int_x^b \exp\left\{-\int_a^z \frac{2\mu(y)}{\sigma^2(y)}\, dy\right\} dz}{\displaystyle\int_a^b \exp\left\{-\int_a^z \frac{2\mu(y)}{\sigma^2(y)}\, dy\right\} dz}, \qquad (a \leqslant x \leqslant b). \qquad (7.26)$$

Finally,

$$\tilde{p}(t; x, \{b\}) = P_x(\tau \leqslant t) - \tilde{p}(t; x, \{a\}) = 1 - \int_{(a,b)} p^0(t; x, y)\, dy - \tilde{p}(t; x, \{a\}). \qquad (7.27)$$

For the case of a Brownian motion on $[a, b]$ with both boundary points a, b absorbing, calculations of $p^0(t; x, y)$, $\psi(x)$ based on eigenfunction expansions and (7.26) are given in Section 8. A purely probabilistic calculation is sketched in Exercises 11.5, 11.6, and 11.10.

7.3 Mixed Two-Point Boundary Case (One Absorbing Boundary Point and One Reflecting)

Let $\{X_t\}$ be a diffusion on $[a, \infty)$ with a reflecting boundary a, having drift and diffusion coefficients $\mu(\cdot)$, $\sigma^2(\cdot)$. For $b > a$, define

$$\tilde{X}_t := \begin{cases} X_t & \text{if } t < \tau_b, \\ b & \text{if } t \geqslant \tau_b. \end{cases} \qquad (7.28)$$

If $\{X_t\}$ starts at $x \in [a, b]$, then $\{\tilde{X}_t\}$ is a Markov process on $[a, b]$, starting at x, which is called a *diffusion on* $[a, b]$ *having coefficients* $\mu(\cdot)$, $\sigma^2(\cdot)$, *and a reflecting boundary point a and an absorbing boundary point b.*

8 CALCULATION OF TRANSITION PROBABILITIES BY SPECTRAL METHODS

Consider an arbitrary diffusion on an interval S with drift $\mu(x)$ and diffusion coefficient $\sigma^2(x)$. If an end point is accessible then choose either a Neumann (reflecting) or a Dirichlet (absorbing) boundary condition. Let S^0 denote the *interior* of S. Define the function

$$\pi(x) = \frac{2a}{\sigma^2(x)} \exp\{I(x_0, x)\}, \qquad x \in S, \tag{8.1}$$

where a is an arbitrary positive constant, x_0 is an arbitrarily chosen state, and

$$I(x_0, x) := \int_{x_0}^{x} \frac{2\mu(z)}{\sigma^2(z)}\, dz. \tag{8.2}$$

Notice that $\pi(x)$ is proportional to the limiting measure obtained from Eq. 4.1 of Chapter III, in the diffusion limit of birth–death chains with the parameters given by (4.2) of the present chapter.

Consider the space $L^2(S, \pi)$ of real-valued functions on S that are square integrable with respect to the density π. Let f, g be twice continuously differentiable functions that satisfy the backward Neumann or Dirichlet boundary condition(s) imposed at the boundary point(s), and that vanish outside a finite interval. Then, upon integration by parts, one obtains the following property for **A** (Exercise 1)

$$\langle \mathbf{A}f, g\rangle_\pi = \langle f, \mathbf{A}g\rangle_\pi, \tag{8.3}$$

where

$$\mathbf{A}f(x) = \tfrac{1}{2}\sigma^2(x)f''(x) + \mu(x)f'(x), \qquad x \in S^0, \tag{8.4}$$

and the inner product $\langle\ ,\ \rangle_\pi$ is defined by

$$\langle f, h\rangle_\pi = \int_S f(x)h(x)\pi(x)\, dx. \tag{8.5}$$

In view of the symmetry of **A** reflected in (8.3), one may expect that there is a spectral representation of the transition probability analogous to that for discrete state spaces (see Section 4 of Chapter III and Section 9 of Chapter IV). That this is indeed the case will be illustrated in forthcoming examples of this section (also see theoretical complement 2).

Consider the case in which S is a closed and bounded interval. The idea behind this method is that if ψ is an eigenfunction of **A** (including boundary conditions) corresponding to an eigenvalue α, i.e.,

$$\mathbf{A}\psi = \alpha\psi \tag{8.6}$$

then $u(t, x) = e^{\alpha t}\psi(x)$ solves the backward equation

$$\frac{\partial u}{\partial t} = e^{\alpha t}\alpha\psi(x) = e^{\alpha t}A\psi(x) = Au(t, x) \tag{8.7}$$

and u satisfies the boundary conditions. Likewise, if $u(t, x)$ is a superposition (linear combination) of such functions then the same will be true. Suppose that the set of eigenvalues (counting multiplicities) is countable, say $\alpha_0, \alpha_1, \alpha_2, \ldots$ with corresponding eigenfunctions $\psi_0, \psi_1, \ldots$ of unit length, i.e., $\|\psi_m\| \equiv \langle\psi_m, \psi_m\rangle^{1/2} = 1$. It is simple to check that eigenfunctions corresponding to distinct eigenvalues are orthogonal with respect to the inner product (8.5) (Exercise 2). Also if there are more than one linearly independent eigenfunctions for a single eigenvalue, then these can be orthogonalized by the Gram–Schmidt procedure (Exercise 2). So $\psi_0, \psi_1, \ldots$ can be taken to be orthonormal. If the set of finite linear combinations of the eigenfunctions is dense in $L^2(S, \pi)$, then the eigenfunctions are said to be *complete*. In this case each $f \in L^2(S, \pi)$ has a Fourier expansion of the form

$$f = \sum_{n=0}^{\infty} \langle f, \psi_n\rangle_\pi \psi_n. \tag{8.8}$$

Consider the superposition defined by

$$u(t, x) = \sum_{n=0}^{\infty} e^{\alpha_n t}\langle f, \psi_n\rangle_\pi \psi_n(x). \tag{8.9}$$

Then u satisfies the backward equation and the boundary condition(s) together with the initial condition

$$u(0, x) = f(x). \tag{8.10}$$

However, the function

$$\mathbf{T}_t f(x) = E_x f(X_t) = \int_S f(y)p(t; x, dy) \tag{8.11}$$

also satisfies the same backward equation, boundary condition and initial condition. So if there is *uniqueness* for a sufficiently large class of initial functions f (see theoretical complement 2) then we get

$$\mathbf{T}_t f(x) = u(t, x), \tag{8.12}$$

which means

$$\int_S f(y)p(t; x, dy) = \int_S f(y)\left(\sum_{n=0}^{\infty} e^{\alpha_n t}\psi_n(x)\psi_n(y) \right)\pi(y)\, dy. \tag{8.13}$$

In such cases, therefore, $p(t; x, dy)$ *has a density* $p(t; x, y)$ and it is given by

$$p(t; x, y) = \sum_{n=0}^{\infty} e^{\alpha_n t}\psi_n(x)\psi_n(y)\pi(y). \tag{8.14}$$

In the present section this method of computation of transition probabilities is illustrated in the case that $\mu(x)$ and $\sigma^2(x)$ are constants.

In Examples 1 and 2 below, $\mu = 0$ and $\sigma^2(x) \equiv \sigma^2 > 0$, so that $\pi(x)$ is a constant that may be taken to be 1.

Explicit computations of eigenvalues and eigenfunctions are possible only in very special cases. There are, however, effective numerical procedures for their approximate computation.

Example 1. (*Brownian Motion on* $[0, d]$ *with Two Reflecting Boundaries*). In this case, $S = [0, d]$, and *Kolmogorov's backward equation* for the transition probability density function $p(t; x, y)$ is

$$\frac{\partial p}{\partial t} = \tfrac{1}{2}\sigma^2 \frac{\partial^2 p}{\partial x^2} \qquad (t > 0, 0 < x < d;\ y \in S), \tag{8.15}$$

with the *backward boundary condition*

$$\left.\frac{\partial p}{\partial x}\right|_{x=0,d} = 0 \qquad (t > 0;\ y \in S). \tag{8.16}$$

We seek eigenvalues α of $\mathbf{A}$ and corresponding eigenfunctions ψ. That is, consider solutions of

$$\tfrac{1}{2}\sigma^2\psi''(x) = \alpha\psi(x), \tag{8.17}$$

$$\psi'(0) = \psi'(d) = 0. \tag{8.18}$$

Check that the functions

$$\psi_m(x) = b_m \cos\left(m\pi \frac{x}{d}\right) \qquad (m = 0, 1, 2, \ldots), \tag{8.19}$$

satisfy (8.17), (8.18), with α given by

$$\alpha = \alpha_m = -\frac{m^2\pi^2\sigma^2}{2d^2} \qquad (m = 0, 1, 2, \ldots). \tag{8.20}$$

Now the function (8.1) is here $\pi(x) = 1$. Since

$$\int_0^d 1^2\,dx = d, \qquad \int_0^d \cos^2\left(\frac{m\pi x}{d}\right)dx = \frac{d}{2}, \tag{8.21}$$

the normalizing constant b_m is given by

$$b_m = \sqrt{\frac{2}{d}} \quad (m \geqslant 1); \qquad b_0 = \sqrt{\frac{1}{d}}. \tag{8.22}$$

To see whether (8.19) and (8.20) provide all the eigenvalues and eigenfunctions, one needs to check the *completeness* of $\{\psi_m : m = 0, 1, 2, \ldots\}$ in $L^2([0, d], \pi)$. This may be done by using Fourier series (Exercise 10). Thus,

$$\begin{aligned} p(t; x, y) &= \sum_{m=0}^{\infty} e^{\alpha_m t}\psi_m(x)\psi_m(y) \\ &= \frac{1}{d} + \frac{2}{d}\sum_{m=1}^{\infty} \exp\left\{-\frac{\sigma^2\pi^2 m^2}{2d^2}\,t\right\}\cos\left(\frac{m\pi x}{d}\right)\cos\left(\frac{m\pi y}{d}\right) \\ &\qquad (t > 0, 0 < x, y < d). \end{aligned} \tag{8.23}$$

Notice that

$$\lim_{t\to\infty} p(t; x, y) = \frac{1}{d}, \tag{8.24}$$

the convergence being *exponentially fast*, uniformly with respect to x, y. Thus, as $t \to \infty$, the distribution of the process at time t converges to the uniform distribution on $[0, d]$. In particular, this limiting distribution provides the invariant initial distribution for the process.

Example 2. (*Brownian Motion with One Reflecting Boundary*). Let $S = [0, \infty)$, $\mu(x) \equiv 0$, $\sigma^2(x) \equiv \sigma^2 > 0$. We seek the transition density p that satisfies

$$\begin{aligned} \frac{\partial p}{\partial t} &= \frac{\sigma^2}{2}\frac{\partial^2 p}{\partial x^2} \qquad \text{for } t > 0, \quad x > 0, \quad y \geqslant 0, \\ \left.\frac{\partial p}{\partial x}\right|_{x=0} &= 0 \qquad \text{for } t > 0, \quad y \geqslant 0. \end{aligned} \tag{8.25}$$

This diffusion is the limiting form of that in Example 1 as $d \uparrow \infty$. Letting $d \uparrow \infty$

in (8.23) one has

$$\begin{aligned} p(t; x, y) &= \lim_{d \uparrow \infty} \frac{1}{d} \sum_{m=-\infty}^{\infty} \exp\left\{-\frac{\sigma^2\pi^2(m/d)^2}{2} t\right\} \cos\left(\frac{m\pi x}{d}\right) \cos\left(\frac{m\pi y}{d}\right) \\ &= \int_{-\infty}^{\infty} \exp\left\{-\frac{t\sigma^2\pi^2u^2}{2}\right\} \cos(\pi x u) \cos(\pi y u)\, du \\ &= \frac{1}{2} \int_{-\infty}^{\infty} \exp\left\{-\frac{t\sigma^2\pi^2u^2}{2}\right\} [\cos(\pi(x+y)u) + \cos(\pi(x-y)u)]\, du \\ &= \frac{1}{2\pi} \int_{-\infty}^{\infty} (e^{-i\xi(x+y)} + e^{-i\xi(x-y)}) \exp\left\{-\frac{t\sigma^2\xi^2}{2}\right\} d\xi, \qquad (\xi = \pi u) \\ &= \frac{1}{(2\pi t\sigma^2)^{1/2}} \left(\exp\left\{-\frac{(x+y)^2}{2t\sigma^2}\right\} + \exp\left\{-\frac{(x-y)^2}{2t\sigma^2}\right\}\right) \\ &\qquad (t > 0, 0 \leqslant x, y < \infty). \end{aligned} \tag{8.26}$$

The last equality in (8.26) follows from Fourier inversion and the fact that $\exp\{-t\sigma^2\xi^2/2\}$ is the characteristic function of the normal distribution with mean zero and variance $t\sigma^2$.

Example 3. *(Diffusion on $S = [0, d]$ with Constant Coefficients and Two Absorbing Boundaries).* Here $S = [0, d]$, and the transition probability has a density $p(t; x, y)$ on $0 < x, y < d$ that satisfies the backward equation

$$\frac{\partial p}{\partial t} = \tfrac{1}{2}\sigma^2 \frac{\partial^2 p}{\partial x^2} + \mu \frac{\partial p}{\partial x}, \qquad (t > 0, 0 < x < d; 0 < y < d), \tag{8.27}$$

the backward boundary conditions

$$\lim_{x \downarrow 0} p(t; x, y) = 0, \qquad \lim_{x \uparrow d} p(t; x, y) = 0, \tag{8.28}$$

for $t > 0$, $0 < y < d$. Check that the functions

$$\psi_m(x) = b_m \exp\left\{-\frac{\mu x}{\sigma^2}\right\} \sin\left(\frac{m\pi x}{d}\right) \qquad (m = 1, 2, \ldots) \tag{8.29}$$

are eigenfunctions corresponding to eigenvalues

$$\alpha = \alpha_m = -\frac{m^2\pi^2\sigma^2}{2d^2} - \frac{\mu^2}{2\sigma^2}, \qquad (m = 1, 2, \ldots). \tag{8.30}$$

Since the function in (8.1) is in this case given by

$$\pi(x) = \exp\left\{\frac{2\mu x}{\sigma^2}\right\} \qquad (0 \leqslant x \leqslant d), \tag{8.31}$$

the normalizing constants are

$$b_m = \sqrt{\frac{2}{d}} \qquad (m = 1, 2, \ldots). \tag{8.32}$$

The eigenfunction expansion of p is therefore

$$\begin{aligned} p(t; x, y) &= \sum_{m=1}^{\infty} e^{\alpha_m t}\psi_m(x)\psi_m(y)\pi(y) \\ &= \frac{2}{d}\exp\left\{\frac{\mu(y-x)}{\sigma^2}\right\}\exp\left\{-\frac{\mu^2 t}{2\sigma^2}\right\}\sum_{m=1}^{\infty}\exp\left\{-\frac{m^2\pi^2\sigma^2 t}{2d^2}\right\} \\ &\quad \times \sin\left(\frac{m\pi x}{d}\right)\sin\left(\frac{m\pi y}{d}\right) \qquad (t > 0, 0 < x, y < d). \end{aligned} \tag{8.33}$$

It remains to calculate

$$p(t; x, \{0\}) = P_x(X_t = 0), \qquad p(t; x, \{d\}) = P_x(X_t = d), \qquad 0 < x < d. \tag{8.34}$$

Note that

$$\begin{aligned} p(t; x, \{0\}) + p(t; x, \{d\}) &= 1 - \int_{(0,d)} p(t; x, y)\, dy \\ &= 1 - \frac{2}{d}\exp\left\{-\frac{\mu x}{\sigma^2}\right\}\sum_{m=1}^{\infty} a_m \exp\left\{-\frac{m^2\pi^2\sigma^2 t}{2d^2}\right\}\sin\left(\frac{m\pi x}{d}\right) \end{aligned} \tag{8.35}$$

where

$$a_m = \int_0^d \exp\left\{\frac{\mu y}{\sigma^2}\right\}\sin\left(\frac{m\pi y}{d}\right) dy. \tag{8.36}$$

Thus, it is enough to determine $p(t; x, \{0\})$; the function $p(t; x, \{d\})$ is then determined from (8.35). From (7.25) we get

$$p(t; x, \{0\}) = \psi(x) - \int_{(0,d)} \psi(y)p(t; x, y)\, dy \tag{8.37}$$

where $\psi(x) = P_x(\{X_t\}$ reaches 0 before $d)$. From the calculations of (9.6) in

Chapter I we have

$$\psi(x) = \begin{cases} \dfrac{x}{d} & \text{if } \mu = 0 \\[2ex] \dfrac{1 - \exp\{2\mu(d - x)/\sigma^2\}}{1 - \exp\{2d\mu/\sigma^2\}} & \text{if } \mu \neq 0. \end{cases} \tag{8.38}$$

Substituting (8.33), (8.38) in (8.37) yields $p(t; x, \{0\})$.

Example 4. *(Diffusion on $S = [0, \infty)$ with Constant Coefficients μ, $\sigma^2 > 0$; Zero an Absorbing Boundary).* The transition density function $p(t; x, y)$ for $x, y \in (0, \infty)$ is obtained as a limit of (8.33) as $d \uparrow \infty$. That is,

$$\begin{aligned} p(t; x, y) &= 2 \exp\left\{\frac{\mu(y - x)}{\sigma^2} - \frac{\mu^2 t}{2\sigma^2}\right\} \int_0^\infty \exp\left\{-\frac{\pi^2\sigma^2 t u^2}{2}\right\} \sin(\pi x u) \sin(\pi y u)\, du \\ &= \frac{2}{\pi} \exp\left\{\frac{\mu(y - x)}{\sigma^2} - \frac{\mu^2 t}{2\sigma^2}\right\} \int_0^\infty \exp\left\{-\frac{\sigma^2 t \xi^2}{2}\right\} \sin(\xi x) \sin(\xi y)\, d\xi \\ &= \frac{1}{\pi} \exp\left\{\frac{\mu(y - x)}{\sigma^2} - \frac{\mu^2 t}{2\sigma^2}\right\} \int_0^\infty \exp\left\{-\frac{\sigma^2 t \xi^2}{2}\right\} \\ &\qquad \times [\cos(\xi(x - y)) - \cos(\xi(x + y))]\, d\xi \\ &= \frac{1}{2\pi} \exp\left\{\frac{\mu(y - x)}{\sigma^2} - \frac{\mu^2 t}{2\sigma^2}\right\} \int_{-\infty}^\infty e^{-i\xi(x - y)} - e^{-i\xi(x + y)}) e^{-\sigma^2 t \xi^2/2}\, d\xi \\ &= \exp\left\{\frac{\mu(y - x)}{\sigma^2} - \frac{\mu^2 t}{2\sigma^2}\right\} \frac{1}{(2\pi\sigma^2 t)^{1/2}} \\ &\qquad \times \left[\exp\left\{-\frac{(x - y)^2}{2\sigma^2 t}\right\} - \exp\left\{-\frac{(x + y)^2}{2\sigma^2 t}\right\}\right]. \end{aligned} \tag{8.39}$$

Integration of (8.39) with respect to y over $(0, \infty)$ yields $p(t; x, \{0\}^c)$, which is the probability that, starting from x, the first passage time to zero is greater than t.

Examples 3 and 4 yield the distributions of the maximum and the minimum and the joint distribution of the maximum, minimum and the state at time t of a diffusion with constant coefficients over the time period $[0, t]$ (Exercises 4–6).

9 TRANSIENCE AND RECURRENCE OF DIFFUSIONS

Let $\{X_t^x : t \geqslant 0\}$ be a diffusion on an interval S, with drift and diffusion coefficients $\mu(x)$ and $\sigma^2(x)$, starting at x. Let $[c, d] \subset S$, $c < d$, and let

$$\psi(x) = P(\{X_t^x\} \text{ reaches } c \text{ before } d), \qquad c \leqslant x \leqslant d. \tag{9.1}$$

Assume, as always, that Condition (1.1) holds. Criteria for transience and recurrence of diffusions may be derived from the computation of $\psi(x)$ given in Section 4 (see Eq. (4.17)). A different method of computation based on solving a differential equation governing ψ is given in this section; recall Chapter III, Eqs. 2.4, 2.5 in the case of the birth–death chain for the analogous discrete problem. The present method is similar to that of Proposition 2.5, but does not make use of martingales. It does, however, use the fact derived in Exercise 3.5, namely,

$$P_x\left(\max_{0 \leqslant t \leqslant h} |X_t - x| > \varepsilon\right) = o(h) \qquad \text{as } h \downarrow 0 \qquad \text{for every } \varepsilon > 0.$$

Assume that this last fact holds.

Proposition 9.1. ψ is the solution to the two-point boundary-value problem:

$$\begin{aligned} \tfrac{1}{2}\sigma^2(x)\psi''(x) + \mu(x)\psi'(x) &= 0 \qquad \text{for } c < x < d, \\ \psi(c) = 1, \qquad \psi(d) &= 0. \end{aligned} \tag{9.2}$$

Proof. Let τ denote the first passage to the set $\{c, d\}$ with two elements. The event

$$\{\tau > h\} \equiv \{c < \min\{X_u: 0 \leqslant u \leqslant h\} \leqslant \max\{X_u: 0 \leqslant u \leqslant h\} < d\}$$

is determined by $\{X_u: 0 \leqslant u \leqslant h\}$. Therefore, by the Markov property,

$$\begin{aligned} P_x(\tau > h, X_\tau = c) &= E_x[P(\tau > h, X_\tau = c \mid \{X_u: 0 \leqslant u \leqslant h\})] \\ &= E_x[P(\tau > h, (X_h^+)_{\tau'} = c \mid \{X_u: 0 \leqslant u \leqslant h\})] \\ &= E_x[\mathbf{1}_{\{\tau > h\}} P((X_h^+)_{\tau'} = c \mid \{X_u: 0 \leqslant u \leqslant h\})] \\ &= E_x(\mathbf{1}_{\{\tau > h\}} \psi(X_h)). \end{aligned} \tag{9.3}$$

Here X_h^+ is the shifted process $\{(X_h^+)_t\} := \{X_{h+t}: t \geqslant 0\}$ and $\tau' = \inf\{t: (X_h^+)_t = c\}$ $(= \tau - h$ on $\{\tau > h\})$. Now extend the function $\psi(x)$ over $x < c$ and $x > d$ smoothly so as to vanish outside a compact set in S. Denote this extended function also by ψ. Then (9.3) may be expressed as

$$\begin{aligned} P_x(\tau > h, X_\tau = c) &= E_x\psi(X_h) - E_x(\mathbf{1}_{\{\tau \leqslant h\}}\psi(X_h)) \\ &= \int_S \psi(y)p(h; x, y)\, dy + o(h), \qquad \text{as } h \downarrow 0. \end{aligned} \tag{9.4}$$

The remainder term in (9.4) is $o(h)$ by the third condition in $(1.2)'$. Actually, one needs the stronger property that $P_x(\max_{0\leqslant s\leqslant h}|X_s - x| > \varepsilon) = o(h)$ (see Exercise 3.5). For the same reason,

$$P_x(\tau \leqslant h, X_\tau = c) \leqslant P_x(\tau \leqslant h) = o(h), \qquad \text{as } h \downarrow 0. \tag{9.5}$$

By (9.4) and (9.5)

$$\begin{aligned}\psi(x) &= P_x(\tau > h, X_\tau = c) + P_x(\tau \leqslant h, X_\tau = c)\\ &= (\mathbf{T}_h\psi)(x) + o(h), \qquad \text{as } h \downarrow 0.\end{aligned} \tag{9.6}$$

Hence

$$\lim_{h\downarrow 0} \frac{(\mathbf{T}_h\psi)(x) - \psi(x)}{h} = 0. \tag{9.7}$$

But the limit on the left is $(\mathbf{A}\psi)(x) = \frac{1}{2}\sigma^2(x)\psi''(x) + \mu(x)\psi'(x)$ (see Section 2). ■

In order to solve the two-point boundary-value problem (9.2), write

$$I(x, z) := \int_x^z \frac{2\mu(y)}{\sigma^2(y)}\, dy \qquad (x, z \in S), \tag{9.8}$$

with the usual convention that $I(z, x) = -I(x, z)$. Fix an arbitrary $x_0 \in S$, and define the *scale function*

$$s(x) \equiv s(x_0; x) := \int_{x_0}^x \exp\{-I(x_0, z)\}\, dz, \qquad (x \in S), \tag{9.9}$$

and the *speed function*

$$m(x) \equiv m(x_0; x) := \int_{x_0}^x \frac{2}{\sigma^2(z)} \exp\{I(x_0, z)\}\, dz, \qquad (x \in S). \tag{9.10}$$

In terms of these two functions, the differential operator

$$\mathbf{A} = \tfrac{1}{2}\sigma^2(x)\frac{d^2}{dx^2} + \mu(x)\frac{d}{dx} \tag{9.11}$$

takes the simple form

$$\mathbf{A} = \frac{d}{dm(x)}\frac{d}{ds(x)}. \tag{9.12}$$

In other words, for every twice-differentiable function f on S (Exercise 1)

$$(\mathbf{A}f)(x) = \frac{d}{dm(x)}\left(\frac{df(x)}{ds(x)}\right), \tag{9.13}$$

where, by the chain rule,

$$\frac{df(x)}{ds(x)} = \frac{df(x)}{dx}\frac{dx}{ds(x)} = \frac{1}{s'(x)}f'(x), \qquad \frac{df(x)}{dm(x)} = \frac{1}{m'(x)}f'(x). \tag{9.14}$$

The differential equation in (9.2) may now be expressed as

$$\frac{d}{dm(x)}\left(\frac{d\psi(x)}{ds(x)}\right) = 0, \tag{9.15}$$

which yields

$$\frac{d\psi(x)}{ds(x)} = c_1, \qquad \psi(x) = c_1 s(x) + c_2. \tag{9.16}$$

The boundary conditions in (9.2) now yield,

$$c_1 = -\frac{1}{s(d) - s(c)}, \qquad c_2 = -c_1 s(d) = \frac{s(d)}{s(d) - s(c)}. \tag{9.17}$$

Use these constants in (9.16) to get

$$\psi(x) = \frac{s(d) - s(x)}{s(d) - s(c)} = \frac{s(d; x_0) - s(x; x_0)}{s(d; x_0) - s(c; x_0)}. \tag{9.18}$$

Taking $x_0 = c$ and noting that $s(c; c) = 0$, one has

$$\psi(x) = \frac{\displaystyle\int_x^d \exp\{-I(c, z)\}\, dz}{\displaystyle\int_c^d \exp\{-I(c, z)\}\, dz}. \tag{9.19}$$

The relation (9.18) justifies the *scale function* nomenclature for the function s. When distance is measured in this scale, i.e., when $s(y) - s(x)$ is regarded as the distance between x and y ($x < y$), then the probability of reaching c before d starting at x is proportional to the distance $s(d) - s(x)$ between x and d. This is a property of Brownian motion, for which $s(x; 0) = x$ (also see Eq. 9.6, Chapter I). In particular, starting from the middle of an interval *in this scale*, the probability of reaching the left end point before the right end point is $\frac{1}{2}$, the same as that of reaching the right end point first.

It follows from (9.18) (Exercise 2) that

$$\varphi(x) := P_x(\{X_t\} \text{ reaches } d \text{ before } c) = \frac{s(x) - s(c)}{s(d) - s(c)} = \frac{\displaystyle\int_c^x e^{-I(c,z)}\,dz}{\displaystyle\int_c^d e^{-I(c,z)}\,dz}. \tag{9.20}$$

Of course, (9.20) may be proved directly in the same way as (9.18).

Next write

$$\rho_{xy} = P_x(\{X_t\} \text{ ever reaches } y\}), \qquad (x, y \in S). \tag{9.21}$$

Then the following results hold.

Proposition 9.2. Let $S = (a, b)$. Fix $x_0 \in S$ arbitrarily.

(a) If

$$s(a) \equiv s(x_0; a) = -\infty, \tag{9.22}$$

then $\rho_{xy} = 1$ for all $y > x$. If $s(x_0; a)$ is finite, then

$$\rho_{xy} = \frac{s(x) - s(a)}{s(y) - s(a)} = \frac{\displaystyle\int_a^x \exp\{-I(x_0, z)\}\,dz}{\displaystyle\int_a^y \exp\{-I(x_0, z)\}\,dz}, \qquad (y > x). \tag{9.23}$$

(b) If

$$s(x_0; b) = \infty, \tag{9.24}$$

then $\rho_{xy} = 1$ for all $y < x$. If $s(x_0; b)$ is finite, then

$$\rho_{xy} = \frac{s(b) - s(x)}{s(b) - s(y)} = \frac{\displaystyle\int_x^b \exp\{-I(x_0, z)\}\,dz}{\displaystyle\int_y^b \exp\{-I(x_0, z)\}\,dz}, \qquad (y < x). \tag{9.25}$$

Proof. (a) If $y > x$ then ρ_{xy} is obtained from (9.20) by letting $d = y$, and $c \downarrow a$.

(b) If $y < x$, then use (9.18) with $c = y$, and let $d \uparrow b$. ∎

Definition 9.1. A *state y is recurrent* if $\rho_{xy} = 1$ for all $x \in S$ such that $\rho_{yx} > 0$, and is *transient* otherwise. If all states in S are recurrent, then the *diffusion* is said to be *recurrent.*

The following corollary is a useful consequence of Proposition 9.2.

Corollary 9.3. A diffusion on $S = (a, b)$ with coefficients $\mu(x)$, $\sigma^2(x)$ is recurrent if and only if $s(a) = -\infty$ and $s(b) = \infty$.

If S has one or two boundary points, then ρ_{xy} is given by the following proposition. The modifications of the above to get these conditions are left to exercises.

Proposition 9.4

(a) Suppose $S = [a, b)$ and a is reflecting. Then the diffusion is recurrent if and only if $s(b) = \infty$. If, on the other hand, $s(b) < \infty$ then one has $\rho_{xy} = 1$ for $y > x$ and

$$\rho_{xy} = \frac{s(b) - s(x)}{s(b) - s(y)} = \frac{\int_x^b \exp\{-I(x_0, z)\}\, dz}{\int_y^b \exp\{-I(x_0, z)\}\, dz}, \qquad y < x. \tag{9.26}$$

(b) Suppose $S = [a, b)$ and a is absorbing. Then the only recurrent state is a and

$$\rho_{xy} = \begin{cases} \dfrac{s(x) - s(a)}{s(y) - s(a)} & \text{if } 0 < x \leqslant y, \\ 1 & \text{if } s(b) = \infty \text{ and } 0 \leqslant y < x, \\ \dfrac{s(b) - s(x)}{s(b) - s(y)} & \text{if } s(b) < \infty \text{ and } 0 \leqslant y \leqslant x. \end{cases} \tag{9.27}$$

(c) Suppose $S = [a, b]$ with both boundaries reflecting. Then $\rho_{xy} = 1$ for all $x, y \in S$, and the diffusion is recurrent.

(d) Suppose $S = [a, b]$ with a absorbing and b reflecting. Then the only recurrent state is a, $\rho_{xy} = 1$ for $y \leqslant x$, and

$$\rho_{xy} = \frac{s(x) - s(a)}{s(y) - s(a)}, \qquad y > x. \tag{9.28}$$

(e) Suppose $S = [a, b]$ with both boundaries absorbing. Then all states, other than a and b, are transient and one has

$$\rho_{xy} = \begin{cases} \dfrac{s(b) - s(x)}{s(b) - s(y)} & a \leqslant y \leqslant x < b, \\[2ex] \dfrac{s(x) - s(a)}{s(y) - s(a)} & a < x \leqslant y \leqslant b. \end{cases} \tag{9.29}$$

10 NULL AND POSITIVE RECURRENCE OF DIFFUSIONS

As in the case of Markov chains, the existence of an invariant initial distribution for a diffusion is equivalent to its positive recurrence. The present section is devoted to the derivation of a criterion for positive recurrence.

For a diffusion $\{X_t\}$ let τ_y denote the *first passage time* to a state y defined by

$$\tau_y = \inf\{t \geqslant 0 \colon X_t = y\}, \tag{10.1}$$

with the usual convention that $\tau_y(\omega) = \infty$ for those sample points ω for which $X_t(\omega)$ does not equal y for any t (i.e., if the set on the right side of (10.1) is empty). If $c < d$ are two states, then write

$$\tau := \inf\{t \geqslant 0 \colon X_t \notin (c, d)\} = \tau_c \wedge \tau_d, \tag{10.2}$$

where $\wedge$ denotes the minimum, i.e., $s \wedge t = \min(s, t)$. Then τ is the first time the process is at c or d, provided $X_0 \in [c, d]$.

Let us now calculate the *mean escape time* $M(x)$ of a diffusion from an interval (c, d),

$$M(x) := E_x \tau, \tag{10.3}$$

Note that $M(x) = 0$ if $x \notin (c, d)$. Here $[c, d] \subset S$. Now denoting by X_h^+ the shifted process $\{(X_h^+)_t := X_{h+t} \colon t \geqslant 0\}$, one has

$$\tau = h + \tau(X_h^+) \qquad \text{on the set } \{\tau > h\}, \tag{10.4}$$

since $\tau(X_h^+) \equiv \tau(\{(X_h^+)_t\})$ is precisely the additional time needed by the process $\{X_t\}$, after time h, to escape from (c, d). Because $\mathbf{1}_{\{\tau > h\}}$ is determined by (i.e., measurable with respect to) $\{X_u \colon 0 \leqslant u \leqslant h\}$, it follows by the Markov property that

$$\begin{aligned} E_x(\mathbf{1}_{\{\tau > h\}} \tau(X_h^+)) &= E_x[\mathbf{1}_{\{\tau > h\}} E(\tau(X_h^+) \mid \{X_u \colon 0 \leqslant u \leqslant h\})] = E_x(\mathbf{1}_{\{\tau > h\}} (E_y \tau)_{y = X_h}) \\ &= E_x(\mathbf{1}_{\{\tau > h\}} M(X_h)) = E_x(M(X_h)) - E_x(\mathbf{1}_{\{\tau \leqslant h\}} M(X_h)). \end{aligned} \tag{10.5}$$

By (10.4) and (10.5), for $x \in (c, d)$,

$$\begin{aligned} M(x) \equiv E_x\tau &= hP_x(\tau > h) + E_x(M(X_h)) - E_x(\mathbf{1}_{\{\tau \leqslant h\}}M(X_h)) + E_x(\tau\mathbf{1}_{\{\tau \leqslant h\}}) \\ &= h + E_x(M(X_h)) - hP_x(\tau \leqslant h) - E_x(\mathbf{1}_{\{\tau \leqslant h\}}M(X_h)) + E_x(\tau\mathbf{1}_{\{\tau \leqslant h\}}) \\ &= h + E_x(M(X_h)) + o(h), \qquad (h \downarrow 0). \end{aligned} \tag{10.6}$$

The last relation in (10.6) clearly follows if $E_x(\mathbf{1}_{\{\tau \leqslant h\}}M(X_h)) = o(h)$, as $h \downarrow 0$; a proof of which is sketched in Exercise 1. Now (10.6) leads to

$$\lim_{h \downarrow 0} \frac{1}{h}\{(T_h M)(x) - M(x)\} = -1, \tag{10.7}$$

i.e., $(\mathbf{A}M)(x) = -1$ for $x \in (c, d)$. Therefore the following result is proved.

Proposition 10.1. The mean escape time $M(x)$ from (c, d) satisfies

$$\begin{gathered} \tfrac{1}{2}\sigma^2(x)M''(x) + \mu(x)M'(x) = -1, \qquad c < x < d, \\ M(c) = M(d) = 0. \end{gathered} \tag{10.8}$$

In order to solve the *two-point boundary-value problem* (10.8), express the differential equation as (see Section 9)

$$\frac{d}{dm(x)}\left(\frac{dM(x)}{ds(x)}\right) = -1, \tag{10.9}$$

whose general solution is given by

$$\frac{dM(x)}{ds(x)} = -m(x) + c_1, \qquad M(x) = c_1 s(x) - \int_c^x m(y)\, ds(y) + c_2. \tag{10.10}$$

From the boundary conditions in (10.8) it follows that

$$c_1 = \frac{\displaystyle\int_c^d m(y)\, ds(y)}{s(d) - s(c)}, \qquad c_2 = -c_1 s(c). \tag{10.11}$$

Substituting this in (10.10), one gets

$$M(x) = \frac{s(c; x)}{s(c; d)} \int_c^d m(c; y)s'(c; y)\, dy - \int_c^x m(c; y)s'(c; y)\, dy$$

$$= \frac{\int_c^x e^{-I(c,z)}\,dz}{\int_c^d e^{-I(c,z)}\,dz} \int_c^d \left[\int_c^y \frac{2}{\sigma^2(z)} e^{I(c,z)}\,dz\right] e^{-I(c,y)}\,dy$$

$$- \int_c^x \left[\int_c^y \frac{2}{\sigma^2(z)} e^{I(c,z)}\,dz\right] e^{-I(c,y)}\,dy, \qquad (c \leqslant x \leqslant d). \quad (10.12)$$

Equation (10.9) justifies the nomenclature *speed measure* given to m. For suppose the diffusion is in natural scale, i.e., that it has been relabeled so that $s(x) = x$. Then (10.9) may be expressed as

$$M''(x) = -m'(x). \qquad (10.13)$$

If m_1, m_2 are two speed measures and M_1, M_2 the corresponding expected times to escape from (c, d), then $m_1'(x) \geqslant m_2'(x)$ for all x implies $M_1(x) \geqslant M_2(x)$ for all x (Exercise 8). *Thus, the larger m' is, the slower is the speed of escape.* For a Brownian motion with zero drift and diffusion coefficient σ^2, the speed measure is $m(x) = (2/\sigma^2)x$ (with $x_0 = 0$) and $m'(x) = 2/\sigma^2$.

Suppose the state space S has an *inaccessible* right end point b. Write $\tau_c \wedge \tau_d$ for $\min\{\tau_c, \tau_d\}$, the first passage time to $\{c, d\}$, so that

$$M(x) = E_x(\tau_c \wedge \tau_d), \qquad c \leqslant x \leqslant d. \qquad (10.14)$$

By letting $d \uparrow b$ in (10.14), one obtains $E_x\tau_c$ provided

$$\tau_b := \lim_{d \uparrow b} \tau_d = \infty \qquad \text{with probability 1}, \qquad (10.15)$$

which is what *inaccessibility* of b means. It may be shown that (10.15) holds, in particular, if $s(b) = \infty$ (Exercise 2). In this case, $E_x\tau_c$ is finite if and only if (Exercise 3)

$$m(b) := m(x_0; b) \equiv \int_{x_0}^b \frac{2}{\sigma^2(z)} e^{I(x_0,z)}\,dz < \infty. \qquad (10.16)$$

It now follows from the first equality in (10.12) that (Exercise 3)

$$E_x\tau_c = \int_c^x (m(b) - m(y))\,ds(y), \qquad x > c. \qquad (10.17)$$

Proceeding in the same manner, it follows (Exercise 4) that if S has an inaccessible lower end point a and $s(a) = -\infty$, *then* $E_x\tau_d < \infty$ $(x < d)$ *if and only if*

$$-m(a) := -m(x_0; a) := -\int_{x_0}^{a} \frac{2}{\sigma^2(z)} e^{I(x_0,z)}\,dz \equiv \int_{a}^{x_0} \frac{2}{\sigma^2(z)} e^{-I(z,x_0)}\,dz < \infty. \tag{10.18}$$

Then (Exercise 4)

$$E_x\tau_d = \int_x^d (m(y) - m(a))\,ds(y), \qquad x < d. \tag{10.19}$$

Finiteness (with probability 1) of the passage time τ_y under every initial state is the meaning of recurrence of a diffusion. A stronger property is the finiteness of their expectations.

Definition 10.1. A diffusion on S is *positive recurrent* if $E_x\tau_y < \infty$ for all $x, y \in S$. A recurrent diffusion that is not positive recurrent is *null recurrent*.

The positive recurrence of a diffusion implies its recurrence, although the converse is not true.

The following result has been proved above.

Proposition 10.2. Suppose $S = (a, b)$. Then the diffusion is positive recurrent, if and only if

$$s(a) = -\infty, \qquad m(a) > -\infty, \tag{10.20}$$

and

$$s(b) = +\infty, \qquad m(b) < \infty. \tag{10.21}$$

For intervals S having boundary points one may prove the following result. The modifications that give this result are left to exercises.

Proposition 10.3

(a) Suppose $S = [a, b)$ with a a reflecting boundary. Then the diffusion is positive recurrent if and only if $s(b) = \infty$ and $m(b) < \infty$.

(b) Suppose $S = [a, b]$ with a and b reflecting boundaries. Then the diffusion is positive recurrent.

11 STOPPING TIMES AND THE STRONG MARKOV PROPERTY

Take Ω to be the set of all possible paths, i.e., the set of all continuous functions ω on the time interval $[0, \infty)$ into the interval S (state space). In this case X_t

is the value of the trajectory (function) at time t, $X_t(\omega) = \omega_t$. The sigmafield $\mathscr{F}$ is the smallest sigmafield that includes all finite-dimensional events of the form $\{X_{t_i} \in B_i \text{ for } 1 \leqslant i \leqslant n\}$, where $0 \leqslant t_1 < t_2 < \cdots < t_n$ and $B_1, \ldots, B_n$ are subintervals of the state space S. The probabilities of such finite-dimensional events are specified by the transition probability $p(t; x, dy)$ of the diffusion and an initial distribution π (see Eq. 1.4). This determines P on all of $\mathscr{F}$.

Let $\mathscr{F}_t$ denote the sigmafield of events determined by finite-dimensional events of the form $\{X_{t_i} \in B_i \text{ for } 1 \leqslant i \leqslant n\}$ with $t_i \leqslant t$ for all i. Thus, $\mathscr{F}_t = \sigma\{X_s : 0 \leqslant s \leqslant t\}$ is the class of events determined by the sample path or trajectory of the diffusion *up to time t*.

Definition 11.1. A random variable τ on the probability space $(\Omega, \mathscr{F})$ with values in $[0, \infty]$ is a *stopping time* or a *Markov time* if, for every $t \geqslant 0$, $\{\tau \leqslant t\} \in \mathscr{F}_t$.

This says that, for a stopping time τ, whether or not to stop by time t is determined by the sample path or trajectory up to time t. Check that *constant times* $\tau \equiv t_0$ are stopping times. The *first passage times* τ_y, $\tau_c \wedge \tau_d$ are important examples of stopping times (Exercise 1). By the "*past up to time* τ" we mean the *pre-*τ *sigmafield* of events $\mathscr{F}_\tau$ defined by

$$\mathscr{F}_\tau := \sigma\{X_{t \wedge \tau} : t \geqslant 0\}. \tag{11.1}$$

The stochastic process $\{\tilde{X}_t\} := \{X_{t \wedge \tau}\}$ is referred to as the *process stopped at* τ. Events in $\mathscr{F}_\tau$ depend only on the process stopped at τ. The stopped process contains no further information about the process $\{X_t\}$ beyond the time τ.

An alternative description of the pre-τ sigmafield, which is often useful in checking whether a particular event belongs to it, is as follows,

$$\mathscr{F}_\tau = \{F \in \mathscr{F} : F \cap \{\tau \leqslant t\} \in \mathscr{F}_t \text{ for all } t \geqslant 0\}. \tag{11.2}$$

For example, using this it is simple to check that

$$\{\tau < \infty\} \in \mathscr{F}_\tau. \tag{11.3}$$

The proof of the equivalence of (11.1) and (11.2) is a little long and is omitted (see theoretical complement 1). For our purposes in this section, we take (11.2) as the definition of $\mathscr{F}_\tau$.

It follows from (11.1) or (11.2) that if τ is the constant time $\tau \equiv t_0$, then $\mathscr{F}_\tau = \mathscr{F}_{t_0}$.

As intuition would suggest, if τ_1 and τ_2 are two stopping times such that $\tau_1 \leqslant \tau_2$ then

$$\mathscr{F}_{\tau_1} \subset \mathscr{F}_{\tau_2}. \tag{11.4}$$

For, if $A \in \mathscr{F}_{\tau_1}$, then

$$A \cap \{\tau_2 \leqslant t\} = (A \cap \{\tau_1 \leqslant t\}) \cap (A \cap \{\tau_1 \leqslant t\} \cap \{\tau_2 > t\})^c. \qquad (11.5)$$

Since $A \cap \{\tau_1 \leqslant t\} \in \mathscr{F}_t$ and $\{\tau_2 > t\} = \{\tau_2 \leqslant t\}^c \in \mathscr{F}_t$, the set on the right side of (11.5) is in $\mathscr{F}_t$. Hence $A \in \mathscr{F}_{\tau_2}$.

Another important property of stopping times is that, if τ_1 and τ_2 are stopping times, then $\tau_1 \wedge \tau_2 \equiv \min\{\tau_1, \tau_2\}$ is also a stopping time. This is intuitively clear and simple to prove (Exercise 2). It follows from (11.4) that $\mathscr{F}_{\tau_1 \wedge \tau_2} \subset \mathscr{F}_{\tau_1} \cap \mathscr{F}_{\tau_2}$.

Finite-dimensional events in $\mathscr{F}_\tau$ are those that depend only on finitely many coordinates of the stopped process, i.e., events of the form

$$G = \{(X_{t_1 \wedge \tau}, \ldots, X_{t_m \wedge \tau}) \in B\} \qquad (11.6)$$

where B is a Borel subset of $S^m := S \times S \times \cdots \times S$ (m-fold). Similarly, a finite-dimensional $\mathscr{F}_\tau$-measurable function is of the form

$$f(X_{t_1 \wedge \tau}, \ldots, X_{t_m \wedge \tau}) \qquad (11.7)$$

where f is a Borel-measurable function on S^m. For example, one may take f to be a continuous function on S^m.

More generally, one may consider an arbitrary measurable set F of paths and let G be the event that the stopped process lies in F,

$$G = [\{X_{t \wedge \tau}\} \in F] \qquad (F \in \mathscr{F}). \qquad (11.8)$$

The sigmafield $\mathscr{F}_\tau$ is precisely this class of sets G. The Markov property holds if the "past" and "future" are defined relative to stopping times, and not merely constant times. For this the past relative to a stopping time τ is taken to be the pre-τ sigmafield information contained in the stopped process $\{X_{t \wedge \tau}\}$. The future relative to to τ is the *after-τ process* $X_\tau^+ = \{(X_\tau^+)_t\}$ obtained by viewing $\{X_t\}$ from time $t = \tau$ onwards, for $\tau < \infty$. That is,

$$(X_\tau^+)_t(\omega) = X_{\tau(\omega)+t}(\omega), \qquad t \geqslant 0, \qquad \text{on } \{\omega: \tau(\omega) < \infty\}. \qquad (11.9)$$

Since the after-τ process is defined only on the set $\{\tau < \infty\}$, events based on it are subsets of $\{\tau < \infty\}$. The *after-τ sigmafield* on $\{\tau < \infty\}$ comprises sets of the form

$$H = \{X_\tau^+ \in F\} \cap \{\tau < \infty\}, \qquad (F \in \mathscr{F}). \qquad (11.10)$$

Finite-dimensional events and functions measurable with respect to this sigmafield of subsets on $\{\tau < \infty\}$ may be defined as in (11.6) and (11.7) with $X_{t \wedge \tau}$ replaced by $X_{\tau + t}$.

Theorem 11.1. (*Strong Markov Property*). Let τ be a stopping time. On $\{\tau < \infty\}$, the conditional distribution of X_τ^+ given the past up to the time τ is the same as the distribution of the diffusion $\{X_t\}$ starting at X_τ. In other words, the conditional distribution is P_{X_τ} on $\{\tau < \infty\}$.

Proof. First assume that τ has countably many values ordered as $0 \leqslant s_1 < s_2 < \cdots$. Consider a finite-dimensional function of the after-τ process of the form

$$h(X_{\tau+t'_1}, X_{\tau+t'_2}, \ldots, X_{\tau+t'_r}), \qquad \{\tau < \infty\}, \tag{11.11}$$

where h is a bounded continuous real-valued function on S^r and $0 \leqslant t'_1 < t'_2 < \cdots < t'_r$. It is enough to prove

$$E[h(X_{\tau+t'_1}, \ldots, X_{\tau+t'_r})\mathbf{1}_{\{\tau<\infty\}}|\mathscr{F}_\tau] = (E_y h(X_{t'_1}, \ldots, X_{t'_r}))_{y=X_\tau}\mathbf{1}_{\{\tau<\infty\}}. \tag{11.12}$$

If (11.12) holds, then for every bounded $\mathscr{F}_\tau$-measurable function Z we have

$$\begin{aligned} E(Zh(X_{\tau+t'_1}, \ldots, X_{\tau+t'_r})\mathbf{1}_{\{\tau<\infty\}}) &= E(E[Zh(X_{\tau+t'_1}, \ldots, X_{\tau+t'_r})\mathbf{1}_{\{\tau<\infty\}} \mid \mathscr{F}_\tau]) \\ &= E(ZE[h(X_{\tau+t'_1}, \ldots, X_{\tau+t'_r})\mathbf{1}_{\{\tau<\infty\}} \mid \mathscr{F}_\tau]) \\ &= E(Z[E_y h(X_{t'_1}, \ldots, X_{t'_r})]_{y=X_\tau}\mathbf{1}_{\{\tau<\infty\}}). \end{aligned} \tag{11.13}$$

Conversely, if the equality between the first and last terms in (11.13) holds for all bounded measurable functions Z of the stopped process, then (11.12) holds. Indeed, to verify (11.12) it is enough to check (11.13) for finite-dimensional functions Z of the form (11.7), where f is bounded and continuous on S^m (Exercise 3). Now

$$\begin{aligned} &E(f(X_{\tau\wedge t_1}, \ldots, X_{\tau\wedge t_m})h(X_{\tau+t'_1}, \ldots, X_{\tau+t'_r})\mathbf{1}_{\{\tau=s_j\}}) \\ &\quad = E(f(X_{s_j\wedge t_1}, \ldots, X_{s_j\wedge t_m})h(X_{s_j+t'_1}, \ldots, X_{s_j+t'_r})\mathbf{1}_{\{\tau=s_j\}}) \\ &\quad = E(E[(f(X_{s_j\wedge t_1}, \ldots, X_{s_j\wedge t_m})h(X_{s_j+t'_1}, \ldots, X_{s_j+t'_r})\mathbf{1}_{\{\tau=s_j\}}) \mid \mathscr{F}_{s_j}]) \\ &\quad = E(f(X_{s_j\wedge t_1}, \ldots, X_{s_j\wedge t_m})\mathbf{1}_{\{\tau=s_j\}}E[h(X_{s_j+t'_1}, \ldots, X_{s_j+t'_r}) \mid \mathscr{F}_{s_j}]), \end{aligned} \tag{11.14}$$

since $X_{s_j\wedge t_1}, \ldots, X_{s_j\wedge t_m}$ are determined by $\{X_s: 0 \leqslant s \leqslant s_j\}$, i.e., are measurable with respect to $\mathscr{F}_{s_j}$; also,

$$\{\tau = s_j\} = \{\tau \leqslant s_j\} \cap \{\tau \leqslant s_{j-1}\}^c \in \mathscr{F}_{s_j}.$$

By the Markov property, the last expression in (11.14) equals

$$\begin{aligned} &E(f(X_{s_j\wedge t_1}, \ldots, X_{s_j\wedge t_m})\mathbf{1}_{\{\tau=s_j\}}[E_y h(X_{t'_1}, \ldots, X_{t'_r}\})]_{y=X_{s_j}}) \\ &\qquad = E(f(X_{\tau\wedge t_1}, \ldots, X_{\tau\wedge t_m})\mathbf{1}_{\{\tau=s_j\}}[E_y h(X_{t'_1}, \ldots, X_{t'_r})]_{y=X_\tau}). \end{aligned}$$

Therefore,

$$E(f(X_{\tau\wedge t_1},\dots,X_{\tau\wedge t_m})h(X_{\tau+t'_1},\dots,X_{\tau+t'_r})\mathbf{1}_{\{\tau=s_j\}})$$
$$= E(f(X_{\tau\wedge t_1},\dots,X_{\tau\wedge t_m})\mathbf{1}_{\{\tau=s_j\}}[E_y h(X_{t'_1},\dots,X_{t'_r}\})]_{y=X_\tau}). \quad (11.15)$$

Summing (11.15) over j one gets

$$E(f(X_{\tau\wedge t_1},\dots,X_{\tau\wedge t_m})h(X_{\tau+t'_1},\dots,X_{\tau+t'_r})\mathbf{1}_{\{\tau<\infty\}})$$
$$= E(f(X_{\tau\wedge t_1},\dots,X_{\tau\wedge t_m})[E_y h(X_{t'_1},\dots,X_{t'_r})]_{y=X_\tau}\mathbf{1}_{\{\tau<\infty\}}). \quad (11.16)$$

This completes the proof in the case τ has countably many values $0\leqslant s_1<s_2<\cdots$.

The case of more general τ may be dealt with by approximating it by stopping times assuming countably many values. Specifically, for each positive integer n define

$$\tau_n = \begin{cases} \dfrac{j}{2^n} & \text{if } \dfrac{j-1}{2^n} < \tau \leqslant \dfrac{j}{2^n}, \qquad j = 0, 1, 2, \dots \\ \infty & \text{if } \tau = \infty. \end{cases} \quad (11.17)$$

Since

$$\left\{\tau_n = \frac{j}{2^n}\right\} = \left\{\frac{j-1}{2^n} < \tau \leqslant \frac{j}{2^n}\right\} = \left\{\tau \leqslant \frac{j}{2^n}\right\}\Big\backslash\left\{\tau \leqslant \frac{j-1}{2^n}\right\} \in \mathscr{F}_{j/2^n},$$

$$\{\tau_n \leqslant t\} = \bigcup_{j:\, j/2^n\leqslant t}\left\{\tau_n = \frac{j}{2^n}\right\} \in \mathscr{F}_t \qquad \text{for all } t\geqslant 0.$$

Therefore, τ_n is a stopping time for each n and $\tau_n(\omega)\downarrow\tau(\omega)$ as $n\uparrow\infty$ for each $\omega\in\Omega$. Also, $\mathscr{F}_\tau\subset\mathscr{F}_{\tau_n}$ by (11.4). Let f, h be bounded *continuous* functions on S^m, S^r, respectively. Define

$$\varphi(y) \equiv E_y h(X_{t'_1},\dots,X_{t'_r}). \quad (11.18)$$

Then

$$\varphi(y) = \int g(y_1)p(t'_1; y, y_1)\,dy_1, \quad (11.19)$$

where $g(y_1) = E[h(X_{t'_1},\dots,X_{t'_r}) \mid X_{t'_1} = y_1]$, so that g is bounded. Since $y\to p(t'_1; y, y_1)$ is continuous, φ is continuous (see Exercise 4 and theoretical complement 1). Applying (11.16) to $\tau=\tau_n$ one has

$$E(f(X_{\tau_n\wedge t_1},\dots,X_{\tau_n\wedge t_m})h(X_{\tau_n+t'_1},\dots,X_{\tau_n+t'_r})\mathbf{1}_{\{\tau_n<\infty\}})$$
$$= E(f(X_{\tau_n\wedge t_1},\dots,X_{\tau_n\wedge t_m})\varphi(X_{\tau_n})\mathbf{1}_{\{\tau_n<\infty\}}). \quad (11.20)$$

Since f, h, φ are continuous, $\{X_t\}$ has continuous sample paths, and $\tau_n \downarrow \tau$ as $n \to \infty$, Lebesgue's dominated convergence theorem may be used on both sides of (11.20) to get

$$E(f(X_{\tau \wedge t_1}, \ldots, X_{\tau \wedge t_m})h(X_{\tau + t'_1}, \ldots, X_{\tau + t'_r})\mathbf{1}_{\{\tau < \infty\}})$$
$$= E(f(X_{\tau \wedge t_1}, \ldots, X_{\tau \wedge t_m})\varphi(X_\tau)\mathbf{1}_{\{\tau < \infty\}}). \quad (11.21)$$

This establishes (11.13), and therefore (11.12). ■

Examples 1, 2, 3 below illustrate the importance of Theorem 11.1 in typical computations. In all these examples $\{X_t\}$ is a one-dimensional Brownian motion with zero drift.

Example 1. (*Probability of Reaching One Boundary Point Before Another*). Let $\{X_t\}$ be a one-dimensional Brownian motion with zero drift and diffusion coefficient $\sigma^2 > 0$. Let $c < d$ be two numbers and define

$$\psi(x) = P_x(\{X_t\} \text{ reaches } c \text{ before } d), \qquad (c \leqslant x \leqslant d). \quad (11.22)$$

Fix $x \in (c, d)$ and $h > 0$ such that $[x - h, x + h] \subset (c, d)$. Write $\tau = \tau_{x-h} \wedge \tau_{x+h}$, i.e., τ is the first time $\{X_t\}$ reaches $x - h$ or $x + h$. Then $P_x(\tau < \infty) = 1$, since

$$P_x(\tau > t) \leqslant P_x(x - h < X_t < x + h) = \frac{1}{(2\pi\sigma^2 t)^{1/2}} \int_{x-h}^{x+h} \exp\left\{-\frac{(y - x)^2}{2\sigma^2 t}\right\} dy$$
$$= \frac{1}{(2\pi)^{1/2}} \int_{-h/\sigma\sqrt{t}}^{h/\sigma\sqrt{t}} \exp\left\{-\frac{z^2}{2}\right\} dz \to 0 \qquad \text{as } t \to \infty. \quad (11.23)$$

Now,

$$\psi(x) = P_x(\{X_t\} \text{ reaches } c \text{ before } d) = P_x(\{(X_\tau^+)_t\} \text{ reaches } c \text{ before } d)$$
$$= E_x(P_x(\{(X_\tau^+)_t\} \text{ reaches } c \text{ before } d \mid \{X_{t \wedge \tau}; t \geqslant 0\})). \quad (11.24)$$

The strong Markov property now gives that

$$\psi(x) = E_x(\psi(X_\tau)) \quad (11.25)$$

so that by symmetry of the normal distribution,

$$\psi(x) = \psi(x - h)P_x(X_\tau = x - h) + \psi(x + h)P_x(X_\tau = x + h)$$
$$= \psi(x - h)\tfrac{1}{2} + \psi(x + h)\tfrac{1}{2}. \quad (11.26)$$

Therefore,

$$(\psi(x+h)-\psi(x))-(\psi(x)-\psi(x-h))=0. \tag{11.27}$$

Dividing the second-order difference equation (11.27) by h^2 and letting $h \downarrow 0$, we get

$$\psi''=0, \qquad \psi(c)=1, \qquad \psi(d)=0. \tag{11.28}$$

Therefore,

$$\psi(x)=\frac{d-x}{d-c} \tag{11.29}$$

which is a special case of (2.24), or (4.17) or (9.18); see also Eq. 9.6 of Chapter I.
Now, by (11.23) and (11.29),

$$P_x(\{X_t\} \text{ reaches } d \text{ before } c)=1-\psi(x)=\frac{x-c}{d-c} \tag{11.30}$$

for $c \leqslant x \leqslant d$. It follows, on letting $d \uparrow \infty$ in (11.29) and $c \downarrow -\infty$ in (11.30) that

$$P_x(\tau_y<\infty)=1 \qquad \text{for all } x, y. \tag{11.31}$$

Example 2. (*Independent Increments of the First Passage Time Process*). Let $\{X_t\}$ be as in Example 1, with $X_0=0$ with probability 1. Fix $0<a<b<\infty$. By Theorem 11.1 the conditional distribution of the *after-*τ_a process $X_{\tau_a}^+$ given the past up to time τ_a is $P_{X_{\tau_a}}=P_a$, since $X_{\tau_a}=a$ on $\{\tau_a<\infty\}$, and $P_0(\tau_a<\infty)=1$ by (11.31). Thus $\{X_{t\wedge\tau_a}\}$ and $\{X_{\tau_a+t}\}$ are *independent stochastic process* and the distribution of $X_{\tau_a}^+$ under P_0 is P_a. That is, $X_{\tau_a}^+ := \{(X_{\tau_a}^+)_t\}$ has the same distribution as that of a Brownian motion starting at a.

Also, starting from 0, the state b can be reached only after a has been reached. Hence, with P_0-probability 1,

$$\tau_b=\tau_a+\inf\{t \geqslant 0\colon X_{\tau_a+t}=b\}=\tau_a+\tau_b(X_{\tau_a}^+), \tag{11.32}$$

where $\tau_b(X_{\tau_a}^+)$ is the first hitting time of b by the after-τ_a process $X_{\tau_a}^+$. Since this last hitting time depends only on the after-τ_a process, it is independent of τ_a which is measurable with respect to $\mathscr{F}_{\tau_a}$. Hence $\tau_b-\tau_a \equiv \tau_b(X_{\tau_a}^+)$ and τ_a are independent, and the distribution of $\tau_b-\tau_a$ under P_0 is the distribution of τ_b under P_a. This last distribution is the same as the distribution of τ_{b-a} under P_0. For, if $\{X_t\}$ is a Brownian motion starting at a then $\{X_t-a\}$ is a Brownian motion starting at zero; and if τ_b is the first passage time of $\{X_t\}$ to b then it is also the first passage time of $\{X_t-a\}$ to $b-a$.

If $0<a_1<a_2<\cdots<a_n$ are arbitrary, then the above arguments applied

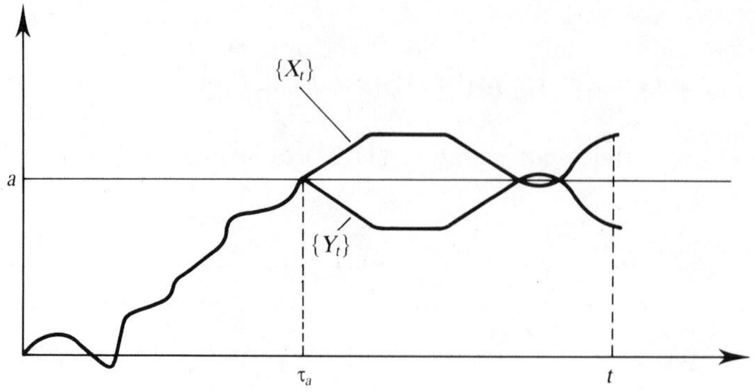

Figure 11.1

to $\tau_{a_{n-1}}$ shows that $\tau_{a_n} - \tau_{a_{n-1}}$ is independent of $\{\tau_{a_i}: 1 \leqslant i \leqslant n-1\}$ and its distribution (under P_0) is the same as the distribution of $\tau_{a_n - a_{n-1}}$ under P_0. We have arrived at the following proposition.

Proposition 11.2. Under P_0, the stochastic process $\{\tau_a: 0 \leqslant a < \infty\}$ is a process with independent increments. Moreover, the increments are homogeneous, i.e., the distribution of $\tau_b - \tau_a$ is the same as that of $\tau_d - \tau_c$ if $d - c = b - a$; both distributions being the same as the distribution of τ_{b-a} under P_0.

Example 3. (*Distribution of the First Passage Time*). Let $\{X_t\}$ be a one-dimensional Brownian motion with zero drift and diffusion coefficient $\sigma^2 > 0$, $X_0 = 0$ with probability 1. Fix $a > 0$. Let $\{Y_t\}$ be the same as $\{X_t\}$ up to time τ_a, and then the reflection of $\{X_t\}$ about the level a after time τ_a (see Figure 11.1 above). That is,

$$Y_t \equiv \begin{cases} X_t & \text{for } t < \tau_a, \\ 2X_{\tau_a} - X_t = 2a - X_t & \text{for } t \geqslant \tau_a. \end{cases} \tag{11.33}$$

Thus, $Y_{\tau_a}^+ := \{Y_{\tau_a + t}\} = a + (a - X_{\tau_a}^+)$. Now the process $a - X_{\tau_a}^+$ is independent of the past up to time τ_a, and has the same distribution as $X_{\tau_a}^+ - a$, namely, that of a Brownian motion starting at zero. Therefore, $Y_{\tau_a}^+ \equiv a + (a - X_{\tau_a}^+)$ is independent of the past up to time τ_a and has the same distribution as that of $a + (X_{\tau_a}^+ - a) = X_{\tau_a}^+$. We have arrived at the following result.

Proposition 11.3. (*The Reflection Principle*). The distribution of $\{Y_t\}$ is the same as that of $\{X_t\}$.

Now for $x \geqslant a$ one has (see Figure 11.1 above)

$$P_0(X_t \geqslant x) = P_0(X_t \geqslant x, \tau_a \leqslant t) = P_0(Y_t \leqslant 2a - x, \tau_a(Y) \leqslant t), \tag{11.34}$$

where $\tau_a(Y)$ is the first passage time to a of $\{Y_t\}$. Note that $\tau_a(Y) = \tau_a$. By Proposition 11.3, the extreme right side of (11.34) equals $P_0(X_t \leqslant 2a - x, \tau_a \leqslant t)$. Hence,

$$P_0(X_t \geqslant x) = P_0(X_t \leqslant 2a - x, \tau_a \leqslant t) = P_0(X_t \leqslant 2a - x, M_t \geqslant a), \quad (11.35)$$

where $M_t = \max\{X_s : 0 \leqslant s \leqslant t\}$. Setting $x = a$ in (11.35) one gets

$$\begin{aligned} P_0(X_t \geqslant a) = P_0(X_t \leqslant a, M_t \geqslant a) &= P_0(M_t \geqslant a) - P_0(M_t \geqslant a, X_t > a) \\ &= P_0(M_t \geqslant a) - P_0(X_t > a) \qquad (11.36) \end{aligned}$$

since $\{X_t > a\} \subset \{M_t \geqslant a\}$. Thus,

$$P_0(M_t \geqslant a) = 2P_0(X_t > a). \quad (11.37)$$

Hence,

$$\begin{aligned} P_0(\tau_a \leqslant t) = P_0(M_t \geqslant a) &= \frac{2}{(2\pi\sigma^2 t)^{1/2}} \int_a^\infty e^{-y^2/(2\sigma^2 t)}\, dy \\ &= \frac{2}{(2\pi)^{1/2}} \int_{a/(\sigma\sqrt{t})}^\infty e^{-z^2/2}\, dz. \qquad (11.38) \end{aligned}$$

Thus, the probability density function (p.d.f.) of τ_a is given by

$$f_a(t) = \frac{a}{\sigma(2\pi)^{1/2}} t^{-3/2} e^{-a^2/(2\sigma^2 t)}, \qquad 0 < t < \infty. \quad (11.39)$$

The p.d.f. of M_t is obtained by differentiating the first integral in (11.38) with respect to a, namely,

$$g_t(a) = \frac{2}{(2\pi\sigma^2 t)^{1/2}} e^{-a^2/(2\sigma^2 t)} = 2\varphi\left(\frac{a}{\sigma\sqrt{t}}\right), \qquad -\infty < a < \infty, \quad (11.40)$$

where φ is the standard normal density, $\varphi(z) = (2\pi)^{1/2} \exp\{-z^2/2\}$.

Changing variables $x \to y = 2a - x$ in (11.35) yields the joint distribution of (X_t, M_t),

$$\begin{aligned} P_0(X_t \leqslant y, M_t \geqslant a) &= P_0(X_t \geqslant 2a - y) \\ &= \int_{(2a-y)/(\sigma\sqrt{t})}^\infty \frac{1}{(2\pi)^{1/2}} e^{-z^2/2}\, dz, \qquad -\infty < y \leqslant a, \quad a \geqslant 0. \end{aligned} \quad (11.41)$$

Differentiating successively with respect to y and a, one gets the joint p.d.f. of

(X_t, M_t) as

$$h_t(y, a) = \frac{2(2a - y)}{\sigma^3 t^{3/2}} \varphi\left(\frac{2a - y}{\sigma\sqrt{t}}\right)$$
$$= \sqrt{\frac{2}{\pi}} \frac{2a - y}{\sigma^3 t^{3/2}} \exp\left\{-\frac{(2a - y)^2}{2\sigma^2 t}\right\}, \qquad -\infty < y \leqslant a, \quad 0 \leqslant a < \infty. \tag{11.42}$$

12 INVARIANT DISTRIBUTIONS AND THE STRONG LAW OF LARGE NUMBERS

In this section it is shown that the existence of an invariant probability is equivalent to positive recurrence. In addition, a computation of the invariant density is provided.

First, an important consequence of the strong Markov property (Theorem 11.1) is the following result.

Proposition 12.1. If the diffusion is recurrent, $\underline{x}, \underline{y} \in S$ $(x < y)$ and f is a real-valued (measurable) function on S, then the sequence of random variables

$$Z_r \equiv \int_{\eta_{2r}}^{\eta_{2r+2}} f(X_s)\, ds \qquad (r = 1, 2, \ldots) \tag{12.1}$$

is i.i.d., where

$$\eta_1 = \inf\{t \geqslant 0\colon X_t = \underline{x}\}, \qquad \eta_{2r} = \inf\{t \geqslant \eta_{2r-1}\colon X_t = \underline{y}\},$$
$$\eta_{2r+1} = \inf\{t \geqslant \eta_{2r}\colon X_t = \underline{x}\} \qquad (r = 1, 2, \ldots). \tag{12.2}$$

Proof. By the strong Markov property, the conditional distribution of Z_r given the past up to time η_{2r} is the same as the distribution of $\int_0^{\eta_2} f(X_s)\, ds$ with initial state y. This last distribution does not change with the sample point ω, and therefore Z_r is independent of the past up to time η_{2r}. In particular, Z_r is independent of $Z_1, Z_2, \ldots, Z_{r-1}$. ∎

The main result may now be derived.

Theorem 12.2. Suppose that the diffusion is positive recurrent on $S = (a, b)$.

(a) Then there exists a unique invariant distribution $\pi(dx)$.

(b) For every real-valued f such that

$$\int_S |f(x)|\pi(dx) < \infty, \tag{12.3}$$

the strong law of large numbers holds, i.e., with probability 1,

$$\lim_{t\to\infty} \frac{1}{t} \int_0^t f(X_s)\, ds = \int_S f(x)\pi(dx), \tag{12.4}$$

no matter what the initial distribution may be.

(c) If the end points a, b, of S are inaccessible or reflecting, then the invariant probability has a density $\pi(x)$, which is the unique normalized integrable solution of $A^*\pi(x) = 0$, i.e.,

$$\frac{1}{2}\frac{d^2}{dx^2}(\sigma^2(x)\pi(x)) - \frac{d}{dx}(\mu(x)\pi(x)) = 0 \qquad \text{for } x \in S, \tag{12.5}$$

or simply of

$$\frac{1}{2}\frac{d}{dx}(\sigma^2(x)\pi(x)) - \mu(x)\pi(x) = 0. \tag{12.6}$$

Indeed, the invariant measure is the normalized speed measure,

$$\pi(x) = \frac{m'(x)}{m(b) - m(a)}. \tag{12.7}$$

Proof. Positive recurrence implies $E_y(\eta_{2r} - \eta_{2r-2}) = E_y\eta_2 < \infty$. Hence, by the classical strong law of large numbers,

$$\frac{\eta_{2r}}{r} = \frac{\sum_{r'=1}^{r} (\eta_{2r'} - \eta_{2(r'-1)})}{r} \to E_y\eta_2 \tag{12.8}$$

with P_y-probability 1, as $r \to \infty$. Let f be a bounded real-valued (measurable) function on S. Then applying the strong law to the sequence Z_r $(r = 0, 1, 2, \ldots; \eta_0 = 0)$ in (12.1) one gets

$$\lim_{r\to\infty} \frac{1}{r} \sum_{r'=1}^{r} Z_{r'-1} = E_y Z_0 = E_y \int_0^{\eta_2} f(X_s)\, ds. \tag{12.9}$$

As in Section 9 of Chapter II (or Section 8 of Chapter IV), one has (Exercise 1)

$$\lim_{r\to\infty} \frac{1}{\eta_{2r}} \int_0^{\eta_{2r}} f(X_s)\, ds = \lim_{t\to\infty} \frac{1}{t} \int_0^t f(X_s)\, ds, \tag{12.10}$$

for every f such that

$$E_y \int_0^{\eta_2} |f(X_s)|\, ds < \infty. \tag{12.11}$$

Combining (12.8)–(12.10), one gets

$$\lim_{t\to\infty} \frac{1}{t} \int_0^t f(X_s)\, ds = \frac{1}{E_y\eta_2} E_y \int_0^{\eta_2} f(X_s)\, ds, \tag{12.12}$$

for all f satisfying (12.11). In the special case $f = \mathbf{1}_B$ with B a Borel subset of S, (12.12) becomes

$$\lim_{t\to\infty} \frac{1}{t} \int_0^t \mathbf{1}_B(X_s)\, ds = \pi(B) \tag{12.13}$$

where

$$\pi(B) := \frac{1}{E_y\eta_2} E_y \int_0^{\eta_2} \mathbf{1}_B(X_s)\, ds. \tag{12.14}$$

Thus, (12.13) says that the limiting proportion of time the process spends in a set B equals the expected amount of time it spends in B during a single cycle relative to (i.e., divided by) the mean length of the cycle. It is simple to check from (12.14) that π is a probability measure on S (Exercise 2). Also, if $f = \sum_1^n a_i \mathbf{1}_{B_i}$, where $a_1, a_2, \ldots, a_n$ are real numbers and $B_1, B_2, \ldots, B_n$ are pairwise disjoint (Borel) subsets of S, then

$$\begin{aligned} \frac{1}{E_y\eta_2} E_y \int_0^{\eta_2} f(X_s)\, ds &= \frac{1}{E_y\eta_2} \sum_{i=1}^n a_i \int_0^{\eta_2} \mathbf{1}_{B_i}(X_s)\, ds \\ &= \sum_{i=1}^n a_i \pi(B_i) = \int_S f(x)\pi(dx). \end{aligned}$$

The equality

$$\frac{1}{E_y\eta_2} E_y \int_0^{\eta_2} f(X_s)\, ds = \int_S f(x)\pi(dx) \tag{12.15}$$

may now be extended to all f satisfying (12.11) (Exercise 2). Combining (12.12) and (12.15), one has

$$\lim_{t\to\infty} \frac{1}{t} \int_0^t f(X_s)\, ds = \int_S f(x)\pi(dx) \tag{12.16}$$

with P_y-probability 1, for all f satisfying (12.11). Taking expectations in (12.16) for *bounded* f one gets, by arguments used in Section 9 of Chapter II,

$$\lim_{t\to\infty} \frac{1}{t} \int_0^t E_y f(X_s)\, ds = \int_S f(x)\pi(dx) \tag{12.17}$$

for *all* y. Writing

$$(\mathbf{T}_s f)(y) = E_y f(X_s) = \int_S f(z)p(s; y, z)\, dz, \tag{12.18}$$

one may express (12.17) as

$$\lim_{t\to\infty} \frac{1}{t} \int_0^t (\mathbf{T}_s f)(y)\, ds = \int_S f(x)\pi(dx) \tag{12.19}$$

for *all* $y \in S$. But the left side also equals, for any given $h > 0$,

$$\lim_{t\to\infty} \frac{1}{t} \int_h^{t+h} (\mathbf{T}_s f)(y)\, ds = \lim_{t\to\infty} \frac{1}{t} \int_0^t (\mathbf{T}_{s+h} f)(y)\, ds = \lim_{t\to\infty} \frac{1}{t} \int_0^t (\mathbf{T}_s \mathbf{T}_h f))(y)\, ds$$
$$= \int (\mathbf{T}_h f)(x)\pi(dx). \tag{12.20}$$

The last equality follows from (12.19) applied to the function $T_h f$. It follows from (12.19) and (12.20) that, at least for all bounded (measurable) f, one has

$$\int_S (\mathbf{T}_h f)(x)\pi(dx) = \int_S f(x)\pi(dx) \qquad (h > 0). \tag{12.21}$$

Specializing this to $f = \mathbf{1}_B$ one has $(\mathbf{T}_h f)(x) = p(h; x, B)$, so that (12.21) yields

$$P_\pi(X_h \in B) = E_\pi P(X_h \in B \mid X_0) = \int_S p(h; x, B)\pi(dx)$$
$$= \int_S \mathbf{1}_B(x)\pi(dx) = \int_B \pi(dx) = \pi(B) = P_\pi(X_0 \in B), \tag{12.22}$$

proving that π is an invariant probability. The proof of parts (a), (b) of Theorem 12.2 is now complete, excepting for uniqueness. Uniqueness may be proved in the same manner as in Section 6 of Chapter II (Exercise 3).

In order to prove (c), first note that $\pi(x)$ given by (12.7) is a probability density function (p.d.f.) which satisfies (12.6) and, therefore, (12.5). To prove its invariance one needs to check (see (12.14)), for π given by (12.7),

$$\frac{1}{E_{\underline{y}}\eta} E_{\underline{y}} \int_0^{\eta_2} f(X_s)\,ds = \int_S f(z)\pi(z)\,dz \equiv \int_S \frac{f(z)\,dm(z)}{m(b) - m(a)} \tag{12.23}$$

where f is an arbitrary bounded measurable function on S, and a, b, are the end points of S. But, as in the proof of (10.17) and (10.18) (Exercise 4)

$$\begin{aligned} E_{\underline{x}} \int_0^{\tau_{\underline{y}}} f(X_s)\,ds &= \int_{\underline{x}}^{\underline{y}} \int_a^z f(u)\,dm(u)\,ds(z), \\ E_{\underline{y}} \int_0^{\tau_{\underline{x}}} f(X_s)\,ds &= \int_{\underline{x}}^{\underline{y}} \int_z^b f(u)\,dm(u)\,ds(z). \end{aligned} \tag{12.24}$$

Hence, by the strong Markov property,

$$\begin{aligned} E_{\underline{y}} \int_0^{\eta_2} f(X_s)\,ds &= E_{\underline{y}} \int_0^{\tau_{\underline{x}}} f(X_s)\,ds + E_{\underline{x}} \int_0^{\tau_{\underline{y}}} f(X_s)\,ds \\ &= \int_{\underline{x}}^{\underline{y}} \left(\int_a^b f(u)\,dm(u) \right) ds(z) = s(\underline{x}; \underline{y}) \int_a^b f(u)\,dm(u). \end{aligned} \tag{12.25}$$

In particular, taking $f \equiv 1$,

$$E_{\underline{y}}\eta_2 = s(\underline{x}; \underline{y})(m(b) - m(a)). \tag{12.26}$$

Dividing (12.25) by (12.26), one arrives at (12.23). ■

The following alternative argument is instructive, and may be justified under appropriate assumptions on the transition probability density $p(t; x, y)$ (Exercise 5). By the backward equation (2.15) and integration by parts one has, for the function $\pi(x)$ given by (12.7),

$$\begin{aligned} \frac{\partial}{\partial t} \int_S p(t; x, y)\pi(x)\,dx &= \int_S \frac{\partial p(t; x, y)}{\partial t} \pi(x)\,dx = \int_S (\mathbf{A}_x p(t; x, y))\pi(x)\,dx \\ &= \int_S p(t; x, y)(\mathbf{A}^*\pi)(x)\,dx = 0, \qquad (t > 0). \end{aligned} \tag{12.27}$$

Since $\int_S p(t; x, y)\pi(x)\,dx$ is the p.d.f. of X_t when the initial p.d.f. is $\pi(x)$, it follows from (12.27) that the distribution of X_t does not change for $t > 0$; but $X_t \to X_0$ a.s. as $t \downarrow 0$, so that X_t converges in distribution to $\pi(x)\,dx$. Therefore, X_t has p.d.f. $\pi(x)$ for every $t > 0$.

Example 1. *(Price Adjustment in a Two-Commodity Model).* The *excess* *demand function* $a(x) = (z_1(x), z_2(x))$ of two commodities is a function of prices

$$x = (x_1, x_2) \in \Delta := \{(x_1, x_2): x_1 > 0, x_2 > 0, x_1 + x_2 = 1\}$$

such that *Walras' law* holds, namely,

$$x_1 z_1(x) + x_2 z_2(x) = 0, \qquad x \in \Delta. \tag{12.28}$$

In view of (12.28) one may concentrate on the behavior of just one excess demand, say $z_1(x)$. Price adjustments in a market usually depend on excess demands. Since $x_2 = 1 - x_1$, one may consider the price $X_1(t)$ of the first commodity as a diffusion on $(0, 1)$ with a drift function determined by the excess demand z_1 at this price. For example, take the generator of such a diffusion to be of the form

$$\tilde{A} = \tfrac{1}{2}\sigma^2(x)\frac{d^2}{dx^2} + z_1(x)\frac{d}{dx}, \qquad (0 < x < 1), \tag{12.29}$$

so that the mean rate of change in price is proportional to the excess demand. The drift function $z_1(x)$ satisfies

$$\lim_{x \downarrow 0} z_1(x) = \infty, \qquad \lim_{x \uparrow 1} z_1(x) = -\infty. \tag{12.30}$$

This says that if the price of commodity 1 is very low (high) its demand is very high (low), so the price will tend to go up (down) fast at the mean rate z_1. Also, assume that there is a constant $\sigma_0^2 > 0$ such that

$$\sigma^2(x) \geqslant \sigma_0^2. \tag{12.31}$$

Then the boundaries 0 and 1 become *inaccessible*. Indeed, it is easy to check in this case that $s(0) = -\infty$, $s(1) = \infty$, and that the diffusion on $S = (0, 1)$ is positive recurrent and, therefore, 0 and 1 cannot be reached from S (Exercise 6). There is, therefore, a unique *steady-state* or *equilibrium distribution* $\pi(dx) = \pi(x)\,dx$, and, no matter where the system starts, the distribution of $X_1(t)$ will approach this steady-state distribution as $t \uparrow \infty$ (see theoretical complement 1).

Example 2. (*Stochastic Changes in the Size of an Industry*). By the "*size*" of a competitive industry is meant its productive capacity. The *profit level* $f(y)$ for a given size y is the rate of additional return (profit) per unit of increase in the industry size from its present size y. The *law of diminishing return* requires that $f'(y) < 0$ for all y. There is a "normal" profit level $\bar{p}$. When the current level f falls below $\bar{p}$, firms tend to drop out or reduce their individual capacities. If the current profit level f is higher than $\bar{p}$, then the productive capacity of the industry is increased owing either to the appearance of new firms or to expansions in existing firms. In a deterministic dynamic model one might consider the industry size y_t to be governed by an equation of the form

$$\frac{dy_t}{dt} = g(f(y_t) - \bar{p}), \tag{12.32}$$

where g is assumed to be *sign-preserving*. That is,

$$g(u)\begin{cases} > 0 & \text{if } u > 0 \\ < 0 & \text{if } u < 0. \end{cases} \tag{12.33}$$

The *deterministic equilibrium* is the value y^* such that $f(y^*) = \bar{p}$. We assume $f(0) > \bar{p} > f(\infty)$. In a stochastic dynamic model one may take the industry size Y_t at time t to be a diffusion on $S = (0, \infty)$ with drift

$$\mu(y) = g(f(y) - \bar{p}), \qquad 0 < y < \infty, \tag{12.34}$$

and a diffusion coefficient $\bar{\sigma}^2(y)$ that represents the (time) rate of local fluctuation (variance) when the process is at y. If one assumes

$$\sigma^2(y) \geqslant \sigma_0^2 > 0, \qquad \overline{\lim_{y\to\infty}}\, g(f(y) - \bar{p}) < 0,$$
$$\underline{\lim_{y\downarrow 0}}\, yg(f(y) - \bar{p}) > 0, \tag{12.35}$$

then the process $\{Y_t\}$ is positive recurrent (and, in particular, 0 and ∞ are inaccessible) (Exercise 7). The *strict* sign-preserving property (12.33) is not needed for this. The stochastic steady-state distribution is approached as time increases, no matter how the process starts. This distribution is sometimes referred to as the *stochastic equilibrium*.

13 THE CENTRAL LIMIT THEOREM FOR DIFFUSIONS

Consider a positive recurrent diffusion on an interval S. The following theorem may be proved in much that same way as described in Section 10 of Chapter II (Exercise 1).

Let $y \in S$, and define η_2 as in (12.2).

Theorem 13.1. Let f be a real-valued (measurable) function on the state space S of a positive recurrent diffusion, satisfying

$$\int_S f^2(x)\pi(x)\,dx < \infty, \qquad \delta^2 := E_y\left(\int_0^{\eta_2} (f(X_s) - E_\pi f)\,ds\right)^2 < \infty, \tag{13.1}$$

where $\pi(dx) = \pi(x)\,dx$ is the invariant probability. Then as $t \to \infty$,

$$\frac{1}{\sqrt{t}} \int_0^t (f(X_s) - E_\pi f)\,ds \tag{13.2}$$

converges in distribution to a Gaussian law with mean zero and variance $\delta^2/(E_y \eta_2)$, whatever the initial distribution.

In order to use Theorem 13.1, one must be able to express the variance parameter in terms of the drift and diffusion coefficients. A derivation of such an expression will be given now.

Assume $E_\pi f = 0$ (i.e., replace f by $f - E_\pi f$). Under the initial distribution π the variance of (13.2) is given by

$$\begin{aligned}
\mathrm{Var}_\pi \frac{1}{\sqrt{t}} \int_0^t f(X_s)\,ds &= E_\pi\left(\frac{1}{\sqrt{t}} \int_0^t f(X_s)\,ds\right)^2 = \frac{1}{t} E_\pi \int_0^t \int_0^t f(X_s) f(X_{s'})\,ds'\,ds \\
&= \frac{2}{t} E_\pi \int_0^t \int_0^t f(X_{s'}) f(X_s)\,ds'\,ds = \frac{2}{t} \int_0^t \int_0^s E_\pi(f(X_{s'}) f(X_s)) \\
&= \frac{2}{t} \int_0^t \int_0^s E_\pi[f(X_{s'}) E(f(X_s) \mid X_{s'})]\,ds'\,ds \\
&= \frac{2}{t} \int_0^t \int_0^s E_\pi(f(X_{s'})(\mathbf{T}_{s-s'} f)(X_{s'}))\,ds'\,ds.
\end{aligned} \tag{13.3}$$

Now

$$E_\pi(f(X_{s'})(\mathbf{T}_{s-s'} f)(X_{s'})) = \int_S f(x)(\mathbf{T}_{s-s'} f)(x)\pi(x)\,dx. \tag{13.4}$$

Therefore, (13.3) may be expressed as

$$\begin{aligned}
\mathrm{Var}_\pi \frac{1}{\sqrt{t}} \int_0^t f(X_s)\,ds &= \int_S \left(\frac{2}{t} \int_0^t \int_0^s (\mathbf{T}_{s-s'} f)(x)\,ds'\,ds\right) f(x)\pi(x)\,dx \\
&= \int_S \left[\frac{2}{t} \int_0^t \left(\int_0^s (\mathbf{T}_u f)(x)\,du\right) ds\right] f(x)\pi(x)\,dx.
\end{aligned} \tag{13.5}$$

Assume that the limit

$$h(x) \equiv \lim_{s \to \infty} \int_0^s (\mathbf{T}_u f)(x)\,du \tag{13.6}$$

exists in $L^2(S, \pi)$ (see remark following proof of Theorem T.13.4, page 515).

Then

$$\lim_{t\to\infty} \frac{1}{t}\int_0^t \int_0^s (\mathbf{T}_u f)(x)\, du\, ds = h(x), \tag{13.7}$$

and one has, using (13.5) and (13.7),

$$\lim_{t\to\infty} \operatorname{Var}_\pi \frac{1}{\sqrt{t}}\int_0^t f(X_s)\, ds = 2\int_S h(x)f(x)\pi(x)\, dx. \tag{13.8}$$

Let us show that the function h satisfies the equation

$$\mathbf{A}h(x) = -f(x), \tag{13.9}$$

where A is the infinitesimal generator of the diffusion, i.e.,

$$\mathbf{A} = \tfrac{1}{2}\sigma^2(x)\frac{d^2}{dx^2} + \mu(x)\frac{d}{dx}, \tag{13.10}$$

together with boundary conditions if S has a boundary. For this, note that, by (13.6), for $\varepsilon > 0$,

$$\begin{aligned}
\mathbf{T}_\varepsilon h(x) &= \lim_{s\to\infty}\int_0^s (\mathbf{T}_\varepsilon \mathbf{T}_u f)(x)\, du = \lim_{s\to\infty}\int_0^s (\mathbf{T}_{\varepsilon+u} f)(x)\, du \\
&= \lim_{s\to\infty}\int_\varepsilon^{s+\varepsilon} (\mathbf{T}_v f)(x)\, dv = \lim_{s'\to\infty}\int_\varepsilon^{s'} (\mathbf{T}_v f)(x)\, dv \\
&= \lim_{s'\to\infty}\int_0^{s'} (\mathbf{T}_v f)(x)\, dv - \int_0^\varepsilon (\mathbf{T}_v f)(x)\, dv = h(x) - \int_0^\varepsilon (\mathbf{T}_v f)(x)\, dv.
\end{aligned} \tag{13.11}$$

Hence,

$$\mathbf{A}h(x) = \lim_{\varepsilon\downarrow 0}\frac{(\mathbf{T}_\varepsilon h)(x) - h(x)}{\varepsilon} = -\lim_{\varepsilon\downarrow 0}\frac{1}{\varepsilon}\int_0^\varepsilon (\mathbf{T}_v f)(x)\, dv = -(\mathbf{T}_0 f)(x) = -f(x). \tag{13.12}$$

These calculations provide the following result.

Proposition 13.2. The variance of the limiting Gaussian distribution in Theorem (13.1) is given by

$$2\langle f, h\rangle := 2\int_S f(x)h(x)\pi(x)\, dx \tag{13.13}$$

where h is a solution to (13.9) in $L^2(S, \pi)$.

It may be shown that if there exists a function h in $L^2(S, \pi)$ satisfying (13.9) for a given f in $L^2(S, \pi)$ with $E_\pi f = 0$, then such an h is unique up to the addition of a constant (see theoretical complement 4).

The condition $E_\pi f = 0$ ensures that in this case the expression (13.13) does not change if a constant is added to h. Also note that, in order that (13.9) may admit a solution h, one must have $E_\pi f = 0$. This is seen by integrating both sides of (13.9) with respect to $\pi(x)\,dx$ and reducing the first integral by integration by parts. That is,

$$\int_S \mathbf{A}h(x)\pi(x)\,dx = \int_S h(x)(\mathbf{A}^*\pi(x))\,dx$$

$$= \int_S h(x)[\tfrac{1}{2}(\sigma^2(x)\pi(x))'' - (\mu(x)\pi(x))']\,dx = 0, \quad (13.14)$$

using (12.5).

14 INTRODUCTION TO MULTIDIMENSIONAL BROWNIAN MOTION AND DIFFUSIONS

A *k-dimensional standard Brownian motion* with initial state $\mathbf{x} = (x^{(1)}, x^{(2)}, \ldots, x^{(k)})$ is the process $\{\mathbf{B}_t^{\mathbf{x}} = (B_t^{x^{(1)}}, \ldots, B_t^{x^{(k)}}): t \geqslant 0\}$ where $\{B_t^{x^{(i)}}\}$, $1 \leqslant i \leqslant k$, are k independent one-dimensional Brownian motions starting at $x^{(i)}$ $(1 \leqslant i \leqslant k)$. It is easily seen to be Markovian and the conditional density of $\mathbf{B}_{t+s}^{\mathbf{x}}$ given $\{\mathbf{B}_h^{\mathbf{x}}: 0 \leqslant u \leqslant s\}$ is a Gaussian (k-dimensional) density with mean vector $\mathbf{B}_s^{\mathbf{x}}$ and variance–covariance matrix tI where I is the $(k \times k)$ *identity matrix*. The transition density function is

$$p(t; \mathbf{x}, \mathbf{y}) = \frac{1}{[(2\pi t)^{1/2}]^k} \exp\left\{-\frac{1}{2t}\sum_{i=1}^{k}(y^{(i)} - x^{(i)})^2\right\}$$

$$= \prod_{i=1}^{k} \frac{1}{(2\pi t)^{1/2}} \exp\left\{-\frac{(y^{(i)} - x^{(i)})^2}{2t}\right\}$$

where

$$\mathbf{x} = (x^{(1)}, \ldots, x^{(k)}), \qquad \mathbf{y} = (y^{(1)}, \ldots, y^{(k)}). \quad (14.1)$$

As in the one-dimensional case, the Markov property also follows from the fact that $\{\mathbf{B}_t^{\mathbf{x}}: t \geqslant 0\}$ is a (vector-valued) *process with independent increments*. It is straightforward to check that p satisfies the *backward equation*

$$\frac{\partial p}{\partial t} = \frac{1}{2}\sum_{i=1}^{k} \frac{\partial^2 p}{(\partial x^{(i)})^2}, \quad (14.2)$$

as well as the *forward equation*

$$\frac{\partial p}{\partial t} = \frac{1}{2} \sum_{i=1}^{k} \frac{\partial^2 p}{(\partial y^{(i)})^2}. \tag{14.3}$$

A Brownian motion $\{\mathbf{X}_t = (X_t^{(1)}, \ldots, X_t^{(k)})\}$ with *drift vector* $\boldsymbol{\mu} = (\mu^{(1)}, \ldots, \mu^{(k)})$ and *diffusion matrix* $\mathbf{D} = ((d_{ij}))$, is defined by

$$\mathbf{X}_t = \mathbf{x}_0 + t\boldsymbol{\mu} + \sigma \mathbf{B}_t, \tag{14.4}$$

where $\mathbf{X}_0 = \mathbf{x}_0 = (x_0^{(1)}, \ldots, x_0^{(k)})$ is the initial state, $\boldsymbol{\sigma}$ is a $k \times k$ matrix satisfying $\boldsymbol{\sigma}\boldsymbol{\sigma}' = \mathbf{D}$, and $\mathbf{B}_t = (B_t^{(1)}, \ldots, B_t^{(k)})$ is a standard Brownian motion with initial state "zero" (vector). For each $t > 0$, $\boldsymbol{\sigma}\mathbf{B}_t$ is a nonsingular linear transformation of a Gaussian vector $\mathbf{B}_t$ whose mean (vector) is zero and variance–covariance matrix $t\mathbf{I}$. Therefore, $\boldsymbol{\sigma}\mathbf{B}_t$ is a Gaussian random vector with mean vector zero and variance–covariance matrix $\boldsymbol{\sigma} t \mathbf{I} \sigma' = t\boldsymbol{\sigma}\mathbf{I}\boldsymbol{\sigma}' = t\mathbf{D} = ((td_{ij}))$. Therefore, $\mathbf{X}_t$ is Gaussian with mean $\mathbf{x}_0 + t\boldsymbol{\mu}$ and variance–covariance matrix $t\mathbf{D}$. Since $\mathbf{X}_{t+s} - \mathbf{X}_t = s\boldsymbol{\mu} + \boldsymbol{\sigma}(\mathbf{B}_{t+s} - \mathbf{B}_t)$, $\{\mathbf{X}_t\}$ is a process with independent increments (and is, therefore, Markovian) having a transition density function

$$p(t; \mathbf{x}, y) = \frac{1}{((2\pi t)^{1/2})^k (\det \mathbf{D})^{1/2}} \times \exp\left\{-\frac{1}{2t} \sum_{i=1}^{k} \sum_{j=1}^{k} d^{ij}(y^{(i)} - x^{(i)} - t\mu^{(i)})(y^{(j)} - x^{(j)} - t\mu^{(j)})\right\}$$

$$(\mathbf{x}, \mathbf{y} \in R^k;\ t \geqslant 0). \tag{14.5}$$

Here $((d^{ij})) = \mathbf{D}^{-1}$. This transition probability density satisfies the backward and forward equations (Exercise 3)

$$\begin{aligned} \frac{\partial p}{\partial t} &= \frac{1}{2} \sum_{i=1}^{k} \sum_{j=1}^{k} d_{ij} \frac{\partial^2 p}{\partial x^{(i)}\, \partial x^{(j)}} + \sum_{i=1}^{k} \mu^{(i)} \frac{\partial p}{\partial x^{(i)}}, \\ \frac{\partial p}{\partial t} &= \frac{1}{2} \sum_{i=1}^{k} \sum_{j=1}^{k} d_{ij} \frac{\partial^2 p}{\partial y^{(i)}\, \partial x^{(j)}} - \sum_{i=1}^{k} \mu^{(i)} \frac{\partial p}{\partial y^{(i)}}. \end{aligned} \tag{14.6}$$

More generally, we have the following.

Definition 14.1. *A k-dimensional diffusion with* (nonconstant) *drift bector* $\boldsymbol{\mu}(\mathbf{x}) = (\mu^{(1)}(\mathbf{x}), \ldots, \mu^{(k)}(\mathbf{x}))$ *and* (nonconstant) *diffusion matrix* $\mathbf{D}(\mathbf{x}) = ((d_{ij}(\mathbf{x})))$ is a Markov process whose transition density function $p(t; \mathbf{x}, \mathbf{y})$ satisfies the

Kolmogorov equations

$$\begin{aligned}\frac{\partial p}{\partial t} &= \frac{1}{2}\sum_{i=1}^{k}\sum_{j=1}^{k} d_{ij}(\mathbf{x})\frac{\partial^2 p}{\partial x^{(i)}\,\partial x^{(j)}} + \sum_{i=1}^{k}\mu^{(i)}(\mathbf{x})\frac{\partial p}{\partial x^{(i)}},\\ \frac{\partial p}{\partial t} &= \frac{1}{2}\sum_{i=1}^{k}\sum_{j=1}^{k}\frac{\partial^2(d_{ij}(\mathbf{y})p)}{\partial y^{(i)}\,\partial y^{(j)}} - \sum_{i=1}^{k}\frac{\partial(\mu^{(i)}(\mathbf{y})p)}{\partial y^{(i)}}.\end{aligned} \tag{14.7}$$

Such a p may be shown to exist and be unique under the assumptions:

1. $((d_{ij}(\mathbf{y})))$ is, for each $\mathbf{y}$, a positive definite matrix.
2. $d_{ij}(\mathbf{y})$ is, for each pair (i, j), twice continuously differentiable.
3. $\mu^{(i)}(\mathbf{y})$ is, for each i, continuously differentiable.
4. $|\mu^{(i)}(\mathbf{y})|$ does not go to infinity faster than $|\mathbf{y}|$, and $|d_{ij}(\mathbf{y})|$ does not go to infinity faster than $|\mathbf{y}|^2$.

An alternative definition of a multidimensional diffusion may be given by requiring that the appropriate analog of (1.2)′ holds (Exercise 2). The state space may be taken to be a suitably "regular" open subset of $\mathbb{R}^k$, e.g., a rectangle, a ball, $\mathbb{R}^k\backslash\{0\}$, etc.

Once p is given, the joint distribution of the stochastic process at any finite set of time points is assigned in the usual manner, e.g., given $\mathbf{X}_0 = \mathbf{x}_0$,

$$p(t_1; \mathbf{x}_0, \mathbf{x}_1)p(t_2 - t_1; \mathbf{x}_1, \mathbf{x}_2)\cdots p(t_n - t_{n-1}; \mathbf{x}_{n-1}, \mathbf{x})$$

is the joint density of $(\mathbf{X}_{t_1}, \mathbf{X}_{t_2}, \ldots, \mathbf{X}_{t_n})$ where $0 < t_1 < t_2 < \cdots < t_n$. The sample paths may also be taken to be continuous (vector-valued functions of t).

An alternative method of constructing the diffusion is to solve *Itô's stochastic differential equations*:

$$d\mathbf{X}_t = \boldsymbol{\mu}(\mathbf{X}_t)\,dt + \boldsymbol{\sigma}(\mathbf{X}_t)\,d\mathbf{B}_t, \qquad \mathbf{X}_0 = \mathbf{x}, \tag{14.8}$$

where $\{\mathbf{B}_t\colon t \geqslant 0\}$ is a standard Brownian motion. The equations (14.8) (there are k equations corresponding to the k components of the vector $\mathbf{X}_t$) may be roughly interpreted as follows: *Given* $\{\mathbf{X}_u\colon 0 \leqslant u \leqslant t\}$, *the conditional distribution of the increment* $d\mathbf{X}_t = \mathbf{X}_{t+dt} - \mathbf{X}_t$ *is approximately Gaussian with mean* $\boldsymbol{\mu}(\mathbf{X}_t)\,dt$ *and variance–covariance matrix* $\mathbf{D}(\mathbf{X}_t)\,dt$. We will study the Itô calculus in detail in Chapter VII.

The operator

$$\mathbf{A} := \frac{1}{2}\sum_{i,j=1}^{k} d_{ij}(\mathbf{x})\frac{\partial^2}{\partial x^{(i)}\,\partial x^{(j)}} + \sum_{i=1}^{k}\mu^{(i)}(\mathbf{x})\frac{\partial}{\partial x^{(i)}} \tag{14.9}$$

is called the (*infinitesimal*) *generator of the diffusion* on $\mathbb{R}^k$ with *drift* $\boldsymbol{\mu}(\cdot)$ and *diffusion matrix* $\mathbf{D}(\cdot)$.

It is sometimes possible to reduce the state space by transformation from multidimensional to one dimensional while still preserving the Markov property. Such reductions often facilitate computation of certain probabilities. Of course, if φ is a one-to-one measurable transformation then $\varphi(\mathbf{x})$ may be thought of only as a relabeling of $\mathbf{x}$, and the Markov property is preserved, since knowing $\varphi(\mathbf{X}_u)$, $0 \leqslant u \leqslant s$, is equivalent to knowing $\mathbf{X}_u$, $0 \leqslant u \leqslant s$. In particular, if φ is a one-to-one twice continuously differentiable map of $\mathbb{R}^k$ onto an open subset S of $\mathbb{R}^k$, then the Markov process $\{\mathbf{Y}_t := \varphi(\mathbf{X}_t)\}$ is also a diffusion whose drift and diffusion coefficients may be expressed in terms of those of $\{\mathbf{X}_t\}$ in a manner analogous to that in Proposition 3.2 (Exercise 2). The Itô calculus of Chapter VII provides another method for calculating the coefficients of the transformed diffusion (Exercise 3).

We are here interested, however, in transformations that are not one-to-one and, in fact, reduce the dimension of the state space.

Example 1. (*The Bessel Process or the Radial Brownian Motion*). Let $\{\mathbf{B}_t\}$ be a k-dimensional standard Brownian motion, $k > 1$. Let

$$\phi(\mathbf{x}) = |\mathbf{x}| = \left(\sum_{i=1}^{k} (x^{(i)})^2\right)^{1/2}, \qquad R_t := \varphi(\mathbf{B}_t) = |\mathbf{B}_t|.$$

Let us show that $\{R_t\}$ is a Markov process on the state space $\bar{S} = [0, \infty)$.

Let f be an arbitrary real-valued, bounded measurable function on $\bar{S}$. Write $g(\mathbf{x}) = (f \circ \varphi)(\mathbf{x}) = f(|\mathbf{x}|)$. Using the Markov property of $\{\mathbf{B}_t\}$ one has

$$\begin{aligned} E[f(R_{t+s}) \mid \{R_u : 0 \leqslant u \leqslant s\}] &= E[E(f(R_{t+s}) \mid \{\mathbf{B}_u : 0 \leqslant u \leqslant s\}) \mid \{R_u : 0 \leqslant u \leqslant s\}] \\ &= E[E(g(\mathbf{B}_{t+s}) \mid \{\mathbf{B}_u : 0 \leqslant u \leqslant s\}) \mid \{R_u : 0 \leqslant u \leqslant s\}] \\ &= E[(\mathbf{T}_t g)(\mathbf{B}_s) \mid \{R_u : 0 \leqslant u \leqslant s\}], \end{aligned} \tag{14.10}$$

where $\{\mathbf{T}_t\}$ is the semigroup of transition operators associated with $\{\mathbf{B}_t\}$, i.e.,

$$\begin{aligned} (\mathbf{T}_t g)(\mathbf{x}) &= \int_{\mathbb{R}^k} g((y)) p(t; \mathbf{x}, \mathbf{y}) \, d\mathbf{y} \\ &= \int_{\mathbb{R}^k} f(|\mathbf{y}|)(2\pi t)^{-k/2} \exp\left\{-\frac{|\mathbf{y} - \mathbf{x}|^2}{2t}\right\} d\mathbf{y}. \end{aligned} \tag{14.11}$$

To reduce the last integral, change to polar coordinates. That is, first integrate over the surface of the sphere $\{|\mathbf{y}| = r\}$ with respect to the normalized surface area measure da_r and then integrate out r after multiplying the integrand by the surface area $\omega_k r^{k-1}$ of $\{|\mathbf{y}| = r\}$. Since

$$|\mathbf{y} - \mathbf{x}|^2 = |\mathbf{y}|^2 + |\mathbf{x}|^2 - 2\sum_{i=1}^{k} x^{(i)}y^{(i)},$$

$$\int_{\{|\mathbf{y}|=r\}} \exp\left\{-\frac{|\mathbf{y}-\mathbf{x}|^2}{2t}\right\} da_r = \exp\left\{-\frac{|\mathbf{x}|^2 + |\mathbf{y}|^2}{2t}\right\} \times \int_{\{|\mathbf{z}|=1\}} \exp\left\{\frac{|\mathbf{x}|\cdot|\mathbf{y}|}{t}\sum_{i=1}^{k} w^{(i)}z^{(i)}\right\} da_1(\mathbf{z}) \tag{14.12}$$

where $\mathbf{w} = \mathbf{x}/|\mathbf{x}|$, $\mathbf{z} = \mathbf{y}/|\mathbf{y}|$. Note that $\sum_{i=1}^{k} w^{(i)}z^{(i)}$ is the cosine of the angle between the unit vectors $\mathbf{w}$ and $\mathbf{z}$, and its distribution under da_1 does not depend on the particular unit vector $\mathbf{w}$. In particular, one may replace $\mathbf{w}$ by $(1, 0, \ldots, 0)$. Hence,

$$\begin{aligned}\int_{\{|\mathbf{y}|=r\}} \exp\left\{-\frac{|\mathbf{y}-\mathbf{x}|^2}{2t}\right\} da_r &= \exp\left\{-\frac{|\mathbf{x}|^2 + r^2}{2t}\right\}\int_{\{|\mathbf{z}|=1\}} \exp\left\{\left(\frac{r|\mathbf{x}|}{t}\right)z^{(1)}\right\} da_1 \\ &= \exp\left\{-\frac{|\mathbf{x}|^2 + r^2}{2t}\right\} h\left(\frac{r|\mathbf{x}|}{t}\right),\end{aligned} \tag{14.13}$$

where (Exercise 9)

$$h(v) := \int_{\{|\mathbf{z}|=1\}} e^{vz^{(1)}}\, da_1 = \frac{\int_{-1}^{1} e^{vu}(1 - u^2)^{(k-2)/2}\, du}{\int_{-1}^{1} (1 - u^2)^{(k-2)/2}\, du}. \tag{14.14}$$

Thus,

$$(\mathbf{T}_t g)(\mathbf{x}) = (2\pi t)^{-k/2}\omega_k e^{-|\mathbf{x}|^2/2t}\int_0^\infty f(r)e^{-r^2/2t}h\left(\frac{r|\mathbf{x}|}{t}\right) r^{k-1}\, dr = \psi(t, |\mathbf{x}|), \tag{14.15}$$

say. Using (14.15) in (14.10) one gets the desired Markov property of $\{R_t\}$,

$$E[f(R_{t+s}) \mid \{R_u : 0 \leqslant u \leqslant s\}] = E[\psi(t, |\mathbf{X}_s|) \mid \{R_u : 0 \leqslant u \leqslant s\}] = \psi(t, R_s). \tag{14.16}$$

Further, it follows from (14.15) that the transition probability density function of $\{R_t\}$ is

$$q(t; r, r') = (2\pi t)^{-k/2}\omega_k \exp\left\{-\frac{r^2 + r'^2}{2t}\right\} h\left(\frac{rr'}{t}\right) r'^{k-1}. \tag{14.17}$$

Write the semigroup of transition operators for $\{R_t\}$ as $\{\bar{\mathbf{T}}_t\}$. Then,

$$(\bar{\mathbf{T}}_t f)(r) = (\mathbf{T}_t g)(\mathbf{x})\big|_{|\mathbf{x}|=r} = \psi(t, r),$$

$$\frac{\partial}{\partial t}(\bar{\mathbf{T}}_t f)(r) = \frac{\partial}{\partial t}\psi(t,r) = \frac{\partial}{\partial t}(\mathbf{T}_t g)(\mathbf{x})\bigg|_{|\mathbf{x}|=r} = \frac{1}{2}\sum_{i=1}^{k}\frac{\partial^2(\mathbf{T}_t g)(\mathbf{x})}{(\partial x^{(i)})^2}\bigg|_{|\mathbf{x}|=r}$$

$$= \frac{1}{2}\sum_{i=1}^{k}\frac{\partial^2\psi(t,|\mathbf{x}|)}{(\partial x^{(i)})^2}\bigg|_{|\mathbf{x}|=r} = \frac{1}{2}\sum_{i=1}^{k}\left[\frac{\partial}{\partial x^{(i)}}\left(\frac{\partial\psi(t,r)}{\partial r}\frac{\partial|\mathbf{x}|}{\partial x^{(i)}}\right)\right]_{|\mathbf{x}|=r}$$

$$= \frac{1}{2}\sum_{i=1}^{k}\left[\frac{\partial}{\partial x^{(i)}}\left(\frac{\partial\psi(t,r)}{\partial r}\frac{x^{(i)}}{|\mathbf{x}|}\right)\right]_{|\mathbf{x}|=r}$$

$$= \frac{1}{2}\sum_{i=1}^{k}\left[\frac{\partial^2\psi(t,r)}{\partial r^2}\frac{(x^{(i)})^2}{r^2} + \frac{\partial\psi(t,r)}{\partial r}\left(\frac{1}{r} - \frac{(x^{(i)})^2}{r^3}\right)\right]_{|\mathbf{x}|=r}$$

$$= \frac{1}{2}\frac{\partial^2\psi(t,r)}{\partial r^2} + \frac{k-1}{2r}\frac{\partial\psi(t,r)}{\partial r} = \frac{1}{2}\frac{\partial^2(\bar{\mathbf{T}}_t f)(r)}{\partial r^2} + \frac{k-1}{2r}\frac{\partial(\bar{\mathbf{T}}_t f)(r)}{\partial r}. \tag{14.18}$$

Thus, the transformed Markov process has as its infinitesimal generator the *Bessel operator*

$$\mathbf{A}_r := \frac{1}{2}\frac{d^2}{dr^2} + \frac{k-1}{2r}\frac{d}{dr}. \tag{14.19}$$

It will be shown later that the state space of $\{R_t\}$ may be restricted to $(0, \infty)$, so that $(k-1)/2r$ in (14.19) is defined.

It may be noted that not only is $\{R_t\}$ Markovian, its distribution under $P_{\mathbf{x}}$ is $Q_{\varphi(\mathbf{x})}$; where $P_{\mathbf{x}}$ is the distribution of the Brownian motion $\{\mathbf{X}_t\}$ when $\mathbf{X}_0 \equiv \mathbf{x}$, and Q_r that of $\{R_t\}$ when $R_0 \equiv r$. This is easily checked by looking at the joint distribution, under $P_{\mathbf{x}}$, of $(R_{t_1}, \ldots, R_{t_n})$ by successive conditioning with respect to $\{\mathbf{X}_u : 0 \leqslant u \leqslant t_i\}$,

$$i = n-1, n-2, \ldots, 1 \qquad (0 < t_1 < t_2 < \cdots < t_n).$$

Let us make use of the above reduction to calculate the probability (for given $0 < c < d < \infty$)

$$P_{\mathbf{x}}(\{\mathbf{X}_t\} \text{ reaches } \partial B(\mathbf{a}:c) \text{ before } \partial B(\mathbf{a}:d)), \tag{14.20}$$

for $c < |\mathbf{x} - \mathbf{a}| < d$. Here $B(\mathbf{a}:r) = \{\mathbf{z} \in \mathbb{R}^k : |\mathbf{z} - \mathbf{a}| < r\}$ is the k-dimensional ball of radius r and center a, and $\partial B(\mathbf{a}:r) = \{|\mathbf{z} - \mathbf{a}| = r\}$ is its boundary (surface). By translation, (14.20) is the same as

$$P_{\mathbf{x}-\mathbf{a}}(\{\mathbf{X}_t\} \text{ reaches } \partial B(\mathbf{0}:c) \text{ before } \partial B(\mathbf{0}:d)), \tag{14.21}$$

for $c < |\mathbf{x} - \mathbf{a}| < d$.

In turn, (14.21) equals

$$P_{\mathbf{x}-\mathbf{a}}(\{R_t\} \text{ reaches } c \text{ before } d) = Q_{|\mathbf{x}-\mathbf{a}|}(\{Y_t\} \text{ reaches } c \text{ before } d), \tag{14.22}$$

where $\{Y_t\}$ is the canonical one-dimensional diffusion (i.e., the coordinate process) on $\bar{S} = [0, \infty)$, having the same distribution as $\{R_t\}$. From (2.24) (or, (4.17), or (9.19)), the last probability is given by

$$\psi(|\mathbf{x} - \mathbf{a}|) = \frac{\displaystyle\int_{|\mathbf{x}-\mathbf{a}|}^{d} e^{-I(c,r)}\, dr}{\displaystyle\int_{c}^{d} e^{-I(c,r)}\, dr} \tag{14.23}$$

where, with $\mu(y) = (k - 1)/2y$, $\sigma^2(y) = 1$ (see (14.19)),

$$I(c, r) = \int_c^r \frac{2\mu(y)}{\sigma^2(y)}\, dy = \int_c^r \frac{k-1}{y}\, dy = \log\left(\frac{r}{c}\right)^{k-1}, \tag{14.24}$$

so that

$$\psi(|\mathbf{x} - \mathbf{a}|) = \begin{cases} \dfrac{\log d - \log|\mathbf{x} - \mathbf{a}|}{\log d - \log c} & \text{if } k = 2, \\[2ex] \dfrac{\left(\dfrac{c}{|\mathbf{x} - \mathbf{a}|}\right)^{k-2} - \left(\dfrac{c}{d}\right)^{k-2}}{1 - \left(\dfrac{c}{d}\right)^{k-2}} & \text{if } k > 2, \end{cases} \tag{14.25}$$

for $c < |\mathbf{x} - \mathbf{a}| < d$.

Further, letting $d \uparrow \infty$ in (14.25) (and (14.21)), one arrives at

$$P_{\mathbf{x}}(\{\mathbf{X}_t\} \text{ ever reaches } \partial B(\mathbf{a}:c)) = \begin{cases} 1 & \text{if } k = 2 \\ \left(\dfrac{c}{|\mathbf{x} - \mathbf{a}|}\right)^{k-2} & \text{if } k > 2, \end{cases} \qquad (c < |\mathbf{x} - \mathbf{a}|). \tag{14.26}$$

In other words, a two-dimensional Brownian motion is *recurrent*, while higher-dimensional Brownian motions are *transient* (Exercises 5 and 6). For $k = 2$, given any ball, $\{\mathbf{X}_t\}$ will reach it (with probability 1) no matter where it starts. For $k > 2$, there is a positive probability that $\{\mathbf{X}_t\}$ will never reach a ball if it starts from a point outside. Recall that an analogous result is true for multidimensional simple symmetric random walks (see Section 5 of Chapter I).

In one respect, however, multidimensional random walks on integer lattices differ from multidimensional Brownian motions. In the former case, in view of the countability of the state space, the random walk reaches every state with positive probability. This is not true for multidimensional Brownian motions. Indeed, by letting $c \downarrow 0$ in (14.25) one gets for $0 < |\mathbf{x} - \mathbf{a}| < d$,

$$P_{\mathbf{x}}(\{\mathbf{X}_t\} \text{ reaches } \mathbf{a} \text{ before } \partial B(\mathbf{a}:d)) = 0. \tag{14.27}$$

Letting $d \uparrow \infty$ in (14.27) it follows that

$$P_{\mathbf{x}}(\{\mathbf{X}_t\} \text{ ever reaches } \mathbf{a}) = 0 \qquad (0 < |\mathbf{x} - \mathbf{a}|). \tag{14.28}$$

In particular, taking $\mathbf{a} = 0$, one gets

$$Q_r(\{Y_t\} \text{ ever reaches } 0) = 0, \qquad (0 < r < \infty). \tag{14.29}$$

In view of (14.29), zero is said to be an *inaccessible boundary* for the Bessel process. The state space of the Bessel process may, therefore, be restricted to $S = (0, \infty)$.

Observe that the Markov property of $\{\varphi(\mathbf{X}_t) := |\mathbf{X}_t| = R_t\}$ in Example 1 follows from the fact that $\mathbf{T}_t$ *transforms* (bounded measurable) *functions* of $\varphi(\mathbf{x})$ *into function of* $\varphi(\mathbf{x})$. This is a general property, as may be seen from the steps (14.10), (14.11), (14.15), and (14.16). In most cases, however, it is difficult, if not impossible, to calculate $\mathbf{T}_t g$, since the transition probability cannot be computed explicitly. A simple check for the process $\{\varphi(\mathbf{X}_t)\}$ to be Markov is furnished by the following.

Proposition 14.1

(a) If $\mathbf{A}$ transforms functions of $\varphi(\mathbf{x})$ into functions of $\varphi(\mathbf{x})$ then $\{\varphi(\mathbf{X}_t)\}$ is a Markov process.

(b) If $P_{\mathbf{x}}$ denotes the distribution of $\{\mathbf{X}_t\}$ when $X_0 = \mathbf{x}$ and Q_u that of $\{\varphi(\mathbf{X}_t)\}$ when $\varphi(\mathbf{X}_0) = u$, then the $P_{\mathbf{x}}$ distribution of $\{\varphi(\mathbf{X}_t)\}$ is $Q_{\varphi(\mathbf{x})}$.

This statement needs some qualification, since not all bounded measurable functions are in the domain (of definition) of $\mathbf{A}$ (see theoretical complement 1).

15 MULTIDIMENSIONAL DIFFUSIONS UNDER ABSORBING BOUNDARY CONDITIONS AND CRITERIA FOR TRANSIENCE AND RECURRENCE

Consider a diffusion $\{\mathbf{X}_t\}$ on $\mathbb{R}^k$ with drift coefficients $\mu^{(i)}(\cdot)$ and diffusion coefficients $d_{ij}(\cdot)$ as described in Section 13. Let G be a proper open subset of $\mathbb{R}^k$ with a "smooth" boundary ∂G (see theoretical complement 1), e.g., $G = H^k := \{\mathbf{x} \in \mathbb{R}^k : x^{(i)} > 0\}$, $G = B(\mathbf{0}:a) := \{\mathbf{x} \in \mathbb{R}^k : |\mathbf{x}| < a\}$. Note that $\partial H^k =$

$\{\mathbf{x} \in \mathbb{R}^k: x^{(1)} = 0\}$, $\partial B(\mathbf{0}:a) = \{\mathbf{x} \in \mathbb{R}^k: |\mathbf{x}| = a\}$. One may "restrict" the diffusion to $\bar{G} \equiv G \cup \partial G$ in various ways, by prescribing what the diffusion must do on reaching the boundary ∂G starting from the interior of G, consistent with the Markov property and the continuity of sample paths. The two most common prescriptions are (i) *absorption* and (ii) *reflection*. This section is devoted to absorption.

In the case of *absorption* the new process $\{\tilde{X}_t\}$ is defined in terms of the original *unrestricted* process $\{\mathbf{X}_t\}$ (which is assumed to start in $S = G \cup \partial G$) by

$$\tilde{\mathbf{X}}_t = \begin{cases} \mathbf{X}_t & \text{if } t < \tau_{\partial G}, \\ \mathbf{X}_{\tau_{\partial G}} & \text{if } t \geqslant \tau_{\partial G}, \end{cases} \tag{15.1}$$

where $\tau_{\partial G}$ is the *first passage time* to ∂G, also called the *first hitting time of* ∂G, defined by

$$\tau_{\partial G} = \inf\{t \geqslant 0: \mathbf{X}_t \in \partial G\}. \tag{15.2}$$

As always, $\tau_{\partial G} = \infty$ if $\{\mathbf{X}_t\}$ never hits ∂G.

Definition 15.1. The process $\{\tilde{\mathbf{X}}_t\}$ defined by (15.1) is called a *diffusion* on $\bar{G} = G \cup \partial G$ *with absorption* on ∂G.

The Markov property of $\{\tilde{\mathbf{X}}_t\}$ follows from calculations that are almost the same as carried out for Theorem 7.1 (Exercise 1). Its transition probability is given by

$$\begin{aligned} \tilde{p}(t; \mathbf{x}, B) &= P(\tilde{\mathbf{X}}_t \in B \mid \tilde{\mathbf{X}}_0 = \mathbf{x}) \\ &= \begin{cases} P(\mathbf{X}_t \in B, \tau_{\partial G} > t \mid \mathbf{X}_0 = \mathbf{x}) & \text{if } B \subset G, \\ P(\tau_{\partial G} \leqslant t, \mathbf{X}_{\tau_{\partial G}} \in B \mid \mathbf{X}_0 = \mathbf{x}) & \text{if } B \subset \partial G. \end{cases} \end{aligned} \tag{15.3}$$

Now $\{\mathbf{X}_t\}$ has a transition probability *density*, say, $p(t; \mathbf{x}, \mathbf{y})$, and

$$P(\mathbf{X}_t \in B, \tau_{\partial G} > t \mid \mathbf{X}_0 = \mathbf{x}) \leqslant P(\mathbf{X}_t \in B \mid \mathbf{X}_0 = \mathbf{x}) = \int_B p(t; \mathbf{x}, \mathbf{y})\, d\mathbf{y} = 0, \tag{15.4}$$

if $B \subset G$ has Lebesgue measure zero. Hence, the measure $B \to \tilde{p}(t; \mathbf{x}, B)$ on the Borel sigmafield of G has a density, say $p^0(t; \mathbf{x}, \mathbf{y})$ (which vanishes outside G), so that (15.3) may be expressed as

$$\tilde{p}(t; \mathbf{x}, B) = \begin{cases} \displaystyle\int_B p^0(t; \mathbf{x}, \mathbf{y})\, d\mathbf{y}, & B \subset G, \quad \mathbf{x} \in G, \\ P_{\mathbf{x}}(\tau_{\partial G} \leqslant t, \mathbf{X}_{\tau_{\partial G}} \in B), & B \subset \partial G, \quad \mathbf{x} \in G, \end{cases} \tag{15.5}$$

where $P_{\mathbf{x}}$ denotes the *distribution of the unrestricted process* starting at $\mathbf{x}$. If $\mathbf{x} \in \partial G$, then of course $p(t; \mathbf{x}, B) = \mathbf{1}_B(\mathbf{x})$ for all $t \geqslant 0$.

Example 1. (*Brownian Motion on a Half-Space with Absorption*). Let $\mu^{(i)}(\cdot) = \mu_i$, $d_{ij}(\cdot) = \sigma_i^2 \delta_{ij}$, where the μ_i, $\sigma_i^2 > 0$ are constants. Let $\{\mathbf{X}_t = (X_t^{(1)}, \ldots, X_t^{(k)})\}$ be an unrestricted Brownian motion with these coefficients, starting at some $\mathbf{x} \in H^k$. First consider the case $k = 1$. For this case the state space is $S = [0, \infty)$, whose boundary is $\{0\}$. Example 8.4 provides the density $p^0 = p_1^0$, say, given by

$$p_1^0(t; u, v) = \exp\left\{\frac{\mu_1(v - u)}{\sigma_1^2} - \frac{\mu_1^2 t}{2\sigma_1^2}\right\}(2\pi\sigma_1^2 t)^{-1/2} \times \left[\exp\left\{\frac{(v - u)^2}{2\sigma_1^2 t}\right\} - \exp\left\{-\frac{(v - u)^2}{2\sigma_1^2 t}\right\}\right], \tag{15.6}$$

for $t > 0, u > 0, v > 0$.

As a special case for $\mu_1 = 0$ one has

$$p_1^0(t; u, v) = p_1(t; u, v) - p_1(t; u, -v) \tag{15.7}$$

for $t > 0, u > 0, v > 0$, where p_1 is the transition probability density of an unrestricted one-dimensional Brownian motion with drift zero and diffusion coefficient σ_1^2.

Also, for the case $k = 1$, one has $\partial G = \{0\}$, so that $X_{\tau_{\partial G}} = 0$

$$P_u(\tau_{\partial G} \leqslant t \text{ and } X_{\tau_{\partial G}} = 0) = P_u(\tau_{\partial G} \leqslant t), \tag{15.8}$$

which is the distribution of the first passage time to zero of a one-dimensional Brownian motion starting at u. By relation (10.15) in Section 10 of Chapter I, we have

$$P_u(\tau_{\partial G} \leqslant t) = \int_0^t f_{\sigma_1^2, \mu_1}(s; u)\, ds \tag{15.9}$$

where

$$f_{\sigma_1^2, \mu_1}(s; u) = u(2\pi\sigma_1^2 s^3)^{-1/2} \exp\left\{-\frac{(u - s\mu_1)^2}{2\sigma_1^2 s}\right\}, \tag{15.10}$$

for $u > 0$.

For $k > 1$, the $(k - 1)$-dimensional Brownian motion $\{(X_t^{(2)}, \ldots, X_t^{(k)})\}$ is independent of $\{X_t^{(1)}\}$ and, therefore, of the first passage time

$$\tau_{\partial G} = \inf\{t \geqslant 0 \colon X_t^{(1)} = 0\}. \tag{15.11}$$

Therefore, for every interval $I_1 \subset (0, \infty)$ and every Borel subset C of $\mathbb{R}^{k-1}$ it follows that

$$\begin{aligned}\tilde{p}(t; \mathbf{x}, I_1 \times C) &= P_{\mathbf{x}}(\{X_t^{(1)} \in I_1, (X_t^{(2)}, \ldots, X_t^{(k)}) \in C\} \cap \{\tau_{\partial G} > t\}) \\ &= P(\{X_t^{(1)} \in I_1\} \cap \{\tau_{\partial G} > t\} \mid X_0^{(1)} = x^{(1)}) \\ &\quad \times P((X_t^{(2)}, \ldots, X_t^{(k)}) \in C \mid X_0^{(j)} = x^{(j)} \text{ for } 2 \leqslant j \leqslant k) \\ &= \left(\int_{I_1} p_1^0(t; x^{(1)}, v)\, dv\right) \int \cdots \int_C \left(\prod_{j=2}^k p_j(t; x^{(j)}, v^{(j)})\right) dv^{(2)} \cdots dv^{(k)}, \\ &\qquad\qquad x^{(1)} > 0, \quad (15.12)\end{aligned}$$

where p_1^0 is given by (15.6) and p_j is the transition probability density of an unrestricted one-dimensional Brownian motion with drift μ_j and diffusion coefficient σ_j^2; namely,

$$p_j(t; u, v) = (2\pi\sigma_j^2 t)^{-1/2} \left\{-\frac{(v - u - t\mu_j)^2}{2\sigma_j^2 t}\right\}, \qquad (u, v \in \mathbb{R}^1). \quad (15.13)$$

Hence, for $\mathbf{x}, \mathbf{y} \in H^k$,

$$p^0(t; \mathbf{x}, \mathbf{y}) = p_1^0(t; x^{(1)}, y^{(1)}) \prod_{j=2}^k p_j(t; x^{(j)}, y^{(j)}). \quad (15.14)$$

Consider a Borel set $B \subset \partial G$. Then $B = \{0\} \times C$ for some Borel subset C of $\mathbb{R}^{k-1}$ and, for $\mathbf{x} \in G$,

$$\begin{aligned}\tilde{p}(t; \mathbf{x}, B) &= P_{\mathbf{x}}(\{\tau_{\partial G} \leqslant t\} \cap \{\mathbf{X}_{\tau_{\partial G}} \in \{0\} \times C\}) \\ &= P_{\mathbf{x}}(\{(X_{\tau_{\partial G}}^{(2)}, \ldots, X_{\tau_{\partial G}}^{(k)}) \in C\} \cap \{\tau_{\partial G} \leqslant t\}) \\ &= \int P_{\mathbf{x}}((X_s^{(2)}, \ldots, X_s^{(k)}) \in C\} \mid \tau_{\partial G} = s) f_{\sigma_1^2, \mu_1}(s; x^{(1)})\, ds \\ &= \int_0^t \left(\int \cdots \int_C \prod_{j=2}^k p_j(s; x^{(j)}, v^{(j)})\, dv^{(2)} \cdots dv^{(k)}\right) f_{\sigma_1^2, \mu_1}(s; x^{(1)})\, ds. \quad (15.15)\end{aligned}$$

The last equality holds owing to the independence of $\tau_{\partial G}$ and the $(k - 1)$-dimensional Brownian motion $\{(X_t^{(2)}, \ldots, X_t^{(k)})\}$. Letting $t \to \infty$ in (15.15) one arrives at the $P_{\mathbf{x}}$-distribution $\psi(\mathbf{x}, d\mathbf{y})$ of $\mathbf{X}_{\tau_{\partial G}}$. Let $\mathbf{y}' = (0, y^{(2)}, \ldots, y^{(k)}) \in \partial G$. It follows from (15.15) that the distribution $\psi(\mathbf{x}, d\mathbf{y})$ has a density $\psi(\mathbf{x}', \mathbf{y}')$ with respect to Lebesgue measure on $\partial G = \{0\} \times \mathbb{R}^{k-1}$, given by

$$\psi(\mathbf{x}, \mathbf{y}') = \int_0^\infty f_{\sigma_1^2, \mu_1}(s; x^{(1)}) \prod_{j=2}^k p_j(s, x^{(j)}, y^{(j)})\, ds \qquad (\mathbf{y}' \in \partial G; \mathbf{x} \in H^k). \quad (15.16)$$

For the special case $\mu_i = 0$, $\sigma_i^2 = \sigma^2$ for $1 \leqslant i \leqslant k$, we have

$$\psi(\mathbf{x}, \mathbf{y}') = \int_0^\infty x^{(1)}(2\pi\sigma^2)^{-k/2} s^{-(k+2)/2} \exp\left\{-\frac{1}{2\sigma^2 s}\left[(x^{(1)})^2 + \sum_2^k (y^{(j)} - x^{(j)})^2\right]\right\} ds$$

$$= \frac{x^{(1)}}{(2\pi)^{k/2}} \frac{2^{k/2}}{\left[(x^{(1)})^2 + \sum_2^k (y^{(j)} - x^{(j)})^2\right]^{k/2}} \int_0^\infty t^{(k/2)-1} e^{-t}\, dt$$

$$= \frac{\Gamma(k/2)}{\pi^{k/2}} \frac{x^{(1)}}{|\mathbf{x} - (0, \mathbf{y}')|^k}. \tag{15.17}$$

The second equality in (15.17) is the result of the change of variables

$$s \to t = \left[(x^{(1)})^2 + \sum_2^k (y^{(j)} - x^{(j)})^2\right] \Big/ 2\sigma^2 s.$$

It is interesting to note that, for $k = 2$, (15.17) is the *Cauchy distribution*.

The explicit calculations carried out for the example above are not possible for general diffusions, or for more general domains G. One may, however, derive the linear second-order equations whose solutions are $p^0(t; \mathbf{x}, d\mathbf{y})$. The function $p^0(t; \mathbf{x}, \mathbf{y})$ satisfies *Kolmogorov's backward equation* (see theoretical complement 2)

$$\frac{\partial p^0(t; \mathbf{x}, \mathbf{y})}{\partial t} = \frac{1}{2} \sum_{i,j=1}^k d_{ij}(\mathbf{x}) \frac{\partial^2 p^0(t; \mathbf{x}, \mathbf{y})}{\partial x^{(i)}\, \partial x^{(j)}} + \sum_{i=1}^k \mu^{(i)}(\mathbf{x}) \frac{\partial p^0(t; \mathbf{x}, \mathbf{y})}{\partial x^{(i)}} = \mathbf{A} p^0,$$

$$(t > 0;\ \mathbf{x}, \mathbf{y} \in G), \tag{15.18}$$

and the *backward boundary condition*

$$\lim_{\mathbf{x} \to \partial G, \mathbf{x} \in G} p^0(t; \mathbf{x}, \mathbf{y}) = 0 \qquad (t > 0, \mathbf{y} \in G). \tag{15.19}$$

Relations (15.18)–(15.19) may be checked directly for Example 1.

Finally, for $B \subset \partial G$ using the same argument as used in Section 7, (see (7.22)–(7.24)) one gets, writing τ for $\tau_{\partial G}$,

$$\begin{aligned}
\tilde{p}(t; \mathbf{x}, B) &= P_{\mathbf{x}}(\tilde{\mathbf{X}}_t \in B) = P_{\mathbf{x}}(\tau \leqslant t, \mathbf{X}_\tau \in B) \\
&= P_{\mathbf{x}}(\tau < \infty, \mathbf{X}_\tau \in B) - P_{\mathbf{x}}(\tau > t, \mathbf{X}_\tau \in B) \\
&= P_{\mathbf{x}}(\tau < \infty \text{ and } \mathbf{X}_\tau \in B) \\
&\quad - E_{\mathbf{x}}[P_{\mathbf{x}}(\{\tau > t\} \cap \{\mathbf{X}_\tau \in B\} \mid \{\mathbf{X}_u : 0 \leqslant u \leqslant t\})]
\end{aligned}$$

$$
\begin{aligned}
&= P_{\mathbf{x}}(\tau < \infty \text{ and } \mathbf{X}_\tau \in B) \\
&\quad - E_{\mathbf{x}}[E_{\mathbf{x}}(\mathbf{1}_{\{\tau > t\}} \mathbf{1}_{\{\mathbf{X}_{\tau(X_t^+)}^+ \in B\}} \mid \{\mathbf{X}_u : 0 \leqslant u \leqslant t\})] \\
&= P_{\mathbf{x}}(\tau < \infty, \mathbf{X}_\tau \in B) - E_{\mathbf{x}}[\mathbf{1}_{\{\tau > t\}} P_{\mathbf{y}}(\tau < \infty, \mathbf{X}_\tau \in B)_{\mathbf{y} = \mathbf{X}_t}] \\
&= \psi(\mathbf{x}; B) - \int \psi(\mathbf{y}; B) p^0(t; \mathbf{x}, \mathbf{y})\, d\mathbf{y} \qquad (t > 0, \mathbf{x} \in G), \qquad (15.20)
\end{aligned}
$$

where

$$
\psi(\mathbf{x}, B) := P_{\mathbf{x}}(\tau < \infty, \mathbf{X}_\tau \in B), \qquad (B \subset \partial G). \qquad (15.21)
$$

Since $p^0(t; \mathbf{x}, \mathbf{y})$ is determined by (15.18) and (15.19) (along with the *initial condition* $\lim_{t \downarrow 0} p^0(t; \mathbf{x}, \mathbf{y})\, d\mathbf{y} = \delta_{\mathbf{x}}$ for $\mathbf{x} \in G$), it remains to determine ψ. Now, for fixed $\mathbf{x}$, the *hitting distribution* $\psi(\mathbf{x}; d\mathbf{y})$ on ∂G is determined by the collection of functions

$$
\psi_f(\mathbf{x}) := \int_{\partial G} f(\mathbf{y}) \psi(\mathbf{x}; d\mathbf{y}) = E_{\mathbf{x}}(f(\mathbf{X}_\tau) \mathbf{1}_{\{\tau < \infty\}}) \qquad (\mathbf{x} \in G), \qquad (15.22)
$$

for arbitrary bounded continuous f on ∂G. Let us show that

$$
\begin{aligned}
\mathbf{A} \psi_f(\mathbf{x}) &= 0 \qquad (\mathbf{x} \in G), \\
\lim_{\mathbf{x} \to \mathbf{a}} \psi_f(\mathbf{x}) &= f(\mathbf{a}) \qquad (\mathbf{a} \in \partial G).
\end{aligned} \qquad (15.23)
$$

Drop the subscript f and assume for simplicity that $\psi(\mathbf{x})$ can be extended to a twice-differentiable function $\bar{\psi}$ on $\mathbb{R}^k$ that belongs to the domain of $\mathbf{A}$, the infinitesimal generator of the unrestricted diffusion $\{\mathbf{X}_t\}$ (theoretical complement 4). Then, for $\mathbf{x} \in G$, assuming $P_{\mathbf{x}}(\tau_{\partial G} \leqslant t) = o(t)$ as $t \downarrow 0$, (see Exercise 14.10), proceed as in the proof of Proposition 9.1 to get, writing τ for $\tau_{\partial G}$

$$
\begin{aligned}
\mathbf{T}_t \bar{\psi}(\mathbf{x}) &\equiv E_{\mathbf{x}}(\bar{\psi}(\mathbf{X}_t)) = E_{\mathbf{x}}(\mathbf{1}_{\{\tau > t\}} \bar{\psi}(\mathbf{X}_t)) + E_{\mathbf{x}}(\mathbf{1}_{\{\tau \leqslant t\}} \bar{\psi}(\mathbf{X}_t)) \\
&= E_{\mathbf{x}}(\mathbf{1}_{\{\tau > t\}} \psi(\mathbf{X}_t)) + o(t) = E_{\mathbf{x}}(\mathbf{1}_{\{\tau > t\}} E_{\mathbf{X}_t}(f(\mathbf{X}_\tau) \mathbf{1}_{\{\tau < \infty\}})) + o(t) \\
&= E_{\mathbf{x}}(\mathbf{1}_{\{\tau > t\}} E(f(\mathbf{X}_{\tau(\mathbf{X}_t^+)}^+) \mathbf{1}_{\{\tau(\mathbf{X}_t^+) < \infty\}} \mid \{\mathbf{X}_u : 0 \leqslant u \leqslant t\})) + o(t) \\
&\qquad\qquad \text{(by the Markov property)} \\
&= E_{\mathbf{x}}(\mathbf{1}_{\{\tau > t\}} f(\mathbf{X}_{\tau(\mathbf{X}_t^+)}) \mathbf{1}_{\{\tau(\mathbf{X}_t^+) < \infty\}}) + o(t) \\
&= E_{\mathbf{x}}(\mathbf{1}_{\{\tau > t\}} f(\mathbf{X}_\tau \mathbf{1}_{\{\tau < \infty\}}) + o(t) \\
&\qquad \times [\mathbf{X}_\tau = \mathbf{X}_{\tau(\mathbf{X}_t^+)} \text{ and } \{\tau < \infty\} = \{\tau\{\mathbf{X}_t^+) < \infty \text{ on } \{\tau > t\}] \\
&= E_{\mathbf{x}}(f(\mathbf{X}_\tau) \mathbf{1}_{\{\tau < \infty\}}) + o(t) \qquad [P_{\mathbf{x}}(\tau \leqslant t) = o(t)] \\
&= \psi(\mathbf{x}) + o(t) = \bar{\psi}(\mathbf{x}) + o(t) \qquad (\mathbf{x} \in G, t > 0). \qquad (15.24)
\end{aligned}
$$

Hence,

$$\mathbf{A}\psi(\mathbf{x}) = \mathbf{A}\bar{\psi}(\mathbf{x}) = \lim_{t \downarrow 0} \frac{T_t\bar{\psi}(\mathbf{x}) - \bar{\psi}(\mathbf{x})}{t} = 0 \qquad (\mathbf{x} \in G), \tag{15.25}$$

establishing the first relation in (15.23). The second relation in (15.23), namely, continuity at the boundary, may also be established under broad assumptions (theoretical complement 2).

If G is *bounded* and ∂G *smooth* then the so-called *Dirichlet problem* (15.23), for a given continuous f specified on ∂G, can be shown to have a *unique* solution by the so-called maximum principle (see theoretical complement 3). A proof of this principle is sketched in Exercises 10–12 for the case $\mathbf{A} = \mathbf{\Delta}$ is the *Laplacian*

$$\mathbf{\Delta} = \sum_{i=1}^{k} \frac{\partial^2}{(\partial x^{(i)})^2}. \tag{15.26}$$

The Maximum Principle. Let u be a continuous function on $\bar{G} = G \cup \partial G$, where G is a bounded, connected, open set, such that $\mathbf{A}u = 0$ on G. Then u attains its maximum and minimum on ∂G, and not on G, unless it is a constant.

To prove uniqueness of the solution of (15.23), suppose ψ_1, ψ_2 are two solutions. Let $u = \psi_1 - \psi_2$, whose value on ∂G is zero. Now apply the maximum principle to u.

We briefly illustrate these ideas by an example.

Example 2. (*Brownian Motion on a Ball with Absorption at the Boundary*). Take $\mu^{(i)}(\cdot) = 0$, $d_{ij}(\cdot) = \sigma^2\delta_{ij}$ for $1 \leqslant i, j \leqslant k$. Let $G = \{\mathbf{x} \in \mathbb{R}^k; |\mathbf{x}| < a\}$, $\partial G = \{|\mathbf{x}| = a\}$ for some $a > 0$. The case $k = 1$ is dealt with in Example 8.3, in detail. Let us consider $k > 1$. The *hitting distribution* $\psi(\mathbf{x}; d\mathbf{y})$, i.e., the $P_{\mathbf{x}}$-distribution of $\mathbf{X}_{\tau_{\partial G}}$ is called the *harmonic measure* on ∂G. Its density $\psi(\mathbf{x}, \mathbf{y})$ with respect to the surface area measure $s(d\mathbf{y})$ on ∂G (or the arc length measure in the case $k = 2$) is the *Poisson kernel*,

$$\psi(\mathbf{x}; \mathbf{y}) = \frac{a^2 - |\mathbf{x}|^2}{\omega_k a |\mathbf{x} - \mathbf{y}|^k}, \qquad (|\mathbf{y}| = a), \tag{15.27}$$

where ω_k is the surface area of the unit sphere (Exercise 7). The solution to (15.23) is

$$\psi_f(\mathbf{x}) = \frac{a^2 - |\mathbf{x}|^2}{\omega_k a} \int_{\{|\mathbf{y}| = a\}} \frac{f(\mathbf{y})}{|\mathbf{x} - \mathbf{y}|^k} s(d\mathbf{y}). \tag{15.28}$$

It is left as an exercise (Exercise 11) to check that $\mathbf{x} \to \psi(\mathbf{x}, \mathbf{y})$ satisfies $\mathbf{\Delta}\psi = 0$ in G for each $\mathbf{x} \in \partial G$, where $\mathbf{\Delta}$ is the *Laplacian* (15.26). One may then differentiate under the integral sign to prove $\mathbf{\Delta}\psi_f = 0$ in G. The checks that

$\int_{\partial G} \psi(\mathbf{x}; \mathbf{y}) s(d\mathbf{y}) = 1$ for each $\mathbf{x} \in G$, and that $\psi(\mathbf{x}, \mathbf{y})\, d\mathbf{y}$ converges weakly to $\delta_{\mathbf{y}_0}(d\mathbf{y})$ as $\mathbf{x} \to \mathbf{y}_0 \in \partial G$ are left to Exercise 11.

Let us now turn to transience and recurrence of unrestricted multidimensional diffusions $\{\mathbf{X}_t\}$.

Now, according to the probabilistic representation (15.22) of the solution to the Dirichlet problem (15.23), the probability

$$\psi(\mathbf{x}) := P_{\mathbf{x}}(\{\mathbf{X}_t\} \text{ reaches } \partial B(\mathbf{0}; c) \text{ before } \partial B(\mathbf{0}; d)), \qquad (0 < c < |\mathbf{x}| < d), \tag{15.29}$$

is the solution to

$$\begin{aligned} \mathbf{A}\psi(\mathbf{x}) &= 0 \qquad \text{for } 0 < c < |\mathbf{x}| < d, \\ \lim_{\mathbf{x}\to\mathbf{y},\, c<|\mathbf{x}|<d} \psi(\mathbf{x}) &= \begin{cases} 1 & \text{if } |\mathbf{y}| = c, \\ 0 & \text{if } |\mathbf{y}| = d. \end{cases} \end{aligned} \tag{15.30}$$

To see this take G in (15.22) and (15.23) to be the annulus $G = \{c < |\mathbf{x}| < d\}$, and f to be 1 on $\{|\mathbf{x}| = c\}$, and 0 on $\{|\mathbf{x}| = d\}$. Recall that the analog (9.2) to (15.30) was solved in Section 9 to compute ψ for the derivation of criteria for transience and recurrence of one-dimensional diffusions. In the case of multidimensional Brownian motion, (15.30) reduces to a two-point boundary value problem, such as (9.2), in a single variable $r = |\mathbf{x}|$. This equation was solved in Section 14 to derive criteria for transience and recurrence of multidimensional Brownian motion. Unfortunately, for general multidimensional diffusions the Dirichlet problem (15.30) cannot be solved explicitly. It is possible, however, to use a multidimensional analog of Corollary 2.4 to derive lower and upper bounds for the probability ψ. Notice that the arguments sketched in Section 2 remain valid for multidimensional diffusions and, hence, Corollary 2.4 is valid with $S = \mathbb{R}^k$. Such a generalization is also derived in Chapter VII, Corollary 3.2, by the method of stochastic differential equations.

In order to derive lower and upper bounds for ψ, some notation is needed. Let F be a given real-valued twice continuously differentiable function on $(0, \infty)$. Consider the *radial function*

$$f(\mathbf{x}) := F(|\mathbf{x}|) \qquad (|\mathbf{x}| > 0). \tag{15.31}$$

Straightforward differentiations yield (also see (14.18))

$$\begin{aligned} \frac{\partial}{\partial x^{(i)}} f(\mathbf{x}) &= \frac{x^{(i)}}{|\mathbf{x}|} F'(|\mathbf{x}|), \\ \frac{\partial^2}{(\partial x^{(i)})^2} f(\mathbf{x}) &= \frac{(x^{(i)})^2}{|\mathbf{x}|^2} F''(|\mathbf{x}|) + \left(\frac{1}{|\mathbf{x}|} - \frac{(x^{(i)})^2}{|\mathbf{x}|^3}\right) F'(|\mathbf{x}|), \\ \frac{\partial^{(2)}}{\partial x^{(i)}\, \partial x^{(j)}} &= \frac{x^{(i)} x^{(j)}}{|\mathbf{x}|^2} F''(|\mathbf{x}|) - \frac{x^{(i)} x^{(j)}}{|\mathbf{x}|^3} F'(|\mathbf{x}|), \qquad (i \neq j). \end{aligned} \tag{15.32}$$

Also write,

$$
\begin{aligned}
d(\mathbf{x}) &:= \sum_{i,j} \frac{d_{ij}(\mathbf{x})x^{(i)}x^{(j)}}{|\mathbf{x}|^2}, \\
B(\mathbf{x}) &:= \text{trace of } \mathbf{D}(\mathbf{x}) = \sum_i d_{ii}(\mathbf{x}), \\
C(\mathbf{x}) &:= 2\sum_i x^{(i)}\mu^{(i)}(\mathbf{x}), \\
\bar{\beta}(r) &:= \max_{|\mathbf{x}|=r} \frac{B(\mathbf{x}) + C(\mathbf{x})}{d(\mathbf{x})} - 1, \\
\underline{\beta}(r) &:= \min_{|\mathbf{x}|=r} \frac{B(\mathbf{x}) + C(\mathbf{x})}{d(\mathbf{x})} - 1, \\
\bar{\alpha}(r) &:= \max_{|\mathbf{x}|=r} d(\mathbf{x}), \qquad \underline{\alpha}(r) := \min_{|\mathbf{x}|=r} d(\mathbf{x}).
\end{aligned}
\tag{15.33}
$$

Finally define, for some $c > 0$,

$$
\bar{I}(r) := \int_c^r \frac{\bar{\beta}(u)}{u}\, du, \qquad \underline{I}(r) := \int_c^r \frac{\underline{\beta}(u)}{u}\, du. \tag{15.34}
$$

Using (15.33) it is simple to check that

$$
2\mathbf{A}f(\mathbf{x}) = d(\mathbf{x})F''(|\mathbf{x}|) + \frac{B(\mathbf{x}) - d(\mathbf{x}) + C(\mathbf{x})}{|\mathbf{x}|} F'(|\mathbf{x}|). \tag{15.35}
$$

Proposition 15.1. Under conditions (1)–(4) in Section 14, the function ψ in (15.29) satisfies

$$
\frac{\int_{|\mathbf{x}|}^d \exp\{-\bar{I}(u)\}\, du}{\int_c^d \exp\{-\bar{I}(u)\}\, du} \leqslant \psi(\mathbf{x}) \leqslant \frac{\int_{|\mathbf{x}|}^d \exp\{-\underline{I}(u)\}\, du}{\int_c^d \exp\{-\underline{I}(u)\}\, du}, \qquad (c \leqslant |\mathbf{x}| \leqslant d) \tag{15.36}
$$

Proof. Let

$$
F(r) := \frac{\int_c^r \exp\{-\bar{I}(u)\}\, du}{\int_c^d \exp\{-\bar{I}(u)\}\, du}, \qquad f(\mathbf{x}) := F(|\mathbf{x}|) \qquad (c \leqslant |\mathbf{x}| \leqslant d). \tag{15.37}
$$

Then $F'(r) > 0$ and $F''(r) + F'(r)\bar{\beta}(r)/r = 0$, for $r \geqslant c$. Hence by (15.35) and the definition of $\bar{\beta}$, writing $r = |\mathbf{x}|$,

$$\frac{\mathbf{A}f(\mathbf{x})}{d(\mathbf{x})} \leqslant \max_{|\mathbf{y}|=r} \frac{\mathbf{A}f(\mathbf{y})}{d(\mathbf{y})} \leqslant F''(r) + \frac{F'(r)\bar{\beta}(r)}{r} = 0, \tag{15.38}$$

so that

$$\mathbf{A}f(\mathbf{x}) \leqslant 0 \qquad \text{for all } c \leqslant |\mathbf{x}| \leqslant d. \tag{15.39}$$

Now extend f to a twice continuously differentiable function on all of $\mathbb{R}^k$ that vanishes outside a compact set (Exercise 15). Denote this extension also by f. By the extension of Corollary 2.4 to $S = \mathbb{R}^k$ mentioned above,

$$Z_t := f(\mathbf{X}_t) - \int_0^t \mathbf{A}f(\mathbf{X}_s)\, ds \qquad (t \geqslant 0) \tag{15.40}$$

is a martingale. By optional stopping (see Proposition 13.9 of Chapter I and Exercise 21),

$$E_{\mathbf{x}} Z_\tau = E_{\mathbf{x}} Z_0 = f(\mathbf{x}), \tag{15.41}$$

with

$$\tau = \tau_{\partial G(\mathbf{0};d)} \wedge \tau_{\partial G(\mathbf{0};d)}. \tag{15.42}$$

In other words, using (15.39),

$$E_{\mathbf{x}} f(\mathbf{X}_\tau) = f(\mathbf{x}) + E_{\mathbf{x}} \int_0^\tau \mathbf{A}f(\mathbf{X}_s)\, ds \leqslant f(\mathbf{x}) \qquad (c \leqslant |\mathbf{x}| \leqslant d). \tag{15.43}$$

On the other hand, $f(\mathbf{X}_\tau)$ is 1 if $\{\mathbf{X}_t\}$ reaches $\partial B(\mathbf{0}; d)$ before $\partial B(\mathbf{0}; c)$, and is 0 otherwise. Therefore, (15.43) becomes

$$P_{\mathbf{x}}(\{\mathbf{X}_t\} \text{ reaches } \partial B(\mathbf{0}; d) \text{ before } \partial B(\mathbf{0}; c)) \leqslant f(\mathbf{x}). \tag{15.44}$$

Subtracting both sides from 1, the first inequality in (15.36) is obtained. For the second inequality in (15.36), replace $\bar{\beta}$ by $\underline{\beta}$ in the definition of f in (15.37) to get $\mathbf{A}f(\mathbf{x}) \geqslant 0$, and $E_{\mathbf{x}} f(\mathbf{X}_\tau) \geqslant f(\mathbf{x})$. ■

The following corollary is immediate on letting $d \uparrow \infty$ in (15.36). Write

$$\rho_c(\mathbf{x}) := P_{\mathbf{x}}(\{\mathbf{X}_t\} \text{ eventually reaches } \partial B(\mathbf{0}; c)). \tag{15.45}$$

Corollary 15.2. If the conditions (1)–(4) of Section 14 hold, then for all $\mathbf{x}$, $|\mathbf{x}| > c$,

$$\rho_c(\mathbf{x}) = 1 \quad \text{if} \int_c^\infty \exp\{-\bar{I}(u)\}\, du = \infty, \tag{15.46}$$

$$\rho_c(\mathbf{x}) < 1 \quad \text{if} \int_c^\infty \exp\{-\underline{I}(u)\}\, du < \infty. \tag{15.47}$$

Note that the convergence or divergence of the integrals in (15.46), (15.47) does not depend on the value of c (Exercise 16).

Theorem 15.3 below says that the divergence of the integral in (15.46) implies recurrence, while the convergence of the integral in (15.47) implies transience. First we must define more precisely transience and recurrence for multidimensional diffusions. Since "point recurrence" cannot be expected to hold in multidimensions, as is illustrated for the case of Brownian motion in (14.28), the definition of recurrence on $\mathbb{R}^k$ has to be modified (in the same way as for Brownian motion).

Definition 15.2. A diffusion on $\mathbb{R}^k$ ($k > 1$) is said to be *recurrent* if for every pair $\mathbf{x} \neq \mathbf{y}$

$$P_{\mathbf{x}}(|\mathbf{X}_t - \mathbf{y}| \leqslant \varepsilon \text{ for some } t) = 1 \tag{15.48}$$

for all $\varepsilon > 0$. A diffusion is *transient* if

$$P_{\mathbf{x}}(|\mathbf{X}_t| \to \infty \text{ as } t \to \infty) = 1 \tag{15.49}$$

for all $\mathbf{x}$.

Theorem 15.3. Assume that the conditions (1)–(4) in Section 14 hold.

(a) If $\int_c^\infty e^{-\bar{I}(u)}\, du = \infty$ for some $c > 0$, then the diffusion is recurrent.
(b) If $\int_c^\infty e^{-\underline{I}(u)}\, du < \infty$ for some $c > 0$, then the diffusion is transient.

Proof. (a) Fix $\mathbf{x}_0, \mathbf{y}$ ($\mathbf{x}_0 \neq \mathbf{y}$) and $\varepsilon > 0$. Let $d > \max\{|\mathbf{x}_0|, |\mathbf{y}| + \varepsilon\}$. Choose c, $0 < c < d$, such that $\{|\mathbf{x}| = c\}$ is disjoint from $B(\mathbf{y}; \varepsilon)$. Define the stopping times

$$\begin{aligned} \tau_1 &:= \inf\{t \geqslant 0\colon |\mathbf{X}_t| = d\}, & \tau_2 &:= \inf\{t \geqslant \tau_1\colon |\mathbf{X}_t| = c\}, \\ \tau_{2n+1} &:= \inf\{t \geqslant \tau_{2n}\colon |\mathbf{X}_t| = d\}, & \tau_{2n} &:= \inf\{t \geqslant \tau_{2n-1}\colon |\mathbf{X}_t| = c\}. \end{aligned} \tag{15.50}$$

Now the function $\mathbf{x} \to P_{\mathbf{x}}(|\mathbf{X}_t| = d \text{ for some } t \geqslant 0)$, $|\mathbf{x}| \leqslant d$, is the solution to the Dirichlet problem (15.23) with $G = B(\mathbf{0}; d)$, and boundary function $f \equiv 1$ on $\{|\mathbf{x}| = d\}$. But $\psi(\mathbf{x}) \equiv 1$ is a solution to this Dirichlet problem. By the uniqueness of the solution,

$$P_{\mathbf{x}}(|\mathbf{X}_t| = d \text{ for some } t \geqslant 0) = 1 \text{ for } |\mathbf{x}| \leqslant d. \tag{15.51}$$

This means

$$P_{\mathbf{x}}(\tau_1 < \infty) = 1, \qquad (|\mathbf{x}| \leqslant d). \tag{15.52}$$

As τ_2 is the first hitting time of $\{|\mathbf{x}| = c\}$ by the *after*-τ process $X_{\tau_1}^+$, it follows from (15.52), the strong Markov property, and (15.46), that

$$P_{\mathbf{x}}(\tau_2 < \infty) = 1, \qquad (|\mathbf{x}| \leqslant d). \tag{15.53}$$

It now follows by induction that

$$P_{\mathbf{x}}(\tau_n < \infty) = 1, \qquad \text{for } |\mathbf{x}| \leqslant d, \quad n \geqslant 1. \tag{15.54}$$

Now define the events

$$A_n := \{\{\mathbf{X}_t\} \text{ does not reach } \partial B(\mathbf{y}; \varepsilon) \text{ during } (\tau_{2n}, \tau_{2n+1}]\} \qquad (n \geqslant 1). \tag{15.55}$$

Then

$$P_{\mathbf{x}_0}(A_n \mid \mathscr{F}_{\tau_{2n}}) = E_{\mathbf{x}_0}(\psi(X_{\tau_{2n}})), \tag{15.56}$$

where

$$\psi(\mathbf{x}) = P_{\mathbf{x}}(A_1). \tag{15.57}$$

But $\psi(\mathbf{x})$ is the solution to the Dirichlet problem

$$\begin{aligned} \mathbf{A}\psi(\mathbf{x}) = 0 \qquad & \text{for } \mathbf{x} \in G := B(\mathbf{y}; d) - \overline{B(\mathbf{y}; \varepsilon)}, \\ \lim_{\mathbf{x} \to \mathbf{z}} \psi(\mathbf{x}) = & \begin{cases} 1 & \text{if } |\mathbf{z}| = d, \\ 0 & \text{if } \mathbf{z} \in \partial B(\mathbf{y}; \varepsilon). \end{cases} \end{aligned} \tag{15.58}$$

By the maximum principle, $0 < \psi(\mathbf{x}) < 1$ for $\mathbf{x} \in G$; in particular,

$$1 - \delta_0 := \max_{|\mathbf{x}| = c} \psi(\mathbf{x}) < 1. \tag{15.59}$$

Therefore, from (15.56),

$$P_{\mathbf{x}_0}(A_n \mid \mathscr{F}_{\tau_{2n}}) \leqslant (1 - \delta_0) \qquad (n \geqslant 1). \tag{15.60}$$

Hence, (Exercise 17)

$$P_{\mathbf{x}_0}(\{\mathbf{X}_t\} \text{ does not reach } \partial B(\mathbf{y};\varepsilon)) \leqslant P_{\mathbf{x}_0}(A_1 \cap \cdots \cap A_n) \leqslant (1-\delta_0)^n \quad \text{for all } n, \tag{15.61}$$

which implies (15.48), with $\mathbf{x} = \mathbf{x}_0$.

(b) The proof is analogous to the proof of the transience of Brownian motion for $k \geqslant 3$, as sketched in Exercise 14.5 and Exercise 18. ■

16 REFLECTING BOUNDARY CONDITIONS FOR MULTIDIMENSIONAL DIFFUSIONS

The probabilistic construction of reflecting diffusions is not quite as simple as that of absorbing diffusions. Let us begin with an example.

Example 1. (*Reflecting Brownian Motion on a Half-Space*). Take $S = H_k \cup \partial H_k = \bar{H}_k$, where

$$H_k = \{\mathbf{x} \in \mathbb{R}^k;\, x^{(1)} > 0\}, \qquad \partial H^k = \{\mathbf{x} \in \mathbb{R}^k;\, x^{(1)} = 0\}.$$

First consider the case $k = 1$. This is dealt with in detail in Subsection 6.2, Example 1. In this case a reflecting Brownian motion $\{X_t\}$ with zero drift and diffusion coefficient $\sigma^2 > 0$ is given by

$$Y_t = |X_t|, \tag{16.1}$$

where $\{X_t\}$ is a one-dimensional (unrestricted) Brownian motion with zero drift and diffusion coefficient $\sigma^2 > 0$. The transition probability density of $\{X_t\}$ is given by

$$q_1(t; x, y) = p_1(t; x, y) + p_1(t; x, -y) \qquad (x, y \geqslant 0), \tag{16.2}$$

where $p_1(t; x, y)$ is the transition probability density of $\{B_t\}$,

$$p_1(t; x, y) = (2\pi\sigma^2 t)^{-1/2} \exp\left\{-\frac{(y-x)^2}{2\sigma^2 t}\right\}, \qquad (x, y \in \mathbb{R}^1). \tag{16.3}$$

For $k > 1$, consider k independent Brownian motions $\{X_t^{(j)}\}$, $1 \leqslant j \leqslant k$, each having drift zero and diffusion coefficient σ^2. Then $\{\mathbf{Y}_t := (|X_t^{(1)}|, X_t^{(2)}, \ldots, X_t^{(k)})\}$ is a Brownian motion in $\bar{H}_k := \{\mathbf{x} \in \mathbb{R}^k;\, x^{(1)} \geqslant 0\} = H_k \cup \partial H_k$ with *normal reflection* at the boundary $\{\mathbf{x} \in \mathbb{R}^k;\, x^{(1)} = 0\} = \partial H_k$. Its transition probability density function q is given by

$$q(t; \mathbf{x}, \mathbf{y}) = q_1(t; x^{(1)}, y^{(1)}) \prod_{j=2}^{k} p_1(t; x^{(j)}, y^{(j)}), \tag{16.4}$$
$$(\mathbf{x} = (x^{(1)}, \ldots, x^{(k)}), \qquad \mathbf{y} = (y^{(1)}, \ldots, y^{(k)}) \in \bar{H}_k).$$

This transition density satisfies Kolmogorov's backward equation

$$\frac{\partial q}{\partial t} = \tfrac{1}{2}\sigma^2 \boldsymbol{\Delta}_{\mathbf{x}} q(t; \mathbf{x}, \mathbf{y}), \qquad (t > 0;\ \mathbf{x} \in H_k,\ \mathbf{y} \in \bar{H}_k), \tag{16.5}$$

where

$$\boldsymbol{\Delta}_{\mathbf{x}} := \sum_{j=1}^{k} \frac{\partial^2}{\partial x^{(j)2}},$$

and the backward boundary condition

$$\frac{\partial q(t; \mathbf{x}, \mathbf{y})}{\partial x^{(1)}} = 0, \qquad (t > 0;\ \mathbf{x} \in \partial H_k,\ \mathbf{y} \in \bar{H}_k). \tag{16.6}$$

The following extension of Theorem 6.1 provides a class of diffusions on $\bar{H}_k$ with normal reflection at the boundary.

Suppose that drift coefficients $\mu^{(i)}(\mathbf{x})$, $1 \leqslant i \leqslant k$, and diffusion coefficients $d_{ij}(\mathbf{x})$ are prescribed on $\bar{H}_k$ satisfying the assumptions (1)–(4) in Section 14. Assume also that

$$\mu^{(1)}(\mathbf{x}) = 0, \qquad d_{1j}(\mathbf{x}) = 0 \qquad (2 \leqslant j \leqslant k;\ \mathbf{x} \in \partial H_k). \tag{16.7}$$

For $\mathbf{x} = (x^{(1)}, \dots, x^{(k)})$ write

$$\bar{\mathbf{x}} = (-x^{(1)}, x^{(2)}, \dots, x^{(k)}). \tag{16.8}$$

Now extend the coefficients $\mu^{(i)}(\cdot)$, $d_{ij}(\cdot)$ on all of $\mathbb{R}^k$ by setting, on $(\bar{H}_k)^c$,

$$\begin{aligned}
\mu^{(1)}(\mathbf{x}) &= -\mu^{(1)}(\bar{\mathbf{x}}), \qquad & \mu^{(j)}(\mathbf{x}) &= \mu^{(j)}(\bar{\mathbf{x}}) \qquad & &(2 \leqslant j \leqslant k),\\
d_{11}(\mathbf{x}) &= d_{11}(\bar{\mathbf{x}}), & d_{1j}\mathbf{x} &= -d_{1j}(\bar{\mathbf{x}}) & &(2 \leqslant j \leqslant k),\\
& & d_{ij}(\mathbf{x}) &= d_{ij}(\bar{\mathbf{x}}) & &(2 \leqslant i, j \leqslant k).
\end{aligned} \tag{16.9}$$

Proposition 16.1. Let $\{\mathbf{X}_t\}$ be a diffusion on $\mathbb{R}^k$ with coefficients $\mu^{(1)}(\cdot)$, $d_{ij}(\cdot)$ defined on $\mathbb{R}^k$ and satisfying (16.7), (16.9). Then the process

$$\{\mathbf{Y}_t := (|X_t^{(1)}|, X_t^{(2)}, \dots, X_t^{(k)})\}$$

is a Markov process on $\bar{H}_k$, whose transition probability density function q is given by

$$\begin{aligned}
q(t; \mathbf{x}, \mathbf{y}) &= p(t; \mathbf{x}, \mathbf{y}) + p(t; \mathbf{x}, \bar{\mathbf{y}})\\
&= p(t; \mathbf{x}, \mathbf{y}) + p(t; \bar{\mathbf{x}}, \mathbf{y}), \qquad (\mathbf{x}, \mathbf{y} \in \bar{H}_k),
\end{aligned} \tag{16.10}$$

where p is the transition probability density of $\{\mathbf{X}_t\}$.

Proof. By a change of variables, the Markov process $\{\bar{\mathbf{X}}_t: (-X_t^{(1)}, X_t^{(2)}, \ldots, X_t^{(k)})\}$ has the transition probability density $\bar{p}$ given by

$$\bar{p}(t; \mathbf{x}, \mathbf{y}) = p(t; \bar{\mathbf{x}}, \bar{\mathbf{y}}). \tag{16.11}$$

Therefore $\bar{p}$ satisfies the backward equation

$$\begin{aligned}\frac{\partial \bar{p}(t; \mathbf{x}, \mathbf{y})}{\partial t} &= \frac{\partial p(t; \bar{\mathbf{x}}, \bar{\mathbf{y}})}{\partial t} \\ &= \frac{1}{2} \sum_{i,j=1}^{k} d_{ij}(\bar{\mathbf{x}})(\partial_i \partial_j p(t; \bar{\mathbf{x}}, \bar{\mathbf{y}}) + \sum_{i=1}^{k} \mu^{(i)}(\bar{\mathbf{x}})(\partial_i p)(t; \bar{\mathbf{x}}, \bar{\mathbf{y}}),\end{aligned} \tag{16.12}$$

where $\partial_i p$ denotes differentiation of p with respect to the ith backward coordinate. Now

$$\begin{aligned}(\partial_1 p)(t; \bar{\mathbf{x}}, \bar{\mathbf{y}}) &= -(\partial_1 \bar{p})(t; \mathbf{x}, \mathbf{y}), \\ (\partial_j p)(t; \bar{\mathbf{x}}, \bar{\mathbf{y}}) &= (\partial_j \bar{p})(t; \mathbf{x}, \mathbf{y}), & (2 \leqslant j \leqslant k), \\ (\partial_1^2 p)(t; \bar{\mathbf{x}}, \bar{\mathbf{y}}) &= (\partial_1^2 \bar{p})(t; \mathbf{x}, \mathbf{y}), \\ (\partial_1 \partial_j p)(t; \bar{\mathbf{x}}, \bar{\mathbf{y}}) &= -(\partial_1 \partial_j \bar{p})(t; \mathbf{x}, \mathbf{y}), & (2 \leqslant j \leqslant k), \\ (\partial_i \partial_j p)(t; \bar{\mathbf{x}}, \bar{\mathbf{y}}) &= (\partial_i \partial_j p)(t; \mathbf{x}, \mathbf{y}), & (2 \leqslant i, j \leqslant k).\end{aligned} \tag{16.13}$$

It follows from (16.7), (16.9), (16.12), and (16.13) that $\{\mathbf{X}_t\}$ and $\{\bar{\mathbf{X}}_t\}$ have the same drift and diffusion coefficients and, therefore, the same transition probability density, i.e.,

$$\bar{p}(t; \mathbf{x}, \mathbf{y}) = p(t; \mathbf{x}, \mathbf{y}). \tag{16.14}$$

It now follows, as in the proof of Theorem 6.1, that

$$\{\mathbf{Y}_t := (|X_t^{(1)}|, X_t^{(2)}, \ldots, X_t^{(k)})\}$$

is a Markov process whose transition probability density is given by $p(t; \mathbf{x}, \mathbf{y}) + p(t; \mathbf{x}, \bar{\mathbf{y}})$. The second equality in (16.10) is obtained using (16.14). ■

The backward equation for p leads, by (16.10), to the *backward equation for* q,

$$\frac{\partial q}{\partial t} = \frac{1}{2} \sum_{i,j=1}^{k} d_{ij}(\mathbf{x}) \frac{\partial^2 q}{\partial x^{(i)} \partial x^{(j)}} + \sum_{i=1}^{k} \mu^{(i)}(\mathbf{x}) \frac{\partial q}{\partial x^{(i)}}, \qquad (\mathbf{x} \in H_k; t > 0). \tag{16.15}$$

The second equation in (16.10), together with the assumption that $p(t; \mathbf{x}, \mathbf{y})$ is differentiable in $\mathbf{x}$, implies the *backward boundary condition*

$$\frac{\partial q(t; \mathbf{x}, \mathbf{y})}{\partial x^{(1)}} = 0 \qquad (t > 0; \mathbf{x} \in \partial H_k, \mathbf{y} \in \bar{H}_k). \tag{16.16}$$

Definition 16.1. A Markov process on the state space $\bar{H}_k$ that has continuous sample paths and whose transition density q satisfies (16.15), (16.16) is called a *reflecting diffusion on $\bar{H}_k$ with drift coefficients $\mu^{(i)}(\mathbf{x})$ and diffusion coefficients $d_{ij}(\mathbf{x})$*.

It may be shown that the first requirement in (16.7), namely,

$$\mu^{(1)}(\mathbf{x}) = 0 \qquad \text{for } \mathbf{x} \in \partial H_k, \tag{16.17}$$

is really not needed for the validity of Proposition 16.1. The condition (16.17) ensures that the extended drift coefficients in (16.9) are continuous. If it does not hold, extend the coefficients as in (16.9) and modify $\mu^{(1)}(\cdot)$ on ∂H_k by setting it zero there. Although the extended $\mu^{(1)}(\cdot)$ is no longer continuous, it is still possible to define a diffusion having these coefficients and Proposition 16.1 goes through (theoretical complement 1).

The second requirement in (16.7), i.e., the condition

$$d_{1j}(\mathbf{x}) = 0 \qquad \text{for } \mathbf{x} \in \partial H_k, \quad 2 \leqslant j \leqslant k, \tag{16.18}$$

is of a different nature. The next proposition shows that the failure of (16.18) means that the direction of reflection is no longer *normal* to the boundary.

First consider a one-dimensional reflecting Brownian motion on $[0, \infty)$ with mean drift $\mu^{(1)}$ and diffusion coefficient 1. It may be constructed by the method described in the proof of Proposition 16.1, after setting the drift $-\mu^{(1)}$ on $(-\infty, 0)$ and modifying the drift at 0 to have the value zero as described above (also see Section 6). Let q_1 denote the transition density of this reflecting Brownian motion. Let p_j denote the transition probability density of a one-dimensional Brownian motion (on $\mathbb{R}^1$) with drift $\mu^{(j)}$ and diffusion coefficient $1, 2 \leqslant j \leqslant k$, i.e.,

$$p_j(t; x, y) := (2\pi t)^{-1/2} \exp\left\{-\frac{(y - x - t\mu^{(j)})^2}{2t}\right\}. \tag{16.19}$$

Then

$$q(t; \mathbf{x}, \mathbf{y}) := q_1(t; x^{(1)}, y^{(1)}) \prod_{j=2}^{k} p_j(t; x^{(j)}, y^{(j)}), \qquad (\mathbf{x}, \mathbf{y} \in \bar{H}_k), \tag{16.20}$$

is the transition probability density of a Markov process on $\bar{H}_k$ satisfying the backward equation

$$\frac{\partial q(t; \mathbf{x}, \mathbf{y})}{\partial t} = \frac{1}{2} \sum_{i=1}^{k} \frac{\partial^2 q}{(\partial x^{(i)})^2} + \sum_{i=1}^{k} \mu^{(i)} \frac{\partial q}{\partial x^{(i)}}, \qquad (t > 0; \mathbf{x} \in H_k, \mathbf{y} \in \bar{H}_k), \quad (16.21)$$

and the backward boundary condition

$$\frac{\partial q(t; \mathbf{x}, \mathbf{y})}{\partial x^{(1)}} = 0 \qquad \text{for } \mathbf{x} \in \partial H_k, \quad (t > 0; \mathbf{y} \in \bar{H}_k). \qquad (16.22)$$

This Markov process will be referred to as a *reflecting Brownian motion on* $\bar{H}_k$ *having drift* $\boldsymbol{\mu} := (\mu^{(1)}, \ldots, \mu^{(k)})$ *and diffusion matrix* $\mathbf{I}$ (or, *diffusion coefficients* δ_{ij} $(1 \leqslant i, j \leqslant k)$).

To construct a reflecting Brownian motion on an arbitrary half-space and having (constant, but) arbitrary drift and diffusion coefficients, first denote by $\bar{H}$ the half-space

$$\bar{H} = \bar{H}\boldsymbol{\gamma} := \{\mathbf{x} \in \mathbb{R}^k; \boldsymbol{\gamma} \cdot \mathbf{x} \geqslant 0\}, \qquad (16.23)$$

where $\boldsymbol{\gamma} = (\gamma_1, \ldots, \gamma_k)$ is a *unit* vector in $\mathbb{R}^k$, and $\boldsymbol{\gamma} \cdot \mathbf{x} := \sum \gamma_i x^{(i)}$ is the Euclidean inner product. The *interior of* $\bar{H}$ is $H := \{\boldsymbol{\gamma} \cdot \mathbf{x} > 0\}$ and its *boundary* is $\partial H := \{\boldsymbol{\gamma} \cdot \mathbf{x} = 0\}$. Let $\boldsymbol{\mu} = (\mu^{(1)}, \ldots, \mu^{(k)})$ be an arbitrary vector in $\mathbb{R}^k$ and $D := ((d_{ij}))$ an arbitrary $k \times k$ positive definite matrix. We write, for every real-valued differentiable function f (on $\mathbb{R}^k$ or $\bar{H}$), $\mathbf{grad}\, f$ for the *gradient* of f, i.e.,

$$(\mathbf{grad}\, f)(\mathbf{x}) = \left(\frac{\partial f(\mathbf{x})}{\partial x^{(1)}}, \ldots, \frac{\partial f(\mathbf{x})}{\partial x^{(k)}} \right). \qquad (16.24)$$

Also write $\mathbf{D}^{1/2}$ for the positive definite matrix such that $\mathbf{D}^{1/2}\mathbf{D}^{1/2} = \mathbf{D}$, and $\mathbf{D}^{-1/2}$ for its inverse.

Proposition 16.2. On the half-space $\bar{H} = \bar{H}\boldsymbol{\gamma}$ there exists a Markov process $\{\mathbf{Z}_t\}$ with continuous sample paths whose transition probability density r satisfies the Kolmogorov backward equation

$$\frac{\partial r(t; \mathbf{z}, \mathbf{w})}{\partial t} = \frac{1}{2} \sum_{i,j=1}^{k} d_{ij} \frac{\partial^2 r}{\partial z^{(i)} \partial z^{(j)}} + \sum_{i=1}^{k} \mu^{(i)} \frac{\partial r}{\partial z^{(i)}}$$

$$(t > 0; \mathbf{z} \in H, \mathbf{w} \in \bar{H}), \quad (16.25)$$

and the backward boundary condition

$$(\mathbf{D}\boldsymbol{\gamma}) \cdot (\mathbf{grad}\, r)(t; \mathbf{z}, \mathbf{w}) = 0 \qquad (t > 0; \mathbf{z} \in \partial H, \mathbf{w} \in \bar{H}). \qquad (16.26)$$

Further, $\{\mathbf{Z}_t\}$ has the representation

$$\mathbf{Z}_t := \mathbf{D}^{1/2}\mathbf{O}\mathbf{Y}_t \qquad (16.27)$$

where $\mathbf{O}$ is an orthogonal transformation such that $\mathbf{O}'$ maps $\mathbf{D}^{1/2}\boldsymbol{\gamma}/|\mathbf{D}^{1/2}\boldsymbol{\gamma}|$ into $\mathbf{e} := (1, 0, \ldots, 0)$, and $\{\mathbf{Y}_t\}$ is a reflecting Brownian motion on $\bar{H}_k$ having drift vector $\mathbf{v} := (\mathbf{D}^{1/2}\mathbf{O})^{-1}\boldsymbol{\mu}$ and diffusion matrix $\mathbf{I}$.

Proof. Let $\{\mathbf{Y}_t\}$ denote a reflecting Brownian motion on the half-space $\bar{H}_k$ and $\mathbf{O}$ an orthogonal matrix as described. Then $\{\mathbf{Z}_t\}$, defined by (16.27), is a Markov process on the state space $\bar{H}$ (Exercise 3). Let q, r denote the transition probability densities of $\{\mathbf{Y}_t\}$, $\{\mathbf{Z}_t\}$, respectively. Then, writing $\mathbf{T} := \mathbf{D}^{1/2}\mathbf{O}$,

$$r(t; \mathbf{z}, \mathbf{w}) = (\det \mathbf{D}^{1/2})^{-1} q(t; \mathbf{T}^{-1}\mathbf{z}, \mathbf{T}^{-1}\mathbf{w}). \tag{16.28}$$

Since whenever $\{\mathbf{B}_t\}$ is a standard Brownian motion, $\{\mathbf{T}\mathbf{B}_t\}$ is a Brownian motion with dispersion matrix $\mathbf{T}\mathbf{T}' = \mathbf{D}$, one may easily guess that r satisfies the backward equation (16.25). To verify (16.25) by direct computation, use the fact that,

$$\frac{\partial q(t; \mathbf{x}, \mathbf{y})}{\partial t} = \tfrac{1}{2}\Delta_{\mathbf{x}} q + \sum_{i=1}^{k} v^{(i)} \frac{\partial q}{\partial x^{(i)}}, \tag{16.29}$$

and that, for every real-valued twice-differentiable function f on $\mathbb{R}^k$ the following standard rules on the differentiation of composite functions apply (Exercise 4):

$$\begin{aligned} \mathbf{T}' \,\mathbf{grad}(f \circ \mathbf{T}^{-1}) &= (\mathbf{grad}\, f) \circ \mathbf{T}^{-1}, \\ \mathbf{T}\mathbf{T}'(((\partial_i \partial_j f) \circ \mathbf{T}^{-1})) &= (((\partial_i \partial_j)(f \circ \mathbf{T}^{-1}))). \end{aligned} \tag{16.30}$$

This is used with f as the function $\mathbf{x} \to q(t; \mathbf{x}, \mathbf{T}^{-1}\mathbf{w})$, for fixed t and $\mathbf{T}^{-1}\mathbf{w}$, to arrive at (16.25) using (16.28) and (16.29).

The boundary condition $\partial q/\partial x^{(1)} \equiv (\mathbf{e} \cdot \mathbf{grad}\, q) = 0$ on ∂H_k becomes, using the first relation in (16.30),

$$\mathbf{e} \cdot (\mathbf{T}' \,\mathbf{grad}\, r)(t; \mathbf{z}, \mathbf{w}) = 0 \qquad (\mathbf{z} \in \partial H \equiv \{\boldsymbol{\gamma} \cdot \mathbf{z} = 0\}). \tag{16.31}$$

Recall that $\mathbf{O}'\mathbf{D}^{1/2}\boldsymbol{\gamma} = |D\boldsymbol{\gamma}|\mathbf{e}$, and $\mathbf{T} = \mathbf{D}^{1/2}\mathbf{O}$, to express (16.31) as

$$(\mathbf{D}\boldsymbol{\gamma}) \cdot (\mathbf{grad}\, r)(t; \mathbf{z}, \mathbf{w}) = 0 \qquad \text{if } \mathbf{z} \in \partial H. \tag{16.32}$$

■

Since a diffusion on $\bar{H}_k$ with *spatially varying* drift coefficients and *constant* diffusion matrix $\mathbf{I}$ may be constructed by the method of Proposition 16.1, the preceding result may be proved with $\{\mathbf{Y}_t\}$ taken as such a diffusion. This leads to a Markov process (diffusion) on $\bar{H}$ whose transition probability density r

satisfies the backward equation

$$\frac{\partial r(t; \mathbf{z}, \mathbf{w})}{\partial t} = \frac{1}{2} \sum_{i,j=1}^{k} d_{ij} \frac{\partial^2 r}{\partial z^{(i)} \partial z^{(j)}} + \sum_{i=1}^{k} \mu^{(i)}(\mathbf{z}) \frac{\partial r}{\partial z^{(i)}} \tag{16.33}$$

and the backward boundary condition (16.26).

To extend the preceding construction to a spatially dependent dispersion matrix $\mathbf{D}(\mathbf{x})$ that does not satisfy (16.18), involves some difficulty, which can in general be resolved only locally; constructions on local pieces must be put together to come up with the desired diffusion on $\bar{H}$ (theoretical complement 2).

In order to define reflecting diffusions on more general domains, we consider domains of the form

$$\bar{G} := \{\mathbf{x} \in \mathbb{R}^k;\ \varphi(\mathbf{x}) \geqslant 0\}, \tag{16.34}$$

where φ is a real-valued continuously differentiable function on $\mathbb{R}^k$ such that $|\mathbf{grad}\ \varphi|$ is bounded away from zero on the boundary, i.e., for some $c > 0$,

$$|\mathbf{grad}\ \varphi(\mathbf{x})| \geqslant c > 0 \qquad \text{for } \mathbf{x} \in \partial G := \{\mathbf{x}: \varphi(\mathbf{x}) = 0\}. \tag{16.35}$$

Write $G := \{\varphi(\mathbf{x}) > 0\}$. Let $\mathbf{D}(\mathbf{x}) := ((d_{ij}(\mathbf{x})))$, and $\mu^{(i)}(\mathbf{x})$ $(1 \leqslant i \leqslant k)$ satisfy the assumptions (1)–(4) of Section 14, on $\bar{G}$.

Definition 16.2. Let $\{\mathbf{X}_t\}$ be a Markov process on the state space $\bar{G}$ in (16.34) that has continuous sample paths and a transition probability density q satisfying the Kolmogorov backward equation

$$\frac{\partial q(t; \mathbf{x}, \mathbf{y})}{\partial t} = \mathbf{A}q := \frac{1}{2} \sum_{i,j=1}^{k} d_{ij}(\mathbf{x}) \frac{\partial^2 q}{\partial x^{(i)} \partial x^{(j)}} + \sum_{i=1}^{k} \mu^{(i)}(\mathbf{x}) \frac{\partial q}{\partial x^{(i)}},$$
$$(t > 0;\ \mathbf{x} \in G, \mathbf{y} \in \bar{G}), \tag{16.36}$$

and the backward boundary condition

$$\mathbf{D}(\mathbf{x})(\text{grad } \varphi)(\mathbf{x}) \cdot \mathbf{grad}\ q = 0 \qquad (t > 0;\ \mathbf{x} \in \partial G, \mathbf{y} \in \bar{G}). \tag{16.37}$$

Then $\{\mathbf{X}_t\}$ is called a *reflecting diffusion* on $\bar{G}$ having *drift coefficients* $\mu^{(i)}(\cdot)$ and *diffusion coefficients* $d_{ij}(\cdot)$ $(1 \leqslant i, j \leqslant k)$. The vector $\mathbf{D}(\mathbf{x})(\mathbf{grad}\ \varphi)(\mathbf{x})$ at $\mathbf{x} \in \partial G$ (or any nonzero multiple of it), is said to be *conormal to the boundary at* $\mathbf{x}$. The *reflection* of $\{\mathbf{X}_t\}$ is said to be *in the direction of the conormal* to the boundary.

Note that, according to standard terminology, $(\mathbf{grad}\ \varphi)(\mathbf{x})$ is *normal to the boundary* of $\bar{G}$ at $\mathbf{x} \in \partial G$.

So far we have considered domains (16.37) with $\varphi(\mathbf{x}) = x^{(1)}$. Now let $\varphi(\mathbf{x}) = 1 - |\mathbf{x}|^2$, so that $\bar{G} \equiv \{\mathbf{x} \in \mathbb{R}^k: |\mathbf{x}| \leqslant 1\}$ is the closed unit ball.

Example 2. The reflecting standard Brownian motion $\{\mathbf{X}_t\}$ on the unit ball $\bar{B}(\mathbf{0}:1) = \{\mathbf{x} \in \mathbb{R}^k : |\mathbf{x}| \leqslant 1\}$ has a transition probability density p satisfying the backward equation

$$\frac{\partial p(t;\mathbf{x},\mathbf{y})}{\partial t} = \tfrac{1}{2}\Delta_{\mathbf{x}}\, p(t;\mathbf{x},\mathbf{y}) \qquad (t > 0;\ |\mathbf{x}| < 1, |\mathbf{y}| \leqslant 1), \tag{16.38}$$

and the boundary condition

$$\sum_{i=1}^{k} x^{(i)} \frac{\partial p(t;\mathbf{x},\mathbf{y})}{\partial x^{(i)}} = 0 \qquad (t > 0;\ |\mathbf{x}| = 1, |\mathbf{y}| \leqslant 1). \tag{16.39}$$

It is useful to note that the radial motion $\{R_t := |\mathbf{X}_t|\}$ is a Markov process on $[0, 1]$. This follows from the fact that the Laplacian $\Delta_{\mathbf{x}}$ transforms radial functions into radial functions (Proposition 14.1) and the boundary condition is radial in nature. Indeed, if f is a twice continuously differentiable function on $[0, 1]$ and $g(\mathbf{x}) = f(|\mathbf{x}|)$, then g is a twice continuously differentiable radial function on $\bar{B}(\mathbf{0}:1)$ and the function

$$T_t g(\mathbf{x}) := E_{\mathbf{x}}\, g(\mathbf{X}_t) = E_{\mathbf{x}} f(|\mathbf{X}_t|), \tag{16.40}$$

is the solution to the *initial-value problem*

$$\begin{aligned} &\frac{\partial u(t,\mathbf{x})}{\partial t} = \tfrac{1}{2}\Delta_{\mathbf{x}} u(t,\mathbf{x}), && (t > 0;\ |\mathbf{x}| < 1),\\ &\sum_{1}^{k} x^{(i)} \frac{\partial u(t,\mathbf{x})}{\partial x^{(i)}} = 0, && (|\mathbf{x}| = 1),\\ &\lim_{t \downarrow 0} u(t,\mathbf{x}) = f(|\mathbf{x}|), && (|\mathbf{x}| \leqslant 1). \end{aligned} \tag{16.41}$$

Let $v(t, r)$ be the unique solution to

$$\begin{aligned} &\frac{\partial v(t,r)}{\partial t} = \frac{1}{2}\frac{\partial^2 v}{\partial r^2} + \frac{k-1}{2r}\frac{\partial v}{\partial r}, && (t > 0;\ 0 \leqslant r < 1),\\ &\frac{\partial v(t,r)}{\partial r} = 0, && (t > 0;\ r = 1),\\ &\lim_{t \downarrow 0} v(t,r) = f(r), && (0 \leqslant r \leqslant 1). \end{aligned} \tag{16.42}$$

It is not difficult to check that the function $v(t, |\mathbf{x}|)$ satisfies (16.41) (Exercise 5). By uniqueness of the solution to (16.41) it follows that $u(t, \mathbf{x}) \equiv v(t, |\mathbf{x}|)$.

Thus, the transition operators $\mathbf{T}_t$ transform radial functions to radial functions. Hence, by Proposition 14.1, $\{R_t\}$ is a Markov process on $[0,1]$ whose infinitesimal generator is

$$A_r := \frac{1}{2}\frac{d^2}{dr^2} + \frac{k-1}{2r}\frac{d}{dr},$$

with boundary condition $d/dr = 0$ at $r = 1$. Recall (see Eq. 14.29, Section 14) that 0 is an *inaccessible* boundary for $\{R_t\}$. Hence, a reflecting boundary condition attached to the accessible boundary $r = 1$ suffices to specify the Markov process $\{R_t\}$.

17 CHAPTER APPLICATION: G. I. TAYLOR'S THEORY OF SOLUTE TRANSPORT IN A CAPILLARY

In a classic study (see theoretical complement 1), G. I. Taylor showed that when a solute in dilute concentration is injected into a liquid flowing with a steady slow velocity through an infinite straight capillary of uniform circular cross section, the concentration along the capillary becomes Gaussian as time increases. Such results have diverse applications. Taylor himself used his theory, along with some meticulous experiments, to determine molecular diffusion coefficients for various substances. Another application is to determine the rate at which a chemical injected into the bloodstream propagates.

Let a denote the radius of the circular cross-section of the capillary whose interior G and boundary ∂G are given by

$$\begin{aligned} G &= \{\mathbf{y} = (y^{(1)}, \mathbf{y}')\colon -\infty < y^{(1)} < \infty, |\mathbf{y}'| < a\}, \\ \partial G &= \{\mathbf{y}\colon |\mathbf{y}'| = a\}, \qquad (\mathbf{y}' = (y^{(2)}, y^{(3)})). \end{aligned} \tag{17.1}$$

Let $c(t; \mathbf{y})$ denote the solute concentration at a point $\mathbf{y}$ at time t. The velocity of the liquid is along the capillary length (i.e., in the $y^{(1)}$ direction) and is given, as the solution of a linearized Navier–Stokes equation governing a steady nonturbulent flow, by

$$F(\mathbf{y}') = U_0\left(1 - \frac{|\mathbf{y}'|^2}{a^2}\right). \tag{17.2}$$

The parameter U_0 is the maximum velocity, attained at the center of the capillary.

As in Einstein's theory of Brownian motion, a solute particle injected at a point $\mathbf{x} = (x^{(1)}, x^{(2)}, x^{(3)}) \in G$ is locally like a three-dimensional Brownian motion with drift $(F(\mathbf{x}'), 0, 0)$ and diffusion matrix $D_0\mathbf{I}$, where D_0 is the *molecular diffusion coefficient* of the solute (relative to the liquid in the capillary), and $\mathbf{I}$

is the 3×3 identity matrix. When this solute particle reaches the boundary ∂G, it is reflected. In other words, the location $\{\mathbf{X}_t = (X_t^{(1)}, X_t^{(2)}, X_t^{(3)}) = (X_t^{(1)}, \mathbf{X}_t')\}$ of the solute particle is a *reflecting diffusion* on $\bar{G} = G \cup \partial G$ having drift $\boldsymbol{\mu}(\mathbf{x}) = (F(\mathbf{x}'), 0, 0)$ and diffusion matrix $D_0\mathbf{I}$. Therefore, its transition probability density $p(t; \mathbf{x}, \mathbf{y})$ satisfies Kolmogorov's *backward equation*,

$$\frac{\partial p}{\partial t} = \tilde{\mathbf{A}}p := \tfrac{1}{2}D_0\mathbf{\Delta}_{\mathbf{x}}\, p + F(\mathbf{x}')\frac{\partial p}{\partial x^{(1)}} \qquad (\mathbf{x} = (x^{(1)}, \mathbf{x}'), \quad \mathbf{x}' = (x^{(2)}, x^{(3)})), \quad (17.3)$$

where $\mathbf{\Delta}_{\mathbf{x}}$ is the *Laplacian* $\sum_i \partial^2/(\partial x^{(i)})^2$. The *backward boundary condition* is that of *normal reflection* (see (16.37) with $\varphi(\mathbf{x}) = a^2 - |\mathbf{x}'|^2$),

$$x^{(2)}\frac{\partial p}{\partial x^{(2)}} + x^{(3)}\frac{\partial p}{\partial x^{(3)}} \equiv \mathbf{n}(\mathbf{x})\cdot\mathbf{grad}\, p = 0 \qquad (\mathbf{x} \in \partial G, t > 0; \mathbf{y} \in \bar{G}). \quad (17.4)$$

Here, $\mathbf{n}(\mathbf{x}) = (0, \mathbf{x}')$ is the normal (of length a) at the boundary point $\mathbf{x} = (x^{(1)}, \mathbf{x}')$.

The *Fokker–Planck* or forward equation may now be shown, by arguments entirely analogous to those given in Sections 2 and 6, to be

$$\frac{\partial p}{\partial t} = \tfrac{1}{2}D_0\mathbf{\Delta}_{\mathbf{y}}\, p - \frac{\partial}{\partial y^{(1)}}(F(\mathbf{y}')p) \qquad (\mathbf{y} \in G, t > 0). \qquad (17.5)$$

The *forward boundary condition* is the no-flux condition (see Section 2)

$$y^{(2)}\frac{\partial p}{\partial y^{(2)}} + y^{(3)}\frac{\partial p}{\partial y^{(3)}} \equiv \mathbf{n}(\mathbf{y})\cdot\mathbf{grad}\, p = 0 \qquad (\mathbf{y} \in \partial G, t > 0). \qquad (17.6)$$

By using the divergence theorem, which is a multidimensional analog of integration by parts, it is simple to check, in the manner of Sections 2 and 6, that (17.5) and (17.6) are indeed the forward (or, adjoint) conditions. Given an initial solute concentration distribution $c_0(d\mathbf{x})$, the solute concentration $c(t; \mathbf{y})$ at $\mathbf{y}$ at time t is then given by

$$c(t; \mathbf{y}) = \int_{\bar{G}} p(t; \mathbf{x}, \mathbf{y})c_0(d\mathbf{x}) \qquad (17.7)$$

and it satisfies the Fokker–Planck equations (17.5) and (17.6).

Since (a) the diffusion matrix is $D_0\mathbf{I}$, (b) the drift velocity is along the $x^{(1)}$-axis and depends only on $x^{(2)}, x^{(3)}$, and (c) the boundary condition only involves $x^{(2)}, x^{(3)}$, it should be at least intuitively clear that $\{\mathbf{X}_t\}$ has the following representation:

1. $\{\mathbf{X}_t' := (X_t^{(2)}, X_t^{(3)})\}$ is a two-dimensional, reflecting Brownian motion on the disc $\bar{B}_a := \{|\mathbf{y}'| \leqslant a\}$ with diffusion matrix $D_0\mathbf{I}'$, where $\mathbf{I}'$ is the 2×2 identity matrix.

2. $X_t^{(1)} = X_0^{(1)} + \int_0^t F(\mathbf{X}_s')\,ds + \sqrt{D_0}\,B_t$, where $\{B_t\}$ is a one-dimensional standard Brownian motion independent of $\{\mathbf{X}_t'\}$ and $X_0^{(1)}$. (17.8)

A proof of this representation is given toward the end of this section.

Now the transition probability density p' of the process $\{\mathbf{X}_t'\}$ is related to p by

$$p'(t; \mathbf{x}', \mathbf{y}') = \int_{-\infty}^{\infty} p(t; \mathbf{x}, \mathbf{y})\,dy^{(1)}. \tag{17.9}$$

By the Markov property of $\{\mathbf{X}_t'\}$, the integral on the right does not involve $x^{(1)}$. Now the disc $\bar{B}_a = \{|\mathbf{y}'| \leqslant a\}$ is compact and $p'(t; \mathbf{x}', \mathbf{y}')$ is positive and continuous in $\mathbf{x}', \mathbf{y}' \in \bar{B}_a$ for each $t > 0$. It follows, as in Proposition 6.3 of Chapter II, that for some positive constants c_1, c_2, one has

$$\max_{\mathbf{x}', \mathbf{y}' \in \bar{B}_a} |p'(t; \mathbf{x}', \mathbf{y}') - \gamma(\mathbf{y}')| \leqslant c_1 e^{-c_2 t} \qquad (t > 0), \tag{17.10}$$

where $\gamma(\mathbf{y}')$ is the unique invariant probability for p'. Let us check that $\gamma(\mathbf{y}')$ is the uniform density,

$$\gamma(\mathbf{y}') = \frac{1}{\pi a^2} \qquad (|\mathbf{y}'| \leqslant a). \tag{17.11}$$

It is enough to show that

$$\frac{\partial}{\partial t} \int_{\{|\mathbf{y}'| \leqslant a\}} p'(t; \mathbf{x}', \mathbf{y}')\gamma(\mathbf{x}')\,d\mathbf{x}' = 0. \tag{17.12}$$

But p' satisfies the backward equation

$$\frac{\partial p'}{\partial t} = \tfrac{1}{2} D_0 \Delta_{\mathbf{x}'} p' \qquad (|\mathbf{x}'| < a), \tag{17.13}$$

along with the boundary condition

$$x^{(2)} \frac{\partial p'}{\partial x^{(2)}} + x^{(3)} \frac{\partial p'}{\partial x^{(3)}} = 0 \qquad (|\mathbf{x}'| = a). \tag{17.14}$$

Interchanging the order of differentiation and integration on the left side of (17.12), and using (17.13) and (17.14) and the divergence theorem, (17.12) is established. In other words, the adjoint operator, which in this symmetric case is the same as the infinitesimal generator, annihilates constants.

As a consequence of (17.10), the concentration $c(t; \mathbf{y})$ gets uniformized, or becomes constant, in the $\mathbf{y}'$-plane. That is,

$$c'(t; \mathbf{y}') := \int_{-\infty}^{\infty} c(t; \mathbf{y})\, dy^{(1)} \to c'_0 := \left(\int_{\bar{G}} c_0(d\mathbf{x})\right)\Big/ \pi a^2 \tag{17.15}$$

exponentially fast as $t \to \infty$. Of much greater interest, however, is the asymptotic behavior of the concentration in the $y^{(1)}$-direction, i.e., of

$$\bar{c}(t; y^{(1)}) := \int_{\{|\mathbf{y}'| \leqslant a\}} c(t; \mathbf{y})\, d\mathbf{y}' = \int_{\bar{G}} c_0(d\mathbf{x}) \left[\int_{\{|\mathbf{y}'| \leqslant a\}} p(t; \mathbf{x}, \mathbf{y})\, d\mathbf{y}'\right]. \tag{17.16}$$

The study of the asymptotics of $\bar{c}(t; y^{(1)})$ is further simplified by the fact that the radial process $\{R'_t := |\mathbf{X}'_t|\}$ is Markovian. This is a consequence of the facts that (1) the backward operator $\frac{1}{2}D_0\boldsymbol{\Delta}_{\mathbf{x}'}$ of $\{\mathbf{X}'_t\}$ transforms all smooth radial functions on the disc $\bar{B}_a$ into radial functions and (2) the boundary condition (17.12) is radial, asserting that the derivative in the radial direction at the boundary vanishes (see Proposition 14.1). Hence $\{R'_t\}$ is a diffusion on $[0, a]$ whose transition probability density is

$$q(t; r, r') := \int_{\{|\mathbf{y}'| = r'\}} p'(t; \mathbf{x}', \mathbf{y}') s_{r'}(d\mathbf{y}') \qquad (|\mathbf{x}'| = r), \tag{17.17}$$

where $s_{r'}(d\mathbf{y}')$ is the arc length measure on the circle $\{|\mathbf{y}'| = r'\}$. The infinitesimal generator of $\{R'_t\}$ is given by the backward operator

$$\mathbf{A}_R := \tfrac{1}{2}D_0\left(\frac{d^2}{dr^2} + \frac{1}{r}\frac{d}{dr}\right), \qquad 0 < r < a \tag{17.18}$$

and the backward reflecting boundary condition

$$\left.\frac{\partial q}{dr}\right|_{r=a} = 0. \tag{17.19}$$

One may arrive at (17.18) and (17.19) from (17.17), (17.13), and (17.14), as in Example 2 of Section 16 (see Eq. 16.42 for $k = 2$). It follows from (17.10), (17.11), and (17.17) that $\{R'_t\}$ has the unique invariant density

$$\pi(r) = \frac{2r}{a^2}, \qquad 0 \leqslant r \leqslant a, \tag{17.20}$$

and that

$$\max_{0 \leqslant r, r' \leqslant a} \left| q(t; r, r') - \frac{2r}{a^2} \right| \leqslant 2\pi a c_1 e^{-c_2 t} \qquad (t > 0). \tag{17.21}$$

For convenience, write

$$f(r) = 1 - \frac{r^2}{a^2} \qquad (0 \leqslant r \leqslant a). \tag{17.22}$$

Then $F(\mathbf{y}') = U_0 f(|\mathbf{y}'|)$.

Now by the central limit theorem (Theorem 13.1, Exercise 13.2. Also see theoretical complement 13.3) it follows that, as $n \to \infty$,

$$Z_t^{(n)} := n^{-1/2} \int_0^{nt} (f(R'_s) - E_\pi f)\, ds \to N(0, \sigma^2 t), \tag{17.23}$$

where the convergence in (17.23) is in distribution, and

$$E_\pi f = \int_0^a f(r)\pi(r)\, dr = \int_0^a \left(1 - \frac{r^2}{a^2}\right)\left(\frac{2r}{a^2}\right) dr = \frac{1}{2}. \tag{17.24}$$

The variance parameter σ^2 of the limiting Gaussian is given by (Proposition 13.2, Exercise 13.2)

$$\sigma^2 = 2 \int_0^a (f(r) - E_\pi f) h(r) \pi(r)\, dr \tag{17.25}$$

where h is a function in $L^2([0, a], \pi)$ satisfying

$$\mathbf{A}_R h(r) = -(f(r) - E_\pi f) = \frac{r^2}{a^2} - \frac{1}{2}. \tag{17.26}$$

Such an h is unique up to the addition of a constant, and is given by

$$h(r) = \frac{1}{8D_0}\left(\frac{r^4}{a^2} - 2r^2\right) + c_3 \tag{17.27}$$

where c_3 is an arbitrary constant, which may be taken to be zero in carrying out the integration in (17.25). One then gets

$$\sigma^2 = \frac{a^2}{48D_0}. \tag{17.28}$$

Finally, since $\{B_t\}$ and $\{X'_t\}$ in (17.8) are independent it follows that, as $n \to \infty$,

$$Y_t^{(n)} := n^{-1/2}(X_{nt}^{(1)} - \tfrac{1}{2}U_0 nt) \to N(0, Dt) \tag{17.29}$$

in distribution. Here

$$D := \frac{a^2 U_0^2}{48D_0} + D_0, \tag{17.30}$$

is the *large scale* or *effective dispersion* along the capillary axis. Equations (17.29) and (17.30) represent Taylor's main result as completed by R. Aris (theoretical complement 1). *The effective dispersion D is larger than the molecular diffusion coefficient D_0 and grows quadratically with velocity.*

Actually, the functional central limit theorem holds, i.e., $\{Y_t^{(n)}\}$ converges in distribution to a Brownian motion with zero drift and variance parameter D.

Also, for each $t > 0$, the convergence in (17.29) is much stronger than in distribution. Indeed, since the distribution of $Y_t^{(n)}$ is the convolution of those of $Z_t^{(n)}$ and Brownian motion $\sqrt{D_0}\, B_t$ (see (17.8(ii))), $Y_t^{(n)}$ has a density that converges to the density of $N(0, Dt)$ as $n \to \infty$. Hence, by Scheffé's theorem (Section 3 of Chapter 0), the density of $Y_t^{(n)}$ converges to that of $N(0, Dt)$ in the L^1-norm. Since the distribution of $X_t^{(1)}$ has the density $\bar{c}(t; \cdot)/C_0$, where $C_0 := \int_{\bar{G}} c_0(d\mathbf{x})$ is the total amount of solute present, the density of $Y_t^{(n)}$ is given by

$$z \to n^{1/2}\bar{c}(nt; n^{1/2}z + \tfrac{1}{2}U_0 nt)/C_0. \tag{17.31}$$

One therefore has, writing $\varepsilon = n^{-1/2}$,

$$\int_{-\infty}^{\infty} \left| \varepsilon^{-1}\bar{c}(\varepsilon^{-2}t; \varepsilon^{-1}z + \tfrac{1}{2}U_0\varepsilon^{-2}t) - \frac{C_0}{(2\pi Dt)^{1/2}} \exp\left\{-\frac{z^2}{2Dt}\right\} \right| dz \to 0 \qquad \text{as } \varepsilon \downarrow 0. \tag{17.32}$$

Another way of expressing (17.32) is by changing variables $z' = \varepsilon^{-1}z$, $t' = \varepsilon^{-2}t$. Then (17.32) becomes

$$\int_{-\infty}^{\infty} \left| \bar{c}(t'; z' + \tfrac{1}{2}U_0 t') - \frac{C_0}{(2\pi Dt')^{1/2}} \exp\left\{-\frac{z'^2}{2Dt'}\right\} \right| dz' \to 0 \qquad \text{as } t' \uparrow \infty. \tag{17.33}$$

From a practical point of view, (17.33) says that at time t' much larger than the *relaxation time* over which the error in (17.10) becomes negligible, the concentration along the capillary axis becomes Gaussian. The center of mass of the solute moves with a velocity $\frac{1}{2}U_0$ along the capillary axis, with a dispersion D per unit of time. The relaxation time in (17.10) is $1/c_2$, and $-c_2$ is estimated by the first (i.e., closest to zero) nonzero eigenvalue of $\frac{1}{2}D_0\Delta_{\mathbf{x}'}$ on the disc $\bar{B}_a$ with the no-flux, or Neumann, boundary condition.

It remains to give a proof of the representation (17.8). First, let us show that $\{\mathbf{X}_t = (X_t^{(1)}, \mathbf{X}_t')\}$ as defined by (17.8) is a time-homogeneous Markov process. Fix s, t positive, and let the initial state be $\mathbf{x}_0 = (x_0^{(1)}, \mathbf{x}_0')$. Write

$$X_{t+s}^{(1)} = X_t^{(1)} + \int_0^s F((\mathbf{X}_t'^+)_u)\, du + \sqrt{D_0}\,(B_{t+s} - B_t), \tag{17.34}$$

where $\mathbf{X}_t'^+$ is the after-t process $\{(\mathbf{X}_t'^+)_u := \mathbf{X}_{t+u}' : u \geqslant 0\}$. Let g be a bounded measurable function on $\bar{G}$. Since $\{B_u : 0 \leqslant u \leqslant t\}$, $\{\mathbf{X}_u' : 0 \leqslant u \leqslant t\}$ determine

$\{\mathbf{X}_u: 0 \leqslant u \leqslant t\}$,

$$E[g(\mathbf{X}_{t+s}) \mid \{\mathbf{X}_u: 0 \leqslant u \leqslant t\}]$$
$$= E[E(g(\mathbf{X}_{t+s}) \mid \{B_u: 0 \leqslant u \leqslant t\}, \{\mathbf{X}'_u: 0 \leqslant u \leqslant t\}) \mid \{\mathbf{X}_u: 0 \leqslant u \leqslant t\}]. \quad (17.35)$$

Now,

$$g(\mathbf{X}_{t+s}) = g(X_t^{(1)} + \int_0^s F((\mathbf{X}_t'^{+})_u)\, du + \sqrt{D_0}\,(B_{t+s} - B_t), \mathbf{X}'_{t+s}). \quad (17.36)$$

Since $B_{t+s} - B_t$ is independent of the conditioning variables and of $X_t^{(1)}$ and $\mathbf{X}_t'^{+}$, the inner conditional expectation on the right in (17.35) may be expressed as

$$E\left(h\left(s, \left(X_t^{(1)} + \int_0^s F((\mathbf{X}_t'^{+})_u)\right) du, \mathbf{X}'_{t+s}\right) \middle| \{B_u: 0 \leqslant u \leqslant t\}, \{\mathbf{X}'_u: 0 \leqslant u \leqslant t\}\right), \quad (17.37)$$

where

$$h(s, v, \mathbf{z}') = Eg(v + \sqrt{D_0}\,(B_{t+s} - B_t), \mathbf{z}'). \quad (17.38)$$

To evaluate (17.37) use the facts that (1) $X_t^{(1)}$ is determined by the conditioning variables, (2) $\{X'_t\}$ is independent of $\{B_t\}$, so that the conditional distribution of $\mathbf{X}_t'^{+}$ given $\{B_u: 0 \leqslant u \leqslant t\}$, $\{\mathbf{X}'_u: 0 \leqslant u \leqslant t\}$, is the same as that given $\{\mathbf{X}'_u: 0 \leqslant u \leqslant t\}$. Then (17.37) becomes

$$\left(E\left(h\left(s, y + \int_0^s F((\mathbf{X}_t'^{+})_u)\, du, \mathbf{X}'_{t+s}\right) \middle| \{\mathbf{X}'_u: 0 \leqslant u \leqslant t\}\right)\right)_{y = X_t^{(1)}}. \quad (17.39)$$

But $\{\mathbf{X}'_t\}$ is a time-homogeneous Markov process. Therefore, (17.39) becomes

$$(h_1(s, y, \mathbf{X}'_t))_{y = X_t^{(1)}} = h_1(s, X_t^{(1)}, \mathbf{X}'_t), \quad (17.40)$$

where h_1 is some function on $[0, \infty) \times \bar{G}$. The expression (17.40) is already determined by $\{\mathbf{X}_u: 0 \leqslant u \leqslant t\}$. Therefore, the outer conditional expectation in (17.35) is the same as (17.40), completing the proof that $\{\mathbf{X}_t\}$ is a time-homogeneous Markov process.

Observe next that by Proposition 14.1, if $\{\mathbf{Z}_t := (Z_t^{(1)}, \mathbf{Z}'_t)\}$ is a Markov process on $\bar{G}$ whose transition probability density satisfies the Kolmogorov equations (17.3) and (17.4), then $\{\mathbf{Z}'_t\}$ is a Markov process on the disc $\bar{B}_a$ whose transition probability density satisfies the Kolmogorov equations (17.13) and (17.14). Since the latter are also satisfied by $\{\mathbf{X}'_t\}$, (17.8(i)) is established. It is enough then to identify the conditional probability density of $X_t^{(1)}$, given $\mathbf{X}_0 = \mathbf{x}$, with that of $Z_t^{(1)}$, given $\mathbf{Z}_0 = \mathbf{x}$. Let g be a smooth bounded function on $\mathbb{R}^1$, and let $\bar{g}$ denote the function on $\bar{G}$ defined by $\bar{g}(\mathbf{x}) = g(x^{(1)})$. Then, denoting by $E_{\mathbf{x}}$

expectation given $\mathbf{X}_0 = \mathbf{x}$,

$$(\mathbf{T}_t \bar{g})(\mathbf{x}) := E_{\mathbf{x}} \bar{g}(\mathbf{X}_t) = E_{\mathbf{x}} g(X_t^{(1)}). \tag{17.41}$$

By (17.8), as $t \downarrow 0$ one has by a Taylor expansion,

$$\begin{aligned}(\mathbf{T}_t \bar{g})(x) &= g(x^{(1)}) + g'(x^{(1)}) E_{\mathbf{x}}\left(\int_0^t F(\mathbf{X}'_s)\, ds + \sqrt{D_0}\, B_t\right) \\ &\quad + \tfrac{1}{2} g''(x^{(1)}) E_{\mathbf{x}}\left(\int_0^t F(\mathbf{X}'_s)\, ds + \sqrt{D_0}\, B_t\right)^2 + \mathrm{o}(t) \\ &= g(x^{(1)}) + g'(x^{(1)}) F(\mathbf{x}') + \tfrac{1}{2} g''(x^{(1)}) D_0 t + \mathrm{o}(t) \end{aligned} \tag{17.42}$$

where g', g'' are the first and second derivatives of g. It follows that

$$\left(\frac{\partial}{\partial t} \mathbf{T}_t \bar{g}(x)\right)_{t=0} = F(\mathbf{x}') g'(x^{(1)}) + \tfrac{1}{2} D_0 g''(x^{(1)}). \tag{17.43}$$

But the right side is $\mathbf{A}\bar{g}$, where $\mathbf{A}$ is as in (17.3). Therefore, the infinitesimal generator of $\{\mathbf{X}_t\}$ agrees with that of $\{\mathbf{Z}_t\}$ for functions depending only on the first coordinate. In particular, the backward equation for $\mathbf{T}_t \bar{g}$ becomes

$$\frac{\partial}{\partial t} \mathbf{T}_t \bar{g}(\mathbf{x}) = \mathbf{A}\mathbf{T}_t \bar{g}(\mathbf{x}), \tag{17.44}$$

so that, by the uniqueness of the solution to the initial-value problem, $E_{\mathbf{x}} g(X_t^{(1)}) \equiv (\mathbf{T}_t \bar{g})(\mathbf{x}) = E(g(Z_t^{(1)}) \mid \mathbf{Z}_0 = \mathbf{x})$. Thus, the conditional distribution of $X_t^{(1)}$, given $\mathbf{X}_0 = \mathbf{x}$, is the same as that of $Z_t^{(1)}$, given $\mathbf{Z}_0 = \mathbf{x}$. This completes the proof of (17.8).

It may be noted that, although kinetic theoretic arguments given earlier provide a justification for the validity of the Fokker–Planck equations (17.5) and (17.6) (with $c(t; \mathbf{y})$ replacing $p(t; \mathbf{x}, \mathbf{y})$), they are not needed for the analysis of the asymptotics of $c(t; \mathbf{y})$ carried out above. We have simply given a probabilistic proof, using the central limit theorem for Markov processes, of an important analytical result.

EXERCISES

Exercises for Section V.1

1. Check (1.2) for
   (i) a Brownian motion with drift μ and diffusion coefficient $\sigma^2 > 0$, and
   (ii) the Ornstein–Uhlenbeck process.

2. Show that if (1.2) holds then so does (1.2)′ for every $\varepsilon > 0$.

3. Suppose that $\{X_t\}$ is a Markov process on $S = (a, b)$, and φ is a continuous strictly monotone function on (a, b) onto (c, d).
   (i) Prove that $\{\varphi(X_t)\}$ is a Markov process.
   (ii) Compute the transition probability density for the process $\{X_t := \exp\{B_t^3\}\}$ in Example 1.

4. Consider the Ornstein–Uhlenbeck velocity process $\{V_t\}$ starting at $V_0 = v$. Let $\gamma = \beta/m$ (Example 2).
   (i) Show that

$$\mathrm{Cov}(V_t, V_{t+s}) = \frac{\sigma^2}{2\gamma}(1 - e^{-2\gamma t})e^{-\gamma s}, \qquad s > 0.$$

   (ii) Calculate the limit of the transition probability density $p(t; x, y)$ in (1.6), as $t \to \infty$, if (a) $\gamma < 0$, or (b) $\gamma > 0$.
   (iii) Define the Ornstein–Uhlenbeck position process by $X_t = x + \int_0^t V_s\,ds$, $t \geqslant 0$.
      (a) Show that $\{X_t\}$ is a Gaussian process.
      (b) Show $EX_t = x + v\gamma^{-1}(1 - e^{-\gamma t})$.
      (c) Show

$$\mathrm{Var}\, X_t = \frac{\sigma^2}{\gamma^2}t - \frac{3}{2}\frac{\sigma^2}{\gamma^3} + \frac{\sigma^2}{2\gamma^2}\left(\frac{4}{\gamma}e^{-\gamma t} - \frac{1}{\gamma}e^{-2\gamma t}\right).$$

      (d) Show

$$\mathrm{Cov}(X_s, X_{s+t} - X_s) = \frac{\sigma^2}{2\gamma^3}(1 - e^{-\gamma t})(1 - e^{-\gamma s})^2.$$

   (iv) Explain why $\{X_t\}$ is not a Markov process.
   (v) Write $\gamma = \beta m^{-1}$ and let $\sigma^2 = \gamma\sigma_0^2$ for some $\sigma_0^2 > 0$. Use (iii) to show that as $\gamma \to \infty$ the process $\{X_t\}$ converges to a Brownian motion with zero drift and diffusion coefficient σ_0^2.

5. In Example 3, take $\{X_t\}$ to be an Ornstein–Uhlenbeck process and $f(t) = ct$ $(c \neq 0)$. Compute the transition probability density of $\{Z_t = X_t + ct\}$.

6. Let $v \in \mathbb{R}^1$, $v \neq 0$, be an arbitrary (nonrandom) constant. Define a deterministic motion by $X_t = X_0 + vt$, $t \geqslant 0$, i.e., randomness only in the initial distribution of X_0. Show that $\{X_t\}$ is a Markov process having continuous sample paths and satisfying (1.2) *but* with $\sigma^2 = 0$. Does the transition probability distribution have a density?

Exercises for Section V.2

1. If $\int_a^b f(x)h_1(x)\,dx = \int_a^b f(x)h_2(x)\,dx$ for all twice continuously differentiable functions f vanishing outside some closed, bounded subinterval of (a, b), then prove that $h_1(x) = h_2(x)$ outside a set of Lebesgue measure zero.

2. Show that the derivations of (2.13) and of the backward equation (2.15) do not require the assumption of the existence of a *density* of the transition probability.

3. Let f be a twice continuously differentiable function on $[c, d]$. Show that given $\varepsilon > 0$ there exists a twice-differentiable g on $\mathbb{R}^1$ such that $g = f$ on $[c, d]$ and g vanishes outside $[c - \varepsilon, d + \varepsilon]$.

4. (i) Show that (1.2) in Section 1 holds uniformly for all x in the following cases:
   (a) $\mu(x) \equiv 0, \sigma^2(x) \equiv \sigma^2$.
   (b) $\mu(x) \equiv \mu, \sigma^2(x) \equiv \sigma^2$.
   (ii) Show that, for the Ornstein–Uhlenbeck process, $P_x(|X_t - x| > \varepsilon) = o(t)$ does not hold uniformly in x.

5. Let $p(t; x, y)$ be the transition probability density of a diffusion on $\mathbb{R}^1$ and λ a real number. Write $q(t; x, y) := \exp\{-\lambda t\} p(t; x, y)$.
   (i) Show that q satisfies the Chapman–Kolmogorov equation (2.2), and the backward and forward equations

$$\frac{\partial q}{\partial t} = \mu(x)\frac{\partial q}{\partial x} + \tfrac{1}{2}\sigma^2(x)\frac{\partial^2 q}{\partial x^2} - \lambda q,$$

$$\frac{\partial q}{\partial t} = -\frac{\partial}{\partial y}(\mu(y)q) + \frac{\partial^2}{\partial y^2}(\tfrac{1}{2}\sigma^2(y)q) - \lambda q.$$

   (ii) (*Initial-Value Problem*) Let f be a bounded continuous function and define $\mathbf{T}_t f := \int f(y)q(t; x, y)\,dy$. Show that the function $u(t, x) := (\mathbf{T}_t f)(x)$ solves the *initial-value problem*

$$\frac{\partial u}{\partial t} = \mu(x)\frac{\partial u}{\partial x} + \tfrac{1}{2}\sigma^2(x)\frac{\partial^2 u}{\partial x^2} - \lambda u \qquad (x \in \mathbb{R}^1, t > 0),$$

$$\lim_{t \downarrow 0} u(t, x) = f(x).$$

   (iii) (*Killing at a Constant Rate*) Let $\{X_t\}$ be a diffusion with transition probability density p, and ξ an exponentially distributed random variable independent of $\{X_t\}$ and having parameter λ. Then q may be interpreted as the (defective) transition probability density of the process $\{X_t\}$ *killed* at time ξ. More specifically, show that the function u in (ii) may be represented as

$$E(f(X_t)\mathbf{1}_{\{\xi > t\}} \mid X_0 = x).$$

   (iv) (*Killing*) Let $\lambda(x)$ be a continuous, nonnegative function on $\mathbb{R}^1$. Define the operators $\mathbf{T}_t$ acting on bounded continuous functions f, by

$$(\mathbf{T}_t f)(x) := E\left(f(X_t)\exp\left\{-\int_0^t \lambda(X_s)\,ds\right\} \,\middle|\, X_0 = x\right),$$

   where $\{X_t\}$ is a diffusion on $\mathbb{R}^1$. Show that the operators $\mathbf{T}_t$ have the semigroup property, and that $u(t, x) := (\mathbf{T}_t f)(x)$ solves the initial-value problem in (ii) with $\lambda = \lambda(x)$.
   (v) Note that $(\mathbf{T}_t f)(x)$ is finite (indeed, $\mathbf{T}_t f$ is bounded if f is bounded), if $\lambda(x) \geqslant 0$. One may also express $\mathbf{T}_t f$ as

$$(\mathbf{T}_t f)(x) = E(f(X_t)\mathbf{1}_{\{\xi\{X_s\} > t\}}),$$

where conditionally given the sample path $\{X_s: s \geqslant 0\}$, the *killing time* $\xi_{\{X_s\}}$ has the (nonhomogeneous) exponential distribution

$$P(\xi_{\{X_s\}} > t \mid \{X_s: s \geqslant 0\}) = \exp\left\{-\int_0^t \lambda(X_s)\,ds\right\}.$$

The killing times $\xi_{\{X_s\}}$ may take the value $+\infty$ with positive probability. If $\{X_t\}$ is defined on a probability space of trajectories $(\Omega_1, \mathscr{F}_1, P_1)$, then show that by enlarging the space appropriately one may define both $\{X_t\}$ and $\xi_{\{X_s\}}$ on a common probability space $(\Omega, \mathscr{F}, P)$. [*Hint*: First construct the product space $(\Omega_2, \mathscr{F}_2, P_2)$ for an independent family of random variables $\{\xi_{\omega_1}\}$ indexed by the set Ω_1 of all trajectories ω_1. That is, $\Omega_2 = \times_{\omega_1 \in \Omega_1} I\omega_1$, where $I\omega_1 = [0, \infty]$ for all $\omega_1 \in \Omega_1$. Then take the product space $(\Omega = \Omega_1 \times \Omega_2$, $\mathscr{F} = \mathscr{F}_1 \otimes \mathscr{F}_2$, $P = P_1 \times P_2)$.]

(vi) (*Feynman–Kac Formula*) If $\lambda(x)$ is bounded below and continuous, and not necessarily nonnegative, show that $\mathbf{T}_t f$ given in (iv) is well defined, defines a semigroup, and solves the initial-value problem (ii) with $\lambda = \lambda(x)$.

6. (*Adjoint Diffusions*) Let $p(t; x, y)$ denote the transition probability density of a diffusion on $\mathbb{R}^1$ with coefficients satisfying Condition (1.1). In addition, assume that

$$\frac{1}{2}\frac{d^2}{dx^2}(\sigma^2(x)) - \frac{d}{dx}\mu(x) = 0, \qquad (x \in \mathbb{R}^1).$$

(i) Show that $\tilde{p}(t; x, y) := p(t; y, x)$ is a transition probability density of a diffusion on $\mathbb{R}^1$ and compute its drift and diffusion coefficients. What is the forward equation for $\tilde{p}$?

(ii) (*Time Reversibility*) Let $\{X_t\}$ and $\{Y_t\}$ be diffusions with transition probability densities p and $\tilde{p}$ as above, with $X_0 \equiv x$, $Y_0 \equiv y$. Prove that for arbitrary time points $0 < t_1 < \cdots < t_n < t$ the conditional p.d.f. of $(X_{t_1}, X_{t_2}, \ldots, X_{t_n})$ given $X_t = y$, evaluated at $(x_1, x_2, \ldots, x_n)$, is the same as the conditional p.d.f. of $(Y_{t-t_n}, Y_{t-t_{n-1}}, \ldots, Y_{t-t_1})$ given $Y_t = x$, evaluated at $(x_n, x_{n-1}, \ldots, x_1)$. Thus, the sample path of one process over any finite time period cannot be distinguished probabilistically from that of the other with the direction of time reversed.

7. (*Sources and Sinks*) Consider the Fokker–Planck equation with a source (or sink) term $h(t, y)$,

$$\frac{\partial c(t, y)}{\partial t} = -\frac{\partial}{\partial y}(\mu(y)c(t, y)) + \frac{\partial^2}{\partial y^2}(\tfrac{1}{2}\sigma^2(y)c(t, y)) + h(t, y),$$

where $h(t, y)$ is a bounded continuous function of t and y, and $\int |h(t, x)| p(t; x, y)\,dx < \infty$. One may interpret $h(t, y)$ (in case $h \geqslant 0$) as the rate at which new solute particles are created at y at time t, providing an additional source for the change in concentration with time other than the flux. The contribution of the initial concentration c_0 to the concentration at y at time t is

$$c_1(t, y) := \int_{\mathbb{R}^1} c_0(x) p(t; x, y)\,dx.$$

The contribution due to the $h(s, x)\,\Delta x\,\Delta s$ new particles created in the region $[x, x + \Delta x]$ during the time $[s, s + \Delta s]$ is $h(s, x)p(t - s; x, y)\,\Delta s\,\Delta x$ $(0 \leqslant s \leqslant t)$. Integrating, the overall contribution from this to the concentration at y at time t is

$$c_2(t, y) = \int_0^t \int_{\mathbb{R}^1} h(s, x)p(t - s; x, y)\,dx\,ds.$$

(i) (*Duhamel's Principle*) Show that c_1 satisfies the *homogeneous* Fokker–Planck equation with $h = 0$ and initial concentration c_0, while c_2 satisfies the *nonhomogeneous* Fokker–Planck equation above with initial concentration zero. Hence $c := c_1 + c_2$ solves the nonhomogeneous Fokker–Planck equation above with initial concentration c_0. Assume here that $c_0(x)$ is integrable, and

$$\lim_{s \downarrow 0} \int_{\mathbb{R}^1} h(t - s, x)p(s; x, y)\,ds = h(t, x).$$

Show that this last condition is satisfied, for example, under the hypothesis of Exercise 6. *Can you make use of Exercise 5 to verify it more generally?

(ii) Apply (i) to the case $\mu(x)$, $\sigma^2(x)$, $h(t, x)$ are constants.

(iii) Solve the corresponding initial-value problem for the backward equation.

Exercises for Section V.3

1. Derive (3.21) using (3.15) and Proposition 3.1.
2. Show that the process $\{Z_t\}$ of Example 4 does not have a finite first moment.
3. By a change of variables directly derive the backward equation satisfied by the transition probability density $\tilde{p}$ of $\{Z_t\} := \{\varphi(X_t)\}$, using the corresponding equation for p. From this read off the drift and diffusion coefficients of $\{Z_t\}$.
4. (*Natural Scale for Diffusions*) For given drift and diffusion coefficients $\mu(\cdot)$, $\sigma^2(\cdot) > 0$ on $S = (a, b)$, define

$$I(x, z) := \int_x^z \frac{2\mu(y)}{\sigma^2(y)}\,dy,$$

with the usual convention $I(z, x) = -I(x, z)$ for $z < x$. Fix an arbitrary $x_0 \in S$, and define the *scale function*

$$s(x) := \int_{x_0}^x \exp\{-I(x_0, z)\}\,dz,$$

again following the convention $\int_{x_0}^x = -\int_x^{x_0}$ for $x < x_0$. If $\{X_t\}$ is a diffusion on S with coefficients $\mu(\cdot)$, $\sigma^2(\cdot)$, find the drift and diffusion coefficients of $\{s(X_t)\}$ (see Section 9 for a justification of the nomenclature).

5. Use Corollary 2.4 to prove that the following stronger version of the last relation

in (1.2)′ holds: for each $x \in \mathbb{R}^1$ and each $\varepsilon > 0$,

$$P_x\left(\max_{0 \leqslant t \leqslant h} |X_t - x| > \varepsilon\right) = o(h) \qquad \text{as } h \downarrow 0.$$

[*Hint*: Let f be twice continuously differentiable, such that (a) $f(y) = 2$ for $|x - y| > 1$, (b) $1 \leqslant f(y) \leqslant 2$ for $|y - x| > \varepsilon$, (c) $f(y) = |y - x|^3/\varepsilon^3$ for $|y - x| \leqslant \varepsilon$. Then

$$\begin{aligned} P_x\left(\max_{0 \leqslant t \leqslant h} |X_t - x| > \varepsilon\right) &\leqslant P_x\left(\max_{0 \leqslant t \leqslant h} f(X_t) \geqslant 1\right) \\ &\leqslant P_x\left(\max_{0 \leqslant t \leqslant h} \left\{f(X_t) - \int_0^t \mathbf{A}f(X_s)\, ds\right\} \geqslant \frac{1}{2}\right), \end{aligned}$$

if h is chosen so small that $\max\{|\mathbf{A}f(y)|: y \in \mathbb{R}^1\} \leqslant 1/2h$. By the maximal inequality (see Chapter I, Theorem 13.6), the last probability is less than

$$E_x\left(f(X_h) - \int_0^h \mathbf{A}f(X_s)\, ds\right)^2 \leqslant 2E_x f^2(X_h) + O(h^2).$$

Now show that

$$E_x f^2(X_h) \leqslant 4P_x(|X_h - x| > \varepsilon) + E_x\left(\frac{|X_h - x|^3}{\varepsilon^3} \mathbf{1}_{\{|X_h - x| \leqslant \varepsilon\}}\right) = o(h),$$

by (1.2)′ and the lemma in Section 3.

6. Show that the geometric Brownian motion $\{X_t\}$ is a process with *independent ratios* $(X_{t_1}/X_{t_0}), \ldots, (X_{t_k}/X_{t_{k-1}})$, for $0 = t_0 < t_1 < \cdots < t_k$, $k \geqslant 2$. In particular, at any time scale, $t_k - k\Delta$, $k = 0, 1, 2, \ldots$, the values $X_{t_0}, X_{t_1}, \ldots$ satisfy the *law of proportionate effect*: $X_{t_{k+1}} = L_k(\Delta)X_{t_k}$, $k = 0, 1, 2, \ldots$, where $L_0(\Delta), L_1(\Delta), \ldots$ are i.i.d. (*load factors*) positive random variables.
7. Let $\{\mathbf{B}(t)\}$ be a standard Brownian motion starting at 0. Let $V_t = e^{-t}\sigma B(e^{2t})$, $t \geqslant 0$. Show that $\{V_t\}$ is an Ornstein–Uhlenbeck process. Note continuity of sample paths from $\{\mathbf{B}(t)\}$.

Exercises for Section V.4

1. Show that (4.4), (4.5) are solved by (4.2).
2. Derive (4.17).
3. (i) Using (4.17) derive the probability ρ_{xc} that, starting at x, $\{X_t\}$ ever reaches c. (ii) Find a necessary and sufficient condition for $\{X_t\}$ to be recurrent.
4. Apply the criterion in Exercise 3(ii) to the Ornstein–Uhlenbeck process. What if $\gamma \equiv \beta/m > 0$?
5. (i) Using the results of Section 3 of Chapter III, give an informal derivation of a necessary and sufficient condition for the existence of a unique invariant probability distribution for $\{X_t\}$ (i.e., of $p(t; x, y)\, dy$).

(ii) Compute this invariant density.
(iii) Show that the Ornstein–Uhlenbeck process $\{V_t\}$ has a unique invariant distribution and compute this distribution (called the *Maxwell–Boltzmann velocity distribution*).
(iv) Show that under the invariant (Maxwell–Boltzmann) distribution,

$$\mathrm{Cov}_\pi(V_t, V_{s+t}) = \frac{\sigma^2}{2\gamma} e^{-\gamma s}, \qquad s > 0.$$

(v) Calculate the average kinetic energy $E(\frac{1}{2}mV_t^2)$ of an Ornstein–Uhlenbeck particle under the invariant initial distribution.
(vi) According to *Bochner's Theorem* (Chapter 0, Section 8), one can check that $\rho(s) = \mathrm{Cov}_\pi(V_t, V_{s+t})$ in (iv) is the Fourier transform of a measure μ (*spectral distribution*). Calculate μ.

6. Briefly indicate how the Ornstein–Uhlenbeck process may be viewed as a limit of the Ehrenfest model (see Section 5 of Chapter III) as the number of balls d goes to infinity.

7. Let

$$s(x) = \int_{x_0}^{x} \exp\left\{-\int_{x_0}^{z} \frac{2\mu(y)}{\sigma^2(y)}\, dy\right\} dz$$

be the so-called scale function on $S = (a, b)$, where x_0 is some point in S. If $\{X_t\}$ is a diffusion on S with coefficients $\mu(\cdot)$, $\sigma^2(\cdot)$, then show that the diffusion $\{Y_t := s(X_t)\}$ on $\tilde{S} = (s(a), s(b))$ has the property

$$P(\{Y_t\} \text{ reaches } c \text{ before } d \mid Y_0 = y) = \frac{d - y}{d - c}, \qquad c \leqslant y \leqslant d.$$

8. Derive the forward equation

$$\frac{\partial p(t; x, y)}{\partial t} = -\frac{\partial}{\partial y}(\mu(y)p(t; x, y)) + \frac{\partial^2}{\partial y^2}(\tfrac{1}{2}\sigma^2(y)p(t; x, y)),$$

$(t > 0, -\infty < x, y < \infty)$, along the lines of the derivation of the backward equation (4.14) given in this section. [*Hint*: $\mathbf{p}^{n+1} = \mathbf{p}^n\mathbf{p}$.]

9. (*Maxwell–Boltzmann Velocities*) A *monatomic* gas consists of a large number N ($\sim 10^{23}$) molecules of identical masses m each. Label the velocity of the ith molecule by $v_i = (v_{3i-2}, v_{3i-1}, v_{3i})$, for $i = 1, 2, \ldots, N$. To say that the gas is an *ideal* gas means that molecular interactions are ignored. In particular, the only energy represented in an ideal monatomic gas is the translational kinetic energy of individual molecular motions. The total energy is therefore

$$T = T(v_1, v_2, \ldots, v_{3N}) = \sum_{j=1}^{3N} \tfrac{1}{2}mv_j^2 = \tfrac{1}{2}m\|\mathbf{v}\|_2^2, \qquad \text{for } \mathbf{v} = (v_1, \ldots, v_{3N}) \in \mathbb{R}^{3N},$$

where $\|\mathbf{v}\|_2^2 := \sum_{j=1}^{3N} v_j^2$. Define *equilibrium* for a gas in a closed system of energy E

to mean that the velocities are (purely random) uniformly distributed over the *energy surface* S given by the surface of the $3N$-dimensional ball of radius $R = (2E/m)^{1/2}$, i.e.,

$$S = \left\{(v_1, v_2, \ldots, v_{3N}) \colon \sum_{j=1}^{3N} v_j^2 = \frac{2E}{m}\right\}.$$

One may also define *temperature* in proportion to E.

(i) Show that the distribution of the jth component of velocity is given by

$$P(a < v_j < b) = \int_a^b [1 - (x/R)^2]^{(3N-3)/2}\, dx \bigg/ \int_{-R}^{R} [1 - (x/R)^2]^{(3N-3)/2}\, dx.$$

(ii) Calculate the limiting distribution as $N \to \infty$ and $E \to \infty$ such that the average energy per particle E/N (density) stabilizes to $\varepsilon > 0$.

(iii) How do the above calculations change when the l_2-norm $\| \ \|_2$ defined above is replaced by the l_p-norm defined by $\|\mathbf{v}\|_p^p := \sum_{j=1}^{3N} |v_j|^p$, where $p \geqslant 1$ is fixed? (This interesting generalization was brought to our attention by Professor Svetlozar T. Rachev.)

Exercises for Section V.5

1. Extend the argument given in Example 1 to compute the transition probability density of a Brownian motion with drift μ and diffusion coefficient $\sigma^2 > 0$.

2. Use the method of Example 2 to compute the transition probability density of a diffusion with drift $\mu(x) = -\gamma x - g$ and diffusion coefficient $\sigma^2 > 0$.

3. (i) Use Kolmogorov's backward equation and the Fourier transform to solve the *initial-value problem*

$$\frac{\partial u(t, x)}{\partial t} = \tfrac{1}{2}\sigma^2 \frac{\partial^2 u}{\partial x^2} + \mu \frac{\partial u}{\partial x} \qquad (t > 0, -\infty < x < \infty),$$

$$\lim_{t \downarrow 0} u(t, x) = f(x),$$

where f is a bounded and continuous function on $(-\infty, \infty)$.

(ii) Use (i) to compute the transition probability density of a Brownian motion.

4. In Exercise 3 take the initial-value problem to be the one corresponding to the Ornstein–Uhlenbeck process.

Exercises for Section V.6

1. Use (1.2)′ to show that the diffusions $\{Y_t := -X_t\}$ and $\{X_t\}$ have the same drift and diffusion coefficients, if (6.7), (6.8) hold.

2. Use (1.2)′ to show that the diffusions $\{Y_t := X_t - md\}$ $(m = 0, \pm 1, \ldots)$ have the same drift and diffusion coefficients, provided $\mu(\cdot)$ and $\sigma^2(\cdot)$ are periodic with period d.

3. Express the transition probability density of the Markov process $\{Z_t\}$ in Theorem 6.4 in terms of that of $\{X_t\}$.

4. Using the construction of reflecting diffusions on $[0, \infty)$, $[0, 1]$, construct reflecting diffusions on $[a, \infty)$, $(-\infty, a]$, $[a, b]$ for arbitrary $a < b$. What are the analogs of (6.7), (6.8) and (6.36)–(6.38) in these cases? Express the transition probability densities for these reflecting diffusions in terms of appropriate unrestricted diffusions.

5. Suppose that φ is a measurable map on an interval S onto an interval S'. Suppose that $\{X_t\}$ is a Markov process having a homogeneous transition probability p. If $T_t(f \circ \varphi)(x) := E[f(\varphi(X_t)) \mid X_0 = x]$ is a function $v_f(t, \varphi(x))$ of $\varphi(x)$, for every real-valued bounded measurable function f on S, then prove that $\{Y_t := \varphi(X_t)\}$ is a homogeneous Markov process on S'.

6. Give a derivation of the forward equation (6.17).

7. Prove that the transition probability density $q(t; x, y)$ of $\{Z_t\}$ in Theorem 6.4 satisfies $(\partial q/\partial x)_{x=1} = 0$.

8. (i) Assume that $p(t; x, y) > 0$ for all $t > 0$, and $x, y \in (-\infty, \infty)$ for every diffusion whose drift and diffusion coefficients satisfy Condition (1.1). Prove that the Markov process $\{Z_t\}$ on the circle in Theorem 6.2 has a unique invariant probability $m(y)\,dy$, and that the transition probability density $q(t; x, y)$ of $\{Z_t\}$ converges to $m(y)$ in the sense that

$$\int_S |q(t; x, y) - m(y)|\, dy \to 0 \qquad \text{as } t \to \infty,$$

the convergence being exponentially fast and uniform in x.

(ii) Compute $m(y)$. [*Hint*: Differentiate $\int T_t f(x) m(x)\, dx = \int f(x) m(x)\, dx$ with respect to t, to get $\int_S (\mathbf{A} f)(x) m(x)\, dx = 0$ for all twice-differentiable f on S with compact support, so that $\mathbf{A}^* m = 0$.]

9. Derive the forward boundary conditions for a reflecting diffusion on $[0, 1]$.

10. Suppose (a) $\{X_t\}$ and $\{-X_t\}$ have the same transition probability density, and (b) $\{1 + X_t\}$ and $\{1 - X_t\}$ have the same transition probability density. Show that the drift and diffusion coefficients of $\{X_t\}$ must be periodic with period 2.

Exercises for Section V.7

1. Use the Radon–Nikodym Theorem (Section 4 of Chapter 0) to prove the existence of a nonnegative function $y \to p^0(t; x, y)$ such that $\int_B p^0(t; x, y)\, dy = \tilde{p}(t; x, B)$ for every Borel subset B of (a, ∞). Show that $p^0(t; x, y) \leqslant p(t; x, y)$ where p is the transition probability density of the unrestricted process $\{X_t\}$.

2. Let $\tilde{\mathbf{T}}_t$ $(t > 0)$ be the semigroup of transition operators for the Markov process $\{\tilde{X}_t\}$ in Theorem 7.1, i.e.,

$$(\tilde{\mathbf{T}}_t f)(x) = E_x f(\tilde{X}_t) = \int_{[a,\infty)} f(y)\tilde{p}(t; x, dy)$$

for all bounded continuous f on $[a, \infty)$.

(i) If $f(a) = 0$, show that

(a) $$(\tilde{\mathbf{T}}_t f)(x) = \int_{(a,\infty)} f(y)p^0(t; x, y)\,dy \qquad (x > a),$$

(b) $(\tilde{\mathbf{T}}_t f)(x) \to 0$ as $x \downarrow a$. [*Hint*: Use (7.11).]

(ii) Use (i) to show that $\mathbf{T}_t^0$ $(t > 0)$, defined by

$$(\mathbf{T}_t^0 f)(x) := \int_{(a,\infty)} f(y)p^0(t; x, y)\,dy,$$

is a *semigroup* of operators on the class C of all bounded continuous functions on $[a, \infty)$ vanishing at a. Thus $\mathbf{T}_t^0$ is the *restriction of* $\tilde{\mathbf{T}}_t$ *to* C (and $\mathbf{T}_t^0 C \subset C$).

(iii) Let $\mathbf{A}^0$ denote the infinitesimal generator of $\mathbf{T}_t^0$, i.e., the restriction of the infinitesimal generator of $\tilde{\mathbf{T}}_t$ to C. Give at least a rough argument to show that (7.10) holds (along with (7.11)).

3. Derive (7.13) and (7.14).

4. (*Method of Images*) Consider the drift and diffusion coefficients $\mu(\cdot), \sigma^2(\cdot)$ defined on $[0, \infty)$. Assume $\mu(0) = 0$. Extend $\mu(\cdot), \sigma^2(\cdot)$ to $\mathbb{R}^1$ as in (6.8) by setting $\mu(-x) = -\mu(x)$, $\sigma^2(-x) = \sigma^2(x)$ for $x > 0$. Let p^0 denote the (defective) transition probability density on $[0, \infty)$ (extended by continuity to 0) defined by (7.9). Let p denote the transition probability density for a diffusion on $\mathbb{R}^1$ having the extended coefficients above. Show that

$$\begin{aligned} p^0(t; x, y) &= p(t; x, y) - p(t; x, -y) \\ &= p(t; x, y) - p(t; -x, y), \qquad (t > 0;\ x, y \geqslant 0). \end{aligned}$$

[*Hint*: Show that p^0 satisfies the Chapman–Kolmogorov equation (2.30) on $S = [0, \infty)$, the backward equation (7.10), the boundary condition (7.11), and the initial condition $p^0(t; x, y)\,dy \to \delta_x(dy)$ as $t \downarrow 0$ (i.e., $(\mathbf{T}_t^0 f)(x) \to f(x)$ as $t \downarrow 0$ for every bounded continuous function f on $[0, \infty)$ vanishing at 0). Then appeal to the uniqueness of such a solution (see theoretical complements 2.3, 8.1). An alternative probabilistic derivation is given in Exercise 11.11.]

5. (*Method of Images*) Let $\mu(\cdot), \sigma^2(\cdot)$ be drift and diffusion coefficients defined on $[0, d]$ with $\mu(0) = \mu(d) = 0$, as in (6.36). Extend $\mu(\cdot), \sigma^2(\cdot)$ to $\mathbb{R}^1$ as in (6.37) and (6.38) (with d in place of 1). Let p denote the transition probability density of a diffusion on $\mathbb{R}^1$ having these extended coefficients. Define

$$q(t; x, y) := \sum_{m=-\infty}^{\infty} p(t; x, y + 2md), \qquad -d \leqslant x, y \leqslant d,$$

$$p^0(t; x, y) := q(t; x, y) - q(t; x, -y), \qquad 0 \leqslant x, y \leqslant d.$$

(i) Prove that p^0 is the density component of the transition probability of the diffusion on $[0, d]$ absorbed at the boundary $\{0, d\}$. [*Hint*: Check (2.30). (7.10), and (7.11) on $[0, d]$, as well as the initial condition: $(\mathbf{T}_t^0 f)(x) \to f(x)$ as $t \downarrow 0$ for every continuous f on $[0, d]$ with $f(0) = f(d) = 0$.]

(ii) For the special case $\mu(\cdot) \equiv 0, \sigma^2(\cdot) \equiv \sigma^2$ compute p^0.

6. (*Duhamel's Principle*)

(i) (*Nonhomogeneous Equation*) Solve the initial-boundary-value problem

$$\frac{\partial}{\partial t} u(t, x) = \mu(x)\frac{\partial u}{\partial x} + \tfrac{1}{2}\sigma^2(x)\frac{\partial^2 u}{\partial x^2} + h(t, x), \qquad (t > 0, a < x < b);$$

$$\lim_{x \downarrow a} u(t, x) = 0, \qquad \lim_{x \uparrow b} u(t, x) = 0, \qquad (t > 0, a \leqslant x \leqslant b);$$

$$\lim_{t \downarrow 0} u(t, x) = f(x), \qquad (a \leqslant x \leqslant b),$$

where h is a bounded continuous function on $[0, \infty) \times [a, b]$, and f is a bounded continuous function on $[a, b]$ vanishing at a, b. [*Hint*: Let p^0 be as in (7.17). Define

$$u_0(t, x) := \int_0^t \int_{[a,b]} h(s, y)p^0(t - s; x, y)\, dy, \qquad u_1(t; x) = \int_{[a,b]} f(y)p^0(t; x, y)\, dy.$$

Check that u_0 satisfies the nonhomogeneous equation and the given boundary conditions, but has zero initial value. Then check that u_1 satisfies the homogeneous differential equation (i.e., with $h = 0$), the given boundary and initial conditions.]

(ii) (*Nonzero Constant Boundary Values*) Solve the initial-boundary-value problem

$$\frac{\partial}{\partial t} u(t, x) = \tfrac{1}{2}\sigma^2 \frac{\partial^2}{\partial x^2} u(t, x), \qquad (t > 0; a < x < b);$$

$$\lim_{x \downarrow a} u(t, x) = \alpha, \qquad \lim_{x \uparrow b} u(t, x) = \beta;$$

$$\lim_{t \downarrow 0} u(t, x) = f(x),$$

where f is a bounded continuous function on $[a, b]$ with $f(a) = \alpha$, $f(b) = \beta$. [*Hint*: Let $u_0(x)$ satisfy the differential equation and the boundary conditions. Find $u_1(t, x)$ which satisfies the differential equation, zero boundary conditions, and the initial condition: $\lim_{t \downarrow 0} u_1(t, x) = f(x) - u_0(x)$. Then consider $u_0(x) + u_1(t, x)$.]

(iii) (*Time Dependent Boundary Values*) In (ii) take $\alpha = \alpha(t)$, $\beta = \beta(t)$ where $\alpha(t)$, $\beta(t)$ are continuous differentiable functions on $[0, \infty)$. [*Hint*: Let $u_0(t, x) := \alpha(t) + ((\beta(t) - \alpha(t))/(b - a))(x - a)$. Find $u_1(t, x)$ solving the nonhomogeneous equation:

$$\frac{\partial u_1}{\partial t} = \tfrac{1}{2}\sigma^2 \frac{\partial^2 u_1}{\partial x^2} - \frac{\partial u_0}{\partial t},$$

with zero boundary conditions and initial condition $\lim_{t \downarrow 0} u_1(t, x) = f(x) - u_0(0, x)$.]

7. (*Maximum Principle*)

(i) Let $u(t, x)$ be twice-differentiable with respect to x and once with respect to t on

$(0, T] \times (a, b)$, continuous on $[0, T] \times [a, b]$, and satisfy $(\partial u/\partial t) - \mathbf{A}u < 0$ on $(0, T] \times (a, b)$, where $\mathbf{A} = \frac{1}{2}\sigma^2(x)\, d^2/dx^2 + \mu(x)\, d/dx$, with $\sigma^2(x) > 0$. Show that the maximum value of u is attained either (a) initially, i.e., at a point $(0, x_0)$, or (b) at a point (t, a) or (t, b) for some $t \in (0, T]$. [*Hint*: Suppose not. Let the maximum value be attained at a point (t_0, x_0) with $0 < t_0 \leqslant T$, and $x_0 \in (a, b)$. Then $(\mathbf{A}u)(t_0, x_0) \leqslant 0$, so that $(\partial u/\partial t)(t_0, x_0) < 0$. But this means $u(t, x_0) > u(t_0, x_0)$ for some $t < t_0$].

(ii) Extend (i) to the case $\partial u/\partial t - \mathbf{A}u \leqslant 0$. [*Hint*: Let θ be a continuous function on $[a, b]$ satisfying $\mathbf{A}\theta = 1$. For each $\varepsilon > 0$, consider $u_\varepsilon(t, u) := u(t, x) + \varepsilon\theta(x)$.]

(iii) Let $u(x)$ be twice-differentiable on (a, b), continuous on $[a, b]$ and satisfy $(\mathbf{A}u)(x) \geqslant 0$ for $a < x < b$. Show that the maximum value of u is attained at $x = a$ or $x = b$. [*Hint*: First assume $\mathbf{A}u > 0$; if the maximum value is attained at $x_0 \in (a, b)$, then $\mathbf{A}u(x_0) \leqslant 0$.]

(iv) Prove that the solutions to the initial-boundary-value problems in Exercise 6 are *unique*.

Exercises for Section V.8

1. Prove that (8.3) holds under Dirichlet or Neumann boundary conditions at the end points of a finite interval S.

2. Prove that
   (i) if α is an eigenvalue of $\mathbf{A}$ (with Dirichlet or Neumann boundary conditions at the end points of S), then $\alpha \leqslant 0$;
   (ii) if α_1, α_2 are two distinct eigenvalues and ψ_1, ψ_2 corresponding eigenfunctions, then ψ_1 and ψ_2 are *orthogonal*, i.e., $\langle \psi_1, \psi_2 \rangle = 0$;
   (iii) if ϕ_1, ϕ_2 are linearly independent, then $\phi_1, \phi_2 - (\langle \phi_1, \phi_2 \rangle / \|\phi_1\|^2)\phi_1$ are orthogonal; if $\phi_1, \phi_2, \ldots, \phi_k$ are linearly independent, and $\phi_1, \ldots, \phi_{k-1}$ are orthogonal to each other, then $\phi_1, \ldots, \phi_{k-1}, \phi_k - (\sum_{j=1}^{k-1} \langle \phi_j, \phi_k \rangle / \|\phi_j \cdot\|^2)\phi_j$ are orthogonal.

3. Assume that the nonzero eigenvalues of $\mathbf{A}$ are bounded away from zero (see theoretical complement 2).
   (i) In the case of two reflecting boundaries prove that

$$\lim_{t \to \infty} p(t; x, y) = m(y) \qquad (x, y \in [0, d])$$

   where $m(y)$ is the normalization of $\pi(y)$ in (8.1). Show that this convergence is exponentially fast, uniformly in x, y.
   (ii) Prove that $m(y)\, dy$ is the unique invariant probability for p.
   (iii) Prove (ii) without the assumption concerning eigenvalues.

4. By a change of scale in Example 3 compute the transition probability of a Brownian motion on $[a, b]$ with both boundaries absorbing.

5. Use Example 4, and a change of scale, to find
   (i) the distribution of the minimum m_t of a Brownian motion $\{X_s\}$ over the time interval $[0, t]$;
   (ii) the distribution of the maximum M_t of $\{X_s\}$ over $[0, t]$;
   (iii) the joint distribution of (X_t, m_t).

6. Use Example 3, and change of scale, to compute the joint distribution of (X_t, m_t, M_t), using the notation of Exercise 5 above.

7. For a Brownian motion with drift μ and diffusion coefficient $\sigma^2 > 0$, compute the p.d.f. of $\tau_a \wedge \tau_b$ when $X_0 = x$ and $a < x < b$.

8. Establish the identities

$$(2\pi\sigma^2 t)^{-1/2} \sum_{m=-\infty}^{\infty} \left[\exp\left\{-\frac{(2md + y - x)^2}{2\sigma^2 t}\right\} + \exp\left\{-\frac{(2md + y + x)^2}{2\sigma^2 t}\right\}\right]$$
$$= \frac{1}{d} + \frac{2}{d} \sum_{m=1}^{\infty} \exp\left\{-\frac{\sigma^2\pi^2 m^2 t}{2d^2}\right\} \cos\left(\frac{m\pi x}{d}\right) \cos\left(\frac{m\pi y}{d}\right),$$

$$(2\pi\sigma^2 t)^{-1/2} \sum_{m=-\infty}^{\infty} \left[\exp\left\{-\frac{(2md + y - x)^2}{2\sigma^2 t}\right\} - \exp\left\{-\frac{(2md + y + x)^2}{2\sigma^2 t}\right\}\right]$$
$$= \frac{2}{d} \sum_{m=1}^{\infty} \exp\left\{-\frac{\sigma^2\pi^2 m^2 t}{2d^2}\right\} \sin\left(\frac{m\pi x}{d}\right) \sin\left(\frac{m\pi y}{d}\right), \qquad (0 \leqslant x, y \leqslant d).$$

Use these to derive *Jacobi's identity for the theta function*,

$$\theta(z) := \sum_{m=-\infty}^{\infty} \exp\{-\pi m^2 z\} = \frac{1}{\sqrt{z}} \sum_{m=-\infty}^{\infty} \exp\{-\pi m^2/z\} \equiv \frac{1}{\sqrt{z}}\,\theta\left(\frac{1}{z}\right), \qquad (z > 0).$$

[*Hint*: Compare (8.23) with (6.31), and (8.33) with Exercise 7.5(ii).]

9. (*Hermite Polynomials and the Ornstein–Uhlenbeck Process*) Consider the generator $\mathbf{A} = \frac{1}{2}\,d^2/dx^2 - x\,d/dx$ on the state space $\mathbb{R}^1$.
   (i) Prove that $\mathbf{A}$ is symmetric on an appropriate dense subspace of $L^2(\mathbb{R}^1, e^{-x^2}\,dx)$.
   (ii) Check that $\mathbf{A}$ has eigenfunctions $H_n(x) := (-1)^n \exp\{x^2\}(d^n/dx^n)\exp\{-x^2\}$, the so-called Hermite polynomials, with corresponding eigenvalues $n = 0, 1, 2, \ldots$.
   (iii) Give some justification for the expansion

$$p(t; x, y) = e^{-y^2} \sum_{n=0}^{\infty} c_n e^{-nt} H_n(x) H_n(y), \qquad c_n := \frac{1}{\sqrt{\pi}\,2^n n!}.$$

10. According to the theory of Fourier series, the functions $\cos nx$ $(n = 0, 1, 2, \ldots)$, $\sin nx$ $(n = 1, 2, \ldots)$ form a complete orthogonal system in $L^2[-\pi, \pi]$. Use this to prove the following:
   (i) The functions $\cos nx$ $(n = 0, 1, 2, \ldots)$ form a complete orthogonal sequence in $L^2[0, \pi]$. [*Hint*: Let $f \in L^2[0, \pi]$. Make an even extension of f to $[-\pi, \pi]$, and show that this f may be expanded in $L^2[-\pi, \pi]$ in terms of $\cos nx$ $(n = 0, 1, \ldots)$.]
   (ii) The functions $\sin x$ $(n = 1, 2, \ldots)$ form a complete orthogonal sequence in $L^2[0, \pi]$. [*Hint*: Extend f to $[-\pi, \pi]$ by setting $f(-x) = -f(x)$ for $x \in [-\pi, 0)$.]

Exercises for Section V.9

1. Check (9.13).
2. Derive (9.20) by solving the appropriate two-point boundary-value problem.

3. Suppose $S = [a, b)$ and a is reflecting. Show that the diffusion is recurrent if and only if $s(b) = \infty$. If, on the other hand, $s(b) < \infty$ then one has $\rho_{xy} = 1$ for $y > x$ and

$$\rho_{xy} = \frac{s(b) - s(x)}{s(b) - s(y)} = \frac{\int_x^b \exp\{-I(x_0, z)\}\, dz}{\int_y^b \exp\{-I(x_0, z)\}\, dz}, \qquad y < x.$$

[*Hint*: For $y > x$, assume the fact that the transition probability density $p(t; x, y)$ is strictly positive and continuous in x, y for each $t > 0$, to show that $\delta := \min\{P_z(X_{t_0} \geq y): a \leq z \leq y\} > 0$ for $t_0 > 0$.]

4. Suppose $S = [a, b)$ and a is absorbing. Then show that no state, other than a, is recurrent and that

$$\rho_{xy} = \begin{cases} \dfrac{s(x) - s(a)}{s(y) - s(a)} & \text{if } a < x \leq y, \\ 1 & \text{if } s(b) = \infty, \text{ and } a \leq y < x, \\ \dfrac{s(b) - s(x)}{s(b) - s(y)} & \text{if } s(b) < \infty, \text{ and } a \leq y \leq x. \end{cases}$$

5. Suppose $S = [a, b]$ with both boundaries reflecting. Then show that $\rho_{xy} = 1$ for all $x, y \in S$, and the diffusion is recurrent.

6. Apply Corollary 9.3 and Exercise 3 to decide which of the following diffusions are recurrent and which are transient:
   (i) $S = (-\infty, \infty)$, $\mu(x) \equiv \mu \neq 0$, $\sigma^2(x) \equiv \sigma^2 > 0$;
   (ii) $S = (-\infty, \infty)$, $\mu(x) \equiv 0$, $\sigma^2(x) \equiv \sigma^2 > 0$;
   (iii) $S = (-\infty, \infty)$, $\mu(x) = -\beta x$, $\sigma^2(x) \equiv \sigma^2 > 0$ (consider separately, $\beta > 0$, $\beta < 0$);
   (iv) $S = [0, \infty)$, $\mu(x) \equiv \mu$, $\sigma^2(x) \equiv \sigma^2 > 0$, 0 reflecting (consider separately the cases $\mu < 0$, $\mu = 0$, $\mu > 0$);
   (v) $S = (0, 1)$, $\mu(x) \to \infty$ as $x \downarrow 0$, $\mu(x) \to -\infty$ as $x \uparrow 1$, $\sigma^2(x) \geq \sigma^2 > 0$ for all x.

7. Suppose $S = [a, b]$ with a absorbing and b reflecting. Show that all states but a are transient, $\rho_{xy} = 1$ for $y \leq x$, and

$$\rho_{xy} = \frac{s(x) - s(a)}{s(y) - s(a)}, \qquad y > x.$$

8. Suppose $S = [a, b]$ with both boundaries absorbing. Show that all states, other than a and b, are transient and

$$\rho_{xy} = \begin{cases} \dfrac{s(b) - s(x)}{s(b) - s(y)} & a \leq y \leq x < b, \\ \dfrac{s(x) - s(a)}{s(y) - s(a)} & a < x \leq y \leq b. \end{cases}$$

9. Let $[c, d] \subset S$. Solve the two-point boundary-value problem: $\mathbf{A}f(x) = 0$ for $c < x < d$, $\lim_{x \downarrow c} f(x) = \alpha$, $\lim_{x \uparrow d} f(x) = \beta$, where α, β are given numbers. Show that the solution equals $\alpha P_x(\tau_c < \tau_d) + \beta P_x(\tau_d < \tau_c)$.

Exercises for Section V.10

1. (i) Assuming that the transition probability density $p(t; x, y)$ is positive and continuous in x, y for each $t > 0$, show that $M(x)$, defined by (10.3), is bounded on $[c, d]$. [*Hint*: Use Proposition 13.5 of Chapter I.]
   (ii) Let τ be as in (10.2), $x \in (c, d)$. Assume that

$$P_x\left(\max_{0 \leqslant t \leqslant h} |X_t - x| > \varepsilon \right) = o(h) \qquad \text{as } h \downarrow 0,$$

   for every $\varepsilon > 0$ (this is proved in Exercise 3.5). Show that $E_x(\mathbf{1}_{\{\tau \leqslant h\}} M(X_h)) = o(h)$ as $h \downarrow 0$. Check the assumption in the case of constant coefficients.

2. Show that (10.15) holds if $s(b) = \infty$.

3. Assume $s(b) = \infty$.
   (i) Show that (10.16) is necessary and sufficient for finiteness of $E_x \tau_c$ $(c < x)$.
   (ii) Prove (10.17), under the assumption (10.16).

4. Suppose $s(a) = -\infty$.
   (i) Prove that, for $x < d$, $E_x \tau_d < \infty$ if and only if (10.18) holds.
   (ii) Derive (10.19).

5. (i) Suppose $S = [a, b)$ with a a reflecting boundary. Show that the diffusion is positive recurrent if and only if $s(b) = \infty$ and $m(b) < \infty$. [*Hint*: Assume that the transition probability density $p(t; x, y)$ is positive and continuous in x, y for each $t > 0$, and proceed as in Exercise 9.3, or use Proposition 13.5 of Chapter I.] Show that (10.19) holds in case of positive recurrence.
   (ii) Suppose $S = [a, b]$ with a and b reflecting boundaries. Show that the diffusion is positive recurrent. Show that (10.17) and (10.19) hold.

6. Apply Proposition 10.2 and Exercise 5 above (as well as Exercise 9.6) to classify the diffusions in Exercise 9.6 into transient, null recurrent and positive recurrent ones.

7. Let $[c, d] \subset S$. Solve the two-point boundary-value problem $\mathbf{A}f(x) = -g(x)$ for $c < x < d$, $\lim_{x \downarrow c} f(x) = \alpha$, $\lim_{x \uparrow d} f(x) = \beta$, where g is a given bounded measurable (or, continuous) function on $[c, d]$ and α, β are given constants. Show that the solution represents

$$E_x \int_0^{\tau_c \wedge \tau_d} g(X_s)\, ds + \alpha P_x(\tau_c < \tau_d) + \beta P_x(\tau_d < \tau_c).$$

8. Show that, under natural scale (i.e., $s(x) = x$), $m_1'(x) \geqslant m_2'(x)$ for all x implies $M_1(x) \geqslant M_2(x)$ for all x.

Exercises for Section V.11

1. Prove that τ_c, $\tau_c \wedge \tau_d$ are stopping times, by showing that they are measurable with respect to $\mathscr{F}_{\tau_c}$, $\mathscr{F}_{\tau_c \wedge \tau_d}$, respectively.

2. Prove that $\tau_1 \wedge \tau_2$ is a stopping time, if τ_1 and τ_2 are.

3. (i) Suppose that

$$E(Zh(X_{\tau+t'_1}, \ldots, X_{\tau+t'_r})\mathbf{1}_{\{\tau<\infty\}}) = E(Z[E_y h(X_{t'_1}, \ldots, X_{t'_r})]_{y=X_\tau}\mathbf{1}_{\{\tau<\infty\}})$$

for all Z of the form (11.7), m being arbitrary and f bounded continuous. Prove that (11.12) holds. [*Hint*: Use (a) the fact that bounded continuous functions on S^m are dense in $L^1(S^m, \mu)$ where μ is any probability measure, (b) Dynkin's Pi-Lambda Theorem (Section 4 of Chapter 0).]

   (ii) Prove that if (11.12) holds for arbitrary finite sets $t'_1 < t'_2 < \cdots < t'_r$ and bounded continuous h, then the conditional distribution of X_τ^+ given $\mathscr{F}_\tau$ is P_{X_τ}, on the set $\{\tau < \infty\}$.

4. Prove that $\varphi(y)$ in (11.18) is continuous for every bounded continuous h on S^r, if $x \to p(t; x, dy)$ is weakly continuous (see Section 5 of Chapter 0, for the definition of weak convergence). This weak continuity, also called the (*weak*) *Feller property*, is proved in Section 3 of Chapter VII.

5. Let $\{X_t\}$ be a Brownian motion with drift μ and diffusion coefficient $\sigma^2 > 0$, starting at x.

   (i) Prove that the conditional distribution of $\{X_u; 0 \leqslant u \leqslant t\}$, given X_t does not depend on μ. [*Hint*: Look at finite-dimensional conditional distributions.]

   (ii) Use (i) and (11.42) to compute the joint p.d.f. of (X_t, M_t), where $M_t = \max\{X_u: 0 \leqslant u \leqslant t\}$.

   (iii) Use (ii) to compute the distributions of (a) M_t, and (b) $\tau_a (a > x)$.

   (iv) Check that the conditional distribution in (i), given $X_t = y$, is the same as the distribution of the process

$$\left\{Y_u - \frac{u}{t}(Y_t - (y - x)) + x: 0 \leqslant u \leqslant t\right\},$$

   where $\{Y_u: u \geqslant 0\}$ is a Brownian motion with zero drift and diffusion coefficient σ^2, starting at zero.

6. Assume the result of Example 8.3, for the case $\mu = 0$. Show how you can use Exercise 5(i) to derive the joint distribution of (X_t, m_t, M_t) where $\{X_t\}$, M_t are as in Exercise 5(i), (ii), and $m_t = \min\{X_u: 0 \leqslant u \leqslant t\}$.

7. Let $\{X_t\}$ be a Brownian motion with drift zero and diffusion coefficient $\sigma^2 > 0$, starting at x. Let $\tau = \tau_c \wedge \tau_d$, where $c < x < d$. Show that the after-τ process X_τ^+ is *not* independent of the pre-τ sigmafield.

8. Let $\{X_t\}$ be a Brownian motion with drift $\mu > 0$ and diffusion coefficient $\sigma^2 > 0$, starting at zero.

   (i) Prove that $\{\tau_a: a \geqslant 0\}$ is a process with independent and homogeneous increments.

   (ii) Find the distribution of τ_a.

   (iii) What can you say concerning (i), (ii), if $\mu < 0$?

9. Show that the proof of the strong Markov property (Theorem 11.1) essentially applies (indeed, more simply) to the countable state space case (see Section 5 of Chapter IV).

10. (i) Use Exercise 5(ii) to derive the transition probability density (component) of a Brownian motion on $(-\infty, 0]$ with absorption at 0.
    (ii) Derive the corresponding transition density on $[0, \infty)$ with absorption at 0.
    (iii) Use Exercise 6 to derive the transition probability density (component) of a Brownian motion on $[0, 1]$ with absorption at both ends.

11. (*Method of Images*) Consider the drift and diffusion coefficient $\mu(\cdot), \sigma^2(\cdot)$ defined on $(-\infty, 0]$. Assume $\mu(0) = 0$. Extend $\mu(\cdot), \sigma^2(\cdot)$ to $\mathbb{R}^1$ as in (6.8) by setting $\mu(-x) = -\mu(x), \sigma^2(-x) = \sigma^2(x)$ for $x < 0$. Let $\{X_t\}$ be a diffusion on $\mathbb{R}^1$ with these coefficients, starting at $x < 0$, and let τ_0 denote the first passage time to 0. Let $\{Y_t\}$ be the process defined by

$$Y_t = \begin{cases} X_t & \text{for } t < \tau_0 \\ -X_t & \text{for } t \geqslant \tau_0. \end{cases}$$

    (i) Show that $\{Y_t\}$ has the same distribution as $\{X_t\}$.
    (ii) Show that $P_x(M_t \geqslant 0) = 2P_x(X_t > 0)$, where $M_t := \max\{X_s : 0 \leqslant s \leqslant t\}$. [*Hint*: Show that (11.34)–(11.36) hold with 0 replaced by x and a replaced by 0.]
    (iii) Show that $P_x(X_t \leqslant y, \tau_0 > t) = P_x(X_t \geqslant -y)$ for all $y \leqslant 0$. [*Hint*: Use the analog of (11.35).]
    (iv) Deduce from (iii) that

$$P_x(X_t \leqslant y, \tau_0 > t) = P_x(X_t \leqslant y) - P_x(X_t \geqslant -y) \qquad \text{for all } x < 0, y \leqslant 0.$$

    (v) Derive the analog of (iv) for a diffusion on $[0, \infty)$.

Exercises for Section V.12

1. Use the strong Markov property to prove (12.10) assuming that the diffusion is recurrent and that (12.11) holds.

2. Assume that the diffusion is positive recurrent.
    (i) Prove that (12.14) defines a probability measure $\pi(dx)$, i.e., prove countable additivity. [*Hint*: Use Monotone Convergence Theorem.]
    (ii) Prove (12.15) for all f satisfying (12.11).

3. Prove the uniqueness of the invariant probability in Theorem 12.2.

4. (i) Prove (12.24).
    (ii) Use the strong Markov property to prove (12.25).

5. Under what assumptions on p (and $\partial p/\partial t$) can you justify the equalities in (12.27)?

6. (i) Prove that the diffusion in Example 1 is positive recurrent.
    (ii) Let $-\infty < a < b < \infty$, and suppose that a diffusion on (a, b) is recurrent. Prove that it cannot reach either a or b with positive probability. [*Hint*: The time to reach $\{a, b\}$ is larger than η_{2r} (see (12.2)) for every r.]

7. Prove that the diffusion in Example 2 is positive recurrent under the assumptions (12.35).

8. Find the invariant distributions of the following diffusions:
   (i) $S = (-\infty, \infty)$, $\mu(x) = -\beta x$ $(\beta < 0)$, $\sigma^2(x) \equiv \sigma^2$.
   (ii) $S = (0, 1]$, $\mu(x) = (k - 1)/x$ $(k > 1)$, $\sigma^2(x) \equiv \sigma^2$, "1" a reflecting boundary.
   (iii) $S = [0, \infty)$, $\mu(x) = -\beta x$ $(\beta > 0)$, $\sigma^2(x) \equiv \sigma^2$, "0" is a reflecting boundary.

Exercises for Section V.13

1. Give a proof of Theorem 13.1 similar to the proof of Proposition 10.1 in Chapter II. (Also, see the proof of Proposition 12.1.)

2. Consider a diffusion $\{X_t\}$ on $S = (0, 1]$, with "1" a reflecting boundary and coefficients $\mu(x) = (k - 1)/x$ $(k > 1)$, $\sigma^2(x) \equiv \sigma^2$. Apply Theorem 13.1 to compute the asymptotic distribution of

$$t^{-1/2} \int_0^t (f(X_s) - E_\pi f)\, ds \qquad \text{as } t \uparrow \infty,$$

where $f(x) = 1 - x^2$. What is $E_\pi f$? [*Hint*: Remember the boundary condition for **A**.]

Exercises for Section V.14

1. Let $\{X_t^{(i)}\}$ $(i = 1, 2, \ldots, k)$ be k independent diffusions on $\mathbb{R}^1$ with drift coefficients $\mu^{(i)}(\cdot)$ and diffusion coefficients $\sigma_i^2(\cdot)$.
   (i) Show that $\{\mathbf{X}_t := (X_t^{(1)}, \ldots, X_t^{(k)})\}$ is a Markov process on $\mathbb{R}^k$, and express its transition probability density in terms of those of $\{X_t^{(i)}\}$. [*Hint*: Look at $P(\{\mathbf{X}_{s+t} \in R\} \cap F)$, where R is a k-dimensional rectangle $(a_1, b_1) \times \cdots \times (a_k, b_k)$, and $F = F_1 \cap F_2 \cap \cdots \cap F_k$ with F_i determined by $\{X_u^{(i)}: 0 \leqslant u \leqslant s\}$ $(1 \leqslant i \leqslant k)$.]
   (ii) Show that the transition probability density $p(t; \mathbf{x}, \mathbf{y})$ of $\{\mathbf{X}_t\}$ satisfies Kolmogorov's backward and forward equations

$$\frac{\partial p}{\partial t} = \frac{1}{2} \sum_{i=1}^k \sigma_i^2(x^{(i)}) \frac{\partial^2 p}{(\partial x^{(i)})^2} + \sum_{i=1}^k \mu^{(i)}(x^{(i)}) \frac{\partial p}{\partial x^{(i)}},$$

$$\frac{\partial p}{\partial t} = \frac{1}{2} \sum_{i=1}^k \frac{\partial^2}{(\partial y^{(i)})^2} (\sigma_i^2(y^{(i)}) p) - \sum_{i=1}^k \frac{\partial}{\partial y^{(i)}} (\mu^{(i)}(y^{(i)}) p).$$

   (iii) Let $\mu(\mathbf{x}) = -\gamma \mathbf{x}$, $\mathbf{D}(\mathbf{x}) = \sigma^2 \mathbf{I}$, with $\gamma > 0$, $\sigma^2 > 0$. Write down its transition probability density $p(t; \mathbf{x}, \mathbf{y})$, and check that it satisfies the Kolmogorov equations. Show $p(t; \mathbf{x}, \mathbf{y})$ converges to a Gaussian density as $t \to \infty$, and deduce that the diffusion has a unique invariant distribution.

2. Write down the analogs of (1.2), (1.2)′ for a multidimensional diffusion.

3. (i) Show that a k-dimensional Brownian motion with drift $\mu = (\mu^{(1)}, \ldots, \mu^{(k)})$ and diffusion matrix $\mathbf{D} = ((d_{ij}))$ is a process with independent increments, and that its transition probability density $p(t; \mathbf{x}, \mathbf{y})$ satisfies Kolmogorov's equations (14.6).

(ii) More generally, let φ be a one-to-one twice continuously differentiable map on $\mathbb{R}^k$ onto an open subset of $\mathbb{R}^k$. If $\{X_t\}$ is a diffusion on $\mathbb{R}^k$ with generator $\mathbf{A}$ given by (14.9), compute the generator of $\{\mathbf{Y}_t := \varphi(\mathbf{X}_t)\}$.

4. Check that the proof of Theorem 11.1 (Strong Markov property) remains unchanged if the state space is taken to be $\mathbb{R}^k$ or a subset of $\mathbb{R}^k$.

5. Let $\{\mathbf{B}_t\}$ be a k-dimensional standard Brownian motion, $k \geqslant 3$, starting at some $\mathbf{x} \in \mathbb{R}^k$.
   (i) Let $d > |\mathbf{x}|$. Prove that $P(|\mathbf{B}_t| > d$ for all sufficiently large $t) = 1$. [*Hint*: Fix $d_1 > d$.
   (a) $P(\{\mathbf{B}_t\}$ ever reaches $\{\mathbf{z}: |\mathbf{z}| = d_1\}) = 1$, by the recurrence of one-dimensional Brownian motions.
   (b) Write

$$\tau_1 := \inf\{t > 0: |\mathbf{B}_t| = d_1\},$$
$$\tau_2 := \inf\{t > \tau_1: |\mathbf{B}_t| = d\}, \qquad \tau_{2n+1} := \inf\{t > \tau_{2n}: |\mathbf{B}_t| = d_1\},$$
$$\tau_{2n} := \inf\{t > \tau_{2n-1}: |\mathbf{B}_t| = d\}.$$

   By the strong Markov property and (14.26), $P(\tau_{2n} < \infty) = (d/d_1)^{(k-2)n}$, and $P(\tau_{2n+1} < \infty \mid \tau_{2n} < \infty) = 1$.
   (c) By (b), $P(\tau_{2n} = \infty$ for some $n) = 1$.]
   (ii) From (i) conclude that

$$P(|\mathbf{B}_t| \to \infty \text{ as } t \to \infty) = 1.$$

6. Let $\{\mathbf{B}_t\}$ be a two-dimensional standard Brownian motion. Let $B = \{\mathbf{z}: |\mathbf{z} - \mathbf{z}_0| \leqslant \varepsilon\}$, with $\mathbf{z}_0$ and $\varepsilon > 0$ arbitrary. Prove that

$$P(\sup\{t \geqslant 0: \mathbf{B}_t \in B\} = \infty) = 1.$$

7. Let $\{\mathbf{X}_t\}$ be a k-dimensional diffusion with periodic drift and diffusion coefficients, the period being $d > 0$ in each coordinate. Write $Z_t^{(i)} := X_t^{(i)} (\operatorname{mod} d)$, $\mathbf{Z}_t := (Z_t^{(1)}, \ldots, Z_t^{(k)})$.
   (i) Use Proposition 14.1 to show that $\{\mathbf{Z}_t\}$ is a Markov process on the *torus* $T = [0, d)^k$—which may be regarded as a Cartesian product of k circles.
   (ii) Assuming that the transition probability density $p(t; \mathbf{x}, \mathbf{y})$ of $\{\mathbf{X}_t\}$ is continuous and positive for all $t > 0$, $\mathbf{x}$, $\mathbf{y}$, prove that the transition probability $q(t; \mathbf{z}, \mathbf{z}')\, d\mathbf{z}'$ of $\{\mathbf{Z}_t\}$ admits a unique invariant probability $\pi(\mathbf{z}')\, d\mathbf{z}'$ and that

$$\max_{\mathbf{z}, \mathbf{z}' \in T} |q(t; \mathbf{z}, \mathbf{z}') - \pi(\mathbf{z}')| \leqslant [1 - \delta d^k]^{[t]-1} \theta,$$

   where $[t]$ is the integer part of t, $\delta := \min\{q(1; \mathbf{z}, \mathbf{z}'): \mathbf{z}, \mathbf{z}' \in T\}$ and

$$\theta = \max\{q(1; \mathbf{y}, \mathbf{z}') - q(1; \mathbf{z}, \mathbf{z}'): y, \mathbf{z}, \mathbf{z}' \in T\}.$$

   [*Hint*: Recall Exercise 6.9 of Chapter II.]

8. Let $\{\mathbf{X}_t\}$ be a k-dimensional diffusion, $k \geqslant 2$.
   (i) Use Proposition 14.1 and computations such as (14.18) to find a necessary and

sufficient condition (in terms of the coefficients of $\{\mathbf{X}_t\}$) for $\{R_t := |\mathbf{X}_t|\}$ to be a Markov process. Compute the generator of $\{R_t\}$ if this condition is met.

(ii) Assume that the sufficient condition in (i) is satisfied. Find necessary and sufficient conditions for $\{\mathbf{X}_t\}$ to be (a) recurrent, (b) transient.

9. Derive (14.14).

10. (i) Sketch an argument, similar to that in Section 2, to show that Corollary 2.4 holds for diffusions on $\mathbb{R}^k$ (under conditions (1)–(4) following Definition 14.1).

 (ii) Use (i) to prove

$$P_{\mathbf{x}}\left(\max_{0 \leqslant t \leqslant h} |\mathbf{X}_t - \mathbf{x}| > \varepsilon\right) = o(h) \qquad \text{as } h \downarrow 0.$$

 [*Hint*: See Exercise 3.5.]

Exercises for Section V.15

1. Prove that the process $\{\tilde{\mathbf{X}}_t\}$ defined by (15.1) is Markovian and has the transition probability (15.3). [*Hint*: Mimic the proof of Theorem 7.1.]

2. Specialize (15.23) to the case $k = 1$ to compute $P_x(\{X_t\}$ reaches c before $d)$, where $\{X_t\}$ is a one-dimensional diffusion and $c \leqslant x \leqslant d$. Check that this agrees with (9.19).

3. Apply (15.23) to compute $P_{\mathbf{x}}(\{\mathbf{B}_t\}$ reaches $\partial B(\mathbf{0}{:}c)$ before $\partial B(\mathbf{0}{:}d))$ where $\{\mathbf{B}_t\}$ is a k-dimensional standard Brownian motion, $k > 1$, $\partial B(\mathbf{0}{:}r) = \{\mathbf{y} \in \mathbb{R}^k : |\mathbf{y}| = r\}$, and $c \leqslant |\mathbf{x}| \leqslant d$. Check that this computation agrees with (14.23).

4. Use Exercise 14.8(i) and (15.23) to compute

$$P_{\mathbf{x}}(\{\mathbf{X}_t\} \text{ reaches } \partial B(\mathbf{0}{:}c) \text{ before } \partial B(\mathbf{0}{:}d)), \qquad c \leqslant |\mathbf{x}| \leqslant d,$$

 for a k-dimensional diffusion $(k > 1)$ $\{\mathbf{X}_t\}$ such that the radial process $\{R_t := |\mathbf{X}_t|\}$ is Markovian.

5. Consider a standard k-dimensional Brownian motion $(k > 1)$ on $\bar{G} = \{\mathbf{x} \in \mathbb{R}^k : x^{(i)} \geqslant 0$ for $1 \leqslant i \leqslant k\}$ with absorption at the boundary.
 (i) Compute $p^0(t; \mathbf{x}, \mathbf{y})$ for $t > 0$; $\mathbf{x}, \mathbf{y} \in G)$.
 (ii) Compute the distribution of $\tau_{\partial G}$, if the process starts at $\mathbf{x} \in G$.
 *(iii) Compute the hitting (or, harmonic) measure $\psi(\mathbf{x}, d\mathbf{y})$ on ∂G, for $\mathbf{x} \in G$. ($\psi(\mathbf{x}, d\mathbf{y})$ is the $P_{\mathbf{x}}$-distribution of $\mathbf{X}_{\tau_{\partial G}}$).
 *(iv) Compute the transition probability $\tilde{p}(t; x, B)$ for $x \in G$, $B \subset \partial G$.

6. Do the analog of Exercise 5 for $\bar{G} = \{\mathbf{x} \in \mathbb{R}^k : 0 \leqslant x^{(i)} \leqslant 1$ for $1 \leqslant i \leqslant k\}$.

7. Prove (15.27).

8. For G in Example 1 and Exercises 5 and 6, show that $P_{\mathbf{x}}(\tau_{\partial G} > t) = o(t)$ as $t \downarrow 0$ if $\mathbf{x} \in G$.

9. Show that the function ψ_f in (15.28) satisfies (15.23) with

$$\mathbf{A} = \tfrac{1}{2}\Delta\left(\Delta := \sum_{i=1}^{k} \frac{\partial^2}{(\partial x^{(i)})^2}\right).$$

10. Let G be an open subset of $\mathbb{R}^k$. A Borel measurable function u on G, which is bounded on compacts, is said to have the *mean value property* in G if $u(\mathbf{x}) = \int_{S^{k-1}} u(\mathbf{x} + r\theta)\, d\theta$, for all $\mathbf{x}$ and $r > 0$ such that the closure of the ball $B(\mathbf{x}:r)$ with center $\mathbf{x}$ and radius r is contained in G. Here S^{k-1} is the unit sphere $\{|\mathbf{y}| = 1\}$ and $d\theta$ denotes integration with respect to the uniform (i.e., rotation invariant) distribution on S^{k-1}.

 (i) Let $\{\mathbf{B}_t\}$ be a standard Brownian motion on $\mathbb{R}^k$ $(k > 1)$ and φ a bounded Borel measurable function on ∂G. Suppose $P_{\mathbf{x}}(\tau_{\partial G} < \infty) = 1$ for all $\mathbf{x} \in G$, where $P_{\mathbf{x}}$ denotes the distribution of $\{\mathbf{B}_t\}$ starting at $\mathbf{x}$. Show that $u(\mathbf{x}) := E_{\mathbf{x}}\varphi(\mathbf{B}_{\tau_{\partial G}})$ has the mean value property in G. [*Hint*: $u(\mathbf{x}) = E_{\mathbf{x}}u(\mathbf{B}_{\tau_{\partial B(\mathbf{x}:r)}})$ if $\{\mathbf{y}: |\mathbf{y} - \mathbf{x}| \leq r\} \subset G$, by the Strong Markov property. Next note that the $P_{\mathbf{0}}$-distribution of $\{\mathbf{B}_t\}$ is the same as that of $\{\mathbf{O}\mathbf{B}_t\}$ for every orthogonal transformation $\mathbf{O}$.]

 (ii) Show that if u has the mean value property in G, then u is infinitely differentiable. [*Hint*: Fix $\mathbf{x} \in G$. Let $B(\mathbf{x}:2r) \subset G$, and let ψ be an infinitely differentiable radially symmetric p.d.f. vanishing outside $B(\mathbf{0}:r)$. Let $\bar{u} = u$ on $B(\mathbf{x}:2r)$ and zero outside. Show that $u = \bar{u} * \psi$ on $B(\mathbf{x}:r)$].

 (iii) Show that if u has the mean value property in G, then it is *harmonic* in G, i.e., $\Delta u = 0$ in G. [*Hint*:

$$u(\mathbf{x}) = u(\mathbf{y}) + \sum (x^{(i)} - y^{(i)}) \frac{\partial u}{\partial x^{(i)}}(\mathbf{y}) + \frac{1}{2} \sum (x^{(i)} - y^{(i)})(x^{(j)} - y^{(j)}) \frac{\partial^2 u}{\partial x^{(i)}\, \partial x^{(j)}}(\mathbf{y})$$
$$+ \mathrm{O}(\rho^3),$$

 where $\rho = |\mathbf{x} - \mathbf{y}|$. Integrate with respect to the uniform distribution on $\{\mathbf{x}: |\mathbf{x} - \mathbf{y}| = \rho\}$. Show that the integral of the first sum is zero, that of the second sum is $(\rho^2/2k)\Delta u(\mathbf{y})$ and that of the remainder is $\mathrm{O}(\rho^3)$, so that $(\rho^2/2k)\Delta u(\mathbf{y}) + \mathrm{O}(\rho^3)$ is zero for small ρ.]

 (iv) Show that a harmonic function in G has the mean value property. [*Hint*: Use the divergence theorem.]

11. Check that

 (i) the function ψ_f in (15.28) is harmonic,

 (ii) $\displaystyle\int_{\partial G} \psi(\mathbf{x}; \mathbf{y}) s(d\mathbf{y}) = 1$ for $\mathbf{x} \in \partial G$, and

 (iii) $\psi(\mathbf{x}, \mathbf{y}) s(d\mathbf{y})$ converges weakly to $\delta_{\mathbf{y}_0}(d\mathbf{y})$ as $\mathbf{x} \to y_0 \in \partial G$.

12. (*Maximum Principle*) Let u be harmonic in a connected open set $G \subset \mathbb{R}^k$. Show that u cannot attain its infimum or supremum in G, unless u is constant in G. [*Hint*: Use the mean value property.]

13. (*Dirichlet Problem*) Let G be a bounded, connected, open subset of $\mathbb{R}^k$. Given a continuous function φ on ∂G, show, using Exercises 10 and 12, that

 (i) $u(\mathbf{x}) := E_{\mathbf{x}}\varphi(\mathbf{B}_{\tau_{\partial G}})$ is a solution of the Dirichlet problem

$$\Delta u(\mathbf{x}) = 0 \qquad \text{for } \mathbf{x} \in G, \qquad u(\mathbf{x}) = \varphi(\mathbf{x}) \qquad \text{for } \mathbf{x} \in \partial G,$$

 and

 (ii) if u is continuous at the boundary, i.e., $u(\mathbf{y}) = \lim u(\mathbf{x})$ as $\mathbf{x}(\in G) \to \mathbf{y} \in \partial G$, then this solution is *unique* in the class of solutions which are continuous on $\bar{G}$.

14. (*Poisson's Equation*) Let G be as in Exercise 13. Give at least an informal proof, akin to that of (15.25), that

$$u(\mathbf{x}) := E_{\mathbf{x}} \int_0^{\tau_{\partial G}} g(\mathbf{X}_s)\, ds + E_{\mathbf{x}} \varphi(\mathbf{X}_{\tau_{\partial G}})$$

satisfies Poisson's equation

$$\tfrac{1}{2}\Delta u(\mathbf{x}) = -g(\mathbf{x}) \qquad \text{for } \mathbf{x} \in G, \qquad u(\mathbf{x}) = \varphi(\mathbf{x}) \qquad \text{for } \mathbf{x} \in \partial G.$$

Here g, φ are given continuous functions on $\bar{G}$ and ∂G, respectively. Assuming u is continuous at the boundary, show that this solution of Poisson's equation is *unique* in the class of all continuous solutions on $\bar{G}$.

15. Extend the function F in (15.37) to a twice continuously differentiable function on $[0, \infty)$ vanishing outside $[c_1, d_1]$, where $0 < c_1 < c < d < d_1$. Show that the corresponding extension of f (in (15.37)) is twice continuously differentiable on $\mathbb{R}^k$ and vanishes outside a compact set.

16. Prove that
    (i) $\int_c^\infty \exp\{-\bar{I}(u)\}\, du$ diverges for some $c > 0$ if and only if it diverges for all $c > 0$;
    (ii) $\int_c^\infty \exp\{-\underline{I}(u)\}\, du$ converges for some $c > 0$ if and only if it converges for all $c > 0$.

17. Prove (15.61) using (15.60).

18. Write out the details of the proof of Theorem 15.3(b). [*Hint*: Follow the steps of Exercise 14.5(i), replacing step (a) by (15.52) with d replaced by d_1.]

19. Consider the translation $\mathbf{x} \to \mathbf{x} + \mathbf{z}$ (for a given $\mathbf{z}$), and the diffusion $\{\mathbf{Y}_t := \mathbf{X}_t + \mathbf{z}\}$.
    (i) Write down the drift and diffusion coefficients of $\{\mathbf{Y}_t\}$.
    (ii) Show that $\{\mathbf{X}_t\}$ is recurrent (or transient) if and only if $\{\mathbf{Y}_t\}$ is recurrent (transient).
    (iii) Write down (15.46) and (15.47) for $\{\mathbf{Y}_t\}$, in terms of the coefficients $\mu^{(i)}(\cdot)$, $d_{ij}(\cdot)$.

20. Let $\{\mathbf{X}_t\}$ be a diffusion on $\mathbb{R}^k$. Assume that
    (a) $((d_{ij}(\mathbf{x}))) \equiv \sigma^2 \mathbf{I}$, where $\sigma^2 > 0$ and $\mathbf{I}$ is the $k \times k$ identity matrix, and
    (b) $\sum_{i=1}^k x^{(i)} \mu^{(i)}(\mathbf{x} + \mathbf{z}) \leqslant -\delta$ for $|\mathbf{x}| \geqslant M$, where $\mathbf{z} \in \mathbb{R}^k$, $\delta > 0$, $M > 0$ are given.
    Apply Theorem 15.3 (and Exercise 19) to decide whether the diffusion is recurrent or transient.

21. For a diffusion having a positive and continuous (in $\mathbf{x}$, $\mathbf{y}$) transition density $p(t; \mathbf{x}, \mathbf{y})$ $(t > 0)$, prove that the $P_{\mathbf{x}}$-distribution of $\tau_{\partial B(\mathbf{0}:d)}$ has a finite m.g.f. in a neighborhood of zero, if $|\mathbf{x}| < d$.

Exercises for Section V.16

1. Compute the transition probability density of a standard Brownian motion on $\bar{G} = \{\mathbf{x} \in \mathbb{R}^k : x^{(i)} \geqslant 0 \text{ for } 1 \leqslant i \leqslant k\}$ with (normal) reflection at the boundary $(k > 1)$.

2. (i) Compute the transition probability density q of a standard Brownian motion on $\bar{G} = \{\mathbf{x} \in \mathbb{R}^k : 0 \leqslant x^{(i)} \leqslant 1 \text{ for } 1 \leqslant i \leqslant k\}$ with (normal) reflection at the boundary.

(ii) Prove that

$$\max_{\mathbf{x},\mathbf{y}\in\bar{G}} |q((t;\mathbf{x},\mathbf{y}) - 1| \leqslant c_1 e^{-c_2 t}, \qquad (t > 0),$$

for some positive constants c_1, c_2.

3. Prove that $\{\mathbf{Z}_t\}$ defined by (16.27) is a Markov process on $\bar{H} = \{\mathbf{x} \in \mathbb{R}^k : \boldsymbol{\gamma}\cdot\mathbf{x} \geqslant 0\}$.
4. Check (16.30).
5. Check that the function $v(t, |\mathbf{x}|)$ given by the solution of (16.42) (with $r = |\mathbf{x}|$) satisfies (16.41).

THEORETICAL COMPLEMENTS

Theoretical Complements to Section V.1

1. A construction of diffusions on $S = \mathbb{R}^1$ by the method of stochastic differential equations is given in Chapter VII (Theorem 2.2), under the assumption that $\mu(\cdot)$, $\sigma(\cdot)$ are Lipschitzian. If it is also assumed that $\sigma(x)$ is nonzero for all x, then it follows from K. Itô and H. P. McKean, Jr. (1965), *Diffusion Processes and Their Sample Paths*, Springer-Verlag, New York, pp. 149–158, that the transition probability $p(t; x, dy)$ has a positive density $p(t; x, y)$ for $t > 0$, which is once continuously differentiable in t and twice in x. Most of the main results of Chapter V have been proved without assuming existence of a smooth transition probability density. Itô and McKean, *loc. cit.*, contains a comprehensive account of one-dimensional diffusions.

Theoretical Complements to Section V.2

1. For the Markov semigroup $\{\mathbf{T}_t\}$ defined by (2.1) in the case $p(t; x, dy)$ is the transition probability of a diffusion $\{X_t\}$, it may be shown that every twice continuously differentiable function f vanishing outside a compact subset of the state space belongs to $\mathscr{D}_{\mathbf{A}}$. A proof of this is sketched in Exercise 3.12 of Chapter VII. As a consequence, by Theorem 2.3, $f(X_t) - \int_0^t \mathbf{A}f(X_s)\,ds$ is a martingale. More generally, it is proved in Chapter VII, Corollary 3.2, that this martingale property holds for every twice continuously differentiable f such that f, f', f'' are bounded. Many important properties of diffusions may be deduced from this martingale property (see Sections 3, 4 of Chapter VII).

2. (*Progressive Measurability*) The assumption of right continuity of $t \to X_t(\omega)$ for every $\omega \in \Omega$ in Theorem 2.3 ensures that $\{X_t\}$ is *progressively measurable* with respect to $\{\mathscr{F}_t\}$; that is, for each $t > 0$ the map $(s, \omega) \to X_s(\omega)$ on $[0, t] \times \Omega$ into S is measurable with respect to the product sigmafield $\mathscr{B}([0, t]) \otimes \mathscr{F}_t$ (on $[0, t] \times \Omega$) and the Borel sigmafield $\mathscr{B}(S)$ on S. Before turning to the proof, note that as a consequence of the progressive measurability of $\{X_t\}$ the integral $\int_0^t \mathbf{A}f(X_s)\,ds$ is well defined, is $\mathscr{F}_t$-measurable and has a finite expectation, by Fubini's Theorem.

Lemma 1. Let $(\Omega, \mathscr{F})$ be a measurable space. Suppose S is a metric space and $s \to X(s, \omega)$ is right continuous on $[0, \infty)$ into S, for every $\omega \in \Omega$. Let $\{\mathscr{F}_t\}$ be an increasing family of sub-sigmafields of $\mathscr{F}$ such that X_t is $\mathscr{F}_t$-measurable for every $t \geqslant 0$. Then $\{X_t\}$ is progressively measurable with respect to $\{\mathscr{F}_t\}$. □

Proof. Fix $t > 0$. Let f be a real-valued bounded continuous function on S. Consider the function $(s, \omega) \to f(X_s(\omega))$ on $[0, t] \times \Omega$. Define

$$g_n(s, \omega) = f(X_{2^{-n}t}(\omega))\mathbf{1}_{[0,2^{-n}t]}(s) + \sum_{i=2}^{2^n} f(X_{i2^{-n}t}(\omega))\mathbf{1}_{((i-1)2^{-n}t, i2^{-n}t]}(s). \quad \text{(T.2.1)}$$

As each summand in (T.2.1) is the product of an $\mathscr{F}_t$-measurable function of ω and a $\mathscr{B}([0, t])$-measurable function of s, g_n is $\mathscr{B}([0, t]) \otimes \mathscr{F}_t$-measurable. Also, for each $(s, \omega) \in [0, t] \times \Omega$, $g_n(s, \omega) \to f(X_s(\omega))$. Therefore, $(s, \omega) \to f(X_s(\omega))$ is $\mathscr{B}([0, t]) \otimes \mathscr{F}_t$-measurable. Since the indicator function of a closed set $F \subset S$ may be expressed as a pointwise limit of continuous functions, it follows that $(s, \omega) \to \mathbf{1}_F(X_s(\omega))$ is $\mathscr{B}([0, t]) \otimes \mathscr{F}_t$-measurable. Finally, $\mathscr{C} := \{B \in \mathscr{B}(S): (s, \omega) \to \mathbf{1}_B(X_s(\omega))$ is $\mathscr{B}([0, t]) \otimes \mathscr{F}_t$-measurable on $[0, t] \times \Omega\}$ equals $\mathscr{B}(S)$, by Dynkin's Pi-Lambda Theorem. ■

Another significant consequence of progressive measurability is the following result, which makes the statement of the strong Markov property of continuous-parameter Markov processes complete (see Chapter IV, Proposition 5.2; Chapter V, Theorem 11.1; and Exercise 11.9).

Lemma 2. Under the hypothesis of Lemma 1, (a) every $\{\mathscr{F}_t\}$-stopping time τ is $\mathscr{F}_\tau$-measurable, where $\mathscr{F}_\tau := \{A \in \mathscr{F} : A \cap \{\tau \leqslant t\} \in \mathscr{F}_t \ \forall t \geqslant 0\}$, and (b) $X_\tau \mathbf{1}_{\{\tau < \infty\}}$ is $\mathscr{F}_\tau$-measurable. □

Proof. The first part is obvious. For part (b), fix $t > 0$. On the set $\Omega_t := \{\tau \leqslant t\}$, X_τ is the composition of the maps (i) $\omega \to (\tau(\omega), \omega)$ (on Ω_t into $[0, t] \times \Omega_t$) and (ii) $(s, \omega) \to X_s(\omega)$ (on $[0, t] \times \Omega_t$ into S).

Let $\mathscr{F}_t|_{\Omega_t} := \{A \cap \Omega_t : A \in \mathscr{F}_t\} = \{A \in \mathscr{F}_t : A \subset \Omega_t\}$, *be the trace of the sigmafield* $\mathscr{F}_t$ *on* Ω_t. If $r \in [0, t]$ and $A \in \mathscr{F}_t|_{\Omega_t}$ one has

$$\{\omega \in \Omega_t : (\tau(\omega), \omega) \in [0, r] \times A\} = \{\tau \leqslant r\} \cap A \in \mathscr{F}_t|_{\Omega_t}.$$

Thus the map (i) is measurable with respect to $\mathscr{F}_t|_{\Omega_t}$ (on the domain Ω_t) and $\mathscr{B}([0, t] \otimes \mathscr{F}_t|_{\Omega_t}$ (on the range $[0, t] \times \Omega_t$).

Next, the map $(s, \omega) \to X_s(\omega)$ is measurable with respect to $\mathscr{B}([0, t]) \otimes \mathscr{F}_t$ (on $[0, t] \times \Omega$) and $\mathscr{B}(S)$ (on S), in view of the progressive measurability of $\{X_s\}$. Since $[0, t] \times \Omega_t \in \mathscr{B}([0, t]) \otimes \mathscr{F}_t$, and the restriction of a measurable function to a measurable set is measurable with respect to the trace sigmafield, the map (ii) is $\mathscr{B}([0, t]) \otimes \mathscr{F}_t|_{\Omega_t}$-measurable. Hence $\omega \to X_{\tau(\omega)}(\omega)$ is $\mathscr{F}_t|_{\Omega_t}$-measurable on Ω_t. Therefore, for every $B \in \mathscr{B}(S)$,

$$\{\omega \in \Omega : X_{\tau(\omega)}(\omega) \in B\} \cap \Omega_t \equiv \{\omega \in \Omega_t : X_{\tau(\omega)}(\omega) \in B\} \in \mathscr{F}_t. \quad ■$$

3. (*Semigroup Theory and Feller's Construction of One-Dimensional Diffusions*) In a series of articles in the 1950s, beginning with "The Parabolic Differential Equation and the Associated Semigroups of Transformations," *Ann. Math.*, **55** (1952), pp. 468–519, W. Feller constructed all nonsingular Markov processes on an interval S having continuous sample paths. *Nonsingularity* here means that, starting from any point in the interior of S, the process can move to the right, as well as to the left with positive probability. If $S = (a, b)$, and a, b cannot be reached from the interior, then such a process is completely characterized by a strictly increasing continuous *scale function* $s(x)$ and a strictly increasing right-continuous *speed function* $m(x)$. The role of $s(x)$ is the determination of the probability $\phi(x)$ of reaching d before c, starting from x, where $a < c < x < d < b$ (for all c, x, d), by the relation $\phi(x) = (s(x) - s(c))/(s(d) - s(c))$. Then the speed function determines the expected time $M(x)$ of reaching c or d, starting from x, by the relation

$$\left(\frac{d}{dm(x)}\frac{d}{ds(x)}\right)M(x) = -1, \qquad c < x < d, \quad M(c) = M(d) = 0.$$

The infinitesimal generator of this process is $\mathbf{A} = (d/dm(x))(d/ds(x))$. It may be noted that, given the functions ϕ, M, the scale function $s(\cdot)$ is determined up to an additive constant, and the speed function $m(\cdot)$ is determined up to the addition of an affine linear function of $s(\cdot)$. One may, therefore, fix $x_0 \in (a, b)$ and let $s(x_0) = 0$, $m(x_0) = 0$. A necessary and sufficient condition that the boundary point a cannot be reached, that is for the *inaccessibility* of a, is

$$\int_a^{x_0} m(x)\, ds(x) = -\infty, \tag{T.2.2}$$

and that for the inaccessibility of b is

$$\int_{x_0}^{b} m(x)\, ds(x) = \infty. \tag{T.2.3}$$

For the class of diffusions considered in this book (see (9.9), (9.10), (9.13)),

$$s(x) = \int_{x_0}^{x} \exp\{-I(x_0, z)\}\, dz, \qquad m(x) = \int_{x_0}^{x} \frac{2}{\sigma^2(z)} \exp\{I(x_0, z)\}\, dz,$$

$$I(x_0, z) = \int_{x_0}^{z} \frac{2\mu(y)}{\sigma^2(y)}\, dy.$$

If any boundary point is *accessible*, a boundary condition has to be specified at that point, in order to indicate what the Markov process does on reaching this point. This is discussed in theoretical complement 7.1. A very readable account of this characterization of nonsingular Markov processes on S having continuous sample paths is given in D. Freedman (1971), *Brownian Motion and Diffusion*, Holden-Day, San Francisco, pp. 102–138. See Theorem T.3.1 of Chapter VII for (T.2.2), (T.2.3).

The second, and more important, part of Feller's theory is the *construction* of Markov processes on S, given a pair of scale and speed functions s and m. This is carried out by the method of semigroups. The rest of this subsection is devoted to a description of this method.

In general semigroup theory one considers a family of bounded linear operators $\{\mathbf{T}_t: t > 0\}$ on a Banach space $\mathbf{B}$ with norm $\|\cdot\|$, satisfying $\mathbf{T}_{t+s} = \mathbf{T}_t\mathbf{T}_s$. It will be assumed that $\|\mathbf{T}_t f - f\| \to 0$ as $t \downarrow 0$, for every $f \in \mathbf{B}$. Since the set $\mathbf{B}_0$ of all $f \in \mathbf{B}$ for which this convergence holds is itself a Banach space, i.e., a closed linear subset of $\mathbf{B}$, this assumption simply means the restrict of $\mathbf{T}_t$ to $\mathbf{B}_0$.

It follows from the semigroup property that $\{\mathbf{T}_t: t > 0\}$ is determined by $\{\mathbf{T}_t: 0 < t < \varepsilon\}$ for every $\varepsilon > 0$. Indeed, the semigroup is determined by its *infinitesimal generator* $\mathbf{A} := (d/dt)\mathbf{T}_t|_{t=0}$. To be precise, let $\mathscr{D} = \mathscr{D}_\mathbf{A}$ denote the set of all $f \in \mathbf{B}$ such that $(\mathbf{T}_t f - f)/t$ converges in norm to some $g \in \mathbf{B}$, as $t \downarrow 0$. For $f \in \mathscr{D}$, define $\mathbf{A}f$ to be $\lim(\mathbf{T}_t f - f)/t$ as $t \downarrow 0$. We will also assume that $\|\mathbf{T}_t\| \leqslant 1$, i.e., $\{\mathbf{T}_t\}$ is a *contraction semigroup*. This is not a serious restriction since, given an arbitrary semigroup $\{\mathbf{T}_t\}$, the semigroup $\exp\{-ct\}\mathbf{T}_t$ $(t > 0)$ defines a contraction semigroup if $c > \log\|\mathbf{T}_1\|$. For $f \in \mathscr{D}_\mathbf{A}$,

$$\|(\mathbf{T}_{t+s} f - \mathbf{T}_t f)/s - \mathbf{T}_t\mathbf{A}f\| \leqslant \|(\mathbf{T}_s f - f)/s - \mathbf{A}f\| \to 0$$

as $s \downarrow 0$. Therefore, $\mathbf{T}_t f \in \mathscr{D}_\mathbf{A}$ for all $t > 0$ if $f \in \mathscr{D}_\mathbf{A}$, and in this case $\mathbf{A}\mathbf{T}_t f = \mathbf{T}_t\mathbf{A}f$. That is, $\mathbf{T}_t$ and $\mathbf{A}$ *commute* on $\mathscr{D}_\mathbf{A}$. By an argument entirely analogous to that used for proving the fundamental theorem of calculus,

$$\mathbf{T}_t f - f = \int_0^t \left(\frac{d}{ds}\mathbf{T}_s f\right) ds = \int_0^t \mathbf{A}\mathbf{T}_s f\, ds = \int_0^t \mathbf{T}_s\mathbf{A}f\, ds.$$

In particular, the function $t \to \mathbf{T}_t f$ on $(0, \infty)$ solves the *initial-value problem*: for $f \in \mathscr{D}_\mathbf{A}$, solve

$$\frac{d}{dt}u(t) = \mathbf{A}u(t) \qquad (t > 0), \qquad \lim_{t \downarrow 0} u(t) = f.$$

The solution $t \to \mathbf{T}_t f$ is also *unique*. To see this, let $u(t)$ be a solution. Fix $t > 0$ and consider the function $s \to \mathbf{T}_{t-s}u(s)$ $(0 \leqslant s \leqslant t)$. Now,

$$\begin{aligned}\frac{d}{ds}\mathbf{T}_{t-s}u(s) &= \lim_{h \downarrow 0}\frac{1}{h}\{(\mathbf{T}_{t-s-h}u(s+h) - \mathbf{T}_{t-s}u(s+h)) + (\mathbf{T}_{t-s}u(s+h) - \mathbf{T}_{t-s}u(s))\}\\ &= \lim_{h \downarrow 0}\frac{1}{h}\left(-\int_{t-s-h}^{t-s} \mathbf{T}_{s'}\mathbf{A}u(s+h)\, ds'\right) + \mathbf{T}_{t-s}\frac{du(s)}{ds}\\ &= -\mathbf{T}_{t-s}\mathbf{A}u(s) + \mathbf{T}_{t-s}\mathbf{A}u(s) = 0.\end{aligned}$$

Hence $\mathbf{T}_{t-s}u(s)$ is independent of s so that setting in turn $s = 0, t$, we get $\mathbf{T}_t u(0) = u(t)$, that is, $\mathbf{T}_t f = u(t)$.

When is an operator $\mathbf{A}$ defined on a linear subspace $\mathscr{D}_\mathbf{A}$ of $\mathbf{B}$ an infinitesimal generator of a contraction semigroup $\{\mathbf{T}_t\}$? The answer is contained in the important *Hille–Yosida Theorem: Suppose $\mathscr{D}_\mathbf{A}$ is dense in* $\mathbf{B}$. *Then* $\mathbf{A}$ *is the infinitesimal generator of a contraction semigroup on* $\mathbf{B}$ *if and only if, for each* $\lambda > 0$, $\lambda - \mathbf{A}$ *is one-to-one (on* $\mathscr{D}_\mathbf{A}$*) and onto* $\mathbf{B}$ *with* $\|(\lambda - \mathbf{A})^{-1}\| \leqslant 1/\lambda$. A simple proof of this theorem for the case $\mathbf{B} = C[a, b]$, the set of all continuous functions on (a, b) with finite limits at a and b, and for closed linear subspaces of $C[a, b]$, may be found in P. Mandl (1968),

Analytical Treatment of One-Dimensional Markov Processes, Springer-Verlag, New York, pp. 2–5. It may be noted that if $\mathbf{A}$ is a bounded operator on $\mathbf{B}$, then $\mathbf{T}_t = \exp\{t\mathbf{A}\}$ (see Section 3 of Chapter IV). But the differential operator $\mathbf{A} = (d/dm(x))(d/ds(x))$ is *unbounded* on $C[a, b]$.

Consider the case $\mathbf{B} = C_0(a, b)$, the set of all continuous functions on (a, b) having limit zero at a and b. $C_0(a, b)$ is given the *supnorm*. Let $\mathbf{A} = (d/dm(x))(d/ds(x))$ with $\mathscr{D}_{\mathbf{A}}$ comprising the set of all $f \in \mathbf{B} = C_0(a, b)$ such that $\mathbf{A}f \in \mathbf{B}$. Suppose $\mathbf{A}$, $\mathscr{D}_{\mathbf{A}}$ satisfy the hypothesis of the Hille–Yosida Theorem, and $\{\mathbf{T}_t\}$ the corresponding contraction semigroup. It is simple to check that $(\lambda - \mathbf{A})^{-1}f$ equals $g := \int_0^\infty e^{-\lambda t}\mathbf{T}_t f\, dt$, for each $\lambda > 0$; that is, $(\lambda - \mathbf{A})g = f$. Suppose that the *resolvent operator* $(\lambda - \mathbf{A})^{-1}$ is *positive*; if $f \geqslant 0$ then $(\lambda - \mathbf{A})^{-1}f \geqslant 0$. Then, by the uniqueness theorem for Laplace transforms, $\mathbf{T}_t f \geqslant 0$ if $f \geqslant 0$. In this case $f \to \mathbf{T}_t f(x)$ is a positive bounded linear functional on $C_0(a, b)$. Therefore, by the *Riesz Representation Theorem* (see H. L. Royden (1968), *Real Analysis*, 2nd ed., Macmillan, New York, pp. 310–11), there exists a finite measure $p(t; x, dy)$ on $S = (a, b)$ such that $\mathbf{T}_t f(x) = \int f(y)p(t; x, dy)$ for every $f \in C_0(a, b)$. In view of the contraction property, $p(t; x, S) \leqslant 1$.

In order to verify the hypothesis of the Hille–Yosida Theorem in the case a, b are inaccessible, construct for each $\lambda > 0$ two positive solutions u_1, u_2 of $(\lambda - \mathbf{A})u = 0$, u_1 increasing and u_2 decreasing. Then define the symmetric *Green's function* $G_\lambda(x, y) = W^{-1}u_1(x)u_2(y)$ for $a < x \leqslant y < b$, extended to $x > y$ by symmetry. Here the *Wronskian* W given by $W := u_2(x)\, du_1(x)/ds(x) - u_1(x)\, du_2(x)/ds(x)$ is independent of x. For $f \in C_0(a, b)$ the function $g(x) = G_\lambda f(x) := \int G_\lambda(x, y)f(y)\, dm(y)$ is the unique solution in $C_0(a, b)$ of $(\lambda - \mathbf{A})g = f$. In other words, $G_\lambda = (\lambda - \mathbf{A})^{-1}$. The positivity of $(\lambda - \mathbf{A})^{-1}$ follows from that of $G_\lambda(x, y)$. Also, one may directly check that $G_\lambda 1 \equiv 1/\lambda$. This implies $p(t; x, (a, b)) = 1$. For details of this, and for the proof that a Markov process on $S = (a, b)$ with this transition probability may be constructed on the space $C([0, \infty):S)$ of continuous trajectories, see Mandl (1968), *loc. cit.*, pp. 14–17, 21–38.

4. For a diffusion on $S = (a, b)$, consider the semigroup $\{\mathbf{T}_t\}$ in (2.1). Under Condition (1.1), it is easy to check on integration by parts that the infinitesimal generator $\mathbf{A}$ is self-adjoint on $L^2(S, \pi(y)\, dy)$, where $\pi(y)\, dy$ is the speed measure whose distribution function is the *speed function* (see Eq. 8.1). That is,

$$\int \mathbf{A}f(y)g(y)\pi(y)\, dy = \int f(y)\mathbf{A}g(y)\pi(y)\, dy$$

for all twice continuously differentiable f, g vanishing outside compacts. It follows that $\mathbf{T}_t$ is self-adjoint and, therefore, the transition probability density $q(t; x, y) := p(t; x, y)/\pi(y)$ with respect to $\pi(y)\, dy$ is symmetric in x and y. That is, $q(t; x, y) = q(t; y, x)$. Since p satisfies the backward equation $\partial p/\partial t = \mathbf{A}_x p$, it follows that so does q: $\partial q/\partial t = \mathbf{A}_x q$. By symmetry of q it now follows that $\partial q/\partial t = \mathbf{A}_y q$. Here $\mathbf{A}_x q$, $\mathbf{A}_y q$ denote the application of $\mathbf{A}$ to $x \to q(t; x, y)$ and $y \to q(t; x, y)$ respectively. The equation $\partial q/\partial t = \mathbf{A}_y q$ easily reduces to $\partial p/\partial t = \mathbf{A}_y^* p$, as given in (2.41). For details see Itô and McKean (1965), *loc. cit.*, pp. 149–158.

For more general treatments of the relations between Markov processes and semigroup theory, see E. B. Dynkin (1965), *Markov Processes*, Vol. I, Springer-Verlag, New York, and S. N. Ethier and T. G. Kurtz (1986), *Markov Processes: Characterization and Convergence*, Wiley, New York.

Theoretical Complements to Section V.4

1. Theorem 4.1 is a special case of a more general result on the approximation of diffusions by discrete-parameter Markov chains, as may be found in D. W. Stroock and S. R. S. Varadhan (1979), *Multidimensional Diffusions*, Springer-Verlag, New York, Theorem 11.2.3.

   From the point of view of numerical analysis, (4.13) is the discretized, or difference-equation, version of the backward equation (4.14), and is usually solved by matrix methods.

Theoretical Complements to Section V.5

1. The general PDE (partial differential equations) method for solving the Kolmogorov equations, or second-order parabolic equations, may be found in A. Friedman (1964), *Partial Differential Equations of Parabolic Type*, Prentice-Hall, Englewood Cliffs. In particular, the existence of a smooth positive fundamental solution, or transition density, is proved there.

Theoretical Complements to Section V.6

1. Suppose two Borel-measurable function $\mu(\cdot), \sigma^2(\cdot) > 0$ are given on $S = (a, b)$ such that (i) $\mu(\cdot)$ is bounded on compact subsets of S, (ii) $\sigma^2(\cdot)$ is bounded away from zero and infinity on compact subsets of S, and (iii) a and b are *inaccessible* (see theoretical complement 2.1 above). Then Feller's construction provides a Markov process on S having continuous sample paths, with a scale function $s(\cdot)$ and speed function $m(\cdot)$ expressed as functions of $\mu(\cdot), \sigma^2(\cdot)$ as described in theoretical complement 2.1. Hence continuity of $\mu(\cdot)$ is not needed for this construction.

2. The general principle Proposition 6.3 is very useful in proving that certain functions $\{\varphi(X_t)\}$ of Markov processes $\{X_t\}$ are also Markov. But how does one find such functions? It turns out that in all the cases considered in this book the function φ is a *maximal invariant* of a group of transformations G under which the transition probability is *invariant*. Here invariance of the transition probability means $p(t; gx, g(B)) = p(t; x, B)$ for all $t > 0$, $x \in S$, $g \in G$, B Borel subset of the metric space S. A function φ is said to be a *maximal invariant* if (i) $\varphi(gx) = \varphi(x)$ for all $g \in G$, $x \in S$, and (ii) every measurable invariant function is a (measurable) function of φ. For each $x \in S$ the *orbit* of x (under G) is the set $o(x) := \{gx: g \in G\}$. Each invariant function is constant on orbits. Let S' be a metric space and φ a measurable function on S onto S' such that (i)′ φ is constant on orbits, and (ii)′ $\varphi(x) \neq \varphi(y)$ if $o(x) \neq o(y)$. In other words $\varphi(x)$ is a relabeling of $o(x)$ and S' may be thought of as (relabeling of) the space of orbits. Then it is simple to see that φ is a maximal invariant.

   Proposition. If φ is a maximal invariant, then $\{\varphi(X_t)\}$ is Markov.

   Proof. Taking the conditional expectation first with respect to $\sigma\{X_u: u \leqslant s\}$ and then with respect to $\sigma\{\varphi(X_u): u \leqslant s\}$,

$$P(\varphi(X_{s+t}) \in B' \mid \sigma\{\varphi(X_u): u \leqslant s\}) = E(p(t; X_s, \varphi^{-1}(B')) \mid \sigma\{\varphi(X_u): u \leqslant s\}).$$

But, by property (i) above,

$$p(t; x, \varphi^{-1}(B')) = p(t; g^{-1}x, g^{-1}(\varphi^{-1}(B'))) = p(t; g^{-1}x, (\varphi \circ g)^{-1}(B'))$$
$$= p(t; g^{-1}x, \varphi^{-1}(B')),$$

since $\varphi \circ g = \varphi$ by invariance of φ. In other words, the function $x \to p(t; x, \varphi^{-1}(B'))$ is invariant. By (ii) it is therefore a function $q(t; \varphi(x), B')$, say, of φ. Thus

$$E(p(t; X_s, \varphi^{-1}(B')) \mid \sigma\{\varphi(X_u): u \leqslant s\}) = E(q(t; \varphi(X_s), B') \mid \sigma\{\varphi(X_u): u \leqslant s\})$$
$$= q(t; \varphi(X_s), B'). \qquad \blacksquare$$

It has been pointed out to us by J. K. Ghosh that the idea underlying the proposition also occurs in the context of sequential analysis in statistics. See W. J. Hall, R. A. Wijsman, and J. K. Ghosh (1965), "The Relationship Between Sufficiency and Invariance with Applications in Sequential Analysis, *Ann. Math. Statist.*, **36**, pp. 575–614. Also see theoretical complement 16.3.

A maximal invariant of the reflection group $G = \{e, -e\}$ $(ex := x, -ex := -x)$ is $\varphi(x) = |x|$. Thus, $\{|X_t|\}$ is Markov if $\{X_t\}$ and $\{-X_t\}$ have the same transition probability, i.e., if $p(t; -x, -B) = p(t; x, B)$. The group G of transformations on $\mathbb{R}^1$ generated by reflections around 0 and 1 (i.e., by $g_0 x = -x$, $g_1 x \equiv g_1(1 + x - 1) = 1 - x + 1 = 2 - x$) is infinite, and

$$o(x) = \{x + 2m: m = 0, \pm 1, \pm 2, \ldots\} \cup \{-x + 2m: m = 0, \pm 1, \ldots\}.$$

Thus, if $p(t; -x, -B) = p(t; x, B)$ and $p(t; x, B) = p(t; x + 2, B + 2)$, then $\{Z_t\}$ in Theorem 6.4 is Markov, since a maximal invariant is $\varphi(x) = |x(\text{mod } 2) - 1|$.

Theoretical Complements to Section V.7

1. It is shown in Example 14.1, that the so-called radial Brownian motion $\{|\mathbf{B}_t|\}$ is a diffusion on $S = [0, \infty)$, such that "0" is *inaccessible* (from the interior of S). On the other hand, if the process starts at "0", it instantaneously enters the interior $(0, \infty)$ and stays there forever. Thus, although the diffusion may be restricted to the state space $(0, \infty)$, "0" may also be included in the state space. The only other way of including "0" in the state space is to make "0" an absorbing boundary, if the process is to have the Markov property and continuous sample paths. The nonabsorbing boundary point "0" in this case is called an *entrance boundary*. In general, an inaccessible lower boundary a is an *entrance boundary* if

$$v_a := \int_a^{x_0} s(x)\, dm(x) > -\infty.$$

Similarly, an upper inaccessible boundary b is *entrance* if

$$v^b := \int_{x_0}^{b} s(x)\, dm(x) < \infty.$$

See Itô and McKean (1965), *loc. cit.*, p. 108.

Suppose a boundary point is *accessible*, but the process starting at this boundary point cannot enter the interior; that is, no boundary condition other than absorption is consistent with the requirement that the process be Markovian and have continuous sample paths. Then the boundary is said to be an *exit boundary*.

In general, an accessible lower boundary a is *exit* if $v_a = -\infty$, and an accessible upper boundary b is *exit* if $v^b = \infty$. For details see Itô and McKean, *loc. cit.*, p. 108. An example of a lower exit boundary is $S = [0, \infty)$, $\mu(x) = \alpha x$, $\sigma^2(x) = \beta x$ with $\beta > 0$, which occurs as a model of growth of an isolated population (see S. Karlin and H. M. Taylor (1981), *A Second Course in Stochastic Processes*, Academic Press, New York, p. 239).

If one allows *jump discontinuities* at a boundary point, then the diffusion on reaching the boundary (or, starting from it) may stay at this point for a random holding time before jumping to another state. The holding time distribution is necessarily exponential (see Chapter IV, Proposition 5.1), with a parameter $\delta > 0$, while the jump distribution $\gamma(dy)$ is arbitrary. The successive holding times at this boundary are i.i.d. and are independent of the successive jump positions, the latter being also i.i.d. See Mandl (1968), *loc. cit.*, pp. 39, 47, 66–69.

Theoretical Complements to Section V.8

1. (*Semigroups under Boundary Conditions*) Let $S = [a, b]$, $\mu(\cdot)$ and $\sigma(\cdot)$ differentiable, and $\sigma^2(x) > 0$ for all $x \in S$. If Dirichlet or Neumann boundary conditions are prescribed at a, b, then the Hille–Yosida theorem may be used to construct a contraction semigroup $\{\mathbf{T}_t\}$ generated by $\mathbf{A}$ on $\mathbf{B} := C[a, b]$, or $C_0(a, b)$. As indicated in theoretical complement 2.3 above, there exists a transition probability, with total mass $\leqslant 1$, such that $\mathbf{T}_t f(x) = \int f(y) p(t; x, dy)$ for all $f \in \mathbf{B}$. To give an indication how the hypothesis of the Hille–Yosida Theorem is verified, consider *Dirichlet boundary* conditions. That is, let $\mathbf{B} = C_0(a, b)$, $\mathscr{D}_{\mathbf{A}}$ the set of all $f \in \mathbf{B}$ such that $\mathbf{A}f := (d/dm(x))(d/ds(x))f(x)$ is in $\mathbf{B}$. Fix $\lambda > 0$, and let u_1, u_2 be the solutions of $(\lambda - \mathbf{A})u = 0$ with $u_1(a) = 0$, $u_1'(a) = 1$, $u_2(b) = 0$, $u_2'(b) = -1$. By the existence and uniqueness theorem for linear ordinary differential equations, these solutions exist and are unique. Define the Green's function $G_\lambda(x, y) = W^{-1} u_1(x) u_2(y)$ for $x \leqslant y$, and extend it symmetrically to $x > y$, with W as the Wronskian (see theoretical complement 2.3). It is not difficult to check that for every $f \in \mathbf{B}$ the function $G_\lambda f(x) := \int G_\lambda(x, y) f(y)\, dm(y)$ is the unique element in $\mathbf{B}$ satisfying $(\lambda - \mathbf{A})G_\lambda f = f$. Since G_λ is positive and $(G_\lambda 1)(x)$ may be shown directly to be less than 1 for all x, the verification is complete, and we have a transition probability $p(t; x, dy)$ with $p(t; x, S) < 1$.

In the case of *Neumann*, or reflecting, *boundary* conditions at a, b, take $\mathbf{B} = C[a, b]$ and $\mathscr{D}_{\mathbf{A}} =$ the set of all f in $\mathbf{B}$ such that $\mathbf{A}f \in \mathbf{B}$ and $f'(a) = 0 = f'(b)$. Then $(\lambda - \mathbf{A})^{-1}$ may be expressed as $(\lambda - \mathbf{A})^{-1} f = G_\lambda f + l_1(f) u_1 + l_2(f) u_2$, where u_1, u_2, G_λ are as above, and $l_1(f), l_2(f)$ are bounded linear functionals of f determined so that $g := (\lambda - \mathbf{A})^{-1} f$ satisfies the boundary conditions $g'(a) = 0 = g'(b)$. The hypothesis of the Hille–Yosida Theorem may be directly verified now. Since constant functions belong to $\mathscr{D}_{\mathbf{A}}$ and $(\lambda - \mathbf{A})(1/\lambda) = 1$, it follows that

$$(\lambda - \mathbf{A})^{-1} 1 = 1/\lambda, \qquad \text{and} \qquad p(t; x, [a, b]) = 1.$$

We have given a brief outline above of Feller's construction of diffusions under Dirichlet and Neumann boundary conditions. Complete details may be found in

Mandl (1968), *loc. cit.*, Chapter II. The direct probabilistic constructions given in the text (see Sections 6 and 7) do not make use of this theory.

2. (*Eigenfunction Expansions*) For a justification of the eigenfunction expansion described in Section 8, consider first $\{\mathbf{T}_t\}$ on $C_0(a, b)$ under Dirichlet boundary conditions. Extend $\mathbf{T}_t$ to $L^2([a, b], \pi(y)\,dy = dm(y))$. Now $\pi(y)\,dy$ is an invariant measure. For

$$\frac{d}{dt}\int \mathbf{T}_t f(y)\pi(y)\,dy = \int \mathbf{A}\mathbf{T}_t f(y)\pi(y)\,dy = \int \mathbf{T}_t f(y)\mathbf{A}^*\pi(y)\,dy = 0,$$

so that

$$\int \mathbf{T}_t f(y)\pi(y)\,dy = \int f(y)\pi(y)\,dy \qquad \text{for } f \in C_0(a, b).$$

From this it is easily shown that $\{\mathbf{T}_t\}$ is a contraction semigroup on L^2, whose infinitesimal generator is the extension of $\mathbf{A}$ to the set of all $f \in C_0(a, b)$ such that $\mathbf{A}f \in L^2([a, b], \pi(y)\,dy)$. We denote this extension also by $\mathbf{A}$ and note that $\mathbf{A}$ is *self-adjoint*. Fix $\lambda > 0$. Then $(\lambda - \mathbf{A})^{-1}$ is compact and self-adjoint on L^2. Indeed, $(\lambda - \mathbf{A})^{-1}$ is the integral operator G_λ, and it follows from a well-known theorem of Riesz (see F. Riesz (1955), *Functional Analysis*, F. Ungar Publishing Co., New York, Chapter VI) that $(\lambda - \mathbf{A})^{-1}$ is a compact self-adjoint operator whose eigenvalues $\mu_n(\lambda)$ are positive and converge to zero as $n \to \infty$, and that the corresponding normalized eigenfunctions ψ_n comprise a complete orthonormal sequence in L^2. As a consequence, the eigenvalues of $\mathbf{A}$ are $\alpha_n := = \lambda - \mu_n^{-1}(\lambda) < 0$, with the corresponding eigenfunctions ψ_n.

Under Neumann boundary conditions, it follows from the representation $(\lambda - \mathbf{A})^{-1}f = G_\lambda f + l_1(f)u_1 + l_2(f)u_2$ that $(\lambda - \mathbf{A})^{-1}$ is again a compact self-adjoint operator on L^2. The eigenvalues α_n of $\mathbf{A}$ are nonnegative, with 0 as one of them since $\mathbf{A}1 = 0$. Again the normalized eigenfunctions comprise a complete orthonormal set.

The uniqueness of the solution to the initial-value problem, under any of these boundary conditions, follows from the uniqueness result proved in theoretical complement 2.1 above. Another proof may be based on the *maximum principle* for parabolic equations (see Exercise 7.7).

Theoretical Complements to Section V.11

1. A proof that the pre-τ sigmafield as defined by (11.2) is the same as the one generated by the stopped process $\{X_{\tau \wedge t}: t \geq 0\}$ is given in Stroock and Varadhan (1979), *loc. cit.*, p. 33.

The continuity of the function $\varphi(y)$ in (11.18) depends only on the fact that $x \to p(t; x, dy)$ is *weakly continuous*. This fact, referred to as the *Feller property*, is proved in Section 3 of Chapter VII, for diffusions on $\mathbb{R}^1$. More generally, as described in theoretical complements to Sections 2, 7, 8, Feller's construction implies that $\mathbf{T}_t f$ is bounded and continuous if f is.

2. The *Brownian meander* and *Brownian excursion* processes considered in Exercises 12.6, 12.7 of Chapter I and theoretical complements to Section 12 of Chapter I and Section 5 of Chapter IV may also be defined as continuous-parameter Markov processes having continuous sample paths but *nonhomogeneous* transition probabilities. In fact, let $\{B_t\}$ denote a standard Brownian motion, and let $\lambda = \sup\{t \leqslant 1: B_t = 0\}$, $\rho = \inf\{t \geqslant 1: B_t = 0\}$. Consider the stochastic processes defined by

$$(meander): \qquad B_t^+ := |B_{(1-t)\lambda+t}|(1-\lambda)^{-1/2}, \qquad 0 \leqslant t \leqslant 1, \tag{T.11.1}$$

$$(excursion): \qquad B_t^{*+} := |B_{(1-t)\lambda+t\rho}|(\rho-\lambda)^{-1/2}, \qquad 0 \leqslant t \leqslant 1. \tag{T.11.2}$$

Then in B. Belkin (1972), "An Invariance Principle for Conditioned Random Walk Attracted to a Stable Law," *Z. Wahrscheinlichkeitstheorie und Verw. Geb.*, **21**, pp. 45–64, it is shown that the meander process defined by (T.11.1) is a continuous-parameter Markov process starting at 0 with transition probabilities given by the following densities

$$p^+(0,0;t,y) = 2t^{-3/2}y\exp\left\{-\frac{y^2}{2t}\right\}\Phi_{1-t}(0,y), \qquad 0 \leqslant t \leqslant 1, \quad y > 0,$$

$$p^+(s,x;t,y) = \{\varphi_{t-s}(y-x) - \varphi_{t-s}(y+x)\}\frac{\Phi_{1-t}(0,y)}{\Phi_{1-s}(0,x)}, \qquad 0 < s < t \leqslant 1, \quad x, y > 0,$$

where

$$\varphi_u(z) = (2\pi u)^{-1/2}\exp\left\{-\frac{z^2}{2u}\right\}, \qquad \text{and} \qquad \Phi_u(a,b) = \int_a^b \varphi_u(z)\,dz.$$

Likewise, for the case of the excursion process defined by (T.11.2), one obtains a continuous-parameter Markov process with nonhomogeneous transition law with the following density (see Itô and McKean (1965), *loc. cit.*, p. 76):

$$p^{*+}(0,0;t,y) = 2y^2(2\pi t^3(1-t)^3)^{-1/2}\exp\{-y^2/2t(1-t)\}, \qquad 0 < t \leqslant 1, \quad y > 0,$$

$$p^{*+}(s,x;t,y) = \{\varphi_{t-s}(y-x) - \varphi_{t-s}(y+x)\}\frac{(1-s)^{3/2}\ y\exp\{-y^2/2(1-t)\}}{(1-t)^{3/2}\ x\exp\{-x^2/2(1-s)\}},$$
$$0 < s < t < 1, \quad x, y > 0.$$

The equivalence of these processes with those of the same name defined in theoretical complements to Section I.12 is the subject of the paper by R. T. Durrett, D. L. Iglehart, and D. R. Miller (1977), "Weak Convergence to Brownian Meander and Brownian Excursion," *Ann. Probab.*, **5**, pp. 117–129.

Theoretical Complements to Section V.12

1. Consider a positive recurrent diffusion on an interval S, having a transition probability $p(t; x, dy)$ and an invariant initial distribution π. We will use a *coupling argument* to show that $\|p(t;x,dy) - \pi(dy)\| := \sup\{|p(t;x,B) - \pi(B)|: B \in \mathscr{B}(S)\} \to 0$ as $t \to \infty$. For this, let $\{X_t\}$, $\{Y_t\}$ be two independent diffusions having transition probability p and with $X_0 \equiv x$, and Y_0 having distribution π. Let $\tau := \inf\{t \geqslant 0: X_t = Y_t\}$. If one

can show that $\tau < \infty$ a.s., then the process $\{Z_t\}$ defined by

$$Z_t := \begin{cases} X_t & \text{for } t < \tau, \\ Y_t & \text{for } t \geqslant \tau, \end{cases} \tag{T.12.1}$$

has, by the strong Markov property for the two-dimensional Markov process $\{(X_t, Y_t)\}$, the same distribution as $\{X_t\}$. In this case,

$$\begin{aligned} |p(t; x, B) - \pi(B)| &= |P(X_t \in B) - P(Y_t \in B)| = |P(Z_t \in B) - P(Y_t \in B)| \\ &\leqslant P(Z_t \neq Y_t) = P(\tau > t) \to 0 \qquad \text{as } t \to \infty. \end{aligned}$$

In order to prove $\tau < \infty$ a.s., it is enough to prove that the process $\{(X_t, Y_t)\}$ reaches every rectangle $[a, b] \times [c, d]$ $(a < b, c < d)$ almost surely. For the process must then go across the diagonal $\{(u, v): u = v\}$ a.s. Let $\varphi(u, v)$ denote the probability that a pair of such independent diffusions starting at (u, v) ever reach the rectangle $[a, b] \times [c, d]$. Note that $\pi \times \pi$ is an invariant initial distribution for the two-dimensional diffusion, and let $\{(U_t, V_t)\}$ denote such a diffusion having the initial distribution $\pi \times \pi$. Let α denote the probability that the latter ever reaches $[a, b] \times [c, d]$. Also write $D_n := \{(U_t, V_t) \in [a, b] \times [c, d]$ for some $t \geqslant n\}$, and $F_n := \{(U_t, V_t) \in [a, b] \times [c, d]$ for some $t \leqslant n\}$. Then $\alpha = P(D_n) + P(F_n \cap D_n^c)$. But, by stationarity, $P(D_n) = \alpha$. Therefore, $P(F_n \cap D_n^c) = 0$. By the Markov property,

$$P(F_n \cap D_n^c) = E[\mathbf{1}_{F_n}(1 - \varphi(U_n, V_n))] = E(P(F_n \mid (U_n, V_n))(1 - \varphi(U_n, V_n)).$$

Now we may use the positivity of the transition probability density to show easily that $P(F_n \mid U_n = u, V_n = v) > 0$ for almost all pairs (u, v) (with respect to Lebesgue measure on $S \times S$). Therefore, $1 - \varphi(u, v) = 0$ for almost every pair (u, v). By using the continuity of $x \to p(t; x, y)$ it is now shown that $(u, v) \to \varphi(u, v)$ is continuous, so that $\varphi(u, v) = 1$ for *all* (u, v). This proves that $\tau < \infty$ a.s.

Theoretical Complements to Section V.13

1. (*Martingale Central Limit Theorem*) Although a proof of the central limit theorem (Theorem 13.1) for positive recurrent one-dimensional diffusions may be patterned after the corresponding proof for Markov chains without any essential change, the proof does not work for Markov processes on general state spaces and multidimensional diffusions. The reason for this failure is that *point recurrence* may not hold. That is, even for Markov processes having a unique invariant distribution, there may not be any point in the state space to which the process returns (infinitely often) with probability 1. A more general approach that applies to all ergodic Markov processes is via martingales. First, let us prove a general martingale central limit theorem. Central limit theorems for Markov processes will then be derived from it.

Let $\{X_{k,n}: 1 \leqslant k \leqslant k_n\}$ be, for each $n \geqslant 1$, a square integrable martingale difference sequence, with respect to an increasing family of sigmafields $\{\mathscr{F}_{k,n}: 0 \leqslant k \leqslant k_n\}$. Write

$$\sigma_{k,n}^2 := E(X_{k,n}^2 \mid \mathscr{F}_{k-1,n}), \qquad s_{k,n}^2 = \sum_{j=1}^{k} \sigma_{j,n}^2,$$

$$\begin{aligned} L_{k,n}(\varepsilon) &:= \sum_{j=1}^{k} E(X_{j,n}^2 \mathbf{1}_{\{|X_{j,n}| > \varepsilon\}} \mid \mathscr{F}_{j-1,n}), \\ M_n &:= \max\{\sigma_{k,n}^2;\ 1 \leqslant k \leqslant k_n\}, \\ S_{n,k_n} &:= \sum_{j=1}^{k_n} X_{j,n}. \end{aligned} \tag{T.13.1}$$

The following result is due to B. M. Brown (1977), "Martingale Central Limit Theorems," *Ann. Math. Statist.*, **42**, pp. 59–66.

Theorem T.13.1. (*Martingale CLT*). Assume that, as $n \to \infty$, (i) $s_{k_n,n}^2 \to 1$ in probability, and (ii) $L_{k,n}(\varepsilon) \to 0$ in probability, for every $\varepsilon > 0$. Then S_{n,k_n} converges in distribution to $N(0, 1)$. □

Proof. Consider the conditional characteristic functions

$$\varphi_{k,n}(\xi) := E(\exp\{i\xi X_{k,n}\} \mid \mathscr{F}_{k-1,n}), \qquad (\xi \in \mathbb{R}^1). \tag{T.13.2}$$

It would be enough to show that

(a) $E\left(\exp\{i\xi S_{n,k_n}\} \Big/ \prod_1^{k_n} \varphi_{k,n}(\xi)\right) = 1,$

provided $|\prod_1^{k_n} \varphi_{k,n}(\xi))^{-1}| \leqslant \delta(\xi)$, a constant, and

(b) $\prod_1^{k_n} \varphi_{k,n}(\xi) \to \exp\{-\xi^2/2\}$ in probability.

Indeed, if $|(\prod \varphi_{k,n}(\xi))^{-1}| \leqslant \delta(\xi)$, then (a), (b) imply

$$\begin{aligned} |E\exp\{i\xi S_{n,k_n}\} - \exp\{-\xi^2/2\}| &= \exp\{-\xi^2/2\}\left|\frac{E\exp\{i\xi S_{n,k_n}\}}{\exp\{-\xi^2/2\}} - E\left(\frac{\exp\{i\xi S_{n,k_n}\}}{\prod \varphi_{k,n}(\xi)}\right)\right| \\ &\leqslant \exp\{-\xi^2/2\} E\left|\frac{1}{\exp\{-\xi^2/2\}} - \frac{1}{\prod \varphi_{k,n}(\xi)}\right| \to 0. \end{aligned} \tag{T.13.3}$$

Now part (a) follows by taking successive conditional expectations given $\mathscr{F}_{k-1,n}$ $(k = k_n, k_n - 1, \ldots, 1)$, if $(\prod \varphi_{k,n}(\xi))^{-1}$ is integrable. Note that the martingale difference property is not needed for this. It turns out, however, that in general $(\prod \varphi_{k,n}(\xi))$ cannot be bounded away from zero. Our first task is then to replace $X_{k,n}$ by new martingale differences $Y_{k,n}$ for which this integrability does hold, and whose sum has the same asymptotic distribution as S_{n,k_n}. To construct $Y_{k,n}$, first use assumption (ii) to check that $M_n \to 0$ in probability. Therefore, there exists a *nonrandom* sequence $\delta_n \downarrow 0$ such that

$$P(M_n \geqslant \delta_n) \to 0 \qquad \text{as } n \to \infty. \tag{T.13.4}$$

Similarly, there exists, for each $\varepsilon > 0$, a *nonrandom* sequence $\Theta_n(\varepsilon) \downarrow 0$ such that

$$P(L_{k_n,n}(\varepsilon) \geqslant \Theta_n(\varepsilon)) \to 0 \qquad \text{as } n \to \infty. \tag{T.13.5}$$

Consider the events

$$A_{k,n}(\varepsilon) := \{\sigma_{k,n}^2 < \delta_n, L_{k,n} < \Theta_n(\varepsilon), s_{k,n}^2 < 2\}, \qquad (1 \leqslant k \leqslant k_n). \tag{T.13.6}$$

Then $A_{k,n}(\varepsilon)$ is $\mathscr{F}_{k-1,n}$-measurable. Therefore, $Y_{k,n}$ defined by

$$Y_{k,n} := X_{k,n}\mathbf{1}_{A_{k,n}(\varepsilon)} \tag{T.13.7}$$

has zero conditional expectation, given $\mathscr{F}_{k-1,n}$. Although $Y_{k,n}$ depends on ε, we will suppress this dependence for notational convenience. Note also that

$$\begin{aligned} P(Y_{k,n} = X_{k,n} \text{ for } 1 \leqslant k \leqslant k_n) &\geqslant P\left(\bigcap_{k=1}^{k_n} A_{k,n}(\varepsilon)\right) \\ &= P(M_n < \delta_n, L_{k_n,n}(\varepsilon) < \Theta_n(\varepsilon), s_{n,k_n}^2 < 2) \to 1. \end{aligned} \tag{T.13.8}$$

We will use the notation (T.13.1–2) with a *prime* (′) to denote the corresponding quantities for $\{Y_{k,n}\}$. For example, using the fact $E(Y_{k,n} \mid \mathscr{F}_{k,-1,n}) = 0$ and a Taylor expansion,

$$\begin{aligned} \left|\varphi'_{k,n}(\xi) - \left(1 - \frac{\xi^2}{2}\sigma'^2_{k,n}\right)\right| &= \left|E\left[\exp(i\xi Y_{k,n}\} - \left(1 + i\xi Y_{k,n} + \frac{(i\xi)^2}{2}Y_{k,n}^2\right) \,\middle|\, \mathscr{F}_{k-1,n}\right]\right| \\ &= \left|E\left[-\xi^2 Y_{k,n}^2 \int_0^1 (1-u)(\exp\{iu\xi Y_{k,n}\} - 1)\,du \,\middle|\, \mathscr{F}_{k-1,n}\right]\right| \\ &\leqslant \varepsilon\frac{|\xi|^3}{2}\sigma'^2_{k,n} + \xi^2 E(Y_{k,n}^2\mathbf{1}_{\{|Y_{k,n}|>\varepsilon\}} \mid \mathscr{F}_{k-1,n}) \\ &\leqslant \varepsilon\frac{|\xi|^3}{2}\sigma'^2_{k,n} + \xi^2 E(X_{k,n}^2\mathbf{1}_{\{|Y_{k,n}|>\varepsilon\}} \mid \mathscr{F}_{k-1,n}). \end{aligned} \tag{T.13.9}$$

Fix $\xi \in \mathbb{R}^1$. Since $M'_n < \delta_n$, $0 \leqslant 1 - (\xi^2/2)\sigma'^2_{k,n} \leqslant 1$ $(1 \leqslant k \leqslant k_n)$ for all large n. Therefore, using (T.13.6, 9),

$$\begin{aligned} \left|\prod \varphi'_{k,n}(\xi) - \prod\left(1 - \frac{\xi^2}{2}\sigma'^2_{k,n}\right)\right| &\leqslant \sum_{k=1}^{k_n}\left|\varphi'_{k,n}(\xi) - \left(1 - \frac{\xi^2}{2}\sigma'^2_{k,n}\right)\right| \\ &\leqslant |\xi|^3\varepsilon + \xi^2\Theta_n(\varepsilon), \end{aligned} \tag{T.13.10}$$

and

$$\left|\prod\left(1-\frac{\xi^2}{2}\sigma'^2_{k,n}\right)-\exp\left\{-\frac{\xi^2}{2}s'^2_{n,k_n}\right\}\right|=\left|\prod\left(1-\frac{\xi^2}{2}\sigma'^2_{k,n}\right)-\prod\exp\left\{-\frac{\xi^2}{2}\sigma'^2_{k,n}\right\}\right|$$
$$\leqslant\frac{\xi^4}{8}\sum\sigma'^4_{k,n}\leqslant\frac{\xi^4}{8}\delta_n s'^2_{n,k_n}\leqslant\frac{\xi^4}{4}\delta_n. \quad \text{(T.13.11)}$$

Therefore,

$$\left|\prod\varphi'_{k,n}(\xi)-\exp\left\{-\frac{\xi^2}{2}s'_{n,k_n}\right\}\right|\leqslant|\xi|^3\varepsilon+\xi^2\Theta_n(\varepsilon)+\frac{\xi^4}{4}\delta_n. \quad \text{(T.13.12)}$$

Moreover, (T.13.12) implies

$$|\prod\varphi'_{k,n}(\xi)|\geqslant\exp\left\{-\frac{\xi^2}{2}s'^2_{n,k_n}\right\}-|\xi|^3\varepsilon+\xi^2\Theta_n(\varepsilon)+\frac{\xi^4}{4}\delta_n$$
$$\geqslant\exp\{-\xi^2\}-|\xi|^3\varepsilon-\left(\xi^2\Theta_n(\varepsilon)+\frac{\xi^4}{4}\delta_n\right). \quad \text{(T.13.13)}$$

By choosing ε sufficiently small, one has for all sufficiently large n (depending on ε), $|\prod\varphi'_{k,n}(\xi)|$ bounded away from zero (uniformly in n). Therefore, (a) holds for $\{Y_{k,n}\}$, for all sufficiently small ε (and all sufficiently large n, depending on ε). By using relations as in (T.13.3) and the inequalities (T.13.12, 13) and the fact that $s'^2_{n,k_n}\to 1$ in probability, we get

$$\overline{\lim_{n\to\infty}}\left|E\exp\{i\xi S'_{n,k_n}\}-\exp\left\{-\frac{\xi^2}{2}\right\}\right|$$
$$\leqslant\exp\left\{-\frac{\xi^2}{2}\right\}\overline{\lim_{n\to\infty}}\,E\left|\left(\exp\left\{-\frac{\xi^2}{2}\right\}\right)^{-1}-(\prod\varphi'_{k,n}(\xi)^{-1})\right|$$
$$\leqslant\exp\left\{-\frac{\xi^2}{2}\right\}\exp\left\{\frac{\xi^2}{2}\right\}\left(\exp\left\{-\frac{\xi^2}{2}\right\}-|\xi|^3\varepsilon\right)^{-1}\overline{\lim_{n\to\infty}}\,E|\prod\varphi'_{k,n}(\xi)-e^{-\xi^2/2}|$$
$$\leqslant\left(\exp\left\{-\frac{\xi^2}{2}\right\}-|\xi|^3\varepsilon\right)^{-1}|\xi|^3\varepsilon. \quad \text{(T.13.14)}$$

Finally,

$$\overline{\lim_{n\to\infty}}\left|E\exp\{i\xi S_{n,k_n}\}-\exp\left\{-\frac{\xi^2}{2}\right\}\right|$$
$$\leqslant\overline{\lim_{n\to\infty}}\,|E\exp\{i\xi S_{n,k_n}\}-E\exp\{i\xi S'_{n,k_n}\}|+\overline{\lim_{n\to\infty}}\left|E\exp\{i\xi S'_{n,k_n}\}-\exp\left\{-\frac{\xi^2}{2}\right\}\right|$$
$$=0+\overline{\lim_{n\to\infty}}\left|E\exp\{i\xi S'_{n,k_n}\}-\exp\left\{-\frac{\xi^2}{2}\right\}\right|\leqslant\left(\exp\left\{-\frac{\xi^2}{2}\right\}-|\xi|^3\varepsilon\right)^{-1}|\xi|^3\varepsilon. \quad \text{(T.13.15)}$$

The extreme right side of (T.13.15) goes to zero as $\varepsilon \downarrow 0$, while the extreme left does not depend on ε. ■

Condition (ii) of the theorem is called the *conditional Lindeberg condition*. For additional information on martingale central limit theorems, see P. G. Hall and C. C. Heyde (1980), *Martingale Limit Theory and Its Applications*, Academic Press, New York.

One may deduce from this theorem a result of P. Billingsley (1961), "The Lindeberg–Lévy Theorem for Martingales," *Proc. Amer. Math. Soc.*, **12**, pp. 788–792, and I. A. Ibragimov (1963), "A Central Limit Theorem for a Class of Dependent Random Variables," *Theor. Probability Appl.*, **8**, pp. 83–89, stated below.

Theorem T.13.2. Let $\{X_n: n \geq 0\}$ be a stationary ergodic sequence of square integrable martingale differences. Then $Z_n/\sqrt{n} := (X_1 + \cdots + X_n)/\sqrt{n}$ converges in distribution to $N(0, \sigma^2)$, where $\sigma^2 = EX_0^2$. □

Proof. It is convenient to construct a doubly stationary sequence

$$\ldots, \tilde{X}_{-n}, \tilde{X}_{-n+1}, \ldots, \tilde{X}_{-1}, \tilde{X}_0, \tilde{X}_1, \ldots, \tilde{X}_n, \ldots$$

such that $\{\tilde{X}_n: n \geq 0\}$ has the same distribution as $\{X_n: n \geq 0\}$ has. This can be accomplished by setting for each m, the finite-dimensional distribution of any m consecutive terms of the sequence $\{\tilde{X}_n: n \in \mathbb{Z}\}$ to be the same as the distribution of $(X_0, \ldots, X_{m-1})$. Since this specification satisfies Kolmogorov's consistency requirements, the construction is complete on the canonical space $\mathbb{R}^{\mathbb{Z}}$. To simplify notation we shall write X_n, instead of $\tilde{X}_n$, for this process.

If $\sigma^2 = 0$, then $X_n = 0$ almost surely, and the desired conclusion holds trivially. Assume $\sigma^2 > 0$.

First, let us show that $\{X_n\}$ so extended is a $\{\mathcal{F}_n\}$-martingale difference sequence, where $\mathcal{F}_n := \sigma\{X_j: -\infty < j \leq n\}$. We need to show that, for each fixed $n > 0$, $E(X_{n+1} \mid \mathcal{F}_n) = 0$, i.e.,

$$E(\mathbf{1}_A X_{n+1}) = 0 \qquad \text{for all } A \in \mathcal{F}_n. \tag{T.13.16}$$

Suppose A is a finite-dimensional event in $\mathcal{F}_n$, say $A \in \sigma\{X_{n-j}: 0 \leq j \leq m\}$. Note that the (joint) distribution of $(X_{n-m}, \ldots, X_n, X_{n+1})$ is the same as that of any $m + 2$ consecutive terms of the sequence $\{X_n: n \geq 0\}$, e.g., $(X_0, X_1, \ldots, X_{m+1})$. Therefore, (T.13.16) becomes, for such a choice of A, $E(\mathbf{1}_A X_{m+1}) = 0$, for all $A \in \sigma\{X_0, \ldots, X_m\}$, which is clearly true by the martingale difference property of $\{X_n: n \geq 0\}$. Now consider the class $\mathscr{C}_n$ of sets A in $\mathcal{F}_n$ such that $E(\mathbf{1}_A X_{n+1}) = 0$. It is simple to check that $\mathscr{C}_n$ is a sigmafield and since $\mathscr{C}_n \supset \sigma\{X_{n-j}: 0 \leq j \leq m\}$ for all $m \geq 0$ one has $\mathscr{C}_n = \mathcal{F}_n$.

Next define $\sigma_n^2 := E(X_n^2 \mid \mathcal{F}_{n-1})$ $(n \in \mathbb{Z})$. Then $\{\sigma_n^2: n \geq 0\}$ is a stationary ergodic sequence, and $E\sigma_n^2 = \sigma^2 > 0$. In particular, $s_n^2/n = \sum_1^n \sigma_m^2/n \to \sigma^2$ a.s., where $s_n^2 = \sum_{m=1}^n \sigma_m^2$. Observe that $s_n^2 \to \infty$ almost surely. Write $X_{k,n} := X_k/\sqrt{n}$, $\mathcal{F}_{k,n} := \mathcal{F}_k$, $k_n = n$. Then $\sigma_{k,n}^2 = \sigma_k^2/n$, and $s_{n,n}^2 = s_n^2/n \to \sigma^2$ a.s., so that condition (i) of Theorem T.13.1 is checked, assuming $\sigma^2 = 1$ without loss of generality. It remains to check the conditional Lindeberg condition (ii). But $L_{n,n}(\varepsilon)$ is nonnegative, and $EL_{n,n}(\varepsilon) = E(X_1^2 \mathbf{1}_{\{|X_1| > \varepsilon\sqrt{n}\}}) \to 0$. Therefore, $L_{n,n}(\varepsilon) \to 0$ in probability. ■

2. We now apply Theorem T.13.2 to Markov processes. Let $\{X_n : n \geq 0\}$ be a Markov process on a state space S (with sigmafield $\mathscr{S}$), having a transition probability $p(x; dy)$. Assume that $p(x; dy)$ admits an *invariant probability* π and that under this initial invariant distribution the stationary process $\{X_n : n \geq 0\}$ is *ergodic*.

Assume X_0 has distribution π. Consider a real-valued function f on S such that $Ef^2(X_0) < \infty$. Write $f'(x) = f(x) - \bar{f}$, where $\bar{f} = \int f\, d\pi$. Write $\mathbf{T}$ for the *transition operator*, $(\mathbf{T}g)(x) := \int g(y)p(x; dy)$. Then $(\mathbf{T}^m f')(x) = (\mathbf{T}^m f)(x) - \bar{f}$ for all $m \geq 0$. Also, $Ef'(X_0) = 0$. Suppose the series $h_n(x) := \sum_0^n (\mathbf{T}^m f')(x)$ converges in $L^2(S, \pi)$ to h, i.e.

$$\int (h_n - h)^2\, d\pi \to 0 \qquad \text{as } n \to \infty. \tag{T.13.17}$$

This is true, e.g., for *all* f in case S is finite. In the case (T.13.17) holds, h satisfies

$$h(x) - (\mathbf{T}h)(x) = f'(x), \qquad \text{or} \qquad (\mathbf{I} - \mathbf{T})h = f', \tag{T.13.18}$$

i.e., f' belongs to the *range* of $\mathbf{I} - \mathbf{T}$ regarded as an operator on $L^2(S, \mathscr{S}, \pi)$. Now it is simple to check that

$$h(X_n) - (\mathbf{T}h)(X_{n-1}) \qquad (n \geq 1) \tag{T.13.19}$$

is a *martingale difference sequence*. It is also stationary and ergodic. Write

$$Z_n := \sum_{m=1}^{n} (h(X_m) - (\mathbf{T}h)(X_{m-1})). \tag{T.13.20}$$

Then, by Theorem T.13.2, $Z_n/\sqrt{n}$ converges in distribution to $N(0, \sigma^2)$, where

$$\sigma^2 = E(h(X_1) - (\mathbf{T}h)(X_0))^2 = Eh^2(X_1) + E(\mathbf{T}h)^2(X_0) - 2E[(\mathbf{T}h)(X_0)h(X_1)]. \tag{T.13.21}$$

Since $E[h(X_1) \mid \{X_0\}] = (\mathbf{T}h)(X_0)$, we have $E[(\mathbf{T}h)(X_0)h(X_1)] = E(\mathbf{T}h)^2(X_0)$, so that (T.13.21) reduces to

$$\sigma^2 = Eh^2(X_1) - E(\mathbf{T}h)^2(X_0) = \int h^2\, d\pi - \int (\mathbf{T}h)^2\, d\pi. \tag{T.13.22}$$

Also, by (T.13.18),

$$\begin{aligned} Z_n = \sum_{m=1}^{n} (h(X_m) - (\mathbf{T}h)(X_{m-1})) &= \sum_{m=0}^{n-1} (h(X_m) - (\mathbf{T}h)(X_m)) + h(X_n) - h(X_0) \\ &= \sum_{m=0}^{n-1} f'(X_m) + h(X_n) - h(X_0). \end{aligned} \tag{T.13.23}$$

Since

$$E[h(X_n) - h(X_0))/\sqrt{n}]^2 \leq \frac{2}{n}(Eh^2(X_n) + Eh^2(X_0)) = \frac{4}{n}\int h^2\, d\pi \to 0,$$

and

$$\frac{1}{\sqrt{n}}\sum_{m=0}^{n-1} f'(X_m) = \frac{1}{\sqrt{n}} Z_n - \frac{1}{\sqrt{n}}(h(X_n) - h(X_0)), \tag{T.13.24}$$

it follows that the left side converges in distribution to $N(0, \sigma^2)$ as $n \to \infty$. We have arrived at a result of M. I. Gordin and B. A. Lifšic (1978), "The Central Limit Theorem for Stationary Ergodic Markov Processes," *Dokl. Akad. Nauk SSSR*, **19**, pp. 392–393.

Theorem T.13.3. (*CLT for Discrete-Parameter Markov Processes*). Assume $p(x, dy)$ admits an invariant probability π and, under the initial distribution π, $\{X_n\}$ is ergodic. Assume also that $f' := f - \bar{f}$ is in the range of $\mathbf{I} - \mathbf{T}$. Then

$$\frac{1}{\sqrt{n}}\sum_{m=0}^{n-1} (f(X_m) - \bar{f}) \to N(0, \sigma^2) \qquad \text{as } n \to \infty, \tag{T.13.25}$$

where the convergence is in distribution, and σ^2 is given by (T.13.22). □

Some applications of this theorem in the context of processes such as considered in Sections 13, 14 of Chapter II may be found in R. N. Bhattacharya and O. Lee (1988), "Asymptotics of a Class of Markov Processes Which Are Not In General Irreducible," *Ann. Probab.*, **16**, pp. 1333–1347, and "Ergodicity and Central Limit Theorems for a Class of Markov Processes," *J. Multivariate Analysis*, **27**, pp. 80–90.

3. For continuous-parameter Markov processes such as diffusions, the following theorem applies (see R. N. Bhattacharya (1982), "On the Functional Central Limit Theorem and the Law of the Iterated Logarithm for Markov Processes," *Z. Wahrscheinlichkeitstheorie und Verw. Geb.*, **60**, pp. 185–201).

Theorem T.13.4. (*CLT for Continuous-Parameter Markov Processes*). Let $\{X_t\}$ be a stationary ergodic Markov process on a metric space S, having right-continuous sample paths. Let π denote the (stationary) distribution of X_t, and $\hat{\mathbf{A}}$ the infinitesimal generator of the process on $L^2(S, \pi)$. If f belongs to the range of $\hat{\mathbf{A}}$, then $t^{-1/2}\int_0^t f(X_s)\,ds$ converges in distribution to $N(0, \sigma^2)$, with $\sigma^2 = -2\langle f, g\rangle \equiv -2\int f(x)g(x)\pi(dx)$, where $g \in \mathscr{D}_{\mathbf{A}}$ and $\hat{\mathbf{A}}g = f$. □

Proof. It follows from Theorem 2.3 that

$$Z_n = g(X_n) - \int_0^n \hat{\mathbf{A}}g(X_s)\,ds = g(X_n) - \int_0^n f(X_s)\,ds \qquad (n = 1, 2, \ldots)$$

is a square integrable martingale. By hypothesis, $Z_n - Z_{n-1}$ $(n \geqslant 1)$, is a stationary ergodic sequence of martingale differences. Therefore, Theorem T.13.2 applies and $n^{-1/2}Z_n$ converges in distribution to $N(0, \sigma^2)$, where $\sigma^2 = E(Z_1 - Z_0)^2$, i.e.,

$$\sigma^2 = E\left[g(X_1) - g(X_0) - \int_0^1 \hat{\mathbf{A}}g(X_s)\,ds\right]^2. \tag{T.13.26}$$

But

$$E(n^{-1/2}g(X_n))^2 = n^{-1}Eg^2(X_1) \to 0 \qquad \text{as } n \to \infty.$$

Therefore,

$$\pm n^{-1/2} \int_0^n f(X_s)\,ds \to N(0, \sigma^2).$$

Also, for each positive integer n, $\{Z_{k/n} - Z_{(k-1)/n}: 1 \leqslant k \leqslant n\}$ are stationary martingale differences, so that

$$\begin{aligned}
\sigma^2 &= E(Z_1 - Z_0)^2 = \sum_{k=1}^{n} E(Z_{k/n} - Z_{(k-1)/n})^2 = nE(Z_{1/n} - Z_0)^2 \\
&= nE\left[g(X_{1/n}) - g(X_0) - \int_0^{1/n} \hat{\mathbf{A}}g(X_s)\,ds\right]^2 \\
&= nE(g(X_{1/n}) - g(X_0))^2 + nE\left(\int_0^{1/n} \hat{\mathbf{A}}g(X_s)\,ds\right)^2 \\
&\quad - 2nE\left\{(g(X_{1/n}) - g(X_0)) \int_0^{1/n} \hat{\mathbf{A}}g(X_s)\,ds\right\}. \qquad \text{(T.13.27)}
\end{aligned}$$

Now,

$$\begin{aligned}
nE(g(X_{1/n}) - g(X_0))^2 &= n[Eg^2(X_{1/n}) + Eg^2(X_0) - 2Eg(X_{1/n})g(X_0)] \\
&= n\left[2Eg^2(X_0) - 2\int g(x)\mathbf{T}_{1/n}g(x)\pi(dx)\right] \\
&= n\left[2Eg^2(X_0) - 2\int g(x)\left\{g(x) + \int_0^{1/n} \mathbf{T}_s\hat{\mathbf{A}}g(x)\,ds\right\}\pi(dx)\right] \\
&= -2n\int g(x)\left\{\int_0^{1/n} \mathbf{T}_s\hat{\mathbf{A}}g(x)\,ds\right\}\pi(dx) \to -2\int g(x)\hat{\mathbf{A}}g(x)\pi(dx),
\end{aligned} \qquad \text{(T.13.28)}$$

as $n \to \infty$, since $\mathbf{T}_s\hat{\mathbf{A}}g \to \hat{\mathbf{A}}g$ in $L^2(S, \pi)$ as $s \downarrow 0$. Also,

$$\begin{aligned}
\left(E\int_0^{1/n} \hat{\mathbf{A}}g(X_s)\,ds\right)^2 &\leqslant E\left(\frac{1}{n}\int_0^{1/n} (\hat{\mathbf{A}}g(X_s))^2\,ds\right) = \frac{1}{n}\int_0^{1/n} E(\hat{\mathbf{A}}g)^2(X_0)\,ds \\
&= \frac{1}{n^2} E(\hat{\mathbf{A}}g)^2(X_0). \qquad \text{(T.13.29)}
\end{aligned}$$

Applying (T.13.28, 29), the product term on the extreme right side of (T.13.27) is seen to go to zero. Therefore, $\sigma^2 = -2\langle g, f\rangle + o(1)$ as $n \to \infty$, which implies $\sigma^2 = -2\langle g, f\rangle$. This proves the result for the sequence $t = n$. But if

$$M_n := \max\left\{n^{-1/2}\int_{k-1}^{k} |f(X_s)|\,ds: 1 \leqslant k \leqslant n\right\},$$

then

$$P(M_n > \varepsilon) \leqslant \sum_{k=1}^{n} P\left(n^{-1/2}\int_{k-1}^{k} |f(X_s)|\,ds > \varepsilon\right) = nP\left(\int_0^1 |f(X_s)|\,ds > \varepsilon\sqrt{n}\right) \to 0$$

as $n \to \infty$, since

$$E\left(\int_0^1 |f(X_s)|\,ds\right)^2 < \infty. \qquad ■$$

In order to check that $f - \bar{f}$ belongs to the range of $\hat{\mathbf{A}}$ it is enough to show that $-\int_0^t \mathbf{T}_s(f - \bar{f})(x)\,ds$ converges in $L^2(S, \pi)$ as $t \to \infty$. For if the convergence is to a function g in L^2, then

$$\frac{1}{h}(\mathbf{T}_h g - g)(x) = \frac{1}{h}\int_0^h \mathbf{T}_s(f - \bar{f})(x)\,ds \to (f - \bar{f})(x)$$

in L^2 as $h \downarrow 0$. In other words, $\hat{\mathbf{A}}g = f - \bar{f}$. Now,

$$\int_0^t \mathbf{T}_s(f - \bar{f})(x)\,ds = \int_0^t \int_S f(y)(p(s; x, dy) - \pi(dy))\,ds. \qquad \text{(T.13.26)}$$

Therefore, one simple sufficient condition that $f - \bar{f}$ belongs to the range of $\hat{\mathbf{A}}$ for *every* $f \in L^2(S, \pi)$ is

$$\sup\{|p(s; x, B) - \pi(B)|: B \in \mathscr{B}(S),\ x \in S\} \to 0$$

exponentially fast as $s \to \infty$. This is, for example, the case with the reflecting r-dimensional Brownian motion on the disc $\{|\mathbf{x}|^2 \leqslant a^2\}$, $k > 1$.

An alternative renewal approach may be found in R. N. Bhattacharya and S. Ramasubramanian (1982), "Recurrence and Ergodicity of Diffusions," *J. Multivariate Analysis*, **12**, pp. 95–122.

4. It is shown in Bhattacharya (1982), *loc. cit.*, that ergodicity of the Markov process $\{X_t\}$ under the stationary initial distribution π is equivalent to the null space of $\hat{\mathbf{A}}$ being the space of constants.

Theoretical Complements to Section V.14

1. To state Proposition 14.1 more precisely, suppose $\{X_t\}$ is a Markov process on a metric space S having an infinitesimal generator $\mathbf{A}$ on a domain $\mathscr{D}_{\mathbf{A}}$ and transition operators $\mathbf{T}_t$ on the Banach space $C_b(S)$ of all real-valued bounded continuous functions f on S, with the sup norm $\|f\|$. Let φ be a continuous function on S onto a metric space S'. *Let $\mathscr{G}$ be a class of real-valued, bounded and continuous functions g on S' having the following two properties:*

(i) *If $g \in \mathscr{G}$ then $g \circ \varphi \in \mathscr{D}_{\mathbf{A}}$, and $\mathbf{A}(g \circ \varphi) = h \circ \varphi$ for some bounded continuous h on S';*

(ii) *$\mathscr{G}$ is a determining class for probability measures on $(S', \mathscr{B}(S'))$, that is, if μ, ν are two probability measures such that $\int g\,d\mu = \int g\,d\nu$ for all $g \in \mathscr{G}$ then $\mu = \nu$.*

Then $\{\varphi(X_t)\}$ is a Markov process on S'.

To see this consider the subspace $\mathbb{B}$ of $C_b(S)$ comprising all functions $g \circ \varphi$, g in the closure of the linear span of $\mathscr{L}(\mathscr{G})$ of $\mathscr{G}$. Then $\mathbb{B}$ is a Banach space, and $\mathbf{A}$ is the generator of a semigroup on $\mathbb{B}$ by the Hille–Yosida Theorem and property (i). This implies in particular that $\mathbf{T}_t$ maps $\mathbb{B}$ into itself, so that $\mathbf{T}_t(g \circ \varphi)$ is of the form $h_t \circ \varphi$ for some bounded continuous h_t on S'. Since $\mathscr{G}$ is a determining class, it follows that $\mathbf{T}_t(g \circ \varphi)$ is of the form $h_t \circ \varphi$ for every bounded measurable g on S'. Therefore, Proposition 6.3 applies, showing that $\{\varphi(X_t)\}$ is a Markov process on S'.

Theoretical Complements to Section V.15

1. The Markov property of $\{\tilde{\mathbf{X}}_t\}$ as defined by (15.1) does not require any assumption on ∂G. But for the continuity at the boundary of $\mathbf{x} \to \mathbf{T}_t g(\mathbf{x}) := E_{\mathbf{x}} g(\tilde{\mathbf{X}}_t)$ for bounded continuous g on $\bar{G}$, or of the solution ψ_f of the Dirichlet problem (15.23) for continuous bounded f on ∂G, some "smoothness" of ∂G is needed. The minimal such requirement is that every point $\mathbf{b}$ of ∂G be *regular*, that is, for every $t > 0$, $P_{\mathbf{x}}(\tau_{\partial G} > t) \to 0$ as $\mathbf{x} \to \mathbf{b}$, $\mathbf{x} \in G$. Here $\tau_{\partial G}$ is as in (15.2). A simple sufficient condition for the regularity of $\mathbf{b}$ is that it be *Poincaré point*, that is, there exists a truncated cone contained in $\mathbb{R}^k \backslash G$ with vertex at $\mathbf{b}$. For the case of a standard Brownian motion on $\mathbb{R}^k$, simple proofs of these facts may be found in E. B. Dynkin, and A. A. Yushkevich (1969), *Markov Processes: Theorems and Problems*, Plenum Press, New York, pp. 51–62.

   Analytical proofs for general diffusions may be found in D. Gilbarg and N. S. Trudinger (1977), *Elliptic Partial Differential Equations of Second Order*, Springer-Verlag, New York, p. 196, for the Dirichlet problem, and A. Friedman (1964), *Partial Differential Equations of Parabolic Type*, Prentice-Hall, Englewood Cliffs.

2. Let f be a bounded continuous function that vanishes outside G. Define $\mathbf{T}_t^0 f(\mathbf{x}) := E(f(\mathbf{X}_t)\mathbf{1}_{\{\tau_{\partial G} > t\}} \mid \mathbf{X}_0 = \mathbf{x})$. If $\mathbf{b}$ is a regular boundary point then as $\mathbf{x} \to \mathbf{b}$ $(\mathbf{x} \in G)$, $\mathbf{T}_t^0 f(\mathbf{x}) \to 0$. In other words, $\mathbf{T}_t^0 C_0 \subset C_0$, where C_0 is the class of all bounded continuous functions on $\bar{G}$ vanishing at the boundary. If f is a twice continuously differentiable function in C_0 with compact support then one may show that $f \in \mathscr{D}_{\mathbf{A}^0}$, the domain of the infinitesimal generator $\mathbf{A}^0$ of the semigroup $\{\mathbf{T}_t^0\}$, and that $(\mathbf{A}^0 f)(\mathbf{x}) = (\mathbf{A}f)(\mathbf{x})$ for all $\mathbf{x} \in G$. For this one needs the estimate $P_{\mathbf{x}}(\tau_{\partial G} \leqslant t) \to 0$ as $t \to 0$, uniformly for all $\mathbf{x}$ in the support of f (see Exercise 3.11 of Chapter VII). Also see (15.24), (15.25), as well as Exercise 3.12 of Chapter VII.

   As indicated in Section 7 following (7.12), if ∂G is assumed regular and $p^0(t; \mathbf{x}, \mathbf{y})$ has a limit as $\mathbf{x}$ approaches ∂G, then this limit must be zero. For the existence of a smooth p^0 continuous on $\bar{G}$, and vanishing on ∂G, see A. Friedman, *loc. cit.*, p. 82.

3. (*Maximum Principle for Elliptic Equations*) Suppose first that $\mathbf{A}u(\mathbf{x}) > 0$ for all $\mathbf{x} \in G$, where G is a bounded open set. Let us show that the maximum of u cannot be attained in the interior. For if it did at $\mathbf{x}_0 \in G$, then $(\partial u/\partial x^{(i)})_{\mathbf{x}_0} = 0$, $1 \leqslant i \leqslant k$, and $\mathbf{B} := ((\partial^2 u/\partial x^{(i)}\, dx^{(j)}))_{\mathbf{x}_0}$ is negative semidefinite. This implies that the eigenvalues of $\mathbf{CB}$ are all real nonpositive, where $\mathbf{C} := ((d_{ij}(\mathbf{x}_0))$. For if λ is an eigenvalue of $\mathbf{CB}$, then $\mathbf{CBz} = \lambda\mathbf{z}$ for some nonzero $\mathbf{z} \in \mathbb{C}^k$ and, using the inner product $\langle\ ,\ \rangle$ on $\mathbb{C}^k$, $0 \leqslant \langle \mathbf{CBz}, \mathbf{Bz} \rangle = \langle \lambda\mathbf{z}, \mathbf{Bz} \rangle = \lambda \langle \mathbf{z}, \mathbf{Bz} \rangle$. Therefore, $\lambda \leqslant 0$. It follows now that $\sum d_{ij}(\mathbf{x}_0)(\partial^2 u/\partial x^{(i)}\, \partial x^{(j)})\mathbf{x}_0 = \text{trace of } \mathbf{CB}$ is $\leqslant 0$. Therefore, $\mathbf{A}u(\mathbf{x}_0) \leqslant 0$, contradicting $\mathbf{A}u(\mathbf{x}_0) > 0$.

   Now consider u such that $\mathbf{A}u(\mathbf{x}) \geqslant 0$ for all $\mathbf{x} \in G$. For $\varepsilon > 0$ consider the function $u_\varepsilon(\mathbf{x}) := u(\mathbf{x}) + \varepsilon \exp\{\gamma x^{(1)}\}$. Then $\mathbf{A}u_\varepsilon(\mathbf{x}) = \mathbf{A}u(\mathbf{x}) + \varepsilon[\frac{1}{2}d_{11}(\mathbf{x})\gamma^2 + \mu^{(1)}(\mathbf{x})\gamma] \exp\{\gamma x^{(1)}\}$.

Choose γ sufficiently large that the expression within square brackets is strictly positive for all $\mathbf{x} \in \bar{G}$. Then $\mathbf{A}u_\varepsilon(\mathbf{x}) > 0$ for $\mathbf{x} \in G$, for all $\varepsilon > 0$. Applying the conclusion of the preceding paragraph conclude that the maximum of u_ε is attained on ∂G, say at $\mathbf{x}_\varepsilon$. Since ∂G is compact, choosing a convergent subsequence of $\mathbf{x}_\varepsilon$ as $\varepsilon \downarrow 0$ through a sequence, it follows that the maximum of u is attained on ∂G.

Finally, if $\mathbf{A}u(x) = 0$ for all $\mathbf{x} \in G$, then applying the above result to both u and $-u$ one gets: *u attains its maximum and minimum on ∂G*. Note that for the above proof it is enough to require that $\mu^{(i)}(\mathbf{x})$, $d_{ij}(\mathbf{x})$, $1 \leqslant i, j \leqslant k$, are continuous on $\bar{G}$, and $d_{ii}(\mathbf{x}) > 0$ on $\bar{G}$ for some i. Under the additional hypothesis that $((d_{ij}(\mathbf{x})))$ is positive definite on G, one can prove the *strong maximum principle*: *Suppose G is a bounded and connected open set. If $\mathbf{A}u(\mathbf{x}) = 0$ for all $\mathbf{x} \in G$, and u is continuous on $\bar{G}$, then u cannot attain its maximum or minimum in G, unless u is constant.*

The probabilistic argument for the strong maximum principle is illuminating. Under the conditions (1)–(4) of Section 14 it may be shown that the *support* of the probability measure $P_\mathbf{x}$ is the set of *all* continuous functions on $[0, \infty)$ into $\mathbb{R}^k$, starting at $\mathbf{x}$ (see theoretical complements to Section VII.3). Therefore, for any ball $B \ni \mathbf{x}$, the $P_\mathbf{x}$-distribution $\psi(\mathbf{x}; dy)$ of $\mathbf{X}_{\tau_{\partial B}}$ has support ∂B. Now suppose u is continuous on $\bar{G}$ and $\mathbf{A}u(\mathbf{x}) = 0$ for all $\mathbf{x} \in G$. Suppose $u(\mathbf{x}_0) = \max\{u(\mathbf{x}): \mathbf{x} \in \bar{G}\}$ for some $\mathbf{x}_0 \in G$. Then for every closed ball $B_\varepsilon := \{\mathbf{x}: |\mathbf{x} - \mathbf{x}_0| \leqslant \varepsilon\} \subset G$, the strong Markov property yields: $u(\mathbf{x}_0) = E_{\mathbf{x}_0} u(\mathbf{X}_{\tau_{\partial B_\varepsilon}})$, in view of the representation (15.22). (Also see Exercise 3.14 of Chapter VII). That is,

$$u(\mathbf{x}_0) = \int_{\partial B_\varepsilon} u(\mathbf{y})\psi_\varepsilon(\mathbf{x}_0; d\mathbf{y}), \tag{T.15.1}$$

where $\psi_\varepsilon(\mathbf{x}_0; d\mathbf{y})$ is the $P_{\mathbf{x}_0}$-distribution of $\mathbf{X}_{\tau_{\partial B_\varepsilon}}$. Since u is continuous on ∂B_ε and the probability measure $\psi_\varepsilon(\mathbf{x}_0; d\mathbf{y})$ has the full support ∂B_ε, the right side is strictly smaller than the maximum value of u on B_ε, namely $u(\mathbf{x}_0)$, unless $u(\mathbf{y}) = u(\mathbf{x}_0)$ for all $\mathbf{y} \in \partial B_\varepsilon$. By letting ε vary, this constancy extends to the maximal closed ball with center $\mathbf{x}_0$ contained in $\bar{G}$. By connectedness of G, the proof is completed.

4. Let G be a bounded open set such that there exists a twice continuously differentiable real-valued function φ on $\mathbb{R}^k$ such that $G = \{\mathbf{x}: \varphi(\mathbf{x}) < c\}$, $\partial G = \{\mathbf{x}: \varphi(\mathbf{x}) = c\}$, for some $c \in \mathbb{R}^1$. Assume also that $|\mathbf{grad}\ \varphi|$ is bounded away from zero on ∂G. If u is a twice continuously differentiable function in G such that u, $\partial u/\partial x^{(i)}$, $\partial^2 u/\partial x^{(i)}\,\partial x^{(j)}$ $(1 \leqslant i, j \leqslant k)$ have continuous extensions to $\bar{G}$, then there exists a twice continuously differentiable function q on $\mathbb{R}^k$ with compact support such that $q = u$ on $\bar{G}$ (see Gilbarg and Trudinger, *loc. cit.*, p. 130).

Theoretical Complements to Section V.16

1. D. W. Stroock and S. R. S. Varadhan (1979), *Multidimensional Diffusions*, Springer-Verlag, New York, Chapter 10, have constructed diffusions on $\mathbb{R}^k$ under the assumptions (i) $((d_{ij}(\mathbf{x})))$ is continuous and positive definite, (ii) $\mu^{(i)}(\mathbf{x})$ $(1 \leqslant i \leqslant k)$ are Borel-measurable and bounded on compacts, and (iii) an appropriate condition for nonexplosion holds. These conditions apply under our assumptions in Section 16.

2. The present method of constructing reflecting diffusions in one and multidimensions may be found in M. Friedlin (1985), *Functional Integration and Partial Differential*

Equations, Princeton University Press, Princeton, pp. 86, 87, and in R. N. Bhattacharya and E. C. Waymire (1989), "An Extension of the Classical Method of Images for the Construction of Reflecting Diffusions," *Proceedings of the R. C. Bose Symposium on Probability, Statistics and Design of Experiments* (ed. R. R. Bahadur, J. K. Ghosh, K. Sen), Eka Press, Calcutta.

3. The transformation $g(\mathbf{x}) = \bar{\mathbf{x}}$ (see (16.9)) generates a group of transformations $G = \{g_0, g\}$ with g_0 as the identity. A maximal invariant under G is $\varphi(\mathbf{x}) = (|x^{(1)}|, x^{(2)}, \ldots, x^{(k)})$. By the proposition in theoretical complement 6.2, $\{\varphi(\mathbf{X}_t)\}$ is Markov if $p(t; \bar{\mathbf{x}}, \bar{\mathbf{y}}) = p(t; \mathbf{x}, \mathbf{y})$. The transition density of the unrestricted diffusion $\{\mathbf{X}_t\}$ in Proposition 16.1 has this invariance property. As further applications of the proposition, consider the following examples.

Example 1. (*Radial Diffusions*). Let $\{X_t\}$ be a diffusion on $\mathbb{R}^k$ with drift $\boldsymbol{\mu}(\cdot)$, and diffusion matrix $((d_{ij}(\cdot)))$. If $\sum d_{ij}(\mathbf{x})x^{(i)}x^{(j)}$ and $2\sum x^{(i)}\mu^{(i)}(\mathbf{x}) + \sum d_{ii}(\mathbf{x})$ are radial functions (i.e., functions of $|\mathbf{x}|$), then $\{|X_t|\}$ is a Markov process. This follows from the fact that if f is a smooth radial function then $\mathbf{A}f$ is also radial (see (15.35), and theoretical complement 14.1). The underlying group G in this case is the group of all orthogonal matrices.

Example 2. (*Diffusions on the Torus*). Let $\{\mathbf{X}_t\}$ be a diffusion on $\mathbb{R}^k$ such that $\mu^{(i)}(\mathbf{x})$ and $d_{ij}(\mathbf{x})$ are invariant under the group G generated by translations by a set of k linearly independent vectors $\xi_1, \xi_2, \ldots, \xi_k$. In other words,

$$\mu^{(i)}\left(\mathbf{x} + \sum_1^k m_r \xi_r\right) = \mu^{(i)}(\mathbf{x}), \qquad d_{ij}\left(\mathbf{x} + \sum_1^k m_r \xi_r\right) = d_{ij}(\mathbf{x}) \qquad (1 \leqslant i, j \leqslant k),$$

for all $\mathbf{x}$ and all k-tuples of integers $(m_1, m_2, \ldots, m_k)$. Now express $\mathbf{x}$ (uniquely) as $\sum_1^k \delta_r(\mathbf{x})\xi_r$, and let $\mathring{\delta}_r(\mathbf{x}) = \delta_r(\mathbf{x}) (\text{mod } 1) \in [0, 1)$, $1 \leqslant r \leqslant k$. Then the maximal invariant under G is $\varphi(\mathbf{x}) := (\mathring{\delta}_1(\mathbf{x}), \ldots, \mathring{\delta}_k(\mathbf{x}))$, and $\{\varphi(\mathbf{X}_t)\}$ is a Markov process on the torus $[0, 1)^k$.

Theoretical Complements to Section V.17

1. Taylor's article appeared in G. I. Taylor (1953), "Dispersion of Soluble Matter in a Solvent Flowing Through a Tube," *Proc. Roy. Soc. Ser. A*, **219**, pp. 186–203. It was shown there that asymptotically for large U_0, $D \sim a^2 U_0^2/48 D_0$. The explicit computation (17.30) was given by R. Aris (1956), "On the Dispersion of a Solute in a Fluid Flowing Through a Tube," *Proc. Roy. Soc. Ser. A*, **235**, pp. 67–77, using the method of moments and eigenfunction expansions.

   The treatment of the Taylor–Aris theory given in this section follows R. N. Bhattacharya and V. K. Gupta (1984), "On the Taylor–Aris Theory of Solute Transport in a Capillary," *SIAM J. Appl. Math.*, **44**, 33–39.

CHAPTER VI

Dynamic Programming and Stochastic Optimization

1 FINITE-HORIZON OPTIMIZATION

Consider a *finite* set S, called the *state space*, a *finite* set A, called the *action space*, and, for each pair $(x, a) \in S \times A$, a probability function $p(y; x, a)$ on S:

$$p(y; x, a) \geqslant 0, \qquad \sum_{y \in S} p(y; x, a) = 1. \tag{1.1}$$

The function $p(y; x, a)$ denotes the probability that the state in the next period will be y, given that the present state is x and an action a has been taken.

A *policy* (or, a *feasible policy*) is a sequence of functions $(f_0, f_1, \ldots)$ defined as follows. f_0 is a function on S into A. If the state in period $k = 0$ is x_0 then an action $f_0(x_0) = a_0$ is taken. Given the state x_0 and the action $a_0 = f_0(x_0)$, the state in period $k = 1$ is x_1 with probability $p(x_1; x_0, f_0(x_0))$. Now f_1 is a function on $S \times A \times S$ into A. Given the triple $x_0, f_0(x_0), x_1$, an action $f_1(x_0, f_0(x_0), x_1) = a_1$ is taken. Given x_0, a_0 and the state x_1 and action a_1, the probability that the state in period $k = 2$ is x_2 is $p(x_2; x_1, a_1)$. Similarly, f_2 is a function on $S \times A \times S \times A \times S$ into A; given x_0, a_0, x_1, a_1, x_2 an action $a_2 = f_2(x_0, a_0, x_1, a_1, x_2)$ is taken. Given all the states and actions up to period $k = 2$ (namely $x_0, a_0 = f_0(x_0), x_1, a_1 = f_1(x_0, a_0, x_1), x_2, a_2 = f_2(x_0, a_0, x_1, a_1, x_2)$), the probability that a state x_3 occurs during period $k = 3$ is $p(x_3; x_2, a_2)$, and so on. In general, a policy is a sequence of functions $\{f_k: k = 0, 1, 2, \ldots\}$ such that f_k is a function on $(S \times A)^k \times S$ into A $(k = 1, 2, \ldots)$, with f_0 a function on S into A. The sequence may be finite $\{f_k: k = 0, 1, \ldots, N\}$, called *the finite-horizon case*, or it may be infinite, *the infinite-horizon case*.

Consider first the case of a *finite horizon*. Given a policy $\mathbf{f} = (f_0, f_1, \ldots, f_N)$, $N + 1$ random variables $X_0, X_1, \ldots, X_N$ are defined, X_k being the state at time

k. The initial state X_0 may be either fixed or randomly chosen according to a probability $\boldsymbol{\pi}_0 = \{\pi_0(x): x \in S\}$. Given $X_0 = x_0$, the conditional probability distribution of X_1 is given by

$$P(X_1 = x_1 \mid X_0 = x_0) = p(x_1; x_0, f_0(x_0)). \tag{1.2}$$

Given $X_0 = x_0, X_1 = x_1$, the conditional probability distribution of X_2 is given by

$$P(X_2 = x_2 \mid X_0 = x_0, X_1 = x_1) = p(x_2; x_1, a_1), \qquad a_1 = f_1(x_0, f_0(x_0), x_1). \tag{1.3}$$

In general, given $X_0 = x_0, X_1 = x_1, \dots, X_{k-1} = x_{k-1}$, the conditional distribution of X_k is given by

$$P(X_k = x_k \mid X_0 = x_0, X_1 = x_1, \dots, X_{k-1} = x_{k-1}) = p(x_k; x_{k-1}, a_{k-1}), \tag{1.4}$$

where the actions a_k are defined *recursively* by

$$\begin{aligned} a_k &= f_k(x_0, a_0, x_1, a_1, \dots, x_{k-1}, a_{k-1}, x_k) \qquad (k = 1, 2, \dots, N), \\ a_0 &= f_0(x_0). \end{aligned} \tag{1.5}$$

The joint distribution of $X_0, X_1, \dots, X_k$ is then given by

$$\begin{aligned} \boldsymbol{\pi}^{\mathbf{f}}_{0,k}(x_0, x_1, \dots, x_k) &:= P(X_0 = x_0, X_1 = x_1, \dots, X_k = x_k) \\ &= \pi_0(x_0)p(x_1; x_0, a_0)p(x_2; x_1, a_1)\cdots p(x_k; x_{k-1}, a_{k-1}), \\ &\qquad (k = 1, 2, \dots, N) \end{aligned} \tag{1.6}$$

with the states $a_0, a_1, \dots$ defined by (1.5). Expectations with respect to $\boldsymbol{\pi}^{\mathbf{f}}_{0,N}$ will be denoted by $E^{\mathbf{f}}_{\boldsymbol{\pi}_0}$.

The kth period *reward function* is a real-valued function g_k on $S \times A$ such that if the state at time k is x_k and the action at time k is a_k, then a reward $g_k(x_k, a_k)$ accrues. The *total expected return* for a policy $\mathbf{f}$ is given by

$$J^{\mathbf{f}}_{\boldsymbol{\pi}_0} := E^{\mathbf{f}}_{\boldsymbol{\pi}_0} \sum_{k=0}^{N} g_k(X_k, a_k), \tag{1.7}$$

where a_k denotes the (random) action at time k determined by the states $X_0, X_1, \dots, X_{k-1}$ according to (1.5). In case $\boldsymbol{\pi}_0 = \delta_x$ write $J^{\mathbf{f}}_x$ for $J^{\mathbf{f}}_{\boldsymbol{\pi}_0}$.

Let $\mathbf{F}$ denote the set of all policies. The object is to find an *optimal policy*, i.e., a policy $\mathbf{f}^* = (f^*_0, \dots, f^*_N)$ such that

$$J^{\mathbf{f}^*}_{\boldsymbol{\pi}_0} \geqslant J^{\mathbf{f}}_{\boldsymbol{\pi}_0} \qquad \text{for all } \mathbf{f} \in \mathbf{F}, \tag{1.8}$$

for *all* initial distributions $\boldsymbol{\pi}_0$.

In view of the fact that the probability of transition to a state x_{k+1} in period $k+1$ depends only on the state x_k and action a_k in period k, it is plausible that the search for an optimal policy may be confined to the class of policies $\mathbf{f} = (f_0, f_1, \ldots, f_N)$ such that the action a_k depends only on the state x_k in period k. In other words, f_k is a function on S into A $(k = 0, 1, \ldots, N)$. Such policies are called *Markovian*. The reason for this nomenclature is the following proposition.

Proposition 1.1. If $\mathbf{f} = (f_1, f_1, \ldots, f_N)$ is a Markovian policy, then $X_0, X_1, \ldots, X_N$ is a Markov chain (which is, in general, time-inhomogeneous).

Proof. By (1.6) one has

$$\begin{aligned} P(X_0 = x_0, X_1 = x_1, \ldots, X_N = x_N) &\\ = \pi_0(x_0)p(x_1; x_0, f_0(x_0))p(x_2; x_1, f_1(x_1))\cdots p(x_N; x_{N-1}, f_{N-1}(x_{N-1})) &\\ = \pi_0(x_0)p_1(x_1; x_0)p_2(x_2; x_1)\cdots p_N(x_N; x_{N-1}), & \quad (1.9) \end{aligned}$$

where

$$p_k(x_k; x_{k-1}) = p(x_k; x_{k-1}, f_{k-1}(x_{k-1})). \tag{1.10}$$

Hence $X_0, X_1, \ldots, X_N$ forms a finite Markov chain with successive one-step transition probabilities $p_1, p_2, \ldots, p_N$. ■

Let us show that a Markovian optimal policy exists. For each $y \in S$ pick an element $f_N^*(y)$ (perhaps not unique) of A such that

$$\max_{a \in A} g_N(y, a) = g_N(y, f_N^*(y)). \tag{1.11}$$

Write

$$h_N(y) = g_N(y, f_N^*(y)). \tag{1.12}$$

Next, for each $y \in S$ find an element $f_{N-1}^*(y)$ of A such that

$$\begin{aligned} \max_{a \in A}\left[g_{N-1}(y, a) + \sum_{z \in S} h_N(z)p(z; y, a)\right] &\\ = g_{N-1}(y, f_{N-1}^*(y)) + \sum_{z \in S} h_N(z)p(z; y, f_{N-1}^*(y)) = h_{N-1}(y), & \quad (1.13) \end{aligned}$$

say. In general, let f_k^* be a function (on S into A) such that

$$\max_{a \in A}\left[g_k(y, a) + \sum_{z \in S} h_{k+1}(z)p(z; y, a)\right] = g_k(y, f_k^*(y)) + \sum_{z \in S} h_{k+1}(z)p(z; y, f_k^*(y))$$
$$= h_k(y), \qquad (k = N-1, N-2, \ldots, 0) \tag{1.14}$$

say, obtained successively (i.e., by *backward recursion*) starting from (1.11) and (1.12).

Theorem 1.2. The Markovian policy $\mathbf{f}^* := (f_0^*, f_1^*, \ldots, f_N^*)$ is optimal.

Proof. Fix an initial distribution π_0. Let $\mathbf{f} = (f_0, f_1, \ldots, f_N)$ be any given policy. Define the policies

$$\mathbf{f}^{(k)} = (f_0, f_1, \ldots, f_{k-1}, f_k^*, f_{k+1}^*, \ldots, f_N^*) \qquad (k = 1, \ldots, N),$$
$$\mathbf{f}^{(0)} = \mathbf{f}^*. \tag{1.15}$$

We will show that

$$J_{\pi_0}^{\mathbf{f}} \leqslant J_{\pi_0}^{\mathbf{f}^{(k)}} \qquad (k = 0, 1, 2, \ldots, N). \tag{1.16}$$

First note that the joint distribution of $X_0, X_1, \ldots, X_k$ is the same under $\mathbf{f}$ and under $\mathbf{f}^{(k)}$ (i.e., $\pi_{0,k}^{\mathbf{f}} = \pi_{0,k}^{\mathbf{f}^{(k)}}$). In particular,

$$E_{\pi_0}^{\mathbf{f}} \sum_{k'=0}^{k-1} g_{k'}(X_{k'}, a_{k'}) = E_{\pi_0}^{\mathbf{f}^{(k)}} \sum_{k'=0}^{k-1} g_{k'}(X_{k'}, a_{k'}). \tag{1.17}$$

Now,

$$E_{\pi_0}^{\mathbf{f}}[g_N(X_N, a_N) \mid X_0 = x_0, X_1 = x_1, \ldots, X_N = x_N]$$
$$= g_N(x_N, a_N) \leqslant g_N(x_N, f_N^*(x_N)) = h_N(x_N)$$
$$= E_{\pi_0}^{\mathbf{f}^{(N)}}[g_N(X_N, a_N) \mid X_0 = x_0, \ldots, X_N = x_N]. \tag{1.18}$$

Since $\pi_{0,N}^{\mathbf{f}} = \pi_{0,N}^{\mathbf{f}^{(N)}}$, averaging over $x_0, x_1, \ldots, x_N$, one gets

$$E_{\pi_0}^{\mathbf{f}} g_N(X_N, a_N) \leqslant E_{\pi_0}^{\mathbf{f}^{(N)}} g_N(X_N, a_N). \tag{1.19}$$

Together with (1.17) with $k = N$, this proves (1.16) for $k = N$. To prove (1.16) for $k = N-1$, note that

$$E_{\pi_0}^{\mathbf{f}}[g_{N-1}(X_{N-1}, a_{N-1}) + g_N(X_N, a_N) \mid X_0 = x_0, \ldots, X_{N-1} = x_{N-1}]$$
$$= g_{N-1}(x_{N-1}, a_{N-1}) + \sum_{z \in S} g_N(z, a_N(z))p(z; x_{N-1}, a_{N-1}) \tag{1.20}$$

where $a_{N-1}, a_N(z)$ are defined by (1.5) with x_N replaced by z. By (1.12) and (1.13), the right side of (1.20) does not exceed

$$h_{N-1}(x_{N-1}) = E_{\pi_0}^{\mathbf{f}^{(N-1)}}[g_{N-1}(X_{N-1}, a_{N-1}) + g_N(X_N, a_N) \mid X_0 = x_0, \ldots, X_{N-1} = x_{N-1}]. \quad (1.21)$$

Since $\pi_{0,N-1}^{\mathbf{f}} = \pi_{0,N-1}^{\mathbf{f}^{(N-1)}}$, it follows by averaging over $x_0, x_1, \ldots, x_{N-1}$ that

$$E_{\pi_0}^{\mathbf{f}}(g_{N-1}(X_{N-1}, a_{N-1}) + g_N(X_N, a_N)) \leqslant E_{\pi_0}^{\mathbf{f}^{(N-1)}}(g_{N-1}(X_{N-1}, a_{N-1}) + g_N(X_N, a_N)). \quad (1.22)$$

Together with (1.17) for $k = N - 1$, (1.22) yields (1.16) for $k = N - 1$.

To complete the proof we use *backward induction.* Assume that

$$E_{\pi_0}^{\mathbf{f}}\left[\sum_{k'=k+1}^{N} g_{k'}(X_{k'}, a_{k'}) \mid X_0 = x_0, \ldots, X_{k+1} = x_{k+1}\right] \leqslant h_{k+1}(x_{k+1}) \quad (1.23)$$

for all $x_0, x_1, \ldots, x_{k+1} \in S$ and

$$E_{\pi_0}^{\mathbf{f}^{(k+1)}}\left[\sum_{k'=k+1}^{N} g_{k'}(X_{k'}, a_{k'}) \mid X_{k+1} = x_{k+1}\right] = h_{k+1}(x_{k+1}) \quad (1.24)$$

for all $x_{k+1} \in S$. Then

$$\begin{aligned} E_{\pi_0}^{\mathbf{f}}&\left[\sum_{k'=k}^{N} g_{k'}(X_{k'}, a_{k'}) \mid X_0 = x_0, \ldots, X_k = x_k\right] \\ &= g_k(x_k, a_k) + E_{\pi_0}^{\mathbf{f}}\left\{E_{\pi_0}^{\mathbf{f}}\left[\sum_{k'=k+1}^{N} g_{k'}(X_{k'}, a_{k'}) \mid X_0, \ldots, X_{k+1}\right] \Bigg| X_0 = x_0, X_1 = x_1, \ldots, X_k = x_k\right\} \\ &\leqslant g_k(x_k, a_k) + E_{\pi_0}^{\mathbf{f}}[h_{k+1}(X_{k+1}) \mid X_0 = x_0, \ldots, X_k = x_k] \\ &= g_k(x_k, a_k) + \sum_{z \in S} h_{k+1}(z) p(z; x_k, a_k) \leqslant h_k(x_k). \end{aligned} \quad (1.25)$$

The first inequality in (1.25) follows from (1.23), while the last inequality follows from (1.14). Hence (1.23) holds for k. Also,

$$\begin{aligned} h_k(x_k) &= g_k(x_k, f_k^*(x_k)) + \sum_{z \in S} h_{k+1}(z) p(z; x_k, f_k^*(x_k)) \\ &= E_{\pi_0}^{\mathbf{f}^{(k)}}[g_k(X_k, a_k) + h_{k+1}(X_{k+1}) \mid X_k = x_k] \\ &= E_{\pi_0}^{\mathbf{f}^{(k)}}\left[g_k(X_k, a_k) + \sum_{k'=k+1}^{N} g_{k'}(X_{k'}, a_{k'}) \mid X_k = x_k\right], \end{aligned} \quad (1.26)$$

where the last equality follows from the facts that:

(i) conditional distributions of $(X_{k+1}, \dots, X_N)$ given X_{k+1} are the same under $\pi_{0,N}^{\mathbf{f}^{(k)}}$ and $\pi_{0,N}^{\mathbf{f}^{(k+1)}}$, and

(ii) one has

$$E_{\pi_0}^{\mathbf{f}^{(k)}}\left\{E_{\pi_0}^{\mathbf{f}^{(k)}}\left[\sum_{k'=k+1}^{N} g_{k'}(X_{k'}, a_{k'}) \mid X_{k+1}\right]\middle| X_k\right\}$$

$$= E_{\pi_0}^{\mathbf{f}^{(k)}}\left\{E_{\pi_0}^{\mathbf{f}^{(k)}}\left[\sum_{k'=k+1}^{N} g_{k'}(X_{k'}, a_{k'}) \mid X_k, X_{k+1}\right]\middle| X_k\right\}$$

$$= E_{\pi_0}^{\mathbf{f}^{(k)}}\left\{\sum_{k'=k+1}^{N} g_{k'}(X_{k'}, a_{k'}) \mid X_k\right\}. \tag{1.27}$$

Hence, (1.24) holds with $k + 1$ replaced by k, and the induction argument is complete, so that (1.23), (1.24) hold for all $k = 0, 1, \dots, N - 1$. Thus, one has

$$J_{\pi_0}^{\mathbf{f}} = E_{\pi_0}^{\mathbf{f}}\left(\sum_{k'=0}^{N} g_{k'}(X_{k'}, a_{k'})\right)$$

$$= E_{\pi_0}^{\mathbf{f}}\left(\sum_{k'=0}^{k-1} g_{k'}(X_{k'}, a_{k'})\right) + E_{\pi_0}^{\mathbf{f}}\left(\sum_{k'=k}^{N} g_{k'}(X_{k'}, a_{k'})\right)$$

$$= E_{\pi_0}^{\mathbf{f}}\left(\sum_{k'=0}^{k-1} g_{k'}(X_{k'}, a_{k'})\right) + E_{\pi_0}^{\mathbf{f}}\left(E\left[\sum_{k'=k}^{N} g_{k'}(X_{k'}, a_{k'})\middle| X_k\right]\right)$$

$$\leqslant E_{\pi_0}^{\mathbf{f}}\left(\sum_{k'=0}^{k-1} g_{k'}(X_{k'}, a_{k'})\right) + E_{\pi_0}^{\mathbf{f}} h_k(X_k)$$

$$= E_{\pi_0}^{\mathbf{f}^{(k)}}\left(\sum_{k'=0}^{k-1} g_{k'}(X_{k'}, a_{k'})\right) + E_{\pi_0}^{\mathbf{f}^{(k)}} h_k(X_k)$$

$$= E_{\pi_0}^{\mathbf{f}^{(k)}}\left(\sum_{k'=0}^{N} g_{k'}(X_{k'}, a_{k'})\right) = J_{\pi_0}^{\mathbf{f}^{(k)}}. \tag{1.28}$$

Letting $k = 0$, the desired result is obtained. ■

Exactly the same proof works for the more general result below (Exercise 3).

Theorem 1.3. Assume S is countable, A compact metric, and the reward functions $a \to g_k(x, a)$ are bounded and continuous for each $x \in S$. Then $\mathbf{f}^* = (f_0^*, f_1^*, \dots, f_N^*)$ is optimal.

There are several useful extensions of the problem of optimization over a finite horizon considered in this section.

First, the set of all possible actions when in state x may depend on x. Denote this set by A_x $(x \in S)$. In this case one simply maximizes over $a \in A_y$ in (1.11)–(1.14) (Exercises 1, 2).

Secondly, instead of maximizing (rewards), one could minimize (costs). In all the results above, *max* is then replaced by *min*, and inequality signs get reversed (Exercise 1).

One may often derive existence of optimal policies when S is a continuum, g_k are unbounded, and A is noncompact, if certain special structures are guaranteed. For example, if S, A are intervals (finite or infinite) and the functions

$$a \to g_k(x, a), g_k(x, a) + \int h_{k+1}(z)p(dz; x, a) \qquad (0 \leqslant k \leqslant N-1)$$

have unique maxima in A, then all the results carry over. Here (i) $p(dz; x, a)$ is, for each pair (x, a), a probability measure on (the Borel sigmafield of) S, (ii) $h_{k+1}(z)$ is integrable with respect to $p(dz; x, a)$, and its integral is a continuous function of (x, a). See Section 5 for an interesting example.

Finally, the rewards g_k $(0 \leqslant k \leqslant N)$ may depend on the state and action in period k, *and the state in period* $k+1$. This may be reduced to the standard case considered in this section. Suppose $g_k'(x, a, x')$ is the reward in period k if the state and action in period k are x, a, and the state in period $k+1$ is x'. Then define

$$g_k(x, a) := \sum_{z \in S} g_k'(x, a, z)p(z; x, a). \tag{1.29}$$

Then for each policy $\mathbf{f}$,

$$\begin{aligned} E_{\pi_0}^{\mathbf{f}} \sum_{k=0}^{N} g_k'(X_k, a_k, X_{k+1}) &= E_{\pi_0}^{\mathbf{f}} \sum_{k=0}^{N} E(g_k'(x_k, a_k, X_{k+1}) \mid X_0, \ldots, X_k) \\ &= E_{\pi_0}^{\mathbf{f}} \sum_{k=0}^{N} g_k(X_k, a_k). \end{aligned} \tag{1.30}$$

Thus, the maximization of the first expression in (1.30) is equivalent to that of the last (see Exercise 2).

2 THE INFINITE-HORIZON PROBLEM

As in Section 1, consider finite state and action spaces S, A, unless otherwise specified. Assume that

$$\sum_{k=0}^{\infty} \left(\sup_{x \in S, a \in A} |g_k(x, a)| \right) < \infty. \tag{2.1}$$

An important example in which (2.1) is satisfied is the case of *discounted rewards* $g_k(x, a) = \delta^k g_0(x, a)$ for some *discount factor* δ, $0 < \delta < 1$. Those who are interested primarily in this special case may go directly to the *dynamic programming principle* (appearing after (2.13)) and Theorem 2.2.

In view of (2.1), given any $\varepsilon > 0$ there exists an integer N_ε such that

$$\left| \sum_{k=0}^{N_\varepsilon} g_k(x_k, a_k) - \sum_{k=0}^{\infty} g_k(x_k, a_k) \right| < \frac{\varepsilon}{2} \tag{2.2}$$

for all sequences $\{x_k\}$ in S and $\{a_k\}$ in A. Let $(f_0^{*\varepsilon}, f_1^{*\varepsilon}, \ldots, f_N^{*\varepsilon})$ be an optimal Markovian policy over the finite horizon $\{0, 1, 2, \ldots, N_\varepsilon\}$, and $\mathbf{f}^{*\varepsilon} = (f_0^{*\varepsilon}, f_1^{*\varepsilon}, \ldots)$ with $f_k^{*\varepsilon} = f_{N_\varepsilon}^{*\varepsilon}$ for all $k > N_\varepsilon$. Then $\mathbf{f}^{*\varepsilon}$ is a Markovian policy that is ε-*optimal* over the infinite horizon, i.e., given any policy $\mathbf{f}$ one has

$$\begin{aligned} J_{\pi_0}^{\mathbf{f}^{*\varepsilon}} &:= E_{\pi_0}^{\mathbf{f}^{*\varepsilon}} \sum_{k=0}^{\infty} g_k(X_k, a_k) = E_{\pi_0}^{\mathbf{f}^{*\varepsilon}}\left(\sum_{k=0}^{N_\varepsilon} g_k(X_k, a_k) \right) + E_{\pi_0}^{\mathbf{f}^{*\varepsilon}}\left(\sum_{k=N_\varepsilon+1}^{\infty} g_k(X_k, a_k) \right) \\ &\geqslant E_{\pi_0}^{\mathbf{f}}\left(\sum_{k=0}^{N_\varepsilon} g_k(X_k, a_k) \right) - \frac{\varepsilon}{2} \geqslant E_{\pi_0}^{\mathbf{f}}\left(\sum_{k=0}^{\infty} g_k(X_k, a_k) \right) - \varepsilon = J_{\pi_0}^{\mathbf{f}} - \varepsilon. \end{aligned} \tag{2.3}$$

By Cantor's diagonal procedure, one may find a sequence $\delta_n \downarrow 0$ and a sequence of functions $\{f_k^{*\delta_n}\}$ from S into A $(k = 0, 1, 2, \ldots)$ such that (Exercise 1)

$$\lim_{n \to \infty} f_k^{*\delta_n}(x) = f_k^*(x) \qquad \text{for all } k = 0, 1, 2, \ldots, \text{ and for all } x \in S, \tag{2.4}$$

for some sequence of functions f_k^*. Since S and A are finite, (2.4) implies that $f_k^{*\delta_n}(x) = f_k^*(x)$ for all sufficiently large n. From this it follows that given any integer N there exists $n(N)$ such that

$$E_{\pi_0}^{\mathbf{f}^*}\left(\sum_{k=0}^{N} g_k(X_k, a_k) \right) = E_{\pi_0}^{\mathbf{f}^{*\delta_n}}\left(\sum_{k=0}^{N} g_k(X_k, a_k) \right) \qquad \text{for } n > n(N). \tag{2.5}$$

In particular, for fixed $\varepsilon > 0$, (2.5) holds for $N = N_\varepsilon$. In view of this and (2.2), one has for all $n > n(N_\varepsilon)$.

$$J_{\pi_0}^{\mathbf{f}^*} \geqslant E_{\pi_0}^{\mathbf{f}^*}\left(\sum_{k=0}^{N_\varepsilon} g_k(X_k, a_k) \right) - \frac{\varepsilon}{2} = E_{\pi_0}^{\mathbf{f}^{*\delta_n}}\left(\sum_{k=0}^{N_\varepsilon} g_k(X_k, a_k) \right) - \frac{\varepsilon}{2} \geqslant J_{\pi_0}^{\mathbf{f}^{*\delta_n}} - \varepsilon. \tag{2.6}$$

Now apply (2.3) to this to get $J_{\pi_0}^{\mathbf{f}^*} \geqslant J_{\pi_0}^{\mathbf{f}} - \delta_n - \varepsilon$ for all $n > n(N_\varepsilon)$, and then let $n \to \infty$ to obtain

$$J_{\pi_0}^{\mathbf{f}^*} \geqslant J_{\pi_0}^{\mathbf{f}} - \varepsilon.$$

Since this is true for all ε, it follows that

$$J^{\mathbf{f}^*}_{\pi_0} \geqslant J^{\mathbf{f}}_{\pi_0} \qquad \text{for all } \mathbf{f} \in \mathbf{F}. \tag{2.7}$$

In other words, $\mathbf{f}^*$ is an optimal Markovian policy.

We have thus proved the following result.

Theorem 2.1. Under the hypothesis (2.1), there exists an optimal Markovian policy.

We now consider a special case of Theorem 2.1, the case of *discounted dynamic programming*. Assume that

$$g_k(x, a) = \delta^k g_0(x, a) \qquad (k = 0, 1, 2, \ldots), \tag{2.8}$$

where g_0 is a given real-valued function on $S \times A$, and δ is a constant satisfying

$$0 < \delta < 1. \tag{2.9}$$

A Markovian policy $\mathbf{f} = (f_0, f_1, \ldots)$ is said to be *stationary* if $f_k = f_0$ for all k. It will be shown now that for the reward (2.8) there exists a stationary optimal policy.

Let $\mathbf{f}^* = (f_0^*, f_1^*, \ldots)$ be an optimal Markovian policy in this case (which exists by Theorem 2.1). Write $E^{\mathbf{f}}_x$ for $E^{\mathbf{f}}_{\pi_0}$ and $J^{\mathbf{f}}_x$ for $J^{\mathbf{f}}_{\pi_0}$ in the case $\pi(\{x\}) = 1$. Then,

$$\begin{aligned} J^{\mathbf{f}^*}_x &= E^{\mathbf{f}^*}_x \sum_{k=0}^{\infty} \delta^k g_0(X_k, a_k) \\ &= g_0(x, f_0^*(x)) + \delta E^{\mathbf{f}^*}_x \left(E\left[\sum_{k=0}^{\infty} \delta^k g_0(X_{k+1}, a_{k+1}) \,\middle|\, X_1 \right]\right) \\ &= g_0(x, f_0^*(x)) + \delta \sum_{z \in S} J^{\mathbf{f}^{(1)}}_z p(z; x, f_0^*(x)), \end{aligned} \tag{2.10}$$

where $\mathbf{f}^{(1)} = (f_1^*, f_2^*, \ldots)$. Since $\mathbf{f}^*$ is optimal, one has

$$J^{\mathbf{f}^{(1)}}_z \leqslant J^{\mathbf{f}^*}_z \qquad \text{for all } z \in S. \tag{2.11}$$

Therefore,

$$J^{\mathbf{f}^*}_x \leqslant g_0(x, f_0^*(x)) + \delta \sum_{z \in S} J^{\mathbf{f}^*}_z p(z; x, f_0^*(x)) = J^{\mathbf{f}^{*(1)}}_x, \tag{2.12}$$

where $\mathbf{f}^{*(1)} = (f_0^*, f_0^*, f_1^*, f_2^*, \ldots)$. But $\mathbf{f}^*$ is optimal. Therefore, one must have equality in (2.12). Similarly, one shows successively that

$$J_x^{\mathbf{f}^{*(n)}} = J_x^{\mathbf{f}^{*(n-1)}} \qquad (n = 1, 2, \ldots), \qquad \mathbf{f}^{*(0)} = \mathbf{f}^* \tag{2.13}$$

where

$$\mathbf{f}^{*(n-1)} := (f_0, f_0, \ldots, f_0, f_1, f_2, \ldots),$$

with f_0 appearing in the first n positions. By (2.2) it follows that the *stationary policy* $\mathbf{f}^* = (f_0^*, f_0^*, f_0^*, \ldots, f_0^*, \ldots)$ is optimal. We shall denote it by $\mathbf{f}^*$. This proves the first part of Theorem 2.2 below. The dynamic programming equation (2.14) below may be derived by letting $k \uparrow \infty$ in (1.14) with $g_k(x, a) = \delta^k(x, a)$. However, an alternative derivation of Theorem 2.2 is given below.

Let us describe first the so-called *dynamic programming principle*. Suppose there exists a stationary optimal policy $\mathbf{f}^* = (f^*, f^*, \ldots)$ in the case of discounted rewards (2.8) with the discount factor δ satisfying (2.9). Let $J_x := J_x^{\mathbf{f}^*}$ denote the optimal reward, starting at state x. Fix $x \in S$ and $a \in A$. Consider a Markovian policy $\mathbf{f}' = (f_0, f^*, f^*, \ldots)$ for which $f_0(x) = a$, while the optimal policy is used from the next period onward. Since the conditional distribution of $\{X_k : k \geqslant 1\}$, given X_1 under $\mathbf{f}'$ is the same as that under $\mathbf{f}^*$,

$$J_x^{\mathbf{f}'} = g_0(x, a) + E_x^{\mathbf{f}'}\left[E_0^{\mathbf{f}^*}\left(\sum_{k=1}^{\infty} \delta^k g_0(X_k, f^*(x_k)) \mid X_1\right)\right] = g_0(x, a) + \delta E_x^{\mathbf{f}'} J_{X_1}$$
$$= g_0(x, a) + \delta \sum_{z \in S} J_z\, p(z; x, a).$$

The optimal reward J_x must equal the maximum of $J_x^{\mathbf{f}'}$ over all choices of $f_0(x) \equiv a \in A$. Thus, one arrives at the *dynamic programming equation* (2.14).

Theorem 2.2. Let g_0 be bounded and $0 < \delta < 1$. Then for the discounted rewards (2.8) there exists a stationary optimal policy $\mathbf{f}^* = (f^*, f^*, \ldots)$. This policy satisfies the *dynamic programming equation*, or the *Bellman equation*, for each $x \in S$, namely

$$J_x^{\mathbf{f}^*} = \max_{a \in S}\left\{g_0(x, a) + \delta \sum_{z \in S} J_x^{\mathbf{f}^*} p(z; x, a)\right\}. \tag{2.14}$$

Conversely, if a stationary Markovian policy $\mathbf{f}^*$ satisfies (2.14) for all $x \in S$, then it is optimal.

Proof. Denote by $\mathbf{B}(S)$ the set of all real-valued functions on S. For each function f on S into A, define the maps $T, T_f\colon \mathbf{B}(S) \to \mathbf{B}(S)$ by

$$(T_f J)(x) = g_0(x, f(x)) + \delta \sum_{z \in S} J(z) p(z; x, f(x)),$$

$$(TJ)(x) = \max_{a \in A} \left\{ g_0(x, a) + \delta \sum_{z \in S} J(z) p(z; x, a) \right\}, \qquad (x \in S,\ J \in \mathbf{B}(S)). \tag{2.15}$$

For all $J, J' \in \mathbf{B}(S)$ one has

$$\|T_f J - T_f J'\| = \max_{x \in S} |T_f J(x) - T_f J'(x)| = \delta \max_{x \in S} \left| \sum_{z \in S} (J(z) - J'(z)) p(z; x, f(x)) \right|$$

$$\leqslant \delta \|J - J'\|, \tag{2.16}$$

where $\|\cdot\|$ denotes "sup norm", $\|g\| := \max\{|g(x)|\colon x \in S\}$. Turning to T, fix $J \in \mathbf{B}(S)$. Let the maximum in (2.15) be attained at a point $f(x)$ $(x \in S)$. Then,

$$(TJ)(x) = (T_f J)(x). \tag{2.17}$$

Also, note that for all $J' \in \mathbf{B}(S)$ one has

$$(TJ')(x) \geqslant (T_f J')(x) \qquad (x \in S). \tag{2.18}$$

Combining (2.17) and (2.18), one gets, for all J',

$$(TJ)(x) - (TJ')(x) = (T_f J)(x) - (TJ')(x) \leqslant (T_f J)(x) - T_f J'(x). \tag{2.19}$$

By (2.16) and (2.19),

$$\sup_{x \in S} [(TJ)(x) - (TJ')(x)] \leqslant \sup_{x \in S} [(T_f J)(x) - (T_f J')(x)] \leqslant \delta \|J - J'\|. \tag{2.20}$$

This inequality being true for all $J, J' \in \mathbf{B}(S)$, one gets (interchanging the roles of J and J'),

$$\|TJ - TJ'\| \leqslant \delta \|J - J'\|. \tag{2.21}$$

Thus, T is a (uniformly strict) contraction on the space $\mathbf{B}(S)$. Therefore, it has a *unique fixed point* J^*; i.e., $TJ^* = J^*$ (Exercise 2),

$$J^*(x) = \max_{a \in A} \left\{ g_0(x, a) + \delta \sum_{z \in S} J^*(z) p(z; x, a) \right\}. \tag{2.22}$$

Let $a = f^*(x)$ maximize the right side of (2.22) $(x \in S)$. Then,

$$J^*(x) = g_0(x, f^*(x)) + \delta \sum_{z \in S} J^*(z) p(z; x, f^*(x)) = T_{f^*} J^*(x). \tag{2.23}$$

Now let $\mathbf{f}$ be an arbitrary policy. In view of (2.22) one has, for all $k = 0, 1, 2, \ldots$ (and all x),

$$J^*(X_k) \geqslant g_0(X_k, a_k) + \delta E_x^{\mathbf{f}}(J^*(X_{k+1}) \mid \{X_k, a_k\}). \tag{2.24}$$

Multiplying both sides by δ^k and taking expectations one gets

$$\delta^k E_x^{\mathbf{f}} J^*(X_k) \geqslant \delta^k E_x^{\mathbf{f}} g_0(X_k, a_k) + \delta^{k+1} E_x^{\mathbf{f}} J^*(X_{k+1}). \tag{2.25}$$

Summing (2.24) over $k = 0, 1, 2, \ldots$ and canceling common terms from both sides it follows that

$$J^*(x) \geqslant \sum_{k=0}^{\infty} \delta^k E_x^{\mathbf{f}} g_0(X_k, a_k) = J_x^{\mathbf{f}}. \tag{2.26}$$

Now note that one has equality in (2.24)–(2.26) if $\mathbf{f} = \mathbf{f}^* = (f^*, f^*, f^*, \ldots)$ (see (2.23)). Hence $\mathbf{f}^*$ is optimal. This proves that there exists a stationary optimal policy $\mathbf{f}^*$ satisfying (2.14). Conversely, if (2.14) holds (i.e., (2.22) and (2.23) hold) then (2.24)–(2.26) (with equalities in case $\mathbf{f} = \mathbf{f}^*$) show that $\mathbf{f}^*$ is optimal. ■

For the map T defined in (2.15), $(T^N 0)(x)$ is the N-stage optimal reward function for the discounted case, as may be seen from (1.11)–(1.14) and Theorem 1.2. Hence $J^*(x) = \lim_{N\to\infty}(T^N 0)(x)$ is the optimal reward function in the infinite-horizon case. The contraction property of T shows that $T^N J(x)$ converges to the optimal reward function, *no matter what J is.*

The proof of Theorem 2.2 extends word for word to the following more general case.

Theorem 2.3. Let S be countable, A a compact metric, g_0 bounded, and $a \to g_0(x, a)$ continuous for each x. Then the conclusions of Theorem 2.2 hold.

Further extensions are indicated in the theoretical complements.

Example. Let $S = \mathbb{Z}$, the set of all integers, $A = [\alpha, 1 - \alpha]$ with $0 < \alpha < \frac{1}{2}$, and

$$\begin{aligned} &p(x+1; x, a) = a, \qquad p(x-1; x, a) = 1 - a \qquad \text{for } x \neq 0, \\ &p(0; 0, a) = 1 - a, \qquad p(1; 0, a) = \frac{a}{2}, \qquad p(-1; 0, a) = \frac{a}{2}. \end{aligned} \tag{2.27}$$

Take $g_k(x, a) = \delta^k \varphi(x)$ $(0 < \delta < 1)$, where φ is bounded and even (i.e., $\varphi(x) = \varphi(-x)$), and $\varphi(|x|)$ decreases as $|x|$ increases. Then the hypothesis of Theorem 2.3 is satisfied. Under a stationary policy $\mathbf{f} = (f, f, \ldots)$ the stochastic process $\{X_k\}$ is a birth–death chain on $\mathbb{Z}$ with transition probabilities

$$\beta_x := P(X_{k+1} = x+1 \mid X_k = x) = f(x),$$
$$\delta_x := P(X_{k+1} = x-1 \mid X_k = x) = 1 - f(x) \qquad (x \neq 0),$$
$$\beta_0 := P(X_{k+1} = 1 \mid X_k = 0) = \delta_0 = P(X_{k+1} = -1 \mid X_k = 0) = \frac{f(0)}{2},$$
$$1 - \beta_0 - \delta_0 = P(X_{k+1} = 0 \mid X_k = 0) = 1 - f(0). \tag{2.28}$$

Since the maximum of φ is at zero, and $\varphi(|x|)$ decreases as $|x|$ increases, one expects the maximum (discounted) reward to be attained by a policy $\mathbf{f}^* = (f^*, f^*, \ldots)$ which assigns the highest possible probability to move in the direction of zero. In other words, an optimal policy is perhaps given by

$$f^*(x) = \begin{cases} 1-\alpha & \text{if } x < 0, \\ \alpha & \text{if } x > 0, \\ \alpha & \text{if } x = 0. \end{cases} \tag{2.29}$$

Let us check that $\mathbf{f}^*$ is indeed optimal. To simplify notation, write E_x for $E_x^{\mathbf{f}^*}$. Assume, as an *induction hypothesis*, that the following order relations hold.

$$\begin{aligned} &\text{(i) } E_x\varphi(X_k) \geqslant E_{x+1}\varphi(X_k) && \text{for all } x \geqslant 0, \\ &\text{(ii) } E_x\varphi(X_k) \leqslant E_{x+1}\varphi(X_k) && \text{for all } x \leqslant -1, \\ &\text{(iii) } E_x\varphi(X_k) = E_{-x}\varphi(X_k) && \text{for all } x. \end{aligned} \tag{2.30}$$

Clearly, (2.30) holds for $k = 0$ in view of the assumptions on φ. *Suppose it holds for some* k. Then for $x \geqslant 1$ one has, by (i),

$$\begin{aligned} E_x\varphi(X_{k+1}) &= (1-\alpha)E_{x-1}\varphi(X_k) + \alpha E_{x+1}\varphi(X_k) \geqslant (1-\alpha)E_x\varphi(X_k) + \alpha E_{x+2}\varphi(X_k) \\ &= E_{x+1}\varphi(X_{k+1}), \qquad (x \geqslant 1). \end{aligned} \tag{2.31}$$

Similarly, using (ii),

$$E_x\varphi(X_{k+1}) \leqslant E_{x+1}\varphi(X_k), \qquad (x \leqslant -1). \tag{2.32}$$

Also, (iii) is trivially true for all k if $x = 0$. For $x > 0$ one has, by (iii),

$$\begin{aligned} E_x\varphi(X_{k+1}) &= (1-\alpha)E_{x-1}\varphi(X_k) + \alpha E_{x+1}\varphi(X_k) \\ &= (1-\alpha)E_{-x+1}\varphi(X_k) + \alpha E_{-x-1}\varphi(X_k) = E_{-x}\varphi(X_{k+1}). \end{aligned} \tag{2.33}$$

Finally, using (2.31)–(2.33),

$$\begin{aligned} E_0\varphi(X_{k+1}) &= (1-\alpha)E_0\varphi(X_k) + \frac{\alpha}{2}E_{-1}\varphi(X_k) + \frac{\alpha}{2}E_1\varphi(X_k) \\ &= (1-\alpha)E_0\varphi(X_k) + \alpha E_1\varphi(X_k) \\ &\geqslant (1-\alpha)E_0\varphi(X_k) + \alpha E_2\varphi(X_k) = E_1\varphi(X_{k+1}). \end{aligned} \tag{2.34}$$

By (2.31) and (2.34), the relation (2.30(i)) holds with k replaced by $k+1$. By (2.32), (2.30(ii)) holds for $k+1$; in view of (2.33), (2.30(iii)) holds for $k+1$. Hence the relations (2.30) hold for all $k \geqslant 0$. To complete the proof that $\mathbf{f}^*$ is optimal, let us now show that the expected discounted reward J_x under $\mathbf{f}^*$ given by

$$J_x := \sum_{k=0}^{\infty} \delta^k E_x\varphi(X_k) \tag{2.35}$$

satisfies the dynamic programming equation (2.14). That is, one needs to show

$$\begin{aligned} J_x &= \varphi(x) + \delta \max_{a\in A}[(1-a)J_{x-1} + aJ_{x+1}] \qquad (x \neq 0), \\ J_0 &= \varphi(0) + \delta \max_{a\in A}\left[(1-a)J_0 + \frac{a}{2}J_{-1} + \frac{a}{2}J_1\right]. \end{aligned} \tag{2.36}$$

Now, by (2.30(i)),

$$J_{x-1} \geqslant J_{x+1} \qquad \text{for } x \geqslant 1,$$

so that, assigning the largest possible weight to the larger quantity J_{x-1},

$$\max_{a\in A}[(1-a)J_{x-1} + aJ_{x+1}] = (1-\alpha)J_{x-1} + \alpha J_{x+1}, \qquad (x \geqslant 1). \tag{2.37}$$

But J_x satisfies the equation

$$\begin{aligned} J_x &= \varphi(x) + \delta \sum_{k=1}^{\infty} \delta^{k-1}E_x\varphi(X_k) \\ &= \varphi(x) + \delta \sum_{k=1}^{\infty} \delta^{k-1}[(1-\alpha)E_{x-1}\varphi(X_{k-1}) + \alpha E_{x+1}\varphi(X_{k-1})] \\ &= \varphi(x) + \delta[(1-\alpha)J_{x-1} + \alpha J_{x+1}], \qquad (x \geqslant 1). \end{aligned} \tag{2.38}$$

Hence, J_x satisfies (2.36) for $x \geqslant 1$. In exactly the same way one shows, using (2.30(ii)), that J_x satisfies (2.36) for $x \leqslant -1$. Finally, since $J_{-1} = J_1$ by (2.30(iii)), and $J_0 \geqslant J_1 \ (= J_{-1})$,

$$\max_{a \in A} \left[(1-a)J_0 + \frac{a}{2} J_{-1} + \frac{a}{2} J_1 \right] = (1-\alpha)J_0 + \frac{\alpha}{2} J_{-1} + \frac{\alpha}{2} J_1, \qquad (2.39)$$

assigning the largest possible weight to J_0 for the last equality. The proof of (2.36), and therefore of the optimality of $\mathbf{f}^*$, is now complete.

In general, even for the infinite-horizon discounted case, it is difficult to compute explicitly optimal policies and optimal rewards. The approximations $T^N 0$, and the corresponding N-period optimal policy given by backward recursion (1.11)–(1.14), are therefore useful.

3 OPTIMAL CONTROL OF DIFFUSIONS

Let G be an open set $\subset \mathbb{R}^k$, and consider the problem of minimizing the expected penalty $E_{\mathbf{x}} f(\mathbf{X}_{\tau_{\partial G}})$ on reaching the boundary ∂G, starting from a point $\mathbf{x}$ outside of $\bar{G} := G \cup \partial G$. Here $\{\mathbf{X}_t\}$ is a diffusion on $\mathbb{R}^k \setminus G$, with absorption at ∂G. The function f is a continuous bounded function on ∂G. If $f \equiv 1$ on ∂G, the problem is to minimize the probability of ever reaching ∂G. As usual $\tau_{\partial G}$ denotes the first time to reach the boundary ∂G.

The drift vector $\boldsymbol{\mu}(\mathbf{x}, c)$ and diffusion matrix $\mathbf{D}(\mathbf{x}, c)$ of the diffusion are specified. The control variable c, which may depend on the present state $\mathbf{x}$, takes values in some compact set $C \subset \mathbb{R}^m$. The function $c(\cdot)$ (on $\mathbb{R}^k \setminus G$ into C) so chosen is called a *control*, or a *feedback control*. The class of allowable, or *feasible*, controls may vary. For example, it may be the class of all measurable functions, the class of all continuous functions, the class of all continuously differentiable functions, etc.

Let us give a heuristic derivation of the dynamic programming equation, by means of the so-called *dynamic programming principle*. Write $E_{\mathbf{x}}^c$ for the expectation when the diffusion has coefficients $\boldsymbol{\mu}(\cdot, c)$, $\mathbf{D}(\cdot, c)$ with c *constant*, and $\mathbf{x}$ is the initial state. If $c(\cdot)$ is a control (function), this expectation is denoted $E_{\mathbf{x}}^{c(\cdot)}$, the diffusion having drift and diffusion coefficients $\boldsymbol{\mu}(\mathbf{y}, c(\mathbf{y}))$, $\mathbf{D}(\mathbf{y}, c(\mathbf{y}))$. Write

$$J^c(\mathbf{x}) := E_{\mathbf{x}}^c f(\mathbf{X}_{\tau_{\partial G}}), \qquad J^{c(\cdot)}(\mathbf{x}) := E_{\mathbf{x}}^{c(\cdot)} f(\mathbf{X}_{\tau_{\partial G}}), \qquad J(\mathbf{x}) := \inf_{c(\cdot)} J^{c(\cdot)}(\mathbf{x}), \quad (3.1)$$

where the infimum is over the class of all feasible controls. Suppose there exists an *optimal control* $c^*(\cdot)$. Consider the following small perturbation $c_1(\cdot)$ from the optimal policy. $c_1(\cdot)$ takes the constant value c for an initial period $[0, t]$, after which it switches to the optimal control $c^*(\cdot)$. This control is of course time-dependent and not feasible under our description above. But one may actually allow time-dependent controls in the present development (theoretical complement 1). One has, by the Markov property,

$$J^{c_1(\cdot)}(\mathbf{x}) = E_{\mathbf{x}}^{c} J(\mathbf{X}_t) + \mathrm{o}(t), \qquad \text{as } t \downarrow 0. \tag{3.2}$$

Note that if after time t the state is $\mathbf{X}_t \in \mathbb{R}^k \setminus \bar{G}$, then the conditional expectation of $f(X_{\tau_{\partial G}})$ is $J(\mathbf{X}_t)$, as the optimal control $c^*(\cdot)$ is now in effect. Also, the probability that $t \geqslant \tau_{\partial G}$ is $\mathrm{o}(t)$ for $\mathbf{x} \in \mathbb{R}^k \setminus \bar{G}$ (see Exercise 1 for an indication of this). As $J^{c_1(\cdot)}(\mathbf{x}) \geqslant J(\mathbf{x})$, one gets

$$E_{\mathbf{x}}^{c} J(\mathbf{X}_t) - J(\mathbf{x}) \geqslant \mathrm{o}(t), \qquad \text{as } t \downarrow 0. \tag{3.3}$$

But the left side is $\mathbf{T}_t^c J(\mathbf{x}) - J(\mathbf{x})$, where $\{\mathbf{T}_t^c\}$ is the semigroup of transition operators generated by

$$\mathbf{A}_c := \frac{1}{2} \sum_{i,j} d_{ij}(\mathbf{x}, c) \frac{\partial^2}{\partial x^{(i)} \partial x^{(j)}} + \sum_i \mu^{(i)}(\mathbf{x}, c) \frac{\partial}{\partial x^{(i)}} \tag{3.4}$$

with absorbing boundary condition on ∂G. Hence, dividing both sides of (3.3) by t and letting $t \downarrow 0$, one gets

$$\mathbf{A}_c J(\mathbf{x}) \geqslant 0. \tag{3.5}$$

By minimizing the left side of (3.3) with respect to c, one expects an equality in (3.3) and, therefore, in (3.5); i.e., the *dynamic programming equation* is

$$\inf_{c \in C} \mathbf{A}_c J(\mathbf{x}) = 0 \qquad \text{for } \mathbf{x} \in \mathbb{R}^k \setminus \bar{G}, \qquad J(\mathbf{x}) = f(x) \qquad \text{for } \mathbf{x} \in \partial G. \tag{3.6}$$

We will illustrate the use of this rigorously in Example 1 below. Before this is undertaken, a simple proposition in the case $k = 1, f \equiv 1$, directly shows how to minimize the expected penalty when the diffusion coefficient is fixed. Let $\mu^{(i)}(\cdot)$ $(i = 1, 2)$ be two drift functions, $\sigma^2(\cdot) > 0$ a diffusion coefficient specified on $(0, \infty)$, satisfying appropriate smoothness conditions. Let $\psi^{(i)}(z)$ denote the probability that a diffusion with coefficients $\mu^{(i)}(\cdot), \sigma^2(\cdot)$ reaches 0 before reaching d, starting at z. Here $d > 0$ is a fixed number.

It is intuitively clear that, for the same diffusion coefficient $\sigma^2(\cdot)$, larger the drift $\mu(\cdot)$ better the chance of staying away from zero. Here is a proof.

Proposition 3.1. Consider two diffusions on $[0, \infty)$, with absorption at 0, having a common diffusion coefficient $\sigma^2(\cdot)$ but different drifts $\mu^{(i)}(\cdot)$ $(i = 1, 2)$ satisfying $\mu^{(1)}(z) \leqslant \mu^{(2)}(z)$ for every $z > 0$. Then $\psi^{(2)}(z) \leqslant \psi^{(1)}(z)$ for all $z > 0$.

Proof. By relation (9.19) in Chapter V,

$$\psi^{(i)}(z) = \int_z^d \exp\left\{-\int_0^u \frac{2\mu^{(i)}(y)}{\sigma^2(y)}\right\} du \bigg/ \int_0^d \exp\left\{-\int_0^u \frac{2\mu^{(i)}(y)}{\sigma^2(y)}\,dy\right\} du. \tag{3.7}$$

Write

$$\eta(z) := \mu^{(2)}(z) - \mu^{(1)}(z), \qquad \mu_\varepsilon(z) := \mu^{(1)}(z) + \varepsilon\eta(z),$$

and let $F(\varepsilon; z)$ denote the probability of reaching 0 before d starting at z, for a diffusion with drift coefficient $\mu_\varepsilon(\cdot)$ and diffusion coefficient $\sigma^2(\cdot)$. It is straightforward to check (Exercise 2) that

$$F(\varepsilon; z) = \psi^{(1)}(z)(1 - \varepsilon\gamma(z)) + O(\varepsilon^2) \qquad \text{as } \varepsilon \to 0, \tag{3.8}$$

where

$$\gamma(z) := \int_z^d \exp\left\{-\int_0^u \frac{2\mu^{(1)}(y)}{\sigma^2(y)}\,dy\right\}\left(\int_0^u \frac{2\eta(y)}{\sigma^2(y)}\,dy\right)du \Big/ \int_z^d \exp\left\{-\int_0^u \frac{2\mu^{(1)}(y)}{\sigma^2(y)}\,dy\right\}du. \tag{3.9}$$

Since $\eta(z) \geqslant 0$ for all z, it follows that

$$\gamma(z) > 0 \tag{3.10}$$

unless $\mu^{(1)}(\cdot) \equiv \mu^{(2)}(\cdot)$. Therefore, barring the case of identical drifts, one has

$$\frac{d}{d\varepsilon} F(\varepsilon; z)\bigg|_{\varepsilon=0} = -\psi^{(1)}(z)\gamma(z) < 0. \tag{3.11}$$

Hence, $F(\varepsilon; z)$ is strictly decreasing in ε in a neighborhood of $\varepsilon = 0$, say on $[0, \varepsilon_0)$. Since one can continue beyond ε_0, by replacing $\mu^{(1)}(\cdot)$ by $\mu_{\varepsilon_0}(\cdot)$, the supremum of all such ε_0 is ∞. In particular, one can take $\varepsilon_0 = 1$. ■

Example 1. (*Survival Under Uncertainty*). Consider first the following discrete-time model, in which h denotes the length of time between successive periods. Let Z_t denote an agent's capital at time t. His capital at time $t + h$ is given by

$$\begin{aligned} Z_0 &= z > 0, \\ Z_{t+h} &= \exp\{W_{t+h}\}(Z_t - ah), \qquad (t = nh;\ n = 0, 1, \ldots) \end{aligned} \tag{3.12}$$

where a is a positive constant and $\{W_{nh}: n = 1, 2, \ldots\}$ is a sequence of i.i.d. Gaussian random variables with mean mh and variance vh. The constant a denotes a fixed consumption rate per unit of time, so that $Z_t - ah$ is the amount available for investment in the next period. The random exponential term represents the uncertain rate of return. For the Markov process $\{Z_{nh}\}$, it is easy to check (Exercise 3) that

$$\begin{aligned}
E(Z_{t+h} \mid Z_t) &= \exp\{mh + \tfrac{1}{2}vh\}(Z_t - ah),\\
E(Z_{t+h}^2 \mid Z_t) &= \exp\{2mh + 2vh\}(Z_t - ah)^2,\\
E(Z_{t+h} - Z_t \mid Z_t) &= [(m + \tfrac{1}{2}v)Z_t - a]h + o(h),\\
E((Z_{t+h} - Z_t)^2 \mid Z_t) &= vZ_t^2 h + o(h),\\
E(|Z_{t+h} - Z_t|^3 \mid Z_t) &= o(h), \qquad \text{as } h \downarrow 0.
\end{aligned} \tag{3.13}$$

Therefore, one may, in the limit as $h \downarrow 0$, model Z_t by a diffusion with drift $\mu(\cdot)$ and diffusion coefficient $\sigma^2(\cdot)$ given by

$$\mu(z) = (m + \tfrac{1}{2}v)z - a, \qquad \sigma^2(z) = vz^2. \tag{3.14}$$

The state space of this diffusion should be taken to be $[0, \infty)$, since negative capital is not allowed. One takes "0" as an absorbing state, which when reached indicates the agent's economic ruin.

In the discrete-time model, suppose that the agent is free to choose (m, v) for the next period depending on the capital in the current period. This choice is restricted to a set $C \subset \mathbb{R}^1 \times (0, \infty)$. In the diffusion approximation, this amounts to a choice of drift and diffusion coefficients,

$$\mu(z) = (m(z) + \tfrac{1}{2}v(z))z - a, \qquad \sigma^2(z) = v(z)z^2, \tag{3.15}$$

such that

$$(m(z), v(z)) \in C \qquad \text{for every } z \in (0, \infty). \tag{3.16}$$

The object is to choose $m(z)$, $v(z)$ in such a way as to minimize the probability of ruin. The following assumptions on C are needed:

$$\begin{aligned}
&\text{(i) } 0 < v_* := \inf\{v: (m, v) \in C\} \leq v^* := \sup\{v: (m, v) \in C\} < \infty,\\
&\text{(ii) } f(v) := \sup\{m: (m, v) \in C\} < \infty, \text{ and } (f(v), v) \in C \text{ for each } v \in [v_*, v^*],\\
&\text{(iii) } v \to f(v) \textit{ is twice continuously differentiable and concave on } (v_*, v^*),\\
&\text{(iv) } \lim_{v \downarrow v_*} f'(v) = \infty, \ \lim_{v \uparrow v^*} f'(v) = -\infty.
\end{aligned} \tag{3.17}$$

First fix a $d > 0$. Suppose that the agent wants *to quit while ahead*, i.e., when a capital d is reached. The goal is to maximize the probability of reaching d before 0. This is equivalent to minimizing the probability $\psi(z)$ of reaching 0 before d, starting at z.

It follows from Proposition 3.1 that an optimal choice of $(m(z), v(z))$ should be of the form

$$(m(z) = f(v(z)), v(z)). \tag{3.18}$$

The problem of optimization has now been reduced to a choice of $v(z)$. The dynamic programming equation for this choice is contained in the following proposition. To state it define, for each constant $v \in [v_*, v^*]$, the infinitesimal generator $\mathbf{A}_v$ by

$$(\mathbf{A}_v g)(z) := \tfrac{1}{2}vz^2 g''(z) + \{(f(v) + \tfrac{1}{2}v)z - a\}g'(z). \tag{3.19}$$

For a given measurable function $v(\cdot)$ on $(0, \infty)$ into $[v_*, v^*]$ define the infinitesimal generator $\mathbf{A}_{v(\cdot)}$ by

$$(\mathbf{A}_{v(\cdot)} g)(z) := \tfrac{1}{2}v(z)z^2 g''(z) + \{(f(v(z)) + \tfrac{1}{2}v(z))z - a\}g'(z). \tag{3.20}$$

Fix $d > 0$. Let $\psi_{v(\cdot)}(z)$ denote the probability that a diffusion having generator $\mathbf{A}_{v(\cdot)}$, starting at $z \in [0, d]$, reaches 0 before d. For simplicity consider $v(\cdot)$ to be differentiable, although the arguments are valid for all measurable $v(\cdot)$ (see theoretical complement 2). Write

$$\bar{\psi}(z) := \inf_{v(\cdot)} \psi_{v(\cdot)}(z). \tag{3.21}$$

Proposition 3.2. Assume the hypothesis (3.17) on C. Then the dynamic programming equation

$$\min_{v \in [v_*, v^*]} \mathbf{A}_v \psi(z) = 0 \qquad \text{for } 0 < z < d; \qquad \lim_{z \downarrow 0} \psi(z) = 1, \qquad \lim_{z \uparrow d} \psi(z) = 0, \tag{3.22}$$

has a solution ψ that is the minimal probability of ruin $\bar{\psi}$. For each $z \in (0, d)$ there exists a unique $\bar{v}(z) \in [v_*, v^*]$ such that the minimum is attained at $v = \bar{v}(z)$. The function $\bar{v}$ is strictly decreasing on $(0, \infty)$ and is optimal.

Proof. In the case $v_* = v^*$, Proposition 3.1 already provides the optimum choice. We therefore assume $v_* < v^*$. Let $\psi(z)$ be a twice continuously differentiable function satisfying (3.22), and $\bar{v}(\cdot)$ a (measurable) function such that the minimum in (3.22) is attained at $v = \bar{v}(z)$. Then,

$$\mathbf{A}_{\bar{v}(\cdot)} \psi(z) = 0 \qquad \text{for } 0 < z < d; \qquad \lim_{z \downarrow 0} \psi(z) = 1, \qquad \lim_{z \uparrow d} \psi(z) = 0. \tag{3.23}$$

Then (see (3.20)),

$$z^2 \psi''(z) = -\frac{2}{\bar{v}(z)} \{(f(\bar{v}(z)) + \tfrac{1}{2}\bar{v}(z))z - a\}\psi'(z). \tag{3.24}$$

Using this in (3.22), one gets

$$\min_{v \in [v_*, v^*]} [-v\{(f(\bar{v}(z)) + \tfrac{1}{2}\bar{v}(z))z - a\}/\bar{v}(z) + (f(v) + \tfrac{1}{2}v)z - a]\psi'(z) = 0. \tag{3.25}$$

It follows from (3.23), since its solution is unique (Exercise 7.7(iii), (iv) of Chapter V), that $\psi(z)$ is the probability of reaching zero before d, starting at z, for the diffusion generated by $\mathbf{A}_{\bar{v}(\cdot)}$. In particular, $\psi'(z) < 0$. One may actually explicitly calculate $\psi'(z)$ from (9.19) in Chapter V. Therefore, to determine $\bar{v}(z)$ we find the critical point(s) of the expression within the square brackets in (3.25) by setting its derivative equal to zero. This leads to

$$\frac{(f(\bar{v}(z)) + \frac{1}{2}\bar{v}(z))z - a}{\bar{v}(z)} = (f'(v) + \tfrac{1}{2})z, \tag{3.26}$$

or,

$$f(\bar{v}(z)) - \bar{v}(z)f'(v) = \frac{a}{z}.$$

Since this must hold for $v = \bar{v}(z)$, the equation to solve is

$$f(v) - vf'(v) = \frac{a}{z}. \tag{3.27}$$

As the derivative of the left side is $-vf''(v) > 0$, the left side increases from $-\infty$ to $+\infty$ as v increases from v_* to v^*. Therefore, for each $z > 0$ there exists a unique solution of (3.27), say $v = \bar{v}(z) \in (v_*, v^*)$. Since the right side of (3.27) decreases as z increases, it follows that the function $\bar{v}(\cdot)$ is strictly decreasing.

It remains to show that ψ is minimal. For this, consider an arbitrary (measurable) choice $v(\cdot)$. It follows from (3.22) that

$$(\mathbf{A}_{v(\cdot)}\psi)(z) \geq 0 \qquad \text{for all } z \in (0, d). \tag{3.28}$$

Since $(\mathbf{A}_{v(\cdot)}\psi_{v(\cdot)})(z) = 0$ and ψ and $\psi_{v(\cdot)}$ satisfy the same boundary conditions, letting $u(z) := \psi(z) - \psi_{v(\cdot)}(z)$, one gets

$$(\mathbf{A}_{v(\cdot)}u)(z) \geq 0 \qquad \text{for all } z \in (0, d), \qquad \lim_{z \downarrow 0} u(z) = 0, \qquad \lim_{z \uparrow d} u(z) = 0. \tag{3.29}$$

By the maximum principle (see Exercise 7.7(iii) of Chapter V) it follows that $u(z) \leq 0$ for all z, i.e.,

$$\psi(z) \leq \psi_{v(\cdot)}(z) \qquad \text{for all } z \in [0, d]. \tag{3.30}$$

■

It is important to note that the optimal choice

$$\bar{\mu}(y) := (f(\bar{v}(y)) + \tfrac{1}{2}\bar{v}(y))y - a, \bar{v}(y), \tag{3.31}$$

does not depend on d. Therefore, with this choice one also minimizes the probability ρ_{z0} of ever reaching 0, starting from z. For this, simply let $d \uparrow \infty$ in the inequality

$$\psi_{\bar{v}(\cdot)}(z) \leqslant \psi_{v(\cdot)}(z). \tag{3.32}$$

An example of a C satisfying (3.17) is the following set bounded by the lines $v = v_0 \pm \beta$, and a semi-elliptical cap (bounding m away from $+\infty$),

$$C = \left\{(m, v)\colon v_0 - \beta \leqslant v \leqslant v_0 + \beta, m \leqslant m_0 + \alpha\left(1 - \frac{(v - v_0)^2}{\beta^2}\right)^{1/2}\right\}. \tag{3.33}$$

Here α, β,v_0 are positive constants, $v_0 > \beta$, and m_0 is an arbitrary constant.

Next consider the problem of maximizing the expected discounted reward

$$J^{c(\cdot)}(\mathbf{x}) := E_{\mathbf{x}}^{c(\cdot)} \int_0^\infty e^{-\beta s} r(\mathbf{X}_s, c(\mathbf{X}_s))\, ds, \tag{3.34}$$

where the state space of the diffusion $\{\mathbf{X}_t\}$ is an open set $G \subset \mathbb{R}^k$. The diffusion has drift $\boldsymbol{\mu}(\mathbf{y}, c(\mathbf{y}))$ and diffusion matrix $\mathbf{D}(\mathbf{y}, c(\mathbf{y}))$; $r(\mathbf{y}, c)$ is the *reward rate* when in state $\mathbf{y}$ and an *action* c belonging to a compact set $C \subset \mathbb{R}^m$ is taken, and $\beta > 0$ is the *discount rate*.

The given functions $\mu^{(i)}(\mathbf{y}, c)$ $(1 \leqslant i \leqslant k)$ are appropriately smooth on $G \times C$, as are $d_{ij}(\mathbf{y}, c)$ $(1 \leqslant i, j \leqslant k)$, and $D(\mathbf{y}, c)$ is positive definite. Once again, the set of feasible controls $c(\cdot)$ may vary. For example, it may be the class of all continuously differentiable functions on G into C. The function $r(\mathbf{y}, c)$ is continuous and bounded on $G \times C$.

If, under some feasible control $c(\cdot)$, the diffusion $\{\mathbf{X}_t\}$ can reach ∂G with positive probability, then in (3.34) the upper limit of integration should be replaced by $\tau_{\partial G}$. Let

$$J(\mathbf{x}) := \sup_{c(\cdot)} J^{c(\cdot)}(\mathbf{x}). \tag{3.35}$$

In order to derive the dynamic programming equation informally, let $c^*(\cdot)$ be an optimal control. Consider a control $c_1(\cdot)$ which, starting at $\mathbf{x}$, takes the action c initially over the period $[0, t]$ and from then on uses the optimal control $c^*(\cdot)$. Then, as in the derivation of (3.2), (3.3),

$$\begin{aligned}
J^{c_1(\cdot)}(\mathbf{x}) &= E_{\mathbf{x}}^c \int_0^t e^{-\beta s} r(\mathbf{X}_s, c)\, ds + E_{\mathbf{x}}^{c(\cdot)} \int_t^\infty e^{-\beta s} r(\mathbf{X}_s, c(\mathbf{X}_s))\, ds \\
&= E_{\mathbf{x}}^c \int_0^t e^{-\beta s} r(\mathbf{X}_s, c)\, ds + E_{\mathbf{x}}^{c(\cdot)} e^{-\beta t} \int_0^\infty e^{-\beta s'} r(\mathbf{X}_{t+s'}, c(\mathbf{X}_{t+s'}))\, ds' \\
&= E_{\mathbf{x}}^c \int_0^t e^{-\beta s} r(\mathbf{X}_s, c)\, ds + e^{-\beta t} E_{\mathbf{x}}^c J(\mathbf{X}_t) \\
&= t(r(\mathbf{x}, c)) + o(t) + e^{-\beta t} \mathbf{T}_t^c J(\mathbf{x}) \leqslant J(\mathbf{x}),
\end{aligned}$$

or,

$$\frac{(1 - e^{-\beta t})}{t} J(\mathbf{x}) \geqslant r(\mathbf{x}, c) + e^{-\beta t} \frac{\mathbf{T}_t^c J(\mathbf{x}) - J(\mathbf{x})}{t} + o(1). \tag{3.36}$$

Letting $t \downarrow 0$ in (3.36) one arrives at

$$\beta J(\mathbf{x}) \geqslant r(\mathbf{x}, c) + \mathbf{A}_c J(\mathbf{x}). \tag{3.37}$$

One expects equality if the right side of (3.37) (or (3.36)) is maximized with respect to c, leading to the *dynamic programming equation*

$$\beta J(\mathbf{x}) = \sup_{c \in C} \{r(\mathbf{x}, c) + \mathbf{A}_c J(\mathbf{x})\}. \tag{3.38}$$

In the example below, $k = 1$.

Example 2. The state space is $G = (0, \infty)$. Let c denote the consumption rate of an economic agent, i.e., the "fraction" of stock consumed per unit time. The *utility*, or reward rate, is

$$r(x, c) = \frac{(cx)^\gamma}{\gamma}, \tag{3.39}$$

where γ is a constant, $0 < \gamma < 1$. The stock X_t corresponding to a constant control c is a diffusion with drift and diffusion coefficients

$$\mu(x, c) = \delta x - cx, \qquad \sigma^2(x, c) = \sigma^2 x^2, \tag{3.40}$$

where $\delta > 1$ is a given constant representing growth rate of capital, while c is the depletion rate due to consumption. For any given feasible control $c(\cdot)$ one replaces c by $c(x)$ in (3.40), giving rise to a corresponding diffusion $\{X_t\}$. It is simple to check that such a diffusion never reaches the boundary (Exercise 4).

The dynamic programming equation (3.38) becomes

$$\beta J(x) = \max_{c \in [0,1]} \left\{ \frac{c^\gamma}{\gamma} x^\gamma + \tfrac{1}{2}\sigma^2 x^2 J''(x) + (\delta x - cx) J'(x) \right\}. \tag{3.41}$$

Let us for the moment assume that the maximum in (3.41) is attained in the interior so that, differentiating the expression within braces with respect to c one gets

$$c^{\gamma - 1} x^\gamma = x J'(x),$$

or,

$$c^*(x) = \frac{1}{x} (J'(x))^{1/(\gamma - 1)}. \tag{3.42}$$

Since the function within braces in (3.41) is strictly *concave* it follows that (3.42) is the unique maximum, provided it lies in (0, 1). Substituting (3.42) in (3.41) one gets

$$\tfrac{1}{2}\sigma^2 x^2 J''(x) + \delta x J'(x) - \left(1 - \frac{1}{\gamma}\right)(J'(x))^{\gamma/(\gamma-1)} - \beta J(x) = 0. \tag{3.43}$$

Try the "trial solution"

$$J(x) = dx^{\gamma}. \tag{3.44}$$

Then (3.43) becomes, as x^{γ} is a factor in (3.43),

$$\tfrac{1}{2}\sigma^2 d\gamma(\gamma - 1) + d\delta\gamma - \left(1 - \frac{1}{\gamma}\right)(d\gamma)^{\gamma/(\gamma-1)} - \beta d = 0, \tag{3.45}$$

i.e.,

$$d = \frac{1}{\gamma}\left[\frac{1-\gamma}{\beta + \frac{1}{2}\sigma^2\gamma(1-\gamma) - \delta\gamma}\right]^{1-\gamma}. \tag{3.46}$$

Hence, from (3.42), (3.44), and (3.46),

$$c^* := c^*(x) = (\gamma d)^{-1/(1-\gamma)} = \frac{\beta + \frac{1}{2}\sigma^2\gamma(1-\gamma) - \delta\gamma}{1-\gamma}. \tag{3.47}$$

Thus $c^*(x)$ is *independent of* x. Of course one needs to assume here that this constant is positive, since a zero value leads to the *minimum* expected discounted reward zero. Hence,

$$\beta + \tfrac{1}{2}\sigma^2\gamma(1-\gamma) - \delta\gamma > 0. \tag{3.48}$$

For *feasibility*, one must also require that $c^* \leqslant 1$; i.e., if the expression on the extreme right in (3.47) is greater than 1, then the maximum in (3.41) is attained always at $c^* = 1$ (Exercise 5). Therefore,

$$c^* = \max\{1, d'\}, \tag{3.49}$$

where d' is the extreme right side of (3.47). We have arrived at the following.

Proposition 3.3. Assume $\delta > 1$ and (3.48). Then the control $c^*(\cdot) \equiv \min\{1, d'\}$ is optimal in the class of all continuously differentiable controls.

Proof. It has been shown above that $c^*(\cdot)$ is the unique solution to the dynamic programming equation (3.38) with $J = J^{c^*}$. Let $c(\cdot)$ be any continuously differentiable control. Then it is simple to check (see Exercise 6) that

$$\beta J^{c(\cdot)}(x) = r(x, c(x)) + \mathbf{A}_{c(\cdot)} J^{c(\cdot)}(x). \tag{3.50}$$

Using (3.38) with $J = J^{c^*}$, one then has for the function

$$h(x) := J(x) - J^{c(\cdot)}(x), \tag{3.51}$$

the inequality

$$\begin{aligned}\beta h(x) &= r(x, c^*) + \mathbf{A}_{c^*} J(x) - \{r(x, c(x)) + \mathbf{A}_{c(\cdot)} J^{c(\cdot)}(x)\} \\ &\geqslant r(x, c(x)) + \mathbf{A}_{c(\cdot)} J(x) - \{r(x, c(x)) + \mathbf{A}_{c(\cdot)} J^{c(\cdot)}(x)\} = \mathbf{A}_{c(\cdot)} h(x),\end{aligned} \tag{3.52}$$

or,

$$g(x) := (\beta - \mathbf{A}_{c(\cdot)}) h(x) \geqslant 0. \tag{3.53}$$

Since the nonnegative function

$$h_1(x) := E_x^{c(\cdot)}\left(\int_0^\infty e^{-\beta s} g(X_s)\, ds\right), \tag{3.54}$$

satisfies the equation (Exercise 6)

$$(\beta - \mathbf{A}_{c(\cdot)}) h_1(x) = g(x), \tag{3.55}$$

one has, assuming *uniqueness* of the solution to (3.55) (see theoretical complement 3), $h(x) = h_1(x)$. Therefore, $h(x) \geqslant 0$. ■

Under the optimal policy c^* the growth of the capital (or stock) X_t is that of a diffusion on $(0, \infty)$ with drift and diffusion coefficients

$$\mu(x) := (\delta - c^*)x, \qquad \sigma^2(x) := \sigma^2 x^2. \tag{3.56}$$

By a change of variables (see Example 3.1 of Chapter V), the process $\{Y_t := \log X_t\}$ is a Brownian motion on $\mathbb{R}^1$ with drift $\delta - c^* - \frac{1}{2}\sigma^2$ and diffusion coefficient σ^2. Thus, $\{X_t\}$ is a geometric Brownian motion.

4 OPTIMAL STOPPING AND THE SECRETARY PROBLEM

On a probability space $(\Omega, \mathscr{F}, P)$ an increasing sequence $\{\mathscr{F}_n: n \geqslant 0\}$ of sub-sigmafields of $\mathscr{F}$ are given. For example, one may have $\mathscr{F}_n = \sigma\{X_0, X_1, \ldots, X_n\}$, where $\{X_n\}$ is a sequence of random variables. Also given are real-valued *integrable* random variables $\{Y_n: n \geqslant 0\}$, Y_n being $\mathscr{F}_n$-measurable. The objective is to find a $\{\mathscr{F}_n\}$-stopping time τ_m^* that *minimizes* EY_τ in the class $\mathscr{T}_m$ of all $\{\mathscr{F}_n\}$-stopping times $\tau \leqslant m$. One may think of Y_n as the *loss incurred by stopping at time n*, and m as the maximum number of observations allowed.

This problem is solved by *backward recursion* in much the same way as the finite-horizon dynamic programming problem was solved in Section 1.

We first give a somewhat heuristic derivation of τ_m^*. If $\mathscr{F}_n = \sigma\{X_0, \dots, X_n\}$ $(n \geq 0)$, and $X_0, X_1, \dots, X_{m-1}$ have been observed, then by stopping at time $m-1$ the loss incurred would be Y_{m-1}. On the other hand, if one decided to continue sampling then the loss would be Y_m. But Y_m is not known yet, since X_m has not been observed at the time the decision is made to stop or not to stop sampling. The (conditional) expected value of Y_m, given $X_0, \dots, X_{m-1}$, must then be compared to Y_{m-1}. In other words, τ_m^* is given, *on the set* $\{\tau_m^* \geq m-1\}$, by

$$\tau_m^* = \begin{cases} m-1 & \text{if } Y_{m-1} \leq E(Y_m \mid \mathscr{F}_{m-1}), \\ m & \text{if } Y_{m-1} > E(Y_m \mid \mathscr{F}_{m-1}). \end{cases} \tag{4.1}$$

As a consequence of such a stopping rule, one's expected loss, given $\{X_0, \dots, X_{m-1}\}$, is

$$V_{m-1} := \min\{Y_{m-1}, E(Y_m \mid \mathscr{F}_{m-1})\} \qquad \text{on } \{\tau_m^* \geq m-1\}. \tag{4.2}$$

Similarly, suppose one has already observed $X_0, X_1, \dots, X_{m-2}$ (so that $\tau_m^* \geq m-2$). Then one should continue sampling only if Y_{m-2} is greater than the conditional expectation (given $\{X_0, \dots, X_{m-2}\}$) of the loss that would result from continued sampling. That is,

$$\tau_m^* \begin{cases} = m-2 & \text{if } Y_{m-2} \leq E(V_{m-1} \mid \mathscr{F}_{m-2}), \\ \geq m-1 & \text{if } Y_{m-2} > E(V_{m-1} \mid \mathscr{F}_{m-2}) \text{ on } \{\tau_m^* \geq m-2\}. \end{cases} \tag{4.3}$$

The conditional expected loss, given $\{X_0, \dots, X_{m-2}\}$, is then

$$V_{m-2} := \min\{Y_{m-2}, E(V_{m-1} \mid \mathscr{F}_{m-2})\} \qquad \text{on } \{\tau_m^* \geq m-2\}. \tag{4.4}$$

Proceeding backward in this manner one finally arrives at

$$\tau_m^* \begin{cases} = 0 & \text{if } Y_0 \leq E(V_1 \mid \mathscr{F}_0), \\ \geq 1 & \text{if } Y_0 > E(V_1 \mid \mathscr{F}_0) \qquad \text{on } \{\tau_m^* \geq 0\} \equiv \Omega. \end{cases} \tag{4.5}$$

The conditional expectation of the loss, given $\mathscr{F}_0$, is then

$$V_0 := \min\{Y_0, E(V_1 \mid \mathscr{F}_0)\}. \tag{4.6}$$

More precisely, V_j are defined by *backward recursion*,

$$V_m := Y_m, \qquad V_j := \min\{Y_j, E(V_{j+1} \mid \mathscr{F}_j)\} \qquad (j = m-1, m-2, \dots, 0), \tag{4.7}$$

and the stopping time τ_m^* is defined by

$$\tau_m^* := \min\{j: 0 \leqslant j \leqslant m,\ Y_j = V_j\}. \tag{4.8}$$

Although the optimality of τ_m^* is intuitively clear, a formal proof is worthwhile. First we need the following extension of the Optional Stopping Theorem (Chapter I, Theorem 13.3). A sequence $\{S_j: j \geqslant 0\}$ is an $\{\mathscr{F}_j\}$-*submartingale* if $E(S_j \mid \mathscr{F}_{j-1}) \geqslant S_{j-1}$ a.s. for all $j \geqslant 1$.

Theorem 4.1. (*Optional Stopping Theorem for Submartingales*). Let $\{S_j: j \geqslant 0\}$ be an $\{\mathscr{F}_j\}$-submartingale, τ a stopping time such that (i) $P(\tau < \infty) = 1$, (ii) $E|S_\tau| < \infty$, and (iii) $\lim_{m\to\infty} E(S_m \mathbf{1}_{\{\tau > m\}}) = 0$. Then

$$ES_\tau \geqslant ES_0,$$

with equality in the case $\{S_j\}$ is a $\{\mathscr{F}_j\}$-martingale.

Proof. The proof is almost the same as that of Theorem 13.1 of Chapter I if we write $X_j := S_j - S_{j-1}$ $(j \geqslant 1)$, $X_0 = S_0$, and note that $E(X_j \mid \mathscr{F}_{j-1}) \geqslant 0$ and accordingly change (13.8) of Chapter I to

$$E(X_j \mathbf{1}_{\{\tau \geqslant j\}}) = E[\mathbf{1}_{\{\tau \geqslant j\}} E(X_j \mid \mathscr{F}_{j-1})] \geqslant 0,$$

and replace the equalities in (13.9) and (13.11) by the corresponding inequalities. ■

The main theorem of this section may be proved now. Theorem 4.1 is needed for the proof of part (c), and we use it only for $\tau \leqslant m$.

Theorem 4.2. (a) The sequence $\{V_j: 0 \leqslant j \leqslant m\}$ is an $\{\mathscr{F}_j\}$-submartingale. (b) The sequence $\{V_{\tau_m^* \wedge j}: 0 \leqslant j \leqslant m\}$ is an $\{\mathscr{F}_j\}$-martingale. (c) One has

$$E(Y_\tau) \geqslant E(V_0) \qquad \forall \tau \in \mathscr{F}_m, \qquad E(Y_{\tau_m^*}) = E(V_0). \tag{4.9}$$

That is, τ_m^* is optimal in the class $\mathscr{T}_m$.

Proof. (a) By (4.7), $V_j \leqslant E(V_{j+1} \mid \mathscr{F}_j)$.

(b) Let Z be an arbitrary $\mathscr{F}_j$-measurable bounded real-valued random variable. We need to prove (see Chapter 0, relation (4.14))

$$E(ZV_{\tau_m^* \wedge j}) = E(ZV_{\tau_m^* \wedge (j+1)}) \qquad (0 \leqslant j \leqslant m-1). \tag{4.10}$$

For this, write

$$\begin{aligned} E(ZV_{\tau_m^* \wedge j}) &= E(ZV_{\tau_m^* \wedge j} \mathbf{1}_{\{\tau_m^* \leqslant j\}}) + E(ZV_{\tau_m^* \wedge j} \mathbf{1}_{\{\tau_m^* > j\}}) \\ &= E(ZV_{\tau_m^* \wedge j} \mathbf{1}_{\{\tau_m^* \leqslant j\}}) + E(ZV_j \mathbf{1}_{\{\tau_m^* > j\}}). \end{aligned} \tag{4.11}$$

But, on $\{\tau_m^* > j\}$, $V_j = E(V_{j+1} \mid \mathscr{F}_j)$. Also, $\{\tau_m^* > j\} \in \mathscr{F}_j$. Therefore,

$$\begin{aligned} E(ZV_j \mathbf{1}_{\{\tau_m^* > j\}}) &= E(Z\mathbf{1}_{\{\tau_m^* > j\}} E(V_{j+1} \mid \mathscr{F}_j)] \\ &= E(ZV_{j+1}\mathbf{1}_{\{\tau_m^* > j\}}) = E(ZV_{\tau_m^* \wedge (j+1)}\mathbf{1}_{\{\tau_m^* > j\}}). \end{aligned} \tag{4.12}$$

Using (4.12) in (4.11) one gets (4.10), since $\tau_m^* \wedge j = \tau_m^* \wedge (j+1)$ on $\{\tau_m^* \leqslant j\}$.

(c) Let $\tau \in \mathscr{T}_m$. Since $Y_j \geqslant V_j$ for all j (see (4.7)), one has $Y_\tau \geqslant V_\tau$. By (4.7), Theorem 4.1, and the submartingale property of $\{V_j\}$ it now follows that

$$E(Y_\tau) \geqslant E(V_\tau) \geqslant E(V_0). \tag{4.13}$$

This gives the first relation in (4.9). The second relation in (4.9) follows by the martingale property of $\{V_{\tau_m^* \wedge j}\}$ (and Theorem 4.1). Note $Y_{\tau_m^*} = V_{\tau_m^*}$. ■

In the minimization of EY_τ over $\mathscr{T}_m$, Y_n need not be $\mathscr{F}_n$-measurable. In such cases one may replace Y_n by $E(Y_n \mid \mathscr{F}_n) = U_n$, say, and note that, for every $\tau \in \mathscr{T}_m$,

$$E(Y_\tau) = \sum_{j=0}^{m} E(Y_j \mathbf{1}_{\{\tau = j\}}) = \sum_{j=0}^{m} E[E(Y_j \mathbf{1}_{\{\tau = j\}} \mid \mathscr{F}_j)] = \sum_{j=0}^{m} E[\mathbf{1}_{\{\tau = j\}} U_j) = EU_\tau.$$

Hence the minimization of EY_τ reduces to the minimization of EU_τ over $\mathscr{T}_m$.

Also, instead of minimization one could as easily maximize EY_τ over $\mathscr{T}_m$. Simply replace *min* by *max* in (4.7), and replace "$\geqslant$" by "$\leqslant$" in (4.9). This is the case in the following example, in which the indexing of the random variables and sigmafields starts at $j = 1$ (instead of $j = 0$).

Example 1. (*The Secretary Problem or the Search for the Best*). Let $m > 1$ labels carrying m distinct numbers be in a box. A person takes out one label at a time at random, observes the number on it and sets it aside before the next draw. This continues until he stops after the τth draw. The objective is to *maximize the probability* that this τth number is the *maximum* M in the box. Thus, if $W_j = 1$ or 0 according as the jth number X_j drawn is the maximum or not, one would like to maximize EW_τ over the class $\mathscr{T}_m$ of all stopping times $\tau(\leqslant m)$ relative to the sigmafields $\mathscr{F}_j := \sigma\{X_1, \ldots, X_j\}$, $1 \leqslant j \leqslant m$.

Define the $\mathscr{F}_j$-measurable random variables

$$Y_j := E(W_j \mid \mathscr{F}_j) = P(X_j = M \mid \mathscr{F}_j). \tag{4.14}$$

If X_j is the maximum, say M_j, among $X_1, \ldots, X_j$, then the condition probability (given $X_1, \ldots, X_j$) that it is the maximum in the whole box is j/m; if $X_j < M_j$ then of course this conditional probability is zero. Therefore,

$$Y_j = \frac{j}{m} \mathbf{1}_{\{X_j = M_j\}}. \tag{4.15}$$

Also, for every $\tau \in \mathcal{T}_m$, as explained above,

$$EY_\tau = \sum_{j=1}^{m} E(Y_j \mathbf{1}_{\{\tau=j\}}) = \sum_{j=1}^{m} E[E(W_j \mid \mathcal{F}_j)\mathbf{1}_{\{\tau=j\}}] = \sum_{j=1}^{m} E[E(\mathbf{1}_{\{\tau=j\}} W_j \mid \mathcal{F}_j)]$$
$$= \sum_{j=1}^{m} E(\mathbf{1}_{\{\tau=j\}} W_j) = EW_\tau .$$

Hence the maximum of EW_τ over $\mathcal{T}_m$ is also the maximum of EY_τ over $\mathcal{T}_m$. In order to use Theorem 4.2 with *min* replaced by *max* in (4.7) and "$\geqslant$" replaced by "$\leqslant$" in (4.9), we need to calculate V_j $(1 \leqslant j \leqslant m)$. Now $V_m = Y_m$ and (see Eq. 4.15)

$$E(Y_m \mid \mathcal{F}_{m-1}) = \frac{m}{m} P(X_m = M_m \mid \mathcal{F}_{m-1}) = \frac{1}{m} \qquad (M_m = M).$$

Since $(m-1)/m \geqslant 1/m$, it then follows that

$$V_{m-1} := \max\{Y_{m-1}, E(Y_m \mid \mathcal{F}_{m-1})\} = \frac{m-1}{m} \mathbf{1}_{\{X_{m-1} = M_{m-1}\}} + \frac{1}{m} \mathbf{1}_{\{X_{m-1} < M_{m-1}\}}.$$

Then,

$$E(V_{m-1} \mid \mathcal{F}_{m-2}) = \frac{m-1}{m} P(X_{m-1} = M_{m-1} \mid \mathcal{F}_{m-2}) + \frac{1}{m} P(X_{m-1} < M_{m-1} \mid \mathcal{F}_{m-2})$$
$$= \frac{m-1}{m} \frac{1}{m-1} + \frac{1}{m} \frac{m-2}{m-1} = \frac{m-2}{m} \left(\frac{1}{m-2} + \frac{1}{m-1} \right). \tag{4.16}$$

To evaluate

$$V_{m-2} := \max\{Y_{m-2}, E(V_{m-1} \mid \mathcal{F}_{m-2})\}$$

note that if $(m-2)^{-1} + (m-1)^{-1} \leqslant 1$ then (see Eqs. 4.14 and 4.16)

$$V_{m-2} = \frac{m-2}{m} \mathbf{1}_{\{X_{m-2} = M_{m-2}\}} + \frac{m-2}{m} \left(\frac{1}{m-2} + \frac{1}{m-1} \right) \mathbf{1}_{\{X_{m-2} < M_{m-2}\}} \tag{4.17}$$

so that a calculation akin to (4.16) yields

$$E(V_{m-2} \mid \mathcal{F}_{m-3}) = \frac{m-3}{m} \left(\frac{1}{m-3} + \frac{1}{m-2} + \frac{1}{m-1} \right).$$

Assume, as a backward induction hypothesis, that

$$E(V_{j+1} \mid \mathscr{F}_j) = \frac{j}{m}\left(\frac{1}{j} + \frac{1}{j+1} + \cdots + \frac{1}{m-1}\right), \tag{4.18}$$

for some j such that

$$a_j := \frac{1}{j} + \frac{1}{j+1} + \cdots + \frac{1}{m-1} \leqslant 1. \tag{4.19}$$

Then,

$$V_j := \max\{Y_j, E(V_{j+1} \mid \mathscr{F}_j)\} = \max\left\{\frac{j}{m}\mathbf{1}_{\{X_j = M_j\}}, \frac{j}{m} a_j\right\}$$

$$= \frac{j}{m}\mathbf{1}_{\{X_j = M_j\}} + \frac{j}{m} a_j \mathbf{1}_{\{X_j < M_j\}},$$

and, since $P(X_j = M_j \mid \mathscr{F}_{j-1}) = 1/j$, it follows that

$$E(V_j \mid \mathscr{F}_{j-1}) = \frac{j-1}{m}\left(\frac{1}{j-1} + \frac{1}{j} + \cdots + \frac{1}{m-1}\right) = \frac{j-1}{m} a_{j-1}.$$

The induction is complete, i.e., (4.18) holds for all $j \geqslant j^*$, where

$$j^* := \max\{j\colon 1 \leqslant j \leqslant m, a_j > 1\}. \tag{4.20}$$

In other words, one gets

$$E(V_{j+1} \mid \mathscr{F}_j) = \frac{j}{m} a_j \qquad \text{for } j^* \leqslant j \leqslant m. \tag{4.21}$$

Also,

$$V_{j^*} \equiv \max\{Y_{j^*}, E(V_{j^*+1} \mid \mathscr{F}_{j^*})\} = \frac{j^*}{m} a_{j^*}, \tag{4.22}$$

since $a_{j^*} > 1$. In particular, V_{j^*} is *nonrandom*, which implies

$$E(V_{j^*} \mid \mathscr{F}_{j^*-1}) = \frac{j^*}{m} a_{j^*},$$

which in turn leads to

$$V_{j^*-1} \equiv \max\{Y_{j^*-1}, E(V_{j^*} \mid \mathscr{F}_{j^*-1})\} = \frac{j^*}{m} a_{j^*}.$$

Continuing in this manner one gets

$$V_j = \frac{j^*}{m} a_{j^*} \qquad \text{for } 1 \leqslant j \leqslant j^*. \tag{4.23}$$

The optimal stopping rule is then given by (see Eq. 4.8)

$$\tau_m^* = \begin{cases} \min\{j: j \geqslant j^* + 1, X_j = M_j\}, & \text{if } X_j = M_j \text{ for some } j \geqslant j^* + 1, \\ m, & \text{if } X_j \neq M_j \text{ for all } j \geqslant j^* + 1. \end{cases} \tag{4.24}$$

For, if $j \leqslant j^*$ then $Y_j < E(V_{j+1} \mid \mathscr{F}_j) = (j^*/m)a_{j^*}$. On the other hand, if $j > j^*$ then $a_j \leqslant 1$ so that (i) $Y_j \geqslant E(V_{j+1} \mid \mathscr{F}_j) = (j/m)a_j$ on $\{X_j = M_j\}$ and (ii) $0 = Y_j < E(V_{j+1} \mid \mathscr{F}_j)$ on $\{X_j < M_j\}$.

Simply stated, the optimal stopping rule is to *draw j^* observations and then continue sampling until an observation larger than all the preceding shows up* (and if this does not happen, stop after the last observation has been drawn). The maximal probability of stopping at the maximum value is then

$$E(V_1) = V_1 = \frac{j^*}{m} a_{j^*} = \frac{j^*}{m}\left(\frac{1}{j^*} + \frac{1}{j^*+1} + \cdots + \frac{1}{m-1}\right). \tag{4.25}$$

Finally, note that, as $m \to \infty$,

$$a_{j^*} \equiv \frac{1}{j^*} + \frac{1}{j^*+1} + \cdots + \frac{1}{m-1} \approx 1, \tag{4.26}$$

where the difference between the two sides of the relation "$\approx$" goes to zero. This follows since j^* must go to infinity (as the series $\sum_1^\infty (1/j)$ diverges and j^* is defined by (4.20)) and $a_{j^*} > 1$, $a_{j^*+1} \leqslant 1$. Now,

$$a_{j^*} = \frac{1}{m} \sum_{j=j^*}^{m-1} \frac{1}{j/m} \approx \int_{j^*/m}^{1} \frac{1}{x}\, dx = -\log(j^*/m). \tag{4.27}$$

Combining (4.26) and (4.27) one gets

$$-\log \frac{j^*}{m} \approx 1, \qquad \frac{j^*}{m} \sim e^{-1}, \tag{4.28}$$

where the ratio of the two sides of "$\sim$" goes to one, as $m \to \infty$. Thus,

$$\lim_{m \to \infty} \frac{j^*}{m} = e^{-1}, \qquad \lim_{m \to \infty} E(V_1) = e^{-1}. \tag{4.29}$$

5 CHAPTER APPLICATION: OPTIMALITY OF (S, s) POLICIES IN INVENTORY PROBLEMS

Suppose a company has an amount x of a commodity on stock at the beginning of a period. The problem is to determine the amount a of additional stock that should be ordered to meet a *random demand* W at the end of this period. The cost of ordering a units is

$$\mathscr{C}(a) := \begin{cases} 0 & \text{if } a = 0, \\ K + ca & \text{if } a > 0, \end{cases} \tag{5.1}$$

where $K \geqslant 0$ is the *reorder cost* and $c > 0$ is the cost per unit. There is a *holding cost* $h > 0$ per unit of stock left unsold, and a *depletion cost* $d > 0$ per unit of unmet demand. Thus the *expected total cost* is

$$I(x, a) := \mathscr{C}(a) + L(x + a), \tag{5.2}$$

where

$$L(y) := E(h \max\{0, y - W\} + d \max\{0, W - y\}). \tag{5.3}$$

In this model x may be negative, but $a \geqslant 0$. The objective is to find, for each x, the value $a = f^*(x)$ that minimizes (5.2).

Assume

$$EW < \infty, \qquad d > c > 0, \qquad h \geqslant 0. \tag{5.4}$$

Consider the function G on $\mathbb{R}^1$,

$$G(y) := cy + L(y). \tag{5.5}$$

Then,

$$I(x, a) = \begin{cases} G(x) - cx & \text{if } a = 0, \\ G(x + a) + K - cx & \text{if } a > 0. \end{cases} \tag{5.6}$$

Minimizing $I(x, a)$ over $a \geqslant 0$ is equivalent to minimizing $I(x, a) + cx$ over $a \geqslant 0$, or over $x + a \geqslant x$. But

$$I(x, a) + cx = \begin{cases} G(x) & \text{if } a = 0, \\ G(x + a) + K & \text{if } a > 0. \end{cases} \tag{5.7}$$

Thus, the optimum $a = f^*(x)$ equals $y^*(x) - x$, where $y = y^*(x)$ minimizes the function (on $[x, \infty)$)

$$\bar{G}(y; x) := \begin{cases} G(x) & \text{if } y = x, \\ G(y) + K & \text{if } y > x, \end{cases} \tag{5.8}$$

over $y \geqslant x$.

Now the function $y \to L(y)$ is convex, since the functions $y \to \max\{0, y - w\}$, $\max\{0, w - y\}$ are convex for each w. Therefore, $G(y)$ is convex. Clearly, $G(y)$ goes to ∞ as $y \to \infty$. Also, for $y < 0$, $G(y) \geqslant cy + d|y|$. Hence $G(y) \to \infty$ as $y \to -\infty$, as $d > c > 0$. Thus, $G(y)$ has a minimum at $y = S$, say. Let $s \leqslant S$ be such that

$$G(s) = K + G(S). \tag{5.9}$$

If $K = 0$, one may take $s = S$. If $K > 0$ then such an $s < S$ exists, since $G(y) \to \infty$ as $y \to -\infty$ and G is continuous. Observe also that G decreases on $(-\infty, S]$ and increases on (S, ∞), by convexity. Therefore, for $x \in (-\infty, s]$, $G(x) \geqslant G(s) = K + G(S)$, and the minimum of $\bar{G}$ in (5.8) is attained at $y = y^*(x) = S$. On the other hand, for $x \in (s, S]$,

$$G(x) \leqslant G(s) = K + G(S) \leqslant K + G(y) \qquad \text{for all } y \geqslant x.$$

Therefore, $y^*(x) = x$ for $x \in (s, S]$. Finally, if $x \in (S, \infty)$ then $G(x) \leqslant G(y)$ for all $y \geqslant x$, since G is increasing on (S, ∞). Hence $y^*(x) = x$ for $x \in (S, \infty)$.

Since $f^*(x) = y^*(x) - x$, we have proved the following.

Proposition 5.1. Assume (5.4). There exist two numbers $s \leqslant S$ such that the optimal policy is to order

$$f^*(x) = \begin{cases} S - x & \text{if } x \leqslant s, \\ 0 & \text{if } x > s. \end{cases} \tag{5.10}$$

The minimum cost function is

$$\begin{aligned} I(x) := I(x, f^*(x)) &= \mathscr{C}(f^*(x)) + L(x + f^*(x)) \\ &= \begin{cases} K + G(S) - cx & \text{if } x \leqslant s, \\ G(x) - cx & \text{if } x > s. \end{cases} \end{aligned} \tag{5.11}$$

Instead of fixed costs per unit for holding and depletion, one may assume that $L(y)$ is given by

$$L(y) := E(H(\max\{0, y - W\}) + D(\max\{0, W - y\})). \tag{5.12}$$

Here H and D are convex and increasing on $[0, \infty)$, $H(0) = 0 = D(0)$. The assumption (5.4) may now be relaxed to

$$L(y) < \infty \qquad \text{for all } y, \qquad \lim_{x \to -\infty} (cx + L(x)) > K + cS + L(S), \quad (5.13)$$

where S is a point where G is minimum.

Next consider an $(N + 1)$-*period dynamic inventory problem.* If the initial stock is $x = X_0$ (*state*) and $a = a_0$ units (*action*) are ordered, then the expected total cost in period 0 is

$$g_0(x, a) := E(\mathscr{C}(a) + h \max\{0, x + a - W_1\} + d \max\{0, W_1 - x - a\}) = I(x, a). \quad (5.14)$$

We denote by $W_1, W_2, \ldots, W_N$ i.i.d. random demands arising at the end of periods $0, 1, \ldots, N - 1$. Assumption (5.4) is still in force with W a generic random variable having the same distribution as the W_i. The state X_1 in period 1 is given by $X_1 = x + a - W_1 = X_0 + a_0 - W_1$ and in general

$$X_k = X_{k-1} + a_{k-1} - W_k \quad (5.15)$$

where X_{k-1} is the state in period $k - 1$, and a_{k-1} is the action taken in that period. Thus, the *transition probability* law $p(dz; x, a)$ is the distribution of $x + z - W$. The conditional expectation of the cost for period k, given $X_{k-1} = x$, $a_{k-1} = a$, is

$$g_k(x, a) := \delta^k I(x, a) \qquad (k = 0, 1, \ldots, N), \quad (5.16)$$

where δ is the *discount factor*, $0 < \delta < \infty$. The objective is to *minimize* the total $(N + 1)$-period expected discounted cost

$$J_x^{\mathbf{f}} := E_x^{\mathbf{f}} \sum_{k=0}^{N} \delta^k I(X_k, a_k), \quad (5.17)$$

over all policies $\mathbf{f} = (f_0, f_1, \ldots, f_N)$.

By Theorem 1.2 (changing *max* to *min*), an optimal policy is $\mathbf{f} = (f_0^*, f_1^*, \ldots, f_N^*)$, which is Markovian and is given by backward recursion (see Eqs. 1.11–1.14). In other words, f_N^* is given by (5.10),

$$f_N^*(x) = \begin{cases} S - x & \text{if } x \leqslant s, \\ 0 & \text{if } x > s, \end{cases} \quad (5.18)$$

and f_k^* $(0 \leqslant k \leqslant N - 1)$ are given recursively by (1.14). Let us determine f_{N-1}^*. For this one minimizes

$$g_{N-1}(x, a) + Eh_N(x + a - W), \quad (5.19)$$

where g_{N-1} is given by (5.16) and $h_N(x)$ is the minimum of $g_N(x, a) = \delta^N I(x, a)$, i.e., (see Eq. 5.11)

$$h_N(x) = \delta^N I(x).$$

Thus, (5.19) becomes

$$\delta^{N-1}[I(x, a) + \delta EI(x + a - W)], \tag{5.20}$$

and its minimization is equivalent to that of

$$I_{N-1}(x, a) := I(x, a) + \delta EI(x + a - W). \tag{5.21}$$

Write

$$G_{N-1}(y) := cy + L(y) + \delta EI(y - W). \tag{5.22}$$

Then (see the analogous relations (5.5)–(5.8)),

$$I_{N-1}(x, a) + cx = \begin{cases} G_{N-1}(x) & \text{if } a = 0, \\ G_{N-1}(x + a) + K & \text{if } a > 0. \end{cases} \tag{5.23}$$

Hence, the minimization of $I_{N-1}(x, a)$ over $a \geqslant 0$ is equivalent to the minimization of the function (on $[x, \infty)$),

$$\bar{G}_{N-1}(y; x) = \begin{cases} G_{N-1}(x) & \text{if } y = x, \\ G_{N-1}(y) + K & \text{if } y > x, \end{cases} \tag{5.24}$$

over $y \geqslant x$. The corresponding points $f^*_{N-1}(x)$, $y^*_{N-1}(x)$, where these minima are achieved, are related by

$$f^*_{N-1}(x) = y^*_{N-1}(x) - x. \tag{5.25}$$

The main difficulty in extending the one-period argument here is that the function G_{N-1} is not in general convex, since I is not in general convex. Indeed (see (5.11)), to the left of the point s it is linear with a slope $-c$, while the right-hand derivative at s is $G'(s) - c < -c$ since $G'(s) < 0$ if $K > 0$. One can easily check now that I is not convex, if $K > 0$.

It may be shown, however, that G_{N-1} is *K-convex* in the following sense.

Definition 5.1. Let $K \geqslant 0$. A function g on an interval $\mathscr{I}$ is said to be *K-convex* if, for all $y_1 < y_2 < y_3$ in $\mathscr{I}$,

$$K + g(y_3) \geqslant g(y_2) + \frac{y_3 - y_2}{y_2 - y_1}(g(y_2) - g(y_1)). \tag{5.26}$$

Thus 0-convexity is the same as convexity, and a convex function is K-convex for all $K \geqslant 0$.

We will show a little later that

(i) G_{N-1} is K-convex on $\mathbb{R}^1$,

(ii) $G_{N-1}(y) \to \infty$ as $|y| \to \infty$.

Assuming (i), (ii), we now prove the existence of two numbers $s_{N-1} \leqslant S_{N-1}$ such that

$$y^*_{N-1}(x) := \begin{cases} S_{N-1} & \text{if } x \leqslant s_{N-1}, \\ x & \text{if } x > s_{N-1}, \end{cases} \tag{5.27}$$

minimizes $\bar{G}_{N-1}(y; x)$ over the interval $[x, \infty)$. For this let S_{N-1} be a point where G_{N-1} attains its *minimum* value, and let s_{N-1} be the *smallest number* $\leqslant S_{N-1}$ such that

$$G_{N-1}(s_{N-1}) = K + G_{N-1}(S_{N-1}). \tag{5.28}$$

Such a number s_{N-1} exists, since G_{N-1} is continuous and $G_{N-1}(x) \to \infty$ as $x \to -\infty$.

Now G_{N-1} is *decreasing* on $(-\infty, s_{N-1}]$. To see this, let $y_1 < y_2 < s_{N-1}$ and apply (5.26) with $y_3 = S_{N-1}$,

$$K + G_{N-1}(S_{N-1}) \geqslant G_{N-1}(y_2) + \frac{S_{N-1} - y_2}{y_2 - y_1}(G_{N-1}(y_2) - G_{N-1}(y_1)). \tag{5.29}$$

Also, $G_{N-1}(y_2) > K + G_{N-1}(S_{N-1})$, since $y_2 < s_{N-1}$ and (5.28) holds for the smallest possible s_{N-1}. Using this in (5.29), one gets

$$K + G_{N-1}(S_{N-1}) > K + G_{N-1}(S_{N-1}) + \frac{S_{N-1} - y_2}{y_2 - y_1}(G_{N-1}(y_2) - G_{N-1}(y_1)),$$

so that $G_{N-1}(y_2) < G_{N-1}(y_1)$. Therefore, if $x \leqslant s_{N-1}$, then $G_{N-1}(x) \geqslant G_{N-1}(s_N) = G_{N-1}(S_{N-1}) + K$ and, for all $y > x$, $\bar{G}_{N-1}(y; x) = G_{N-1}(y) + K \geqslant G_{N-1}(S_{N-1}) + K$. Hence, the minimum of $\bar{G}_{N-1}(\cdot\,; x)$ on $[x, \infty)$ is attained at $y = S_{N-1}$, proving the first half of (5.27). In order to prove the second half of (5.27) it is enough to show that

$$G_{N-1}(x) \leqslant G_{N-1}(y) + K = \bar{G}_{N-1}(y; x) \qquad \text{if } s_{N-1} \leqslant x < y. \tag{5.30}$$

If $s_{N-1} < x \leqslant S_{N-1}$ then by (5.28) and K-convexity,

$$G_{N-1}(s_{N-1}) = K + G_{N-1}(S_{N-1}) \geqslant G_{N-1}(x) + \frac{S_{N-1} - x}{x - s_{N-1}}(G_{N-1}(x) - G_{N-1}(s_{N-1})),$$

or,

$$\frac{S_{N-1} - s_{N-1}}{x - s_{N-1}} G_{N-1}(s_{N-1}) \geqslant \frac{S_{N-1} - s_{N-1}}{x - s_{N-1}} G_{N-1}(x),$$

i.e., $G_{N-1}(x) \leqslant G_{N-1}(s_{N-1}) = G_{N-1}(S_{N-1}) + K \leqslant G_{N-1}(y) + K$ for all y. Since (5.30) clearly holds for $x = s_{N-1}$, it remains to prove it in the case $S_{N-1} < x < y$. By K-convexity one has

$$K + G_{N-1}(y) \geqslant G_{N-1}(x) + \frac{y - x}{x - S_{N-1}}(G_{N-1}(x) - G_{N-1}(S_{N-1})).$$

Thus (5.30) is proved, establishing the second half of (5.27).

Finally, let us show that G_{N-1} *is K-convex*. Recall (see Eqs. 5.22, 5.5) that

$$G_{N-1}(y) = G(y) + \delta EI(y - W). \tag{5.31}$$

Since the function G is convex (i.e., 0-convex), it is enough to prove that I is K-convex. For it is easy to check from the definition, on taking expectations, that the K-convexity of I implies that of $y \to EI(y - W)$. It is also easy to show that if J_1 is K_1-convex, J_2 is K_2-convex for some $K_1 \geqslant 0, K_2 \geqslant 0$, then $\delta_1 J_1 + \delta_2 J_2$ is $(\delta_1 K_1 + \delta_2 K_2)$-convex for all $\delta_1 \geqslant 0, \delta_2 \geqslant 0$. For positive δ it now follows from (5.31) that G_{N-1} is K-convex if I is K-convex. Recall the definition of I (see Eq. 5.11),

$$I(x) = \begin{cases} K + G(S) - cx & \text{if } x \leqslant s, \\ G(x) - cx & \text{if } x > s. \end{cases} \tag{5.32}$$

We want to show

$$K + I(x_3) \geqslant I(x_2) + \frac{x_3 - x_2}{x_2 - x_1}(I(x_2) - I(x_1)) \quad \text{for all } x_1 < x_2 < x_3. \tag{5.33}$$

If $x_3 \leqslant s$, then linearity of I in $(-\infty, s]$ implies convexity and, therefore, K-convexity in $(-\infty, s]$. Similarly, if $x_1 \geqslant s$, (5.33) is trivially true since $G(x) - cx$ is convex, and, therefore, K-convex. Consider then $x_1 < s < x_3$. Distinguish two cases.

CASE I. $x_1 < s < x_2$. In this case (5.33) is equivalent to

$$K + G(x_3) - cx_3 \geqslant G(x_2) - cx_2 + \frac{x_3 - x_2}{x_2 - x_1}(G(x_2) - cx_2 - K - G(S) + cx_1). \tag{5.34}$$

Canceling cx_i $(i = 1, 2, 3)$ from both sides and recalling that $K + G(S) = G(s)$,

(5.34) becomes equivalent to

$$K + G(x_3) \geqslant G(x_2) + \frac{x_3 - x_2}{x_2 - x_1}(G(x_2) - G(s)). \tag{5.35}$$

If $G(x_2) \geqslant G(s)$, then the right side of (5.35) is no more than

$$G(x_2) + \frac{x_3 - x_2}{x_2 - s}(G(x_2) - G(s)),$$

which is no more than $G(x_3)$ by convexity of G. Hence (5.35) holds. Suppose $G(x_2) < G(s)$. The left side of (5.35) is no less than $K + G(S) = G(s)$, and (5.35) will be proved if it holds with the left side replaced by $G(s)$, i.e., if

$$G(s) \geqslant G(x_2) + \frac{x_3 - x_2}{x_2 - x_1}(G(x_2) - G(s)). \tag{5.36}$$

By simple algebra, (5.36) is equivalent to $G(s) \geqslant G(x_2)$, which has been assumed to be the case.

CASE II. $x_1 < x_2 \leqslant s < x_3$. In this case (5.33) is equivalent to

$$\begin{aligned} K + G(x_3) - cx_3 &\geqslant K + G(S) - cx_2 + \frac{x_3 - x_2}{x_2 - x_1}(-cx_2 + cx_1) \\ &= K + G(S) - cx_3, \end{aligned}$$

which is obviously true.

Thus, I is K-convex, so that G_{N-1} is K-convex, and the proof of (5.27) is complete.

The minimum value of (5.19), or (5.20), is

$$h_{N-1}(x) := \delta^{N-1} I_{N-1}(x), \tag{5.37}$$

where (see Eqs. 5.23, 5.25, 5.27)

$$I_{N-1}(x) = \min_{a \geqslant 0} I_{N-1}(x, a) = \begin{cases} K + G_{N-1}(S_{N-1}) - cx & \text{if } x \leqslant s_{N-1}, \\ G_{N-1}(x) - cx & \text{if } x > s_{N-1}. \end{cases} \tag{5.38}$$

The equation for the determination of f^*_{N-2} is then (see Eq. 1.14)

$$\begin{aligned} &\min_{a \geqslant 0} [\delta^{N-2} I(x, a) + \delta^{N-1} E I_{N-1}(x + a - W)] \\ &\qquad = \delta^{N-2} \min_{a \geqslant 0} [I(x, a) + \delta E I_{N-1}(x + a - W)]. \end{aligned} \tag{5.39}$$

This problem is mathematically equivalent to the one just considered since (5.32) and (5.38) are of the same structure, except for the fact that G is convex and G_{N-1} is K-convex. But only K-convexity of G was used in the proof of K-convexity of $I(x)$. Proceeding iteratively, we arrive at the following.

Proposition 5.2. Assume (5.4). Then there exists an optimal Markovian policy $\mathbf{f}^* = (f_0^*, \ldots, f_N^*)$ of the (S, s) type for the $(N+1)$-period dynamic inventory model, i.e., there exist $s_k \leqslant S_k$ $(k = 0, 1, \ldots, N)$ such that during period k it is optimal to order $S_k - x$ if the stock x is $\leqslant s_k$, and to order nothing if $x > s_k$. Also $s_k < S_k$ for all k if $K > 0$, and $s_k = S_k$ if $K = 0$.

Finally, consider the *infinite-horizon problem* of minimizing, for a given δ, $0 < \delta < 1$,

$$J_x^{\mathbf{f}} := E_x^{\mathbf{f}} \sum_{k=0}^{\infty} \delta^k I(X_k, a_k), \tag{5.40}$$

over the class of all (measurable) policies $\mathbf{f} = (f_0, f_1, \ldots)$. Write this infimum as

$$J_x := \inf_{\mathbf{f}} J_x^{\mathbf{f}}. \tag{5.41}$$

To show that J_x is finite, consider the policy $\mathbf{f}^0 := (0, 0, \ldots)$, i.e., the stationary policy that orders zero in every period. It is straightforward to check that

$$E_x^{\mathbf{f}^0} I(X_k, a_k) \leqslant (h + d) E_x^{\mathbf{f}^0}(|X_{k-1}| + W_k) = (h + d)(E_x^{\mathbf{f}^0}|X_{k-1}| + EW), \tag{5.42}$$

where $\{W_k\}$ is an i.i.d. sequence, W_k having the same distribution as W considered before. Also, for each k, W_k is independent of $X_0, \ldots, X_{k-1}$. Now,

$$E_x^{\mathbf{f}^0}|X_k| = E_x^{\mathbf{f}^0}|X_{k-1} - W_k| \leqslant EW + E_x^{\mathbf{f}^0}|X_{k-1}|,$$

and iterating one gets

$$E_x^{\mathbf{f}^0}|X_k| \leqslant kEW + |x|. \tag{5.43}$$

Substituting (5.43) in (5.42), (5.40) one gets

$$J_x^{\mathbf{f}^0} \leqslant \frac{(h + d)EW}{(1 - \delta)^2} + \frac{(h + d)|x|}{1 - \delta} < \infty. \tag{5.44}$$

Hence $J_x < \infty$. The dynamic programming equation (2.14) becomes

$$\bar{J}_x = \min_{a \geqslant 0} \{I(x, a) + \delta E(\bar{J}_{x+a-W})\}. \tag{5.45}$$

We will *assume* $K = 0$. Now the $(N+1)$-period optimal costs $(T^N 0)(x)$ are increasing with N. Since these costs are convex, the limit J_x is also easily seen to be convex and, in particular, continuous. It follows that the convergence of

$$I(x, a) + \delta E(T^N 0)(x + a - W) \tag{5.46}$$

to the expression within braces in (5.45) with $\bar{J} = J_x$ is uniform on compact subsets of x and a, at least if W is bounded,

$$P(0 \leqslant W \leqslant M) = 1 \qquad \text{for some } M < \infty. \tag{5.47}$$

The dynamic programming equation, therefore, holds for J_x. This also implies that the convex function

$$G_\infty(y) := G(y) + \delta E J_{y-W} = cy + L(y) + \delta E J_{y-W} \tag{5.48}$$

goes to infinity as $|y| \to \infty$. Thus, by the same argument as above, an optimal policy exists with $S = s$. We have the following.

Proposition 5.3. Assume (5.4) and (5.47). Let $K = 0$, $0 < \delta < 1$.

(a) Then the dynamic programming equation (5.45) holds for the minimum cost J_x in the infinite-horizon discounted dynamic programming problem.
(b) There exists a stationary optimal policy $\mathbf{f}^* = (f^*, f^*, \ldots)$ such that $a = f^*(x)$ minimizes the right-hand side of (5.45) with $\bar{J}_x = J_x$, and f^* is given by

$$f^*(x) = \begin{cases} S - x & \text{if } x \leqslant S, \\ 0 & \text{if } x > S. \end{cases}$$

Here S is the point where the function (5.48) attains its minimum.

EXERCISES

Exercises for Section VI.1

1. Consider the following inventory control problem. Let the possible amounts of a commodity that a business can stock be $0, 1, 2$ (*states*). In each period $k = 0, 1, 2$ $(N = 2)$ the stock is replenished by ordering 0, 1, or 2 units (*actions*), but not more than 2 units can be stored. There is a random demand that arises at the end of each period, demands over different periods being independent of each other and identically distributed. The possible values of demand are 0, 1, 2, which occur with probabilities 0.2, 0.5, and 0.3, respectively. Thus the transition probability $p(z; x, a)$ is the distribution of $\max\{0, x + a - W\} \wedge 2$, where W is the random demand. Find an optimal policy to minimize the total expected cost if in each period the cost is the sum of (a) the cost of ordering a units at the rate of \$1 per unit, (b) the cost of storing

the excess supply $\max\{0, x + a - W\}$ at the rate of \$1 per unit, and (c) a penalty of \$1 per unit for excess demand $\max\{0, W - x - a\}$.

2. (*Taxicab Operation*) The area of operation of a cab driver comprises three towns, T_1, T_2, T_3 (*states*). To get a fare, the driver may follow one of three courses (*actions*): pull over and wait for a radio call (a_1), go to the nearest taxi stand and wait in line (a_2), or go on cruising until hailed by a passenger (a_3). The action a_1 is not available if the driver is in town T_1, as there is no radio service in this town. There is a known probability $p(T_j; T_i, a_k)$ that the next trip will be to town T_j, if the cab is in town T_i and follows the course of action a_k. The corresponding reward is $g(T_i, a_k, T_j)$.
   (i) Write down the backward recursion relations for maximizing the total expected reward over a finite horizon.
   (ii) Find the optimal policy for the case $N = 2$, if the transition probabilities and rewards are as follows:

State	Action	Probability of transition to state			Reward if trip is to state		
		T_1	T_2	T_3	T_1	T_2	T_3
T_1	a_2	$\frac{1}{2}$	0	$\frac{1}{2}$	2	0	4
	a_3	$\frac{1}{4}$	$\frac{1}{2}$	$\frac{1}{4}$	4	2	4
T_2	a_1	$\frac{1}{4}$	$\frac{1}{2}$	$\frac{1}{4}$	2	2	4
	a_2	$\frac{1}{4}$	$\frac{1}{4}$	$\frac{1}{2}$	2	4	2
	a_3	$\frac{1}{4}$	$\frac{1}{2}$	$\frac{1}{4}$	2	2	2
T_3	a_1	$\frac{1}{4}$	$\frac{1}{4}$	$\frac{1}{2}$	4	4	2
	a_2	$\frac{1}{2}$	$\frac{1}{4}$	$\frac{1}{4}$	2	2	2
	a_3	$\frac{1}{4}$	$\frac{1}{4}$	$\frac{1}{2}$	2	4	2

3. Prove Theorem 1.3 along the lines of the proof of Theorem 1.2.

Exercises for Section VI.2

1. Suppose $S = \{x_1, x_2, \ldots\}$ is countable and A is a compact metric space. For each $\varepsilon \in (0, \delta)$ let $\{f_k^\varepsilon : k = 0, 1, \ldots\}$ be a sequence of functions on S into A.
   (i) Show that there exists a sequence $\varepsilon_{n,1} \downarrow 0$ such that $f_0^{\varepsilon_{n,1}}(x_1)$ converges to some point $f_0^*(x_1)$, say, in A. [*Hint*: A is a compact metric space.]
   (ii) Show that there exists a subsequence $\{\varepsilon_{n,2} : n = 1, 2, \ldots\}$ of $\{\varepsilon_{n,1} : n = 1, 2, \ldots\}$ such that $f_0^{\varepsilon_{n,2}}(x_2)$ converges to some point $f_0^*(x_2)$ in A.
   (iii) Having constructed $\{\varepsilon_{n,j} : n = 1, 2, \ldots\}$ in this manner find a subsequence $\{\varepsilon_{n,j+1} : n = 1, 2, \ldots\}$ of $\{\varepsilon_{n,j} : n = 1, 2, \ldots\}$ such that $f_0^{\varepsilon_{n,j+1}}(x_{j+1})$ converges to some point $f_0^*(x_{j+1})$, say.
   (iv) Define $\delta_{n,0} = \varepsilon_{n,n}$, and show that $f_0^{\delta_{n,0}}(x_j) \to f_0^*(x_j)$ for all $j = 1, 2, \ldots$, as $n \to \infty$.
   (v) Use (i)–(iv) to find a subsequence $\{\delta_{n,1} : n \geq 1\}$ of $\{\delta_{n,0}\}$ such that $f_1^{\delta_{n,1}}(x)$ converges to $f_1^*(x)$, say, for every $x \in S$, as $n \to \infty$.
   (vi) Proceeding in this manner find a subsequence $\{\delta_{n,k+1} : n \geq 1\} \subset \{\delta_{n,k} : n \geq 1\}$ such that $f_k^{\delta_{n,k}}(x) \to f_k^*(x)$, say, for every $x \in S$.

(vii) Let $\delta_n := \delta_{n,n}$. Show that $f_k^{\delta_n}(x) \to f_k^*(x)$ for every $k \geqslant 0$ and every $x \in S$, as $n \to \infty$.

2. Let the elements of a finite set S be labeled $1, 2, \ldots, m$.
   (i) Show that the set $\mathbf{B}(S)$ of all real-valued functions on S may be identified with $\mathbb{R}^m$, and the "sup norm" on $\mathbf{B}(S)$ corresponds to the norm $|\mathbf{x}| := \max\{|x^{(i)}|: 1 \leqslant i \leqslant m\}$ on $\mathbb{R}^m$.
   (ii) Let T be a (strict) contraction on S, i.e., $\|Tf - Tg\| \leqslant \delta \|f - g\|$ for some $\delta \in [0, 1)$. Show that this may be viewed as a contraction, also denoted T, on $\mathbb{R}^m$, i.e., $|T\mathbf{x} - T\mathbf{y}| \leqslant \delta |\mathbf{x} - \mathbf{y}|$.
   (iii) Show that $T^n\mathbf{0}$ converges to a point, say $\mathbf{x}^*$, in S. [*Hint*: $\{T^n\mathbf{0}: n \geqslant 1\}$ is a Cauchy sequence in $\mathbb{R}^m$.]
   (iv) Show that $T\mathbf{x}^* = \mathbf{x}^*$.
   (v) Show that $T^n\mathbf{y} \to \mathbf{x}^*$ whatever be $\mathbf{y} \in \mathbb{R}^m$. [*Hint*: If $T\mathbf{x}^* = \mathbf{x}^*$, $T\mathbf{y}^* = \mathbf{y}^*$, then $\mathbf{x}^* = \mathbf{y}^*$.]

3. Write out the dynamic programming equation for the optimal infinite-horizon discounted reward version of part (i) of the taxicab operation problem described in Example 1.2.

4. (i) Prove Theorem 2.3.
   (ii) Under the hypothesis of Theorem 2.3 show that $(T^N 0)(x)$ is the N-stage optimal reward, where T is defined by (2.15).

Exercises for Section VI.3

1. Suppose $\{X_t\}$ is a diffusion on $\mathbb{R}^k \setminus G$ with absorption at ∂G, starting at $\mathbf{x} \notin \partial G$. Give an argument analogous to that given in Exercise 3.5 of Chapter V to prove that $P_{\mathbf{x}}(\tau_{\partial G} \leqslant t) = o(t)$, as $t \downarrow 0$.

2. Check (3.8).

3. Check (3.13).

4. Show that the probability that the diffusion with coefficients (3.40), with c replaced by $c(x)$ (continuously differentiable) ever reaches 0, starting at $x > 0$, is zero.

5. Suppose that the expression on the extreme right in (3.47) is greater than or equal to 1, then the maximum in (3.41) is attained at $c^* = 1$. [*Hint*: $c = 0$ gives a minimum, and the function is strictly concave on $[0, 1]$.]

6. Suppose $\{X_t\}$ is a diffusion on (a, b) with infinitesimal generator $\mathbf{A} = \frac{1}{2}\sigma^2(x)\, d^2/dx^2 + \mu(x)\, d/dx$. Let f be in the domain of $\mathbf{A}$, i.e., $(\mathbf{T}_t f - f)/t \to \mathbf{A}f$ as $t \downarrow 0$. Prove that the function $g(x) := E_x \int_0^\infty e^{-\beta s}(\mathbf{T}_s f)(x)\, ds$ satisfies the *resolvent equation*: $(\beta - \mathbf{A})g(x) = f(x)$. [*Hint*: $(\mathbf{T}_t g)(x) = e^{\beta t}g(x) - e^{\beta t}\int_0^t e^{-\beta s}(\mathbf{T}_s f)(x)\, ds$.]

THEORETICAL COMPLEMENTS

Theoretical Complements to Section VI.1

1. (*Extensions to More General State Spaces: Finite-Horizon Case*) Let S be a complete separable metric space, A a compact metric space, g_k $(0 \leqslant k \leqslant N)$ continuous

real-valued functions on $S \times A$, $p(dz; z, a)$ weakly continuous on $S \times A$ into $\mathscr{P}(S)$, where $\mathscr{P}(S)$ is the set of all probability measures on the Borel sigmafield of S. Then the maxima in (1.11), (1.13), and (1.14) are attained, and the maximal functions h_k are continuous on S. Since there may not be a unique point in A where (1.14) is maximized, one needs to use a *measurable selection theorem* to obtain measurable functions $f_N^*, f_{N-1}^*, \ldots, f_0^*$ on S into A in order to define a Markovian policy $\mathbf{f}^* = (f_0^*, f_1^*, \ldots, f_N^*)$. A proof of the existence of measurable selections may be found in A. Maitra (1968), "Discounted Dynamic Programming on Compact Metric Spaces," *Sankhyā, Ser. A*, **30**, pp. 211–216. The proof of optimality remains unchanged.

The above assumptions can be further relaxed. Assume (**A**):

1. S is a nonempty Borel subset of a complete separable metric space.
2. A is a nonempty compact metric space.
3. g_k $(k = 0, 1, \ldots, N)$ are bounded upper semicontinuous real-valued functions on $S \times A$.
4. $p(dz; x, a)$ is weakly continuous on $S \times A$ into $\mathscr{P}(S)$.

Under (**A**) the following hold:

(i) $h_N(x) := \sup_{a \in A} g_N(x, a)$ is attained (i.e., $h_N(x) = g_N(x, a_N'(x))$ for some $a_N(x) \in A$) and is upper semicontinuous on S.

(ii) Given that h_{k+1} is upper semicontinuous on S:

(a) The function $(x, a) \to \int h_{k+1}(z)p(dz; x, a)$ is upper semicontinuous on $S \times A$.

(b) $h_k(x) \equiv \sup_{a \in A}\{g_k(x, a) + \int h_{k+1}(z)p(dz; x, a)\}$ is upper semicontinuous on S (and the supremum is attained).

(c) There is a Borel-measurable function f_k^* on S into A such that $h_k(x) = g_k(x, f_k^*(x)) + \int h_{k+1}(z)p(dz; x, f_k^*(x))$.

See, Maitra, *loc. cit.*, for this generalization.

Therefore, Proposition 1.1 and Theorem 1.2 go over under (**A**).

Theoretical Complements to Section VI.2

1. (*Infinite-Horizon Discounted Problem*) Consider the assumption (**A**′): Conditions (**A**) above, with (3) replaced by (3)′: g_0 is bounded and upper semicontinuous on $S \times A$.

   Theorem 2.2 holds under (**A**′). The proof goes over word for word if one takes $\mathbf{B}(S)$ to be the set of *all* real-valued bounded Borel measurable functions on S.

2. (*Semi-Markov Models*) All the results of Sections 1 and 2 have extensions to *semi-Markov models*, which include, as special cases, the discrete-time models of Sections 1 and 2, as well as continuous-time jump Markov-type models. In order to describe this extension let (1) the *state space* S be a Borel subset of a complete separable metric space, (2) the *action space* A be compact metric; (3) a reward rate $r(x, a)$ be given, which is bounded and upper semicontinuous on $S \times A$. (4) In addition, one is given the *holding time distribution* $\gamma(du; x, a)$ in state x when an action a is taken, and the *probability distribution of transition* $p(dz; x, a)$ to a new state z from

the present state x when an action a is taken; $(x, a) \to \gamma(du; x, a)$ and $(x, a) \to p(dz; x, a)$ are assumed to be weakly continuous.

A *policy* $\mathbf{f}$ is a sequence of functions $\mathbf{f} = (f_0, f_1, f_2, \ldots)$, where f_0 is a measurable function on S into A, f_1 is a measurable function on $S \times A \times S$ into $A, \ldots, f_k$ is a measurable function on $(S \times A)^k \times S$ into A. Given such a policy and an initial state x, an action $f_0(x)$ is taken. Then the process remains in state x for a random time T_0 having distribution $\gamma(du; x, f_0(x))$. At the end of this time, the state changes to X_1 with distribution $p(dz; x, f_0(x))$. Then an action $\alpha_1 = f_1(x, f_0(x), X_1)$ is taken. The process remains in state X_1 for a random time T_1 having distribution $\gamma(du; X_1, \alpha_1)$, conditionally given X_1. At the end of this period the state changes to X_2 having distribution $p(dz; X_1, \alpha_1)$, conditionally given X_1. This goes on indefinitely. The *expected discounted reward* under the policy is

$$J_x^{\mathbf{f}} := E_x^{\mathbf{f}} \int_0^\infty e^{-\beta t} r(Y_t, a_t)\, dt$$

where $\beta > 0$ is the *discount rate*, Y_t is the state at time t, and a_t is the action at time t. A policy $\mathbf{f}$ is *(semi-)Markov* if, for all $k \geqslant 1$, f_k depends only on the last coordinate among its arguments, i.e., if f_k is a function on S into A. Under such a policy, the stochastic process $\{Y_t: t \geqslant 0\}$ is *semi-Markov*. In other words, although the embedded process $\{X_k: k \geqslant 0\}$ is (nonhomogeneous) Markov having transition probability $p(dz; x, f_{k-1}(x))$ at the kth step, and the holding times $T_0, T_1, \ldots$ are conditionally independent given $\{X_k: k \geqslant 0\}$, the latter (conditional) distributions $\gamma(du; x, f_k(x))$ are *not* in general *exponential*. Hence, $\{Y_t: t \geqslant 0\}$ is not Markov in general. A policy $\mathbf{f} \equiv f^{(\infty)}$ is said to be *stationary* if it is (semi-)Markov and if $f_k = f$ for all k, where f is a fixed Borel-measurable function on S into A. Under a mild additional assumption that (5) $\delta_\beta(x, a) := \int \exp\{-\beta u\}\gamma(du; x, a\}$ is bounded away from 1, it may be shown that *the optimal discounted reward J_x is the unique upper semicontinuous solution to the dynamic programming equation*

$$J_x = \max_a \left\{ r(x, a)\tau_\beta(x, a) + \delta_\beta(x, a) \int J_z\, p(dz; x, a) \right\}.$$

Here $\tau_\beta(x, a) := (1 - \delta_\beta(x, a))/\beta$. In addition, there exists a Borel-measurable function f^* on S into A such that $a = f^*(x)$ maximizes the right side above. For each such f^* the stationary policy $f^{*(\infty)}$ is optimal. For details and references to the literature, see R. N. Bhattacharya and M. Majumdar (1989), "Controlled Semi-Markov Models: The Discounted Case," *J. Statist. Plan. Inference*, **21**, pp. 365–381. The following lists only a few of the significant earlier works on the subject. R. Howard (1960), *Dynamic Programming and Markov Processes*, MIT Press, Cambridge, Mass.; D. Blackwell (1965), "Discounted Dynamic Programming," *Ann. Math. Statist.*, **36**, pp. 226–235; A. Maitra (1968), *loc. cit.*; S. M. Ross (1970), "Average Cost Semi-Markov Decision Processes," *J. Appl. Probability*, **7**, pp. 656–659.

The dynamic programming equations of this chapter (for example, (2.14)) originated in Bellman's pioneering work. See R. Bellman (1957), *Dynamic Programming*, Princeton University Press, Princeton.

Theoretical Complements to Section VI.3

1. The policies that are shown in this section to be optimal in the class of all stationary feasible policies are actually optimal in a much larger class of policies, namely, the class of all nonanticipative feasible policies, which may depend on the entire past. For details, see W. H. Fleming and R. W. Rishel (1975), *Deterministic and Stochastic Optimal Control*, Springer-Verlag, New York, Chapters V and VI. Apart from some technical measurability problems, the ideas involved in this extension of the class of feasible policies are similar to those already considered in Sections 1 and 2.

2. Feller's construction of one-dimensional diffusions (see theoretical complement 2.3 to Chapter V) allows discontinuous drift and diffusion coefficients. Hence, in Example 1 one may allow all measurable functions v with values in $[v_*, v^*]$. This example is due to M. Majumdar and R. Radner (1990), "Linear Models of Economic Survival under Production Uncertainty," Working Paper 427, Dept. Econ., Cornell University. Example 2 is a one-dimensional specialization of a result of R. C. Merton (1971), "Optimal Consumption and Portfolio Rules in a Continuous-Time Model," *J. Economic Theory*, **3**, pp. 373–413.

3. The uniqueness of the solution to (3.55) is a little difficult to check, since $r(x, c)$ is unbounded as a function of x. One way to circumvent this problem is to fix $T > 0$ and consider the problem of maximizing the discounted reward over the finite horizon $[0, T]$. This is carried out in Fleming and Rishel, *loc. cit.*, pp. 160, 161. One may then let $T \to \infty$.

Theoretical Complement to Section VI.4

1. A readable account of the general theory of optimal stopping rules in discrete time may be found in Y. S. Chow, H. Robbins, and D. Siegmund (1971), *Great Expectations: The Theory of Optimal Stopping*, Houghton Mifflin, Boston.

Theoretical Complement to Section VI.5

1. The chapter application is based on H. Scarf (1960), "The Optimality of (S, s) Policies in the Dynamic Inventory Problem," *Mathematical Methods in the Social Sciences*, Stanford University Press, Stanford, pp. 96–202.

CHAPTER VII

An Introduction to Stochastic Differential Equations

1 THE STOCHASTIC INTEGRAL

A diffusion $\{X_t\}$ on $\mathbb{R}^1$ may be thought of as a Markov process that is locally like a Brownian motion. That is, in some sense the following relation holds,

$$dX_t = \mu(X_t)\,dt + \sigma(X_t)\,dB_t, \tag{1.1}$$

where $\mu(\cdot), \sigma(\cdot)$ are given functions on $\mathbb{R}^1$, and $\{B_t\}$ is a standard one-dimensional Brownian motion. In other words, conditionally given $\{X_s: 0 \leqslant s \leqslant t\}$, in a small time interval $(t, t + dt]$ the displacement $dX_t := X_{t+dt} - X_t$ is approximately the Gaussian random variable $\mu(X_t)\,dt + \sigma(X_t)(B_{t+dt} - B_t)$, having mean $\mu(X_t)\,dt$ and variance $\sigma^2(X_t)\,dt$. Observe, however, that (1.1) cannot be regarded as an ordinary differential equation such as $dX_t/dt = \mu(X_t) + \sigma(X_t)\,dB_t/dt$. For, outside a set of probability 0, a Brownian path is nowhere differentiable (Exercise 1; or see Exercises 7.8, 9.3 of Chapter I). The precise sense in which (1.1) is true, and may be solved to construct a diffusion with coefficients $\mu(\cdot), \sigma^2(\cdot)$, is described in this section and the next.

The present section is devoted to the definition of the integral version of (1.1):

$$X_t = x + \int_0^t \mu(X_s)\,ds + \int_0^t \sigma(X_s)\,dB_s. \tag{1.2}$$

The first integral is an ordinary Riemann integral, which is defined if $\mu(\cdot)$ is continuous and $s \to X_s$ is continuous. But the second integral cannot be defined as a Riemann–Stieltjes integral, as the Brownian paths are of unbounded variation on $[0, t]$ for every $t > 0$ (Exercise 1). On the other hand, for a constant function $\sigma(x) \equiv \sigma$, the second integral has the obvious meaning as $\sigma(B_t - B_0)$, so that (1.2) becomes

$$X_t = x + \int_0^t \mu(X_s)\,ds + \sigma(B_t - B_0). \tag{1.3}$$

It turns out that (1.3) may be solved, more or less by Picard's well-known method of iteration for ordinary differential equations, if $\mu(x)$ is Lipschitzian (Exercise 2). This definition of the (stochastic) integral with respect to the Brownian increments dB_s easily extends to the case of an integrand that is a step function. In order that $\{X_t\}$ may have the Markov property, it is necessary to restrict attention to step functions that do not anticipate the future. This motivates the following development.

Let $(\Omega, \mathscr{F}, P)$ be a probability space on which a standard one-dimensional Brownian motion $\{B_t : t \geqslant 0\}$ is defined. Suppose $\{\mathscr{F}_t : t \geqslant 0\}$ is an increasing family of sub-sigmafields of $\mathscr{F}$ such that

$$\begin{aligned} &\text{(i) } B_s \text{ is } \mathscr{F}_s\text{-measurable } (s \geqslant 0),\\ &\text{(ii) } \{B_t - B_s : t \geqslant s\} \text{ is independent of (events in) } \mathscr{F}_s\ (s \geqslant 0). \end{aligned} \tag{1.4}$$

For example, one may take $\mathscr{F}_t = \sigma\{B_s : 0 \leqslant s \leqslant t\}$. This is the smallest $\mathscr{F}_t$ one may take in view of (i). Often it is important to take larger $\mathscr{F}_t$. As an example, let $\mathscr{F}_t = \sigma[\{B_s : 0 \leqslant s \leqslant t\}, \{Z_\lambda : \lambda \in \Lambda\}]$, where $\{Z_\lambda : \lambda \in \Lambda\}$ is a family of random variables independent of $\{B_t : t \geqslant 0\}$. For technical convenience, also assume that $\mathscr{F}_t$, $\mathscr{F}$ are *P-complete*, i.e., if $N \in \mathscr{F}_t$ and $P(N) = 0$, then all subsets of N are in $\mathscr{F}_t$: this can easily be ensured (theoretical complement 1).

Next fix two time points $0 \leqslant \alpha < \beta < \infty$. A real-valued stochastic process $\{f(t) : \alpha \leqslant t \leqslant \beta\}$ is said to be a *nonanticipative step functional on* $[\alpha, \beta]$ if there exists a finite set of time points $t_0 = \alpha < t_1 < \cdots < t_m = \beta$ and random variables f_i $(0 \leqslant i \leqslant m)$ such that

$$\begin{aligned} &\text{(i) } f_i \text{ is } \mathscr{F}_{t_i}\text{-measurable } (0 \leqslant i \leqslant m),\\ &\text{(ii) } f(t) = f_i \text{ for } t_i \leqslant t < t_{i+1}\ (0 \leqslant i \leqslant m-1),\ f(\beta) = f_m. \end{aligned} \tag{1.5}$$

Definition 1.1. The *stochastic integral*, or the *Itô integral*, of the nonanticipative step functional $f = \{f(t) : \alpha \leqslant t \leqslant \beta\}$ in (1.5) is the stochastic process

$$\int_\alpha^t f(s)\,dB_s := \begin{cases} f(t_0)(B_t - B_{t_0}) \equiv f(\alpha)(B_t - B_\alpha) & \text{for } t \in [\alpha, t_1],\\ \displaystyle\sum_{j=1}^{i} f(t_{j-1})(B_{t_j} - B_{t_{j-1}}) + f(t_i)(B_t - B_{t_i}) & \text{for } t \in (t_i, t_{i+1}], \end{cases}$$

$$(\alpha \leqslant t \leqslant \beta). \tag{1.6}$$

Observe that the Riemann-type sum (1.6) is obtained by taking the value of the integrand at the *left end point* of a time interval $(t_{j-1}, t_j]$. As a consequence, for each $t \in [\alpha, \beta]$ the Itô integral $\int_\alpha^t f(s)\,dB_s$ is *$\mathscr{F}_t$-measurable*, i.e., it is nonanticipative. Some other important properties of this integral are contained in the proposition below.

Proposition 1.1

(a) If f is a nonanticipative step functional on $[\alpha, \beta]$, then $t \to \int_\alpha^t f(s)\,dB_s$ is continuous for every $\omega \in \Omega$; it is also additive, i.e.,

$$\int_\alpha^s f(u)\,dB_u + \int_s^t f(u)\,dB_u = \int_\alpha^t f(u)\,dB_u, \qquad (\alpha \leqslant s < t \leqslant \beta). \quad (1.7)$$

(b) If f, g are nonanticipative step functionals on $[\alpha, \beta]$, then

$$\int_\alpha^t (f(s) + g(s))\,dB_s = \int_\alpha^t f(s)\,dB_s + \int_\alpha^t g(s)\,dB_s. \quad (1.8)$$

(c) Suppose f is a nonanticipative step functional on $[\alpha, \beta]$ such that

$$E\int_\alpha^\beta f^2(t)\,dB_t \equiv \int_\alpha^\beta (Ef^2(t))\,dt < \infty. \quad (1.9)$$

Then $\{\int_\alpha^t f(s)\,dB_s \colon \alpha \leqslant t \leqslant \beta\}$ is a square integrable $\{\mathscr{F}_t\}$-martingale, i.e.,

$$E\left(\int_\alpha^t f(u)\,dB_u \,\middle|\, \mathscr{F}_s\right) = \int_\alpha^s f(u)\,dB_u \qquad (\alpha \leqslant s < t \leqslant \beta). \quad (1.10)$$

Also,

$$E\left(\left(\int_\alpha^t f(s)\,dB_s\right)^2 \middle|\, \mathscr{F}_\alpha\right) = E\left(\int_\alpha^t f^2(s)\,ds \,\middle|\, \mathscr{F}_\alpha\right) \quad (1.11)$$

and

$$E\int_\alpha^t f(s)\,dB_s = 0, \qquad E\left(\int_\alpha^t f(s)\,dB_s\right)^2 = E\int_\alpha^t f^2(s)\,ds, \qquad (\alpha \leqslant t \leqslant \beta). \quad (1.12)$$

Proof. (a), (b) follow from Definition 1.1.

(c) Let f be given by (1.5). As $\int_\alpha^s f(u)\,dB_u$ is $\mathscr{F}_s$-measurable, it follows from (1.7) that

$$E\left(\int_\alpha^t f(u)\,dB_u \,\middle|\, \mathscr{F}_s\right) = \int_\alpha^s f(u)\,dB_u + E\left(\int_s^t f(u)\,dB_u \,\middle|\, \mathscr{F}_s\right). \quad (1.13)$$

Now, by (1.6), if $t_{j-1} < s \leqslant t_j$ and $t_i < t \leqslant t_{i+1}$, then

$$\int_s^t f(u)\,dB_u = f(s)(B_{t_j} - B_s) + f(t_j)(B_{t_{j+1}} - B_{t_j}) + \cdots + f(t_{i-1})(B_{t_i} - B_{t_{i-1}}) + f(t_i)(B_t - B_{t_i}). \tag{1.14}$$

Observe that, for $s' < t'$, $B_{t'} - B_{s'}$ is independent of $\mathscr{F}_{s'}$ (property (1.4(ii)), so that

$$E(B_{t'} - B_{s'} \mid \mathscr{F}_{s'}) = 0 \qquad (s' < t'). \tag{1.15}$$

Applying this to (1.14),

$$\begin{aligned} E(f(s)(B_{t_j} - B_s) \mid \mathscr{F}_s) &= f(s)E(B_{t_j} - B_s \mid \mathscr{F}_s) = 0, \\ E(f(t_j)(B_{t_{j+1}} - B_{t_j}) \mid \mathscr{F}_s) &= E[E(f(t_j)(B_{t_{j+1}} - B_{t_j}) \mid \mathscr{F}_{t_j}) \mid \mathscr{F}_s] \\ &= E[f(t_j)E(B_{t_{j+1}} - B_{t_j} \mid \mathscr{F}_{t_j}) \mid \mathscr{F}_s] = 0, \ldots, \\ E(f(t_i)(B_t - B_{t_i}) \mid \mathscr{F}_s) &= E[f(t_i)E(B_t - B_{t_i} \mid \mathscr{F}_s)] = 0. \end{aligned} \tag{1.16}$$

Therefore,

$$E\left(\int_s^t f(u)\,dB_u \mid \mathscr{F}_s\right) = 0 \qquad (s < t). \tag{1.17}$$

From this and (1.7), the martingale property (1.10) follows. In order to prove (1.11), first let $t \in [\alpha, t_1]$. Then

$$\begin{aligned} E\left(\left(\int_\alpha^t f(s)\,dB_s\right)^2 \middle| \mathscr{F}_\alpha\right) &= E(f^2(\alpha)(B_t - B_\alpha)^2 \mid \mathscr{F}_\alpha) = f^2(\alpha)\mathrm{E}((B_t - B_\alpha)^2 \mid \mathscr{F}_\alpha) \\ &= f^2(\alpha)(\mathrm{t} - \alpha) = E\left(\int_\alpha^t f^2(s)\,ds \mid \mathscr{F}_s\right), \end{aligned} \tag{1.18}$$

by independence of $\mathscr{F}_\alpha$ and $B_t - B_\alpha$. If $t \in (t_i, t_{i+1}]$ then, by (1.6),

$$E\left(\left(\int_\alpha^t f(s)\,dB_s\right)^2 \middle| \mathscr{F}_\alpha\right) = E\left[\left\{\sum_{j=1}^{i} f(t_{j-1})(B_{t_j} - B_{t_{j-1}}) + f(t_i)(B_t - B_{t_i})\right\}^2 \middle| \mathscr{F}_\alpha\right]. \tag{1.19}$$

Now the contribution of the product terms in (1.19) is zero. For, if $j < k$, then

$$\begin{aligned} &E[f(t_{j-1})(B_{t_j} - B_{t_{j-1}})f(t_{k-1})(B_{t_k} - B_{t_{k-1}}) \mid \mathscr{F}_\alpha] \\ &\qquad = E[E(\cdots \mid \mathscr{F}_{t_{k-1}}) \mid \mathscr{F}_\alpha] \\ &\qquad = E[f(t_{j-1})(B_{t_j} - B_{t_{j-1}})f(t_{k-1})E(B_{t_k} - B_{t_{k-1}} \mid \mathscr{F}_{t_{k-1}}) \mid \mathscr{F}_\alpha] \\ &\qquad = 0. \end{aligned} \tag{1.20}$$

Therefore, (1.19) reduces to

$$
\begin{aligned}
E\left[\left(\int_\alpha^t f(s)\,dB_s\right)^2 \,\Big|\, \mathscr{F}_\alpha\right] &= \sum_{j=1}^{i} E(f^2(t_{j-1})(B_{t_j} - B_{t_{j-1}})^2 \mid \mathscr{F}_\alpha) \\
&\quad + E(f^2(t_i)(B_t - B_{t_i})^2 \mid \mathscr{F}_\alpha) \\
&= \sum_{j=1}^{i} E[f^2(t_{j-1})E((B_{t_j} - B_{t_{j-1}})^2 \mid \mathscr{F}_{t_{j-1}}) \mid \mathscr{F}_\alpha] \\
&\quad + E[f^2(t_i)E((B_t - B_{t_i})^2 \mid \mathscr{F}_{t_i}) \mid \mathscr{F}_\alpha] \\
&= \sum_{j=1}^{i} E(f^2(t_{j-1})(t_j - t_{j-1}) \mid \mathscr{F}_\alpha) + E(f^2(t_i)(t - t_i) \mid \mathscr{F}_\alpha) \\
&= E\left(\sum_{j=1}^{i} (t_j - t_{j-1})f^2(t_{j-1}) + (t - t_i)f^2(t_i) \,\Big|\, \mathscr{F}_\alpha\right) \\
&= E\left(\int_\alpha^t f^2(s)\,ds \,\Big|\, \mathscr{F}_\alpha\right). \qquad (1.21)
\end{aligned}
$$

The relations (1.12) are immediate consequences of (1.17) and (1.21). ∎

The next task is to extend the definition of the stochastic integral to a larger class of functionals, by approximating these by step functionals.

Definition 1.2. A right-continuous stochastic process $f = \{f(t): \alpha \leqslant t \leqslant \beta\}$, such that $f(t)$ is $\mathscr{F}_t$-measurable, is said to be a *nonanticipative functional on* $[\alpha, \beta]$. Such an f is said to *belong to* $\mathscr{M}[\alpha, \beta]$ if

$$E\int_\alpha^\beta f^2(t)\,dt < \infty. \qquad (1.22)$$

If $f \in \mathscr{M}[\alpha, \beta]$ for all $\beta > \alpha$, then f is said to *belong to* $\mathscr{M}[\alpha, \infty)$.

Proposition 1.2. Let $f \in \mathscr{M}[\alpha, \beta]$. Then there exists a sequence $\{f_n\}$ of nonanticipative step functionals belonging to $\mathscr{M}[\alpha, \beta]$ such that

$$E\int_\alpha^\beta (f_n(t) - f(t))^2\,dt \to 0 \qquad \text{as } n \to \infty. \qquad (1.23)$$

Proof. Extend $f(t)$ to $(-\infty, \infty)$ by setting it zero outside $[\alpha, \beta]$. For $\varepsilon > 0$ write $g_\varepsilon(t) := f(\mathrm{t} - \varepsilon)$. Let ψ be a symmetric, continuous, nonnegative (nonrandom) function that vanishes outside $(-1, 1)$ and satisfies $\int_{-1}^{1} \psi(x)\,dx = 1$. Write $\psi_\varepsilon(x) := \psi(x/\varepsilon)/\varepsilon$. Then ψ_ε vanishes outside $(-\varepsilon, \varepsilon)$ and satisfies

$\int_{-\varepsilon}^{\varepsilon} \psi_\varepsilon(x)\,dx = 1$. Now define $g^\varepsilon := g_\varepsilon * \psi_\varepsilon$, i.e.,

$$g^\varepsilon(t) = \int_{-\varepsilon}^{\varepsilon} g_\varepsilon(t-x)\psi_\varepsilon(x)\,dx = \int_{-\varepsilon}^{\varepsilon} f(t-\varepsilon-x)\psi_\varepsilon(x)\,dx$$
$$= \int_{t-2\varepsilon}^{t} f(y)\psi_\varepsilon(y-t+\varepsilon)\,dy \qquad (y = t-\varepsilon-x). \tag{1.24}$$

Note that, for each $\omega \in \Omega$, the Fourier transform of g^ε is

$$\hat{g}^\varepsilon(\xi) = \hat{g}_\varepsilon(\xi)\hat{\psi}_\varepsilon(\xi) = e^{i\xi\varepsilon}\hat{f}(\xi)\hat{\psi}_\varepsilon(\xi),$$

so that, by the Plancherel identity (Chapter 0, Eq. 8.45),

$$\int_{-\infty}^{\infty} |g^\varepsilon(t) - f(t)|^2\,dt = \frac{1}{2\pi}\int_{-\infty}^{\infty} |\hat{g}^\varepsilon(\xi) - \hat{f}(\xi)|^2\,d\xi$$
$$= \frac{1}{2\pi}\int_{-\infty}^{\infty} |\hat{f}(\xi)|^2 |e^{i\xi\varepsilon}\hat{\psi}_\varepsilon(\xi) - 1|^2)\,d\xi. \tag{1.25}$$

The last integrand is bounded by $2|\hat{f}(\xi)|^2$. Since

$$E\int |\hat{f}(\xi)|^2\,d\xi \equiv \int_\Omega \int_{-\infty}^{\infty} |\hat{f}(\xi)|^2\,d\xi\,dP = 2\pi \int_\Omega \int_{-\infty}^{\infty} f^2(t)\,dt\,dP$$
$$= 2\pi E \int_{[\alpha,\beta]} f^2(t)\,dt < \infty, \tag{1.26}$$

it follows, by Lebesgue's Dominated Convergence Theorem, that

$$E\int_{-\infty}^{\infty} |g^\varepsilon(t) - f(t)|^2\,dt \to 0 \qquad \text{as } \varepsilon \to 0. \tag{1.27}$$

This proves that there exists a sequence $\{g_n\}$ of continuous nonanticipative functionals in $\mathcal{M}[\alpha, \beta]$ such that

$$E\int_\alpha^\beta (g_n(s) - f(s))^2\,ds \to 0. \tag{1.28}$$

It is now enough to prove that if g is a continuous element of $\mathcal{M}[\alpha, \beta]$, then there exists a sequence $\{h_n\}$ of nonanticipative step functionals in $\mathcal{M}[\alpha, \beta]$ such that

$$E\int_\alpha^\beta (h_n(s) - g(s))^2\,ds \to 0. \tag{1.29}$$

For this, first assume that g is also bounded, $|g| \leqslant M$. Define

$$h_n(t) = g\left(\alpha + \frac{k}{n}(\beta - \alpha)\right) \qquad \text{if } \alpha + \frac{k}{n} \leqslant t < \alpha + \frac{k+1}{n}(\beta - \alpha),\ (0 \leqslant k \leqslant n-1),$$

$$h_n(\beta) = g(\beta).$$

Use Lebesgue's Dominated Convergence Theorem to prove (1.29). If g is unbounded, then apply (1.29) with g replaced by its truncation g^M, which agrees with g on $\{t \in [\alpha, \beta]: |g(t)| \leqslant M\}$ and equals $-M$ on the set $\{g(t) < -M\}$ and M on $\{g(t) > M\}$. Note that $g^M = (g \wedge M) \vee (-M)$ is a continuous and bounded element of $\mathscr{M}[\alpha, \beta]$, and apply Lebesgue's Dominated Convergence Theorem to get

$$E\int_\alpha^\beta (g^M(s) - g(s))^2\, ds \to 0 \qquad \text{as } M \to \infty. \qquad \blacksquare$$

We are now ready to define the stochastic integral of an arbitrary element f in $\mathscr{M}[\alpha, \beta]$. Given such an f, let $\{f_n\}$ be a sequence of step functionals satisfying (1.23). For each pair of positive integers n, m, the stochastic integral $\int_\alpha^t (f_n - f_m)(s)\, dB_s$ $(\alpha \leqslant t \leqslant \beta)$ is a square integrable martingale, by Proposition 1.1(c). Therefore, by the Maximal Inequality (see Chapter I, Eq. 13.56) and the second relation in (1.12), for all $\varepsilon > 0$,

$$P(M_{n,m} > \varepsilon) \leqslant \int_\alpha^\beta E(f_n(s) - f_m(s))^2\, ds/\varepsilon^2, \tag{1.30}$$

where

$$M_{n,m} := \max\left\{\left|\int_\alpha^t (f_n(s) - f_m(s))\, dB_s\right| : \alpha \leqslant t \leqslant \beta\right\}. \tag{1.31}$$

Choose an increasing sequence $\{n_k\}$ of positive integers such that the right side of (1.30) is less than $1/k^2$ for $\varepsilon = 1/k^2$, if $n, m \geqslant n_k$. Denote by A the set

$$A := \{\omega \in \Omega: M_{n_k, n_{k+1}}(\omega) \leqslant 1/k^2 \text{ for all sufficiently large } k\}. \tag{1.32}$$

By the Borel–Cantelli Lemma (Section 6 of Chapter 0), $P(A) = 1$. Now on A the series $\sum_k M_{n_k, n_{k+1}} < \infty$. Thus, given any $\delta > 0$ there exists a positive integer $k(\delta)$ (depending on $\omega \in A$) such that

$$M_{n_j, n_l} \leqslant \sum_{k \geqslant k(\delta)} M_{n_k, n_{k+1}} < \delta \qquad \forall j, l \geqslant k(\delta).$$

In other words, on the set A the sequence $\{\int_\alpha^t f_{n_k}(s)\, dB_s: \alpha \leqslant t \leqslant \beta\}$ is Cauchy in the supremum distance (1.31). Therefore, $\int_\alpha^t f_{n_k}(s)\,dB_s$ $(\alpha \leqslant t \leqslant \beta)$ converges uniformly to a continuous function, outside a set of probability zero.

Definition 1.3. For $f \in \mathcal{M}[\alpha, \beta]$, the uniform limit of $\int_\alpha^t f_{n_k}(s)\,dB_s$ $(\alpha \leqslant t \leqslant \beta)$ on A as constructed above is called the *stochastic integral*, or the *Itô integral, of f*, and is denoted by $\int_\alpha^t f(s)\,dB_s$ $(\alpha \leqslant t \leqslant \beta)$. It will be assumed that $t \to \int_\alpha^t f(s)\,dB_s$ is continuous for all $\omega \in \Omega$, with an arbitrary specification on A^c. If $f \in \mathcal{M}[\alpha, \infty)$, then such a continuous version exists for $\int_\alpha^t f(s)\,dB_s$ on the infinite interval $[\alpha, \infty)$.

It should be noted that the Itô integral of f is well defined up to a null set. For if $\{f_n\}, \{g_n\}$ are two sequences of nonanticipative step functionals both satisfying (1.23), then

$$E\int_\alpha^\beta (f_n(s) - g_n(s))^2\,ds \to 0. \tag{1.33}$$

If $\{f_{n_k}\}, \{g_{m_k}\}$ are subsequences of $\{f_n\}, \{g_n\}$ such that $\int_\alpha^t f_{n_k}(s)\,dB_s$ and $\int_\alpha^t g_{m_k}(s)\,dB_s$ both converge uniformly to some processes $\{Y_t\}, \{Z_t\}$, respectively, outside a set of zero probability, then

$$E\int_\alpha^\beta (Y_s - Z_s)^2\,ds \leqslant \varliminf_{k\to\infty} E\int_\alpha^\beta (f_{n_k}(s) - g_{m_k}(s))^2\,ds = 0,$$

by (1.33) and Fatou's Lemma (Section 3 of Chapter 0). As $\{Y_s\}, \{Z_s\}$ are a.s. continuous, one must have $Y_s = Z_s$ for $\alpha \leqslant s \leqslant \beta$, outside a P-null set.

Note also that $\int_\alpha^t f(s)\,dB_s$ is $\mathcal{F}_t$-measurable for $\alpha \leqslant t \leqslant \beta$, $f \in \mathcal{M}[\alpha, \beta]$. The following generalization of Proposition 1.1 is almost immediate.

Theorem 1.3. Properties (a)–(c) of Proposition 1.1 hold if f (and g) $\in \mathcal{M}[\alpha, \beta]$.

Proof. This follows from the corresponding properties of the approximating step functionals $\{f_{n_k}\}$, on taking limits. ■

Example. Let $f(s) = B_s$. Then $f \in \mathcal{M}[\alpha, \infty)$. Fix $t > 0$. Define $f_n(t) = B_t$ and

$$f_n(s) := B_{r2^{-n}t} \qquad \text{if } r2^{-n}t \leqslant s < (r+1)2^{-n}t \quad (0 \leqslant r \leqslant 2^n - 1).$$

Then, writing $B_{r,n} := B_{r2^{-n}t}$,

$$\int_0^t f_n(s)\,dB_s = \sum_{r=0}^{2^n-1} B_{r,n}(B_{r+1,n} - B_{r,n}) = -\frac{1}{2}\sum_{r=0}^{2^n-1}(B_{r+1,n} - B_{r,n})^2 - \tfrac{1}{2}B_0^2 + \tfrac{1}{2}B_t^2$$
$$\to -\tfrac{1}{2}t - \tfrac{1}{2}B_0^2 + \tfrac{1}{2}B_t^2 \quad \text{a.s.} \qquad (\text{as } n \to \infty). \tag{1.34}$$

The last convergence follows from an application of the Borel–Cantelli Lemma (Exercise 4). Hence,

$$\int_0^t B_s\,dB_s = -\tfrac{1}{2}(B_t^2 - B_0^2) - \tfrac{1}{2}t. \tag{1.35}$$

Notice that a formal application of ordinary calculus would yield $\int_0^t B_s\,dB_s = \frac{1}{2}(B_t^2 - B_0^2)$. Also, if one replaced the values of f at the *left end points* of the subintervals by those at the *right end points*, then one would get, in place of the first sum in (1.34),

$$\sum_{r=0}^{2^n-1} B_{r+1,n}(B_{r+1,n} - B_{r,n}) = \frac{1}{2}\sum_{r=0}^{2^n-1} (B_{r+1,n} - B_{r,n})^2 + \tfrac{1}{2}(B_t^2 - B_0^2),$$

which converges a.s. to $\frac{1}{2}t + \frac{1}{2}(B_t^2 - B_0^2)$.

Observe finally that the stochastic integral in (1.2) is well defined if $\{\sigma(X_s)\}$ belongs to $\mathcal{M}[0, \infty)$. In the next section, it is shown that there exists a unique continuous nonanticipative solution of (1.2) if $\mu(\cdot)$ and $\sigma(\cdot)$ are Lipschitzian and, in particular, $\{\sigma(X_s)\} \in \mathcal{M}[0, \infty)$.

2 CONSTRUCTION OF DIFFUSIONS AS SOLUTIONS OF STOCHASTIC DIFFERENTIAL EQUATIONS

In the last section, a precise meaning was given to the stochastic differential equation (1.1) in terms of its integral version (1.2). The present section is devoted to the solution of (1.1) (or (1.2)).

2.1 Construction of One-Dimensional Diffusions

Let $\mu(x)$ and $\sigma(x)$ be two real-valued functions on $\mathbb{R}^1$ that are Lipschitzian, i.e., there exists a constant $M > 0$ such that

$$|\mu(x) - \mu(y)| \leqslant M|x - y|, \qquad |\sigma(x) - \sigma(y)| \leqslant M|x - y|, \qquad \forall x, y. \tag{2.1}$$

The first order of business is to show that equation (1.1) is valid as a *stochastic integral equation* in the sense of Itô, i.e.,

$$X_t = X_\alpha + \int_0^t \mu(X_s)\,ds + \int_\alpha^t \sigma(X_s)\,dB_s, \qquad (t \geqslant \alpha). \tag{2.2}$$

Theorem 2.2. Suppose $\mu(\cdot), \sigma(\cdot)$ satisfy (2.1). Then, for each $\mathcal{F}_\alpha$-measurable square integrable random variable X_α, there exists a unique (except on a set of zero probability) continuous nonanticipative functional $\{X_t : t \geqslant \alpha\}$ belonging to $\mathcal{M}[\alpha, \infty)$ that satisfies (2.2).

Proof. We prove "existence" by the method of iterations. Fix $T > \alpha$. Let

$$X_t^{(0)} := X_\alpha, \qquad \alpha \leqslant t \leqslant T. \tag{2.3}$$

Define, recursively,

$$X_t^{(n+1)} := X_\alpha + \int_\alpha^t \mu(X_s^{(n)})\, ds + \int_\alpha^t \sigma(X_s^{(n)})\, dB_s, \qquad \alpha \leqslant t \leqslant T. \tag{2.4}$$

For example,

$$\begin{aligned} X_t^{(1)} &= X_\alpha + \mu(X_\alpha)(t - \alpha) + \sigma(X_\alpha)(B_t - B_\alpha), \\ X_t^{(2)} &= X_\alpha + \int_\alpha^t \mu([X_\alpha + \mu(X_\alpha)(s - \alpha) + \sigma(X_\alpha)(B_s - B_\alpha)])\, ds \\ &\quad + \int_\alpha^t \sigma([X_\alpha + \mu(X_\alpha)(s - \alpha) + \sigma(X_\alpha)(B_s - B_\alpha)])\, dB_s, \qquad \alpha \leqslant t \leqslant T. \end{aligned} \tag{2.5}$$

Note that, for each n, $\{X_t^{(n)}: \alpha \leqslant t \leqslant T\}$ is a continuous nonanticipative functional on $[\alpha, T]$. Also,

$$\begin{aligned} (X_t^{(n+1)} - X_t^{(n)})^2 &= \left[\int_\alpha^t (\mu(X_s^{(n)}) - \mu(X_s^{(n-1)}))\, ds + \int_\alpha^t (\sigma(X_s^{(n)}) - \sigma(X_s^{(n-1)}))\, dB_s\right]^2 \\ &\leqslant 2\left(\int_\alpha^t (\mu(X_s^{(n)}) - \mu(X_s^{(n-1)}))\, ds\right)^2 \\ &\quad + 2\left(\int_\alpha^t (\sigma(X_s^{(n)}) - \sigma(X_s^{(n-1)}))\, dB_s\right)^2, \qquad (n \geqslant 1). \end{aligned} \tag{2.6}$$

Write

$$D_t^{(n)} := E\left(\max_{\alpha \leqslant s \leqslant t} (X_s^{(n)} - X_s^{(n-1)})^2\right). \tag{2.7}$$

Taking expectations of the maximum in (2.6), over $0 \leqslant t \leqslant T$, and using (2.1) we get

$$\begin{aligned} D_T^{(n+1)} &\leqslant 2EM^2\left(\int_\alpha^T |X_s^{(n)} - X_s^{(n-1)}|\, ds\right)^2 \\ &\quad + 2E \max_{\alpha \leqslant t \leqslant T} \left(\int_\alpha^t (\sigma(X_s^{(n)}) - \sigma(X_s^{(n-1)}))\, dB_s\right)^2. \end{aligned} \tag{2.8}$$

Now,

$$E\left(\int_\alpha^T |X_s^{(n)} - X_s^{(n-1)}|\, ds\right)^2 \leqslant (T-\alpha)E\int_\alpha^T (X_s^{(n)} - X_s^{(n-1)})^2\, ds$$

$$\leqslant (T-\alpha)\int_\alpha^T D_s^{(n)}\, ds. \tag{2.9}$$

Also, by the Maximal Inequality (Chapter I, Theorem 13.6) and Theorem 1.3 (see the second relation in (1.12)),

$$E \max_{\alpha \leqslant t \leqslant T}\left(\int_\alpha^t (\sigma(X_s^{(n)}) - \sigma(X_s^{(n-1)}))\, dB_s\right)^2$$

$$\leqslant 4E\left(\int_\alpha^T (\sigma(X_s^{(n)}) - \sigma(X_s^{(n-1)}))\, dB_s\right)^2 = 4\int_\alpha^T E(\sigma(X_s^{(n)}) - \sigma(X_s^{(n-1)}))^2\, ds$$

$$\leqslant 4M^2\int_\alpha^T E(X_s^{(n)} - X_s^{(n-1)})^2\, ds \leqslant 4M^2\int_\alpha^T D_s^{(n)}\, ds. \tag{2.10}$$

Using (2.9) and (2.10) in (2.8), obtain

$$D_T^{(n+1)} \leqslant (2M^2(T-\alpha) + 8M^2)\int_\alpha^T D_s^{(n)}\, ds = c_1\int_\alpha^T D_s^{(n)}\, ds \qquad (n \geqslant 1), \tag{2.11}$$

say. An analogous, but simpler, calculation gives

$$D_T^{(1)} \leqslant 2(T-\alpha)^2 E\mu^2(X_\alpha) + 8(T-\alpha)E\sigma^2(X_\alpha) = c_2, \tag{2.12}$$

where c_2 is a finite positive number (Exercise 1). It follows from (2.11), (2.12), and induction that

$$D_T^{(n+1)} \leqslant c_2\frac{(T-\alpha)^n c_1^n}{n!} \qquad (n \geqslant 0). \tag{2.13}$$

By Chebyshev's Inequality,

$$P\left(\max_{\alpha \leqslant t \leqslant T} |X_t^{(n+1)} - X_t^{(n)}| > 2^{-n}\right) \leqslant \frac{2^{2n}c_2(T-\alpha)^n c_1^n}{n!}. \tag{2.14}$$

Since the right side is summable in n, it follows from the Borel–Cantelli Lemma that

$$P\left(\max_{\alpha \leqslant t \leqslant T} |X_t^{(n+1)} - X_t^{(n)}| > 2^{-n} \text{ for infinitely many } n\right) = 0. \tag{2.15}$$

Let N denote the set within parentheses in (2.15). Outside N, the series

$$X_t^{(0)} + \sum_{n=0}^{\infty} (X_t^{(n+1)} - X_t^{(n)}) = \lim_{n\to\infty} X_t^{(n)}$$

converges absolutely, uniformly for $\alpha \leqslant t \leqslant T$, to a nonanticipative functional $\{X_t : \alpha \leqslant t \leqslant T\}$, say. Since each $X_t^{(n)}$ is continuous and the convergence is uniform, the limit is continuous. Also, by Fatou's Lemma,

$$\int_\alpha^T E(X_t^{(n)} - X_t)^2\, dt \leqslant \lim_{m\to\infty} \int_\alpha^T E(X_t^{(n)} - X_t^{(m)})^2\, dt. \tag{2.16}$$

By the triangle inequality for L^2-norms, and (2.13), if $m > n$ then

$$\left(\int_\alpha^T E(X_t^{(n)} - X_t^{(m)})^2\, dt\right)^{1/2} \leqslant \sum_{r=n+1}^{m} \left(\int_\alpha^T E(X_t^{(r-1)} - X_t^{(r)})^2\, dt\right)^{1/2}$$
$$\leqslant \sum_{r=n+1}^{m} (D_T^{(r)})^{1/2} \leqslant \sum_{r=n+1}^{\infty} (D_T^{(r)})^{1/2} \to 0, \tag{2.17}$$

as $n \to \infty$ (Exercise 2). Therefore, $\{X_t : \alpha \leqslant t \leqslant T\}$ is in $\mathcal{M}[\alpha, T]$, and

$$\int_\alpha^T E(X_t^{(n)} - X_t)^2\, dt \to 0, \qquad \text{as } n \to \infty. \tag{2.18}$$

To prove that X_t satisfies (2.2), note that

$$\max_{\alpha\leqslant t\leqslant T} \left| \int_\alpha^t \mu(X_s^{(n)})\, ds + \int_\alpha^t \sigma(X_s^{(n)})\, dB_s - \int_\alpha^t \mu(X_s)\, ds - \int_\alpha^t \sigma(X_s)\, dB_s \right|$$
$$\leqslant M \int_\alpha^T |X_s^{(n)} - X_s|\, ds + \max_{\alpha\leqslant t\leqslant T} \left| \int_\alpha^t (\sigma(X_s^{(n)}) - \sigma(X_s))\, dB_s \right|. \tag{2.19}$$

The first term on the right side goes to zero, as proved above. For the second term, use the Maximal Inequality (Chapter I, Theorem 13.6) to get

$$P\left(\max_{\alpha\leqslant t\leqslant T} \left| \int_\alpha^t (\sigma(X_s^{(n)}) - \sigma(X_s))\, dB_s \right| > \frac{1}{k}\right) \leqslant k^2 E \int_\alpha^T (\sigma(X_s^{(n)}) - \sigma(X_s))^2\, ds$$
$$\leqslant k^2 \left(\sum_{r=n+1}^{\infty} (D_T^{(r)})^{1/2}\right)^2 M^2. \tag{2.20}$$

Since the last expression in (2.20) is summable in n (Exercise 2), it follows from the Borel–Cantelli Lemma that $\max\{|\int_\alpha^t (\sigma(X_s^{(n)}) - \sigma(X_s))\, dB_s| : \alpha \leqslant t \leqslant \beta\} \leqslant 1/k$

for all sufficiently large n, outside a set N_k of zero probability. Let $N_0 := \bigcup \{N_k: 1 \leqslant k < \infty\}$. Then N_0 has zero probability and, outside N_0, the second term on the right side of (2.19) goes to zero. Thus, the right side of (2.4) converges uniformly on $[\alpha, T]$ to the right side of (2.2) as $n \to \infty$, outside a set of zero probability. Since $X_t^{(n)}$ converges to X_t uniformly on $[\alpha, T]$ outside a set of zero probability, (2.2) holds on $[\alpha, T]$ outside a set of zero probability.

It remains to prove the *uniqueness* to the solution to (2.2). Let $\{Y_t: \alpha \leqslant t \leqslant T\}$ be another solution. Then, write $\varphi_t := E(\max\{|X_s - Y_s|: \alpha \leqslant s \leqslant t\})^2$ and use (2.7)–(2.11) with φ_t in place of $D_t^{(n)}$, X_t in place of $X_t^{(n)}$ and Y_t in place of $X_t^{(n-1)}$, to get

$$\varphi_T \leqslant c_1 \int_\alpha^T \varphi_s\, ds. \tag{2.21}$$

Since $t \to \varphi_t$ is nondecreasing, iteration of (2.21) leads, just as in (2.12), (2.13), to

$$\varphi_T \leqslant \frac{c_2(T-\alpha)^n c_1^n}{n!} \to 0 \qquad \text{as } n \to \infty. \tag{2.22}$$

Hence, $\varphi_T = 0$, i.e., $X_t = Y_t$ on $[\alpha, T]$ outside a set of zero probability. ■

The next result identifies the stochastic process $\{X_t: t \geqslant 0\}$ solving (2.2), in the case $\alpha = 0$, as a diffusion with drift $\mu(\cdot)$, diffusion coefficient $\sigma^2(\cdot)$, and initial distribution as the distribution of the given random variable X_0.

Theorem 2.2. Assume (2.1). For each $x \in \mathbb{R}^1$ let $\{X_t^x: t \geqslant 0\}$ denote the unique continuous solution in $\mathcal{M}[0, \infty)$ of Itô's integral equation

$$X_t^x = x + \int_0^t \mu(X_s^x)\, ds + \int_0^t \sigma(X_s^x)\, dB_s \qquad (t \geqslant 0). \tag{2.23}$$

Then $\{X_t^x\}$ is a diffusion on $\mathbb{R}^1$ with drift $\mu(\cdot)$ and diffusion coefficient $\sigma^2(\cdot)$, starting at x.

Proof. By the additivity of the Riemann and stochastic integrals,

$$X_t^x = X_s^x + \int_s^t \mu(X_u^x)\, du + \int_s^t \sigma(X_u^x)\, dB_u, \qquad (t \geqslant s). \tag{2.24}$$

Consider also the equation, for $z \in \mathbb{R}^1$,

$$X_t = z + \int_s^t \mu(X_u)\, du + \int_s^t \sigma(X_u)\, dB_u, \qquad (t \geqslant s). \tag{2.25}$$

Let us write the solution to (2.25) (in $\mathscr{M}[s, \infty)$) as $\theta(s, t; z, B_s^t)$, where $B_s^t := \{B_u - B_s : s \leqslant u \leqslant t\}$. It may be seen from the successive approximation scheme (2.3)–(2.5) that $\theta(s, t; z, B_s^t)$ is *measurable* in (z, B_s^t) (theoretical complement 1). As $\{X_u^x : u \geqslant s\}$ is a continuous stochastic process in $\mathscr{M}[s, \infty)$ and is, by (2.24), a solution to (2.25) with $z = X_s^x$, it follows from the uniqueness of this solution that

$$X_t^x = \theta(s, t; X_s^x, B_s^t), \qquad (t \geqslant s). \tag{2.26}$$

Since X_s^x is $\mathscr{F}_s$-measurable and $\mathscr{F}_s$ and B_s^t are independent, (2.26) implies (see Section 4 of Chapter 0, part (b) of the theorem on Independence and Conditional Expectation) that the conditional distribution of X_t^x given $\mathscr{F}_s$ is the distribution of $\theta(s, t; z, B_s^t)$, say $q(s, t; z, dy)$, evaluated at $z = X_s^x$. Since $\sigma\{X_u^x : 0 \leqslant u \leqslant s\} \subset \mathscr{F}_s$, the Markov property is proved. To prove *homogeneity* of this Markov process, notice that for every $h > 0$, the solution $\theta(s + h, t + h; z, B_{s+h}^{t+h})$ of

$$X_{t+h} = z + \int_{s+h}^{t+h} \mu(X_u)\, du + \int_{s+h}^{t+h} \sigma(X_u)\, dB_u, \qquad (t \geqslant s) \tag{2.27}$$

has the same distribution as that of (2.25). This fact is verified by noting that the successive approximations (2.3)–(2.5) yield the same functions in the two cases except that, in the case of (2.27), B_s^t is replaced by B_{s+h}^{t+h}. But B_s^t and B_{s+h}^{t+h} have the same distribution, so that $q(s + h, t + h; z, dy) = q(s, t; z, dy)$. This proves homogeneity.

To prove that $\{X_t^x\}$ is a diffusion in the sense of (1.2) or (1.2)′ of Chapter V, assume for the sake of simplicity that $\mu(\cdot)$ and $\sigma^2(\cdot)$ are bounded (see Exercise 7 for the general case). Then

$$\begin{aligned} E(X_t^x - x) &= E \int_0^t \mu(X_s^x)\, ds + E \int_0^t \sigma(X_s^x)\, dB_s = E \int_0^t \mu(X_s^x)\, ds \\ &= t\mu(x) + \int_0^t E(\mu(X_s^x) - \mu(x))\, \mathrm{ds} = t\mu(x) + \mathrm{o}(t), \end{aligned} \tag{2.28}$$

since $E\mu(X_s^x) - \mu(x) \to 0$ as $s \downarrow 0$. Next,

$$\begin{aligned} E(X_t^x - x)^2 &= E\left(\int_0^t \mu(X_s^x)\, ds\right)^2 + E\left(\int_0^t \sigma(X_s^x)\, dB_s\right)^2 \\ &\quad + 2E\left[\left(\int_0^t \mu(X_s^x)\, ds\right)\left(\int_0^t \sigma(X_s^x)\, dB_s\right)\right] \\ &= \mathrm{O}(t^2) + \int_0^t E\sigma^2(X_s^x)\, ds + \mathrm{o}(t) = \mathrm{o}(t) \qquad \text{as } t \downarrow 0. \end{aligned} \tag{2.29}$$

For, by the Schwarz Inequality,

$$\left|E\left[\left(\int_0^t \mu(X_s^x)\,ds\right)\left(\int_0^t \sigma(X_s^x)\,dB_s\right)\right]\right|$$
$$\leqslant \left[E\left(\int_0^t \mu(X_s^x)\,ds\right)^2\right]^{1/2}\left[E\left(\int_0^t \sigma(X_s^x)\,dB_s\right)^2\right]^{1/2}$$
$$= \mathrm{O}(t)\left[\int_0^t E\sigma^2(X_s^x)\,ds\right]^{1/2} = \mathrm{O}(t)\mathrm{O}(t^{1/2}) = \mathrm{o}(t) \qquad \text{as } t \downarrow 0,$$

leading to

$$E(X_t^x - x)^2 = \int_0^t \sigma^2(X_s^x)\,ds + \mathrm{o}(t) \qquad \text{as } t \downarrow 0. \tag{2.30}$$

Now, as in (2.28),

$$\int_0^t E\sigma^2(X_s^x)\,ds = \int_0^t \sigma^2(x)\,dx + \int_0^t E(\sigma^2(X_s^x) - \sigma^2(x))\,ds$$
$$= t\sigma^2(x) + \mathrm{o}(t) \qquad \text{as } t \downarrow 0. \tag{2.31}$$

The last condition (1.2)′ of Chapter V is checked in Section 3, Corollary 3.5. ■

2.2 Construction of Multidimensional Diffusions

In order to construct multidimensional diffusions by the method of Itô it is necessary to define stochastic integrals for vector-valued integrands with respect to the increments of a multidimensional Brownian motion. This turns out to be rather straightforward.

Let $\{\mathbf{B}_t = (B_t^{(1)}, \ldots, B_t^{(k)})\}$ be a standard k-dimensional Brownian motion. *Assume* (1.4) *for* $\{\mathbf{B}_t\}$. Define a vector-valued stochastic process

$$\mathbf{f} = \{\mathbf{f}(t) = (f^{(1)}(t), \ldots, f^{(k)}(t)) : \alpha \leqslant t \leqslant \beta\}$$

to be a *nonanticipative step functional on* $[\alpha, \beta]$ if (1.5) holds for some finite set of time points $t_0 = \alpha < t_1 < \cdots < t_m = \beta$ and some random vectors $\mathbf{f}_i$ $(0 \leqslant i \leqslant m)$ with values in $\mathbb{R}^k$.

Definition 2.1. The *stochastic integral*, or the *Itô integral* of a nonanticipative step functional $\mathbf{f}$ on $[\alpha, \beta]$ is defined by

$$\int_\alpha^t \mathbf{f}(s)\cdot d\mathbf{B}_s := \begin{cases} \mathbf{f}(t_0)\cdot(\mathbf{B}_t - \mathbf{B}_{t_0}) & \text{for } t \in [\alpha, t_1], \\ \sum_{j=1}^{r} \mathbf{f}(t_{j-1})\cdot(\mathbf{B}_{t_j} - \mathbf{B}_{t_{j-1}}) + \mathbf{f}(t_r)\cdot(\mathbf{B}_t - \mathbf{B}_{t_r}) & \text{for } t \in (t_r, t_{r+1}]. \end{cases} \tag{2.32}$$

Here $\cdot$ (*dot*) denotes euclidean inner product,

$$\mathbf{f}(s)\cdot(\mathbf{B}_t - \mathbf{B}_s) = \sum_{i=1}^{k} f^{(i)}(s)(B_t^{(i)} - B_s^{(i)}). \tag{2.33}$$

It follows from this definition that the stochastic integral (2.32) is the sum of k one-dimensional stochastic integrals,

$$\int_\alpha^t \mathbf{f}(s)\cdot d\mathbf{B}_s = \sum_{i=1}^{k} f^{(i)}(s)\, dB_s^{(i)}. \tag{2.34}$$

Parts (a), (b) of Proposition 1.1 extend immediately. In order to extend part (c) assume, as in (1.9),

$$\int_\alpha^\beta E|\mathbf{f}(u)|^2\, du \equiv \sum_{i=1}^{k} \int_\alpha^\beta E(f^{(i)}(u))^2\, du < \infty. \tag{2.35}$$

The square integrability and the martingale property of the stochastic integral follow from those of each of its k summands in (2.34). Similarly, one has

$$E\left(\int_\alpha^t \mathbf{f}(s)\cdot d\mathbf{B}_s\right) = 0. \tag{2.36}$$

It remains to prove the analog of (1.11), namely,

$$E\left(\left(\int_\alpha^t \mathbf{f}(u)\cdot dB_u\right)^2 \middle| \mathscr{F}_\alpha\right) = E\left(\int_\alpha^t |\mathbf{f}(u)|^2\, du \middle| \mathscr{F}_\alpha\right). \tag{2.37}$$

In view of (1.11) applied to $\int_\alpha^t f^{(i)}(u)\, dB_u^{(i)}$ $(1 \leqslant i \leqslant k)$, it is enough to prove that the product terms vanish, i.e.,

$$E\left[\left(\int_\alpha^t f^{(i)}(u)\, dB_u^{(i)}\right)\left(\int_\alpha^t f^{(j)}(u)\, dB_u^{(j)}\right) \middle| \mathscr{F}_\alpha\right] = 0 \qquad \text{for } i \neq j. \tag{2.38}$$

To see this, note that, for $s < s' < u < u'$,

$$E[f^{(i)}(s)(B_{s'}^{(i)} - B_s^{(i)}) f^{(j)}(u)(B_{u'}^{(j)} - B_u^{(j)}) \mid \mathscr{F}_\alpha]$$

$$= E[E(\cdots \mid \mathscr{F}_u) \mid \mathscr{F}_\alpha] = E[f^{(i)}(s)(B_{s'}^{(i)} - B_s^{(i)}) f^{(j)}(u) E(B_{u'}^{(j)} - B_u^{(j)} \mid \mathscr{F}_u) \mid \mathscr{F}_\alpha] = 0. \tag{2.39}$$

Also, by the indepencence of $\mathscr{F}_s$ and $\{\mathbf{B}_{s'} - \mathbf{B}_s : s' \geqslant s\}$, and by the independence of $\{B_t^{(i)}\}$ and $\{B_t^{(j)}\}$ for $i \neq j$,

$$
\begin{aligned}
&E[f^{(i)}(s)(B_{s'}^{(i)} - B_s^{(i)})f^{(j)}(s)(B_{s'}^{(j)} - B_s^{(j)}) \mid \mathscr{F}_\alpha] \\
&\quad = E[E(\cdots \mid \mathscr{F}_s) \mid \mathscr{F}_\alpha] = E[f^{(i)}(s)f^{(j)}(s)\{E((B_{s'}^{(i)} - B_s^{(i)})(B_{s'}^{(j)} - B_s^{(j)}) \mid \mathscr{F}_s)\} \mid \mathscr{F}_\alpha] \\
&\quad = 0.
\end{aligned}
$$

Thus, Proposition 1.1 is fully extended to k-dimension. For the extension to more general nonanticipative functionals, consider the following analog of Definition 1.2.

Definition 2.2. If $\mathbf{f} = \{\mathbf{f}(t) = (f^{(1)}(t), \ldots, f^{(m)}(t)): \alpha \leqslant t \leqslant \beta\}$ is a right-continuous stochastic process with values in $\mathbb{R}^m$ (for some m) such that $\mathbf{f}(t)$ is $\mathscr{F}_t$-measurable for all $t \in [\alpha, \beta]$, then $\mathbf{f}$ is called a *nonanticipative functional on* $[\alpha, \beta]$ *with values in* $\mathbb{R}^m$. If, in addition,

$$
E\int_\alpha^\beta |\mathbf{f}(t)|^2\, dt < \infty, \tag{2.40}
$$

then $\mathbf{f}$ is said to *belong to* $\mathscr{M}[\alpha, \beta]$. If $\mathbf{f} \in \mathscr{M}[\alpha, \beta]$ for all $\beta > \alpha$, then $\mathbf{f}$ *belongs to* $\mathscr{M}[\alpha, \infty)$.

For $\mathbf{f} \in \mathscr{M}[\alpha, \beta]$ with values in $\mathbb{R}^k$ one may apply the results in Section 1 to each of the k components $\int_\alpha^t f^{(i)}(u)\, dB_u^{(i)}$ $(1 \leqslant i \leqslant k)$ to extend Proposition 1.2 to k-dimension and to define the stochastic integral $\int_\alpha^t \mathbf{f}(u)\, d\mathbf{B}_u$ for a k-dimensional $\mathbf{f} \in \mathscr{M}[\alpha, \beta]$. The following extension of Theorem 1.3 is now immediate.

Theorem 2.3. Suppose $\mathbf{f} \in \mathscr{M}[\alpha, \beta]$ with values in $\mathbb{R}^k$. Then

(a) $t \to \displaystyle\int_\alpha^t \mathbf{f}(u)\, d\mathbf{B}_u$ is continuous and additive; also

(b) $\left\{\displaystyle\int_\alpha^t \mathbf{f}(u)\, d\mathbf{B}_u: \alpha \leqslant t \leqslant \beta\right\}$ is a square integrable $\{\mathscr{F}_t\}$-martingale and

$$
E\left(\left(\int_\alpha^t \mathbf{f}(u)\, d\mathbf{B}_u\right)^2 \,\middle|\, \mathscr{F}_\alpha\right) = E\left(\int_\alpha^t |\mathbf{f}(u)|^2\, du \,\middle|\, \mathscr{F}_\alpha\right). \tag{2.41}
$$

The final task is to construct multidimensional diffusions. For this let $\mu^{(i)}(\mathbf{x}), \sigma_{ij}(\mathbf{x})$ $(1 \leqslant i, j \leqslant k)$ be real-valued Lipschitzian functions on $\mathbb{R}^k$. Then, writing, $\boldsymbol{\sigma}(\mathbf{x})$ for the $k \times k$ matrix $((\sigma_{ij}(\mathbf{x})))$,

$$
|\boldsymbol{\mu}(\mathbf{x}) - \boldsymbol{\mu}(\mathbf{y})| \leqslant M|\mathbf{x} - \mathbf{y}|, \qquad \|\boldsymbol{\sigma}(\mathbf{x}) - \boldsymbol{\sigma}(\mathbf{y})\| \leqslant M|\mathbf{x} - \mathbf{y}|, \qquad (\mathbf{x}, \mathbf{y} \in \mathbb{R}^k), \tag{2.42}
$$

for some positive constant M. Here $\|\cdot\|$ denotes the *matrix norm*, $\|\mathbf{D}\| := \sup\{|\mathbf{Dz}|: \mathbf{z} \in \mathbb{R}^k\}$.

Consider the vector stochastic integral equation,

$$\mathbf{X}_t = \mathbf{X}_\alpha + \int_\alpha^t \boldsymbol{\mu}(\mathbf{X}_s)\,ds + \int_\alpha^t \boldsymbol{\sigma}(\mathbf{X}_s)\,d\mathbf{B}_s, \qquad (t \geqslant s), \tag{2.43}$$

which is shorthand for the system of k equations

$$X_t^{(i)} = X_\alpha^{(i)} + \int_\alpha^t \mu^{(i)}(\mathbf{X}_s)\,ds + \int_\alpha^t \boldsymbol{\sigma}_i(\mathbf{X}_s)\cdot d\mathbf{B}_s, \qquad (1 \leqslant i \leqslant k). \tag{2.44}$$

Here $\boldsymbol{\sigma}_i(\mathbf{x})$ is the k-dimensional vector, $(\sigma_{i1}(\mathbf{x}), \ldots, \sigma_{ik}(\mathbf{x}))$. The proofs of the following theorems are entirely analogous to those of Theorems 2.1 and 2.2 (Exercises 4 and 5). Basically, one only needs to write

$$|\mathbf{X}_{n+1}(t) - \mathbf{X}_n(t)|^2 \quad \text{instead of} \quad (X_{n+1}(t) - X_n(t))^2,$$

$$\left|\int_\alpha^t (\boldsymbol{\mu}(\mathbf{X}_n(s)) - \boldsymbol{\mu}(\mathbf{X}_{n-1}(s)))\,ds\right|^2 \quad \text{in place of} \quad \left(\int_\alpha^t (\mu(X_n(s)) - \mu(X_{n-1}(s)))\,ds\right)^2,$$

$$\|\boldsymbol{\sigma}(\mathbf{X}_n(s)) - \boldsymbol{\sigma}(\mathbf{X}_{n-1}(s))\|^2 \quad \text{for} \quad (\sigma(X_n(s)) - \sigma(X_{n-1}(s))^2, \qquad \text{etc.}$$

Theorem 2.4. Suppose $\boldsymbol{\mu}(\cdot)$, $\boldsymbol{\sigma}(\cdot)$ satisfy (2.42). Then, for each $\mathscr{F}_\alpha$-measurable square integrable random vector $\mathbf{X}_\alpha$, there exists a unique (up to a P-null set) continuous element $\{\mathbf{X}_t: t \geqslant \alpha\}$ of $\mathscr{M}[\alpha, \infty)$ such that (2.43) holds.

Theorem 2.5. Suppose $\boldsymbol{\mu}(\cdot)$, $\boldsymbol{\sigma}(\cdot)$ satisfy (2.42), and let $\{\mathbf{X}_t^\mathbf{x}: t \geqslant 0\}$ denote the unique (up to a P-null set) continuous nonanticipative functional in $\mathscr{M}[0, \infty)$ satisfying Itô's integral equation

$$\mathbf{X}_t = \mathbf{x} + \int_0^t \boldsymbol{\mu}(\mathbf{X}_s)\,ds + \int_0^t \boldsymbol{\sigma}(\mathbf{X}_s)\,d\mathbf{B}_s, \qquad (t \geqslant 0). \tag{2.45}$$

Then $\{\mathbf{X}_t^\mathbf{x}\}$ is a diffusion on $\mathbb{R}^k$ with drift $\boldsymbol{\mu}(\cdot)$ and diffusion matrix $\boldsymbol{\sigma}(\cdot)\boldsymbol{\sigma}'(\cdot)$, $\boldsymbol{\sigma}'(\cdot)$ being the transpose $\boldsymbol{\sigma}(\cdot)$.

It may be noted that in Theorems 2.4 and 2.5 it is not assumed that $\boldsymbol{\sigma}(\mathbf{x})$ is positive definite. The positive definiteness guarantees the existence of a *density* for the transition probability and its *smoothness* (see theoretical complement 5.1 of Chapter V), but is not needed for the Markov property.

Example. Let $k = 1$, $\mu(x) = -\gamma x$, $\sigma(x) = \sigma$. Then the successive approximations

$X_t^{(n)}$ (see Eqs. 2.3–2.5) are given by (assuming $B_0 = 0$)

$$
\begin{aligned}
X_t^{(0)} &\equiv X_0, \\
X_t^{(1)} &= X_0 + \int_0^t - \gamma X_0\, ds + \sigma B_t = X_0(1 - t\gamma) + \sigma B_t, \\
X_t^{(2)} &= X_0 + \int_0^t -\gamma\{X_0(1 - s\gamma) + \sigma B_s\}\, ds + \sigma B_t \\
&= \left(1 - t\gamma + \frac{t^2\gamma^2}{2!}\right) X_0 - \gamma\sigma \int_0^t B_s\, ds + \sigma B_t, \\
X_t^{(3)} &= X_0 + \int_0^t -\gamma\left\{\left(1 - s\gamma + \frac{s^2\gamma^2}{2!}\right) X_0 - \gamma\sigma \int_0^s B_u\, du + \sigma B_s\right\} ds + \sigma B_t \\
&= \left(1 - t\gamma + \frac{t^2\gamma^2}{2!} - \frac{t^3\gamma^3}{3!}\right) X_0 + \gamma^2\sigma \int_0^t \int_0^s B_u\, du\, ds - \gamma\sigma \int_0^t B_s\, ds + \sigma B_t \\
&= \left(1 - t\gamma + \frac{t^2\gamma^2}{2!} - \frac{t^3\gamma^3}{3!}\right) X_0 + \gamma^2\sigma \int_0^t \left(\int_u^t ds\right) B_u\, du - \gamma\sigma \int_0^t B_s\, ds + \sigma B_t \\
&= \left(1 - t\gamma + \frac{t^2\gamma^2}{2!} - \frac{t^3\gamma^3}{3!}\right) X_0 + \gamma^2\sigma \int_0^t (t - u) B_u\, du - \gamma\sigma \int_0^t B_s\, ds + \sigma B_t.
\end{aligned}
\tag{2.46}
$$

Assume, as an induction hypothesis,

$$
X_t^{(n)} = \left(\sum_{m=0}^{n} \frac{(-t\gamma)^m}{m!}\right) X_0 + \sum_{m=1}^{n-1} (-\gamma)^m \sigma \int_0^t \frac{(t - s)^{m-1}}{(m - 1)!} B_s\, ds + \sigma B_t. \tag{2.47}
$$

Now use (2.4) to check that (2.47) holds for $n + 1$, replacing n. Therefore, (2.47) holds for all n. But as $n \to \infty$, the right side converges (for every $\omega \in \Omega$) to

$$
e^{-t\gamma} X_0 - \gamma\sigma \int_0^t e^{-\gamma(t-s)} B_s\, ds + \sigma B_t. \tag{2.48}
$$

Hence, X_t equals (2.48). In particular, with $X_0 = x$,

$$
X_t^x = e^{-t\gamma} x - \gamma\sigma \int_0^t e^{-\gamma(t-s)} B_s\, ds + \sigma B_t. \tag{2.49}
$$

As a special case, for $\gamma > 0$, $\sigma \neq 0$, (2.49) gives a representation of an Ornstein–Uhlenbeck process as a functional of a Brownian motion.

3 ITÔ'S LEMMA

Brownian paths $s \to B_s$ have finite quadratic variation on every finite time interval $[0, t]$ in the sense that

$$\max_{1 \leqslant N \leqslant 2^n} \left| \sum_{m=0}^{N-1} (B_{(m+1)2^{-n}t} - B_{m2^{-n}t})^2 - N2^{-n}t \right| \to 0 \quad \text{a.s.} \qquad \text{as } n \to \infty. \tag{3.1}$$

This is easily checked by recognizing that the expression $Z_{N,n}$, say, within the absolute signs is, for each n, a martingale ($1 \leqslant N \leqslant 2^n$), so that the Maximal Inequality (Chapter I, Eq. 13.56) may be used to prove (3.1) (Exercise 1). Notice that the quadratic variation of $\{B_s\}$ over an interval equals the length of the interval. A consequence of (3.1) is a curious and extremely important "chain rule" for the stochastic calculus, called *Itô's Lemma.* The present section is devoted to a derivation of this chain rule and some applications.

For an intuitive understanding of this chain rule, consider a nonanticipative functional of the form

$$Y(t) = Y(0) + \int_0^t f(s)\, ds + \int_0^t g(s)\, dB_s, \tag{3.2}$$

where $f, g \in \mathscr{M}[0, T]$, and $Y(0)$ is $\mathscr{F}_0$-measurable and square integrable. One may express (3.2) in the differential form

$$dY(t) = f(t)\, dt + g(t)\, dB_t. \tag{3.3}$$

Suppose that φ is a real-valued twice continuously differentiable function on $\mathbb{R}^1$, say with bounded derivatives φ', φ''. Then Itô's Lemma says

$$\begin{aligned} d\varphi(Y(t)) &= \varphi'(Y(t))\, dY(t) + \tfrac{1}{2}\varphi''(Y(t))g^2(t)\, dt \\ &= \{\varphi'(Y(t))f(t) + \tfrac{1}{2}\varphi''(Y(t))g^2(t)\}\, dt + \varphi'(Y(t))g(t)\, dB_t. \end{aligned} \tag{3.4}$$

In other words,

$$\varphi(Y(t)) = \varphi(Y(0)) + \int_0^t \{\varphi'(Y(s))f(s) + \tfrac{1}{2}\varphi''(Y(s))g^2(s)\}\, ds + \int_0^t \varphi'(Y(s))g(s)\, dB_s. \tag{3.5}$$

Observe that a formal application of ordinary calculus would give

$$d\varphi(Y(t)) = \varphi'(Y(t))\, dY(t) = \varphi'(Y(t))f(t)\, dt + \varphi'(Y(t))g(t)\, dB_t. \tag{3.6}$$

The extra term $\frac{1}{2}\varphi''(Y(t))g^2(t)\, dt$ appearing in (3.4) arises because

$$(dY(t))^2 = g^2(t)(dB_t)^2 + o(dt) = g^2(t)\, dt + o(dt).$$

Since the term $g^2(t)\,dt$ cannot be neglected in computing the differential $d\varphi(Y(t))$, one must expand $\varphi(Y(t+dt))$ around $Y(t)$ in a Taylor expansion including the second derivative of φ.

The same argument as above applied to a function $\varphi(t, y)$ on $[0, T] \times \mathbb{R}^1$, such that $\varphi_0 := \partial\varphi/\partial t$, φ', φ'' are continuous and bounded, leads to

$$d\varphi(t, Y(t)) = \{\varphi_0(t, Y(t)) + \varphi'(t, Y(t))f(t) + \tfrac{1}{2}\varphi''(t, Y(t))g^2(t)\}\,dt + \varphi'(t, Y(t))g(t)\,dB_t. \tag{3.7}$$

To state an extension to multidimensions, let $\varphi(t, \mathbf{y})$ be a function on $[0, T] \times \mathbb{R}^m$ that is once continuously differentiable in t and twice in $\mathbf{y}$. Write

$$\partial_0\varphi(t, \mathbf{y}) := \frac{\partial\varphi(t, \mathbf{y})}{\partial t}, \qquad \partial_r\varphi(t, \mathbf{y}) := \frac{\partial\varphi(t, \mathbf{y})}{\partial y^{(r)}} \qquad (1 \leqslant r \leqslant m). \tag{3.8}$$

Let $\{\mathbf{B}_t\}$ be a k-dimensional standard Brownian motion satisfying conditions (1.4(i), (ii)) (with B_s replaced by $\mathbf{B}_s$). Suppose $\mathbf{Y}(t)$ is a vector of m processes of the form

$$\begin{aligned} \mathbf{Y}(t) &= (Y^{(1)}(t), \ldots, Y^{(m)}(t)), \\ Y^{(r)}(t) &= Y^{(r)}(0) + \int_0^t f_r(s)\,ds + \int_0^t \mathbf{g}_r(s)\cdot d\mathbf{B}_s \qquad (1 \leqslant r \leqslant m). \end{aligned} \tag{3.9}$$

Here $f_1, \ldots, f_m$ are real-valued and $\mathbf{g}_1, \ldots, \mathbf{g}_m$ vector-valued (with values in $\mathbb{R}^k$) nonanticipative functionals belonging to $\mathscr{M}[0, T]$. Also $Y^{(r)}(0)$ are $\mathscr{F}_0$-measurable square integrable random variables. One may express (3.9) in the differential form

$$dY^{(r)}(t) = f_r(t)\,dt + \mathbf{g}_r(t)\cdot d\mathbf{B}_t \qquad (1 \leqslant r \leqslant m). \tag{3.10}$$

Itô's lemma says,

$$\begin{aligned} d\varphi(t, \mathbf{Y}(t)) = \Bigg\{&\partial_0\varphi(t, \mathbf{Y}(t)) + \sum_{r=1}^{m} \partial_r\varphi(t, \mathbf{Y}(t))f_r(t) \\ &+ \frac{1}{2!}\sum_{1 \leqslant r, r' \leqslant m} \partial_r\partial_{r'}\varphi(t, \mathbf{Y}(t))(\mathbf{g}_r(t)\cdot\mathbf{g}_{r'}(t))\Bigg\}\,dt \\ &+ \sum_{r=1}^{m} \partial_r\varphi(t, \mathbf{Y}(t))\mathbf{g}_r(t)\cdot d\mathbf{B}_t. \end{aligned} \tag{3.11}$$

In order to arrive at this, write

$$\begin{aligned} d\varphi(t, \mathbf{Y}(t)) &= \varphi(t+dt, \mathbf{Y}(t+dt)) - \varphi(t, \mathbf{Y}(t)) = \varphi(t+dt, \mathbf{Y}(t+dt)) \\ &\quad - \varphi(t, \mathbf{Y}(t+dt)) + \varphi(t, \mathbf{Y}(t+dt)) - \varphi(t, \mathbf{Y}(t)) \\ &= \partial_0 \varphi(t, \mathbf{Y}(t))\, dt + \sum_{r=1}^{m} \partial_r \varphi(t, \mathbf{Y}(t))\, dY^{(r)}(t) \\ &\quad + \frac{1}{2!} \sum_{1 \leqslant r, r' \leqslant m} \partial_r \partial_{r'} \varphi(t, \mathbf{Y}(t))\, dY^{(r)}(t)\, dY^{(r')}(t). \end{aligned} \tag{3.12}$$

In Newtonian calculus, of course, the contribution of the last sum to the differential would be zero. But there is one term in the product $dY^{(r)}(t)\, dY^{(r')}(t)$ which is of the order dt and, therefore, must be retained in computing the stochastic differential. This term is (see Eq. 3.10)

$$\begin{aligned} (\mathbf{g}_r(t) \cdot d\mathbf{B}_t)(\mathbf{g}_{r'}(t) \cdot d\mathbf{B}_t) &= \left(\sum_{i=1}^{k} g_r^{(i)}(t)\, dB_t^{(i)} \right) \left(\sum_{j=1}^{k} g_{r'}^{(j)}(t)\, dB_t^{(j)} \right) \\ &= \sum_{i=1}^{k} g_r^{(i)}(t) g_{r'}^{(i)}(t) (dB_t^{(i)})^2 + \sum_{i \neq j} g_r^{(i)}(t) g_{r'}^{(j)}(t)\, dB_t^{(i)}\, dB_t^{(j)}. \end{aligned} \tag{3.13}$$

Now as seen above (see Eq. 3.1) the first sum in (3.13) equals

$$\left(\sum_{i=1}^{k} g_r^{(i)}(t) g_{r'}^{(i)}(t) \right) dt = \mathbf{g}_r(t) \cdot \mathbf{g}_{r'}(t)\, dt. \tag{3.14}$$

To show that the contribution of the second term in (3.13) to $\varphi(t, \mathbf{Y}(t)) - \varphi(0, \mathbf{Y}(0))$ over any interval $[0, t]$ is zero note that, for $i \neq j$,

$$Z'_{N,n} := \sum_{m=0}^{N-1} g_r^{(i)}(m2^{-n}t) g_{r'}^{(j)}(m2^{-n}t) (B_{(m+1)2^{-n}t}^{(i)} - B_{m2^{-n}t}^{(i)})(B_{(m+1)2^{-n}t}^{(j)} - B_{m2^{-n}t}^{(j)}),$$
$$1 \leqslant N \leqslant 2^n, \tag{3.15}$$

is a martingale. This is true as the conditional expectation of each summand, given all the preceding, is zero. Therefore, as in (3.1),

$$\max_{1 \leqslant N \leqslant 2^n} |Z'_{N,n}| \to 0 \quad \text{a.s.} \qquad \text{as } n \to \infty. \tag{3.16}$$

Thus,

$$\sum_{i \neq j} g_r^{(i)}(t) g_{r'}^{(j)}(t)\, dB_t^{(i)}\, dB_t^{(j)} = 0 \qquad (i \neq j). \tag{3.17}$$

Using (3.14), (3.17) in (3.12), Itô's lemma (3.11) is obtained. A more elaborate argument is given in theoretical complement 3.1. For ease of reference, here is a statement of Itô's Lemma.

Theorem 3.1. (*Itô's Lemma*). Assume $f_1, \ldots, f_m$, $\mathbf{g}_1, \ldots, \mathbf{g}_m$ belong to $\mathcal{M}[0, T]$, $Y^{(r)}(0)$ $(1 \leqslant r \leqslant m)$ $\mathcal{F}_0$-measurable and square integrable. Let $\varphi(t, \mathbf{y})$ be a real-valued function on $[0, T] \times \mathbb{R}^m$ which is once continuously differentiable in t and twice in $\mathbf{y}$. Assume that

$$E \int_0^T (\partial_r \varphi(s, \mathbf{Y}(s))^2 |\mathbf{g}_r(s)|^2 \, ds < \infty, \qquad (1 \leqslant r \leqslant m), \tag{3.18}$$

i.e., $\partial_r \varphi \mathbf{g}_r \in \mathcal{M}[0, T]$ $(1 \leqslant r \leqslant m)$. Then (3.11) holds, i.e.,

$$\begin{aligned} \varphi(t, \mathbf{Y}(t)) - \varphi(s, \mathbf{Y}(s)) = \int_s^t \Bigg\{ & \partial_0 \varphi(u, \mathbf{Y}(u)) + \sum_{r=1}^m \partial_r \varphi(u, \mathbf{Y}(u)) f_r(u) \\ & + \frac{1}{2!} \sum_{1 \leqslant r, r' \leqslant m} \partial_r \partial_{r'} \varphi(u, \mathbf{Y}(u)) \mathbf{g}_r(u) \cdot \mathbf{g}_{r'}(u) \Bigg\} \, du \\ & + \sum_{r=1}^m \int_s^t \partial_r \varphi(u, \mathbf{Y}(u)) \mathbf{g}_r(u) \cdot d\mathbf{B}_u \qquad (0 \leqslant s < t \leqslant T). \end{aligned} \tag{3.19}$$

Applying Itô's Lemma to the diffusion $\mathbf{Y}(t) = \mathbf{X}_t$ $(t \geqslant 0)$ constructed in Section 2, the following corollary is obtained immediately.

Corollary 3.2. Let $\{\mathbf{X}_t\}$ be a diffusion given by the solution to the stochastic differential equation (2.43), with $\alpha = 0$ and $\boldsymbol{\mu}(\cdot), \boldsymbol{\sigma}(\cdot)$ Lipschitzian. Assume $\varphi(t, \mathbf{y})$ satisfies the hypothesis of Theorem 3.1 with $m = k$, and (3.18) replaced by

$$E \int_0^T (\partial_r \varphi)^2 (u, \mathbf{X}_u) |\boldsymbol{\sigma}_r(\mathbf{X}_u)|^2 \, du < \infty, \qquad (1 \leqslant r \leqslant k).$$

(a) Then one has the relation

$$\begin{aligned} \varphi(t, \mathbf{X}_t) = \varphi(s, \mathbf{X}_s) + & \int_s^t \{\partial_0 \varphi(u, \mathbf{X}_u) + (\mathbf{A}\varphi)(u, \mathbf{X}_u)\} \, du \\ & + \sum_{r=1}^k \int_s^t \partial_r \varphi(u, \mathbf{X}_u) \boldsymbol{\sigma}_r(\mathbf{X}_u) \cdot d\mathbf{B}_u \qquad (s < t \leqslant T), \end{aligned} \tag{3.20}$$

where $\boldsymbol{\sigma}_r(\mathbf{x})$ is the rth row vector of $\boldsymbol{\sigma}(\mathbf{x})$, and $\mathbf{A}$ is the differential operator

(14.9) of Chapter V with $\mathbf{D}(\mathbf{x}) = \boldsymbol{\sigma}(\mathbf{x})\boldsymbol{\sigma}'(\mathbf{x})$,

$$(\mathbf{A}\varphi)(u, \mathbf{x}) := \frac{1}{2} \sum_{1 \leqslant r, r' \leqslant k} d_{rr'}(\mathbf{x}) \partial_r \partial_{r'} \varphi(u, \mathbf{x}) + \sum_{r=1}^{k} \mu^{(r)}(\mathbf{x}) \partial_r \varphi(u, \mathbf{x}),$$
$$((d_{rr'}(\mathbf{x}))) := \boldsymbol{\sigma}(\mathbf{x})\boldsymbol{\sigma}'(\mathbf{x}). \tag{3.21}$$

(b) In particular, if $\partial_0 \varphi$, $\partial_{rr'} \varphi$ are bounded then

$$Z_t := \varphi(t, \mathbf{X}_t) - \int_0^t \{\partial_0 \varphi(u, \mathbf{X}_u) + (\mathbf{A}\varphi)(u, \mathbf{X}_u)\}\, du \qquad (0 \leqslant t \leqslant T), \tag{3.22}$$

is a $\{\mathscr{F}_t\}$-martingale.

Note that part (b) is an immediate consequence of part (a), since $Z_t - Z_s$ equals the stochastic integral in (3.20), whose conditional expectation, given $\mathscr{F}_s$, is zero.

Corollary 3.2 generalizes Corollary 2.4 of Chapter V to a wider class of functions and to multidimension. Therefore, one may provide derivations of Propositions 2.5 and 15.1 of Chapter V, as well as criteria for transience and recurrence based on them, using Itô's Lemma instead of Corollary 2.4 of Chapter V. The next result similarly provides a criterion for positive recurrence. Recall the notation (see Eqs. 15.33 and 15.34 of Chapter V)

$$\mathbf{D}(\mathbf{x}) \equiv ((d_{ij}(\mathbf{x}))) := \boldsymbol{\sigma}(\mathbf{x})\boldsymbol{\sigma}'(\mathbf{x}), \qquad d(\mathbf{x}) := \sum_{i,j} d_{ij}(\mathbf{x}) x^{(i)} x^{(j)} / |\mathbf{x}|^2,$$

$$B(\mathbf{x}) := \sum_i d_{ii}(\mathbf{x}), \qquad C(\mathbf{x}) := 2 \sum_i x^{(i)} \mu^{(i)}(\mathbf{x}),$$

$$\bar{\beta}(r) := \max_{|\mathbf{x}| = r} \frac{B(\mathbf{x}) + C(\mathbf{x})}{d(\mathbf{x})} - 1, \qquad \underline{\beta}(r) := \min_{|\mathbf{x}| = r} \frac{B(\mathbf{x}) + C(\mathbf{x})}{d(\mathbf{x})} - 1, \tag{3.23}$$

$$\bar{\alpha}(r) := \max_{|\mathbf{x}| = r} d(\mathbf{x}), \qquad \underline{\alpha}(r) := \min_{|\mathbf{x}| = r} d(\mathbf{x}),$$

$$\bar{I}(r) := \int_c^r \frac{\bar{\beta}(u)}{u}\, du, \qquad \underline{I}(r) := \int_c^r \frac{\underline{\beta}(u)}{u}\, du,$$

where $c > 0$ is a given constant. Also note that (see Eq. 15.35 of Chapter V) for every F that is twice differentiable on $(0, \infty)$, $\mathbf{A}\varphi$ for $\varphi(\mathbf{x}) := F(|\mathbf{x}|)$ is given by

$$2(\mathbf{A}\varphi)(\mathbf{x}) = (d(\mathbf{x}))F''(|\mathbf{x}|) + \frac{B(\mathbf{x}) + C(\mathbf{x}) - d(\mathbf{x})}{|\mathbf{x}|} F'(|\mathbf{x}|) \qquad (|\mathbf{x}| > 0). \tag{3.24}$$

Proposition 3.3. Let $\boldsymbol{\mu}(\cdot)$, $\boldsymbol{\sigma}(\cdot)$ be Lipschitzian, and $\boldsymbol{\sigma}(\mathbf{x})$ nonsingular for all $\mathbf{x}$.

Suppose that, for some $c > 0$,

$$\int_c^\infty \exp\{-\bar{I}(u)\}\,du = \infty, \qquad \int_c^\infty \frac{1}{\underline{\alpha}(u)} \exp\{\bar{I}(u)\}\,du < \infty. \tag{3.25}$$

Then

$$E(\tau_{\partial B(\mathbf{0}:r_0)} \mid \mathbf{X}_0 = \mathbf{x}) < \infty \qquad (|\mathbf{x}| > r_0 > 0), \tag{3.26}$$

where

$$\tau_{\partial B(0:r)} := \inf\{t \geqslant 0: |\mathbf{X}_t| = r\}. \tag{3.27}$$

Proof. First note that if (3.25) holds for some $c > 0$ then it holds for all $c > 0$ (Exercise 2). Let $c = r_0 > 0$. Define

$$F(r) := \int_{r_0}^r \exp\{-\bar{I}(u)\}\left(\int_u^\infty \frac{1}{\underline{\alpha}(v)} \exp\{\bar{I}(v)\}\,dv\right) du, \qquad (r \geqslant r_0) \tag{3.28}$$

and

$$\varphi(\mathbf{x}) = F(|\mathbf{x}|), \qquad |\mathbf{x}| \geqslant r_0. \tag{3.29}$$

Note that $F'(r) > 0$ and $F''(r) + (\bar{\beta}(r)/r)F'(r) = -1/\underline{\alpha}(r)$ for $r \geqslant r_0$. Hence, by (3.24),

$$2(\mathbf{A}\varphi)(\mathbf{x}) \leqslant -d(\mathbf{x})/\underline{\alpha}((\mathbf{x})) \leqslant -1 \qquad \text{for } |\mathbf{x}| \geqslant r_0. \tag{3.30}$$

Fix $\mathbf{x}$ such that $|\mathbf{x}| > r_0$. By Corollary 3.2(b), and optional stopping (Proposition 13.9 of Chapter I; also see Exercises 3, 4),

$$EZ_{\eta_N} = EZ_0 = \varphi(\mathbf{x}) = F(|\mathbf{x}|) \tag{3.31}$$

where Z_t is as in (3.22) with $\varphi(t, \mathbf{y}) = \varphi(\mathbf{y})$ and $\varphi_0(t, \mathbf{y}) = 0$, and η_N is the $\{\mathscr{F}_t\}$-stopping time

$$\eta_N := \inf\{t \geqslant 0: |\mathbf{X}_t^{\mathbf{x}}| = r_0 \text{ or } N\}, \qquad (r_0 < |\mathbf{x}| < N). \tag{3.32}$$

Using (3.30) in (3.31) the following inequality is obtained,

$$2EF(|\mathbf{X}_{\eta_N}^{\mathbf{x}}|) - 2F(|\mathbf{x}|) = E\int_0^{\eta_N} 2(\mathbf{A}\varphi)(\mathbf{X}_s^{\mathbf{x}})\,ds \leqslant -E(\eta_N). \tag{3.33}$$

Now the first relation in (3.25) implies that $\tau_{\partial B(\mathbf{0}:r_0)} < \infty$ a.s. (see Corollary 15.2 of Chapter V, or Exercise 5). Therefore, $\eta_N \to \tau_{\partial B(\mathbf{0}:r_0)}$ a.s. as $N \to \infty$, so that

(3.33) yields

$$E(\tau_{\partial B(\mathbf{0}:r_0)} \mid \mathbf{X}_0 = \mathbf{x}) \leqslant 2F(|\mathbf{x}|). \tag{3.34}$$

■

As in the one-dimensional case (see Section 12 of Chapter V), it may be shown that (3.26) implies the existence of a unique invariant distribution of the diffusion (theoretical complement 5).

As a second application of Itô's Lemma, let us obtain an estimate of the fourth moment of a stochastic integral.

Proposition 3.4. Let $\mathbf{f} = \{(f_1(t), \dots, f_k(t)) : 0 \leqslant t \leqslant T\}$ be a nonanticipative functional on $[0, T]$ satisfying

$$E \int_0^T f_i^4(t)\, dt < \infty \qquad (i = 1, \dots, k). \tag{3.35}$$

Then

$$E\left(\int_0^T \mathbf{f}(t) \cdot d\mathbf{B}_t\right)^4 \leqslant 9k^3 T \sum_{i=1}^k \int_0^T \{E f_i^4(t)\}\, dt. \tag{3.36}$$

Proof. Since $\mathbf{f}(t) \cdot d\mathbf{B}_t = \sum_{i=1}^k f_i(t)\, dB_t^{(i)}$,

$$\begin{aligned} E\left(\int_0^T \mathbf{f}(t) \cdot d\mathbf{B}_t\right)^4 &= k^4 E\left(\frac{1}{k} \sum_{i=1}^k \int_0^T f_i(t)\, dB_t^{(i)}\right)^4 \\ &\leqslant k^4 E \frac{1}{k} \sum_{i=1}^k \left(\int_0^T f_i(t)\, dB_t^{(i)}\right)^4 = k^3 \sum_{i=1}^k E\left(\int_0^T f_i(t)\, dB_t^{(i)}\right)^4. \end{aligned} \tag{3.37}$$

Hence, it is enough to prove

$$E\left(\int_0^T g(t)\, dB_t\right)^4 \leqslant 9T \int_0^T \{E g^4(t)\}\, dt \tag{3.38}$$

for a real-valued nonanticipative functional g satisfying

$$E \int_0^T g^4(t)\, dt < \infty. \tag{3.39}$$

First suppose g is a bounded nonanticipative step functional. Then by an explicit computation (Exercise 6) the nonanticipative functional $s \to (\int_0^s g(u)\, dB_u)^3 g(s)$ belongs to $\mathscr{M}[0, T]$. One may then apply Itô's Lemma (see Eq. 3.5, or Eq.

3.20 with $k = 1$) to $\varphi(y) = y^4$ and $Y(t) = \int_0^t g(s)\,dB_s$ to get

$$\left(\int_0^t g(s)\,dB_s\right)^4 = 6\int_0^t \left(\int_0^s g(u)\,dB_u\right)^2 g^2(s)\,ds + 4\int_0^t \left(\int_0^s g(u)\,dB_u\right)^3 g(s)\,dB_s,$$

$$0 \leqslant t \leqslant T. \quad (3.40)$$

Taking expectations,

$$J_t := E\left(\int_0^t g(s)\,dB_s\right)^4 = 6\int_0^t E\left\{\left(\int_0^s g(u)\,dB_u\right)^2 g^2(s)\right\} ds. \quad (3.41)$$

This shows, in particular, that J_t is absolutely continuous with a density satisfying

$$\frac{dJ_t}{dt} = 6E\left\{\left(\int_0^t g(u)\,dB_u\right)^2 g^2(t)\right\} \leqslant 6\left\{E\left(\int_0^t g(u)\,dB_u\right)^4\right\}^{1/2} \{Eg^4(t)\}^{1/2}$$

$$= 6J_t^{1/2}\{Eg^4(t)\}^{1/2}. \quad (3.42)$$

Therefore, unless $J_t = 0$ (which implies $J_s = 0$ for $0 \leqslant s \leqslant t$),

$$\frac{dJ_t^{1/2}}{dt} = \tfrac{1}{2}J_t^{-1/2}\frac{dJ_t}{dt} \leqslant 3\{Eg^4(t)\}^{1/2},$$

or,

$$J_t^{1/2} \leqslant 3\int_0^t \{Eg^4(s)\}^{1/2}\,ds, \qquad J_t \leqslant 9t\int_0^t (Eg^4(s))\,ds.$$

This proves (3.38) for bounded nonanticipative step functionals. For the case of an arbitrary nonanticipative functional g satisfying (3.39), observe that such a g belongs to $\mathscr{M}[0, T]$. Following the proof of Proposition 1.2, there exists a sequence of bounded step functionals g_n that converge almost surely to g and for which

$$\int_0^T g_n^i(s)\,dB_s \to \int_0^T g^i(s)\,dB_s \quad \text{a.s.} \qquad (i = 1, 2),$$

$$E\int_0^T (g_n^2(s) - g^2(s))^2\,ds \to 0.$$

By Fatou's Lemma,

$$E\left(\int_0^T g(t)\,dB_t\right)^4 \leqslant \varliminf_{n\to\infty} E\left(\int_0^T g_n(t)\,dB_t\right)^4 \leqslant 9T \varliminf_{n\to\infty} E\int_0^T g_n^4(t)\,dt$$
$$= 9TE\int_0^T g^4(t)\,dt. \qquad \blacksquare$$

As a corollary, the property (1.2) of $\{X_t^x\}$ in Section 1 of Chapter V follows.

Corollary 3.5. Let $\{X_t^x\}$ be a diffusion with drift coefficient $\mu(\cdot)$ and diffusion coefficient $\sigma^2(\cdot)$, both Lipschitzian and bounded. Then, for every $\varepsilon > 0$,

$$P(|X_t^x - x| > \varepsilon) = o(t) \qquad \text{as } t \downarrow 0. \tag{3.43}$$

Proof. By Chebyshev's Inequality,

$$P(|X_t^x - x|) \leqslant \frac{1}{\varepsilon^4} E(X_t^x - x)^4.$$

But if $\mu(\cdot)$ and $\sigma(\cdot)$ are both bounded by c, then

$$\begin{aligned} E(X_t^x - x)^4 &= E\left(\int_0^t \mu(X_s^x)\,ds + \int_0^t \sigma(X_s^x)\,dB_s\right)^4 \\ &\leqslant 2^3\left\{E\left(\int_0^t \mu(X_s^x)\,ds\right)^4 + E\left(\int_0^t \sigma(X_s^x)\,dB_s\right)^4\right\} \\ &\leqslant 8\{(ct)^4 + 9t^2c^4\} = O(t^2) = o(t) \qquad \text{as } t \downarrow 0. \qquad \blacksquare \end{aligned}$$

For another application of Itô's Lemma (see Eq. 3.19), let

$$\mathbf{Y}(t) := \mathbf{X}_t^{\mathbf{x}} - \mathbf{X}_t^{\mathbf{y}} = \mathbf{x} - \mathbf{y} + \int_0^t (\boldsymbol{\mu}(\mathbf{X}_s^{\mathbf{x}}) - \boldsymbol{\mu}(\mathbf{X}_s^{\mathbf{y}}))\,ds + \int_0^t (\boldsymbol{\sigma}(\mathbf{X}_s^{\mathbf{x}}) - \boldsymbol{\sigma}(\mathbf{X}_s^{\mathbf{y}}))\,d\mathbf{B}_s, \tag{3.44}$$

and $\varphi(\mathbf{z}) = |\mathbf{z}|^2$, to get

$$\begin{aligned} |\mathbf{X}_t^{\mathbf{x}} - \mathbf{X}_s^{\mathbf{y}}|^2 = |\mathbf{x} - \mathbf{y}|^2 &+ \int_0^t \{2(\mathbf{X}_s^{\mathbf{x}} - \mathbf{X}_s^{\mathbf{y}})\cdot(\boldsymbol{\mu}(\mathbf{X}_s^{\mathbf{x}}) - \boldsymbol{\mu}(\mathbf{X}_s^{\mathbf{y}})) \\ &\qquad + \sum_{r=1}^k |\boldsymbol{\sigma}_r(\mathbf{X}_s^{\mathbf{x}}) - \boldsymbol{\sigma}_r(\mathbf{X}_s^{\mathbf{y}})|^2\}\,ds \\ &+ \int_0^t 2(\mathbf{X}_s^{\mathbf{x}} - \mathbf{X}_s^{\mathbf{y}})\cdot(\boldsymbol{\sigma}(\mathbf{X}_s^{\mathbf{x}}) - \boldsymbol{\sigma}(\mathbf{X}_s^{\mathbf{y}}))\,d\mathbf{B}_s. \end{aligned} \tag{3.45}$$

Note that, since $\mathbf{X}_t^{\mathbf{x}}, \mathbf{X}_t^{\mathbf{y}} \in \mathcal{M}[0, \infty)$ and $\boldsymbol{\mu}(\cdot)$ and $\boldsymbol{\sigma}(\cdot)$ are Lipschitzian, the expectation of the stochastic integral is zero, so that

$$E|\mathbf{X}_t^{\mathbf{x}} - \mathbf{X}_t^{\mathbf{y}}|^2 = |\mathbf{x} - \mathbf{y}|^2 + \int_0^t \left\{ 2E(\mathbf{X}_s^{\mathbf{x}} - \mathbf{X}_s^{\mathbf{y}}) \cdot (\boldsymbol{\mu}(\mathbf{X}_s^{\mathbf{x}}) - \boldsymbol{\mu}(\mathbf{X}_s^{\mathbf{y}})) + \sum_{r=1}^{k} E|\boldsymbol{\sigma}_r(\mathbf{X}_s^{\mathbf{x}}) - \boldsymbol{\sigma}_r(\mathbf{X}_s^{\mathbf{y}})|^2 \right\} ds. \tag{3.46}$$

In particular, the left side is an absolutely continuous function of t with a density

$$\frac{d}{dt} E|\mathbf{X}_t^{\mathbf{x}} - \mathbf{X}_t^{\mathbf{y}}|^2 \leqslant 2ME|\mathbf{X}_t^{\mathbf{x}} - \mathbf{X}_t^{\mathbf{y}}|^2 + kM^2 E|\mathbf{X}_t^{\mathbf{x}} - \mathbf{X}_t^{\mathbf{y}}|^2. \tag{3.47}$$

Integrate (3.47) to obtain

$$E|\mathbf{X}_t^{\mathbf{x}} - \mathbf{X}_t^{\mathbf{y}}|^2 \leqslant |\mathbf{x} - \mathbf{y}|^2 e^{(2M + kM^2)t}. \tag{3.48}$$

As a consequence, the diffusion has the *Feller property*: *If* $\mathbf{y} \to \mathbf{x}$, *then* $p(t; \mathbf{y}, d\mathbf{z})$ *converges weakly to* $p(t; \mathbf{x}, d\mathbf{z})$ *for every* $t \geqslant 0$. To deduce this property, note that (3.48) implies that, for every bounded Lipschitzian function h on $\mathbb{R}^k$ with $|h(\mathbf{z}) - h(\mathbf{z}')| \leqslant c|\mathbf{z} - \mathbf{z}'|$,

$$\begin{aligned} |Eh(\mathbf{X}_t^{\mathbf{y}}) - Eh(\mathbf{X}_t^{\mathbf{x}})| &= |E(h(\mathbf{X}_t^{\mathbf{y}}) - h(\mathbf{X}_t^{\mathbf{x}}))| \leqslant cE|\mathbf{X}_t^{\mathbf{y}} - \mathbf{X}_t^{\mathbf{x}}| \leqslant c(E|\mathbf{X}_t^{\mathbf{y}} - \mathbf{X}_t^{\mathbf{x}}|^2)^{1/2} \\ &\leqslant c|\mathbf{x} - \mathbf{y}| e^{(M + kM^2/2)t} \to 0 \qquad \text{as } \mathbf{y} \to \mathbf{x}. \end{aligned} \tag{3.49}$$

Now apply Theorem 5.1 of Chapter 0. To state the Feller property another way, write $\mathbf{T}_t$ for the transition operator

$$(\mathbf{T}_t f)(\mathbf{x}) = Ef(\mathbf{X}_t^{\mathbf{x}}) = \int f(\mathbf{y}) p(t; \mathbf{x}, d\mathbf{y}). \tag{3.50}$$

The Feller property says that *if f is bounded and continuous, so is* $\mathbf{T}_t f$.

A number of other applications of Itô's Lemma are sketched in the Exercises.

4 CHAPTER APPLICATION: ASYMPTOTICS OF SINGULAR DIFFUSIONS

A significant advantage of the theory of stochastic differential equations over other methods of construction of diffusions lies in its ability to construct and analyze with relative ease those diffusions whose diffusion matrices $\mathbf{D}(\mathbf{x}) := \boldsymbol{\sigma}(\mathbf{x})\boldsymbol{\sigma}'(\mathbf{x})$ are singular. Such diffusions are known as *singular*, or *degenerate, diffusions*. Observe that in Section 2 the only assumption made on the coefficients is that they are Lipschitzian. As we shall see in this section, the stochastic integral representation (2.45) and Itô's Lemma are effective tools in analyzing the asymptotic behavior of these diffusions. Notice, on the other hand, that the method of Section 15 of Chapter V (also see Section 3 of the present chapter) does not work for analyzing transience, recurrence, etc., as quantities such as $\bar{\beta}(r)$ may not be finite.

Singular diffusions arise in many different contexts. Suppose that the velocity $\mathbf{V}_t$ of a particle satisfies the stochastic differential equation

$$d\mathbf{V}_t = \boldsymbol{\mu}_0(\mathbf{V}_t)\,dt + \boldsymbol{\sigma}^{(1)}(\mathbf{V}_t)\,d\mathbf{B}_t^{(1)}, \tag{4.1}$$

where $\{\mathbf{B}_t^{(1)}\}$ is a standard three-dimensional Brownian motion. The position $\mathbf{Y}_t$ of the particle satisfies

$$d\mathbf{Y}_t = \mathbf{V}_t\,dt. \tag{4.2}$$

The process $\{\mathbf{X}_t := (\mathbf{V}_t, \mathbf{Y}_t)\}$ is then a six-dimensional singular diffusion governed by the stochastic differential equation

$$d\mathbf{X}_t = \boldsymbol{\mu}(\mathbf{X}_t)\,dt + \boldsymbol{\sigma}(\mathbf{X}_t)\,d\mathbf{B}_t, \tag{4.3}$$

where

$$\boldsymbol{\mu}(\mathbf{x}) = (\boldsymbol{\mu}_0(\mathbf{v}), \mathbf{v})', \qquad \boldsymbol{\sigma}(\mathbf{x}) = \begin{bmatrix} \boldsymbol{\sigma}^{(1)}(\mathbf{v}) & \mathbf{0} \\ \mathbf{0} & \mathbf{0} \end{bmatrix},$$

writing $\mathbf{x} = (\mathbf{v}, \mathbf{y})$ and $\mathbf{0}$ for a 3×3 null matrix. Here $\{\mathbf{B}_t\}$ is a six-dimensional standard Brownian motion whose first three coordinates comprise $\{\mathbf{B}_t^{(1)}\}$. Often, as in the case of the Ornstein–Uhlenbeck process (see Example 1.2 in Chapter V, and Exercise 14.1(iii) of Chapter V), the velocity process has a unique invariant distribution. The position process, being an integral of velocity, is usually asymptotically Gaussian by an application of the central limit theorem. Thus, the asymptotic properties of $\mathbf{X}_t = (\mathbf{V}_t, \mathbf{Y}_t)$ may in this case be deduced from those of the nonsingular diffusion $\{\mathbf{V}_t\}$. On the other hand, there are many problems arising in applications in which the analysis is not as straightforward. One may think of a deterministic process $\mathbf{U}_t = (\mathbf{U}_t^{(1)}, \mathbf{U}_t^{(2)})$ governed by a system of ordinary differential equations $d\mathbf{U}_t = \boldsymbol{\mu}(\mathbf{U}_t)\,dt$, having a unique fixed point $\mathbf{u}^*$ such that $\mathbf{U}_t \to \mathbf{u}^*$ as $t \to \infty$, no matter what the initial value $\mathbf{U}_0$ may be. Suppose now that a noise is superimposed that affects only the second component $\mathbf{U}_t^{(2)}$ directly. The perturbed system $\mathbf{X}_t = (\mathbf{X}_t^{(1)}, \mathbf{X}_t^{(2)})$ may then be governed by an equation of the form (4.3) with

$$\boldsymbol{\sigma}(\mathbf{x}) = \begin{pmatrix} \mathbf{0} & \mathbf{0} \\ \mathbf{0} & \boldsymbol{\sigma}^{(1)}(\mathbf{x}) \end{pmatrix},$$

where $\boldsymbol{\sigma}^{(1)}(\mathbf{x})$ is nonsingular. The following result may be thought to deal with this kind of phenomenon, although $\boldsymbol{\sigma}(\mathbf{x})$ need not be singular. For its statement, use the notation

$$J_{ij}(\mathbf{x}) := \frac{\partial}{\partial x^{(j)}} \mu^{(i)}(\mathbf{x}), \qquad \mathbf{J}(\mathbf{x}) := ((J_{ij}(\mathbf{x}))). \tag{4.4}$$

Theorem 4.1. Let $\boldsymbol{\mu}(\cdot)$, $\boldsymbol{\sigma}(\cdot)$ be Lipschitzian on $\mathbb{R}^k$,

$$\|\boldsymbol{\sigma}(\mathbf{x}) - \boldsymbol{\sigma}(\mathbf{y})\| \leqslant \lambda_0 |\mathbf{x} - \mathbf{y}| \qquad \text{for all } \mathbf{x}, \mathbf{y} \in \mathbb{R}^k. \tag{4.5}$$

Assume that, for all $\mathbf{x}$, the eigenvalues of the symmetric matrix $\frac{1}{2}(\mathbf{J}(\mathbf{x}) + \mathbf{J}'(\mathbf{x}))$ are all less than or equal to $-\lambda_1 < 0$, where $k\lambda_0^2 < 2\lambda_1$. Then there exists a unique invariant distribution $\pi(d\mathbf{z})$ for the diffusion, and $p(t; \mathbf{x}, d\mathbf{z})$ converges weakly to $\pi(d\mathbf{y})$ as $t \to \infty$, for every $\mathbf{x} \in \mathbb{R}^k$.

Proof. The first step is to show that, for some $\mathbf{x}$, the family $\{p(t; \mathbf{x}, d\mathbf{y}): t \geqslant 0\}$ is *tight*, i.e., given $\varepsilon > 0$ there exists $c_\varepsilon < \infty$ such that

$$p(t; \mathbf{x}, \{\mathbf{y}: |\mathbf{y}| > c_\varepsilon\}) \leqslant \varepsilon \qquad (t \geqslant 0). \tag{4.6}$$

It will then follow (see Chapter 0, Theorem 5.1) that there exist $t_n \to \infty$ and a probability measure π, perhaps depending on $\mathbf{x}$, such that

$$p(t_n; \mathbf{x}, d\mathbf{y}) \xrightarrow{\text{weakly}} \pi(d\mathbf{y}) \qquad \text{as } n \to \infty. \tag{4.7}$$

We will actually prove that

$$\sup_{t \geqslant 0} E|\mathbf{X}_t^{\mathbf{x}}|^2 < \infty, \tag{4.8}$$

where $\{\mathbf{X}_t^{\mathbf{x}}\}$ is the solution to (4.3) with $\mathbf{X}_0 = \mathbf{x}$. Clearly, (4.6) follows from (4.7) by means of Chebyshev's Inequality,

$$p(t; \mathbf{x}, \{|\mathbf{y}| > c\}) = P(|\mathbf{X}_t^{\mathbf{x}}| > c) \leqslant \frac{1}{c^2} E|\mathbf{X}_t^{\mathbf{x}}|^2.$$

In order to prove (4.8), apply Itô's Lemma (see Eq. 3.20) to the function $\varphi(\mathbf{y}) = |\mathbf{y}|^2$ to get

$$|\mathbf{X}_t^{\mathbf{x}}|^2 = |\mathbf{x}|^2 + \int_0^t (\mathbf{A}\varphi)(\mathbf{X}_s^{\mathbf{x}})\, ds + \sum_{r=1}^{k} \int_0^t \partial_r \varphi(\mathbf{X}_s^{\mathbf{x}}) \boldsymbol{\sigma}_r(\mathbf{X}_s^{\mathbf{x}}) \cdot d\mathbf{B}_s. \tag{4.9}$$

Now check that

$$(\mathbf{A}\varphi)(\mathbf{y}) = 2\mathbf{y}\cdot\boldsymbol{\mu}(\mathbf{y}) + \sum_{r=1}^{k} |\boldsymbol{\sigma}_r(\mathbf{y})|^2 = 2\mathbf{y}\cdot\boldsymbol{\mu}(\mathbf{y}) + \operatorname{tr}(\boldsymbol{\sigma}(\mathbf{y})\boldsymbol{\sigma}'(\mathbf{y})), \tag{4.10}$$

where tr **D** denotes the *trace* of **D**, i.e., the sum of the diagonal elements of **D**. Now, by a one-term Taylor expansion,

$$\mu^{(r)}(\mathbf{y}) - \mu^{(r)}(\mathbf{0}) = \int_0^1 \mathbf{y}\cdot\mathbf{grad}\,\mu^{(r)}(\theta\mathbf{y})\,d\theta,$$

$$\boldsymbol{\mu}(\mathbf{y}) - \boldsymbol{\mu}(\mathbf{0}) = \int_0^1 \mathbf{y}'\mathbf{J}(\theta\mathbf{y})\,d\theta, \tag{4.11}$$

$$\mathbf{y}\cdot\boldsymbol{\mu}(\mathbf{y}) = \mathbf{y}\cdot(\boldsymbol{\mu}(\mathbf{y}) - \boldsymbol{\mu}(\mathbf{0})) + \mathbf{y}\cdot\boldsymbol{\mu}(\mathbf{0}) = \int_0^1 (\mathbf{y}'\mathbf{J}(\theta\mathbf{y})\mathbf{y})\,d\theta + \mathbf{y}\cdot\boldsymbol{\mu}(\mathbf{0}).$$

Now,

$$\mathbf{y}'\mathbf{J}(\theta\mathbf{y})\mathbf{y} = \mathbf{y}'\mathbf{J}'(\theta\mathbf{y})\mathbf{y} = \tfrac{1}{2}\mathbf{y}'(\mathbf{J}(\theta\mathbf{y}) + \mathbf{J}'(\theta\mathbf{y}))\mathbf{y} \leqslant -\lambda_1|\mathbf{y}|^2. \tag{4.12}$$

Using this in (4.11),

$$2\mathbf{y}\cdot\boldsymbol{\mu}(\mathbf{y}) \leqslant -2\lambda_1|\mathbf{y}|^2 + 2(|\boldsymbol{\mu}(\mathbf{0})|)|\mathbf{y}|. \tag{4.13}$$

Also,

$$\begin{aligned}\operatorname{tr}(\boldsymbol{\sigma}(\mathbf{y})\boldsymbol{\sigma}'(\mathbf{y})) = \operatorname{tr}\{&(\boldsymbol{\sigma}(\mathbf{y}) - \boldsymbol{\sigma}(\mathbf{0}))(\boldsymbol{\sigma}(\mathbf{y}) - \boldsymbol{\sigma}(\mathbf{0}))' + \boldsymbol{\sigma}(\mathbf{0})(\boldsymbol{\sigma}(\mathbf{y}) - \boldsymbol{\sigma}(\mathbf{0}))' \\ &+ (\boldsymbol{\sigma}(\mathbf{y}) - \boldsymbol{\sigma}(\mathbf{0}))\boldsymbol{\sigma}'(\mathbf{0}) + \boldsymbol{\sigma}(\mathbf{0})\boldsymbol{\sigma}'(\mathbf{0})\}.\end{aligned} \tag{4.14}$$

Since every element of a matrix is bounded by its norm, the trace of a $k \times k$ matrix is no more than k times its norm. Therefore, (4.14) leads to

$$\operatorname{tr}(\boldsymbol{\sigma}(\mathbf{y})\boldsymbol{\sigma}'(\mathbf{y})) \leqslant k\lambda_0^2|\mathbf{y}|^2 + 2k|\boldsymbol{\sigma}(\mathbf{0})|\lambda_0|\mathbf{y}| + k|\boldsymbol{\sigma}(\mathbf{0})|^2. \tag{4.15}$$

Substituting (4.13) and (4.15) into (4.10), and using the fact that $|\mathbf{y}| \leqslant \delta|\mathbf{y}|^2 + 1/\delta$ for all $\delta > 0$, there exist $\delta_1 > 0$, $\delta_2 > 0$, such that $2\lambda_1 - k\lambda_0^2 - \delta_1 > 0$ and

$$(\mathbf{A}\varphi)(\mathbf{y}) \leqslant -(2\lambda_1 - k\lambda_0^2 - \delta_1)|\mathbf{y}|^2 + \delta_2. \tag{4.16}$$

Now take expectations in (4.9) to get

$$E|\mathbf{X}_t^{\mathbf{x}}|^2 = |\mathbf{x}|^2 + \int_0^t E(\mathbf{A}\varphi(\mathbf{X}_s^{\mathbf{x}}))\,ds. \tag{4.17}$$

In arriving at (4.17), use the facts (i) $|\mathbf{A}\varphi(\mathbf{y})| \leqslant c|\mathbf{y}|^2 + c'$ for some c, c' positive, so that $\int_0^t E|\mathbf{A}\varphi(\mathbf{X}_s^{\mathbf{x}})|\, ds < \infty$, and (ii) the integrand of the stochastic integral in (4.9) is in $\mathscr{M}[0, \infty)$. Now (4.17) implies that $t \to E|\mathbf{X}_t^{\mathbf{x}}|^2$ is absolutely continuous with a density satisfying, by (4.16),

$$\frac{d}{dt} E|\mathbf{X}_t^{\mathbf{x}}|^2 = E\mathbf{A}\varphi(\mathbf{X}_t^{\mathbf{x}}) \leqslant -(2\lambda_1 - k\lambda_0^2 - \delta_1)E|\mathbf{X}_t^{\mathbf{x}}|^2 + \delta_2. \tag{4.18}$$

In other words, writing $\delta' := 2\lambda_1 - k\lambda_0^2 - \delta_1 > 0$, $\theta(t) := E|\mathbf{X}_t^{\mathbf{x}}|^2$,

$$\begin{aligned} &\frac{d}{dt}(e^{\delta' t}\theta(t)) \leqslant \delta_2 e^{\delta' t}, \\ &\theta(t) \leqslant \{\theta(0) + (\delta_2/\delta')(e^{\delta' t} - 1)\}e^{-\delta' t} \leqslant |\mathbf{x}|^2 e^{-\delta' t} + \delta_2/\delta'. \end{aligned} \tag{4.19}$$

This proves (4.8) and therefore (4.6) for all $\mathbf{x}$.

The next task is to prove that the limit in (4.7) does not depend on $\mathbf{x}$. For this it is enough to show

$$\lim_{t \to \infty} E|\mathbf{X}_t^{\mathbf{x}} - \mathbf{X}_t^{\mathbf{y}}|^2 = 0 \qquad \forall \mathbf{x}, \mathbf{y} \in \mathbb{R}^k. \tag{4.20}$$

For, if (4.7) holds for some $\mathbf{x}$, and (4.20) holds for this $\mathbf{x}$ and all $\mathbf{y}$ then for every Lipschitzian and bounded f, writing t for t_n,

$$\begin{aligned} \left|Ef(\mathbf{X}_t^{\mathbf{y}}) - \int f(\mathbf{z})\pi(d\mathbf{z})\right| &\leqslant |Ef(\mathbf{X}_t^{\mathbf{y}}) - Ef(\mathbf{X}_t^{\mathbf{x}})| + \left|Ef(\mathbf{X}_t^{\mathbf{x}}) - \int f(\mathbf{z})\pi(d\mathbf{z})\right| \\ &\leqslant cE|\mathbf{X}_t^{\mathbf{y}} - \mathbf{X}_t^{\mathbf{x}}| + \left|Ef(\mathbf{X}_t^{\mathbf{x}}) - \int f(\mathbf{z})\pi(d\mathbf{z})\right| \\ &\leqslant c(E|\mathbf{X}_t^{\mathbf{y}} - \mathbf{X}_t^{\mathbf{x}}|^2)^{1/2} + \left|Ef(\mathbf{X}_t^{\mathbf{x}}) - \int f(\mathbf{z})\pi(d\mathbf{z})\right| \to 0, \end{aligned}$$

as $t_n \to \infty$. It follows (Chapter 0, Theorem 5.1) that if (4.7) holds, then

$$p(t_n; \mathbf{y}, d\mathbf{z}) \to \pi(d\mathbf{z}) \qquad \forall \mathbf{y} \in \mathbb{R}^k. \tag{4.21}$$

To prove (4.20), proceed as in the first step, replacing $\mathbf{X}_t^{\mathbf{x}}$ by

$$\mathbf{Z}_t := \mathbf{X}_t^{\mathbf{x}} - \mathbf{X}_t^{\mathbf{y}} = \mathbf{x} - \mathbf{y} + \int_0^t \mathbf{f}(s)\, ds + \int_0^t \boldsymbol{\gamma}(s)\, d\mathbf{B}_s, \tag{4.22}$$

where

$$\begin{aligned} \mathbf{f}(s) &:= \boldsymbol{\mu}(\mathbf{X}_s^{\mathbf{x}}) - \boldsymbol{\mu}(\mathbf{X}_s^{\mathbf{y}}) = \int_0^1 (\mathbf{X}_s^{\mathbf{x}} - \mathbf{X}_s^{\mathbf{y}})'\mathbf{J}(\mathbf{X}_s^{\mathbf{y}} + \theta(\mathbf{X}_s^{\mathbf{x}} - \mathbf{X}_s^{\mathbf{y}}))\, d\theta, \\ \boldsymbol{\gamma}(s) &:= \boldsymbol{\sigma}(\mathbf{X}_s^{\mathbf{x}}) - \boldsymbol{\sigma}(\mathbf{X}_s^{\mathbf{y}}). \end{aligned} \tag{4.23}$$

Letting $\mathbf{Y}_t = \mathbf{Z}_t$ and $\varphi(\mathbf{z}) = |\mathbf{z}|^2$ in Itô's Lemma (see Eqs. 3.44–3.46), obtain

$$E|\mathbf{Z}_t|^2 = |\mathbf{x} - \mathbf{y}|^2 + E\int_0^t \{2\mathbf{Z}_s \cdot \mathbf{f}(s) + \operatorname{tr} \gamma(s)\gamma'(s)\}\, ds. \tag{4.24}$$

But, by (4.23) and the fact that $\mathbf{z}'J(\mathbf{w})\mathbf{z} \leqslant -\lambda_1|\mathbf{z}|^2$ for all $\mathbf{w}, \mathbf{z}$ (see Eq. 4.12),

$$\mathbf{Z}_s \cdot \mathbf{f}_s = \int_0^1 \mathbf{Z}_s'\mathbf{J}(\mathbf{X}_s^{\mathbf{y}} + \theta\mathbf{Z}_s)\mathbf{Z}_s\, d\theta \leqslant -\lambda_1|\mathbf{Z}_s|^2. \tag{4.25}$$

Also,

$$\|\gamma(s)\gamma'(s)\| \leqslant \|\gamma(s)\|\,\|\gamma'(s)\| \leqslant \lambda_0^2|\mathbf{Z}_s|^2,$$

so that

$$\operatorname{tr} \gamma(s)\gamma'(s) \leqslant k\lambda_0^2|\mathbf{Z}_s|^2. \tag{4.26}$$

Using (4.25) and (4.26) in (4.24), obtain

$$\begin{aligned}\frac{d}{dt}E|\mathbf{Z}_t|^2 &= E\{2\mathbf{Z}_t \cdot \mathbf{f}(t) + \operatorname{tr} \gamma(t)\gamma'(t)\}\\ &\leqslant -(2\lambda_1 - k\lambda_0^2)E|\mathbf{Z}_t|^2, \qquad (E|\mathbf{Z}_0|^2 = |\mathbf{x} - \mathbf{y}|^2).\end{aligned} \tag{4.28}$$

Hence (4.20) holds and, therefore, the limit in (4.21) is independent of $\mathbf{y} \in \mathbb{R}^k$ (i.e., the limit in (4.7) is independent of $\mathbf{x}$).

The final step is to show that (4.21) implies that π is the unique invariant probability. Let $\mathbf{T}_t$ denote the transition operator for the diffusion (see Eq. 3.50). Then for every bounded continuous function f, (4.21) implies

$$(\mathbf{T}_{t_n}f)(\mathbf{y}) \to \bar{f} := \int f(\mathbf{z})\pi(d\mathbf{z}). \tag{4.29}$$

By Lebesgue's Dominated Convergence Theorem, applied to the sequence $\{\mathbf{T}_{t_n}f\}$ and the measure $p(t; \mathbf{x}, d\mathbf{y})$,

$$\mathbf{T}_t(\mathbf{T}_{t_n}f)(\mathbf{x}) = \int (\mathbf{T}_{t_n}f)(\mathbf{y})p(t; \mathbf{x}, d\mathbf{y}) \to \mathbf{T}_t\bar{f} = \bar{f}. \tag{4.30}$$

But $\mathbf{T}_t f$ is a bounded continuous function, by the *Feller property* (see end of Section 3). Therefore, applying (4.29) to $\mathbf{T}_t f$,

$$\mathbf{T}_{t_n}(\mathbf{T}_t f)(\mathbf{x}) \to \int (\mathbf{T}_t f)(\mathbf{z})\pi(d\mathbf{z}). \tag{4.31}$$

As $\mathbf{T}_t(\mathbf{T}_{t_n}f) = \mathbf{T}_{t_n}(\mathbf{T}_t f) = \mathbf{T}_{t+t_n}f$, the limits in (4.30) and (4.31) coincide,

$$\int (\mathbf{T}_t f)(\mathbf{z})\pi(d\mathbf{z}) = \int f(\mathbf{z})\pi(d\mathbf{z}), \tag{4.32}$$

i.e., if $\mathbf{X}_0$ has distribution π, then $Ef(\mathbf{X}_t) = Ef(X_0)$ for all $t \geqslant 0$. In other words, π is an invariant (initial) distribution. To prove uniqueness, let π' be any invariant probability. Then for all bounded continuous f,

$$\int (T_{t_n} f)(\mathbf{z})\pi'(d\mathbf{z}) = \int f(\mathbf{z})\pi'(d\mathbf{z}) \qquad \forall n. \tag{4.33}$$

But, by (4.29), the left side converges to $\bar{f}$. Therefore,

$$\bar{f} \equiv \int f(\mathbf{z})\pi(d\mathbf{z}) = \int f(\mathbf{z})\pi'(d\mathbf{z}), \tag{4.34}$$

implying $\pi' = \pi$. ■

Consider next *linear stochastic differential equations* of the form

$$d\mathbf{X}_t = \mathbf{C}\mathbf{X}_t\, dt + \boldsymbol{\sigma}\, d\mathbf{B}_t \tag{4.35}$$

where $\mathbf{C}$ and $\boldsymbol{\sigma}$ are constant $k \times k$ matrices. This is a continuous-time analog of the difference equation (13.10) of Chapter II for linear time series models. It is simple to check that (Exercise 1)

$$\mathbf{X}_t^{\mathbf{x}} := e^{t\mathbf{C}}\mathbf{x} + \int_0^t e^{(t-s)\mathbf{C}}\boldsymbol{\sigma}\, d\mathbf{B}_s \tag{4.36}$$

is a solution to (4.35) with initial state $\mathbf{x}$. By uniqueness, it is *the* solution with $\mathbf{X}_0 = \mathbf{x}$. Observe that, being a limit of linear combinations of Gaussians, $\mathbf{X}_t^{\mathbf{x}}$ is Gaussian with mean vector $\exp\{t\mathbf{C}\}\mathbf{x}$ and dispersion matrix (Exercise 2)

$$\begin{aligned}\boldsymbol{\Sigma}(t) = E\left(\int_0^t e^{(t-s)\mathbf{C}}\boldsymbol{\sigma}\, d\mathbf{B}_s\right)\left(\int_0^t e^{(t-s)\mathbf{C}}\boldsymbol{\sigma}\, d\mathbf{B}_s\right)' &= \int_0^t e^{(t-s)\mathbf{C}}\boldsymbol{\sigma}\boldsymbol{\sigma}' e^{(t-s)\mathbf{C}'}\, ds \\ &= \int_0^t e^{u\mathbf{C}}\boldsymbol{\sigma\sigma} e^{u\mathbf{C}'}\, du. \end{aligned} \tag{4.37}$$

Suppose that the real parts of the eigenvalues $\lambda_1, \ldots, \lambda_k$, say, of $\mathbf{C}$ are negative. This does not necessarily imply that the eigenvalues of $\mathbf{C} + \mathbf{C}'$ are all negative (Exercise 3). Therefore, Theorem 4.1 does not quite apply. But, writing

$[t]$ for the integer part of t and $\mathbf{B} = \exp\{\mathbf{C}\}$,

$$\|e^{t\mathbf{C}}\| \leqslant \|e^{[t]\mathbf{C}}\| \, \|e^{(t-[t])\mathbf{C}}\| \leqslant \|\mathbf{B}^{[t]}\| \max\{\|e^{s\mathbf{C}}\| : 0 \leqslant s \leqslant 1\} \to 0 \tag{4.38}$$

exponentially fast, as $t \to \infty$, by the Lemma in Section 13 of Chapter II. For the eigenvalues of $\mathbf{B}$ are $\exp\{\lambda_i\}$ $(1 \leqslant i \leqslant k)$, all smaller than 1 in modulus. It follows from the exponential convergence (4.38) that $\exp\{t\mathbf{C}\}\mathbf{x} \to 0$ as $t \to \infty$, and the integral in (4.37) converges to

$$\mathbf{\Sigma} := \int_0^\infty e^{u\mathbf{C}} \boldsymbol{\sigma\sigma} e^{u\mathbf{C}} \, du. \tag{4.39}$$

The following result has, therefore, been proved.

Proposition 4.2. The diffusion governed by (4.35) has a unique invariant distribution if all the eigenvalues of $\mathbf{C}$ have negative real parts. The invariant distribution is Gaussian with zero mean vector and dispersion matrix $\mathbf{\Sigma}$ given by (4.39).

Under the hypothesis of Proposition 4.2, the diffusion governed by (4.35) may be thought of as the multivariate Ornstein–Uhlenbeck process.

EXERCISES

Exercises for Section VII.1

1. Let $s = t_{0,n} < t_{1,n} < \cdots < t_{k_n,n} = t$ be, for each n, a partition of the interval $[s, t]$, such that $\delta_n := \max\{t_{i+1,n} - t_{i,n} : 0 \leqslant i \leqslant k_n - 1\} \to 0$ as $n \to \infty$.
   (i) Prove that

$$s_n^2 := \sum_{i=0}^{n} (B_{t_{i+1,n}} - B_{t_{i,n}})^2$$

   converges to $t - s$ in probability as $n \to \infty$. [*Hint*: $E(s_n^2 - (t-s))^2 \to 0$.]
   (ii) Prove that

$$\gamma_n := \sum_{i=0}^{n} |B_{t_{i+1,n}} - B_{t_{i,n}}| \to \infty$$

   in probability as $n \to \infty$. [*Hint*: $\gamma_n \geqslant s_n^2/\max|B_{t_{i+1,n}} - B_{t_{i,n}}|$.]
   (iii) Let π denote an arbitrary subdivision $s = t_0 < t_1 < \cdots < t_m = t$ of $[s, t]$. Prove that the supremum of $\gamma(\pi) := \sum_{i=0}^{m-1} |B_{t_{i+1}} - B_{t_i}|$ over all partitions π is ∞ for all Brownian paths, outside a set of zero probability. [*Hint*: Choose a subsequence in (ii) such that $\gamma_n \to \infty$ a.s.]

(iv) Prove that, outside a set of zero probability, the Brownian paths are of unbounded variation over every finite interval $[a, b]$, $a < b$. [*Hint*: Use (iii) for every interval $[a, b]$ with rational end points.]

(v) Use (iv) to prove that, outside a set of zero probability, the Brownian paths are nowhere differentiable on $(0, \infty)$. [*Hint*: If f is differentiable at x, then there exists an interval $[x - h, x + h]$ such that $|f(y) - f(z)| < (2|f'(x)| + 1)|y - z|$ $\forall y, z \in [x - h, x + h]$, so that f is of variation less than $(2|f'(x)| + 1)2h$ on $[x - h, x + h)$.]

2. Consider the stochastic integral equation (1.3) with $\mu(\cdot)$ Lipschitzian, $|\mu(x) - \mu(y)| \leqslant M|x - y|$. Solve this equation by the method of successive approximations, with the nth approximation $X_t^{(n)}$ given by

$$X_t^{(n)} = x + \int_0^t \mu(X_s^{(n-1)})\,ds + \sigma(B_t - B_0) \qquad (n \geqslant 1), \qquad X_t^{(0)} \equiv x.$$

[*Hint*: For each $t > 0$ write $\delta_n(t) := \max\{|X_s^{(n)} - X_s^{(n-1)}|: 0 \leqslant s \leqslant t\}$. Fix $T > 0$. Then

$$\delta_n(T) \leqslant M \int_0^T \delta_{n-1}(s)\,ds \leqslant M^2 \int_0^T \int_0^{t_{n-1}} \delta_{n-2}(s)\,ds\,dt_{n-1} \leqslant \cdots$$

$$\leqslant M^n \int_0^T \int_0^{t_{n-1}} \cdots \int_0^{t_2} \delta_1(s)\,ds\,dt_2 \cdots dt_{n-1} \leqslant M^n \delta_1(T) T^{n-1} (n-1)!$$

Hence, $\sum_n \delta_n(T)$ converges to a finite limit, which implies the uniform convergence of $\{X_s^{(n)}: 0 \leqslant s \leqslant T\}$ to a finite and continuous limit $\{X_s: 0 \leqslant s \leqslant T\}$.]

3. Suppose $f_n, f \in \mathscr{M}[\alpha, \beta]$ and $E \int_\alpha^\beta (f_n(s) - f(s))^2\,ds \to 0$ as $n \to \infty$.

(i) Prove that

$$U_n := \max\left\{\left|\int_\alpha^t f_n(s)\,dB_s - \int_\alpha^t f(s)\,dB_s\right| : \alpha \leqslant t \leqslant \beta\right\} \to 0$$

in probability.

(ii) If $\{Y(t): \alpha \leqslant t \leqslant \beta\}$ is a stochastic process with continuous sample paths and $\int_\alpha^t f_n(s)\,dB_s \to Y(t)$ in probability for all $t \in [\alpha, \beta]$, then prove that

$$P\left(Y(t) = \int_\alpha^t f(s)\,dB_s, \forall t \in [\alpha, \beta]\right) = 1.$$

4. Prove the convergence in (1.34).

5. Let $f(t)$ $(t_0 \leqslant t \leqslant t_1)$ be a nonrandom continuously differentiable function. Prove that

$$\int_{t_0}^{t_1} f(s)\,dB_s = f(t_1)B_{t_1} - f(t_0)B_{t_0} - \int_{t_0}^{t_1} B_s f'(s)\,ds.$$

Exercises for Section VII.2

1. If h is a Lipschitzian function on $\mathbb{R}^1$ and X is a square integrable random variable, then prove that $Eh^2(X) < \infty$. [*Hint*:*Look at* $h(X) - h(0)$.]

2. Let $D_T^{(r)}$ be as in the proof of Theorem 2.1. Prove that the sum $\sum_{r=n+1}^{\infty} (D_T^{(r)})^{1/2}$ occurring in (2.17) goes to zero as $n \to \infty$. [*Hint*: Use Stirling's approximation (10.3) in Chapter I.]

3. Assume the hypothesis of Theorem 2.1, with $\alpha = 0$.
   (ii) Prove that
$$E(X_t - X_0)^2 \leqslant 4M^2(t+1)\int_0^t E(X_s - X_0)^2\,ds + 4t^2E\mu^2(X_0) + 4tE\sigma^2(X_0).$$
   (ii) Write $D_t := E(\max\{(X_s - X_0)^2 : 0 \leqslant s \leqslant t\})$, $0 \leqslant t \leqslant T$. Deduce from (i) that
   (a) $D_t \leqslant d_1 \int_0^t D_s\,ds + d_2$, where $d_1 = 4M^2(T+1)$, $d_2 = 4T(TE\mu^2(X_0) + E\sigma^2(X_0))$;
   (b) $D_T \leqslant d_2 e^{d_1 T}$.

4. Write out a proof of Theorem 2.4 by extending step by step the proof of Theorem 2.1.

5. Write out a proof of Theorem 2.5 along the lines of that of Theorem 2.2.

6. Consider the Gaussian diffusion $\{X_t^x\}$ given by (2.49).
   (i) Show that $EX_t^x = e^{-t\gamma}x$.
   (ii) Compute $\mathrm{Cov}(X_t^x, X_{t+h}^x)$.
   (iii) For the case $\gamma > 0$, $\sigma \neq 0$, prove that there exists a unique invariant probability distribution, and specify this distribution.

7. (i) Prove (2.28), (2.30), and (2.31) under the assumption (2.1).
   (ii) Write out a corresponding proof for the multidimensional case.

Exercises for Section VII.3

1. (i) Prove (3.1). [*Hint*: For each n, the finite sequence $(B_{(m+1)2^{-n}t} - B_{m2^{-n}t})^2 - 2^{-n}t$ $(m = 0, 1, \ldots, 2^n - 1)$ is i.i.d. with mean zero and variance $2(4^{-n})t^2$. Use (13.56) of Chapter I to estimate
$$P\left(\max_{1 \leqslant N \leqslant 2^n} |Z_{N,n}| > \frac{1}{n}\right),$$
   and then apply the Borel–Cantelli Lemma.]
   (ii) Prove (3.16).

2. Let $\boldsymbol{\mu}(\cdot)$, $\boldsymbol{\sigma}(\cdot)$ be Lipschitzian and $\boldsymbol{\sigma}(\mathbf{x})$ nonsingular for all $\mathbf{x}$.
   (i) Show that if
$$\int_c^{\infty} \exp\{-\bar{I}(u)\}\,du = \infty \qquad \text{for } some\ c > 0,$$
   then the same holds for *all* $c > 0$.

(ii) Show that if

$$\int_c^\infty \underline{\alpha}^{-1}(u)\exp\{\bar{I}(u)\}\,du < \infty \qquad \text{for } \textit{some } c > 0,$$

then the same holds for *all* $c > 0$.

3. Let $\boldsymbol{\mu}(\cdot)$, $\boldsymbol{\sigma}(\cdot)$ be Lipschitzian and $d_{11}(\mathbf{x}) := |\boldsymbol{\sigma}_1(\mathbf{x})|^2 > 0$ for all $\mathbf{x}$. Prove that $E\tau < \infty$, where $\tau := \inf\{t \geqslant 0\colon |\mathbf{X}_t^{\mathbf{x}}| = d\}$ and $|\mathbf{x}| \leqslant d$. [*Hint*: For a sufficiently large c, the function $\varphi(\mathbf{x}) := -\exp\{cx^{(1)}\}$ in $\{|\mathbf{x}| \leqslant d\}$ (extended suitably) satisfies $\mathbf{A}\varphi(\mathbf{x}) < 0$ in $\{|\mathbf{x}| \leqslant d\}$. Let $-\delta := \max\{\mathbf{A}\varphi(\mathbf{x})\colon |\mathbf{x}| \leqslant d\}$. Apply Itô's Lemma to φ and optional stopping (Proposition 13.9 of Chapter I) with stopping time $\tau \wedge n$ to get $E(\tau \wedge n) \leqslant (2/\delta)\max\{|\varphi(\mathbf{x})|\colon |\mathbf{x}| \leqslant d\}$. Now let $n \uparrow \infty$.]

4. Let F be the twice continuously differentiable function on $[r_0, N]$, $0 < r_0 < N < \infty$, defined by (3.28). Find a twice continuously differentiable extension of F on $[0, \infty)$ that vanishes outside $[r_0 - \varepsilon, N + \varepsilon]$, where $\varepsilon > 0$ is chosen so that $r_0 - \varepsilon > 0$. Show that $\varphi(\mathbf{x}) := F(|\mathbf{x}|)$ is twice continuously differentiable on $\mathbb{R}^k$, vanishing outside a compact set. Apply Itô's Lemma to φ to derive (3.31).

5. Let

$$F(r) := \int_c^r \exp\{-\bar{I}(u))\}\,du, \qquad c \leqslant r \leqslant d,$$

where $\bar{I}(u)$ is defined by (3.23).

(i) Define $\varphi(\mathbf{x}) := F(|\mathbf{x}|)$ for $c \leqslant |\mathbf{x}| \leqslant d$ and obtain a twice continuously differentiable extension of φ vanishing outside a compact.

(ii) For the function φ in (i) check that $\mathbf{A}\varphi(\mathbf{x}) \leqslant 0$ for $c \leqslant |\mathbf{x}| \leqslant d$, and use Itô's Lemma to derive a lower bound for $P(\tau_{\partial B(0:c)} < \tau_{\partial B(0:d)})$, where $\tau_{\partial B(\mathbf{0}:r)} := \inf\{t \geqslant 0\colon |\mathbf{X}_t^{\mathbf{x}}| = r\}$.

(iii) Use (ii) to prove that

$$P(\tau_{\partial B(\mathbf{0}:r_0)} < \infty) = 1 \qquad \text{for } |\mathbf{x}| > r_0,$$

provided

$$\int_c^\infty \exp\{-\bar{I}(u)\}\,du = \infty \qquad \text{for some } c > 0.$$

(iv) Use an argument similar to that outlined in (i)–(iii) to prove that

$$P(\tau_{\partial B(\mathbf{0}:r_0)} < \infty) < 1 \qquad \text{if} \int_c^\infty \exp\{-\underline{I}(u)\}\,du < \infty.$$

6. Suppose that $g(\cdot)$ is a real-valued, bounded, nonanticipative step functional on $[0, T]$. Prove that

$$t \to \left(\int_0^t g(u)\,dB_u\right)^3 g(t) \qquad \text{belongs to} \qquad \mathcal{M}[0, T].$$

[*Hint*: Use (1.6), with g in place of f, to get $E\{(\int_0^t g(u)\,dB_u)^6 g^2(t)\} \leqslant ct^3$ for an appropriate constant c.]

7. Let $\{X_t^x\}$ be a diffusion on $\mathbb{R}^1$ having Lipschitzian coefficients $\mu(\cdot), \sigma(\cdot)$, with $\sigma^2(x) > 0$ for all x.
   (i) Use Itô's Lemma to compute (a) $\psi(x) := P(\{X_t^x\}$ reaches c before $d)$, $c < x < d$; (b) $\rho_{xc} := P(\{X_t^x\}$ ever reaches $c)$, $x > c$; (c) $\rho_{xd} := P(\{X_t^x\}$ ever reaches $d)$, $x < d$. [*Hint*: Consider $\varphi(y) := \int_c^y \exp\{-I(c, r)\}\,dr$ for $c \leqslant y \leqslant d$, where $I(c, r) := \int_c^r (2\mu(z)/\sigma^2(z))\,dz$.]
   (ii) Compute (a) $E\tau_c \wedge \tau_d$, where $\tau_r := \inf\{t \geqslant 0 : X_t^x = r\}$, and $c < x < d$; (b) $E\tau_c$ $(x > c)$; (c) $E\tau_d$ $(x < d)$. [*Hint*: Let φ be the solution of $\mathbf{A}\varphi(y) = -1$ for $c < y < d$, $\varphi(c) = \varphi(d) = 0$; use Itô's Lemma.]

8. (i) Let m be a positive integer, and g a nonanticipative functional on $[0, T]$ such that $E \int_0^T g^{2m}(t)\,dt < \infty$. Prove that

$$E\left(\int_0^T g(t)\,dB_t\right)^{2m} \leqslant (2m-1)^m T^{m-1} \int_0^T E g^{2m}(t)\,dt.$$

[*Hint*: For bounded nonanticipative step functionals g, use Itô's Lemma to get

$$J_t := E\left(\int_0^t g(s)\,dB_s\right)^{2m} = m(2m-1)\int_0^t E\left\{\left(\int_0^s g(u)\,dB_u\right)^{2m-2} g^2(s)\right\} ds,$$

so that

$$\frac{dJ_t}{dt} = m(2m-1)E\left\{\left(\int_0^t g(u)\,dB_u\right)^{2m-2} g^2(t)\right\}$$
$$\leqslant m(2m-1)E\left\{\left(\int_0^t g(u)\,dB_u\right)^{2m}\right\}^{(2m-2)/2m} (Eg^{2m}(t))^{1/m}$$

(by Hölder's Inequality).]

   (ii) Extend (i) to multidimension, i.e., for nonanticipative functionals $\mathbf{f} = \{(f_1(t), \ldots, f_k(t)) : 0 \leqslant t \leqslant T\}$ satisfying $E \int_0^T f_i^{2m}(t)\,dt < \infty$ $(1 \leqslant i \leqslant k)$, prove that

$$E\left(\int_0^T \mathbf{f}(t)\cdot d\mathbf{B}_t\right)^{2m} \leqslant k^{2m-1}(2m-1)^m T^{m-1} \sum_{i=1}^k E \int_0^T f_i^{2m}(t)\,dt.$$

9. (*L^p-Maximal Inequalities*)
   (i) Let $\{Z_n : n = 0, 1, \ldots\}$ be an $\{\mathscr{F}_n\}$-martingale, such that $E|Z_n|^p < \infty$ for all n and for some $p \geqslant 1$. Prove that

$$P\left(\max_{1 \leqslant m \leqslant n} |Z_m| \geqslant \lambda\right) \leqslant \lambda^{-p} E|Z_n|^p \qquad \forall \lambda > 0.$$

[*Hint*: Note that $\{|Z_n|^p\}$ is a submartingale, and see Exercise 13.10 of Chapter I.]

(ii) Assume the hypothesis of (i) for some $p > 1$. Prove that

$$E\left(\max_{0 \leqslant m \leqslant n} |Z_n|^p\right) \leqslant (p/(p-1))^p E|Z_n|^p.$$

[*Hint*: Write $M_n := \max\{|Z_m|: 0 \leqslant m \leqslant n\}$. Then

$$\lambda P(M_n \geqslant \lambda) \leqslant \int_\Omega |Z_n| \mathbf{1}_{\{M_m \geqslant \lambda\}} \, dP,$$

by (13.55) of Chapter I. Therefore,

$$EM_n^p = E\left(p \int_0^{M_n} \lambda^{p-1} \, d\lambda\right) = p \int_0^\infty \lambda^{p-1} P(M_n \geqslant \lambda) \, d\lambda \quad \text{(by Fubini)}$$

$$\leqslant p \int_0^\infty \lambda^{p-2}\left(\int_\Omega |Z_n| \mathbf{1}_{\{M_n \geqslant \lambda\}} \, dP\right) d\lambda = p \int_\Omega |Z_n| \left(\int_0^{M_n} \lambda^{p-2} \, d\lambda\right) dP$$

$$= (p/(p-1)) \int_\Omega |Z_n| M_n^{p-1} \, dP$$

$$\leqslant (p/(p-1))(EM_n^p)^{(p-1)/p}(E|Z_n|^p)^{1/p} \quad \text{(by Hölder's Inequality)} \quad .]$$

(iii) Let $\{Z_t: t \geqslant 0\}$ be a continuous-parameter $\{\mathscr{F}_t\}$-martingale, having a.s. continuous sample paths. If $E|Z_t|^p < \infty$ for all $t \geqslant 0$, and for some $p \geqslant 1$, then show that

$$P\left(\max_{0 \leqslant t \leqslant T} |Z_t| \geqslant \lambda\right) \leqslant \lambda^{-p} E|Z_T|^p \qquad \forall \lambda > 0,$$

and, if $p > 1$, show that

$$E\left(\max_{0 \leqslant t \leqslant T} |Z_t|\right)^p \leqslant (p/(p-1))^p E|Z_T|^p.$$

10. (i) In addition to the hypothesis of Theorem 2.1 assume that $EX_\alpha^{2m} < \infty$, where m is a positive integer. Prove that for $0 \leqslant t - \alpha \leqslant 1$,

$$\|X_t - X_\alpha\|_{2m} \leqslant c_1(m, M)\left[(t-\alpha)\|\mu(X_\alpha)\|_{2m} + (t-\alpha)^{1/2}\|\sigma(X_\alpha)\|_{2m}\left(\frac{(2m)!}{m!\, 2^m}\right)^{1/2m}\right]$$

$$= c_1(m, M)\varphi(t-\alpha),$$

say. Here $\|\cdot\|_{2m}$ is the L^{2m}-norm for random variables, and $c_1(m, M)$ depends only on m and M. [*Hint*: Write $\delta_t^{(n)} := \sup\{E(X_s^{(n)} - X_s^{(n-1)})^{2m}: \alpha \leqslant s \leqslant t\}$. Use the triangle inequality for $\|\cdot\|_{2m}$, and Exercise 8(i), to show that

(a) $\|X_t^{(1)} - X_\alpha\|_{2m} \leqslant \varphi(t-\alpha), \qquad (\delta_t^{(1)})^{1/2m} \leqslant \varphi(t-\alpha);$

(b) $\delta_t^{(n)} \leqslant 2^{2m-1}M^{2m}\{(t-\alpha)^{2m-1} + (2m-1)^m(t-\alpha)^{m-1}\} \int_\alpha^t \delta_s^{(n-1)}\,ds$ for $n \geqslant 2$.

Observe that

$$\|X_t - X_\alpha\|_{2m} \leqslant \sum_{n=1}^{\infty} (\delta_t^{(n)})^{1/2m} \quad .]$$

(ii) Deduce from (i) that $\sup_{\alpha \leqslant s \leqslant t} EX_s^{2m} < \infty$ for all $t \geqslant \alpha$.
(iii) Extend (i), (ii) to multidimension.

11. (i) Use the result in 10(i) to prove, for every $\varepsilon > 0$, that $P(|X_t^x - x| \geqslant \varepsilon) = o(t)$, as $t \downarrow 0$, (a) uniformly for all x in a bounded interval, if $\mu(\cdot)$, $\sigma(\cdot)$ are Lipschitzian, and (b) uniformly for all x in $\mathbb{R}^1$ if $\mu(\cdot)$, $\sigma(\cdot)$ are bounded as well as Lipschitzian.
(ii) Extend (i) to multidimension.
(iii) Let $\boldsymbol{\mu}(\cdot)$, $\boldsymbol{\sigma}(\cdot)$ be Lipschitzian. Prove that

$$P\left(\max_{0 \leqslant s \leqslant t} |\mathbf{X}_s^{\mathbf{x}} - \mathbf{x}| \geqslant \varepsilon\right) = o(t) \qquad \text{as } t \downarrow 0,$$

uniformly for every bounded set of $\mathbf{x}$'s. Show that the convergence is uniform for all $\mathbf{x} \in \mathbb{R}^k$ if $\boldsymbol{\mu}(\cdot)$, $\boldsymbol{\sigma}(\cdot)$ are bounded as well as Lipschitzian. [*Hint*:

$$P\left(\max_{0 \leqslant s \leqslant t} |\mathbf{X}_s^{\mathbf{x}} - \mathbf{x}| \geqslant \varepsilon\right) \leqslant P\left(\int_0^t |\boldsymbol{\mu}(\mathbf{X}_s^{\mathbf{x}})|\,ds \geqslant \frac{\varepsilon}{2}\right) + P\left(\max_{0 \leqslant s \leqslant t} \left|\int_0^s \boldsymbol{\sigma}(\mathbf{X}_u^{\mathbf{x}})\,d\mathbf{B}_u\right| \geqslant \frac{\varepsilon}{2}\right)$$
$$= I_1 + I_2,$$

say. To estimate I_1 note that $|\boldsymbol{\mu}(\mathbf{X}_s^{\mathbf{x}})| \leqslant |\boldsymbol{\mu}(\mathbf{x})| + M|\mathbf{X}_s^{\mathbf{x}} - \mathbf{x}|$, and use Chebyshev's Inequality for the fourth moment and Exercise 10(iii). To estimate I_2 use Exercise 9(iii) with $p = 4$, Proposition 3.4 (or Exercise 8(ii) with $m = 2$) and Exercise 10(iii).]

12. (*The Semigroup* $\{\mathbf{T}_t\}$) Let $\boldsymbol{\mu}(\cdot)$, $\boldsymbol{\sigma}(\cdot)$ be Lipschitzian.
(i) Show that $\|\mathbf{T}_t f - f\| \to 0$ as $t \downarrow 0$, for every real-valued Lipschitzian f on $\mathbb{R}^k$ vanishing outside a bounded set. Here $\|\cdot\|$ denotes the "sup norm." [*Hint*: Use Exercise 10.]
(ii) Let f be a real-valued twice continuously differentiable function on $\mathbb{R}^k$.
(a) If $\mathbf{A}f$ and $\mathbf{grad}\,f$ are polynomially bounded (i.e., $|\mathbf{A}f(\mathbf{x})| \leqslant c_1(1 + |\mathbf{x}|^{m_1})$, $|\mathbf{grad}\,f(\mathbf{x})| \leqslant c_2(1 + |\mathbf{x}|^{m_2})$, show that $\mathbf{T}_t f(\mathbf{x}) \to f(\mathbf{x})$ as $t \downarrow 0$, uniformly on every bounded set of $\mathbf{x}$'s.
(b) If $\mathbf{A}f$ is bounded and $\mathbf{grad}\,f$ polynomially bounded, show that $\|\mathbf{T}_t f - f\| \to 0$. [*Hint*: Use Itô's Lemma.]
(iii) (*Infinitesimal Generator*) Let $\mathscr{D}$ denote the set of all real-valued bounded continuous functions f on $\mathbb{R}^k$ such that $\|t^{-1}(\mathbf{T}_t f - f) - g\| \to 0$ as $t \downarrow 0$, for some bounded continuous g (i.e., $g = ((d/dt)\mathbf{T}_t f)_{t=0}$, where the convergence of the difference quotient to the derivative is uniform on $\mathbb{R}^k$). Write $g = \hat{\mathbf{A}}f$ for such f, and call $\hat{\mathbf{A}}$ the *infinitesimal generator of* $\{\mathbf{T}_t\}$, and $\mathscr{D}$ the *domain of* $\hat{\mathbf{A}}$. Show that every twice continuously differentiable f, which vanishes outside some bounded set, belongs to $\mathscr{D}$ and that for such f, $\hat{\mathbf{A}}f = \mathbf{A}f$ where $\mathbf{A}$ is given

by (3.21). [*Hint*: Suppose $f(\mathbf{x}) = 0$ for $|\mathbf{x}| \geqslant N$. For each $\varepsilon > 0$ and $\mathbf{x} \in \mathbb{R}^k$,

$$|\mathbf{T}_t f(\mathbf{x}) - f(\mathbf{x}) - t\mathbf{A}f(\mathbf{x})| \leqslant t\theta(\mathbf{A}f:\varepsilon) + 2t\|\mathbf{A}f\| P\left(\max_{0 \leqslant s \leqslant t} |\mathbf{X}_s^{\mathbf{x}} - \mathbf{x}| \geqslant \varepsilon\right),$$

where $\theta(h:\varepsilon) = \sup\{|h(\mathbf{y}) - h(\mathbf{z})|: \mathbf{y}, \mathbf{z} \in \mathbb{R}^k, |\mathbf{y} - \mathbf{z}| < \varepsilon\}$. As $\varepsilon \downarrow 0$, $\theta(\mathbf{A}f:\varepsilon) \to 0$, and for each $\varepsilon > 0$,

$$P\left(\max_{0 \leqslant s \leqslant t} |\mathbf{X}_s^{\mathbf{x}} - \mathbf{x}| \geqslant \varepsilon\right) \to 0 \qquad \text{as } t \downarrow 0$$

uniformly for all $\mathbf{x}$ in $N_\varepsilon := \{\mathbf{y}: |\mathbf{y}| \leqslant N + \varepsilon\}$ (see Exercise 11(iii)). For $\mathbf{x} \in N_\varepsilon^c$,

$$|\mathbf{T}_t f(\mathbf{x}) - f(\mathbf{x}) - t\mathbf{A}f(\mathbf{x})| = \left|E \int_0^t \mathbf{A}f(\mathbf{X}_s^{\mathbf{x}})\, ds\right| \leqslant t\|\mathbf{A}f\| P(\tau^{\mathbf{x}} \leqslant t),$$

where $\tau^{\mathbf{x}} := \inf\{t \geqslant 0: |\mathbf{X}_t^{\mathbf{x}}| = N\}$. But $P(\tau^{\mathbf{x}} \leqslant t) \leqslant \sup\{P(\tau^{\mathbf{y}} \leqslant t): |\mathbf{y}| = N + \varepsilon\}$, by the strong Markov property applied to the stopping time $\tau := \inf\{t \geqslant 0: |\mathbf{X}_t^{\mathbf{x}}| = N + \varepsilon\}$. Now

$$P(\tau^{\mathbf{y}} \leqslant t) \leqslant P\left(\max_{0 \leqslant s \leqslant t} |\mathbf{X}_s^{\mathbf{y}} - \mathbf{y}| \geqslant \varepsilon\right) = o(t) \qquad \text{as } t \downarrow 0,$$

uniformly for all $\mathbf{y}$ satisfying $|\mathbf{y}| = N + \varepsilon$.]

(iv) Let f be a real-valued, bounded, twice continuously differentiable function on $\mathbb{R}^k$ such that **grad** f is polynomially bounded, $\mathbf{A}f$ is uniformly continuous, and $\mathbf{A}f(\mathbf{x}) \to 0$ as $|\mathbf{x}| \to \infty$. Show that $f \in \mathscr{D}$. [*Hint*: Extend the argument in (iii).]

(v) (*Local Property of* $\hat{\mathbf{A}}$) Suppose $f, g \in \mathscr{D}$ are such that $f(\mathbf{y}) = g(\mathbf{y})$ in a neighborhood $B(\mathbf{x}:\varepsilon) \equiv \{|\mathbf{y} - \mathbf{x}| < \varepsilon\}$ of $\mathbf{x}$. Then $\hat{\mathbf{A}}f(\mathbf{x}) = \hat{\mathbf{A}}g(\mathbf{x})$. [*Hint*: $\mathbf{T}_t f(\mathbf{x}) - \mathbf{T}_t g(\mathbf{x}) = o(t)$ as $t \downarrow 0$, by Exercise 11(iii).]

13. (*Initial-Value Problem*) Let $\boldsymbol{\mu}(\cdot), \boldsymbol{\sigma}(\cdot)$ be Lipschitzian. Adopt the notation of Exercise 12.

(i) If $f \in \mathscr{D}$, then $\mathbf{T}_t f \in \mathscr{D}$ for all $t \geqslant 0$, and $\hat{\mathbf{A}}\mathbf{T}_t f = \mathbf{T}_t \hat{\mathbf{A}} f$. [*Hint*:

$$\frac{1}{h}(\mathbf{T}_h(\mathbf{T}_t f) - \mathbf{T}_t f) = \frac{\mathbf{T}_t(\mathbf{T}_h f - f)}{h} \to \mathbf{T}_t \hat{\mathbf{A}} f$$

in "sup norm," since $\|\mathbf{T}_t g\| \leqslant \|g\|$ for all bounded measurable g.]

(ii) (*Existence of a Solution*) Let f satisfy the hypothesis of Exercise 12(iv). Show that the function $u(t, \mathbf{x}) := \mathbf{T}_t f(\mathbf{x})$ satisfies *Kolmogorov's backward equation*

$$\frac{\partial u(t, \mathbf{x})}{\partial t} = \hat{\mathbf{A}} u(t, \mathbf{x}) \qquad (t > 0, \mathbf{x} \in \mathbb{R}^k),$$

and the *initial condition*

$$u(t, x) \to f(x) \qquad \text{as } t \downarrow 0, \text{ uniformly on } \mathbb{R}^k.$$

If, in addition, $u(t, \mathbf{x})$ is twice continuously differentiable in x, for $t > 0$, then $\hat{\mathbf{A}}u(t, x) = \mathbf{A}u(t, x)$. [*Hint*: The first part is a restatement of (i). For the second part use Exercise 12(iii), (v).]

(iii) (*Uniqueness*) Suppose $v(t, x)$ is continuous on $\{t \geq 0, \mathbf{x} \in \mathbb{R}^k\}$, once continuously differentiable in $t > 0$ and twice continuously differentiable in $\mathbf{x} \in \mathbb{R}^k$, and satisfies

$$\frac{\partial v(t, \mathbf{x})}{\partial t} = \mathbf{A}v(t, \mathbf{x}) \qquad (t > 0, \mathbf{x} \in \mathbb{R}^k),$$

$$\lim_{t \downarrow 0} v(t, \mathbf{x}) = f(\mathbf{x}) \qquad (\mathbf{x} \in \mathbb{R}^k),$$

where f is continuous and polynomially bounded. If $\mathbf{A}v(t, \mathbf{x})$ and $|\mathbf{grad}\, v(t, \mathbf{x})| = |(\partial/\partial x^{(1)}, \dots, \partial/\partial x^{(k)})v(t, \mathbf{x})|$ are polynomially bounded uniformly for every compact set of time points in $(0, \infty)$, then $v(t, \mathbf{x}) = u(t, \mathbf{x}) := Ef(\mathbf{X}_t^{\mathbf{x}})$. [*Hint*: For each $t > 0$, use Itô's Lemma to the function $w(s, \mathbf{x}) = v(t - s, \mathbf{x})$ to get

$$Ev(\varepsilon, \mathbf{X}_{t-\varepsilon}^{\mathbf{x}}) - v(t, \mathbf{x}) = E\int_{\varepsilon}^{t} \{-v_0(r, \mathbf{X}_{t-r}^{\mathbf{x}}) + \mathbf{A}v(r, \mathbf{X}_{t-r}^{\mathbf{x}})\}\, dr = 0,$$

where $v_0(t, \mathbf{y}) := (\partial v/\partial t)(t, \mathbf{y})$. Let $\varepsilon \downarrow 0$.]

14. (*Dirichlet Problem*) Let G be a bounded open subset of $\mathbb{R}^k$. Assume that $\boldsymbol{\mu}(\cdot), \boldsymbol{\sigma}(\cdot)$ are Lipschitzian and that, for some i, $d_{ii}(\mathbf{x}) := |\boldsymbol{\sigma}_i(\mathbf{x})|^2 > 0$ for $\mathbf{x} \in \bar{G}$. Suppose v is a twice continuously differentiable function on G, continuous on $\bar{G}$, satisfying

$$\mathbf{A}v(\mathbf{x}) = -g(\mathbf{x}) \qquad (\mathbf{x} \in G),$$

$$v(\mathbf{x}) = f(\mathbf{x}) \qquad (\mathbf{x} \in \partial G),$$

where f and g are (given) continuous functions on ∂G and $\bar{G}$, respectively. Assume that v can be extended to a twice continuously differentiable function on $\mathbb{R}^k$. Then

$$v(\mathbf{x}) = Ef(\mathbf{X}_\tau^{\mathbf{x}}) + E\int_0^\tau g(\mathbf{X}_s^{\mathbf{x}})\, ds \qquad (\mathbf{x} \in \bar{G}),$$

where $\tau := \inf\{t \geq 0 : \mathbf{X}_t^{\mathbf{x}} \in \partial G\}$. [*Hint*: v can be taken to be twice continuously differentiable on $\mathbb{R}^k$ with compact support. Apply Itô's Lemma to $\{v(\mathbf{X}_t^{\mathbf{x}})\}$, and then use the Optional Stopping Theorem (Proposition 13.9, Chapter I). Also see Exercise 3.]

15. (*Feynman–Kac Formula*) Let $\boldsymbol{\mu}(\cdot), \boldsymbol{\sigma}(\cdot)$ be Lipschitzian. Suppose $u(t, \mathbf{x})$ is a continuous function on $[0, \infty) \times \mathbb{R}^k$, once continuously differentiable in t for $t > 0$, and twice continuously differentiable in $\mathbf{x}$ on $\mathbb{R}^k$, satisfying

$$\frac{\partial u(t, \mathbf{x})}{\partial t} = \mathbf{A}u(t, \mathbf{x}) + V(\mathbf{x})u(t, \mathbf{x}) \qquad (t > 0, \mathbf{x} \in \mathbb{R}^k), \qquad u(0, \mathbf{x}) = f(\mathbf{x}),$$

where f is a polynomially bounded continuous function on $\mathbb{R}^k$, and V is a continuous function on $\mathbb{R}^k$ that is bounded above. If $\mathbf{grad}\, u(t, \mathbf{x})$ and $\mathbf{A}u(t, \mathbf{x})$ are polynomially bounded in $\mathbf{x}$ uniformly for every compact set of t in $(0, \infty)$, then

$$u(t, \mathbf{x}) = E\left(f(\mathbf{X}_t^{\mathbf{x}}) \exp\left\{\int_0^t V(\mathbf{X}_s^{\mathbf{x}})\, ds\right\}\right) \qquad (t \geqslant 0, \mathbf{x} \in \mathbb{R}^k).$$

[*Hint*: For each $t > 0$, apply Itô's Lemma to

$$\mathbf{Y}(s) = (\mathbf{Y}_1(s), \mathbf{Y}_2(s)) = \left(\mathbf{X}_s^{\mathbf{x}}, \int_0^s V(\mathbf{X}_{s'}^{\mathbf{x}})\, ds'\right) \quad \text{and} \quad \varphi(s, \mathbf{y}) = u(t - s, \mathbf{y}_1),$$

where $\mathbf{y} = (\mathbf{y}_1, y_2)$.]

Exercises for Section VII.4

1. (i) Use the method of successive approximation to show that (4.36) is the solution to (4.35).
   (ii) Use Itô's Lemma to check that (4.36) solves (4.35).
2. Check that (4.37) is the dispersion matrix of the solution $\{\mathbf{X}_t^{\mathbf{x}}\}$ of the linear stochastic differential equation (4.35).
3. (i) Let $\mathbf{C}$ be a $k \times k$ matrix such that $\mathbf{C} + \mathbf{C}'$ is *negative definite* (i.e., the eigenvalues of $\mathbf{C} + \mathbf{C}'$ are all negative). Prove that the real parts of all eigenvalues of $\mathbf{C}$ are negative.
   (ii) Give an example of a 2×2 matrix $\mathbf{C}$ whose eigenvalues have negative real parts, but $\mathbf{C} + \mathbf{C}'$ is not negative definite.
4. In (4.35) let $\mathbf{C} = -c\mathbf{I}$ where $c > 0$ and $\mathbf{I}$ is the $k \times k$ identity matrix. Compute the dispersion matrices $\mathbf{\Sigma}(t)$, $\mathbf{\Sigma}$ explicitly in this case, and show that they are singular if and only if $\boldsymbol{\sigma}$ is singular.

THEORETICAL COMPLEMENTS

Theoretical Complements to Section VII.1

1. (*P-Completion of Sigmafields*) Let $(\Omega, \mathcal{F}, P)$ be a probability space, $\mathcal{G}$ a subsigmafield of $\mathcal{F}$. The P-completion $\bar{\mathcal{G}}$ of $\mathcal{G}$ is the sigmafield $\{G \cup N: G \in \mathcal{G}, N \subset M \in \mathcal{F}$ such that $P(M) = 0\}$. The proof of Lemma 1 of theoretical complement V.1.1 shows that if $\mathcal{F}_t$ is P-complete for all $t \geqslant 0$ then the stochastic process $\{f(t): t \geqslant 0\}$ is progressively measurable provided (1) $f(t)$ is $\mathcal{F}_t$-measurable for all t, and (2) $t \to f(t)$ is *almost surely* right-continuous. Several times in Section 1 we have constructed processes $\{f(t)\}$ that have continuous sample paths almost surely, and which have the property that $f(t)$ is $\mathcal{F}_t$-measurable. These are nonanticipative in view of the fact that $\mathcal{F}_t$ is P-complete.

2. (*Extension of the Itô Integral to the Class* $\mathcal{L}[\alpha, \beta]$) Notice that the Itô integral (1.6) is well defined for *all* nonanticipative step functionals, and not merely those that are in $\mathcal{M}[\alpha, \beta]$. It turns out that the stochastic integral of a right-continuous nonanticipative functional $\{f(t): \alpha \leqslant t \leqslant \beta\}$ exists as an almost surely continuous limit of those of an appropriate sequence of step functionals, provided

$$\int_\alpha^\beta f^2(s)\,ds < \infty \quad \text{a.s.} \tag{T.1.1}$$

The *class* of all nonanticipative right-continuous functionals $\{f(t): \alpha \leqslant t \leqslant \beta\}$ satisfying (T.1.1) *is denoted by* $\mathscr{L}[\alpha, \beta]$. Given an $f \in \mathscr{L}[\alpha, \beta]$ define, for each positive integer n,

$$A(t; n) := \left\{\omega: \int_\alpha^t f^2(s)\,ds < n\right\}, \qquad f_n(t) = f(t)\mathbf{1}_{A(t;n)}.$$

Then $f_n \in \mathscr{M}[\alpha, \beta]$. Write

$$A_n := \left\{\omega: \int_\alpha^\beta f^2(t, \omega)\,dt < n\right\} \equiv A(\beta; n).$$

If $m > n$ then on A_n we have $f_m(t, \omega) = f_n(t, \omega)$ for $\alpha \leqslant t \leqslant \beta$. It follows that

$$\int_\alpha^t f_m(s)\,dB_s = \int_\alpha^t f_n(s)\,dB_s \qquad \text{for } \alpha \leqslant t \leqslant \beta, \text{ for almost all } \omega \in A_n.$$

To see this, consider $g, h \in \mathscr{M}[\alpha, \beta]$ such that $g(t, \omega) = h(t, \omega)$ for $\alpha \leqslant t \leqslant \beta$, $\omega \in A \in \mathscr{F}$. If g, h are step functionals, then it follows from the definition (1.6) of the stochastic integral that

$$\int_\alpha^t g(s)\,dB_s = \int_\alpha^t h(s)\,dB_s \qquad \text{for } \alpha \leqslant t \leqslant \beta, \text{ for } all\ \omega \in A.$$

In the general case, let g_n, h_n be nonanticipative step functionals such that

$$\int_\alpha^t g_n(s)\,dB_s \to \int_\alpha^t g(s)\,dB_s, \qquad \int_\alpha^t h_n(s)\,dB_s \to \int_\alpha^t h(s)\,dB_s \qquad \text{for } \alpha \leqslant t \leqslant \beta, \text{ a.s.}$$

Without loss of generality one may take, for each n, the same partition $\{t_i(n): 0 \leqslant i \leqslant N_n\}$ of $[\alpha, \beta]$ for both g_n, h_n. Modify h_n, if necessary, so that $h_n(t_i^{(n)}, \omega) = g_n(t_i^{(n)}, \omega)$ for $\omega \in \{h(s) = g(s) \text{ for } \alpha \leqslant s \leqslant t_i^{(n)}\}$ $(0 \leqslant i \leqslant N_n)$. Then

$$\int_\alpha^t h_n(s)\,dB_s = \int_\alpha^t g_n(s)\,dB_s \qquad \text{for } \alpha \leqslant t \leqslant \beta, \text{ if } \omega \in A,$$

so that, in the limit,

$$\int_\alpha^t h(s)\,dB_s = \int_\alpha^t g(s)\,dB_s, \qquad \alpha \leqslant t \leqslant \beta, \text{ almost everywhere on } A.$$

Thus, we get

$$\int_\alpha^t f_m(s)\,dB_s = \int_\alpha^t f_n(s)\,dB, \qquad \alpha \leqslant t \leqslant \beta, \text{ almost everywhere on } A_n.$$

Therefore,

$$\lim_{m\to\infty}\int_\alpha^t f_m(s)\,dB_s = \int_s^t f_n(s)\,dB, \qquad \alpha \leqslant t \leqslant \beta, \text{ almost everywhere on } A_n.$$

A stochastic process that equals $\int_\alpha^t f_n(s)\,dB_s$ a.s. on A_n $(n \geqslant 1)$ will be called the *Itô integral of f* and denoted by $\int_\alpha^t f(s)\,dB_s$. Since $P(A_n) \to 1$ in view of (T.1.1), on the complement of $\bigcup_n A_n$ the process may be defined arbitrarily.

The proof of Proposition 1.2 may be adapted for $f \in \mathscr{L}[\alpha, \beta]$ to show that (see (1.25) and the definition of h_n after (1.29)) there exists a sequence of nonanticipative step functionals h_n such that

$$\int_\alpha^\beta (h_n(s) - f(s))^2\,ds \to 0 \qquad \text{in probability.} \tag{T.1.2}$$

From this one may show that $\int_\alpha^\beta h_n(s)\,dB_s$ converges in probability to $\int_\alpha^\beta f(s)\,dB_s$. For this we first derive the inequality

$$P\left(\left|\int_\alpha^\beta f(s)\,dB_s\right| > \varepsilon\right) \leqslant P\left(\int_\alpha^\beta f^2(s)\,ds \geqslant c\right) + \frac{c}{\varepsilon^2} \tag{T.1.3}$$

for all $c > 0$, $\varepsilon > 0$. To see this define

$$g(t) = f(t)\mathbf{1}_{A(t;c)}.$$

Since $g \in \mathscr{M}[\alpha, \beta]$, and $f(t) = g(t)$ for $\alpha \leqslant t \leqslant \beta$ on the set $\{\int_\alpha^\beta f^2(s)\,ds < c\}$, it follows that

$$\begin{aligned} P\left(\left|\int_\alpha^\beta f(s)\,dB_s\right| > \varepsilon\right) &\leqslant P\left(\int_\alpha^\beta f^2(s)\,ds \geqslant c\right) + P\left(\left|\int_\alpha^\beta g(s)\,dB_s\right| > \varepsilon\right) \\ &\leqslant P\left(\int_\alpha^\beta f^2(s)\,ds \geqslant c\right) + E\left(\int_\alpha^\beta g(s)\,dB_s\right)^2\Big/\varepsilon^2 \\ &\leqslant P\left(\int_\alpha^\beta f^2(s)\,ds \geqslant c\right) + \left(E\int_\alpha^\beta g^2(s)\,ds\right)\Big/\varepsilon^2 \\ &\leqslant P\left(\int_\alpha^\beta f^2(s)\,ds \geqslant c\right) + \frac{c}{\varepsilon^2}. \end{aligned} \tag{T.1.4}$$

Applying (T.1.3) to $h_n - f$ one gets, taking $c = \varepsilon^3$,

$$\overline{\lim_{n\to\infty}}\, P\left(\left|\int_\alpha^\beta h_n(s)\,dB_s - \int_\alpha^\beta f(s)\,dB_s\right| > \varepsilon\right) \leqslant \varepsilon,$$

which is the desired result. Properties (a), (b) of Proposition 1.1 hold for $f \in \mathscr{L}[\alpha, \beta]$.

Theoretical Complements to Section VII.2

1. (*The Markov Property of* $\{X_t^x\}$) Consider the equation (2.25). Let $\mathscr{G}_{s,t}$ denote the P-completion of the sigmafield $\sigma\{B_u - B_s: s \leqslant u \leqslant t\}$. The successive approximations (2.3)–(2.5), with $\alpha = s$, $X_\alpha = z$, may be shown to be measurable with respect to $\mathscr{B}(\mathbb{R}^1) \otimes \mathscr{G}_{s,t}$. To see this, note that $(z, u, \omega) \to X_u^{(1)}(\omega)$ is measurable on $(\mathbb{R}^1 \times [s, t] \times \Omega, \mathscr{B}(\mathbb{R}^1) \otimes \mathscr{B}([s, t]) \otimes \mathscr{G}_{s,t})$ for every $t \geqslant s$. Also, if $(z, u, \omega) \to X_u^{(n)}(\omega)$ is measurable with respect to this product sigmafield, then so is the right side of (2.4). This is easily checked by expressing the time integral as a limit of Riemann sums, and the stochastic integral as a limit of stochastic integrals of nonanticipative (with respect to $\{\mathscr{G}_{s,u}: u \geqslant s\}$) step functionals. It then follows that the almost sure limit X_t of $X_t^{(n)}$ (as $n \to \infty$) is measurable as a function of (z, ω) (with respect to the sigmafield $\mathscr{B}(\mathbb{R}^1) \otimes \mathscr{G}_{s,t}$ on $\mathbb{R}^1 \times \Omega$). Write the solution of (2.25) as $\varphi(s, t; z)$. Then $X_t^x = \varphi(s, t; X_s^x)$, by (2.24) and the uniqueness of the solution of (2.25) with $z = X_s^x$. Since X_s^x is $\mathscr{F}_s$-measurable, and is independent of $\mathscr{G}_{s,t}$, it follows that for every bounded Borel measurable function g on $\mathbb{R}^1$, $E(g(X_t^x) \mid \mathscr{F}_s) = [Eg(\varphi(s, t; z))]_{z = X_s^x}$ (see Section 4 of Chapter 0, Theorem on Independence and Conditional Expectation). This proves the Markov property of $\{X_t^x\}$. The multidimensional case is entirely analogous.

2. (*Nonhomogeneous Diffusions*) Suppose that we are given functions $\mu^{(i)}(t, \mathbf{x})$, $\sigma_{ij}(t, \mathbf{x})$ on $[0, \infty) \times \mathbb{R}^k$ into $\mathbb{R}^1$. Write $\boldsymbol{\mu}(t, \mathbf{x})$ for the vector whose ith component is $\mu^{(i)}(t, \mathbf{x})$, $\boldsymbol{\sigma}(t, \mathbf{x})$ for the $k \times k$ matrix whose (i, j) element is $\sigma_{ij}(t, \mathbf{x})$. Given $s \geqslant 0$, we would like to solve the stochastic differential equation

$$d\mathbf{X}_t = \boldsymbol{\mu}(\mathbf{X}_t, t)\, dt + \boldsymbol{\sigma}(\mathbf{X}_t, t)\, d\mathbf{B}_t \qquad (t \geqslant s), \qquad \mathbf{X}_s = \mathbf{Z}, \tag{T.2.1}$$

where $\mathbf{Z}$ is $\mathscr{F}_s$-measurable, $E|\mathbf{Z}|^2 < \infty$. It is not difficult to extend the proofs of Theorems 2,1, 2.2 (also see Theorems 2.4, 2.5) to prove the following result.

Theorem T.2.1. Assume

$$\begin{aligned} |\boldsymbol{\mu}(t, \mathbf{x}) - \boldsymbol{\mu}(s, \mathbf{y})| &\leqslant M(|\mathbf{x} - \mathbf{y}| + |t - s|), \\ \|\boldsymbol{\sigma}(t, \mathbf{x}) - \boldsymbol{\sigma}(s, \mathbf{y})\| &\leqslant M(|\mathbf{x} - \mathbf{y}| + |t - s|), \end{aligned} \tag{T.2.2}$$

for some constant M.

(a) Then for every $\mathscr{F}_s$-measurable k-dimensional square integrable $\mathbf{Z}$ there exists a unique nonanticipative continuous solution $\{\mathbf{X}_t: t \geqslant s\}$ in $\mathscr{M}[s, \infty)$ of

$$\mathbf{X}_t = \mathbf{Z} + \int_s^t \boldsymbol{\mu}(\mathbf{X}_u, u)\, du + \int_s^t \boldsymbol{\sigma}(\mathbf{X}_u, u)\, d\mathbf{B}_u \qquad (t \geqslant s). \tag{T.2.3}$$

(b) Let $\{\mathbf{X}_t^{s,\mathbf{x}}: t \geqslant s\}$ denote the solution of (T.2.3) with $\mathbf{Z} = \mathbf{x} \in \mathbb{R}^k$. Then it is a (possibly nonhomogeneous) Markov process with (initial) state $\mathbf{x}$ at time s and transition probability

$$p(s', t; \mathbf{z}, B) = P(\mathbf{X}_t^{s',\mathbf{z}} \in B) \qquad (s \leqslant s' \leqslant t). \tag{T.2.4}$$

□

Theoretical Complements for Section VII.3

1. (*Itô's Lemma*) To prove Theorem 3.1, assume first that f_r, $\mathbf{g}_r$ are nonanticipative step functionals. The general case follows by approximating f_r, $\mathbf{g}_r$ by step functionals and taking limits in probability.

Fix $s < t$ $(0 \leqslant s < t \leqslant T)$. In view of the additivity of the (Riemann and stochastic) integrals, it is enough to consider the case $f_r(s') = f_r(s)$, $\mathbf{g}_r(s') = g_r(s)$ for $s \leqslant s' \leqslant t$. Let $t_0^{(n)} = s < t_1^{(n)} < \cdots < t_{N_n}^{(n)} = t$ be a sequence of partitions such that

$$\delta_n := \max\{t_{i+1}^{(n)} - t_i^{(n)} := 0 \leqslant i \leqslant N_n - 1\} \to 0 \qquad \text{as } n \to \infty.$$

Write

$$\varphi(t, \mathbf{Y}(t)) - \varphi(s, \mathbf{Y}(s)) = \sum_{i=0}^{N_n-1} [\varphi(t_{i+1}^{(n)}, \mathbf{Y}(t_{i+1}^{(n)})) - \varphi(t_i^{(n)}, \mathbf{Y}(t_{i+1}^{(n)}))]$$
$$+ \sum_{i=0}^{N_n-1} [\varphi(t_i^{(n)}, \mathbf{Y}(t_{i+1}^{(n)})) - \varphi(t_i^{(n)}, \mathbf{Y}(t_i^{(n)}))]. \qquad \text{(T.3.1)}$$

The *first sum* on the right may be expressed as

$$\sum_{i=0}^{N_n-1} (t_{i+1}^{(n)} - t_i^{(n)})\{\partial_0\varphi(t_i^{(n)}, \mathbf{Y}(t_{i+1}^{(n)})) + R_{n,i}^{(1)}\} \qquad \text{(T.3.2)}$$

where

$$|R_{n,i}^{(1)}| \leqslant \max\{|\partial_0\varphi(u, \mathbf{Y}(t_{i+1}^{(n)})) - \partial_0\varphi(u', \mathbf{Y}(t_{i+1}^{(n)}))| : t_i^{(n)} \leqslant u \leqslant u' \leqslant t_{i+1}^{(n)}\} \to 0$$

uniformly in i (for each ω), as $n \to \infty$, because $\partial_0\varphi$ is uniformly continuous on $[s, t] \times \{Y(u, \omega): s \leqslant u \leqslant t\}$. Hence, (T.3.2) converges, for all ω, to

$$J_0 := \int_s^t \partial_0\varphi(u, \mathbf{Y}(u))\, du. \qquad \text{(T.3.3)}$$

By a Taylor expansion, the *second sum* on the right in (T.3.1) may be expressed as

$$\sum_{i=0}^{N_n-1} \sum_{r=1}^{m} (Y^{(r)}(t_{i+1}^{(n)}) - Y^{(r)}(t_i^{(n)}))\partial_r\varphi(t_i^{(n)}, \mathbf{Y}(t_i^{(n)}))$$
$$+ \frac{1}{2!} \sum_{1 \leqslant r,r' \leqslant m} \sum_{i=0}^{N_n-1} (Y^{(r)}(t_{i+1}^{(n)}) - Y^{(r)}(t_i^{(n)}))(Y^{(r')}(t_{i+1}^{(n)}) - Y^{(r')}(t_i^{(n)}))$$
$$\times \{\partial_r\partial_{r'}\varphi(t_i^{(n)}, \mathbf{Y}(t_i^{(n)})) + R_{r,r',n,i}^{(2)}\}, \qquad \text{(T.3.4)}$$

where $R_{r,r',n,i}^{(2)} \to 0$ uniformly in i (for each ω), because $(u, \mathbf{y}) \to \partial_r\partial_{r'}\varphi(u, \mathbf{y})$ and $t \to \mathbf{Y}(t)$ are continuous. Using (3.9) the first sum in (T.3.4) is expressed as

$$\sum_{r=1}^{m} \sum_{i=0}^{N_n-1} \{(t_{i+1}^{(n)} - t_i^{(n)})f_r(s) + \mathbf{g}_r(s)\cdot(\mathbf{B}_{t_{i+1}^{(n)}} - \mathbf{B}_{t_i^{(n)}})\}\partial_r\varphi(t_i^{(n)}, \mathbf{Y}(t_i^{(n)}))$$
$$\to J_{11} := \sum_{r=1}^{m} \left[\int_s^t f_r(u)\partial_r\varphi(u, \mathbf{Y}(u))\, du + \int_s^t \partial_r\varphi(u, \mathbf{Y}(u))\mathbf{g}_r(u)\cdot d\mathbf{B}_u\right],$$
$$\text{in probability}, \qquad \text{(T.3.5)}$$

since $\partial_r \varphi$ is continuous and (see Exercise 1.3)

$$\sum_{i=0}^{N_n-1} \int_{t_i^{(n)}}^{t_{i+1}^{(n)}} (\partial_r \varphi(t_i^{(n)}, \mathbf{Y}(t_i^{(n)})) - \partial_r \varphi(u, \mathbf{Y}(u)))\mathbf{g}_r(s) \cdot d\mathbf{B}_u = \int_s^t h_n(u) \cdot d\mathbf{B}_u,$$

say, where $\int_s^t |h_n^2(u)|\, du \to 0$ a.s. It remains to find the limiting value of the second sum in (T.3.4) (excluding the remainder). For this, write

$$\begin{aligned}
&\sum_{i=0}^{N_n-1} (Y^{(r)}(t_{i+1}^{(n)}) - Y^{(r)}(t_i^{(n)}))(Y^{(r')}(t_{i+1}^{(n)}) - Y^{(r')}(t_i^{(n)}))\partial_r \partial_{r'} \varphi(t_i^{(n)}, \mathbf{Y}(t_i^{(n)})) \\
&\quad = \sum_{i=0}^{N_n-1} \{(t_{i+1}^{(n)} - t_i^{(n)})f_r(s) + \mathbf{g}_r(s) \cdot (\mathbf{B}_{t_{i+1}^{(n)}} - \mathbf{B}_{t_i^{(n)}})\} \\
&\quad \times \{(t_{i+1}^{(n)} - t_i^{(n)})f_{r'}(s) + \mathbf{g}_{r'}(s) \cdot (\mathbf{B}_{t_{i+1}^{(n)}} - \mathbf{B}_{t_i^{(n)}})\}\partial_r \partial_{r'} \varphi(t_i^{(n)}, \mathbf{Y}(t_i^{(n)})). \qquad \text{(T.3.6)}
\end{aligned}$$

By (3.1), (3.15) and (3.16), (T.3.6) converges a.s. (for a suitable sequence of partitions) to

$$\begin{aligned}
&\lim \sum_{i=0}^{N_n-1} \left\{\sum_{j=1}^{k} g_r^{(j)}(s)(B_{t_{i+1}^{(n)}}^{(j)} - B_{t_i^{(n)}}^{(j)})\right\}\left\{\sum_{j=1}^{k} g_{r'}^{(j)}(s)(B_{t_{i+1}^{(n)}}^{(j)} - B_{t_i^{(n)}}^{(j)})\right\}\partial_r \partial_{r'} \varphi(t_i^{(n)}, \mathbf{Y}(t_i^{(n)})) \\
&\quad = \lim \sum_{j=1}^{k} g_r^{(j)}(s) g_{r'}^{(j)}(s) \sum_{i=0}^{N_n-1} (B_{t_{i+1}^{(n)}}^{(j)} - B_{t_i^{(n)}}^{(j)})^2 \partial_r \partial_{r'} \varphi(t_i^{(n)}, \mathbf{Y}(t_i^{(n)})) \\
&\quad = \sum_{j=1}^{k} g_r^{(j)}(s) g_{r'}^{(j)}(s) \lim \sum_{i=0}^{N_n-1} (t_{i+1}^{(n)} - t_i^{(n)}) \partial_r \partial_{r'} \varphi(t_i^{(n)}, \mathbf{Y}(t_i^{(n)})) \\
&\quad = J_{12} := \sum_{j=1}^{k} \int_s^t g_r^{(j)}(u) g_{r'}^{(j)}(u) \partial_r \partial_{r'} \varphi(u, \mathbf{Y}(u))\, du. \qquad \text{(T.3.7)}
\end{aligned}$$

The proof is completed by adding J_0, J_{11}, and J_{12}. ■

It may be noted that the proof goes through for $f_r, \mathbf{g}_r \in \mathscr{L}[0, T]$, $1 \leqslant r \leqslant k$, $\varphi(t, \mathbf{y})$ twice continuously differentiable in $\mathbf{y}$ and once in t.

2. (*Explosion*) Assume that $\boldsymbol{\mu}(\cdot)$, $\boldsymbol{\sigma}(\cdot)$ are *locally Lipschitzian*, that is, (2.42) holds for $|\mathbf{x}| \leqslant n$, $|\mathbf{y}| \leqslant n$, with $M = M_n$, where M_n may go to infinity as $n \to \infty$. One may still construct a diffusion on $\mathbb{R}^k$ satisfying (2.43) up to an *explosion time* ζ. For this, let $\{\mathbf{X}_{t,n}^{\mathbf{x}}: t \geqslant 0\}$ be the solution of (2.43) with *globally Lipschitzian* coefficients $\boldsymbol{\mu}_n(\cdot)$, $\boldsymbol{\sigma}_n(\cdot)$ satisfying

$$\boldsymbol{\mu}_n(\mathbf{y}) = \boldsymbol{\mu}(\mathbf{y}) \quad \text{and} \quad \boldsymbol{\sigma}_n(\mathbf{y}) = \boldsymbol{\sigma}(\mathbf{y}) \quad \text{for } |\mathbf{y}| \leqslant n. \qquad \text{(T.3.8)}$$

We may, for example, let $\boldsymbol{\mu}_n(\mathbf{y}) = \boldsymbol{\mu}(n\mathbf{y}/|\mathbf{y}|)$ and $\boldsymbol{\sigma}_n(\mathbf{y}) = \boldsymbol{\sigma}(n\mathbf{y}/|\mathbf{y}|)$ for $|\mathbf{y}| > n$, so that (2.42) holds for *all* $\mathbf{x}, \mathbf{y}$ with $M = M_n$. Let us show that, if $|\mathbf{x}| \leqslant n$,

$$\mathbf{X}_{t,m}^{\mathbf{x}}(\omega) = \mathbf{X}_{t,n}^{\mathbf{x}}(\omega) \quad \text{for } 0 \leqslant t < \zeta_n(\omega) \quad (m \geqslant n) \qquad \text{(T.3.9)}$$

outside a set of zero probability, where

$$\zeta_n(\omega) := \inf\{t \geqslant 0 \colon |\mathbf{X}^{\mathbf{x}}_{t,n}| = n\}. \tag{T.3.10}$$

To prove (T.3.9), first note that if $m \geqslant n$ then $\boldsymbol{\mu}_m(\mathbf{y}) = \boldsymbol{\mu}_n(\mathbf{y})$ and $\boldsymbol{\sigma}_m(\mathbf{y}) = \boldsymbol{\sigma}_n(\mathbf{y})$ for $|\mathbf{y}| \leqslant n$, so that

$$\boldsymbol{\mu}_m(\mathbf{X}^{\mathbf{x}}_{t,n}) = \boldsymbol{\mu}_n(\mathbf{X}^{\mathbf{x}}_{t,n}) \qquad \text{and} \qquad \boldsymbol{\sigma}_m(\mathbf{X}^{\mathbf{x}}_{t,n}) = \boldsymbol{\sigma}_n(\mathbf{X}^{\mathbf{x}}_{t,n}) \qquad \text{for } 0 \leqslant t < \zeta_n(\omega). \tag{T.3.11}$$

It follows from (T.3.11) (see the argument in theoretical complement 1.2) that

$$\int_0^t \boldsymbol{\mu}_m(\mathbf{X}^{\mathbf{x}}_{s,n})\,ds + \int_0^t \boldsymbol{\sigma}_m(\mathbf{X}^{\mathbf{x}}_{s,n})\,d\mathbf{B}_u = \int_0^t \boldsymbol{\mu}_n(\mathbf{X}^{\mathbf{x}}_{s,n})\,ds + \int_0^t \boldsymbol{\sigma}_n(\mathbf{X}^{\mathbf{x}}_{s,n})\,d\mathbf{B}_u \qquad \text{for } 0 \leqslant t < \zeta_n \tag{T.3.12}$$

almost surely. Therefore, a.s.,

$$\mathbf{X}^{\mathbf{x}}_{t,m} - \mathbf{X}^{\mathbf{x}}_{t,n} = \int_0^t [\boldsymbol{\mu}_m(\mathbf{X}^{\mathbf{x}}_{s,m}) - \boldsymbol{\mu}_m(\mathbf{X}^{\mathbf{x}}_{s,n})]\,ds + \int_0^t [\boldsymbol{\sigma}_m(\mathbf{X}^{\mathbf{x}}_{s,m}) - \boldsymbol{\sigma}_m(\mathbf{X}^{\mathbf{x}}_{s,n})]\,d\mathbf{B}_u \qquad \text{for } 0 \leqslant t < \zeta_n. \tag{T.3.13}$$

This implies

$$\begin{aligned}\varphi(t) := E(|\mathbf{X}^{\mathbf{x}}_{t,m} - \mathbf{X}^{\mathbf{x}}_{t,n}|^2 \mathbf{1}_{\{\zeta_n > t\}}) &\leqslant 2tM_m^2 \int_0^t \varphi(s)\,ds + 2M_m^2 \int_0^t \varphi(s)\,ds \\ &= 2M_m^2(t+1) \int_0^t \varphi(s)\,ds. \end{aligned} \tag{T.3.14}$$

By iteration (see (2.21) and (2.22)), $\varphi(t) = 0$. In other words, on $\{\zeta_n > t\}$, $\mathbf{X}^{\mathbf{x}}_{t,m} = \mathbf{X}^{\mathbf{x}}_{t,n}$ a.s., for each $t \geqslant 0$. Since $t \to \mathbf{X}^{\mathbf{x}}_{t,m}, \mathbf{X}^{\mathbf{x}}_{t,n}$ are continuous, it follows that, outside a set of probability 0, $\mathbf{X}^{\mathbf{x}}_{s,m} = \mathbf{X}^{\mathbf{x}}_{s,n}$ for $0 \leqslant s < \zeta_n$, $m \geqslant n$. In particular, $\zeta_n \uparrow$ a.s. to (the *explosion time*) ζ, say, as $n \uparrow \infty$, and

$$\mathbf{X}^{\mathbf{x}}_t(\omega) = \lim_{n \to \infty} \mathbf{X}^{\mathbf{x}}_{t,n}(\omega), \qquad t < \zeta(\omega), \tag{T.13.15}$$

exists a.s., and has a.s. continuous sample paths on $[0, \zeta(\omega))$. If $P(\zeta < \infty) > 0$, then we say that *explosion occurs*. In this case one may continue $\{\mathbf{X}^{\mathbf{x}}_t\}$ for $t \geqslant \zeta(\omega)$ by setting

$$\mathbf{X}^{\mathbf{x}}_t(\omega) = \text{“}\infty\text{”}, \qquad t \geqslant \zeta(\omega), \tag{T.13.16}$$

where "∞" is a new state, usually taken to be the *point at infinity* in the *one-point compactification* of $\mathbb{R}^k$ (see H. L. Royden (1963), *Real Analysis*, 2nd ed., Macmillan, New York, p. 168). Using the Markov property of $\{\mathbf{X}^{\mathbf{x}}_{t,n}\}$ for each n, it is not difficult to show that $\{\mathbf{X}^{\mathbf{x}}_t \colon t \geqslant 0\}$ defined by (T.13.15) and (T.13.16) is a Markov process on $\mathbb{R}^k \cup \{\text{“}\infty\text{”}\}$, called the *minimal diffusion generated by*

$$\mathbf{A} = \frac{1}{2} \sum d_{ij}(\mathbf{x}) \frac{\partial^2}{\partial x^{(i)}\,\partial x^{(j)}} + \sum \mu^{(i)}(\mathbf{x}) \frac{\partial}{\partial x^{(i)}}.$$

If $k = 1$ one often takes, instead of (T.13.16),

$$X_t^x = \begin{cases} -\infty & \text{for } t \geqslant \zeta(\omega) \text{ on } \left(\lim\limits_{t\to\infty} X_t^x = -\infty\right), \\ +\infty & \text{for } t \geqslant \xi(\omega) \text{ on } \left(\lim\limits_{t\to\infty} X_t^x = +\infty\right). \end{cases} \tag{T.3.17}$$

Write $\zeta = \tau_{-\infty}$ on $\{\lim_{t\to\infty} X_t^x = -\infty\}$, $\zeta = \tau_{+\infty}$ on $\{\lim_{t\to\infty} X_t^x = +\infty\}$. Write $\tau_z := \inf\{t \geqslant 0: X_t^x = z\}$.

Theorem T.3.1. (*Feller's Test for Explosion*). Let $k = 1$ and $I(z) := \int_0^z (2\mu(z')/\sigma^2(z'))\, dz'$. Then

(a) $P(\tau_{+\infty} < \infty) = 0$ iff $\displaystyle\int_0^\infty \exp\{-I(y)\}\left[\int_0^y (\exp\{I(z)\}/\sigma^2(z))\, dz\right] dy = \infty,$

(b) $P(\tau_{-\infty} < \infty) = 0$ iff $\displaystyle\int_{-\infty}^0 \exp\{-I(y)\}\left[\int_y^0 (\exp\{I(z)\}/\sigma^2(z))\, dz\right] dy = \infty.$

□

Proof. (a) Assuming that the integral on the right in (a) diverges we will construct a nonnegative, increasing and twice continuously differentiable function φ on $[0, \infty)$ such that (i) $\mathbf{A}\varphi(y) \leqslant \varphi(y)$ for all y and (ii) $\varphi(y) \to \infty$ as $y \to \infty$. Granting the existence of such a function φ for the moment (and extending it to a twice continuously differentiable function on $(-\infty, 0]$), apply Itô's Lemma to the function $\varphi(t, y) := \exp\{-t\}\varphi(y)$ to get, for all $x > 0$,

$$E\varphi(\tau_n \wedge \tau_0 \wedge t, X^x_{\tau_n \wedge \tau_0 \wedge t}) - \varphi(x) = E\int_0^{\tau_n \wedge \tau_0 \wedge t} (-\exp\{-s\})[\varphi(X_s^x) - \mathbf{A}\varphi(X_s^x)]\, ds$$
$$\leqslant 0.$$

This means $\varphi(x) \geqslant E\varphi(\tau_n \wedge \tau_0 \wedge t, X^x_{\tau_n \wedge \tau_0 \wedge t})$, so that

$$\varphi(x) \geqslant e^{-t}\varphi(n)P(\tau_n \leqslant \tau_0 \wedge t).$$

Letting $n \to \infty$ one has, for all $t \geqslant 0$,

$$P(\tau_{+\infty} \leqslant \tau_0 \wedge t) \leqslant e^t\varphi(x) \lim_{n\to\infty} \varphi^{-1}(n) = 0. \tag{T.3.18}$$

Thus, $P(\tau_{+\infty} \leqslant t, \tau_0 = \infty) = 0$, for all $t \geqslant 0$, so that $P(\tau_{+\infty} < \infty, \tau_0 = \infty) = 0$. But on the set $\{\tau_0 < \infty\}$, $\tau_{+\infty} > \tau_0$ by definition of $\tau_{+\infty}$. Therefore, $P(\tau_{+\infty} \leqslant t, \tau_0 < \infty) = 0$. This implies $P(\tau_{+\infty} < \infty) = 0$. It remains to construct φ. Let $\varphi_0(z) \equiv 1$ $(z \geqslant 0)$ and define, recursively,

$$\varphi_n(z) = 2\int_0^z \exp\{-I(y)\}\left[\int_0^y \frac{\exp\{I(z')\}\varphi_{n-1}(z')}{\sigma^2(z')}\, dz'\right] dy \qquad (z \geqslant 0;\ n \geqslant 1). \tag{T.3.19}$$

Then $\varphi_n \geqslant 0$, $\varphi_n(z)\uparrow$ as $z\uparrow$, and

$$\varphi_n(z) \leqslant \frac{2^n}{n!}\left(\int_0^z \exp\{-I(y)\}\left[\int_0^y \frac{\exp\{I(z')\}}{\sigma^2(z')}\,dz'\right]dy\right)^n \qquad (n \geqslant 0). \quad \text{(T.3.20)}$$

To prove (T.3.20), note that it holds for $n = 0$ and that $\varphi_n(z) \leqslant \varphi_{n-1}(z)\varphi_1(z)$ $(n \geqslant 1)$. Now use induction on n. It follows from (T.3.20) that the series $\varphi(z) := \sum_{n=0}^{\infty} \varphi_n(z)$ converges uniformly on compacts. Also, it is simple to check that $\mathbf{A}\varphi_n(z) = \varphi_{n-1}(z)$ for all $n \geqslant 1$, so that $\mathbf{A}$ may be applied term by term to yield $\mathbf{A}\varphi(z) = \varphi(z)$ $(z \geqslant 0)$. By assumption, $\varphi_1(z) \to \infty$ as $z \to \infty$. Therefore, $\varphi(z) \to \infty$ as $z \to \infty$. Next assume that the series on the right in (a) converges. Again, applying Itô's Lemma to $\varphi(t, y) := \exp\{-t\}\varphi(y)$, we get

$$\begin{aligned}\varphi(x) &= E\varphi(\tau_n \wedge \tau_0 \wedge t, X^x_{\tau_n \wedge \tau_0 \wedge t})\\ &\leqslant \varphi(n)P(\tau_n \leqslant \tau_0 \wedge t) + \varphi(0)P(\tau_0 \leqslant \tau_n \wedge t) + e^{-t}\varphi(n)P(t < \tau_n \wedge \tau_0)\\ &= \varphi(n)\{P(\tau_n \leqslant \tau_0 \wedge t) + e^{-t}P(\tau_n \wedge \tau_0 > t)\}. \qquad \text{(T.3.21)}\end{aligned}$$

Letting $t \to \infty$ we get

$$\varphi(x) \leqslant \varphi(n)\{P(\tau_n \leqslant \tau_0, \tau_0 < \infty) + P(\tau_n < \tau_\infty, \tau_0 = \infty)\} \leqslant \varphi(n)P(\tau_n < \infty),$$

so that

$$P(\tau_{+\infty} < \infty) \geqslant \varphi(x)/\varphi(\infty) > 0.$$

The proof of (b) is essentially analogous. ■

REMARK. It should be noted that $P(\tau_{+\infty} < \infty) > 0$ means $+\infty$ *is accessible*. For diffusions on $S = (a, b)$ with a and/or b finite, the criterion of accessibility mentioned in theoretical complement V.2.3 may be derived in exactly the same manner.

In multidimension $(k > 1)$ the following criteria may be derived. We use the notation (3.23) for the following statement.

Theorem T.3.2. (*Has'minskii's Test for Explosion*).

(a) $P(\zeta < \infty) = 0$ if $\displaystyle\int_1^\infty e^{-\bar{I}(r)}\left(\int_1^r \frac{\exp\{\bar{I}(u)\}}{\bar{\alpha}(u)}\,du\right)dr = \infty$,

(b) $P(\zeta < \infty) > 0$ if $\displaystyle\int_1^\infty e^{-\underline{I}(r)}\left(\int_1^r \frac{\exp\{\underline{I}(u)\}}{\underline{\alpha}(u)}\,du\right)dr < \infty$. □

Proof. (a) Assume that the right side of (a) diverges. Fix $r_0 \in [0, |\mathbf{x}|)$. Define the following functions on (r_0, ∞),

$$\varphi_0(v) \equiv 1, \qquad \varphi_n(v) = \int_{r_0}^v e^{-\bar{I}(r)}\left(\int_{r_0}^r \frac{\exp\{\bar{I}(u)\}\varphi_{n-1}(u)}{\underline{\alpha}(u)}\,du\right)dr. \quad \text{(T.3.22)}$$

Then the radial functions $\varphi_n(|\mathbf{y}|)$ on $(r_0 \leqslant |\mathbf{y}| < \infty)$ satisfy: $\mathbf{A}\varphi_n(|\mathbf{y}|) \leqslant \varphi_{n-1}(|\mathbf{y}|)$. Now, as in the proof of Theorem T.3.1, apply Itô's Lemma to the function

$$\varphi(t, \mathbf{y}) := \exp\{-t\}\varphi(|\mathbf{y}|), \qquad \text{where } \varphi(|\mathbf{y}|) = \sum_{n=0}^{\infty} \varphi_n(|\mathbf{y}|),$$

extended to all of $\mathbb{R}^k$ as a twice continuously differentiable function. Then, writing $\tau_R := \inf\{t \geqslant 0: |\mathbf{X}_t^{\mathbf{x}}| = R\}$,

$$E\varphi(\tau_{r_0} \wedge \tau_R \wedge t, \mathbf{X}^{\mathbf{x}}_{\tau_{r_0} \wedge \tau_R \wedge t}) \leqslant \varphi(|\mathbf{x}|),$$

or,

$$e^{-t}\varphi(R)P(\tau_R \leqslant \tau_{r_0} \wedge t) \leqslant \varphi(|\mathbf{x}|), \qquad P(\tau_R \leqslant \tau_{r_0} \wedge t) \leqslant e^t \frac{\varphi(|\mathbf{x}|)}{\varphi(R)}.$$

Letting $R \to \infty$, we get $P(\zeta \leqslant \tau_{r_0} \wedge t) = 0$ for all $t \geqslant 0$. It follows that $P(\zeta < \infty) = 0$.

(b) This follows by replacing *upper bars* by *lower bars* in the definition (T.3.22) and noting that for the resulting functions, $\mathbf{A}\varphi_n(|\mathbf{y}|) \geqslant \varphi_{n-1}(|\mathbf{y}|)$. The rest of the proof follows the proof of the second half of part (a) of Theorem T.3.1. ■

Theorem T.3.1 is due to W. Feller (1952), "The Parabolic Differential Equations and the Associated Semigroups of Transformations," *Ann. of Math.*, **55**, pp. 468–519. Theorem T.3.2 is due to R. Z. Has'minskii (1960), "Ergodic Properties of Recurrent Diffusion Processes and Stabilization of the Solution of the Cauchy Problem for Parabolic Equations," *Theor. Probability Appl.*, **5**, pp. 196–214.

3. (*Cameron–Martin–Girsanov Theorem*) Let $\boldsymbol{\mu}(\cdot)$, $\boldsymbol{\sigma}(\cdot)$ be Lipschitzian, $\boldsymbol{\sigma}(\cdot)$ nonsingular and $\boldsymbol{\sigma}^{-1}(\cdot)$ Lipschitzian. Let $\{\mathbf{X}_t^{\mathbf{x}}\}$ be the solution of

$$\mathbf{X}_t^{\mathbf{x}} = \mathbf{x} + \int_0^t \boldsymbol{\sigma}(\mathbf{X}_s^{\mathbf{x}})\, d\mathbf{B}_s \qquad (t \geqslant 0). \tag{T.3.23}$$

Define

$$Z_t^{s,\mathbf{x}} := \exp\left\{\int_s^t \boldsymbol{\sigma}^{-1}(\mathbf{X}_{s'}^{\mathbf{x}})\boldsymbol{\mu}(\mathbf{X}_{s'}^{\mathbf{x}})\cdot d\mathbf{B}_{s'} - \frac{1}{2}\int_s^t |\boldsymbol{\sigma}^{-1}(\mathbf{X}_{s'}^{\mathbf{x}})\boldsymbol{\mu}(\mathbf{X}_{s'}^{\mathbf{x}})|^2\, ds'\right\}$$
$$(0 \leqslant s \leqslant t < \infty). \tag{T.3.24}$$

By Itô's Lemma,

$$dZ_t^{s,\mathbf{x}} = Z_t^{s,\mathbf{x}}\boldsymbol{\sigma}^{-1}(\mathbf{X}_t^{\mathbf{x}})\boldsymbol{\mu}(\mathbf{X}_t^{\mathbf{x}})\cdot d\mathbf{B}_t,$$

or,

$$Z_t^{0,\mathbf{x}} = 1 + \int_0^t Z_s^{0,\mathbf{x}}\boldsymbol{\sigma}^{-1}(\mathbf{X}_s^{\mathbf{x}})\boldsymbol{\mu}(\mathbf{X}_s^{\mathbf{x}})\cdot d\mathbf{B}_s. \tag{T.3.25}$$

One may show, using Exercises 3.8, 3.10, that the integrand in (T.3.25) is in $\mathcal{M}[0, \infty)$. Hence $\{Z_t^{0,\mathbf{x}}\}$ is a nonnegative $\{\mathcal{F}_t\}$-martingale and $EZ_t^{0,\mathbf{x}} = 1$. Define the set function Q on $\bigcup_{t \geq 0} \mathcal{F}_t$ by

$$Q(A) := E(\mathbf{1}_A Z_t^{0,\mathbf{x}}) = \int_A Z_t^{0,\mathbf{x}}\, dP, \qquad A \in \mathcal{F}_t. \tag{T.3.26}$$

Note that if $A \in \mathcal{F}_t$ then $A \in \mathcal{F}_{t'}$ for all $t' > t$, and

$$E(\mathbf{1}_A Z_{t'}^{0,\mathbf{x}}) = E[\mathbf{1}_A E(Z_{t'}^{0,\mathbf{x}} \mid \mathcal{F}_t)] = E(\mathbf{1}_A Z_t^{0,\mathbf{x}}).$$

Hence Q is a *well-defined* nonnegative and countably additive set function on the field $\bigcup_{t \geq 0} \mathcal{F}_t$. It therefore has a unique extension to the smallest sigmafield $\mathcal{G}$ containing $\bigcup_{t \geq 0} \mathcal{F}_t$. Also, $Q(\Omega) = E(Z_t^{0,\mathbf{x}}) = 1$. Thus, Q is a probability measure on $(\Omega, \mathcal{F})$. On $(\Omega, \mathcal{F}_t)$, Q is absolutely continuous with respect to P, with a Radon–Nikodym derivative $Z_t^{0,\mathbf{x}}$.

We will show that, *under* Q, $\{\mathbf{X}_t^{\mathbf{x}}\}$ is a diffusion with drift $\boldsymbol{\mu}(\cdot)$ and diffusion matrix $\boldsymbol{\sigma}(\cdot)\boldsymbol{\sigma}'(\cdot)$, starting at $\mathbf{x}$. Denote by E_Q and E, expectations under Q and P, respectively. Let g be an arbitrary real-valued bounded Borel measurable function on $\mathbb{R}^k$. Then

$$E(g(\mathbf{X}_{s+t}^{\mathbf{x}}) Z_{s+t}^{0,\mathbf{x}} \mid \mathcal{F}_s) = Z_s^{0,\mathbf{x}} \psi(t, \mathbf{X}_s^{\mathbf{x}}), \tag{T.3.27}$$

where

$$\psi(t, \mathbf{y}) := E(g(\mathbf{X}_t^{\mathbf{y}}) Z_t^{0,\mathbf{y}}). \tag{T.3.28}$$

Therefore, for every real bounded $\mathcal{F}_s$-measurable h,

$$E_Q(hg(\mathbf{X}_{s+t}^{\mathbf{x}})) = E(hZ_s^{0,\mathbf{x}} \psi(t, \mathbf{X}_s^{\mathbf{x}})) = E_Q(h\psi(t, \mathbf{X}_s^{\mathbf{x}})), \tag{T.3.29}$$

which proves the desired Markov property. Finally, for all twice continuously differentiable g with compart support, Itô's Lemma yields, with $\mathbf{f} = \boldsymbol{\sigma}^{-1}\boldsymbol{\mu}$,

$$\begin{aligned} d(g(\mathbf{X}_t^{\mathbf{x}}) Z_t^{0,\mathbf{x}}) &= (\mathbf{A}_0 g)(\mathbf{X}_t^{\mathbf{x}}) Z_t^{0,\mathbf{x}}\, dt + Z_t^{0,\mathbf{x}} (\mathbf{grad}\, g)(\mathbf{X}_t^{\mathbf{x}}) \cdot \boldsymbol{\sigma}(\mathbf{X}_t^{\mathbf{x}})\, d\mathbf{B}_t \\ &\quad + g(\mathbf{X}_t^{\mathbf{x}}) Z_t^{0,\mathbf{x}} \mathbf{f}(\mathbf{X}_t^{\mathbf{x}}) \cdot d\mathbf{B}_t + Z_t^{0,\mathbf{x}} (\mathbf{grad}\, g)(\mathbf{X}_t^{\mathbf{x}}) \boldsymbol{\mu}(\mathbf{X}_t^{\mathbf{x}})\, dt \\ &= (\mathbf{A} g)(\mathbf{X}_t^{\mathbf{x}}) Z_t^{0,\mathbf{x}}\, dt + Z_t^{0,\mathbf{x}} (\mathbf{grad}\, g)(\mathbf{X}_t^{\mathbf{x}}) + g(\mathbf{X}_t^{\mathbf{x}}) f(\mathbf{X}_t^{\mathbf{x}})) \cdot d\mathbf{B}_t, \end{aligned} \tag{T.3.30}$$

where, writing $d_{ij}(\mathbf{y}) := (\boldsymbol{\sigma}(\mathbf{y})\boldsymbol{\sigma}'(\mathbf{y}))_{ij}$,

$$\mathbf{A}_0 g(\mathbf{y}) := \frac{1}{2} \sum d_{ij}(\mathbf{y}) \frac{\partial^2 g(\mathbf{y})}{\partial y^{(i)} \partial y^{(j)}}, \qquad \mathbf{A} g(\mathbf{y}) := \mathbf{A}_0 g(\mathbf{y}) + \sum \mu^{(i)}(\mathbf{y}) \frac{\partial g}{\partial y^{(i)}}. \tag{T.3.31}$$

Taking expectations in (T.3.30) we get

$$E_Q\, g(\mathbf{X}_t^{\mathbf{x}}) - g(\mathbf{x}) = \int_0^t E_Q \mathbf{A} g(\mathbf{X}_s^{\mathbf{x}})\, ds. \tag{T.3.32}$$

Dividing by t and letting $t \downarrow 0$, it follows that the infinitesimal generator of the Q-distribution of $\{\mathbf{X}_t^{\mathbf{x}}\}$ is $\mathbf{A}$. We have thus proved the following result.

Theorem T.3.3. (*Cameron–Martin–Girsanov Theorem*). Suppose $\boldsymbol{\mu}(\cdot), \boldsymbol{\sigma}(\cdot), \boldsymbol{\sigma}^{-1}(\cdot)$ are Lipschitzian, $\{\mathbf{X}_t^{\mathbf{x}}\}$ the nonanticipative solution of (T.3.23).

(a) Then for every $t \geqslant 0$, the probability measure Q defined by (T.3.26) is absolutely continuous with respect to P on $(\Omega, \mathscr{F}_t)$, with the Radon–Nikodym derivative $Z_t^{0,\mathbf{x}}$.
(b) Under Q, $\{\mathbf{X}_t^{\mathbf{x}}\}$ is a diffusion with drift $\boldsymbol{\mu}(\cdot)$ and diffusion matrix $\boldsymbol{\sigma}(\cdot)\boldsymbol{\sigma}'(\cdot)$. □

An immediate corollary is the following result. Let $\mathscr{B}_t$ denote the sigmafield on $\Omega' := C([0, \infty): \mathbb{R}^k)$ generated by the coordinate projections $\omega' \to \omega'_s$, $0 \leqslant s \leqslant t$. Let $\mathscr{B} = \sigma(\bigcup_{t \geqslant 0} \mathscr{B}_t)$. Denote by $P_{\mathbf{x}}^{(i)}$ the distribution (on $(\Omega', \mathscr{B})$) of a diffusion with drift $\boldsymbol{\mu}_i(\cdot)$ and diffusion matrix $((d_{ij}(\cdot))) = \boldsymbol{\sigma}(\cdot)\boldsymbol{\sigma}(\cdot)'$, starting at $\mathbf{x}$ $(i = 1, 2)$.

Corollary T.3.4. Suppose $\boldsymbol{\mu}_i(\cdot)$ $(i = 1, 2)$, $\boldsymbol{\sigma}(\cdot), \boldsymbol{\sigma}^{-1}(\cdot)$ are Lipschitzian. Then, for every $t \geqslant 0$, $P_{\mathbf{x}}^{(2)}$ is absolutely continuous with respect to $P_{\mathbf{x}}^{(1)}$ on $(\Omega', \mathscr{B}_t)$. □

The preceding results extend easily to nonhomogeneous diffusions on $\mathbb{R}^k$. Recall the notation of Theorem T.2.1, and define

$$\bar{Z}_t^{0,\mathbf{x}} := \exp\left\{\int_0^t \boldsymbol{\sigma}^{-1}(s, \mathbf{X}_s^{0,\mathbf{x}})\boldsymbol{\mu}(s, \mathbf{X}_s^{0,\mathbf{x}}) \cdot d\mathbf{B}_s - \frac{1}{2}\int_0^t |\boldsymbol{\sigma}^{-1}(s, \mathbf{X}_s^{0,\mathbf{x}})\boldsymbol{\mu}(s, \mathbf{X}_s^{0,\mathbf{x}})|^2 \, ds\right\}. \tag{T.3.33}$$

Theorem T.3.5

(a) Suppose $(t, \mathbf{y}) \to \boldsymbol{\mu}(t, \mathbf{y}), \boldsymbol{\sigma}(t, \mathbf{y}), \boldsymbol{\sigma}^{-1}(t, \mathbf{y})$ are Lipschitzian. Let $\{\mathbf{X}_t^{0,\mathbf{x}}\}$ denote the nonanticipative solution of

$$\mathbf{X}_t^{0,\mathbf{x}} = \mathbf{x} + \int_0^t \boldsymbol{\sigma}(s, \mathbf{X}_s^{0,\mathbf{x}}) \, d\mathbf{B}_s, \tag{T.3.34}$$

and Q a probability measure on $(\Omega, \sigma(\bigcup_{t \geqslant 0} \mathscr{F}_t))$ defined by

$$Q(A) = E(\mathbf{1}_A \bar{Z}_t^{0,\mathbf{x}}), \qquad A \in \mathscr{F}_t. \tag{T.3.35}$$

Then Q is absolutely continuous with respect to P on $(\Omega, \mathscr{F}_t)$, with a Radon–Nikodym derivative $\bar{Z}_t^{0,\mathbf{x}}$.
(b) Suppose $(t, (y)) \to \boldsymbol{\mu}_i(t, \mathbf{y})$ $(i = 1, 2)$, $\boldsymbol{\sigma}(t, \mathbf{y}), \boldsymbol{\sigma}^{-1}(t, \mathbf{y})$ are Lipschitzian. Let $P_{0,\mathbf{x}}^{(i)}$ $(i = 1, 2)$ denote the distribution (on $(C([0, \infty): \mathbb{R}^k), \mathscr{B})$) of a diffusion with drift $\boldsymbol{\mu}_i(\cdot,\cdot)$ and diffusion matrix $\boldsymbol{\sigma}(\cdot,\cdot)\boldsymbol{\sigma}'(\cdot,\cdot)$. Then $P_{0,\mathbf{x}}^{(2)}$ is absolutely continuous with respect to $P_{0,\mathbf{x}}^{(1)}$ on $\mathscr{B}_t$. □

The assumption that $\boldsymbol{\mu}, \boldsymbol{\sigma}, \boldsymbol{\sigma}^{-1}$ be Lipschitzian may be relaxed to the assumption that these may be locally Lipschitzian and that the diffusions, with and without the drift $\boldsymbol{\mu}$, be nonexplosive. This may be established by approximating $\boldsymbol{\mu}, \boldsymbol{\sigma}$ by $\boldsymbol{\mu}_n, \boldsymbol{\sigma}_n$ as in (T.3.8), and then letting $n \to \infty$.

R. H. Cameron and W. T. Martin (1944), "Transformation of Wiener Integrals Under Translations," *Ann. of Math.*, **45**, pp. 386–396, proved the above results for the case $\boldsymbol{\sigma}(\cdot) = I$. These were generalized by I. V. Girsanov (1960), "On Transforming a Class of Stochastic Processes by Absolutely Continuous Substitution of Measures," *Theor. Probability Appl.*, **5**, pp. 314–330.

4. (*The Support Theorem and the Maximum Principle*) We want to establish the following result. The notation is the same as in the statement of Corollary T.3.4.

Theorem T.3.6. (*The Support Theorem*). Let $\boldsymbol{\mu}(\cdot)$, $\boldsymbol{\sigma}(\cdot)$, $\boldsymbol{\sigma}^{-1}(\cdot)$ be Lipschitzian. Then the support of the distribution $P_{\mathbf{x}}$ (of a diffusion with drift $\boldsymbol{\mu}(\cdot)$ and diffusion matrix $\boldsymbol{\sigma}(\cdot)\boldsymbol{\sigma}'(\cdot)$) is $C_{\mathbf{x}} := \{\omega' \in C([0, \infty): \mathbb{R}^k): \omega'(0) = x\}$. □

Proof. It is enough to show that all sufficiently smooth functions in $C_{\mathbf{x}}$ are in the support of $P_{\mathbf{x}}$. Let $t \to \mathbf{u}(t) = (u_1(t), \ldots, u_k(t))$ be a twice continuously differentiable function in $C_{\mathbf{x}}$. Let $\varepsilon > 0$, $T > 0$ be given. Define a continuously differentiable (in t) function $\mathbf{c}(t) = (c_1(t), \ldots, c_k(t))$ that vanishes for $t > 2T$ and such that $c_i(t) = (d/dt)u_i(t)$, $1 \leqslant i \leqslant k$, $t \leqslant T$. Consider the diffusion $\{\mathbf{X}_t^{0,\mathbf{x}}\}$ that is the solution of

$$\mathbf{X}_t^{0,\mathbf{x}} = \mathbf{x} + \int_0^t \mathbf{c}(s)\,ds + \int_0^t \boldsymbol{\sigma}(\mathbf{X}_s^{0,\mathbf{x}})\,d\mathbf{B}_s \qquad (t \geqslant 0). \tag{T.3.36}$$

In view of Theorem T.3.5, it is enough to show

$$P\left(\left|\mathbf{X}_t^{0,\mathbf{x}} - \mathbf{x} - \int_0^t \mathbf{c}(s)\,ds\right| < \varepsilon \text{ for all } t \in [0, T]\right) > 0. \tag{T.3.37}$$

In turn, (T.3.37) follows if we can find a Lipschitzian vector field $\mathbf{b}(\cdot)$ such that the solution $\{\mathbf{Y}_t\}$ of

$$\mathbf{Y}_t = \mathbf{x} + \int_0^t \mathbf{c}(s)\,ds + \int_0^t \mathbf{b}(\mathbf{Y}_s)\,ds + \int_0^t \boldsymbol{\sigma}(\mathbf{Y}_s)\,d\mathbf{B}_s \qquad (t \geqslant 0) \tag{T.3.38}$$

satisfies

$$P\left(\left|\mathbf{Y}_t - \mathbf{x} - \int_0^t \mathbf{c}(s)\,ds\right| < \varepsilon \text{ for all } t \in [0, T]\right) > 0. \tag{T.3.39}$$

But (T.3.39) will follow if one may find $\mathbf{b}(\cdot)$ such that

$$E(\tau_{\partial B(\mathbf{x}:\varepsilon)}) > T, \qquad \text{where } \tau_{\partial B(\mathbf{x}:\varepsilon)} := \inf\left\{t \geqslant 0 : \left|\mathbf{Y}_t - \mathbf{x} - \int_0^t \mathbf{c}(s)\,ds\right| \geqslant \varepsilon\right\}.$$

The following lemma shows that such a $\mathbf{b}(\cdot)$ exists. ■

Lemma. Let $\boldsymbol{\sigma}(\cdot)$ be nonsingular and Lipschitzian. Then, for every $\varepsilon > 0$,

$$\sup_{\mathbf{b}(\cdot)} E\tau_{\partial B(\mathbf{x}:\varepsilon)} = \infty, \tag{T.3.40}$$

where the supremum is over the class of all Lipschitzian vector fields $\mathbf{b}(\cdot)$. □

Proof. Without loss of generality, take $\mathbf{x} = \mathbf{0}$. (This amounts to a translation of the coefficients by $\mathbf{x}$.) Write $\mathbf{Z}_t = \mathbf{Y}_t - \int_0^t \mathbf{c}(s)\,ds$. For any given pair $\boldsymbol{\mu}(\cdot) \equiv \mathbf{b}(\cdot)$, $\boldsymbol{\sigma}(\cdot)$

define

$$d(t, \mathbf{z}) := \sum_{i,j} d_{ij}\left(\mathbf{z} + \int_0^t \mathbf{c}(s)\,ds\right)\frac{z^{(i)}z^{(j)}}{|\mathbf{z}|^2}, \qquad B(t, \mathbf{z}) = \sum_i d_{ii}\left(\mathbf{z} + \int_0^t \mathbf{c}(s)\,ds\right),$$

$$C(t, \mathbf{z}) = 2\sum_i z^{(i)}b^{(i)}\left(\mathbf{z} + \int_0^t c(s)\,ds\right),$$

$$\bar{\beta} := \sup_{t \geqslant 0, |\mathbf{z}| = r} \frac{B(t, \mathbf{z}) + C(t, \mathbf{z})}{d(t, \mathbf{z})} - 1, \qquad \bar{\alpha}(r) := \sup_{t \geqslant 0, |\mathbf{z}| = r} d(t, \mathbf{z}).$$

Similarly $\underline{\beta}(r)$, $\underline{\alpha}(r)$ are defined by replacing "sup" by "inf" in the last two expressions. Finally, let

$$\bar{I}(r) := \int_a^r \frac{\bar{\beta}(u)}{u}\,du, \qquad \underline{I}(r) := \int_a^r \frac{\underline{\beta}(u)}{u}\,du,$$

where a is an arbitrary positive constant. Note that $\{\mathbf{Z}_t\}$ is a (nonhomogeneous) diffusion on $\mathbb{R}^k$ whose drift coefficient is $\bar{\mathbf{b}}(t, \mathbf{z}) = \mathbf{b}(\mathbf{z} + \int_0^t \mathbf{c}(s)\,ds)$, and whose diffusion matrix is $\mathbf{D}(t, \mathbf{z}) := \bar{\boldsymbol{\sigma}}(t, \mathbf{z})\bar{\boldsymbol{\sigma}}'(t, \mathbf{z})$ where $\bar{\boldsymbol{\sigma}}(t, \mathbf{z}) = \boldsymbol{\sigma}(\mathbf{z} + \int_0^t \mathbf{c}(s)\,ds)$. The infinitesimal generator of this diffusion is denoted $\mathbf{A}_t$.

Now check that

$$\lim_{r \downarrow 0} \bar{\beta}(r) \geqslant k - 1. \tag{T.3.41}$$

Define

$$F(r) := \int_r^\varepsilon \left(\int_0^u \frac{1}{\underline{\alpha}(v)} \exp\left\{-\int_v^u \frac{\bar{\beta}(v')}{v'}\,dv'\right\}dv\right)du. \tag{T.3.42}$$

Using (T.3.41) it is simple to check that $F(r)$ is finite for $0 \leqslant r \leqslant \varepsilon$. Also, for $r > 0$,

$$\begin{aligned} F'(r) &= -\int_0^r \frac{1}{\underline{\alpha}(v)} \exp\left\{-\int_v^r \frac{\bar{\beta}(v')}{v'}\,dv'\right\}dv < 0, \\ F''(r) &= -\frac{1}{\underline{\alpha}(r)} - \frac{\bar{\beta}(r)}{r}F'(r). \end{aligned} \tag{T.3.43}$$

Define $\varphi(\mathbf{z}) = F(|\mathbf{z}|)$. By (3.24) and (T.3.43) it follows that

$$2\mathbf{A}_t\varphi(\mathbf{z}) \geqslant -1 \qquad \text{for } 0 < |\mathbf{z}| \leqslant \varepsilon.$$

Therefore, by Itô's Lemma applied to $\varphi(\mathbf{Z}_t)$ (and using the fact that $\varphi = 0$ on $\partial B(\mathbf{0}:\varepsilon)$),

$$0 - \varphi(\mathbf{0}) \geqslant -\tfrac{1}{2}E\tau_{\partial B(\mathbf{0}:\varepsilon)},$$

so that

$$E\tau_{\partial B(\mathbf{0}:\varepsilon)} \geqslant 2\varphi(\mathbf{0}) = 2F(0), \qquad 0 \leqslant |\mathbf{z}| \leqslant \varepsilon. \tag{T.3.44}$$

Choose $\mathbf{b}(\mathbf{z}) = -M\mathbf{z}$ $(M > 0)$. Then it is simple to check that as $M \to \infty$, $F(0) \to \infty$.

■

Corollary T.3.7. (*Maximum Principle for Elliptic Equations*). Suppose $\boldsymbol{\mu}(\cdot)$ and $\boldsymbol{\sigma}(\cdot)$ are locally Lipschitzian and $\boldsymbol{\sigma}(\cdot)$ nonsingular. Let $\mathbf{A}$ be as in (T.3.31). Let G be a bounded, connected, open set, and $u(\cdot)$ a continuous function on $\bar{G}$, twice continuously differentiable in G, such that $\mathbf{A}u(\mathbf{x}) = 0$ for all $\mathbf{x} \in G$. Then $u(\cdot)$ cannot attain its maximum or minimum in G, unless u is constant on $\bar{G}$. □

Proof. If possible, let $\mathbf{x}_0 \in G$ be such that $u(\mathbf{x}) \leqslant u(\mathbf{x}_0)$ for all $\mathbf{x} \in G$. Let $\delta > 0$ be such that $B(\mathbf{x}_0:2\delta) \subset G$. Let φ be a bounded, twice continuously differentiable function on $\mathbb{R}^k$ with bounded first and second derivatives, satisfying $\varphi(\mathbf{x}) = u(\mathbf{x})$ on $B(\mathbf{x}_0:\delta)$. Let $\{\mathbf{X}_t\}$ be a diffusion with coefficients $\boldsymbol{\mu}(\cdot)$, $\boldsymbol{\sigma}(\cdot)$, starting at $\mathbf{x}_0$. Apply Itô's Lemma to get

$$E\varphi(\mathbf{X}_{\tau_{\partial B(\mathbf{x}_0:\delta)}}) = \varphi(\mathbf{x}_0),$$

or, denoting by π the distribution of $\mathbf{X}_{\tau_{\partial B(\mathbf{x}_0:\delta)}}$,

$$\int_{\partial B(\mathbf{x}_0:\delta)} u(\mathbf{y})\pi(d\mathbf{y}) = u(\mathbf{x}_0). \tag{T.3.45}$$

But $u(\mathbf{y}) \leqslant u(\mathbf{x}_0)$ for all $\mathbf{y} \in \partial B(\mathbf{x}_0:\delta)$. If $u(\mathbf{y}_0) < u(\mathbf{x}_0)$ for some $\mathbf{y}_0 \in \partial B(\mathbf{x}_0:\delta)$ then, by the continuity of u, there is a neighborhood of $\mathbf{y}_0$ on $\partial B(\mathbf{x}_0:\delta)$ in which $u(\mathbf{y}) < u(\mathbf{x}_0)$. But Theorem T.3.6 implies that π assigns positive probability to this neighborhood, which would imply that the left side of (T.3.45) is smaller than its right side. Thus $u(\mathbf{y}) = u(\mathbf{x}_0)$ on $\partial B(\mathbf{x}_0:\delta)$. By letting δ vary, it follows that u is constant on every open ball B contained in G centered at $\mathbf{x}_0$. Let $D = \{\mathbf{y} \in G: u(\mathbf{y}) = u(\mathbf{x}_0)\}$. For each $\mathbf{y} \in D$ there exists, by the same argument as above, an open ball centered at $\mathbf{y}$ on which u equals $u(\mathbf{x}_0)$. Thus D is open. But D is also closed (in G), since u is continuous. By connectedness of G, $D = G$. ■

D. W. Stroock and S. R. S. Varadhan (1970), "On the Support of Diffusion Processes with Applications to the Strong Maximum Principle," *Proc. Sixth Berkeley Symp. on Math. Statist. and Probability*, Vol. III, pp. 333–360, contains much more than the results described above.

5. (*Positive Recurrence and the Existence of a Unique Invariant Probability*) Consider a positive recurrent diffusion $\{\mathbf{X}_t^{\mathbf{x}}\}$ on $\mathbb{R}^k$. Fix two positive numbers $r_1 < r_2$. Define

$$\begin{gathered}\eta_1 := \inf\{t \geqslant 0: |\mathbf{X}_t^{\mathbf{x}}| = r_1\}, \qquad \eta_{2i} := \inf\{t \geqslant \eta_{2i-1}: |\mathbf{X}_t^{\mathbf{x}}| = r_2\}, \\ \eta_{2i+1} := \inf\{t \geqslant \eta_{2i}: |\mathbf{X}_t^{\mathbf{x}}| = r_1\} \qquad (i \geqslant 1).\end{gathered} \tag{T.3.46}$$

The random variables η_i are a.s. finite. By the strong Markov property $\{\mathbf{X}_{\eta_{2i-1}}^{\mathbf{x}}: i \geqslant 1\}$ is a Markov chain on the state space $\partial B(\mathbf{0}:r_1) = \{|\mathbf{y}| = r_1\}$, having a (one-step) transition probability

$$\begin{gathered}\pi_1(\mathbf{y}, B) := P(\mathbf{X}_{\eta_{2i+1}}^{\mathbf{x}} \in B \mid \mathscr{F}_{\eta_{2i-1}})_{\mathbf{X}_{\eta_{2i-1}}^{\mathbf{x}} = \mathbf{y}} = \int_{\{|\mathbf{z}| = r_2\}} H_1(\mathbf{z}, B) H_2(\mathbf{y}, d\mathbf{z}), \\ (|\mathbf{y}| = r_1,\ B \in \mathscr{B}(\partial B(\mathbf{0}:r_1))),\end{gathered} \tag{T.3.47}$$

where $\mathscr{F}_{\eta_i}$ is the pre-η_i sigmafield, and $H_i(\mathbf{y}, d\mathbf{z})$ is the distribution of $\mathbf{X}^{\mathbf{y}}_{\tau_{\partial B(\mathbf{0}:r_i)}}$—the random point where $\{\mathbf{X}^{\mathbf{y}}_t\}$ hits $\partial B(\mathbf{0}:r_i)$ at the time of its first passage to $\partial B(\mathbf{0}:r_i)$.

Now for all $\mathbf{z}_1, \mathbf{z}_2 \in \partial B(\mathbf{0}:r_1)$, $H_2(\mathbf{z}_1, d\mathbf{y})$, $H_2(\mathbf{z}_2, d\mathbf{y})$ are absolutely continuous with respect to each other and there exists a positive constant c (independent of $\mathbf{z}_1, \mathbf{z}_2$) such that

$$\frac{dH_2(\mathbf{z}_1, d\mathbf{y})}{dH_2(\mathbf{z}_2, d\mathbf{y})} \geqslant c \qquad (|\mathbf{z}_i| = r_1 \text{ for } i = 1, 2). \tag{T.3.48}$$

This is essentially *Harnack's inequality* (see D. Gilbarg and N. S. Trudinger (1977), *Elliptic Partial Differential Equations of Second Order*, Springer-Verlag, New York, p. 189). It follows from (T.3.47) and (T.3.48) that

$$\frac{d\pi_1(\mathbf{y}, d\mathbf{z})}{d\pi_1(\mathbf{y}_1, d\mathbf{z})} \geqslant c \qquad (\mathbf{y}, \mathbf{y}_1 \in \partial B(\mathbf{0}:r_1)). \tag{T.3.49}$$

This implies (see theoretical complements II.6) that there exists a unique invariant probability μ_1 for the Markov chain $\{\mathbf{X}^{\mathbf{x}}_{\eta_{2i-1}}: i \geqslant 1\}$ and the n-step transition probability $\pi_1^{(n)}(\mathbf{y}, B)$ converges exponentially fast to $\mu_1(B)$ as $n \to \infty$, uniformly for all $\mathbf{y} \in \partial B(\mathbf{0}:r_1)$ and all $B \in \mathscr{B}(\partial B(\mathbf{0}:r_1))$.

Now let $\{\mathbf{X}_t\}$ be the above diffusion with initial distribution μ_1, so that $\{\mathbf{X}_{\eta_{2i-1}}: i \geqslant 1\}$ is a *stationary process*, as is the process

$$Z_i(f) := \int_{\eta_{2i-1}}^{\eta_{2i+1}} f(\mathbf{X}_s)\, ds \qquad (i \geqslant 1),$$

where f is a real-valued bounded Borel-measurable function on $\mathbb{R}^k$.

Lemma. The tail sigmafield of $\{Z_i(f): i \geqslant 1\}$ is trivial. □

Proof. Let $A \in \mathscr{T} := \bigcap_{n \geqslant 1} \sigma\{Z_i(f): i \geqslant n\}$—the *tail sigmafield*, and $B \in \sigma\{Z_i(f): 1 \leqslant i \leqslant n\} \subset \mathscr{F}_{\eta_{2n+1}}$ (the *pre-η_{2n+1} sigmafield* of $\{\mathbf{X}_t\}$). For every positive integer m, A belongs to the *after-$\eta_{2(n+m)+1}$ process* $\mathbf{X}^+_{\eta_{2(n+m)+1}}$ and, therefore, may be expressed as $\{\mathbf{X}^+_{\eta_{2(n+m)+1}} \in A_{n+m}\}$ a.s., for some Borel set A_{n+m} of $C([0, \infty): \mathbb{R}^k)$. By the strong Markov property,

$$\begin{aligned} P(A \mid \mathscr{F}_{\eta_{2(n+m)+1}}) &= E(\mathbf{1}_{A_{n+m}}(\mathbf{X}^+_{\eta_{2(n+m)+1}}) \mid \mathscr{F}_{\eta_{2(n+m)+1}}) \\ &= (P_{\mathbf{x}}(A_{n+m}))_{\mathbf{x} = \mathbf{X}_{\eta_{2(n+m)+1}}} = \varphi_{n+m}(\mathbf{X}_{\eta_{2(n+m)+1}}), \end{aligned}$$

say. Hence,

$$P(A \mid \mathscr{F}_{\eta_{2n+1}}) = E(\varphi_{n+m}(\mathbf{X}_{\eta_{2(n+m)+1}}) \mid \mathscr{F}_{\eta_{2n+1}}) = \left(\int \varphi_{n+m}(\mathbf{z}) \pi_1^{(m)}(\mathbf{x}, d\mathbf{z})\right)_{\mathbf{x} = \mathbf{X}_{\eta_{2n+1}}}.$$

But $\pi_1^{(m)}(\mathbf{x}, B) \to \mu_1(B)$ as $m \to \infty$, uniformly for all $\mathbf{x}$ and B. Therefore

$$P(A \mid \mathscr{F}_{\eta_{2n+1}}) - \int \varphi_{n+m}(\mathbf{z}) \mu_1(d\mathbf{z}) \to 0 \quad \text{a.s.} \qquad \text{as } m \to \infty. \tag{T.3.50}$$

But

$$\int \varphi_{n+m}(\mathbf{z})\mu_1(d\mathbf{z}) = E\varphi_{n+m}(\mathbf{X}_{\eta_{2(n+m)+1}}) = P(A).$$

Therefore, (T.3.50) implies, for every $B \in \mathscr{F}_{\eta_{2n+1}}$,

$$P(B \cap A) - P(B)P(A) \to 0 \qquad \text{as } m \to \infty. \tag{T.3.51}$$

But the left side of (T.3.51) does not depend on m. Therefore, $P(B \cap A) = P(B)P(A)$. Hence $\sigma\{Z_i(f): 1 \leqslant i \leqslant n\}$ and the tail sigmafield $\mathscr{T}$ are independent for every $n \geqslant 1$. This implies that $\mathscr{T}$ is independent of $\sigma\{Z_i(f): i \geqslant 1\}$. In particular, $\mathscr{T}$ is independent of itself, so that $P(A) \equiv P(A \cap A) = P(A)P(A)$. Thus, $P(A) = 0$ or 1. ■

Using the above lemma and the Ergodic Theorem (theoretical complements II.9), instead of the classical strong law of large numbers, we may show, as in Section 9 of Chapter II, or Section 12 of Chapter V, that

$$\lim_{t \to \infty} \frac{1}{t} \int_0^t f(\mathbf{X}_s)\, ds \xrightarrow{\text{a.s.}} \frac{1}{E(\eta_3 - \eta_1)} E \int_{\eta_1}^{\eta_3} f(\mathbf{X}_s)\, ds = \int f(x)m(dx), \tag{T.3.52}$$

say, where m is the probability measure on $(\mathbb{R}^k, \mathscr{B}^k)$ with $m(B)$ equal to the average amount of time the process $\{\mathbf{X}_s\}$ spends in the set B during a cycle $[\eta_{2n+1}, \eta_{2n+3}]$. It follows from (T.3.52) that m is the unique invariant probability for the diffusion.

The criterion (3.25) (Proposition 3.3) for positive recurrence is due to R. Z. Has'minskii (1960), "Ergodic Properties of Recurrent Diffusion Processes and Stabilization of the Solution to the Cauchy Problem for Parabolic Equations," *Theor. Probability Appl.*, **5**, pp. 196–214. A proof and some extensions may be found in R. N. Bhattacharya (1978), "Criteria for Recurrence and Existence of Invariant Measures for Multidimensional Diffusions," *Ann. Probab.*, **6**, pp. 541–553; Correction Note (1980), *Ann. Probab.*, **8**. The proof in the last article does not require Harnack's inequality.

Theoretical Complements to Section VII.4

1. Theorem 4.1 and Proposition 4.2 are special cases of more general results contained in G. Basak (1989), "Stability and Functional Central Limit Theorems for Degenerate Diffusions,)) Ph.D. Dissertation, Indiana University.

Two comprehensive accounts of the theory of stochastic differential equations are N. Ikeda and S. Watanabe (1981), *Stochastic Differential Equations and Diffusion Processes*, North-Holland, Amsterdam, and I. Karatzas and S. E. Shreve (1988), *Brownian Motion and Stochastic Calculus*, Springer-Verlag, New York.

CHAPTER 0

A Probability and Measure Theory Overview

1 PROBABILITY SPACES

Underlying the mathematical description of random variables and events is the notion of a *probability space* $(\Omega, \mathscr{F}, P)$. The *sample space* Ω is a nonempty set that represents the collection of all possible outcomes of an experiment. The elements of Ω are called *sample points*. The *sigmafield* $\mathscr{F}$ is a collection of subsets of Ω that includes the empty set $\varnothing$ (the *"impossible event"*) as well as the set Ω (the *"sure event"*) and is closed under the set operations of complements and finite or denumerable unions and intersections. The elements of $\mathscr{F}$ are called *measurable events*, or simply *events*. The *probability measure* P is an assignment of probabilities to events (sets) in $\mathscr{F}$ that is subject to the conditions that (i) $0 \leqslant P(F) \leqslant 1$, for each $F \in \mathscr{F}$, (ii) $P(\varnothing) = 0$, $P(\Omega) = 1$, and (iii) $P(\bigcup_i F_i) = \sum_i P(F_i)$ for any finite or denumerable sequence of *mutually exclusive* (*pairwise disjoint*) events F_i, $i = 1, 2, \dots$, belonging to $\mathscr{F}$. The closure properties of $\mathscr{F}$ ensure that the usual applications of set operations in representing events do not lead to *nonmeasurable* events for which no (consistent) assignment of probability is possible.

The required *countable additivity* property (iii) gives probabilities a sufficiently rich structure for doing calculations and approximations involving limits. Two immediate consequences of (iii) are the following so-called *continuity properties*: if $A_1 \subset A_2 \subset \cdots$ is a *nondecreasing* sequence of events in $\mathscr{F}$ then, thinking of $\bigcup_{n=1}^{\infty} A_n$ as the *"limiting event"* for such sequences,

$$P\left(\bigcup_{n=1}^{\infty} A_n\right) = \lim_n P(A_n). \tag{1.1}$$

To prove this, disjointify $\{A_n\}$ by $B_n = A_n - A_{n-1}$, $n \geqslant 1$, $A_0 = \varnothing$, and apply (iii) to $\bigcup_{n=1}^{\infty} B_n = \bigcup_{n=1}^{\infty} A_n$. By considering complements, one gets for decreasing measurable events $A_1 \supset A_2 \supset \cdots$ that

$$P\left(\bigcap_{n=1}^{\infty} A_n\right) = \lim_n P(A_n). \tag{1.2}$$

While (1.1) holds for all countably additive set functions μ (in place of P) on $\mathscr{F}$, finite or not, (1.2) *does not* in general hold if $\mu(A_n)$ is not finite for at least some n (onwards).

If Ω is a finite or denumerable set, then probabilities are defined for all subsets F of Ω once they are specified for singletons, so $\mathscr{F}$ is the collection of all subsets of Ω. Thus, if f is a *probability mass function* (p.m.f.) for singletons, i.e., $f(\omega) \geqslant 0$ for all $\omega \in \Omega$ and $\sum_\omega f(\omega) = 1$, then one may define $P(F) = \sum_{\omega \in F} f(\omega)$. The function P so defined on the class of all subsets of Ω is *countably additive*, i.e., P satisfies (iii). So $(\Omega, \mathscr{F}, P)$ is easily seen to be a probability space. In this case the probability measure P is determined by the probabilities of singletons $\{\omega\}$.

In the case Ω is not finite or denumerable, e.g., when Ω is the real line or the space of all infinite sequences of 0's and 1's, then the above formulation is no longer possible in general. Instead, for example in the case $\Omega = \mathbb{R}^1$, one is often given a piecewise continuous *probability density function* (p.d.f.) f, i.e., f is nonnegative, integrable, and $\int_{-\infty}^{\infty} f(x)\,dx = 1$. For an interval $I = (a, b)$ or (b, ∞), $-\infty \leqslant a < b \leqslant \infty$, one then assigns the probability $P(I) = \int_a^b f(x)\,dx$, by a Riemann integral. This set function P may be extended to the class $\mathscr{C}$ comprising all finite unions $F = \bigcup_j I_j$ of pairwise disjoint intervals I_j by setting $P(F) = \sum_j P(I_j)$. The class $\mathscr{C}$ is a *field*, i.e., $\varnothing$ and Ω belong to $\mathscr{C}$ and it is closed under complements and finite intersection (and therefore finite unions). But, since $\mathscr{C}$ is not a sigmafield, usual sequentially applied operations on events may lead to events outside of $\mathscr{C}$ for which probabilities have not been defined. But a theorem from measure theory, the *Carathéodory Extension Theorem*, asserts that there is a unique countably additive extension of P from a field $\mathscr{C}$ to the smallest sigmafield that contains $\mathscr{C}$. In the case of $\mathscr{C}$ above, this sigmafield is called the *Borel sigmafield* $\mathscr{B}^1$ on $\mathbb{R}^1$ and its sets are called *Borel sets* of $\mathbb{R}^1$. In general, such an extension of P to the *power set* sigmafield, that is the collection of all subsets of $\mathbb{R}^1$, is not possible. The same considerations apply to all *measures* (i.e., countably additive nonnegative set functions μ defined on a sigmafield with $\mu(\varnothing) = 0$), whether the measure of Ω is 1 or not. The measure $\mu = m$, which is defined first for each interval I as the *length* of the interval, and then extended uniquely to $\mathscr{B}^1$, is called the *Lebesgue measure* on $\mathbb{R}^1$. Similarly, one defines the *Lebesgue measure* on $\mathbb{R}^k$ $(k \geqslant 2)$ whose Borel sigmafield $\mathscr{B}^k$ is the smallest sigmafield that contains all k-dimensional rectangles $I = I_1 \times I_2 \times \cdots \times I_k$, with I_j a one-dimensional rectangle (interval) of the previous type. The Lebesgue measure of a rectangle is the product of the lengths of its sides, i.e., its volume. Lebesgue measure on $\mathbb{R}^k$ has the property that the space can be decomposed into a countable union of measurable sets of finite Lebesgue measure; such measures are said to be *sigma-finite*. All measures referred to in this book are sigma-finite.

The extension of a measure μ from a field $\mathscr{C}$, as provided by the Carathéodory Extension Theorem stated above, is *unique* and may be expressed by the formula

$$\mu(F) = \inf \sum_n \mu(C_n) \qquad (F \in \mathscr{F}), \tag{1.3}$$

where the summation is over a finite collection $C_1, C_2, \ldots$ of sets in $\mathscr{C}$ whose union contains F and the infimum is over all such collections.

As suggested by the construction of measures on $\mathscr{B}^k$ outlined above, starting from their specifications on a class of rectangles, if two measures μ_1 and μ_2 on a sigmafield $\mathscr{F}$ agree on a subclass $\mathscr{A} \subset \mathscr{F}$ closed under finite intersections and $\Omega \in \mathscr{A}$, then they agree on the *smallest sigmafield*, denoted $\sigma(\mathscr{A})$, that contains $\mathscr{A}$. The sigmafield $\sigma(\mathscr{A})$ is called the *sigmafield generated by* $\mathscr{A}$. On a metric space S the sigmafield $\mathscr{B} = \mathscr{B}(S)$ generated by the class of all open sets is called the *Borel sigmafield.*

In case a probability measure P is given on $\mathscr{B}^k$, by specifying a p.d.f. f on $\mathbb{R}^k$, for example, one defines the *(cumulative) distribution function* (c.d.f.) F associated with P as

$$F(x) = P(\{(y_1, \ldots, y_k) \in \mathbb{R}^k \colon y_i \leqslant x_i \text{ for } 1 \leqslant i \leqslant k\}), \qquad x = (x_1, \ldots, x_k) \in \mathbb{R}^k. \tag{1.4}$$

It is simple to show, using the *continuity* (additivity) properties of P that F is *right-continuous* and *coordinatewise nondecreasing*. By the usual inclusion–exclusion procedure one may compute probabilities of rectangles $(a_1, b_1] \times \cdots \times (a_k, b_k]$ in terms of the distribution function.

2 RANDOM VARIABLES AND INTEGRATION

A real-valued function X defined on Ω is a *random variable* provided that events of the form $\{X \in I\} := \{\omega \in \Omega \colon X(\omega) \in I\}$ are in $\mathscr{F}$ for all intervals I; i.e., X is a *measurable function* on Ω with respect to $\mathscr{F}$. This definition makes it possible to assign probabilities to events $\{X \in I\}$. A similar definition can be made for k-dimensional vector-valued random variables by taking I to be an arbitrary open rectangle in $\mathbb{R}^k$. From this definition it follows that if X is a random variable then the event $\{X \in B\} \in \mathscr{F}$ for every Borel set B. The *distribution* of X is the probability measure Q on $(\mathbb{R}^k, \mathscr{B}^k)$ given by

$$Q(B) = P(X \in B), \qquad B \in \mathscr{B}^k. \tag{2.1}$$

Often one writes P_X for Q. We sometimes write $Q(dx) = f(x)\,dx$ to signify that the distribution of X has p.d.f. f with respect to Lebesgue measure.

Suppose that $(\Omega, \mathscr{F}, \mu)$ is an arbitrary measure space and g is a real-valued function on Ω that is measurable with respect to $\mathscr{F}$. For the simplest case, suppose that g takes on only finitely many distinct nonzero values $y_1, y_2, \ldots, y_k$ on the respective sets $A_1, A_2, \ldots, A_k$; such a function g is called a *simple function*. In the case μ is a probability measure, simple functions are discrete random variables. If $\mu(A_i)$ is finite for each $i \geqslant 1$ then the *Lebesgue integral* of g is defined as the sum of the values of g weighted by the μ-measures of the sets on which it takes its values. That is,

$$\int_\Omega g\, d\mu := \sum y_i\, \mu(A_i), \tag{2.2}$$

where $A_i = \{\omega \in \Omega \colon g(\omega) = y_i\}, i = 1, 2, \ldots, k$. If g is a bounded nonnegative measurable function on Ω, then it is possible to approximate g by a sequence of simple functions g_n, $n = 1, 2, \ldots,$ where g_n has values $i/2^n$ on $A_i(n) = \{\omega \colon i/2^n \leqslant g(\omega) < (i+1)/2^n\}$, respectively, for $i = 1, 2, \ldots, k_n = N2^n$ with N so large that $g(\omega) < N$ for all ω. If each $A_i(n)$ has finite measure, then $\int_\Omega g_n\, d\mu$ is a nondecreasing sequence of numbers whose limit therefore exists but may be infinite. In the case that the limit exists and is finite, g is said to be *integrable* and its Lebesgue integral is the limiting value. In the case that

g is an unbounded nonnegative measurable function, first approximate g by $g_N(\omega) = \min\{g(\omega), N - 1\}$, $N = 1, 2, \ldots$. Then each g_N is a bounded nonnegative measurable function. If each g_N is integrable, then the sequence $\int_\Omega g_N \, d\mu$ is a nondecreasing sequence of numbers. If the limit of this sequence is finite, then g is integrable and its integral is the limiting value. Finally, if g is an arbitrary measurable function then write $g = g^+ - g^-$ as a difference of nonnegative measurable functions, where $g^+(\omega) = g(\omega)$ on $A^+ = \{\omega \in \Omega: g(\omega) > 0\}$ and $g^+(\omega) = 0$ on $A^- = \{\omega \in \Omega: g(\omega) < 0\}$, and $g^-(\omega) = 0$ on A^+ and $g^-(\omega) = -g(\omega)$ on A^-. If each of g^+ and g^- is integrable, then g is integrable and its integral is the difference of the two integrals. A basic simple property of the integral is *linearity*. If f, g are integrable and c, d are reals, then $cf + dg$ is integrable and $\int_\Omega (cf + dg)\, d\mu = c \int_\Omega f \, d\mu + d \int_\Omega g \, d\mu$.

In the case $\mu = P$ is a probability measure, the integral of g is called the *expected value* of g. The notation for the expected value (integral) takes various forms, e.g., $EX = \int_\Omega X \, dP = \int_\Omega X(\omega)P(d\omega)$. Sometimes the domain of integration is omitted, and one simply writes $\int X \, dP$ for $\int_\Omega X \, dP$. In addition, one may apply a *change of variable* $\omega \to X(\omega)$ from Ω to $\mathbb{R}^1$ to obtain the integral with respect to the distribution Q of X on $\mathbb{R}^1$ given by

$$EX = \int_{\mathbb{R}^1} xQ(dx).$$

More generally, if φ is a measurable function on $\mathbb{R}^1$ and $\varphi(X)$ integrable, then

$$E\varphi(X) = \int_\Omega \varphi(X)\, dP = \int_{\mathbb{R}^1} \varphi(x)Q(dx). \tag{2.3}$$

To verify the change-of-variable formula (2.3) one begins with simple functions φ and proceeds by limits as described after the definition (2.2).

Estimates and bounds on expected values and other integrals are often obtained by applications of the *basic inequalities for integrals* summarized below. Let $(\Omega, \mathcal{F}, \mu)$ be an arbitrary measure space and let f and g be integrable functions defined on Ω. A property is said to hold μ *almost everywhere* (a.e.) or *almost surely* (a.s.) if the set of points where it fails has μ-measure zero. Then, it is simple to check from the definition of the integral that we have the following.

Proposition 2.1. (*Domination Inequality*). If $f \leqslant g$ μ-a.e. then

$$\int_\Omega f \, d\mu \leqslant \int_\Omega g \, d\mu. \tag{2.4}$$

A real-valued differentiable function φ defined on an interval I whose graph is supported below by its tangent lines is called *convex function*, e.g., $\varphi(x) = e^x$, $x \in \mathbb{R}^1$, $\varphi(x) = x^p$, $x \geqslant 0$ $(p \geqslant 1)$. In other words, a differentiable φ is convex on I if

$$\varphi(y) \geqslant \varphi(x) + m(x)(y - x) \qquad (x, y \in I), \tag{2.5}$$

where $m(x) = \varphi'(x)$. The function φ is *strictly convex* if the inequality in (2.5) is strict except when $x = y$. For twice continuously differentiable functions this means a positive second derivative. Observe that if φ is convex on I, and X and $\varphi(X)$ have finite expectations, then letting $y = X$, $x = EX$ in (2.5) we have $\varphi(X) \geqslant m(EX)(X - EX) + \varphi(EX)$. Therefore, using the above domination inequality (2.6) and linearity of the integral, we have

$$E\varphi(X) \geqslant m(EX)E(X - EX) + E\varphi(EX) = \varphi(EX). \tag{2.6}$$

This inequality is strict if φ is strictly convex and X is not degenerate. More generally, convexity means that for any $x \neq y \in I$, $\varphi(tx + (1-t)y) \leqslant t\varphi(x) + (1-t)\varphi(y)$, $0 < t < 1$; likewise *strict convexity* means strict inequality here. One may show that (2.5) still holds if $m(x)$ is taken to be the right-hand derivative of φ at x. The general inequality (2.6) may then be stated as follows.

Proposition 2.2. (*Jensen's Inequality*). If φ is a convex function on the interval I and if X is a random variable taking its values in I, then

$$\varphi(EX) \leqslant E\varphi(X) \tag{2.7}$$

provided that the indicated expected values exist. Moreover, if φ is strictly convex, then equality holds if and only if the distribution of X is concentrated at a point (degenerate).

From Jensen's inequality we see that if X is a random variable with finite pth absolute moment then for $0 < r < p$, writing $|X|^p = (|X|^r)^{p/r}$, we get $E|X|^p \geqslant (E|X|^r)^{p/r}$. That is, taking the pth root,

$$(E|X|^r)^{1/r} \leqslant (E|X|^p)^{1/p} \qquad (0 < r < p). \tag{2.8}$$

In particular, all moments of lower order than p also exist. The inequality (2.8) is known as *Liapounov's Inequality*.

The next inequality is the *Hölder Inequality*, which can also be viewed as a convexity type inequality as follows. Let $p > 1, q > 1$ and let f and g be functions such that $|f|^p$ and $|g|^q$ are both μ-integrable. If p and q are *conjugate* in the sense $1/p + 1/q = 1$, then using convexity of the function e^x we obtain

$$|f|\cdot|g| = e^{(1/p)\log|f|^p + (1/q)\log|g|^q} \leqslant \frac{1}{p}e^{\log|f|^p} + \frac{1}{q}e^{\log|g|^q} = \frac{1}{p}|f|^p + \frac{1}{q}|g|^q.$$

Thus,

$$|f||g| \leqslant \frac{1}{p}|f|^p + \frac{1}{q}|g|^q. \tag{2.9}$$

Now define the *L^p-norm* by

$$\|f\|_p := \left(\int_\Omega |f|^p\,d\mu\right)^{1/p}. \tag{2.10}$$

Replacing f and g by $f/\|f\|_p$ and $g/\|g\|_q$, respectively, we have upon integration that

$$\frac{\int |fg|\,d\mu}{\|f\|_p\|g\|_q} \leqslant \frac{1}{p}\frac{\int |f|^p\,d\mu}{(\|f\|_p)^p} + \frac{1}{q}\frac{\int |g|^q\,d\mu}{(\|g\|_q)^q} = \frac{1}{p} + \frac{1}{q} = 1. \tag{2.11}$$

Thus, we have proved the following.

Proposition 2.3. (*Hölder Inequality*). If $|f|^p$ and $|g|^q$ are integrable, where $1 < p < \infty$, $1/p + 1/q = 1$, then fg is integrable and

$$\left|\int fg\,d\mu\right| \leqslant \int |fg|\,d\mu \leqslant \left\{\int |f|^p\,d\mu\right\}^{1/p}\left\{\int |g|^q\,d\mu\right\}^{1/q}. \tag{2.12}$$

The case $p = 2$ (and therefore $q = 2$) is called the *Cauchy–Schwarz Inequality*, or the *Schwarz Inequality*. The first inequality in (2.12) is a consequence of the Domination Inequality, since $fg \leqslant |fg|$ and $-fg \leqslant |fg|$.

The next convexity-type inequality furnishes a triangle inequality for the L^p*-norm* $(p \geqslant 1)$.

Proposition 2.4. (*Minkowski Inequality*). If $|f|^p$ and $|g|^p$ are μ-integrable where $1 \leqslant p < \infty$, then

$$\left\{\int |f + g|^p\,d\mu\right\}^{1/p} \leqslant \left\{\int |f|^p\,d\mu\right\}^{1/p} + \left\{\int |g|^p\,d\mu\right\}^{1/p}. \tag{2.13}$$

To prove (2.13) first notice that since,

$$|f + g|^p \leqslant (|f| + |g|)^p \leqslant (2\max(|f|, |g|))^p \leqslant 2^p(|f|^p + |g|^p), \tag{2.14}$$

the integrability of $|f + g|^p$ follows from that of $|f|^p$ and $|g|^p$. Let $q = p/(p - 1)$ be the conjugate to p, i.e., $1/q + 1/p = 1$, barring the trivial case $p = 1$. Then,

$$\int |f + g|^p\,d\mu \leqslant \int (|f| + |g|)^p\,d\mu. \tag{2.15}$$

The trick is to write

$$\int (|f| + |g|)^p\,d\mu = \int |f|(|f| + |g|)^{p-1}\,d\mu + \int |g|(|f| + |g|)^{p-1}\,d\mu.$$

Now we can apply Hölder's inequality to get, since $qp - q = p$,

$$\begin{aligned}\int (|f| + |g|)^p\,d\mu &\leqslant \|f\|_p\left(\int [(|f| + |g|)^{p-1}]^q\,d\mu\right)^{1/q} + \|g\|_p\left(\int [(|f| + |g|)^{p-1}]^q\,d\mu\right)^{1/q}\\ &= (\|f\|_p + \|g\|_p)\left(\int_\Omega (|f| + |g|)^p\right)^{1/q}.\end{aligned}$$

Dividing by $(\int (|f| + |g|)^p)^{1/q}$ and again using conjugacy $1 - 1/q = 1/p$ gives the desired inequality by taking $(1/p)$th roots in (2.15), i.e., $\|f + g\|_p \leqslant \|f\|_p + \|g\|_p$.

Finally, if X is a random variable on a probability space $(\Omega, \mathcal{F}, P)$ then one has *Chebyshev's Inequality*

$$P(|X| \geqslant \varepsilon) \leqslant \frac{1}{\varepsilon^p} E|X|^p \qquad (\varepsilon > 0, p > 0), \tag{2.16}$$

which follows from $E|X|^p \geqslant E(\mathbf{1}_{\{|X| \geqslant \varepsilon\}}|X|^p) \geqslant \varepsilon^p P(|X| \geqslant \varepsilon)$.

3 LIMITS AND INTEGRATION

A sequence of measurable functions $\{f_n: n \geqslant 1\}$ on a measure space $(\Omega, \mathscr{F}, \mu)$ is said to *converge in measure* (or *in probability* in case μ is a probability) *to* f if, for every $\varepsilon > 0$,

$$\mu(\{|f_n - f| > \varepsilon\}) \to 0 \qquad \text{as } n \to \infty. \tag{3.1}$$

A sequence $\{f_n\}$ is said to *converge almost everywhere* (abbreviated a.e.) to f, if

$$\mu(\{f_n \nrightarrow f\}) = 0. \tag{3.2}$$

If μ is a probability measure, and (3.2) holds, one says that $\{f_n\}$ *converges almost surely* (a.s.) *to* f.

A sequence $\{f_n\}$ is said to be *Cauchy in measure* if, for every $\varepsilon > 0$,

$$\mu(\{|f_n - f_m| > \varepsilon\}) \to 0 \qquad \text{as } n, m \to \infty. \tag{3.3}$$

Given such a sequence, one may find an increasing sequence of positive integers $\{n_k: k \geqslant 1\}$ such that

$$\mu(\{|f_{n_k} - f_{n_{k+1}}| > (\tfrac{1}{2})^k\}) < (\tfrac{1}{2})^k \qquad (k = 1, 2, \ldots). \tag{3.4}$$

The sets $B_m := \bigcup_{k=m}^{\infty} \{|f_{n_k} - f_{n_{k+1}}| > 2^{-k}\}$ form a decreasing sequence converging to $B := \{|f_{n_k} - f_{n_{k+1}}| > 2^{-k}$ for infinitely many integers $k\}$. Now $\mu(B_m) \leqslant 2^{-m+1}$. Therefore, $\mu(B) = 0$. Since $B^c = \{|f_{n_k} - f_{n_{k+1}}| \leqslant 2^{-k}$ for all sufficiently large $k\}$, on B^c the sequence $\{f_{n_k}: k \geqslant 1\}$ converges to a function f. Therefore, $\{f_{n_k}\}$ *converges to* f *a.e.* Further, for any $\varepsilon > 0$ and all m,

$$\mu(\{|f_n - f| > \varepsilon\}) \leqslant \mu(\{|f_n - f_{n_m}| > \frac{\varepsilon}{2}\}) + \mu(\{|f_{n_m} - f| > \frac{\varepsilon}{2}\}). \tag{3.5}$$

The first term on the right goes to zero as n and n_m go to infinity. Also,

$$\left\{|f_{n_m} - f| > \frac{\varepsilon}{2}\right\} \subset B_m \qquad \text{if } 2^{-m+1} < \frac{\varepsilon}{2}.$$

As $\mu(B_m) \to 0$ as $m \to \infty$, so does the second term on the right in (3.5). Therefore, $\{f_n: n \geqslant 1\}$ *converges to* f *in measure.* Conversely, if $\{f_n\}$ converges to f in measure then, for every $\varepsilon > 0$,

$$\mu(\{|f_n - f_m| > \varepsilon\}) \leqslant \mu(\{|f_n - f| > \frac{\varepsilon}{2}\}) + \mu(\{|f_m - f| > \frac{\varepsilon}{2}\}) \to 0$$

as $n, m \to \infty$. That is, $\{f_n\}$ is Cauchy in measure. We have proved parts (a), (b) of the following result.

Proposition 3.1. Let $(\Omega, \mathscr{F}, \mu)$ be a measure space on which is defined a sequence $\{f_n\}$ of measurable functions.

(a) If $\{f_n\}$ converges in measure to f then $\{f_n\}$ is Cauchy in measure and there exists a subsequence $\{f_{n_k}: k \geqslant 1\}$ which converges a.e. to f.

(b) If $\{f_n\}$ is Cauchy in measure then there exists f such that $\{f_n\}$ converges to f in measure.

(c) If μ is finite and $\{f_n\}$ converges to f a.e., then $\{f_n\}$ converges to f in measure.

Part (c) follows from the relations

$$\mu(\{|f_n - f| > \varepsilon\}) \leqslant \mu\left(\bigcup_{m=n}^{\infty} \{|f_m - f| > \varepsilon\}\right) \to 0 \qquad \text{as } n \to \infty. \tag{3.6}$$

Note that the sets A_n, say, within parentheses on the right side are decreasing to $A := \{|f_m - f| > \varepsilon$ for infinitely many $m\}$. As remarked in Section 1, following (1.2), the convergence of $\mu(A_m)$ to $\mu(A)$ holds because $\mu(A_m)$ is finite for all m.

The first important result on interchanging the order of limit and integration is the following.

Theorem 3.2. (*The Monotone Convergence Theorem*). If $\{f_n\}$ is an increasing sequence of nonnegative measurable functions on a measure space $(\Omega, \mathscr{F}, \mu)$, then

$$\lim_n \int f_n \, d\mu = \int f \, d\mu, \tag{3.7}$$

where $f = \lim_n f_n$.

Proof. If $\{f_n\}$ is a sequence of simple functions then (3.7) is simply the definition of $\int f \, d\mu$. In the general case, for each n let $\{f_{n,k}: k \geqslant 1\}$ be an increasing sequence of nonnegative simple functions converging a.e. to f_n (as $k \to \infty$). Then $g_k := \max\{f_{n,k}: 1 \leqslant n \leqslant k\}$ $(k \geqslant 1)$ is an increasing sequence of simple functions and, for $k \geqslant n$,

$$f_{n,k} \leqslant g_k \leqslant f_k \tag{3.8}$$

as $f_{n,k} \leqslant f_n \leqslant f_k$ for $1 \leqslant n \leqslant k$. By the Domination Inequality

$$\int f_{n,k} \, d\mu \leqslant \int g_k \, d\mu \leqslant \int f_k \, d\mu. \tag{3.9}$$

By taking limits in (3.8) as $k \to \infty$ one gets

$$f_n \leqslant g \leqslant f \tag{3.10}$$

where $g = \lim_k g_k$. By taking limits in (3.10) as $n \to \infty$, one gets $f \leqslant g \leqslant f$, that is, $g = f$. Now, by the definition of the integral, on taking limits in (3.9) as $k \to \infty$ one gets

$$\int f_n \, d\mu \leqslant \int g \, d\mu = \int f \, d\mu \leqslant \lim_k \int f_k \, d\mu. \tag{3.11}$$

Taking limits, as $n \to \infty$, in (3.11) one gets (3.7). ■

A useful consequence of the Monotone Convergence Theorem is the following theorem.

Theorem 3.3. (*Fatou's Lemma*). If $\{f_n\}$ is a sequence of nonnegative integrable functions, then

$$\int (\liminf f_n)\, d\mu \leqslant \liminf \int f_n\, d\mu. \tag{3.12}$$

Proof. Write $g_n := \inf\{f_k: k \geqslant n\}$, $g := \liminf f_n$. Then $0 \leqslant g_n \uparrow g$. Therefore, by the Monotone Convergence Theorem, $\int g_n\, d\mu \to \int g\, d\mu$. But $g_n \leqslant f_n$ so that $\int g_n\, d\mu \leqslant \int f_n\, d\mu$ for all n. Therefore,

$$\int g\, d\mu = \lim \int g_n\, d\mu = \liminf \int g_n\, d\mu \leqslant \liminf \int f_n\, d\mu. \qquad \blacksquare$$

The final result on the interchange of limits and integration is as follows.

Theorem 3.4. (*Lebesgue's Dominated Convergence Theorem*). Let $\{f_n\}$ be a sequence of measurable functions on a measure space $(\Omega, \mathscr{F}, \mu)$ such that

(i) $f_n \to f$ a.e. or in measure, and
(ii) $|f_n| \leqslant g$ for all n, where $\int g\, d\mu < \infty$. Then

$$\int |f_n - f|\, d\mu \to 0 \qquad \text{as } n \to \infty. \tag{3.13}$$

In particular, $\int f_n\, d\mu \to \int f\, d\mu$.

Proof. Assume $f_n \to f$ a.e. By Fatou's Lemma applied to the sequences $\{g + f_n\}$, $\{g - f_n\}$ one gets $\int f\, d\mu \leqslant \liminf \int f_n\, d\mu$ and $\int f\, d\mu \geqslant \limsup \int f_n\, d\mu$. Therefore, $\lim \int f_n\, d\mu = \int f\, d\mu$. To derive (3.13), apply the above result to the sequence $\{|f_n - f|\}$, noting that $|f_n - f| \leqslant 2g$.

Now assume $\{f_n\}$ converges to f in measure. By Proposition 3.1, for every subsequence $\{f_{n'}\}$ of $\{f_n\}$ there exists a further subsequence $\{f_{n_k}: k \geqslant 1\}$ such that $\{f_{n_k}\}$ converges a.e. to f. By the above, $\int f_{n_k}\, d\mu \to \int f\, d\mu$ as $k \to \infty$. Since the limit is the same for all subsequences, the proof is complete. $\blacksquare$

We now turn to the L^p-spaces. Consider the set of all measurable functions f on a measure space $(\Omega, \mathscr{F}, \mu)$ such that $\int |f|\, d\mu < \infty$, where p is a given number in $[1, \infty)$. For each such f, consider the class of all g such that $f = g$ a.e. Since the relation "$f = g$ a.e." is an equivalence relation (i.e., it is reflexive, symmetric, and transitive), the set of all such f splits into disjoint equivalence classes. The set of all these equivalence classes is denoted $L^p(\Omega, \mathscr{F}, \mu)$ or $L^p(\Omega, \mu)$ or simply L^p if the underlying measure space is obvious. It is customary and less cumbersome to write $f \in L^p$, rather than {equivalence class of $f\} \in L^p$. For $f \in L^p$, the L^p-norm $\|f\|_p$ is defined by (2.10). By Minkowskii's inequality (2.13), L^p is a *linear space*. That is, $f, g \in L^p$ implies $cf + dg \in L^p$ for all reals c, d. Also, under the L^p-norm, L^p is a *normed linear space*. This means (i) $\|f\|_p \geqslant 0$ for all $f \in L^p$, with equality if and only if $f = 0$ (a.e.); (ii) $\|cf\|_p = |c|\,\|f\|_p$ for every real c, every $f \in L^p$;

and (iii) the *triangle inequality* $\|f + g\|_p \leqslant \|f\|_p + \|g\|_p$ holds for all $f, g \in L^p$ (by (2.13)). The space L^p is also *complete*, i.e., if $f_n \in L^p$ ($n \geqslant 1$) is a Cauchy sequence (i.e., $\|f_n - f_m\|_p \to 0$ as $n, m \to \infty$), then there exists f in L^p such that $\|f_n - f\|_p \to 0$. For this last important fact, note first that $\|f_n - f_m\|_p \to 0$ easily implies that $\{f_n\}$ is Cauchy in measure and therefore, by Proposition 3.1, converges to some f in measure. As a consequence, $\{|f_n|^p\}$ converges to $|f|^p$ in measure. It then follows by Fatou's Lemma applied to $\{|f_n|^p\}$ that $\int |f|^p \, d\mu < \infty$, i.e., $f \in L^p$. Applying Fatou's Lemma to the sequence $\{|f_n - f_m|^p : m \geqslant n\}$ one gets

$$\int |f_n - f|^p \, d\mu \leqslant \liminf_m \int |f_n - f_m|^p \, d\mu. \tag{3.14}$$

But $\int |f_n - f_m|^p \, d\mu \equiv (\|f_n - f_m\|_p)^p$. Therefore, the right side goes to zero as $n \to \infty$, proving that $\|f_n - f\|_p \to 0$. A complete normed linear space is called a *Banach space*. Thus, the L^p-spaces are Banach spaces ($1 \leqslant p < \infty$). If μ is a probability measure (more generally, a finite measure), then $p < p'$ implies $L^p \subset L^{p'}$, in view of Liapounov's inequality (2.8). This is not true if $\mu(\Omega) = \infty$.

When does a sequence $\{f_n\}$ in L^p converge to some $f \in L^p$ in L^p-norm? Clearly, $\{f_n\}$ must converge to f in measure. A useful sufficient condition is provided by Lebesgue's Dominated Convergence Theorem if one assumes that the dominating function g is in L^p. For then $|f_n - f|^p \to 0$ in measure, and $|f_n - f|^p \leqslant 2^p(|f_n|^p + |f|^p)$ (see Eq. 2.14) $\leqslant 2^{p+1}|g|^p = h$, say, so that the conditions of the theorem apply to $\{|f_n - f|^p\}$.

If μ is a probability measure, then one may obtain a necessary and sufficient condition for L^p-convergence. For this purpose, define a sequence of random variables $\{X_n\}$ on a probability space $(\Omega, \mathscr{F}, P)$ to be *uniformly integrable* if

$$\sup_n \int_{\{|X_n| \geqslant \lambda\}} |X_n| \, dP \to 0 \qquad \text{as } \lambda \to \infty. \tag{3.15}$$

One then has the following.

Theorem 3.5. (*L^1-Convergence Criterion*). Let $(\Omega, \mathscr{F}, P)$ be a probability space, and $\{X_n\}$ a sequence of integrable random variables. Then X_n converges in L^1 to some random variable X if and only if (i) $X_n \to X$ in probability, and (ii) $\{X_n\}$ is uniformly integrable.

Proof. (*Sufficiency*). Under the hypotheses (i), (ii), given $\varepsilon > 0$ one has for all n,

$$\int |X_n| \, dP \leqslant \int_{\{|X_n| > \lambda\}} |X_n| \, dP + \lambda \leqslant \varepsilon + \lambda(\varepsilon) \tag{3.16}$$

if $\lambda(\varepsilon)$ is sufficiently large. In particular, $\{\int |X_n| \, dP\}$ is a bounded sequence. Then, by Fatou's Lemma, $\int |X| \, dP < \infty$. Now

$$\begin{aligned} \int_{\{|X_n - X| \geqslant \lambda\}} |X_n - X| \, dP &\leqslant \int_{\{|X_n| \geqslant \lambda/2\}} |X_n - X| \, dP + \int_{\{|X_n| < \lambda/2, |X_n - X| \geqslant \lambda\}} |X_n - X| \, dP \\ &\leqslant \int_{\{|X_n| \geqslant \lambda/2\}} |X_n| \, dP + \int_{\{|X_n| \geqslant \lambda/2\}} |X| \, dP \\ &\quad + \int_{\{|X_n| < \lambda/2, |X_n - X| \geqslant \lambda\}} |X_n - X| \, dP. \end{aligned} \tag{3.17}$$

Now, given $\varepsilon > 0$, choose $\lambda(\varepsilon) > 0$ such that the first and second terms of the last sum are less than ε for $\lambda = \lambda(\varepsilon)$ (use hypothesis (ii)). With this value of λ, the third term goes to zero, as $n \to \infty$, by Lebesgue's Dominated Convergence Theorem. Hence

$$\limsup_{n\to\infty} \int_{\{|X_n - X| \geqslant \lambda(\varepsilon)\}} |X_n - X| \, dP \leqslant 2\varepsilon.$$

But, again by the Dominated Convergence Theorem,

$$\limsup_{n\to\infty} \int_{\{|X_n - X| < \lambda(\varepsilon)\}} |X_n - X| \, dP = 0.$$

(*Necessity*). If $X_n \to X$ in L^1, clearly $X_n \to X$ in probability. Also,

$$\begin{aligned}\int_{\{|X_n| \geqslant \lambda\}} |X_n| \, dP &\leqslant \int_{\{|X_n| \geqslant \lambda\}} |X_n - X| \, dP + \int_{\{|X_n| \geqslant \lambda\}} |X| \, dP \\ &\leqslant \int |X_n - X| \, dP + \int_{\{|X| \geqslant \lambda/2\}} |X| \, dP + \int_{\{|X| < \lambda/2, |X_n - X| \geqslant \lambda/2\}} |X| \, dP.\end{aligned} \tag{3.18}$$

The first term of the last sum goes to zero, as $n \to \infty$, by hypothesis. For each $\lambda > 0$ the third term goes to zero by the Dominated Convergence Theorem, as $n \to \infty$. The same convergence theorem implies that the second term goes to zero as $\lambda \to \infty$. Therefore, given $\varepsilon > 0$ there exists $\lambda(\varepsilon)$, $n(\varepsilon)$, such that

$$\sup_{n \geqslant n(\varepsilon)} \int_{\{|X_n| \geqslant \lambda\}} |X_n| \, dP \leqslant \varepsilon \qquad \text{if } \lambda \geqslant \lambda(\varepsilon). \tag{3.19}$$

Since a *finite* sequence $\{X_n : 1 \leqslant n < n(\varepsilon)\}$ of integrable random variables is always uniformly integrable, it follows that $\{X_n\}$ is uniformly integrable. ■

A corollary is the following version. As always, $p \geqslant 1$.

Theorem 3.6. (*L^p-Convergence Criterion*). Let $(\Omega, \mathscr{F}, P)$ be a probability space, and $X_n \in L^p$ $(n \geqslant 1)$. Then $X_n \to X$ in L^p if and only if (i) $X_n \to X$ in probability, and (ii) $\{|X_n^p|\}$ is uniformly integrable.

Proof. Apply the above result to the sequence $\{|X_n - X|^p\}$. For sufficiency, note, as in (3.16), that (i), (ii) imply $X \in L^p$, and then argue as in (3.17) that the uniform integrability of $\{|X_n|^p\}$ implies that of $\{|X_n - X|^p\}$. The proof of necessity is analogous to (3.18) and (3.19), using (2.14). ■

It is simple to check by a Chebyshev-type inequality (see Eq. 2.16) that if $\{E|X_n|^{p'}\}$ is a bounded sequence for some $p' > p$, then $\{|X_n|^p\}$ is uniformly integrable.

There is one important case where convergence a.e. implies L^1-convergence. This is as follows.

Theorem 3.7. (*Scheffé's Theorem*). Let $(\Omega, \mathscr{F}, \mu)$ be a measure space. Let $f_n (n \geqslant 1)$, f be p.d.f.'s with respect to μ, i.e., f_n, f are nonnegative, and $\int f_n \, d\mu = 1$ for all n, $\int f \, d\mu = 1$. If f_n converges a.e. to f, then $f_n \to f$ in L^1.

Proof. Recall that for every real-valued function g on Ω, one has

$$g = g^+ - g^-, \qquad |g| = g^+ + g^-, \qquad \text{where } g^+ = \max\{g, 0\}, g^- = -\min\{g, 0\}.$$

One has

$$\int (f - f_n)\, d\mu = 0 = \int (f - f_n)^+\, d\mu - \int (f - f_n)^-\, d\mu,$$

so that

$$\int (f - f_n)^-\, d\mu = \int (f - f_n)^+\, d\mu, \qquad \int |f - f_n|\, d\mu = 2 \int (f - f_n)^+\, d\mu. \tag{3.20}$$

Now $0 \leqslant (f - f_n)^+ \leqslant f$, and $(f - f_n)^+ \to 0$ a.e. as $n \to \infty$. Therefore, by (3.20) and Lebesgue's Dominated Convergence Theorem,

$$\int |f - f_n|\, d\mu = 2 \int (f - f_n)^+\, d\mu \to 0 \qquad \text{as } n \to \infty. \qquad ■$$

Among the L^p-spaces the space $L^2 \equiv L^2(\Omega, \mathscr{F}, \mu)$ has a particularly rich structure. It is a *Hilbert space*. That is, it is a Banach space with an *inner product* $\langle\ ,\ \rangle$ defined by

$$\langle f, g \rangle := \int fg\, d\mu \qquad (f, g \in L^2). \tag{3.21}$$

The inner product is bilinear, i.e., linear in each argument, and the L^2-norm is given by

$$\|f\|_2 = \langle f, f \rangle^{1/2}, \tag{3.22}$$

and the Schwarz Inequality may be expressed as

$$|\langle f, g \rangle| \leqslant \|f\|_2 \|g\|_2. \tag{3.23}$$

4 PRODUCT MEASURES AND INDEPENDENCE, RADON–NIKODYM THEOREM AND CONDITIONAL PROBABILITY

If $(S_1, \mathscr{S}_1, \mu_1)$, $(S_2, \mathscr{S}_2, \mu_2)$ are two measure spaces, then the *product space* $(S, \mathscr{S}, \mu)$ is a measure space where (i) S is the Cartesian product $S_1 \times S_2$; (ii) $\mathscr{S} = \mathscr{S}_1 \otimes \mathscr{S}_2$ is the smallest sigmafield containing the class $\mathscr{R}$ of all *measurable rectangles*, $\mathscr{R} := \{B_1 \times B_2 : B_1 \in \mathscr{S}_1, B_2 \in \mathscr{S}_2\}$; and (iii) μ is the *product measure* $\mu_1 \times \mu_2$ on $\mathscr{S}$ determined by the requirement

$$\mu(B_1 \times B_2) = \mu_1(B_1)\mu_2(B_2) \qquad (B_1 \in \mathscr{S}_1, B_2 \in \mathscr{S}_2). \tag{4.1}$$

As the intersection of two measurable rectangles is a measurable rectangle, $\mathscr{R}$ is closed under finite intersections. Then the class $\mathscr{C}$ of all finite disjoint unions of sets in $\mathscr{R}$ is a field. By finite additivity and (4.1), μ extends to $\mathscr{C}$ as a countably additive set function. Finally, Carathéodory's Extension Theorem extends μ uniquely to a measure on $\mathscr{S} = \sigma(\mathscr{C})$, the smallest sigmafield containing $\mathscr{C}$.

For each $B \in \mathscr{S}$, every *x-section* $B_{(x,\cdot)} := \{y: (x, y) \in B\}$ is in $\mathscr{S}_1$. This is clearly true for measurable rectangles $F = B_1 \times B_2$. The class $\mathscr{A}$ of all sets in $\mathscr{S}$ for which the assertion is true is a $\mathscr{L}$-class (or, a *lambda class*). That is, (i) $\Omega \in \mathscr{L}$, (ii) $A, B \in \mathscr{L}$, *and* $A \subset B$ *imply* $B \backslash A \in \mathscr{L}$, *and* (iii) $A_n (n \geqslant 1) \in \mathscr{L}$, $A_n \uparrow A$ *imply* $A \in \mathscr{L}$. We state without proof the following useful result from which the *measurable-sections* property asserted above for all $B \in \mathscr{S}$ follows.

Theorem 4.1. (*Dynkin's Pi-Lambda Theorem*).[1] Suppose a class $\mathscr{B}$ is closed under finite intersections, a class $\mathscr{A}$ is a lambda class, and $\mathscr{B} \subset \mathscr{A}$. Then $\sigma(\mathscr{B}) \subset \mathscr{A}$, where $\sigma(\mathscr{B})$ is the smallest sigmafield containing $\mathscr{B}$.

In view of the measurable-section property, for each $B \in \mathscr{S}$ one may define the functions $x \to \mu_2(B_{(x,\cdot)})$, $y \to \mu_1(B_{(\cdot,y)})$. These are measurable functions on $(S_1, \mathscr{S}_1, \mu_1)$ and $(S_2, \mathscr{S}_2, \mu_2)$, respectively, and one has

$$\mu(B) = \int_{S_1} \mu_2(B_{(x,\cdot)})\mu_1(dx) = \int_{S_2} \mu_1(B_{(\cdot,y)})\mu_2(dy) \qquad (B \in \mathscr{S}). \tag{4.2}$$

This last assertion holds for $B = B_1 \times B_2 \in \mathscr{R}$, by (4.2) and the relations $\mu_2(B_{(x,\cdot)}) = \mu_2(B_2)\mathbf{1}_{B_1}(x)$, $\mu_1(B_{(\cdot,y)}) = \mu_1(B_1)\mathbf{1}_{B_2}(y)$. The proof is now completed for all $B \in \mathscr{S}$ by the Pi-Lambda Theorem. It follows that if $f(x, y)$ is a simple function on $(S, \mathscr{S}, \mu)$, then for each $x \in S_1$ the function $y \to f(x, y)$ on $(S_2, \mathscr{S}_2, \mu_2)$ is measurable, and for each $y \in S_2$ the function $x \to f(x, y)$ on $(S_1, \mathscr{S}_1, \mu_1)$ is measurable. Further for all such f

$$\int_S f\,d\mu = \int_{S_1}\left(\int_{S_2} f(x, y)\mu_2(dy)\right)\mu_1(dx) = \int_{S_2}\left(\int_{S_1} f(x, y)\mu_1(dx)\right)\mu_2(dy). \tag{4.3}$$

By the usual approximation of measurable functions by simple functions one arrives at the following theorem.

Theorem 4.2. (*Fubini's Theorem*). (a) If f is integrable on the product space $(S_1 \times S_2, \mathscr{S}_1 \otimes \mathscr{S}_2, \mu_1 \times \mu_2) \equiv (S, \mathscr{S}, \mu)$, then

(i) $x \to \int_{S_2} f(x, y)\mu_2(dy)$ is measurable and integrable on $(S_1, \mathscr{S}_1, \mu_1)$.
(ii) $y \to \int_{S_1} f(x, y)\mu_1(dx)$ is measurable and integrable on $(S_2, \mathscr{S}_2, \mu_2)$.
(iii) One has the equalities (4.3).

(b) If f is nonnegative measurable on $(S, \mathscr{S}, \mu)$ then (4.3) holds, whether the integrals are finite or not.

The concept of a product space extends to an arbitrary but finite number of components $(S_i, \mathscr{S}_i, \mu_i)$ $(1 \leqslant i \leqslant k)$. In this case $S = S_1 \times S_2 \times \cdots \times S_k$, $\mathscr{S} = \mathscr{S}_1 \otimes \mathscr{S}_2 \otimes \cdots \otimes \mathscr{S}_k$ is the smallest sigmafield containing the class $\mathscr{R}$ of all measurable rectangles

[1] P. Billingsley (1986), *Probability and Measure*, 2nd ed., Wiley, New York, p. 36.

$B = B_1 \times B_2 \times \cdots \times B_k$ ($B_i \in \mathscr{S}_i$, $1 \leqslant i \leqslant k$). The sigmafield $\mathscr{S}$ is called the *product sigmafield*, while $\mu = \mu_1 \times \cdots \times \mu_k$ is the *product measure* determined by

$$\mu(B_1 \times B_2 \times \cdots \times B_k) = \mu_1(B_1)\mu_2(B_2)\cdots\mu_k(B_k) \tag{4.4}$$

for elements in $\mathscr{R}$. Fubini's theorem extends in a straightforward manner, integrating f first with respect to one coordinate keeping the $k-1$ others fixed, then integrating the resulting function of the other variables with respect to a second coordinate keeping the remaining $k-2$ fixed, and so on until the function of a single variable is integrated with respect to the last and remaining coordinate. The order in which the variables are integrated is immaterial, the final result equals $\int_S f\,d\mu$.

Product probabilities arise as joint distributions of independent random variables. Let $(\Omega, \mathscr{F}, P)$ be a probability space on which are defined measurable functions X_i ($1 \leqslant i \leqslant k$), with X_i taking values in S_i, which is endowed with a sigmafield $\mathscr{S}_i$ ($1 \leqslant i \leqslant k$). The measurable functions (or random variables) $X_1, X_2, \ldots, X_n$ are said to be *independent* if

$$P(X_1 \in B_1, X_2 \in B_2, \ldots, X_k \in B_k) = P(X_1 \in B_1)\cdots P(X_k \in B_k)$$

$$\text{for all } B_1 \in \mathscr{S}_1, \ldots, B_k \in \mathscr{S}_k. \tag{4.5}$$

In other words the (joint) distribution of $(X_1, \ldots, X_k)$ is a product measure. If f_i is an integrable function on $(S_i, \mathscr{S}_i, \mu_i)$ ($1 \leqslant i \leqslant k$), then the function

$$f\colon (x_1, x_2, \ldots, x_k) \to f_1(x_1)f_2(x_2)\cdots f_k(x_k)$$

is integrable on $(S, \mathscr{S}, \mu)$, and one has

$$\int_S f\,d\mu = \prod_{i=1}^{k}\left(\int_{S_i} f_i\,d\mu_i\right), \tag{4.6}$$

or

$$E\left(\prod_{i=1}^{k} f_i(X_i)\right) = \prod_{i=1}^{k} (Ef_i(X_i)). \tag{4.7}$$

A *sequence* $\{X_n\}$ of random variables *is* said to be *independent* if every finite subcollection is. *Two sigmafields* $\mathscr{F}_1, \mathscr{F}_2$ *are independent* if $P(F_1 \cap F_2) = P(F_1)P(F_2)$ for all $F_1 \in \mathscr{F}_1$, $F_2 \in \mathscr{F}_2$. *Two families* of random variables $\{X_\lambda\colon \lambda \in \Lambda_1\}$ and $\{Y_\lambda\colon \lambda \in \Lambda_2\}$ *are independent of each other* if $\sigma\{X_\lambda\colon \lambda \in \Lambda_1\}$ and $\sigma\{Y_\lambda\colon \lambda \in \Lambda_2\}$ are independent. Here $\sigma\{X_\lambda\colon \lambda \in \Lambda\}$ is the smallest sigmafield with respect to which all the X_λ are measurable.

Events $A_1, A_2, \ldots, A_k$ *are independent* if the corresponding indicator functions $\mathbf{1}_{A_1}, \mathbf{1}_{A_2}, \ldots, \mathbf{1}_{A_k}$ are independent. This is equivalent to requiring $P(B_1 \cap B_2 \cap \cdots \cap B_k) = P(B_1)P(B_2)\cdots P(B_k)$ for all choices $B_1, \ldots, B_k$ with $B_i = A_i$ or A_i^c.

Before turning to Kolmogorov's definition of conditional probabilities, it is necessary to state an important result from measure theory. To motivate it, let $(\Omega, \mathscr{F}, \mu)$ be a measure space and let f be an integrable function on it. Then the set function ν defined by

$$\nu(F) := \int_F f\,d\mu \qquad (F \in \mathscr{F}), \tag{4.8}$$

is a countable additive set function, or a *signed measure*, on $\mathscr{F}$ with the property

$$\nu(F) = 0 \qquad \text{if } \mu(F) = 0. \tag{4.9}$$

A signed measure ν on $\mathscr{F}$ is said to be *absolutely continuous with respect to* μ, denoted $\nu \ll \mu$, if (4.9) holds. The theorem below says that, conversely, $\nu \ll \mu$ implies the existence of an f such that (4.8) holds, if μ, ν are *sigmafinite*. A countably additive set function ν is *sigmafinite* if there exist B_n $(n \geqslant 1)$ such that (i) $\bigcup_{n \geqslant 1} B_n = \Omega$, (ii) $|\nu(B_n)| < \infty$ for all n. All measures in this book are assumed to be sigmafinite.

Theorem 4.3. (*Radon–Nikodym Theorem*).[2] Let $(\Omega, \mathscr{F}, \mu)$ be a measure space. If a finite signed measure ν on $\mathscr{F}$ is absolutely continuous with respect to μ then there exists an a.e. unique integrable function f, called the Radon–Nikodym derivative of ν with respect to μ, such that ν has the representation (4.8).

Next, the *conditional probability* $P(A \mid B)$, *of an event* A *given another event* B, is defined in classical probability as

$$P(A \mid B) := \frac{P(A \cap B)}{P(B)}, \tag{4.10}$$

provided $P(B) > 0$. To introduce Kolmogorov's extension of this classical notion, let $(\Omega, \mathscr{F}, P)$ be a probability space and $\{B_n\}$ a countable partition of Ω by sets B_n in $\mathscr{F}$. Let $\mathscr{D}$ denote the sigmafield generated by this partition, $\mathscr{D} = \sigma\{B_n\}$. That is, $\mathscr{D}$ comprises all countable disjoint unions of sets in $\{B_n\}$. Given an event $A \in \mathscr{F}$, one defines $P(A \mid \mathscr{D})$, the *conditional probability of* A *given* $\mathscr{D}$, by the $\mathscr{D}$-measurable random variable

$$P(A \mid \mathscr{D})(\omega) := \frac{P(A \cap B_n)}{P(B_n)} \qquad \text{for } \omega \in B_n, \tag{4.11}$$

if $P(B_n) > 0$, and an arbitrary constant c_n, say, if $P(B_n) = 0$. Check that

$$\int_D P(A \mid \mathscr{D})\, dP = P(A \cap D) \qquad \text{for all } D \in \mathscr{D}. \tag{4.12}$$

If X is a random variable, $E|X| < \infty$, then one defines $E(X \mid \mathscr{D})$, the *conditional expectation of* X *given* $\mathscr{D}$, as the $\mathscr{D}$-measurable random variable

$$\begin{aligned} E(X \mid \mathscr{D})(\omega) &:= \frac{1}{P(B_n)} \int_{B_n} X\, dP \qquad && \text{for } \omega \in B_n, \text{ if } P(B_n) > 0, \\ &= c_n && \text{for } \omega \in B_n, \text{ if } P(B_n) = 0, \end{aligned} \tag{4.13}$$

where c_n are arbitrarily chosen constants (e.g., $c_n = 0$ for all n). From (4.13) one easily verifies the equality

$$\int_D E(X \mid \mathscr{D})\, dP = \int_D X\, dP \qquad \text{for all } D \in \mathscr{D}. \tag{4.14}$$

Note that $P(A \mid \mathscr{D}) = E(\mathbf{1}_A \mid \mathscr{D})$, so that (4.12) is a special case of (4.14).

[2] P. Billingsley (1986), *loc. cit.*, p. 443.

One may express (4.14) by saying that $E(X \mid \mathscr{D})$ is a *$\mathscr{D}$-measurable random variable whose integral over each $D \in \mathscr{D}$ equals the integral of X over D*. By taking $D = B_n$ in (4.14), on the other hand, one derives (4.13). Thus, the italicized statement above may be taken to be the definition of $E(X \mid \mathscr{D})$. This is Kolmogorov's definition of $E(X \mid \mathscr{D})$, which, however, holds for any subsigmafield $\mathscr{D}$ of $\mathscr{F}$, whether generated by a countable partition or not. To see that a $\mathscr{D}$-measurable function $E(X \mid \mathscr{D})$ satisfying (4.14) exists and is unique, no matter what sigmafield $\mathscr{D} \subset \mathscr{F}$ is given, consider the set function ν defined on $\mathscr{D}$ by

$$\nu(D) = \int_D X\, dP \qquad (D \in \mathscr{D}). \tag{4.15}$$

Then $\nu(D) = 0$ if $P(D) = 0$. Consider the restriction of P to $\mathscr{D}$. Then one has $\nu \ll P$ on $\mathscr{D}$. By the Radon–Nikodym Theorem, there exists a unique (up to a P-null set) $\mathscr{D}$-measurable function, denoted $E(X \mid \mathscr{D})$, such that (4.14) holds.

The simple interpretation of $E(X \mid \mathscr{D})$ in (4.13) as the average of X on each member of the partition generating $\mathscr{D}$ is lost in this abstract definition for more general sigmafields $\mathscr{D}$. But the italicized definition of $E(X \mid \mathscr{D})$ above still retains the intuitive idea that given the information embodied in $\mathscr{D}$, the reassessed (or conditional) probability of A, or expectation of X, must depend only on this information, i.e., must be $\mathscr{D}$-measurable, and must give the correct probabilities, or expectations, when integrated over events in $\mathscr{D}$.

Here is a list of some of the commonly used properties of conditional expectations.

Theorem 4.4. (*Basic Properties of Conditional Expectations*). Let $(\Omega, \mathscr{F}, P)$ be a probability space, $\mathscr{D}$ a subsigmafield of $\mathscr{F}$.

(a) If X is $\mathscr{D}$-measurable and integrable then $E(X \mid \mathscr{D}) = X$.

(b) (*Linearity*) If X, Y are integrable and c, d constants, then $E(cX + dY \mid \mathscr{D}) = cE(X \mid \mathscr{D}) + dE(Y \mid \mathscr{D})$.

(c) (*Order*) If X, Y are integrable and $X \leqslant Y$ a.s., then $E(X \mid \mathscr{D}) \leqslant E(Y \mid \mathscr{D})$ a.s.

(d) If Y and XY are integrable, and X is $\mathscr{D}$-measurable then $E(XY \mid \mathscr{D}) = XE(Y \mid \mathscr{D})$.

(e) (*Successive Smoothing*) If $\mathscr{G}$ is a subsigmafield of $\mathscr{F}$, $\mathscr{G} \subset \mathscr{D}$, and X is integrable, then $E(X \mid \mathscr{G}) = E[E(X \mid \mathscr{D}) \mid \mathscr{G}] = E[E(X \mid \mathscr{G}) \mid \mathscr{D}]$.

(f) (*Convergence*) Let $\{X_n\}$ be a sequence of random variables such that, for all n, $|X_n| \leqslant Z$ where Z is integrable. If $X_n \to X$ a.s., then $E(X_n \mid \mathscr{D}) \to E(X \mid \mathscr{D})$ a.s. and in L^1.

All the properties (a)–(f) are fairly straightforward consequences of the definition of conditional expectations, and are therefore not proved here. The interplay between conditional expectations and independence is described by the following result.

Theorem 4.5. (*Independence and Conditional Expectation*). Let $(\Omega, \mathscr{F}, P)$ be a probability space and $\mathscr{D}$ a subsigmafield of $\mathscr{F}$. Then the following hold.

(a) If X is integrable, and $\sigma\{X\}$ and $\mathscr{D}$ are independent, then $E(X \mid \mathscr{D}) = EX$.

(b) Suppose X and Y are measurable maps on $(\Omega, \mathscr{F})$ into $(S_1, \mathscr{S}_1)$ and $(S_2, \mathscr{S}_2)$, respectively. Let φ be a real-valued measurable function on $(S_1 \times S_2, \mathscr{S}_1 \otimes \mathscr{S}_2)$, and $\varphi(X, Y)$ integrable. Assume X and Y are independent, i.e., $\sigma\{X\}$ and $\sigma\{Y\}$ are independent. Then,

$$E(\varphi(X, Y) \mid \sigma\{Y\}) = [E\varphi(X, y)]_{y=Y}. \tag{4.16}$$

(c) Let X, Y, Z be measurable maps on $(\Omega, \mathscr{F}, P)$ into $(S_1, \mathscr{S}_1)$, $(S_2, \mathscr{S}_2)$, and $(S_3, \mathscr{S}_3)$, respectively. Let φ be a real-valued measurable function on $(S_1, \mathscr{S}_1)$. Assume $\varphi(X)$ is integrable, and $\sigma\{X\}$ and $\sigma\{Y\}$ both independent of $\sigma\{Z\}$. Then, writing $\sigma\{Y, Z\}$ for the smallest sigmafield $\subset \mathscr{F}$ with respect to which both Y and Z are measurable, one has

$$E(\varphi(X)) \mid \sigma\{Y, Z\}) = E(\varphi(X) \mid \sigma\{Y\}). \tag{4.17}$$

Proof. (a) EX is a constant and, therefore, $\mathscr{D}$-measurable. Also, relation (4.14) holds with EX in place of $E(X \mid \mathscr{D})$ by independence.

(b) Let P_X, P_Y denote the distributions of X and Y, respectively. Then the product probability $P_X \times P_Y$ on $(S_1 \times S_2, \mathscr{S}_1 \otimes \mathscr{S}_2)$ is the (joint) distribution of (X, Y). Let $D \in \sigma\{Y\}$. This means there exists $B \in \mathscr{S}_2$ such that $D = \{\omega \in \Omega : Y(\omega) \in B\} \equiv Y^{-1}(B)$. By the change-of-variables formula and Fubini's Theorem,

$$\begin{aligned}
\int_D \varphi(X, Y)\, dP &= \int_\Omega \mathbf{1}_B(Y)\varphi(X, Y)\, dP = \int_{S_1 \times S_2} \mathbf{1}_B(y)\varphi(x, y) P_X(dx) P_Y(dy) \\
&= \int_{S_2} \mathbf{1}_B(y)\left(\int_{S_1} \varphi(x, y) P_X(dx)\right) P_Y(dy) = \int_{S_2} \mathbf{1}_B(y)(E\varphi(X, y)) P_Y(dy) \\
&= \int_\Omega \mathbf{1}_B(Y(\omega))(E\varphi(X, y))_{y = Y(\omega)}\, dP(\omega) = \int_\Omega \mathbf{1}_D(\omega)(E\varphi(X, y))_{y = Y(\omega)}\, dP(\omega) \\
&= \int_D (E\varphi(X, y))_{y = Y}\, dP.
\end{aligned}$$

Also, $[E\varphi(X, y)]_{y = Y}$ is $\sigma\{Y\}$-measurable.

(c) Let $D_1 \in \sigma\{Y\}$, $D_2 \in \sigma\{Z\}$. Then there exist $B_1 \in \mathscr{S}_2$, $B_2 \in \mathscr{S}_3$ such that $D_1 = Y^{-1}(B_1)$, $D_2 = Z^{-1}(B_2)$. With $D = D_1 \cap D_2$ one has

$$\begin{aligned}
\int_D \varphi(X)\, dP &= \int_\Omega \mathbf{1}_{D_1}\mathbf{1}_{D_2}\varphi(X)\, dP = \int_\Omega \mathbf{1}_{B_1}(Y)\mathbf{1}_{B_2}(Z)\varphi(X)\, dP \\
&= E(\mathbf{1}_{B_1}(Y)\varphi(X))E(\mathbf{1}_{B_2}(Z)) = E(\mathbf{1}_{B_1}(Y)\varphi(X))P(Z \in B_2) \\
&= P(Z \in B_2)E[E(\mathbf{1}_{B_1}(Y)\varphi(X) \mid \sigma\{Y\})] = P(Z \in B_2)E(\mathbf{1}_{B_1}(Y)E(\varphi(X) \mid \sigma\{Y\})] \\
&= E(\mathbf{1}_{B_2}(Z)\mathbf{1}_{B_1}(Y)E(\varphi(X) \mid \sigma\{Y\})) = E(\mathbf{1}_{D_1}\mathbf{1}_{D_2}E(\varphi(X) \mid \sigma\{Y\})) \\
&= \int_{D_1 \cap D_2} E(\varphi(X) \mid \sigma\{Y\})\, dP.
\end{aligned}$$

Thus, the desired relation (of the type) (4.14) holds for sets $D = D_1 \cap D_2 \in \sigma\{Y, Z\}$. The class $\mathscr{B}$ of all such sets is closed under finite intersection. Also, the class $\mathscr{A}$ of all sets $D \in \sigma\{Y, Z\}$ for which $\int_D \varphi(X)\, dP = \int_D E(\varphi(X) \mid \sigma\{Y\})\, dP$ holds is a lambda class. Therefore, by Dynkin's Pi-Lambda Theorem, $\sigma(\mathscr{B}) \subset \mathscr{A}$. But $\sigma(\mathscr{B}) = \sigma\{Y, Z\}$. ■

There is an extension of Jensen's inequality (2.7) for conditional expectations that is useful.

Proposition 4.6. (*Conditional Jensen's Inequality*). Let φ be a convex function on an interval J. If X is an integrable random variable such that $P(X \in J) = 1$, and $\varphi(X)$ is integrable, then

$$E[\varphi(X) \mid \mathscr{D}] \geqslant \varphi(E(X \mid \mathscr{D})). \tag{4.18}$$

Proof. In (2.5) take $y = X$, $x = E(X \mid \mathscr{D})$, to get

$$\varphi(X) \geqslant \varphi(E(X \mid \mathscr{D})) + m(E(X \mid \mathscr{D}))(X - E(X \mid \mathscr{D})).$$

Now take conditional expectations, given $\mathscr{D}$, on both sides. ■

As an immediate corollary to (4.18) it follows that the operation of taking conditional expectation is a *contraction* on $L^p(\Omega, \mathscr{F}, P)$. That is, if $X \in L^p$ for some $p \geqslant 1$, then

$$\|E(X \mid \mathscr{D})\|_p \leqslant \|X\|_p. \tag{4.19}$$

If $X \in L^2$, then in fact $E(X \mid \mathscr{D})$ is the orthogonal projection of X onto $L^2(\Omega, \mathscr{D}, P) \subset L^2(\Omega, \mathscr{F}, P)$. For if Y is an arbitrary element of $L^2(\Omega, \mathscr{D}, P)$, then

$$\begin{aligned} E(X - Y)^2 = E(X - E(X \mid \mathscr{D}) + E(X \mid \mathscr{D}) - Y)^2 &= E(X - E(X \mid \mathscr{D}))^2 \\ &+ E(E(X \mid \mathscr{D}) - Y)^2 + 2E[(E(X \mid \mathscr{D}) - Y)(X - E(X \mid \mathscr{D}))]. \end{aligned}$$

The last term on the right side vanishes, on first taking conditional expectation given $\mathscr{D}$ (see Basic Property (e)). Hence, for all $Y \in L^2(\Omega, \mathscr{D}, P)$,

$$E(X - Y)^2 = E(X - E(X \mid \mathscr{D}))^2 + E(E(X \mid \mathscr{D}) - Y)^2 \geqslant E(X - E(X \mid \mathscr{D}))^2. \tag{4.20}$$

Note that X has the orthogonal decomposition: $X = E(X \mid \mathscr{D}) + (X - E(X \mid \mathscr{D}))$.

Finally, the classical notion of conditional probability as a reassessed probability measure may be recovered under fairly general conditions. The technical difficulty lies in the fact that for every given pairwise disjoint sequence of events $\{A_n\}$ one may assert the equality $P(\bigcup A_n \mid \mathscr{D})(\omega) = \sum P(A_n \mid \mathscr{D})(\omega)$ for all ω outside a P-null set (Basic Properties (b), (f)). Since in general there are uncountably many such sequences, there may not exist any choice of versions of $P(A \mid \mathscr{D})$ for all $A \in \mathscr{F}$ such that for each ω, outside a set of zero probability, $A \to P(A \mid \mathscr{D})(\omega)$ is a probability measure on $\mathscr{F}$. When such a choice is possible, the corresponding family $\{P(A \mid \mathscr{D})\colon A \in \mathscr{F}\}$ is called a *regular conditional probability*. The problem becomes somewhat simpler if one does not ask for a conditional probability measure on $\mathscr{F}$, but on a smaller sigmafield. For example, one may consider the sigmafield $\mathscr{G} = \{Y^{-1}(B)\colon B \in \mathscr{S}\}$ where Y is a measurable function on $(\Omega, \mathscr{F}, P)$ into $(S, \mathscr{S})$. A function $(\omega, B) \to Q_\omega(B \mid \mathscr{D})$ on $\Omega \times \mathscr{S}$ into $[0, 1]$ is said to be a *conditional distribution of Y given $\mathscr{D}$* if (i) for each $B \in \mathscr{S}$, $Q_\omega(B \mid \mathscr{D}) = P(\{Y \in B\} \mid \mathscr{D})(\omega) \equiv E(\mathbf{1}_B(Y) \mid \mathscr{D})(\omega)$, for all ω outside a P-null set, and (ii) for each $\omega \in \Omega$, $B \to Q_\omega(B \mid \mathscr{D})$ is a probability measure on $(S, \mathscr{S})$. Note that (i), (ii) say that there is a regular conditional probability on $\mathscr{G}$.

If there exists a conditional distribution $Q_\omega(B \mid \mathscr{D})$ of Y given $\mathscr{D}$, then it is simple to check that

$$E(\varphi(Y) \mid \mathscr{D})(\omega) = \int_S \varphi(y) Q_\omega(dy \mid \mathscr{D}) \quad \text{a.s.} \tag{4.21}$$

for every measurable φ on $(S, \mathscr{S})$ such that $\varphi(Y)$ is integrable. Conditional distributions do exist if S is a (Borel subset of a) complete separable metric space and $\mathscr{S}$ its Borel sigmafield.

In this book we often write $E(Z \mid \{X_\lambda: \lambda \in \Lambda\})$ in place of $E(Z \mid \sigma\{X_\lambda: \lambda \in \Lambda\})$ for simplicity.

5 CONVERGENCE IN DISTRIBUTION IN FINITE DIMENSIONS

A sequence of *probability measures* $\{P_n: n = 1, 2, \ldots\}$ on $(\mathbb{R}^1, \mathscr{B}^1)$ is said to *converge weakly* to a probability measure P (on $\mathbb{R}^1$) if

$$\lim_{n\to\infty} \int_{\mathbb{R}^1} \phi(x)\, dP_n(x) = \int_{\mathbb{R}^1} \phi(x)\, dP(x) \tag{5.1}$$

holds for all bounded continuous functions ϕ on $\mathbb{R}^1$. *It is* actually *sufficient to verify (5.1) for those continuous functions ϕ that vanish outside some finite interval.* For suppose (5.1) holds for all such functions. Let ϕ be an arbitrary bounded continuous function, $|\phi(x)| \leqslant c$ for all x. For notational convenience write $\{x \in \mathbb{R}^1: |x| \geqslant N\} = \{|x| \geqslant N\}$, etc. Given $\varepsilon > 0$ there exists N such that $P(\{|x| \geqslant N\}) < \varepsilon/2c$. Let θ_N, θ'_N be as in Figure 5.1(a), (b). Then,

$$\begin{aligned}\underline{\lim}_{n\to\infty} P_n(\{|x| \leqslant N+1\}) &\geqslant \underline{\lim}_{n\to\infty} \int \theta_N(x)\, dP_n(x) = \int \theta_N(x)\, dP(x)\\ &\geqslant P(\{|x| \leqslant N\}) > 1 - \frac{\varepsilon}{2c},\end{aligned}$$

so that

$$\overline{\lim}_{n\to\infty} P_n(\{|x| > N+1\}) \equiv 1 - \underline{\lim}_{n\to\infty} P_n(\{|x| \leqslant N+1\}) < \frac{\varepsilon}{2c}. \tag{5.2}$$

Hence, writing $\phi_N = \phi\theta'_N$ and noting that $\phi = \phi_N$ on $\{|x| \leqslant N+1\}$ and that on $\{|x| > N+1\}$ one has $|\phi(x)| \leqslant c$, we have

$$\begin{aligned}\overline{\lim}_{n\to\infty} \left| \int_{\mathbb{R}^1} \phi\, dP_n - \int_{\mathbb{R}^1} \phi\, dP \right| &\leqslant \overline{\lim}_{n\to\infty} \left| \int_{\mathbb{R}^1} \phi_N\, dP_n - \int_{\mathbb{R}^1} \phi_N\, dP \right|\\ &\quad + \overline{\lim}_{n\to\infty} (cP_n(\{|x| > N+1\}) + cP(\{|x| > N+1\}))\\ &= \overline{\lim}_{n\to\infty} cP_n(\{|x| > N+1\}) + cP(\{|x| > N+1\}))\\ &< c\frac{\varepsilon}{2c} + c\frac{\varepsilon}{2c} = \varepsilon.\end{aligned}$$

Since $\varepsilon > 0$ is arbitrary, $\int_{\mathbb{R}^1} \phi\, dP_n \to \int_{\mathbb{R}^1} \phi\, dP$, and the proof of the italicized statement above is complete. Let us now show that it is enough to verify (5.1) for *every infinitely*

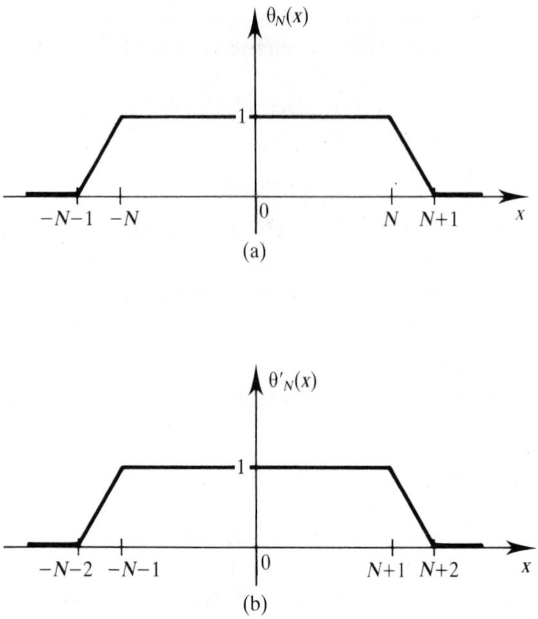

Figure 5.1

differentiable function that vanishes outside a finite interval. For each $\varepsilon > 0$ define the function

$$\rho_\varepsilon(x) = d(\varepsilon)\exp\left\{-\frac{1}{(1 - x^2/\varepsilon^2)}\right\} \qquad \text{for } |x| < \varepsilon,$$
$$= 0 \qquad \text{for } |x| \geqslant \varepsilon, \tag{5.3}$$

where $d(\varepsilon)$ is so chosen as to make $\int \rho_\varepsilon(x)\,dx = 1$. One may check that $\rho_\varepsilon(x)$ is infinitely differentiable in x. Now let ϕ be a continuous function that vanishes outside a finite interval. Then ϕ is uniformly continuous and, therefore, $\delta(\varepsilon) = \sup\{|\phi(x) - \phi(y)|: |x - y| \leqslant \varepsilon\} \to 0$ as $\varepsilon \downarrow 0$. Define

$$\phi^\varepsilon(x) = \phi * \rho_\varepsilon(x) := \int_{-\varepsilon}^{\varepsilon} \phi(x - y)\rho_\varepsilon(y)\,dy, \tag{5.4}$$

and note that, since $\phi^\varepsilon(x)$ is an average over values of ϕ within the interval $(x - \varepsilon, x + \varepsilon)$, $|\phi^\varepsilon(x) - \phi(x)| \leqslant \delta(\varepsilon)$ for all ε. Hence,

$$\left|\int_{\mathbb{R}^1} \phi\,dP_n - \int_{\mathbb{R}^1} \phi^\varepsilon\,dP_n\right| \leqslant \delta(\varepsilon) \qquad \text{for all } n, \qquad \left|\int_{\mathbb{R}^1} \phi\,dP - \int_{\mathbb{R}^1} \phi^\varepsilon\,dP\right| \leqslant \delta(\varepsilon),$$

$$\left|\int_{\mathbb{R}^1} \phi\, dP_n - \int_{\mathbb{R}^1} \phi\, dP\right| \leqslant \left|\int_{\mathbb{R}^1} \phi\, dP_n - \int_{\mathbb{R}^1} \phi^\varepsilon\, dP_n\right| + \left|\int_{\mathbb{R}^1} \phi^\varepsilon\, dP_n - \int_{\mathbb{R}^1} \phi^\varepsilon\, dP\right|$$
$$+ \left|\int_{\mathbb{R}^1} \phi^\varepsilon\, dP - \int_{\mathbb{R}^1} \phi\, dP\right|$$
$$\leqslant 2\delta(\varepsilon) + \left|\int_{\mathbb{R}^1} \phi^\varepsilon\, dP_n - \int_{\mathbb{R}^1} \phi^\varepsilon\, dP\right| \to 2\delta(\varepsilon) \qquad \text{as } n \to \infty.$$

Since $\delta(\varepsilon) \to 0$ as $\varepsilon \to 0$ it follows that $\int_{\mathbb{R}^1} \phi\, dP_n \to \int_{\mathbb{R}^1} \phi\, dP$, as claimed.

Next let F_n, F be the distribution functions of P_n, P, respectively ($n = 1, 2, \ldots$). We want to show that if *$\{P_n: n = 1, 2, \ldots\}$ converges weakly to P, then $F_n(x) \to F(x)$ as $n \to \infty$ for all points of continuity x of F.* For this, fix a point of continuity x_0 of F. Given $\varepsilon > 0$ there exists $\eta(\varepsilon) > 0$ such that $|F(x) - F(x_0)| < \varepsilon$ for $|x - x_0| \leqslant \eta(\varepsilon)$. Let $\psi_\varepsilon^+(x) = 1$ for $x \leqslant x_0$, $= 0$ for $x \geqslant x_0 + \eta(\varepsilon)$, and $\psi_\varepsilon^+(x)$ be linearly interpolated for $x_0 < x < x_0 + \eta(\varepsilon)$. Similarly, define $\psi_\varepsilon^-(x) = 1$ for $x \leqslant x_0 - \eta(\varepsilon)$, $\psi_\varepsilon^-(x) = 0$ for $x \geqslant x_0$, and linearly interpolated in the interval $(x_0 - \eta(\varepsilon), x_0)$. Then, using (5.1),

$$\begin{aligned}
\overline{\lim_{n\to\infty}}\, F_n(x_0) &\leqslant \overline{\lim_{n\to\infty}} \int \psi_\varepsilon^+(x)\, dP_n(x) = \int \psi_\varepsilon^+(x)\, dP(x) \leqslant F(x_0 + \eta(\varepsilon)) < F(x_0) + \varepsilon,\\
\underline{\lim_{n\to\infty}}\, F_n(x_0) &\geqslant \underline{\lim_{n\to\infty}} \int \psi_\varepsilon^-(x)\, dP_n(x) = \int \psi_\varepsilon^-\, dP(x) \geqslant F(x_0 - \eta(\varepsilon)) > F(x_0) - \varepsilon.
\end{aligned} \tag{5.5}$$

Since $\varepsilon > 0$ is arbitrary, $\overline{\lim}_{n\to\infty} F_n(x_0) \leqslant F(x_0) \leqslant \underline{\lim}_{n\to\infty} F_n(x_0)$, showing that $F_n(x_0) \to F(x_0)$ as $n \to \infty$. One may show that the converse is also true. That is, *if $F_n(x) \to F(x)$ at all points of continuity of a distribution function* (d.f.) *F, then $\{P_n: n = 1, 2, \ldots\}$ converges weakly to P.* For this take a continuous function f that vanishes outside $[a, b]$ where a, b are points of continuity of F. One may divide $[a, b]$ into a finite number of small subintervals whose end points are all points of continuity of F, and approximate f by a step function constant over each subinterval. The integral of this step function with respect to P_n converges to that with respect to P. All the above results easily extend to multidimensions and we have the following theorem.

Theorem 5.1. Let $P_1, P_2, \ldots, P$ be probability measures on $\mathbb{R}^k$. The following are equivalent statements.

(a) $\{P_n: n = 1, 2, \ldots\}$ converge weakly to P.

(b) Eq. 5.1 holds for all infinitely differentiable functions vanishing outside a bounded set.

(c) $F_n(\mathbf{x}) \to F(\mathbf{x})$ as $n \to \infty$, for every point of continuity $\mathbf{x}$ of F.

Often the probability measures P_n ($n \geqslant 1$) arise as distributions of given random variables X_n ($n \geqslant 1$). In this situation, if $\{P_n\}$ converges weakly to P, one says $\{X_n\}$ *converges in distribution*, or *in law*, to P.

It is simple to check that *if $\{X_n\}$ converges in probability to a random variable X, then $\{X_n\}$ converges in distribution to the distribution of X.*

In general, if $\{F_n\}$ is a sequence of distribution functions of probability measures P_n that converges to a right-continuous, nondecreasing function F at all points of continuity

x of $F(x)$, it need not be the case that $F(x)$ is the distribution function of a *probability* measure. While there will be a positive measure μ such that $F(x) = \mu(-\infty, x]$, $x \in \mathbb{R}^1$, it can happen that $\mu(\mathbb{R}^1) < 1$; for example, if $P_n = \delta_n$, i.e., $P_n(\{n\}) = 1$,$F_n(x) = 0$, $x < n$ and $F_n(x) = 1$ for $x \geqslant n$, then $F \equiv 0$. Situations such as this are described by the expression, "probability mass has escaped to infinity." A property that prevents this from happening for a sequence $\{P_n\}$ is the so-called *tightness criterion. Namely, a family* $\{P_n\}$ *of probability measures on* $\mathbb{R}^1$ (*with* d.f.'s $\{F_n\}$), is said to be *tight* if for each $\varepsilon > 0$ there is an interval $(a, b]$ in $\mathbb{R}^1$ such that $P_n((a, b]) = F_n(b) - F_n(a) > 1 - \varepsilon$ for each $n \geqslant 1$.

It is clear that if a sequence of probability measures on $\mathbb{R}^1$ (or $\mathbb{R}^k$) converges weakly to a probability measure P, then $\{P_n\}$ is tight. Conversely, one has the following result.

Theorem 5.2. Suppose a sequence of probability measures $\{P_n\}$ on $\mathbb{R}^1$ is tight. Then it has a subsequence converging weakly to a probability measure P.

Proof. Let F_n be the d.f. of P_n. Let $\{r_1, r_2, \ldots\}$ be an enumeration of the rationals. Since $\{F_n(r_1)\}$ is bounded, there exists a subsequence $\{n_1\}$ (i.e., $\{1_1, 2_1, \ldots, n_1, \ldots\}$) of the positive integers such that $F_{n_1}(r_1)$ is convergent. Since $\{F_{n_1}(r_2)\}$ is bounded, there exists a subsequence $\{n_2\}$ of $\{n_1\}$ such that $F_{n_2}(r_2)$ is convergent. Then $\{F_{n_2}(r_i)\}$ $(i = 1, 2)$ converges. Continue in this manner. Consider the "diagonal" sequence $1_1, 2_2, \ldots, k_k, \ldots$. Then $\{F_{n_n}(r_i)\}$ converges, as $n \to \infty$, for every i, as $\{n_n : n \geqslant i\}$ is a subsequence of $\{n_i : n \geqslant 1\}$. Let us write $n_n = n'$. Define

$$\begin{aligned} G(r_i) &:= \lim F_{n'}(r_i) \qquad (i = 1, 2, \ldots), \\ F(x) &:= \inf\{G(r_i): r_i > x\} \qquad (x \in \mathbb{R}^1). \end{aligned} \tag{5.6}$$

Then F is a nondecreasing and right-continuous function on $\mathbb{R}^1$, $0 \leqslant F \leqslant 1$. Let x be a point of continuity of F. Let $x' < y' < x < y'' < x''$ with y', y'' rational. Then

$$F(x') \leqslant G(y') = \lim F_{n'}(y') \leqslant \underline{\lim}\, F_{n'}(x) \leqslant \overline{\lim}\, F_{n'}(x) \leqslant \lim F_{n'}(y'') = G(y'') \leqslant F(x'').$$

Now let $x' \uparrow x$, $x'' \downarrow x$, and use continuity of F at x to deduce

$$\lim_{n'} F_{n'}(x) = F(x). \tag{5.7}$$

It remains to prove that F is the d.f. of a probability measure. By tightness of $\{P_n\}$ (and, therefore, of $\{P_{n'}\}$), given $\varepsilon > 0$, there exist $x_\varepsilon, y_\varepsilon$ such that $F_{n'}(x_\varepsilon) < \varepsilon$, $F_{n'}(y_\varepsilon) > 1 - \varepsilon$. Let $x'_\varepsilon, y'_\varepsilon$ be points of continuity of F such that $x'_\varepsilon < x_\varepsilon$ and $y'_\varepsilon > y_\varepsilon$. Then

$$F(x'_\varepsilon) = \lim F_{n'}(x'_\varepsilon) \leqslant \varepsilon, \qquad F(y'_\varepsilon) = \lim F_{n'}(y'_\varepsilon) \geqslant 1 - \varepsilon.$$

This shows $\lim_{x \downarrow -\infty} F(x) = 0$, $\lim_{x \uparrow \infty} F(x) = 1$. ■

A similar proof applies to P_n, P on $\mathbb{R}^k$.

6 CLASSICAL LAWS OF LARGE NUMBERS

Theorem 6.1. (*Strong Law of Large Numbers*). Let $\{X_n\}$ be a sequence of pairwise independent and identically distributed random variables defined on a probability space

$(\Omega, \mathcal{F}, P)$. If $E|X_1| < \infty$ then with probability 1,

$$\lim_{n\to\infty} \frac{X_1 + \cdots + X_n}{n} = EX_1. \tag{6.1}$$

The proof we present here is due to Etemadi.[3] It is based on Part 1 of the following *Borel–Cantelli Lemmas*. Part 2 is also important and it is cited here for completeness. However, it is *not* used in Etemadi's proof of the SLLN.

Lemma 6.1. (*Borel–Cantelli*). Let $\{A_n\}$ be any sequence of events in $\mathcal{F}$.

Part 1. If $\sum_{n=1}^{\infty} P(A_n) < \infty$ then

$$P(A_n \text{ i.o.}) := P\left(\bigcap_{n=1}^{\infty} \bigcup_{k=n}^{\infty} A_k\right) = 0.$$

Part 2. If $A_1, A_2, \ldots$ are independent events and if $\sum_{n=1}^{\infty} P(A_n)$ diverges then $P(A_n \text{ i.o.}) = 1$.

Proof. For Part 1 observe that the sequence of events $B_n = \bigcup_{k=n}^{\infty} A_k$, $n = 1, 2, \ldots$, is a decreasing sequence. Therefore, we have by the continuity property (1.2),

$$P\left(\bigcap_{n=1}^{\infty} \bigcup_{k=n}^{\infty} A_k\right) = \lim_{n\to\infty} P\left(\bigcup_{k=n}^{\infty} A_k\right) \leqslant \lim_{n\to\infty} \sum_{k=n}^{\infty} P(A_k) = 0.$$

For Part 2 note that

$$P(\{A_n \text{ i.o.}\}^c) = P\left(\bigcup_{n=1}^{\infty} \bigcap_{k=n}^{\infty} A_k^c\right) = \lim_{n\to\infty} P\left(\bigcap_{k=n}^{\infty} A_k^c\right) = \lim_{n\to\infty} \prod_{k=n}^{\infty} P(A_k^c).$$

But

$$\prod_{k=n}^{\infty} P(A_k^c) = \lim_{m\to\infty} \prod_{k=n}^{m} (1 - P(A_k))$$

$$\leqslant \lim_{m\to\infty} \exp\left\{-\sum_{k=n}^{m} P(A_k)\right\} = 0. \quad \blacksquare$$

Without loss of generality we may assume for the proof of the SLLN that the random variables X_n are nonnegative, since otherwise we can write $X_n = X_n^+ - X_n^-$, where $X_n^+ = \max(X_n, 0)$ and $X_n^- = -\min(X_n, 0)$ are both nonnegative random variables, and then the result in the nonnegative case yields that

$$\frac{S_n}{n} = \frac{1}{n}\sum_{k=1}^{n} X_k^+ - \frac{1}{n}\sum_{k=1}^{n} X_k^-$$

converges to $EX_1^+ - EX_1^- = EX_1$ with probability 1.

[3] N. Etemadi (1983), 'On the Laws of Large Numbers for Nonnegative Random Variables," *J. Multivariate Analysis*, **13**, pp. 187–193.

Truncate the variables X_n by $Y_n = X_n \mathbf{1}_{\{X_n \leqslant n\}}$. Then Y_n has moments of all orders. Let $T_n = \sum_{k=1}^n Y_k$ and consider the sequence $\{T_n\}$ on the "fast" time scale $\tau_n = [\alpha^n]$, for a fixed $\alpha > 1$, where brackets [ ] denote the integer part. Let $\varepsilon > 0$. Then by Chebyshev's Inequality and pairwise independence,

$$P\left(\left|\frac{T_{\tau_n} - ET_{\tau_n}}{\tau_n}\right| > \varepsilon\right) \leqslant \frac{\operatorname{Var}(T_{\tau_n})}{\varepsilon^2\tau_n^2} = \frac{1}{\varepsilon^2\tau_n^2}\sum_{k=1}^{\tau_n} \operatorname{Var} Y_k \leqslant \frac{1}{\varepsilon^2\tau_n^2}\sum_{k=1}^{\tau_n} EY_k^2$$

$$= \frac{1}{\varepsilon^2\tau_n^2}\sum_{k=1}^{\tau_n} E\{X_k^2\mathbf{1}_{\{X_k \leqslant k\}}\} = \frac{1}{\varepsilon^2\tau_n^2}\sum_{k=1}^{\tau_n} E\{X_1^2\mathbf{1}_{\{X_1 \leqslant k\}}\}$$

$$\leqslant \frac{1}{\varepsilon^2\tau_n^2}\sum_{k=1}^{\tau_n} E\{X_1^2\mathbf{1}_{\{X_1 \leqslant \tau_n\}}\} = \frac{1}{\varepsilon^2\tau_n^2}\tau_n E\{X_1^2\mathbf{1}_{\{X_1 \leqslant \tau_n\}}\}. \quad (6.2)$$

Therefore,

$$\sum_{n=1}^{\infty} P\left(\left|\frac{T_{\tau_n} - ET_{\tau_n}}{\tau_n}\right| > \varepsilon\right) \leqslant \sum_{n=1}^{\infty}\frac{1}{\varepsilon^2\tau_n} E\{X_1^2\mathbf{1}_{\{X_1 \leqslant \tau_n\}}\} = \frac{1}{\varepsilon^2} E\left\{X_1^2 \sum_{n=1}^{\infty}\frac{1}{\tau_n}\mathbf{1}_{\{X_1 \leqslant \tau_n\}}\right\}. \quad (6.3)$$

Let $x > 0$ and let $N = \min\{n \geqslant 1 : \tau_n \geqslant x\}$. Then $\alpha^N \geqslant x$ and, since $y \leqslant 2[y]$ for any $y \geqslant 1$,

$$\sum_{n=1}^{\infty}\frac{1}{\tau_n}\mathbf{1}_{\{x \leqslant \tau_n\}} = \sum_{\tau_n \geqslant x}\frac{1}{\tau_n} \leqslant 2\sum_{n \geqslant N}\alpha^{-n} = \frac{2}{a-1}\alpha^{-N} = k\alpha^{-N} \leqslant \frac{k}{x},$$

where $k = 2/(\alpha - 1)$. Therefore,

$$\sum_{n=1}^{\infty}\frac{1}{\tau_n}\mathbf{1}_{\{X_1 \leqslant \tau_n\}} \leqslant \frac{k}{X_1} \qquad \text{for } X_1 > 0.$$

So

$$\sum_{n=1}^{\infty} P\left(\left|\frac{T_{\tau_n} - ET_{\tau_n}}{\tau_n}\right| > \varepsilon\right) \leqslant k\frac{E[X_1]}{\varepsilon^2} < \infty. \quad (6.4)$$

By the Borel–Cantelli Lemma (Part 1), taking a union over positive rational values of ε, with probability 1, $(T_{\tau_n} - ET_{\tau_n})/\tau_n \to 0$ as $n \to \infty$. Therefore,

$$\frac{T_{\tau_n}}{\tau_n} \to EX_1 \quad (6.5)$$

since

$$\lim_{n\to\infty}\frac{1}{\tau_n}ET_{\tau_n} = \lim_{n\to\infty} EY_{\tau_n} = EX_1.$$

Since

$$\sum_{n=1}^{\infty} P(X_n \neq Y_n) = \sum_{n=1}^{\infty} P(X_1 > n) \leqslant \int_0^{\infty} P(X_1 > u)\,du = EX_1 < \infty$$

we get by another application of the Borel–Cantelli Lemma that, with probability 1,

$$\frac{S_n - T_n}{n} \to 0 \qquad \text{as } n \to \infty. \tag{6.6}$$

Therefore, the previous results about $\{T_n\}$ give for $\{S_n\}$ that

$$\frac{S_{\tau_n}}{\tau_n} \to EX_1 \qquad \text{as } n \to \infty \tag{6.7}$$

with probability 1. If $\tau_n \leqslant k \leqslant \tau_{n+1}$ then, since $X_i \geqslant 0$,

$$\frac{\tau_n}{\tau_{n+1}} \frac{S_{\tau_n}}{\tau_n} \leqslant \frac{S_k}{k} \leqslant \frac{\tau_{n+1}}{\tau_n} \frac{S_{\tau_{n+1}}}{\tau_{n+1}}. \tag{6.8}$$

But $\tau_{n+1}/\tau_n \to \alpha$, so that now we get with probability 1,

$$\frac{1}{\alpha} EX_1 \leqslant \liminf_k \frac{S_k}{k} \leqslant \limsup_k \frac{S_k}{k} \leqslant \alpha EX_1. \tag{6.9}$$

Take the intersection of all such events for rational $\alpha > 1$ to get $\lim_{k\to\infty} S_k/k = EX_1$ with probability 1. This is the *Strong Law of Large Numbers* (SLLN). ■

The above proof of the SLLN is really quite remarkable, as the following observations show. First, *pairwise independence* is only used to make sure that the positive and negative parts of X_n, and their truncations, remain pairwise *uncorrelated* for the calculation of the variance of T_k as the sum of the variances. Observe that if the random variables are all of the same sign, say nonnegative, and bounded, then it suffices to require that they merely be *uncorrelated* for the same proof to go through. However, this means that, if the random variables are bounded, then it suffices that they be uncorrelated to get the SLLN; for one may simply add a sufficiently large constant to make them all positive. Thus, we have the following.

Corollary 6.2. If $X_1, X_2, \ldots,$ is a sequence of mean zero uncorrelated random variables that are uniformly bounded, then with probability 1,

$$\frac{X_1 + \cdots + X_n}{n} \to 0 \qquad \text{as } n \to \infty.$$

7 CLASSICAL CENTRAL LIMIT THEOREMS

In view of the great importance of the *central limit theorem* (CLT) we shall give a general but self-contained version due to Lindeberg. This version is applicable to non-identically distributed summands.

Theorem 7.1. (*Lindeberg's CLT*). For each n, let $X_{1,n}, \ldots, X_{k_n,n}$ be independent random variables satisfying

$$EX_{j,n} = 0, \qquad \sigma_{j,n} := (EX_{j,n}^2)^{1/2} < \infty, \qquad \sum_{j=1}^{k_n} \sigma_{j,n}^2 = 1, \tag{7.1}$$

and, for each $\varepsilon > 0$,

$$\text{(Lindeberg condition)} \qquad \lim_{n\to\infty} \sum_{j=1}^{k_n} E(X_{j,n}^2 \mathbf{1}_{\{|X_{j,n}| > \varepsilon\}}) = 0. \tag{7.2}$$

Then $\sum_{j=1}^{k_n} X_{j,n}$ converges in distribution to the standard normal law $N(0,1)$.

Proof. Let $\{Z_j : j \geqslant 1\}$ be a sequence of i.i.d. $N(0,1)$ random variables, independent of $\{X_{j,n} : 1 \leqslant j \leqslant k_n\}$. Write

$$Z_{j,n} := \sigma_{j,n} Z_j \qquad (1 \leqslant j \leqslant k_n), \tag{7.3}$$

so that $EZ_{j,n} = 0 = EX_{j,n}$, $EZ_{j,n}^2 = \sigma_{j,n}^2 = EX_{j,n}^2$. Define

$$\begin{aligned} U_{m,n} &:= \sum_{j=1}^{m} X_{j,n} + \sum_{j=m+1}^{k_n} Z_{j,n} \qquad (1 \leqslant m \leqslant k_n - 1), \\ U_{0,n} &:= \sum_{j=1}^{k_n} Z_{j,n}, \qquad U_{k_n,n} := \sum_{j=1}^{k_n} X_{j,n}, \\ V_{m,n} &:= U_{m,n} - X_{m,n} \qquad (1 \leqslant m \leqslant k_n). \end{aligned} \tag{7.4}$$

Let f be a real-valued function on $\mathbb{R}^1$ such that f, f', f'', f''' are bounded. Recall the following version of the Taylor expansion, which is easy to check by integration by parts,

$$f(x+h) = f(x) + hf'(x) + \frac{h^2}{2!} f''(x) + h^2 \int_0^1 (1-\theta)\{f''(x+\theta h) - f''(x)\}\, d\theta \qquad (x, h \in \mathbb{R}^1). \tag{7.5}$$

Taking $x = V_{m,n}$, $h = X_{m,n}$ in (7.5), one gets

$$Ef(U_{m,n}) \equiv Ef(V_{m,n} + X_{m,n}) = Ef(V_{m,n}) + E(X_{m,n} f'(V_{m,n})) + \tfrac{1}{2} E(X_{m,n}^2 f''(V_{m,n})) + E(R_{m,n}), \tag{7.6}$$

where

$$R_{m,n} := X_{m,n}^2 \int_0^1 (1-\theta)\{f''(V_{m,n} + \theta X_{m,n}) - f''(V_{m,n})\}\, d\theta. \tag{7.7}$$

As $X_{m,n}$ and $V_{m,n}$ are independent, and $EX_{m,n} = 0$, $EX_{m,n}^2 = \sigma_{m,n}^2$, (7.6) reduces to

$$Ef(U_{m,n}) = Ef(V_{m,n}) + \frac{\sigma_{m,n}^2}{2} Ef''(V_{m,n}) + E(R_{m,n}). \tag{7.8}$$

Also $U_{m-1,n} = V_{m,n} + Z_{m,n}$, and $V_{m,n}$ and $Z_{m,n}$ are independent. Therefore, exactly as above one gets, using $EZ_{m,n} = 0$, $EZ^2_{m,n} = \sigma^2_{m,n}$,

$$Ef(U_{m-1,n}) = Ef(V_{m,n}) + \frac{\sigma^2_{m,n}}{2} Ef''(V_{m,n}) + E(R'_{m,n}), \tag{7.9}$$

where

$$R'_{m,n} := Z^2_{m,n} \int_0^1 (1-\theta)\{f''(V_{m,n} + \theta Z_{m,n}) - f''(V_{m,n})\}\, d\theta. \tag{7.10}$$

Hence,

$$|Ef(U_{m,n}) - Ef(U_{m-1,n})| \leqslant E|R_{m,n}| + E|R'_{m,n}| \qquad (1 \leqslant m \leqslant k_n). \tag{7.11}$$

Now, given an arbitrary $\varepsilon > 0$,

$$\begin{aligned} E|R_{m,n}| &= E(|R_{m,n}| \mathbf{1}_{\{|X_{m,n}| > \varepsilon\}}) + E(|R_{m,n}| \mathbf{1}_{\{|X_{m,n}| \leqslant \varepsilon\}}) \\ &\leqslant E\left[X^2_{m,n} \mathbf{1}_{\{|X_{m,n}| > \varepsilon\}} \int_0^1 (1-\theta) 2 \|f''\|_\infty \, d\theta \right] \\ &\quad + E\left[X^2_{m,n} \mathbf{1}_{\{|X_{m,n}| \leqslant \varepsilon\}} \int_0^1 (1-\theta)|X_{m,n}| \, \|f'''\|_\infty \, d\theta \right] \\ &\leqslant \|f''\|_\infty E(X^2_{m,n} \mathbf{1}_{\{|X_{m,n}| > \varepsilon\}}) + \tfrac{1}{2}\varepsilon \sigma^2_{m,n} \|f'''\|_\infty. \end{aligned} \tag{7.12}$$

We have used the notation $\|g\|_\infty := \sup\{|g(x)| : x \in \mathbb{R}^1\}$. By (7.1), (7.2), and (7.12),

$$\overline{\lim} \sum_{m=1}^{k_n} E|R_{m,n}| \leqslant \tfrac{1}{2}\varepsilon \|f'''\|_\infty.$$

As $\varepsilon > 0$ is arbitrary,

$$\lim \sum_{m=1}^{k_n} E|R_{m,n}| = 0. \tag{7.13}$$

Also,

$$\begin{aligned} E|R'_{m,n}| &\leqslant E\left[Z^2_{m,n} \int_0^1 (1-\theta) \|f'''\|_\infty |Z_{m,n}| \, d\theta \right] = \tfrac{1}{2} \|f'''\|_\infty E|Z_{m,n}|^3 = \tfrac{1}{2} \|f'''\|_\infty \sigma^3_{m,n} E|Z_1|^3 \\ &\leqslant c\sigma^3_{m,n} \leqslant c\left(\max_{1 \leqslant m \leqslant k_n} \sigma_{m,n} \right) \sigma^2_{m,n}, \end{aligned} \tag{7.14}$$

where $c = \frac{1}{2}\|f''\|_\infty E|Z_1|^3$. Now, for each $\delta > 0$,

$$\sigma^2_{m,n} = E(X^2_{m,n} \mathbf{1}_{\{|X_{m,n}| > \delta\}}) + E(X^2_{m,n} \mathbf{1}_{\{|X_{m,n}| \leqslant \delta\}}) \leqslant E(X^2_{m,n} \mathbf{1}_{\{|X_{m,n}| > \delta\}}) + \delta^2,$$

which implies that

$$\max_{1 \leqslant m \leqslant k_n} \sigma^2_{m,n} \leqslant \sum_{m=1}^{k_n} E(X^2_{m,n} \mathbf{1}_{\{|X_{m,n}| > \delta\}}) + \delta^2.$$

Therefore, by (7.2),

$$\max_{1 \leqslant m \leqslant k_n} \sigma_{m,n} \to 0 \qquad \text{as } n \to \infty. \tag{7.15}$$

From (7.14) and (7.15) one gets

$$\sum_{m=1}^{k_n} E|R'_{m,n}| \leqslant c\left(\max_{1 \leqslant m \leqslant k_n} \sigma_{m,n}\right) \to 0 \qquad \text{as } n \to \infty. \tag{7.16}$$

Combining (7.13) and (7.16), one finally gets

$$|Ef(U_{k_n,n}) - Ef(U_{0,n})| \leqslant \sum_{m=1}^{k_n} (E|R_{m,n}| + E|R'_{m,n}|) \to 0 \qquad \text{as } n \to \infty. \tag{7.17}$$

But $U_{0,n}$ is a standard normal random variable. Hence,

$$Ef\left(\sum_{j=1}^{k_n} X_{j,n}\right) - \int_{\mathbb{R}^1} f(y)(2\pi)^{-1/2} \exp\{-\tfrac{1}{2}y^2\}\, dy \to 0 \qquad \text{as } n \to \infty.$$

By Theorem 5.1, the proof is complete. ∎

It has been shown by Feller[4] that in the presence of the *uniform asymptotic negligibility* condition (7.15), the Lindeberg condition is also *necessary* for the CLT to hold.

Corollary 7.2. (*The Classical CLT*). Let $\{X_j: j \geqslant 1\}$ be i.i.d. $EX_j = \mu$, $0 < \sigma^2 :=$ Var $X_j < \infty$. Then $\sum_{j=1}^{n} (X_j - \mu)/(\sigma\sqrt{n})$ converges in distribution to $N(0, 1)$.

Proof. Let $X_{j,n} = (X_j - \mu)/(\sigma\sqrt{n})$, $k_n = n$, and apply Theorem 7.1. ∎

Corollary 7.3. (*Liapounov's CLT*). For each n let $X_{1,n}, X_{2,n}, \ldots, X_{n,k_n}$ be k_n independent random variables such that

$$\sum_{j=1}^{k_n} EX_{j,n} = \mu, \qquad \sum_{j=1}^{k_n} \operatorname{Var} X_{j,n} = \sigma^2 > 0,$$

(Liapounov condition) $$\lim_{n \to \infty} \sum_{j=1}^{k_n} E|X_{j,n} - EX_{j,n}|^{2+\delta} = 0 \tag{7.18}$$

for some $\delta > 0$. Then $\sum_{j=1}^{k_n} X_{j,n}$ converges in distribution to the Gaussian law with mean μ and variance σ^2.

Proof. By normalizing one may assume, without loss of generality, that

$$EX_{j,n} = 0, \qquad \sum_{j=1}^{k_n} EX_{j,n}^2 = 1.$$

[4] P. Billingsley (1986), *loc. cit.*, p. 373.

It then remains to show that the hypothesis of the corollary implies the Lindeberg condition (7.2). This is true since, for every $\varepsilon > 0$,

$$\sum_{j=1}^{k_n} E(X_{j,n}^2 \mathbf{1}_{\{|X_{j,n}| > \varepsilon\}}) \leqslant \sum_{j=1}^{k_n} E \frac{|X_{j,n}|^{2+\delta}}{\varepsilon^\delta} \to 0, \tag{7.19}$$

as $n \to \infty$, by (7.18). ■

Observe that the most crucial property of the normal distribution used in the proof of Theorem 7.1 is that the sum of independent normal random variables is normal. In other words, the normal distribution is *infinitely divisible*. In fact, the normal distribution $N(0, 1)$ may be realized as the distribution of the sum of independent normal random variables having zero means and variances σ_i^2 for any arbitrarily specified set of nonnegative numbers σ_i^2 adding up to 1. Another well-known infinitely divisible distribution is the *Poisson distribution*.

The following multidimensional version of Corollary 7.2 may be proved along the lines of the proof of Theorem 7.1.

Theorem 7.4. (*Multivariate Classical CLT*). Let $\{\mathbf{X}_n: n = 1, 2, \ldots\}$ be a sequence of i.i.d. random vectors with values in $\mathbb{R}^k$. Let $E\mathbf{X}_1 = \mu$ and assume that the dispersion matrix (i.e., variance–covariance matrix) $\mathbf{D}$ of $\mathbf{X}_1$ is nonsingular. Then as $n \to \infty$, $n^{-1/2}(\mathbf{X}_1 + \cdots + \mathbf{X}_n - n\boldsymbol{\mu})$ converges in distribution to the Gaussian probability measure with mean zero and dispersion matrix $\mathbf{D}$.

8 FOURIER SERIES AND THE FOURIER TRANSFORM

Consider a real- or complex-valued periodic function on the real line. By changing the scale, if necessary, one may take the period to be 2π. Is it possible to represent f as a superposition of the periodic functions ("waves") $\cos nx, \sin nx$ of *frequency* n $(n = 0, 1, 2, \ldots)$? The *Weierstrass approximation theorem* (Theorem 8.1) says that every continuous periodic function f of period 2π is the limit (in the sense of *uniform convergence* of functions) of a sequence of trigonometric polynomials, i.e., functions of the form

$$\sum_{n=-T}^{T} c_n e^{inx} = c_0 + \sum_{n=1}^{T} (a_n \cos nx + b_n \sin nx).$$

The theory of Fourier series says, among other things, that with the weaker notion of L^2-convergence the approximation holds for a wider class of functions, namely for all square integrable functions f on $[-\pi, \pi]$; here square integrability means that $|f|^2$ is measurable and that $\int_{-\pi}^{\pi} |f(x)|^2\, dx < \infty$. This class of functions is denoted by $L^2[-\pi, \pi]$. The successive coefficients c_n for this approximation are the so-called *Fourier coefficients*:

$$c_n = \frac{1}{2\pi} \int_{-\pi}^{\pi} f(x) e^{-inx}\, dx \qquad (n = 0, \pm 1, \pm 2, \ldots). \tag{8.1}$$

The functions $\exp\{inx\}$ $(n = 0, \pm 1, \pm 2, \ldots)$ form an *orthonormal set*:

$$\frac{1}{2\pi}\int_{-\pi}^{\pi} e^{inx}e^{-imx}\,dx = 0 \qquad \text{for } n \neq m,$$
$$= 1 \qquad \text{for } n = m, \tag{8.2}$$

so that the *Fourier series of f* written formally, without regard to convergence for the time being, as

$$\sum_{n=-\infty}^{\infty} c_n e^{inx} \tag{8.3}$$

is a representation of f as a superposition of orthogonal components. To make matters precise we first prove the following theorem.

Theorem 8.1. Let f be a continuous periodic function of period 2π. Then, given $\delta > 0$, there exists a trigonometric polynomial $\sum_{n=-N}^{N} d_n \exp\{inx\}$ such that

$$\sup_{x\in\mathbb{R}_1}\left| f(x) - \sum_{n=-N}^{N} d_n \exp\{inx\}\right| < \delta.$$

Proof. For each positive integer N, introduce the *Féjer kernel*

$$k_N(x) := \sum_{n=-N}^{N}\left(1 - \frac{|n|}{N+1}\right)\exp\{inx\}. \tag{8.4}$$

This may also be expressed as

$$(N+1)k_N(x) = \sum_{0\leq j,k\leq N} \exp\{i(j-k)x\} = \left|\sum_{j=0}^{N}\exp\{ijx\}\right|^2 = \left|\frac{\exp\{i(N+1)x\} - 1}{\exp\{ix\} - 1}\right|^2$$
$$= \frac{2\{1 - (\cos(N+1)x\}}{2(1 - \cos x)} = \left(\frac{\sin\{\frac{1}{2}(N+1)x\}}{\sin\frac{1}{2}x}\right)^2. \tag{8.5}$$

The first equality in (8.5) follows from the fact that there are $N + 1 - |n|$ pairs (j, k) such that $j - k = n$. It follows from (8.5) that k_N is a positive continuous periodic function with period 2π. Also, k_N is a p.d.f. on $[-\pi, \pi]$, as follows from (8.4) on integration. For every $\varepsilon > 0$ it follows from (8.5) that $k_N(x)$ goes to zero uniformly on $[-\pi, -\varepsilon] \cup [\varepsilon, \pi]$ so that

$$\int_{[-\pi,-\varepsilon]\cup[\varepsilon,\pi]} k_N(x)\,dx \to 0 \qquad \text{as } N \to \infty. \tag{8.6}$$

In other words, $k_N(x)\,dx$ converges weakly to $\delta_0(dx)$, the point mass at 0, as $N \to \infty$.

Consider now the approximation f_N of f defined by

$$f_N(x) := \int_{-\pi}^{\pi} f(y)k_N(x-y)\,dy = \sum_{n=-N}^{N}\left(1 - \frac{|n|}{N+1}\right)c_n \exp\{inx\}, \tag{8.7}$$

where c_n is the nth Fourier coefficient of f. By changing variables and using the periodicity of f and k_N, one may express f_N as

$$f_N(x) = \int_{-\pi}^{\pi} f(x - y)k_N(y)\, dy.$$

Therefore, writing $M = \sup\{|f(x)|: x \in \mathbb{R}^k\}$, and $\delta_\varepsilon = \sup\{|f(y) - f(y')|: |y - y'| < \varepsilon\}$, one has

$$|f(x) - f_N(x)| \leqslant \int_{-\pi}^{\pi} |f(x - y) - f(x)|k_N(y)\, dy \leqslant 2M \int_{[-\pi,-\varepsilon]\cup[\varepsilon,\pi]} k_N(y)\, dy + \delta_\varepsilon. \tag{8.8}$$

It now follows from (8.6) that $f - f_N$ converges to zero uniformly as $N \to \infty$. Now write $d_n = (1 - |n|/(N + 1))c_n$. ■

The next task is to establish the convergence of the Fourier series (8.3) to f in L^2. For this note that for every square integrable f and all positive integers N,

$$\frac{1}{2\pi}\int_{-\pi}^{\pi}\left(f(x) - \sum_{-N}^{N} c_n e^{inx}\right)e^{-imx}\, dx = c_m - c_m = 0 \qquad (m = 0, \pm 1, \ldots, \pm N). \tag{8.9}$$

Therefore, if one defines the *norm* (or "length") of a function g in $L^2[-\pi, \pi]$ by

$$\|g\| = \left(\frac{1}{2\pi}\int_{-\pi}^{\pi} |g(x)|^2\, dx\right)^{1/2} \equiv \|g\|_2, \tag{8.10}$$

then, writing $\bar{z}$ for the complex conjugate of z,

$$\begin{aligned} 0 \leqslant \left\| f(x) - \sum_{-N}^{N} c_n e^{inx}\right\|^2 &= \frac{1}{2\pi}\int_{-\pi}^{\pi}\left(f(x) - \sum_{-N}^{N} c_n e^{inx}\right)\left(\bar{f}(x) - \sum_{-N}^{N} \bar{c}_n e^{-inx}\right) dx \\ &= \frac{1}{2\pi}\int_{-\pi}^{\pi} |f(x)|^2\, dx - \sum_{-N}^{N}\left(\frac{c_n}{2\pi}\int_{-\pi}^{\pi} e^{inx}\bar{f}(x)\, dx + \frac{\bar{c}_n}{2\pi}\int_{-\pi}^{\pi} e^{-inx} f(x)\, dx - c_n\bar{c}_n\right) \\ &= \|f\|^2 - \sum_{-N}^{N} c_n\bar{c}_n = \|f\|^2 - \sum_{-N}^{N} |c_n|^2. \end{aligned} \tag{8.11}$$

This shows that $\|f(x) - \sum_{-N}^{N} c_n \exp\{inx\}\|^2$ *decreases* as N increases and that

$$\lim_{N\to\infty}\left\| f(x) - \sum_{-N}^{N} c_n e^{inx}\right\|^2 = \|f\|^2 - \sum_{-\infty}^{\infty} |c_n|^2. \tag{8.12}$$

To prove that the right side of (8.12) vanishes, first assume that f is continuous and $f(-\pi) = f(\pi)$. Given $\varepsilon > 0$ there exists, by Theorem 8.1, a trigonometric polynomial $\sum_{-N_0}^{N_0} d_n e^{inx}$ such that

$$\max_x \left| f(x) - \sum_{-N_0}^{N_0} d_n e^{inx}\right| < \varepsilon.$$

This implies

$$\frac{1}{2\pi}\int_{-\pi}^{\pi}\left|f(x)-\sum_{-N_0}^{N_0} d_n e^{inx}\right|^2 dx < \varepsilon^2. \tag{8.13}$$

But, by (8.9), $f(x)-\sum_{-N_0}^{N_0} c_n \exp\{inx\}$ is orthogonal to $\exp\{imx\}$ $(m = 0, \pm 1, \ldots, \pm N_0)$ so that

$$\begin{aligned}\frac{1}{2\pi}\int_{-\pi}^{\pi}\left|f(x)-\sum_{-N_0}^{N_0} d_n e^{inx}\right|^2 dx &= \frac{1}{2\pi}\int_{-\pi}^{\pi}\left|f(x)-\sum_{-N_0}^{N_0} c_n e^{inx}+\sum_{-N_0}^{N_0}(c_n-d_n)e^{inx}\right|^2 dx\\ &= \frac{1}{2\pi}\int_{-\pi}^{\pi}\left|f(x)-\sum_{-N_0}^{N_0} c_n e^{inx}\right|^2 dx\\ &\quad+\frac{1}{2\pi}\int_{-\pi}^{\pi}\left|\sum_{-N_0}^{N_0}(c_n-d_n)e^{inx}\right|^2 dx.\end{aligned} \tag{8.14}$$

Hence, by (8.13) and (8.14),

$$\frac{1}{2\pi}\int_{-\pi}^{\pi}\left|f(x)-\sum_{-N_0}^{N_0} c_n e^{inx}\right|^2 dx < \varepsilon^2, \qquad \lim_{N\to\infty}\left\|f(x)-\sum_{-N}^{N} c_n e^{inx}\right\|^2 \leqslant \varepsilon^2. \tag{8.15}$$

Since $\varepsilon > 0$ is arbitrary, it follows that

$$\lim_{N\to\infty}\left\|f(x)-\sum_{-N}^{N} c_n e^{inx}\right\| = 0, \tag{8.16}$$

and, by (8.12),

$$\|f\|^2 = \sum_{-\infty}^{\infty} |c_n|^2. \tag{8.17}$$

This completes the proof of convergence for continuous periodic f. Now it may be shown that given a square integrable f and $\varepsilon > 0$, there exists a continuous periodic g such that $\|f - g\| < \varepsilon/2$. Also, letting $\sum d_n \exp\{inx\}$, $\sum c_n \exp\{inx\}$ be the Fourier series of g, f, respectively, there exists N_1 such that

$$\left\|g(x)-\sum_{-N_1}^{N_1} d_n \exp\{inx\}\right\| < \frac{\varepsilon}{2}.$$

Hence (see (8.14))

$$\begin{aligned}\left\|f(x)-\sum_{-N_1}^{N_1} c_n e^{inx}\right\| &\leqslant \left\|f(x)-\sum_{-N_1}^{N_1} d_n e^{inx}\right\| \leqslant \|f-g\| + \left\|g(x)-\sum_{-N_1}^{N_1} d_n e^{inx}\right\|\\ &< \frac{\varepsilon}{2}+\frac{\varepsilon}{2} = \varepsilon.\end{aligned} \tag{8.18}$$

Since $\varepsilon > 0$ is arbitrary and $\|f(x) - \sum_{-N}^{N} c_n e^{inx}\|^2$ decreases to $\|f\|^2 - \sum_{-\infty}^{\infty} |c_n|^2$ as $N \uparrow \infty$ (see Eq. 8.12), one has

$$\lim_{N\to\infty} \left\| f(x) - \sum_{-N}^{N} c_n e^{inx} \right\| = 0, \qquad \|f\|^2 = \sum_{-\infty}^{\infty} |c_n|^2. \tag{8.19}$$

Thus, we have proved the first part of the following theorem.

Theorem 8.2

(a) For every f in $L^2[-\pi, \pi]$, the Fourier series of f converges to f in L^2-norm, and the identity $\|f\| = (\sum_{-\infty}^{\infty} |c_n|^2)^{1/2}$ holds for its Fourier coefficients c_n.

(b) If (i) f is differentiable, (ii) $f(-\pi) = f(\pi)$, and (iii) f' is square integrable, then the Fourier series of f also converges uniformly to f on $[-\pi, \pi]$.

Proof. To prove part (b), let f be as specified. Let $\sum c_n \exp\{inx\}$ be the Fourier series of f, and $\sum c_n^{(1)} \exp\{inx\}$ that of f'. Then

$$c_n^{(1)} = \frac{1}{2\pi} \int_{-\pi}^{\pi} f'(x) e^{-inx}\, dx = \frac{1}{2\pi} f(x) e^{-inx} \Big|_{-\pi}^{\pi} + \frac{in}{2\pi} \int_{-\pi}^{\pi} f(x) e^{-inx}\, dx = 0 + inc_n = inc_n. \tag{8.20}$$

Since f' is square integrable,

$$\sum_{-\infty}^{\infty} |nc_n|^2 = \sum_{-\infty}^{\infty} |c_n^{(1)}|^2 < \infty. \tag{8.21}$$

Therefore, by the Cauchy–Schwarz Inequality,

$$\sum_{-\infty}^{\infty} |c_n| = |c_0| + \sum_{n\neq 0} \frac{1}{|n|} |nc_n| \leqslant |c_0| + \left(\sum_{n\neq 0} \frac{1}{n^2} \right)^{1/2} \left(\sum_{n\neq 0} |nc_n|^2 \right)^{1/2} < \infty. \tag{8.22}$$

But this means that $\sum c_n \exp\{inx\}$ is uniformly absolutely convergent, since

$$\max_x \left| \sum_{|n|>N} c_n \exp\{inx\} \right| \leqslant \sum_{|n|>N} |c_n| \to 0 \qquad \text{as } N \to \infty.$$

Since the continuous functions $\sum_{-N}^{N} c_n \exp\{inx\}$ converge uniformly (as $N \to \infty$) to $\sum_{-\infty}^{\infty} c_n \exp\{inx\}$, the latter must be a continuous function, say h. Uniform convergence to h also implies convergence in norm to h. Since $\sum_{-N}^{N} c_n \exp\{inx\}$ also converges in norm to f, $f(x) = h(x)$ for all x. For if the two continuous functions f and h are not identically equal, then

$$\int_{-\pi}^{\pi} |f(x) - g(x)|^2\, dx > 0. \qquad \blacksquare$$

For a finite measure (or a finite signed measure) μ on the circle $[-\pi, \pi)$ (identifying

$-\pi$ and π), the nth *Fourier coefficient of* μ is defined by

$$c_n = \frac{1}{2\pi}\int_{[-\pi,\pi)} \exp\{-inx\}\mu(dx) \qquad (n = 0, \pm 1, \ldots). \tag{8.23}$$

If μ has a density f, then (8.23) is the same as the nth Fourier coefficient of f given by (8.1).

Proposition 8.3. A finite measure μ on the circle is determined by its Fourier coefficients.

Proof. Approximate the measure $\mu(dx)$ by $g_N(x)\,dx$, where

$$g_N(x) := \int_{-\pi}^{\pi} k_N(x-y)\mu(dy) = \sum_{-N}^{N}\left(1 - \frac{|n|}{N+1}\right)c_n \exp\{inx\}, \tag{8.24}$$

with c_n defined by (8.23). For every continuous periodic function h (i.e., for every continuous function on the circle),

$$\int_{[-\pi,\pi)} h(x)g_N(x)\,dx = \int_{[-\pi,\pi)}\left(\int_{[-\pi,\pi)} h(x)k_N(x-y)\,dx\right)\mu(dy). \tag{8.25}$$

As $N \to \infty$, the probability measure $k_N(x-y)\,dx = k_N(y-x)\,dx$ on the circle converges weakly to $\delta_y(dx)$. Hence, the inner integral on the right side of (8.25) converges to $h(y)$. Since the inner integral is bounded by $\sup\{|h(y)|: y \in \mathbb{R}^1\}$, Lebesgue's Dominated Convergence Theorem implies that

$$\lim_{N\to\infty}\int_{[-\pi,\pi)} h(x)g_N(x)\,dx = \int_{[-\pi,\pi)} h(y)\mu(dy). \tag{8.26}$$

This means that μ is determined by $\{g_N: N \geq 1\}$ The latter in turn are determined by $\{c_n\}$. ■

We are now ready to answer an important question: When is a given sequence $\{c_n: n = 0, \pm 1, \ldots\}$, the sequence of Fourier coefficients of a finite measure on the circle?

A sequence of complex numbers $\{c_n: n = 0, \pm 1, \pm 2, \ldots\}$ is said to be *positive definite* if for any finite sequence of complex numbers $\{z_j: 1 \leq j \leq N\}$, one has

$$\sum_{1\leq j,k\leq N} c_{j-k}z_j\bar{z}_k \geq 0. \tag{8.27}$$

Theorem 8.4. (*Herglotz's Theorem*). $\{c_n: n = 0, \pm 1, \ldots\}$ is the sequence of Fourier coefficients of a probability measure on the circle if and only if it is positive definite, and $c_0 = 1$.

Proof. (*Necessity*). If μ is a probability measure on the circle, and $\{z_j: 1 \leq j \leq N\}$ a given finite sequence of complex numbers, then

$$\sum_{1 \leqslant j,k \leqslant N} c_{j-k} z_j \bar{z}_k = \frac{1}{2\pi} \sum_{1 \leqslant j,k \leqslant N} z_j \bar{z}_k \int_{[-\pi,\pi)} \exp\{-i(j-k)x\} \mu(dx)$$

$$= \frac{1}{2\pi} \int \left(\sum_1^N z_j \exp\{-ijx\} \right) \left(\sum_1^N \bar{z}_k \exp\{ikx\} \right) \mu(dx)$$

$$= \frac{1}{2\pi} \int \left| \sum_1^N z_j \exp\{-ijk\} \right|^2 \mu(dx) \geqslant 0. \tag{8.28}$$

Also,

$$c_0 = \int_{[-\pi,\pi)} \mu(dx) = 1.$$

(*Sufficiency*). Take $z_j = \exp\{i(j-1)x\}$, $j = 1, 2, \ldots, N+1$ in (8.27) to get

$$g_N(x) := \frac{1}{N+1} \sum_{0 \leqslant j,k \leqslant N} c_{j-k} \exp\{i(j-k)x\} \geqslant 0. \tag{8.29}$$

Again, as there are $N + 1 - |n|$ pairs (j, k) such that $j - k = n$ $(-N \leqslant n \leqslant N)$, (8.29) becomes

$$0 \leqslant g_N(x) = \sum_{-N}^{N} \left(1 - \frac{|n|}{N+1} \right) \exp\{inx\} c_n. \tag{8.30}$$

In particular,

$$\frac{1}{2\pi} \int_{[-\pi,\pi)} g_N(x)\, dx = c_0 = 1. \tag{8.31}$$

Hence $(1/2\pi) g_N$ is a p.d.f. on $[-\pi, \pi]$. By Theorem 5.2, there exists a subsequence $\{g_{N'}\}$ such that $(1/2\pi) g_{N'}(x)\, dx$ converges weakly to a probability measure $\mu(dx)$ as $N' \to \infty$. Also, integration yields

$$\frac{1}{2\pi} \int_{[-\pi,\pi]} \exp\{-inx\} g_N(x)\, dx = \left(1 - \frac{|n|}{N+1} \right) c_n \qquad (n = 0, \pm 1, \ldots, \pm N). \tag{8.32}$$

For each fixed n, take $N = N'$ in (8.32) and let $N' \to \infty$. Then

$$c_n = \lim_{N' \to \infty} \left(1 - \frac{|n|}{N'+1} \right) c_n = \int_{[-\pi,\pi]} \exp\{-inx\} \mu(dx) \qquad (n = 0, \pm 1, \ldots). \tag{8.33}$$

In other words, c_n is the nth Fourier coefficient of μ. If $\mu(\{\pi\}) > 0$, one may change μ to μ' where $\mu'(\{-\pi\}) = \mu(\{-\pi\}) + \mu(\{\pi\})$, and $\mu' = \mu$ on $(-\pi, \pi)$ to get a probability measure μ' on the circle whose Fourier coefficients are c_n. Note that (8.33) holds with μ replaced by μ', because $\exp\{-in\pi\} = \exp\{in\pi\} = 1$. ■

Corollary 8.5. A sequence $\{c_n\}$ of complex numbers is the sequence of Fourier coefficients of a finite measure on the circle $[-\pi, \pi)$ if and only if $\{c_n\}$ is positive definite.

Proof. Since the measure $\mu = 0$ has Fourier coefficients $c_n = 0$ for all n, and the latter trivially comprise a positive definite sequence, it is enough to prove the correspondence between nonzero positive definite sequences and nonzero finite measures. It follows from Theorem 8.4, by normalization, that this correspondence is 1–1 between positive definite sequences $\{c_n\}$ with $c_0 = c > 0$, and measures on the circle having total mass c. ∎

The *Fourier transform* of an integrable (real- or complex-valued) function f on $(-\infty, \infty)$ is the function $\hat{f}$ on $(-\infty, \infty)$ defined by

$$\hat{f}(\xi) = \int_{-\infty}^{\infty} e^{i\xi y} f(y)\, dy, \qquad -\infty < \xi < \infty. \tag{8.34}$$

As a special case take $f = \mathbf{1}_{(c,d]}$. Then,

$$\hat{f}(\xi) = \frac{\exp\{i\xi d\} - \exp\{i\xi c\}}{i\xi} \tag{8.35}$$

so that $\hat{f}(\xi) \to 0$ as $|\xi| \to \infty$. This convergence to zero as $\xi \to \pm\infty$ is clearly valid for arbitrary *step functions*, i.e., finite linear combinations of indicator functions of finite intervals. Now let f be an arbitrary integrable function. Given $\varepsilon > 0$ there exists a step function f_ε such that

$$\| f_\varepsilon - f \|_1 := \int_{-\infty}^{\infty} |f_\varepsilon(y) - f(y)|\, dy < \varepsilon. \tag{8.36}$$

Now it follows from (8.34) that $|\hat{f}_\varepsilon(\xi) - \hat{f}(\xi)| \leqslant \| f_\varepsilon - f \|_1$ for all ξ. Since $\hat{f}_\varepsilon(\xi) \to 0$ as $\xi \to \pm\infty$, one has $\overline{\lim}_{|\xi| \to \infty} |\hat{f}(\xi)| \leqslant \varepsilon$. Since $\varepsilon > 0$ is arbitrary,

$$\hat{f}(\xi) \to 0 \qquad \text{as } |\xi| \to \infty. \tag{8.37}$$

The property (8.37) is generally referred to as the *Riemann–Lebesgue Lemma*.

If f is continuously differentiable and f, f' are both integrable, then integration by parts yields

$$\hat{f}'(\xi) = -i\xi \hat{f}(\xi). \tag{8.38}$$

The boundary terms in deriving (8.38) vanish, for if f' is integrable (as well as f) then $f(x) \to 0$ as $x \to \pm\infty$. More generally, if f is r-times continuously differentiable and $f^{(j)}$, $0 \leqslant j \leqslant r$, are all integrable, then one may repeat the relation (8.38) to get

$$\hat{f}^{(r)}(\xi) = (-i\xi)^r \hat{f}(\xi). \tag{8.39}$$

In particular, (8.39) implies that if f, f', f'' are integrable then $\hat{f}$ is integrable.

It is instructive to consider the Fourier transform as a limiting version of a Fourier series. Consider for this purpose that f is differentiable and vanishes outside a finite interval, and that f' is square integrable. Then, for all sufficiently large integers N, the function

$$g_N(x) := f(Nx) \tag{8.40}$$

vanishes outside $(-\pi, \pi)$. Let $\sum c_{n,N} e^{inx}$, $\sum c^{(1)}_{n,N} e^{inx}$ be the Fourier series of g_N and its derivative g'_N, respectively. Then

$$\begin{aligned} c_{n,N} &= \frac{1}{2\pi}\int_{-\pi}^{\pi} g_N(x)e^{-inx}\,dx = \frac{1}{2\pi}\int_{-\pi}^{\pi} f(Nx)e^{-inx}\,dx = \frac{1}{2N\pi}\int_{-\infty}^{\infty} f(y)e^{-iny/N}\,dy \\ &= \frac{1}{2N\pi}\hat{f}\left(-\frac{n}{N}\right). \end{aligned} \tag{8.41}$$

Now writing $A = (2\sum_{n=1}^{\infty} n^{-2})^{1/2}$,

$$\begin{aligned} \sum_{n=-\infty}^{\infty} |c_{n,N}| &= |c_{0,N}| + \sum_{n\neq 0} \frac{1}{|n|}(|nc_{n,N}|) \leqslant |c_{0,N}| + \left(\sum_{n\neq 0}\frac{1}{n^2}\right)^{1/2}\left(\sum_{n\neq 0} |nc_{n,N}|^2\right)^{1/2} \\ &\leqslant |c_{0,N}| + A\left(\sum_{n=-\infty}^{\infty} |c^{(1)}_{n,N}|^2\right)^{1/2} \\ &= \frac{1}{2\pi}\left|\int_{-\infty}^{\infty} g_N(x)\,dx\right| + A\left(\frac{1}{2\pi}\int_{-\pi}^{\pi} |g'_N(x)|^2\,dx\right)^{1/2} < \infty. \end{aligned}$$

Therefore, for all sufficiently large N, the following convergence is uniform:

$$f(z) = g_N\left(\frac{z}{N}\right) = \sum_{n=-\infty}^{\infty} c_{n,N} e^{inz/N} = \sum_{n=-\infty}^{\infty} \frac{1}{2N\pi}\hat{f}\left(-\frac{n}{N}\right)e^{inz/N}. \tag{8.42}$$

Letting $N \to \infty$ in (8.42), if $\hat{f} \in L^1(\mathbb{R}^1, dx)$, one gets the *Fourier inversion formula*,

$$f(z) = \frac{1}{2\pi}\int_{-\infty}^{\infty} \hat{f}(-y)e^{izy}\,dy = \frac{1}{2\pi}\int_{-\infty}^{\infty} \hat{f}(\xi)e^{-iz\xi}\,d\xi. \tag{8.43}$$

One may show that this formula holds for *all* f such that both f and $\hat{f}$ are integrable. Next, any f that vanishes outside a finite interval and is square integrable is automatically integrable, and for such an f one has, for all sufficiently large N,

$$\frac{1}{2\pi}\int_{-\pi}^{\pi} |g_N(x)|^2\,dx = \frac{1}{2N\pi}\int_{-\infty}^{\infty} |f(y)|^2\,dy,$$

$$\frac{1}{2\pi}\int_{-\pi}^{\pi} |g_N(x)|^2\,dx = \sum_{n=-\infty}^{\infty} |c_{n,N}|^2 = \frac{1}{4N^2\pi^2}\sum_{n=-\infty}^{\infty}\left|\hat{f}\left(-\frac{n}{N}\right)\right|^2,$$

so that

$$\frac{1}{N}\sum_{n=-\infty}^{\infty}\left|\hat{f}\left(-\frac{n}{N}\right)\right|^2 = 2\pi\int_{-\infty}^{\infty} |f(y)|^2\,dy. \tag{8.44}$$

Therefore, letting $N \uparrow \infty$, one has the *Plancherel identity*

$$\int_{-\infty}^{\infty} |\hat{f}(\xi)|^2 \, d\xi = 2\pi \int_{-\infty}^{\infty} |f(y)|^2 \, dy. \tag{8.45}$$

Now every square integrable f on $(-\infty, \infty)$ may be approximated in norm arbitrarily closely by a continuous function that vanishes outside a finite interval. Since (8.45) holds for the latter, it also holds for all f that is integrable as well as square integrable. We have, therefore, the following.

Theorem 8.6

(a) If f and $\hat{f}$ are both integrable, then the Fourier inversion formula (8.43) holds.

(b) If f is integrable as well as square integrable, then the Plancherel identity (8.45) holds.

Next define the *Fourier transform* $\hat{\mu}$ *of a finite measure* μ by setting

$$\hat{\mu}(\xi) = \int_{-\infty}^{\infty} e^{i\xi x} \, d\mu(x). \tag{8.46}$$

If μ is a *finite signed measure*, i.e., $\mu = \mu_1 - \mu_2$ where μ_1, μ_2 are finite measures, then also one defines $\hat{\mu}$ by (8.46) directly, or by setting $\hat{\mu} = \hat{\mu}_1 - \hat{\mu}_2$. In particular, if $\mu(dx) = f(x)\, dx$, where f is real-valued and integrable, then $\hat{\mu} = \hat{f}$. If μ is a probability measure, then $\hat{\mu}$ is also called the *characteristic function* of μ (or of any random variable whose distribution is μ).

We next consider the *convolution* of two integrable functions f, g:

$$f * g(x) = \int_{-\infty}^{\infty} f(x-y)g(y)\, dy \qquad (-\infty < x < \infty). \tag{8.47}$$

Since

$$\int_{-\infty}^{\infty} |f * g(x)| \, dx = \int_{-\infty}^{\infty} \int_{-\infty}^{\infty} |f(x-y)||g(y)| \, dy \, dx$$
$$= \int_{-\infty}^{\infty} |f(x)| \, dx \int_{-\infty}^{\infty} |g(y)| \, dy, \tag{8.48}$$

$f * g$ is integrable. Its Fourier transform is

$$(f * g)\hat{}\,(\xi) = \int_{-\infty}^{\infty} e^{i\xi x}\left(\int_{-\infty}^{\infty} f(x-y)g(y)\, dy\right) dx$$
$$= \int_{-\infty}^{\infty} \int_{-\infty}^{\infty} e^{i\xi(x-y)} e^{i\xi y} f(x-y) g(y) \, dy \, dx$$
$$= \int_{-\infty}^{\infty} \int_{-\infty}^{\infty} e^{i\xi z} e^{i\xi y} f(z) g(y) \, dy \, dz = \hat{f}(\xi)\hat{g}(\xi), \tag{8.49}$$

a result of importance in probability and analysis. By iteration, one defines the *n-fold convolution* $f_1 * \cdots * f_n$ of n integrable functions $f_1, \ldots, f_n$ and it follows from (8.49) that $(f_1 * \cdots * f_n)\hat{} = \hat{f}_1\hat{f}_2 \cdots \hat{f}_n$. Note also that if f, g are real-valued integrable functions and one defines the measures μ, ν by $\mu(dx) = f(x)\,dx$, $\nu(dx) = g(x)\,dx$, and $\mu * \nu$ by $(f * g)(x)\,dx$, then

$$(\mu * \nu)(B) = \int_B (f * g)(x)\,dx = \int_{-\infty}^{\infty}\left(\int_B f(x - y)\,dx\right)g(y)\,dy$$
$$= \int_{-\infty}^{\infty} \mu(B - y)g(y)\,dy = \int_{-\infty}^{\infty} \mu(B - y)\,d\nu(y), \tag{8.50}$$

for every interval (or, more generally, for every Borel set) B. Here $B - y$ is the *translate* of B by $-y$, obtained by subtracting from each point in B the number y. Also, $(\mu * \nu)\hat{} = (f * g)\hat{} = \hat{f}\hat{g} = \hat{\mu}\hat{\nu}$. In general (i.e., whether or not μ and/or ν have densities), the last expression in (8.50) defines the *convolution* $\mu * \nu$ of finite signed measures μ and ν. The Fourier transform of this finite signed measure is still given by $(\mu * \nu)\hat{} = \hat{\mu}\hat{\nu}$. Recall that if X_1, X_2 are independent random variables on some probability space $(\Omega, \mathscr{A}, P)$ and have distributions Q_1, Q_2, respectively, then the distribution of $X_1 + X_2$ is $Q_1 * Q_2$ whose characteristic function (i.e., Fourier transform) may also be computed from

$$(Q_1 * Q_2)\hat{}(\xi) = Ee^{i\xi(X_1 + X_2)} = Ee^{i\xi X_1}Ee^{i\xi X_2} = \hat{Q}_1(\xi)\hat{Q}_2(\xi). \tag{8.51}$$

This argument extends to finite signed measures, and is an alternative way of thinking about (or deriving) the result $(\mu * \nu)\hat{} = \hat{\mu}\hat{\nu}$.

Theorem 8.4 and Corollary 8.5 may be extended to a correspondence between finite (probability) measures on $\mathbb{R}^1$ and positive definite functions on $\mathbb{R}^1$ defined as follows.

A complex-valued function φ on $\mathbb{R}^1$ is said to be *positive definite* if for every positive integer N and finite sequences $\{\xi_1, \xi_2, \ldots, \xi_N\} \subset \mathbb{R}^1$ and $\{z_1, z_2, \ldots, z_n\} \subset \mathbb{C}$ (the set of complex numbers), one has

$$\sum_{1 \leqslant j,k \leqslant N} z_j\bar{z}_k\varphi(\xi_j - \xi_k) \geqslant 0. \tag{8.52}$$

Theorem 8.7. (*Bochner's Theorem*). A function φ on $\mathbb{R}^1$ is the Fourier transform of a finite measure on $\mathbb{R}^1$ if and only if it is positive definite and continuous.

The proof of necessity is entirely analogous to (8.28). The proof of sufficiency may be viewed as a limiting version of that for Fourier series, as the period increases to infinity. We omit this proof.

The above results and notions may be extended to higher dimensions $\mathbb{R}^k$. This extension is straightforward. The Fourier series of a square integrable function f on $[-\pi, \pi) \times [-\pi, \pi) \times \cdots \times [-\pi, \pi) = [-\pi, \pi)^k$ is defined by $\sum_{\mathbf{v}} c_{\mathbf{v}} \exp\{i\mathbf{v}\cdot\mathbf{x}\}$ where the summation is over all *integral vectors* (or *multi-indices*) $\mathbf{v} = (v^{(1)}, v^{(2)}, \ldots, v^{(k)})$, each $v^{(i)}$ being an integer. Also $\mathbf{v}\cdot\mathbf{x} = \sum_{i=1}^k v^{(i)}x^{(i)}$—the usual euclidean inner product on $\mathbb{R}^k$ between two vectors $\mathbf{v} = (v^{(1)}, \ldots, v^{(k)})$, and $\mathbf{x} = (x^{(1)}, x^{(2)}, \ldots, x^{(k)})$. The *Fourier coefficients* are given by

$$c_{\mathbf{v}} = \frac{1}{(2\pi)^k}\int_{-\pi}^{\pi} \cdots \int_{-\pi}^{\pi} f(\mathbf{x})e^{i\mathbf{v}\cdot\mathbf{x}}\,d\mathbf{x}. \tag{8.53}$$

The extensions of Theorems 8.1–8.4 are fairly obvious. Similarly, the *Fourier transform* of an integrable function (with respect to Lebesgue measure on $\mathbb{R}^k$) f is defined by

$$\hat{f}(\xi) = \int_{-\infty}^{\infty} \cdots \int_{-\infty}^{\infty} e^{i\xi\cdot\mathbf{y}} f(\mathbf{y})\, d\mathbf{y} \qquad (\xi \in \mathbb{R}^k), \tag{8.54}$$

and the Fourier inversion formula becomes

$$f(\mathbf{z}) = \frac{1}{(2\pi)^k} \int_{-\infty}^{\infty} \cdots \int_{-\infty}^{\infty} \hat{f}(\xi) e^{-i\mathbf{z}\cdot\xi}\, d\xi, \tag{8.55}$$

which holds when $f(\mathbf{x})$ and $\hat{f}(\xi)$ are integrable. The Plancherel identity (8.45) becomes

$$\int_{-\infty}^{\infty} \cdots \int_{-\infty}^{\infty} |\hat{f}(\xi)|^2\, d\xi = (2\pi)^k \int_{-\infty}^{\infty} \cdots \int_{-\infty}^{\infty} |f(y)|^2\, dy, \tag{8.56}$$

which holds whenever f is integrable and square integrable. Theorem 8.6 now extends in an obvious manner. The definitions of the Fourier transform and convolution of finite signed measures on $\mathbb{R}^k$ are as in (8.46) and (8.50) with integrals over $(-\infty, \infty)$ being replaced by integrals over $\mathbb{R}^k$. The proof of the property $(\mu_1 * \mu_2)\hat{} = \hat{\mu}_1\hat{\mu}_2$ is unchanged.

Our final result says that the correspondence $P \to \hat{P}$, on the set of probability measures onto the set of characteristic functions, is *continuous*.

Theorem 8.8. (*Cramér–Lévy Continuity Theorem*). Let $P_n (n \geqslant 1), P$ be probability measures on $(\mathbb{R}^k, \mathscr{B}^k)$.

(a) If P_n converges weakly to P, then $\hat{P}_n(\xi)$ converges to $\hat{P}(\xi)$ for every $\xi \in \mathbb{R}^k$.
(b) If $\hat{P}_n(\xi) \to \hat{P}(\xi)$ for every ξ, then P_n converges weakly to P.

Proof. (a) Since $\hat{P}_n(\xi), \hat{P}(\xi)$ are the integrals of the bounded continuous function $\exp\{i\xi\cdot\mathbf{x}\}$ with respect to P_n and P, it follows from the definition of weak convergence that $\hat{P}_n(\xi) \to \hat{P}(\xi)$.

(b) Let f be an infinitely differentiable function on $\mathbb{R}^k$ having compact support. Define $\tilde{f}(\mathbf{x}) = f(-\mathbf{x})$. Then $\tilde{f}$ is also infinitely differentiable and has compact support. Now,

$$\int f(\mathbf{x}) P_n(d\mathbf{x}) = (\tilde{f} * P_n)(\mathbf{0}), \qquad \int f(\mathbf{x}) P(d\mathbf{x}) = (\tilde{f} * P)(\mathbf{0}). \tag{8.57}$$

As $\hat{\tilde{f}}$ is integrable, $(\tilde{f} * P_n)\hat{} = \hat{\tilde{f}}\hat{P}_n$ and $(\tilde{f} * P)\hat{} = \hat{\tilde{f}}\hat{P}$ are integrable. By Fourier inversion and Lebesgue's Dominated Convergence Theorem,

$$(\tilde{f} * P_n)(\mathbf{0}) = \frac{1}{(2\pi)^k} \int_{\mathbb{R}^k} \exp\{-i\mathbf{0}\cdot\mathbf{x}\} \hat{\tilde{f}}(\xi) \hat{P}_n(\xi)\, d\xi \to (\tilde{f} * P)(\mathbf{0}). \tag{8.58}$$

Hence, $\int f\, dP_n \to \int f\, dP$ for all such f. The proof is complete by Theorem 5.1. ∎

Author Index

Subject Index

GNANADESIKAN • Methods for Statistical Data Analysis of Multivariate Observations
GREENBERG and WEBSTER • Advanced Econometrics: A Bridge to the Literature
GROSS and HARRIS • Fundamentals of Queueing Theory, *Second Edition*
GROVES • Survey Errors and Survey Costs
GROVES, BIEMER, LYBERG, MASSEY, NICHOLLS, and WAKSBERG • Telephone Survey Methodology
GUPTA and PANCHAPAKESAN • Multiple Decision Procedures: Theory and Methodology of Selecting and Ranking Populations
GUTTMAN, WILKS, and HUNTER • Introductory Engineering Statistics, *Third Edition*
HAHN and SHAPIRO • Statistical Models in Engineering
HALD • Statistical Tables and Formulas
HALD • Statistical Theory with Engineering Applications
HAND • Discrimination and Classification
HEIBERGER • Computation for the Analysis of Designed Experiments
HOAGLIN, MOSTELLER and TUKEY • Exploring Data Tables, Trends and Shapes
HOAGLIN, MOSTELLER, and TUKEY • Understanding Robust and Exploratory Data Analysis
HOCHBERG and TAMHANE • Multiple Comparison Procedures
HOEL • Elementary Statistics, *Fourth Edition*
HOEL and JESSEN • Basic Statistics for Business and Economics, *Third Edition*
HOGG and KLUGMAN • Loss Distributions
HOLLANDER and WOLFE • Nonparametric Statistical Methods
HOSMER and LEMESHOW • Applied Logistic Regression
IMAN and CONOVER • Modern Business Statistics
JESSEN • Statistical Survey Techniques
JOHNSON • Multivariate Statistical Simulation
JOHNSON and KOTZ • Distributions in Statistics
Discrete Distributions
Continuous Univariate Distributions—1
Continuous Univariate Distributions—2
Continuous Multivariate Distributions
JUDGE, GRIFFITHS, HILL, LÜTKEPOHL and LEE • The Theory and Practice of Econometrics, *Second Edition*
JUDGE, HILL, GRIFFITHS, LÜTKEPOHL and LEE • Introduction to the Theory and Practice of Econometrics, *Second Edition*
KALBFLEISCH and PRENTICE • The Statistical Analysis of Failure Time Data
KASPRZYK, DUNCAN, KALTON, and SINGH • Panel Surveys
KAUFMAN and ROUSSEEUW • Finding Groups in Data: An Introduction to Cluster Analysis
KEENEY and RAIFFA • Decisions with Multiple Objectives
KISH • Statistical Design for Research
KISH • Survey Sampling
KUH, NEESE, and HOLLINGER • Structural Sensitivity in Econometric Models
LAWLESS • Statistical Models and Methods for Lifetime Data
LEAMER • Specification Searches: Ad Hoc Inference with Nonexperimental Data
LEBART, MORINEAU, and WARWICK • Multivariate Descriptive Statistical Analysis: Correspondence Analysis and Related Techniques for Large Matrices
LINHART and ZUCCHINI • Model Selection
LITTLE and RUBIN • Statistical Analysis with Missing Data
McNEIL • Interactive Data Analysis
MAGNUS and NEUDECKER • Matrix Differential Calculus with Applications in Statistics and Econometrics
MAINDONALD • Statistical Computation

Applied Probability and Statistics (Continued)

MALLOWS • Design, Data, and Analysis by Some Friends of Cuthbert Daniel

MANN, SCHAFER and SINGPURWALLA • Methods for Statistical Analysis of Reliability and Life Data

MARTZ and WALLER • Bayesian Reliability Analysis

MASON, GUNST, and HESS • Statistical Design and Analysis of Experiments with Applications to Engineering and Science

MIKÉ and STANLEY • Statistics in Medical Research: Methods and Issues with Applications in Cancer Research

MILLER • Beyond ANOVA, Basics of Applied Statistics

MILLER • Survival Analysis

MILLER, EFRON, BROWN, and MOSES • Biostatistics Casebook

MONTGOMERY and PECK • Introduction to Linear Regression Analysis

NELSON • Accelerated Testing, Statistical Models, Test Plans, and Data Analyses

NELSON • Applied Life Data Analysis

OCHI • Applied Probability and Stochastic Processes in Engineering and Physical Sciences

OSBORNE • Finite Algorithms in Optimization and Data Analysis

OTNES and ENOCHSON • Applied Time Series Analysis: Volume I, Basic Techniques

OTNES and ENOCHSON • Digital Time Series Analysis

PANKRATZ • Forecasting with Univariate Box-Jenkins Models: Concepts and Cases

PLATEK, RAO, SARNDAL and SINGH • Small Area Statistics: An International Symposium

POLLOCK • The Algebra of Econometrics

RAO and MITRA • Generalized Inverse of Matrices and Its Applications

RÉNYI • A Diary on Information Theory

RIPLEY • Spatial Statistics

RIPLEY • Stochastic Simulation

ROSS • Introduction to Probability and Statistics for Engineers and Scientists

ROUSSEEUW and LEROY • Robust Regression and Outlier Detection

RUBIN • Multiple Imputation for Nonresponse in Surveys

RUBINSTEIN • Monte Carlo Optimization, Simulation, and Sensitivity of Queueing Networks

RYAN • Statistical Methods for Quality Improvement

SCHUSS • Theory and Applications of Stochastic Differential Equations

SEARLE • Linear Models

SEARLE • Linear Models for Unbalanced Data

SEARLE • Matrix Algebra Useful for Statistics

SKINNER, HOLT, and SMITH • Analysis of Complex Surveys

SPRINGER • The Algebra of Random Variables

STAUDTE and SHEATHER • Robust Estimation and Testing

STEUER • Multiple Criteria Optimization

STOYAN • Comparison Methods for Queues and Other Stochastic Models

STOYAN, KENDALL and MECKE • Stochastic Geometry and Its Applications

THOMPSON • Empirical Model Building

TIJMS • Stochastic Modeling and Analysis: A Computational Approach

TITTERINGTON, SMITH, and MAKOV • Statistical Analysis of Finite Mixture Distributions

UPTON and FINGLETON • Spatial Data Analysis by Example, Volume I: Point Pattern and Quantitative Data

UPTON and FINGLETON • Spatial Data Analysis by Example, Volume II: Categorical and Directional Data

VAN RIJCKEVORSEL and DE LEEUW • Component and Correspondence Analysis

WEISBERG • Applied Linear Regression, *Second Edition*

WHITTLE • Optimization Over Time: Dynamic Programming and Stochastic Control, Volume I and Volume II

WHITTLE • Systems in Stochastic Equilibrium